Recent Advances in Materials Manufacturing and Machine Learning

Compendium of 2nd International Conference (RAMMML-23)

T0256297

About the conference

The international conference on recent advances in materials manufacturing and machine learning-2023 (RAMMML-23), is organized by Yeshwantrao Chavan College of Engineering, Nagpur, Maharashtra, India during February 02-04, 2023. The conference received more than 530 papers. More than 120 papers were accepted and orally presented in the conference.

The role of manufacturing in the country's economy and societal development has long been established through their wealth generating capabilities. To enhance and widen our knowledge of materials and to increase innovation and responsiveness to ever-increasing international needs, more in-depth studies of functionally graded materials/ tailor- made materials, recent advancements in manufacturing processes and new design philosophies are needed at present. The objective of this conference is to bring together experts from academic institutions, industries and research organizations and professional engineers for sharing of knowledge, expertise and experience in the emerging trends related to design, advanced materials processing and characterization, advanced manufacturing processes.

The conference is structured with plenary lectures followed by parallel sessions. The plenary lectures introduces the theme of the conference delivered by eminent personalities of international repute .Each parallel session starts with an invited talk on specific topic followed by contributed papers. Papers are invited from the prospective authors from industries, a cademic institutions and R&D organizations and from professional engineers. This conference brings academicians, industrial experts, researchers, and scholars together from areas of Mechanical Design Engineering, Materials Engineering and Manufacturing Processes. The topics of interest includes Design, Materials and Manufacturing engineering and other related areas such as Mechatronics, Prosthetic design and Bio inspired design and Smart materials. This conference is to provide a platform for learning, exchange of ideas and networking with fellow colleagues and participants across the globe in the field of Mechanical Design, Materials and Manufacture.

This conference RAMMML-2023 have paved the way to understand the latest technological and innovative advancements especially in the fields of manufacturing, design, materials, machine learning and interdisciplinary sciences. The conference has developed the solutions to physical problems, questions how things work, make things work better, and create ideas for doing things in new and different ways in the manufacturing, design and materials engineering. The conference focuses on the frontier themes of recent advances in manufacturing, design and materials engineering, as applied to multiple disciplines of engineering. Researchers, Academicians, Industrialist and Students has benefitted with the latest trends and developments in design, manufacturing and materials engineering applied to various disciplines of engineering. The objective of the conference is to have the orientation of research and practice of professionals towards attaining global supremacy in manufacturing, design and materials engineering. Also this conference aims in understanding the recent trends in manufacturing, design and materials engineering including optimization and innovation. This conference RAMMML-2023 also aims in improving Research culture in the minds of faculty in exploring the knowledge base, establishing better insights and maintaining dynamism in the teaching - learning process

Compendium of 2nd international conference on recent advances in materials manufacturing and machine learning-2023 (RAMMML – 23), February 02-04, 2023, Nagpur, Maharashtra, India

Recent Advances in Materials Manufacturing and Machine Learning

Edited By

Dr Bjorn Schuller
Imperial College London & University of Augsburg CSO, audEERING
0000-0002-6478-8699

Dr. Rajiv Gupta
North Carolina State University, Raleigh, NC, United States
0000-0003-2684-1994

Dr. Rakesh Mote
Indian Institute of Technology, Bombay, India
0000-0001-7853-5150

Dr. Abhishek Sharma
Osaka University, Japan
0000-0002-3384-861X

Dr. J.P. Giri
Yeshwantrao Chavan College Of Engineering, Nagpur
0000-0003-4438-2613

Dr. R.B. Chadge
Yeshwantrao Chavan College Of Engineering, Nagpur
0000-0001-9072-0607

CRC Press
Taylor & Francis Group
Boca Raton London New York

CRC Press is an imprint of the
Taylor & Francis Group, an **informa** business

First edition published 2024
by CRC Press
4 Park Square, Milton Park, Abingdon, Oxon, OX14 4RN

and by CRC Press
2385 NW Executive Center Drive, Suite 320, Boca Raton FL 33431

CRC Press is an imprint of Informa UK Limited

British Library Cataloguing-in-Publication Data
A catalogue record for this book is available from the British Library

ISBN: 9781032584799 (pbk)
ISBN: 9781003450252 (ebk)

DOI: 10.1201/9781003450252

Typeset in Sabon LT Std
by HBK Digital

Contents

List of Figures

List of Tables

Foreword

India as a growing economy has made its significant impact on global affairs in the last few decades. The country has survived many economic, social and political challenges in the past and is determined to become a stronger economy in coming future. The path of this growth is full of challenges of Material research, manufacturing technologies, design and development while ensuring inclusive growth for every citizen of the country. The role of manufacturing in the country's economy and societal development has long been established through their wealth generating capabilities. To enhance and widen our knowledge of materials and to increase innovation and responsiveness to ever-increasing international needs, more in-depth studies of functionally graded materials/ tailor- made materials, recent advancements in manufacturing processes and new design philosophies are needed at present. The objective of this conference is to bring together experts from academic institutions, industries and research organizations and professional engineers for sharing of knowledge, expertise and experience in the emerging trends related to design, advanced materials processing and characterization, advanced manufacturing processes. Since its inception, Yeshwantrao Chavan College of Engineering, Nagpur has been working as an institution committed to contribute in the field of Materials, design, automation and sustainable development through its various academic and non-academic activities. Different departments of YCCE have been conducting various academic conferences, seminars and workshops to discuss on contemporary issues being faced by Design, development and material research sectors and bring together academicians, researchers, technologists, industry experts and policy-makers together on common platforms.

Yeshwantrao Chavan College of Engineering, Nagpur is the pioneer in organizing academic conferences to share and disseminate academic research in the field of engineering & technology with the industry and policy makers since its establishment (1984).The department of mechanical engineering, YCCE is organizing its 2nd international conference on recent advances in materials manufacturing and machine learning-2023 (RAMMML-23), February 02-04, 2023 with an objective & scope to deliberate, discuss and document the latest technological and innovative advancements especially in the fields of manufacturing, design and materials engineering.

It was good news to hear from the Conference Organizing Committee that they have received more than 640 papers on different topics related to different subthemes of the conference. These papers cover topics related to manufacturing, design, materials engineering, machine learning, simulation, civil engineering and many more interdisciplinary topics. We believe that this conference will become a common platform to disseminate new researches done by various researchers from different universities/ institutes before industry professionals and policy makers in the government. We hope that these new researches will suggest new directions to innovations in Material research, design, development, manufacturing industries and government policies pertaining to these sectors.

We wish all the best to our conference participants who are the real knowledge champions of their universities/ institutes/ organizations. We strongly believe that all of us at YCCE will make this conference a good experience for every participant of the conference and this conference will achieve its objectives effectively.

We welcome you all to RAMMML-2023.

Preface

The main aim of the 2nd international conference on recent advances in materials manufacturing and machine learning-2023 (RAMMML-23) is to bring together all interested academic researchers, scientists, engineers, and technocrats and provide a platform for continuous improvement of manufacturing, machine learning, design and materials engineering research. RAMMML 2023 received an overwhelming response with more than 530 full paper submissions. After due and careful scrutiny, about 120 of them have been selected for presentation. The papers submitted have been reviewed by experts from renowned institutions, and subsequently, the authors have revised the papers, duly incorporating the suggestions of the reviewers. This has led to significant improvement in the quality of the contributions, Taylor & Francis publications, CRC Press have agreed to publish the selected proceedings of the conference in their book series of Advances in Mechanical Engineering and Interdisciplinary Sciences. This enables fast dissemination of the papers worldwide and increases the scope of visibility for the research contributions of the authors.

This book comprises four parts, viz. Materials, Manufacturing, Machine learning and interdisciplinary sciences. Each part consists of relevant full papers in the form of chapters. The Materials part consists of chapters on research related to Advanced Materials, Ceramics, Shape Memory Alloys and Nano materials, Materials for Aerospace applications, Polymers and Polymer Composites, Glasses and Amorphous Systems, Material characterization and testing, MEMS/NEMS, Bio Materials, Optical/Electronic Materials, Magnetic Materials, 3D Materials, Cryogenic Materials, Materials applications, performance and life cycle etc. The Manufacturing part consists of chapters on Micro/Nano Machining, Metal Forming, Green Manufacturing, Non-Conventional Machining Processes, Additive Manufacturing, Subtractive Manufacturing, Industry 4.0, Sustainable Manufacturing Technologies, Casting Technology, Joining Technology, Plastic processing technology, CAD/CAM/CAE/CIM/HVAC, Product Design and Development, Multi Objective Optimization, Modelling, Analysis and Simulation, Process Monitoring and Control, Vibration Noise Analysis and Control, Thermal Optimization, Energy Analysis etc. The Machine learning part consists of chapters on Machine learning, knowledge discovery, and data mining, Artificial intelligence in biomedical engineering and informatics, Artificial neural networks and algorithms, Knowledge acquisition, representation and reasoning methodologies, Genetic algorithms, Probability-based systems and fuzzy systems, Healthcare process management, Imaging, signal processing and text analysis, Bioinformatics and neurosciences. And the Interdisciplinary part consists of chapters on Condition Monitoring, NDT, Soft Computing, VLSI, Embedded System, Computer Vision, Environment Sustainability, Water Management, Advanced Mechatronics System and Control, Structural and Geo-technical Engineering areas. This book provides a snapshot of the current research in the field of Materials, Manufacturing, Machine learning and interdisciplinary sciences and hence will serve as valuable reference material for the research community.

Details of programme committee

International Advisory Committee

S. No	Name	Details
1.	Dr. Schuller Bjoern	Imperial college of London
2.	Dr. Rajeev Gupta	North carolina state university
3.	Dr. Avishek Dey	University of Delaware, USA
4.	Dr. Meysam Heydari Gharahcheshmeh	Massachusetts Institute of Technology, USA
5.	Dr. Vipindev Adat Vasudevan	Massachusetts Institute of Technology, USA
6.	Er. Amol Borkar	Ford Motor, USA
7.	Dr. Shukui Liu	Nanyang Technological University, Singapore
8.	Dr. Pritom Jyoti Bora	National University of Singapore
9.	Dr. Sreeprasad Sreenivasan	The University of Texas, USA
10.	Dr. Alessandro Ruggiero	University of Salerno, Italy
11.	Dr. Peng Cao	University of Auckland, NZ
12.	Dr. Nur Hassan	Central Queensland University, Australia
13.	Dr. Saurav Goel	London South Bank University ,UK
14.	Dr. Deepak Selvakumar R	Chung-Ang University, South Korea
15.	Dr. Arcot A Somashekar	The University of Auckland, NZ
16.	Dr. Maziar Ramezani	Auckland University of Technology, NZ
17.	Dr. Fangchen Liu	Nanyang Technological University, Singapore
18.	Dr. Yeliz Yoldas	Kayseri University, Turkey
19.	Dr. Naveen Shrivastava	BITS, Dubai Campus

National Advisory Committee

S.No	Name	Details
1.	Dr. Dilip Pratihar	IIT, Kharagpur
2.	Dr. Milind Atre	IIT, Bombay
3.	Dr. Sudarsan Ghosh	IIT, Delhi
4.	Er. Milind Pathak	Chairman IEI, Nagpur Local Centre
5.	Dr. Seshadri Sekhar, A.	IIT, Madras
6.	Dr. Chandramouli, P.	IIT, Madras
7.	Dr. Sujeet Kumar Sinha	IIT, Delhi
8.	Dr. Mallikarjuna, J.M.	IIT, Chennai
9.	Dr. Rakesh G. Mote	IIT, Bombay
10.	Dr. P.P. Datey	IIT, Bombay
11.	Dr. Nalinaksh Vyas	IIT, Kanpur
12.	Dr. Mukul S. Sutaone	Officiating Director, COEP, Pune
13.	Dr. Mohan P. Khond	COEP, Pune
14.	Dr. Rajiv B.	COEP, Pune
15.	Dr. R. L. Shrivastava	Institute of Engineers, India
16.	Dr. Satish V. Kailas	IISC, Banglore
17.	Er. Nitin Balsaraf	DRDO, Bangalore
18.	Dr. Pramod Padole	Director, VNIT
19.	Dr. V. R. Kalamkar	VNIT, Nagpur
20.	Dr. Awanikumar Patil	VNIT, Nagpur
21.	Dr. Y. M. Puri	VNIT, Nagpur

S.No	Name	Details
22.	Dr. G. S. Dangayach	MNIT, Jaipur
23.	Dr. R. V. Rao	SVNIT, Surat
24.	Dr. Sunil Bhirud	Director, VJTI, Mumbai
25.	Dr. R. N. Awale	VJTI, Mumbai
26.	Dr. D. M. Kulkarni	BITS, Goa
27.	Dr. Sachin Waigaokar	BITS, Goa
28.	Dr. Anupan Agnihotri	JNARDDC, Nagpur
29.	Mr. Pramod Patwardhan	SVP Mining Pvt. Ltd. Raipur C.G.
30.	Mr. Shree Jamdar	Kinetic Gears, Nagpur, India

Conference Chair & Organizing Secretary

S.No	Commitee	Name	Details
1.	Conference Chair	Dr. J.P. Giri	Head of The Department, Department of Mechanical Engineering, Yeshwantrao Chavan College of Engineering, Nagpur
2.	Conference Chair	Dr. R.B. Chadge	Department of Mechanical Engineering, Yeshwantrao Chavan College of Engineering, Nagpur
3.	Convenor	Prof. V. M. Korde	Department of Mechanical Engineering, Yeshwantrao Chavan College of Engineering, Nagpur
4.	Organizing Secretary	Prof. Neeraj Sunheriya	Department of Mechanical Engineering, Yeshwantrao Chavan College of Engineering, Nagpur
5.	Organizing Secretary	Prof. C. A. Mahatme	Department of Mechanical Engineering, Yeshwantrao Chavan College of Engineering, Nagpur
6.	Organizing Secretary	Prof. A. P. Edlabadkar	Department of Mechanical Engineering, Yeshwantrao Chavan College of Engineering, Nagpur
7.	Organizing Secretary	Prof. G. H. Waghmare	Department of Mechanical Engineering, Yeshwantrao Chavan College of Engineering, Nagpur

Organizing Committee Members

Prof. D. I. Sangotra	Prof. A. B. Amale	Prof. M. M. Dakhore
Prof. N. J. Giradkar	Dr S R Jachak	Prof. P. S. Barve
Dr S P Ambade	Prof. R. G. Bodkhe	Prof. N. D. Gedam
Prof. A. S. Bonde	Prof. D. Y. Shahare	Prof. P. V. Lande
Dr. S. T. Bagde	Dr. S. S. Khedkar	Prof. G. M. Dhote
Dr. S. S. Chaudhari	Dr. P. D. Kamble	Dr. V. R. Khawale
Dr. S. V. Prayagi	Prof. R. V. Adakane	Prof. P. A. Hatwalne
Dr. A. P. Kedar	Prof. D. N. Kashyap	Prof. N. P. Mungle
Prof. V. G. Thakre	Prof. A. R. Narkhede	Prof. Praful Shirpurkar
Prof. M. S. Tufail	Prof. S. P. Kamble	Prof. Ritu Shrivastava
Prof. P. N. Shende	Prof. Y. Y. Nandurkar	

Chapter 1

A review of shape memory alloys: origins, occurrence, properties, and their applications

K Shreyas Suvarna[a], Ram Rohit V[b], H R Prakash[c], Kruttik R[d], Niketh S R[e] and Rahul D[f]

Department of Mechanical Engineering, B.M.S. College of Engineering, Bengaluru, Karnataka, India

Abstract

Shape memory alloys (SMAs) are an important set of materials that find widespread applications due to their 'shape retaining' abilities when heated above a certain critical temperature in the fields of automotive engineering, aerospace, medical, robotics, structural engineering, spintronics, non-conventional energy resources, etc. The present paper discusses SMAs, detailing their origin and their classifications. Furthermore, the various properties of SMAs are discussed along with a study of nitinol, an extremely popular SMA used in many applications. A detailed study of SMA applications in the domains of aerospace and robotics has been presented in this paper. The motivation for this review work on SMAs is to understand the properties, characteristics and behavior of SMAs and utilize it in developing mechanisms for a diverse range of applications. SMAs are one of the most prominent research fields for engineers and scientists across the globe and significant progress has been made in recent years.

Keywords: Aerospace, nitinol, robotics, shape memory alloys (SMAs), shape-retaining ability

Introduction

Materials that are called shape memory alloys (SMAs) can return to their previous shape when heated. At temperatures below its transformation temperature, it can be deformed with ease and the new shape will be retained because of the low yield strength. A change in the crystal structure occurs when the SMA is heated above the given transformation temperature. This rise in temperature causes the alloy to return to its original, predetermined structure. SMEs provide a unique mechanism for remote actuation. This is based on the phenomenon that extremely large forces are generated when they come across any resistance. Nitinol is an SMA that is typically used widely. Nitinol is an alloy of nickel (49%) and titanium (51%). It has excellent electromechanical properties and can withstand extreme environments and exceptionally long fatigue life. If used as an actuator, it can provide a restoration stress of 344.737 MPa and a strain recovery of around 5% (for many repeated cycles). Approximately, 7.25 kg can be pulled by a wire made of nitinol that is 0.508 mm in diameter against gravity [1]. Nitinol can also be actuated by the Joule heating effect by passing electric currents through it. A transformation of phase occurs when a current is passed through the wire, causing sufficient heat generation. The chosen transition temperature is usually much higher than the room temperature. The reason for this is to ensure that the SMA actuates only with the intentional addition of heat. Essentially, nitinol functions as a sensor, an actuator, and a heater. Even so, SMAs cannot be used for all applications. In the case of large mechanisms, solenoids, electromagnets, and motors become better options as actuators. However, in cases where they cannot be utilised, SMAs become an excellent alternative. Nitinol produces high amounts of work per unit volume, and very few mechanisms produce more work per unit volume than nitinol. Rod and bar stock, wire, and thin films are the forms in which nitinol can be made available [1].

Origin of SMA

Milestones in the research of SMAs

The martensite phase, in a non-cubic state, is an unconventional crystalline state (named after Adolf Martens) and is a hard metastable non-equilibrium state of a material that originates from quenching [2]. The austenite, named after Charles Austen in its hot phase, has a centered cubic structure and was first used to describe a nonmagnetic form operating at high temperatures [3].

Read discovered phase transition and its link to strain memory and shape memory in a study based on alloys of Au and Cd in 1951 [4]. In 1932, Olander found pseudo elasticity in the AuCd alloy and defined it as a "rubber-like effect" [5]. During the examination of CuZn and CuSn [6], Greninger and Mooradian discovered that a martensitic phase formed and vanished as a function of temperature, demonstrating clear thermoelastic effects in CuZn. Kurdyumov conducted a comprehensive investigation of thermos elasticity

[a]shreyas.k.suvarna@gmail.com, [b]ramrohit.mech@bmsce.ac.in, [c]prakash.mech@bmsce.ac.in, [d]ktuttik.me18@bmsce.ac.in, [e]niketh.me18@bmsce.ac.in

DOI: 10.1201/9781003450252-1

in CuZn along with L.G. Khandros [7]. Olander and others [8,9] described the material which has elastic properties similar to materials such as rubber when subject to strain at lower temperatures, emphasising the material's thermoelastic properties.

Nevertheless, shape memory has been found in a variety of different alloys. For specific applications, copper-based compositions with Al and Zn [10,11], as well as iron-based compositions with metals like Pt, Pd, and Mn [12-14], have got a lot of applications in medical fields. Copper-based materials have higher transition temperatures and are effective in preventing unregulated actuation, and Fe-based systems are relatively much less expensive and more appealing for extensive large-scale deployment. Nitinol is currently one of the alloys with the top general shape memory, superb corrosion resistance, highly stable configuration, and nearly ideal biocompatibility. As a result, nitinol is a highly suitable material for implants that are biologically compatible. Efforts are currently being made to improve its quality [15]. However, such materials, on the other hand, are costly and complex to melt and treat [16]. Figure 1.1 depicts the phase transformation of SMAs with respect to temperature.

Nitinol

The discovery of SME in nitinol was an accidental discovery by its inventors William J. Buehler and Frederick Wang. The alloy retained a memory of its previous state and reverted to its original shape when accidentally exposed to the flame [17]. Figure 1.2 is an image of a nitinol wire of 0.5 mm thickness.

The most well-known properties of nitinol alloys are their super elasticity and thermal shape memory. Super elasticity refers to that property of the alloys, which makes them ten times more elastic than the best stainless steel now used. Shape memory is the phenomenon of recovering a predefined shape through heating after having "plastically" deformed that shape [18,19].

Applications of Shape Memory Alloys

It has been a long-standing engineering challenge and engineers and inventors across the globe have been involved in research activities to convert thermal energy into mechanical work through the phase change

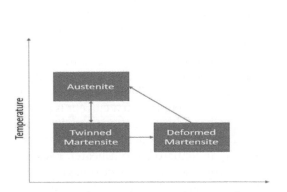

Figure 1.1 Transformation between phases in SMA [2, 3]
Source: Author

Figure 1.2 Nitinol wire of 0.5 mm thickness [2]
Source: Author

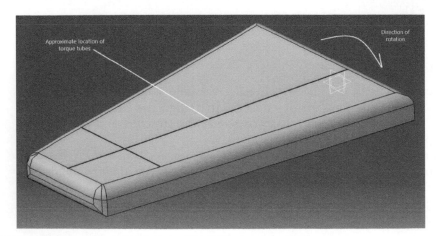

Figure 1.3 A representation of the SmartWing [23]
Source: Author

in the crystal structure of SMAs [20]; these methods have found many applications in the real world satisfactorily from the past several decades. The hydraulic tubing coupling used in the F-14 in 1971 is one of the most renowned of these early applications [21]. Since then, designers in aerospace and other industries have continued to use the shape memory and pseudo-elastic effects of SMAs to solve engineering problems. SMA technology implementations have included fixed-wing aircraft, rotorcraft, and spacecraft, actuators, and electromechanical devices [22].

Smart wing program and smart aircraft and marine propulsion system

The Smart Wing program and the Smart Aircraft and Marine Propulsion System (SAMP-SON) demonstration are two such projects where shape memory was extensively utilised for applications in fixed-wing aircraft [23,24]. The Smart Wing program was created to demonstrate and implement the use of various active materials like SMAs, to improve the lift and control surface performance [25-27]. The project consisted of two phases, the first of which required the maximum quantity of SMA. SMA wire fibers were utilised in the actuation of hinge-less ailerons, with a torque tube in place to activate wing twist across the span on a scaled-down model of the F-18. These applications, the actuation is induced by means of shape recovery which occurs under non-zero conditions of stress. Figure 1.3 depicts the torque tube installation as tested. This work was done in the United States' Virginia headquartered Defense Advanced Research Projects Agency (DARPA) contract with aerospace and advanced military tech giant Northrop Grumman and was overseen by the Air Force Research Lab (AFRL) based out of Ohio founded in October 1997 [28].

The primary objective of the SAMPSON program [29] was to show the value of shape memory ability of materials in optimizing the geometry of inlets and to experiment with orientations and combinations of multiple propulsion elements for applications primarily in military and combat aircraft. The experimental validation utilized a full-sized inlet from the F-15 platform. The SMAs were extensively used to control and modify the inlet cowl and vary the area of cross section to obtain the required characteristics. A pair of dissimilar SMA bundles was designed to initiate movement in directions opposite to each other, with shape recovery coming into effect with the heating of one bundle and the bundle in the cold state undergoing detwinning. For actuation in the opposite direction the bundle in the cold state was subjected to heat and the previously heated bundle underwent detwinning.

The SMA bundles were in most cases having more than 30 wires or rods generating about 27 kN of force to control the motion of the inlet cowl. Development activities also focused on achieving much more complex actuation including the shaping of the inlet lip.

Shape memory alloys as mini-actuators

Mechanical performance is increased by miniaturising the actuators and consuming less space; as a result, compactness and reduction in weight are the important features among the actuator elements. From Figure. 1.4 observing the power delivered per unit weight in various types of actuators, SMA actuators are better than lightweight technologies, implying that they have a significant potential for downsizing [30,31].

There are numerous other benefits to using SMA materials in mini actuators, the most important of which are mechanism simplicity, eco-friendly nature, remote sensing and operations capability, and low driving voltage. The disadvantages include low energy efficiency (<10%), low relative feedback, and non-uniform behavior [30].

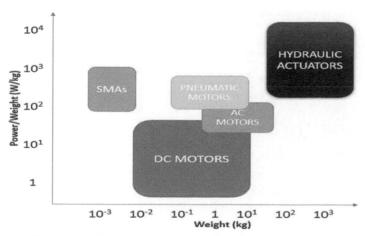

Figure 1.4 Weight vs Power-to-weight ratio of various actuators [30,31]
Source: Author

The SMA element is usually shaped like a helix or a linear wire, and it is usually paired with an element which is biased, which might also be an SMA constituent. In most cases, the SMA constituent is connected to an output shaft which is utilised to transmit motion to external systems. There is vast research that proposes utilisation of materials that exhibit shape memory as components for actuators in micro and mini mechanical systems that can be used extensively in applications demanding modularity [32-34].

Rotational shape memory alloy mini-actuators

An actuator was proposed by Jansen for applications requiring angular positioning in constrained spaces made up of many modules [34]: two SMA wire-activated drive modules as shown in Figure 1.5, a gear module is coupled with output shaft and in turn coupled to drive modules. SMA wires actuate a brake module and a snap module which is used as a braking mechanism. It also consists of a casing module and a module to control and regulate the sensors.

Each driving module with dimensions $70 \times 67 \times 40$ mm^3 can deliver an utmost displacement of around 25 mm-35 mm and an utmost force of till 4 N making use of a SMA wire about 0.2 mm in diameter. Both drive modules' output SMA wires are coupled inside the gear module to a pulley, causing the output shaft to rotate. The output shaft is coupled with a gear pulley. The highest output torque is 40N mm, and the maximum angular range is 180°.

Pöhlau and Meier [35] proposed to use SMA wires in a high-torque drive (Figure 1.5) which consists of an internal gear wheel named flex-ring which is flexible. SMA wires attached radially within the flex-ring, and an exterior rigid gear wheel make up the device. The flex-ring has exterior teeth, whereas the exterior gear wheel consists of interior teeth; as the teeth mate with one another on both sides with, both gear wheels being interconnected to each other. SMA wires which are actuated cause flex-ring deformation; the force generated in the radial direction created by the wires is converted into a force along the tangent at the tooth flanks, resulting in torque generation.

SMA wires are actuated one by one due to which the external gear wheel rotates intermittently. Figure 1.5 depicts the operational principle. 27.5 mm is the external radius of the actuator with 15 mm as its length. The diameter of SMA wires is 0.3 mm. Approximately 2.5 Nm is the maximum driving torque.

Using SMA wires, Park et al. created numerous miniature rotary actuators [36]. A simple joint actuator can be used to accomplish more complicated mechanisms to perform continuous rotational motion from

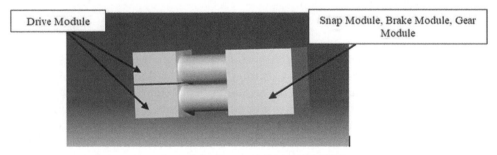

Figure 1.5 A representation of the Jansen drive module [34]
Source: Author

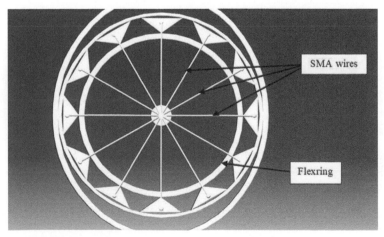

Figure 1.6 A simplified representation of the Pöhlau and Meier drive system [35]
Source: Author

step wise rotational motion accomplishing 60° of angular bidirectional deflection where dual SMA wires are implanted with radius of 50 μm. The continual rotational motion of a ratchet wheel is caused by the unidirectional rotary joint, the torque measured at the ratchet wheel is 0.08 N mm in one direction only.

Sharma et al. investigated a poly-phase motor utilising wires made of SMA [36,37]. The three-phase one is 150 × 150 × 20 mm in size and manufactured using SMA wires having a diameter of about 0.35 mm and a length of 12 cm, having a linear spring connected serially, in turn fixed with cam and an anchor which can be adjusted as required. The SMA are folded back by means of pulleys in order to minimise the size. The SMA elements use an active cooling system during activation to prevent strong heating around the link between them and the other portions of the motor. Spin of 72 degree per second is made for a force of 19.6 N mm.

A Japanese company named Toki Corporation manufactured a Biometal SMA rotary actuator, the SmartServo RC-1, which can span 60 degrees in 0.25 seconds (+/-30 degrees). The element is attached to a bias spring and has a wire-like shape with a length of 28 mm and a diameter of 60 μm [38,39]. The torque achieved is within the range of 1.5 Nmm while using an average power of approximately 0.15W in a temperature range of 0 to 40°C. The actuator is 38 × 9 × 3 mm^3 in size.

Ferromagnetic shape memory alloy actuators

When mechanical stresses with or without a magnetic field are applied to a ferromagnetic SMA (FSMA), they alter the structure/shape [40-42]. FSMA is ideal for producing actuators because of its quick response time (within the range of a milli second) and great strain (above 10%). Ni/Mn rich alloys are the most effective and well-known FSMA.

FSMA element activated mechanical devices are relatively rare compared to SMA element-based devices which find extensive application.

AdaptaMat's most recent actuator is made up of two NiMnGa components measuring 20 × 2.5 × 1.0 mm^3 each and is pre-stressed by a spring (that can be changed to suit the desired application). It has a maximum stroke of 0.7–0.8 mm and forces output of 5–7 N. The dimensions are approximately 80 mm × 80 mm × 70 mm.

AdaptaMat manufactures a variety of actuators. The A06-3 actuator has a blocking or stopping force of 2.5 N and a maximum stroke of 0.6 mm for a frequency of 200 Hz. It reaches a stroke of around 3% in approximately 0.2 ms. With an iron core, the A-1 2000 actuator can achieve stresses of 2.8% at 1.25 MPa. Linear actuator is the final device that tends to generate a linear motion with approximately 40 mm/s average shaft speed and 1 N of force [43]. Du Plessis et al and Murray [44] created two more FSMA actuator prototypes. Du Plessis' actuator tends to control the position of a valve spool using two FSMA opposing elements.

Recent Developments

In the Mars Curiosity Vehicle, NASA (National Aeronautics and Space Administration) and Goodyear [45] found that the vehicle's tires were subject to damage due to heavy wear and tear and unexpected harsh terrain. Therefore, they have chosen SMA (Nickel-Titanium wire mesh) for their tires as they retain their original shape (super elastic Ni-Ti alloy) despite being subject to severe deformation. Testing a lightweight SMA which allows the aircraft to fold their wings during flight, a part of the wing material is controlled by temperature, basically, an actuator in the form of a tube when subjected to heat undergoes twisting thereby moving the external portions of the wing in either directions by up to almost 70 degree. NASA's objective is to make the aircraft lighter, more fuel-efficient, and quieter. Flight wings must be rigid, to adapt to the conditions of space, and speed and to support heavy loads during turbulent environments. However, the wing's fixed position is not optimal for every aspect of flight, including take-off, cruising, and landing phases. Folding the wing's outer portion during flight at specific times offers directional stability and reduces drag during harsh flight conditions. This wing folding will also give it greater controllability, which would lead to reduced dependency on the heavier components of the aircraft, including components like the tail rudder. Folding the wings downwards, not only reduces drag but tends to increase performance as the supersonic aircraft transitions from subsonic speeds to supersonic speeds. Currently, in the F-18, the folding mechanism of the wing section has a heavy hydraulic system with an electric motor and gear transmission. This got replaced with high torque actuators based on SMA which made it lightweight and easy to use [46]. In the medical field, nitinol is used as a stent to treat blocked arteries. Here, it is compressed and sent through blood vessels, and on reaching the blocked site, it expands [47]. Due to its super elastic behavior, some of the most successful applications are eyeglass frames, and toys that are movable and flexible. Another development that has come up in recent times for SMAs is that they are being used to open greenhouse windows based on temperature. In this application, the SMA acts as a sensor and an actuator based on the outside temperature. It is also being used to restrict the flow of hot water in shower heads beyond a certain temperature to prevent scalding. They are also being used in satellites and space vehicles for deploying antennae to moving solar panel covers. The antennae while being sent up are coiled to save space and avoid damage and on reaching their destination, it is uncoiled by applying the necessary amount of heat. In the same way, solar panels are opened using shape memory alloys. To preserve heritage buildings and important structures, an SMA wire, bar, or strip in martensitic condition, can be placed around the structural member. It is placed transversely to existing shear cracks or even before the shear cracks appear. Here, Nickel-Titanium-Niobium alloys or low-cost Fe-based SMAs are used. These elements must be anchored carefully onto the structural member with nails, screws, or rivets for durability and long-term behavior. By heating the reinforcing element, causing its transformation from a martensitic phase into an austenitic phase, this tends to shorten, which closes the existing cracks and increases the shear strength. The high ductility of the materials capable of shape memory used will ensure an increase in the ductility of the shear failure. This low-cost iron-based shape memory alloy has 10% chromium and a high amount of silicon. This SMA also exhibits similar corrosion resistance as conventional stainless steel. With the addition of Hf and Pd to a Ni-rich nitinol, these types of alloys will tend to exhibit ideal super elasticity over a wider temperature gradient when compared to Nitinol, depending on various conditions, primarily the constituting elements, their composition and material treatment such as aging. The stress holding capability of concentrated Ni TiNiHf alloys is far greater compared to nitinol alloys. These new higher functionalities will open doors for innovative usage and applications.

Conclusions

Shape memory alloy (SMAs) demand will reach upwards of 20 billion USD by the year 2026. Considering SMA's unique features of super elasticity and shape retention and will replace the traditional heavy sensor and actuator mechanisms. Its unique feature allows it to behave as a sensor and actuator simultaneously. It is being widely deployed in applications like motors, actuators, transducers, structural materials, and sensors. Early commercial development started from pipe couplings to orthodontic wires many decades ago and now they are used across industries from aerospace to automobile, manufacturing, telecom, marine, and construction. These alloys will replace traditional equipment and bring about more efficiency, reliability, and robustness in operation. Research into doping the base alloy of Nitinol with other elements like hafnium and palladium and other new alloys to make them even more elastic and robust in operation is underway. This will go a long way in reducing operational complexity across industry verticals. This effort is being generously funded by industry market leaders. In the medical industry, reliability is the key that makes SMA's elasticity the default choice from bone repair to stents, catheters, medical guide wires, and dentistry. Initial costs are high, however, in time with an increase in volume, prices will become more affordable. The aircraft, space, and satellite industries are working towards deploying this material wherever possible. Smart wings, engine stablisers, landing gear for aircraft, tires for space vehicles, antennae, and solar panel deployment in satellites are just the beginning of what promises to be a replacement for many complex circuits using this wonder-alloy. The civil industry is also deploying SMA to reinforce heritage buildings, damping devices, structural connections, isolation devices, frames, and others. The automotive industry is deploying this alloy for the body frame, engine, and batteries. Evolving consumer electronics, rapid industrialisation, and automation increasing robotics demand for mundane and complex operations, and increased defense budgets, especially in India and China are spurring research and deployment of this material.

References

[1] *Tinialloy.com* (2022). [Online]. Available from: http://www.tinialloy.com/pdf/introductiontosma.pdf. [Accessed 15 May 2022].

[2] Gümpel, P. (2004), Formgedächtnislegierungen, expert verlag, Renningen [Accessed 24 May 2022].

[3] Callister, W. D. (2020). Materials science and engineering: an introduction. *John Wiley and Sons Inc., Hoboken.* [Accessed 28 May 2022].

[4] Mouritz, A. P. (2012). Introduction to aerospace materials, Woodhead Publishing, Sawston. [Accessed 3 June 2022].

[5] Chang, L. and Read, T. (1951). Plastic deformation and diffusionless phase changes in metals — the gold-cadmium beta phase. JOM. 3(1), 47-52, 1951.

[6] Ölander, A. (1932). AN electrochemical investigation of solid cadmium-gold alloys. Journal of the American Chemical Society. 54(10), 3819-33.

[7] Greninger, A. and Mooradian, V. (1938). Train transformation in metastable beta copper-zinc and beta copper-tin alloys. AIME Transactions. 138, 337-68.

[8] Kurdyumov, and Khandros, L. (1949). On the thermoelastic equilibrium on martensitic transformations. Dokl. Akad. Nauk SSSR. 66(2), 211-14.

[9] Basinski, Z. S. and Christian, J. (1954). Crystallography of deformation by twin boundary movements in indium-thallium alloys. Acta Metallurgica. 2(1), 101-16. Available from: 10.1016/0001-6160(54)90100-5.

[10] Asanovic, V. and Delijic, K. (2007). The mechanical behavior and shape memory recovery of Cu-Zn-Al alloys. Metalurgija-Journal of Metallurgy. 13.

[11] Hamdaoui, K. (2008). Experimental applications on cu-based shape memory alloys: Retrofitting of historical monuments and base isolation (Ph.D. thesis), University of Pavia.

[12] Khalil, W., Sulpice, L. S., Chirani, S. A., Bouby, C. Mikolajczak, A., and Zineb, T. B. (2013). Experimental analysis of Fe-based shape memory alloy behavior under thermomechanical cyclic loading. Mechanics of Materials. 631-11. Available from: 10.1016/j.mechmat.2013.04.002.

[13] Yang, G. S., Jonasson, R., Baek, S. N., Murata, K., Inoue, S. and Koterazawa, K. (2003). Phase transformations of ferromagnetic FePdPt-based shape memory alloys. Material Research Society Proceedings, Materials, and Devices for Smart Systems. 1, e5, 2003.

[14] Hsu, T. (2002). Fe-Mn-Si based shape memory alloys. Materials Science Forum. (327-328), 199-206. Available from: 10.4028/www.scientific.net/msf.327-328.199.

[15] Kurtoğlu, S. Yağcı, M. Uzun, A. Ünal, U. and Canadinc, D. (2020). Enhancing biocompatibility of NiTi shape memory alloys by simple NH3 treatments. Applied Surface Science. 525, 146547. Available from: 10.1016/j.apsusc.2020.146547.

[16] van der Wijst, M. W. M. (1992). Shape memory alloys featuring nitinol (Ph.D. thesis), TU Eindhoven, Faculteit der Werktuigbouwkunde, Eindhoven (NL).

[17] Corneliu Cismasiu (2010), Shape memory alloys / monograph. Rijeka, Crotia: Sciyo. [Accessed: 26 May 2022].

[18] Duerig, T. Pelton, T., and Stöckel, D. (1996). The utility of superelasticity in medicine. Bio-Medical Materials and Engineering. 6(4), 255-66. Available from: 10.3233/bme-1996-6404.

[19] Antonio, C. and Lecce, L. (2014). Shape memory alloy engineering: for aerospace. Structural and Biomedical Applications. [Accessed: 26 May 2022]

[20] Renner, E. (1976). Thermal engine. United States Patent. 3(937), 19.

[21] Melton, K. R. (1999). General applications of shape memory alloys and smart materials. In Shape memory materials (Eds K. Otsuka and C. M. Wayman). 10, 220–239

[22] Kudva, J. (2004). Overview of the DARPA Smart wing project. Journal of Intelligent Material Systems and Structures. 15(4), 261-267. Available from: 10.1177/1045389x04042796.

[23] Garcia, E. (2002). Smart structures and actuators: past, present, and future. In Proceedings of SPIE, Smart Structures and Materials. San Diego, CA. 17–21, 1–12.

[24] Sanders, B., Crowe, R., and Garcia, E. (2004). Defense advanced research projects agency – smart materials and structures demonstration program overview. Journal of Intelligent Material Systems and Structures. 15(4), 227-33. 2004. Available from: 10.1177/1045389x04042793.

[25] Kudva, J., Appa, K., Martin, C., and Jardine, A. (1997). Design, fabrication, and testing of the DARPA/Wright lab 'smart wing' wind tunnel model. Structures, Structural Dynamics, and Materials Conference and Exhibit. Kissimmee, FL. pp. 1–6.

[26] Stroud, H. and Hartl, D. (2020). Shape memory alloy NiTi actuators for twist control of smart designs. Smart Materials and Structures. [Accessed: 1 August 2022].

[27] Jardine, A. P., Flanagan, J. S., Martin, C. A., and Carpenter, B. F. (1997). Smart wing shape memory alloy actuator design and performance. Proceedings of SPIE - The International Society for Optical Engineering. DOI: 10.1117/12.2746853044, 48-55.

[28] Hartl, D. and Lagoudas, D. (2007). Aerospace applications of shape memory alloys. Journal of Aerospace Engineering. 221(4), 535-52. Available from: 10.1243/09544100jaero211. [Accessed 28 August 2022].

[29] Pitt, D. M., Dunne, J., White, E. V., and Garcia, E. (2001). SAMPSON smart inlet SMA powered adaptive lip design and static test. Conference: 19th AIAA Applied Aerodynamics Conference. 1-11.

[30] IKUTA, D. (1990). Miniature gripper and micro actuator using shape memory alloy. Journal of the Robotics Society of Japan. 8(4), 489-91.

[31] Mavroidis, C. (2002). Development of advanced actuators using shape memory alloys and electrorheological fluids. Research in Nondestructive Evaluation. 14(1), 1-32. Available from: 10.1080/09349840209409701.

[32] Nespoli,V. Besseghini, S., Pittaccio, S. Villa, E., and Viscuso, S. (2010). The high potential of shape memory alloys in developing miniature mechanical devices: A review on shape memory alloy mini-actuators. Sensors and Actuators A: Physical. 158(1), 149-60. Available from: 10.1016/j.sna.2009.12.020.

[33] Pöhlau, F. and Meier, H. (2004). Extremely compact high-torque drive with shape memory actuators and strain wave gear wave drive. International Conference on New Actuators, 2004, International Exhibition on Smart Actuators and Drive Systems. 3, 98-102.

[34] Sharma, S., Nayak, M. and Dinesh, N. (2008). Modelling, design and characterization of shape memory alloy-based poly phase motor. Sensors and Actuators A: Physical. 147(2), 583-92. Available from: 10.1016/j.sna.2008.05.021.

[35] Sharma, S., Nayak, M., and Dinesh, N. (2008). Shape memory alloy-based motor. Sadhana. 33(5), 699-712. Available from: 10.1007/s12046-008-0052-z.

[36] Homma, D., Uemura, S., and Nakazawa, F. (2007). Functional anisotropic shape memory alloy fiber and differential servo actuator. Proceedings of the International Conference on Shape Memory and Superelastic Technologies. [Accessed: 15 Aug 2022].

[37] Adaptamat.com (2022). Available from: http://www.adaptamat.com. [Accessed 28- Aug- 2022].

[38] Tellinen, J., Suorsa, I., Jääskeläinen, A., Aaltio, I., and Ullakko, K. (2002). Basic properties of magnetic shape memory actuators. 8, 566.

[39] Chernenko, V. and Besseghini, S. (2008). Ferromagnetic shape memory alloys: Scientific and applied aspects. Sensors and Actuators A: Physical. 142(2), 542-8. Available from: 10.1016/j.sna.2007.05.023.

[40] Murray, S. (2004). Ferromagnetic shape memory alloys, principles and applications.

[41] Plessis, A. D., Jessiman, A., Muller, G., and Schoor, M. V. (2003). Latching valve control using ferromagnetic shape memory alloy actuators. SPIE Smart Structures and Materials. 5054. [Accessed 28 August 2022].

[42] Youtube.com, 2022. [Online]. Available: https://www.youtube.com/watch?v=wI-qAxKJoSU. [Accessed 28- Aug- 2022].

[43] Mishra, S. (2022). Applications of shape memory alloys: a review. National Institute of Technology, Rourkela, India. [Accessed 28 August 2022].

[44] Youtube.com, 2022. [Online]. Available: https://www.youtube.com/watch?v=NRqG-o_DNnA. [Accessed 28- Aug- 2022].

[45] Tong, Y., Shuitcev, A., and Zheng, Y. (2020). Recent development of TiNi☐based shape memory alloys with high cycle stability and high transformation temperature. Advanced Engineering Materials. 22(4), 1900496. Available from: 10.1002/adem.201900496.

[46] Youtube.com, 2022. [Online]. Available from: https://www.youtube.com/watch?v=Pn-6bGORy0U. [Accessed 28- Aug- 2022].

[47] GlobeNewswire News Room, (2022). [Online]. Shape Memory Alloy (SMAs): Shaping technology for the future. https://www.globenewswire.com/news-release/2019/05/20/1827636/0/en/Shape-Memory-Alloy-SMAs-Shaping-technology-for-the-future.html. [Accessed 28- Aug- 2022].

Chapter 2

Optimisation of vibration in boring operation to obtain required surface finish using 45 degree carbon fiber orientation

Dr. H.P. Ghongade[a] and Dr. A.A. Bhadre

Brahma Valley College of Engineering and Research Institute

Abstract

The boring procedure is widely utilised worldwide to enlarge the specified existing holes of machine components. The surface quality and precision of the hole are more likely to be harmed by static deflections or self-excited palaver vibrations whenever the boring machine tool is long and thin. Additionally, it causes the tools' rapid wear and chipping. In order to operate within a cavity during an inside turning operation, a long, thin, boring tool is usually required. One of the boring tool's primary bending kinds is typically highly correlated with vibrations. There are several ways to eliminate vibrations, but selecting the most effective and stable ones requires a thorough comprehension of the dynamic categorisation of the tooling system. The contact between the clamping housing and the tool considerably impacts the functional classification of the securely secured boring tool. In order to achieve the needed surface smoothness in microns, 45 degree carbon fibre orientation tests were conducted in the lab along with different tool lengths to dampen the vibration with respect to the natural frequency. This study focuses on the functionality of a boring tool used in the boring operation under varied overhang lengths of carbon fibre tools at a 45 degree angle.

Keywords: Surface vibration, damping, fiber orientations, boring, natural frequency

Introduction

In the boring operation the dynamic excitation occurred in the boring tool, this is because of the material plastic deformation occurred during in process of cutting operation. This may bring ups a time relevant deflection of a boring tool. In the boring tool operation if its natural frequencies match with anyoneofthe-excitedfrequenciesthenthesituationofresonanceoccurred. Under such types conditions the maximum vibration occurred, accordingly the mathematical formulation of the natural frequencies play major part of the significance in the investigation of vibration sin machining process. In the process of material cutting, forces produce the maximum bending vibration in boring tool operation. The force that is delivered to the cutting tool during a cutting operation comes from the chip deformation process. Almost all significant structural systems are taken into consideration based on their capacity to sustain transverse shear in addition to having internal stiffness.

The quality needed for surface finishes must be controlled by choosing cutting settings that are optimised. Sadly, tool nose radius, feed rate, and cutting speed are not the only aspects that affect surface roughness. Other elements that may contribute to ruffling include inappropriate tool vibrations, friction between the cut surface and the cutting tip, and the embedding of the surface-level substance particles [1]. Consequently, the applied forces, which may be thought of as the sum of constant forces, harmonic forces, and random forces, act on the cutting tool and alter its stiffness and damping, hence altering the tool's vibration response. The causes of these differences in damping and stiffness are due to factors that are difficult to foresee in real-world scenarios. In addition, the transformation in the model tool parameters likewise impacted by the impacts of cutting parameters on the cutting tool, which makes this approach more helpful in managing tool vibration. This research focuses on gathering and analysing data on cutting force, tool vibration, and tool modal parameters produced by dry-turning mild carbon steel samples on a lathe at various speeds feeds, cut depths, tool nose radius, tool hanging lengths, and work-piece diameter and lengths. With the two-level interactions between the independent variables in mind, a full factorial experimental design was implemented. With the two-level interactions between the independent variables in mind, a full factorial experimental design was implemented. Fibre-reinforced composites provide a superior combination of strength and modulus than many conventional metals. Given their low specific gravities, these composite materials' strength-to-weight ratios and modulus-to-weight ratios surpass those of metallic materials. Composite materials will conform to the special needs of each design. Possible design factors include the choice of materials, the percentage by volume of fibre and matrix, the manufacturing technique, the number of layers in a particular direction, the thickness of individual layers, the kind of layer, and the sequence of layer stacking. The expense of the materials is the most significant downside of composites. In modern strength design, fibre-reinforced composite constructions such as aeroplanes, spacecraft, and automobiles predominate. Compared to many traditional metals, fiber-reinforced composites provide

---

[a]ghongade@gmail.com

DOI: 10.1201/9781003450252-2

a better mix of strength and modulus. Their low specific gravities make them, the strength-to-weight ratios and modulus-to-weight ratios of these composite materials surpass those of metallic materials. Composite materials will conform to the special needs of each design. Possible design factors include the choice of materials, the percentage by volume of fibre and matrix, the manufacturing technique, the number of layers in a particular direction, the thickness of individual layers, the kind of layer, and the sequence of layer stacking. The expense of the materials is the most significant downside of composites. In modern strength design, fiber-reinforced composite constructions such as aeroplanes, spacecraft, and automobiles predominate [2]. The rubber was either free or restricted by a rigid outer layer, known as the constrained configuration, in the study that was done today. The rubber layer was compelled to follow the surrounding material when it was constrained, causing it to undergo significant shear deformations. Shear deformation of the rubber provided the most damping for the current load instance since it was more important than axial or bending deformation. According to research, the quantity of shear decreases as rubber thickness increases while the amount of material with strong damping capabilities rises at a fixed thickness of the restricting layer. Studies reveal that these two effects essentially cancel each other out and that rubber thickness adjustments have little influence on the damping level [3]. Nevertheless, there are significant fluctuations in the damping depending on the thickness and stiffness of the constraining layer, and configurations with high damping are discovered for thick constraining layers with acceptable stiffness. So that the impact of the rubber stiffness may be ignored, the material of the beam and confining layer should be much stiffer than the rubber. The impact of incorrectly placed cuts in the constraining layer—which often occur while applying loads and supports to the beam—is also covered in the current study. While considerable dampening losses were seen for layer discontinuities, the significance of a continuous constraining layer is shown to be important [21].

Research Gap

In the boring operation, there are many other ways to damp the vibration but the using the carbon fiber orientation at 45 degree in the machining process with respect to the boring bar length while performing operation is unique idea. To testify the results there are many experiments are carried out in the laboratory and results are noted.

Objective

During the time spent on deep boring operations, typically when the length/diameter ratio (overhang length of tool/diameter of the boring tool) is larger, then excessive vibration occurs at the tip of the boring tool, which limits the final quality of the product and the surface finish of the product. In addition, it lowers the life of the cutting tool. Henceforth, minimise and improve the vibration of the boring tool by covering a composite carbon fibre layer as an inert damper with a varied combination of fibres.

1. Create a novel method for reducing the vibration in boring machine operation, i.e., damping effect optimisation over the boring tool.
2. Detect the dynamic condition classes of the boring tool operation under the numerous cutting parameters used during a boring process, such as feed rate and cut depth.
3. To examines the impact of carbon fibre covering with dissimilar optimise vibration and angle of coating – to increase or decrease the size of carbon fibre coating over the tool to get optimum results.
4. To examine the vibration behaviour of a coated tool by experimental results the number of tests done on the different cutting parameters to figure out the influence of the carbon fibre on the acceleration amplitude.

SCOPE

This article employed the finite element method (FEM) for the boring tool operation without and with the damper and via different layer of testing setup will be carried out. To research the impact of many cutting parameters on the boring tool operation the investigational set-up will be constructed. It is anticipated that the polymer-based composite material should be configured in a way that produces a certain dampening effect. Different conclusions will be reached using the finite element method, and they will be validated using the incremental result analysis.

Methodology

An experimental setup is prepared to testify to the obtained results from the various combination of carbon fiber with different degrees of angles to improve the vibration-damping effect and get the required overall surface finish in the machining process. Thoroughly to minimise the vibration impact on the boring tool, the damping and stiffness of the tool should be increased. This paper mainly focuses on the boring tool laminated with carbon fiber and different fiber orientations to achieve the maximum damping effect. In testing, four different types of boring tools were laminated with various carbon fiber combination, and which is listed below:

1.   10 degree fibre positioning
2.   Cross 10 degree fibre positioning
3.   45 degree fibre positioning
4.   Cross 45 degree fiber placement

Various Number of experimentations is performed with various cutting parameters to figure out the overall effect of carbon fiber on the acceleration amplitude.

Construction of the Damped Boring Tool

A diameter of 100 mm of boring tool is used in boring operation mild steel workpiece. A 16 mm diameter steel boring tool with a carbide tip is used. Because it creates vibrations during the boring process, boring with an excessive length divided by the radius of gyration ratio is especially challenging. This is due to the intrinsic frequency and dynamic stiffness of high-speed steel. The whole boring operation is hampered by lows peed, low feed rate, excessive depth of cut, and poor tool characteristics. Compared to steel, composite materials are more rigid. The shank of the boring tool was subsequently coated with carbon fibre in different orientations (100, cross 10, 450, and cross 450) to increase bending and longitudinal stiffness. Epoxy resin is a powerful adhesive that is used to laminate carbon fibre. The epoxy resin works well as a bonding agent and enhances the structure's rigidity. Figure 2.1 shows the image of a normal laminated tool.

Experimental Setup

The experiment is conducted using a lathe machine with a maximum diameter of 120 mm and a capacity of 2.2 KW based on vibration analysis of the boring tool with and without lamination. The experiment is conducted in the workshop of the college. The conventional boring tool configuration used in mild steel boring operation is shown in Figure 2.2. The mild steel work pieces utilised in this operation have an inner

Figure 2.1 Normal tool vs laminated tool
Source: Author

Table 2.1 Standard for photographs and laminated boring tools

		Material properties		
Part of boring tool	Material	Young's modulus (MPa)	Density (Kg/m³)	Poission's ratio
Shank	High speed steel	2.84 E5	7850	0.3
Tip	Cemented carbide	6.25 E5	1495	0.22
Lamination	Carbon fiber	3.5 E5	1800	0.4

Source: Author

Figure 2.2 Configuration of the standard boring tool
Source: Author

Table 2.2 Provides the specs of the boring tool utilised in the cutting test

Tool number	S16Q-SCLCR-09T3WIDAX
Material of the tool	Steel
Length of the tool (mm)	180
Diameter of the tool (mm)	16
Nose radius of the toll (mm)	0.4

Source: Author

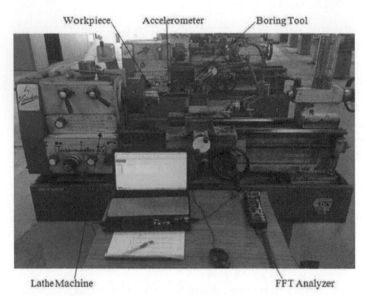

Figure 2.3 Experimental configuration
Source: Author

diameter of 80 mm. Utilising a WIDAX S16Q SCLCR 09T3 Boring Bar Holder, the experiment is run (Table 2.2).

The Fast Fourier Transform spectrum analyser (FFT) for various combinations of cutting settings, has been utilised to calculate the displacement of the boring tool under and the magnitude of the vibration acceleration. In order to deliver more accurate results for the tip under varied circumstances, an accelerometer has been put at the tip of the boring tool. To get the experimental findings as graphics, a laptop and FFT analyser are linked through a USB connection. Figure 2.3 illustrates the whole experiment's setup.

Results

Table 2.3 Findings from harmonic analysis

Acceleration/Amplitude (m/s²)			
Overhanging length (mm)	Standardise tool	Cross 45° fiber orientation	In percentage reduction in acceleration amplitude
96	102.7	89	15.23002
	214.88	191.92	11.75354
104	163.5	79.26	42.1232
	173.6	118.91	39.6843
112	225.54	69.88	69.01658
	132.85	46.08	65.31426
120	260.875	72.33	71.88549
	202.300	79.98	61.70405
128	296.21	74.78	74.7544
	271.75	113.88	58.09384

Source: Author

Conclussion

Boring operation vibration optimization to achieve the required surface finish four tools were built using 45 degree carbon fiber orientation to dampen the boring tool. Initial trial tests were done throughout the execution stage to evaluate each boring tool's performance separately. Ten measurements (96 mm, 112 mm, and 128 mm) of the starting depth of cut and the overhang tool length were made while the feed rate and spindle speed remained constant at 0.05 mm and 715 R P M per rotation, respectively (0.1 mm, 0.2 mm, 0.3 mm). In the initial trial experiments, it was found that the tool with cross-fibre orientation 10 degree lamination and cross-fibre orientation is more effective than the other two tools are 45 degree lamination as the acceleration amplitude keeps decreasing when the lamination or coating affects the overall damping effect in boring tool vibration (Table 2.3).

References

[1] González, H. L. W., Ahmed, S. Y., Rodríguez, P. R., Robledo, Z. P. D. C., and Mata, G. M. P. (2018). Selection of machining parameters using a correlative study of cutting tool wear in high-speed turning of aisi 1045 steel. Journal of Manufacturing and Materials Processing. 2, 66. https://doi.org/10.3390/jmmp2040066

[2] Lee, D. G., Hwang, H. Y., and Kim, J. K. (2003). Design and manufacture of a carbon fiber epoxy rotating boring bar composite structures. Composite Structures. 60(1), 115-24. https://doi.org/10.1016/S0263-8223(02)00287-8.

[3] Kristensen, R. F., Nielsen, K. L., and Mikkelsen, L. P. (2008). Numerical studies of shear damped composite beams using a constrained damping layer. Composite Structures. 83, 304-311.

[4] Houck, L. III, Tony, L., Schmitz, K. Scott Smith, (2011). A tuned holder for increased boring tool dynamic stiffness. Journal of Manufacturing Processes. 13, (24–29.

[5] Dutta, G. S. and Venkatesan, C. (2013). Analytical and empirical modelling of multilayered elastomeric isolators from damping experiments. Journal of Sound and Vibration. 332, 6913-23.

[6] Sortino, M., Totis, G., and Prosperi, F. (2013). Modelling the dynamic properties of conventional and high-damping boring tools. Mechanical Systems and Signal Processing. 343, 40-52.

[7] Rubio, L., Loya, J. A., Miguélez, M. H., and Fernández, S. J. (2013). Optimization of passive vibration absorbers to reduce chatter in boring. Mechanical Systems and Signal Processing. 4, 691-704.

[8] Esakkiraja, K., Antony, B. S., and Kannan, A. (2014). Investigation of chatter stability in boring operations using polymers as impact dampers. International Journal of Research in Aeronautical and Mechanical Engineering. 2321, 3051.

[9] Kriby, E. D., Zhe, Z., and Chen, J. C. (2004). Development of an accelerometer-based surface roughness prediction system in turning operations using multiple regression techniques. Journal of Industrial Technology. 20(4), 1-8

[10] Dimla, D. E., Lister, P. M., and Leighton, N. J. (2000). Neural networks solutions to the tool condition monitoring problem in metal cutting: a critical review of methods. International Journal of Machine Tools and Manufacture. 40, 1219-1241

[11] Dimla, D. E. (2000) Sensors signals for tool-wear monitoring in metal cutting operations - a review of methods. International Journal of Machine Tools and Manufacture. 40,1073-1098

[12] Xiaoli, L. I. (2002). A brief review: acoustic emission method for tool wear monitoring during turning. International Journal of Machine Tools and Manufacture. 42, 157-65

[13] Jemielniak, K. (1999). Commercial tool condition monitoring systems. International Journal of Advanced Manufacturing Technology. 15(4), 711-721

[14] Ghasempoor, A., Jeswiet, J., and Moore, T. N. (1999). Real time implementation of online tool condition monitoring in turning.

[15] International Journal of Advanced Manufacturing Technology. 39, 1883-1902

[16] Byrne, G., Dornfeld, D., Insaki, I., Ketteler, G., König, W., and Teti, R. (1995). Tool condition monitoring (TCM), the status of research and industrial application. Annals of CIRP. 44(2), 541-67

[17] Sick, B. (2002) Online and indirect tool wear monitoring in turning with artificial neural networks: a review of more than a decade of research. Mechanical Systems and Signal Processing. 16(4), 487-546

[18] Scheffer, C., Kratz, H., Heyns, P. S., and Klocke, F. (2003). Development of tool wear monitoring system for hard turning. International Journal of Machine Tools and Manufacture. 43, 973-85

[19] Kurada, S. and Bradley, C. (1997). Are view of machine vision sensors for tool condition monitoring. Computing. 34, 55-72

[20] Kakade, S. K., Raghavan, V. L., and Murthy, K. R. (1995). Monitoring of tool status using intelligent acoustic emission sensing and decision based neural network. Institute of Electrical and Electronics Engineers. 25-29

[21] Sen, G. C. and Bhattacharyya, A. (1969). Principles of Metal Cutting. India, New Delhi: New Central Book Agency.

[22] Youssef and El- Hofy H. (2008). Machining Technology: Machine tools and operations. pp. 8-10. New York, NY: CRC Press.

[23] Ghongade, H. P. (2022). Investigation of vibration in boring operation to improve machining process to get required surface finish. 62(8), 5392-95

[24] Thomas, M. and Beauchamp, Y., (2003) Statistical investigation of modal parameters of cutting tools in dry turning. International Journal of Machine Tools and Manufacture. 43, 1093-1106.

[25] Kurt, A., Sürücüler, S., and Ali, K., (2010). Developing a mathematical model for the cutting forces prediction depending on the cutting parameters. Journal of Engineering Science and Technology. 13(1), 23-30.

[26] Axinte, D. A., Belluco, W., and De Chiffre. L., (2001). Evaluation of cutting force uncertainty components in turning. International Journal of Machine Tools & Manufacture. 41, 719-730.

[27] Ezeanyagu, P. I. (2015). Cutting force coefficients estimation and tool dynamics of turning process: an unpublished. Master of Engineering Thesis, Nnamdi Azikiwe University Awka.

Chapter 3

Metal inert gas cladding simulations and residual stress analysis of mild steel

Arka Banerjee[a], Rajeev Ranjan[b], Manoj Kundu[c] and Subhas Chandra Moi[d]

Department of Mechanical Engineering, Dr. B. C. Roy Engineering College, Durgapur, West Bengal, India

Abstract

In this article, residual stress analysis and mathematical modeling for mild steel cladding applied by metal inert gas welding are presented. To determine the features of residual stresses, several cladding materials and varying process parameters are applied on mild steel. Simulating the cladding process involves using element birth and death function of ANSYS software. Firstly, ANSYS top-down technique is used to build geometry. To make brick meshing of the geometry easier, the geometry is separated. Thermal analysis is initially performed using the proper elements to determine the temperature distribution during the cladding process. In order to determine the nature of the residual stress, coupled field analysis is performed after the structure has been cooled to its original temperature (no load condition). The authors noted that several clad materials were proposed by earlier research to reduce the coating's residual stress. This study primarily emphasises the impact of clad thickness on residual stress for the substrate of mild steel and recommends the appropriate settings for the least amount of residual stress. The findings show that the residual stress decreases with decreasing clad material thickness.

Keywords: ANSYS, finite element analysis, MIG cladding, residual stresses

Introduction

Obtaining a material with the desired qualities is challenging. Hardness, yield strength, corrosion, and wear resistance of a material play significant roles in structural design and the lifespan of a product. Corrosion in machine components leads to considerable reduction in their service life [1, 2]. Corrosion is chemical hazard to a material due to unavoidable environmental conditions, which is very common in industries like nuclear power plant, petrochemical industries, fertiliser industries, food processing and chemical, industries etc [3-6]. However, materials with the qualities are pricey As a result, their price will increase if they are used to make components. The procedure of surface treatment is utilised to get around this sort of issue so that the service life of a component may be improved while keeping the cost of the material acceptable. Consequently, cladding is carried out to enhance the necessary properties on the surface of the base metal. In cladding, an inexpensive substrate is covered with a layer of a hard or corrosion-resistant alloy to boost a component's hardness, corrosion resistance and wear resistance with the goal of extending the component's service life [7-8]. Welding is a potential candidate for the cladding process because of the availability wide range of processes to cater the needs of the different industries [9]. Diverse welding techniques, such as TIG, MIG, SMAW, SAW, FCAW, PTAW, and laser deposition can be used to apply cladding [10]. During the MIG welding procedure, which joins the two base materials, welding cannon feeds a continuous solid wire electrode into the weld pool. To prevent contamination of the weld pool, a shielding gas is further provided through the welding gun. The term MIG really stands for metal inert gas. Its vernacular name is wire welding, while its scientific name is gas metal arc welding (GMAW) [11-12]. Manual MIG welding is frequently referred to as a semi-automatic process since the arc length and wire feed are controlled by the power source, but the wire position and travel speed are manually controlled. When none of the process parameters are directly under the control of a welder, the process can also be automated; welding may still need manual adjustments. Welding can be considered as automated when no manual assistance is required. [13-14]. In order for the process to function, the wire must be positively charged and connected to a power source that offers a constant voltage. The choice of wire feed rate and wire diameter determine the welding current since the burn-off rate of the wire will stabilise with the feed rate [15-18]. Finite element studies are frequently used as the basis for numerical assessments of cladding residual stresses [19]. The authors note that earlier research on the metal inert gas cladding approach recommended different coating materials to reduce residual stress. However, the present article investigated the variation of residual stresses on the clad thickness.

Cladding residual stresses

Even in the absence of external loads or heat gradients, residual stresses that persist, especially in welded components, cause considerable plastic deformation, which causes warping and distortion of an item.

[a]arka.banerjee@bcrec.ac.in, [b]rajeevranjan.br@gmail.com, [c]manoj.kundu@bcrec.ac.in, [d]subhas.moi@bcrec.ac.in

DOI: 10.1201/9781003450252-3

Temperature affects the residual strains on the cladding, which decrease with rising temperatures. The amount and distribution of residual stresses in the cladding are influenced by the cladding's thickness, geometry, base metal and cladding material characteristics, heat and pressure treatments used before and after the cladding process, and geometry of the cladded component. These residual stresses are intricate and frequently assessed using computer modelling and tests [20-21]. The basis for experimental assessments frequently stems from tests carried out in the lab on specimens that were fabricated from cladded parts. To determine the through-thickness residual stresses, experimental methods that entail cutting and machining the plate must be destructive. They are based on relaxation technique. Surface strain gauges are used to keep track of the deformations that occur during each cutting and machining phase. These deformations serve as the input for a computational process that determines the residual stresses present in the cladded plate prior to cutting. The size and distribution of cladding residual stresses are affected by several temperature dependent characteristics, though; therefore, computer simulations are often carried out under certain over-simplified assumptions.

Numerical methods

Residual stress and temperature distribution in the clad and parent materials are to be determined. For this analysis, initially the temperature is at 200°C. Bottom side of the base material is considered to be insulated, at the top surface convection is considered. To find out the temperature distribution, a transient thermal analysis is required. As the initial condition for the geometry in the thermal analysis, a uniform temperature of 1800°C.is used. Step size is chosen as 0.5 sec for the optimisation of the calculation time. To get the residual stress, the geometry has been cooled down to the initial-K temperature.

FEM based analysis has been carried out using ANSYS thermal-structural coupled analysis [22]. The material is considered as homogeneous for the structural analysis. Other material properties for parent and clad material are given in Table 3.1.

Brick mesh has been used to discretise entire geometry; total numbers of cells are 4500. In the thermal-structural coupled analysis calculated thermal results are used by the solver for structural analysis to give stress and deformation results. In this study, ANSYS top-down technique is used to predict distribution of residual stress. The substrate had the following measurements: 100 mm in length, 100 mm in width, and 10 mm in height. The clad measured 2.4 mm in height and 4.1 mm in breadth. The geometry for the remaining half of the 3D meshing process was determined by considering the uniformity along the weld line (Z-direction). The 3D Solid 120 hexahedron element, which has eight nodes and one degree of freedom i.e., temperature at each node, was employed for thermal analysis. At each node, a mechanical analogue of the 120 solid components has three degrees of freedom (translations in X, Y, and Z directions). For thermo-mechanical modelling, the solid 120 element was changed to the solid 205 element throughout the modelling process, and thermal analysis and structural analysis were combined. The deposit of clad material on the substrate was imitated using the element birth and kill method. A moving consumable electrode that served as the process' heat source deposited heat on the substrate. The components of the armour were destroyed and rendered inactive in the model when a heat source was applied. The clad zone and a portion of the substrate towards the interface are where the heating source's heat/temperature distribution is most constrained. As a result, this section has more finely meshed than other areas. Coarser elements were gradually messed into the remaining substrate region. This kind of non-uniform mesh distribution is known as the transitional mapped mesh approach [23, 24], where a ratio between fine and coarse element size is followed. The transitory nature of the MIGC process prompted the model to do a transient coupled thermal and structural study. The simulated results are compared with the experimental work of Lai et al. [25], and a good agreement is found in the trend of residual stress values.

Results and Discussion

The residual stresses and temperature distribution resulting from the cladding have been analysed using the component birth and death feature. Convection loads are given to the top of the base material, while the

Table 3.1 Properties of material

	Density	Poison's ratio	Young's modulus	Thermal conductivity	Thermal expansion coefficient	Specific heat
Base metal	7800kg/m^3	0.3	200GPa	42W/m-K	11.6/ °C	365W
clad material	7800kg/m^3	0.3	190GPa	15W/m –K	18.6/ °C	465W

Source: Author

bottom is thermally insulated. The body is first heated to 180°C. Around 1800°C is the maximum temperature, while 180°C is the lowest. Elements that are clad are made visible by applying component birth and death charactcristics. Other components are eliminated and kept out of sight. The temperature increases at the top and decreases as it moves towards the base metal. A temperature of about 180°C can be seen at the base of the foundation material. The temperatures are at their lowest outside of the cladding area.

The temperature is shown in Figure 3.1 from the top clad layer to the base material. The temperature from top to bottom on the graph is decreasing parabolic. Temperatures are highest at the top and lowest at the bottom.

Figure 3.2 depicts the growth of residual stress in the clad material. The midsection of the materials contained the most tension. Due to the differing materials, this can be a result of differential expansion and distortion. There is virtually any discernible difference in tension between the top and bottom. The analysis is carried out using a variety of thicknesses in order to examine the impact of clad layer thickness on residual stress produced.

Figure 3.3 illustrates the relationship between von Mises stress variation and cladding material thickness. With thickness, the von Mises stress is consistently rising. To discover the optimal clad material, more study is done using several materials. For comparison, aluminium, brass, and zerconium are considered. The characteristics of the clad material are altered, and analysis is done to determine the generation of residual stress.

Figure 3.4 displays the geometry-wide von Mises stress plot. Since the stiffness modulus of base metals is substantially higher than the young's modulus, the aluminium material is under the least stress.

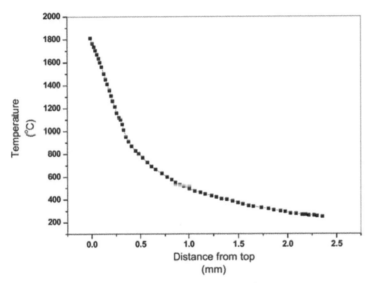

Figure 3.1 Temperature across the entire thickness
Source: Author

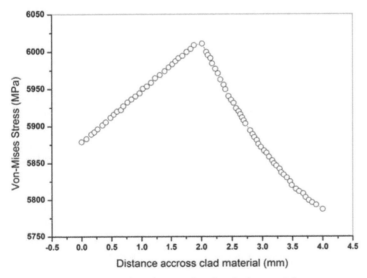

Figure 3.2 Stress distribution for the clad material
Source: Author

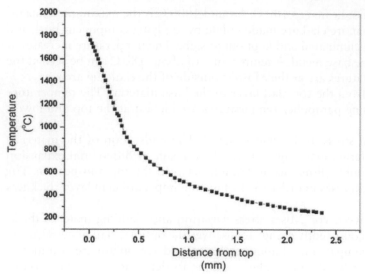

Figure 3.3 Clad thickness vs von Mises stress
Source: Author

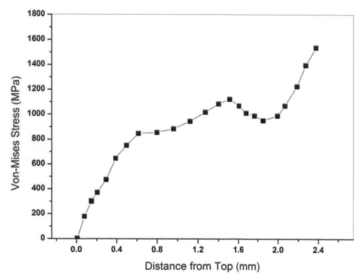

Figure 3.4 von Mises stress graph
Source: Author

von Mises stress tension is seen in the brass cladding in Figure 3.5. Once again, the stresses in the brass material are lower than in the base material. This is mostly caused by steel having a greater Young's modulus than brass. The structure experiences additional stress production as a result of the larger thermal expansion coefficient differences. The optimum material for cladding is zirconium since it has the lowest thermal expansion coefficient among engineering materials with good mechanical qualities.

Figure 3.6 depicts residual stress in the zirconium-clad structure. The base metal has the highest stress levels, whereas zirconium exhibits the lowest. This is primarily explained by the structure's reduced stresses caused by heating, which results in a decreased thermal expansion coefficient of zirconium.

The cladding process is simulated in the current study using ANSYS and finite element analysis. To make map meshing easier, the initial clad geometry and base metal are modelled and divided. Relevant attributes are given to the mesh geometry. The whole body is first heated to 180°C. Block by block simulation of the cladding is used for further investigation. Plots of temperature are shown. The region distant from the cladding process is exposed to least temperature, while the top clad geometry is subjected to highest temperature. After 2000 sec of cooling time it returns to its starting point. By relating thermal loads to structural conditions, the study is transformed into a structural analysis. The coupled field analysis demonstrates that residual stress has built up throughout the cladding process because it shows a gradient of stress, with the lowest stress at the top and the maximum stress in the region where it is restricted. The investigation and analysis are carried out for various thicknesses in order to identify the effect of cladding material thickness on the formation of residual stress. Less clad material thickness is associated with lower residual stress.

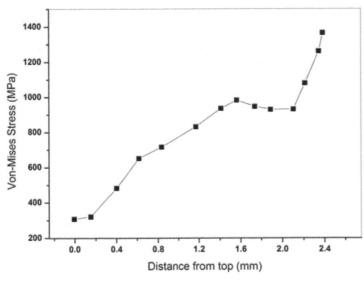

Figure 3.5 Stress diagram of brass
Source: Author

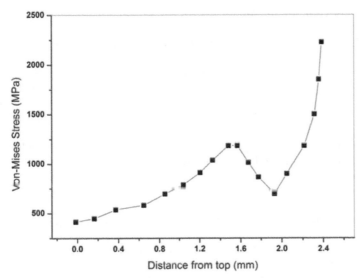

Figure 3.6 Residual stress graphic plot
Source: Author

Lower heat input to the structure may be too responsible for this. Various cladding materials are used in further analysis. The findings demonstrate that materials with higher thermal expansion coefficients result in greater strains than those with lower thermal expansion coefficients. The most effectively constructed cladding material with the least amount of residual stress development in the structure is zirconium, which has a reduced thermal expansion. The amount of stress that the materials can sustain is also significantly influenced by Young's modulus.

Conclusions

The ANSYS is used to model the thermal cladding process. The distribution of temperatures throughout the cladding process, the use of component birth and death, the impact of clad material thickness, and the role of various clad materials in producing residual stress are all discussed.

- To facilitate brick meshing of the structure, the initial clad geometry and base material geometry are constructed and divided.
- Thermal boundary conditions are implemented initially with a 180° initial preheat temperature.
- The component birth and death feature together with the Newton Raphson iterative approach are used to mimic the cladding process.
- A plot is made showing the section's temperature distribution and variance.

- After being cooled to room temperature, the structure exhibits residual stress. The constraint zone exhibits the highest stresses, while the outside surface exhibits lower stresses.
- Results for residual stress formation are further analysis by altering the thickness of the clad material. The findings show that the residual stress decreases with decreasing clad material thickness. This might be explained by the cladding's thinner thickness having lower heat content.
- According to the research, zirconium is the best cladding material since it has the fewest residual stresses.

References

[1] Chang, Y. -N. and Wei, F. I. (1991). High-temperature chlorine corrosion of metals and alloys. Journal of materials science. 26(14), 3693–3698.

[2] Melchers, R. E. (2018). A review of trends for corrosion loss and pit depth in longer-term exposures. Corrosion and Materials Degradation. 1(1), 42–58.

[3] Melchers, R., E. (2019). Predicting long-term corrosion of metal alloys in physical infrastructure. Materials Degradation. 3(1), 1–7.

[4] Melchers, R., E. (2005). The effect of corrosion on the structural reliability of steel offshore structures. Corrosion Science. 47(10), 2391–2410.

[5] Cragnolino, G. (1994). Application of accelerated corrosion tests to service life prediction of materials. United States, West Conshohocken, Pennsylvania: ASTM International.

[6] Albrecht, P. and Hall, T. T. (2003). Atmospheric corrosion resistance of structural steels. Journal of Materials in Civil Engineering. 15(1), 2–24.

[7] WiseGEEK, (2016). WiseGEEK clear answers for common questions, What is Cladding? http://www.wisegeek. com/what-is cladding.htm.

[8] Lee, J. W., Nishio, K., Katoh, M., Yamaguchi, T., and Mishima, K. (2005). The performance of wear resistance cladding layer on a mild steel plate by electric resistance welding, welding in the world. 49(9/10), 94–101.

[9] Palani, P. K. and Murugan, N. (2006). Development of mathematical models for prediction of weld bead geometry in cladding by flux cored arc welding. International Journal of Advanced Manufacturing Technology. 30, 669–6.

[10] Saha, M. K. and Das, S.A. (2016). Review on different cladding techniques employed to resist corrosion. Journal of the Association of Engineers India. 86(1-2), 51–63.

[11] Chakrabarti, B., Das, S., Das, H. and Pal, T. K. (2013). Effect of process parameters on clad quality of Duplex Stainless steel using GMAW process. Transactions of the Indian Institute of Metals. 66(3), 221–230

[12] Dreilich, T., V., Assis, K. S., de Sousa, F. V. V., and de Mattos, O. R. (2014). Influence of multipass pulsed gas metal arc welding on corrosion behaviour of a duplex stainless steel. Corrosion Science. 86, 268–74. doi: https://doi.org/10.1016/j.co rsci.2014.06.004.

[13] Elango, P. and Balaguru, S. (2015). Welding parameters for inconel 625 overlay on carbon steel using GMAW. Indian Journal of Science & Technology. 8(31), 1–5.

[14] Ghosh, P. K., Gupta, P.C. and Goyal, V. K. (1998).Stainless steel cladding of structural steel plate using pulsed current GMAW process. Welding Journal. 7(7), 307–12

[15] Ibrahim, T., Yawas, D. S., and Aku, S. Y. (2013). Effects of gas metal arc welding techniques on the mechanical properties of duplex stainless steel. Journal of Minerals and Materials Characterization and Engineering. 1, 222–30.

[16] Kannan, T. and Yoganandh, J. (2010). Effect of process parameters on clad bead geometry and its shape relationships of stainless steel claddings deposited by GMAW. International Journal for Advanced Manufacturing Technology, 47, 1083–1095.

[17] Kumar, V., Singh, G., and Yusufzai, M. Z. K. (2012). Effects of process parameters of gas metal arc welding on dilution in cladding of stainless steel on mild steel. MIT International Journal of Mechanical Engineering. 2(2), 127–31

[18] Murugan, N. and Parmar, R. S. (1997). Stainless steel cladding deposited by automatic gas metal arc welding. Welding Journal. 76, 391–403

[19] Zhang, Q., Xu, P., Zha, G., Ouyang, Z., and He, D. (2021). Numerical simulations of temperature and stress field of Fe-Mn-Si-Cr-Ni shape memory alloy coating synthesized by laser cladding. *Optik, 242*, 167079.

[20] Makhnenko, O.(2019) Influence of residual stresses in the cladding zones of rpv wwer-1000 on integrity assessment, proceedings of the second international Conference on Theoretical, Applied and Experimental Mechanics (pp.341–347).

[21] Tamanna, N., I. R. Kabir, and Naher, S. (2022). "Thermo-mechanical modelling to evaluate residual stress and material compatibility of laser cladding process depositing similar and dissimilar material on Ti6Al4V alloy." Thermal Science and Engineering Progress. 31, 101283.

[22] ANSYS analysis user's manual, Version 10. Theory reference.

[23] Kabir, I. R., D. Yin, and Naher, S. (2017). 3D thermal model of laser surface glazing for H13 tool steel. In AIP Conference Proceedings. 1896(1), 130003. AIP Publishing LLC.

[24] Kabir, I. R., Yin, D., Tamanna, N., and Naher, S. (2018). Thermomechanicalmodelling of laser surface glazing for H13 tool steel. Applied Physics. A 124(3), 1–9.

[24] Lai, Y., Yue, X., and Yue, W., (2022). A study on the residual stress of the co-based alloy plasma Cladding Layer. Materials. 15(15), 5143.

Chapter 4

Solution of goal programming using alternative approach of dual simplex method

Monali Dhote and Girish Dhote

Yeshwant Rao Chavan College of Engineering, Nagpur, India

Abstract

Charnes and Cooper discovered that goal programming (GP) is a generalisation of linear programming which solves mathematical programming with multiple objectives. There are many methods to solve Goal Programming Problem, but it is found that, among all the methods, the simplex algorithm is immensely used and proficient algorithm ever invented and shown extremely accurate in the formulation of optimisation problems. In this paper to solve GP problem, the dual simplex method is studied, which is most powerful method gives out iterative process, which first designing a basic feasible solution then proceeds towards the optimal solution. We use some modification to existing dual simplex method with the new approach of modified elimination method, hence the method become more efficient in reaching solution. Through this method, we are able to solve GPP with better solution which takes less iteration or sometimes equal iterations, save many efforts and time by neglecting calculations of net evaluation. The results of calculation for solved problem confirm that our method generates a smaller number of iterations as other methods.

Keywords: Goal programming (GP), linear programming (LP), optimal solution, dual simplex method, less or same iterations

Introduction

Basically, linear programming is one of the techniques which are applicable only for only one goal, which is minimizing the profit or minimising the loss. But sometimes a number of circumstances come wherever the single goal is not sufficient and system needs more than one goal i.e., multiple goals. Basically, many industries do have a bunch of goals, which contain stability of employment, product quality should high, minimisation of earnings, minimising overtime or cost, and then for these circumstances, one special practice must have required which seek a compromise solution where all objectives are relatively important. Hence goal programming (GP) was introduced, which started with the important goal furthermore continued up to the attainment of less important goal and proposes at minimizing the deviations from the goals.

Significantly in GP, the interest has been improved in the latest past, as has its concrete execution. Charnes and Cooper [1] had developed the initial progress of the goal programming, done in 1961 during one argument. They found that GP is an extension of linear programming (LP) which originated and arrived in the same way, the dissimilarity was the total description of the entire objective functions which approached as per preferred priority. GP problem has quite similar formulation as that of LP problems (LPP), but in 1961, as per Charnes and Cooper, the linear programming formulation has been extended by GP so that mathematical programming with multiple objectives is accommodated. Also, many of their students extended this tool; again it was enhanced by them. GP had an extra approach that defines all goals the same as constraints as well as the normal constraints and is used in optimization of multiple objective goals by minimising the deviation for each of the objectives from the desired target. Many algorithms have been applied to solve GP problem based on maximising the profit and minimising the cost.

Initially, in 1948 simplex method has been developed by Dantzig [2] to solve LPP. Later this simplex algorithm was modified by Lee [3] to solve a GPP. He handles the entire tables, extending the evaluation z_jc_j for all preemptive priority. At that time, it was found that the simplex method is immensely used and proficient algorithm ever invented and is still the standard technique to solve optimisation problems. Then afterwards while solving LPP to decrease the calculation time, Schniederjans and Kwaks [4] produced an alternative simplex method which eliminates half of the columns in the deviation table. Moreover, using Schniederjans and Kwaks's [4] procedure, Olson [5] represents the revised method to resolve GP problem which help to calculate a new element in simplex table and this result shows that dual simplex method is better as compare to revised simplex method. In 1954, when dual simplex method was presented by Lemke [6], for nearly forty years nobody has assumed that it will be considered like a competitive alternative to the primal simplex method. But finally in early 1990s due to the contributions of Forrest and Goldfarb, this fact has changed, where they have developed a comparatively in expensive dual version of the highest edge pricing rule. Enormous improvement has been made by commercial solvers in last decade to launch dual simplex method. Many inventors extended this tool and again it was enhanced by them. Koberstein [7] proposed dual simplex method for fast and stable implementation in his dissertation report in 2005. Khobragade [8] and Lokhande [12] recommended new approach to Wolfe's modified simplex method for

DOI: 10.1201/9781003450252-4

quadratic programming problems. Ghadle et al. [9] discussed alternative simplex method to solve game theory problems. Moorthy and Kalyani [11] give the intelligence of dual simplex method to solve linear fractional fuzzy transportation problem, in 2015. An alternative method for dual simplex method has been mentioned by Safitri et al [10] in 2015. Again in 2020, Goli and Nasseri [13] recommended an extended dual simplex method for LPP. All the above inventions regarding dual simplex method was very helpful to all researchers while solving GPP.

The purpose of this paper is to present dual simplex method with new approach to get better solution which takes less iteration, save many efforts and time. Safitri et al [10] in his the previous study update the new tables using use the Gauss-Jordan elimination which need more iterations to get optimal solution, while authors apply the Modified Gauss-Jordan elimination to get new element for new table by dividing it by pivot element and convert remaining elements in the column to zero and finally changed the resulting sign and solve GPP with better solution which takes less iteration. The principle of the present article resolves the GP problem by means of new approach of elimination method has been mentioned which framed so as to find a fundamental or key solution for GP problem based on maximising the profit along with minimising the cost.

Formulation to solve GPP

In this paper the most usually applied type, pre-emptive weighted priority GP is discussed whose model is referred by Schniederjans and Kwaks in 1982 as follows:

$$\min z = \sum_{k=1}^{m} w_k P_k (d_k^- + d_k^+) \tag{1}$$

$$\text{Subject to } \sum_{l=1}^{n} s_{kl} x_l + d_k^- + d_k^+ = r_k, \quad k = 1,2,3,......m \tag{2}$$

$$\text{And } x_l, d_k^-, d_k^+ \geq 0; \quad k = 1,2,......m; l = 1,2,3,......n \tag{3}$$

Here
z - Addition of deviations of entirely essential goals with m goal constraints and n decision variables.
w_k - The relative non -ve weight allotted to deviational variables d_k^- and d_k^+ for all goal constraint.
P_k - Pre-emptive priorities allotted to bunch of goals in rank order assembled with each other in formulation of GPP.
d_k^- - variable with –ve deviation for k[th] goal
d_k^+ - variable with +ve deviation for k[th] goal
x_l - The jth decision variable
S_{kl} - Constant involved to each decision variable
r_k - Values at right-hand-side.

Here initial table for this GP problem has been set up in the way that of the table of LPP.
So, from equation (1), we write the deviation variable d_k^+ in the following way,

$$d_k^+ = -r_k + \sum_{l=1}^{n} s_{kl} x_l + d_k^-, \quad k = 1,2,3,......m \tag{4}$$

In this paper we used d_k^+ as initial basic variables and in the absence of a variable d_k^+, an artificial zero priority is given to a goal constraint to formulate the initial table, Table 4.1 shows initial presentation. First row of initial table contain basic variables and non-basic variables (i.e. decision variables x_j and negative deviational variables d_k^-). In column 1 priority with weights are included for all positive deviational variables. Column 3 and column 4 contain all values at right-hand-side r_k and coefficients of decision variable S_{kj}. Last column 5 carries an identity matrix which represents the inclusion of d_k^-. In row 2, column 5, total absolute deviation is added. Row 2, column 4 contains zero values which represents the addition of all decision variables in the calculation procedure. And in row 4, column 5, proper weights are listed for all negative deviational variables.

Proposed Alternative Approach for GPP

Proposed alternative method adds following stages to find GPP:

Table 4.1 Initial table

Initial iteration	Formulation of initial table				
	Column (1)	Column (2)	Column (3)	Column (4)	Column (5)
Row (1)	Formulation table	Basic variables	RHS values r_k	Decision variables x_l	-ve deviational variables d_k^-
Row (2)	priorities with Weight	z	$\sum_{k=1}^{m} \|w_k r_k\|$	$0.....0$	$w_1.....w_m$
Row (3)	$w_1 P_1$	d_1^+	$-r_1$	$S_{11}.....S_{1n}$	1
	$w_2 P_2$	d_2^+	$-r_2$	$S_{21}.....S_{2n}$	1
	$\vdots$	$\vdots$	$\vdots$	$\vdots$	$\vdots$
	$w_1 P_m$	d_m^+	$-r_m$	$S_{m1}.....S_{mn}$	1

Source: Author

Stage (1) State the given GPP in standard format.

Stage (2) Prefer the variable with higher priority such as $(P_0 > P_1 > P_2 > P_3 .... > P_n)$ to leave the variable.

a) For more than one variable with same priority, the variable having highest weighs leave first.

b) And for only one variable with higher priority, select the pivotal row which is a row with leaving variable.

Stage (3) Prefer highest coefficient in pivot row to enter the variable. Again, select the column corresponding to this highest coefficient of pivot row which is also known as pivot column. And the point where pivot row and pivot column intersect, gives us pivot element.

Stage (4) Exchange all coefficients of pivot row with pivot column, then we get new element in new table which obtained by dividing by pivot element. Convert remaining elements in the column to zero and finally changed the resulting sign.

Stage (5). Again revise the new table by using the modified Gauss-Jordan elimination method. Find the fresh fixed deviation by the formula given below:

$$z = \sum_{k=1}^{m} \left| w_k r_k \right|$$

Stage (6) Optimal solution is existing when all basic variables are positive and pre-emptive priority is fulfilled. If some basic variables are negative, replicate above stages. But if all basic variables are positive and the pre-emptive priority is unfulfilled, the solution is not optimal. Then follow the following steps.

Stage (7) Prefer the highest positive element from column 4 of initial table which also have highest priority level and find the variable which exits the solution basis. This row also gives the pivot element.

Stage (8) If negative coefficients of pivot row are split into their corresponding positive elements in row 2 changing the resultant sign, prefer the column with smallest resulting ratio and enter the variable to the solution basis. Repeat steps 3-8.

Stage (9) Optimum solution exits if all basic variables are positive and one or more objective function in row 2 are negative.

An Ilustrative Example

Example 1. Bharat Architecture and decorator's firm works on many areas out of which it has two most demanded departments named as planning and decorating. Each unit of planning and decorating department requires 1 hr. per room equally. It is assumed that only 350 hrs. are available for each unit of both departments weekly. The managing director of the firm has conveyed the wish that each member of planning department must achieve profit of Rs. 200 weekly also each member of decorating department Rs. 300 weekly. The management of Bharat firm wants to get the following goals and priorities:

P_1: Achieve a weekly profit of Rs. 350 as possible.

P_2: Neglect over time of all departments and make full use of available hours of all departments.

P_3: Minimise underutilisation of the daily operation hours of both departments.

Solve given GPP by alternative approach to attain all the goals and gives maximum profit on weekly production.

Solution

Assume that x_1 and x_2 denotes unit of planning and decorating department correspondingly.
    Following equations shows, the constraints and goals:

$$\text{Min } Z = P_1 d_1^- + 2P_2 d_2^- + P_2 d_3^- + P_3 d_1^+ \tag{5}$$

Subject to

$$x_1 + x_2 + d_1^- - d_1^+ = 350 \quad \text{(Profit)} \tag{6}$$

$$x_1 + d_2^- - d_2^+ = 200 \quad \text{(Overtime)} \tag{7}$$

$$x_2 + d_3^- - d_3^+ = 300 \quad \text{(Under-utilization)} \tag{8}$$

$$x_1, x_2, d_1^-, d_1^+, d_2^-, d_3^- \geq 0 \quad \text{(Non-Negativity)} \tag{9}$$

Now, the Dual form of above GPP is as:

$$\text{Min } Z = P_1 d_1^- + 2P_2 d_2^- + P_2 d_3^- + P_3 d_1^+$$

Subject to:

$$x_1 + x_2 + d_1^- - d_1^+ = 350 \rightarrow d_1^+ = x_1 + x_2 + d_1^- - 350 \tag{10}$$

$$x_1 + d_2^- - d_2^+ = 200 \rightarrow d_2^+ = x_1 + d_2^- - 200 \tag{11}$$

$$x_2 + d_3^- - d_3^+ = 300 \rightarrow d_3^+ = x_2 + d_3^- - 300 \tag{12}$$

We make required calculations of the given example by the following tables. Let's start with, Table 4.1.
    As largest element in pivot row d_2^+ correspond with pivot column x_1, so exchange pivot row d_2^+ with pivot column x_1. Then we get new element in, Table 4.3, obtained by dividing by pivot element, remaining elements in the pivot column are converted to zero and finally changed the resulting sign.
    The first iteration is shown in, Table 4.3.
    The second iteration in, Table 4.4 shows the exchange of pivot row d_3^+ with pivot column x_2, as largest element in d_3^+ correspond with x_2.
    The third iteration in, Table 4.5 shows the exchange of pivot row d_1^+ with pivot column d_1^-, as largest element in d_1^+ correspond with d_1^-.
    After getting all positive variable and all fulfilled pre-emptive priority, from, Table 4.5 it is concluded that we get optimal solution as

$$x_1 = 200, x_2 = 300, d_1^- = 850, d_1^+ = 0, d_2^+ = 0, d_3^+ = 0$$

Table 4.2 First table

First iteration			First iteration table					
			0	0	P_1	$2P_2$	P_2	P_3
C_B	y_B	x_B	x_1	x_2	d_1^-	d_2^-	d_3^-	d_1^+
P_1	d_1^+	-350	1	1	1	0	0	-1
$2P_2$	d_2^+	-200	1	0	0	1	0	0
P_2	d_3^+	-300	0	1	0	0	1	0
$Z_j - C_j$	P_1	0	0	0	0	0	0	-1
	P_2	-700	2	1	0	0	0	0
	P_3	-350	1	1	0	0	0	-1

Source: Author

Table 4.3 Enter x_1 and Exchange d_2^+

Second iteration			Exchange variable from pivot row and pivot column					
			0	0	P_1	$2P_2$	P_2	P_3
C_B	y_B	x_B	d_2^+	x_2	d_1^-	d_2^-	d_3^-	d_1^+
P_1	d_1^+	-550	0	1	1	-1	0	-1
$2P_2$	x_1	200	1	0	0	-1	0	0
P_2	d_3^+	-300	0	1	0	0	1	0
$Z_j - C_j$	P_1	0	0	0	0	0	0	-1
	P_2	-1100	0	1	0	-2	0	0
	P_3	-550	0	1	0	-1	0	-1

Source: Author

Table 4.4 Enter x_2 And Exchange d_3^+

Third iteration			Exchange variable from pivot row and pivot column					
			$2P_2$	P_2	P_1	$2P_2$	P_2	P_3
C_B	y_B	x_B	d_2^+	d_3^+	d_1^-	d_2^-	d_3^-	d_1^+
P_1	d_1^+	-850	0	0	1	-1	1	-1
0	x_1	200	1	0	0	-1	0	0
0	x_2	300	0	1	0	0	-1	0
$Z_j - C_j$	P_1	0	0	0	0	0	0	-1
	P_2	-1400	0	0	0	-2	1	0
	P_3	-850	0	0	0	-1	1	-1

Source: Author

Table 4.5 Enter d_1^- and exchange d_1^+

Final result			Exchange variable from pivot row and pivot column					
			$2P_2$	P_2	P_1	$2P_2$	P_2	P_3
C_B	y_B	x_B	d_2^+	d_3^+	d_1^+	d_2^-	d_3^-	d_1^+
P_1	d_1^+	850	0	0	1	1	-1	1
0	x_1	200	1	0	0	-1	0	0
0	x_2	300	0	1	0	0	-1	0
$Z_j - C_j$	P_1	0	0	0	0	0	0	-1
	P_2	-1400	0	0	0	-2	1	0
	P_3	-850	0	0	0	-1	1	-1

Source: Author

As per above table, Planning and Decorating department have achieved the profit goal of Rs 350 weekly by neglecting overtime which is ample to get the first, second goals, also minimum under-utilisation of daily hours for both departments is achieved.

The above result can also be represented by graphical presentation Figure 4.1, which makes data easy to understand.

Result and Discussion

In this paper, it is shown that by using the proposed methods with suggested modifications at stage (4) and stage (5), described better solution to the GP problems, occurring in real-life situations as compared to the existing method. Also correspondingly the time required in the solution process of the problems by proposed methods is less than those in other method.

GRAPHICAL VIEW

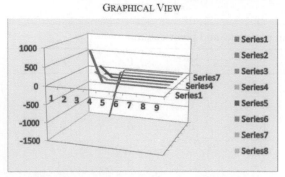

Figure 4.1 Optimal solution of Table 4.5
Source: Author

Table 4.6 Summary table of result

Sources	Number of variables	Number of goals	Number of iterations
[10]	3	3	5
Author	3	3	4

Source: Author

Conclusion

This paper presents the mathematical model, computational technique and implementation detail of propose dual simplex method which perform the best existing result. We have done an attempt to explain GPP by means of dual simplex method. This method is going to be a new alternative approach. Through this method, we are able to solve GPP with better solution which takes less iteration or sometimes equal iterations, save many efforts and time by neglecting calculations of net evaluation $Z_j - C_j$.

Acknowledgment

My sincere gratitude goes to the mysterious referees for their constructive comments and suggestions in anonymous ways in the improvement of this paper.

References

[1] Charnes A. and Cooper W. W.(1961) A book on management models and industrial applications of linear programming. New York, NY: John Wiley.

[2] Dantzig, G. B. (1948). Programming a linear structure. Report of the meeting in Madison, Econometrica. 9, 73–74.

[3] Lee, S. M. (1972). Goal programming for decision analysis. Auerbach, Philadelphia.

[4] Schniederjans, M. J and Kwak, N. K. (1982). An alternative solution method for goal programming problems: a tutorial. Journal of Operation Research Society. 33, 247–52.

[5] Olson, D. L. (1984). Comparison of four goal programming algorithm. Journal of the Operational Research Society. 35, 347–54.

[6] Lemke, Y C. E. (1953). The Dual method of solving the linear programming problem. Carnegie institute of Technology, Department of Mathematics. Technical Report 29.

[7] Koberstein, A. (2005). The dual simplex method, techniques for a fast and stable implementation. Dissertation report, Paderborn.

[8] Khobragade N. W.(2012). Alternative approach to wolfe's modified simplex method for quadratic programming problems. International Journal of Latest Trends Mathematics. 2(1), 1–18.

[9] Ghadle, K. P. and Pawar, T. S. (2014). Game theory problems by an alternative simplex method. International Journal of Research in Engineering and Technology. 3(5), 900–05.

[10] Safitri, E., Saleh, H. and Gamal, M. D. H. (2015). Solving Goal programming using dual simplex method. Bulletin of Mathematics. 7(15), 1–9.

[11] Moorthy, S. N. and Kalyani, S. (2015). The intelligence of dual simplex method to solve linear fractional fuzzy transportation problem. Computational Intelligence and Neuroscience. 1–7.

[12] Lokhande, K., Khot, P. G., and Khobragade N. W. (2017). Optimum solution of quadratic programming problem bywolfe's modified simplex method. International Journal of Latest Technology in Engineering, Management and AppliedScience. 6(3), 11–19.

[13] Goli, M. and Nasseri, S. H. (2020). Extension of duality results and a dual simplex method forlinear programming problems with intuitionistic fuzzyvariables. Fuzzy Information and Engineering. 12(3), 392–411.

Chapter 5

Experimental investigation on the use of new generation waste material in concrete for M-50 Grade

Pawan K. Hinge[1] and Tushar G. Shende[2]

Assistant Professor, Department of Civil Engineering, Yeshwantrao Chavan College of Engineering, Nagpur, India

Associate Professor, Department of Civil Engineering, Raisoni Centre of research and Innovation, G. H. Raisoni University, Amravati, Maharashtra, India

Abstract

The journey of concrete began around 600 BC, when the ancient Romans began widely using concrete for construction by mixing volcanic ash, lime, and seawater. Centuries later, it was in 1824 that Joseph Aspdin invented Portland cement by burning finely crushed chalk and clay until the carbon dioxide was eliminated. Between 1850 and 1880, Francois Coignet used Portland cement extensively for the first time in the construction of homes in England and France. Since then, though there have been vast innovative leaps, improvements and inventions in the material used in construction, the construction industry is still heavily reliant on Portland cement and aggregates. This excessive use and mass dependence on concrete is now a concern for the construction industry as the natural resources used in concrete like sand and gravel are slowly being depleted. In addition, the environmental concerns relating to CO_2 emissions from concrete have forced the industry to start looking for alternatives for the ingredients used in concrete.

On the other hand, electronic waste, also known as electronic waste, is considered one of the harmful types of electronic waste in today's new age world. Drinking water supplies are becoming polluted, and the environment and ecosystems are being impacted all around the world because of the unceasing increase of these technological wastes. Both current environmental issues can be mitigated to some extent by experimenting with and utilising electronic waste materials in the construction industry. Along with traditional aggregate alternatives such as fly ash and rice husk ash, the industry is now experimenting with E-waste to determine if it can be a viable material for improving concrete strength and longevity.

The primary objective of this research is to better understand the behavior of M-50 grade high strength concrete when coarse aggregate is replaced by E-waste plastics in staggered percentages. The strength of concrete containing E-waste plastic was tested at different time intervals after coarse aggregates were replaced with varying percentages of E plastic waste by the volume range from 0-30%. When compared to normal concrete, concrete with E-waste had comparable compressive, flexural, and tensile strength for 28 days up to a specified percentage of replacement. When the E-waste plastic component percentage is considerably high, however, there is a decline in strength after a certain point.

Introduction

The massive leap that technology has taken over the past few decades have proven to be both a boon and a curse for our world. This rapidly advancing society is demanding and causing a massive growth in electronic waste creation. According to the global E-waste monitor, India is one of the countries which is leading in E-waste generation. People are being pushed to acquire new technology as it becomes available, and old devices are being discarded as they become obsolete. This trend, combined with a lack of awareness about safe electronic item disposal and a lack of infrastructure to manage these massive amounts of E-waste, has resulted in most of the E-waste being dumped directly into the ground, resulting in hazardous solid waste. The chemical "lead" present in E-waste is almost certain to impair human health which has researchers all over the world concerned about the management, disposal, and handling of E-plastic waste. Research activities are now being focused on adopting numerous strategies to manage E-waste plastics, one of which is utilising this E-waste plastic in the construction industry. This serves a greater purpose as today, natural resources utilised in construction materials are depleting at an alarming rate, necessitating the search for a replacement.

During 2007, the Manufacturers' Association for Information Technology (MAIT) in India conducted an e-waste inventory based on three products: mobile phones, laptops, and TV. In India, the total amount of e-waste generated in year 2007 was 3, 32,979 (MT) (computer: 56324MT, mobile phones: 1655 MT, and televisions: 275000 MT). According to this estimate, the volume of E-waste reached about 0.7 million MT in year 2015, and is expected to reach 2 million MT by 2025 (Figure 5.1) [14].

An experiment with E-waste as a fine aggregate replacement was done [7]. According to their research, the CS, FS, TS of concrete reduced when check with a control mix. Lakshmi and Nagan conducted research on the partial replacement of CA with E-waste particles in proportions range from 0-3% [8]. According to research, when the E-plastic component was increased by 20%, the strength fell. Shamili et al. gave a comprehensive overview of E-waste composition, preparation, characteristics, and classification [6].

DOI: 10.1201/9781003450252-5

Arora and Dave (2013) examined the strength of concrete when fine aggregate was replaced with E-waste and plastic waste in percentages of 0%, 2%, and 4%. Their research revealed that the strength of concrete rose by 5% while the cost of concrete manufacture decreased by 7% [9].

The objective of this experiment is to determine if using E-waste plastic as a partial replacement for coarse aggregate in M-50 grade concrete with cement OPC 43 grade complying to IS 8112 in the proportions of 0 to 30% is feasible. On the age of 7,14,28 days of curing, compression strength is evaluated and compared, while tensile strength and flexural strength are compared on the 28th, 56th and 90th day of curing. The purpose of the experimental work was to use E-waste components as a partial replacement for coarse aggregate. The results from the study revealed that mix types exhibit good compressive strength in concrete with 15% replacement in M50 grade concrete. Crushed E-waste plastic material could be used as an alternative material in the construction industry, thereby lowering the cost of concrete construction and manufacturing while also helping with the environmental concerns with respect to concrete.

Materials

1.1. E-plastic waste: Every day, tons of e-waste is generated across the country and around the world. The E-plastic waste used in this study comes from discarded electronic equipment, household electronic goods, etc. After being cut into 10 mm pieces, the discarded and crushed circuit boards and chips are used as a replacement for coarse aggregate in varied amounts.

The classification of the electronic waste material used in this investigation are listed in Table 5.1, and Figure 5.2 depicts the E-waste material collected for this study.

1.2. Cement: The tests were conducted on OPC of 53 grade and all properties of cement were examined according to IS 2386 (PART-1) 1963. Table 5.2 lists the properties of cement that were examined and used in the experimental study.

1.3. Fine aggregate: The sieve size is used for test is 4.75 mm for fine aggregates, removing undesirable stones and impurities. Fine aggregate is aggregate that has passed through a 4.75 mm sieve. IS-2386 (PART-1) 1963 I Is used to characterize fine aggregate properties. Table 5.3 depicts the Fineness Modulus, bulking, Water Absorption, and Specific Gravity values of Fine Aggregate.

1.4. Course aggregate: For the experiment, coarse aggregates with a size of 20 mm were used. The investigation was carried out in accordance with IS 2386 (PART-1) 1963 criteria. Table 5.4 shows the properties of coarse Aggregate, including water absorption, crushing value, impact value and specific gravity.

Figure 5.1 Growth of E-waste in India
Source: Author

Table 5.1 Classification of E-waste material

Combination no.	P.C.B. (gm)	Kit material. (gm)	Steel. (gm)	Plastic. (gm)	Other combined materials	Total wt.
1st	1360 gm	439 gm	78 gm	105 gm	18 gm	2000 gm
2nd	1540 gm	390 gm	30 gm	33 gm	7 gm	2000 gm
3rd	1460 gm	408 gm	62 gm	60 gm	10 gm	2000 gm
4th	1520 gm	410 gm	42 gm	22 gm	6 gm	2000 gm
Total (%)	73.5%	20.5875%	2.65%	2.755%	0.512%	100%

Source: Author

Figure 5.2 Crushed E-waste material collected for replacement
Source: Author

Table 5.2 Test on Cement

Sr. No.	Characteristics	Value
1	Consistency	31%
2	Initial setting time	87 min
3	Final setting time	350 min
4	Fineness	94%
5	Soundness	2 mm

Source: Author

Table 5.3 Test on Aggregates

Sr. No.	Characteristics	Value
1	FM of fine aggregate	4.96
2	Bulking (%)	3
3	Water absorption (%)	1.2
4	Specific gravity	2.74

Source: Author

Methodology

2.1 Concrete design: (IS-10262-2019)

Material (kg/m30)	0%	5%	10%	15%	20%	25%	30%
Cement	427	427	427	427	427	427	427
Water	192	192	192	192	192	192	192
Fine aggregate	701	701	701	701	701	701	701
Coarse aggregate 20 mm	572	572	572	572	572	572	572
Coarse aggregate 10 mm	572	535	516	480	461	425	406
E waste	0	12	36	48	72	84	108
Water-cement ratio	0.5	0.5	0.5	0.5	0.5	0.5	0.5

Source: Author

2.2. Replacement: Course aggregate are replaced in percentages of 0, 5, 10, 15, 20, 25, and 30% by E-waste material which has undergone various tests as per Table 5.5.

2.3. Casting: Cubes of dimensions 15 × 15 × 15 cm were cast to test the compressive strength of the various mixes of concrete at 7th, 14th and 28th days. Similarly, 15 × 30 cm cylinders are cast to test the tensile strength of concrete, and 10 × 10 × 50 cm beams are cast to test the flexural strength of concrete.

Testing for tensile strength and flexural strength was done at 28, 56 and 90 days. Figure 5.5 depicts the casting of moulds that will be cured and evaluated.

2.4. Curing⊗IS: 516): Cubes, cylinders, and beams cast were cured in water tanks under appropriate conditions and tested at designated days of curing. Figure 5.6 shows the curing process for the samples.

2.5. Testing of material: Tests were conducted on E-waste, as per respective IS codes. Table 5.5 represents respective data for the same.

Tests conducted on materials:

Slump cone method (IS: 1199)

The workability of concrete at varying percentages of replacement was tested using SCT. Compressive strength (CS), FST, and STT were used to determine the strength of the concrete after 7, 14, and 28 days of curing when partial replacement of coarse aggregate by E-waste material was in the proportions of 0, 5, 10, 15, 20, 25, and 30 percentage. The slump values for M-50 grade concrete are tested when CA is replaced

Table 5.4 Properties of Course Aggregate

Sr. No.	Characteristics	Value
1	FM of course aggregate	8.69
2	Specific gravity of course aggregate	2.74
3	Impact value of course aggregate	10.52%
4	Crushing value of course aggregate	16.90%
5	Abrasion value of course aggregate	16.76%
6	Water absorption of course aggregate	Nil

Source: Author

Table 5.5 Properties of E-waste

Sr. No.	Characteristics	Value
1	Specific gravity	1.77
2	Impact value	12.03%
3	Crushing value	17.74%
4	Abrasion value	2.258
5	Water absorption	Nil

Source: Author

Figure 5.3 Casting of Cubes
Source: Author

Figure 5.4 Casting of Cylinder
Source: Author

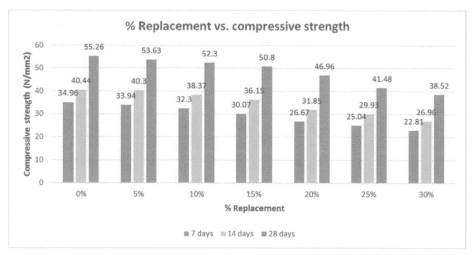

Figure 5.5 CS test results
Source: Author

with E-waste in proportions of 0%, 5%, 10%, 15%, 20%, 25%, and 30%. Figure 5.7 depicts the slump cone test in action with results noted down in Table 5.6.

Results and Discussion

The different tests conducted to find the different properties of concrete after partially substituting CA with E-waste plastics of varied percentages, namely 0%, 5%, 10%, 15%, 20%, 25% and 30%, and the acquired results are shown below.

Test result data for CS test: The test CS and % replacement results of 7, 14, 28 days are shown in Figure 5.5., which shows upto 15% replacement getting positive results in compression at the age of 28 days curing after 15% replacement results goes on decreasing.

Figure 5.5 shows, the early age strength goes on increasing upto 14 days after that rate of gain of strength is slower.

The test CS vs age in days results of 7, 14, and 28 days are shown in Figure 5.6.

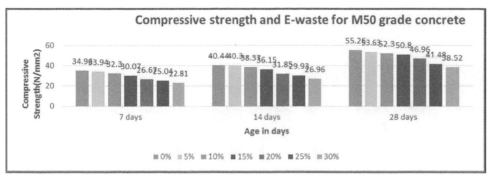

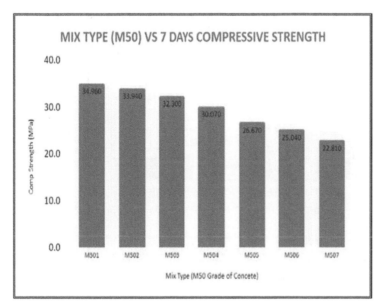

Figure 5.6 Compressive strength results at various days of curing
Source: Author

Figure 5.7 7 days curing
Source: Author

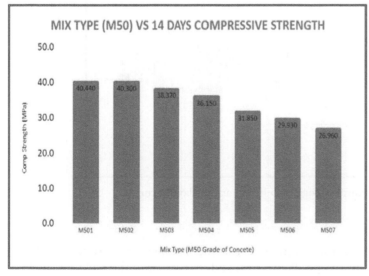

Figure 5.8 14 days curing
Source: Author

The test CS vs days curing results of 7, 14, and 28 days are shown in Figure 5.7, 5.8, 5.9, as per the results, shown below 7 days curing gain the strength upto 68% for control mix and for 30% replacement it is 44%. Similarly for 14 days and 28 days the curing strength continues to decrease.

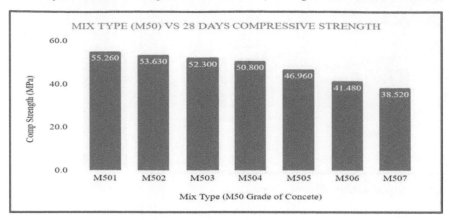

Figure 5.9 28 days curing
Source: Author

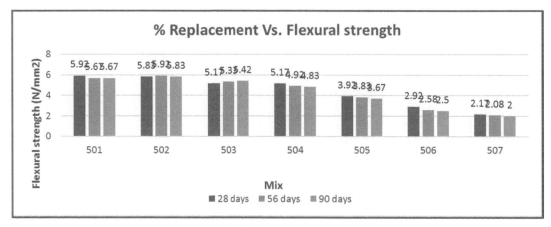

Figure 5.10 Graphical representation of flexural strength test results
Source: Author

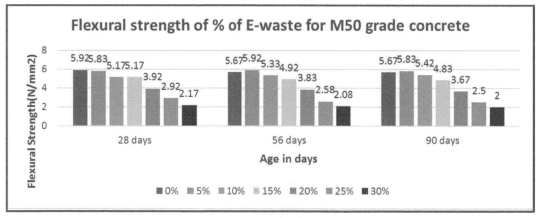

Figure 5.11 Graphical representation of Flexural strength results at various days of curing
Source: Author

Test result data for flexural strength test: The test flexural strength and % replacement results of 28, 56, 90 days are shown in Figure 5.10 in flexural test as per IS code strength is same upto the replacement 15% and after this it is decrease with an increase in partial replacement.

The test flexural strength Vs Age in days results of 28, 56, 90 days are shown in Figure 5.11 as per the results, the shown in below at the early age flexural strength upto 28 days after replacement of 15% it is decrease by 12.67%.

The test flexural strength Vs days curing results of 28, 56 and 90 are shown in Figure 5.12, 5.13, 5.14 as per the results, the shown in below at the early age flexural strength upto 56 and 90 days after replacement of 15% it is decrease by 13% and 14%.

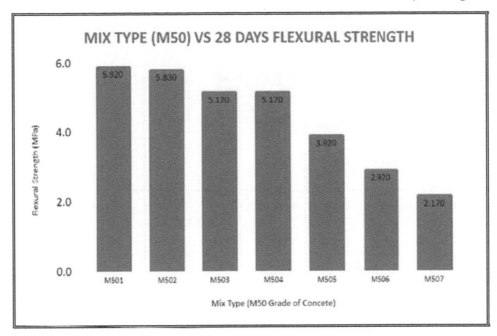

Figure 5.12 FT results for 28 days curing
Source: Author

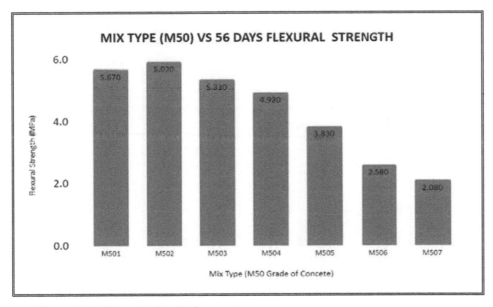

Figure 5.13 FT results for 56 days curing
Source: Author

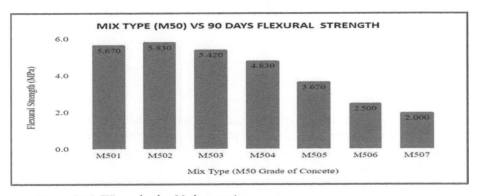

Figure 5.14 FT results for 90 days curing
Source: Author

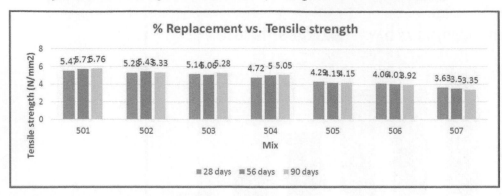

Figure 5.15 Graphical representation of tensile strength test results
Source: Author

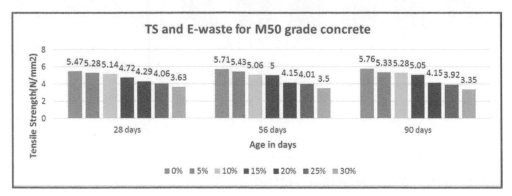

Figure 5.16 Graphical representation of split tensile strength results at various days of curing
Source: Author

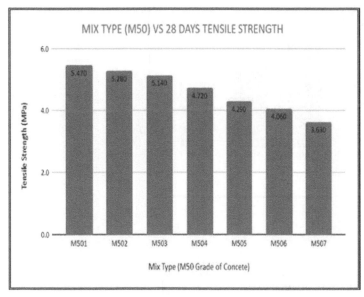

Figure 5.17 Tensile strength test results for 28 days
Source: Author

Split tensile test

Test result data for tensile strength test: The test tensile strength and % replacement results of 28, 56, 90 days are shown in Figure. 5.15. The test tensile strength and % replacement results of 28, 56, 90 days are shown in Figure. 5.10 in tensile test as per IS code strength is same upto the replacement 15% and after this it is decrease with an increase in partial replacement.

The test tensile strength vs Age in days results of 28, 56, 90 days are shown in Figure 5.16 as per the results, the shown in below at the early age tensile strength upto 28 days after replacement of 15% it is decrease by 13.71% at the age of 28 days curing.

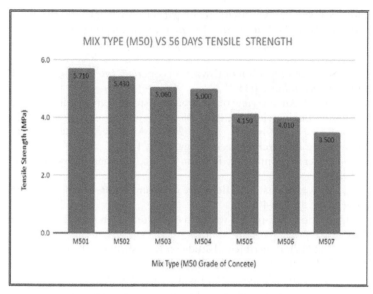

Figure 5.18 Tensile strength test results for 56 days
Source: Author

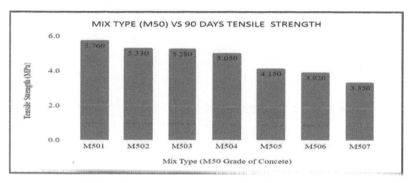

Figure 5.19 Tensile strength test results for 56 days
Source: Author

The test tensile strength vs days curing results of 28, 56 and 90 are shown in Figure 5.17, 5.18, 5.19 as per the results, the shown in below at the early age flexural strength upto 56 and 90 days after replacement of 15% it is decrease by 12% and 13%.

Conclusion

Experiments on M50 grade concrete using E-waste as a partial replacement for coarse aggregate showed that at the age of 28 days for curing period, the concrete with 15% E-waste replacement had compressive strength (CS) nearly equal to that of concrete without any replacement. Thus, 15% of E-waste can be used as a partial replacement for coarse aggregate. At the age of 28 days for curing period, the concrete with 15% E-waste replacement had flexural strength of 5.17 N/mm² and tensile strength of 4.72 N/mm², respectively.

E-waste reuse and recycling can potentially help in reducing E-waste and solid waste while also helping to conserve the environment to some extent. The utilization of e-waste in construction sites can lower construction costs, and additional research in this field around the world can yield greater outcomes.

References

[1] Takahashi, H., Satomi, T., and Bui, K. (2017). Improvement of mechanical properties of recycled aggregate concrete based on anew combination method between recycled aggregate and natural aggregate. Construction and Building Materials. 148.

[2] Kurama, C., Weldon, D., Rosa, A., Davis, A., and Mcginnis, J. (2017). Strength and stiffness of concrete with recycled concrete aggregates. Construction and Building Materials. 5, 25–30.

[3] Balasubramanian, B., Krishna, G. V. T. G., and Saraswathy, V. (2016). Investigation on partial replacement of coarse aggregate using E-waste in concrete ISSN 0974-5904. International Journal of Engineering Science. 9(3).

[4] Borthakur, A. and Singh, P. (2012). Electronic waste in India: problems and policies. International Journal of Environmental Sciences. 3(1).

[5] Siddique,S., Shakil, S. andSiddiqui, M. S. (2015). Scope of utilization of E-waste in concrete. *International Journal of Advanced Research in Science, Engineering and Technology.* 4(1)

[6] Shamili, S. R., Natarajan, C., and Karthikeyan, J. (2017) .An overview of electronic waste as aggregate in concrete. International Journal of Structural and Construction Engineering. 11, 1444-48.

[7] Alagusankareshwari, K., kumar, S. S., Vignesh, K. B., and Niyas, K. A. H. (2016). An experimental study on E-waste concrete. International Journal of Engineering Research & Technology. 9 (2), 1–5.

[8] Lakshmi, R. and Nagan, S. (2010). Studies on concrete containing E-waste plastic, International Journal of Environmental Science and Technology. 1, 270-281.

[9] Arora, A., and Dave, U. (2013) Utilization of E-waste and plastic bottle waste in concrete. *International Journal of Students' Research in Technology & Management..* 1, 398-406.

[10] Soni, A. S. and Sutar, D., and Patel, P. (2016). Utilization of E-plastic waste in concrete. International Journal of Engineering Research & Technology. 5, 594-601.

[11] Manikandan, M., Prakash, A., and Manikandan, P. (2017). Experimental study on E-waste concrete and comparing with conventional concrete. Journal of Industrial Pollution Control. 33(S3), 1490-95.

[12] Mohd, S. and Kaushal, V. (2018). E-waste management in India: current practices and challenges.

[13] Fu, J., Zhang, H. Zhang, A., and Jiang, G. (2018). E-waste recycling in China. Environmental Science & Technology. 52(12), 6727–28.

[14] Report on E-waste Inventorisation in India, MAIT-GTZ Study, 2007.

[15] Ahirwar, S., Malviya, P., Patidar, V., and Singh, V. K.(2016). An experimental study on concrete by using E-waste as partial replacement for coarse aggregate. International Journal of Engineering, Science and Technology. 3(4).

Chapter 6

An overview of post-processing techniques for ss316l processed by direct metal laser sintering

Purushottam Balaso Pawar[a] and Swanand G Kulkarni[b]

[1]SKNs Sinhgad College of Engineering Pandharpur, India

[2]Department of Mechanical Engineering, SKNs Sinhgad College of Engineering Pandharpur, India

Abstract

Additive manufacturing (AM) is preferred over conventional machining due to its ability to process various materials and fabricate complex geometries with ease directly from a prepared CAD model. AM processed products are currently being used in aviation, marine, and the biomedical field. Direct metal laser sintering (DMLS) is the most used AM technique due to its ability to process material ranging from polymers to metal powders by sintering mechanism. In spite of optimal parameter selection, certain defects are observed in DMLS processed products. Some of the defects recorded in research include porosity, residual stresses, stair stepping defect, balling defect, poor surface finish, cracks, etc., leading to reduced strength of the component, reduced corrosion resistance, and fatigue strength. These defects are inevitable, though their effect on product quality can be minimised by post-processing. A reliable analytical model is not available that can fully explain and predict this process. A reliable analytical model is not available that can fully explain and predict this process The purpose of this study is to provide an overview of various post-processing techniques used for improving the quality of DMLS processed products, along with their effect on product quality. The post-processing techniques adopted for enhancing mechanical and tribological properties include shot peening, polishing, laser shock peening, vibratory bowl abrasion, and conventional machining methods. Post-processing methods have successfully reduced surface roughness and increased corrosion and wear resistance, which has resulted in the elimination of defects. A 99.99% reduction in porosity and a 95% reduction in surface roughness has been observed compared to printed samples. Changing the microstructure orientation with heat and surface treatment also improves mechanical and surface properties. The specific process can be selected based on the product improvement needed. An overview such as this can be useful when selecting a particular process based on the user's requirements.

Keywords: DMLS, ss316l, defects in AM, post processing, property enhancement

Introduction

Additive manufacturing (AM) is a layer-by-layer manufacturing technique for making a solid object. The geometry of a component is used to generate a 3D CAD model, which is then used to directly print the prototype based on the CAD model. The technology has proven to be the best option for dealing with complex geometries, intricate parts, and product development with less tooling and waste [1]. It has also helped in reducing overall cycle time. Now days AM finding its applications in aerospace, marine, automobile, and the biomedical field. Technology has developed to the point that it can process a wide variety of materials, such as polymers and metal powders. Additive manufacturing methods mainly used are categorized as laser powder bed fusion (LPBF), direct energy deposition, binder jetting, VAT polymerisation, Sheet Lamination, etc. Laser Powder Bed Fusion is mostly the preferred technique for processing materials. Major LPBF techniques include: 1. Direct metal laser sintering (DMLS), 2. selective laser melting (SLM), 3. selective laser sintering (SLS), 4. electron beam melting (EBM) [2]. The SLS and DMLS processes operate similarly in concept, but DMLS sinters pre-alloyed metal powders, rather than polymers or coated metal powders. Over the past few years, DMLS has proved its capability to process products with varying alloying contents, and it has been used to develop products commercially.

The strength and other mechanical properties of the DMLS product are comparable to those of conventional machining methods. Due to the lack of special tooling required, DMLS is suitable for short production runs, unlike castings. Materials processed by DMLS include stainless steel (304L, 316L), titanium, CoCrMo, alloys of aluminium. The SS316L material is widely used in a variety of engineering applications that require high strength and corrosion resistance, such as food processing, medical, aerospace, and many others. Furthermore, surgical tools and medical implants can be made with 316L because it is biocompatible. The quality of DMLS product depends on process parameters involved in process viz. Laser power, hatch spacing, layer thickness, scan speed, build orientation, laser spot diameter, etc. There are some parameters related to material also viz. powder morphology, powder composition, etc. Despite optimal process parameters, research has demonstrated that these products still suffer from defects related to tribology, strength of the component, or microstructure. For minimising these defects' detrimental effects and enhancing the product's properties, post-processing is required. Some of the defects observed

[a]purushottampawar_07@rediffmail.com, [b]swanand.kulkarni@sknscoe.ac.in

DOI: 10.1201/9781003450252-6

in DMLS products include porosity, residual stresses, balling defect, stair stepping defect, microstructural defect, etc. [3].

In current study an overview has been taken related to DMLS process defects, and posts processing techniques used for curing these defects. According to the study on post-heat treatment of AM components, T6 heat treatment enhances their fatigue resistance. Moreover, heat treatments can cause a reduction in material hardness as well as the formation of residual tensile stresses. The as-built AM parts are then subjected to processes such as HIP, shot peening, and FSP to improve their surface and mechanical properties. These techniques need to be further investigated in order to determine how they affect the porosity, residual stress, fatigue life, or any other mechanical properties of parts built by AM. Thus, heat treatment and other post-processing techniques must be combined in the right way to improve AM components. The particular material, AM method, and application must also be understood along with the suitable post-processing technique. The selection of the optimal post-processing strategy requires continuing research.

DMLS Process And Process Parameters

DMLS is a form of AM or rapid prototyping (RP), where metal powder and a high-power laser are used to produce usable parts. A high-power laser beam is directed over a powder layer and sinters the powder into a solid. As with other laser machining processes such as laser cutting and laser welding, DMLS systems used either CO_2 or Nd: YAG lasers. The important parts of DMLS process are dispenser chamber, building chamber, collector chamber, computer with process software, Optic lens and high-power laser. In accordance with the CAD model, a high-powered laser beam moved over a thin powder layer. The entire process is carried out in closed chamber with argon gas environment. With powder from the dispenser platform, successive powder layers are spread by the recoater blade. During the recoater operation, the dispenser moves up so that sufficient powder is taken by the recoater to spread a new layer over the layer that has already been sintered. Once a powder layer has been sintered excess powder accumulates in collector platform. This process is repeated until final geometry is formed [4]. Schematic constructional diagram of DMLS process has been represented in Figure 6.1. It is possible to produce near-net shape products with an almost 90% density using this process. The steps involved in the DMLS process have been summarized in Figure 6.2. The final quality of DMLS processed product depends on process parameters involved in the process. Major parameters influencing product quality include:

a.　Laser power (P), W
b.　Layer thickness (t), mm
c.　Scan speed (v), mm/sec
d.　Hatch spacing (h), mm
e.　Laser spot diameter (d), mm
f.　Scan pattern
g.　Build direction

And parameters related to material influencing product quality includes

a.　Powder morphology
b.　Powder material alloying

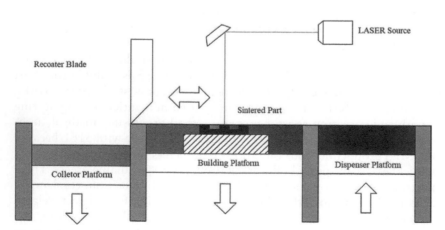

Figure 6.1 Direct metal laser sintering (DMLS) process
Source: Author

While AM is still in its infancy, it exhibits a number of process problems, which should be understood when designing a part for AM, specifically under DMLS. One major drawback associated with DMLS process is slow build rate. There is still no satisfactory correlation between printing parameters and the resulting microstructure and surface roughness. Microstructure and mechanical properties need to be evaluated in relation to varying print parameters so that parameters can be validated and specified for industrial adoption [5].

Defects in DMLS Process

There have been several challenges encountered during the operation of the DMLS process. In manufacturing, defects can occur if process parameters are not predicted and chosen appropriately, powders spread randomly, etc. Manufacturing uncertainty also contributes to defects. Compared to conventional manufacturing processes, AM processes are plagued by problems such as poor surface finish, unwanted microstructure phases, defects, and wear tracks, as well as reduced corrosion resistance and fatigue life. Porosity, balling phenomenon, delamination, undesired microstructure formation, residual stresses, and distortions are common defects found during direct metal laser sintering. There are many defects in metal-based additive manufacturing processes related to solidification, such as porosity and hot cracking. These defects are mainly associated with DMLS product irrespective of geometry. While stair stepping defect, surface roughness have been found more predominant with inclined surfaces. A printed part's mechanical and material properties are very difficult to estimate due to non-equilibrium phase changes caused by irregular and repetitive thermal cycles during printing. Figure 6.3 summarises major defects observed in DMLS that influence mechanical and tribological properties [6].

Multiple factors influence surface roughness and surface morphology of manufactured parts, including scan strategy and laser specifications. It is possible to minimise defects during the manufacturing process by using high-quality powder and optimising the process parameters, such as layer thickness, energy input, build direction, scanning strategy, hatch spacing, and scanning speed. In addition to the energy density, the powder morphology and scan speed also affect roughness of the surface [7].

Several factors may contribute to porous material formation, including process parameters, if impurities are present in the solid material, its low laser energy absorption, and wettability of the material. A variable process of heating, melting, and cooling occurs in the printing chamber and material during a printing process, generating residual stresses. A support structure that is higher in temperature than the substrate is provided to reduce residual stress. As a result of different scanning strategies like meandering and chessboard patterns, parts develop low residual stresses due to uniform temperature distribution. Printed parts

Figure 6.2 Steps in product development with DMLS
Source: Author

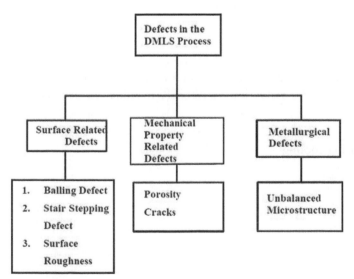

Figure 6.3 Defects in the DMLS process
Source: Author

are susceptible to crack propagation, distortion, and layer delamination due to residual stresses. Printed parts with a long length will experience residual stress that can result in warping and delamination. Figure 6.4 illustrates the mechanism of residual stress formation [8].

Post Processing Techniques

While metal AM has several benefits, the process still requires post processing before the component can be used. For the product to achieve optimal material performance and visual appeal, it must undergo a number of treatments and finishes. In many cases, a support structure is needed to print the product, so it needs to be removed before it can undergo heat treatment or finishing. An initial heat treatment is performed to relieve the metal of accumulated stresses [9]. During the printing process, the material is rapidly heated and cooled, which creates residual stress. Additional heat treatment procedures relieve this residual stress. A heat treatment improves the strength, ductility, and hardness of a material. It is necessary to apply surface treatments to solve problems caused by powder accumulation (partially melted powder) and stair casing or layering defects [10].

Shot peening

As a result of shot peening, components are induced to experience compressive residual stress, thereby extending their fatigue life. Researchers have found that printed products exhibit higher surface roughness values, which are not recommended for many applications. As a result of shot peening, the surface morphology improvements, which results in reduced roughness parameters and the emergence of trough profiles. By shot peening, surface layers also become harder without the formation of any secondary phases, such as martensitic phases from austenite. There are several advantages to shot peening, including its cost-effectiveness, ease of handling, and viability in the industry [11].

Electrochemical polishing

During EP treatment, a metallic piece is polished by removing material by electrochemical means. With EP, irregularities at large scales can be effectively reduced, or surfaces can be brightened. In comparison to mechanical finishing, EP allows achieving a surface free of stresses and deformations. Deburring, polishing, and passivating metal parts are all electrochemical processes that involve electro polishing. A bright, uniform finish can be achieved with this process instead of abrasive fine polishing for microstructural preparation in DMLS projects that do not rely on tolerances. A surface's preparation determines how successful electro polishing will be [12].

Abrasive flow machining (extrude hone) polishing

To achieve better surface integrity and to reduce balling effect and powder adhesion, the DMLS product can be polished using AFM polishing. DMLS projects that do not require tolerance and need a more uniform surface roughness can benefit from this inexpensive option. In particular, the process is useful for difficult-to-reach inner passageways, bends, cavities, and edges. In AFM, an abrasive-laden semisolid media flows through or across a workpiece to remove small amounts of material.

Friction stir processing (FSP)

This technique has been used to improve the surface properties of AM parts by a number of researchers. Compared to as-built, friction stir processing (FSP) contributes to grain refinement, porosity reduction, micro-hardness improvement, and residual stress reduction. With FSP, metallic surfaces can be

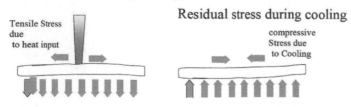

Figure 6.4 Residual stress developed due to thermal cycling
Source: Author

Table 6.1 Surface roughness after different mechanical post processing [13].

Type of the post process	SR Ra in µm	SR Rq in µm
As printed	15.03 ± 0.06	23.04 ± 0.10
Grinded	0.96 ± 0.05	1.15 ± 0.04
Mechanical polished	0.12 ± 0.02	0.13 ± 0.01
Blasted with sand (25–50 µm)	4.88 ± 0.06	5.74 ± 0.08

Source: Author

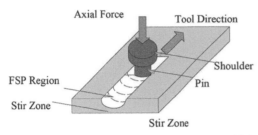

Figure 6.5 Schematic representation of friction stir processing
Source: Author

microstructurally modified by rotating tools with traverse speeds inserted into the work-pieces. Figure 6.5 illustrates the mechanism of FSP [14].

Micro machining process (MMP)

An item is treated with micro machining process (MMP) in a tank featuring a mechanical, physical, and chemical process, providing highly accurate selective surface finishing. MMP is used only on areas where a particular surface finish is required to achieve the desired result. This process is ideal for a large number of parts that require precision tolerance finishing, as well as parts that have intricate internal passages that cannot be reached in another way [15].

Vibratory bowl abrasion

It is achieved through vibrating at a constant speed in order to improve surface finish of the product. The abrasive material and the model are vibrated in a 'U'-shaped bowl at a constant speed during this process. Abrasion takes place as the media is re-circulated around the bowl. There is less aggression involved in this process than in earlier studies when tumbling caused damage to the models [16].

Laser shock peening (LSP)

The purpose of laser peening is usually to reduce micro defects and improve the quality of surfaces. In laser shock peening (LSP), the material is compressed perpendicular to the surface in order to expand laterally. A significant difference between LSP and shot peening was the level of surface roughness after LSP. In contrast to shot peening, LSP did not significantly change the surface roughness. When a laser beam is focused on a metallic surface for 30 ns, the heated zone reaches 10,000° C, creating plasma. Materials are subjected to pressure generated by shock waves. A critical parameter in shock peening is laser energy. Figure 6.6 shows an illustration of LSP [17].

Laser polishing (LP)

The Laser polishing (LP) process involves re-melting to alter the surface morphology without affecting the bulk behavior. Laser rapid melting and re-solidification improves pitting corrosion resistance by causing microstructural changes. It is during LP that laser energy is irradiated onto the material surface that morphological apexes quickly reach melting temperatures. Following the melt pool generation, gravity and surface tension cause the liquefied material to reorganise at the same level as shown in Figure 6.7.

Conventional machining methods (CMP)

The conventional machining methods (CMP) process enhances the surface quality and dimensional accuracy of manufactured parts. It is possible to reduce the surface roughness of a specimen by grinding and abrasion by removing burrs and pores within the provided specimen. It is commonly used to enhance

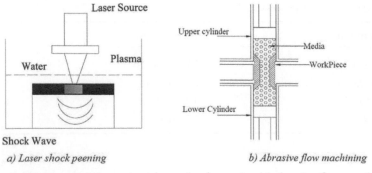

Figure 6.6 Schematic a) laser shock peening b) abrasive flow machining
Source: Author

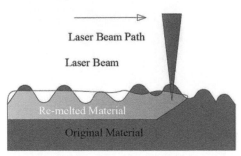

Figure 6.7 Laser polishing
Source: Author

surface characteristics such as surface roughness and skewness. Rolled surfaces have shown to have significant improvements in hardness and roughness.

Thermal processes

It is inevitable that as-built AM metals will show residual stresses, segregation, and non-equilibrium phases due to the highly localized heat input and the high thermal gradient of laser-based AM processes. An effective fatigue performance is believed to be achieved through stress relief heat treatment, which can decrease residual tensile stresses. The microstructure can be altered by heat treatment, resulting in increased fatigue crack resistance. It is possible to alter the mechanical characteristics of additively manufactured SS316L functional components through heat treatment [18-20].

Heat treatment

A study found that the yield strength of HIP + annealed samples decreased as a result of the better ductility compared to other samples. As a result of the combination of HIP and annealing, better properties are achieved compared with other heat treatment methods. Reducing residual stress and homogenizing microstructure are two benefits of heat treating print samples. In addition to altering the microstructure of the material, an increase in ductility may lead to a loss of strength if the material is treated at high temperatures. In comparison with printed samples, heat-treated samples have been found to have higher corrosion resistance and wear resistance due to the reduction in the ferrite phase [21]. Yield strength of a component depends on grain size as evident from Hall-Petch equation,

$$\sigma_y = \sigma_0 + \frac{k_y}{\sqrt{d}} \qquad\qquad 1$$

Where σ_y represents yield strength of material, σ_0 is material constant for starting stress of dislocation movement, k_y is strengthening coefficient and d is average grain size.

Hot isostatic pressing

The combination of high heat and pressure enhances part solidity and density, diminishes residual stress, and extends fatigue life by a great deal. Temperature and pressure are applied to a surface for a certain period of time during HIP. Metal parts are heated and pressed simultaneously (typically around 100 MPa), which collapses the pores and seals them up. Improvements in diffusion and better metallurgical bonding

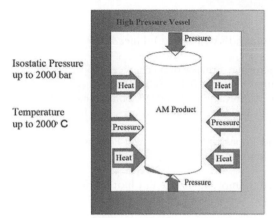

Figure 6.8 Schematic diagram of hip
Source: Author

Table 6.2 Comparison of as printed, annealing ht and hip+ annealed samples [22]

Mechanical property	AP (BD)	AP (TD)	AHT (BD)	AHT (TD)	HIP+AHT (BD)	HIP+AHT (TD)
Tensile strength, Mpa	579 ± 10	585 ± 20	582± 05	595 ± 10	592 ± 05	611 ± 05
Tensile strength, Mpa	439 ± 5	445 ± 20	365 ± 10	370 ± 10	257 ± 05	263 ± 05
Yield strength, Mpa	21 ± 2	21 ± 2	31 ± 03	29 ± 02	47 ± 03	48 ± 03

Source: Author

have led to a significant improvement in tensile strength after HIP. Process duration, pressure, and temperature are three crucial parameters in the HIP process. HIP can also homogenise microstructures and relieve residual stresses because both high pressure and high temperatures are simultaneously applied. In this process, the average temperature ranges from 1000 to 2000 degrees Celsius. Enhanced mechanical properties can be achieved when parts are treated with HIP because the porosity and un-melted material are reduced, as well as the microstructure is coarsened. The mechanical properties of AM products subjected to HIP and annealing were comparable to those of wrought products, as shown in Table 6.2.

Conclusions

Tooling costs are lower and more time-efficient with powder-based additive manufacturing (AM). A reliable analytical model is not available that can fully explain and predict Direct metal laser sintering (DMLS) process. In order to improve mechanical properties, we can control the contributing parameters that affect microstructure. There are some defects that occur during this process that result in weakened mechanical properties. It is due to this limitation that AM technology based on metal is not widely used, resulting in limited repeatability and precision [23-25]. In conclusion, the study can be summarised as follows:

- In view of the fact that as-built AM components produce many defects, it is imperative that the optimum processing parameters are selected and the right combination of post-processing techniques with heat treatment is used to ensure that the as-fabricated component's surface integrity and mechanical properties remain intact and won't fail in service.
- Energy density and the combination of machining parameters can also regulate the mechanical properties of a particular material product.
- The final step in the production process - or set of steps - is post-processing. Post-processing ensures that the final product meets its structural, material, and aesthetic requirements.
- By considering its effects on other mechanical properties, a particular heat treatment process should be suggested. Although heat treatment reduces porosity, it has also been reported to reduce strength in some cases.
- Research needs to be conducted to identify the right post-processing strategy and optimise its processing parameters for standardising AM.

Acknowledgment

Authors acknowledge support from all those who directly or indirectly extended their support in preparing this manuscript.

References

[1] Duda, T. and Raghavan, L. V. (2016). 3D metal printing technology. IFAC-Papers OnLine. 49(29), 103–10. doi: 10.1016/j.ifacol.2016.11.111.

[2] Lewandowski, J. J. and Seifi, M. (2016). Metal additive manufacturing: a review of mechanical properties. Annual Review of Materials Research. 46(14), 1–14. doi 10.1146/annurev-matsci-070115-032024.

[3] Sugavaneswaran, M., Jebaraj, A. V., Kumar, M. D. B., Lokesh, K., and Rajan, A. J. (2018). Enhancement of surface characteristics of direct metal laser sintered stainless steel 316L by shot peening. Surfaces and Interfaces. doi: 10.1016/j.surÞn.2018.04.010.

[4] Mahmood, M. A., Chioibasu, D., Rehman, U., Mihai, A. and Popescu, S. A.C. (2022). Post-processing techniques to enhance the quality of metallic parts produced by additive manufacturing. Metals. 12, 77. https://doi.org/10.3390/met12010077.

[5] Kumbhar, N. N. and Mulay, A. V. Post processing methods used to improve surface finish of products which are manufactured by additive manufacturing technologies: a review. Journal of The Institution of Engineers. doi: 10.1007/s40032-016-0340-z.

[6] Malekipour, E. and -Mounayri, H. E. Common defects and contributing parameters in powder bed fusion AM process and their classification for online monitoring and control: a review. International Journal of Advanced Manufacturing Technology. https://doi.org/10.1007/s00170-017-1172-6.

[7] Mower TM, Long MJ. Mechanical behavior of additive manufactured, powder-bed laser-fused materials. Materials Science and Engineering: A. 2016 Jan 10; 651: 198–213.

[8] Santa-aho, S., Kiviluoma, M., Jokiaho, T., Gundgire, T., Honkanen, M., Lindgren, M., and Vippola, M. (2021). Additive manufactured 316l stainless-steel samples: microstructure, residual stress and corrosion characteristics after post-processing. Metals. 11, 182. https://doi.org/10.3390/met11020182.

[9] Shiyas, K. A. and Ramanujam, R. (2021). A review on post processing techniques of additively manufactured metal parts for improving the material properties. Materials Today: Proceedings. 46 1429–36, https://doi.org/10.1016/j.matpr.2021.03.016.

[10] Duleba B, Greškovič F, Sikora JW. Materials and finishing methods of DMLS manufactured parts. Transfer inovácií. 2011; 21: 143–8.

[11] Beretta, S. and Romano, S. (2017). A comparison of fatigue strength sensitivity to defects for materials manufactured by AM or traditional processes. International Journal of Fatigue. 94, 178–191. http://dx.doi.org/10.1016/j.ijfatigue.2016.06.020.

[12] Kim, K. T. (2022). Mechanical performance of additively manufactured austenitic 316L stainless steel. Nuclear Engineering and Technology. 54, 244e254. https://doi.org/10.1016/j.net.2021.07.041.

[13] Löber, L., Flache, C., Petters, R., Kühn, U., and Eckert, J. (2013). Comparison of different post processing technologies for SLM generated 316l steel parts. Rapid Prototyping Journal. 19(3), 173–179. doi:10.1108/13552541311312166.

[14] Li, K. Liu, X. and Zhao, Y. (2019). Research status and prospect of friction stir processing technology, Coatings 2019, 9, 129; doi:10.3390/coatings9020129.

[15] Matthieu Rauch, Jean Yves Hascoet, (2022). A comparison of post-processing techniques for additive manufacturing. Procedia CIRP. 108, 442–7. doi: 10.1016/j.procir.2022.03.069.

[16] Eleonora Atzeni, et al. (2020). Performance assessment of vibro finishing technology for additively manufactured components. 13th CIRP Conference on Intelligent Computation in Manufacturing Engineering, CIRP ICME '19, Procedia CIRP. 88, 427–432. doi 10.1016/j.procir.2020.05.074.

[17] Gupta RK, Pant BK, Kain V, Kaul R, Bindra KS. Laser shock peening and its applications: a review. Lasers in Manufacturing and Materials Processing. 2019 Dec; 6(4): 424–63.

[18] Mohyla, P., Hajnys, J.,Gembalová, L., Zapletalová, A. and Krpec, P. (2022). Influence of heat treatment of steel AISI316L produced by the selective laser melting method on the properties of welded joint. Materials. 15, 1690. https://doi.org/10.3390/ma15051690.

[19] Tommasi, A., Maillol, N., Bertinetti, A., Penchev, P., Bajolet, J., Gili, F., Pullini, D., and Mataix, D. B. (2021). Influence of surface preparation and heat treatment on mechanical behavior of hybrid aluminum parts manufactured by a combination of laser powder bed fusion and conventional manufacturing processes. Metals. 11, 522. https://doi.org/10.3390/met11030522.

[20] Zeng Q, Gan K, Wang Y. Effect of heat treatment on microstructures and mechanical behaviors of 316L stainless steels synthesized by selective laser melting. Journal of Materials Engineering and Performance. 2021 Jan; 30: 409–22.

[21] Pala, S., Tiyyaguraa, H. R., Drstvenšek, I. and Kumar, C. S. (2016). The effect of post-processing and machining process parameters on properties of Stainless Steel PH1 product produced by direct metal laser sintering. Procedia Engineering. 149, 359–65. doi: 10.1016/j.proeng.2016.06.679.

[22] Chadha, K., Tian, Y., Spray, J. G., and Aranas, C. (2020). Jr, effect of annealing heat treatment on the microstructural evolution and mechanical properties of hot isostatic pressed 316l stainless steel fabricated by laser powder bed fusion. Metals. 10, 753. doi:10.3390/met10060753.

[23] Yuan, L. (2019). Solidification defects in additive manufactured materials. JOM. 71, 9. https://doi.org/10.1007/s11837-019-03662-x

[24] Gu, D. and Shen, Y. (2009). Balling phenomena in direct laser sintering of stainless steel powder: Metallurgical mechanisms and control methods. Materials and Design. 30, 2903–2910. doi:10.1016/j.matdes.2009.01.013.

[25] Anand, M. and Das, A .K. (2021). Issues in fabrication of 3D components through DMLS technique: A review. Optics & Laser Technology. 139, 106914. https://doi.org/10.1016/j.optlastec.2021.106914.

Chapter 7

Investigation of mechanical properties of nitrided-microblasted and ticn coated D3 tool steel

Santosh Bhaskar[a], Dhiraj Bhaskar[b], Amjad Shaikh[c], Pankaj Patil[d], Jalees Ahemad[e] and Mahesh Nagarkar[f]

Mechanical Engineering Department, Sanjivani College of Engineering, Kopargaon, Kopargaon, India

Abstract

Manufacturing of parts requires increasing productivity and higher productivity can be achieved by improvements in properties of tools. The properties of tools can be improved by the application of thin coatings which can be extended to numerous sectors of industry in order to improve the life of tools from failure by wear, corrosion etc. To improve the surface hardness, decrease friction, and improve wear resistance of tool steel substrates, surface engineering techniques such as surface treatment, coating, and surface modification are applied which is very useful in various industries for improvement in life of tool in various industries. However, in the forming industry, it is still more of an exception than a norm to find surface treated forming tools. The aim of this work is to investigate the performance of D3 tool steel with plasma nitriding, micro-blasting and different coatings.

The influence of pretreatment such as micro-blasting, nitriding and TiCN coating on the behavior of AISI D3 tool steels are experimentally investigated. The specimen of D3 material were nitride with plasma nitriding process at 500^0C. The mico-blasting process is applied on the specimen at pressure of 3 bar with Al_2O_3 particles. Also, the TiCN coating is applied on D3 steel with monolayer and multilayer at 450°C after the micro-blasting and nitriding process. A comparison between the mechanical properties such as yield strength, ultimate strength, and compressive strength were made. After conduction of tests, it is observed that due to various combinations of micro-blasting nitriding and TiCN coating, the mechanical properties of D3 tool steels are improved.

Keywords: Coating, forming tool, mechanical characterization, micro-blasting, nitriding

Introduction

Conditions of sliding contact could be applied to all forming tools, with the material being formed, and, hence, subjected to wear. It also possesses strong adhesion when it is sliding against itself and other metals [1]. Such processes necessitate pretreatment before being put to service. Surface modifications are generally employed to improve wear resistance of tool substrates by increasing surface hardness and minimizing adhesion (reduce friction) [2]. One of the strategies the industry has used to lessen wear and enhance life of the tool is the use of physical vapor deposition (PVD) technique [3]. PVD hard coatings provide engineering surfaces with high surface hardness and improved tribological properties. The use of PVD hard coatings on substrate materials, however, is a well-known fact that does not always result in the best tribological performance, if an appropriate pretreatment is not applied to the substrate material [4]. It is brought on by the substrate's plastic deformation, which could lead to the failure of the coating [1].

Forming processes have their own importance in mechanical industries where surface treated forming tools have not been used yet. The D3 steel has various applications such as blanking, stamping, and cold forming dies and punches for long runs, lamination dies, bending, forming, and seaming rolls. In this work, the forming tool application has been considered and D3 tool steel has been selected which is widely used for making forming dies.

In metal forming process, different types of operations are used to achieve a finished component. At one point in the manufacturing process, a toughened tool steel punch is utilised in the blanking operation to remove material from a blank. Throughout the entire process, the load on the punch and die is quite high due to the rapid impact and to support the applied loads in such applications, one needs a strong tool. Also tolerating tool surface wear is dependent on the tool having a hard surface. Throughout the entire process, heavy normal and shearing loads will be imparted to the surfaces of the punch and the die. Therefore, the tool needs to be able to endure several impacts and the ensuing increase in internal stresses [5]. The growth and characteristics of the hard coating are greatly impacted by the coating deposition, if done after nitriding.[6]. Surface hardness is significantly increased by the development of the hardened surface layer [7]. Moreover, nitriding results in the development of compressive stresses on the substrate's surface [8] reducing the difference between the stress environments in the coating and the steel substrate, as a result [9].

[a]bhaskarsantoshmech@sanjivani.org.in, [b]bhaskardhirajmech@sanjivani.org.in, [c]shaikhamjadmech@sanjivani.org.in, [d]patilpankajhmech@sanjivani.org.in, [e]jaleesahemadmech@sanjivani.org.in, [f]nagarkarmaheshmech@sanjivani.org.in

DOI: 10.1201/9781003450252-7

The punch moves downward with force, thus resulting in shearing of the blank material. These contact surfaces will be formed due to shear failure of the material, hence, the contact surfaces of tool and material are also important factors in forming process [10].

Generally, in metal forming processes, if we use untreated tools, there are different drawbacks such as, degradation of material and wear which occurs due to the frequently coming into contact with the tool surfaces and the untreated, softer punch material. A tribological pair's wear might differ based on the materials in contact, the speed, temperature, and loads being applied, as well as the surface conditions (roughness, area of contact etc) [11]. Machine parts become damaged due to material deterioration and wear, necessitating repairs and operational downtime. Wear and tear on materials can also have a significant impact on personal and occupational safety [12]. Najari et. al. [13] investigated the possibility of forming a chromium carbide layer on the surface of an AISI W1 cold work tool steel and found that increasing the temperature and immersion time of the TRD process increases the thickness and hardness of coatings, resulting in low friction coefficient values and improvement in the wear resistance up to six times compared to the untreated AISI W1 sample. Taheri et. al. [14] examined the effect of coating tungsten carbide drill bits with graphene in order to examine the effect of graphene on the wear, as well as the rate of penetration of the drilling bit. Study revealed slight increase in deformation, while two times increase in the maximum shear elastic strain which ultimately indicate that the bit's wear was significantly reduced after the coating. Dementyev and Ivanova [15], in their work discussed the results of the study of coated and uncoated drills. The study revealed strengthening of the corners of the drill and reduction of coefficient of chip friction during its moving along the tool grooves. Vereschaka et. al. [16] compared performance of uncoated tools with Ti-(Ti,Al) N coated AISI 321 steel and found better performance properties with the coated tools. Uhlmann et. al. [17] studied ultrasonic assisted drilling of cemented carbide and found reduction in cutting forces and found increase in productivity of the drilling process.

In this work, an attempt has been made to overcome these drawbacks with surface treatments such as micro-blasting, nitriding and coating which is applied on forming tool material D3 steel. Investigating how D3 tool steel performs post plasma nitriding and micro-blasting as pre-treatments with TiCN coating and effect of these processes on mechanical and tribological properties of the tool steel is the objective of this research.

Materials and Methodology

Material

The D3 tool is made of a high-carbon, high-chromium tool steel that is air hardened. It possesses outstanding dimensional stability and excellent compressive strength. It has chemical composition which is presented in Table 7.1.

The ends of the specimens were polished to a roughness of around 0.4 mm. Following surface treatments and materials have been used in investigation.

Micro-blasting

It is the operation of applying a stream of abrasive material such as Al_2O_3 with high force on a required surface under intense pressure to smooth out a rough surface, shape a surface, roughen a smooth surface or remove surface contaminants. It is crucial to take air pressure, powder flow rate, nozzle size, and angle of contact into account during the micro-blasting process in order to achieve the best outcomes [18]. In this

Table 7.1 Chemical Composition of D3 Steel [4]

Element	Content (%)	Actual value
C	2.00-2.35	2.09
Mn	0.60	0.39
Si	0.60	0.35
Cr	11.00-13.50	11.22
Ni	0.30	0.018
W	1.00	0.49
V	1.00	0.60
P	0.03	0.019
S	0.03	0.016

Source: Author

Table 7.2 Surface treatments of D3 steel

Material	Notation
Untreated D3 tool steel	D3
Blasting +TiCN coating D3 tool steel	B-TiCN-D3
Nitriding + TiCN coating D3 tool steel	N-TiCN-D3
Blasting + nitriding +TiCN coating D3 tool steel	B-N-TiCN D3
Blasting + nitriding +TiCN (multilayer) coating D3 tool steel	B-N-multi TiCN D3

Source: Author

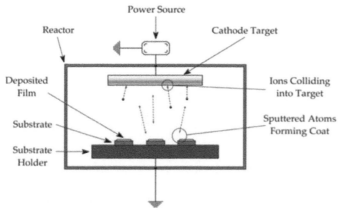

Figure 7.1 Physical vapour deposition process [18]
Source: Author

work, the abrasive micro-blasting process is used by applying a stream of abrasive material Al_2O_3 with size of 13 μ onto a surface. Pressurized fluid, typically air, is used to propel the media. The micro-blasting treatment was conducted for 30 seconds to confirm full coverage of the surfaces. The pressure varied within the range from 0.05 to 0.3 MPa in steps of 0.05 MPa

Nitriding and pvd coating

This material is selected because it could be nitrided without compromising with its strength. The ends of the specimens were polished to a roughness of around 0.4 mm.

The samples were polished with alumina before being cleaned with acetone and then nitrided. The specimens were nitrided using the plasma nitriding (PN) technique for approximately 16 hours at 540°C in a 75% H_2-25%N_2 environment.

After nitriding and microblasting, the specimens were machined and polished again in order to eliminate the white layer. Coatings were deposited on nitrided-microblasted surfaces, with the help of PVD process, the technique developed by 'Oerilikon Balzers Coating Pvt. Ltd', at a substrate temperature between 400°C and 450°C. A set of five samples each was coated with TiCN. Experiments were conducted on all the five samples and the average of the five observations is reported in the results.

Results and Discussion

Tesion test

The most frequently specified qualities of materials are their tensile strength, elastic modulus, and percentage elongation at break, which are three of the most significant indicators of a material's strength. A controlled amount of tension is applied to a sample during a tensile test till failure. Properties which are measured using tensile test involve yield strength, ultimate tensile strength, maximum elongation etc. The universal testing machine is the most often utilized tensile testing setup. The schematic of the tension and compression test and the photograph of the actual setup are shown in Figure 7.2 and Figure 7.3, respectively.

The tensile test specimens are prepared as per ASTM standard ASTM E8-04 [19] as shown in Figure 7.4 having length of 300 mm, neck diameter 12.5 mm and gauge length of 60 mm.

The prepared specimens as per ASTM standard ASTM E8-04 of D3 tool steel are then micro blasted, nitrited and PVD Coated with TiCN Coating with thickness 4 micron are shown in Figure 7.5 below.

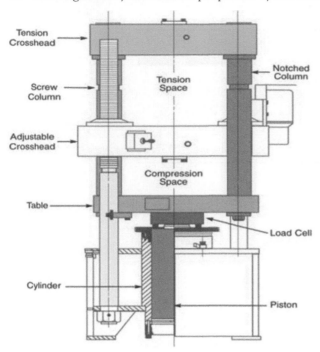

Figure 7.2 Schematic of the tension and compression test setup [14]
Source: Author

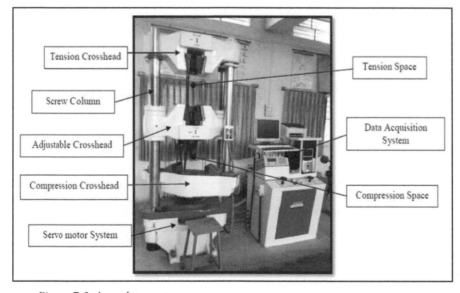

Figure 7.3 Actual test setup
Source: Author

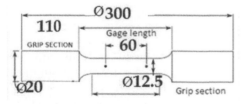

Figure 7.4 Tensile test specimen dimensions [19]
Source: Author

The results of yield, ultimate strength and % elongation for treated D3 material under investigation are presented in table 7.III below.

Comparison of yield strength and ultimate strength of various surface treated D3 steel is shown in Figure 7.6.

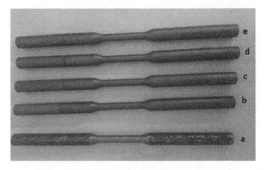

Figure 7.5 (a) Untreated D3 tool steel specimen (b) Micro-blasted and coated D3 steel (c) Nitrided and TiCN coated D3 steel (d) Microblasted, nitrided and TiCN monolayer coated D3 steel (e) Microblasted, nitrided and TiCN multilayer coated D3 steel

Source: Author

Table 7.3 Yeild strength, ultimate strength and percentage elongation of D3 steel

Materials	Yield strength	Ultimate strength	% Elongation
D3	261	398.1	17.50
B-TiCN-D3	521.1	740.7	8.33
N-TiCN-D3	469.7	598.5	2.50
B-N-TiCN D3	526	668.6	19.17
B-N-multi TiCN D3	530	693.8	7.50

Source: Author

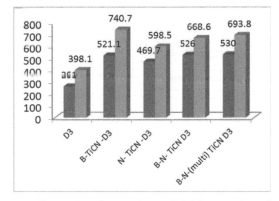

Figure 7.6 Comparison of yield strength and ultimate strength of various surface treated D3 steel

Source: Author

Comparison of percentage increase in yield strength w.r.t. untreated D3 steel specimen is shown Figure 7.7.

From the above tension test results, it can be concluded that when the various surface treatments such as nitriding, micro-blasting followed by TiCN coating are applied on D3 tool steel, its yield strength increases considerably. The yield strength of untreated D3 tool steel is low as compared to treated D3 steel specimens. The maximum yield strength is obtained with the combination of blasting, nitriding and TiCN multilayer coating process which is up to 103%. Improvement in yield strength can be attributed to an increase in adhesion force between the substrate and the coating and to the diffusion of nitrogen into the substrate due to nitriding and microblasting. Previous research [21] suggested that TiCN film deposited by plasma assisted physical vapor deposition (PAPVD) resulted in significant increase in fatigue life. This increase in fatigue life was attributed to the high mechanical strength of the film, its compressione residual stress state and the excellent adhesion of the coat to the steel substrate.

The surface treatments also affect the ultimate strength of D3 steel. The ultimate strength of D3 steel reaches its maximum value with the combination of micro-blasting and TiCN coating process which increase up to 86.05%. The ultimate strength of untreated D3 steel is low as compared to treated D3 specimens. Improvement in ultimate strength can be attributed to the high mechanical strength of the film and excellent adhesion of the coat to the steel substrate [21].

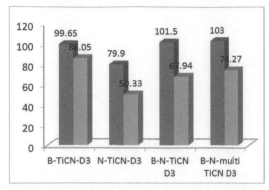

Figure 7.7 Percentage rise in yield strength w.r.t untreated D3 steel specimen
Source: Author

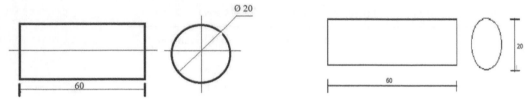

Figure 7.8 Standard specimen for compression test [20]
Source: Author

Figure 7.9 (a) Compression test specimen D3 tool steel (b) Micro-blasted &coated D3 steel (c) Nitrided and TiCN coated D3 steel (d) Microblasted, nitrided and TiCN monolayer coated D3 steel (e) Microblasted, nitrided and TiCN multilayer coated D3 steel
Source: Author

Compression test

A crucial engineering factor is the compressive strength of materials and structures. Applying balanced inward stresses to various locations on a material result in compression, which reduces the size of the object in one or more directions. In uniaxial compression, the forces are only applied in one direction, helping to shorten the object's length in that direction. On the same apparatus used to conduct tension tests, compression tests are also carried out.

The compression test samples are made in accordance with ASTM standard ASTM E9 89a [20] by machining to length of 60 mm and diameter 20 mm with the L/D ratio equal to 3 as shown in Figure 7.8.

The prepared specimens of D3 Tool steel are then micro-blasted, nitrided and PVD Coated with TiCN Coating with thickness 4 micron as shown in Figure 7.9.

The Results of compressive strength for the D3 material under investigation are presented in Table 7.4 below.

Comparison of Compressive Strength of various surface treated D3 Steel is presented in Figure. 7.10.

Discussion

From the above compression test results, it can conclude that if specimen undergoes various surface treatments such as nitriding, micro-blasting and coating, its compressive strength increases considerably. Increased compressive strength can be attributed to high mechanical strength of the film, its compressive residual stress state, and excellent adhesion of the coat to the steel substrate [21], due to nitriding and microblasting and also to the diffusion of nitrogen into the substrate. The compressive strength of

Table 7.4 Compressive strength of tested specimen

Material	Compressive load (KN)	Compressive strength (N/mm^2)
D3	296.45	943.62
B-TiCN-D3	300.8	957.5
N-TiCN-D3	286	910.36
B-N-TiCN D3	302.7	963.52
B-N-multi TiCN D3	307.15	977.68

Source: Author

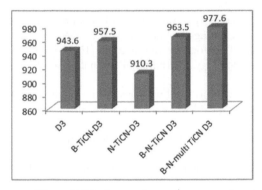

Figure 7.10 Comparison of compressive strength of various surface treated D3 steel.
Source: Author

untreated D3 tool steel is low as compared to treated D3 steel specimens. The maximum compressive strength is obtained with the combination of blasting, nitriding and TiCN multilayer coating process. The compressive strength of D3 steel with nitriding& coating is low as compared to other specimens.

Conclusion

The mechanical characteristics of D3 tool steel were investigated and following conclusions can be drawn.

- The yield strength of untreated D3 steel specimen is low as compared to treated D3 specimens.
- Yield strength of blasting, nitriding and TiCN multilayer coating was found to be maximum qual to 530 N/mm^2 among all other surface treated D3 steel specimens.
- The D3 tool steel specimen has maximum % rise in yield strength i.e 103% in case of blasting, nitriding and TiCN multilayer coating with respect to untreated D3 tool steel specimen.
- The ultimate strength of untreated D3 steel specimen is low as compared to treated D3 specimens.
- Ultimate strength of blasting, and TiCN coating was found to be maximum equal to 740.7 N/mm^2 among all other surface treated D3 tool steel specimens.
- The D3 tool steel specimen has maximum % rise in ultimate strength i.e. 86.05% in case of blasting, and TiCN coating with respect to untreated D3 specimen.
- The compressive strength of untreated D3 steel specimen is low as compared to treated D3 specimens.
- Compressive strength of blasting, nitriding and TiCN multilayer coating was found to be maximum equal to 977.6 N/mm^2 among all other surface treated D3 steel specimens. The corresponding percentage rise is 3.6%.
- Improvement in mechanical properties can be attributed to high mechanical strength of the film, its compressive residual stress state, and excellent adhesion of the coat to the steel substrate.

Hence, from above results, it can be concluded that due to various surface treatments such as micro-blasting, nitriding and TiCN coating, the mechanical properties of D3 tool steel increase considerably.

References

[1] Perumal, D. A., .Elaya, A., Kailas, A. J., Satish, V., and Venugopal, S. (2012). Sliding wear behaviour of plasma nitrided austenitic stainless steel type AISI 316LN in the temperature range from 25 to 400°C at 10^{-4} bar. Wear, 288, 17–26.

[2] Zeghni , A. E. and Hashmi, M. S. J. (2004). Comparative wear characteristics of TiN and TiC coated and uncoated tool steel abrasive wear behaviour of a 0.19 wt% C dual phase steel. 155-156, 1923–6.

[3] Luo, D. B., Fridrici, V., and. Kapsa, P. H. A systematic approach for the selection of tribological coatings. Wear. 271, 2132–2143.

[4] Zeghni, A. E. and Hashmi, M. S. J. (2004). The effect of coating and nitriding on the wear behaviour of tool steels. Journal of Materials Processing Technology. 155-156, 1918–22.

[5] Gates, J. D. (1998). Two-body and three-body abrasion: a critical discussion. Wear. 214, 139–146.

[6] Yilbas, B. S. and Nizam, S. M. (2000). Wear behaviour of TiN coated AISI H11 and AISI M7 twist drills prior to plasma nitriding. Journal of Materials Processing Technology. 105, 352–8.

[7] Sharma, A. and. Swami, K. C. (2014). A study of plasma nitriding process on the AISI 4140 steel Journal of Materials Science and Surface Engineering. 1(3), 81–83.

[8] Yildiz, F., Yetim, A. F., Alsaran, A., Celik, A., and Kaymaz. (2011). Fretting fatigue properties of plasma nitrided AISI 316L stainless steel: Experiments and finite element analysis. Tribology International. 44, 1979–86.

[9] Yilbas, B. S. and Nizam, S. M. (2000). Wear behaviour of TiN coated AISI H11 and AISI M7 twist drills prior to plasma nitriding. Journal of Materials Processing Technology 105, 352–358.

[10] Liu, Y. and. Fischer, T. E. (2003). Comparison of HVOF and plasma-sprayed alumina/titania coatings--microstructure, mechanical properties and abrasion behavior. Surface and Coatings Technology. 167, 68–76.

[11] Williams, J. A. (2005).Wear and wear particles--some fundamentals. Tribology International 38 (10), 863–870.

[12] Tichy, J. A. and Meyer, D. M. (2000). Review of solid mechanics in tribology. International Journal of Solids and Structures. 37, 391–400.

[13] Najari, M. R., Sajjadi, S. A., and Ganji, O. (2022). Microstructural evolution and wear properties of chromium carbide coatingformed by thermo-reactive diffusion (TRD) process on a cold-work tool steel. Results in Surfaces and Interfaces. 8, 1–14.

[14] Taheri, R., Jalali, M., Yaseri, A. A., and Yabesh, G. (2022). Improving TC drill bit's efficiency and resistance to wear by grapheme coating. Pet. Res. 7, 430–36.

[15] Dementyev, V. B. and Ivanova, T. N. (2021). Research on efficiency of high-strength coating during drilling. Procedia Structural Integrity. 32, 291–94.

[16] Vereschaka, A. A., Grigoriev, S., Sitnikov, N. N. Bublikov, J. I., and Batako, A. D. L. (2018). Effect produced by thickness of nanolayers of multilayer composite wear resistant coating on tool life of metal-cutting tool in turning of steel AISI 321. Procedia CIRP. 77, 549–52.

[17] Uhlmann, E., Protz, F., and Sassi, N. (2021). Ultrasonic assisted drilling of cemented carbide. Procedia CIRP. 101, 222–5.

[18] Andres, B. (2006). Investigation on nitriding with emphasis in plasma nitriding process. Materials Science.

[19] ASTM E8-04, American Society for Testing and Materials, Standard test methods for tension testing of metallic materials. ASTM International.

[20] ASTM E9-89a, American Society for Testing and Materials, Standard test methods of compression testing of metallic materials at room temperature. ASTM International.

[21] Cabrera, E. S. P., Staia, M. H. Quinto, D. T., Gutierrez, C. V., and Perez, E. O. (2007). Fatigue properties of a SAE 4340 steel coated with TiCN by PAPVD. International Journal of Fatigue. 29, 471–480.

Chapter 8

Ergonomic arm rest for comfortable drive in a passenger car

Rajesh R[a], Vijay S[b], Aswin T M[c], Dhanush Prabakaran S.[d], Dharshika S[e] and Dhanushsree B[f]

Department of Production Engineering, PSG College of Technology, Coimbatore, Tamil Nadu, India

Abstract

Car driving has become one of the essential activities of human life due to increasing number of cars. The driving posture often changes without the inherent knowledge of the car driver during long distance journey. The driver experiences shoulder, neck pain due to positioning of hand continuously on the steering wheel. Eventually, the driver changes the position to rest or hold the gear stick for a better feel. However, this also develops the pain in the upper limb in the body which leads to muscular skeletal disorder (MSD) injuries. In this work, the various height adjustment mechanism concepts were developed in the armrest using free hand sketching and the best sketch was selected based on concept evaluation tools such as Pugh matrix and weighted assessment method. The prototype armrest developed was positioned in a car and tested for ergonomic comfort. The new skeletal computer-aided design (CAD) models are developed with armrest was compared with the actual experiment posture angle and it was found that driver comfort level improved with the addition of the armrest prototype in the car.

Keywords: Car, armrest, concept evaluation, ergonomics, RULA analysis, prototype

Introduction

Nowadays, driving is one of the main activities of our day-to-day life. The destination spot for travel could be reached safely and comfortably using a car. The passenger car is the only option for a commuter to reach the desired place as per convenience as most of the places in the universe are connected only through roads. A passenger car is used predominantly over public transport due to easy access and availability for traveling in the universe. The commuters either use their own car or a taxi to travel to the desired location for purchase and other essential works.

A commuter is while driving car for a longer distance experience muscular skeletal disorder (MSD) injuries due to positioning of hand continuously on steering wheel [1, 3]. The problem definition was quantified from customer feedback survey and literature study. The real identification of the problem was based on the customer feedback survey to identify the problem faced bycommuters while driving a car. The survey results indicated that 74% of commuters experience the pain at the neck, upper and lower arm and torso area. The above-mentioned problem is caused due to the following reasons.

1. The driver continuously sitting in the same posture during a long-distance journey due to road condition the body posture gets tilted to the left side from the initial sitting position.
2. The driver both hands are continuously positioned on the steering wheel.
3. The driver positions the one hand on the gear stick knob and another hand on the steering wheel to overcome the discomfort while driving a car [1].

The above three points were considered as the important reasons for the MSD problems. Musculoskeletal disorders are the include damages that to tendon sheaths, tendons, and synovial lubrication of tendon sheaths, and muscles, and nerves of hands, related to bones, elbows, wrists, shoulders, neck and back. These musculoskeletal disorders belong to a collection of health problems that are more prevalent among the workingclass than the general population [2, 3]. The solution offered in this work is the development of armrest for improving the comfort level of driver in a passenger car. The armrest is not included in the driver seating position in most of the current car models in the market.

In some cars, armrests are attached only in front and rear seats. The design of the armrest plays a major role for drivers to achieve a better driving experience and reduce MSD injuries. Good ergonomic armrest should consider the ergonomic criteria with safety features for the driver while journey from one place to another in a passenger car [3].

[a]rajeshpsgtech2012@gmail.com, [b]vijaymerni@gmail.com, [c]20P101@psgtech.ac.in, [d]20P103@psgtech.ac.in, [e]21P105@psgtech.ac.in, [f]21P104@psgtech.ac.in

DOI: 10.1201/9781003450252-8

Customer feedback survey

The voice of the customers is registered by preparing a set of questions related to the problem concerned in this work. The questionnaires are designed to measure the commuter problems while driving the car for a longer distance. A survey was conducted among the professional cab drivers and the drivers who had their own car. The survey locations were the car showroom, service centers and through online mode. Thefeedback was collected from the 500 car drivers.

Questionnaires designed to measure the driver problems are mentioned below.

1. Are you interested in driving a Car?
2. Purpose of driving the car?
3. Approximate distance travelled in a day?
4. While driving a car for 50 km and above, do you feel any discomforts?
5. If yes means mention your discomfort/pain?
6. If you feel pain mention the action, you do get relaxed?
7. How often you take rest while driving?
8. During long drive whether you feel discomfort-able?
9. Any suggestions for avoiding the discomfort caused?

Concept Development

The concept development is the initial activity in the sequence of productdevelopment process. The starting point of the concept development is creating sketches where the ideas are given a definite shape. More concept sketches are developed, and a best concept is selected based on a set of criteria for prototype development [4, 14].

The concept selection method is used to reduce the time-consumption in the engineering design process because it involves the decision making and consideration of multiple factors. Initially, the sketch ideation of four concepts was developed for an ergonomic armrest that is used for the comfortable dive based on problem identification through the customer feedback survey. This process is used to identify the functional requirements, cost, and manufacturability. The concept screening method is an important process to evaluate the generated concepts against the criteria and helps to determine the suitable criteria, comparing the relative strength and weakness of the concepts. The concept screening is done using weighted decision matrix method [14].

Weighted Decision Matrix Method

The weightage rating used in weighted matrix method Table 8.1 is mentioned as follows.
  0: Not satisfied, 1: Just tolerable, 2: Adequate, 3: Good, 4: Satisfied, 5: Very satisfied
  Concept 4 has been selected as it outperforms the concept 1 based on ease of use, swivel adjustment and ease of assembly.

Table 8.1 Weighted matrix

WEIGHTED decision matrix selection					
	Weight (%)	Weightage in Rating	Concept - 1	Weightage in Rating	Conce pt - 4
Defined criteria					
Ease of use	15	3	0.45	4	0.6
Ergonomic assessment	20	4	0.8	5	1
Serviceability	10	3	0.3	3	0.3
Manufacturing feasibility	10	4	0.4	4	0.4
Height adjustment	15	3	0.45	3	0.45
Swivel adjustment	15	3	0.45	4	0.6
Easy to assemble	10	2	0.2	2	0.2
Customisation	5	1	0.05	3	0.15
Total score	100		0.387		0.462

Source: Author

Concept 1

In concept 1, two types of mechanisms are proposed. It has a height adjustment mechanism and swivel mechanism.

Height adjustment mechanism

In this working method, lever engages in the ratchet slot with the help of the spring force shown in the left side of Figure 8.6. As the button is pressed, the lever disengages from the ratchet slot and the armrest is moved upward or downward for positioning at convenient position. As the button is released the lever gets engaged in the ratchet slot. Once the lever gets engaged in the slot the movement is restricted, it cannot move without releasing the button. The detail sketch is shown in Figure 8.1.

Swivel mechanism

This mechanism consists of rotating the arm top assembly to the convenient position for the user as shown in Figure 8.2. The swivel adjustment mechanism consists of the cam, ratchet slot, mounting plate and spring shown in the right bottom of Figure 8.2. The mounting plate is mounted on the arm stem assembly. The cam with a cam follower supports the arm movement after positioning the armrest on the arm top plate

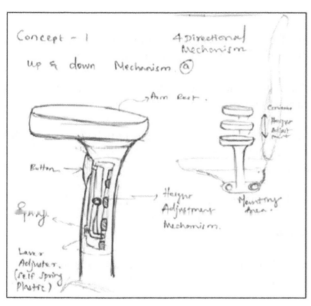

Figure 8.1 Height adjustment mechanism concept 1
Source: Author

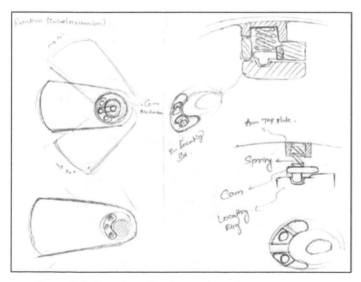

Figure 8.2 Swivel mechanism concept 1
Source: Author

and it could be adjusted suitably by the user. The cam is attached with the spring and is inserted to the top mounting plate. The overall mechanism sub assembly is between the arm and arm rest assembly. The functions and details are shown in Figure 8.2.

Concept 4

Concept 4 was developed to minimise the usage of springs in the mechanism. It has height adjustment, swivel mechanism, and similar functions of concept 1.

Height adjustment mechanism

The height adjustment mechanism functions, assembly arrangement is same as the concept 1. The only difference is that use of rubber spring instead of a spring. The concept functions are shown in the Figure 8.3.

Swivel mechanism

Concept 4 of swivel mechanism works based on cam mechanism. A cam movement happens with a rubber spring arrangement instead of the spring components. The rotation movement consists of single side movement to rest the arm for user feasibility. This mechanismarrangement is shown in Figure 8.4.

Passenger car interior modeling

The construction of the car cabin was developed using the above-mentioned dimensions data as shown in Figure 8.4. The model of car cabin with the exact dimension of the car interior dimensions is shown in Figure 8.5. The elements are seat, cabin, dashboard and steering wheels.

The computer-aided design (CAD) design was completed with the help of collected data from the journal paper and some dimension taken from the manual measurement of the car [1, 2].

a)   The distance between the steering wheel and the sitting driver: the distance will be measured the dimension from the center of the steering wheel to the chest – dimension 1.
b)   Distance of the chair: the distance will be measured from the chair's edge to the floor under pedals – dimension 2, (X, Y plane)
c)   To the height of the chair in the distance from the chair to the floor – dimension 3 (X, Y plane).
d)   The dimensions of the length and width of placing the gear shift stick the distance measured from both edges of the floor – manual measurement dimension 4.
e)   Steering wheel angle of dimension 5,
f)   A driver's chair leaning angle – the angle between the backrest and the seat surface – dimension 6.
g)   Width of the seat and position, gear stick position dimension taken by manually measured from the car [1, 4].

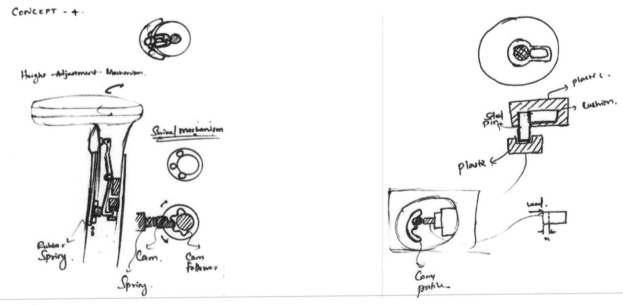

Figure 8.3 Height adjustment mechanism concept 4
Source: Author

Figure 8.4 Swivel mechanism concept 4
Source: Author

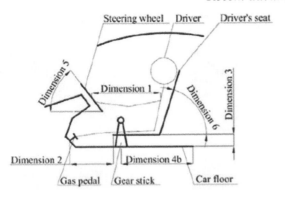

Dimension number		Value of dimension
1		430 mm
2		548 mm
3		190 mm
4	a	767 mm
	b	877 mm
5		85°
6		80°

Figure 8.5 Passenger car interior dimension
Source: Author

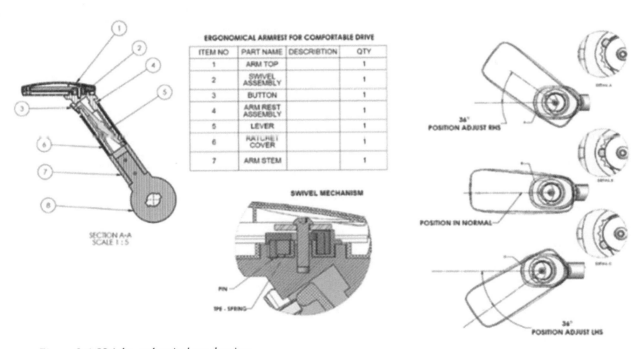

Figure 8.6 Height and swivel mechanism
Source: Author

Spring lever

The spring lever is the main component of the height adjustment mechanism. It's shown in Figure 8.7. The material used in spring lever is polypropylene because of the high elasticity and load resistance. The center of the part has pivot circle, based on the pivot circle it is attached with the armrest stem components. From the pivot point, one end connects the button and other end connects the ratchet slot. The force formula is shown in Equation 1.

$$F1 \times \text{distance } 1 = F2 \times \text{distance } 2 \tag{1}$$

Initial force required for arm rest is 40 N. From the pivot point the lever upside dimension has the 93.2 mm bottom side has the 76.5 mm. This value applied in equation 1.

Posture "a"   **Posture "b"**

Figure 8.7 Armrest with car cabin assembly and BOM
Source: Author

Figure 8.8 Typical driving postures
Source: Author

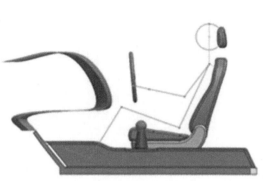

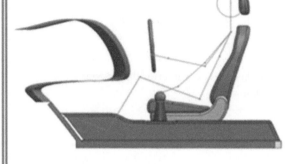

Figure 8.9 Posture 'a' and 'b'.
Source: Author

$$40N \times 93.2 \text{ mm} = F2 \times 76.5 \text{ mm}$$

48.73 N is required force to release from the slot in worst case.40 N is the BIFMA 5x standard case from ASTM.

Arm rest assembled in car cabin

The entire arm rest assembly was developed using the Solid work's software. The components are seat, cabin, dashboard, and steering wheels. The design developed with the help of collected data from the journal paper and some dimensions taken from the manual measurement of the car. The components are mentioned in BOM as shown in Figure 8.7.

Experiments conducted under different postures

Experiment test was conducted under posture a) and posture b) shown in Figure 8.8. The posture surveillance test was conducted among 50 peoples who visited the car showroom. A sitting position with hands on steering wheel and a sitting position with a right-hand holding gear shift stick [1, 2]. It was shown in which areas of the cabin of the passenger car changes in dimensions are necessary and then driver's comfort improvement was shown in percentage and taken dimensions from SAE Technical paper series and Indian anthropometry data [1, 13].

The figure indicates the typical posture like 'a' and 'b'. The problem was identified from the experiment test in these postures and checked for uncomfortable feel/pain by driving of 200 kms in a single stretch.

Based on that posture the dimensions taken from the professional drivers with from above mentioned reference 50th percentile population male skeleton model was developed. The skeleton model of posture 'a' was created based on anthropometry data. Posture 'a' is both hands holding the steering wheel. Hand position dimension measurements were done based on experiments method with devices such as ruled scale and ergon point compass. Posture 'b' manikin left hand resting is resting on the gear knob shown in Figure 8.9. The model for posture 'b' is developed similar to posture 'a'. The armrest attached model is shown as Posture 'c'. The new posture skeleton model was developed based on the attached arm rest assembly. It is shown in Figure 8.9 and 10 [7, 9, 13].

Rapid Upper Limp Analysis Results

Posture 'A'

Rapid Upper Limp (RULA) analysis is done by the RULA standard work sheet in manual calculation methods. The RULA score was developed based on manikin sitting position angle that is both hands holding the steering wheel. [8]

From the H point, the torso angle is 65°, with reference to the torso, the upper arm angle is 12°, the lower arm is 12° from the floor, and the wrist angle is 1° tilt upside of the elbow is 13°. Finally, the score results obtained was 4, it is shown in Figure 8.13 [7, 10, 12, 13].

Posture 'B'

Posture 'b' consists of left hand resting on the gear knob. The evaluated angles from the H point of the torso angle shoulder have 65°, With reference to shoulder, the upper arm angle is 15°, the lower arm is 41° down from the floor, and wrist angle is 20°. Finally, the results score was 3. It is shown in Figure 8.10. Driving position and angles are same as posture 'a' only the upper arm and lower arm position have been changed. The final score calculated as '3' which is better than posture 'a'.

Posture 'C'

The posture 'c' consists of armrest included in the car cabin. The RULA score was obtained based on the manikin sitting position angle left hand arm positioned on the arm rest. The upper arm angle is 5°, the lower arm 5° from the floor and a wrist angle is 1° tilt upside of the elbow. RULA score was 2 is shown in the Figure 8.11, indicating the result as a safe ergonomic condition. All posture is considered in static mode at 55 min timing.

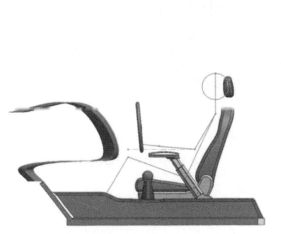

Figure 8.9 Posture 'c'
Source: Author

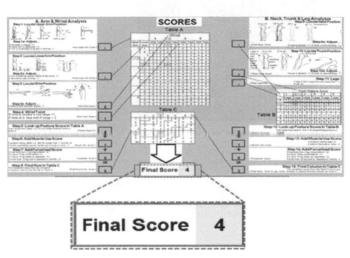

Figure 8.10 RULA score of posture
Source: Author

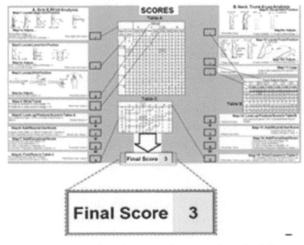

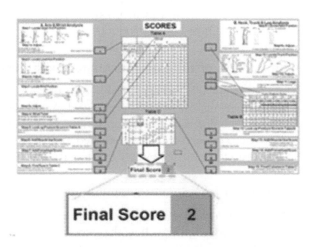

Figure 8.11 RULA score of posture b and c [5]
Source: Author

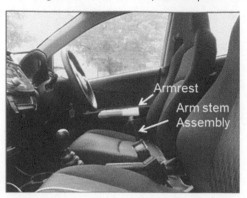

Figure 8.12 Armrest assembly assembled with CAR [6]
Source: Author

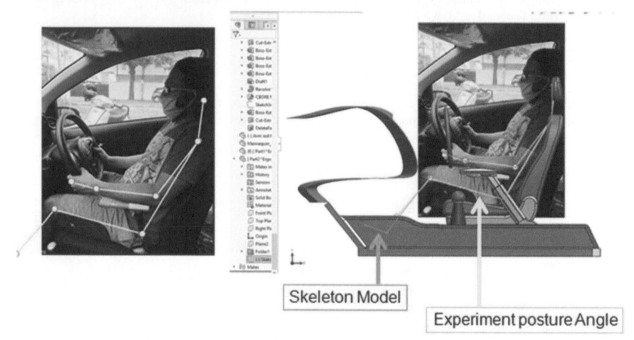

Figure 8.13 Result comparison of skeleton model with actual posture
Source: Author

Prototype development and experiments analysis

The final prototype model attached to the car driver seat is shown in Figure 8.12. The armrest assembly consists of arm stem and armrest top. The prototype was made using available resources like a plastic container and household product.

Comparison of the experiments test with the design skeleton model

The prototype armrest was attached to the car cabin and the posture based on design skeleton model and actual experiment posture is compared. As shown in Figure 8.13.

The blue color represents the design model, and the yellow color indicates the experiment's actual posture angle. The posture was predicted almost similar to the CAD design model. From the analysis results it was identified that the newly designed armrest will provide the best sitting posture for the long drive [1, 3].

Conclusion

The developed new arm rest has improved the comfort level of drivers. The ergonomic analysis was performed by virtually as well as the experiment methods. In this work, product design and development steps have been used from concept sketching to prototype development. The outcomes are discussed in the following points. The armrest helps to reduce driving fatigue. it has enhanced the comfort zone and provides a better experience for the drivers. The armrest designed with height and swivel adjustment mechanism was compatible for all professional drivers. The armrest was designed for easy of assembly, consist of less no of components, and the mechanism was also simple to achieve the function. The armrest will provide the best sitting posture for the long drive. Initially the rapid upper limb assessment analysis results were validated by experiment test.

Acknowledgements

I acknowledge the PSG management for supporting with facilities and teaching assistants for this work.

References

[1]   Cichański, A. and Wirwicki M. (2014). Faculty of ocean engineering hip technology. Journal of POLISH CIMAC. 8(2).

[2]   Tamene, A., Mulugeta, H., Ashenafi, T., and Thygerson, S. M. (2020). Musculoskeletal disorders and associated factors among vehicle repair workers in awassa city, southern Ethiopia. Journal of Environmental and Public Health Volume. doi:10.1155/2020/9472357

[3]   Mozafari, A., Vahedian, M., Mohebi, S., and Birru, M. N. (2015). Work-related musculoskeletal disorders in truck drivers and official workers. Acta Medica Iranica. 53(7), 432–8.

[4]   Arunachalam M., Arun Prakash R, Rajesh R (2014), A typical approach in conceptual and embodiment design of foldable bicycle. International Journal of Computer Applications. 87(19). doi: 10.5120/15458-4031.

[5]   Agarwal, A., Nair, S. K., Chada, V. K. K., Pardeshi, A., and Sarawade, S. S. (2016). Ergonomic Evaluation to improve work posture. International Journal of Engineering Research & Technology. 5(3).

[6]   Chakrabarti, D. K. (1999). Indian anthropometric dimensions for ergonomics design practice India. India: National Institute of Design.

[7]   Wei, D. (2017). Base on RULA analysis comfort evaluation tractor river method. IEEE 978-1-5090-6414-4/17.

[8]   Manary, M. A., Reed, M. P. Flannagan, C. A. C., and Schneider, L. W. (1998). ATD positioning based on driver posture and position. United States: University of Michigan Transportation Research Institute.

[9]   Yusop, M. S. M., Mat, S., Ramli, F. R., Dullah, A. R. S. Khalil, K. Case (2018). Design of welding armrest based on ergonomics analysis: case study at educational institution In Johor Bahru, Malaysia. ARPN Journal of Engineering and Applied Sciences ISSN 1819-6608.

[10] Mat, S., Abdullah, M. A., Dullah, A. R., Shamsudin, S. A., and Hussin, M. F. (2017). Car seat design using RULA analysis. Proceedings of Mechanical Engineering Research Day. 1–2.

[11] kadir, S. A. A., Dodo, S. M., and Vandi, L. T. (2018). Design of an ergonomic chair with headrest and armrest using anthropometric data. Journal of Engineering Science and Technology Research. 4(2), 33–43.

[12] Yasobant, S., Chandran, M., and Reddy, E. M. (2015). Are bus drivers at an increased risk for developing musculoskeletal disorders: an ergonomic risk assessment study. Journal of Ergonomics. S3:011.

[13] Schmidt, S., Amereller, M., Franz, M., Kaiser, R., and Schwirtz, A. (2014) A literature review on optimum and preferred joint angles in automotive sitting posture. Journal, Applied Ergonomic. 45, 247–60.

Chapter 9

Explicit finite difference approach on multi-dimensional wave equation with constant propagation

Malabika Adak[1,a], Akshaykumar Meshram[1,b] and Anirban Mandal[2,c]

[1]Yeshwantrao Chavan College of Engineering, Nagpur, India

[2]Visvesvaraya National Institute of Technology, Nagpur, India

Abstract

Different types of wave motion like light wave motion, sound wave motion, water wave motion and seismic wave motion arises in the field of vibrating string (musical instrument), electromagnetic field and fluid flow etc. Wave equation is a second-order linear hyperbolic partial differential equation (PDE). In engineering and physical science problems, solution of partial differential equations with complicated boundary condition (BCS) as well as irregular domain is a challenging work particularly with analytical method. Therefore, numerical method is one of the best choices to solve the boundary value problems (BVP) involving partial differential equation and suitable boundary conditions in more efficient way. In present study, one- and two-dimensional wave equations have been considered with constant propagation for solving BVP using explicit finite difference method (FDM). To illustrate error calculation, stability and convergence, two examples have been considered. Numerical solutions have been verified with analytical solution to test the accuracy and subsequently error has been calculated at each node for different mesh grid size in the direction of space and time. The numerical results for very small mesh grid size and small-time space will give the better accuracy. In this study two different mesh size along with space and two different mesh size along with time have been taken to verify the accuracy of numerical technique. While comparing with existing analytical solution it is observed that the percentage error for present solution technique is within the range of 7.45% for average mesh size. In case of finer mesh and small time step the error is further reduced to 4.16 to 5.84 %.

Keywords: Explicit finite difference method, hyperbolic partial differential equation. initial boundary conditions, wave equation

Introduction

Multi-dimensional wave equations with constant propagation (hyperbolic partial differential equation) are solved using explicit finite difference method (FDM). Different types of wave motion are very important in engineering and physical science problems like light wave, sound wave, water wave and seismic wave motion in the vibrating fields. Solution of partial differential equations with complicated boundary condition (BCS) with irregular domain is a challenging work particularly with analytical method. So, numerical method is more convenient to solve problem.

Many scientific and engineering problems can be mathematically modeled with the help of PDE. Particularly, wave equation or motion equation normally represented by the hyperbolic type PDE. PDE contains partial derivatives with respect to at least two independent variables. The general form of second order and first degree linear PDE is given by

$$a\frac{\partial^2 g}{\partial x^2} + b\frac{\partial^2 g}{\partial x \partial y} + c\frac{\partial^2 g}{\partial y^2} + d\frac{\partial g}{\partial x} + e\frac{\partial g}{\partial y} + pg = q \tag{1}$$

where a, b, c, d, e, p, q are function of x and y only.

PDE (1) can be three types like parabolic, hyperbolic and elliptic as $b^2 - 4ac = 0$ $b^2 - 4ac > 0$ and $b^2 - 4ac < 0$ respectively. In this study hyperbolic PDE has been considered to solve the equation. In very early decades researcher worked with 1D wave equation, and after 10 years researcher used 3D wave equation. Several researchers Balagurusamy [7], Jain [8], Sastry [13, 14], Scheid [15] worked with the boundary values problems by using FDM. Lakshmi [10] used the finite difference method to solve the second order boundary value problem with ordinary differential equation for Dirichlel boundary condition. Muhammad [11] computed approximate solution of a third order boundary value problem (BVP) using new two stage finite difference method. Siddiqi [16] used a tool homotopy analysis method (HAM) to solve higher order BVP and Chebyshev wavelets by Xu [17] solved eighth-order two-point BVPs using numerical method. Adak [1-6] provides numerical results for elliptic and diffusion partial differential equation using FDM with convergency study. Muhammad [12] investigated the initial boundary hyperbolic equation using implicit FDM. Jaakko [9] presented the FEM and FDM for solution of wave equation.

---

[a]malabikaadak@yahoo.co.in, [b]ajm.ycce@gmail.com, [c]amandalthesis@yahoo.com

DOI: 10.1201/9781003450252-9

For clear understanding on diffusion, vibration system, wave mechanics etc, the numerical studies are very necessary. Hence, this is very necessary to study on how to calculate wave motion from hyperbolic partial differential equation using explicit finite difference scheme.

Multidimensional problem is handled in this article which is lacking in previous literatures particularly for complicated boundary conditions. More importantly present work will help to build up the confidence among researcher to use this solution technique in different engineering problems.

Problem Identification

Consider a string of length l is bobbing up and down in which a simple harmonic excitation is created as shown in Figure 9.1. Since the tension acts necessarily along the line of the string, the tension F is a small angle $dg(x)/dx$ to the horizontal line at point x on the string fragment. Due to pulling to the left, there creates downward force component which is $F (dg(x + dx))/dx$. There is upward force component $F (dg(x + dx))/dx$ at the right-hand end of the string fragment.

Putting $g (x + dx) = g(x) + (dg/dx)dx$, and adding algebraically the upwards and downwards forces together, a total force is acting $F \left(\frac{d^2 g}{dx^2}\right) dx$ on the bit of string. The string mass is ρdx, so force $= ma$ becomes $F \left(\frac{d^2 g}{dx^2}\right) dx = \rho dx$ giving the standard wave equation

$$\left(\frac{d^2 g}{dt^2}\right) = c^2 \left(\frac{d^2 g}{dx^2}\right) \quad 0 \leq x \leq l \tag{2}$$

where $c^2 = F/\rho$ is the wave velocity which is considered as constant.

Constant C is the propagation speed of vibration. It gives the solution of wave motion function $g(x, t)$ defined in the interval x Î [0, l] and t Î [0, ∞), satisfy the initial boundary conditions (IBC) $g(x, 0) = h(x)$, $g_t(x, 0) = r(x)$, $g(0, t) = s1(t)$, $g(1, t) = s2(t)$, for $0 \leq t \leq T$

In 2D problem, consider a rectangular structure with dimension m × n ,is surrounded by tightly stretched elastic membrane and edges are kept fixed with structure. To develop the mathematically model of the membrane surface is the goal. Suppose $g(x, y, t)$ is the deflection of membrane from equilibrium at position (x, y) at a time t. Surface $Z = g(x, y, t)$ gives the shape of the structure at time t. Considering assumption like uniform density, uniform tension, no resistance to motion, small deflection etc., one can model 2D wave equation in the following form,

$$\left(\frac{d^2 g}{dt^2}\right) = c^2 \left(\frac{d^2 g}{dx^2} + \frac{d^2 g}{dy^2}\right) \quad m \leq x, y \leq n \tag{3}$$

Explicit finite difference approximation

For 1D wave equation

In finite difference method, second order partial derivatives are replaced by centre difference formula in the given differential equation as well as boundary conditions. At every unknown point arise one equation. After putting initial conditions and boundary conditions, solution is obtained at every unknown point from each equation.

For solving BVP defined by equation (2), the interval or range [m, n] or $[x_0, x_r]$ along x axis is divided intor numbers of sub-intervals with length Δx, therefore $x_i = x_0 + ih$, where, $i = 1, 2, \ldots \ldots \ldots r$.

The g value along y axis at x points are given by $g(x_i) = g_i = g(x_0 + ih)$, $i = 0, 1, 2, \ldots \ldots \ldots \ldots r$.

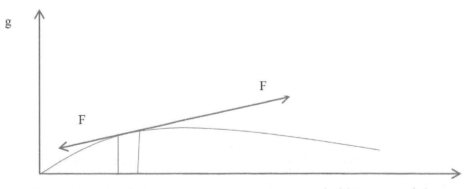

Figure 9.1 A simple harmonic motion in a string as bobbing up and down
Source: Author

Using Taylor's expansion, central difference approximations of $\frac{d^2g}{dx^2}$ and $\frac{d^2g}{dt^2}$ from equation (2) at the point $x = x_i$ are given by

$$\frac{d^2g}{dx^2} = \frac{1}{\Delta x^2}(g_{i-1}^r - 2g_i^r + g_{i+1}^r)$$
$$\frac{d^2g}{dt^2} = \frac{1}{\Delta t^2}(g_i^{r-1} - 2g_i^r + g_i^{r+1})$$

where Δx and Δt are small step sizes along space and time domain. Now, the wave equation. (2) becomes the following difference equation

$$\frac{1}{\Delta t^2}(g_i^{r-1} - 2g_i^r + g_i^{r+1}) = \frac{c^2}{\Delta x^2}(g_{i-1}^r - 2g_i^r + g_{i+1}^r)$$

Taking $\lambda^2 = \frac{c^2 \Delta t^2}{\Delta x^2}$

Here, $c\frac{\Delta t}{\Delta x}$ is called the CFL number (Courant Friedrichs Lewy number). This number is selected in such a way that numerical speed $\frac{\Delta x}{\Delta t}$ is less than the propagation constant, ie, $\frac{\Delta x}{\Delta t} < c$

$$\blacktriangleright \quad g_i^{r+1} = -g_i^{r-1} + \lambda^2(g_{i-1}^r + g_{i+1}^r) + 2(1 - \lambda^2)g_i^r \qquad (4)$$

In equation (4), to determine $(r+1)^{th}$ time level function values, r^{th} and $(r-1)^{th}$ time level are required. Since three-time level is present, so difference schemes are said to be three level finite difference schemes. Equation (4) is also called the **explicit scheme**. It is stable for the condition $\lambda^2 \le 1$.

For 2D wave equation

Use the central difference approximations for 2nd order derivatives in two-dimensional wave equation:

$$g_{xx} = \frac{1}{\Delta x^2}(g_{i-1,j}^r - 2g_{i,j}^r + g_{i+1,j}^r)$$
$$g_{yy} = \frac{1}{\Delta y^2}(g_{i,j-1}^r - 2g_{i,j}^r + g_{i,j+1}^r)$$
$$g_{tt} = \frac{1}{\Delta t^2}(g_{i,j}^{r-1} - 2g_{i,j}^r + g_{i,j}^{r+1})$$

2D wave equation (3) reduces to the following difference equation:

$$\frac{1}{\Delta t^2}(g_{i,j}^{r-1} - 2g_{i,j}^r + g_{i,j}^{r+1}) = \frac{c^2}{\Delta x^2}[(g_{i,j-1}^r - 2g_{i,j}^r + g_{i,j+1}^r) + (g_{i,j-1}^r - 2g_{i,j}^r + g_{i,j+1}^r)]$$

Taking $\lambda^2 = \frac{c^2 \Delta t^2}{\Delta x^2}$

$$g_{i,j}^{r+1} = -g_{i,j}^{r-1} + \lambda^2(g_{i-1,j}^r + g_{i+1,j}^r + g_{i,j-1}^r + g_{i,j+1}^r) + 2(1 - 2\lambda^2)g_{i,j}^r$$

which is called the explicit scheme for two-dimensional wave problem.

Numerical illustration and verification

A. Example 1

Consider 2nd order one dimensional wave equation $g_{tt} = g_{xx}$ with IBC $g(0, t) = 0$, $g(1, t) > 0$ and $g_t(x, 0) = 0$, $g(x, 0) = sin^3(\pi x)$, $0 \le x \le 1$.

Solve the IBV problems using explicit method with taking $\Delta x = \frac{1}{3}$, $\Delta t = 0.2$. Analytical solution of the given BVP is given by $g(x,t) = \frac{3}{4}\sin \pi x \cos \pi t - \frac{1}{4}\sin 3\pi x \cos 3\pi t$.

Solution: In the given problem, x range is defined by $0 \le x \le 1$, ie, [0, 1]. Consider mesh size is $\Delta x = 1/3$ along length, $\Delta t = 1/5$ is the mesh size along time domain. Since, $c = 1$, then $\lambda = c \frac{\Delta t}{\Delta x} = \frac{1/5}{1/3} = 0.6$, so $\lambda^2 = 0.36$ < 1 (Stability condition is satisfied).

Here $0 \le x \le 1$ if $\Delta x = 1/3$, then along x domain there are four nodes like $x_0 = 0$, $x_1 = 1/3$, $x_2 = 2/3$, $x_3 = \frac{3}{3} = 1$, The given BCS are

$$g(0,t) = g_0^r = 0, \quad g(1,t) = g_3^r = 0,$$

The given IC is $g(x, 0) = g_x^0 = sin^3(\pi x)$, obtain

$$g(\tfrac{1}{3}, 0) = g_1^0 = sin^3\left(\tfrac{\pi}{3}\right) = 0.6495, \quad g(\tfrac{2}{3}, 0) = g_2^0 = sin^3\left(\tfrac{2\pi}{3}\right) = 0.6495,$$

Other IC is $f_t(x, 0) = 0$

➤ $\left(\frac{g_i^{n+1} - g_i^{n-1}}{2k}\right)_{n=0} = 0$

➤ $g_i^{-1} = g_i^1$

Taking $\lambda^2 = 0.36$, the explicit formula (4) is reduced by

$$g_i^{n+1} = -g_i^{n-1} + 0.36(g_{i-1}^n + g_{i+1}^n) + 2(0.64)g_i^n$$

Putting $n = 0$, 1st iteration is

$$g_i^1 = -g_i^{-1} + 0.36(g_{i-1}^0 + g_{i+1}^0) + 1.28g_i^0$$

➤ $g_i^1 = -g_i^{-1} + 0.36(g_{i-1}^0 + y_{i+1}^0) + 1.28g_i^0$ (since $g_i^{-1} = g_i^1$)

➤ $g_i^1 = 0.18(g_{i-1}^0 + g_{i+1}^0) + 0.64g_i^0$ (5)

Putting $i = 1, 2, 3$ in equation (5)

$$g_1^1 = 0.18(g_0^0 + g_2^0) + 0.64g_1^0$$
$$= 0.18 \,(0 + 0.6495) + 0.64(0.6495)$$
$$= 0.53259$$

Similarly,

$$g_2^1 = 0.18(g_1^0 + g_3^0) + 0.64g_2^0 = 0.53259$$

The exact solution of the above IBV problem is:

$$g(x,t) = \frac{3}{4} sin\pi x \, cos\pi t - \frac{1}{4} sin\,3\pi x \, cos\,3\pi t$$

Comparison of numerical solution and analytical solution are shown in Table 9.1, 9.2 and 9.3 for different conditions.

B. *Example 2*

Consider the two-dimensional boundary value problem $g_{tt} = g_{xx} + g_{yy}$, subject to condition $g(x, y, t) = 0$, $t > 0$ on the boundary and shown in Figure 9.2 with $g_t(x, y, 0) = 0$, $g(x, y, 0) = sin\,\pi x\,sin\,\pi y$ $0 < x, y < 1$ shown in Figure 9.2 with $\Delta x = \frac{1}{3}$, $\lambda = \frac{1}{3}$. Find the numerical solution using Explicit finite difference method.

Solution:
From the boundary condition $g(x, y, t) = 0$, we have $g(0, 0, t) = g_{0,0}^r = 0$, $g_{1,0}^r = g_{2,0}^r = g_{3,0}^r = 0$, $g_{0,1}^r = g_{0,2}^r = g_{0,3}^r = 0$, $g_{3,1}^r = g_{3,2}^r = g_{3,3}^r = 0$, $g_{0,3}^r = g_{1,3}^r = g_{2,3}^r = 0$.
From the initial condition $g(x, y, 0) = sin\,\pi x\,sin\,\pi y$

Table 9.1 Comparison of numerical solution and analytical solution when $\Delta x = \frac{1}{3}$, $\Delta t = 0.2$

x	Analytical solution	Numerical solution (FDM)	Error	Percentage error
0	0	0	0	0
0.3333	0.57547	0.53259	0.04288	7.45
0.6667	0.57547	0.53259	0.04288	7.45
1	0	0	0	0

Source: Author

Table 9.2 Comparison of numerical solution and analytical solution when $\Delta x = \frac{1}{4}$, $\Delta t = 0.2$

x	Analytical solution	Numerical solution (FDM)	Error	Percentage error
0	0	0	0	0
0.25	0.4836	0.4473	0.0363	7.50
0.5	0.5295	0.5867	0.0568	10.72
0.75	0.4836	0.4473	0.0363	7.50
1	0	0	0	0

Source: Author

Table 9.3 Comparison of numerical solution and analytical solution when $\Delta x = \frac{1}{4}$, $\Delta t = 0.1$

x	Analytical solution	Numerical solution (FDM)	Error	Percentage error
0	0	0	0	0
0.25	0.4004	0.3770	0.0234	5.84
0.5	0.8602	0.8960	0.0358	4.16
0.75	0.4004	0.3770	0.0234	5.84
1	0	0	0	0

Source: Author

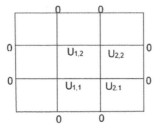

Figure 9.2 Specified boundary conditions in a square domain for 2D problem
Source: Author

➤ $g_{i,j}^r = sin\,(\pi i \Delta x)\,sin\,(\pi j \Delta x)$
➤ $g_{1,1}^0 = u\left(\frac{1}{3},\frac{1}{3},0\right) = sin\left(\frac{\pi}{3}\right) sin\left(\frac{\pi}{3}\right) = 0.75,$
  $g_{2,1}^0 = u\left(\frac{2}{3},\frac{1}{3},0\right) = sin\left(\frac{2\pi}{3}\right) sin\left(\frac{\pi}{3}\right) = 0.75,$
  $g_{1,2}^0 = u\left(\frac{1}{3},\frac{2}{3},0\right) = sin\left(\frac{\pi}{3}\right) sin\left(\frac{2\pi}{3}\right) = 0.75$
  $g_{2,2}^0 = u\left(\frac{2}{3},\frac{2}{3},0\right) = sin\left(\frac{2\pi}{3}\right) sin\left(\frac{2\pi}{3}\right) = 0.75$

From the condition $g_t(x, y, 0) = 0$, we have $\frac{g_{i,j}^{r+1} - g_{i,j}^{r-1}}{2k} = 0$

➤ $g_{i,j}^{r+1} = g_{i,j}^{r-1}$

➤ For $r = 0$, $g_{i,j}^1 = g_{i,j}^{-1}$

Using explicit scheme for two-dimensional wave equation

$$g_{i,j}^{r+1} = -g_{i,j}^{r-1} + \lambda^2 (g_{i-1,j}^r + g_{i+1,j}^r + g_{i,j-1}^r + g_{i,j+1}^r) + 2(1 - 2\lambda^2)g_{i,j}^r$$

Putting $r = 0$, $i = 1, 2$ and $j = 1$

$$g_{1,1}^1 = -g_{1,1}^{-1} + \frac{1}{9}(g_{0,1}^0 + g_{2,1}^0 + g_{1,0}^0 + g_{1,2}^0) + 2(1 - 2\lambda^2)g_{1,1}^0$$

> $2g_{1,1}^1 = \frac{1}{9}(0 + 0.75 + 0 + 0.75) + 2.\frac{7}{9}(0.75)$
> $g_{1,1}^1 = \frac{2}{3}$

Similarly, $g_{2,1}^1 = \frac{2}{3}$

Again Putting $r = 0$, $i = 1, 2$ and $j = 2$, we obtain,

$$g_{1,2}^1 = \frac{2}{3} = 0.667, \quad g_{2,2}^1 = \frac{2}{3} = 0.667$$

For the second iteration Putting $r = 1$, we have

$$g_{1,1}^2 = \frac{47}{108}, \quad g_{2,1}^2 = \frac{47}{108}, \quad g_{1,2}^2 = \frac{47}{108}, \quad g_{2,2}^2 = \frac{47}{108}$$

Table 9.1, 9.2 and 9.3 show the comparison of numerical and analytical results at each node for different mesh grid size. If points are increased, then mesh size will be decreased, and error should be reduced. For better accuracy, one can consider very small mesh size. If mesh size along time domain is decreased results will be changed. Therefore, good accuracy small step size is required. In this study, the second order wave equation (boundary value problem) which arise in simple harmonic motion is solved by using approximate numerical explicit FDM. Linear and nonhomogeneous problems are solved, and results are displayed in table 9.1 and table 9.2 and table 9.3. Numerical stability and accuracy depend on Δx and Δt, that is related through $\lambda = \frac{c\Delta t}{\Delta x}$. Large value of Δx and Δt will give the large numerical error or unstable solution. Very small value of Δx and Δt will give stable numerical solution with taking more computing time.

Conclusion

Numerical solutions are verified with analytical solutions to test the accuracy and error has been calculated at each node for different mesh grid size in the direction of space and time. It is observed that better accuracy was obtained for finer mesh size with fine step size. Explicit finite difference method is proved to be easy and convenient way to solve complicated hyperbolic partial differential equation. Therefore, it will encourage to use such solution technique among researchers for fast and accurate solution. While comparing with existing analytical solution it is observed that the percentage error for present solution technique is within the range of 7.45% for average mesh size. In the case of finer mesh and small time step the error is further reduced to 4.16-5.84%. Explicit finite difference method is proved to be an easy and convenient way to solve the complicated hyperbolic partial differential (wave) equation. Therefore, this method could be easily applicable to solve the multi-dimensional problems which arise in the field of vibration system, motion, and diffusion problems. The initial and boundary conditions are required to solve the hyperbolic partial differential equation using explicit Finite difference method. This method is conditionally stable if $\lambda \leq 1$. Numerical results are compared with analytical results that showed the error is very reasonable. Error is to be reduced by increasing the iterations. It will give better accuracy if the mesh size along space and time is small. This method is suitable for the complicated domain. Hence, it will be encouraged to use such solution technique among researchers for fast and accurate solution. Present work will help to build up the confidence among researcher to use this solution technique in different engineering problems.

Reference

[1] Adak, M. and Mandal, N. R. (2010). Numerical and experimental study of mitigation of welding distortion. Applied Mathematical Modelling. 34, 146–158. doi:10.1016/j.apm.2009.03.035.

[2] Adak, M. and Soares Guedes, C. (2014). Effects of different restraints on the weld-induced residual deformations and stresses in a steel plate. International Journal of Advanced Manufacturing Technology. 71, 699–710. doi:10.1007/s00170-013-5521-9

[3] Adak, M. and Mandal, A. (2021). Numerical solution of fourth-order boundary value problems for euler-bernoulli beam equation using FDM. Journal of Physics Conference Series. 2070(1), 012052. doi:10.1088/1742-6596/2070/1/012052.

[4] Adak, M. (2020). Comparison of Explicit and Implicit Finite difference Schemes on Diffusion Equation, In book: Mathematical Modeling and Computational Tools. 227-238. doi:10.1007/978-981-15-3615-1_15.

[5] Adak, M. (2021). Electrostatic potential distribution by numerical method alternating direction implicit method. Proceedings of the 2021 1st International Conference on Advances in Electrical, Computing, Communications and Sustainable Technologies, ICAECT 2021 9392562.| doi: 10.1109/ICAECT49130.2021.9392562.

[6] Adak, M. (2022). Numerical solution of laplace and poisson equations for regular and irregular domain using five-point formula. Lecture Notes in Electrical Engineering. 897, pp. 271–281.

[7] Balagurusamy, E. (1999). Numerical Methods. India, New Delhi: McGraw Hill.,

[8] Jain, M. K., Iyengar, S. R., and Jain, R. K. (1985). Numerical methods for scientific and engineering computation, India, New Delhi: Wiley Eastern Limited.

[9] Lihtinen, J. (2003). Time domain numerical solution of the wave equation, thesis. Corpus ID: 17799283, pp. 1–17, 2003

[10] Lakshmi R., Muthuselvi M. (2013). Numerical solution for boundary value problem using finite difference method. International Journal of Innovative Research in Science, Engineering and Technology. 2, 10. http://www.ijirset.com/upload/october/26_NUMERICAL.pdf.

[11] Muhammad A. N., Eisa Al-Said, and Khalida I. N. (2012). Finite difference method for solving a system of third order boundary value problems. Journal of Applied Mathematics. https://doi.org/10.1155/2012/351764.

[12] Muhammad A. (2011). The Solution of the hyperbolic differential equation by finite difference method using the implicit scheme. Jurnal Rekayasa Struktur & Infrastruktur. 5(1), 33.

[13] Sastry, S. S. (2004). Engineering Mathematics. India, New Delhi: Prentice-Hall of India.

[14] Sastry, S. S. (2012). Introductory methods of numerical analysis. India, New Delhi: PHI Learning Private Limited.

[15] Scheid, F. (1968). Theory and problems of numerical analysis, schaum series. New York, NY: McGraw Hill.

[16] Siddiqi, S. S. and Iftikhar, M. (2013). Numerical solution of higher order boundary value problems, abstract and applied analysis. 2013, 1–12. doi.org/10.1155/2013/427521.

[17] Xu, X. and Zhou, F. (2015). Numerical solutions for the eighth-order initial and boundary value problems using the second kind chebyshev wavelets, advances in mathematical physics. 2015, 964623, 1–9. https://doi.org/10.1155/2015/964623.

Chapter 10

Experimental study of copper slag as a natural sand replacement

Amruta A. Yadav[1,a], Ajay R. Gajbhiye[2,b] and Pranita S. Bhandari[3,c]

[1]Research Scholar, Department of Civil Engineering, Yeshwantrao Chavan College of Engineering, Nagpur, India

[2]Professor, Department of Civil Engineering, Yeshwantrao Chavan College of Engineering, Nagpur, India

[3]Assistant Professor, Department of Civil Engineering, Priyadarshini College of Engineering, Nagpur, India

Abstract

This experimental research work examined the impact of incorporating copper slag (CS) as a substitute material for natural sand (NS) in concrete mix design on material's mechanical qualities. Eight concrete mixtures with CS concentrations for the control mix i. e 0-100% were casted. Compressive, split tensile, flexural strengths and workability of concrete mixes were investigated. According to the findings, workability significantly increased as the CS content increased. NS replaced by CS up to 60% produced strength comparable to the control mix. The mixes that replaced CS by 80% and 100% had the lowest compressive strength values. According to the findings, CS gives NS a 30% replacement rate that maximises flexural, split tensile and compressive strength. For concrete to be made with good strength, it is recommended to use 30% of CS as NS.

Keywords: Copper slag, strength, workability

Introduction

Cement, sand and aggregate are necessary ingredients for the construction industry. Cement and natural sand (NS) are two essential ingredients in the preparation of concrete, and they have a significant impact on mix design. Industrialisation is essential for the development of our nation. However, numerous tonnes of industrial waste are released annually from industries. Utilising industrial waste materials in concrete can assist in compensating for the shortage of natural resources, solve the waste disposal issue, and develop alternative methods of protecting the environment. The primary objectives of the government and environmental protection organisations are to find solutions to the disposal issues and health risks associated with by-products. In the construction industry, some industrial waste items have been effectively used to produce concrete. The quantity and variety of waste produced have increased along with the increase in global population. Copper slag (CS) is a mining waste material obtained during the production of copper. It is produced in large quantities during the extraction of copper [7]. For every million tonnes of extracted copper ore, around 2.2- million tonnes of CS are produced [1]. The disposal of such a large amount of waste poses a significant environmental challenge. The recyclable materials or the waste material of mining industry can be used as an artificial aggregate for providing a viable solution because of a new challenge imposed caused by the depletion of natural aggregates in the near future [3]. Because of its unique mechanical and chemical properties, CS can be utilised in concrete as an aggregate or as a partial replacement for Portland cement. When CS usage as aggregate, it possesses a number of advantageous mechanical features, including good soundness characteristics, good abrasion resistance and good stability [5]. Also, CS exhibits pozzolanic properties since it contains a low CaO content and other oxides such as Al_2O_3, SiO_2 and Fe_2O_3 [5]. When incorporated as an ingredient in concrete, CS, which has a glossy shine and is granular in nature, enhances the strength of the material, durability, workability, etc. [7-8]. Concrete is made more affordable by using waste materials, and recycling material is considered to be the most environment friendly solution to the problem of waste disposal [2]. As a result, the current study attempts to mitigate the potential environmental impact by substituting CS for NS.

Materials and methods

Cement

In this investigation, regular Portland cement of Ultra tech 53 grade is used. The construction industry in India uses this cement the most. The table below shows cement's various characteristics. According to IS guidelines, it conforms number of standard tests.

Fine aggregates

Sand should be hard and durable and free of clay or inorganic materials. Fine aggregate is the material that passes through the 4.75 mm BIS sieve.

---

[a]amruta1227@gmail.com, [b]argajbhiye@ycce.edu, [c]pranita.bhandari11@gmail.com

DOI: 10.1201/9781003450252-10

Copper slag

In this study the CS that is produced from Birla Copper is used. Because of its high specific gravity, the material is heavier.

Coarse aggregate

The coarse aggregate used in this research are locally available and retained on 4.75 mm IS sieve.

Table 10.1 Physical properties of cement

Property	Experimental value	Standard value for OPC
Fineness of cement (%)	1.5	–
Consistency (%)	30	–
Initial setting time (min)	40	>60
Final setting time (min)	225	<600

Source: Author

Table 10.2 Physical properties of sand

Property	Natural sand
Specific gravity (gm/cc)	2.64
Water absorption (%)	1.06
Fineness modulus	2.84
Bulking of sand (%)	28

Source: Author

Table 10.3 Physical properties of CS

Property	Copper slag
Specific gravity (gm/cc)	3.31
Water absorption (%)	0.87
Fineness modulus	2.86

Source: Author

Table 10.4 Chemical properties of natural and copper slag

Component	Natural sand	Copper slag
SiO_2	73.79	27.94
K_2O	3.46	0.98
CaO	1.79	3.55
MgO	0.47	0.99
P_2O_5	0.13	0.1
Fe_2O_3	2.69	56.75
Al_2O_3	15.39	3.75
Na_2O	0.59	0.52
TiO_2	0.34	0.5

Source: Author

Table 10.5 Physical properties

Property	Coarse aggregate
Specific gravity (gm/cc)	2.89
Water Absorption (%)	0.8

Source: Author

Plasticiser

Viscofluz 5507 is a high performance superplasticiser designed for applications that necessitate significant water reduction and prolonged workability retention. A plasticiser weighing 1.0% of the cement was added to the mixture.

Fineness modulus

The fineness modulus is an empirical number that may be calculated by summing the percentages of aggregate retained on each sieve in the sample and dividing the result by 100. The value of the fineness modulus is determined using the Sieve analysis method. A finer aggregate is indicated by the lower value. Sieve analysis method is conducted on fine aggregate for various combination of NS and CS as per IS 2386 part 1 1963.

The value of fineness modulus is found to be 2.8 in most cases. It means that the average value of the aggregate falls between the second and third sieves. According to the Figure 10.7, the typical aggregate size ranges from 0.3–0.6 mm.

Gradation

According to IS 2386 part I, the grading of the fine aggregate was determined. Figure 10.1 shows that the NS lies in zone II, as per the grading limit provide in IS 383-2016. The graph is plotted between sieve size and passing percentage.

Table 10.6 Fineness modulus

S.N	IS Sieve	Cumulative passing (%)							
		CS(0) + NS(100)	CS(10) + NS(90)	CS(20) + NS(80)	CS(30) + NS(70)	CS(40) + NS(60)	CS(60) + NS(40)	CS(80) + NS(20)	CS(100) + NS(0)
1	10 mm	100	100	100	100	100	100	100	100
2	4.75 mm	93.75	94.6	95.2	96.25	97.325	96.565	96.55	100
3	2.36 mm	89.75	91.7	92.4	94.475	95.675	94.615	94.2	100
4	1 18 mm	70.25	75.25	75.1	77.55	77.65	77.29	75.965	84.875
5	600 um	48.25	48.05	48.3	45.525	43.15	43.19	42.125	26.675
6	300 um	12.25	10.15	11.35	9.625	10.6	9.675	10.685	1.525
7	150 um	1.25	0.75	1.4	0.7	0.675	0.95	1.535	0.425
8	10 mm	–	–	–	–	–	–	–	–
Fineness Modulus		2.84	2.79	2.762	2.75	2.74	2.77	2.78	2.86

Source: Author

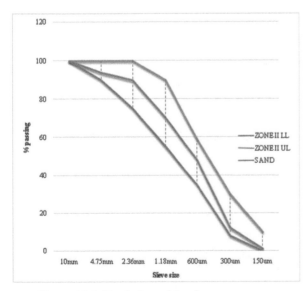

Figure 10.1 Gradation of Sand

Source: Author

The perfect grading or perfect particle size distribution is observed for CS(30)+NS(70) mix since particle size distribution curve of zone II nearly matches with the combination between the distribution of particle sizes of 30% CS as shown in Figure 10.2

Experimetal procedure

In the initial state, workability parameters like the slump value were investigated. Strength tests such as flexural strength, split tensile as well as compressive strength were examined in the hardened stage. By preparing 150 mm × 150 mm x 150 mm cube specimens in moulds to measure the value of compressive strength for a curing period of 7, 14 and 28 days. A 100 mm by 100 mm by 500 mm beam specimen was cast for 28 days to determine its flexural strength. Cylindrical specimens with 150 mm diameters and 300 mm heights were casted for 28 days curing to measure their split tensile strength.

Mix proportions

The following Table 10.7 shows the mix proportion used in this study. A mix proportion of M-30 grade were used for this work. For the comparative study, eight mixes were casted using varied amounts of CS as a partial replacement for sand, ranging from 0-100% as shown in Table 10.8

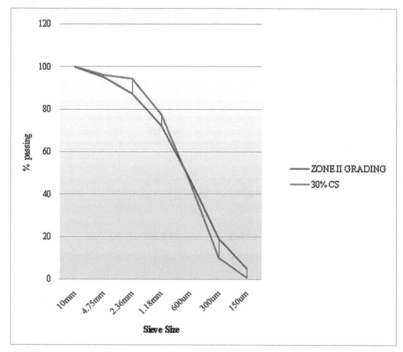

Figure 10.2 Gradation of 30% Copper Slag
Source: Author

Table 10.7 Various mix replacements (%)

S.N	Mix	Proportion			
		Cement	Fine Aggregate		Coarse Aggregate
			CS	NS	
1	CS(0)+NS(100)	100	0	100	100
2	CS(10)+NS(90)	100	10	90	100
3	CS(20)+NS(80	100	20	80	100
4	CS(30)+NS(70)	100	30	70	100
5	CS(40)+NS(60)	100	40	60	100
6	CS(60)+NS(40)	100	60	40	100
7	CS(80)+NS(20)	100	80	20	100
8	CS(100)+NS(0)	100	100	0	100

Source: Author

Table 10.8 Mix proportions (kg/m³)

S.N	Mix	Cement	Fine aggregate		Coarse aggregate	Water
			Natural sand	Copper slag		
1	CS(0)+NS(100)	350	701.18	–	1368.17	158.53
2	CS(10)+NS(90)	350	631.06	88.08	1368.17	158.55
3	CS(20)+NS(80	350	560.94	176.16	1368.17	158.57
4	CS(30)+NS(70)	350	490.83	264.25	1368.17	158.59
5	CS(40)+NS(60)	350	420.71	352.33	1368.17	158.62
6	CS(60)+NS(40)	350	280.47	528.49	1368.17	158.66
7	CS(80)+NS(20)	350	140.24	704.66	1368.17	158.70
8	CS(100)+NS(0)	350	–	880.82	1368.17	158.75

Source: Author

Table 10.9 Slump values for various mixes

Mix	Slump value in mm
CS(0)+NS(100)	90
CS(10)+NS(90)	100
CS(20)+NS(80	105
CS(30)+NS(70)	110
CS(40)+NS(60)	112
CS(60)+NS(40)	115
CS(80)+NS(20)	120
CS(100)+NS(0)	130

Source: Author

Table 10.10 Compressive strength

Mix	7 Days (MPa)	14 Days (MPa)	28 Days (MPa)
CS(0)+NS(100)	36.69	37.78	39.33
CS(10)+NS(90)	39.80	42.67	46.07
CS(20)+NS(80	41.78	44.44	48.44
CS(30)+NS(70)	43.56	45.78	50.22
CS(40)+NS(60)	37.78	40.00	48.89
CS(60)+NS(40)	36.89	37.78	41.33
CS(80)+NS(20)	32.44	33.33	35.56
CS(100)+NS(0)	28.00	29.33	31.11

Source: Author

Results and Discussion

Workability

Concrete was prepared for various mixes. Fresh concrete was used for the slump cone test. The outcome of the slump test, in accordance with IS 1199, demonstrates the behaviour of a compacted, inverted cone of concrete when subjected to gravity. The concrete's consistency is assessed by slump cone test.

According to the findings, increasing the CS content in the mix increases the concrete's workability. The value of the slump for CS(100)+NS(0) was 130 mm while for control mix it is 90 mm. This high increase was brought on by the excess free water that was left over after the absorption and hydration processes were complete because CS has a glassy or smooth surface and has a lower water absorption rate than NS [8]. It is observed that mixes with high levels of CS shows the bleeding and segregation problems, which adversely affect the performance concrete.

Compressive strength

According to Figure 10.3 and 10.4, the mix CS(30)+NS(70) outperformed all other mixes in terms of compressive strength. For each curing period, the strength was gradually increased. This can be because of the CS's higher iron oxide (Fe_2O_3) content as well as the material's toughness. The control mix, CS(0)+NS(100) had a compressive strength of 36.69 N/mm², 37.78 N/mm² and 39.33 N/mm², whereas CS(30)+NS(70) had a compressive strength of 43.56 N/mm², 45.78 N/mm² and 50.22 N/mm² after 7, 14 and 28 days of curing, respectively.

Figure 10.5 shows the percentage increase in compressive strength of the cubes over the control mixture concrete after 28 days of curing. For concrete mixes CS(10) +NS(90), CS(20) +NS(80), CS(30) +NS(70), CS(40) +NS(60), and CS(60) +NS(40), respectively, the strength increased by 17.14%, 23.16%, 27.68%, 24.29% and 5.08%.

Flexural strength

The beam specimen was casted (500 mm × 100 mm × 100 mm) and allowed to cure for 28 days before being put through the flexural test apparatus.

For CS replacement of 30% as NS the maximum flexural strength was achieved i.e. for CS(30) +NS(70) mix, as shown in Figure 10.6. After that, the mix's strength decreases as the copper slag content increased. The mix CS(30) +NS(70) produced the highest flexural strength of 9.4 N/mm². Flexural strength increased

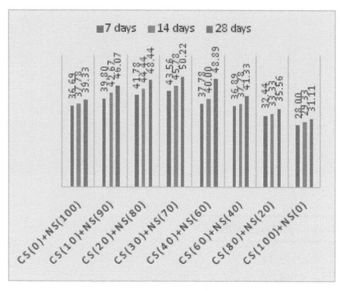

Figure 10.3 Compressive strength
Source: Author

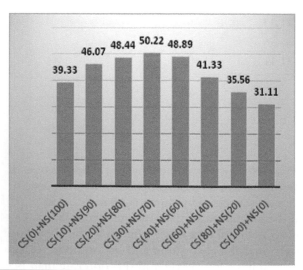

Figure 10.4 Compressive strength at 28 days
Source: Author

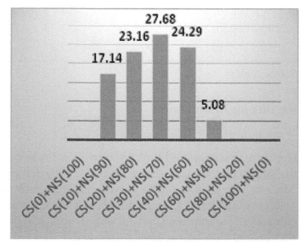

Figure 10.5 % Increase in compressive strength w.r.t control mix
Source: Author

Table 10.11 Flexural strength

Mix	28 Days (MPa)
	7.87
CS(10)+NS(90)	8.80
CS(20)+NS(80)	9.20
CS(30)+NS(70)	9.40
CS(40)+NS(60)	8.60
CS(60)+NS(40)	8.00
CS(80)+NS(20)	6.48
CS(100)+NS(0)	5.72

Source: Author

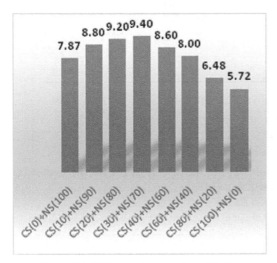

Figure 10.6 Flexural strength at 28 days
Source: Author

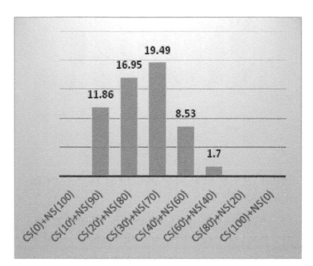

Figure 10.7 % Increase in flexural strength
Source: Author

Table 10.12 Split tensile strength

Mix	28 Days (MPa)
CS(0)+NS(100)	2.87
CS(10)+NS(90)	2.97
CS(20)+NS(80	3.11
CS(30)+NS(70)	3.47
CS(40)+NS(60)	3.26
CS(60)+NS(40)	3.04
CS(80)+NS(20)	2.83
CS(100)+NS(0)	2.12

Source: Author

by 11.86%, 16.95%, 19.49%, 8.53%, and 1.7% for the CS(10) + NS(90), CS(20) + NS(80), CS(30) + NS(70), CS(40) + NS(60) and CS (60) + NS (40) mixes, respectively as seen in Figure 10.7

Spilt tensile strength

The cylinder is casted with dimensions 300 mm in height and 150 mm in diameter for 28 day of curing to evaluatethe spiltlt tensile strength.

These results are also similar to the other strength test, as indicated in Table 10.12. The maximum spilt tensile strength was found when natural sand was replaced by 30% CS, as shown in Figure 10.8. After that, the mix's strength decreases as the copper slag content increased. CS(30) + NS(70) mix produced the highest spilt tensile strength of 3.47 N/mm².

According to Figure 10.9, the spilt tensile strength for the mixes CS(10) + NS(90), CS(20) +NS(80), CS(30) + NS(70), CS (40) + NS(60), and CS(60) + NS(40), increased by 3.7%, 8.64%,20.99%, and 13.58% and 6.17%, respectively as compared to control mix.

Density

The results shown in Table 10.13, prove that as the content of CS increases simultaneously the density of concrete increases this is because of higher specific gravity of CS which was 3.31 as compared to NS value i.e. 2.64.

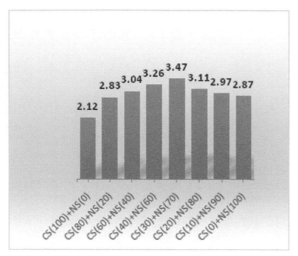

Figure 10.8 Split tensile strength at 28 days (MPa)
Source: Author

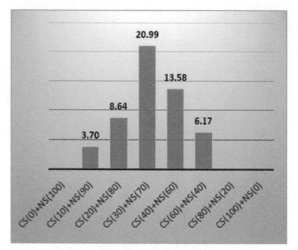

Figure 10.9 % Increase in split tensile strength w.r.t control mix (MPa)
Source: Author

Table 10.13 Density

Mix	Density (kg/m³)
CS(0)+NS(100)	2630.37
CS(10)+NS(90)	2634.07
CS(20)+NS(80	2654.81
CS(30)+NS(70)	2723.70
CS(40)+NS(60)	2752.59
CS(60)+NS(40)	2794.07
CS(80)+NS(20)	2844.44
CS(100)+NS(0)	2897.78

Source: Author

Conclusion

As the content of copper slag (CS) increases simultaneously the workability of concrete mix increases. The compressive strength of concrete increases by 27.68% at a replacement level of 30% CS. The addition of 30% of CS in concrete, the flexural strength increased by 19.49%. There was an increase of 20.99% split tensile strength at 30% replacement of CS in concrete. Strength of specimen tested was found to be higher upto 60% replacement level of CS but further addition reduces the strength. Concrete's density increases as the amount of CS in it increased.

References

[1] Al-Jabri, K. S., Hisada, M., Al-Oraimi, S. K., and Al-Saidy, A. H. (209). Copper slag as sand replacement for high performance concrete. ELSEVEIR. 31(7). doi:10.1016/j.cemconcomp.2009.04.007

[2] Dash, M. K. and Patro, S. K., and Rath, A. K. (2016). Sustainable use of industrial-waste as partial replacement of fine aggregate for preparation of concrete – A review. International Journal of Sustainable Built Environment. 5(2), 484-516. ISSN 2212-6090. https://doi.org/10.1016/j.ijsbe.2016.04.006

[3] Edwin, R. S., Gruyaert, E., and Belie, N. D. (2022). Valorization of secondary copper slag as aggregate and cement replacement in ultra-high performance concrete. Journal of Building Engineering. 5, 104567. https://doi.org/10.1016/j.jobe.2022.104567

[4] Al Jabri, K. S., Al-Saidy, A. H., and Taha, R. (2011). Effect of copper slag as a fine aggregate on the properties of cement mortars and concrete. Construction and Building Materials. 25, 933–938. doi:10.1016/j.conbuildmat.2010.06.090

[5] Ameri, F., Shoaei, P., Musaeei, H. R., Zareei, S. A., and Ban, C. C. Partial replacement of copper slag with treated crumb rubber aggregates in alkali-activated slag mortar. Construction and Building Materials. 256, 119468. https://doi.org/10.1016/j.conbuildmat.2020.119468

[6] Patil, M. P. and Patil, Y. D. (2020). Effect of copper slag and granite dust as sand replacement on the properties of concrete. Materials Today: Proceedings. 43. Part 2, 2021, 1666–1677. https://doi.org/10.1016/j.matpr.2020.10.029

Chapter 11

Characterisation of thin sandwich panels made of glass fibre and polyester foam

V. B. Ugale and P. A. Thakare

Mechanical Department, College of Military Engineering, Pune, India

Abstract

Three kinds of glass fibre reinforced (GFRP) thin sandwich panels were studied. A thin layer of glass mat Glass chopped strand mat (CSM) was added between core and face sheet to boost interlaminar shear strength in two kinds of panels manufactured by hand layup process, while the third panel was manufactured without inserting CSM but the vacuum bagging technique was employed for fabrication of panel. The polyester foam was offered in two totally different forms, XM and Xi, which were used as core material to fabricate the panels. There were two different kinds of plate tests performed: (i) low speed impact (ii) gradually applied transverse loading through Universal Testing Machine. The properties of sandwich panels were compared with that of MS sheet. The weight of 3 mm thick GFRP sandwich panels was less than 50% of that of 1 mm thick MS sheet. The damage area, due to separation between face sheet and core, was spread just around 36 mm diameter and permanent depression was less than 1 mm under 18 J impact. In the panels fabricated by hand layup process, the damage was due to weak interface between face sheet and core. However, in the panels fabricated by vacuum bagging technique, interface was stronger and delamination was due to failure of the core. In addition, permanent deformation in MS sheet was much higher up to 10 mm and spread over a large area. Thin GFRPs and which panel is the suitable alternative material to MS sheet for various applications.

Keywords: GFRP, thin sandwich panel, characterisation, polyester foam

Introduction

Fibre reinforced polymer (FRP) materials are getting used in several applications in the monocoque kind (Figure 11.1) where the fibre strengthened chemical compound laminas made of same reinforcement material are used. During this kind, the fibre becomes valuable and the layers in the central region near the neutral plane aren't utilised to their full potential.

To form higher use of fibre and to possess light-weight, sandwich panels (Figure 11.2) are being developed where the light weight and low-cost core material is used at the central region of the FRP laminate. For sandwich panels, the moment-of-inertia of the structure increases by separating the face sheets by the core. On being subjected to flexural loading, the face sheets are under maximum tension/compression and largely contribute to the strength of the sandwich panel [19,20].

The polyester foam is offered as the sheets of 1-3 mm thickness [1]. Hence, it's able to fabricate sandwich panels of thickness less than three millimetres. The polyester foam is offered in two totally different forms, XM and Xi. It is appropriate for usage as a core material because of its low weight. Thick sandwich panels are studied by several investigators [2-14]. In addition, the study on the FRP panels where all the layers are made of same reinforcement material is done by researchers [15-27]. The impact harm on FRP composite

Figure 11.1 Monocoque structure made of FRP laminas
Source: Author

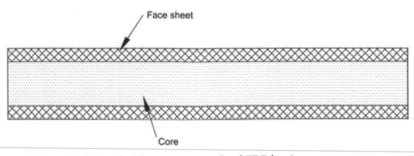

Figure 11.2 Sandwich structure made of FRP laminas
Source: Author

DOI: 10.1201/9781003450252-11

plates has drawn the attention of researchers in a big waysince the layered structure of FRP laminate is prone to delamination under impact loading and there are numerous researchers' findings in the literature. However, there is need to characterise thin FRP sandwich panels since very few literatures are available on it. FRP material is very attractive for use in structural applications due to its high strength-to-weight and stiffness-to-weight ratios, corrosion resistance, light weight, potentially high durability and less prone to environmental degradation [19-22].

The target of this work is to characterise the thin sandwich FRP laminate of glass fibre and polyester foam of thickness close to three mm for low-rate impact and gradually applied load. The properties are compared with the conventionally used material, MS sheet.

Experimental Procedure

Types of FRP Laminates

For the current investigation, three different types of panels were created: (i) XMC/CSM TSP (ii) XiC/CSM TSP (iii) XMC/VB TSP. In all these three panels, outer layers were made of glass fabric (GF).

XMC/CSM TSP was prepared by using core material made of polyester foam XM (Figure 11.3a). A layer of glass mat (CSM) was used at the interfaces between GF and Polyester foam XM to boost the strength in shear. This was fabricated through the hand layup technique.

XiC/CSM TSP was similar to XMC/CSM TSP except with the difference that polyester foam XM was replaced by polyester foam Xi (Figure 11.3b).

XMC/VB TSP was prepared by using core material made of polyester foam XM without CSM at the interfaces between face sheet and core. This was fabricated using vacuum bagging technique.

Fabrication of Sandwich Panels

When producing XMC/CSM TSP and XiC/CSM TSP by hand layup, a moderate pressure of about 0.04 bar was applied using dead weights. These TSPs were 3.00 mm thick, with a 0.35 fibre volume fraction. Vacuum bagging was used to create XMC/VB TSP. Nearly 2.5 mm of thickness and 0.45 of fibre volume fraction were found in the panel.

Mass of Panels

XMC/CSM TSP and XiC/CSM TSP each had a mass of 3004 grams per square meter (gsm) and 3028 gsm, respectively. The mass of XMC/VB TSP was, however, significantly lowered to 1890 gsm. Mass of MS sheet and FRP panels are depicted in Figure 11.4. However, the mass of XMC/CSM TSP and XiC/CSM TSP was around 50% of mass of MS sheet that is 1 mm thick. The mass of XMC/VB TSP was approximately 33% of that of MS sheet.

Experimental Technique

Impact loading: There were three different energy levels used in the impact test, 6 J, 12 J, and 18 J. A plate specimen with a 150 mm diameter was firmly fastened around its perimeter by the device. At the centre of the specimen, an indenter was dropped vertically. The mass of impactor was 1 kg. It was dropped from the height depending on impact energy. The damage on the specimen was observed after impact. The schematic layout of drop weight impact test setup and specimen of sandwich panel for impact test are shown in Figure 11.5 [19, 20].

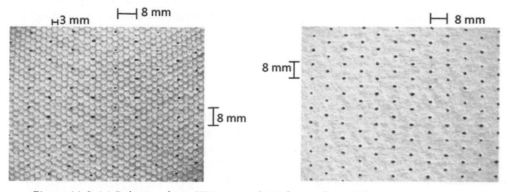

Figure 11.3 (a) Polyester foam XM core   (b) *Polyester* foam Xi core
Source: Author

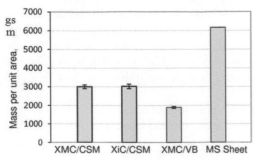

Figure 11.4 Mass of panels and MS sheet
Source: Author

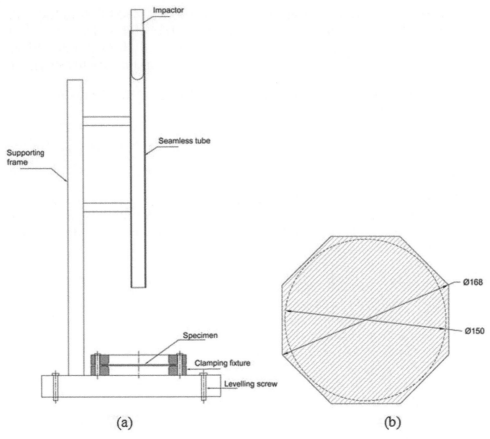

Figure 11.5 (a) Schematic layout of drop weight impact test set up, (b) Specimen of sandwich panel for impact test
Source: Author

Static load test: The transverse load was applied gradually under UTM at low speed of 0.1 mm/min. The plate specimen of diameter 150 mm with rigid clamping at the circumference was used to carry out the test. The load versus deflection graph and strength of the specimen were observed. Schematic and photograph of plate test set up are shown in Figure 11.6 [21, 22].

Numerical simulation

Impact loading

LS-DYNA software was used to perform the detail analysis of impacts at low velocity. The material of impactor was steel which was meshed by solid elements by assigning material properties. The incident velocity was given to the impactor. Constituent layers of the panel were discretised with the shell elements and material attributes were given to it (Figure 11.7). Table 11.1 displays the attributes of the constituent materials. To imitate the situation of rigid clamping of the specimen, all boundary elements' translation and rotation were set to zero. The sandwich panel's adjacent layers were characterised as having surface-to-surface contact with one another.

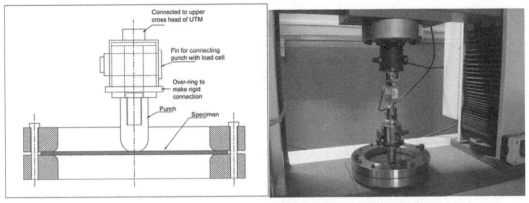

Figure 11.6 (a)Schematic of plate test, (b) Photograph of plate test set up of UTM
Source: Author

Figure 11.7 Meshed model of sandwich panel with impactor
Source: Author

Table 11.1 Mechanical properties

Material	E_x (MPa)	E_y (MPa)	E_z (MPa)	v_{xy}	v_{yz}	v_{zx}	G_{xy} (MPa)	G_{yz} (MPa)	G_{zx} (MPa)
GF/resin layer	19100	19100	6720	0.17	0.28	0.28	3500	2.06	2.06
CSM/resin layer	7120	7120	4810	0.17	0.28	0.28	2100	1550	1550
Polyester foam XM/ resin layer	1100	-	-	0.30	-	-	-	-	-
Polyester foam Xi/ resin layer	1000	-	-	0.30	-	-	-	-	-

Source: Author

The face sheet and core separated upon being loaded by impact mostly because of the stresses τ_{zx} and τ_{zy} at the interface (thickness of panel is along z-axis). By using a numerical simulation, they were ascertained. The degree of delamination damage was assessed by comparing the resulting shear stress with the interface's shear strength.

Results and Discussion

Damage under impact loading

It was investigated how sandwich panels were damaged under impact load through drop weight impact test set up (Figure 11.12). The XMC/CSM TSP, XiC/CSM TSP, and XMC/VB TSP have average total damage areas of 418 mm², 650 mm², and 201 mm² at 6J impact, respectively. It was raised for the XMC/CSM TSP, XiC/CSM TSP, and XMC/VB TSP at 12 J of impact to 807 mm², 843 mm², and 531 mm², respectively. The average damage area increased to 1030 mm², 1021 mm², and 991 mm² at 18 J of impact, respectively. At a high impact level of 18 J, it was discovered that the damage of XMC/VB TSP was nearly identical to that of XMC/CSM TSP and XiC/CSM TSP. However, the overall XMC/VB TSP damage decreased at 6 J and 12 J impact.

This study also created a GFRP monocoque laminate with a 3 mm thickness using the same glass fibers and epoxy in order to evaluate its impact resistance to that of sandwich panels. The impact damage on the GFRP monocoque laminate was less severe. Sandwich panel damage was not noticeably worse than monocoque structure damage. The damage area with an 18 J high impact is just around 36 mm in diameter, which is tolerable. Additionally, there is barely any panel deflection.

Through numerical simulation by LS-DYNA, the delamination was estimated for all three types of panels. At the back interface of the XMC/CSM TSP (z-axis was normal to the specimen plate), the resultant of the shear stresses τ_{zx} and τ_{zy}, $\tau_R = \sqrt{\tau_{zx}^2 + \tau_{zy}^2}$ was calculated. It was calculated at the elements along the 0^0 direction from the impact point (Figure 11.8 (a)) and along the 45^0 direction (Figure 11.8 (b)). The interlaminar shear strength of 5.2 MPa determined through experimentation was compared to the produced resultant shear stress at the rear interface. A short beam test was used to experimentally assess the shear strength of the interface of FRP laminas. Figures 11.8 displays the shear strength as a horizontal dotted line. According to the figures, the amount of delamination for impacts of 6 J, 12 J, and 18 J was 9 mm, 12 mm, and 18 mm along 0^0 direction. Delamination damage for impacts of 6 J, 12 J, and 18 J measured 9 mm, 12 mm, and 16 mm in the 45^0 direction. They were compared with the experimentally observed overall damage area of the XMC/CSM TSP (Figure 11.12). Experimental results matched well with the numerical results.

Through numerical simulation, it was possible to calculate the resultant of shear stresses, τ_R at the rear interface of XiC/CSM TSP panels at the elements along the 0^0 direction (Figure 11.9a) and along the 45^0 direction from the impact point (Figure 11.9b). The experimentally determined interlaminar shear strength of 4.9 MPa was compared to the produced resultant shear stress at the rear interface. Figures 11.9 depicts the interlaminar shear strength as a horizontal dotted line. According to the figures, the delamination was 10 mm, 13.2 mm, and 18.1 mm in the $0°$ direction for impacts of 6 J, 12 J, and 18 J. 10.2 mm, 12.1 mm, and 16 mm of delamination damage were present in the 45^0 direction. The predicted damage in XiC/CSM TSP panels was observed to be rather similar to the experimentally determined damage region (Figure 11.12).

Because the interface in XMC/VB TSP was stronger, the breakdown of the core caused the face sheet and core to separate. Figure 11.10 displays cross-sectional photographs of the panel taken under an optical microscope. It indicates that with 18 Jimpact, the rear face sheet of the panel separated with a failure of the core close to the interface. Within the damaged area, the back face sheet moved off its plane during impact pressure, and the core failed close to the rear interface.

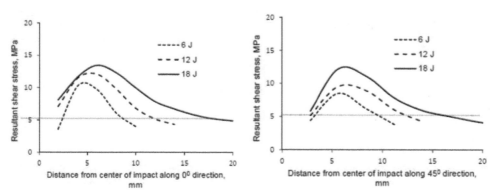

Figure 11.8 Resultant of shear stresses, τ_R at rear interface (a) along 0^0 direction (b) along 45^0 direction of XMC/CSM TSP

Source: Author

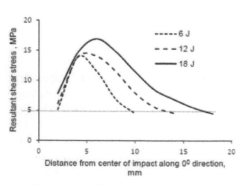 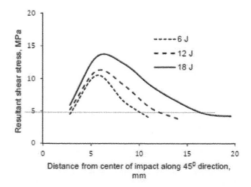

Figure 11.9 Resultant of shear stresses, τ_R at rear interface (a) along 0^0 direction, (b) along $45°$ direction of XiC/CSM TSP

Source: Author

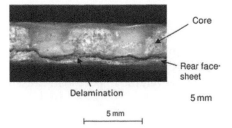

Figure 11.10 Cross section of XMC/VB TSP panel after impact at 18 J
Source: Author

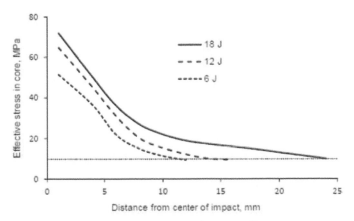

Figure 11.11 Under impact loading, an effective stress experienced by the core material of XMC/VB TSP panel
Source: Author

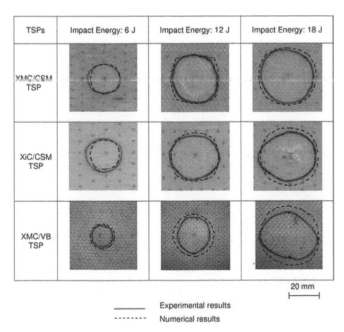

Figure 11.12 Overall area of damage after impact
Source: Author

The tensile stress caused in the lowest part of the core under impact loading was determined by LS-DYNA. It was compared with the tensile strength of 9.6 MPa of the core (polyester foam XM/epoxy layer) to estimate the degree of core failure. Under impact loading of 6 J, 12 J, and 18 J, Figure 11.11 depicts the tensile stress created in the lower side of the XMC/VB TSP in relation to the distance from the point of impact. Also, Figure 11.11 depicts the experimentally determined tensile strength of the polyester foam, XM/epoxy layer core through a horizontal dotted line. The illustration depicts the degree of core failure near the back interface which predicts the separation of the core from the back face sheet was 11 mm, 16 mm and 24 mm at 6 J, 12 J and 18 J impact respectively due to core failure. The predicted damage in XMC/VB TSP was observed to be rather similar to the experimentally determined damage region (Figure 11.12).

Deflection versus time curve under impact loading

Figure 11.13 displays numerically computed graph of deflection versus time. Because the thickness of the XMC/VB TSP was significantly less than that of the other two panels, the amplitude of deflection and time of the deflection mode were bigger for the XMC/VB TSP than for the XMC/CSM and XiC/CSM TSPs. As anticipated, the deflection mode's amplitude grew as impact energy rose. However, the impact load caused the MS sheet to deflect with significant plastic deformation of around 10 mm.

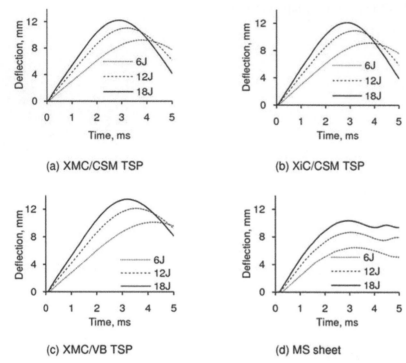

Figure 11.13 Graph of deflection versus time within impact force
Source: Author

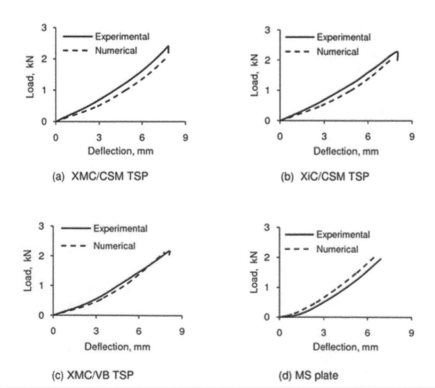

Figure 11.14 Graph of load versus deflection under gradually applied transverse loading
Source: Author

XMC/VB TSP was found to be superior to XMC/CSM TSP and XiC/CSM TSP because its mass was substantially less, and impact damage resistance of XMC/VB TSP was found to be similar to XMC/CSM TSP and XiC/CSM TSP. However, the elastic deflection of XMC/VB TSP was more because it is much thinner. But the permanent deflection was less than 1 mm and almost same in all the three kinds of panels.

Deflection versus time curve under static load test

Figure 11.14 displays the graph of load versus deflection obtained through experimentation for each type of panel, XMC/CSM TSP, XiC/CSM TSP, and XMC/VB TSP, including thin MS sheet when a plate specimen is statically loaded with a centre punch. The greatest load that XMC/VB TSP could withstand was 2122.95 N. However, the highest load that XMC/CSM TSP and XiC/CSM TSP could withstand was 2340 N and 2317 N, respectively.

Conclusions

Three different types of glass fibre and polyester foam reinforced thin FRP panels were created, two utilizing the hand layup method and one using the vacuum bagging method. They underwent impact and static loading studies.

XMC/VB TSP panel fabricated by vacuum bagging technique had a mass of 33% of 1 mm thick MS sheet, whereas the panels fabricated by hand layup process, XMC/CSM TSP and XiC/CSM TSP, had a mass of around 50% of 1 mm thick MS sheet.

Sandwich thin FRP panels' overall damage from the separation of the core and face sheet during low-velocity impact was not much higher than that of monocoque FRP panels. At a collision of 18 J, the damage spread outward just by radial distance of 18 mm. Additionally, the FRP panel's permanent deformation is hardly noticeable. However, permanent depression is upto 10 mm in 1 mm thick MS sheet and spread over a large distance.

In case of panels manufactured using the hand layup process, the damage due to delamination is caused by a weak interface between face sheet and core. However, core breaking in the vacuum bagged panels is what caused the delamination damage but the interface between the layers is stronger. Thin panels made of glass fabric and polyester foam XM, fabricated by vacuum bagging technique, is the most suitable alternative to MS sheets for various applications.

Acknowledgment

Authors acknowledge to College of Military Engineering, Pune for the support and helping in successful completion of fabrication, testing of the research work.

References

[1] Lantor Composites. https://www.lantorcomposites.com/
[2] Ahmed, K. S. and Vijayarangan, S. (2008). Tensile, flexural and interlaminar shear properties of woven jute and jute glass fabric reinforced polyester composites. Journal of Materials Processing Technology. 207(1-3). 330–5
[3] Nangia, S. and Biswas, S. (2000). Jute composite: technology and business opportunities. Proceedings of International Conference on Advances in Composites at IISc and HAL. pp. 295–305.
[4] Ning, H., Janowski, G. M., Vaidya, U. K., and Husman, G. (2007). Thermoplastic sandwich structure design and manufacturing for the body panel of mass transit vehicle. Composite Structures. 80(1), 82–91.
[5] Gaudenzi, P., Pascucci, A., Barboni, R., and Horoschenkoff, A. (1997). Analysis of a glass-fibre sandwich panel for car body constructions. Composite Structures. 38(1-4), 421–33.
[6] Park, J. H. and Ha, S. K. (2008). Impact damage resistance of sandwich structure subjected to low velocity impact. Journal of Material Processing Technology. 425–30.
[7] Imieliska, K., Guillaumat, L., Wojtyra, R., and Castaings, M. (2008). Effects of manufacturing and face/core bonding on impact damage in glass/polyester-PVC foam core sandwich panels. Composites: Part B. 39(6), 1034–41.
[8] Corigliano, A., Rizzi, E., and Papa, E. (2000). Experimental characterization and numerical simulations of a syntactic-foam/glass-fibre composite sandwich. Composite Science and Technology. 60, 2169–80.
[9] Qiao, P. and Yang, M. (2007). Impact analysis of fibre reinforced polymer honeycomb Composite sandwich beams. Composites: Part B. 38, 739–50.
[10] Scarponi, C., Briotti, G., Barboni, R., Marcone, A., and Iannone, M. (1996). Impact testing on composites and sandwich panels. Journal of Composite Materials. 30(17), 1873-1911.
[11] Mines, R. A. W. and Jones, N. (1995). Approximate elastic plastic analysis of static and impact behaviour of polymer composite sandwich beams. Journal of Composites. 26(12), 803–14.
[12] Anderson, T. and Madenci, E. (2000). Experimental investigation of low velocity impact characteristics of sandwich composites. Composite Structures. 50, 239–47.

[13] Lim, T. S., Lee, C. S., and Lee, D. G. (2004). Failure modes of foam core sandwich beams under static and impact loads. Journal of Composite Materials. 38, 1639–62.

[14] Schubel, P. M., Luo, J. J., and Daniel, I. M. (2005). Low velocity impact behaviour of composite sandwich panels. Composites: Part A. 36, 1389–96.

[15] Caprino, G. and Lopresto, V. (1999). Influence of material thickness on the response carbon fabric / epoxy panels to low velocity impact. Composite Science and Technology. 59, 2279–86.

[16] Lopez-puente, J., Zaera, R., and Navarro, C. (2002). The effect of low temperatures on the intermediate and high velocity impact response of CFRPs. Composites Part B. 33, 559–66.

[17] Cantwell, W. J. and Morton, J. (1989). Comparison of the low and high velocity impact response of CFRP. Composites. 20, 545–51.

[18] Dorey, G., Sidey, G. R., and Hutchings, J. (1978). Impact properties of carbon fibre/Kevlar 49 fibre hybrid composites. Composites. 1, 25–32.

[19] Kumar, P. and Rai, B. (1993). Delaminations of barely visible impact damage in CFRP laminates. Composite Structures. 23(4), 313–18.

[20] Kumar, P. and Rai, B. (1991). Reduction of Impact Damage in KFRP through Replacement of Surface Plies with Glass Fabric Plies, Composite Materials. 25(4), 694–702.

[21] Naik, N. K. and Nemani, B. (2001). Initiation of Damage in Composite Plates under Transverse Central Static Loading, Journal of Composite Structures, 52(2), 167–72.

[22] Naik, N. K., Chandra Sekher, Y., and Meduri, S. (2000). Damage in woven-fabric composites subjected to low-velocity impact. Journal of Composites Science and Technology. 60, 731–44.

[23] Tiberkak, R., Bachene, M., Rechak, S., and Necib B. (2008). Damage prediction in composite plates subjected to low velocity impact. Composite Structures, 83, 73–82.

[24] Feraboli, P. and Masini, A. (2004). Development of carbon/epoxy structural components for a high performance vehicle. Journal of Composites, Part B: Engineering. 35(4), 323–30.

[25] Singh, K. K., Singh, R. K., and Kumar, P. (2009). Toughness of adhesive bonded interface under static and dynamic loads — an experimental study. Journal of reinforced plastics and composites. 28(5), 601.

[26] Alexander, V., Alexander, S., Fausto, T., and Pierpaolo, C. (2020). Pultruded materials and structures: A review, Journal of Composite Materials, Sage publication, 54 (26).

[27] Abbood, I. S.,Odaa, S., Hasan, K. F., and Jasim M. A. (2021). Properties evaluation of fiber reinforced polymers and their constituent materials used in structures. A review, Materials Today Proceedings. 43(2), 1003–8.

Chapter 12

Effect of chemical admixtures on the hardened and durability properties of concrete

Kuruba Chandraprakash[1,a], Muhammed Zain Kangda[1,b], Mohammed Aslam[2,c], Divyarajsinh M. Solanki[3,d] and Sandeep Sathe[3,e]

[1]REVA University, India

[2]Changwon National University, South Korea

[3]MIT World Peace University, India

Abstract

Due to considerable traffic congestion in places such as Mumbai, particularly during weekdays, the time necessary to transport RMC from plant to building site is quite long. Keeping concrete workable for so long without sacrificing essential strength has become one of the most difficult problems in the RMC business. Furthermore, high ambient temperatures in the summer intensify the problem because high temperatures decrease the workability of fresh concrete. In this sense, the appropriate type and dosage of chemical admixture can play a vital impact. In light of the foregoing discussion, a thorough experimental examination was conducted with certain chemical admixtures often used in India's ready-mix concrete industry. The effects of their doses on fresh (workability) and hardened (compressive strength, splitting tensile strength, and ultrasonic pulse velocity properties of concrete have been examined in order to determine the optimum content of chemical admixture. Taking a look into concrete technology discipline, many minerals as well as chemical admixtures have been selected to upgrade the physical and chemical properties of the normally manufactured concrete. In the present study chemical additives namely sodium bicarbonate and aluminium nitrate have been selected to compose and enhance the performance of concrete. These selected admixtures are added in the proportions of 0.25%, 0.5%, 0.75% and 1% by weight of cement. This study conducts the compressive strength test, splitting tensile strength tests after 7, 14 and 28 days of curing the specimens to evaluate the efficiency of chemical admixtures in improving the performance of freshly prepared concrete. In addition, durability test is also performed using sulphuric acid and sodium sulphate solutions. The observation that was made depicts that sodium bicarbonate yields the highest compressive strength amongst the selected admixtures. It also upgrades the porosity of cement paste and reduces the expansion of mortar. The chemical agent in the form of aluminium nitrate acts as a strong oxidizing and corrosion inhibitor agent that prevents the corrosion of reinforcement It also enhance the compressive strength of concrete in comparison to the traditional concrete specimens and it is also non-flammable. The durability test results confirm the improved resistance against both alkaline and acidic attack of the normal concrete which happened due to addition of the selected admixtures without any reduction in strength. The addition of sodium bicarbonate and aluminium nitrate in the range of 0.25-0.5% generates low workable concrete and 0.75-1.00% dose leads to medium workable concrete. It is observed that in order to attain maximum strength, the optimum percentage of admixture to be added was 0.5% of the weight of cement. The minimum loss in the concrete sample was also achieved at 0.5% dose of chemical admixture, during acid attack test. The study observes that an increases in the cube compressive strength by 3.70% and 6.02% at 0.25% and 0.50% admixtures as compared to normal concrete without admixture at 28 days curing. An increase in the cylindrical compressive strength by 4.54% and 7.35% at 0.25% and 0.50% admixtures at 28 days curing is also observed. The split tensile strength is increased by 3.70% and 6.06% at 0.25% and 0.50% admixture as compared to normal concrete without admixture at 28 days curing.

Keywords: Chemical admixtures, concrete, aluminium nitrate, sodium bicarbonate, durability and compressive strength test

Introduction

The American Concrete Institute (ACI) committee 212 has defined admixtures as the unique ingredients that are added to the basic composition of concrete prepared by an appropriate proportioning of cement, sand, aggregates and water. In the concrete industry, the term additives are used when special raw materials namely alumina, lime, iron oxide and silica are added to cement to improve its properties during the phase of manufacturing. The primary objective of adding these substances is to improve and escalate the characteristics of the concrete which is developed in fresh or hardened state. Admixtures added to concrete in the past have played the role of accelerators, retarders and wear resistance. The other advantages of admixtures also include reduction in parameters like bleeding, shrinkage, permeability as well as corrosion of concrete reinforcement. The presence of admixtures also supplemented numerous properties like workability, strength, durability and pozzolanic properties of the concrete. The history of admixtures reverts back to the Roman era when animal blood, milk and lard produced the specimens of durable concrete. The other civilizations,

[a]k.chandraprakash@reva.edu.in, [b]zainkangda@gmail.com, [c]aslam@changwon.ac.kr, [d]divyarajsinhs911@gmail.com, [e]sandeepsatheresearch@gmail.com

DOI: 10.1201/9781003450252-12

that is to say the Chinese and Mayans added boiled bananas and bark extracts to ameliorate the performance of concrete which was prepared. The employment of substances including rice paste, oil, cactus juice, latex and molasses have also behaved as admixtures in the medieval period to upgrade the quality of building materials in past. Admixtures are broadly classified as mineral and chemical and the comparison between them are tabulated in Table 12.1. The review presented by Nagataki [1] recommended that the primary aim of the mineral admixtures is to reduce price of cementitious mix and enhance the strength characteristics of concrete. The study predicted the use of rice husk and ultra-low heat blends as admixtures along with improving the skills of engineers to use them in the construction sites. The researches which are governing the ability to improve the performance of the concrete along with addition of various mineral admixtures are tabulated in Table 12.2. The study gives an overview of the character of several mineral admixtures used in the past and their efficiency in improving the quality of the developed concrete for numerous civil engineering projects. The concrete that is supplied for the construction of civil engineering structures has advanced with

Table 12.1 Comparison between mineral and chemical admixtures.

Sr. No.	Mineral admixtures	Chemical admixtures
1	Mineral admixtures are fine solid waste that are produced from industries.	Chemical admixtures are the compounds which are water soluble.
2	The admixtures are added in larger quantity, and they act as replacement for cement.	These admixtures are added in certain proportions to concrete.
3	Mineral based admixtures help in improving the durability, serviceability of concrete and reduces the cost of construction.	Chemical based admixtures help in constructing structure that is durable, strong and waterproof.
4	Examples are fly ash, silica fume and slag	Classified as accelerators (calcium chloride), retarders (calcium sulphate), plasticizers (calcium ligno-sulphonates), super-plasticizers (acrylic polymer) and water proofing (aluminum sulfate).

Table 12.2 Effectiveness of mineral admixtures in improving concrete performance.

Sr. No.	Researcher	Mineral admixtures	Tests conducted/discussed	Highlights
1	Kanamarlapudi et al. [11]	Silica fume (SF), GGBS, fly-ash (FA), rice husk (RH), metakaolin and palm oil fuel ash.	Tests performed includes split tensile, flexural strength test, durability, workability and compressive strength test.	The review highlights the implication of numerous mineral admixtures in concrete.
2	Gonen and Yazicioglu [12]	FA and SF	The list of tests performed includes compressive strength test, capillary absorption test, porosity, wet-dry cycle and accelerated carbonation test.	This study highlights improvement in results yielded when the ratio of cement content to fly-ash as well as silica fumes was 10–15%.
3	Uysal and Sumer [13]	GGBS, FA, Basalt, limestone and marble powder.	Tests performed on concrete workability test, ultrasonic pulse velocity test, compressive strength test, sulphate and density resistance test.	The study focuses on how fly-ash and GGBS helped in increasing the compressive strength whereas basalt, limestone and marble powder not only acts as filler material but also provides resistance against sulphate attack.
4	Sulapha et al. [14]	GGBS, SF and FA.	Carbonation resistance test is performed.	The study highlights the relation between carbonation and compressive strength. The carbonation coefficient is inversely proportional to compressive strength. It also highlights that few favoring admixtures along with adequate curing can help in improving the ability to resist carbonation in concrete.

Sr. No.	Researcher	Mineral admixtures	Tests conducted/discussed	Highlights
5	Khan et al. [15]	SF, GGBS, FA, RA and metakaolin.	Test mainly including workability, heat of hydration, setting time, bleeding and reactivity were performed.	This review highlights the classification of mineral based admixtures into chemically active and micro-filler admixtures. It also discusses merits and demerits of the aforementioned classification.
6	Shannag [16]	FA and SF.	Density test, workability along with compressive strength test and split tensile strength test entailed with stress-strain diagram are a part of this study	Concrete cubes and cylinders using chemical admixtures as well as lightweight aggregates are casted in this study. With partially replacing the cement by adding 10% of silica fume and 5% Fly-ash helped in improving the stiffness and strength of samples.
7	Ramanathan et al. [17]	GGBS, SF and FA.	Primary tests including compressive strength test, workability test and split tensile strength test were performed on the sample.	Partial replacement of cement with fly-ash, silica fume and GGBS by 30%, 40%, and 50% by weight respectively, while keeping the waterpower ratio of 0.35 constant. Another important factor of how particle shape, size distribution and surface characteristics affected the performance of the admixtures incorporated in cementitious matrix was also observed.
8	Ping et al. [18]	Metakaolin, SF and slag.	Microstructure characteristics analysis along with pore structure and interfacial transition zone were considered for analysis	It highlights the fact that the descending order (metakaolin > silica fume > slag) of mineral admixtures helps to refine the microstructure behavior of concrete. Another observation made was, higher pozzolanic activity and surface area of metakaolin yielded higher compressive strength of concrete specimens.
9	Memon et al. [19]	FA, GGBS, SF and superplasticizer.	Firstly, porosity of the samples was tested, followed by pore size distribution and compressive strength test were performed.	The study highlights tests being conducted for obtaining high-strength concrete under seawater condition exposed to tidal zone. 30% and 70% of admixtures replaced cement which yielded a high-strength concrete with fine pore distribution while compared to natal Portland cement concrete.
10	Nehdi et al. [20]	RH	Numerous tests include carbon content analysis, oxide analysis. X-ray diffraction test was the part of experimentation. Along with the above tests grind ability and water demanding property were also taken into consideration.	The study highlights the fact that Torbed reactor technique was used to produce rice husk yielded low carbon content as compared to the conventional fluidized bed technology. The proposed technology tends to improve the properties like compressive strength, surface and chloride resistance of concrete specimens when compared to the same content of silica fume and rice husk produced using conventional methods.

the help of incorporating chemical and natural admixtures. It is no longer restricted to just mixing of cement, aggregate and water [2]. The addition of admixtures in the cementitious matrix has led to an improved production of an effective construction material. It has also been observed that the role of chemical admixtures is becoming crucial in enhancing the attributes of both fresh and hardened concrete [3]. Considering few physical and mechanical properties which are improved due to addition of chemical admixtures which are helping in facilitating the activities like characteristics setting, entrain air, reducing water content, increasing cohesiveness, enhancing the flow, introducing self-levelling properties, improving the durability, appearance and lastly enhancing the strength parameters. The new admixtures have proved that they play a crucial role in concrete as compared to new cement versions [4]. Other advantages of chemical admixtures also include perks like reduction in cost of the concrete mix by altering the plastic and hardened properties which are required to maintain the quality at the time of mixing the raw ingredients, transporting, placing, and curing [5, 6]. There are numerous disadvantages associated with admixtures which includes stiffer concrete gets produced, getting variations in initial slump, slump loss and large disparity in the flow characteristics using a blend of different admixtures [7, 8]. Thus, it is essential to ensure that the compatibility of admixtures along with cement and other additives before implementing the applications in the concrete mix. These chemical admixtures can be further classified under inorganic and organic materials [9]. The review by Francisca et al. [10] highlighted the demerits of chemical admixtures when added into an alkali activated concrete mix. The commonly used chemical admixtures in the previous years helped in improving the performance of concrete are tabulated in Table 12.3. Although several research with various types of chemical admixture were conducted, only few of them addressed the influence of chemical admixtures on the properties of concrete. There is a lot of data on old-generation plasticizers, but there is very little research on the chemical admixtures used in the RMC industries right now. The novelty of the present study is to investigate the role of chemical admixtures on the mechanical and durability properties of prolonged mixed concrete. The qualitative and quantitative impact of the selected chemical admixtures on the mechanical and durability properties of concrete at various dosages and mix design parameters are also explored in the present study. The influence of chemical admixtures on mechanical and durability properties of concrete is not addressed in old literatures. So that the effects of chemical admixtures on mechanical and durability properties of concrete at different dosages and for different mix design parameters are essential to be investigated.

Thus, the above-mentioned literature submits a proof that sodium bicarbonate has been extensively used as a chemical admixture in enhancing the compressive strength and workability characteristics of prepared concrete. The present research is directed towards enhancing the quality of concrete using two chemical admixtures namely sodium bicarbonate and aluminium nitrate. In the present study, chemical admixtures have been added in small proportions of 0.25%, 0.5%, 0.75%, and 1% to understand the performance of prepared concrete. Cement of OPC grade 53 is selected to produce M30 grade concrete in the mix proportions of 1:1.6:2.3. The effect of the admixtures and their dosages on fresh, hardened and durability properties of concrete is also investigated. The main objective of the present study is to investigate the role of chemical admixtures on the mechanical and durability properties of prolonged mixed concrete. The qualitative and quantitative impact of the selected chemical admixtures on the mechanical and durability properties of concrete at various dosages and mix design parameters are also explored in the present study. The qualitative impact of the concrete specimens after adding the necessary chemical admixture dosage are studied in the present study. This study investigates the influence of chemical admixtures on the concrete mix design which is different from old literature. Comprehensive description of the methodology, dosage selection and observing various changes effecting the mechanical and durability properties of mix design are given for better understanding of the study. The study likewise also entails qualitative impact of chemical admixtures on the mechanical and durability properties of the concrete mix design.

Experimental Program

Material Used

OPC grade 53 with a specific gravity of 3.15 is used as a binder for preparing the cementitious mix. The coarse aggregates used were having the maximum size of 12 mm and manufactured sand (M-Sand) which was used was passing through 4.75mm sieve which is retained on a 600-micron sieve. M-Sand generally can be used as a substitute for river sand. The M-sand used for the current study is produced using hard granite stone by crushing, with a cubical shape along with grounded edges. After crushing the M-sand is washed and graded into as a type of construction material. The sand is free from impurities like clay silt and organic impurities and also conforms to zone III [28]. Specific gravity of coarse and fine aggregate was obtained as 2.56 and 2.53 respectively. The fineness moduli for coarse and fine aggregates were also calculated, with values as 6.8 and 2.75, respectively [29]. The two types of chemical admixtures which are used viz. aluminium nitrate - Al (NO3)3 and Sodium bicarbonate NaHCO3. Aluminium nitrate being an admixture is generally used as a strong oxidizing agent. It depicts properties like high solubility and is also

Table 12.3 Significance of chemical admixtures in concrete.

Sr. No.	Researcher	Chemical admixtures	Tests conducted/discussed	Highlights
1	Collepardi [21]	Superplasticizers	The tests include slump test, shrinkage test and placing characteristics.	The study highlights merits and demerits of superplasticizers in the form of sulphonated and acrylic polymers, and their role in improving workability of cementitious mix at lower w/c content.
2	Yasuhide et al. [22]	Baking soda (sodium bicarbonate)	Electrical resistivity and compressive strength are the two tests performed on the specimens for desired investigation.	study highlights mathematical relationship between compressive strength of the concrete specimen and electrical resistivity property from the experimental analysis.
3	Almedia and Haselbach [23, 24]	Sodium bicarbonate	Compressive strength and porosity of the specimens were analyzed.	This study is based on mitigating the magnesium chloride attacks in concrete specimens without hampering the strength test results.
4	Nugroho [25]	baking powder	Workability and compressive strength are the prime tests performed to find out results supporting the study.	This study focuses that an optimum dose of 0.45% of sodium carbonate incorporated in concrete samples increases the compressive strength of specimen by 6.43%.
5	Gopakumar et al. [26]	There are numerous chemical admixtures incorporated in this study. They include Calcium acetate $(1H_2O)$, Magnesium acetate $(4H_2O)$, Calcium format, Calcium bromide $(2H_2O)$, Magnesium bromide $(6H_2O)$, Calcium nitrate $(4H_2O)$ and lastly Magnesium nitrate $(6H_2O)$.	The study includincludedes essential "ests like workability, mechanical properties tests, setting time and durability.	The study highlights development of new alkali silica reaction powder or liquid chemical admixture for the concrete mix.
6	Korhonen [27]	Calcium chloride and sodium chloride	Freeze-thaw durability test is considered to analyze the samples to support the bestowed study.	The study reviewed the performance of air entrained agents and high dose of chemicals admixtures and de-icing agents who provide resistance against complications like thawing and freezing, also improve the durability of concrete mixes in the snowy geographical locations.

used as a corrosion inhibitor to prevent steel from getting corroded. It is odourless and non-flammable in nature. The density of aluminum nitrate is 1.401 g/cm3and molecular weight is 375.13 g/mol. Sodium bicarbonate - NaHCO3 is added in the specimens to facilitate the progress of chemical interactions in the middle of carbonate components and free hydroxides present inside the cement paste. Sodium bicarbonate is odourless, white crystalline powder and it increases the rate of concrete carbonation. The density of sodium carbonate is 2.2 g/cm3 and molecular weight of sodium bicarbonate is and 84 g/mol, respectively.

Mix Proportioning

The concrete mix proportioning that was made for the purpose of experiment was designed according to M30 grade (30MPa) standards. The cementitious mix prepared was based on the guidelines provided by the Indian standard code i.e. (IS Code) [30]. The concrete mix design (reference) comprised a total 1080.3

kg/m³ of coarse aggregate, 717.9 kg/m³ fine aggregate and 444.4 kg/m³ of cement respectively. The w/c ratio was fixed at 0.45 to achieve reasonable workability of concrete. The cement powder and fine aggregates were first mixed together inside a pan mixer, and after initial mixing was done, coarse aggregates were added in it. Later the mixing was continued for further 1 to 2min to attain a uniform mixture. The chemical admixtures were also added in concrete mix during the time of mixing. Table 12.4 presents the quantity of materials required for preparing concrete along with admixtures.

Results and Discussion

Workability

The determination of workability of concrete mix was done as per the procedure given in IS code [31]. Workability is generally measured in terms of slump by a 300-mm-high slump cone with top diameter of 100mm and bottom diameter of 200 mm. The height difference between the mould and the highest point of concrete is considered as slump value which is tabulated in Table 12.5 accordingly. Another observation made shows that the slump increases along with an increase in the admixture content in the cementitious matrix.

Hardened Properties

To study and analyse the effect of admixtures on compressive strength of concrete samples, cube and cylindrical shaped moulds were incorporated having size 150mm diameter and 300mm height respectively. The specimens are prepared as per the guidelines given in IS 516:1959 [32]. Initially, the cube and cylinder mould were cleansed and brushed with oil on all the internal walls for easy removal of concrete samples after casting them. Curing of specimens was done under ambient condition where average room temperature was 23–28°C and then they were demoulded after 24 hrs. of casting. After 7, 14, and 28 days of curing, these specimens were taken for testing in a load-controlled compression-testing machine i.e., CTM having a capacity of 2000kN and at the rate of 400 kg/min. The test setup for cube and cylinder specimens are shown in Figure 12.1a and 12.1b, respectively, while Figure 12.2 shows the variation in compressive strength of concrete for different mixes respectively. From Figure 12.4, it is being confirmed that compressive strength of these specimens increases with the inclusion of admixture up to 0.5% of the weight of cement. Further addition of admixtures decreases the strength of cubes as well as cylindrical specimens. The trend is similar for all the specimens which are cured for 7, 14, as well as 28 days.

In order to evaluate tensile strength of the concrete specimens, split tensile strength was performed. Split tensile strength of concrete can be evaluated by conducting the tests according to IS 5816:1999 [33] on cylinder samples with a diameter of about 150 mm and a height of 300 mm respectively. The split tensile test's experimental setup is shown in Figure 12.3. Certain variation in split tensile strength along with addition of admixture is plotted in Figure 12.4. It was observed that in order to attain maximum tensile strength, the optimum percentage of admixture to be added was 0.5% of the weight of cement.

Table 12.4 Quantity of materials for concrete mix.

Mineral	Cement (kg/m³)	Fine aggregate (kg/m³)	Coarse aggregate (kg/m³)	Water (kg/m³)	Sodium bicarbonate (by wt. of cement)	Aluminium nitrate (By wt. of cement)
	444.40	717.9	1080.3	200	0%	0%
M-0.25	442.18	717.9	1080.3	200	0.25%	0.25%
M-0.50	439.96	717.9	1080.3	200	0.50%	0.50%
M-0.75	437.73	717.9	1080.3	200	0.75%	0.75%
M-1.00	435.51	717.9	1080.3	200	1.00%	1.00%

Table 12.5 Slump and compaction factor obtained for different mixes.

Mix	Slump (mm)
Reference	65
M-0.25	60
M-0.50	55
M-0.75	55
M-1.00	50

(a)   (b)

Figure 12.1 Test setup for compressive strength (a) Cube strength (b) Cylinder strength

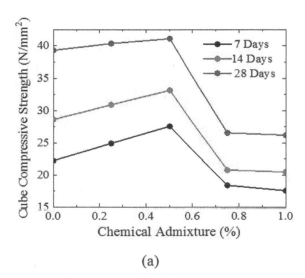
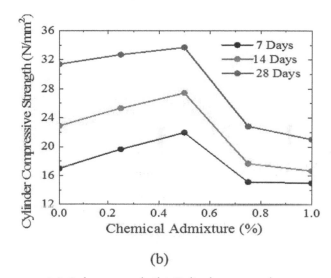

(a)   (b)

Figure 12.2 Compressive strength variation of concrete (a) Cube strength (b) Cylinder strength

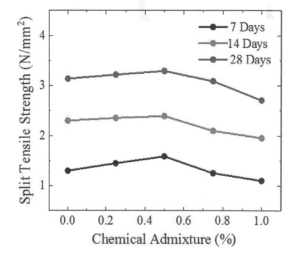

Figure 12.4 Split tensile strength variations

Figure 12.3 Setup for splitting tensile test

Durability Characteristics

Acid Attack

The cube samples being 150 mm × 150 mm in size were oven-dried for 24 hrs. and dipped in 3% sulfuric acid (H_2SO_4) solution (10 N) for about 180 days which is a 3-month time period in other words. After 180 days, the % loss in weight as well as compressive strength of the same cube were measured. Table 12.6 shows the variation in both weight loss as well as compressive strength for various mixes under acid attack for a time period of 180 days. The weight loss and compressive strength for the M-0.50 mix was 1.38 times and 1.4 times lower than that of the reference mix was also observed. Specimens with admixtures showed a higher resistance towards the acid attack while compared to the ordinary concrete specimens.

Sulfate Attack

The cube specimens with similar size to that of used in acid attack were oven-dried for 24 hrs. Once drying process got over, they were then immersed in a sodium sulphate solution (3%) for around 180 days. Solution was agitated every day as well as solution was changed once in a month for providing uniformity to the cementitious mix. Table 12.7 shows reduction in weight loss and compressive strength of the specimens after 180 days of keeping the samples immersed in the above-mentioned sodium sulphate solution. The loss in percentage which occurred in compressive strength for the M-0.50 mix during sulphate attack was 1.3 times less than the reference specimen, which signifies that the concrete mix which incorporated the chosen admixtures has a slightly higher resistance against sulphate attack.

Ultra-sonic pulse velocity Test

The ultra-sonic pulse velocity (UPV) test is performed to assess the quality of concrete specimens along with varying admixture contents. The fundamental principle of undergoing ultrasonic pulse velocity is to obtain comparatively higher velocities generally when the concrete's quality in terms of density, homogeneity, and uniformity is good. An UPV test is conducted on a cylindrical concrete specimen and as per IS 13311 (Part 1): 1992. From Figure 12.5, it is observed that for the specimens to attain good health after 28 days of curing, there is a slight increase in UPVs for concrete along with admixtures. It is also noted that mix- M-0.50 has a 2.9% increase in UPV value while compared to the reference mix. When observing the specimens under the criteria of acid and sulphate attack, mix M − 0.50 showed 64.7% and 27.7% increase in UPV values while compared to the reference mix. Lower velocity indicates more transit time which in turn denotes that the integrity of concrete is more disturbed.

The chemical admixture used in present study, namely sodium bicarbonate helped in enhancing the performance of the concrete, while several other literature works mentioned above also showed the same observation in their study. Studies which not only incorporated sodium bicarbonate but also, admixtures like fly-ash, silica fumes, GGBS are all the examples resembling the results. One of the most common tests out of several other tests mentioned in both the present study as well as old literature work is the

Table 12.6 Weight and compressive strength losses due to acid attack.

Mix	% Weight loss	% Strength loss
Reference	15.96	36.54
M-0.25	12.33	30.27
M-0.50	11.56	26.10
M-0.75	11.69	27.15
M-1.00	12.13	28.81

Table 12.7 Percentage weight and strength loss due to sulphate attack.

Mix	% Weight loss	% Strength loss
Reference	0.71	18.37
M-0.25	0.56	16.29
M-0.50	0.44	14.15
M-0.75	0.48	15.05
M-1.00	0.51	16.63

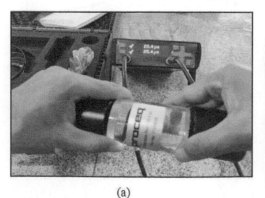

(a)   (b)

Figure 12.5 UPV test (a) Calibration (b) testing on cylindrical specimens

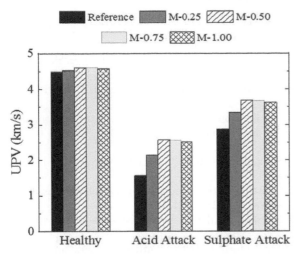

Figure 12.6 UPV values for different specimens subjected to acid and sulphate attack

workability test. Conclusion for the workability test is like that discussed in old studies that is upon increasing the dosage of chemical admixtures workability of concrete also increases significantly. Superplasticizers are one such type which enhances the workability of the mix. Similarly, baking powder and admixtures like calcium acetate, magnesium acetate, calcium format, etc are used in various studies to study the development of liquid chemical admixtures which enhances or likewise improves the workability of the specimens When the concrete specimens were subjected to acid attack showed that specimens casted with chemical admixtures in it survived the chemical attack which is similar to that as described in old literature. Similar result where admixtures like GGBS, fly-ash, basalt, marble powder, silica fume and superplasticizers added in concrete specimens helped them survive in tidal as well as acidic conditions and prevent loss in the overall strength of the specimen. Another study by Shannag, (2011).[16] described how partially replacing the cement by adding 10% of silica fume and 5% Fly-ash helped in improving the stiffness and strength of samples. Another study [18] shows higher pozzolanic activity and surface area of metakaolin yielded higher compressive strength of concrete specimens. Similarly Nehdi et al (2003) [20] also described an increase in the strength parameters of the concrete by using the Torbed reactor technology which upon implementation improved the compressive strength and resistivity with respect to surface and chloride resistance.

Conclusions

Present experimental investigation advocates the effectiveness of chemical admixtures in improving certain parameters like workability, strength and resistance against chemical attack. The experimental conclusions which are procured from the experimental study done above are summarized below:

1. In the current study, chemical admixtures which were used includes sodium bicarbonate and aluminium nitrate. They are mixed to the conventional concrete to enhance its performance.
2. The test results from workability test concludes about how increase in chemical admixture content also increases the workability of concrete.
3. The maximum compressive and split tensile strength is achieved at an optimum dose of 0.5% by weight of cement when testing is done for 7th, 14th and 28th days. Further increase in the dose of chemical admixture results in decrement of the concrete strength.
4. Present study also conducted chemical attack test using sulphuric acid solution and it was observed that upon adding chemical admixture in the concrete specimens helps in protecting the conventional concrete sample from degradation. The minimum loss in the concrete sample was also achieved at 0.5% dose of chemical admixture.

References

[1] Nagataki, S. (1994). Mineral admixtures in concrete: state of the art and trends. ACI Special Publications. 144, 447-447.
[2] Sabet, F. A., Libre, N. A., and M. Shekarchi (2013). Mechanical and durability properties of self consolidating high performance concrete incorporating natural zeolite, silica fume and fly ash. Construction and Building Materials. 44, 184.
[3] Shah, D. S., Shah, M. P. and Pitroda, J. (2014). Chemical admixtures: a major role in modern concrete materials and technologies. In National conference on trends and challenges of civil engineering in today's transforming world.

[4] Shunsuke, H. and Yamada, K.(1999). Interaction between cement and chemical admixture from the point of cement hydration, absorption behaviour of admixture, and paste rheology. Cement and Concrete Research. 29(8), 1159-1165.

[5] Kosmatka, S. H., Panarese, W. C., and Kerkhoff, B. (2002). Design and control of concrete mixtures. Skokie, IL: Portland Cement Association.

[6] Mehta, P. K. and Monteiro, P. (2014). Concrete: microstructure, properties, and materials. McGraw-Hill Education.

[7] Mario, C. (2005). Admixtures: Enhancing concrete performance. In Admixtures-Enhancing Concrete Performance: Proceedings of the International Conference, University of Dundee, Scotland, UK., pp. 217-230, Thomas Telford Publishing, 2005.

[8] Shah, A. A. and Ribakov, Y. (2011). Recent trends in steel fibered high-strength concrete. Materials & Design. 32(8-9), 4122-4151.

[9] Sari, M, . Prat, E., and Labastire, J. F. (1999). High strength self-compacting concrete original solutions associating organic and inorganic admixtures. Cement and Concrete Research. 29(6), 813-818.

[10] Francisca, P., Palacios, M., and Provis, J. L. (2014). Admixtures. In Alkali Activated Materials, pp. 145-156. Springer, Dordrecht.

[11] Kanamarlapudi, L., Jonalagadda, K. B., Jagarapu, D. C. K., and Eluru, A. (2020). Different mineral admixtures in concrete: a review. SN Applied Sciences. 2(4), 1-10.

[12] Gonen, T. and Yazicioglu, S. (2007). The influence of mineral admixtures on the short and long-term performance of concrete. Building and Environment. 42(8), 3080-3085.

[13] Uysal, M. and Sumer, M. (2011). Performance of self-compacting concrete containing different mineral admixtures. Construction and Building Materials. 25(11), 4112-4120.

[14] Sulapha, P., Wong, S. F., Wee T. H., and Swaddiwudhipong, S. (2003). Carbonation of concrete containing mineral admixtures. Journal of Materials In Civil Engineering. 15(2), 134-143.

[15] Khan, S. U., Nuruddin, M. F., Ayub, T., and Shafiq, N. (2014). Effects of different mineral admixtures on the properties of fresh concrete. The Scientific World Journal.

[16] Shannag, M. J. (2011). Characteristics of lightweight concrete containing mineral admixtures. Construction and Building Materials. 25(2), 658-662.

[17] Ramanathan, P., Baskar, I., Muthupriya, P., and Venkatasubramani, R. (2023). Performance of self-compacting concrete containing different mineral admixtures. KSCE journal of Civil Engineering. 17(2), 465-472.

[18] Ping, D., Shui, Z., Chen, W., and Shen, C. (2013). Efficiency of mineral admixtures in concrete: Microstructure, compressive strength and stability of hydrate phases. Applied Clay Science. 83, 115-121.

[19] Memon, A. H., Radin, S. S., Zain, M. F. M., and Trottier, J.-F. (2002). Effects of mineral and chemical admixtures on high-strength concrete in seawater," Cement and Concrete Research. 32(3), 373-377.

[20] Nehdi, M., Duquette, J., and A. El Damatty (2003). Performance of rice husk ash produced using a new technology as a mineral admixture in concrete. Cement and Concrete Research. 33(8), 1203-1210.

[21] Collepardi, M. (2005). Chemical admixtures today. In Proceedings of Second International Symposium on Concrete Technology for Sustainable, Development with Emphasis on Infrastructure, Hyderabad, India, pp. 527-541.

[22] Yasuhide, M., Olabamiji, O. J., and Kasahara, K. (2020). Study on the effect of baking soda on bleeding and compressive strength of cement milk. GEOMATE Journal. 19(73), 64-69.

[23] Almeida, N. and Haselbach, L. (2021). Proposed method to reduce magnesium chloride deicer damage to pervious concrete. Journal of Cold Regions Engineering. 35(3).

[24] Almeida, N. and Haselbach, L. (2021). Time impacts of treating pervious concrete with sodium bicarbonate. Journal of Infrastructure Preservation and Resilience. 2(1), 1-8.

[25] Nugroho, F. B., Sumarni, S., Thamrin, A. G., and Isnantyo, F. D. (2020). Using baking powder as additional concrete material. In IOP Conference Series: Materials Science and Engineering. 858(1).

[26] Gopakumar, K., Szeles, T., Stoffels, S. M., and Rajabipour, F. (2021). Novel admixtures for mitigation of alkali-silica reaction in concrete. Cement and Concrete Composites. 120, 104028. doi:10.1016/j.cemconcomp.2021.104028

[27] Korhonen, C. (2002). Effect of high doses of chemical admixtures on the freeze-thaw durability of portland cement concrete. Technical Report, ERDC/CRREL TR-02-5, US Army Corps of Engineers Engineer Research and Development Center.

[28] IS (Indian Standards). 2016. Coarse and Fine Aggregate for Concrete Specification. IS 383. New Delhi, India: IS

[29] IS (Indian Standards). 1963. Methods of test for aggregate for concrete, Part 3–Specific gravity, density, voids, absorption and bulking. IS 2386 (Part III). New Delhi, India: IS.

[30] IS (Indian Standards). 2019. Concrete Mix Proportioning – Guidelines. IS 10262. New Delhi, India: IS.

[31] IS (Indian Standards). 1959. Methods of Sampling and Analysis of Concrete. IS 10262. New Delhi, India: IS.

[32] IS (Indian Standards). 1959. Methods of Tests for Strength of Concrete. IS 516. New Delhi, India: IS

[33] IS (Indian Standards). 1999. Splitting Tensile Strength of Concrete - Method of Test. IS 5816. New Delhi, India: IS.

Chapter 13

A mini-review on electro discharge machining of aluminium metal matrix composite materials

Prajwal Chamat[a], Tushar Motghare[b], Shreyash Borghare[c],
Aaradhya Wange[d] and Md. Sharique Tufail[e]
Yeshwantrao Chavan College of Engineering, Nagpur, India

Abstract

Spark erosion machining, also known as electro-discharge machining, is commonly used to machine hard materials, which is hard to do by using the traditional machining process. Electrical discharge machining (EDM) is prominently used for machining advanced materials like ceramic, composite, metal matrix composite, etc. which is most popular in various fields of application like automobiles, aerospace, defense, etc. EDM machines advanced materials effectively, therefore electro-discharge machining is most popular among non-traditional machining processes. Properties of aluminum-based MMC like wear resistance, low density, high tensile strength, etc, make aluminum-based MMC very difficult to machine effectively even with EDM. Therefore, this paper focuses on the electric discharge machining of the aluminum-based MMC due to the number of applications it has and its attractive properties. This paper also reviews the effect of process variables like pulse-on time (T-on), pulse-off time (T-off), voltage (V), and current (I) on MMR, SR, and EWR, in addition to this various methodology of design of experimentation of aluminum-based MMC has been discussed. Future research trends which can be investigated in the future in the same field are also discussed. This study revealed that flushing pressure T-on and pulse current rise cause MMR to climb and reduce as T-off decreases, while SR rises with growing I and lowers with growing T-on. TWR increases with rising current and decreases with an increasing pulse on time.

Keywords: EDM, aluminum-based MMC, machining parameters

Introduction

Nowadays the world is always in quest of materials that exhibit properties like enhanced rigidity, strong wear resistance, low density, excellent thermal properties, and restricted thermal expansion. Aluminum-based metal matrix composite (MMC) is an advanced material that exhibited all these properties and can be further improved by combining different reinforced materials such as SiC, TiC, B_4C, Al_2O_3, etc [1]. Therefore, making it popular in various fields of application like automobiles, aerospace, defense, etc.

Aluminum-based MMC demonstrated improved chemical, thermal, and improved mechanical characteristics, because of the addition of strong reinforcing components in MMC its wear resistance improved [2]. Because of this, it is more challenging to machine using a traditional machining method. Therefore, several forms of non-convectional machine processes, including electrical discharge machining (EDM), laser, and other techniques, were progressively employed to successfully and efficiently machine MMC materials. Because of Al-based MMC abrasiveness, hardness, and brittleness [3]. EDM is more suitable for machine Al-based MMC because of its capacity to work with very tough material precisely and accurately. There is no friction between the electrode as well as the work surface. Thus, it undeniably resolves the issues with vibration, noise, and mechanical stress [4].

It also removes material by anodic dissolution. High material removal rates and application regardless of material hardness are benefits of EDM [3]. EDM has input process variables that influence the output response. The properties of the aluminum-based MMC change with reinforced material (SiC, TiC, B_4C, etc.). Therefore, the process controls variables for getting the appropriate output values for the aluminum-based MMC were quite uncertain [5]. In order to achieve the appropriate output response, and support evaluating the efficacy of EDM, therefore it is crucial to tune the EDM process variables of Al-based MMC. Past literature is more focused on the optimization of the process parameter and analysis of their effects on the output responses and very less articles review the current research in EDM of aluminum-based MMC [6-9, 14]. Therefore, this work concentrates on the current research in EDM of Al-based MMC. This study also discusses the effects of process variables such as V, I, T-on, T-off, dielectric fluid, electrode material on the MMR, surface quality, and TWR. Additionally, different design methodologies for experiments with aluminum-based MMC have been discussed.

[a]prajwalchamat@gmail.com.iq, [b]tusharmotghare06@gmail.com, [c]shreyashborghare6@gmail.com,
[d]wangeaaradhya@gmail.com, [e]mstufail@rediffmail.com

DOI: 10.1201/9781003450252-13

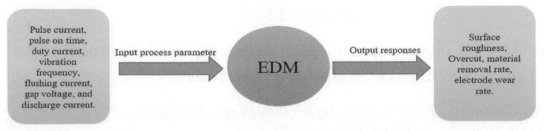

Figure 13.1 EDM I/P and O/P responses
Source: Author

Literature Review

Gopalakannan et al. [6] did research to determine the significant process variables influencing the output characteristics while EDM of Al 7075-B_4C through RSM. The authors used copper electrodes of 10 mm diameter and PC, GV, T-on, and T-off were input responses, and MMR, EWR, and SR as output parameters considered. This study revealed an increase in MMR with a rise in pulse current (PC) and Ton. A higher Toff gives a lower value of the EWR, and SR rises with the rise in Ton and pulses current. Voltage 49.02-volt, Pulse current 14 Amps, T-on 7.77μs, and T-off 5 μs for maximising MRR, minimising TWR, and SR have suggested combinations of optimum process parameter setting in this study.

Naik et al. [7] conducted experimental research on Al-SiC MMC EDM with brass electrode and dielectric fluid based on vegetable oil. Based on box-design Behnken's experiments, 46 sets of experimental trials were undertaken by taking into account the five cutting parameters (T-on, T-off, DC, GV, and flushing pressure (BBDOEs). Additionally, for experimental analysis of various techniques like RSM and genetic algorithms were used. The confirmation test is then used to assess the performance of the two suggested (RSM, GA) multi-objective optimisation strategies. The outcome demonstrates that discharge current significantly improves material removal rate while degrading surface quality and hole diameter dimensional deviation, particularly overcut.

Palanisamy et al. [8] investigated an experiment to enhance the input process responses while LM6-Alumina MMC EDM. To achieve the optimal value of MMR, SR, and EWR, the input process responses such as T-off, T-on, and discharge current, were optimised using grey relation analysis. In this study, MRR is increased as increased in discharge current, Ton, and Toff. TWR rise as a rise in applied current, T-on, and T-off. With a rise in I, T-on, and T-off surface roughness increased. The parameter that most significantly influences the SR and MMR, as this experiment has shown, is discharge current.

Kandpal et al. [9] optimised the process parameter such as gap current (GC), Ton, and duty current of EDM of AA 6061/10% Al_2O_3 aluminum-based MMC. The Taguchi approach was employed to optimize the process parameters with the aid of analysis of variance and S/N ratio. Die-sinking electric discharge machining with an electrode of copper was used in the study. This study found that an increase in MMR is caused by a greater pulse current and T-on.

Singh et. al [10] investigate the result of machining process variables such as T-on, T-off, V, and I on MRR during EDM of Al-based MMC reinforced with Sic and Grp with an electrode of copper of 12 mm diameter. In this study mathematical model was developed by RSM and analysis of variance is used to examine the significance of the model. T-on and I are found to be the most significant parameters which help to increase the MMR and the high value of T-off gives a minimum MMR.

Srivastava et al. [11] optimised machining parameter during EDM of Al 6061/SiC by using an electrode of copper of 20mm diameter. Analysis of variance was used to assess the importance of process factors, and RSM was utilised to construct a statistical model and optimise process parameters. In this study T-on, I_p, and duty cycle were input responses, and MMR, EWR, and SR were considered output responses. I_p affects MMR, T-on affects TWR, and I_p and T-on affect SR.

Kumar et al. [14] examined the impact of input process variables such as I, T-on, T-off, and electrode material on MRR, TWR, and SR during Al-B_4C MMC EDM. In this study, Taguchi is used for the DOE, and analysis of variance is utilised to analyze experimental results. A rise in current causes a rise in MMR and initially increase in T-on causes a decrease in MMR then it increases. The highest value of MMR was obtained with a graphite electrode. An increase in I causes increases in TWR. An increase in T-on causes a decrease in TWR; later, it increases with a T-on. Similarly, EWR decreases with a rise in T-off but increases with a further increase. The highest value of TWR was obtained with the E-19 electrode. I and T-on were the most affected parameter on SR and the low SR was obtained with a graphite material electrode.

Shyn et al. [13] investigated the process variables during A6061/6%B_4C MMC EDM. This research considers flushing pressure, spark-off, T-on, gap V, duty factor, and as input process variables and SR, TWR, and MRR were output parameters. Input parameters optimized using CCD and RSM techniques. A rise in current, spark-on time, T-on, gap voltage, and duty factor causes to increase in MMR and MMR decreases with an increase in spark-off time. An increase in I, T-on, gap V, and duty factor help to rise in TWR and a decrease with a rise in flushing pressure and on time of spark. SR rise as a rise in I, on time of spark, and T-on, and decreases as a rise of flushing pressure. Other parameters did not make any major impact on surface roughness.

Kumar et al. [14] compared the surface roughness, MRR and TWR of the AA2024 base material, AA2024/2wt%SiO$_2$, and AA2024/3wt%SiO$_2$ using EDM output responses with the copper electrode and transformer oil as dielectric fluid. It was discovered that the EDM process properties were considerably impacted by the weight of the reinforcing agent SiO$_2$. In comparison to the base aluminum MMC AA2024, the MRR (gm/min) decreases as the weight percentage of SiO$_2$ in the composite increases. While the TWR increases by 2 weight percent SiO2, it then decreases by 3 weight percent SiO$_2$. It can be seen from the measured SR that the SiO2 reinforcement in the aluminum MMC had a significant impact on the machining SF/roughness. It has been observed that the SR rises with the reinforcement particles in the aluminum metal matrix.

Mohantya et al. [15] investigated to optimise the process parameters in EDM of Al-SiC12% MMC like Ton, Toff, and peak current on MRR, TWR, and SR. The copper electrode and spark gap between the tool were increased using the powder which is a mixture of nanoparticles in dielectric fluid. This is done to increase efficiency. Mathematical modeling and correlating the EDM parameters were done by RSM. To examine the significance of the model ANOVA is used. In this study, the MRR has increased because of the use of nano powders in EDM which results in a wide gap between tool and workpiece. It also found that the surface roughness has reduced compared to traditional techniques to the EDM because of uniform spark. Along that with a rise in current tool wear rate rises and decreases with an increase in pulse rate.

Kar et al. [16] conducted an experiment to determine the influence process variables in EDM of red mud MMC by using a brass electrode. For that the open circuit voltage Ton, and discharge current took an input and TWR, radial overcut and MRR was taken as the output. Examine the significance of this model using the ANOVA. MMR and EWR were increased an increase in pulse on time and pick current. ROC within the Increased peak current, Ton and value of gap voltage. Peak currents were having a significant effect on output responses and in comparison, to input parameters. such as pulse on time and gap voltage.

Choudhary et al. [17] investigated the process parameters of Al6061/14%wt fly-ash composite using EDM. Using multiple electrodes like copper and brass. For the determination of optimal parameter setting, analyses of the S/N signal noise ratio are used. and quality parameters like SR, MRR, and TWR were selected. Examine the significance of this mode using the ANOVA Method. MRR gets decreased with an increase in voltage with an increase the shift of tool electrode, duty cycle, current, and Ton. TWR acted the same as MRR. Brass electrodes were seen to be more significant than copper electrode because of the unavailability of a carbon layer on them.

Singh et al. [18] conducted to determine the MRR and TWR of Al6061/SiC metal matrix composite material by using multiple electrodes like brass and copper. To achieve the optimal value of MRR and TWR, the input parameters like T-on, T-off, and voltage gap were optimised using the ANOVA technique. In this study, it is observed that the copper electrode achieves less MRR rate than that of brass, both reached a maximum MRR of 15A I$_p$. Also brass has higher TWR compared to the copper electrode, TWR was high for both electrodes at a lower I$_p$ value is 5A.

Mahanta et al. [19] investigated the process parameters of hybrid MMCs using EDM with the minimum powder consumption quality jobs in the Production. For that gap voltage, I$_p$, Pulse Duty, and T-on were taken as input to see the optimal value of SR and P using the ANOVA technique. It was found that the pulse current has dominating influence over different cutting conditions for the SR and P.

Phate et al. [20] analysed the TWR of Al-Cu-Ni alloy in electro-discharge machining using ANFIS Method. In For that T-on, T-off, and input current were taken as input and analysed over Al alloy. It was found that TWR depends upon the mixture of elements in the alloy as TWR rises as a rise in copper and falls as a fall in aluminum. In ANFIS using a high degree of precision is found.

Sajeevan et al. [21] analysed the parameters of aluminum-based MMC on die sink EDM. This study was conducted to determine the MRR and surface finish of MMC by the various input parameters. Electric Discharge Machining of Aluminum Titanium Di-Boride was done using copper electrodes using the Taguchi method. It's observed from the investigation that MRR got rise as a rise in I and T-on and better surface finishing was found with the lower value of these two.

Dar et al. [22] investigated optimising the process parameter in electric discharge machining (EDM) Al-7%SiC3% Using copper electrodes. The impact of EDM I/P process variables I$_p$, applied voltage, T-on, and T-off on the rate of material removal and EWR was investigated. Analyzing the relevant parameters and their quantitative impact on chosen process parameters using ANOVA and the Taguchi method. The impacts of all input factors on MRR are considerable, according to experimental findings, I$_p$, which accounts for 88.84% of the factors influencing TWR, is followed by the applied voltage. IP and V are important process variables that add values of 52.17% and 26.17%, respectively, of diametral overcut, whereas other parameters are minor.

Markopoulos et al. [23] examined ways to improve the process parameters for Al5052 EDM in MMC. By using a Cu electrode and hydrocarbon oil as a dielectric fluid. SR, MRR, the formation of white layers, and the microhardness of the heat-affected zone were taken into consideration as an output response. T-on and T-off as input responses. Whereas the SR is mostly influenced by the T-on, depending on the present pulse. The interplay and fusion of the machining parameters, notably the Pulse on current and the T-on, have a sizable influence on the process efficiency. The existence of a HAZ with reduced microhardness was finally established.

Srivastava [24] utilised a Cu electrode to carry out the research on improved machining parameters during EDM of Al 6063/SiC. Combining this substance with the matrix material is preferable. Utilizing SiC as a reinforced material. They constructed a statistical model with the RSM Technique, optimized the process parameters, and utilized ANOVA to assess the importance of the process parameters. An RHS test revealed that the composite material had better tensile properties than the basic metal, Al6063. The findings of the machining show that all the process variables significantly affect the MMR. Using ideal set of process parameters resulted in a 37.5% improvement.

Jayendra et al. [25] investigated the impact of process variables including T-on, gap voltage, Duty factor, I, and spark-on as input parameters, and SR, TWR, and MRR were output parameters by EDM of Al-7075/B4C/Graphite Reinforced Hybrid MMC. Utilizing the stir-casting manufacturing method, graphite, and B4C content were produced for the experiment. An increase in graphite and B_4C contain in MMC lead to an increase in its hardness as well as impact strength.

Chandramoulia et. al [26] optimised the process parameters during the Al 6061/MoS2 metal matrix composite EDM. RSM and GR Analysis Techniques were used. SR, TWR, and MMR were thought to study output parameters. The input parameters included I, T-on, T-off, gap voltage, Duty factor, and flush pressure. The central composite design served as the foundation for the statistical model, and RSM was used for optimization. In his most recent work, he integrated the Grey-RSM technique for EDM process parameters for Al6061/MoS2 composite materials: Ip of 10 A, T-on 100 s, T-off of 60 s, and V of 60 volts.

The EDM of the aluminum-based MMCs had very unpredictable process control parameters that provided the best output variables. Each individual composite component can have a very distinct surface roughness and material removal rate. The optimisation technique utilised to improve the input parameter is therefore crucial. ANOVA, Taguchi, and response surface technique has all been the subject of extensive investigation as shown in Figure 13.2.

Impact of Input Parameters on Output Responses

The speed at which material is taken out of the workpiece is an MRR. MRR in EDM is more closely tied to the work sample's thermal qualities than to its mechanical attributes. T-on and pulse current have the biggest effects on the MMR as compared to the other electrical process parameters, followed by peak current and peak current shown in Figure 13.3. The desirable increase in MMR will occur if we raise T-on. Higher

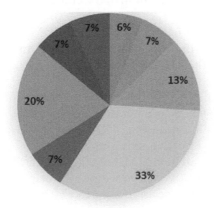

■ Taguchi and Gauss Elimination Method ■ Genetic Algorithm

■ ANOVA   ■ Response Surface Method

■ Taguchi   ■ Taguchi and Grey Relation Analysis

■ Artificial Bee Colony   ■ Lexicographic Goal Programming

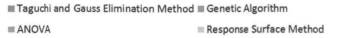

Figure 13.2 Various employed optimisation methods to produce the optimum EDM responses
Source: Author

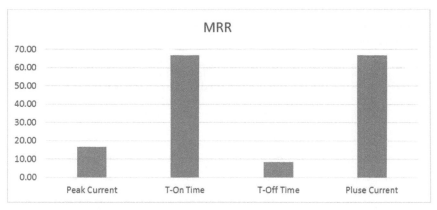

Figure 13.3 Impact of input parameters on MMR
Source: Author

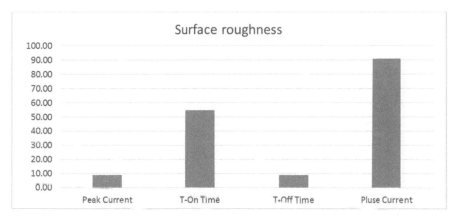

Figure 13.4 Impact of input parameters on surface roughness
Source: Author

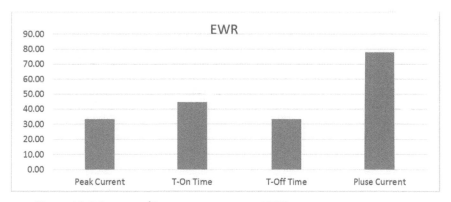

Figure 13.5 Impact of input parameters on EWR
Source: Author

T-off decreases MRR, which is not desired. The low T-on, MRR is quite dependent on shifts in T-off. With an increase in pulse current, MRR also rises [6, 7]. Surface roughness is the irregularities on the machine's surface. Which is not desired for effective machining. Figure 13.4 demonstrates that the 90% plus current impacted the surface roughness followed by T-on with 55%, T-off and peak current has a nearly 10% impact each. Some researchers have investigated that increasing T-on causes an increase in surface roughness which is not desired. An increase in I also increases SR [13]. The shape and accuracy of the dimensions of the work specimen are significantly influenced by the tool wear rate, which is associated with the melting point of the electrode material [34]. Figure 13.5 showed that the TWR is significantly impacted by pulse current, which is close to 80%. The fact that a higher current generates more heat between the workpiece and the tool leads to tool wear. Peak current comes in second with 33%, followed by T-off with a near 45% contribution.

Table 13.1 Research Advancements in EDM of Al-based MMC

Reference	Material	Electrode material	Findings
[6]	Al 7075-B_4C	Copper	The study found that pulse current and T-on rise were associated with an increase in MMR. A lower EWR rating is the result of a longer T-off. Surface quality decreases as T-on and pulse current rise.
[7]	Al-SiC	Brass	Discharge current significantly lowers the dimensional accuracy and surface quality of holes, especially overcut holes, while raising MRR.
[34]	LM6-Alumina	Copper	The parameter that has the greatest impact on the SR and MMR is discharge current.
[9]	AA 6061/10% Al_2O_3	Copper	Rises in pulse current and T-on result in a rise in MMR.
[10]	Al-20%SiC-8%Grp	Copper	The T-on and I are shown to be the most significant factors that raise the MMR, whereas a high T-off value results in the lowest MMR.
[11]	Al 6061/SiC	Copper	SR is affected by both I_p and T-on, whereas I_p also influences MMR and EWR.
[12]	Al-B_4C	Graphite, EN-19, Copper, and Brass	Graphite electrode help to produce lower values of SR and increase MMR. The E-19 electrode produced the greatest value of EWR.
[13]	A6061-6%B_4C	Copper	A high current gives a higher value of SR.
[14]	AA2024/2wt%SiO_2, and AA2024/3wt%SiO_2	Copper	The amount of SiO_2 used as reinforcement in AA 2024 causes a raise in surface roughness, a raise in tool wear rate, and a decrease in material removal rate.
[15]	Al-SiC12%	Copper	The use of nanoparticles in spark erosion machining improves surface qualities and enhances the MRR.
[16]	Al7075-red mud	Brass	I_p was discovered as the most important factor.
[17]	Al6061/14% fly-ash	Copper and Brass	The key constants that influence SR, MMR, and TWR were duty cycle, T-on, and I.
[18]	Al6061/SiC	Copper and Brass	MMR is increased with brass electrodes as opposed to Cu electrodes. Compared to a brass electrode, a Cu electrode has a reduced TWR.
[19]	Al6351- 5%SiC-10%B4C	Copper	A higher MMR score and a less-than-smooth surface are produced by an increase in pulse rate with time.
[20]	Aluminum/Copper/Nickel Alloy	copper	The most important characteristic impacting TWR was the percentage change in copper and aluminum. Changes in T-on, and I_p.

Source: Author

Conclusion

Electrical discharge machining metal matrix composites (MMCs) of the aluminum-based MMCs, the process control parameters that produced the best output variables was extremely unpredictable. Surface roughness and material removal rate for each different component of the composites can vary greatly. Therefore, the optimization method which is used for the optimization of the input parameter is vital. So much research has been done on Taguchi, Response surface methodology, and ANOVA. When used in optimization, ANN with a genetic algorithm needs to be explored more as this approach produces better results than RSM.

Tool wear rate is not only affected by the input responses but also by the tool (electrode) material. Therefore, selecting the proper tool material is vital for getting the ideal output responses. Copper, brass, and graphite are mostly used electrode materials for EDM of Al-MMC. Among graphite, copper, and brass. Graphite electrode gives a lower value of SR and increases MMR therefore it's found to be more effective as compared to copper and brass.

SR rises with increased I and falls with a growing T-on, and flushing pressure. Pulse current is the most affecting process parameter followed by t-on, t-off, and peak current. T-on and pulse current rise cause MMR rise and falls as T-off falls. TWR rises as the current rises and drops as a pulse on time increases.

Nano powder mixed with dielectric fluid EDM produces superior results compared to non-nano powder combined with EDM.

The research field of electro-discharge machining of metal matrix composites is very emerging. This study will be beneficial to researchers who are new to the field of electro-discharge machining and EDM of aluminum-based MMCs.

References

[1] Senthil, S., Raguraman, M., and Manalan, D. T. (2021). Manufacturing processes & recent applications of aluminium metal matrix composite materials: A review. Materials Today: Proceedings. 45, 5934–38.

[2] Bharat, N. and Bose, P. S. C. (2021). An overview on the effect of reinforcement and wear behaviour of metal matrix composites. Materials Today: Proceedings. 46, 707–13.

[3] Rizwee, M., Rao, P. S., and Khan, M. Y. (2021). Recent advancement in electric discharge machining of metal matrix composite materials. Materials Today: Proceedings. 37, 2829–36.

[4] Gouda, D., Panda, A., Nanda, B. K., Kumar, R., Sahoo, A. K., and Routara, B. C. (2021). Recently evaluated Electrical Discharge Machining (EDM) process performances: A research perspective. Materials Today: Proceedings. 44, 2087–92.

[5] Devi, M. B., Birru, A. K., and Bannaravuri, P. K. (2021). The recent trends of EDM applications and its relevance in the machining of aluminium MMCs: A comprehensive review. Materials Today: Proceedings. 47, 6870–73.

[6] Gopalakannan, S., Senthilvelan, T., and Ranganathan, S. (2012). Modeling and optimization of EDM process parameters on machining of Al 7075-B4C MMC using RSM. Procedia Engineering. 38, 685–90.

[7] Naik, S., Das, S. R., and Dhupal, D. (2021). The experimental investigation, predictive modeling, parametric optimization and cost analysis in electrical discharge machining of Al-SiC metal matrix composite. Silicon. 13(4), 1017–40.

[8] Paganism, D., Devaraju, A., Manikandan, N., Balasubramanian, K., and Arulkirubakaran, D. (2020). Experimental investigation and optimization of process parameters in EDM of aluminium metal matrix composites. Materials Today: Proceedings. 22, 525–30.

[9] Kandpal, B. C., Kumar, J., and Singh, H. (2018). Optimization Of Process Parameters Of Electrical Discharge Machining Of Fabricated AA 6061/10% Al2 O3 Aluminium Based Metal Matrix Composite. Materials Today: Proceedings. 5(2), 4413–20.

[10] Singh, M. an Maharana, S. (2020). Investigating the EDM parameter effects on aluminium based metal matrix composite for high MRR. Materials Today: Proceedings. 33, 3858–63.

[11] Srivastava, A., Yadav, S. K., and Singh, D. K. (2021). Modeling and optimization of electric discharge machining process parameters in machining of Al 6061/SiCp metal matrix composite. Materials Today: Proceedings. 44, 1169–74.

[12] Kumar, P. and Parkash, R. (2016). Experimental investigation and optimization of EDM process parameters for machining of aluminum boron carbide (Al–B4C) composite. Machining Science and Technology. 20(2), 330–48.

[13] Shyn, C. S., Rajesh, R., and Anand, M. D. (2021). Modeling and prediction of die sinking EDM process parameters for A6061/6% B4C metal matrix composite material. Materials Today: Proceedings. 42, 677–85.

[14] Kumar, V., Singh, B., Chandel, S., and Singhal, P. (2020). Evaluation of EDM characteristics of synthesised AA2024-2 & 3 wt% SiO2 metal matrix nanocomposite (MMNC). Materials Today: Proceedings. 26, 1449–54.

[15] Mohanty, S., Singh, S. S., Routara, B. C., Nanda, B. K., and Nayak, R. K. (2019). A Comparative study on machining of AlSiCp metal matrix composite using Electrical discharge machine with and without nano powder suspension in dielectric. Materials Today: Proceedings. 18, 4281–89.

[16] Kar, C., Surekha, B., Jena, H., and Choudhury, S. D. (2018). Study of influence of process parameters in electric discharge machining of aluminum–red mud metal matrix composite. Procedia Manufacturing. 20, 392–99.

[17] Choudhary, R., Singh, G., Kumar, K., Bharti, P., Kumar, R., and Kumar, V. (2018). Investigations of electrical discharge machining of Al6061/14% wt fly-ash composite with different tool electrodes. Materials Today: Proceedings. 5(9), 19923–32.

[18] Singh, H., Singh, J., Sharma, S., and Chohan, J. S. (2022). Parametric optimization of MRR & TWR of the Al6061/SiC MMCs processed during die-sinking EDM using different electrodes. Materials Today: Proceedings. 48, 1001–1008.

[19] Mahanta, S., Chandrasekaran, M., and Samanta, S. (2018). GA based optimization for the production of quality jobs with minimum power consumption in EDM of hybrid MMCs. Materials Today: Proceedings. 5(2), 7788–96.

[20] Phate, M., Bendale, A., Toney, S., and Phate, V. (2020). Prediction and optimization of tool wear rate during electric discharge machining of Al/Cu/Ni alloy using adaptive neuro-fuzzy inference system. Heliyon. 6(10), e05308.

[21] Sajeevan, R. and Dubey, A. K. (2021). Parametric study of die-sinking electric discharge machining on aluminium based metal matrix composite. Materials Today: Proceedings. 44, 930–34.

[22] Dar, S. A., Kumar, J., Sharma, S., Singh, G., Singh, J., Aggarwal, V., … Obaid, A. J. (2021). Investigations on the effect of electrical discharge machining process parameters on the machining behavior of aluminium matrix composites. Materials Today: Proceedings. doi:10.1016/j.matpr.2021.07.126

[23] Markopoulos, A. P., Papazoglou, E. L., Svarnias, P., and Karmiris-Obratański, P. (2019). An Experimental Investigation of Machining Aluminum Alloy Al5052 with EDM. Procedia Manufacturing. 41, 787–94.

[24] Srivastava, A. K. (2019). Assessment of mechanical properties and EDM machinability on Al6063/SiC MMC produced by stir casting. Materials Today: Proceedings. doi:10.1016/j.matpr.2019.07.429

[25] Jayendra, B., Sumanth, D., Dinesh, G., & Rao, M. V. (2020). Mechanical Characterization of Stir Cast Al-7075/B4C/Graphite Reinforced Hybrid Metal Matrix Composites. Materials Today: Proceedings. 21, 1104–10.

[26] Chandramouli, A., CHVS, R., and Kumar, M. (2022). Multi Response Optimization of Electric Discharge Machining (EDM) of Al 6061/MoS2 MMC through Response Surface Methodology and Grey Relational Analysis Technique.

[27] Palanisamy, D., Devaraju, A., Manikandan, N., Balasubramanian, K., and Arulkirubakaran, D. (2020). Experimental investigation and optimization of process parameters in EDM of aluminium metal matrix composites. Materials Today: Proceedings. 22, 525–30.

Chapter 14

Design and fabrication of modular, light weight bridge structure for crevasse crossing in high altitude areas

Suresh Madhavan[a] and Vinay Ugale[b]

College of Military Engineering, India

Abstract

In high altitude areas like Siachen-Glaceir, the movement of soldiers is mainly affected due to the presence of crevasse. The existing resources like extendable ladders along with ropes are extremely dangerous to cross the crevasse. Hence, modular, light weight, high strength and high stiff crevasse crossing bridge structure is designed and fabricated. The span of bridge is 9.1 m and width is 0.4 m. Total weight is 125 kg and can be assembled in 30 min. It is made of FRP material except joints. 20% carbon fibers and 80% glass fibers are reinforced in epoxy to fabricate the structural members. The joints are made of high-grade stainless steel (SS304) material. The bridge is designed for 120 kg load using ANSYS software. The dimensions of the structural members are optimised through simulation. It consists of seven ladder type structures, each of length 1.3 m and weighs around 5 kg. The under slung reinforcement of steel wire, GFRP decking panels, handrails and trolley are provided for additional safety. The project is fabricated in collaboration with the firm S. V. Composites, Pune. Suitable pallet has been designed for transportation of the parts of bridge.

Keywords: Crevasse, modular, FRP, bridge

Introduction

Glaciated terrain mostly found in the northern part of Indian Himalayas of Ladakh is characterised by temperatures within a range of -50 to -10°C around the year. Unlike any normally ice-capped mountain, glaciers have crevasses which create disruption in movement of troops by foot and vehicles. Crevasse crossing with existing non-conventional technique (Figure 14.1) such as aluminium ladders tied with rope proves to be extremely challenging. It provides least safety to people crossing over due to non-availability of handrails and results in most of the personnel getting misbalanced due to sagging and swaying of the ladder. In 85% of the cases when a person falls off the harness/safety rope, he can never be rescued, and the person loses life. Similar cases arise due to poor availability of light weight equipment for free movement in highly undulated glacier terrain. So, it becomes imminent to provide an effective life-saving solution for those personnel living in Glaciated terrain.

In this paper, design and fabrication of light weight (man portable), high strength and stiff structure for crossing the crevasse has been discussed. The glass fibres along with the carbon fibres are reinforced in epoxy to fabricate the structural members of bridge through pultrusion process. Extensive literature survey has been carried out to select suitable material for the required application. The study on the fibre reinforced polymer (FRP) laminate where all the layers are made of same reinforcement material or combination of different materials has drawn the attention of researchers in a big way and there are numerous researchers' findings related to material characterisation like low velocity impact strength, high velocity impact strength, flexural strength, compressive strength, tensile strength, low temperature tolerance are available in the literature. The analysis of impact strength of FRP is carried by the researchers [1-4]. No failure in carbon fabric/epoxy panels under low velocity impact was found till 18J [1,2]. The impact behaviour of plain weave E-glass/epoxy and twill weave T300 carbon/epoxy composites has been compared by the investigators [3,4] and good impact strength is observed in glass fibre/ epoxy composite plate due to its low stiffness. The analysis of flexural strength of Glass fibre reinforced polymer (GFRP) and Carbon fibre reinforced polymer (CFRP) composites have been carried out by the researchers [4,8] and the excellent strength to weight ratiois observed. FRP material is very attractive for use in structural applications due to their high strength-to-weight and stiffness-to-weight ratios, corrosion resistance, light weight, potentially high durability and less prone to environmental degradation [9-11]. FRP bars have high tensile strength, but their compressive strength is relatively low and often neglected [12], which can be improved by suitable hybrid combination of glass and carbon. In the comparative study of different FRP composites like CFRP, GFRP, aramid fibre reinforced polymer (AFRP), boron fibre reinforced polymer (BFRP) along with steel, it is observed that GFRP is the most economical material for most of the structural applications [13].

[a]sureshmadhavan82@gmail.com, [b]vinayugale@gmail.com

DOI: 10.1201/9781003450252-14

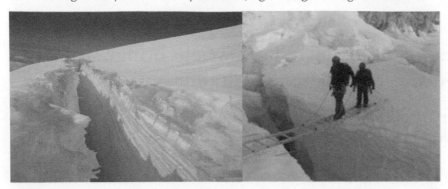

Figure 14.1 Crevasse crossing using extendable aluminium ladders
Source: Author

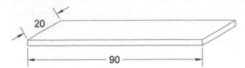

Figure 14.2 Strip specimen for 3-point flexural test
Source: Author

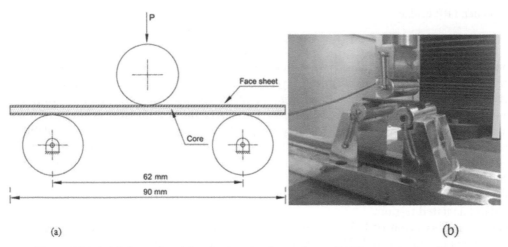

(a)   (b)

Figure 14.3 (a) Schematic of the three-point flexural test, (b) Three-point bending test set up of UTM
Source: Author

Experimental Investigation

Constituent materials

The hybrid composite with a configuration of 80% glass mat/glass rovings and 20% carbon mat were used to fabricate the bridge members, where outer layers are of glass and carbon fibre mat and inner core is made of glass fibre rovings. The glass mat and glass roving used were manufactured by Owens Cornings, USA and carbon fibre mat used was manufactured by Zoltek, USA. The resin used was industrial grade epoxy resin manufactured by Lapox, India. Other additives such as hardener, accelerators were also used. The high-grade stainless steel (SS304) was used for the joining members.

Experimental technique

The structural members of 4 mm thickness were used for making the crevasse bridge structure. The strip specimens (4 mm thickness) having dimensions 90 mm × 20 mm were used for the test to check the strength (Figure 14.3). The span length (L) between the supporting rollers was 62 mm. The span to thickness ratio (L/t) was controlled to be close to 16. The schematic of the three-point flexural test is shown in Figure 14.3(a). The load was applied at the loading rate of 1 mm/min by using a load cell of 2 kN. Figure 14.3(b) shows the three-point bending test set up of UTM. Max load sustained by the specimen was observed to be 1562 N which gives rise to the flexural strength of 454 MPa. The test was conducted as per ASTM D 790 on a UTM.

Design of Bridge through Numerical Simulation

Numerical simulation of the bridge structure was carried out to design and optimise the dimensions of the structural members. ANSYS APDL and CATIA software were used for the simulation.

Validation of numerical method

The finite element modelingof strip specimen (Figure 14.2) of size 90 mm × 20 × 4 mm is as shown in figure 14.4 was done in ANSYS APDL. The different layers of the specimen were modelled and the material properties were assigned to the layers as shown in Table 14.1. The solid elements (20 node 186) were used for meshing the model. The experimentally determined critical load 1562 N was applied at the line contact of loading roller. The translation along thickness, of the nodes at the line contact of the supporting rollers was equated to zero. The induced stress and deflection of panel were determined through the numerical evaluation.

The experimentally observed load-deflection relation and numerically determined load-deflection relation are shown in Figure 14.5. The experimental load-deflection curve matched reasonably well with the numerical values, which proves the accuracy of numerical method.

Modelling of crevasse crossing bridge

A detailed survey on available bridging options and latest trends was carried out after which a design was conceptualised for going ahead. It was found that the most stable and light structure was possible with ladder-shaped structure. When a ladder is laid across a gap as a bridge, major problems encountered

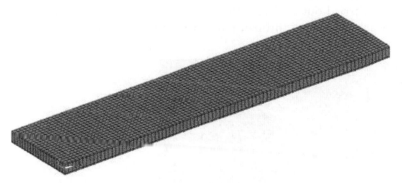

Figure 14.4 Finite element model of strip specimen
Source: Author

Table 14.1 Mechanical properties [14]

Material	E_x(MPa)	E_y(MPa)	E_z (MPa)	τ_{xy}	τ_{yz}	τ_{zx}	G_{xy}(MPa)	G_{yz} (MPa)	G_{zx} (MPa)
Unidirectional glass fibre mat/resin layer	41000	10400	10400	0.28	0.50	0.28	4300	3500	4300
Unidirectional carbon fibre mat/resin layer	147000	10300	10300	0.27	0.54	0.27	7000	3700	7000

Source: Author

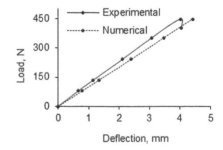

Figure 14.5 Load versus deflection under three-point bending test
Source: Author

where downward sag and lateral sway when it is dynamically loaded i.e., when the load is moving on the bridge. This problem could be solved by various methods like providing support from bottom or top on the ground, however reliability of such a support becomes a question considering the amount of undulations found in the designated glaciated terrain. The idea of reinforcing the structure through external reinforcements was analysed. The bridge can be reinforced either from the top of ladder as a superstructure or by underslung manner. Superstructure type reinforcements were analysed and found that a separate composite section has to be designed and fitted thereby increasing the weight of the overall bridge which would continuously remain in compression. The other disadvantage was that the ladder member always remains on tensile load. In the other case of underslung reinforcements, the reinforcements can be of much slender sizes like a steel wire rope which will take complete tensile load and keep the ladder member in a stable condition. The other criteria of rejecting superstructure reinforcements were that the structural stability gets compromised due to higher position of centre of gravity. The bridge will tend to topple when loaded in such a case. After a lot of analysis, it was found that underslung reinforcements be provided as an integral part of the bridge which can be connected during assembly of the bridge at site. This can be provided with two steel wire ropes connected between two end sections and passed through two compression struts fitted on the lower side.

The next major part to be analysed was length of each ladder member. After a lot of deliberation, the ladder member was designed for 1300 mm length so that it could be man portable with an overall width of 400 mm (Figure 14.6) to accommodate a person comfortably.

The model of bridge structure was prepared by using CATIA software (Figure 14.6). Box type section (50 mm × 50 mm × 4 mm) was considered for longitudinal member and hollow circular section (outer radius = 16 mm, inner radius = 14 mm) was considered for cross members and solid circular section (Ø 6 mm) for wire members (figure 14.4). Solid (20 node 186) elements were used for meshing the complete model. Mesh independence study has been carried out to decide the optimum element size. The model is discretised with optimum element size, which is received through number of iterations. The element edge

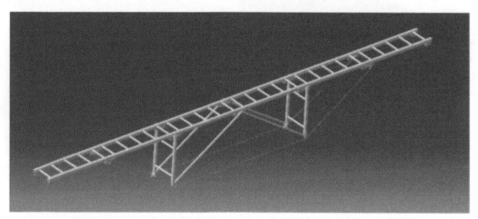

Figure 14.6 Bridge of span 9.1 m made of seven ladder type members each of 1.3 mm length
Source: Author

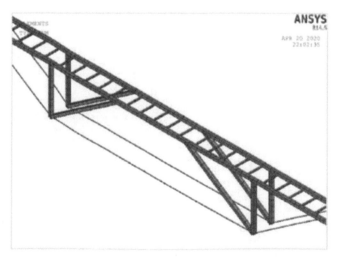

Figure 14.7 Meshed model of bridge
Source: Author

length is 1 mm along thickness and, 4 mm along the length and width. There is negligible change in results with further decreasing element size. The meshed model is shown in figure 14.7. The orthotropic material properties were supplied to the model (Table 14.1). The hybrid FRP composite material with a configuration of 80% glass fibre mat/ glass fibre rovings and 20% unidirectional carbon fibre mat reinforced in epoxy, was used for the structural members. Steel material (E = 210 GPa, U = 0.3) was defined for the wire used in underslung reinforcement.

Boundary conditions

As a worst-case scenario, the load analysis was carried out. The maximum concentrated load of 1200 N was applied at the centre of the ladder. All the degrees of freedom was made zero for the nodes at both the ends of ladder to simulate the actual condition, since the ladder structure was supported at the ends. Joints of the structure were not considered in the analysis, however higher factor of safety six was considered due to presence of joints, associated stress concentrations and safety of soldiers.

Results and discussion

In the ladder type structure, major load was taken up by the longitudinal members. The hollow box type section was selected for longitudinal members since it has high flexural/torsional stiffness and strength, and greater aerodynamic stability. The simulation was carried out for various square and rectangular box type sections with varying widths and breadths. Various sections analysed in ANSYS software are as shown in Figure 14.8. All screen shots have the corresponding size of section in the inset (Table 14.2). Maximum stress induced and maximum deflection of the structure under loading condition were determined. It is observed that the square cross section of size 50 mm × 50 mm is stiffer, hence this cross section is used for further analysis and fabrication of the required structure.

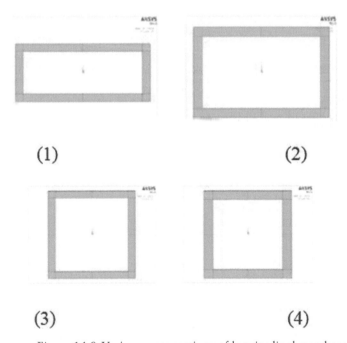

(1)   (2)

(3)   (4)

Figure 14.8 Various cross sections of longitudinal members analyzed in ANSYS software.
Source: Author

Table 14.2 Dimensions of cross section of longitudinal members

Sr No	Size of section (mm)			Max stress induced (MPa)	Max deflection (mm)
	Width (w)	Height (h)	Thickness (t)		
1	70	30	4	50.8	6.85
2	60	40	4	50.4	4.95
3	50	50	4	49.9	3.86
4	40	40	5	50.6	5.26

Source: Author

The maximum stress induced was observed to be 49.89 MPa at the steel wire of underslung reinforcement (Figure 14.9a). However, for ladder structure, maximum stress induced was around 25 MPa near the ladder end supported on ground (Figure 14.9b). Since the strength of steel wire was above 500 MPa and for FRP composite ladder structure was 454 MPa, design was very much safe. Maximum deflection was 3.86 mm at the mid of ladder structure, which was permissible since it would maintain the stability of structure.

Fabrication

Fabrication of longitudinal members of bridge

A set of 125 glass rovings, two glass mats and two carbon mats were used in manufacturing of the hollow square tube. The roving set was dipped in the matrix made of epoxy resin with accelerator and hardener. As per the project requirement orange colour was mixed in the matrix. The colour was preferred due to fact that orange colour enhances visible identification of bridge/ individual components when used in permafrost icy conditions. The impregnated glass roving, glass mats and carbon mats were fed into the forming and curing die which was maintained at 140⁰C (Figure 14.10). The ends of the complete set of roving and mats were tied to the puller on the other end and pultruded to obtain the finished or cured product of size 50 mm × 50 mm × 4 mm.

Fabrication of cross members of bridge

A set of 75 glass rovings, two glass mats and two carbon mats were used in manufacturing of the hollow circular tube. The roving set was dipped in the matrix made of epoxy resin with accelerator and hardener. As per the project requirement orange colour was mixed in the matrix. The impregnated glass roving, glass mats and carbon mats were fed into the forming and curing die which was maintained at 140°C. The ends of the complete set of roving and mats were tied to the puller on the other end and pultruded to obtain the hollow circular tube of size Ø 32 mm × 4mm.

Fabrication of ladder structure

The pultruded sections were cut to required sizes such as for manufacturing of ladder structure, cutting of 1300 mm length for square tubes to be utilised as long member of the ladder and 390 mm length for circular tubes to be utilised as cross member (Figure 14.11). The long member has four cross members at distance of 325 mm each starting at 165 mm from the end of the square tube.

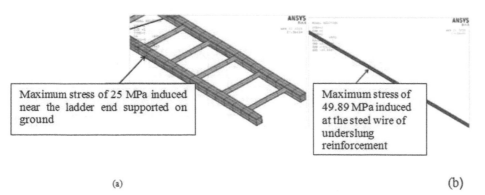

Maximum stress of 25 MPa induced near the ladder end supported on ground

Maximum stress of 49.89 MPa induced at the steel wire of underslung reinforcement

(a)

(b)

Figure 14.9 (a) Stress induced in the ladder structural of bridge, (b) Stress induced in the underslung reinforcement of steel wire
Source: Author

Figure 14.10 Hollow square tube of size 50 mm × 50 mm × 4 mm obtained after pultrusion process
Source: Author

Joining blocks

Joining blocks along with the cover plate were provided to join two ladder sections. The joining blocks were designed using high grade stainless steel (SS304) square tubes. The joining blocks were fitted into the ends of the ladder sections using specially designed stainless-steel pins. The pins were provided with the thumb ring for ease of handling in ice cold conditions. The joining blocks had provision for fitting of handrail member which could be directly fitted after launching of the bridge.

Underslung reinforcements

The bridge was provided with underslung reinforcement to support the tensile load while it is in loaded condition (Figure 14.10). It also provided rigidity to the bridge and avoided sagging and swaying while the loads were being crossed. The reinforcements were based on two steel wire ropes connected from first ladder to the last ladder through struts provided at 2nd and 5th ladder section. These struts were H-shaped sections of 900 mm length fitted under the joining blocks of respective sections.

Bridge Trials and Observations

The manufacturing process was completed, and the prototype bridge had been made ready for subjecting it to loading trials as per the envisaged working standards. The aim of the trials was to ascertain the capability of the material i.e., the hybrid composite to withstand requisite load and terrain conditions. The bridge was completely assembled and launched for the first time at Integrated training area at College of Military Engineering Pune across a clear gap of 8 m. The stores were laid out and the launching crew was briefed about the equipment. As per the planned procedure, inverted launch was carried out with trolley used as launching roller (Figure 14.11). The bridge was first tested for move of prescribed loads without provision of Handrail and Decking, i.e., by use of trolley for movement of personnel. It remained stable during move of trolley loaded with one man and two men simultaneously. No visible fatigue cracks were observed in the ladder or reinforcement struts.

The bridge was again assembled and launched at Integrated Training Area inside College of Military Engineering. This time the launch was carried out with handrail and decking. These manual passes were executed by walking, first one person at a time, then two simultaneously and then followed by three persons simultaneously. With an average weight of a person to be 70 kg, the bridge was subjected to dynamic

Figure 14.11 Ladder structure
Source: Author

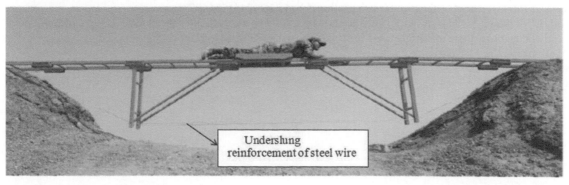

Figure 14.12 Bridge trial
Source: Author

load of 210 kg. Total of 10 passes were conducted. No visible fatigue cracks were observed in the ladder or reinforcement struts.

All components of bridge could be bag packed so that the person can easily carry it. Total nine bags were required to accommodate all the bridge components. Total weight of bridge was 125 kg. The weight of each bag was around 15 kg after keeping the bridge parts.

Conclusions

The light weight, man portable crevasse crossing bridge structure have been designed and fabricated using FRP material. The salient features are as follows:

- Light weight for easy carriage and handling. Easy transportation to site by all avail means (Manpack/ animal transport/snow sled/snow mobile/helicopter). Strong and sturdy parts which require low maintenance. Easy assembling and dismantling by least manpower. Launching and retrieving from both sides. Reduced launching time. Approximate span of nine meters to cover maximum available gaps in mountains and glacier. Designed for 120 kg load with factor of safety as six.

Acknowledgment

Authors acknowledge to College of Military Engineering, Pune for the support and helping in successful completion of fabrication, testing of the research work.

References

[1] Caprinoa, G., Loprestoa, V., Scarponib, C., and Briottib, G. (1999). Influence of material thickness on the response of carbon-fabric/epoxy panels to low velocity impact. Composites Science and Technology. 9(15), 2279-86.

[2] Puente, J. L., Zaera, R., and Navarro, C. (2002). The effect of low temperatures on the intermediate and high velocity impact response of CFRPs. Composites Part B: Engineering. 33(8), 559-66.

[3] Naik, N. K. and Shrirao, P. (2004). Composite structures under ballistic impact. Composite Structures. 66, 579-90.

[4] Naik, N. K., Joglekar, M. N., Arya, H., Borade, S. V. and Ramakrishna, K. N. (2004). Impact and compression after impact characteristics of plain weave fabric composites; effect of plate thickness, Advanced Composite Materials. 12, 261-80.

[5] Shimpi, R. P., Arya, H., and Naik, N. K. (2003). A higher order displacement model for plate analysis. Journal of Reinforced Plastics and Composites. 22, 1667-88

[6] Naik, N. K., Tiwari, S. I., and Kumar, R. S. (2003). An analytical model for compressive strength of plain weave fabric composites. Composites Science and Technology. 63, 609--25.

[7] Naik, N. K., Reddy, K. S., and Raju, N. B. (2003). Damage evolution in woven fabric composites: transverse static loading, Journal of Composite Materials. 37, 21-34.

[8] Naik, N. K., Azad, Sk. N. M., and Prasad, P. D. (2002). Stress and failure analysis of 3D angle interlock woven composites. Journal of Composite Materials. 36, 93-123.

[9] Einde, L. V. D. and Friedereible, L. J. (2003). Use of FRP composites in civil structural applications. Journal of Construction and Building Materials. 17(6–7). doi: 10.1016/S0950-0618

[10] YuBai, H. F., Liu, W., Qi, Y., and Wang, J. (2003). Connections and structural applications of fibre reinforced polymer composites for civil infrastructure in aggressive environments. Composites Part B: Engineering. 164, 129-43. doi: 10.1016/J.COMPOSITESB.2018.11.047

[11] Lochan, P. P. and Polak, M. A. (2022). Determination of tensile strength of GFRP bars using flexure tests. International journal of Construction and Building Materials. 314, Part A, 3.

[12] Liu, Y., Zhang, H. T., Zhao, H. H., Lu, L., Han, M. Y., Wang, J. C., and Shuai. (2021). Experimental study on mechanical properties of novel FRP bars with hoop winding layer. International Journal of Advances in Materials Science and Engineering. Hindawi publications. doi: 10.1155/2021/9554687

[13] Abbood, I. S., Odaa, S., Hasan, K. F., and Jasim, M. A. (2021). Properties evaluation of fiber reinforced polymers and their constituent materials used in structures – A review. Materials Today: Proceedings, Elsevier. doi: 10.1016/j.matpr.2020.07.636

[14] Daniel, I. M. and Ishai, O. (2006). Engineering Mechanics of Composite Materials. New York, Oxford: Oxford University Press.

Chapter 15

Effect of confinement on tension stiffening effect of RC beams subjected to flexure

Sumant Kulkarni[1,a], Mukund Shiyekar[2,b], Sandip Shiyekar[3,c] and Zain Kangda[1,d]

[1]School of Civil Engineering, REVA University, Bangalore, Karnataka, India

[2]Government College of Engineering, Karad, Maharashtra, India

[3]D. Y. Patil College of Engineering, Akurdi, Pune, Maharashtra, India

Abstract

In structural analysis, it is important to recognize the material as well as geometric properties of structural elements in indeterminate structures. Concrete and steel elastic properties recommended by codal provisions are usually accurate enough. The tension stiffening effect is fairly addressed by many international codes. The studies have revealed that tension stiffening effect varies with parameters like concrete grade and area of steel in tension zone. However, Indian Codal provisions remain silent about tension stiffening effect. This intension of this research work is to evaluate the influence of spacing of stirrups on the tension stiffening developed in members subjected to bending. The experimental study under consideration has included testing beams with various percentages of reinforcement to see most appropriate effect in accordance with existing code provisions for flexural members. The model size of beams used in the study is 150 × 150 × 700 mm. The study considers the effect of confinement on the results. The experimental results are confirmed using three-dimensional finite element analysis. The stress-strain values for steel bars obtained from FE analysis and those calculated by Eurocode2 are compared.

Keywords: Geometric properties, material properties, tension stiffening, confinement effect

Introduction

Reinforced concrete members consist of two materials namely steel and concrete. The response of concrete is influenced by stress-strain relationship of the same constituents. The ideal modelling of non-linear response of these materials proves to be a difficult task. In order to overcome this issue, many approximations are taken into account to study appropriate nonlinear aspects of reinforced concrete material. However, weak tension carrying capacity of concrete is major influential parameter for causing the non-linearity. This limitation is overcome by adding tension reinforcement in concrete specimen. The ultimate load carrying capacity of reinforced concrete member is not much under the effect of the lower magnitudes of tensile strength. It results in improved stiffness of the element subjected to higher magnitude of external loads owing to tension stiffening. The whole area between adjacent cracks in reinforced concrete can support tensile stresses after cracks are induced. This concept is regarded as tension stiffening (Figure 15.1).

The tension stiffening phenomenon has been modelled by various codal provisions as well as numerous studies in the past decade. Tension stiffening effect proposed by Eurocode 2 [1] and CEB-fib Model Code 2010 [2] takes into account strain, curvature, or deflection. These factors are interpolated and computed parameters are on the plain sections and sections with crack using the equations mentioned below.

$$\alpha = \zeta\alpha_2 + (1-\zeta)\alpha_1 \qquad (1) \qquad \text{and} \qquad \zeta = 1 - \beta\left(\frac{\sigma_{SF}}{\sigma_S}\right) \qquad (2)$$

where α = average value of strain, curvature, or deflection of the section between two successive cracks; α_1 = value for uncracked section; α_2 = value for sections with fully developed cracks; ζ = the coefficient of distribution; β = 1.0 for short term loading; β = 0.5 for long-term loading; σ_S = tensile stress in steel bar calculated on the cracked section; and σ_{SF} = stress corresponding to first crack as per the situation of load (short-term or long-term).

Eurocode 2 [1] also recommended a relation for the estimation of the crack's interaxis average value, s_{rm}:

$$s_{rm} = 50 + 0.25K_1K_2\frac{\emptyset_s}{\rho_{P,eff}} \text{ (in mm)} \quad (3)$$

where: ϕ_s = average bar diameter; K_1 = factor for bond reinforcement; K_2 = factor corresponding to strain distribution; and $\rho_{p,eff}$ = effective reinforcement ratio

[a]sumantk4@yahoo.co.in, [b]mukundshiyekar@gmail.com, [c]shiyekar@gmail.com, [d]zainkangda@gmail.com

DOI: 10.1201/9781003450252-15

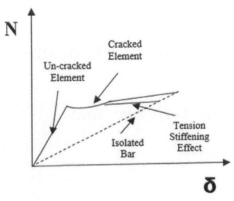

Figure 15.1 Tension stiffening effect on bare reinforcement bar
Source: Author

The flexural response of reinforced concrete members under tension stiffening has been studies by past several researchers. In preliminary stage, work was carried out by Vecchio and Collins [3], Collins and Mitchell [4], and Belarbi and Hsu [5,6]. Bentz [7] compared outcomess of the investigations and differentiated them through well documented research. The study could evaluate effectively the crack width and stiffness of reinforced concrete members under serviceability conditions. Salvatore et al. [8] studied its effect on reinforced concrete beams subjected to flexure. The study was further undertaken to evaluate the ductility of concrete sections rectangular in shape reinforced with particular, two-phase [9], steel bars. Shukri et al. [10] studied the behavior of reinforced concrete beams strengthened with carbon fiber reinforced polymer (CFRP) and subsequent effect of tension stiffening. A mechanical model was established by Shukri et al. [11] to introduce the effect of tension stiffening by simulating reinforced concrete hinges subjected to reversed periodic loading. The investigation of reinforced concrete members bonded externally with fiber-reinforced polymers was further extended by Sato et al. [12] to provide models for the evaluation of spacing of cracks and tension stiffening effects. A stress-strain model was developed by Stramandinoli et al. [13] to express exponential decay beyond cracking. It was interpreted by a specific factor that is a function of the reinforcement ratio and along with steel-to-concrete modular ratio. The rectangular cross-section beam specimen under study was subjected to four point loading. In order to calculate mean stresses in tension in steel bar of a reinforced concrete member post yielding, a tension stiffening model was proposed by Lee et al. [14]. Rectangular sections with symmetric as well as unsymmetric reinforcement were studied [15]. A relationship of tension stiffening without shrinkage effect was developed for the sections under study.

Tension stiffening effect is emphasised by various codes for estimation of the crack width of members subjected to bending [16-19]. However, tension stiffening due to confinement effect is neglected by few codes including IS: 456-2000 [20] and by the researchers earlier. A model was developed to calculate flexural deflection of reinforced ultra-high performance beams [22]. It was observed that the developed model could predict successfully the deflection of cracked section beam under service loads. An attempt was made [23] to assess the effect of tension stiffening on the flexural stiffness of RC circular sections to arrive at a conclusion that tension stiffening effect increases with the increase in percentage of tension reinforcement (p_t). Axial force does not influence tension stiffening except for lightly reinforced sections. Finally, it was observed that tension stiffening was higher in magnitude for smaller sections. A one-dimensional finite element model was developed to simulate the tension stiffening developed in concrete having cracks [24]. The model was used to predict the stress, strain, and displacement fields in the concrete along the rebar. Many researchers have proposed tension stiffening value empirically [25]. In present work, an attempt has been made to address this issue through experimental study. The experimental program has included testing of beams with various percentages of reinforcement to see the most appropriate effect in accordance with existing Codal provisions for flexural members. As far as the scope of this research is concerned, the study deals with different grades of normal concrete, variation in spacing of stirrups and the analysis is done for post-cracking stage of the beam. The details of the experimental program, results and discussion and validation by FE analysis are discussed in the subsequent sections along with the conclusions.

Experimental Program

Selection of Model Size

The experimental program consisted of flexure test for reinforced concrete beams of grades M20, M25 and M30. For each concrete grade, percentage of tension reinforcement (p_t) was decided as per IS:456-2000 [20]. The specimen were of size 150 × 150 × 700 mm in accordance with Clause 7.3, IS 516:1959

(Reaffirmed 2018) Methods of Tests for Strength of Concrete [21]. Three number of specimen were cast for each combination of reinforcement and average results were presented.

Area of Tension Steel bars in percentage (pt)

Minimum and maximum area of reinforcement as per IS: 456-2000 [20] provisions is $A_{st\,min} = \frac{0.85bd}{fy}$ (Clause No. 26.5.1.1 a and b), $A_{st}\,max = 0.04bD$, where b = width of beam, d = effective depth of beam, D = overall depth of beam and f_y = Stress in steel bats at yield stage.

In present scenario, calculation of percentage of reinforcement for a cross-section shown in Figure 15.2 is explained below.

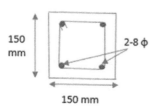

Figure 15.2 Beam cross-section
Source: Author

Effective depth of section= d= D-Effective cover to reinforcement= (150-24) =126mm.

According to IS:456-2000, $A_{st\,min} = \frac{0.85bd}{fy} = \frac{0.85 \times 150 \times 126}{415} = 38.71$ mm² and $A_{st}\,max = 0.04 \times 150 \times 150 = 900$mm². It is not possible to provide this small area as tension reinforcement in models since minimum bar diameter to be provided is of 6 mm. Hence, $A_{st\,provided} = 2 \times \frac{\pi}{4} \times (8)^2 = 100.53$ mm². Therefore, percentage of tension reinforcement pt= $\frac{100A_{st}}{bd} = \frac{100 \times 100.53}{150 \times 126} = 0.54\%$

Likely groupings of reinforcement for selected concrete grades and Fe415 grade of steel are shown in Table 15.1 [26, 27]. The experimental program has also considered the spacing of stirrups. The spacing of stirrups adopted was 50 mm, 10 0mm and 200 mm for all the specimens. Overall, there were four models for M20, five models for M25 and six models for M30 grade of concrete along with one plain concrete model for each concrete grade. The RC specimen were also provided with 2-8 ϕ bars at top to hold tension reinforcement with the help of stirrups.

Results and Discussion

Load deflection behaviour of plain concrete specimen of beam is shown in Figure 15.3. It can be observed that the model collapsed corresponding to a load of 77.67 kN, 78.53 kN and 79.43 kN respectively for M20, M25 and M30 grades of concrete. It was noted that hairline cracks developed in the model corresponding to a load of 68.65 kN, 62.75 kN and 69.34 kN for all the three concrete grades in the investigation. The recorded first crack deflections were 0.26 mm, 0.22 mm and 0.23 mm respectively. Due to absence of tension reinforcement in the section, all specimen failed without any warning. The response of RC sections for models 2–7 are plotted in Figures 15.4–15.6 for all the concrete grades selected in study.

Loads with respect to initial crack, yielding and failure were noted for all the specimen tested under flexure. It was observed that performance of these specimens was improved due to provision of reinforcement

Table 15.1 Steel reinforcement for concrete grades

Grade of concrete	Bar diameter and percentage of steel (p$_t$)						
	Model 1	Model 2	Model 3	Model 4	Model 5	Model 6	Model 7
M20	0%	2-8 ϕ 0.54%	2-8 ϕ+1-6 ϕ 0.69%	3-8 ϕ 0.80%	2-10 ϕ 0.84%	-----	------
M25	0%	2-8 ϕ 0.54%	2-8 ϕ+1-6 ϕ 0.69%	3-8 ϕ 0.80%	2-10 ϕ 0.84%	2-12 ϕ 1.21%	------
M30	0%	2-8 ϕ 0.54%	2-8 ϕ+1-6 ϕ 0.69%	3-8 ϕ 0.80%	2-10 ϕ 0.84%	2-12 ϕ 1.21%	2-12 ϕ+1-10 ϕ 1.63%

Source: Author

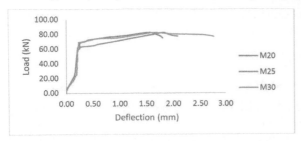

Figure 15.3 Load deflection response for model 1
Source: Author

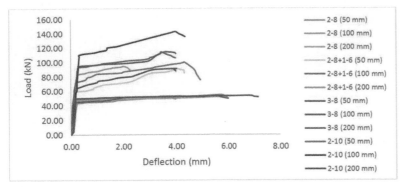

Figure 15.4 Load deflection response for models 2-5 (M20)
Source: Author

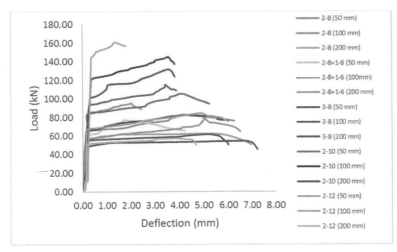

Figure 15.5 Load deflection response for models 2-6 (M25)
Source: Author

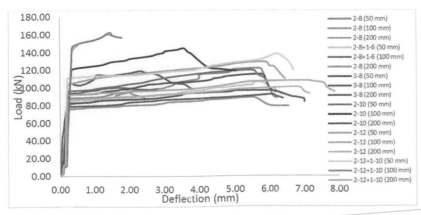

Figure 15.6 Load deflection response for models 2-7 (M30)
Source: Author

and variation in stirrup spacing. Central deflection was recorded for all the specimens using linear variable displacement transducer (LVDT). The experimental results of flexural member model for 50 to 200 mm apart stirrups for all the three concrete grades are discussed in subsequent sections below:

Beam models with M20 Concrete Grade

In case of RC models 2-5 having reinforcement range from 0.54 to 0.84%, it is noteworthy that there was a gradual rise in load at different stages (first crack, yielding and failure) with escalation in percentage of tension steel. Model 2 in 50 mm spacing group was observed to have developed first crack at 44 kN and corresponding deflection of 0.17 mm. At yielding stage, the mid-span deflection was recorded as 1.80 mm at a load of 47.96 kN. An unexpected rise in deflection was detected owing to de-bonding of the steel bars with adjacent concrete.

The failure load for model 2 was 55 kN and deflection was noted as 5.70 mm. A defection of 0.20 mm was recorded for RC models 3-5 with load at first crack as 59.32 kN, 64.12 kN, and 72.99 kN. A closer spacing of confinement of 50 mm resulted in same deflection for the models but with enhanced loading. Yield loads for RC models 3-5 were 66.06 kN, 79.20 kN, and 85.81 kN with deflections of 0.94 mm, 1.83 mm and 1.90mm respectively. At failure, these models were subjected to a deflection range of 4mm to 4.3 mm. It is evident that a fair warning was given by models from yield stage to failure stage.

Models with same combination reinforcements with 100 mm stirrup spacing were subjected to two-point loading in universal testing machine. The loads and corresponding deflections at different stages were noted for all the specimens. Models 2-5 recorded first crack loads as 83 kN, 92 kN, 95.30 kN and 110.49 kN with corresponding defections as 0.27 mm, 0.29 mm, 0.28 mm and 0.31 mm respectively. These models reached yield stage with a little increment in loads as 94 kN, 101.43 kN, 98.92 kN and 124.17 kN and recorded deflection deflections were 1.90 mm, 1.90 mm, 1.98 mm and 1.93 mm respectively. Along with a fair warning for failure, models 2-5 failed at higher magnitudes of loads of 95 kN, 114.83 kN, 114.65 kN and 142.98 kN with deflection values in the range of 2 mm to 3.98 mm.

The models with stirrup spacing of 200mm showed relatively lesser loads at all the stages i.e. first crack, yielding stage and failure of specimen. Owing to increased spacing of stirrups, models 2-5 developed first crack at 47.34 kN, 49.31 kN, 45.11 kN, and 50.31 kN with deflections of 0.19 mm, 0.19 mm, 0.17 mm and 0.17 mm respectively. The yield loads were in the range of 50.11 kN to 52.47 kN whereas failure loads were in the cluster of 52.34 kN to 54.00 kN. Increased spacing of stirrups from 100mm to 200mm resulted in more deflection of the specimen at the same time, loads recorded were of lower magnitude.

Table 15.2 Load deflection at different stages for m20 concrete

Stages of loading	Load (kN) and deflection (mm) values					
	Model 1 0%	Model 2 2-8ϕ 0.54%	Model 3 2-8 ϕ+1-6 ϕ 0.69%	Model 4 3-8 ϕ 0.80%	Model 5 2-10 ϕ 0.84%	Stirrups spacing (mm)
First crack	68.65 0.26	44.00 0.17	54.32 0.20	64.12 0.20	72.99 0.20	50
Yield stage	78.49 1.13	47.96 1.80	56.06 0.94	79.20 1.83	85.81 1.90	
Failure stage	77.67 1.76	55.00 5.70	57.58 6.50	83.78 4.00	100.36 4.30	
First crack	68.65 0.26	83.00 0.27	92.00 0.29	95.30 0.28	110.49 0.31	100
Yield stage	78.49 1.13	94.00 1.90	101.43 1.90	98.92 1.98	124.17 1.93	
Failure stage	77.67 1.76	95.00 2.00	114.83 2.68	114.65 3.25	142.98 3.98	
First crack	68.65 0.26	47.34 0.19	49.31 0.19	45.11 0.17	50.31 0.17	200
Yield stage	78.49 1.13	50.11 2.90	51.25 2.60	51.55 2.98	52.47 2.93	
Failure stage	77.67 1.76	52.34 4.95	53.47 5.34	53.92 6.18	54.00 6.80	

Source: Author

Table 15.3 Load deflection at different stages for M25 concrete (50 mm spacing)

Stages of loading	Load (kN) and deflection (mm) values						
	Model 1 0%	Model 2 2-8 φ 0.54%	Model 3 2-8 φ+1-6 φ 0.69%	Model 4 3-8 φ 0.80%	Model 5 2-10 φ 0.84%	Model 6 2-12 φ 1.21%	Stirrups spacing (mm)
First crack	62.75 0.22	66.12 0.21	61.60 0.20	65.60 0.20	84.00 0.20	67.20 0.16	50
Yield stage	76.52 1.37	71.50 2.22	71.89 1.10	75.48 2.19	89.71 1.81	72.97 2.00	
Failure stage	78.53 1.98	82.65 4.20	77.00 1.70	82.00 4.30	105.00 4.00	84.00 4.80	
First crack	62.75 0.22	83.00 0.27	93.67 0.30	101.09 0.31	120.75 0.35	144.20 0.37	100
Yield stage	76.52 1.37	90.00 1.21	102.77 2.21	116.43 1.98	132.72 2.01	157.28 1.23	
Failure stage	78.53 1.98	95.00 2.00	114.83 3.43	131.44 3.54	144.69 3.55	158.79 1.54	
First crack	62.75 0.22	55.70 0.20	57.13 0.20	55.17 0.18	48.60 0.15	51.30 0.15	200
Yield stage	76.52 1.37	60.39 1.23	60.25 1.24	56.93 1.62	52.49 1.15	53.22 1.15	
Failure stage	78.53 1.98	62.00 5.30	79.35 5.00	61.30 5.20	54.00 6.80	57.00 3.90	

Source: Author

Beam models with M25 Concrete Grade

For M25 grade of concrete, maximum compression reinforcement increases to 1.21%. All the models have exhibited enhanced performance in terms of load and deflection which is evident from Figure 15.5 as well as Table 15.3. For all the models with 50 mm spacing of stirrups, load at first crack stage is in the range of 62.75 kN to 84 kN with deflection range of 0.16 mm to 0.22 mm. There is a gradual increase in load at yield stage in the range of 71.50 kN to 89.7kN and deflection in the range from 1.10 mm to 2.00 mm. There is a considerable rise in load at failure stage from 77 kN to 105kN and deflection in the range of 1.70 mm to 4.80 mm. All the models failed with a fair warning giving sufficient time. The models with 100mm spacing of stirrups were observed to have higher loads at different stages of loading in comparison to 50 mm spacing of stirrups. The models with 200 mm spacing of stirrups failed at relative lower range of loads and higher deflection owing to increased spacing of stirrups.

Beam models with M30 Concrete Grade

For M30 grade of concrete, maximum compression reinforcement is 1.65% as per IS Codal provisions. All the models were observed to have increased load and deflection as reflected Figure 15.6 and Table 15.4. For all the models with 50 mm spacing of stirrups, load at first crack stage is in the range of 69.34 kN to 110.94 kN with deflection range of 0.23 mm to 0.25 mm. There is a gradual increase in load at yield stage in the range of 80.56 kN to 113.64 kN and deflection in the range from 1.06 mm to 1.45 mm. There is a substantial increase in load at failure stage from 79.43 kN to 138.67 kN and deflection in the range of 2.56 mm to 6.20 mm. All the models failed with a reasonable notice giving adequate time. The models with 100 mm spacing of stirrups were observed to have higher loads at different stages of loading in comparison to 50 mm spacing of stirrups. The models with 200mm spacing of stirrups failed at relative lower range of loads and higher deflection owing to increased spacing of stirrups.

Investigation of Tension Stiffening

Tension stiffening effect can be evaluated by various codal provisions. Some of the tension stiffening equations recommended by different codes around the globe are as mentioned below:

Egyptian Code

The Egyptian Code ECP203-2007 recommends equation (4) for the mean strain ε_{sm} as

Table 15.4 Load deflection at different stages for m30 concrete (50 mm spacing)

Stages of loading	Load (kN) and deflection (mm) values							
	Model 1 0%	Model 2 2-8 φ 0.54%	Model 3 2-8 φ+1-6 φ 0.69%	Model 4 3-8 φ 0.80%	Model 5 2-10 φ 0.84%	Model 6 2-12 φ 1.21%	Model 7 2-12 φ+1-10 φ 1.63%	Stirrups spacing (mm)
First crack	69.34 0.23	88.20 0.25	85.60 0.24	92.27 0.24	96.14 0.24	103.47 0.23	110.94 0.23	50
Yield stage	80.56 1.45	90.00 1.22	87.70 1.69	93.60 1.06	99.76 1.15	112.11 1.24	113.64 1.24	
Failure stage	79.43 2.56	122.15 5.70	107.00 5.60	115.34 5.80	120.17 5.50	129.34 5.90	138.67 6.20	
First crack	69.34 0.23	82.00 0.37	93.67 0.30	105.42 0.31	120.57 0.34	140.93 0.36	144.20 0.36	100
Yield stage	80.56 1.45	86.00 1.13	97.41 1.10	114.65 1.06	125.88 1.20	157.28 1.23	158.29 1.26	
Failure stage	79.43 2.56	95.00 2.00	100.09 1.60	118.93 2.30	144.69 3.55	160.55 1.33	162.55 1.43	
First crack	69.34 0.23	75.91 0.24	77.47 0.24	78.74 0.24	82.65 0.23	86.05 0.23	92.11 0.23	200
Yield stage	80.56 1.45	78.85 1.64	79.03 1.22	80.87 1.65	85.64 1.25	88.35 1.26	94.15 1.15	
Failure stage	79.43 2.56	89.31 5.50	91.14 5.90	92.64 6.10	97.21 6.00	101.24 6.80	108.36 7.10	

Source: Author

$$\varepsilon_{sm} \text{ as } \varepsilon_{sm} = \frac{f_s}{E_s}\left(1 - \beta_1\beta_2\left(\frac{f_{scr2}}{f_s}\right)^2\right) \tag{4}$$

where fs is the value of stress developed in the tension steel. It can be evaluated on the basis of a cracked section in N/mm². Longitudinal stress in N/mm² due to tension in reinforcement fscr2 is calculated based on a cracked section. The load to be considered for this calculation is load at the first crack.

β1 = Coefficient for bar bond characteristics. Its value is 0.5 for plain bars and 0.8 for deformed bars.

For sustained loading, β_2 is the coefficient. It is taken as 1 for short-term loading whereas 0.5 for sustained/cyclic loading.

E_s is the Young's Modulus of steel, N/mm². Tension stiffening is evaluated by the term $\beta_1\beta_2\left(\frac{f_{scr2}}{f_s}\right)^2$ in equation (4).

British standards BS8110-1997

As per BS8110-1997, when cracking in the section is taken into account, the average strain is,

$$\varepsilon_m = \varepsilon_1 - \frac{b(h-x)(a'-x)}{3E_sA_s(d-x)} \tag{5}$$

where ε1 is the strain at the level considered. It is estimated by overlooking the result of stiffening of concrete in tension region. b = The cross-section width at CG of reinforcement in zone below neutral axis (NA), a'= The distance from extreme compression fiber to the location of crack. In Equation 5, tension stiffening is evaluated by the term $\frac{b(h-x)(a'-x)}{3E_sA_s(d-x)}$.

Eurocode2- 2004

As IS:456-2000 [20] does not have any equation for the calculation of effect of tension stiffening, Eurocode2 equation was assumed for the same. The value of mean tensile strain ($\varepsilon_{sm} - \varepsilon_{cm}$) as per Eurocode2-2004 is given by equation (6). This equation is utilized for the evaluation of the crack width of a member subjected to bending.

$$(\varepsilon_{sm} - \varepsilon_{cm}) = \frac{\left(f_s - K_t\left(\frac{f_{cteff}(1+\eta\rho_{eff})}{\rho_{eff}}\right)\right)}{E_s} \geq 0.6\frac{f_s}{E_s} \tag{6}$$

where K_t is the factor used for the duration of loading. Its value is 0.6 for short-term and 0.4 for long-term. f_s = Steel stress under tension. It is estimated based on section with developed cracks, η = modular ratio E_s/E_c and f_{cteff} = The average value of concrete in tension corresponding to first crack,

$$\rho_{eff} = \frac{A_s}{A_{ceff}} \tag{7}$$

A_{ceff} = Net area in tension. In Equation 6, tension stiffening is evaluated by the term $\frac{\left(K_t\left(\frac{f_{cteff}(1+\eta\rho_{eff})}{\rho_{eff}}\right)\right)}{E_s}$.

The estimation of tension stiffening in steel bars for M20 concrete is done using equations recommended by Eurocode 2- 2004 and FE simulation. These values are presented graphically in Figures 15.7–15.10 for M20 grade of concrete.

FE Analysis

The results of models considered in experimental study are validated by FE analysis. The elements used for different materials of RC models are SOLID65 for concrete material and LINK8 for main reinforcement and stirrups. The FE model of the concrete beam and reinforcement created are presented in Figures 15.11 and 15.12 respectively.

Bilinear Kimematic properties and the typical Multilinear Isotropic have been represented in Figures 15.13 and 15.14 respectively. The values of stress-strain induced in tension reinforcement bars are utilized for analytical evaluation of tension stiffening. Each FE model is subjected to incremental load in a total of sixty one number of sub-steps. It was noted that the gain of strain in tension reinforcement for each model is higher with the increase in load for each sub-step. Initial stages of loading (sub-step 1) for model 1 revealed that initial strain range was from 3.77×10^{-6} to 1.21×10^{-5}. This strain was consistent for later sub-step 15 from 3.25×10^{-4} to 3.98×10^{-4}. It was prominently observed that in the initial sub-steps, cracks were totally absent. This behavior may be attributed to the presence of reinforcement in tension region. It ultimately

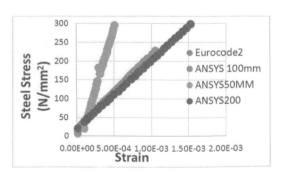

Figure 15.7 Steel strain for beam model 2
Source: Author

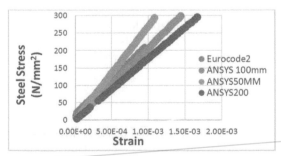

Figure 15.8 Steel strain for beam model 3
Source: Author

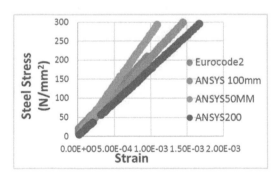

Figure 15.9 Steel strain for beam model 4
Source: Author

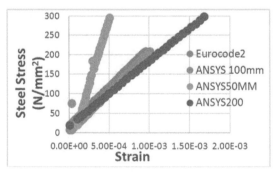

Figure 15.10 Steel strain for beam model 5
Source: Author

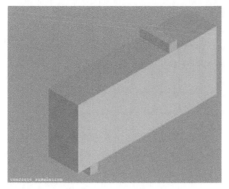

Figure 15.11 Concrete beam model
Source: Author

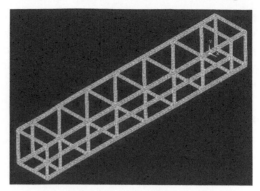

Figure 15.12 Steel bar and stirrup model
Source: Author

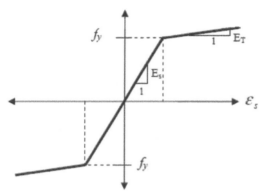

Figure 15.13 BKIN simulation for steel
Source: Author

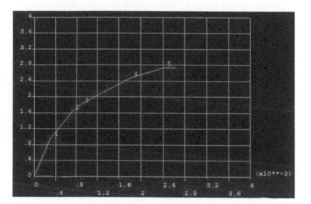

Figure 15.14 MISO simulation for model 2 of M20
Source: Author

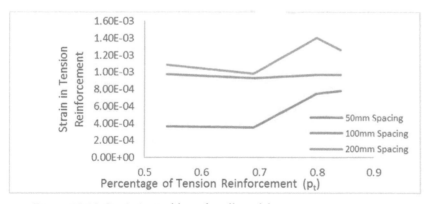

Figure 15.15 Strain in steel bars for all models
Source: Author

resulted in improvement in the models resisting the considerably. When the cracks were induced in the FE models of the beam, there was some part of concrete below neutral axis which was undisturbed. The strain values induced in tension reinforcement with respect to the loads corresponding to first crack may be owing to this fact. Evaluation of strain due to tension by Eurocode2 and FE software were found to be in fairly good agreement. Figures 15.7-15.10 reflect that Eurocode2 values are nearly same for FE models having a stirrup spacing of 100mm only. Eurocode2 values differ from FE results with a difference of as good as ±7%. However, RC models confined with 50mm and 200mm stirrup spacing show a substantial difference in case of strain values of steel in the tension zone. Figure 15.15 represents variation of strain in tension steel bars. For evaluation of average strain in tension steel bars, stress considered in steel is 200 N/mm^2. It is witnessed that with the increase in confinement spacing from 50 mm to 200 mm, strain values induced in steel bars under tension also tend to increase. With the increase in spacing of stirrups, area of concrete between two consecutive stirrups also increases. Lesser the spacing of confinement, less is the crack developed in concrete zone. It was also observed that the concrete region remains relatively stable in between two consecutive stirrups even first crack is developed in tension zone of the beam. This is useful to extend the study the effect of confinement on tension stiffening also. Nevertheless, it demands an exhaustive investigational study to establish an empirical relation for evaluation of tension stiffening effect as per IS:456-200 which is missing.

Conclusions

The experimental study was undertaken to estimate the tension stiffening influence of RC members with steel ratios of 0.54% to 0.84% for M20 concrete, 0.54-1.21% for M25 concrete and 0.54-1.65% for M30 concrete. Results and deductions from this study are summarized as follows:

- Evaluation of tension strain by Eurocode2 and FE software were found to be in fairly good agreement for models with a stirrup spacing of 100 mm. The effect of confinement does affect the average strain values. More the confinement, lesser is the strain value in tension reinforcement.
- Initiation of cracks in stabilised cracking stage and the failure load stage were observed with naked eye. It was deduced that crack numbers increase with the increment in area of tension steel bars in percentage (pt).
- The tension stiffening is a function of amount of tension steel bars in percentage (pt) along with equi-spaced stirrups. With increase in p_t, the effect of tension stiffening reduces.
- For models with 0.80% and 0.84% tension reinforcement, tension stiffening value evaluated by Eurocode2 is slightly on higher side in comparison to other lightly reinforced sections (0.54-1.65%).
- Experimental program has revealed that tension stiffening effect reduces considerably owing to higher magnitude of strain rate. However, concrete is able sustain tensile stresses at higher levels of strain.

References

[1] EN 1992, Eurocode 2: Design of Concrete Structures, Part 1–1: General Rules and Rules for Buildings, European Committee for Standardization (CEN): Brussels, Belgium, 2005.
[2] FIB-Special Activity Group 5. FIB Bulletin 65: Model Code 2010—Final Draft; International Federation for Structural Concrete (FIB): Lausanne, Switzerland, 2012.
[3] Vecchio, F. and Collins, M. (1982). Response of Reinforced Concrete to In-Plane Shear and Normal Stresses. University of Toronto: Toronto, ON, Canada.
[4] Collins M. and Mitchell, D. (1987). Prestressed Concrete Basics; Canadian Prestressed Concrete Institute: Ottawa, ON, Canada.
[5] Belarbi, A. and Hsu, T. (1994). Constitutive Laws of Concrete in Tension and Reinforcing Bars Stiffened by Concrete. Journal of Structural. 91, 465–74.
[6] Belarbi, A. and Hsu, T. (1995). Constitutive laws of softened concrete in biaxial tension-compression. Journal of Structural. 92, 562–73.
[7] Bentz, E. (2005). Explaining the riddle of tension stiffening models for shear panel experiments. Journal of Structural Engineering. 131, 1422–25.
[8] Salvatore, W., Buratti, G., Maffei, B., and Valentini, R. (2007). Dual-phase steel re-bars for high-ductile r.c. structures, Part 2: Rotational capacity of beams. Engineering Structures. 29, 3333–41.
[9] Maffei, B., Salvatore, W., and Valentini, R. (2007). Dual-phase steel rebars for high-ductile RC structures, Part 1: Microstructural and mechanical characterization of steel rebars. Journal of Structural Engineering. 29, 3325–32.
[10] Shukri, A., Darain, K., and Jumaat, M. (2015). The tension-stiffening contribution of NSM CFRP to the behavior of strengthened RC beams. Materials. 8, 4131–46.
[11] Shukri, A., Visintin, P., Oehlers, D., and Jumaat, M. (2016). Mechanics model for simulating rc hinges under reversed cyclic loading. Materials. 9, 305.
[12] Sato, Y. and Vecchio, F. (2003). Tension stiffening and crack formation in reinforced concrete members with fiber-reinforced polymer sheets. Journal of Structural Engineering. 129, 717–24
[13] Stramandinoli, R. and La Rovere, H. (2008). An efficient tension-stiffening model for nonlinear analysis of reinforced concrete members. Engineering Structures. 30, 2069–80.
[14] Lee, S., Cho, J., and Vecchio, F. (2011). Model for post-yield tension stiffening and rebar rupture in concrete members. Engineering Structures. 33, 1723–33.
[15] Kaklauskas, G., Gribniak, V., Bacinskas, D., and Vainiunas, P. (2009). Shrinkage influence on tension stiffening in concrete members. Engineering Structures. 31, 1305–12.
[16] ACI Committee 224R-01, Control of Cracking in Concrete Structures, ACI Report 224R-01, American Concrete Institute, Farmington Hills, MI, 2001, pp. 46.
[17] BS 8110: Part 2:1997: Structural Use of Concrete, Part 2, Code of Practice for Special Circumstances. British Standard Institution, London.
[18] Eurocode 2: Design of Concrete Structures – Part 1: General Rules and Rules for Buildings; The European Standard EN1992-1-1.

[19] ECP 203-2007; The Egyptian Code for Design and Construction of Reinforced Concrete Structures, Ministry of Housing, Egypt.

[20] IS 456:2000: Code of practice for plain and reinforced concrete, New Delhi, India.

[21] IS 516:1959 (Reaffirmed 2018): Methods of tests for strength of concrete, New Delhi, India.

[22] Teng, L., Zhang, R., and Khayat, K. H. (2022). Tension stiffening effect consideration for modeling deflection of cracked reinforced UHPC beams. MDPI Sustainability. 14, 1-21.

[23] Morelli, F., Amico, C., Salvatore, W., Squeglia, N., and Stacul, S. (2017). Influence of tension stiffening on the flexural stiffness of reinforced concrete circular sections. MDPI Materials. 10, 1-16.

[24] Yankelevsky, D. and Jabareen, M. (2008). One-dimensional analysis of tension stiffening in reinforced concrete with discrete cracks. Engineering Structures. 30, 206-217.

[25] Welch, G. and Janjua, M. (1971). Width and spacing of tensile cracks in reinforced concrete. UNICIV Report No R76, University of NSW, Kensington.

[26] Kulkarni, S. K., Shiyekar, M. R., Shiyekar, S. M., and Wagh, B. (2014). Elastic properties of RCC under flexural loading-experimental and analytical approach. Sadhana - Academy Proceedings in Engineering Sciences. 39, 677-697.

[27] Kulkarni, S. K., Shiyekar, M. R., and Shiyekar, S. M. (2017). Confinement effect on material properties of rc beams under flexure. Journal of The Institution of Engineers (India): Series A. 98, 413-428.

Chapter 16

Overview of gravity die casting process parameters affecting product quality

Vishal Dhore[1,a] and Lalit Toke[2,b]

[1]MET Bhujbal Knowledge City, Institute of Engineering, Nashik, India

[2]Sandip Institute of Engineering and Management, India

Abstract

The automotive industry faces many challenges in the process of die casting which is specially used for manufacturing of complex shaped components. One of the industries located in Maharashtra state is involved in manufacturing automotive brake parts used in verity of sport bikes. Recently the industry has identified the failure of the main component used in brake assembly. Basically, component is manufactured using die casting technique. The preliminary analysis reveals that there are many parameters that affect the product quality. The objective of this study is to review the gravity flow die casting process parameters that has an influence on casting component. The literature review considers the use of a variety of major index publications by renowned publishers. We observed that process validation was necessary to examine the effects of different gravity flow die casting process parameters. The gravity flow die casting technique where reviewed. It was observed that temperature, pouring time and chemical constitutes have significant impact on product quality. To reduce the defects of casting components, a pouring temperature, die temperature and molten metal velocity flow rate, these process parameters are needed to study experimentally and simulation technique. This paper represents the effect of these gravity flow die casting process parameters and suggested remedies to minimize the rejection percentage and increase casting component quality.

Keywords: Casting, mould filling, mould temperature, pouring time, simulation

Introduction

Automotive components like brake linear, brake calliper, engine valves, cylinder head etc. are complex in shape. These components are difficult to manufacture with traditional machining processes. All such small and complex shaped components are manufactured with various casting processes. Automotive industries uses these components in various automotive assemblies like braking system, engine system, transmission system, body etc. Among the available manufacturing processes, industry prefers casting processes. The advantage of casting process is ability of producing complex shaped components with higher accuracy [1-3]. In India around 14,000 automotive industries are established till the year 2021[4].There are many small scale and medium scale industries who are the suppliers for automotive industries. These small scale industries are involved in manufacturing of automotive components of complex shaped. Different types of casting processes adopted by the industry as shown in Figure 16.1.

Currently automotive industries are adopting innovative casting method for the production of these complex shaped components. However, for the various requirements, most small-scale casting industries face challenges to increase the production of goods on a budget. During the casting process, many defects are occur, which shows in poor quality product. It demonstrates the significance of the casting process and, as a result, has become an area of attention in terms of defect reduction and process improvement.

An industry located in Pune MIDC of the Maharashtra state is manufacturer and supplier of braking system. The industry has established in the year 2012. Nearly 20,000 components was supplied by the industry, presently the industry, facing the problem of higher rejection due to casting defects. The industry uses gravity die casting process for the manufacturing of these component. The author was motivated to take up this as research work. This paper considers the review of various peer review articles for study and analysis of the casting process, the process parameters, effects of these parameters and defects.

Total 34 research publication in the related field was reviewed. These articles are reviewed that focuses on the various casting methodology and technical advances. A total of 60% of the articles was considered for the study of experimental and simulation processes. Remaining articles are considered for the study of analysis techniques used by the researchers. All publications under review process are peer publications from various conferences and journals.

Die Casting Materials

Mixing the more than one alloying elements together to form a metals and other time setting materials are used as casting materials in various automotive component. Many automotive industries are relies on

[a]dhore.vishal@gmail.com, [b]lalittoke2010@gmail.com

DOI: 10.1201/9781003450252-16

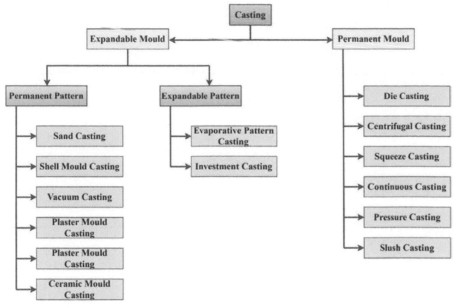

Figure 16.1 Types of castings [5-7]
Source: Author

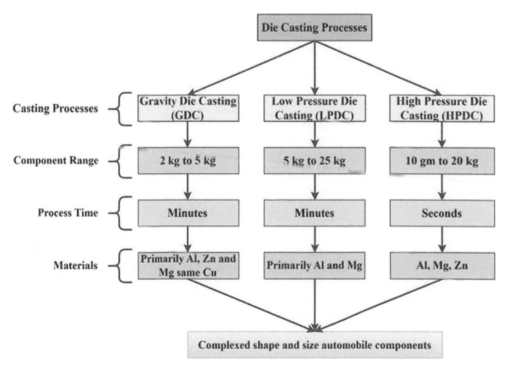

Figure 16.2 Die casting processes
Source: Author

casting industry for manufacturing the complex shapes and sizes component with very high dimensional accuracy. As per the Foundry Informatics Center and Statista Research Department, information the overall casting production in Worldwide is 105.5 million metric tons and in India is 11.31 million metric tons during 2020-21. The automobile sector consumes 32% share of casting products [8-9].

Automobile sector is enormous market for casting products using various non-ferrous metals and its alloys. Aluminium, Copper, Zinc, Magnesium and their alloys are most common casting materials are used. These materials has excellent mechanical properties are low density, light in weight, higher strength to density ratio, non-magnetic, higher conductivity, resistance to corrosion [10].

Many automotive industry recommend the die casting processes as gravity, low and high pressure. For excellent surface finishing, strength, high dimensional accuracy, lightweight, complex in shape and size, large number in quantity for manufacturing. Figure 16.2 represents die casting processes with range of component, process time and materials used in it.

Table 16.1 Chemical composition (in %) of AC4C (Al-Si-Mg) material [13]

Al	Cr	Cu	Fe	Mg	Mn	Si	Ti	Zn
97.5-99	< = 0.10	< = 0.10	< = 0.35	0.40-0.60	< = 0.10	0.60-0.90	< = 0.10	< = 0.10

Source: Author

Table 16.2 Chemical composition (in%) of AC7A (Al-Mg 5000) material [13]

Al	Mg	Cr	Cu	Fe	Mn	Si	Zn	Ni	Li	B
97.5	0.5-13	<= 0.5	< = 0.2	< = 0.8	< = 2	< = 2	< = 3	< = 0.5	< = 3	< = 0.05

Source: Author

Table 16.3 Summary of Al-alloys casting materials

Sr. No.	Year	Casting metals	Properties	Applications	References
1	2011	AC4C (Al-Si-Mg)	Thermal expansion coefficient 21.5×10^{-6} Density 2680 kg/m^3 Specific heat 0.96 kJ/kg.K	Tier mould	[11-12]
2	2011	AC7A (Al-Mg 5000)	Thermal expansion coefficient 23.6×10^{-6} Density 2670 kg/m^3 Specific heat 0.88 kJ/kg.K	Tier mould	[12]
3	2015	Al-Zn-Mg-Cu alloy	High tensile strength, good elongation, good corrosion resistance	Transportation and Racing Car	[15]
4	2016	AlSi7Cu3Mg	High-temperature fatigue, yield strength 364 MPa	Automotive gasoline V16 cylinder head	[16,19]
5	2017	Al–5.5Mg–1.5Li–0.5Zn–0.07Sc–0.07Zr alloy	Lightweight, good strength,	Structural purpose in automobile	[17]
6	2018	Al-8.1Mg-2.6Si alloy	Excellent thermal conductivity, high tensile strength, high yield strength	Bearings	[18]

Source: Author

The review is totally focus on various casting materials, simulation software, experimentation and numerical methods are utilized for complex shape geometry components of automobile by using die casting (gravity) process.

Die casting (gravity)

Die casting (gravity) is process developed for manufacturing the complex geometrical shapes of automobile components. It has the vertical and horizontal mould opening and tilt between 0 to 90° and 0 to 120° angle with the tilting arrangement. The molten metal are directly poured into the mould opening and filling molten metal into the mould cavity with help of tilting arrangement of die at certain tilting angle. By using the die casting (gravity) process high mechanical properties like strength, stiffness and surface finishing are easily obtained. Due to that reason it is strongly recommended for automotive application like engine cylinder head, brake calliper, engine blocks and piston, brake bracket and so many.

Reddy and Rajanna, conducted the experiment on Al-Si-Mg alloy samples. It was observed that solidification time decrease when Si (%) increase from 2-4 and increase when Si (%) varies from 4-9. To remove the dissolved gasses from melt 1% tetrachloremethane is sufficient [11]. Yoon et. al, reported that tier mould productivity of AC7A material was better than AC4C material [12]. Table 16.1 represents the chemical composition of the AC4C (Al-Si-Mg) casting material and Table 16.2 represents chemical composition of AC7A (Al-Mg 5000) casting materials.

Rahul et al. reported the experimental study on the Al-6wt%Cu alloy. The various samples was prepared with frequency of 0 Hz, 40 Hz, 80 Hz, and 150 Hz by using mould vibration. They measured the hardness number of sample by using Brinell hardness test and observed the microstructure of each sample at top, centre and bottom location. If vibration frequency varies from 40 Hz to 180 Hz refinements grains structures are observed and hardness is increases at bottom location this will effect on the solidification of casting due to high cooling rate [14]. Saikawa et. al conclude from there experimental study that fine crystallised alpha-Al grain structure of (5Zn-M,6.5Zn-M and 8Zn-M) was observed in microstructure with increasing the Zn content in Al-Zn-Mg-Cu alloy in die casting (gravity). It will impact the solidification of casting due to high cooling rate. The tensile test result of Al-Zn-Mg-Cu alloys in the 6.5Zn-M alloy cast are 537 MPa, 519 MPa and 1.3 % elongation [15]. Camicia and Timelli, reported the effect of addition of AlTi5B1 elements in the AlSi7Cu3Mg alloy improve the uniform and fine grain structure in throughout the casting. Due to that refinement of grain structure it will improve the cooling rate and mechanical property [16]. Chunchang Shi et. al, reported the heat treatment and die casting (gravity) process. With this analysis they concluded that 8.7% elongation, 270.5 MPa yield strength and 435.5 MPa ultimate tensile strength after the solid solution heat treated (500°C/10h) with subsequent aging process in the cast Al–5.5Mg–1.5Li–0.5Zn–0.07Sc–0.07Zr alloy [17]. Longfei Li et. al, varying the Zn content 0.08 % to 4.62 % in the Al-8.1Mg-2.6Si alloy. By using the heat treatment process 3.59 % Zn content added in the Al-8.1Mg-2.6Si alloy. It was observed that 1.16 % and 2.91% increase in tensile and yield strength as well as 5.72% decrease in elongation [18]. Table 16.3 shows the various Al alloys with their properties and their application used in the automobile components.

From the review of various Al-alloy materials, it was observed that if changing the Si, Zn and adding Ti content in the materials refine grain structure has observed in throughout the casting and it has the impact on the solidification due to high cooling rate and improve the mechanical properties of the casting. If pouring the molten metal in cavity by using the vibration frequency a fine grain structure observed in casting and it has impact on the solidification due to high cooling rate and improve the hardness property of die casting (gravity).

Casting simulation

A simulation is the act of imitating the behaviour of situation of present or future system, providing indication for decision making. This provide 3D environment for more understanding. Simulation allows to take number of test runs for various combination of situations. It also helps in designing the process, identification of defects and its location, prediction purpose, development of product to reduce time and cost of casting industry. Due to that casting industries increase the productivity of casting product [20].

There are many casting simulation software packages are working in industry right now. AutoCAST, ProCAST, MAGMASoft, SolidCAST, ANYCAST, CastCAE, Nova-Solid flow, ANSYS these are the casting simulation software's are used by casting industry. Fig. 16.3 represents the various simulation software with their add-on modules, solution methods, defects predicted and materials database information.

Hussainy et. al, reported the simulation findings for the casting defects using SolidCAST simulation software. Simulation was performed using the process parameters, pouring temperature 720°C, pouring time

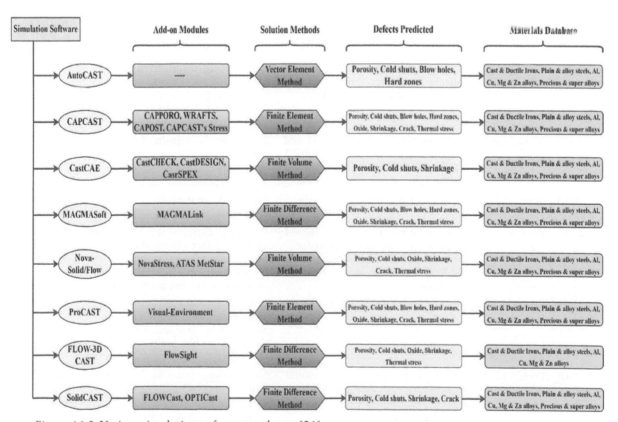

Figure 16.3 Various simulation software packages [21]
Source: Author

3 sec, mould temperature 200°C and 250°C, casting material LM6 and LM25. Simulation result shows the LM25 material has better result than LM6 material at 250°C mould temperature. In simulation they varying the riser diameter from 12 mm to 35 mm for reducing the defect near riser region. The simulation result of 35 mm riser diameter has no defect in casting [22]. Patil et. al, reported the simulation findings for casting using MAGMA software. Simulation was performed using the process parameters, pouring temperature 730°C, mould temperature 350°C, pouring time 7 sec, solidification time 200 sec, molten metal velocity 1 m/s. The simulation result show the no shrinkage porosity defect in the casting components. Simulation software also helped to save the time and cost of the industry for development of new component design [23]. Kumar and Reddy, used ProCAST simulation software to identify the casting defect in wheel rim component. The shrinkage porosity defect was observed at various points in gating system and high turbulence flow by selecting the pouring temperature as 680°C with 60% yield. After modifying the gating system and selected the pouring temperature 680°C with 80% yield, no shrinkage porosity defect was observed in the component [24]. Yu Fu et. al used the ProCAST software to identify the shrinkage porosity region. Simulation was simulated using pouring temperature 730°C, die temperature 200°C, heat transfer coefficient of casting mould 500 W/m²k, casting velocity 0.5 m/s. Simulating solidification and molten metal filling processes in the bottom gating system and slitting (vertical) gating system at intermediate time intervals. The shrinkage porosity region was observed in the top riser, tank and near the slag tank of bottom gating system. In slitting (vertical) gating system no shrinkage porosity defects was observed in the casting [25]. Kamal et. al used ANYCAST software for optimizing the design and simulation of cavity mould. 2.1857 sec in horizontal design and 1.6546 sec of vertical design of filling time was observed while pouring temperature of molten metal at 740°C. Longer solidification time of 504.7066 sec was observed in vertical design as compared to horizontal design. In horizontal design more shrinkage defect was observed as compare to the vertical design in the mould [26]. Li et. al reported that the shrinkage porosity defect in the component by experimentally and ProCAST simulation software. They varying the mould temperature from 150, 250 and 350°C and keep the pouring temperature at 750°C. If mould temperature at 150°C, no shrinkage porosity defect was observed. The shrinkage porosity defect was observed along the transverse and longitudinal direction, when increase the mould temperature from 150°C-350°C [27]. Šabík et. al reported use of NovaFlow and Solid simulation software for designing the gating system for an automotive clutch wheel. Simulation was performed using the process parameter pouring temperature 1410°C and molten metal flow rate 11 kg/sec. Initial casting design requires 3.460 sec time for the 49% filling the

Table 16.4 Summary of Al-alloys casting materials

Sr. No.	Year	Simulation software	Process parameters	Key findings	Reference
1	2015	SolidCAST	Pouring temperature 720°C, Pouring Time 3 sec, mould temperature 200°C and 250°C	Better result of LM25 material casting at 250°C mould temperature. 35 mm riser diameter has shown no defects in casting.	[22]
2	2015	MAGMA	Pouring temperature 730°C, Mould temperature 350°C, Pouring time 7 seconds, Solidification time 200 sec and Molten metal velocity 1 m/s	No shrinkage porosity defect in casting	[23]
3	2015	ProCAST	Pouring temperature 680°C with 80% yield	No shrinkage porosity defect in casting	[24]
4	2017	ProCAST	Pouring temperature 730°C, mould temperature 200°C, heat transfer coefficient of casting mould 500 W/m²k, casting velocity 0.5 m/sec.	In bottom gating system shrinkage porosity defect observed in top riser, tank and near the slag tank. No shrinkage porosity defect was observed in slitting gating system.	[25]
5	2018	ANYCAST	Pouring temperature 740°C	In vertical design filling time is 1.6546 sec, solidification time is 504.7066 sec and less shrinkage defects in mould cavity was observed.	[26]
6	2018	ProCAST	Pouring temperature 750°C, mould temperature 150, 250 and 350°C	No shrinkage porosity defect at 150°C mould temperature.	[27]
7	2021	NovaFlow	Pouring temperature 1410°C, velocity of molten metal flow 11 kg/sec	Modified design of casting requires 3.168 sec to fill the 49% mould.	[28]

Source: Author

mould whereas the modified casting design takes only 3.168 sec [28]. Table 16.4 represents the summary of various simulation software, process parameters and key finding of simulation result.

By review the various simulation software, it observe that so many researchers are only focus on the pouring temperature, die temperature and molten metal velocity process parameters to reduce the casting defects as well as casting design. However, in gravity die casting process tilting angle, tilting speed and tilting time are not analysed by using the simulation software. It was also observed that very few researchers were reported the analysis of solidification time.

Experimental Studies

In a casting process experimental studies was performed to identify the relationship between the various process parameters and casting defects. With the help of experimental studies we identify the causes of various defects in the components. Mathematical model and quality control techniques are implemented to minimise the casting defects.

Wong and Pao has reported computer experimentation for thermal control of uniform solidification in GDC by using the genetic algorithms approach. In this experimentation they optimised the interfacial heat transfer coefficient at 5000 W/K/m^2. It was observed that the left section cool faster than the right section. By implementing the GA optimization algorithms, 5°C temperature difference was observed between the right and left section after 50 generation. So, by using the GA optimization algorithms uniform solidification was achieved [29]. Mehmood et. al. reported the experimental findings for grain refinement using sloping plate process. The molten metal poured at 800°C temperature on 800 mm length slopping channel. The specimens was casted at 15°, 30°, 45°, 60°, and 75° slopping plate angles. High heat transfer rate was observed at 60° slopping plate angle which leads to refine grain structure. Comparatively highest ultimate tensile strength 205.9 Mpa and 2.81 % elongation was observed at 60° slopping plate angle [30]. The refined grains structure improve the solidification time as well as mechanical properties of the casting. Barot and Ayar has reported the experimental findings for shrinkage porosity defect depth by varying the thickness of plate. The experiment was conducted on three different cavity size of 130 × 105 × 8 mm, 130 × 105 x 12 mm, and 130 × 105 × 16 mm. The molten metal was poured into the cavity at 785°C and 550°C die temperature. The experimental result was compared with approximate simulation result the minimum 4.34% shrinkage porosity depth defect was observed in 16 mm thick plate [31]. Wolff et. al reported the influences of process parameters on heat transfer coefficient using the statistical ANOVA methods. The experimental analysis was conducted for a pouring temperature of 720°C. During the experimentation cooling oil temperature was set to 30°C, 100°C, 200°C and 300°C. Mold inserts was made of steel and copper materials. Cooling channel distances was varied from 10 to 15 mm. At cooling oil temperature of 100°C it was observed that copper mold material has the highest impact of heat transfer coefficient [32].

Table 16.5 Various experimental studies on GDC

Sr. No.	Year	Methodology	Process parameters	Key findings	Reference
1	2011	Genetic algorithm (GA)	Solidification temperature difference	Uniform solidification temperature achieved after 50 generation of GA.	[29]
2	2016	Slopping plate angle	Pouring temperature 800°C, Slopping plate angles 15°, 30°, 45°, 60° and 75°	High heat transfer rate, 205.9 MPa ultimate tensile strength and 2.81% elongation at 60° slopping plate angle	[30]
3	2020	Simulation and geometry varied plate	Pouring temperature 785°C, Die temperature 550°C	Minimum 4.34% shrinkage porosity depth defect in 16 mm thick plate	[31]
4	2020	ANOVA	Pouring temperature 720°C, cooling oil temperature as 30, 100, 200 and 300°C	Highest impact of heat transfer coefficient at 100°C cooling oil temperature (copper mold material)	[32]
5	2021	CNN	Pouring temperature 725°C, die temperature 400°C	Oxide related defects were occurred at a cross-section close to the ingate	[33]
6	2021	ANN	Pouring temperature 550°C to 650°C, mould temperature 150°C to 350°C, pouring time 3-12 sec	Higher shrinkage defect in upper and lateral feeder section at 150°C mould temperature ANN model revealed 99% prediction	[34]

Source: Author

Scampone et. al. reported the experimental findings for oxide related defects using convolutional neural network deep learning algorithm technique. The experiment was conducted at pouring temperature 725°C, die temperature 400°C and high speed resolution camera was used for capturing the images of molten metal flow path. It was reported that greater concentration of oxide related defects has obtained near the ingate section [33]. Kumruoglu, reported the use of experimental findings for predicting the casting defects using ANN prediction model. The experiments was conducted for a pouring temperature range 550°C-650°C, mould temperature range 150°C-350°C, pouring time 3-12 sec. Higher shrinkage defect was observed in lateral and upper surface of feeder at 150°C mould temperature. At 350°C mould temperature lowest shrinkage defects was observed. ANN model was designed with (5-14-7-1) arrangement to train and test model with sigmoid function. The ANN model has R2 Coefficient 0.99 and the model has good prediction ability [34]. Table 16.5 represents the summary of various experimental studies on GDC.

After reviewing various experimental studies, it was observed that pouring temperature and die temperature has significant role in reduction of casting defects as well as improving the mechanical properties of casting product. Limited use of ANN, DoE, ANOVA, and CNN was reported for prediction and minimising the casting defect by varying/optimising material composition.

Conclusion

This paper presents an overview of gravity die casting process parameters affecting product quality. It was observed that temperature, pouring time, and chemical constitutes has a significant impact on product quality. This paper represents the effect of these gravity flow die casting process parameters and suggested remedies to minimise the rejection percentage and increase casting component quality.

- Changing the Si and Zn content in Al-alloy, a uniform refine grain structure was observed throughout casting. It improved the mechanical properties and cooling rate of casting product.
- Adding Ti content in Al-alloy, a uniform refine grain structure was observed throughout the casting. It improved the cooling rate of solidification and mechanical properties of casting product.
- Pouring the molten metal in cavity by using vibration frequency, fine grain structure was observed in casting. It improved the cooling rate of solidification and hardness of casting product.
- To reduce the defects of casting components, a pouring temperature, die temperature, and molten metal velocity flow rate, process parameters, were simulated using various simulation software's.
- Simulation software results save the time and cost for the casting industries.
- Various experimental studies shows that pouring temperature, die temperature, and molten metal flow rate has a significant role towards the casting defects.
- It was observed that some process parameters as angle of tilting, tilting time, and tilting speed not analysed through the simulation software.
- For predicting and minimising the casting defects in gravity die casting, the limited use of ANN, DoE, ANOVA and CNN was observed.

References

[1] Goede, M., Stehlin, M., Rafflenbeul, L., Kopp, G., and Beeh, E. (2009). Super light car—lightweight construction thanks to a multi-material design and function integration. European Transport Research Review. 1, 5-10.

[2] Campbell, J. (2015). Complete Casting Handbook: Metal Casting Processes, Metallurgy, Techniques and Design, Second edition. UK, Complete Casting Handbook: Metal Casting Processes, Metallurgy, Techniques and Design, Second edition. UK,Oxford: Butterworth-Heinemann.

[3] Ravi, B. (2005). Metal casting: Computer-aided design and analysis, sixth Edition. India, New Delhi: PHI Learning. Pvt. Ltd, 2005. Automotive Industry in India, https://www.ibef.org/industry/india-automobiles

[4] Daws, K. M., AL-Dawwod, Z. I., and AL-Kabi, S. H. (2008). Selection of metal casting processes: a fuzzy approach. Jordan Journal of Mechanical and Industrial Engineering,. 2(1), 45-52.

[5] Shi, C., Wu, G., Zhang, L., and Zhang, X. (2018). Al–5.5Mg–1.5Li–0.5Zn–0.07Sc–0.07Zr alloy produced by gravity casting and heat treatment processing. Materials and Manufacturing Processes. 33(8), 891-7.

[6] Doan, B. Q., Nguyen, D. T., Nguyen, M. N., Le, T. H., Dong, T. M. H., and Duong, L. H. (2021). A review on properties and casting technologies of aluminum alloy in the machinery manufacturing. Journal of Mechanical Engineering Research and Developments. 44(8), 204-17. Foundry Informatics Centre. http://foundryinfo-india.org/profile_of_indian.aspx, Statista Research Department. https://www.statista.com/statistics/237526/casting-production-worldwide-by-country/

[7] Goenka, M., Nihal, C., Ramanathan, R., Gupta, P., Parashar, A., and Joel, J. (2020). Automobile parts casting-method and materials used: a review. Materials Today Proceedings. 22, 2525-31.

[8] Reddy, A. C. and Rajanna, C. H. (2009). Design of gravity die casting process parameters of Al-Si-Mg alloys. Journal of Machining and Forming Technologies. 1, 1-25.

[9]   Yoon, H. -S., Yang, H. -D., and Oh,Y. -K. (2011). A study on cooling characteristics of ac4c and AC7A cast-ing material for manufacturing tire mold. Advanced Materials Research. 264-265, 379-83. MatWeb Material Property Data sheet. https://www.matweb.com/

[10] Kumar, R., Ansari, M. S., Mishra, S. S., and Kumar, A. (2014). Effect of mould vibration on microstructure and mechanical properties of casting during solidification. International Journal of Engineering Research & Technology. 3(4), 90-2.

[11] Saikawa, S., Aoshima, G., Ikeno, S., Morita, K., Sunayama, N., and Komai, K. (2015). Microstructure and mechanical properties of an Al-Zn-Mg-Cu alloy produced by gravity casting process. Archives of Metallurgy and Materials. 60, 871-4.

[12] Camicia, G. and Timelli, G. (2016). Grain refinement of gravity die cast secondary AlSi7Cu3Mg alloys for auto-motive cylinder heads. Transactions of Nonferrous Metals Society of China,. 26(5), 1211-21.

[13] Shi, C., Wu, G., Zhang, L., and Zhang, X. (2018). Al–5.5Mg–1.5Li–0.5Zn–0.07Sc–0.07Zr alloy produced by gravity casting and heat treatment processing. Materials and Manufacturing Processes. 33(18), 891-7.

[14] Li, L., Ji, S., Zhu, Q., Wang, Y., Dong, X., Yang, W., Midson, M., and Kang, Y. (2018). Effect of Zn concentra-tion on the microstructure and mechanical properties of Al-Mg-Si-Zn alloys processed by gravity die casting. Metallurgical and Materials Transactions A. 49, 3247-56.

[15] Moria, A. D., Timelli, G., Berto, F., and Fabrizi, A. (2020). High temperature fatigue of heat treated secondary AlSi7Cu3Mg alloys. International Journal of Fatigue. 138, 1-10.

[16] Behera, R. Das, S., Dutta, A., Chatterjee, D., Sutradhar, G. (2010). Comparative evaluation of usability of FEM- and VEM based casting simulation software, 58th Indian Foundry Congress. 31-39.

[17] Khan, M.A.A. and Shaikh, A. K. (2018). A comparative study of simulation software for modelling metal casting processes. International Journal of Simulation Modelling. 17(2), 197-209.

[18] Hussainy, S. F., Mohiuddin, M. V., Laxminarayana, P., Krishnaiah, A., and Sundarrajan, S. (2015). A practical approach to eliminate defects in gravity die cast al-alloy casting using simulation software. International Journal of Research in Engineering and Technology. 4(1), 114-23.

[19] Patil, R. T., Metri, V. S., and Tambore, S. S. (2015). Analysis and simulation of die filling in gravity die casting using MAGMA software. International Journal of Engineering Research & Technology. 4(1), 556-9.

[20] Kumar, S. K. and Reddy, K. S. (2015). Casting simulation of automotive wheel rim using aluminium alloy mate-rial. International Journal of Manufacturing and Mechanical Engineering. 1(1), 41-46.

[21] Fu, Y. Wang, H., Zhang, C. and Hao, H. (2018). Numerical simulation and experimental investigation of a thin-wall magnesium alloy casting based on a rapid prototyping core making method. International Journal of Cast Metals Research. 31, 37-46.

[22] Kamal, M. R. M., Tahir, H.A.M., Salleh, M. S., Bazri, H. N. H., Musa, M. K., and Bazilah, N. F. (2018). Design and thermal analysis simulation of gravity die casting on aluminium alloy (ADC12). Journal of Advanced Manufacturing Technology. 12, 271-88.

[23] Li, L., Li, D., Gao, J., Zhang, Y., and Kang, Y. (2018) Influence of mold temperature on microstructure and shrinkage porosity of the A357 alloys in gravity die casting. Advances in Materials Processing, pp. 793-801. Lecture Notes in Mechanical Engineering. .

[24] Vladimír Šabíka, Peter Futáš , Alena Pribulová , Petra Delimanová, (2021). Optimization of a gating system by means of simulation software to eliminate cold shut defects in casting. Journal of Casting and Materials Engineering. 5(1), 1-4.

[25] Wong, M. L. D. and ennis Pao, W. K. S. (2011). A genetic algorithm for optimizing gravity die casting's heat trans-fer coefficients. Expert Systems with Applications. 38, 7076-80.

[26] Mehmood, A., Shah, M., Sheikh, N. A., Qayyum, J. A., and Khushnood, S. (2016). Grain refinement of ASTM A356 aluminum alloy using sloping plate process through gravity die casting. Alexandria Engineering Journal. 55, 2431-38.

[27] Barot, R. P. and Ayar, V.S. (2020). Casting simulation and defect identification of geometry varied plates with experimental validation. Materials Today Proceedings. 26, 2754-62.

[28] Wolff, N., Zimmermann, G., Vroomen, U., and Bührig-Polaczek, A. (2020). A statistical evaluation of the influ-ence of different material and process parameters on the heat transfer coefficient in gravity die casting. Advances in Metal Casting Technology. 10(10), 1-12.

[29] Scampone, G., Pirovano R., Mascetti, S., and Timelli, G. (2021). Experimental and numerical investigations of oxide-related defects in Al alloy gravity die castings. The International Journal of Advanced Manufacturing Technology. 117, 1765-80.

[30] Kumruoglu, L. C. (2021). Prediction of shrinkage ratio of ZA-27 die casting alloy using artificial neural network, computer aided simulation, and comparison with experimental studies. Scientia Iranica B. 28(5), 2684-2700.

Chapter 17

Numerical modal and harmonic analysis using viscoelastic material

Gaurav Sharma[1,a], Adepu Kumaraswamy[1,b], Sangram K Rath[2,c] and Debdatta Ratna[2,d]

[1]Department of Mechanical Engineering, Defence Institute of Advanced Technology, Pune, India

[2]Polymer Science and Technology Directorate, Naval Materials Research Laboratory, Ambernath, India

Abstract

A numerical study was conducted with the help of the finite element method to minimise the vibration generating in the structures using free layer damping and constrained layer damping. Viscoelastic materials (VEMs) are extensively used for vibration damping due to their excellent dynamic properties. In literature, characterisation of the efficacy of VEMs for vibration damping applications is mainly through measurement of material loss factor (tan δ), system loss factor, or by simulation using FEM. It is posited that a combination of all three tools is pertinent for robust characterisation; however, such reports are rather scanty. To this end, the VEM was characterised for its viscoelastic properties in terms of storage modulus and loss factor using dynamic mechanical analysis (DMA). A Prony series expansion of the relaxation modulus was used to model the material behavior and evaluate the coefficients. The experimental data was adapted for the ANSYS mechanical finite element method (FEA) through a numerical algorithm for the evaluation of frequency response function in both free layer damping (FLD) and constrained layer damping (CLD) mode. The results revealed pronounced amplitude attenuation in constrained layer mode compared to free layer mode.

Keywords: CLD, DMA, FLD, modal analysis, Prony series, viscoelastic material

Introduction

Viscoelastic materials (VEM) are extensively used for the attenuation of the machine and structure-borne vibrations. Two approaches are distinguished: free layer damping (FLD) and constrained layer damping (CLD). Viscoelastic layers applied on the structure with a constrained layer are known as CLD while without a constraining layer just one single layer of VEM over the structures is called FLD [1]. In the case of the former, alternate compression and extension of the VEM leads to the dissipation of mechanical vibrations. In the latter case, on the other hand, shearing off the VEM imposed by the stiff constraining layer induces vibration damping. The evaluation of the dynamic properties of VEMs is a prerequisite for understanding their vibration-damping efficacies. Modal analysis is a dynamic study of a structure that undergoes vibrational excitation. A structure's dynamic properties include natural frequencies, mode shaper, and damping which can be identified using the modal analysis [2]. The modal analysis helps to find out the resonance frequency which is critical in the case of dynamic types of machinery and suggests the optimised frequency which helps vibration engineers and to identify the dynamics of the structure and the response of the structure to different types of excitations. Harmonic analysis is the forward study of modal analysis in which the response of the structure in the terms of displacement, velocity, and acceleration concerning frequency can be identified also known as frequency response or bode diagram [3].

The viscoelastic behavior of polymers depends on several factors such as molecular weight, glass transition temperature, degree of cross-linking, and stress relaxation characteristics due to the re arrangements of the molecular chain with time [4]. There are generally two basic mechanical models in practice to identify the properties of viscoelastic materials named Voigt's model and Maxwell elements, which can be represented as a mathematical model where springs and dashpots are connected in parallel and series connection. Voigt model describes the parallel connection between spring and dashpot while in the maxwell model both are connected series connection. The complex behavior of viscoelastic materials can be explained by combining Maxwell and Voigt's models based on that one simple model is there which is known as the generalized Maxwell model which is the parallel connection o of Maxwell elements [5-7].

There is a few mathematical modelling of viscoelastic materials found in the literature the Prony series is one of the effective modelings which can be used in FEM software for vibration analysis of the structure [8]. The Prony series is one of the most common and efficient numerical methods in the time domain which shows the relation of the relaxation and creep functions of VEMs, this method has been widely used by researchers in an attempt to achieve the modeling of viscoelastic materials which were verified by experimental data from some polymeric materials. An optimised numerical method to calculated the Prony series coefficients of viscoelastic materials was developed by Ferry and Tschoegl [9]. At temperatures above the

[a]akswamy@gmail.com, [b]sharmag603@gmail.com, [c]akswamy@gmail.com, [d]sangramkrath@gmail.com

DOI: 10.1201/9781003450252-17

glass transition, the materials show noticeable viscoelastic behavior. Hu et al., conducted a study on viscoelastic behavior of an epoxy thermoset by tensile relaxation test and determined the relaxation modulus of the material as a function of time [10-14].

Prony series can be expressed assuming linear viscoelastic behavior by a generalised maxwell model: -

$$E(t) = E_\alpha + \sum_{i=1}^{n} E_i \cdot \exp\left(-\frac{t}{\tau_i}\right) \tag{1}$$

Where E_α is the instantaneous elastic modulus
$\quad$ E_i is the modulus of ith element
$\quad$ t is time
$\quad$ τ_i is the relaxation time of the respected element.
$\quad$ E(t) is the relaxation modulus.

Relaxation modulus and relaxation time are the two parameters we are required to add in the FEM software as an important property of viscoelastic material.

The motive behind the present study is to identify the resonant frequencies and harmonic response of structural steel beams subjected to free layer as well constraint layer damping treatment. Using viscoelastic material as a damping layer and structural steel as a constraint layer in the frequency range from 0 – 3000 Hz and studying its response in terms of acceleration. A literature survey was done on the procedure of modal analysis and harmonic analysis. The finite element method (FEM) is one of the most useful and efficient options in analysing modal characteristics as its quick and very informative. We employ numerical modal and harmonic analysis of free layer and constrained layer damping using viscoelastic material in ANSYS will be explained.

Dynamic Mechanical Analysis

A dynamic mechanical analyser is a device that is useful to do characterisation of polymeric materials. The working principle of the DMA is that it has two clamps one is movable another one is fixed. The fixed clamps remain static while the movable clamp help to deforming the material in an oscillating pattern by moving up and down. The frequency and amplitude of the oscillations are programmed based on the properties of each material displayed. The results are often frequency response diagrams and temperature response diagrams called frequency sweep and temperature sweep. where the values of the storage modulus E′ and the loss modulus E″ are shown on the vertical axis and the frequencies in the frequency sweep where the analysis was performed are shown on the horizontal axis. Most of the time, the horizontal axis is presented in a logarithmic scale because it is important to show high and low-frequency values. We can get the loss factor or known as tan delta through DMA data because it is the ratio of loss modulus to storage modulus and the sum of storage modulus and loss modulus is known as complex modulus [15].

$$E^* = E′ + iE″ \tag{2}$$

$$\tan\delta = E″ / E′ \tag{3}$$

Using DMA data to obtain Prony series for VEM

The values of storage modulus E′ and loss modulus E″ as the function of the frequency can be determined by performing DMA on a sample of material of a specified size. Also, as explained by Ferry and Tschoegl these two storage and loss moduli can be represented as functions of frequency in the following equations.

$$E′(w) = E_\infty + \Sigma((E_i * w^2 * \tau_i^2/(1 + w^2 * \tau_i^2)) \text{ .. storage modulus.} \tag{4}$$

$$E″(w) = \Sigma((E_i * w * \tau_i^2/(1 + w^2 * \tau_i^2)) \text{ .... Loss modulus...} \tag{5}$$

Here both storage as well loss modulus can be found with these mathematical equations. In this study storage modulus was used for the calculation as it gives more accurate results due to higher values compared to loss modulus.

The dynamic analysis of the VEM was done in the on DMA machine Gabbro in tensile mode. And the required inputs for the test were given as: -

• Strain: -0.05%
• Frequency: -0.5–50 Hz
• Temperature Range: -20-40°C

- Sample size: -25 *mm* × 10 *mm* × 2 *mm*
- Reference temperature: 25°C

To get the results in extended frequency range WLF equation was used and the master curve was plotted at the reference temperature.

Data obtained by DMA

Figure 17.1 shows the data obtained by the DMA in frequency sweep. Figure 17.1(a) shows the storage modulus concerning the frequency and the value of the storage modulus is increasing with frequency. While Figure 17.2(b) shows the loss modulus concerning the frequency and the value of loss modulus is increasing till a point after it decreases with frequency. And Figure 17.1(c) indicates the loss factor or tan delta concerning the frequency and we can observe the highest value of tan delta is 0.35 which is good for damping and the lowest value of tan delta is 0.08 at a very high frequency.

Equation 4 of storage modulus was used till the 4th terms.

$$E'(w) = E\infty + ((E1 * w^2 * \tau1^2/(1 + w^2 * \tau1^2)) + (E2 * w^2 * \tau2^\wedge2/(1 + w^2 * \tau2^2)$$
$$+ (E3 * w^2 * \tau3^2/(1 + w^2 * \tau3^2)) + (E4 * w^2 * \tau4^2/(1 + w^2 * \tau4^2)) \tag{6}$$

The above Equation 6 was used to curve fit the DMA data to get the unknows
The approximate values of storage modulus $E'(w)$ from curve fitting approximation are the following:

$E_\infty = 7.239, E_1 = 2.886, E_2 = 2.595, E_3 = 2.941, E_4 = 3.611, \tau_1 = 1.664, \tau_2 = 0.322, \tau_3 = 23.33, \tau_4 = 0.1686.$
putting the above values in Equation 6 we got the storage modulus as a function of frequency(w) only, as is shown in Equation 7:

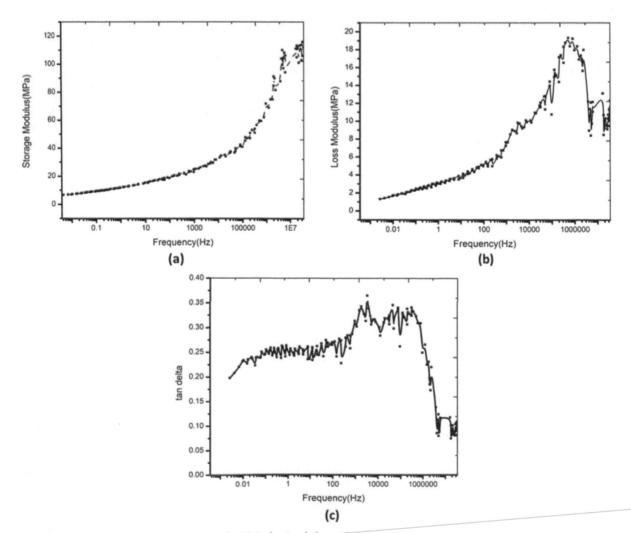

Figure 17.1 Dynamic properties of VEM obtained through DMA
Source: Author

$$E'(w) = 7.239 + ((2.886 * w^2 * 1.664^2/(1 + w^2 * 1.664^2)) + (2.595 * w^2 *$$
$$0.322^2/(1 + w^2 * 0.322^2) + (2.941 * w^2 * 23.22^2/(1 + w^2 * 23.22^2)) +$$
$$(3.611 * w^2 * 0.1686^2/(1 + w^2 * 0.1686^2)) \tag{7}$$

Through the equation, we have calculated the value of storage modulus concerning the frequency w and compared the calculated value with the experimentally measured value from DMA. Table 17.1 shows the measured and calculated storage modulus and the error for storage modulus and Figure 17.2 shows the graphical comparison of both calculated and measured storage modulus.

The accuracy of the numerical modal performed can be determined by comparing the calculated values of $E'(w)$ with Equation 7 and it has a difference in values maximum of 2% with reference to the data measured by experimental measurement through the DMA. These results are justifying the use of this Prony

Table 17.1 Error calculation of values measured with DMA vs. values calculated with the Prony Series

Frequency	Measured storage modulus (DMA data)	Calculated storage modulus (Prony series)	| Error |
Hz	MPa	MPa	%
0.010447	7.34705	7.384086	0.504092
0.037929	8.53241	8.384166	1.737425
0.054824	8.80851	8.867886	0.674074
0.08689	9.19158	9.382112	2.072899
0.104465	9.60067	9.542819	0.60257
0.218258	10.1695	10.07802	0.899583
0.239314	10.4631	10.15357	2.958261
0.262404	10.3553	10.23522	1.159579
0.315478	10.5079	10.42037	0.832999
0.5	10.9998	11.02132	0.195669
0.548239	11.3355	11.16003	1.547942
0.601133	11.4836	11.30146	1.58613
0.792447	11.6877	11.72294	0.301542
0.868898	11.874	11.85688	0.144175
0.95273	11.9043	11.98554	0.682474
1.25594	12.223	12.33677	0.930825
1.37712	12.5413	12.44353	0.77958
1.99054	12.7917	12.84958	0.452503
2.18258	13.0518	12.95388	0.750224
3.15479	13.5043	13.43065	0.545394
5	14.1361	14.21836	0.581937
5.00001	14.4572	14.21837	1.652
7.92447	14.8543	15.11523	1.756573
9.5273	15.4205	15.44959	0.188674
12.5594	15.8254	15.8988	0.463833
15.0998	16.2795	16.16892	0.679257
19.9054	16.6346	16.55637	0.47026
23.9314	17.0506	16.82002	1.352316
31.5479	17.2349	17.2412	0.036569
37.929	17.8764	17.53687	1.899319
49.9999	18.2482	17.9769	1.486745
60.1133	18.7135	18.24673	2.494277
79.2448	19.1451	18.58691	2.915564
95.273	18.765	18.76357	0.007615
100	18.7151	18.80342	0.471945

Source: Author

series model as it is showing a good prediction for E' (w) since the maximum difference in the values is 2%, and this also confirms that equation 1 of relaxation modulus will show a good correlation with real behavior of the material.

Modal and Harmonic Analysis Using the Prony Series

Prony series constants obtained from Equation 7 used in Prony series Equation 1 relaxation modulus.

$$E(t) = 7.239 + 2.886. \exp\left(-\frac{t}{1.664}\right) + 2.595. \exp\left(-\frac{t}{0.322}\right) + 2.941. \exp\left(-\frac{t}{23.22}\right) + 3.611. \exp\left(-\frac{t}{0.1686}\right) \quad (8)$$

The Prony series characteristics are different for each FEM software. In the case of the ANSYS workbench in the section of engineering material, for the addition of new material we need to put the properties which describe the material fully in the case of viscoelastic material software gives the option to add the damping property in the terms of the Prony series, where the requested parameters are: *ei*, and τ_i. The *ei* is the normalized Prony coefficients which represent the relaxation moduli and the τ_i is the relaxation times of the Prony series. relaxation modulus is defined as the ratio of the respective Prony series coefficient of modulus to the cumulative relaxation modulus.

$$e_i = \frac{E_i}{E_o} \quad (9)$$

where E_i is the coefficient of the Prony series and E_o is the relation modulus E(t) at time t=0.

Here Replacing the value of E_o = 19.266 in the equation we get the value of relaxation modulus.

Table 17.2 represents the required relaxation modulus and relaxation time as input of viscoelastic material properties for modal and harmonic analysis.

In ANSYS harmonic analysis followed by modal analysis was done for the bare beam of structural steel subtracting of dimension (280*30*5) mm. Then viscoelastic material which was added in the software was

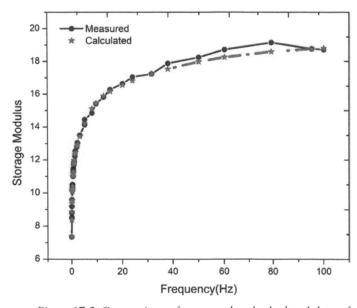

Figure 17.2 Comparison of measured and calculated data of storage modulus vs frequency
Source: Author

Table 17.2 Prony series coefficients

Relaxation modulus(e$_i$)	Relaxation time(τ_i)
0.1497	1.664
0.1346	0.322
0.4062	23.22
0.1874	0.1686

Source: Author

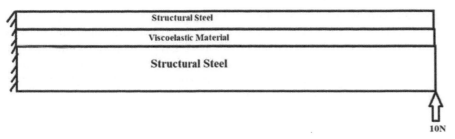

Figure 17.3 Cantilever beam sample of CLD for modal and harmonic analysis
Source: Author

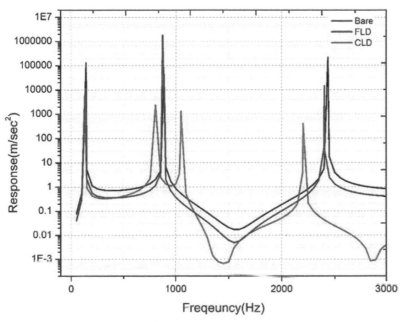

Figure 17.4 Bode diagram of bare beam, FLD, and CLD treatment
Source: Author

applied on the structural steel substrate as a free layer damping treatment for modal and harmonic analysis. And last, a damping layer (VEM) and a constrained layer (structural steel) were applied on the substrate for modal and harmonic analysis of constrained layer damping. The thickness of the damping layer and the constrained layer was kept at 2 mm each.

The sample was used as a cantilever beam fixed at one end and free at another end. For modal analysis, the required inputs are material selection and the fixed or free beam. For harmonic analysis excitation of the force of 10N was applied at the free end as shown in Figure 17.3.

As a result of modal analysis natural frequencies and mode shapes of all three samples was generated and after that harmonic analysis by applying the excitation of 10N at the free end was done and as result, we got the response of the structures in the terms of acceleration concerning the frequency shown in Figure 17.4.

Figure 17.4 shows the frequency response or bode diagram in term of acceleration for all three samples. Here we can observe that for CLD which is shown as the red curve in the graph, the amplitude is very low compared to the bare beam, which is shown as black, and the FLD treatment which is shown as the blue curve in the graph. This difference in the amplitude showing which good damping of the CLD treatment. Along with the response, we can observe a slight shifting of resonance frequency for the CLD treatment in the above graph. Whereas if we observe the FLD treatment the damping is almost nothing as its response is almost near to the Bare beam which indicates the less efficiency of FLD treatment in the case of viscoelastic material as a damping layer.

Conclusion

In this study, the efficiency of viscoelastic material for damping has been shown and the ability of constrained layer damping (CLD) treatment in damping using viscoelastic material has been described in the study we found that CLD gives better treatment than the free layer damping (FLD)treatment while using the viscoelastic material further studies can be done by varying the thickness of damping and constraining layers. The method of adding the viscoelastic material in ANSYS software for modal and harmonic analysis using the Prony series and the calculation of Prony series coefficients using DMA data was studied in this article. Calculated data of storage modulus using the Prony series and experimental storage modulus data were compared and found an error of less than 2% which justify the Prony series method to be used as a study of structural vibration analysis using the viscoelastic material.

References

[1] Moreira, R. and Rodrigues, J. D. (2004). Constrained damping layer treatments: Finite element modelling. Journal of Vibration and Control. 10(4), 575-95.

[2] Ewins, D. J. (2000). Modal testing: theory, practice, and applications, 2nd edition. United Kingdom, Hertfordshire: Research Studies Press.

[3] Ferhat, Ç. and Bekir, A. (2021). Modal and harmonic response analysis of new CFRP laminate reinforced concrete railway sleepers. Engineering Failure Analysis. 127.

[4] Barrientos, E., Pelayo, F., Noriega, Á. (2019). Optimal discrete-time Prony series fitting method for viscoelastic materials. *Mech Time-Depend Mater.* 23, 193–206.

[5] Deng, R., Davies, P., and Bajaj, A. K. (2003). Flexible polyurethane foam modeling and identification of viscoelastic parameters for automotive seating applications. Journal of Sound and Vibration. 262(3), 391-417

[6] Brinson, H. L. and Brinson, L. C. (2008). Polymer engineering science and viscoelasticity, New York, NY: Springer Verlag.

[7] Sánchez, E., Nájera, A., and Sotomayor, O. (2021). Numerical study of the viscoelastic mechanical response of polystyrene in the process of thermoforming through the generalized Maxwell model. *Materials Today: Proceedings.* 49, 107–114.

[8] Roy, S. and Reddy, J. N. (1988). Finite element models of viscoelasticity and diffusion in adhesively bonded joints. International Journal for Numerical Methods in Engineering. 26, 11.

[9] Ferry, J. D. (1980). Viscoelastic Properties of Polymers. 3rd edition. United States, NY: John Wiley & Sons.

[10] Chen, T. (2000). Determining a Prony Series for a viscoelastic material from time varying strain data. Internal report. NASA - National Technical Information Service.

[11] Christensen, R. M. (1971). Theory of Viscoelasticity. New York, NY: Academic Press.

[12] Felhos, D., XU, D., Schlare, A. K., Varadi, K., and Goda T. (2008). Viscoelastic characterization of an EPDM rubber and finite element simulation of its dry rolling friction. Express Polymer Letters. 2(3), 157-64.

[13] Hu, G., Tay, A. A. O., Zhang Y., Zhu, W., and Chew, S. (2006). Characterization of viscoelastic behaviour of a molding compound with application to delamination analysis in IC packages. Electronics Packaging Technology Conference. 53-59.

[14] Huang, G., Wang, B., and Lu, H. (2004). Measurements of viscoelastic functions of polymers in the frequency-domain using nanoindentation. Mechanics of Time-Dependent Materials. 8(4), 345-64.

[15] Menardn, K. P. (1999). Dynamic Mechanical Analysis – a Practical Introduction. CRC Press.

Chapter 18

Characteristics and performance analysis for diesel engines using cerium oxide nanoparticles blended with biodiesel

Prakash Kadam[1,a], D. R. Dolas[2,b], Ravikant K. Nanwatkar[3,c] and Hitendra Bhusare[1,d]

[1]JSPM Imperial College of Engineering and Reserch, Wagholi, Pune, India

[2]Jawaharlal Nehru Engineering College, MGM University, Aurangabad, India

[3]Assistant Professor, Department of Mechanical Engineering, NBN Sinhgad Technical Institutes Campus, Ambegaon, Pune, SPPU, India

Abstract

The proposed work basically aims to recognise the application of nanoparticles as biodiesel fuel. Biodiesel is an unsurpassed alternative fuel source aiming to increase the value of fossil fuels also increase the cleanliness of diesel engine and reduces greenhouse gas emission. Nanoparticles are mixed with biodiesel to increase their efficiency, decrease emissions, and improve fuel properties. A mixture of soybean biodiesel and cerium oxide nanoparticles is used in the mass fraction of 25, 50, and 100 ppm, with an assessment of mechanically operated conventional homogeniser and an ultrasonicate. Test reading shows B60C25 fuel reduced smoke by 4%, emission of hydrocarbon and carbon monoxide reduced by 15%, and NOx by 7% specific fuel consumption by 4% when compared to diesel.

Keywords: Biodiesel, CeO_2 nanoparticles, engine performance

Introduction

Globally depletion of fossil fuels is a serious issue for the environment and human beings. The supply of conventional fuels is limited, and it also produces greenhouse gas emissions. A lot of work is being performed to identify alternative fuels that can fulfill the current and future needs of energy storage, without causing further global-warming effects. The challenges of unwanted climate change and the dependency on fossil fuels can be solved by suitable alternatives for petroleum fuels and it's also reported by the International Energy Agency (IEA). One of the best sources of alternative fuel is biodiesel as it is biodegradable, harmless, and beneficial for the environment. Further, it is used in a diesel engine without any diminishing. As a result of the recent volatility in the price of oil and the severe limitation of pollutant emissions, interest in renewable fuels. Biodiesel is made from various natural sources, including animal fats, cooking oils, soybean oil, jatropha oil, and vegetable oil. Utilising biodiesel results in greater combustion because it contains more oxygen. As fossil fuel demand is increasing, oil supplies tend to deplete. According to survey results for the past 30 years between 1990 and 2020, there will be a three-time increase in the demand for fossil fuels in the upcoming future to meet energy demands. To increase the thermal and physical properties of the fuel, different additives like nanoparticles and water emulsions are added and an investigation is performed on how engine performance and emission parameters varies over a period of time.

Literature Review

Kathikuan et. al. studied Alvarez brown algae biodiesel used in CI engines to analyse Al_2O_3 and $C_{18}H_{34}O_3$. Fuel is prepared by using correlations of colloidal particles and additives of 10, 20, and 50 ppm results show a reduction in the cylinder temperature and an improvement in ignition delay time [1]. Satish et. al. studied in combustion behavior of diesel engines in dual fuel mode. In the first mode, nanoparticles are created by using an ultrasonicate and converted homogenizer mixed with diary scum oil methyl ester in a mass fraction of 10 to 30 ppm. The outcome demonstrates reduced brake thermal efficiency by 11.5% hydrocarbon by 23.2% and carbon monoxide by 32.6% during a load of 80% [2]. Syed et. al. worked on the effect of nanoparticles distributed in biodiesel made up of waste cooking oil on the thermal parameters of a variable compression ratio engine with the blending of 20%, 40%, and 60% by mixing with cerium oxide, and magnesium oxide nanoparticles to decrease the emission [3]. Hossain et. al. showed the consequences of nano additives on neat jatropha biodiesel. Cerium oxide nanoparticle on neat jatropha biodiesel. Cerium oxide nanoparticle at a concentration of 100 ppm and 50 ppm is added separately also 100 ppm aluminum oxide nanoparticles (J100A100) were selected for engine experimentation. the results indicate that the

[a]kadamprakash.20@gmail.com, [b]dhananjaydolas@jnec.ac.in, [c]ravikant.nanwatkar@sinhgad.edu, [d]hitendra14594@gmail.com

DOI: 10.1201/9781003450252-18

brake thermal efficiency of J100A100 fuel was 3% higher and at fuel load J100A100 gasoline was shown to have 6% lower and 5% higher BSFC consumption [4]. Jayakumar et. al. studied the solution of B20 was combined with concentrations of 50, 75, and 100 ppm of the Cr_2O_3 nanoparticle addition and dispersion (QPAN 80) at a persistent speed of 1500 rpm during the experimentation. Using gasoline that has been treated with nanoparticles significantly enhanced cylinder pressure (CP) and net heat release rate (NHRR). Additionally, thermal efficiency was rise and brake-specific fuel consumption was reduced [5]. Pourhoseini et. al. studied the effect of Al_2O_3 nanoparticles on flame characteristics, pollutant emissions, and thermal performance. Biodiesel prepared by blending 20% palm oil with diesel fuel and adding Al_2O_3 nanoparticles, affects flame properties, temperature radiation, and emissions of pollutants in the burner. To make a solution, 500 ppm of B20-blend biodiesel fuel was mixed with Al_2O_3 nanoparticles. Instead of absorbing heat, Al2O3 nanoparticles prefer to scatter heat [6]. Ghanmari et. al. studied using the response surface method adding alumina nanoparticles to diesel blends affected the engine performance and emission characteristics nanoparticles of alumina of 40, 80, 120, and 160 ppm at an engine speed of 800, 850, 900, 950, and 1000 rpm finding results maximum as 11.76% and 1899 ppm respectively. Maximum values of torque and brake power were measured as 402.8 NM and 42.82 kW. The minimum BSFC, CO_2, and HC emissions were calculated to be 207.21 gr/kw 1.15% and 9%, respectively [7]. Srinivas et. al. investigated the direct-injection diesel engine biodiesel blend nanoparticles that may have an effect (B20), which was supplemented with $CuCl_2$ and $COCl_2$ nanoparticle concentrations of 50, 75, and 100 ppm, respectively. Nanoparticles were given a 100 ppm dose of QPAN 80 dispersion, which was then ultrasonically processed. Day 1 and day 15 were chosen specifically for the stability test [8]. Rastogi et. al experimented with the effect of the concentration of cerium-oxide nanoparticles by using jojoba biodiesel. CuO nanoparticles in various concentrations were added to the JB20 fuel 25, 50, and 75 ppm. experimental results show that JB20CN50 fuel has a greater BTE than other samples of jojoba biodiesel fuel [9]. Hoseini et. al. showed the impact of fuel additives graphene oxide (GO) nanoparticles, on diesel engines' performance and emission. The biodiesel blends of Ailanthus altissima (B0, B10, and B20) were combined with the GO nanoparticles. the graphene oxide nanoparticles each fuel blend was done at concentrations of 30, 60, and 90 ppm. At a speed of 2100 rpm, testing of the engine was done with load 100%, 75%, 50%, 25%, and 0%. Investigations were made into NOx emissions [10]. Gad et. al studied the characteristics and fatty acid composition of Egyptian jatropha seeds oil, B100, B80, B60, B40, and B20 were the volumetric blends of biodiesel and diesel oil. At a 75% engine load, tests were conducted at various engine speeds. The biggest drops in volumetric efficiency and output braking power for B100 were 27% and 9%, respectively, even though B100's thermal efficiency was reduced by a maximum of 33% when compared to diesel. NOx emissions increased by 47% and 22% reduction in smoke emissions. The cylinder pressures and heat release rates of biodiesel blends were lower than those of pure diesel [11]. Prabu et. al. studied alumina and nanoparticles of cerium oxide were examined as an emission control approach in biodiesel. Jatropha biodiesel is blended with alumina and cerium oxide nanoparticles to create 10, 30, and 60 ppm. Results show that carbon dioxide, carbon monoxide, unburned hydrocarbons, and smoke emissions are reduced by 13%, 60%, 33%, and 32% respectively for the nanoparticle blended test fuels, with the brake thermal efficiency improving significantly [12].

From the literature review it can be seen that the use of cerium oxide nanoparticles improves the performance of the four-stroke single-cylinder diesel engine various nanoparticle dosages must be investigated.

Methodology

Before carrying out the experiment the fuel blends were made with soybean biodiesel and cerium oxide an ultrasonicate and placed for experimental validation. Blends are labeled as B100C100 (contains 100% diesel and cerium oxide 100 pmm), B20C25 (contains 20% biodiesel, and 25% cerium oxide), B40C100 (contains 40% biodiesel and 100% cerium oxide), B60C25 (cerium oxide 25 ppm, 40% biodiesel).

The engine setup has a four strokes single-cylinder diesel engine attached to a dynamometer. The signals are linked to a computer for PV diagram through an engine indication. Panel box with an engine indicator, an air-box, a tank of fuel, a manometer, transmitters used for measurement of the flow of both fuel and air, and rotameters for measuring the flow of water calorimeter and cooling water.

Result and Discussion

Performance results

The specific fuel consumption (SFC) versus load. At 0.150 kg shown in the Figure 18.3. Load D100C100 has 50% more fuel consumption than other biodiesel blends. B20C25 has the lowest fuel consumption. At 3, 6, and 12 kg loads all samples have nearly the same specific fuel consumption.

Brake thermal efficiency measures how energy is converted into mechanical output. Figure 18.4 depicts the change of BTE in relation to load for fuel and CeO_2 nanoparticle mixtures.

Figure 18.1 Testing rig setup
Source: Author

Figure 18.2 Emission gas analyser (AVL437 smoke meter)
Source: Author

Table 18.1 Specification of engine

Cylinder	1
Stroke length	110 mm
Compression ratio	18.1
Orifice diameter	20 mm
Connecting rod length	234
Power	3.5 kW
Speed	1500 RPM
Cylinder diameter	87.5 mm
Swept volume	661.45 (cc)
Length of Dynamometer	185 mm

Source: Author

Emission parameter results

The Co, NOx, and smoke emissions for various fuel samples with additives are shown in the emission characteristics for distinct samples. These additives are effective for lowering CO, NOx, and HC emissions and enhancing fuel combustion in diesel engines. At a 6 kg load, B40 C100 has more CO emission and at full load, hence it cannot be the best fuel blend.

Fuel nanoparticle oxides increase the amount of oxygen used during combustion. Compared to other fuels, those that contain CeO_2 nanoparticles have lower CO emissions compared to B20C25 shown

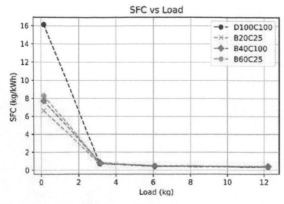

Figure 18.3 SFC vs load
Source: Author

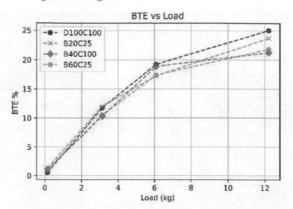

Figure 18.4 Brake thermal efficiency (BTE) vs load
Source: Author

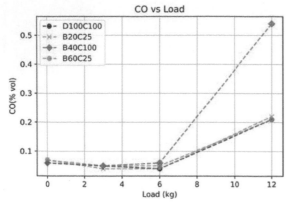

Figure 18.5 CO (% volume) emission vs load
Source: Author

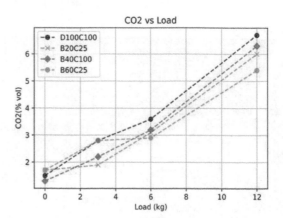

Figure 18.6 CO_2 emission vs load
Source: Author

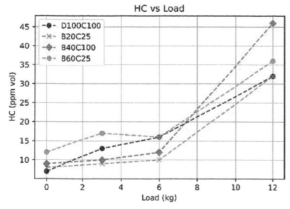

Figure 18.7 HC (ppm volume) vs load
Source: Author

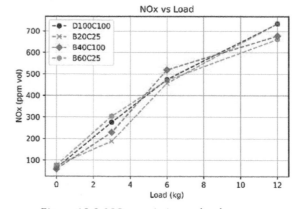

Figure 18.8 NO x emission vs load
Source: Author

in Figure 18.6 CO_2 emission graph, in Figure 18.6, we see that diesel has the largest CO_2 emission volume percent. B40C100 also have more emissions similar to diesel, B60C25 has the lowest emission percentage.

As result, B40C100 has 20% more unburnt hydrocarbon emission than other fuel blends. Figure 18.7 reveals that UHC has an impact on engine performance. Biodiesel contains a higher amount of oxygen and therefore the UHC has been reduced for B20.

Due to the nanoparticle NOx converting to N_2 at higher adiabatic flame temperatures, eventually, NOx emission decreases shown in Figure 18.8.

Nanoparticles improve combustion characteristics and fuel quality, resulting in reduced smoke emissions when used with biodiesel blends. Smoke opacity is reduced by increasing oxygen and sulfur levels in the fuel. This will also result in better combustion and less smoke.

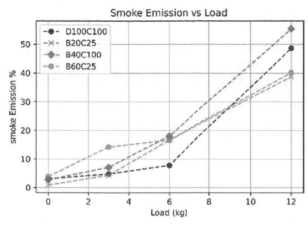

Figure 18.9 Smoke emission percent vs load
Source: Author

Conclusion

In the present study on the performance of four strokes single-cylinder diesel engine at a constant speed of 1500 rpm. Biodiesels made from soybeans with the addition of nanoparticles were tried. It is found that When the engine run on B60C25 fuel, better combustion brought in lower smoke opacity values due to the nano-additives. The minimum values of SFC were recorded as 6.64 kg/kW-hr. mainly biodiesel blends have lower SFC values, hence it is preferable to use biodiesel blends. The parameters of emission were also noticed lesser with B60 + 25 ppm CeO2 at 100% load, the emissions of the smoke by 4%, hydro-carbon by 15%, and NOx by 7%. Compared to diesel and other samples B60C25 biodiesel-blend improved. Further It can be decided that the addition of cerium oxide nanoparticles improvise the performance of the four stroke single-cylinder diesel engine. For this study, various dosage of nanoparticles needs to be studied.

References

[1] S. Karthikeyan,A. Elango &A. Prathima (2016). The effect of cerium oxide additive on the performance and emission characteristics of a CI engine operated with rice bran biodiesel and its blends. http://doi.org/10.1080/15435 075.2014.932419.

[2] K. A. Sateesh, V. S. Yaliwal, Manzoore Elahi M. Soudagar, N. R. Banapurmath, H. Fayaz, Mohammad Reza Safaei, Ashraf Elfasakhany & Ahmed I. EL-Seesy 2021). Utilization of biodiesel Al2O3 nanoparticles for combustion behavior enhancement of a diesel engine operated on duel fuel mode. 10.1007010973-021-1098-7.

[3] Dr-Syed Abbas,Shirajahammad Hunagund, Syed Sameer Hussain,Altaf Hussain Bagwan (2020). The effect of nanoparticle dispersed in waste cooking oil (WCO) biodiesel on thermal performance characteristics of VCR engine. 101016.

[4] Abul Kalam Hossain; Hussain, Abdul. (2019). Impact of nano additives on the performance and combustion characteristics of neat jatropha biodiesel. doi 10.3390?en12050921.

[5] S. Jaikumar, V. Srinivas, M. Rajasekhar (2021). Infl uence of dispersant of added nanoparticle additives with diesel-biodiesel blend on diesel injection compression ignition engine. 10.1016/j.energy2021.120197.

[6] S.H. Pourhoseini , Maryam Ghodrat (2021). Experimental investigation of the effect of AL2O3 nanoparticles as additives to B20 blended biodiesel fuel Flame characteristics thermal performance and pollutant emission. 10.1016/j.esite.2021.101292.

[7] Mani Ghanbari , Lotfali Mozafari-Vanani , Masoud Dehghani-Soufi , Ahmad Jahanbakhshi (2021). Effect of alumina nanoparticles as additives with diesel -blends on performance and emission characteristics of a six-cylinder diesel engine using response surface methodology (RSM).

[8] S. Jaikumar, V. Srinivas, V. V. S. Prasad, G. Susmitha, P. Sravya, A. Sajala & L. Jaswitha (2021). Experimental studies on the performance and emission parameters of a direct injection diesel engine fueled with nanoparticle-dispersed biodiesel blend. 10, 1007.

[9] Pankaj Mohan Rastogi , Abhishek Sharma , Naveen Kumar (2021) Effect of Cuo nanoparticles concentration on the performance and emission characteristics of the diesel engine running on jojoba biodiesel (2021). 10.1016/j. fuel. 2020.

[10] S.S. Hoseini , G. Najafi , B. Ghobadian , R. Mamat , M.T. Ebadi , Talal Yusaf (2018). Characteristics and optimization of ultrasonication assisted biodiesel production.

[11] M.S. Gad , A.S. El-Shafay , H.M. Abu Hashish (2020). Assessment of diesel engine performance, emission and combustion characteristics burning biodiesel blends from jatropha seeds.

[12] Sivakumar Sivalingam,Ponnusamy Palanisamy,Anbarasan Baluchamy (2016). Experimental investigation on jatropha oil methyl ester fi lled CI engine using oxide nanoparticle in biodiesel. 10.1016/j.joei.2015.03.00

Chapter 19

Synthesis and characterisation of TiO_2-$Cu_3(BTC)_2$ metal organic framework based advanced materials for photocatalytic CO_2 reduction

Ramyashree M S[a], Veekshit Udayakumar Ail[b] and Shanmuga Priya S[c]

Manipal Institute of Technology, India

Abstract

Global warming has caused a serious threat on the environment. The emission of harmful gases like CO_2 has been the major contributor for global warming. Besides that, the exhaustion of fossil fuels has created major energy barriers. Photocatalytic CO_2 reduction to fuels using metal-organic framework (MOF) based nanocomposites will help in addressing the above issues. MOF has emerged as a novel class of nanomaterials that has seen diverse applications. TiO_2 is considered one of the most impressive materials for photon absorption application. Thus, incorporating the two superior materials to form a nano composite will broaden its application. The work projects the synthesis of TiO_2@$Cu_3(BTC)_2$ (HKUST-1) through the hydrothermal method and solvothermal techniques. The synthesized MOF composite has been studied for structural, textural, morphological, and optical behavioral changes after incorporating TiO_2 in $Cu_3(BTC)_2$. The characteristics of the catalysts were determined using various characterisation techniques. The structural properties were studied through XRD and FTIR analysis. The prominent peaks in XRD of the composite match with the reference data of the individual components. The peaks at different wave numbers in FTIR shows the formation of different bonds in the composite. Thus, both methods confirm the formation of the composite. The surface area along with pore volume analysis, was incorporated to quantify the textural properties. The BET of 38.75 m^2/gm with pore volume of 0.0054 cc/gm was obtained for the composite. The STEM and FESEM revealed the morphological properties of the composite showing their nano structures. The optical property was studied using UV Vis spectrophotometry. The composite showed an increase in absorbance in the visible region after the incorporation of TiO_2. The studied properties suggest the potential of the synthesised nano composite in tuning better photo absorption capacity of any nano material for photo catalytic CO_2 reduction.

Keywords: Metal organic frameworks, CO_2, photocatalytic reduction

Introduction

Overpopulation and rapid industrialisation have led to the widespread use of fossil fuels for energy production [1]. The average CO_2 level over time was 408 ppm, whereas the maximum allowed level is only 350 ppm. Hence, sea level rise and global warming are witnessed [2]. Photocatalytic CO_2 reduction to solar fuels like methanol and methane has seen greater scope in recent years since it can work on natural photosynthesis mechanisms. Porous materials are known as solid groups with voids which are seized within a fluid [3]. Charcoals, polymer foams, zeolites, porous metal, and activated porous ceramics these materials are collectively called porous materials [4]. Porous materials are of prime importance because of their capability to interact with the ions and molecules at the external level and through the volume of the material [5]. In the class of porous materials, zeolites are most widely used microporous material over the years, which can be found naturally and synthesised. They have high thermal stability, excellent durability, and low-cost commercially proven technology [6]. Zeolites are used in photocatalysis, gas adsorption and separation applications. However, there were few difficulties like turning the shapes and size in activated carbon, hydrophilicity of zeolites, low surface area, and limited number of structures with the existing materials, thus gave rise to other porous solids like metal organic-frameworks (MOFs) [7].

MOFs are crystalline polymers with metallic nodes and linkers with extremely superior surface area properties, and chemical tunability [8, 9]. Figure 19.1 represents different synthesized MOFs for over the years for varied applications like gas storage, catalysis, and adsorption etc[10, 11].

Combining MOFs with different functional materials has shown to enhance the performance of the composite. MOFs have varied functional materials like metal-carbon nanotubes, graphene etc. TiO_2 is one such functional material which features high thermal and chemical stabilities[12]. Incorporating TiO_2 will eventually broaden the scope of application of the synthesised nanocomposite exhibiting favorable properties.

The objective of the work includes structural, textural, morphological, and optical studies on synthesised MOFs, $Cu_3(BTC)_2$ and TiO_2@$Cu_3(BTC)_2$ nanocomposite through XRD, FTIR, FESEM, STEM, BET and UV Vis analysis to study the scope of the material in CO_2 photoreduction application.

[a]ramyashree.s1@learner.manipal.edu, [b]veekshit100@gmail.com, [c]shan.priya@manipal.edu

DOI: 10.1201/9781003450252-19

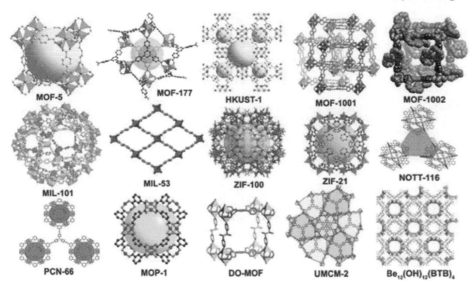

Figure 19.1 Different metal organic framework structures [10]
Source: Author

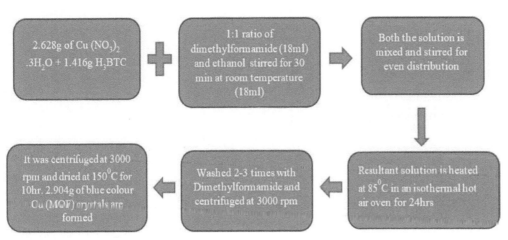

Figure 19.2 Synthesis of HKUST-1
Source: Author

Materials and Methods

Materials required

Copper (II) nitrate trihydrate, zinc (II) nitrate hexahydrate [merck], 1,4-benzene dicarboxylic acid, 1, 3, 5-benzene tricarboxylic acid [alfa aesar], N,N-dimethylformamide [loba], ethanol, methanol, sodium nitrate (99% loba), H_2SO_4, copper (II) acetate monohydrate (98%), titanium isopropoxide, acetylaceton.

Synthesis Procedure

The synthesis procedure of the nanocomposites is shown in following flow sheet.

Synthesis of HKUST-1
The step-by-step synthesis of HKUST-1 is shown in Figure 19.2.
Synthesis of $TiO_2@Cu_3(BTC)_2$

The step-by-step synthesis of $TiO_2@Cu_3(BTC)_2$(HKUST-1) is represented in Figure 19.3.

Results and Discussion

Structural Properties

X-ray diffraction (XRD) characterization of Cu_3BTC_2 and $TiO_2@Cu_3(BTC)_2$ was analysed. $Cu_3(BTC)_2$ powdered XRD pattern fits the pattern reported in the literature, emphasising the substance's crystalline

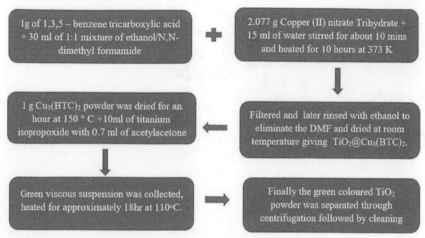

Figure 19.3 Synthesis of $TiO_2@Cu_3(BTC)_2$(HKUST-1)
Source: Author

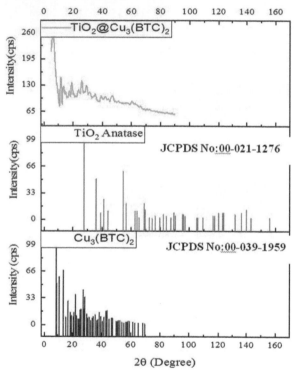

Figure 19.4 XRD pattern of synthesized components $TiO_2@Cu_3(BTC)_2$, TiO_2 and $Cu_3(BTC)_2$
Source: Author

form. From the Figure 19.4, the typical peaks observed at 2θ, 6.10°, 11.49°, 13.02° matches with the JCDPS No. 00-039-1959 of $Cu_3(BTC)_2$ which is in accordance to literature and corresponds well with the data [13]. The deformities at the end are due to the formation of CuO and some trapped solvents that haven't been washed away.

The peaks at 2θ, 24.25°, 34.95°, 35.33°, 61.10°, 70.33° and 87.64° matches with the matches with the JCDPS No. 00-021-1276 of TiO_2 in anatase phase. The obtained XRD pattern in Figure 2 almost resembles with the peaks given in the work reported by Credico et. al. [14]. The characteristic peaks between 15° and 40° also match with that of literature patterns of TiO_2 and HKUST-1($TiO_2@Cu_3(BTC)_2$) have been given individually in the literature as shown in Figure 19.3. However, both constituents have less crystallinity because of a higher fraction of the amorphous phase. $Cu_3(BTC)_2$ core-shell in $TiO_2@Cu_3(BTC)_2$ did not have expected crystallinity as indicated by obtained powder XRD pattern. The compound obtained was amorphous in nature which may be due to insufficient dispersion of TiO_2. The average crystalline size was found to be 3.48 nm.

The Fourier transform infrared spectra (FTIR) of the composite is shown in Figure 19.5. $Cu_3(BTC)_2$ peaks at 1630 cm⁻¹ represent the O-C=O vibration, and peak value at 1230 cm⁻¹ represents O-H vibrations

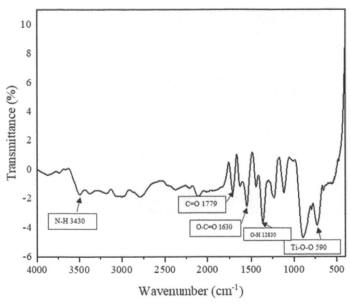

Figure 19.5 FTIR of TiO₂@Cu₃(BTC)₂

Source: Author

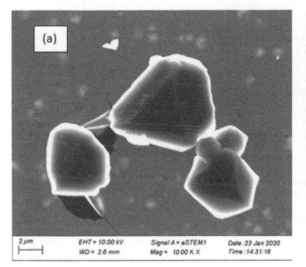

Figure 19.6 (a-b) STEM images of Cu₃(BTC)₂

Source: Author

[15]. The Ti-O-O bond's vibration is responsible for the FTIR signal at 590 cm. The N-H stretching is visible in the spectra as a broad, strong band at 3430 cm1. Asymmetrical C=O linked systems have a significant absorption peak at 1779 cm⁻¹ [16].

Textural property

Measurement of Brunauer-Emmett-Teller surface area

The surface area of the measured sample of Cu₃(BTC)₂ was 819 m²/gm with pore volume 0.025 cm³/gm. The surface area indicated in the literature is much higher than the observed surface area. This discrepancy in results may be ascribed for the solvent which is trapped in the pores of the prepared sample. The surface area of TiO₂@Cu₃(BTC)₂ was measured to be 38.75 m²/gm with pore volume of 0.0054 cm³/gm. Because of the incorporation of TiO₂, the composite shows a reduced surface area value.

Morphological properties

Scanning transmission electron microscope analysis of Cu₃BTC₂

The synthesised samples show octahedral morphology Figure 19.6 as characterised by STEM. The monodispersed solid octahedron crystals have been formed at a fairly large scale, and the surfaces of these specified micro-octahedrons are very smooth [17].

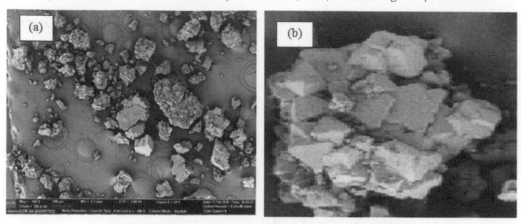

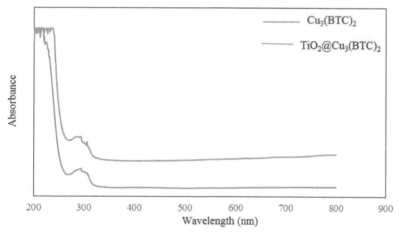

Figure 19.7 (a) FESEM of $TiO_2@Cu_3(BTC)_2$ (b) At 400 nm (c) TiO_2 particles dispersed on the surface of the MOF
Source: Author

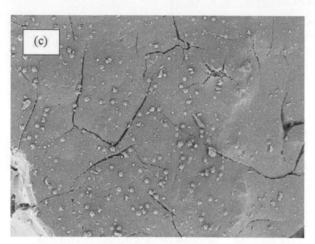

Figure 19.8 Absorption spectra of HKUST-1 and $TiO_2@Cu_3(BTC)_2$
Source: Author

Field emission scanning electron microscopy of $TiO_2@Cu_3BTC_2$

FESEM analysis was done to analyse the morphology of the composite. The $TiO_2@Cu_3(BTC)_2$ with Octahedron structure is shown in Figure 19.7(a-b). The TiO_2 particles are seen embedded on this octahedron structure, as seen in Figure 19.7(c). However, the synthesized catalyst lacked a long-range octahedron lattice as it was amorphous with some crystal defects. The average diameter of the composite is 434.54 μm.

Optical property

Utilising UV-visible spectrophotometry, the composite's optical characteristics were examined. From the Figure 19.8 we can observe that the absorbance of the TiO_2 incorporated MOF is higher than the pristine

MOF in the visible region. The peak value at 300 nm wavelength shows the highest absorbance for both the samples. Thus, the optical behaviour of the composite helps to improve the light harvesting capacity which has more scope in catalysis application.

The overall observation of the all the analysis shows us the importance of the synthesised material as a potential nano composite which can be used for various applications. The characterisation techniques helped us to draw inferences on various parameters of shape, size, surface area and absorbance.

Conclusion

- In this work, different categories of metal-organic framework (MOF) based photocatalyst were synthesised successfully. Cu_3BTC_2 metal organic framework was synthesised using a hydrothermal method whereas $TiO_2@Cu_3(BTC)_2$ was synthesised using solvothermal method.
- Pure TiO_2 offers good performance but fabricating it with other transition metal oxide compounds and even MOFs such as $Cu_3(BTC)_2$ have been reported to enhance the photon absorption and conversion properties to a great extent.
- The nanocomposites were then characterised using various techniques. The purpose of synthesising and characterising these coupled and fabricated MOF based nanocomposite was to study its potential for photon reduction application.
- Extensive structural and optical analysis of other mixed photocatalyst such as transition oxide and graphene-based substituents will be helpful for further research in this field.

Future Scope of Work

As we saw in our studies the importance of studying the structural aspects of metal-organic frameworks (MOFs) can really pave the path for successful incorporation of MOFs and its composites in photon absorption application. Further the same composite can be studied for gas storage application like hydrogen gas which is the known as upcoming green fuel. The light harvesting ability of the composite will attract catalytic applications. In the further studies it is vital to focus on bringing different ways of incorporating the same MOFs and semiconductors with advanced structural properties.

Acknowledgement

The corresponding author would like to express their gratitude to Vision Group of Science and Technology (VGST), Department of Science and Technology (DST), Government of Karnataka, India for the project grant titled "Photocatalytic conversion of carbon dioxide to methanol using ZIF-8/BiVO$_4$/GO and rGO/CuOnanocomposites as an adsorbent material" (2020) (Ref No: VGST/RGS-F/GRD-918/2019e20/2020-21/198). The authors would also like to thank National Institute of Technology, Tiruchirappalli and National Institute of Technology, Surathkal, for supporting with the characterization of MOF based nanocomposites.

References

[1] Shet, S. P., Shanmuga, S. P, Sudhakar, K., and Tahir, M. (2021). A review on current trends in potential use of metal-organic framework for hydrogen storage. International Journal Hydrogen Energy. 46, 11782–11803. https://doi.org/10.1016/j.ijhydene.2021.01.020.

[2] Ramyashree, M. S., Shanmuga, S. P, Freudenberg, N. C., Sudhakar, K., and Tahir, M. (2021). Metal-organic framework-based photocatalysts for carbon dioxide reduction to methanol: A review on progress and application. Journal of CO$_2$ Utilisation. 43 101374. https://doi.org/10.1016/j.jcou.2020.101374.

[3] Marcos-Hernández, M. M., and Villagrán, D. (2019). Mesoporous composite nanomaterials for dye removal and other applications. Composite Nanoadsorbents, 265–93.

[4] Yap, M. H., Fow, K. L., and Chen, G. Z. (2017). Synthesis and applications of MOF-derived porous nanostructures. Green Energy and Environment. 2, 218–245. https://doi.org/10.1016/j.gee.2017.05.003.

[5] Davis, M. E. (2002). Ordered porous materials for emerging applications. Nature. 417, 813–21. https://doi.org/10.1038/nature00785.

[6] Slater, A. G. and Cooper, A. I. (2015). Function-led design of new porous materials. Science (1979). 348.

[7] Makal, T. A., Li, J. R. Lu, W., and Zhou, H. C. (2012). Methane storage in advanced porous materials. Chemical Society Reviews. 41, 7761–79. https://doi.org/10.1039/C2CS35251F.

[8] Wang, Q., Gao, Q., Al-Enizi, A. M., Nafady, A, and Ma, S. (2020). Recent advances in MOF-based photocatalysis: Environmental remediation under visible light. Inorganic Chemistry Frontiers. 7, 300–39. https://doi.org/10.1039/c9qi01120j.

[9] Adhikari, A. K. and Lin, K. S. (2016). Improving CO_2 adsorption capacities and CO_2/N_2 separation efficiencies of MOF-74(Ni, Co) by doping palladium-containing activated carbon. Chemical Engineering Journal. 284, 1348–1360. https://doi.org/10.1016/j.cej.2015.09.086.

[10] Kampouraki, Z. C., Giannakoudakis, D. A. Nair, V. Bandegharaei, A. H., Colmenares, J. C., and Deliyanni, E. A. (2019). Metal organic frameworks as desulfurization adsorbents of DBT and 4,6-DMDBT from fuels. Molecules. 24, 1–23. https://doi.org/10.3390/molecules24244525.

[11] Kaur, R., Kaur, A., Umar, A., Anderson, W. A., and Kansal, S. K. (2019). Metal organic framework (MOF) porous octahedral nanocrystals of Cu-BTC: Synthesis, properties and enhanced absorption properties. Materials Research Bulletin. 109, 124–33. https://doi.org/10.1016/j.materresbull.2018.07.025.

[12] Wang, L., Jin, P., Duan, S., She, H., Huang, J., and Wang, Q. (2019). In-situ incorporation of Copper (II) porphyrin functionalised zirconium MOF and TiO2 for efficient photocatalytic CO_2 reduction. Science Bulletin. (Beijing). 64 926–933. https://doi.org/10.1016/j.scib.2019.05.012.

[13] Wang, Q. M., Shen, D., Bülow, M Lau, M. L., Deng, S., Fitch, F. R., Lemcoff, N. O., and Semanscin, J. (2002). Metallo-organic molecular sieve for gas separation and purification. Microporous and Mesoporous Materials. 55, 217–230. https://doi.org/https://doi.org/10.1016/S1387-1811(02)00405-5.

[14] Ryder, M. R. and Tan, J. C (2014). Nanoporous metal organic framework materials for smart applications. Materials Science and Technology. 30, 1598–1612. https://doi.org/10.1179/1743284714Y.0000000550.

[15] Phuong, N. T., Herman, C. B., Thom, N. T., Nam, P T., Lam, T. D., and Thanh, D. T. M. (2016). Synthesis of Cu-BTC, from Cu and benzene-1,3,5-tricarboxylic acid (H3BTC), by a green electrochemical method. Green Processing and Synthesis. 5, 537–547. https://doi.org/10.1515/gps-2016-0096.

[16] Rajakumar, G., Rahuman, A. A., Roopan, S. M., Khanna, V. G., Elango, G., Kamaraj, C., Zahir, A. A., and Velayutham, K. (2012). Fungus-mediated biosynthesis and characterisation of TiO_2 nanoparticles and their activity against pathogenic bacteria. *Spectrochimica Acta*, Part A: Molecular and *Biomolecular Spectroscopy*. 91, 23–29. https://doi.org/10.1016/j.saa.2012.01.011.

[17] Hu, X., Li, C., Lou, X., Yang, Q., and Hu, B. Hierarchical CuOoctahedra inherited from copper metal-organic frameworks: High-rate and high-capacity lithium-ion storage materials stimulated by pseudocapacitance. Journal of Materials Chemistry: A Material. 5, 12828–12837. https://doi.org/10.1039/c7ta02953e.

Chapter 20

Direct liquid fuel cell: challenges and advances

Ashish P. Umarkar[1], Dr. Rajkumar Chadge[1], Dr. Jayant Giri[1] and Dr. Naveen Shriwastava[2]

[1]Yeshwantrao Chavan College of Engineering, Nagpur, India

[2]Georgia Institute of Technology, Atlanta, USA

Abstract

Global energy security is put into doubt by the paucity of fossil fuels. Researchers from all around the world are attempting to find a solution to the issue of rising energy consumption and simultaneous mitigation of environmental damage. Batteries, fuel cells and renewable energy sources (solar, wind, hydro, bio, etc.) technologies are being considered as viable solutions, but each one has its own drawbacks and prospects.

The fuel cell is a energy convertor that converts chemical energy of fuel into electrical energy directly without any moving part. A fuel cell has potential to replace the conventional power generators because of certain advantages-high energy conversion efficiency, non-polluting, noiseless, uses renewable fuel, and so on.

Hydrogen as a fuel has advantageous for fuel cells but the storage, transportation, and handling is difficult because of its gaseous nature and high inflammability. So, H_2 rich liquid fuels area probable solution for the issue, as usage of the liquid fuel even does not demand change in supply chain infrastructure that is already available for the liquid fuels we are using currently (gasoline). In this paper, the focus is on significance of H_2 rich liquid fuels in fuel cell is discussed alongwith their current status and futurescope.

Keyword: DLFC, DMFC, fuel cell

Introduction

Both developed and emerging nations today face a variety of difficulties. Population expansion, food security, energy security, resource depletion, global warming, and other major issues demand more attention. Scientists are motivated to find environmentally friendly solutions to the issue of sustainable energy sources by the century-long rise in global energy demand. Fuel cell has potential to be the world's fuel of the future. Without combustion or moving parts, the fuel cell is an energy converter that transforms the chemical energy of fuel into electrical energy [1]. The pie chart shows the installed apacity of various power sources in India (CEA, MNRE, Mercom India Solar Project Tracker 31 Dec, 2020) and about 60% of energy is generated by the hydrocarbon based fuel and only 20% supplied by renewable energy sourses.

Renewable energy generators

Renewable energy sources, such as solar, wind, biopower, and others, are contributing more and more each year. These sources offer very lucrative advantages, including free availability, a straightforward system, no environmental impact, the ability to be installed at the location of energy use (solar), and others. However, there are some drawbacks to the use of renewable energy sources, including dependence on the occurrence of natural phenomena, concentration that varies from day to night and location to location, the justification of windmills for only hilly or coastal regions, and the reduced efficiency of solar panels in the rainy or winter months. As a result, this problem restricts the adoption of renewable energy sources as primary energy producers [2].

Conventional power generators

Energy demand has been steadily rising over the past century, and conventional energy generators (ICEs, thermal, gas, and diesel power plants, etc.) are able to provide it by burning hydrocarbon-based fuels like coal, oil, and natural gas. The previous century has seen a sharp growth in the use of fossil fuels, which has had a significant impact on the ecosystem of our planet by causing different pollutions (air, land, noise), the emission of greenhouse gases, and an increase in carbon footprints [3]. Furthermore, the usage of fossil fuels is limited after 3050 years due to their scarcity. Therefore, for the nation to continue growing, it is crucial to have an energy-secured society that demands future-ready technology that uses renewable yet readily available fuel and does not simultaneously harm the environment.

Battery

Batteries need a power source to store the energy from because they are energy-storing devices rather than generators. Around the world, lithium-ion batteries are currently used extensively in a wide range

DOI: 10.1201/9781003450252-20

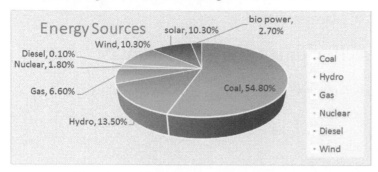

Figure 20.1 Share of energy sources in India Renewable energy generators
Source: India Energy Data, Statistics and Analysis-Oil, Gas, Electricity, Coal

of equipment, including laptops, cell phones, automobiles, power tools, cameras, medical devices, etc. Lithium-ion batteries are the most effective batteries in the last 10 years because of a variety of advantages. For example, having a high energy density yet being lightweight, less charge loss than other batteries, on average, significant power density. It is apparent that Li-ion batteries are popular given the dearth of nearby alternatives. Most likely, scientists were motivated to develop more powerful batteries by the oil crisis of the 1970s. Lithium is used in place of lead, which makes sense given the demands for weight and power density. Furthermore, studies are being conducted for development of certain parameters, such as quick charging, cost optimisation, safety, and operational range [4].

As more companies around the world add products powered by lithium-ion batteries to their product lines, the system is swiftly absorbing them due to their capacity to store electricity. We need more powerful batteries that can run for at least 1012 hours and charge in a couple of minutes, thus this cannot be the substitute for the current energy producing technology. The environmental impact of battery disposal in the afterlife and the requirement for intensive waste management are also prominent. The energy required to produce batteries greatly increases greenhouse gas emissions [5]. Hence, this issue poses over the future of batteries.

Fuel cell technology

The fuel cells are primarily an open thermodynamic system. The electrochemical reaction is the basis for the fuel cell's operation, and it uses the reactants that are supplied to produce an output of electric charge, heat, and water [6]. There are certain limitations to the hydrogen feed fuel cell, such as the need for large storage areas to hold sizable quantities of hydrogen gas as well as high-pressure storage and transportation. The direct liquid fuel cell, which potentially replace conventional liquid-fueled engines, has received comparable attention. Additionally, the fuel cell is better than ICE since it is more efficient, produces heat and pure water as its output, emits very little pollution, is silent, has fewer moving parts and components, and is also efficient [7].

Among various types of fuel cell polymer electrolyte membrane fuel cell (PEMFC) are the most popular choice due to low operating temperature, quick start and stop, high efficiency [1, 8]. The research data on PEMFC suggested different liquid fuels such as methanol, ethanol, propanol, ethylene glycol, formic acid, ammonia, glycerol, hydrazine, etc. A typical polymer electrolyte fuel cell with an anode, cathode, and proton exchange membrane is depicted in Figure 20.2. The electrochemical reaction at the anode causes the hydrogen molecules to split into protons (H^+) and electrons (e). The protons exit the proton exchange membrane and are oxidised to produce water at the cathode [9, 10].

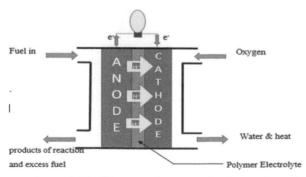

Figure 20.2 Schematic diagram of polymer electrolyte fuel cell
Source: Author

There are two types of PEMFC: active and passive. A pressurised oxygen tank, heat exchanger, gas liquid separator, and pump for pumping fuel and reactant are only a few of the other components used in an active fuel cell. As opposed to active DLFC, passive DLFC relies solely on natural phenomena including capillarity, natural convection, and gravity to work [6, 11].

Fuel Selection Criteria

Table 20.1 above provide a detailed overlooked of some of the fuels which are developed or under research. Analysis of different fuel is needed while selecting fuel for the fuel cell that comprised of the various parameters that affect the fuel cell ecosystem. This parameters are shown in Figure 20.3.

Methanol

Methanol is a readily available, inexpensive fuel that may be produced from virtually any hydrocarbon, including coal, biomass, methane, natural gas, etc. The DMFC has the potential to be used in a wide range of applications, including mobile electronics devices and automobiles. The portability, ambient operating temperature, small weight, and simple fuel storage mechanism of DMFC are its key advantage [2830]. Despite its advantages, DMFC still has a long way to go before it can be widely used. These barriers include slow anode kinetics, methanol crossover, species management, heat management, expensive electrode material, durability, and stability [31, 32].

Ethanol

Ethanol is seen as a desired and promising fuel for fuel cells due to benefits such being non-toxic, naturally occurring, renewable, having a higher power density, and not emitting greenhouse gases. Additionally, ethanol can be fed directly into a fuel cell's anode, where it will be oxidised at the electrolyte interface with an incredibly high level of efficiency in either a liquid or a vapour feed. Ethanol is a renewable fuel source for DLFC since it can be produced from agricultural products, biomass, and raw materials like molasses. Ethanol is the fuel for DLFC that has received the second-most investigation since it has a higher power density and less toxicity than methanol [33, 34].

Propanol

Propanol provides a number of significant benefits when used in DLFCs, including safety while handling and transportation, the ability to be quickly identified in the event of a leak or other malfunction, and a high energy density. The disadvantage of propanol is that several unwanted byproducts are produced during reaction, which considerably lowers DPFC performance. Propanol, including 1-propanol, which is a higher primary alcohol, is not regarded as a useful fuel due to slower kinetics in acidic environments. Additionally, anode catalyst poisoning results from the creation of a number of intermediate species during propanol oxidation. Therefore, regular electrode cleaning is required, which requires a lot of time and effort [35, 36].

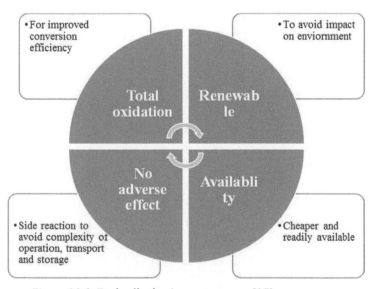

Figure 20.3 Fuel cell selection parameters [27]
Source: Author

Table 20.1 Characteristic of different DLFC fuel cell

Sr. No.	Fuel cell	Name of the fuel used	Nature of the fuel	Molecular Formula	Synonym	Energy Density	Flammable	Toxic	Health Hazard	Corrosive	environmental hazard	Molicularwt	Pka	Renewabilty	cost (Rs./Unit)	Reference
1	DMFC	Methanol	The most common kind of alcohol is methanol, which has the appearance of a colorless, volatile liquid with a faint aroma of sweetness.	CH_3OH or CH_4O	wood alcohol, Methyl alcohol, 67-56-1, carbonil,		4820	Highly	acute	Damages organ	-	32.042	15.3	Natural gas is the main fuel used in industry, however biomass can also be gasified.	30pl	[12]
2	DEFC	Ethanol	A primary alcohol with a clear colour and a distinctively vinous aroma and flavor is ethanol. Additionally, it vapourises more strongly than air.	CH_3CH_2OH or C_2H_6O	Alcohol, ethyl alcohol, 64-17-5		6280	Highly	-	-	-	46.07	15.9 @25 c	On large scale ethanol is reduced by fermentation of molasses.	60pl	[13]
3	DPFC	Propanol	Itsclear, colorless liquid	$CH_3CH_2CH_2OH$ or C_3H_8O	Propylalcohol, 1-propanol, n-propanol		7080	Highly	-	Serious eye and organ	Corrosive	60.1	16.1	For industrial urose it is roduced from natural gas.	120 pl	[14]

Sr. No.	Fuel cell	Name of the fuel used	Nature of the fuel	Molecular Formula	Synonym	Energy Density	Flammable	Toxic	Health Hazard	Corrosive	environmental hazard	Molicularwt	Pka	Renewability	cost (Rs./Unit)	Reference
4	DEGFC	Ethylene glycol	Its an antifreeze element (for car and airoplane) and clear, colorless, odourless and syrupy liquid with sweet taste.	CH$_2$OHCH$_2$OH or C$_2$H$_6$O$_2$ or HOCH$_2$CH$_2$OH	Ethane-1, 107-21-1	5870	-	-	acute	harmful if swalloed	easily cotaminate ground water	62.02	15.1	Glycerol, a waste product of the biodiesel industry, is a renewable bioresource that can be used to generate ethylene glycol in a sustainable manner.	50 pl	[15]
5	DGFC	Glycerol	Its atrihydroxyalcohol	CH$_2$OH-CHOH-CH$_2$OH or C$_3$H$_8$O$_3$	56-81-5, glycerine	6400	-	-	-	Severe eye damage	dangourous to aquatic enivornment	92.09	14.4	During roduction of fatty ester, fatty acid otsoa from oils and fats glycerine is obtained naturally as a primarily co-product.	85 pk	[16]
6	DFAFC	Formic acid	Pungent odor and colorless are character of Formic acid which is corrosive to tissuesand metals.	HCOOH or CH$_2$O$_2$	Aminic acid, 64-18-6, methanoic acid	1750	-	-	-	Damages eye and causes skin burn.	Corrosive	46.025	3.75 @25c	Methanol can be used for production of formic acid	85pk	[17]

Sr. No.	Fuel cell	Name of the fuel used	Nature of the fuel	Molecular Formula	Synonym	Energy Density	Flammable	Toxic	Health Hazard	Corrosive	enviornmental hazard	Moliculwt	Pka	Renewablity	cost (Rs./Unit)	Reference
7	DDEFC	Dimethyl ether	With a very light ether smell DME is almost an odourless gas. During transport it is in liquefied gas form vapor pressure.	CH_3OCH_3 or C_2H_6O	Methyl ether, dimethyl oxide, methoxymethane	5610	Etremely flammable	-	Skin and eyes irritation	-	-	46.07	-	Lignocellulosic biomass can be utilise to reduce DME.	180 pk	[18]
8	DHFC	Hydrazine	Its anhydrous, colorless, fluming oily liquid with anodour like ammonia.	N_2H_4 or H_4N_2	Diamine, 302-01-2, Livoxine	5400	Highly	acute	Severe skin burn and damage to eye damage, carciogenic	Corrosive	Highly toxic to aquatic animals	32.046	7.96	During nitrogen fixation of azotobacter hydrazine is naturally produced.	130 pk	[19]
9	DBFC	Borane	Borane is present in Erysimumin conspucuum naturally. Due to high calorific value its used in some high energy fuel.	BH_3	Trihydridoboron, 13283-31-3	8701	-	-	-	-	-	13.84	-	-	12500 pk	[20][21]

Sr. No.	Fuel cell	Name of the fuel used	Nature of the fuel used	Molecular Formula	Synonym	Energy Density	Flammable	Toxic	Health Hazard	Corrosive	environmental hazard	Molecularwt	Pka	Renewablity	cost (Rs./Unit)	Reference
10	DAFC	Ascorbate	1 Ascorbic acid which is almost odorless is a crystalline powder that ranges in colour from white to extremely pale yellow and has a deliciously acidic flavor.	$C_6H_8O_6$ or $HC_6H_7O_6$	Ascorbic acid, Ascorbic acid, 50-81-7, L vitamin C	-	-	-	-	-	-	176.12	4.17 @ 10C	Basically vitamin C, and produced organically by almost all plants.	650 pk	[22] food, and pharmaceutical sector as nutritional supplement and preservative, making use of its antioxidative properties. Until recently, the Reichstein-Grüssner process, designed in 1933, was the main industrial route. Here, D-sorbitol is converted to L-ascorbic acid via 2-keto-L-gulonic acid (2KGA [23]
11	DABFC	Ammonia borane	-	BH_3N	13774-81-7	6100	Highly	acute	Serious eye irritation, harmful if inhaled, respiration irritation	-	-	-	-	-	27.84	-

Sr. No.	Fuel cell	Name of the fuel used	Nature of the fuel	Molecular Formula	Synonym	Energy Density	Flammable	Toxic	Health Hazard	Corrosive	enviornmental hazard	Molicularwt	Pka	Renewablity	cost (Rs./Unit)	Reference
12	DAFC	Ammonia	ammonia gas	NH_3	7664-41-7, Azane	6250	Highly	acute	Causes severe skin burn and eye damage	Corrosive	Very toxic to aquatic animals	17.031	9.25	-	200 pl	[24]
13	DUFC	Urea	It is odorless, white crystal and does not combust.	CH_4N_2O or NH_2CONH_2	Carbamide, 57-13-6	4694	-	-	-	-	-	60.056	0.1 @2.5	Urea is an organic compound of carbon, nitrogen, oxygen, and hydrogen.	-	[25]
14	DGFC	Glucose	Its primary source of power for life. It is naturally present in free form in fruits and plants.	$C_6H_{12}O_6$	D- GLC, D-glucopyranose	4430	-	-	-	-	-	180.16	-	-	-	[26]

Source: Author

Ethylene glycol

Every year, millions of tonnes of the alcohol ethylene glycol are produced for use as antifreeze in the automotive industry, which has well established supply chain [29]. Since it loses less fuel to evaporation because of its higher vapour pressure than a fuel with a single hydroxyl group, it has better properties. Alkaline direct ethylene glycol fuel cells are an option for powering applications like portable and stationary electronics since they use a sustainable fuel and have a large number of inexpensive parts [37].

Glycerol

Glycerol is a suitable fuel for DLFCs since it is less expensive, nonvolatile, nontoxic, and nonflammable. It is a byproduct of the transesterification process used to produce biodiesel from plant and animal fats [16]. Glycerol is regarded by the bidiesel industry as a waste product. As a result, its price is frequently low. Since glycerol is less expensive than methanol and ethanol, using it in commercial applications will be more financially feasible [35].

Formic acid

Due to safety issues, formic acid is an ideal fuel competitor in the commercialisation of fuel cells because it is non-flammable, non-toxic, and naturally occurring. Low doses of it can be added to food as an additive. Formic acid has a significant disadvantage as a fuel since it has a lower volumetric energy density than unprocessed methanol. Performance for smaller systems is expected to be better with DFAFC than DMFC [38].

Dimythel ether

Dimethyl ether is a gas that naturally occurs. DME may be shipped more easily and safely than hydrogen because the main components of liquefied petroleum gas, propane and butane, have similar vapor pressures to DME [39].

Hydrazine

Hydrazine is a desirable fuel for DLFC since it may be entirely oxidised to nitrogen and water. It smells like ammonia and is a white liquid. The electro-oxidation reaction of hydrazine does not result in the formation of carbon dioxide since hydrazine is entirely devoid of carbon atoms [35]. Hydrazine has been shown to be the riskiest fuel of those being researched for PEMFCs. It can have mild to severe impacts on human organs such the neurological system, lungs, and kidneys. It also contains cancer-causing elements [35].

Borohydride

The hydrogen content in borhydrides compounds is high. Due to its high hydrogen content and simplicity of H_2 release, it was extensively investigated for the purpose of hydrogen storage and generation. Simple cooling systems, refuelling, and quick and secure handling are other advantages provided by DBFCs [20, 40].

Foramte

Among the different alternative fuels, formate is a desirable fuel. Compared to methanol, it produces more output, and its salts are ease to store and transport. It appears to be a potential supply of electricity for mobile equipment such as laptops, drones, and cell phones [41, 42].

Ascorbic acid

Ascorbic acid is a fuel that is both chemically stable and safe for the environment. It's an inexpensive, risk-free powder that might dissolve in water. It is manufactured in factories by chemically converting glucose, despite the fact that it originates from plants. Low-power electronics and medically safe components are both compatible with it. Ascorbic acid is said to work better in alkaline media than it does in acidic media [43]. According to research by Fujiwara et al., the crossover of AA over a Nafion membrane was almost 100 times lower than that of methanol [44].

Ammonia Borane

At rom temperature, ammonia borane is chemically stableIt is a non-toxic, safe for the environment, highly water soluble solid with a high hydrogen content that is transportable and only moderately unstable during hydrolysis in the absence of a catalyst. When compared to the significantly less complex borohydride fuel cells, which even have a substantially higher energy density, the author raises questions about the viability of AB fuel cells [45].

Ammonia

Ammoia is an excellent choice for liquid fuel since it can be completely oxidised to produce water and nitrogen. It also has a large amount of hydrogen. When employing PEM fuel cells directly in acidic conditions, the main problems are membrane conductivity and catalyst poisoning [46, 47].

Urea

A sizable volume of waste water with varying urea concentrations is produced during the commercial manufacture of urea. Urea can be found in a variety of materials, including industrial waste that contains different levels of urea and human urine. Furthermore, its electrooxidation results in the non-toxic gases CO_2, N_2, and H_2 [48].

Glucose

A cost-effective, widely available, nontoxic, inert, easy to store, and transport biofuel, glucose is also readily available. Since the direct oxidation of glucose on a metal electrode has a slow kinetics at room temperature, extremely active catalysts must be used. DGFCs are increasingly being used in medical implants like pacemakers and glucose sensors [44, 49].

Future Scope

The fuel cell technology is on the verge of implementation and need governmental contraints on certain area for usage of fuel cell technology. However, "the journey is never ending" and has scope for improvement.

Due to the usage of noble materials (platinum, gold) electrocatalyst costs account for around 51% of fuel cell costs [50]. This opens up numerous opportunities for the development of non-noble electrocatalyst materials and cutting-edge production techniques to enable the commercialisation of fuel cells. Furthermore, durability and reliability are the matter of research as the degradation of the different component reported at about 3500 hours [51].

This paper is an attempt to develop criteria for selection of fuel for fuel cell. Fuel such as methanol and ethanol are very popular in the race but other fuels like ascorbate, glycerol has potential as power source for portable application. Despite high energy density of hydrazine it stand as most hazardous fuel in group. Moreover, search for inexpensive, organic, easily available, electrochemically active, high power density fuel is on going. The passive fuel cell has a lot of potential for the various fuels indicated in this research and has found tremendous use in portable devices.

Conclusion

Fuel cell is undoubtedly a strong contender in competition of energy generator. However, direct liquid fuel cell can replace ICE's in near future considering an aspect of transporation and storage of the fuel. In addition, DLFC has negligible impact on environment that makes it more lucarativesubject for researcher. Although, certain issues such that- durability, expensive electrode material, relaibility are hurdles in fuel cell development and can be addressed through consistent research activities and commercialisation of fuel cell.

Despite of various liquid fuel non of the fuel is considered non toxic except ascorbic acid which is edible in pure form. The most studied fuel cell is DMFC, and developed form of it is in market as well.Hydrazine forming the most hazardous among the group yet having maximum energy density. Most of the fuel are not tested for passive fuel cell that generate scope for further research in this area.

This evaluation represents an effort to create criteria for choosing liquid fuel for use in fuel cells. Additionally, compare various liquid fuels for additional study on particular fuel feed DLFCs. Some of the fuels have only recently begun their voyage, therefore the research data is not yet publicly available.

References

[1] Alias, Kamarudin, S. K., Zainoodin,A. M., and Masdar, M. S. (2020). Active direct methanol fuel cell. An overview. International Journal Hydrogen Energy. 45,(38), 19620–641. doi: 10.1016/j.ijhydene.2020.04.202.

[2] Chu, S. and Majumdar, A. (2012). Opportunities and challenges for a sustainable energy future. Nature 488, 294–303.

[3] Du, H. et al. (2018). Carbon nanomaterials in direct liquid fuel cells. Chemical Record. 18(9), 136572. doi:

[4] Li, M., Lu,Chen, ,and Amine, K. (2018). 30 years of lithium-ion batteries. 1800561, 124. doi: 10.1002/adma.201800561.

[5] Ashok, K., Babu,,Jula, and Mullai, (2021). Impact of used battery disposal in the environment. Linguistic Cultural Review. 5(1), 127686. doi: 10.21744/lingcure.v5ns1.1598.

[6] Mekhilef, S.,Saidur, R., and Safari, A. (2012). Comparative study of different fuel cell technologies. Renewable and Sustainable Energy Reviews. 6(1), 98189. doi: 10.1016/j.rser.2011.09.020.

[7] Nacef, M. and Affoune, (2010). Comparison between direct small molecular weight alcohols fuel cells' and hydrogen fuel cell's parameters at low and high temperature thermodynamic study. *International Journal* of Hydrogen Energy. 36(6), 4208–19 doi: 10.1016/j.ijhydene.2010.06.075.

[8] Wang, Y., Chen, K. S., Mishler, J., Cho, S. C., and Adroher, X. C. (2011). A review of polymer electrolyte membrane fuel cells: technology, applications, and needs on fundamental research. Applied Energy. 88(4), 981–1007. doi: 10.1016/j.apenergy.2010.09.030.

[9] Chadge, R. B. (2016). Effect of ethanol concentration and cell orientation on the performance of passive direct ethanol fuel cell. 11th International Conference on Industrial and Information Systems (ICIIS).

[10] Kothekar, K. P., Shrivastava, N. K., and Thombre, S. B. (2020). Gas diffusion layers for direct methanol fuel cells. INC, 2020.

[11] Munjewar, S. S., Thombre, S. B. and Mallick, R. K. (2017). Approaches to overcome the barrier issues of passive direct methanol fuel cell: review. Renewable and Sustainable Energy Reviews. 67, 1087–1104. doi: 10.1016/j.rser.2016.09.002.

[12] Jorg Ott, Veronika G, Florian P, Eckhard F, Grorg G, D B K, Gunther W, Claus W (2012). Methanol. Ullmann's encyclopedia of industrial chemistry. doi:10.1002/14356007.a16_465.pub3

[13] Hansen, A. C., Goering, C. E. and Ramadhas, A. S. (2016). Ethanol. Alternative Fuels for Transportation. 24, 129–166, doi: 10.1016/b978-0-12-819286-3.00021-x.

[14] Walther T. and François, J. M. (2016). Microbial production of propanol. Biotechnology. Advance. 34. (5), 984–996, doi: 10.1016/j.biotechadv.2016.05.011.

[15] Yue, H., Zhao, Y., Ma, X. and Gong, J. (2012). Ethylene glycol: Properties, synthesis, and applications. Chemical Society Reviews. 41(11), 4218–44. doi: 10.1039/c2cs15359a.

[16] Wernke, M. J. (2014). Glycerol. Encyclopedia of Toxicology: Third Edition., pp. 754–756, doi: 10.1016/B978-0-12-386454-3.00510-8.

[17] Jukka H, Antti V, Pekka J, Ilkka P, Werner R, Heinz K (2016). Formic acid. Ullmann's encyclopedia of industrial. Doi: 10.1002/14356007.a12_013.pub3

[18] Semelsberger, T. A., R. L. Borup, R. L., and Greene, H. L. (2006). Dimethyl ether (DME) as an alternative fuel. 156, 497–511. doi: 10.1016/j.jpowsour.2005.05.082.

[19] (2023). Hydrazine-Production import use and disposal. Harvard University.

[20] Santos, D. M. F. (2021). Direct borohydride fuel cells (DBFCs). 203–32. doi: 10.1016/B978-0-12-818624-4.00010-8.

[21] Leon, C. P. De., Walsh, F. C., Pletcher, D., Browning, D. J., and Lakeman, J. B. (2006) Direct borohydride fuel cells. 155, 172–81. doi: 10.1016/j.jpowsour.2006.01.011.

[22] Pappenberger, G. and Hohmann, H. P. (2013). Industrial production of L-ascorbic acid (Vitamin C) and D-isoascorbic acid. Advances in Biochemical Engineering/Biotechnology. 143, 143–88. doi: 10.1007/10_2013_243.

[23] Marder, T. B. (2007). Will we soon be fueling our automobiles with ammonia-borane?. Angewandte Chemie International Edition. 46(43), 8116–18. doi: 10.1002/anie.200703150.

[24] Medina, A. V., Xiao, H. Jones, M. O., David, W. I. F. and Bowen, P. J. (2018). Ammonia for power. Progress in Energy and Combustion Science. 69, 63–102, doi: 10.1016/j.pecs.2018.07.001.

[25] Sayed, E. T. et al. (2018). Direct urea fuel cells: Challenges and opportunities. Journal. Power Sources. 417, 159–75. doi: 10.1016/j.jpowsour.2018.12.024.

[26] Song, B., He, Y., He, Y., Huang, D., and Zhang, Y. (2019). Experimental study on anode components optimization for direct glucose fuel cells. *Energy*, 176, 15–22. doi: 10.1016/j.energy.2019.03.169.

[27] Demirci, U. B. (2009). How green are the chemicals used as liquid fuels in direct liquid-feed fuel cells? Environment International 35(3), 626–31. doi: 10.1016/j.envint.2008.09.007.

[28] Shrivastava, N. K., Thombre, S. B., and Chadge, R. B. (2016). Liquid feed passive direct methanol fuel cell: challenges and recent advances. pp. 123. doi: 10.1007/s11581-015-1589-6.

[29] Demirci U. B. (2007). Direct liquid-feed fuel cells: Thermodynamic and environmental concerns.169, 239–46. doi: 10.1016/j.jpowsour.2007.03.050.

[30] Gong, L., Yang, Z., Li, K., Ge, J., Liu, C., and Xing, W. (2018). Journal of Energy Chemistry. doi: 10.1016/j.jechem.2018.01.029.

[31] Shrivastava, N. and Thombre, S. B. (2011). Barriers to commercialisation of passive direct methanol fuel cells: a reveiw. International Journal of Engineering Science and Technology.

[32] Apanel. G. and Johnson, E. (2004). Direct methanol fuel cells - Ready to go commercial? Fuel Cells Bull. (11), 12–17. doi: 10.1016/S1464-2859(04)00410-9.

[33] Shrivastava, N. K., Chadge, R. B., Ahire, P., and Giri, J. B. (2018). Experimental investigation of a passive direct ethanol fuel cell. Springer. Ionics 25(2). doi:10.1007/s11581-018-2797-7

[34] Saisirirat, P. (2018). The passive direct ethanol fuel cell performance investigation for applying with the portable electronic devices. IOSR Journal of Engineering. 8(7), 57–68.

[35] Ong, B. C., Kamarudin, S. K., and Basri, S. (2017). Science direct direct liquid fuel cells: a review. International Journal Hydrogen Energy. 42(15), 10142–157. doi: 10.1016/j.ijhydene.2017.01.117.

[36] Qi Z. and Kaufman, A. (2002). Performance of 2-propanol in direct-oxidation fuel cells. Journal. Power Sources. 112(1), 121–29. doi: 10.1016/S0378-7753(02)00357-9.

[37] An, L. and Chen, R. (2016). Recent progress in alkaline direct ethylene glycol fuel cells for sustainable energy production. 329, 484–501. doi: 10.1016/j.jpowsour.2016.08.105.

[38] Miesse, C. M. et al. (2006). Direct formic acid fuel cell portable power system for the operation of a laptop computer. 162, 532–40. doi: 10.1016/j.jpowsour.2006.07.013.

[39] Xing, L., Yin, G., Wang, Z., Zhang, S., Gao, Y. and Du, C. (2012). Investigation on the durability of direct dimethyl ether fuel cell. Part I: Anode degradation. Journal. Power Sources, 198 pp. 170–175. doi: 10.1016/j.jpowsour.2011.09.090.

[40] Li, Z. P., Liu, B. H., Arai, K., and Suda, S. (2005). Development of the direct borohydride fuel cell. 406, 648–52. doi: 10.1016/j.jallcom.2005.01.130.

[41] Su, X., Pan, Z., and An, L. (2019). Performance characteristics of a passive direct formate fuel cell. 111. doi: 10.1002/er.4775.

[42] An, L. and Chen, R. (2016). Direct formate fuel cells : a review. 320. doi: 10.1016/j.jpowsour.2016.04.082.

[43] Muneeb, O. et al. (2017). A direct ascorbate fuel cell with an anion exchange membrane. Journal Power Sources. 351, 74–78. doi: 10.1016/j.jpowsour.2017.03.068.

[44] Fujiwara, N., Yamazaki, S. I., Siroma, Z., Ioroi, T., and Yasuda, K. (2007). l-Ascorbic acid as an alternative fuel for direct oxidation fuel cells. Journal. Power Sources. 167(1), 32–38. doi: 10.1016/j.jpowsour.2007.02.023.

[45] Nagle, L. C. and Rohan, J. F. Nanoporous gold catalyst for direct ammonia borane fuel cells. 158(7), 772–778. doi: 10.1149/1.3583637.

[46] Siddiqui, O. and Dincer, I. (2018). A review and comparative assessment of direct ammonia fuel cells. Thermal Science and Engineering Progress. 5, 568–78. doi: 10.1016/j.tsep.2018.02.011.

[47] Soloveichik, G. L. (2014). Liquid fuel cells. 1399–1418. doi: 10.3762/bjnano.5.153.

[48] Lan, R. and Irvine, J. T. S. (2010). A direct urea fuel cell – power from fertiliser and waste. (3), 438–41. doi: 10.1039/b924786f.

[49] Basu, D. and Basu, S. (2010). Electrochimica Acta A study on direct glucose and fructose alkaline fuel cell. Electrochim Acta. 55(20), 5775–79. doi: 10.1016/j.electacta.2010.05.016.

[50] Corti, H. R. and Gonzalez, E. R. (2014). Introduction to direct alcohol fuel cells. doi: 10.1007/978-94-007-7708-8

[51] Chen, M., Wang, M., Yang, Z., Ding, X., and Wang, X. (2018). Long-term degradation behaviors research on a direct methanol fuel cell with more than 3000h lifetime. Electrochimica Acta. 282, 702–10. doi: 10.1016/j.electacta.2018.06.116.

Chapter 21

Role of biopolymers in drag reduction enhancement of flow – a review

Prashant Baghele[1,a] and Pramod Pachghare[2,b]

[1]Government College of Engineering, Amravati, India

[2]Government College of Engineering, Yavatmal, India

Abstract

Fluid transportation through pipelines generally faces the problem of pressure loss, leading to pumping power loss and lowering of flow efficiency. The drag is responsible for pressure loss, so there is a need of drag reduction. The fluid flow through a pipeline can be either a single phase or multiphase flow. The drag reduction can be achieved through addition of drag reducing agent or additive to the flow. A minute concentration of additive can achieve considerable amount of drag reduction. The widely applied drag reducing agents are synthetic polymers. The high molecular weight synthetic polymers have outstanding drag reduction capabilities, but they can create environment issues and safety problems. Therefore, to eradicate such problems, biopolymers are the good choices over synthetic polymers. Biopolymers are easily available, stable, cheaper, biodegradable, biocompatible and non-toxic. Many biopolymers have also shown good drag reducing abilities. This article describes about different biopolymers and their significant characteristics for achieving drag reduction. The different factors affecting drag reduction performance of biopolymers are also mentioned.

Keywords: Drag reduction, biopolymers, two-phase flow, guar gum, xanthan gum

Introduction

The discovery of drag reduction (DR) phenomenon was accidentally made by Toms in 1949. The DR has advantages for pipeline systems such as, pumping power reduction, flow rate improvement or pump size reduction in turbulent flows [1]. It can be applied to various areas for energy savings purposes and the promising areas of applications are disposal of flood water, irrigation in the fields, oil pipelines, functioning of oil wells, fire-extinguishing purpose, slurries and suspensions transportation, water-based cooling and heating systems, wastewater collection system, filling of airplane tank, oceanic systems and bioengineering [2]. The DR can be achieved by lowering pressure drop due to friction across a channel or pipe. DR can be obtained by using different techniques, which include additive and non-additive methods. The additive method is the chief method of DR, where drag reducing agents (DRAs) are added to the flow. Generally, high molecular weight DRAs such as suspended solids, polymers or surfactants are preferred for DR and they have potential to reduce drag up to 80%. These DRAs are effective and simple to use. Non-additive methods of DR do not involve the use of additives or agents. The five most famous non-additive methods of DR are micro-bubbles, compliant surfaces, wavy and oscillating walls, riblets and dimples. These are also known as passive DR techniques. Non-additive methods can be preferred in the areas such as beverage, food, chemical and pharmaceutical industries where fluid parameters (such as viscosity, density and specific heat capacity) are of great importance and no chance of alteration of physical and chemical makeup of fluid flow as additives are not used [3].

Drag-reducing polymers (DRPs) fall under the category of macromolecules with longer chains. They have slow rate of degradation, fast rate of dissolution and high molecular weight (MW). The performance of DR relies on the factors such as MW and concentration of polymer, phase fraction and mixture velocity (U_m). DRPs with magnificent solubility, greatest extensity and linear structure at a particular MW are most efficient. The interaction between additive and fluid molecules results in DR. The most famous DR mechanism theory is polymer elongation theory. The additive molecules play the role of suppressing turbulence formation in buffer layer, resulting in prevention of turbulent eddy formation [4]. There are two types of DRPs, artificial (or synthetic) and natural (or bio-based) polymers. Artificial polymers can be obtained from petroleum oil, whereas biopolymers can be acquired from natural resources. The artificial polymers as DRAs have good thermal stability and mechanical properties but they have slow biodegradation rate. Furthermore, for identical molecular weights (MWs), artificial polymers cost more than biopolymers. Beside this, biopolymers are biodegradable in nature and easily produced in the form of polysaccharides or carboxy methyl cellulose (CMC) by plants and microorganisms. The very familiar natural polymer is a cellulose and which has vast appearance in fruits and plants. Therefore, utilising it from organic waste material helps to prevent water pollution and save the environment [5].

[a]baghele.prashant@gmail.com, [b]pramod.pachghare@gmail.com

DOI: 10.1201/9781003450252-21

Drag Reduction Using Synthetic Polymers for Gas-Liquid and Liquid-Liquid Flows

Gas-liquid flows

Scott and Rhodes [6] carried out investigations on slug flow of gas and liquid using Polyhall 295 as a DRP. The highest recorded DR was 33% at 68 wppm (weight ppm) DRP concentration. The DRP indicated shear degradability and it lost its DR potential after six residence times. Mowla and Naderi [7] further investigated air-crude oil slug flow using polyalpha-olefin (polyisobutylene) as DRP through three horizontal pipes, one polycarbonate material smooth pipe with 2.54 cm Id and two galvanised iron rough pipes with 2.54 cm and 1.27 cm inner diameters (IDs). The findings showed that DR was achieved even at very low concentrations of polymer addition. Also, the increase in % DR was noticed with increasing polymer concentration up to the optimum value (18 ppm) and after that no DR was noticed. The optimum polymer concentration was independent of pipe type or diameter. The roughness of pipe also affected DR and it was noticed that DR was more for rough pipe as compared to smooth pipe, because pipe roughness is responsible for flow turbulence enhancement. Moreover, in case of rough pipes, 1.27 cm ID pipe indicated more DR than 2.54 cm ID pipe. Al-Sarkhi and Hanratty [8] investigated air-water annular flows to determine the influence of a pipe diameter on DR potential of a DRP, and the pipe diameters were 2.54 cm and 9.53 cm. The 2.54 cm inner diameter (ID) pipe showed drag reductions (DRs) up to 63%, while in their previous investigations the 9.53 cm ID pipe indicated DRs up to 48%. Al-Sarkhi and Abu-Nada [9] also investigated air-water horizontal annular flows through a small diameter pipe using a polyacrylamide (PAM), Magnafloc 101l. The pipe with 7 m length and 0.0127 m ID was utilised for the investigation. The PAM at 1000 wppm concentration was added to water in a 150 litre tank. The polymer solution injection did not use a pump, and 47% DR was achieved at just 40 wppm concentration.

Liquid-liquid flows

Al-Wahaibi et al. [10] carried out investigations on oil-water liquid-liquid flow through a smaller pipe (acrylic material) of 14 mm ID using Magnafloc 1011 (a copolymer of PAM and sodium acrylate) as a DRP. The DRP addition to annular flow led to a maximum DR of about 50%. Al-Yaari et al. [11] determined the influence of water-soluble polymers, Magnafloc 1011 and polyethylene oxide (PEO) on DR in horizontal oil and water flow through 0.0254 m ID acrylic pipe. The remarkable amount of DR was achieved. The effect of salt content on oil-water polymer flow was also determined at 1.5 and 3 m/s mixture velocities, and the results indicated that DRP performance decreases with salt addition. Furthermore, the DR performance increased with increasing MW of PEO and it had MWs of 4×10^6 g/mol and 8×10^6 g/mol. Rise of MW leads to enhancement in polymer entanglement resulting in polymer aggregates formation which has a role in DR. Yusuf et al. [12] tested high MW anionic PAM (Magnafloc 1035) with average MW of 15×10^6 g/mol for horizontal flow of oil and water through 25.4 mm ID acrylic pipe. The significant amount of DR was just possible at around 2 ppm concentration. It was observed that when polymer concentration was increased, the DR amount also increased, and it approached highest plateau value at 10 ppm. The maximum DRs were 45 % and 60 % at superficial oil velocities (U_{so}) of 0.52 and 0.14 m/s respectively. DR increased with increasing superficial water velocity (U_{sw}) up to 1.3 m/s, whereas DR declined with increment in U_{so} for $U_{sw} > 1.0$ m/s. Eshrati et al. [13] conducted an experimental analysis on the flow of oil and water through 30.6 mm ID horizontal pipe to determine DR effect of five different high MW sulfonated polyacrylamides (PAMs) with their commercial names as AN125SH, AN125, AN125VLM, AN113SH and AN105SH (water soluble and linear polymers). The highest DR was achieved beyond 20 ppm polymer concentration. The DR performance reduced with rising oil fraction (α) during continuous phase of oil. Furthermore, increase in DR was seen with the rising U_m particularly over 1.0 m/s for more than 10 ppm polymer concentration. Application of these polymers inferred that polymer DR effectiveness depends upon polymer chain rigidity. More rigid polymer chains showed higher DR, since polymer rigidity increases with MW and lower charge density. Eshrati et al. [46] further investigated oil-in-water dispersed flow using four PAMs (cationic and linear) as DRPs in a 30.6 mm ID pipe. Three polymers had dissimilar MWs and charge density of 10%, and remaining one polymer had charge density of 45%. The test section flow was adjusted at α of 0.3 and U_m of 0.8, 1.0, 1.2 and 1.5 m/s ($Re > 34,400$) for 10 – 40 ppm DRP concentrations. The maximum DR was achieved beyond 20 ppm DRP concentrations. The DR advanced with polymer MW and flow velocity. It was also found that polymer elasticity increased with increase of MW and charge density. The maximum DR improved with rise of MW and charge density, indicating a relationship between effectiveness and elasticity of a polymer. Furthermore, DR improved with U_m (or Re) increase.

Different Factors Affecting the Drag Reduction Performance of Polymers

Edomwonyi-Otu and Angeli [14] investigated DR performance of two kinds of PEO polymers having MWs of 5×10^6 g/mol and 8×10^6 g/mol. Results showed that increment in DR was noticed with the increase of

MW of polymer, but it also relied on polymers ionic strength and their mechanical degradation at higher Re. Liu et al. [15] carried out experiments and simulations to investigate the effects of PAM (a polymer) and cetyltrimethyl ammonium chloride (CTAC) (a cationic surfactant) interactions on DR. The polymer-surfactant interactions can help to enhance the anti-shear ability of the surfactant micelles. Moreover, polymers can also play the role of balancing the energy distribution of micelles, delaying the rupture of micelles and postponing the energy extreme point appearance. It was concluded that the destruction of network structure of micelle in polymer-surfactant combination is not easy under the shear conditions, which indicates enhancement in DR performance. Mahmood et al. [16] tested a biopolymer–surfactant combination (chitosan-sodium laurel ether sulfate (SLES)) as drag reducing agent(DRA) in rotating disk apparatus (RDA). The solutions of individual DRAs (Chitosan and SLES) and their combinations showed a non-Newtonian behavior. The viscosity of chitosan-SLES combination was more than that of surfactant SLES but less than the high concentration of biopolymer chitosan (beyond 300 ppm). The chitosan and SLES combination (at 300 and 400 ppm respectively), attained a maximum DR of 47.75% at 3000 rpm, which was higher than DRs achieved by individual DRAs. This combination also yielded outstanding viscoelastic properties, showing its better DR ability. Furthermore, the combination also showed higher shear stability than individual DRAs. Mohammadtabar et al. [17] carried out investigations on flexible and rigid polymers. The flexible polymers were three grades of PAM (Magnafloc 5250, Superfloc A-150 and Superfloc A-110) and PEO, and rigid polymers were xanthan gum (XG) and CMC. At similar high shear rate viscosities, initial DRs achieved by flexible PAM and PEO solutions were 50–58% and 44% respectively, while 12 % DR was initially achieved by rigid polymers. The DR degraded negligibly for rigid polymers over a 2 hour period. The DR moderately degraded for a flexible polymer PAM, though PEO showed an abrupt fall of DR after 20 min. The investigations by Soares [18] indicated that the DR ability of a polymer is lowered by its mechanical degradation and which limits practical applications of flexible polymers. The variables such as concentration, MW and chemical structure of polymer, temperature, linear polymer configuration, turbulence intensity, solvent quality, Reynolds number (Re), relaxation and residence time, affect the mechanical scission or degradation of flexible polymers. Also, DR ability can be lowered by de-aggregation. In de-aggregation, inter molecular associations undergo splitting rather than intramolecular splitting [18].

The biopolymer can face some problems during its application as DRA and biodegradability is one of the problems. No doubt that polymers must be biodegradable, but biodegradability has a disadvantage of shelf-life reduction of biopolymer, so it needs a proper control. The biodegradability can be reduced by a technique such as grafting of biopolymer. This technique combines the properties of artificial and natural polymers. The biodegradability will be reduced due to increased content of artificial polymer in the biopolymer. Besides this, grafting leads to improved DR performance of biopolymer. Along with grafting, cross linking can also be used to improve the DR capabilities of polymers. Polymers should also be shear resistant for retaining their DR performance. It was found that grafting can make polymers shear stable and efficient [19].

Biopolymers and Their Applications in Drag Reduction

The biopolymers are natural substances which are usually created by living organisms such as polysaccharides, mucilage, natural gums, and chitosan. They have environment friendly nature and can sustain significant amount of mechanical degradation. The chemical compounds of polysaccharides are joined through glycosidic bonds, and they form ultimate complex structures. The mostly applied natural polysaccharide is galactomannan. Tara, cassia, locust bean and guar are the primary sources of galactomannan [20]. Polysaccharide is a carbohydrate. The significant types of polysaccharides include xylan, guar gum (GG), XG and chitin. The plant-based polysaccharides can be used as a natural substitutes to synthetic polymers. The natural gums such as okra, karaya, tragacanth, locust bean and guar work as a DRAs in both brine and freshwater solutions. The plant mucilage is a substance which has sticky, viscus and water-soluble nature. It is created by some microorganisms and most types of plants including aloe vera and okra. The mucilage present in plants has vital role in thickening membranes, seed germination and, water and food storage [19]. A review on biopolymers such as GG, XG, okra mucilage (OM) and aloe vera mucilage (AVM)is conducted for describing their DR abilities and factors affecting their DR performance:

Guar gum

The GG is extracted from the guar plant seeds, which is a galactomannan polysaccharide. It easily dissolves in water [21]. The abundance of hydroxyl groups across the long molecular chain of GG helps to form hydrogen bonds in the GG water solution and it gives remarkable viscosity and thickening to the solution [22]. It covers wide application range including cosmetics, paper, agriculture, food, pharmaceuticals, bioremediation and hydraulic fracturing due to its distinctive features such as biodegradability, emulsifying and thickening agent, quick dissolution rate in cold water, wide pH stability, etc. [23].

Sokhal et al. [21] used GG as a flow improver (or DRA) and determined its effect on DR in one-phase water flow (turbulent) through 19.05 mm ID galvanized iron pipe. The KCl salt was also added to the flow for investigating its effects on DR and shear degradation of biopolymer. The maximum DR was found to be 71.4% at 3000 ppm polymer concentration. The %DR increased with the polymer concentration. The boundary layer injection of polymer resulted to improved DR as compared to pre-mixed injection used in the literature. The shear stability of GG was improved up to 47% after KCl salt addition for 45000 Re but %DR decreased with increasing salt concentration. Sokhal et al. [24] further applied GG as a flow improver to water-oil horizontal flow through 19.5 mm ID pipe. The flow improver GG was heterogeneously injected to the boundary layer for determining its effect on DR. The highest DR was found to be 50% at 50 ppm GG concentration and $\alpha = 0.1$. Dosumu et al. [25] also determined the effects of a flow improver GG on one-phase water flow and oil-water two-phase flow through pipes having 12 and 20 mm IDs. Investigations showed that the maximum DR was 45% for one-phase water flow through 12 mm ID pipe at GG concentration of 200 ppm and 69,000 Re. In case of oil-water flow, the maximum DRs were achieved at zero oil fractions. At $\alpha = 0.25$, highest DR was 23% for GG concentration of 200 ppm. It was also investigated that increasing oil fraction led to decreased DR and increasing pipe diameter resulted to increased DR. Attempts were also made to evaluate the MW of a GG and it was found out to be 994035 ± 1.9% g/mol. Edomwonyi-Otu et al. [26] also attempted to investigate the effects of GG addition to both one-phase water flow and oil-water flow through 12 mm ID conduit. It was found that GG with 200 ppm concentration achieved 39% DR in one-phase water flow at 59000 Re and 19 % DR in two-phase oil and water flow. It indicated that one-phase water flow had more DR than two-phase oil-water flow. Motta et al. [27]obtained DR of 45% using GG in one-phase turbulent flow of water, and the maximum DR effect was seen within 30 min after its injection. Following 30 min in the experimental run, GG started degradation because it experienced the shear stresses and powerful stretching. The degradation of GG led to reduction in its DR ability. Ma et al. [47] investigated different properties (such as damage rate, temperature resistance and apparent viscosity) of GMN fracturing fluid which was produced by mechanical nano hybridisation of GG and inorganic nano layered montmorillonite (MMT). The GMN (GG and MMT nanocomposite) showed improved viscoelasticity and apparent viscosity as compared to GG because of extreme cross-linked 3D nexus of GG and MMT. The GMN also showed excellent thermal stability and it indicated 40 °C improvement in resistant temperature as that of GG fracturing fluid. Furthermore, a nanocomposite GMN was smoothly broken and showed damage rate of 10.34% for imparting suitable fracture conductivity.

Singh et al. [28] carried out investigations on grafting biopolymers with synthetic polymers. Investigations showed that grafting a polysaccharide GG with PAM enhanced the DR efficiency of GG. This grafting method combines the strength of the polysaccharide main chain and effectiveness of PAM. Crosslinking is another method to grafting and it was reported that cross linking of GG with borax beneath the gelation level improved the DR performance of GG. In crosslinking, the polymer molecules' dimensions increase due to intermolecular crosslinks and results showed DR improvement up to 35% above 500 ppm concentrations. Though, the shear degradation resistance of a biopolymer was not affected by crosslinking. The further results also showed that the grafted GG had superior DR performance as compared to both commercial and purified GG. Also, the purified GG was better in achieving DR than commercial one. Deshmukh and Singh [29] synthesised seven copolymers of a PAM and GG by grafting method, and their biodegradation rate, shear stability and DR capabilities were investigated. The findings showed that the shear stability and DR capability of graft copolymers depended upon the number of grafts in the molecule and graft length. The shear stability of all the grafted copolymers was found to be lower than the GG, but the grafting improved DR performance and minimised biodegradation rate. Phukan et al. [30] applied commercial and purified GG to sprinkler irrigation system for achieving DR and finally lowering the power requirement. The commercial GG was utilised to obtain the purified GG. Extracting fat and protein impurities from the commercial GG yields to purified GG, which showed improved DR performance. The highest DR was 40% for purified GG at 500 ppm and commercial GG at 1000 ppm.

Xanthan Gum

The XG is a member of microbial polysaccharide family and it is obtained from fermentation process conducted by xanthomonas campestris (bacterium). It has better rheological properties and therefore, it finds wide applications in industries. For aqueous systems, it is an effective stabiliser. At the low shear forces, it is also capable of forming high viscosity solutions and such a property makes it valuable for the petroleum industry, including enhanced oil recovery, oil drilling fluids, pipeline cleaning and fracturing [23].

Edomwonyi-Otu et al. [26] investigated effects of XG addition to both one-phase water flow and oil-water flow through 12 mm ID conduit. It was found that XG with 200 ppm concentration achieved 44% DR in one-phase water flow at 59000 Re and 32% DR in oil-water flow, showing more DR for one-phase water flow than two-phase oil-water flow. Their investigations also included determining the impact of mixture velocity (U_m) on DR with biopolymer XG in case of oil-water flow. Improvement in DR was seen with increasing U_m for 0, 0.25, 0.5 and 0.75 oil fractions. Besides this, DR was seen to be decreasing with increasing oil fraction(α)for 2.22, 3.33 and 4.67 m/s mixture velocities. This all can be depicted in Figure 21.1.

Sohn et al. [31] investigated the aqueous solutions of XG using RDA and tested different MWs of XG such as XGU (3.61 X 10^6g/mol), XG15 (3.41 X 10^6 g/mol), XG30 (3.25 X 10^6 g/mol) ,and XG60 (2.80 X 10^6 g/mol) for DR analysis, and it was found that DR increased with biopolymer concentration. The maximum DRs for XGU, XG15, XG30 and XG60 were 33.0%, 32.3%, 32.1%, and 31% respectively. The maximum DR values increased with increasing MW. The DR for XG solution in deionised water at 50-60 ^{0}C temperature was higher than at room temperature. At the higher rotation speed, the DR was also higher but after some time, significant polymer degradation was noticed. The findings also suggested that XG showed more shear stability in both deionised water and the salt solution than most of the flexible artificial polymers. Kim et al. [32],also tested XG as a DRA in water for RDA. Results showed that XG was more resistant to mechanical degradation in the salt and deionised water solution as compared to many of the artificial polymers (flexible type). Hong et al. [33] determined effects of XG and KCl salt concentrations on DR in turbulent flow of water using RDA. Results showed that DR capacity of XG increased with increasing its concentration, but it declined with the rising KCl salt concentration. At the lower salt concentrations, the %DR decreased with time, but at higher salt concentrations %DR was not significantly affected by time and it seemed to be constant. The mechanical degradation resistance of XG increased with the rising concentration of KCl salt.,Tian et al. [34]carried out experimental studies on hydroxypropyl xanthan gum (HXG) and XG. The XG and propylene oxide mixture under alkaline conditions gives HXG. The DR and rheological properties of XG and HXG solutions in aqueous media at various concentrations were studied. The XG and HXG solutions achieved maximum DRs of 68.1% and 72.8% respectively at 1000 ppm concentration in smooth tube (Figure. 21.2). It was seen that HXG achieved more DR than XG. Also, %DR increases with the concentration of DRAs (XG and HXG) (Figure 21.2).

The investigations on the apparent viscosity of XG and HXG solutions indicated that along with their increasing concentrations viscosity also increased, but the viscosity increase for HXG was remarkably higher than XG. At 6000 ppm, XG and HXG solutions had viscosity values of 74.3 and 167.1 mPa·s respectively. HXG also showed significant improvement in dissolution rate and water solubility due to grafting of hydroxyl to XG. HXG showed the greater values of G' (loss modulus) and G' (storage modulus) as compared to XG. Also, HXG solutions indicated remarkable thixotropic and viscoelastic properties.

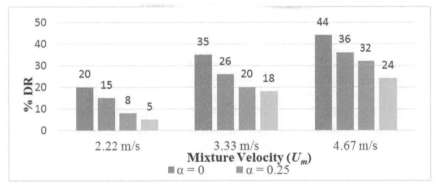

Figure 21.1 The effect of mixture velocities on DR at various oil fractions ($0 \leq \alpha \leq 0.75$) for biopolymer XG [26]
Source: Edomwonyi-Otu et al. [26]

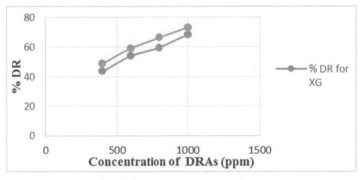

Figure 21.2 The DRs for XG and HXG at different concentrations in ppm [34]
Source: Tian et al. [34]

Cryo-FESEM technique was used to reveal the microstructures of HXG and XG solutions. Microstructures were seen for 6000 ppm of XG and HXG solutions and these were appeared to be honeycomb network. The HXG microstructure was much tighter as compared to XG. This may show the better drag reducing abilities of HXG solutions over XG. Yousif [35] experimentally tested a surfactant-biopolymer mixture of poly diallyldimethyl ammonium chloride (PDDAC) and XG as a DRA for the turbulent flow in pipeline and RDA. The XG biopolymer alone achieved 51% DR in RDA and 58% DR in the pipe. Also, the PDDAC surfactant alone achieved 32% DR in RDA and 36% DR in the pipe. Though, a combination of surfactant and biopolymer (PDDAC-XG) yielded about 62% DR. This showed that the surfactant-biopolymer combination was more effective in reducing drag as compared to individual ones and this duo can be referred as a new type of DRA. Investigations also showed that biopolymer XG was more effective DRA over surfactant PDDAC but in regard of stability (mechanical degradation resistance), surfactant was superior to biopolymer. This is due to the fact that polymer molecules generally cannot withstand high shear forces, break down permanently and lose their DR ability. Whereas, in case of surfactant molecules, their aggregates (micelles) are broken under high shear conditions but the aggregates reform and re-gain their DR ability. This newly formed surfactant-polymer combination gathered properties of individual materials and emerged with good DR abilities as well as mechanical stability. Zhao et al. [36] carried out investigations on turbulent DR features of rigid (XG) and flexible (partially hydrolysed PAM) polymers. Their rheological measurements indicated that both polymers had shear thinning characteristics and this extent of shear thinning increased with the increasing polymer concentration. Their findings also showed that DR effect for both polymers increased by increasing the concentration. The XG showed better shear stability than PAM. At lower Re, XG showed better DR performance; but at higher Re, the highest DR achieved by XG was lower than PAM. Their measurements on flow field indicated that for both the polymers, DR mechanism was similar. Both polymers were also able to supress the turbulence. Li et al. [48] investigated different properties (such as DR, rheological and shear resistance properties) of XG, hydrolyzed polyacrylamide (HPAM) and XG-HPAM composite solutions for exploring the synergy effect between rigid and flexible polymers. It was found that the XG-HPAM solution showed enhanced shear thinning and visco elasticity as compared to one-component solutions (XG and HPAM), indicating that XG (a rigid polymer) affected the visco elasticity of HPAM (flexible polymer). Furthermore, XG-HPAM solution achieved more DR and showed better shear resistance than one-component solutions.

Okra mucilage

Okra plant belongs to mellow or hibiscus family. Abelmoschus esculentus or hibiscus esculentus is the scientific name of okra. It is found in many countries. It is biodegradable, inexpensive, hydrophilic, renewable and stable. Its MW is very high, which goes up to 2×10^5 g/mol and more [37]. The OM is obtained from the leaves of hibiscus plant [39].

Abdulbari et al. [37] used OM as DRA in one-phase water flow through 0.015 and 0.025 m ID galvanised iron pipes. The flow system had a testing length of 1.5 m. The DRA was injected to the flow at various concentrations ranging from 100 to 1000 ppm. The flow rate effect on DR was investigated for flow rates of 0.5, 1, 1.5, 2, 2.5 and 3 m³/hr and it was noticed that %DR lowered with increasing flow rates. The experimental results also showed that %DR was raised by raising biopolymer concentration. The water flow system had a maximum DR of 71% at 1000 ppm biopolymer concentration and 11788 Re for pipe ID of 0.015 m, whereas for the pipe ID of 0.025 m, the highest DR was 57% at 1000 ppm biopolymer concentration and 14145 Re. This showed that the amount of DR was lowered by increasing internal pipe diameter at all flow conditions. Abdulbari et al. [38] further carried out investigations on biopolymer OM as DRA in turbulent one-phase water flow through horizontal pipes. Investigations suggested that with the increase of pipe diameter, pipe length and fluid velocity, the %DR increased. Near about at all Re, with the rising concentration of biopolymer (100-500 ppm) the DR also increased. For a pipe with 0.0254 m ID, the maximum DR was 31.81% at DRA concentration of 200 ppm, while for a pipe with 0.0381 m ID, the maximum DR was 60.7% at DRA concentration of 200 ppm.

Abdulbari et al. [39] investigated water and diesel (gas-oil) flows through pipes of IDs 0.0381 and 0.0127 m. The new type of DRA was made from grafting OM with a monomer acrylonitrile. The OM is a water-soluble DRA. This water soluble mucilage has glycoprotein with 1.7 million MW, which is responsible for making viscoelastic, shear thinning and viscous solutions in water. After grafting, the solubility of mucilage transitioned to hydrocarbon media. Investigations showed that this new grafted DRA indicated high DR performance with maximum DR of 60%. When DRA concentration was increased, the DR performance was also seen to be increasing. Moreover, DR performance improved with increasing Re when DRA added to water media whereas DR decreased with increasing Re when DRA added to oil media. The pipe diameter also influenced DR and DR amount was found to be more in 0.0381 m pipe ID. OM and grafted polymer (mucilage-acrylonitrile) were shear stable up to 200 s of continuous circulation through closed-loop system. Coelho et al. [40] applied a mixture of OM and fiber as a DRA to water flow through a pipeline at high Re. The results showed that the maximum DR was obtained at 1600 ppm of DRA, which was near to maximum DR asymptote. In this case of DRA, the loss of flow efficiency or DR was assigned to de-aggregation or biodegradation rather than mechanical degradation. This is due to the fact that, OM and fiber both belong to rigid materials family like GG and XG. So, they do not undergo mechanical degradation. After 24 hrs, abrupt fall in DR was noticed and this may indicate the biological degradation of DRA. The bactericides can be utilised to minimise the biological degradation. The findings may suggest that this DRA (OMand fiber) can be applied as an alternative over artificial polymers or other bio-based polymers due to its availability and low cost.

Aloe vera mucilage

An AVM is extracted from hydroparenchyma cells of the succulent leaves of aloe vera plant. The mucilage has water content of nearly 98.5 – 99.5% and remaining solids largely made of polysaccharides almost 60% w/w, typically pectic substances and acemannan. The fruitful effects associated with the AVM mostly rely on the chemical features of these polysaccharides (pectic substances and acemannan), such as the acetylation pattern and MW. The studies have shown that the mode of irrigation affects mucilage composition, mostly affecting the mannose-rich polymers. It is also found that the polysaccharides of aloevera especially the mannose-rich polymers, which remarkably contribute to AVM's rheological behavior that is responsible for their viscoelastic properties and flow [41].

Abdulbari et al. [42] used AVM as a DRA in turbulent one-phase water flow through 2.54 cm ID horizontal pipe having 2 m total length. The pipe was divided into four sections with testing lengths of 0.5, 1.0, 1.5 and 2.0 m. This biopolymer was tested at four different concentrations of 100, 200, 300 and 400 ppm. A closed loop system was designed for liquid circulation. In the experiment, the highest DR was found to be 63% at 400 ppm DRA concentration. The findings showed that the experimental parameters such as biopolymer concentration, Re and pipe length affected the DR performance. For some cases DR increased with increasing Re but for other cases DR decreased with increasing Re at all polymer concentrations. The DR increased with increasing biopolymer concentration. DR increased with increase of pipe length up to 1 m but after that DR decreased with increase of pipe length at all polymer concentrations. For a pipe length of 1 m, the optimum DR was 30% at 400 ppm. For longer pipe distances DR decreased and it showed the appearance of shear degradation effect. Abdallah et al. [43] also applied AVM as DRA to both one-phase water flow and oil-water flow for testing its performance of DR.Both flows were tested through a horizontal pipe of 1.2 cm ID. The biopolymer was injected to the flows in the concentration range of 50 – 500 ppm. The oil fraction (α)and mixture velocity (U_m) were selected in range of 0 to 1 and 1.67 to 1.11 m/s respectively for the multiphase flow system. The one-phase water flow showed the highest DR of 64 % at velocity (U, single-phase velocity) of 4.67 m/s, whereas the oil-water flow showed the highest DR of 53.80% at $U_m$of 4.67 m/s and α of 0.25. The maximum DRs for both the flow systems (one-phase and two-phase flows) were obtained at 400 ppm biopolymer concentration (optimum concentration). For one-phase flow, the amount of DR increased along with increasing biopolymer (AVM) concentration and Re until threshold value of 62,949. After that Re, %DR decreased remarkably at 83,690 Re. The decrease of DR after Re of 62,949, may indicate the mechanical degradation of biopolymer. For oil-water flow, the maximum DR occurred at $\alpha = 0$ (only water flow) and it reduced significantly with increasing α. At pure oil flow ($\alpha = 1$), the %DR was zero. Also, %DR increased with increasing U_m at a particular α. The influence of pipe inclination on DR was also investigated and it was found that inclination angle slightly affected DR. Soares et al. [49] investigated the consequences of AVM aging on its DR ability. The outcomes from ^{1}H nuclear magnetic resonance revealed that young and mature aloe vera leaves have different compositions and such a variation affects their DR efficiency. In the tests, DR was analysed in a pipeline system and rotating apparatus. It was found that the samples of young AVM (with low acid contents and plentiful complex polysaccharides) showed more DRs than samples of mature AVM.

Gimba et al. [44] tested the DR performance of Magnafloc 1011 (which is partially hydrolysed polyacrylamide, HPAM), PEO, AVM, and the mixture of PEO-AVM and HPAM-AVM for one-phase water flow through two horizontal unplasticised PVC pipes with IDs of 0.012 and 0.02 m at different flow rates. The individual polymers AVM, HPAM and PEO achieved DRs of 64%, 73.6% and 76% respectively, while HPAM-AVM combination and PEO-AVM combination achieved DRs of 80% and 84%, and 81.6% and 84.8% at 3:1 and 1:19 mixing ratios respectively for a pipe with ID of 0.012 m. The pipe with larger ID of 0.02 m showed lower DRs than pipe ID of 0.012 m at the same conditions (flow rate, polymer type and concentration). Edomwonyi-Otu et al. [45] also investigated the effects of polymers such as HPAM, PEO, AVM and their combinations such as HPAM – AVM and PEO-AVM on DR for one-phase water flow and two-phase oil-water flow through 0.02 m ID horizontal pipe (unplasticised PVC). For both types of flow, the polymer combinations showed higher amounts of DR than individual polymers due to synergistic effect of both polymer molecules. This synergistic effect may enhance the rigidity of the biopolymer molecules present in the mixture. For oil-water flows, DR diminished with rising oil fraction. The maximum DRs for HPAM-AVM combination at 30 and 400 ppm were 62% and 67% respectively and for PEO-AVM combination at 30 and 400 ppm were 63% and 68% respectively at $\alpha = 0.25$.

Conclusions

The following conclusions can be made based on reviewing the synthetic polymers, biopolymers and polymer combinations for DR in the fluid flows:

- Nearly most of the researchers inferred that DR increased with the increasing concentration of DRAs such as synthetic polymers, biopolymers and polymer mixtures (combinations) for one-phase or two-phase flows. The salt addition to one or two-phase flows using DRAresulted into fall of DR with rising salt concentration, but shear or mechanical stability of DRAwas improved with increasing salt concentration [11, 21, 23, 33]. The pipe roughness is also an important factor which affects the DR. The rough pipe attained higher DR than that of smooth pipe [7]. In case of oil-in-water dispersed flow using PAMs, it can be concluded that DR increases with the increase of polymer elasticity [46]. The biopolymers such as GG, XG and AVM for the same concentration and pipe diameter, achieved more DR in one-phase water flow than two-phase oil-water flow [25, 26], [43].

- For oil-water flows, %DR increased with increasing MW of a DRA (synthetic polymer) [11-14, 46]. The MW of a polymer also influenced its rigidity and it was found that polymer chain rigidity increased with the increasing MW of a polymer [13]. Also, for biopolymer XG in aqueous media using RDA, the increase of DR was noticed along with rising MW of XG [31]. Investigations on oil-water flows using DRAs such as high MW sulfonated PAMs, GG, XG and AVM suggested that DR decreased with increasing oil fraction and it was maximum at $\alpha = 0$ (pure water flow) and it was zero at $\alpha = 1$ (pure oil flow) [13, 25, 26, 43]. For oil-water flows using PAMs, XG and AVM as DRAs, increase in %DR was seen with increasing U_m [26, 43, 46]. For water flow using OM as DRA, %DR increased with increasing fluid velocity [38]. Whereas, for further studies on water flow using OM as DRA, the %DR reduced with rising flow rates [37].

- The declining DR was noticedwith respect to increasing pipe diameter for the cases such as air-crude oil flow through rough pipes using Polyalpha-olefin as DRA,air-water annular flow using DRP, water flow using OM as DRA and water flow through PVC pipes using DRPs such as HPAM, PEO, AVM,PEO-AVMand HPAM-AVM [7, 8, 37, 44]. Whereas, DR increased with increasing pipe diameter for the cases such as biopolymer GG as a DRA in oil-water flows,OM as a DRA in one-phase water flow and grafted OM (mucilage-acrylonitrile) as DRA [25, 38, 39]. For AVM as a biopolymeric DRA in one-phase water flow, DR was affected by pipe length and it declined for longer pipe lengths due to emergence of shear degradation effect [42].

- The polymer-surfactant (PAM- CTAC) combination showed enhanced anti-shear capability, which led to improved DR performance [15]. The otherbiopolymer-surfactant (Chitosan and SLES) combination as DRA showed higher DR and shear stability than individual DRAs. It also indicated improved visco-elastic properties [16]. The surfactant-biopolymer (PDDAC- XG) combination showed higher DR than individual DRAs such as PDDAC and XG. The XG was better in DR performance than PDDAC but in respect of mechanical stability PDDAC was supreme over XG. So, this new DRA (PDDAC- XG) came out with better DR potential and mechanical stability [35]. The rigid and flexible polymer combination (XG-HPAM) indicated superior viscoelasticity, shear thinning and shear resistance than individual polymers (XG and HPAM). Also, XG-HPAM attained higher DR than single XG and HPAM [48]. For water flow using OM-fiber combination as DRA, the fall of DR may be due to de-aggregation or biodegradation regardless of mechanical degradation. The OM and fiber belong to rigid polymers and therefore they are mechanically stable [40]. For water and oil-water flows, the polymer combinations such as PEO-AVM and HPAM-AVM showed higher DRs than individual polymers such as AVM, HPAM and PEO because of the synergistic effect between two polymer molecules [44, 45]. The aging of AVM affected its DR capacity and the results showed that young AVM had superior DR potential than mature AVM [49].

- The grafted GG with PAM showed improved DR performance of GG. Also, grafted GG was ahead of both purified and commercial GG in DR performance [28]. Moreover, purification of commercial GG resulted into improved DR [28, 30]. An alternate method to grafting is crosslinking, and crosslinking of GG enhanced its DR performance [28]. Further investigations on seven grafted guar gums with PAM also showed enhanced DR and minimum biodegradation rate, but grafted GGs indicated lower shear stability than pure GG [29]. The nanocomposite GMN showed enhanced properties (such apparent viscosity, viscoelasticity, and thermal stability) as compared to GG [47]. For XG as DRA in aqueous solution using RDA, XG indicated better shear stability and mechanical degradation resistance in salt and deionized water solution than many artificial polymers (flexible) [31, 32]. The grafted XG (HXG) showed higher DR than XG. Also, the rheological properties of HXG were improved over XG [34]. The grafted OM (mucilage-acrylonitrile) as DRA also showed better DR [39].

References

[1] Pouranfard, A. R., Mowla, D., and Esmaeilzadeh, F. (2015). An experimental study of drag reduction by nanofluids in slug two-phase flow of air and water through horizontal pipes. Chinese Journal of Chemical Engineering. 23(3), 471-75.

[2] Brostow, W. (2008). Drag reduction in flow: Review of applications, mechanism and prediction. Journal of Industrial and Engineering Chemistry. 14(4), 409–16.

[3] Abdulbari, H. A., Yunus, R. M., Abdurahman, N. H., and Charles, A. (2013). Going against the flow - A review of non-additive means of drag reduction. Journal of Industrial and Engineering Chemistry. 19(1), 27-36.

[4] Asidin, M. A., Suali, E., Jusnukin, T., and Lahin, F. A. (2019). Review on the applications and developments of drag reducing polymer in turbulent pipe flow. Chinese Journal of Chemical Engineering. 27(8), 1921-32.

[5] Kaur, H. and Jaafar, A. (2018). The effect of sodium hydroxide on drag reduction using banana peel as a drag reduction agent. In: International Conference on Engineering and Technology. AIP Conference Proceedings 1930. 020031-1–020031-7.

[6] Scott, D. and Rhodes, E. (1972). Gas–liquid slug flow with drag reducing polymer solutions. AIChE Journal. 18, 744–50.

[7] Mowla, D. and Naderi, A. (2006). Experimental study of drag reduction by a polymeric additive in slug two-phase flowof crude oil and air in horizontal pipes. Chemical Engineering Science. 61(5), 1549-54.

[8] Sarkhi, A. A. L. and Hanratty, T. J. (2001). Effect of pipe diameter on the performance of drag-reducing polymers in annular gas–liquid flows. Chemical Engineering Research and Design. 79 (4), 402-8.

[9] Al-sarkhi, A. M. and Ab-Nada, E. (2005). Effect of drag reducing polymer on annular flow patterns of air and water in a small horizontal pipeline. In: Twelfth International Conference on Multiphase Production Technology, Barcelona, Spain.

[10] Wahaibi, T. A. L., Smith, M., and Angeli, P. (2007). Effect of drag-reducing polymers on horizontal oil–water flows. Journal of Petroleum Science and Engineering. 57(3–4), 334-46.

[11] Al-soleimani, Y. M., Sharkh, B. A., Mubaiyedh, U, and Al-sarkhi, A. (2009). Effect of drag reducing polymers on oil–water flow in a horizontal pipe. International Journal of Multiphase Flow. 35(6), 516-24.

[12] Yusuf, N., Wahaibi, T. A. L., Al-wahaibi, Y., Al-ajmi, A., Al-hashmi, A. R., Olawale, A. S., and Mohammed, I. A. (2012). Experimental study on the effect of drag reducing polymer on flow patterns and drag reduction in a horizontal oil–water flow. International Journal of Heat and Fluid Flow. 37, 74-80.

[13] Eshrati, M. Al-hashmi, A. R., Al-wahaibi, T., Wahaibi, Y., Ajmi, A. and Abubakar, A. (2015). Drag reduction using high molecular weight polyacrylamides during multiphase flow of oil and water: A parametric study. Journal of Petroleum Science and Engineering. 135, 403-9.

[14] Edomwonyi, L. C. and Angeli, P. (2019). Separated oil-water flows with drag reducing polymers experimental. Thermal and Fluid Science. 102, 467-78.

[15] Liu, D., Wang, S., Ivitskiy, I., Wei, J., Tsui, O. K. C., and Chen, F. (2021). Enhanced drag reduction performance by interactions of surfactants and polymers. Chemical Engineering Science. 232, 116336.

[16] Mahmood, W. K., Khadum, W. A,. Eman E., and Abdulbari, H. A. (2019). Biopolymer–surfactant complexes as flow enhancers: Characterization and performance evaluation. Applied Rheology. 29 (1), 12-20.

[17] Mohammadtabar, M., Sanders, R. S., and Ghaemi, S. (2020). Viscoelastic properties of flexible and rigid polymers for turbulent drag reduction. Journal of Non-Newtonian Fluid Mechanics. 283, 104347.

[18] Soares, E. J. (2020). Review of mechanical degradation and de-aggregation of drag reducing polymers in turbulent flows. Journal of Non-Newtonian Fluid Mechanics. 276, 104225.

[19] Abdulbari, H. A. Shabirin, A., and Abdurrahman, H. N. (2014). Bio-polymers for improving liquid flow in pipelines - A review and future work opportunities. Journal of Industrial and Engineering Chemistry. 20, 1157-70.

[20] Baghele, P. and Pachghare, P. (2022). Review on methods of drag reduction for two-phase horizontal flows. In: 2nd International Conference on Advanced Research in Mechanical Engineering, AIP Conference Proceedings. 2421.

[21] Sokhal, K. S., Gangacharyulu, D., and Bulasara, V. K. (2018). Effect of guar gum and salt concentrations on drag reduction and shear degradation properties of turbulent flow of water in a pipe. Carbohydrate Polymers. 181, 1017-25.

[22] Gong, H.,Liu, M., Chen, J. Han, F. Gao, C., and Zhang, B. (2012). Synthesis and characterisation of carboxymethyl guar gum and rheological properties of its solutions. Carbohydrate Polymers. 88(3), 1015-22.

[23] Han, W. J. and Choi, H. J. (2017). Role of bio-based polymers on improving turbulent flow characteristics: Materials and application.Polymers. 9 (6), 209.

[24] Sokhal, K. S. Gangacharyulu, D., and Bulasara, V. K. (2019). An experimental investigation of heterogeneous injection of biopolymer (guar gum) on the flow patterns and drag reduction percentage for two phase (water-oil mixture) flow. Experimental Thermal and Fluid Science. 102, 342-50.

[25] Dosumu, A. I., Edomwonyi-Otu, L. C., Yusuf, N. and Abubakar, A. (2020). Guar gum as flow improver in single-phase water and liquid–liquid flows Arabian Journal for Science and Engineering 45,, pp. 7267-7273.

[26] Edomwonyi-Otu, L. C. Dosumu, A. I., and Yusuf, N. (2021). Effect of oil on the performance of biopolymers as drag reducers in fresh water flow. Heliyon. 7(3), E06535.

[27] Motta, M. V. L., Ribeiro de Castro, E. V. Muri, E. J. B. Costalonga, M. L. Loureiro, B. V., and Filgueiras, P. R. (2019). Study of the mechanical degradation mechanism of guar gum in turbulent flow by FTIR. International Journal of Biological Macromolecules. 121, 23-28.

[28] Singh, R. P., Pal, S., Krishnamoorthy, S., Adhikary, P., and Ali, S. A. (2009). High-technology materials based on modified polysaccharides. Pure and Applied Chemistry. 81(3), 525-47.

[29] Deshmukh, S. R. and Singh, R. P. (1987). Drag reduction effectiveness, shear stability and biodegradation resistance of guargum-based graft copolymers. Journal of Applied Polymer Science. 33, 1963-75.

[30] Phukan, S., Kumar, P., Panda, J., Nayak, B. R., Tiwari, K. N., and Singh, R. P. (2001). Application of drag reducing commercial and purified guar gum for reduction of energy requirement of sprinkler irrigation and percolation rate of the soil. Agricultural Water Management. 47(2), 101-18.

[31] Sohn, J. I., Kim, C. A., Choi, H. J., and Jhon, M. S. (2001). Drag-reduction effectiveness of xanthan gum in a rotating disk apparatus. Carbohydrate Polymers. 45(1), 61-68.

[32] Kim, C. A., Choi, H. J., Kim, C. B., and Jhon, M. S. (1998). Drag reduction characteristics of polysaccharide xanthan gum. Macromolecular Rapid Communication. 19, 419-22.

[33] Hong, C. H., Choi, H. J. Zhang, K., Renou, F., and Grisel, M. (2015). Effect of salt on turbulent drag reduction of xanthan gum. Carbohydrate Polymers 121, 342-47.

[34] Tian, M., Fang, B., Jin, L., Lu, Y., Qiu, X., Jin, H., and Li, K. (2015). Rheological and drag reduction properties of hydroxypropyl xanthan gum solutions Chinese Journal of Chemical Engineering. 23 (9), 1440-46.

[35] Yousif, Z. (2018). Drag reduction study of xathan gum with polydiallyldimethylammonium chloride (PDDAC) solutions in turbulent flow. Engineering and Technology Journal. 36(8A), 891-99.

[36] Zhao, S., Li, E., Yang, H., Fan, L., Zheng, L., and Zhang, Y. (2022). Comparative study on turbulent drag reduction characteristics of flexible polymer and rigid polymer. International Journal of Fluid Mechanics Research. 49(1), 1-18.

[37] Abdulbari, H. A., Ahmad, M. A., and Yunus, R. B. M. (2010). Formulation of okra-natural mucilage as drag reducing agent in different size of galvanized iron pipes in turbulent water flowing system. Journal of Applied Sciences. 10(23), 3105-10.

[38] Abdulbari, H. A., Izam, F., and Man, R. C. (2011). Investigating drag reduction characteristic using okra mucilage as new drag reduction agent. Journal of Applied Sciences. 11(14), 2554-61.

[39] Abdulbari, H. A., Kamarulizam,N. S., and Nour, A. H. (2012). Grafted natural polymer as new drag reducing agent: An experimental approach. Chemical Industry & Chemical Engineering Quarterly. 18(3), 361-71.

[40] Coelho, E. C., Barbosa, K. C. O., Soares, E. J., Siqueira, R. N., and Freitas, J. C. C. (2016). Okra as a drag reducer for high Reynolds numbers water flows. Rheologica Acta. 55, 983-91.

[41] -Fuentes, R. M., Torres, L. M., Laredo, R. F. G. González, V. M. R., Eim, V., and Femenia, A. (2017). Influence of water deficit on the main polysaccharides and the rheological properties of Aloe vera (Aloe barbadensis Miller) mucilage. Industrial Crops and Products. 109, 644-53.

[42] Abdulbari, H. A., Letchmanan, K., and Yunus, R. M. (2011). Drag reduction characteristics using aloe vera natural mucilage: An experimental study. Journal of Applied Sciences. 11(6), 1039-43.

[43] Abdallah, M. N., out, L. C. E., Yusuf, N., and Baba, A. (2019). Aloe vera mucilage as drag reducing agent in oil-water flow Arid zone. Journal of Engineering, Technology & Environment. 15(2), 248-58.

[44] Gimba, M. M., -Otu, L. C. E.,Yusuf, N., and Abubakar, A. (2020). Synergistic effect of natural and synthetic polymers as drag reducing agents in water flow: Effect of pipe diameter. FUPRE Journal of Scientific and Industrial Research. 4(2).

[45] -Otu, L. C. E., Gimba, M. M., and Yusuf, N. (2020). Drag reduction with biopolymer-synthetic polymer mixtures in oil-water flows: effect of synergy. Engineering Journal. 24(6), 1-10.

[46] Eshrati, M., Wahaibi, T. AL., Hashmi, A. R. AL., Wahaibi, Y. AL., Ajmi, A., and Abubakar, A. (2022). Significance of polymer elasticity on drag reduction performance in dispersed oil-in-water pipe flow. Chemical Engineering Research and Design. 182, 571-79.

[47] Ma, Y. X., Du, Y. R., Zou, C. H., Lai, J., Ma, L. Y., and Guo, J. C. (2022). A high-temperature-resistant and metallic-crosslinker-free fracturing fluid based on guar gum/montmorillonite nanocomposite. Journal of Natural Gas Science and Engineering. 105, 104712.

[48] Li, E., Zheng, L., Li, Y.M, Fan, L., Zhao, S., and Liu, S. (2022). Investigation of the drag reduction of hydrolyzed polyacrylamide–xanthan gum composite solution in turbulent flow. Asia-Pacific Journal of Chemical Engineering.17(5), e2791.

[49] Soares, E. J., Siqueira, R. N., Leal, L. M., Barbosa, K. C. O., Cipriano, D. F., and Freitas, J. C. C. (2019). The role played by the aging of aloe vera on its drag reduction properties in turbulent flows. Journal of Non-Newtonian Fluid Mechanics. 265, 1–10.

Chapter 22

Experimental analysis of compound material combination of concrete-steel beams using non-symmetrical and symmetrical castellated beams structures

H. P. Ghongade[a] and A. A. Bhadre[b]

Brahma Valley College of Engineering And Research Institute, Nashik, India

Abstract

This paper aims to identify the characteristic performance of cement slabs stand-in with commonly non-symmetrical and symmetrical castellated beams structures. The connecting stud connection is used to connect cement slab steel structures. The common practice in building construction is that castellated beams of compound concrete-steel beams are used. The five compound beams with simple support are investigated in these experiments with two-point loading conditions. To carry out testing experiments, the two beams specimen are prepaid using steel only, and three specimens are prepaid using rolled steel beams. One specimen is prepared using a non-symmetrical cross section made up of two dis-similar standard sections are, HEA120 and IPE120. The cement slab for all testing specimens has the same characteristics and dimensions. The testing results indicate that the rigidity and strength are significantly notable for steel compound castellated beams compared to the compound castellated beams from the main section. The final load-carrying capacity for the compound rolled steel beams are fabricated from an IPE-120 section are 47% more than that of the compound castellated beams, and the final load-carrying capacity of compound castellated beam made-up of from the wide flanged section HEA120, which resulted in a sudden rise of 22% compare the main beam control sample. The final load-carrying capacity of the compound material sample beam manufactured using the non-symmetrical rolled steel beam (HEA-120 and IPE-120) gained a rise-up of 13% and 70%, correspondingly associated with the command over the specimen set up from graded sections.

Keywords: Buckling, compound material beam, castellated beam, Vierendeel mechanism

Introduction

In the modern design era, the long span compound roofs are commonly found. The rolled steel beams are premeditated to toil combinedly with the cement roof slab, therefore it indicates a notable increase in their final load carrying magnitude. The compound castellated beams containing there in forcing cements lab associated to the rolled steel beam is commonly used in civil common constructions all over all countries. The chemical bonding between cements lab and the beams are giving better member stiffness and increases the final load carrying capacity. There are some new sites specimen of projects using the rolled steel beam for compound material roof system as marked in Figures 22.1 and 22.2 [1].

Litrature Review

A rolled steel beam is manufactured by processing standard steel web by cutting operation and with welding process they have joined together with two half sections formed by the separating to create the skeleton structure member through the large dimension's depth than the main beam [1]. In the fabrication process induces a mechanism it combines an enhancement in membrane strength by overall decrease in members own weight after it's equated to standard beams of same strength. In on additional benefit of the rolled steel beams are allow service pipes, electric cables and ducts to go through the hexagonal web openings to removing the essential passes through the such facilities inside the beams. Opening in the web also looks beautiful when steel rolled beams are erected in civil constructions with visibility of them [2]. The use of non-symmetrical concrete-steel compound steel rolled beams has massively accepted in civil construction standard practices because of its weight to strength ratio. This paper lightens the conclusion of the by means of non-symmetrical steel rolled beams to compile compound the beams as shown in Figure 22.3 [3]. Hosain and Speirs [4] tried 12 samples of the rolled steel beams to examine the results of opening dimensions on failure mode. Outcome of these are optimal sizes for the opening are necessities a least neck thickness to minimalise the possibilities of not getting successful arises because of Bending of Vierendeel. The Nethercot- Kerdal [5] studied the failure modes of rolled steel beams besides determined that the greatest probably modes of failure occurred where a Vierendeel bending, the bending mechanism, welded rupture, buckling of web post in joint then torsional lateral buckling. Another study approved out examinations

[a]ghongade@gmail.com, [b]ghongade@gmail.com

DOI: 10.1201/9781003450252-22

Figure 22.1 Rolled steel beams in parking area
Source: Author

Figure 22.2 Rolled steel beams in factory erection
Source: Author

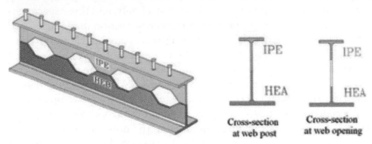

Figure 22.3 Non-symmetrical rolled steel beam
Source: Author

proceeding compound halve rolled steel beams, these are manufactured by one halve of a rolled steel beam of the flange of flat plate fused on the way to upper part of the post of web and the studies on shear force component fused to the steel plate [6]. After they detected web-post buckling break in in the section of steel beam. Later on, over all achievement of the halve-rolled steel beam with-out a cement slab was reviewed for assessment resolutions. These sample fails because of lateral torsional buckling occurred, re-presenting that of addition a cement slab to the beam not able arises the amount final load carrying capacity. Lawson et al. [1] investigated case study on four symmetrical compound beams with cells has been processed and made available (made-up of IPE-400), in on adding to the five sample with on-symmetrical cross-section manufactured it from IPE-400 and HEB-340 to examine the consequences a non-symmetrical cross section takes on modes of failures. The results displayed that non-symmetry beams are on dangerous aspect for the reason that the subject causes to escalate the moment of bending in the post of webs. Sheehan et al. [3] shown that sample experiments on two sets of complete gage non-symmetrical compound rolled steel beams with extended openings at middle of the span to examine the change of the compound action on the bending resistance of Vierendeel at the web opening. The shear load on the beam 46% opposes than that of steel cross section of shear resistance. Al-Thabhawee and Mohammad [7] revises focuses on consolidation of rolled steel beams by shape of octagonal opening thru fusing a rounded circle ring in-side of the opening of web. The solidification method directed in the direction of reinforcing the web exposed area besides by-passing failure buckling in web post. Final readings have it has been listed to that, final loading by strengthening the entrance of the web via circle-ring improved by extent 289% comparatively by the main beam. Al-Thabhawee and Alhasan [9] focuses on improving the beam by placing a steel ring inside the octagonal openings to strengthen the weakest part, the web-post. To increase the strength of castellated beams with expansion plates by using different types of stiffeners around web opening [10]. It found that using circular ring stiffeners was more efficient than using varios ring stiffeners, and can be used as additional steel material to increase load capacity to the original beam [10]. The Abaqus program was used to conduct a finite element (FE) analysis of TCBs, which showed excellent agreement with the experimental results in terms of ultimate load capacity vs. mid-span deflection response and failure mechanisms. TCBs can be used for increasing the strength and stiffness of the I-section parent beam with adding expansion plates.

In this paper focuses on overview of the present information in compound concrete-steel members and roofing system in addition to that its reveal modern trends for upcoming study. The primary aim of this

Table 22.1 Compounds ample–details of steel beam

Number of sets of series	Test Samples	Steel Sections/beam	Rolled steel beams		Dimension in mm	
			Upper T- Section	Lower T- Section	Depth	Height
Set 1	CNB-I type	IPE-120 section	-	-	121	191
	CNB-H Type	HEA-120 section	-	-	121	191
Set 2	CCB-I Section	Rolled steel beam symmetrical	IPE-120	IPE-120	171	241
	CCB-H Section	Rolled steel beam symmetrical	HEA-120	HEA-120	171	241
Set 3	CCB-IH	Rolled steel beam-non- symmetrical	IPE-120	HEA-120	171	241

Note– "d" stand for depth beam and "H" stand for height of the beam

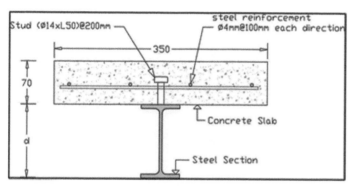

Figure 22.4 Standard sample of cement flange(compound beam)
Source: Author

research study is to find out novel innovative construction procedures and foremost usages of construction materials, on the top of finical constraints towards resilience and sustainability.

Experimental Conduction

The number of sets are organised in the college laboratory to carry the testing and result of the compound concrete-steel beams. The testing specimen of each dimension are shown in Table 22.1. The Span of each specimen is 1800 mm and cement slab thickness is 70 mm and 350 mm in wide. To carry the testing the minimum standard of steel using ACI code using were placed in respectively direction in the cement's lab. In figure the standard sample of cements lab with dimension and reinforcement details are provided.

In set 1 – 2 testing sample are built up using standard procedure of hot-rolling section. The first sample CNB-I with specification IPE-120 I-section having flange width 64 mm, depth 120 mm, flange thickness 6.3 mm and thickness of web 4.4 mm and second sample consist of CNB-H remained made up of HEA-120 wide flanges sections of width of 120 mm, total depth of 120 mm, flange thickness 8 mm including web thickness of 5 mm. Figure 22.5a show that the detailing and dimension of CNB-H and CNB-I.

In Set 2 – 2 testing samples were considered of CCB-H and CCB-I built up using symmetrical rolled steel beams. The primary sampleof CCB-I is utilised for rolled steel beams fabricated from the IPE-120 section, subsequently in depth of the beam 170 mm with-out of any additional weight. As the increases in the depth-ness of rolled steel beam with corresponding to main section mentioned as the ratio of expansion was it 42%. Coming to second sample of CCB-H utilized for rolled steel beam fabricated from HEA-120 wide-ranging flange section, similarly the proceeding in the deepness of the beam is 170 mm. Both rolled steel beam set of 2 are having same expansion ratio and the dimensional of hexagonal opening as shown in Figure 22.5b.

In set 3 – only one sample consist of CCB-IH made up of a non-symmetrical rolled steel beam. This sample is used for rolled steel beam made-up from two dissimilar steel section. The first upper section T was cut from an IPE-120 and the second lower section T was cut from HEA-120 wide flange and both beams consist of depth-ness of 170 mm. For simple judgement altogether non- symmetrical and symmetrical rolled steel beams utilised in the examination of same expansion ratio of 0.43. In Figure 22.5c indicate that the particulars of CCB-IH specimen and dimensions. Table 22.1 indicates particulars of each verified compound specimen are recorded in table.

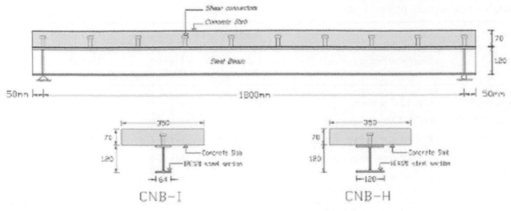

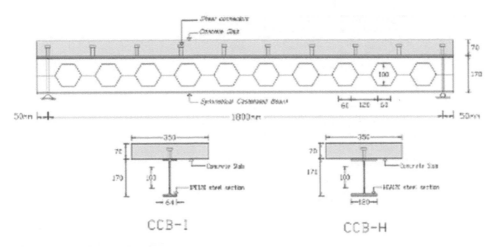

(a) Series 1: Standard beams

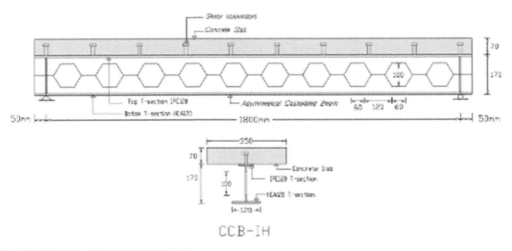

(b) Series 2: Symmetrical castellated beams

(c) Series 3: Asymmetrical castellated beam

Figure 22.5 Detailing and dimension of compound concrete –steel specimens
Source: Author

Properties of Material

Example All compound samples were prepared from the same steel bar to ensure that the materials properties of the testing specimens remain constant and for cement flanges were prepared from the same concentrated mixture to ensure the properties remain same. Natural fine aggregate is mixed with Portland concrete with crushed coarse to harvest this mixture. Set 3, the cement cubes of dimensions length 150 mm x breadth 150 mm x height 150 mm was established to examination the compressive strength of it and it is found average strength of 30.5 MPa. The strength of slabs come from 4 mm wired mesh with material

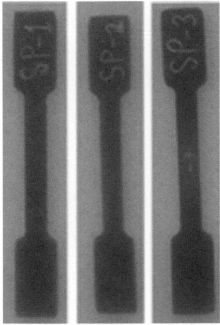

a. A steel samples   b. Materials testing machine

Figure 22.6 Material tensile testing for sample steels
Source: Author

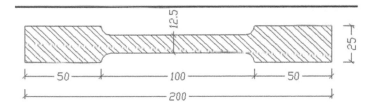

Table 22.2 Steel properties and testing sample result

Sample	$f_u$inMPa	$f_y$inMPa	EinMPa
Sample 1	556.1	305.7	2.01×105
Sample 2	551.7	305.1	2.03×105
Sample 3	543.2	304.4	2.07×105
Average	550.4	305.1	2.03×105

Source: Author

of yield strength is 410 MPa. Three tensile samples were cutted from the flange and its web of IPE-120 with simultaneously web of HEA-120 section was tested. The steel specimen consists of standard dimensions' as per specified in A.S.T.M. E-8M. As per the A.S.T.M. A 370-17 [8], the samples of steel undergone testing of materials testing machine having the capacity of loading of 590 KN as mentioned of in Figure 22.6. Young's Modulus, Yield stress and Final Stress results are noted down from materials testing machine which is recorded in Table 22.2.

The bonding between the steel beams and cement slab are throughout constant in all compound specimen by shear connectors of height of 50 mm and 14 mm in diameter connected using welding to the flanges of the beam steel material as of shown in Figure 22.7. A distribution of studs of shear is in a one side aligned by side wise the distance of beam steel of at distance 200 mm.

Instruments: Loading and Testing

Altogether the compound samples are put on the simply supported position at 50 mm apart from both side ends besides that were testing under condition of two - point load as displayed in Figure 22.8. For testing the load over the beams, a hydraulic Jack of capacity 1000 KN is applying single point load to the cement

Figure 22.7 Studs welding over the steel flange of the beam
Source: Author

Figure 22.8 CCB-H test setup of samples
Source: Author

Table 22.3 Compound testing samples-experimental results plotted

Set of series	Test samples	Types of the steel beam	Load for crack propagation (KN)	Final load for breaking (KN)	Slip of the beam (mm)	Types/modes of failure
Set 1st	CNB-H	Std section	35.0	165.0	9.3	Due to compression failure occurred in cement slab
	CNB-I	Std sectio n	25.0	110.0	10.3	Due to compression failure occurred in cement slab
Set 22nd	CCB-H	Symmetrical	45.0	200.0	10.2	Steel beam – web post buckling occurred
	CCB-I	Symmetrical	30.0	160.0	14.2	Steel beam Vierendeel bending occurred
Set 3rd	CCB-IH	Non-symmetrical	40.0	186.0	13.10	Upper T-section of steel beam Vierendeel ending occurred

Source: Author

block over the compound beam. Electronics dial and LVDT gauge are used to measure the samples vertical displacement at the middle of span. To get the pure bending interesting samples in the middle of beam at distance 600 mm portion of each sample the two-point loading was adopted.

Results

In this experimental investigation, it is out lined to analyse the effect rolled steel beams utilised in compound steel cement beams along with to examine the effect of rolled steel beams over the non-symmetrical

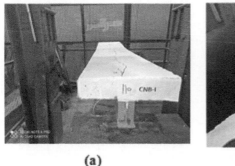
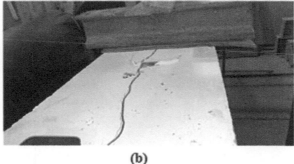

(a)   **(b)**

Figure 22.9 Specimen CNB-I-failure mode
Source: Author

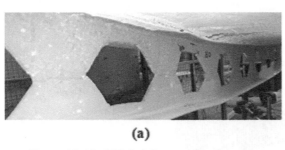
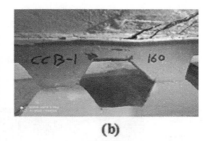

(a)   **(b)**

Figure 22.10 CCB-I–failure mode of sample
Source: Author

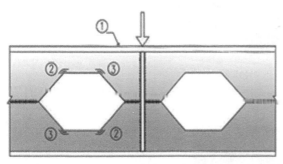

Figure 22.11 Vierendeel bending sequence of yielding
Source: Author

cross section. These investigational results for failure modes, midspan deflection, maximum slipping and final load carrying capacity are reflected in the next paragraph. As the testing results are listed in Table 22.3.

Compound Beams with IPE-120 Sections

The samples CCB-I together with CNB-I be presently inspected to observe the result to make use of rolled steel beams over the performance of compound beams. Material of IPE-120 steel section was utilised to shape up sample of CNB-I and IPE-120 section was shape extent to rolled steel beam of expansion ratio of 0.42 at sample CCB-I.In The listed Figure 22.9 indicate that sample CNB-I is having not succeeded in the bending failure mode. Crack of starting happened in the cement slab on load of 25 KN and as the bend cracking increases as of testing load increases to till break down occurs at final loading carrying capacity of 110 KN because of breaking slab of cement.

As of acting loading implemented for testing sample CCB-I it's surpassed the failure loading of sample CNB-I of 110 KN in addition its continue up till failure started as on final load atf 160 KN because of Vierendeel bending failure of rolled steel beams as listed in Figure 22.10. Figure 22.11 demonstrates the order of yielding point in the failure regions. Based upon the plotted results the starting indication showing yield was noted at the Point 1 where an applied loading at 153 KN. By way of the implemented load is increased to 157 KN, then the yielding remained cited at the side edges of hexagonal-openings at Point-2, afterwards opposite side of coroner Point-3. The yielding continue still up to failure of the beams occurred because of Vierendeel bending as of final load at 160 KN.

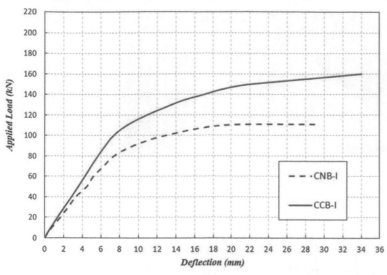

Figure 22.12 Curve of load vs deflection for sample CCB-Iand CNB-I
Source: Author

(a)   (b)

Figure 22.13 Modes of failure for sample CNB-H
Source: Author

(a)   (b)

Figure 22.14 Modes of failure for sample CCB-H. compound beam with HEA-120 section
Source: Author

Compound Beam was made up of as of the rolled steel beams of CCB-I which had and final load carrying capacity of 46% greater than the existing compound beam made up of from main beam of CNB-I.A so fit increases the final load carrying capacity because of axis of neutral compound cross sectional at CNB-I taking place near to cement slab than that of CCB-I.

Curve of Load vs Deflection for sample CCB-Iand CNB-I samples as listed in Figure 22.12. As the sample example of CCB-I, the bending stiffness has enhanced remarkably since the rolled steel beam is in depth of that the main section.

The sample CCB-H and CNB-H was tried to observe the result of utilising wide flange rolled steel beam on the behavior of compound cement-steel beam. AHEA-120 section were used to manufactured sample CNB-H, then HEA-120 section was used to create the sample CCB-H. The sample CNB-H the starting of crack occurred at load 35 KN and the bending failure be fallen at functional loading on it 165-KN is listed in Figure 22.13.

In the beam CCB-H samples breakdown outstanding to the web-post-buckling that happened on the beam of rolled steel by the side of a final loading 200 KN as per listed in Figure 22.14. The mechanism of break down arisen because in web post the high-shear forces of twisting of beam occurs. Because of the bending that occurs towards the bottom and top of web, as shown in Figure 22.15, its indicate horizontal

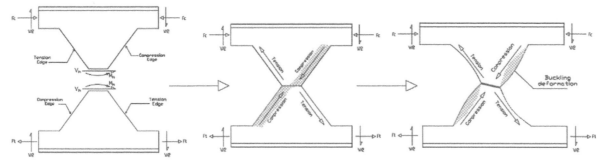

Figure 22.15 Buckling of post of web occurred because of force of shearing
Source: Author

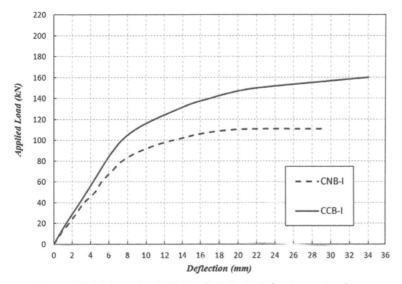

Figure 22.16 Samples CCB-hand CNB-H-deflection vs load curve
Source: Author

Figure 22.17 Sample CCB-IH–failure mode
Source: Author

shear force on the middle depth, or Vh, results in tensile and compressive stresses. Different types of stresses from point to point on opening of hexagonal shape and the angel of max stress differs depends upon web post width. Sfd and Bmd curves of samples CCB-Hand CNB-Has revealed in Figure 22.16 which indicates the enhanced performance of the compound rolled steel beams. The ability of sample beam CCB-Hon 21% more than it CNB-H with respect to final load carrying capacity.

Compound beams with an non-symmetrical rolled steel beams

The CCB-IH sample was made up of using non-symmetrical rolled steel beam fabricated from IPE-120 section. While testing of this beam to investigate the influence using non-symmetrical rolled steel beam on the behavior of compound beams. When Yielding Start, it was identified in the corner at the hexagonal surface of web openings of an implemented load of 180 KN besides yielding continue still final break out occurs at a final load of 186KN because of Vierendeel shear in upper surface T-section of IPE-120 together with web post buckling as shown in picture 17. On the upper T-Section the shear force was opposed by bending of Vierendeel and rest of the force of share was opposed by heavy weight lower surface T-section. Sample of CCB-I His undergone load-deflection giving reading is joined by those of the additional sample as listed in

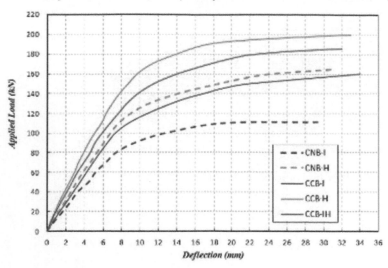

Figure 22.18 Load-deflection curves of all specimens
Source: Author

Figure 22.18. Final capacity of load carrying of sample beam CCB-IH is superior than that of CNB-H and CNB-I as a result of 12% and 69% respectively.

Conclussion

Overall, the strength of the beam (cum rigidity) of compound concrete-steel beam improved noticeably when using rolled steel beam associated total though its utilising the standard main section for the reason that average deepness of the beam of steel was increases on the outside of any additional material. As the compound cement-steel beam remained formed utilising a rolled beam of steel made-up from IPE-120I-section, its final load carrying capacity increases by 46% on over compound be am made up of standard IPE-120 beam. As the side-by-side increases in final load carrying capacity when utilizing the section HEA-120 H by way of main beam remained 21%. Considering the results for investigational revision it is observed especially a non- symmetrical rolled steel beams manufactured from HEA-120 and IPE-120 to enhance of the compound concrete-steel inveigled to rises in the final load carrying capability of 12% and 69% correspondingly, as of that of compound beams was made up of using standard HEA-120 and IPE-120 sections.

References

[1] Lawson,R. M., Lim, J., Hicks, S. J. and Simms, W. I. (2006). Design of composite asymmetric cellular beams and beams with large web. Journal of Constructional Steel Research. 62(6), 614–29.
[2] Al-Thabhawee HW, andAl-Hassan A. (2021). Experimental study for improving behavior of castellated steel beam using steel rings. Pollack Period. 16(1), 45–51.
[3] Sheehan, T., Dai, X., Lam, D., Aggelopoulos,. E., Lawson, M. and Obiala, R. (2016). Experimental study on long spanning composite cellular beamunder flexure and shear. Journal of Constructional Steel Research. 116(1), 40–54.
[4] Hosain, M. U. and Speirs, W. G. (1973). Experiments on castellated SteelBeams. 1973, 329–42.
[5] Nethercot, D.A. and Kerdal, D.(1982). Lateral torsional buckling of castellated beams. Structural Engineering. 60B(3), 53–61.
[6] Hartono, W.and Chiew, S. P. (1996). Composite behavior of half castellated beam with concrete top slab. In: Chan SL, Teng JG, editors. Ad-vances in Steel Structures (ICASS '96). Oxford: Pergamon Press.. p. 437-42.
[7] Al-Thabhawee HW and Mohammed, A. (2019). Experimental study for strengthening octagonal castellated steel beams using circular and octagonal ring stiffeners. https://doi.org/10.1088/1757-899X/584/1/012063.
[8] ASTM E8M. Standard Test Methods for Tension Testing of Metal- lic Materials (Metric). 1999 ed. ASTM; 1999.
[9] Thabhawee, A. L. and Abbas, H. A. (2021). Experimental study for improving behavior of castellated steel beam using steel rings. Pollack Periodica. 16. 10.1556/606.2020.00215.
[10] Thabhawee, A. L. and Abbas, H. A. (2019). Experimental study for strengthening octagonal castellated steel beams using circular and octagonal ring stiffeners. IOP Conference Series: Materials Science and Engineering. 584. 012063. 10.1088/1757-899X/584/1/012063.
[11] Kannoon, A. L . M. and Al-Thabhawee, H. (2022). Investigation of flexural and shear failure modes of tapered castellated steel beams using expansion plates. Eastern-European Journal of Enterprise Technologies. 4. 6-13. 10.15587/1729-4061.2022.262558.

Chapter 23

Analysis of impact of high velocity projectile on layered configuration of armor materials

Divyanshu S. Morghode[a] and D.G. Thakur

Department of Mechanical Engineering, Defence Institute of Advanced Technology, Pune, India

Abstract

Armed forces use commercial vehicles to transport men in highly active counter insurgency environment. These vehicles are vulnerable to attacks by the militants, who generally use AK 47 rifles. Thus to prevent loss of precious lives of personals of our armed forces, need arises to fix an add on armour of optimized thickness on these commercial vehicles so that perforation of 7.62 mm steel core bullets of AK 47 rifle can be arrested within the armour material itself. In the paper, thickness optimisation of layered configuration of armor materials is attempted to prevent the perforation of 7.62 mm bullets having steel core and velocity of impact up to 850 m/s. Numerical analysis is utilised to study the ballistic impact on multilayer configuration of Al 7075-T651 and Al_2O_3. A numerical model was constructed using LS DYNA software and it was validated using experimental results available in the literature. The validated numerical model was further used to conduct ballistic impact simulations on layered configuration of Al 7075-T651 and Al_2O_3 using LS DYNA. During the numerical simulations, the thickness of each material of the layered configuration is varied. The optimum thickness of each material in the layered configuration was predicted based on safety of personals and the total weight of the armour. The numerical simulations results shows that the layered configuration of 20 mm Al_2O_3 and 10 mm Al 7075-T651 can be used as add on armour as its weight penalty is minimum and it prevents perforation of 7.62 mm steel core bullets.

Keywords: Ballistic impact, 7.62 mm steel core projectile, Al 7075-T651, Al_2O_3, layered configuration

Introduction

In India, security forces use commercial vehicles to transport men in highly active counter insurgency environment. These vehicles are vulnerable to small arm attacks by the militants. Thus, to prevent loss of precious lives, the need arises to fix optimised thickness of add on armor to these vehicles so that perforation of 7.62 mm steel core bullets of AK 47 rifle can be prevented.

Study of research papers reveals that under the effect of high velocity impact, materials show nonlinear and dynamic behavior which includes thermal softening, fracture and strain rate hardening. High velocity impact simulations of bullets on different types of armor materials are done using explicit dynamic FE analysis. During such analysis, different contact algorithms and material models are used. Johnson et al [1] performed numerical impact simulations on single layer and multilayers of steel and aluminium, using 7.62 mm armor piercing bullet having impact velocity of 770-950 m/s. LS DYNA software was used for numerical simulations. It was shown that multi layered plates having different materials show greater resistance at same areal density. Rahman et al [2] conducted similar simulations on high strength steel and Al 7075-T6. In this study, velocity of ballistic limit, process of penetration and permanent deformation was studied. Results showed that triple layered configuration achieves maximum weight reduction, without compromising the performance. Appropriate material model for steel core bullet, Al 2024-T351 and ceramic was given by Turhan et al [3]. They used plastic kinematic hardening model for steel core projectile and Al 2024-T351 and Johnson Holmquist model for ceramics. Optimum thickness of adhesive layer of toughened epoxy resin for alumina-aluminium armors was given by Lopez-Puente et al [4]. Mazaheri et al [5] studied the effect on velocity of ballistic limit and energy absorption after wrapping Al foil on the impact face of ceramic tiles. The study shows that there is an increase of 13% in ballistic limit velocity and 11% increase in energy absorption by just 2.4% increase in weight.

A numerical study conducted for development of projectiles and attempted effective penetration of target by using ANSYS explicit dynamics/Autodyn software [6-8]. On referring the research papers, it was observed that minimal study has been done to find out optimum thickness of materials used as armor, so that it can prevent bullet of 7.62 mm caliber from perforating it, at maximum impact velocity. In this paper, an attempt is made to optimise thickness of materials, to prevent the perforation of 7.62 mm bullet having steel core and 850 m/s impact velocity, such that less weight armor and protection against bullets can be achieved simultaneously. In order to achieve this aim, numerical method is used to analyse impact on layered configurations of Al 7075-T651 and Al_2O_3. Following sections explain in detail the numerical approach followed by results observed.

[a]divyanshu25@gmail.com

DOI: 10.1201/9781003450252-23

Numerical Analysis

Geometry

Most researchers, working with standard NATO ammunitions are using 7.62 mm steel core bullets for conducting the impact analysis on various target materials. 7.62 mm steel core bullets are given in Figure 23.1. These types of bullets consist of an inner steel core and a protective outer jacket. This jacket is generally made of brass and is used to engage with the lands of the barrel so that spin can be provided to the bullet during its travel inside the barrel. But this brass jacket doesn't affect the collision between the target material and the bullet. Therefore, the jacket is not considered during the simulations to reduce the time required for computation. Senthil et al [9] also used only steel core shown in Figure 23.2 and compared his results with experimental data obtained using projectile with jacket. The results obtained were matching with the experimental results.

Modeling

LS DYNA software was used in this study. Dimensions of the projectile used in this paper are given in Figure 23.2. The target was designed as a circular plate having diameter equivalent to 152 mm. The materials, used as target were Al 7075-T651 and Al_2O_3. Johnson-Cook (JC) strength, failure models are used for metals and for ceramic Johnson Holmquist model is used. Material properties that were used for projectile and target are given in Table 23.1 and in Table 23.2. To validate the designed model, experimental results available in the literature were utilised.

Meshing

To cover maximum volume and curvature of the bullet and target plate, hexahedron elements are used for meshing. The target was divided into three different regions having three different element sizes. The impact region at the center of the circular target was designed in form of a square having each side of 10 mm. The second and the third region are of circular shape having radius 15 mm and 25 mm respectively. Mesh convergence study was conducted to select the optimised element size of the projectile and the impact region

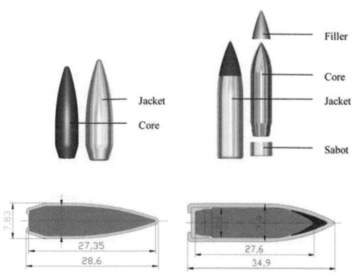

Figure 23.1 Standard NATO 7.622 projectiles (a) lead core, (b) steel core [1]

Source: https://www.researchgate.net/figure/Standard-NATO-762mm-Ball-and-API-Bullets-Depiction_fig3_274257056

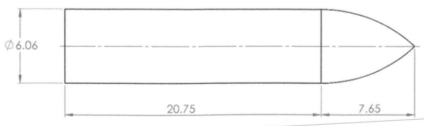

Figure 23.2 7.62 mm AP projectile (equivalent hard core only)

Source: Author

Table 23.1 Material properties

Parameter	Units	Al 7075-T651 [10]	Projectile [9]
Young's modulus E	GPa	71.7	202
Poisson's ratio u	-	0.33	0.32
Density ρ	$\frac{kg}{m^3}$	2810	7850
Johnson Cook strength model			
Yield strength A	MPa	520	2700
Strain hardening parameter B	MPa	477	211
Strain hardening parameter n	-	0.52	0.065
Reference strain rate ε_0	S^{-1}	5e-4	1e-4
Strain rate constant C	-	0.0025	0.005
Reference temperature	K	293	293
Melting temperature	K	893	1800
Thermal softening parameter m	-	1.61	1.17
Specific heat capacity C_P	$\frac{J}{kg.K}$	910	452
Thermal expansion coefficient α	1/K	2.3e-5	1.2e-5
Johnson Cook failure model			
Failure parameter D_1	-	0.096	0.4
Failure parameter D_2	-	0.049	0
Failure parameter D_3	-	3.465	0
Failure parameter D_4	-	0.016	0
Failure parameter D_5	-	1.099	0

Source: Author

Table 23.2 Values of parameters in the Johnson Holmquist ceramic model [11]

Parameter	Units	Al2O3
Density ρ	g/cm^3	3.84
Bulk modulus G	GPa	93
Yield strength A	-	0.93
Strain hardening parameter B	-	0.31
Strain rate constant C	-	0.007
Thermal softening parameter m	-	0.6
Strain hardening parameter n	-	0.64
EPSI	-	1
Reference temperature	K	262
SFMAX	-	1
HEL	MPa	8000
PHEL	MPa	1460
Beta	-	1
Johnson Cook failure model		
Failure parameter D_1	-	0.01
Failure parameter D_2	-	0.7
Failure parameter K_1	-	131
Failure parameter K_2	-	0
Failure parameter K_3	-	0

Source: Author

of target, in order to minimise the computation time. On analysing the results obtained by changing the meshing element size and comparing these results with that given in the literature, it was decided that the size of the elements in the square impact region will be 0.25 mm. Similarly, the size of the elements in the circular portion of radius 15 mm and 25 mm will be 0.5 mm and 1 mm respectively. Element size beyond 25 mm radius will be 2 mm. The final meshing is shown in Figure 23.3.

Impact Analysis

For the present analysis, impact velocity considered is 850 m/s. Rotational velocity of projectile is neglected. Initially, with validated model conditions, the depth of penetration of projectile for 850 m/s is computed by using 50 mm thick target. It is found that depth of penetration is 38.7 mm with failure having ductile hole formation. The penetration simulation image is shown in Figure 23.4. A 20 mm ceramic plate was considered and impacted with same velocity. It is found that single ceramic plate was unable to stop the projectile. The target has experienced failure having brittle fracture, with projectile having 790 m/s residual velocity. Simulation image is shown in Figure 23.5. So, it can be concluded that monolithic targets need more thickness for protection and thus increases the weight of target. So, in this analysis, combination of ceramic front plate and ductile back plate is used.

Layered Configuration

From the simulation conducted on monolithic plates of Al 7075-T651 and ceramic, it is observed that ceramic can absorb more energy from projectile compared to Al 7075-T651. It is also observed that Al

Figure 23.3 Meshing (a) Projectile (b) Target
Source: Author

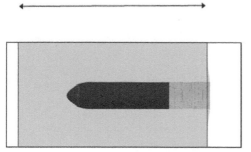

Figure 23.4 Depth of penetration for Al 7075-T651 target
Source: Author

Figure 23.5 Brittle fracture of ceramic material
Source: Author

7075-T651 failure is ductile hole enlargement type while ceramic fails by brittle fracture as it has more toughness and absorb more energy from projectile. For better protection against the projectiles, it is required to use a target which can absorb the more energy as well as, it shouldn't transmit any vibration to rear face of armor i.e., armor should be toughed from front side and should be ductile enough to absorb vibration without transmitting it to back surface of armor. So, combination of these materials in layered form can be useful for better protection against AK 47 projectiles (i.e., 7.62 mm steel core). So, front plate is considered as ceramic and back plate is considered of Al 7075 T651. Initially, both plates are considered as 20 mm thickness each. It is observed that the projectile was successfully stopped at ceramic plate itself and only few stress waves are experienced by Al 7075-T651 plate and projectile has also experienced large damage. The simulation images are shown in Figure 23.6.

   Weight was also one of the parameters of study. From the above study, it can be concluded that, ceramic-Al 7075-T651 layer configuration is good enough to damage the projectile and to stop the perforation. To reduce the weight, there are two options available, either by decreasing the thickness of Ceramic or by decreasing the thickness of Al 7075-T651. From observing the densities of materials, reducing the thickness will reduce the weight of armor effectively, but it will drastically decrease the ballistic resistant properties of armor. So, the thickness of Al 7075-T651 plate has been decreased. For current study, the optimized condition observed for 7.62 mm Steel core projectile is found to be 20 mm front plate of ceramic and 10 mm back plate of Al 7075-T651. The simulations are shown in Figure 23.7.

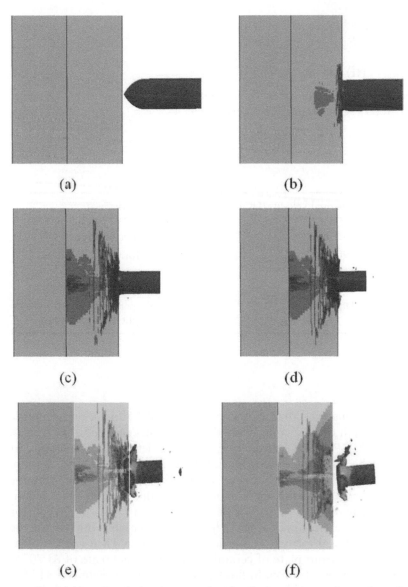

Figure 23.6 Simulation images of projectile impacting on layered configuration of 20 mm front plate of ceramic and Al 7075-T651 back plate of 20 mm. At (a) 0 msec (b) 59e-3 msec (c) 17e-2 msec (d) 26e-2 msec (e) 44e-2 msec (f) 89e-2 msec
Source: Author

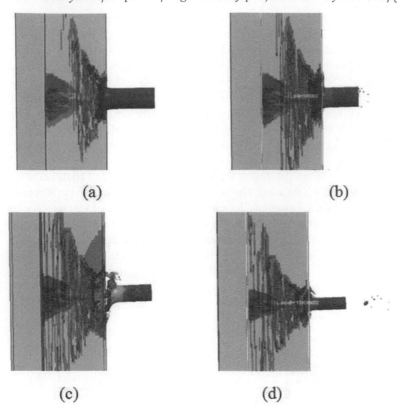

Figure 23.7 Simulation images of projectile impacting on layered configuration of 20 mm front plate of ceramic and 10 mm Al 7075-T651 back plate. At (a) 59e-3 msec (b) 178e-2 msec (c) 435e-2 msec (d) 89e-2 msec
Source: Author

Table 23.4 Results

S.No.	Target	Thickness	Residual velocity/DOP	Weight	Remarks
1.	Al 7075-T651	50 mm	38.7 mm	2.449 kg	No perforation
2.	Ceramic	20 mm	790 m/s	1.433 kg	Perforation
3.	Ceramic + Al 7075-T651	40 mm (20 mm+20 mm)	15.63 mm	2.413 kg	No perforation, Complete protection
4.	Ceramic + Al 7075-T651	30 mm (20 mm+10 mm)	16.08 mm	1.923 kg	No perforation, Optimized weight with protection

Source: Author

Results and Discussion

In the initial study of monolithic plates, it is observed that at impact velocity of 850 m/s, the projectile can penetrate the Al 7075-T651 target to depth of 38.7 mm, while the target thickness is 50 mm. From considered sample geometry, the weight is 2.449 kg. By considering the small deformation tolerance at back surface a 40 mm thickness, it might be sufficient to stop projectile. It is proposed to use 20 mm ceramic front plate, and 20 mm Al 7075-T651 back plate configuration as initial point for analysing the layered configuration. To understand the behavior of ceramic plate, only monolithic ceramic plate is considered, which failed with brittle fracture. Here residual velocity is higher compared to Al 7075-T651 for normalised analysis, this is because monolithic ceramic plate doesn't have any backplate, which made the cracks to be formed and propagate easily in conical formation in thickness direction. Then Layered configuration is analysed, it is observed that 20 mm front plate of ceramic and 20mm backplate of Al 7075-T651 configuration was successfully able to stop penetration. In this configuration the weight is 2.413 kg. From the simulations it is observed that, projectile was able to penetrate the ceramic target upto 15.63 mm only, this is because of support of back plate. In next step, the backplate thickness is reduced to 10 mm, it is observed that penetration depth is increased to 16.08mm, but the projectile was unable to perforate. In this configuration the weight is 1.923kg. The summarised results are given in Table 23.4.

Conclusion

In the study, it is attempted to study the layered armor configuration to protect soldiers from 7.62 mm steel core projectiles. It is observed that monolithic targets of ceramic and Al 7075-T651 can perform ballistic resistance action, but these monolithic plates needed higher thickness, which is directly proportional to weight of armor. So, it is proposed to use layered formation with ceramic front plate and Al 7075-T651 back plate. By considering weight optimization as well as ballistic protection, it is proposed to use 20 mm front plate of Ceramic and 10 mm back plate of Al 7075-T651. This configuration has weight of 1.923kg only, and complete protection against the projectile is achieved.

Acknowledgment

Authors want to thank Defence Institute of Advanced Technology (DU), Pune for providing the support, resources and guidance for the conduct of the study.

References

[1] Flores-Johnson, E., Saleh, M., and Edwards, L. (2011). Ballistic performance of multi-layered mettalic plates impacted by a 7.62-mm APM2 projectile. International Journal of Impact Engineering. 38, 1022-32. doi : https://doi.org/10.1016/j.ijimpeng.2011.08.005.

[2] Rahman, N. A., Abdullah, S., Zamri, W. F. H., Abdullah, M. F., Omar, M. Z., and Sajuri, Z. (2016). Ballistic limit of high strength steel and Al 7075-T6 multil-Layered plates under 7.62-mm Armour Piercing projectile impact. Latin Americal Journal of Solids and Structures. 13(9), 1658-76. doi: https://doi.org/10.1590/1679-78252657.

[3] Turhan,L., Eksik, Ö., Yalçın, E. B., Demirural, A., Baykara, T., and Günay, V. (2008). Comutational simulations and ballistic verification tests for 7.62mm AP and 12.7mm AP bullet impact against ceramic metal composite armours. Structures Under Shock and Impact X. 98, 379-88.

[4] López-Puente, J., Arias, A., Zaera, R., and Navarro, C. (2005). The effect of the thickness of the adhesive layer on the ballistic limit of ceramic/metal armours. An experimental and numerical study," International Journal of Impact Engineering. 32(1-4). 321-36. doi: https://doi.org/10.1016/j.ijimpeng.2005.07.014.

[5] Mazaheri, H., Naghdabadi, R., and Arghavani, J. (2017). Experimental and numerical study on the effect of aluminum foil wrapping on penetration resistance of ceramic tiles. Scientia Iranica. 24(3), 1126-35. doi: https://doi.org/10.24200/sci.2017.4094..

[6] Gálvez, F., Chocron, S., and Cendón, D. (2005). Numerical simulation of tumbling of kinetic energy projectiles after impact on ceramic/metal armours. WIT transactions on modelling and simuations. 40, 9.

[7] Pranay, V. and Panigrahi, S. K. (2022). Effects of spinning on residual velocity of ogive-nosed projectile undergoing ordnance velocity impact. Proceedings of the Institution of Mechanical Engineers, Part C: Journal of Mechanical Engineering Science. 36(2), 1685-97. doi: https://doi.org/10.1177/09544062211020030.

[8] Pranay, V. and Panigrahi, S. K. (2022). Design and development of new spiral head projectiles undergoing ballistics impact. International Journal of Structural Integrity. 13(3), 490-510. doi: https://doi.org/10.1108/IJSI-01-2022-0008.

[9] Senthil, K. Iqbal, M. A. (2021). Prediction of superior target layer configuration of armour steel, mild steel and aluminium 7075-T651 alloy against 7.62 AP projectile. Structures. l(29), 2109-19. doi: https://doi.org/10.1016/j.istruc.2020.06.010.

[10] Jørgensen, K. C. and Swan, V. (2014). Modeling of Armour-piercing Projectile perforation of thick Aluminium plates. In proceedings of the 13th international LS-DYNA Users conference.

[11] Zochowski, P., Bajkowski, M., Grygoruk, R., Magier, M., Burian, W., Pyka, D., Bocian, M., and Jamroziak, K. (2021). Comparison of numerical simulation techniques of ballistic ceramics under projectile impact conditions. Materials. 15(1), 18. doi: https://doi.org/10.3390/ma15010018.

Chapter 24

A finite element analysis of ball type linear motion guide ways subjected to vertical, horizontal and combined loading

Sayaji Patil[1,a], Dr. Sanjay Sawant[2,b] and Dr. Prashant Powar[3,c]

[1]Department of Technology, Shivaji University Kolhapur Institute of Technology (Autonomous), COEK, India

[2]Sant Gajanan Maharaj College of Engineering, Mahagaon Kolhapur, India

[3]Kolhapur Institute of Technology (Autonomous), College of Engineering, Kolhapur, India

Abstract

Linear motion guide ways have drastically reduced frictional resistance and have provided reliable rigidity, accuracy, and repeatability. Machines are widely used in manufacturing of components required for automotive, medical, printing, automation, ship building and aerospace industries demanding higher production rates with cost effective processes. These all aspects increase the need for investigating the guide ways in many depths. Major studies have focused on vertical loading, while few have nominally considered horizontal loading. In this paper linear motion guide way model has been developed and analysed for vertical, horizontal, and combined loading. The analysis provides researchers computed values for deflection and stiffness in vertical, horizontal and combined (resultant) conditions ranging from 1KN till 10 KN force using finite element analysis. The proposed work can benefit designers, machine tool manufacturers, research scholars to have an insight on different sizes, loading, deflection, stiffness, and applications of linear motion guide ways.

Keywords: Vertical load, horizontal load, combined load, deformation, combined stiffness, guide ways, finite element analyses

Introduction

Ball type linear motion guide ways are extensively used for different applications as it overcomes limitations of sliding type guide ways. It has high accuracy, repeatability, acceleration, and decelerations rates and is cost effective with good life. Our work is focused on studying its different loading, deflection, and stiffness characteristics.

Three-parameter life equation can be used to estimate linear motion ball guide system dynamic load rating [1]. The machine tool properties are dependent on stiffness of fixed and movable joints, it has special attribution in constructional system of machines. Finite element analysis is used for calculating static properties of guide ways [2]. Stiffness and damping properties are relatively compared for conventional high precision rolling-element bearing [3]. Optimum algorithm is developed by taking contact stiffness coefficient as design variable. The main aim to minimise sum of square of Eigen frequency difference between experimental modes and finite element required for identification of mode shape [4]. Ball bearings supported by rigid shaft in radial and axial vibrations are studied , the nature of ball and race contact are nonlinear spring type and hertz elastic contact deformation theory is used to obtain stiffness [5]. Load distribution, co-ordinate transformations are used to obtain values of stiffness and damping matrices while evaluation of compatibility relations is done numerically [6]. Stiffness and damping are equivalent dynamic quantities in linear rolling elements exhibiting hysteretic friction and it is examined experimentally by two set -ups [7]. Method of superposition is used for deriving stiffness equation of linear motion guide ways having crowned profile, different profiles such as fourth order power, quadratic, cubic, circular and exponential are investigated [8]. Stiffness and deformations in outward carriages can't be explained with rigid model of conventional type so to overcome this difficulty flexible model of carriage and rail are discussed and explained by using finite element analysis [9]. Finite element model for vertical column-spindle is developed with contact stiffness rolling interface to assess dynamic characteristics of linear guides [10]. Ball type linear motion guide ways subjected to different loading and preloading conditions results in changing contact angle which affects the life and accuracy of guide ways. Study reveals the factors which effect linear guide ways angle of contact [11]. Finite element model and analysis is done on ball type linear motion guide ways for improving the stiffness which is important for rigidity and precision of machines. Preload also plays an important role in improving stiffness of machines reducing deviation in position under different loading conditions. Experiments are performed to prove the correctness of FEA model [12]. Static stiffness results are obtained by simulating slice of full model and merits and demerits are discussed for equivalent models. General analytical solution using hertz is used for validating numerical models [13]. Stiffness and damping are the joint parameters and are complex structures for modeling

[a]patil.sayaji@kitcoek.in, [b]sanjaysawant2010@gmail.com, [c]powar.prashant@kitcoek.in

DOI: 10.1201/9781003450252-24

, finite element analysis are used to evaluate configurations of different mechanical systems for dynamic characteristics [14]. Finite element analysis and experiments are performed to evaluate stiffness of linear guide ways[15]. Investigation proposes to determine static and dynamic behavior along with rigidity of guide ways in the stage of design. Theoretical and experimental work for stiffness investigation is carried out on non-preload guide ways [16]. The design and performance of linear guides can be improved by investigating stiffness and wear characteristics. A model is proposed through experimentation and simulations [17]. Static and dynamic characteristics of guide ways are studied through 8-spring equivalent stiffness model [18]. A numerical analysis is done to present different models of guide ways having preloads in carriage and profiled guide rails [19]. Numerical investigations of machine tool ball bearing linear guides is done for machine tools ball bearing linear guides using finite element method using ANSYS software [20]. Manufacturer's catalogue does not provided much data on friction and stick slip behavior of linear ball guide bearings, investigates it in different working conditions [21]. Estimations for stiffness and damping is done with machine drives. Simulated and experimental results are useful for adjusting parameters for CNC machine tool axis drive [22]. Machining efficiency and accuracy of CNC machines is affected directly by static and dynamic loads coming on guide ways as it exhibits complex mechanical behavior due to more rolling interfaces [23].

Stiffness greatly influences accuracy and performance of different manufacturing systems. Stiffness is investigated by experimental and analytical approach [24]. Stiffness is important in dynamic behavior and service life of linear guide ways. Finite element analysis and experimental work is carried out for stiffness investigations [25]. Theoretical analysis and experiment are combined for obtaining equivalent dynamic parameters of guide ways and it is done in four steps [26]. Reviewing of analysis and modeling of ball and roller bearings is done focusing on design and application aspects [27]. A investigations are carried out for newly designed oil groove system in guide ways for understanding damping characteristics [28]. Different Static and dynamic behavior of spindle system having precision ball bearing is investigated by using response impact [29]. Finite element method is used for investigating stiffness of load bearing machine tool system [30]. Static analysis for linear rolling bearing is done by novel five degrees-of-freedom model subjected to external loading [31]. Machine tool performance is influenced directly by guide ways , an attempt is made to establish for single ball–raceway joint surface contact model [32]. Dynamic characteristics of linear guide ways enhance machine tool performance. Investigations are done to develop analytical model design optimization of high-precision linear guide way [33]. The review paper addresses load and deflection behave in nonlinear manner and stiffness characteristics of linear motion guide ways and propose to build mathematical model for combined stiffness in order to optimize stiffness values [34].

Extensive review done clearly reveals that past work has focused on vertical and to some extent horizontal loading considering different performance characteristics of linear motion guide ways using theoretical, experimental, numerical analysis. Our work is focused on applying vertical and horizontal load simultaneously giving a rise to resultant force coming on to the linear motion ball type of guide ways. The condition has been experienced by guide ways on many machines performing operations. The load ranges selected for study are varied in the range of 1 KN till 10 KN and the testing is done through building model and analysing the guide ways by applying boundary conditions and finding the values for deflection using ANSYS to compute the values of stiffness's.

Finite Element Model and Analysis Approach

Linear guide THK HSR 25 is selected for research work. This series is widely used pairs in market. Angle of contact is 45° making load equally distributed in directions such as vertically up and down, horizontally right and left. Four rows of recirculating balls transport linear motion carriage Figure 24.1 shows the geometry of linear guide THK HSR 25. It consists of grease/oil nipple, end cap, retainers, side seals, recirculating balls, rails etc.

Pre-process performed

The manufacturers catalogue was referred for detailing the model of ball type linear motion guide ways. The model was drafted in CATIA V5/R16. The model consisted of three main parts namely carriage, rail and ball elements drafted separately and then assembled. The model was developed ignoring screws, chamfers, fillets, radius, tapped holes, grease fittings, cap, and end seals. The main purpose behind simplifying the model is to reduce the computational time that will be required for analysis purpose. Figure 24.3 and Table 24.1 shows the detailed geometric parameters of ball type linear motion guide ways. Part drawings are drawn and assembly is done as shown in Figures 24.4-24.7 using CATIA V5/R16.

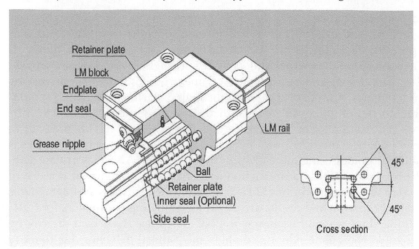

Figure 24.1 Structure of linear motion ball guideway along with contact angle details (Courtesy THK Catalogue)

Source: Author

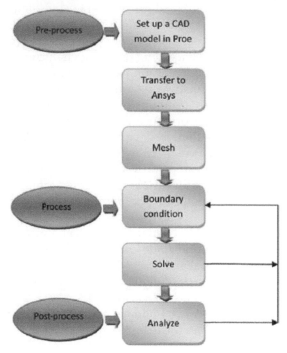

Figure 24.2 Flow chart for finite element analysis (Courtesy: Azhar Shaukharova, Yi Liang, Hutian Feng, Bin Xu (2016)

Source: Author

Table 24.1 Details of HSR 25 (Courtesy THK Catalogue)

Description	Details
Rail length	450
Rail height	22
Rail width (W1) +/- 0.05	23
Block length (L1)	59.5
Block height (K)	30.5
Block width (W)	70
Ball radius	4.760
Number of balls in one row of block	13
Total number of balls	52

Source: Author

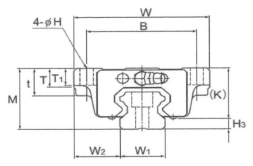

Figure 24.3 Details of HSR 25 (Courtesy THK catalogue)
Source: Author

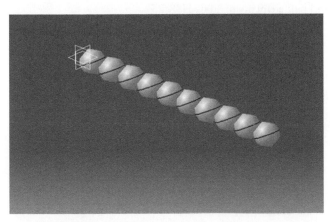

Figure 24.4 Assembly comprising of ball elements of THK HSR 25
Source: Author

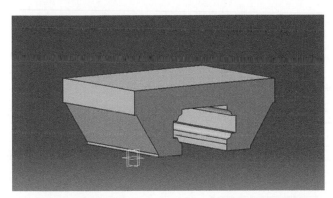

Figure 24.5 CATIA model of THK HSR 25 carriage
Source: Author

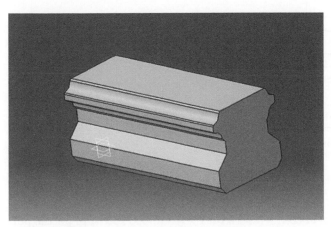

Figure 24.6 CATIA model of THK HSR 25 rail
Source: Author

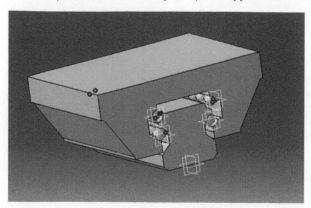

Figure 24.7 Assembly comprising of carriage, rail, and ball elements of THK HSR 25
Source: Author

Table 24.2 Properties of material

Young's modulus(E)	210 GPA
Poisson's ratio	0.3
Density	7800 kg/m³
Frictional co-efficient	0.3

Source: Author

Table 24.4 Component constraints

Components	Constraints
Rail	dof 1,2
Block	dof 1,2,3,4,5,6
Balls	dof 1,2,3,4,5,6

Source: Author

Table 24.3 Meshing properties

Components	Block
Rail	Steel ball
Element type	Tetrahedral
Element	12833
Nodes	34471
Aspect ratio	<3.5
Warpage	<1.25
Jacobian	<0.7

Source: Author

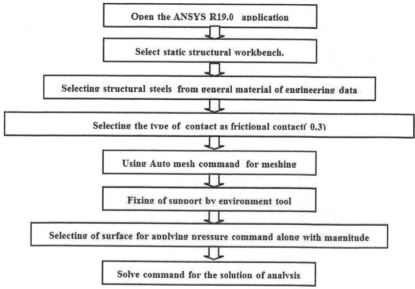

Figure 24.8 Flow chart for process to be followed in ANSYS
Source: Author

Process and post process

The computer aided drafted model is imported from CATIA V5/R16 to ANSYS R19.0 software for the purpose of analysis. For the analysis purpose the boundary conditions are selected for loading steps, constraint conditions, elements, mesh density and material property are finalised (Tables 24.2-24.4). The material

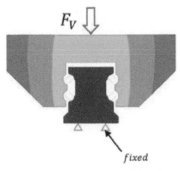

Figure 24.9 Vertical loading conditions to ball type linear motion guideways (Courtesy [30])
Source: Author

Table 24.5 Vertical load and values obtained for deformation and stiffness of THK HSR 25

Load (N) Fv	Total deformation (mm) Δy	Stiffness (N/mm) $Ks = Fv / \delta y$
1000	0.0013475	742115.0278
2000	0.002695	742115.0278
3000	0.0040424	742133.3861
4000	0.0053899	742128.7965
5000	0.0067374	742126.0427
6000	0.0080849	742124.2069
7000	0.0094324	742122.8956
8000	0.01078	742115.0278
9000	0.012127	742145.6255
10000	0.013475	742115.0278

Source: Author

selected for carriage, rail and ball elements is structural steel. The carriage, rail and ball elements are modeled as solid elements. The stiffness of the guide is served by four rows of ball rolling elements. There are 13 rolling ball type elements in each row so in total 52 balls sustain the exerted load. As there are number of ball elements contacting each other and to the carriage and rail considerable computational time will be required.

We have developed, modeled, and analysed full finite element model consisting of 52 parts and 104 contacting surfaces. The contact of surfaces is in between balls and rail, balls, and carriage grooves. The model developed is subjected to loading conditions in vertical, horizontal and combined (resultant) loads varying from 1KN till 10 KN and the cases are explained in Figure 24.9, 24.12, 24.15 along with fixing surfaces. The loads are increased step by step by 1KN. The deformation undergone by the assembly is done by the analysis software by increasing the loads as specified and then computing the values for vertical, horizontal, and combined stiffness for loading conditions. Flow of process for obtaining results for ball type linear motion guide ways subjected to vertical, horizontal and combined (resultant) is shown in Figure 24.8. Results can be obtained for total deformation, directional deformation, and equivalent stress. We have focused on total deformation considering static structural analysis.

ANSYS results

Vertical loading

As shown in Figure 24.9 vertical load (Fv) is applied externally and is transmitted to the ball type linear motion guide ways causing it to deflect (δy). The Vertical load and deflection are considered, the vertical stiffness of ball type linear motion guide ways, Kv, as Kv = Fv / δy. Table 24.5 provides us with the steps in which vertical loads are applied on ball type linear motion guide ways and subsequent values obtained by application of load.

The values for stiffness are tabulated for respective loads. Figure 24.10 shows the graph plotted for vertical loads applied on ball type linear motion guide ways with respect to deformation. The plot exhibits linear trend. Static structural analysis is done in ANSYS for ball type linear motion guide ways and the vertical loads are applied in increasing manner ranging from 1 KN till 10 KN as shown in Figure 24.11.

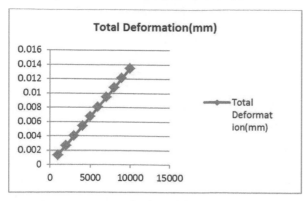

Figure 24.10 Graph plotted for vertical load vs deformation obtained in ANSYS
Source: Author

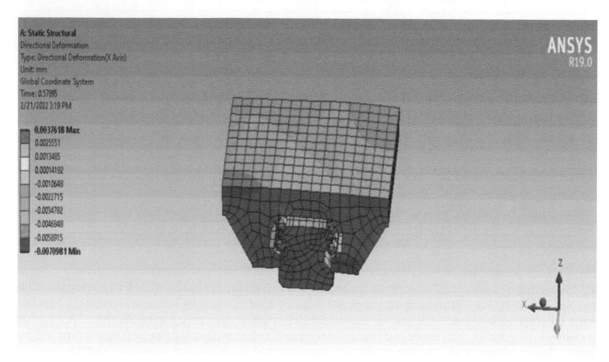

Figure 24.11 Static structural analysis of vertical load applied 1 KN till 10 KN in ANSYS
Source: Author

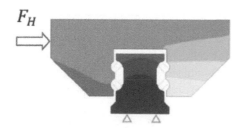

Figure 24.12 Horizontal loading conditions to ball type linear motion guideways (Courtesy Pawel Dunaj, 2019)
Source: Author

Horizontal loading

As shown in Figure 24.12 horizontal load (F_H) is applied externally and is transmitted to the ball type linear motion guide ways causing it to deflect (δx). The horizontal load and deflection are considered, we can express the horizontal stiffness of ball type linear motion guide ways, K_H, as $K_H = F_H / \delta x$

Table 24.6 gives us the steps in which horizontal loads are applied on ball type linear motion guide ways and subsequent values obtained by application of load. The values for stiffness are tabulated for respective loads. Figure 24.13 shows the graph plotted for horizontal loads applied on ball type linear motion guide ways with respect to deformation. The plot exhibits linear trend. Static structural analysis is done in ANSYS

Table 24.6 Horizontal load and values obtained for deformation and stiffness of THK HSR 25

Load (N) F_H	Total deformation (mm) δx	Stiffness (N/mm) $K_H = F_H / δx$
1000	0.0037602	265943.3009
2000	0.0075203	265946.8372
3000	0.01128	265957.4468
4000	0.01504	265957.4468
5000	0.018801	265943.3009
6000	0.022561	265945.6584
7000	0.026321	265947.3424
8000	0.030081	265948.6054
9000	0.033841	265949.5878
10000	0.037602	265943.3009

Source: Author

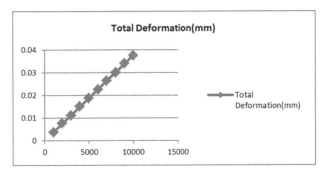

Figure 24.13 Graph plotted for horizontal load vs deformation obtained in ANSYS
Source: Author

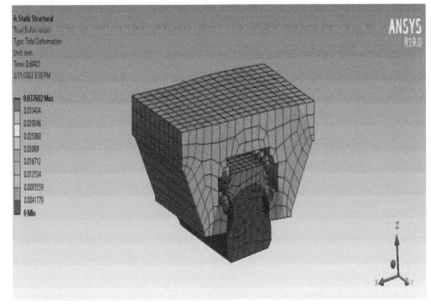

Figure 24.14 Static structural analysis of horizontal load applied 1 KN till 10 KN in ANSYS
Source: Author

for ball type linear motion guide ways and the horizontal loads are applied in increasing manner ranging from 1 KN till 10 KN as shown in Figure 24.14.

Combined Loading (Vertical and horizontal loading)

In our work we have applied vertical and horizontal load simultaneously which results in developing a resultant force on ball type of linear motion guide ways as shown in Figure 24.15. We have considered this

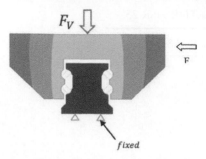

Figure 24.15 Combined loading conditions to ball type linear motion guideways (Courtesy [30])
Source: Author

Table 24.7 Combined loads and values obtained for deformation and stiffness of THK HSR 25

Load (N) Fc	Total Deformation (mm) δc	Stiffness (N/mm) $K_c = F_c / δc$
1000	0.004616	216637.7816
2000	0.0092321	216635.4351
3000	0.013848	216637.7816
4000	0.018464	216637.7816
5000	0.02308	216637.7816
6000	0.027696	216637.7816
7000	0.032312	216637.7816
8000	0.036928	216637.7816
9000	0.041544	216637.7816
10000	0.04616	216637.7816

Source: Author

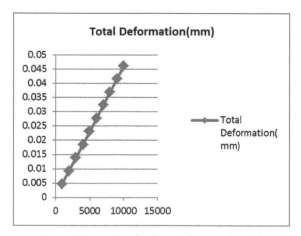

Figure 24.16 Graph plotted for combined loads vs Deformation obtained in ANSYS
Source: Author

situation because till date majority of the work focuses on vertical loads while few have considered horizontal loads to study the static and dynamic characteristics of guide ways. Our attempt is to explore and study the behavioral changes guide ways exhibits when such type of loads is applied. In this paper we have only considered static loads and their behavior. Considering Combined load (δc), deformation in guide way (dc) is obtained and Combined stiffness (Kc), $K_c = F_c / δc$ can be tabulated.

The Table 24.7 gives us the steps in which combined loads are applied on ball type linear motion guide ways and subsequent values obtained by application of load. The values for stiffness are tabulated for respective loads. The Figure 24.16 provides graph plotted is for combined loads applied on ball type linear motion guide ways with respect to deformation. The plot exhibits linear trend. Static structural analysis is done in ANSYS for ball type linear motion guide ways and the combined loads are applied in increasing manner ranging from 1 KN till 10 KN as shown in Figure 24.17.

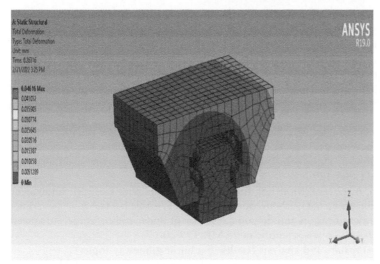

Figure 24.17 Static structural analysis of combined loads applied 1 KN till 10 KN in ANSYS
Source: Author

Conclusion

The results obtained from ANSYS considering boundary and loading conditions are discussed in the conclusion providing us with details that can benefit designers and researchers are put forth below:

The study determines deflection and stiffness of ball type linear motion guide ways subjected to vertical, horizontal, and combined loads. In addition to it, it also provides us with vertical, horizontal, and combined stiffness. The deflection results obtained by ANSYS for vertical, horizontal, and combined loading are 0.0013475 to 0.013475, 0.0037602 to 0.037602, 0.004616 to 0.04616 respectively for loads applied ranging from 1 KN till 10 KN. The stiffness results for vertical, horizontal, and combined loading are tabulated using the formula for THK HSR 25 linear motion ball type guide way. The graph plotted for load versus deflection in all cases vertical, horizontal, and combined exhibit linear trend as obtained through ANSYS results. ANSYS work provided a good see through in the static behavior of ball type of linear motion guide ways. The computational time required for obtaining the solution is ANSYS was less as we have considered a simple geometry of carriage, rail, and ball element, but when it will be applied to actual complex application, the time required will drastically increase. Designers, researchers can use this method for analyzing different make guide ways for obtaining values of deflection and stiffness subjected to various applications. This information is not provided in the manufacturer's catalogue.

In this research the effort was to focus on different loading condition and its results on deflection and stiffness. To complete the research, it is planned to develop experimental set up consisting of zero preload condition and carry out testing to validate the Analysis results and check for nonlinear behavior .We expect that the results obtained from developing experimental set up will exhibit non -linear trend for load versus deflection, change of contact angle and increase in stiffness by reduction in deflection values by employing medium and high preload conditions. The work can be taken further to study the effect of dynamic characteristics of guide ways.

References

[1] Shimizu, S., Saito, E., Uchida, H., Sharma, C. S., and Taki, Y. (1998). Tribological studies of linear motion ball guide systems. Tribology Transactions. 41(1), 49–59. doi: 10.1080/10402009808983721.
[2] Chlebus, E. and Dybala, B. (1999). Modelling and calculation of properties of sliding guideways. International Journal of Machine Tools and Manufacture. 39(12), 1823-39.
[3] Prashad, H. (2002). Relative comparison of stiffness and damping properties of double decker high precision and conventional rolling-element bearings. [Online]. Available: from www.elsevier.com/locate/triboint.
[4] Lin, Y. and Chen, W. (2022). A method of identifying interface characteristic for machine tools design. Journal of Sound and Vibration. 255(3), 481–87. doi: 10.1006/jsvi.2001.4200.
[5] Harsha, S. P., Sandeep, K., and Prakash, R. (2003). Effects of Preload and number of balls on nonlinear dynamic behavior of ball bearing system. International Journal of Nonlinear Sciences and Numerical Simulation. 4(3).
[6] Sarangi, M., Majumdar, B. C., and Sekhar, A. S. (2004). Stiffness and damping characteristics of lubricated ball bearings considering the surface roughness effect. Part 1: Theoretical formulation. Proceedings of the Institution of Mechanical Engineers, Part J: Journal of Engineering Tribology. 218(6), 529–38. doi: 10.1243/1350650042794716.
[7] Al-Bender, F. and Symens, W. (2015). Characterization of frictional hysteresis in ball-bearing guideways. Wear. 258(11–12), 1630–42. doi: 10.1016/j.wear.2004.11.018.

[8] Horng, T. L. (2009). An analytical solution of the stiffness equation for linear guideway type recirculating rollers with arbitrarily crowned profiles. Proceedings of the Institution of Mechanical Engineers, Part C: Journal of Mechanical Engineering Science. 223(6), 1351–58. doi: 10.1243/09544062JMES1208.

[9] Ohta, H. and Tanaka, K. (2010). Vertical stiffnesses of preloaded linear guideway type ball bearings incorporating the flexibility of the carriage and rail. Journal of Tribology, 132(1), 1–9. doi: 10.1115/1.4000277.

[10] Lin, C. Y., Hung, J. P., and Lo, T. L. (2010). Effect of preload of linear guides on dynamic characteristics of a vertical columnspindle system. International Journal of Machine Tools and Manufacture. 50(8), 741–46. doi: 10.1016/j.ijmachtools.2010.04.002.

[11] Shaw, D. (2011). Theoretical Study of contact angles of a linear guideway. Structural Longevity. 5(3), 139-45. https://doi.org/10.3970/sl.2011.005.139

[12] Shaw, D. and Su, W. L. (2011). Study of stiffness of a linear guideway by FEA and experiment. Structural Longevity. 5(3), 129-38. https://doi.org/10.3970/sl.2011.005.129

[13] Dadalau, A., Groh, K., Reuß, M., and Verl, A. (2012). Modeling linear guide systems with CoFEM: Equivalent models for rolling contact. Production Engineering. 6(1), 39–46. doi: 10.1007/s11740-011-0349-3.

[14] Li, L. and Zhang, J. R. (2011). Parameters identification and dynamic analysis of linear rolling guide. Advanced Materials Research. 199–200, 7–12. doi: 10.4028/www.scientific.net/AMR.199-200.7.

[15] Shaw D. and Su, W.-L. (2012). Experimental results and analysis results of a linear guideway topic.

[16] Shaw, D. and Su, W. L. (2013). Stiffness analysis of linear guideways without preload. Journal of Mechanical Sciences. 29(2), 281–86. doi: 10.1017/jmech.2012.136.

[17] Tao, W., Zhong, Y., Feng, H., and Wang, Y. (2013). Model for wear prediction of roller linear guides. Wear. 305(1–2), 260–66. doi: 10.1016/j.wear.2013.01.047.

[18] Jeong, J., Kang, E., and Jeong, J. (2014). Equivalent stiffness modeling of linear motion guideways for stage systems. International Journal of Precision Engineering and Manufacturing. 15(9), 1987–93. doi: 10.1007/s12541-014-0555-y.

[19] Pawełko, P., Berczyński, S., and Grzadziel, Z. (2013). Modeling roller guides with preload. Archives of Civil and Mechanical Engineering. 14(4), 691–99. doi: 10.1016/j.acme.2013.12.002.

[20] Mahdi, R., Stephan, K., and Friedrich. B. (2015). Numerical investigations on the static stiffness of industrial ball bearing linear guides. Annals of DAAAM and Proceedings of the International DAAAM Symposium. 608–13, doi: 10.2507/26th.daaam.proceedings.082.

[21] Mahdi, R., Stephan, K., and Friedrich, B. (2015). Experimental investigations on stick-slip phenomenon and friction characteristics of linear guides. Procedia Engineering. 100, 1023–31. doi: 10.1016/j.proeng.2015.01.462.

[22] Holroyd, G., Pislaru, C., and Ford, D. G. (2015). Determination of stiffness and damping sensitivity for computer numerically controlled machine tool drives. 217(10). https://doi.org/10.1243/095440603322517171

[23] Sun, W., Kong, X., Wang, B., and Li, X. (2015). Statics modeling and analysis of linear rolling guideway considering rolling balls contact. Proceedings of the Institution of Mechanical Engineers, Part C: Journal of Mechanical Engineering Science. 229(1), 168–79. doi: 10.1177/0954406214531943.

[24] Rahmani, M. and Bleicher, F. (2016). Experimental and analytical investigations on normal and angular stiffness of linear guides in manufacturing systems. Procedia CIRP. 41, 795–800. doi: 10.1016/j.procir.2015.12.033.

[25] Shaukharova, A., Liang, Y., Feng, H., and Xu, B. (2016). Study of Stiffness of linear guide pairs by experiment and FEA. World Journal of Engineering Research and Technology. 4(3), 115–28. doi: 10.4236/wjet.2016.43d015.

[26] Xu, D. and Feng, Z. (2016). Research on dynamic modeling and application of kinetic contact interface in machine tool. Shock and Vibration. doi: 10.1155/2016/5658181.

[27] Hong S. W. and Tong, V. C. (2016). Rolling-element bearing modeling: A review. International Journal of Precision Engineering and Manufacturing. 17(12). 1729–49. doi: 10.1007/s12541-016-0200-z.

[28] Ke, N., Feng, H. T., Chen, Z. T., Ou, Y., and Zhou, C. G. (2017). Real contact total length of linear motion roller guide without preload based on Greenwood-Williamson rough contact model. Proceedings of the Institution of Mechanical Engineers, Part C: Journal of Mechanical Engineering Science. 231(22), 4274–84. doi: 10.1177/0954406216664548.

[29] Lee, C. B., Zolfaghari, A., Kim, G. H., and Jeon, S. (2019). An optical measurement technique for dynamic stiffness and damping of precision spindle system. Journal of the International Measurement Confederation. 131, 61–68. doi: 10.1016/j.measurement.2018.08.049.

[30] Dunaj, P., Berczyński, S., Pawełko, P., Grządziel, Z., and Chodźko, M. (2019). Static condensation in modeling roller guides with preload. Archives of Civil and Mechanical Engineering. 19(4), 1072–82. doi: 10.1016/j.acme.2019.06.005.

[31] Kwon, S. W., Tong, V. C., and Hong, S. W. (2019). Five-degrees-of-freedom model for static analysis of linear roller bearing subjected to external loading. Proceedings of the Institution of Mechanical Engineers, Part C: Journal of Mechanical Engineering Science. 233(8), 2920–38. doi: 10.1177/0954406218792573.

[32] Wang, J., Zhang, G., Fan, H., Fan, Z., and Huang, Y. (2021). Effect of off-sized balls on contact stiffness and stress and analysis of the wear prediction model of linear rolling guideways. Advances in Mechanical Engineering. 13(8). doi: 10.1177/16878140211034433.

[33] M. Xu *et al.*, (2021). Model and nonlinear dynamic analysis of linear guideway subjected to external periodic excitation in five directions. Nonlinear Dynamics. 105(4), 3061–92. doi: 10.1007/s11071-021-06796-3.

[34] Patil, S. B. and Sawant, S. H. (2022). A review on non-linear and mathematical modelling for stiffness characteristics of linearmotion guide ways. Journal of Mines, Metals and Fuels. 70(8A), 1-479. doi: 10.18311/jmmf/2022/32013

Chapter 25

Potential of nano-fluid based minimum quantity lubrication in machining difficult-to-cut materials: a review on the perception of sustainable manufacturing

A. P. Vadnere[1,a] and Shyamkumar D. Kalpande[2,b]

Research Scholar, Mechanical Engieering Research Centre, MET's Institute of Engineering, Nashik, (SPPU), India

Professor, Department of Mechanical Engineering, GGSFCOE & RC, Nashik, India

Abstract

The conception of advanced materials reinvented material science and accelerated industrial development. Even though advanced materials have improved properties over most engineering materials, such as high strength, stiffness, heat capacity, toughness, hardness, corrosion, oxidation, fatigue resistance, and so on. These materials are extremely challenging to the machine for the reasons mentioned above, frequently referred to as "difficult-to-cut," "hard-to-machine," or "challenging materials." Obtaining superior quality, precise dimensions, a smooth surface, fast production speeds, and cost reductions are all priorities that contemporary manufacturing companies are currently working to achieve. The cutting fluid is a critical element of any machining process since it cools the tool and work surface by removing the chips from the high-temperature area. With lower use of coolants and the right selection of cutting parameters, MQL machining can achieve better operating conditions while reducing production expenses. This research seeks to assess the prior studies and investigate MQL with nanofluids as potential candidates to increase the effectiveness of various machining techniques on challenging materials. Finally, a conclusion has been attained with some suggestions based on prior studies, which will be helpful for any further research in this area.

Keywords: Difficult-to-cut materials, minimum quantity lubrication, nanofluid, sustainable machining

Introduction

The metal-cutting industry is becoming increasingly interested in the topic of sustainable and eco-friendly machining. The best way to reduce negative environmental and social impacts is to minimise the overuse of cutting fluids. According to researchers, the majority of cutting fluids are obtained from mineral oil; therefore, dumping them into the environment without first recycling them may cause environmental destruction. In addition, the high cost of treating spilled oil increases production costs [8, 11]. Thus, achieving overall sustainability in all industrial sectors is essential due to a wide range of establishing and evolving factors, such as strict environmental and occupational safety and health regulatory frameworks, a growing demand from consumers for environmentally friendly products, and the global depletion of non-renewable resources [3]. The concepts of "sustainability" and "sustainable development" have gained popularity recently. Adaptation to sustainability will promote an effective alignment between the environment, health, safety, and economics in organisations to generate new resources. As a result, it is an effort to provide a concise summary of the extensive studies reported on MQL applications of cutting fluids based on nanofluids for various machining processes [4, 31].

We have covered the advantages of the environmentally friendly MQL technique with nanofluid in this review paper, which is a strong replacement for conventional lubrication methods. As shown in Figure 25.1, the conclusions of the study represent sustainable machining (including environmental, safety, and economic aspects) for the improvement of overall quality when machining hard-to-machine materials.

Machining Challenges

As shown in Figure 25.2, the processing of hard materials is very expensive and labor-intensive. Besides, manufacturing methods that are easy to use for a variety of engineering materials might not always be effective when used on hard materials due to their exceptional characteristics, such as good strength, stiffness, toughness, hardness, corrosion, oxidation, fatigue resistance, heat capacity, etc. One of the most prominent manufacturing processes is machining, which entails removing material from a work piece's surface layer by layer with a wedge-shaped cutting tool as per the industrial requirement. In advanced materials like superalloys and steels, cutting tools quickly degrade, creating a number of challenges for part quality and process efficiency [9, 36]. The heat and pressure produced during the procession are the causes of this rapid deterioration [21]. Rapid tool wear is a challenge for the machining industry because of the

[a]apvadnere@rediffmail.com, [b]shyamkalpande@gmail.com

DOI: 10.1201/9781003450252-25

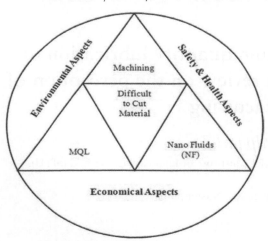

Figure 25.1 Perception of sustainable manufacturing
Source: Author

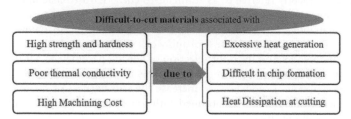

Figure 25.2 Significance of difficult-to-cut (challenging) material
Source: Author

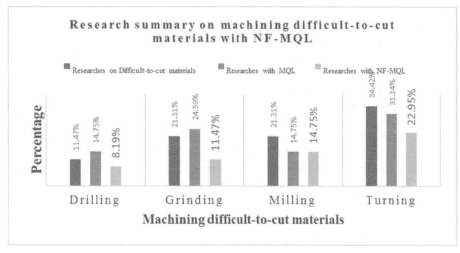

Figure 25.3 Research on machining difficult-to-cut material using NF-MQL
Source: Author

corrosive nature of these materials. Several studies cited in the review focused on machinery for these challenging materials.

As shown in Figure 25.3, the current review offers a clear insight into the potential of MQL and NF-MQL to address the primary issues related to machining challenging materials for environmentally friendly manufacturing processes.

Literature Summary on MQL Assisted Machining

The substantial quantity of heat produced while machining hard-to-cut materials is a major hurdle in identifying cutting tools and cutting conditions while keeping the performance specification, production efficiency, and appropriate production volume. Therefore, the progression of hard machining processes depends heavily on providing the cutting zone with adequate cooling and lubricating effects [26]. Cutting tools are used in machining operations to eliminate surplus material and obtain the desired expected outcome. Tool wear and failure will consequently happen more quickly. Improper heat regulation impacts tool life. The utilisation of cutting fluids is necessary to decrease tool wear and minimise the cutting force while reducing the high heat developed in the process to enhance the surface characteristics [28, 37-39]. The review paper discusses four fundamental machining operations—grinding, turning, drilling, and milling—on which NF-MQL research has been done. Tables 25.1 through IV provide a review of earlier studies on various machining operations on challenging materials under the NF-MQL condition.

MQL Assisted machining:

Numerous literature works have been reviewed in order to gain a general understanding of the influence of various lubrication techniques on a number of cutting parameters. A review of earlier research on various machining operations on challenging materials under the NF-MQL condition is given in Tables 25.1 through IV. Heat production and high temperatures in the tooltip precede deformation in drilling. High temperatures run the risk of damaging the drill and minimising the surface quality. The selection of cutting fluid used has a noticeable influence on how well the drilling tool works [49]. According to research, it may be challenging to control the intense heat generated during grinding. The heat produced during oxidation and metallurgical transformation might be detrimental to the surface's quality [51, 53-57]. Milling is extensively utilised in production areas including the automobile and aviation industries where quality is vital [3, 48]. Cutting fluids are also essential and utilised in milling operations [30]. Studies revealed that the previous researchers worked on machining challenging materials, considering the crucial input and output variables [21-23]. Related to this, in grinding and milling operations, tools wear (T_w) and tool life (T_L) are the factors that receive the least attention. Evidently, there has been little research on how MQL process parameters affect tool life and chip deformation coefficients (C_{DC}) as an output parameter in milling and grinding operations. The line plot of input-output variables that affect the machining of challenging materials is shown in Figure 25.4.

Table 25.1 Review on grinding difficult-to-cut materials under the nf-mql condition

Ref.	Material used	Machining used	Input (cutting) parameters			Output parameters (response)					
			C_S	F_R	DOC	Ra	G_F	T_W	T_L	C_T	C_{DC}
13 (2010)	100Cr6	CNC	√	-	√	√	√	-	-	-	-
28 (2015)	Ti-6A1-4V	Surface Grinder	√	-	√	√		-	-	-	√
51 (2013)	Inconel 751	Surface Grinder	√	-	√	√	√	-	-	√	
52 (2008)	Ti-6Al-4V	Surface Grinder	√	-	√	√	√	-	-	-	-
53 (2013)	AISI 4340	Surface Grinder	√	-	√	√		-	√	-	-
54 (2013)	AA6061	Surface Grinder	√	-	√	√	√				√
55 (2010)	EN8,M2, EN31	Surface Grinder	√	-	√	√	√	-	-	√	-
56 (2010)	100Cr6	Surface Grinder	√	-	√	√	√	-	-	-	-

Source: Author

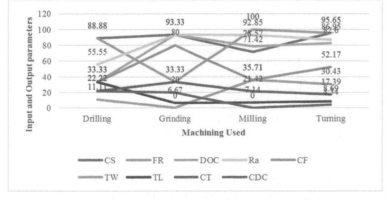

Figure 25.4 Line plot of input-output variables that influence the machining of challenging materials
Source: Author

Literature Review on Sustainable Machining with MQL and Nanofluids

To enhance the lubricating properties in accordance with machining requirements, cutting fluids are formulated using a base fluid and additives. Along with cooling and lubricating, they have been frequently used to improve machine processes' productivity and quality. In order to protect the environment worldwide, cutting fluid usage must now be reduced and chlorine-free cutting fluids must be used. To effectively remove the heat produced by the machining process, a substantial amount of lubricant is applied and saturates the entire machining area in conventional lubricating and cooling methods. Cutting fluid was used in copious amounts when lubricating using the traditional flooding method. The health of the machining center operator(s), environmental risks, excessive cutting fluid waste, and high production expenses are all consequences of this method [6]. Sustainable manufacturing can be accomplished in manufacturing by reducing power consumption, extending tool life, and enhancing workpiece surface quality. Recently, new substitutes have been created to address the principal shortcomings of conventional cutting fluids. MQL has undergone extensive technical evaluation [14]. Numerous issues with conventional (flood) cooling included the reachability of the cutting fluid, as well as numerous other negative effects like environmental pollution, a variety of operator-specific diseases, unsafe working conditions, and problems with waste disposal [19].

MQL is an alternative strategy that involves spraying a combination of air-mist onto the cutting area, which has turned out to be a more effective choice. It has drawn significant recognition from all over the world. On the other hand, using MQL during the machining process can significantly lower the overall cost of manufacturing [6, 10]. In MQL systems, there are two ways to combine air and lubricant: i.e., external feeds and internal feeds. The amount of dangerous and harmful compounds is constantly increasing as companies grow, influencing both the ecological systems and the operator's health. As a result, "sustainable" manufacturing" concepts have become more important in modern industrial operations. The best approaches to improve sustainability performance are reduced energy use, recycling of industrial wastes, efficient use of industrial resources, advancements in lubricating technology, and optimal use of natural resources. As a result, MQL technology effectively eliminates the mist generation issue that is typically associated with flood lubrication, making it a suitable replacement for conventional lubrication [31, 36]. Nanofluids have recently attracted more attention, in a variety of industries, including automobile, farming, pharmaceuticals, machining, electronics, and more. The addition of various nanoparticles to fluids has demonstrated tremendous benefits, subsequently reducing power consumption [40-41].

The unique characteristics and incredibly small sizes of the nanoparticles have provided an important area of study that will undoubtedly lead to the creation of new and more effective systems in the head, which makes them of considerable interest to researchers [26]. In comparison to traditional colloidal suspensions, nanofluids have certain significant advantages, including great stability, less particle clogging, and strong heat transmission capacities [18]. The blockage of tiny channels brought on by the significant agglomeration of solid particles is the biggest limitation of conventional flooded lubrication systems. In addition to these, nanoparticles have an enormous surface area over which a heat transfer process between the particle and its environment occurs. Still, there are some challenges that need to be solved for commercial acceptance of nanofluids like the stability and operational performance of nanoparticles [25-27]. The NF-MQL process parameters that affect the machining of challenging materials are shown in Figure 25.5. It is revealed that earlier researchers studied the machining of challenging materials while considering MQL process variables like nozzle diameter, nozzle angle, nozzle location, the quantity of nozzles, air flow velocity, and pressure, among others. In all machining operations, air flow velocity and pressure have been widely recognised as MQL process parameters by prior researchers. Researchers need to make improvements to and gain a better understanding of a number of factors through commercialising these types of advanced fluids, including chemical stability, thermo-physical properties, toxicity, availability, compatibility with the base fluid, and nanofluid cost [17, 19].

A Brief Review of the Findings from the Previous Research

For the machining of challenging materials, there has been a demand in the past few years to reduce the consumption of cutting oils. However, characteristics like high hardness and low thermal conductivity have a detrimental effect on the induced surface quality and tool life, which have an adverse effect on the general machinability of these materials. From the detailed literature review, the following findings can be summarised:

- The previous researchers revealed their work on various hard-to-machine materials like; nickel-based superalloys (i.e., Inconel 718, Inconel 625, Hastelloy C276), titanium based superalloys (i.e., AA6061-T6), manganese based hardened steels (i.e. 60Si2Mn), medium and hardened alloy steels (i

.e., AISI 1020, AISI 4340, AISI 9310, 90CrSi steel), etc. for different machining operations, especially turning, drilling, and grinding.

- Numerous parameters and how they interact with one another have an impact on cutting performance, particularly when working with challenging materials. Therefore, using the right cutting fluid in machining with the best cutting parameters will enable you to achieve better conditions at the lowest possible cost and with the optimal amount of lubricant or coolant.
- According to studies, earlier researchers had tested MQL machining on challenging materials using techniques like grinding, turning, drilling, and milling with output parameters including surface finish, tool wear, tool life, cutting temperature, chip deformation coefficient, and cutting forces, among others. Surface roughness has been noted as a key output parameter in all machining operations by prior researchers.
- The use of MQL with hybrid nanofluids for the machining of various metals and alloys, can be studied further.
- The literature study also reveals that there are currently few investigations on the use of NF-MQL in machining employing multi-criteria decision making (MCDM) methods like response surface methodology (RSM). Through MCDM techniques like gray relational analysis (GRA), analytical hierarchical process (AHP), TOPSIS, etc., it would be possible to analyse the performance parameters.

Concluding Remarks with Further Research Scope

As has been previously mentioned, low productivity, poor surface integrity, and short tool life are frequently associated with the machining of challenging materials used in the automobile and aviation fields. The use of cutting oils is a common method for enhancing mechanical characteristics; however, the risks to the environment and operator's health posed by conventional cutting fluids, as well as emerging governmental regulations, have led to rising machining expenses. While cutting challenging materials, selecting the best machining parameters for quality, productivity, and profitability is extremely important. In accordance with recent and previous studies, Figure 25.5 illustrates the key machining parameters of NF-MQL-assisted machining processes that affect various machinability characteristics. Tool wear, tool life, cutting temperature, chip deformation coefficient, cutting forces, surface integrity, etc. are a few examples of machinability characteristics that have been examined by previous researchers. The various parameters that affect or to be considered for NF-MQL machining of challenging materials are categorized as MQL process parameters (e.g., method of spraying, nozzle location, nozzle angle, the quantity of nozzles, the diameter of the nozzle, air and fluid flow velocity, air pressure), NF parameters (e. g., size of nanoparticles, the base fluid used, stability), workpiece/specimen (e.g. material, hardness, size), etc. According to published research, the MQL system opens possibilities as a cutting fluid substitute [26, 33-35]. Utilising MQL, which emphasises lubricant properties over coolant properties, heat removal is primarily accomplished by compressed air. The outcomes of this review can be concluded as:

Most of the metal-cutting industries still employ the traditional lubrication method for machining. Only a very small number of contemporary workshops and industries use MQL because it is more complicated and expensive to implement initially than traditional flooded lubrication. This issue might be solved if more work is done in the small- and medium-scale production industries.

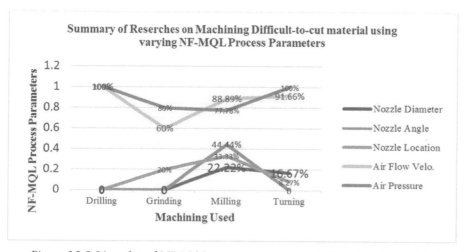

Figure 25.5 Line plot of NF-MQL process parameters influencing the machining challenging materials
Source: Author

- It has been found that applying nanofluids to the machining zone and adding the proper lubrication can enhance the tribological properties of a variety of challenging materials.
- The literature review also reveals that there are currently few studies on the use of MQL with nanofluids for challenging materials. Fewer researchers are studying milling techniques. Therefore, more research is needed to understand how nano-lubricants affect this process.

Table 25.2 Review on turning difficult-to-cut materials under the NF-MQL condition

Ref.	Material used	Machining used	Input (cutting) parameters			Output parameters (response)					
			C_S	F_R	DOC	Ra	G_F	T_W	T_L	C_T	C_{DC}
[1]	Inconel 718	CNC Lathe	√	√	√	-	-	-	-	√	√
[4]	Ductile Iron	Lathe	√	√	√	√	-	√	-	-	√
[5]	Inconel 718	CNC Lathe	√	√	√	√	-	-	-	-	-
[7]	AISI 4340	CNC Lathe	√	√	√	√	-	√	-	-	-
[9]	Inconel 718	CNC Lathe	√	√	√	√	-	-	√	-	-
[10]	AISI 9310	Lathe	√	√	-	√	-	-	√	√	√
[11]	90CrSi	Lathe	√	-	-	√	√	-	-	-	-
[12]	AISI 1020	Lathe	-	√	√	√	-	-	-	√	-
[18]	Inconel 718	CNC	√	√	√	√	-	-	-	-	√
[20]	EN8 Steel	Lathe	√	√	√	√	√	√	√	-	-
[24]	Alloy Steel 4140	CNC Lathe	√	√	√	√	-	-	-	-	-
[29]	NICROFER C263	CNC Lathe	√	√	√	√	-	-	-	√	-
[42]	Ti-6Al-4V	CNC Lathe	√	√	√	√	√	-	-	-	-
[43]	AA6061	CNC Lathe	√	√	√	-	√	-	-	-	-

Source: Author

Table 25.3 Review on drilling difficult-to-cut materials under the Nf-MQL condition

Ref.	Material used	Machining used	Input (cutting) parameters			Output parameters (response)					
			C_S	F_R	DOC	Ra	G_F	T_W	T_L	C_T	C_{DC}
[2]	Stain steel 304	CNC	√	√	√	√	-	-	√	-	√
[44]	Inconel 718	CNC	√	√	-	√	-	-	-	-	-
[45]	Ti-6Al-4V	CNC	-	√	√	√	-	-	√	-
[46]	Al7075	Drilling	√	√	-	√	-	-	-	√	-
[47]	Al Alloy (A390.0)	CNC	√	√	-	-	-	√	-	-	√
[48]	AISI-1040	CNC	√	√	-	√	-	-	-	-	-
[49]	Hardened Steel	CNC	√	√	-	√	√	√	-	-	-
[50]	B 319 Al alloy	CNC	√	√	-	-	-	-	√	-	√

Source: Author

Table 25.4 Review on milling difficult-to-cut materials under the Nf-MQL condition

Ref.	Material used	Machining used	Input (cutting) parameters			Output parameters (response)					
			C_S	F_R	DOC	Ra	G_F	T_W	T_L	C_T	C_{DC}
[3]	Aluminium 6061	VMC	-	√	√	√	-	-	-	-	-
[6]	AA 6061-T6	VMC	-	√	√	√	-	-	-	√	-
[8]	60 Si2Mn	VMC	√	√	-	√	√	-	-	-	-
[14	EN-GSJ 700-02	VMC	√	√	√	√	-	√	-	-	-
[15]	SKD 11	VMC	-	√	-	√	-	-	-	-	-
[27]	Hastelloy C276	VMC	√	√	-	√	-	√	√	-	-

Source: Author

- It is evident that research on the effects of performance evaluation of nanofluid-based MQL process parameters (hybrid lubrication) in the machining of challenging materials is still very restricted, particularly for toughened steels utilised in automobile applications.
- Thus, the efforts of industry practitioners toward sustainable manufacturing will be reduced substantially through detailed research on NF-MQL machining on challenging materials.

References

[1] Kadam, G. and Pawade, R. (2018). Chip deformation aspects in relative eco-friendly HSM of Inconel 718. Procedia Manufacturing. 20, 35-40.

[2] Subhedar, D., Patel, Y., Ramani, B., and Patange, G. (2021). An experimental investigation on the effect of Al_2O_3/ cutting oil based nano coolant for minimum quantity lubrication drilling of SS 304. Cleaner Engineering and Technology. 3, 100104.

[3] Ojolo, S., Money D., and Ismail, S. (2015). Experimental investigation of cutting parameters on surface roughness prediction during end milling of Aluminium 6061 under MQL (minimum quantity lubrication). Journal of Mechanical Engineering and Automation. 5(1), 1-13.

[4] Sakharkar, S. Pawade, R., and Brahmankar, P. (2015). Model development and sustainability assessment of minimum quantity lubrication technique in turning of 700/3 austempered ductile iron. Asian Journal of Convergence in Technology. 2(3).

[5] Kadam, G. and Pawade, R. (2017). Water vapor as eco-friendly cutting fluid – parametric explorations in HSM of Inconel 718. Journal of Materials Science & Surface Engineering. 5(7), 679-84.

[6] Safiei, W., Rahman, M., and Rusdan, S. (2018). Experimental investigation of MQL optimum parameters in end milling of AA6061-T6 using Taguchi method. International Journal of Engineering & Technology. 7(4), 186–186.

[7] Patole, P. and Kulkarni, V. (2017). Experimental investigation and optimization of cutting parameters with multi response characteristics in MQL turning of AISI 4340 using nano fluid. Cogent Engineering. 4(1).

[8] Duc, T., Long, T., and Tuan, N. (2021). Performance investigation of MQL parameters using nano cutting fluids in hard milling. Fluids. 6(7), 248.

[9] Kamata, Y. and Obikawa, T. (2007). High speed MQL finish-turning of Inconel 718 with different coated tools. Journal of Materials Processing Technology. 192-193, 281-6.

[10] Khan, M., Mithu, M., and Dhar, N. (2009). Effects of minimum quantity lubrication on turning AISI 9310 alloy steel using vegetable oil-based cutting fluid. Journal of Materials Processing Technology. 209(15-16), 5573-83.

[11] Duc, T., Long, T., and Chien, T. (2019). Performance evaluation of MQL parameters using Al2O3 and MoS2 nanofluids in hard turning 90CrSi steel. Lubricants. 7(5), 40.

[12] Sonowal, D., Sarma, D., Barua, P., and Nath, T. (2017). Taguchi optimization of cutting parameters in turning AISI 1020 MS with M2 HSS tool. IOP Conference Series: Materials Science and Engineering. 225.

[13] Tawakoli, T., Hadad, M., and Sadeghi, M. (2010). Influence of oil mist parameters on minimum quantity lubrication – MQL grinding process. International Journal of Machine Tools and Manufacture. 50(6), 521-31.

[14] Bülent, Ç. and Alaattin, K. (2020). Experimental investigation on nano MoS2 application in milling of EN-GSJ 700-02 cast iron with minimum quantity lubrication (MQL). Journal of Scientific & Industrial Research. 79, 479-83.

[15] Duc, T. and Long, T. (2020). Effects of MQL and MQCL parameters on surface roughness in hard milling of SKD 11 tool steel. International Journal of Mechanical Engineering. 7(10), 28-31.

[16] Velasquez, M. and Hester, P. (2013). An analysis of multi-criteria decision making methods. International Journal of Operations Research. 10(2), 56-66.

[17] Said, Z., Gupta, M., Hegab, H., Arora, N., Khan, A., Jamil, M., and Bellos, E. (2019). A comprehensive review on minimum quantity lubrication (MQL) in machining processes using nano-cutting fluids. The International Journal of Advanced Manufacturing Technology. 105(5-6), 2057-86.

[18] Ali, M. A. M., Khalil, A. N. M., Azmi, A. I., and Salleh, H. M. (2022). Optimization of cutting parameters for surface roughness under MQL, using Al2O3 nanolubricant, during turning of Inconel 718. Materials Science and Engineering. 226, 012067.

[19] Sharma, V., Singh, G., and Sørby, K. (2014). A review on minimum quantity lubrication for machining processes. Materials and Manufacturing Processes. 30(8), 935-53.

[20] Nageswara Rao, D. and Vamsi Krishna, P. (2008). The influence of solid lubricant particle size on machining parameters in turning. International Journal of Machine Tools and Manufacture. 48(1), 107-11.

[21] Patole, P. and Kulkarni, V. (2018). Optimization of process parameters based on surface roughness and cutting force in MQL turning of AISI 4340 using nano fluid. Materials Today: Proceedings. 5(1), 104-12.

[22] Patole, P., Kulkarni, V., and Bhatwadekar, S. (2021). MQL machining with nano fluid: a review. Manufacturing Review. 8, 13.

[23] Siddharth, J. and Kar, S. (2018). Review on application of minimum quantity lubrication (MQL) in metal turning operations using conventional and nano-lubricants based cutting fluids. International Journal of Advanced Mechanical Engineering. 8, 63-70.

[24] Bhosale, B. and Bhedasgaonkar, R. (2020). Sustainable machining of alloy steel 4140 with minimum quantity lubrication (MQL). International Research Journal of Engineering and Technology. 7(9), 3287-92.

[25] Rifat, M., Rahman, M. H., and Das, D. (2017). A review on application of nanofluid MQL in machining. AIP Conference Proceedings. 1919(1).

[26] Long, T. T. and Du, T. M. (2019). The characteristics and application of nanofluids in MQL and MQCL for sustainable cutting. Advances in microfluidic technologies for energy and environmental applications processes. doi: 10.5772/intechopen.90362

[27] Günan, F., Kıvak, T., Yıldırım, Ç., and Sarıkaya, M. (2020). Performance evaluation of MQL with AL2O3 mixed nanofluids prepared at different concentrations in milling of Hastelloy C276 alloy. Journal of Materials Research and Technology. 9(5), 10386-400.

[28] Bhargavi, A. and Prashantha Kumar, S. (2015). Application of nano cutting fluid under minimum quantity lubrication (MQL) technique to improve grinding of Ti – 6Al – 4V alloy. International Journal of Engineering Research & Technology (IJERT). 3(19), 1-4.

[29] Bose, P. S. C. Rao, C S P., and Jawale, K. (2014). Role of MQL and nano fluids on the machining of NICROFER C263. 5th International & 26th All India Manufacturing Technology, Design and Research Conference (AIMTDR 2014) December 12th–14th, 2014, IIT Guwahati, Assam, India.

[30] Singh, P., Dureja, J., Singh, H., and S. Bhatti, M. (2019). Nanofluid-based minimum quantity lubrication (MQL) face milling of inconel 625. International Journal of Automotive and Mechanical Engineering. 16(3), 6874-88.

[31] Sen, B., Mia, M., Krolczyk, G., Mandal, U., and Mondal, S. (2019). Eco-friendly cutting fluids in minimum quantity lubrication assisted machining: a review on the perception of sustainable manufacturing. International Journal of Precision Engineering and Manufacturing-Green Technology. 8(1), 249-80.

[32] Srikant, R., Prasad, M., Amrita, M., Sitaramaraju, A., and Krishna, P. (2013). Nanofluids as a potential solution for minimum quantity lubrication: a review. Proceedings of the Institution of Mechanical Engineers, Part B: Journal of Engineering Manufacture. 228(1), 3-20.

[33] Benedicto, E., Carou, D., and Rubio, E. (2017). Technical, economic and environmental review of the lubrication/cooling systems used in machining processes. Procedia Engineering. 184, 99-116.

[34] Ali, N., Teixeira, J., and Addali, A. (2018). A Review on nanofluids: fabrication, stability, and thermophysical properties. Journal of Nanomaterials. 1-33.

[35] Reverberi, A., D'Addona, D., Bruzzone, A., Teti, R., and Fabiano, B. (2019). Nanotechnology in machining processes: recent advances. Procedia CIRP. 79, 3-8.

[36] Hegab, H., Darras, B., and Kishawy, H. (2018). sustainability assessment of machining with nano-cutting fluids. Procedia Manufacturing. 26, 245-54.

[37] Saidur, R., Leong, K., and Mohammed, H. (2011). A review on applications and challenges of nanofluids. Renewable and Sustainable Energy Reviews. 15(3), 1646-68.

[38] Kadirgama, K. (2020). Nanofluid as an alternative coolant in machining: a review. Journal of Advanced Research in Fluid Mechanics and Thermal Sciences. 69(1), 163-73.

[39] Shokoohi, Y. and Shekarian, E. (2016). Application of nanofluids in machining processes-a review. Journal of Nanoscience and Technology. 2(1), 59–63.

[40] Wong, K. and De Leon, O. (2010). Applications of nanofluids: current and future. Advances in Mechanical Engineering. 2, 519659.

[41] Deb Majumder, S., and Das, A. (2021). A short review of organic nanofluids: preparation, surfactants, and applications. Frontiers in Materials. 8.

[42] Liu, Z., Xu, J., Han, S., and Chen, M. (2013). A coupling method of response surfaces (CRSM) for cutting parameters optimization in machining titanium alloy under minimum quantity lubrication (MQL) condition. International Journal of Precision Engineering and Manufacturing. 14(5), 693-702.

[43] Shashidhara, Y. and Jayaram, S. (2013). Experimental determination of cutting power for turning and material removal rate for drilling of AA 6061-T6 using vegetable oils as cutting fluid. Advances in Tribology. 1-7.

[44] Rahim, E. and Sasahara, H. (2011). An Analysis of Surface Integrity when drilling Inconel 718 using Palm Oil and Synthetic Ester under MQL condition. Machining Science and Technology. 15(1), 76-90.

[45] Zeilmann, R. and Weingaertner, W. (2006). Analysis of temperature during drilling of Ti6Al4V with minimal quantity of lubricant. Journal of Materials Processing Technology. 179(1-3), 124-27.

[46] Kilickap, E., Huseyinoglu, M., and Ozel, C. (2011). Empirical study regarding the effects of minimum quantity lubricant utilization on performance characteristics in the drilling of Al 7075. Journal of the Brazilian Society of Mechanical Sciences and Engineering. 33(1), 52-57.

[47] Bagawade, A. 2018. Effects of Machining Parameters on Cutting Forces during Turning of Hardened AISI 52100 Steel using PCBN Tooling. Int. Journal of Engineering Research and Application. 8(2), 16-21.

[48] Xiufang, B., Juan, J., Changhe, L., Lan, D. Z. D., and Haiying, Z. (2021). Lubrication Performance of different concentrations of Al_2O_3 nanofluids on minimum quantity lubrication milling.

[49] Tasdelen, B., Wikblom, T., and Ekered, S. (2008). Studies on minimum quantity lubrication (MQL) and air cooling at drilling. Journal of Materials Processing Technology. 200(1-3), 339-46.

[50] Fox-Rabinovich, G., Dasch, J., Wagg, T., Yamamoto, K., Veldhuis, S., Dosbaeva, G., and Tauhiduzzaman, M. (2011). Cutting performance of different coatings during minimum quantity lubrication drilling of aluminum silicon B319 cast alloy. Surface and Coatings Technology. 205(16), 4107-16.

[51] Balan, A., Vijayaraghavan, L., and Krishnamurthy, R. (2013). Minimum quantity lubricated grinding of Inconel 751 alloy. Materials and Manufacturing Processes. 28(4), 430-35.

[52] Sadeghi, M., Haddad, M., Tawakoli, T., and Emami, M. (2008). Minimal quantity lubrication-MQL in grinding of Ti–6Al–4V titanium alloy. The International Journal of Advanced Manufacturing Technology. 44(5-6), 487-500.

[53] Silva, L., Corrêa, E., Brandão, J., and de Ávila, R. (2020). Environmentally friendly manufacturing: Behavior analysis of minimum quantity of lubricant - MQL in grinding process. Journal of Cleaner Production. 256, 103287.

[54] Hadad, M. and Hadi, M. (2013). An investigation on surface grinding of hardened stainless steel S34700 and aluminum alloy AA6061 using minimum quantity of lubrication (MQL) technique. The International Journal of Advanced Manufacturing Technology. 68(9-12), 2145-58.

[55] Barczak, L., Batako, A., and Morgan, M. (2010). A study of plane surface grinding under minimum quantity lubrication (MQL) conditions. International Journal of Machine Tools and Manufacture. 50(11), 977-85.

[56] Tawakoli, T., Hadad, M., and Sadeghi, M. (2010). Investigation on minimum quantity lubricant-MQL grinding of 100Cr6 hardened steel using different abrasive and coolant–lubricant types. International Journal of Machine Tools and Manufacture. 50(8), 698-708.

[57] Tawakoli, T., Hadad, M., Sadeghi, M., Daneshi, A., Stöckert, S. and Rasifard, A. (2009). An experimental investigation of the effects of workpiece and grinding parameters on minimum quantity lubrication—MQL grinding. International Journal of Machine Tools and Manufacture. 49(12-13), 924-32.

Chapter 26

Study and comparison of physical, mechanical properties of polymer-based bio-composites

Hrushikesh Anil Chari[a], Tanmay Kailas Kanmahale[b] and Dr. A.M.Patki[c]

Sinhgad College of Engineering, India

Abstract

Natural fibers polymer composites (NFPC) using sugarcane bagasse (S) and pineapple leaf (P) woven fabrics reinforced with epoxy matrix were prepared using the vacuum infusion method. Study of different combinations of (S) and (P), such as S-S-S-S-S, P-P-P-P-P, P-S-P-S-P, and S-P-S-P-S were carried out since they are the emerging bio-degradable materials of the present time that are recyclable, sustainable, abundantly available and are pollution-free compared to synthetic fibers. Mechanical tests such as tensile, flexural, compression, impact strengths and physical tests such as water absorption were carried out according to ASTM standards. It was observed that the tensile strength and the flexural strength of the P-P-P-P-P composite showed the best results of 41.87 MPa and 81.89 MPa, respectively, compared to the other three composites. Maximum compressive strength was also achieved in the P-P-P-P-P fibers composite (88.45 MPa) as compared to the other composites but P-S-P-S-P stacking sequence showed the highest impact strength of 84.76 J/m. Water absorption percentage was observed to be higher in sugarcane composite (3.05%) than in pineapple composite, due to the hydrophilic nature of sugarcane fiber and its lower cellulose content.

Keywords: NFRC, pineapple fibers, bagasse fibers, composite

Introduction

In today's fast-growing world, researchers have been progressively looking for more advanced lightweight materials which have value effectiveness, and eco-friendliness with comparative strength, as a substitute to synthetic materials. Natural composites are made up using incorporating two or more natural fibers within a common polymeric matrix. Materials are formed of two or more constituent materials that have distinctly dissimilar chemical and physical properties. As a result, the composite material has qualities that are different from those of the constituent materials. Each component remains unique and well-defined inside the final structure, distinguishing it from material combinations and solid solutions. The matrix (blinders) and the reinforcement are the two primary types of constituent materials. To form a composite, it is necessary to use at least one of the members from each category. The matrix phase both embeds and supports the reinforcements by maintaining their relative placements throughout the process. The matrix's qualities are improved because of the reinforcements' contributions, which manifest in the form of their physical and chemical properties. There are two criteria that must be met to classify composite materials. The first type of composite is determined by the matrix material, such as metal, ceramic and polymer matrix composites. The second type of composite is determined by the reinforcement phases, such as fiber reinforced composites, laminar composites, whisker reinforced composites, and particle composites, which are further subdivided into various categories.

Animal fibers comprise proteins extracted from animals like sheep's wool, goat l's hair, horse's hair, etc while mineral fibers occur from modified fibers procured by minerals. But the plant's fibers can be further classified into seed and bast fibers, leaf fibers (pineapple, palm), grass fibers (oat, barley, bagasse, corn, bamboo, and wheat), and fruit fibers (coir, loofa), as well as all other kinds (woods and roots), are most commonly grown crops in the all parts of the world. These plant fibers are immensely available in a wide range leading to thinking of their use as eco-friendly material.

Eco-friendly materials are sometimes known to as natural fibers polymer composites (NFPC). Since most components used to create NFPCs are derived from living plants and even animal skin. Plant fibers natural fibers with high tensile strengths such as bagasse, kenaf, jute, oil palm, and flax can be included into the production of polymer composites. The elements of PMC are thermosets and thermoplastics. Since their qualities and form may be customised to fit the demands of a particular application, polymer matrix composites have the advantages of being stronger, stiffer, and lighter than conventional metals. The characteristics of composite structures are influenced by the fiber reinforcement as well as the polymer matrix, the qualities of the fib-matrix link, and the manufacturing technique used to create the finished structure.

Natural fibers have become an area of interest because of some superior advantages like their renewability, biodegradability, lower cost, lower density, recyclability, low energy consumption, posing no harm to health, comparable specific strength and stiffness properties, corrosive resistance, and flexibility thus reduce

[a]charihrushi08@gmail.com, [b]kanmahaletanmay@gmail.com, [c]ampatki.scoe@sinhgad.edu

DOI: 10.1201/9781003450252-26

machine wear, etc. Despite having good properties, the qualities and performance of NPCs can be impacted by a variety of things. The main drawback of natural fibers is their hydrophilic nature i.e., the presence of moisture/water. Because they contain the hydroxyl group, natural fibers are naturally water- repellent/ hydrophilic. The hydrophilic nature of fibers can cause low interfacial bonding between reinforcement and matrix which can affect negatively on structure and properties of the composite. This can be neglected or minimised by treating a surface of fibers with various chemical treatments. But these chemical processes are costlier and more time-consuming.

Literature Review

Hussin et al. discovered that natural fiber composites are much less expensive than synthetic composites, more environmentally friendly, biodegradable, readily available, renewable, and lightweight. According to his research, natural fibers can be obtained from animals, plants or even minerals. There are more than 2,000 different species of fiber plants, and most of them are composed of cellulose. Common examples include kenaf, sugarcane, corn, bamboo, banana, hay, flax, hemp, jute, henequen, pineapple leaf, cotton, sisal, and ramie [1].

According to research by Dixit et al., cellulose, hemicellulose, lignin, pectins, and waxes are the primary elements of natural fibers. The physical characteristics of cellulose are greatly influenced by hydrogen bonding, which reinforces the fibers [2]. More than 200 million of tons of sugarcane bagasse are generally produced each year in India alone, according to a work by Devadiga et al. His research also identifies the composition of bagasse fibers to determine their strength. About 25% lignin, 25% hemicellulose, and 50% cellulose make up the sugarcane bagasse. The sugarcane bagasse's high cellulose content and crystalline structure make it a fantastic composite reinforcement [3]. According to Reddy et al., pineapple fibers have a high specific strength and stiffness and are hydrophilic by nature as a result of the high cellulose content (about 70–82%). Furthermore, he also concluded that the pineapple leaf fibers have the best mechanical characteristics of all natural fibers [4]. Due to the easy accessibility of natural fibers in nature, which makes them affordable and sustainable materials, Ahmed et al researches found that NFRCs are better than synthetic fiber reinforced composites (SFRCs). Additionally, according to his findings, NFRCs are ideally suited for ballistics applications due to their excellent impact characteristics and extremely minimal environmental impact [5].

Epoxy resins are superior molecular weight polymers that typically contain at least two epoxide groups, according to research by Mohanavel et al. In order to build the matrix, epoxy resin is used because of its excellent binding abilities between the fibre layers. At room temperature, LY556 is the epoxy resin used. To improve the composite's interfacial adhesion and toughness, HY951 hardener is used. The use of an epoxy resin and a hardener mixture of 10:1 result in a stronger matrix composition [6]. Vacuum bag technology, according to Bere et al. experimental' s research, enables the creation of strong and compact composite materials. Using this technique, the composite fibers have high and consistent mechanical characteristics [7]. The best fiber-to-resin ratio is provided by the vacuum infusion technique, according to Majid et al research. Although VIP offers a cleaner, safer, and more congenial working environment, it is still crucial to operate in a space that is properly ventilated and to use safety gear like respirators and other protective clothing. His studies explain that before resin is added, the materials are set out inside the mold dry and a vacuum is applied. When a complete vacuum has been attained, the resin is practically sucked into the laminate using properly positioned tubing. A variety of tools and materials help this procedure [8].

After doing the above literature survey, it was found that microplastics have a huge impact on pollution. To reduce this problem, bio-composite materials are a promising alternative. In combination with biodegradable plastics, fully bio-based and biodegradable solutions are possible. India is the largest prolific of sugarcane and pineapple across the globe. It has been observed that a large amount of sugarcane and pineapple residuals are being wasted. Today the growing industry is looking for ecological lightweight materials with minimal carbon footprint, which can certainly be provided by bio composites. To bridge the research gap, a proper combination of sugarcane (S) and pineapple (P) layers were selected and composites with an appropriate strength-to-weight ratio were manufactured using Epoxy (LY556 grade) resin and Hardener (HY951 grade). It was also found from the above literature study that the vacuum infusion method is the most convenient for manufacturing because the results are the most accurate. Various mechanical and physical properties were tested to find out the best combination of S and P.

Methodology

Material selection

Natural composites are made up of incorporating two or more natural fibers within a common polymeric matrix. These natural fibers are also known as plant or green fibers. The fibers act as the reinforcement and the polymer acts as the matrix. Finding the correct combination of the reinforcement and the polymer can attribute to creating exceptional material.

Reinforcement

Sugarcane/Bagasse fiber

As an agro-residue, bagasse fibers are abundant, and biocomposites created from such renewable resources have the ability for expansion and greater reliance. It is estimated that India produces about 200 million tons of sugarcane bagasse annually. The sugarcane crop is a sustainable agricultural and natural resource that can be harvested repeatedly. Typically, the composition of bagasse is 25% legnin, 25% hemicellulose, and 50% cellulose. Sugarcane bagasse's superior performance as a composite reinforcement is due to the material's high cellulose content and crystalline structure. A-cellulose makes up around 50% of it, followed by pentosanes at 30% and ash at 2%.

Pineapple leaf fiber

Large amounts of pineapple leaf fiber are harvested each year, but only very little amounts are used for feedstock and power generation. It encourages the agricultural sector to enter markets that are not reliant on food. Its medium-length fibers have a good tensile strength and rigidity, are white in color, and are as smooth and shiny as silk. Its surface is finer than that of other natural fibers, and it effectively absorbs and holds color. Because of the high proportion of cellulose in it, it has a hydrophilic character. Mechanical and fermentation methods are used to extract the fibers from pineapple leaf fibers. About 2-3% of fibers are produced by fresh leaves. After mechanically removing the entire top layer of the pineapple after harvesting, the fibrous cell is retrieved, it has a system of vascular bundles in the design of bunches.

There are numerous chemical components found in pineapple fibers. It is a multicellular, natural polymer called a cellulosic fiber. Along with a few other minor substances like fat, wax, pectin, color pigment, anhydride, pentosan, uronic acid, and inorganic material, lignin makes up the majority of the components. A collection of small, multicellular fibers that resemble thread is called fibers. Pectin is present, which helps to firmly connect these cells together. Pineapple fibers make up cellulose (70–82%), and the fiber structure is same to cotton (82.7%). Pineapple leaf fiber has a better chemical makeup and is a more suitable natural fiber resource across all groups.

Bagasse (sugarcane fiber), Pineapple leaf fibers are used for the manufacturing of composite specimens. The bagasse and pineapple fibers are obtained from Go Green Products, Chennai. The natural fibers used for composite manufacturing are the bidirectional woven mat of 300 and 285 GSM respectively with 0-90° orientation. Properties of natural fibers are given below in Table 26.1.

Matrix

Epoxy (LY556 grade) resin and hardener (HY951 grade) are obtained from Herenba Instruments & Engineers, Chennai which is used as the polymer matrix. Hardener HY951 having a viscosity of 10-20 MPa at 25°C is selected as it possesses good mechanical properties as well as increases the interfacial bonding capacity of resin while epoxy LY556 possesses high strength and low molecular weight. It's biocompatible and eco-friendly as well as it is convenient for both hand lay-up and vacuum infusion processes, having good resistance to atmospheric and chemical degradation. The minimum curing cycle time of resin and hardener is 24⁻48 hours at room temperature or cured for 6 hours at room temperature plus 6 hours at 23°C. Resin viscosity at 25°C is about 10350-12000 MPa and hardener viscosity at 25°C is 10-20 MPa as

Table 26.1 Properties of natural fibers [12]

Sr. No.	Properties	Bagasse	Pineapple
Mechanical			
1.	Tensile strength (Mpa)	20 to 289.8	170 to 1672
2.	Young's modulus (Gpa)	19.69	82
3.	Density (g/cm3)	1.2	1.5
Chemical			
1.	Cellulose %	55.2	81
2.	Hemicellulose %	16.8	-
3.	Lignin %	25.3	12.7

per ISO 2555 and ISO 12058 respectively. Figure 26.1 is of the epoxy resin and hardener used in preparation of the sample [9, 10].

Layup selection

Polymer reinforced with pineapple having all five layers of pineapple fibers i. e., (P-P-P-P-P) as shown in Figure 26.2(b). The second combination is of polymer reinforced with bagasse (sugarcane) having all five layers of bagasse fibers i. e., (S-S-S-S-S) as shown in Figure 26.2(a). The third sample taken into consideration is of five alternate layers of pineapple and bagasse fibers having facial layers of pineapple fibers while core layers were of sugarcane fibers (P-S-P-S-P) as shown in Figure 26.2(c). Again, for the fourth sample taken into consideration is of five alternate layers of bagasse and pineapple fibers having facial layers of bagasse fibers while core layers were of pineapple fibers i.e.(S-P-S-P-S) as shown in Figure 26.2(d). Therefore, pineapple to bagasse fibers ratio in third and fourth stack considered as 3:2.

To maintain uniformity in interfacial layers, different stacking sequences were decided based on fiber content. Pineapple fibers content varies from 14.8-36.37% max. Similarly, bagasse fibers content varied from 14.8-36.37% according to weight fraction given in Table 26.2.

Figure 26.1 Resin and fibers
Source: Author

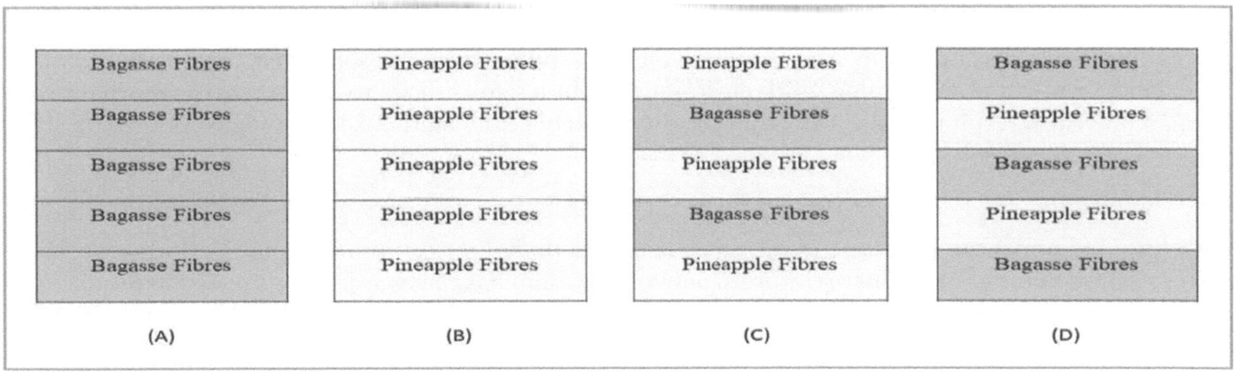

Figure 26.2 Stacking sequence of composite
Source: Author

Table 26.2 Weight fraction of composite laminates

Sr.No.	Stacking sequence	Weight (%)			Weight fraction (%)		
		wf			Wf		
		wp	ws	wm	Wp	Ws	Wm
1.	P-P-P-P-P	40 ± 1	-	63 ± 2	38.83	-	61.16
2.	S-S-S-S-S	-	45 ± 1	70 ± 2	-	39.13	71.5
3.	P-S-P-S-P	24 ± 1	18 ± 1	65 ± 2	22.42	16.82	60.74
4.	S-P-S-P-S	16 ± 1	27 ± 1	62 ± 2	15.22	25.71	59.04

Source: Author

Process selection

Open mold layup

Open mold layup is a cost-efficient and efficient method for producing composites. Both spray-up and hand lay-up are steps in the process. In this procedure, the form and outside surface of the part are provided by a single-sided mold. On the surface of the prepared mold, a gel coat is applied.

Hand lay-up

The most straightforward way to manufacture composites is through hand layup. Usually, fiber preforms are arranged in a mold with a thin anti-adhesive coat to facilitate removal. On a reinforcement material, the resin is either sprayed on or simply applied with a brush. In order to provide a better connection between the subsequent layers of the reinforcement and the matrix materials, the resin is pushed into the fabrics using a roller. In a process known as a "wet layup," each ply is coated with resin and resect or compressed after placement. In a typical hand lay-up, resin often makes up more than 100% of the fabric weight. Any surplus will make the part weaker because resin on its own is exceedingly fragile. Hand lay-up resins typically have higher potential for injury than high molecular weight goods because of their low molecular weights. Due to the resin's lesser viscosity, they also have a higher tendency to pierce skin, clothing, and other materials.

Closed mold layup

Vacuum infusion

The vacuum infusion practice (VIP) is a technique for injecting resin into a laminate using vacuum pressure. It begins with filling the mold with dry components and then vacuuming it before adding resin. After a complete vacuum has been created, the resin is pulled into the laminate through strategically placed tubing. A practical method for producing high- quality composite parts is VIP. Higher quality, improved consistency, good interior finish, shorter cycle times, and cheaper costs are all benefits of VIP. A better ratio of fibers to resin is provided by vacuum infusion. However, vacuum infusion provides limitless setup time. Since a vacuum is applied while the reinforcements are still wet, there is no resin timer. The leaks can be gradually located after applying the bag. Just start the vacuum and move anything that isn't sitting comfortably. Up until the decision is made that it is time to infuse the resin, there are no time restrictions enforced. There is still time to make changes before that moment. The method of vacuum infusion is cleaner. There are no rollers or brushes, so there won't be any spills or splatters. There are also fewer resin fumes to worry about. The only emissions from the resin reservoir are vapors, which are somewhat contained. VIP offers a hygienic, secure, and welcoming work environment, but it's still crucial to operate in a space that's well-ventilated, wear a respirator, and use other necessary safety gear. Figure 26.3 shows how VIP works [11].

Vacuum bagging

With little investment in tooling, the vacuum bagging methodology can produce high-performance, large-scale products made of glass fiber reinforced polymer. Vacuum bagging was initially created to produce big, high-quality composite products. It makes use of the pressure to fulfil special criteria. Vacuum bagging can produce huge and complicated composite components with high quality thanks to its flexible mold tooling

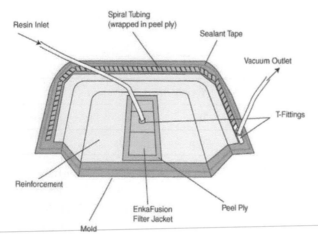

Figure 26.3 Vacuum infusion process [13]

design and material selection. Vacuum bagging may be easily changed to produce various part geometries because it resembles the hand lay-up procedure' open mold. The limitation of vacuum bagging includes vacuum bags, resin tube, flow distribution medium, sealing tape, and peel ply that may not be reusable. The training and experience of the worker have a big impact on how likely it is that air may leak. The injection pressure for the resin has a limit that lies between air pressure and vacuum.

The composites were laminated VIP, which seals laminates together using vacuum pressure to bond the resin matrix. The main motive to choose the vacuum infusion process over the vacuum bagging is that in vacuum bagging we use breather materials which we do not need in vacuum infusion hence making the infusion process much easier. In vacuum infusion, we get a better flow rate.

For polymerisation, the VIP method is used as it offers more benefits than hand lay-up. It reduces void formation, therefore, gets better surface finishing than hand lay-up. Also, it consumes less time for polymerisation, though the setup is a bit more complicated than hand lay-up. In addition, the process allows the resin to flow uniformly over the surface of laminates and least resin is used. The weight ratio of mixing epoxy and hardener is 10: 1 i.e., 1 gm of hardener per 10 gm of resin. Figure 26.4 represents sample preparation using VIP.

The infusion mesh mat of green color acts as flow media which can be used as a channel where extra resin is stored during the process. Pill ply is a fabric material that is set down upon the surface of the composite to absorb additional matrix resin and to help to remove fabricated polymer composite. Two to three layers of high-temperature wax are applied on the glass surface and let dry for some time to ensure easy removal of composite and good surface finishing.

Results

The tests were conducted using specimens of sequential layer P-P-P-P-P, S-S-S-S-S, S-P-S-P-S, P-S-P-S-P. (P: Pineapple and S: Sugarcane). The composite specimen underwent mechanical and physical tests per ASTM standards. Compared to other combinations, the P-S-P-S-P sequence produced the best results overall.

Tensile characteristics of composite specimen

The universal testing machine or UTM was used to perform the tensile testing. The test was carried out at 25°C room temperature with a 5 mm/min loading rate. P-S-P-S-P sample's tensile strength was determined to be 33.62 Mpa. Figure 26.5 represents the P-S-P-S-P sample after tensile test and Figure 26.6 gives the graphical comparison of the different sample. In the P-S-P-S-P type, it has been observed in the results that composite tensile strength has slightly decreased because of less cellulose content. Their chemical composition also influences tensile strength. A higher Strength will be achieved compared to other combinations by reinforcing low strength fibers (bagasse fibers with low% of cellulose) at the facial and core layers with high strength fibers (pineapple fibers with high% of cellulose) at an intermediate level.

Flexure tensile characteristics of composite specimen

To determine the flexure characteristics of the specimen P-S-P-S-P, a three-point bending test was performed. P-S-P-S-P composite's flexural strength was determined to be 65.37 MPa. Figure 26.7 represents the P-S-P-S-P sample after flexure test and Figure 26.8 gives the graphical comparison of the different sample. It has been observed that when we reinforced higher-strength fibers in the facial and core layers of composite laminate, we achieved better bending results than others. As a result, it is obvious that mechanical properties are affected not only by the woven fiber mat but also by the stacking sequence. Thus, the stacking sequence needs to be chosen wisely according to the loading application.

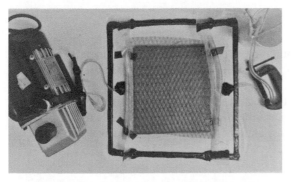

Figure 26.4 Composite manufacturing using vacuum infusion process
Source: Author

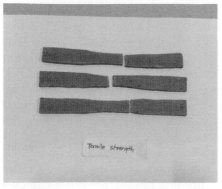

Figure 26.5 P-S-P-S-P samples after testing
Source: Author

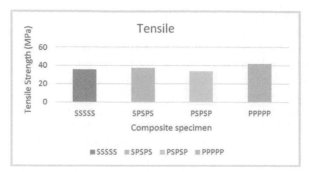

Figure 26.6 Bar diagram of tensile strength of composite specimen
Source: Author

Figure 26.7 P-S-P-S-P samples after testing
Source: Author

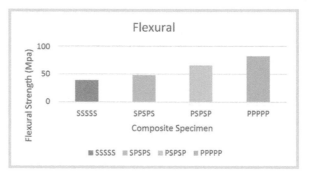

Figure 26.8 Bar diagram of flexural strength of composite specimen
Source: Author

Figure 26.9 P-S-P-S-P samples after testing
Source: Author

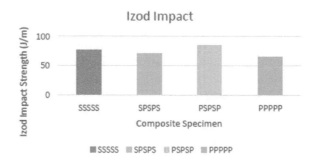

Figure 26.10 Bar diagram of izod impact strength of composite specimen
Source: Author

Impact characteristics of composite specimen

The impact strength of the composite laminate is determined by the energy absorbed during the fracture generated by the impact loads. The impact properties of the specimen P-S-P-S-P were determined using izod impact test equipment, and the stacking of the specimens shows an impact strength of 84.76 J/m. Figure 26.9 represents the P-S-P-S-P sample after the izod impact test, and Figure 26.10 gives the graphical comparison of the different samples. The impact characteristics are mainly determined by the type of fibers, the fiber-matrix interface, the stacking sequence, the dimensions, and the testing apparatus.

Compressive characteristics of composite specimen

The compression strength of specimen P-S-P-S-P was tested using a compression strength testing machine. The maximum compression strength of specimen was found to be 77.66 Mpa. Figure 26.11 represents the

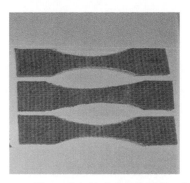

Figure 26.11 P-S-P-S-P samples after testing
Source: Author

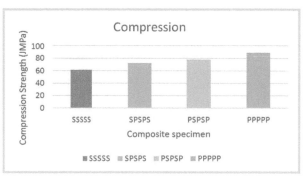

Figure 26.12 Bar diagram of compression strength of composite specimen
Source: Author

Figure 26.13 P-S-P-S-P sample after water-absorption test
Source: Author

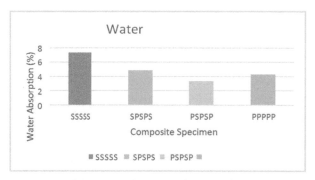

Figure 26.14 Bar diagram of water absorption of composite specimen
Source: Author

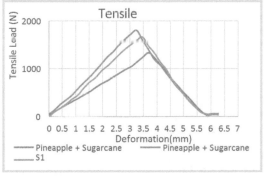

Figure 26.15 Graph of tensile load(N) vs deformation(mm) for P-S-P-S-P samples
Source: Author

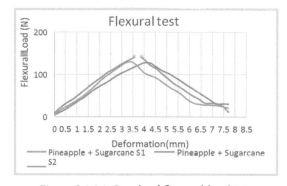

Figure 26.16 Graph of flexural load(N) vs deformation(mm) for P-S-P-S-P samples
Source: Author

P-S-P-S-P sample after compression test and Figure 26.12 gives the graphical comparison of the different sample. Again, using the pineapple fibers layer at the facial layer showed better properties. Correct stacking sequence and fabrication will result in increased resistance, prevention of matrix movement, and proper stress transfer among the fibers and matrix.

Physical characteristics of composite specimen

The presence of irregularities and voids in the composite laminate, which also contributes to the weakening of the mechanical and physical characteristics of the composite, is the fundamental reason why the theoretical and experimental densities are never the same. The vacuum infusion process was used to prepare the composite efficiently and with the fewest voids possible. The experimental density of the P-S-P-S-P specimen was found to be 1.20 g/cc. The composite specimens were immersed in water at 25°C room temperature for 24 hours to evaluate the kinetics of water absorption. Less water absorption was seen in the composite P-S-P-S-P (3.32%). Figure 26.13 represents the P-S-P-S-P sample after water absorption test and Figure 26.14 gives the graphical comparison of the different sample.

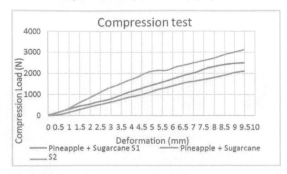

Figure 26.17 Graph of compression load(N) vs deformation(mm) for P-S-P-S-P samples
Source: Author

Conclusion

The natural fibers composite using pineapple and sugarcane bagasse woven fabrics reinforced with epoxy matrix was prepared using the vacuum infusion method. Various kinds of literature were reviewed, and a specific stacking sequence was adopted to achieve maximum strength and modulus. Sugarcane and pineapple fibers were chosen as it is eco-friendly biodegradable materials with abundant availability, recyclability, and sustainability compared to synthetic fibers. The vacuum infusion method is used for manufacturing laminates to achieve low void formation with strong interfacial bonding. Although the tensile strength of the P-S-P-S-P is on the lower side, the flexural and compressive strength of the layer are in the medium range. The impact strength is the best among the other alternatives.

Figures 15-17 show the deformation of the sample against load for different test. The Table 26.3 shows the specification of the P-S-P-S-P for different tests. Even water absorption % was observed less in P-S-P-S-P (3.32%). The hydrophilic nature of the natural fibers was limited to a maximum extent by the P-S-P-S-P. Hence, it can be used as a potential alternative where high-impact strength and moderate flexural and compression strength are required.

References

[1] Hussin, M. H. N., Hambali, A., Yuhazri, Y. M., Taufik, Zolkarnain, M., and Saifuddin, H. Y. (2014). A review of current development in natural fibres composites in automotive applications. Applied Mechanics and Materials. 564.

[2] Rohit, K. and Dixit, S. A. (2016). Review – future aspect of natural fibres reinforced composite. Polymers from Renewable Resources. 7(2), 43-59

[3] Mahesha, G. T., Bhat, K. S., and Devadiga, D. G. (2020). Sugarcane bagasse fibre reinforced composites: recent advances and applications. Cogent Engineering. 7(1).

[4] Reddy, M. I., Kumar, M. A., and Raju, C. R. (2018). Tensile and flexural properties of jute, pineapple leaf and glass fibres reinforced polymer matrix hybrid. Composites Materials Today: Proceedings, 5(1), 458-62.

[5] Khalid, M. Y., Rashid, A. A., Arif, Z. U., Ahmed, W., Arshad, H., and Zaidi, A. A. (2021). Natural fibres reinforced composites: sustainable materials for emerging applications. Results in Engineering. 11.

[6] Mohanavel, V., Ravichandran, M., Sathish, T., Kumar, M. M. R., Ganeshan, P., and Sabbiah, R. (2021). Experimental investigations on mechanical properties of cotton/hemp fibre reinforced epoxy resin hybrid composites. Journal of Physics Conference Series. 2027(1), 12015. doi: 10.1088/1742- 6596/2027/1/012015, 2021

[7] Bere, P. (2014). Experimental research regarding vacuum bag technology for obtaining carbon/epoxi composites. Academic Journal of Manufacturing Engineering. 12, 86-90.

[8] Yidris, N., Majid, D. L., Baitab, D. M., Hashim, N., and Zahari, R. (2019). Tensile properties of woven intraply carbon/kevlar reinforced epoxy hybrid composite at sub-ambient temperature. Encyclopedia of Materials: Composites. 766-73.

[9] Hashim, N., Majid, D. L., Baitab, D. M., Yidris, N., and Zahari, R., (2019). Tensile properties of woven intraply carbon/kevlar reinforced epoxy hybrid composite at sub-ambient temperature. Encyclopedia of Materials: Composites.

[10] Zhang, Z. Y., Dhakal, H. N., Ahmed, M. M., Zahari, R., and Barouni, A. (2021). Enhancement of impact toughness and damage behaviour of natural fibre reinforced composites and their hybrids through novel improvement techniques: a critical review. Composite Structures. 259, 113496.

[11] Mohanavel, V., Jacob, M. S., Raaj, M. N., and Benjamin, T. M. (2020). Mechanical properties of polymer matrix composites prepared through hand lay-up route. International Journal of Advanced Science and Technology. 29(9), 6586-91.

[12] Mohammed, L., Ansari, M. N., Pua, G., Jawaid, M., and Islam, M. S. (2015). A Review on Natural Fibres Reinforced Polymer Composite and Its Applications. International Journal of Polymer Science, 2015, 1–15.

[13] Fibre Glast Developments Corporation, Vacuum Infusion Complete Guide, 2019. [Online]. Available:https://www.fibreglast.com/product/vacuum-infusion Guide/Learning_Center

Chapter 27

Light transmitting concrete: a systematic review of research trends

Manoj Kumar Rajak[1,a], Ravish Kumar[2,b], Rahul Biswas[3,c] and Sinam Hudson Singh[1,d]

[1]Bureau of Indian Standards, India

[2]National Institute of Technology, Patna, India

[3]Visvesvaraya National Institute of Technology, Nagpur, India

Abstract

In recent decade, an innovative, smart, and revolutionary construction material termed light transmitting concrete/translucent concrete/transparent concrete has developed. It is made by embedding translucent materials in concrete. It has the property of allowing light to travel through it. One of the most remarkable qualities of light transmitting concrete (LTC) is its ability to save energy, enhance aesthetics and improve building sunlight inside condition. Translucent materials have a negative impact on durability. Variations in the translucent material content, shape, and size impact the physical, mechanical, durability, and optical transmittance ability of LTC. Orientation of translucent material is also one of the critical parameters which influence the optical transmittance ability of LTC. This systematic review study presents and discusses several perimeters that impact the fundamental features of light transmittance concrete about the most recent research trend. These innovative construction materials have enormous potential to use as an architectural construction material and for infrastructure purposes due to their unique property, which helps the world achieve its sustainability goal.

Keywords: light transmitting concrete, transparent concrete, translucent concrete

Introduction

In 2001, Losonczi, a Hungarian architect, created and produced light transmitting concrete (LTC), commonly known as transparent concrete [1-3]. LTC allows light to pass through the concrete, improving visibility and lowering the building's light energy requirement [4]. According to published data, artificial lighting contributes almost one-fourth of worldwide energy consumption. LTC has a lot of potential in architecture [5, 6]. LTC has been proven to be quite effective in various applications, including partition walls that allow sunlight to penetrate interior spaces, boosting the intensity of illumination in indoor surroundings [7]. Different measures have been taken to cut down the amount of energy used by artificial lighting, involving the development of novel construction materials that can transmit light, such as LTC. Its capability to transmit light not only improves visibility but also reduces the amount of light energy required by structures [4]. The transmission of light by optical fibers is sufficiently bright and efficient that almost no light is lost by the fibers [8–11]. Thus, the invention of LTC is an excellent solution for transmitting light inside structures. LTC is employed in a wide range of applications, including road markings, pavements, walkways, speed humps, stairwells, and tunnels, in addition to architectural features and construction components [12-15]. Apart from that, load or non load bearing walls panel made of LTC may be used to allow sunshine to enter interior spaces. In previous studies, LTC was shown to convey light and reduce light energy consumption by up to 50% while maintaining compressive strength [16]. Global warming is one of the significant forces for the development of LTC, as global energy uses is one of the major causes of global warming in this quickly changing era. Artificial lighting is used to illuminate the living area, which increases energy consumption and carbon emissions. LTC could help lower the carbon footprint. LTC will also improve the structure's safety and aesthetic appeal [12, 17-19]. Transparent or translucent concrete is another name for LTC. It is a new sustainable and environmentally friendly construction material. LTC is made up of light transmitting materials incorporated in the concrete, such as glass and optical fire [12, 20]. By placing optical fibers in a horizontal shape within the mold prior to concrete casting, basic building units may be created. Polymethylmethacrylate as polymer fibers or plastic optical fiber, as well as glass optical fiber, were used in these applications [21, 22]. LTC, on the other hand, is still not commonly accepted and is employ ed as a green construction material in the construction sector. The lack of skilled labour, as well as the uncertainty about its long-term mechanical and durability properties, are two important factors influencing the acceptance of LTC. However, extensive study is being conducted around the world to get widespread acceptance of this unique material. The following subsections will help to the understand the design and the durability properties of LTC concrete.

[a]manoj@bis.gov.in, [b]ravish@nitp.ac.in, [c]rahulbiswas@apm.vnit.ac.in, [d]shudson@bis.org.in

DOI: 10.1201/9781003450252-27

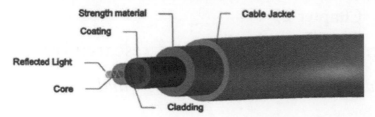

Figure 27.1 Typical structure of optical fiber cable
Source: Author

Effect of Translucent Materials on Manufacturing and Mix Design of Light Transmitting Concrete

The most common translucent materials used to generate LTC are broken glass, optical fiber, and resin. To meet this prerequisite and achieve its value, LTC must be made with materials that have a high translucency or light transmittance. Different types of translucent materials, have different degrees of translucency, characteristics, strength, and thermal conductivity, all these factors influence the overall performance of light-transmitting concrete. Mixing coarse broken glass with self-compacting concrete is a common method for creating light-transmitting concrete. As long as both ends of the waste glass are at the panel's surface, light may travel through transparent broken glass utilised in LTC. Alkaline pore solution combines with particles metastable silica to generate an alkali-silica fluid that expands and destroys concrete [19]. For LTC, additives, such as GGBs, are recommended for preventing or avoiding the development of alkali-silica fluid, fly-ash, silica-fume or metakaolin are also used in the design mix [23]. LTC was found to be reliant on the thickness of the light-transmitting concrete blocks and the shape and size of the waste glass as coarse aggregate for its light translucency and air purification properties [24]. To reduce the alkali-silica reaction, investigator [24] used coarse scrap glass that wasn't alkali reactive. To strengthen the concrete panel, steel fiber was added to the mix design. The researchers found that using LTC panels with an 11% light translucency, it is possible to cut the demand for light energy by up to 20.6%. This indicates that a LTC panel with coarse waste glass inclusions could be employed in building interior walls to minimise the usage of light energy. Manuello et al. [25] investigated the bending of LTC panels with glass inclusions using acoustic emission monitoring. When LTC without steel fiber was compared to the LTC with glass inclusion, the results revealed that LTC with glass inclusion had greater ductile behavior. The overall expenditure with respect to salary for workers and the quantity of material, polymer resin beats glass and optical fiber. A silicone casting was used by the researchers [12] to construct a polymer resin light transmitting objects. The casting mold was then used to place the polyester resin light guide before it was filled with mortar. According to research, LTC is influenced by the aspect ratio of resin used. As a result, they proposed that if the concrete panel thickness is greater than 100 mm, only 60% of light can get through. Polymethylmethacrylate resin was utilised to construct LTC in another experiment by Shen and Zhou [26], which demonstrated that due to its improved light transmittance characteristics, the usages of artificial light energy may be decreased from 72% to 41%. Epoxy resin and acrylic resin were utilised to create light-transmitting concrete [27]. The light-transmitting concrete was built in multiple layers with the use of liquid resin and concrete.

Concrete split at resin-concrete interfacial zone due to a lack of mechanical strength in the solidified mixture. For facades, [28] employed polymethylmethacrylate resins to their precast translucent concrete samples. To strengthen the toughness of the concrete and reduce the risk of cracking, steel and polypropylene fibers were also incorporated. Overall, different casting methods can change the mechanical strength of LTC with polymer resin insertion. When opposed to layer-by-layer casting, a prefabricated polymer resin lighting guidance or isolated components may give greater adhesive at the fiber-matrix interfacial zone. A concrete panel's thickness and angle of incidence restrict the amount of light that can be conveyed, even if polymeric resins have a light transmittance as excellent as optical fiber. As a result of superior properties of light transmittance, optical fiber is the most often and frequently utilised light transmission element in concrete construction. Glass or plastic are the most common materials used in optical fiber. Total internal reflection is used to transmit light through optical fiber. The fundamental rationale for using optical fiber as translucent materials in producing LTC, according to Han et al. [29], is that it has outstanding light transmittance capabilities even when the angle of light incident is greater than 60 degrees. Optical fibers transmit light with such brightness and efficiency that nearly no light is lost in the process [9, 28, 30-32]. Concrete strain, deformation, vibration, and corrosion may be tracked using fiber optic sensors inserted into panels [33]. For the manufacturing of transparent self-healing cementitious composites Snoeck et al. [34] utilise glass fibers. Primarily two mix techniques are used by Snoeck et al [34] first by direct and second by hand insertion. Due to the challenges in handling and positioning large amounts of glass fibers, manual insertion was determined to be impractical. The direct-mixing method, on the other hand, is less labor-intensive.

Huang [35] investigated the light presentation of building envelopes using LTC enclosing optical fibers and discovered that the smoothness of the optical fibers' ends is critical for light transmitting performance. Mosalam and Casquero-Modrego [36] use compound parabolic concentrators and straight cones. To enhance light transmitting, they claimed that compound parabolic concentrators and straight cones have a half acceptance angle that corresponds to the optical fiber's numerical aperture (NA). PVC and glass tubes, glow-in-the-dark powder, are some of the other transparent materials utilised to make light-transmitting concrete. Fiber optics and glass rods were compared to the light transmission of light-transmitting concrete [37]. Light transmitting through glass rods, on the other hand, was not as effective as optical fiber. The use of plastic rods and pipes instead of optical fibers decreased manufacturing costs and increased constructability and production. Glow-in-the-dark powder in light-transmitting concrete can absorb sunlight during the day and emit visible light at night or in dimly illuminated areas. Components including Portland cement, fine aggregate (0.6-4.75 mm), coarse aggregate (10 to 12 mm), and optical fibers have been the most often used. In addition, the spacing between the optical fibers within the mold defines aggregates with a specified particle size [37]. Reasonably of using optical-fibers, light-transmitting concretes make use of broken glasses [20, 24], rods of glass, acrylic [38], epoxy resin [39], and other plastics materials [42]. For LTC, cement, sands, and optical fiber are most typical ingredients in the mix [29]. Several studies have used coarse aggregates with a diameter of 10 mm and below for load bearing structures [19, 40]. The amount of optical fiber used for LTC should be 2.5–5% for efficient transmittance [3, 19]. According to reports, LTC containing more than 4% optical fiber will have a significant reduction in compressive strength and will have a negative impact on structural performance [19].

Factor Affecting Mechanical and Physical Property of Light Transmitting Concrete

As shown by the available research, LTC containing more than 4% optical fiber will have a significant reduction in compressive strength and will have a negative impact on structural performance [19]. The most significant reason for this reduction in the physical and mechanical property is the smooth and hydrophobic character of optical fiber, which produce a weak connection between the fiber-matrix contacts [41]. It is because of a lack of bonding between the glass particles and the cement matrix, the inclusion of recycled glass in LTC has a similar problem [42]. The addition of optical fiber to concrete reduces its strength when compared to normal concrete. For light-transmitting concrete with a volumetric percentage of 4% optical fiber, the compression strength was decreased by roughly 20%, according to Li et al. [22]. The loss in the compressive strength have been observed due to voids present in between concrete and fibre and these voids are also observed when scanning through electron microscopy [22]. According to Sawant et al., [43] as the volume of fiber is increase beyond a certain limit the compressive strength start decreasing.

A LTC system is made up of a variety of components, and the properties of the mixture mainly depend on the interfacial bond strength [44]. However, the flexural toughness of LTC was approximately increased by 10% compared to normal concrete. When Li et al. looked at the light transmittance properties of LTC, they discovered that the fiber degradation caused by the high curing temperature had such a negative impact on those properties [21]. In comparison to the randomly aligned plastic optical fiber LTC, the results reported by Henriques et al. revealed that the regularly aligned plastic optical fiber has no substantial impact on the performance of LTC's transmittance [32, 41, 45, 46]. By raising plastics optical fiber quantity in light-transmitting concrete from 2-3.5%, the enhancement of LTCs transmittance ability is same in both cases of plastic optical fiber alignment. Furthermore, the plastic optical fiber content from 2-5% resulted in a 22% increase in the light transmittance ability of regularly aligned plastic optical fiber LTC, compared to the equal improvement for randomly aligned plastic optical fiber LTC. The researchers concluded from their experiments that the bigger optical fiber diameter improves LTC's optical power capability [21, 22, 47-49]. Concrete with increased compressive and flexural strength and better light transmission that has the ideal diameter is 2 mm diameter plastic optical fiber [50]. The greater the diameter of the fibers, the greater the compressive strength of the concrete. Due to shorter linking lengths, microcracks spread more rapidly inside the matrix, resulting in a decrease in compression strength properties. Increased flexural and compressive strength was achieved by increasing fiber diameter because to the smaller contact area [51]. According to Bashbash et al. [52], larger diameter optical fibers contribute for increasing the compressive strength of LTC, as the higher size diameter fibers are stronger and capable to bear compressive loads than smaller-diameter fibers [50]. The flexural strength of light-transmitting concrete generally falls. Flexural strength decreases at an average rate of LTC for plastic optical fiber content less than 2% was about 10.2% for each 1% increase in plastic optical fiber content, for plastic optical fiber percentage above 2%, a mean reduction in flexural strength for each 1% increase in plastic optical fiber content was 15.5%. Furthermore, the flexural strength of LTC decreased by roughly 6.5 and 16.5% as the plastic optical fiber's diameter changes from 1.5-2 and 3 mm, respectively. Reduced fiber spacing leads in shorter interconnection lengths for microcrack propagation under compressive loads when the volumetric percentage of fibers is more than

2%. However, Momin et al. [53] asserted the increasing the gap between the fibers affects both the strength of concrete and the amount of light that passes through it.

Factor Affecting Light Transmittance of Light Transmitting Concrete

Data and equation reported by Tahwia et al. [54] show that the light transmittance increases as the volume of optical fiber increases, which is displayed in Figure 27.2. However, due to the difficulty in gently filling the spaces between the fibers without breaking the plastic optical fiber, volumetric fiber content greater than 5% cannot be achieved. As a result, when fiber content exceeds 5%, LTC's transmittance performance begins to deteriorate [41]. Optical transmittance reduced as the diameter of the fibers grew [55]. According to Momin et al. [53], increasing fiber spacing reduces both concrete compressive strength and light transmitting. According to Li et al. [22], LTC's light transmittance ability decreases if the gap between the source of light and the panel increases.

Figure 27.3 shows that the light transmittance decreases as the distance of light source are increased on the basis of the data and equation reported by Tahwia et al. [54]. The concrete specimen's light transmission is determined by calculating the difference between the source of light and the light-transmitting object. An incident beam is refracted as it travels through an opaque object. As a result, light will be reflected rather than refracted if its incidence is larger than critical angle. The ideal tilt angle of LTC varies depending on latitude in the area, particularly in low and high latitude areas. The LTC specimen's light transmittance is affected by the distance between the source of light and the concrete object. According to Tuaum et al. [55], increasing the gap between the illumination and the LTC reduces the light transmittance of the LTC.

Factor Affecting Durability of Light Transmitting Concrete

Microstructure study of light transiting concrete was conducted by (Mehta and Monteiro, 2017) [56] and explain the microstructure of light transiting concrete and provide the useful information related to durability and strength. They explain the permeability and porosity characteristics of LTC and its effect on the strength and durability of the concrete. Only a few researchers, employed microscopic research to acquire a more in-depth understanding of these two concrete qualities [12, 21, 22, 41, 42]. The concrete's porosity and water-absorbing capacity were both enhanced by the fiber insertion.

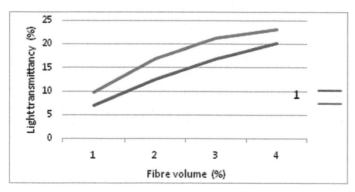

Figure 27.2 Relation between light transmittance (%) and fiber volume (%) [54]
Source: Author

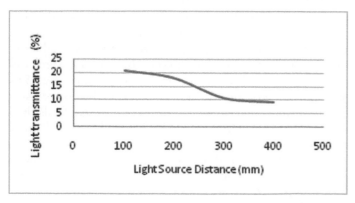

Figure 27.3 Relation between light transmittance (%) and light source distance (mm) [54]
Source: Author

Due to the weak link given by the optical fiber's extraordinarily smooth and slippery surface, they observed holes between the fiber–matrix interface using SEM analysis and a capillary water absorption test [41]. Some researchers have conducted water absorption studies on LTC [41, 45, 57, 64]. As a result, the water absorption capacity of light-transmitting concrete was improved by about two to five times by adding optical fibers between 2% and 5%. This is because the optical fibers are surrounded by a crisp, porous, and brittle matrix with many pores. Other water absorption experiments on LTC, which contains about 4% optical fibers, revealed water absorption and sorptivity were found to have decreased by about 9 and 0.5%, respectively, water absorption and sorptivity fell by 9 and 5%, respectively, representing the insignificance of simple concrete [57]. He et al. [58] examined the imperviousness to determine the long-term durability of the concrete. Final results showed, light-transmitting concrete's permeability is increased when optical fibers are added to the mix.

Factors Influencing the Cost of Light-Transmitting Concrete

While light-transmitting concrete is more expensive to manufacture than regular concrete, its long-term advantages in terms of reduced lighting costs as well as a reduced environmental impact outweigh the greater cost in the short term. The expenses and payback time of light-transmitting concrete were analysed using energy efficiency as a measure [44]. The least expensive transparent substance is glass aggregate [59]. Optical fiber is commonly recognised as the most costly transparent material. Replacement of glass optical fiber with polymer resin and plastic optical fiber has resulted in significant cost savings. Optic fiber architecture and installation challenges prevented large-scale production of light-transmitting concrete, reducing the material's appeal for use in construction [12]. Just one researchers that have employed glow in the dark powder to construct LTC blocks is done by Saleem and Blaisi [14]. The cost of manufacture, however, is three-times that of normal interlocking blocks. In contrast, light-transmitting concrete was used to perform a thorough investigation for maintain a comfort level in the building and its relation to the consumed energy [16]. The heat conductivity of transparent materials allowed light to permeate through the light-transmitting concrete, allowing both light and heat to enter the structure. As a consequence, when lighting and HVAC system costs are taken into account, a light transmitting panel with a 5.6% fiber volume fraction reduced total energy consumption by 18% [16].

Application of Light Transmitting Concrete

Light-transmitting concrete may be employed in infrastructure. This is owing to a lack of thorough and persuasive study into LTC's effectiveness in meeting standard indoor visibility standards rather than focusing on its aesthetic appeal [36, 60, 61]. However, several applications can be used as models for future research and development. The Italian Pavilion at the 2010 Shanghai Expo was built with light-transmitting concrete and polymer resin [12, 41, 62]. Light-transmitting concrete panels have also been employed to construct the Al Aziz Mosque in Abu Dhabi [29], Europe Gate in Hungary [70], transparent facades at Aachen University in Germany, and the walls and partitions of the Bank of Georgia [19]. Lucem Lichbeton, a German company, was the first to produce light-transmitting concrete walkways and sidewalks [63]. Some facilities, including banks and museums, may benefit from light-transmitting concrete, which can reduce light energy usage while also increasing aesthetics [5].

Azambuja and Castro [64] employed LTC to create a prison wall prototype that aimed to improve living conditions while also lowering light [46] energy usage. Smart traffic markings and light-transmitting concrete were also used by Zhu et al. [65] to improve exposure and enhance safety, respectively. In terms of infrastructure, LTC can illuminate dark subway stations and tunnels [17]. From above, light was sent via an optical fiber as lighting pipes to illuminate the highway tunnel's various portions. The optical fiber might be used in the tunnel's construction as it has a special ability of pass the light. Pena-Garcia et al. [65] are implemented the Light-transmitting concrete for tunnel construction for transmitting the daylight into the tunnel. The most significant piece showcasing the LTC idea is the "European Gate," an artistic work created in Hungary in 2004 [3]. Its usage is appreciated in LTC dividers or partition walls in office cabins or residences, constructing appealing furniture and clever light fixtures, illuminating gloomy subway stops, etc. [9, 28, 66]. Huashuyan Tunnel's solar optical transmission systems were studied by Qin et al. [67].

Conclusion

The light transmitting concrete (LTC) is a very innovative material, and it has sufficient potential to use as a state-of-the-art material for improving the architectural view of any structure with lowering electricity consumption. The light transmittance property is mainly dependent upon the spacing, percentage and inherent characteristics of the translucent materials. As per various research, it is found that optical fiber has the advantage over other translucent materials concerning its unparalleled light transmittance property. However, due to the difficulty in gently filling the spaces between the fibers without breaking the plastic optical fiber, volumetric fiber content greater than 5% cannot be achieved. The research is very limited in the case of translucent materials other than optical fiber. To compete with regular concrete, extensive research is required with other translucent materials as they are considerably cheaper than optical fiber. Although, it was studied that a light transmitting panel with a 5.6% fiber volume fraction reduced total energy consumption by 18%. LTC has the potential to replace conventional architectural material to achieve the global sustainability goal. Research data are available for the water absorption and permeability of LTC; however, extensive research is required to evaluate the other durability property, such as weathering action, abrasion, chemical attack, resistance to freezing and thawing, etc. The manufacturing of LTC is very labour intensive. Research is required for the automation of LTC for faster and mass production. This review's conclusion is that LTC is a skillful way to maximize and using the free day-light system in an intelligent way of the "live and go green" concept in the form of a potential, novel, promising, innovative, and bright future with buildings made of materials that improve their aesthetic appeal and save energy, among other things.

References:

[1] Goho, A. (2005). Concrete nation: Bright future for ancient material. Science News. 167. doi: 10.2307/4015970.

[2] Timina, A., Yanova, R., Popov, A., and Sorokoumova, T. (2019). Modern translucent materials and their impact on architectural forming. E3S Web of Conferences. 97. doi: 10.1051/e3sconf/20199701035.

[3] Zielińska, M. and Ciesielski, A. (2017). Analysis of Transparent concrete as an innovative material used in civil engineering. IOP Conference Series: Materials Science and Engineering. 245(2). doi: 10.1088/1757-899X/245/2/022071.

[4] Patel, J. and Goyal, A. (2018). Smart materials in construction technology. doi: 10.1109/ICSCET.2018.8537256.

[5] Tuaum, A., Shitote, S., Oyawa, W., and Biedebrhan, M. (2019). Structural performance of translucent concrete façade panels. Advancements in Civil Engineering. doi: 10.1155/2019/4604132.

[6] Dutta, M., Reeti, R., and Mandal, S. K. (2018). Light transmitting and self cleansing concrete as a smart building material-a review. doi: 10.1109/ICSCET.2018.8537328.

[7] Roye, A. (2013). Intriguing transparency. German Research. 35(3). doi: 10.1002/germ.201490000.

[8] Zhou, Z., Ou, G., Hang, Y., Chen, G., and Ou, J. (2009). Research and development of plastic optical fiber based smart transparent concrete. Smart Sensor Phenomena, Technology Networks, System. 7293, 72930F. doi: 10.1117/12.816638.

[9] Bhushan, M. N. V. P., Johnson, D., Pasha, A. B., and Ms. K. (2013). Optical fibres in the modeling of translucent concrete blocks. International Journal of Engineering Research and Applications. 3(3).

[10] Kashiyani, B. K., Raina, V., Pitroda, J., and Shah, B. K. (2013). A study on transparent concrete: a novel architectural material to explore construction sector. International Journal of Engineering and Technology. 2(8).

[11] Mainini, A. G., Poli, T., Zinzi, M., and Cangiano, S. (2012). Spectral light transmission measure and radiance model validation of an innovative transparent concrete panel for façades. Energy Procedia. 30, 1184–1194. doi: 10.1016/j.egypro.2012.11.131.

[12] Shen, J. and Zhou, Z. (2013). Some progress on smart transparent concrete. Pacific Science Review. 15(1), 51-55

[13] Peña-García, A., Gil-Martín, L. M., and Rabaza, O. (2016). Application of translucent concrete for lighting purposes in civil infrastructures and its optical characterization. Key Engineering Materials. 663. doi: 10.4028/www.scientific.net/KEM.663.148.

[14] Saleem, M. and Blaisi, N. I. (2019). Development, testing, and environmental impact assessment of glow-in-the-dark concrete. Structural Concrete. 20(5). doi: 10.1002/suco.201800221.

[15] Saleem, M., Elshami, M. M., and Najjar, M. (2017). Development, testing, and implementation strategy of a translucent concrete-based smart lane separator for increased traffic safety. Journal of Construction Engineering and Management. 143(5). doi: 10.1061/(asce)co.1943-7862.0001240.

[16] Ahuja, A. and Mosalam, K. M. (2017). Evaluating energy consumption saving from translucent concrete building envelope. Energy Building. 153. 10.1016/j.enbuild.2017.06.062.

[17] Said, S. H. 2020). State-of-the-art developments in light transmitting concrete. Materials Today: Proceedings. 33, 1967–73. doi: 10.1016/j.matpr.2020.06.128.

[18] Paul, S. and Dutta, A.(2013). Translucent concrete. Concrete. 3. International Journal of Scientific and Research Publications. 3(10). doi: 10.11129/detail.9783034614740.26.

[19] Altlomate, A., Alatshan, F., Mashiri, A., and Jadan, M./ (2016). Experimental study of light-transmitting concrete. International Journal of Sustainable Building Technology and Urban Development. 7(3–4). doi: 10.1080/2093761X.2016.1237396.

[20] Pagliolico, S. L., Lo Verso, V. R. M., Torta, A., Giraud, M., Canonico, F., and Ligi, L. (2015). A preliminary study on light transmittance properties of translucent concrete panels with coarse waste glass inclusions. Energy Procedia. 78. doi: 10.1016/j.egypro.2015.11.317.

[21] Li, Y., Li, J., and Guo, H. (2015). Preparation and study of light transmitting properties of sulfoaluminate cement-based materials. Materials & Design. 83. doi: 10.1016/j.matdes.2015.06.021.

[22] Li, Y., Li, J., Wan, Y., and Xu, Z. (2015). Experimental study of light transmitting cement-based material (LTCM). Construction and Building Materials. 96. doi: 10.1016/j.conbuildmat.2015.08.055.

[23] Carles-Gibergues, A., Cyr, M., Moisson, M., and Ringot, E. (2008). A simple way to mitigate alkali-silica reaction. Materials and Structures. 41(1). doi: 10.1617/s11527-006-9220-y.

[24] Spiesz, P., Rouvas, S., and Brouwers, H. J. H. (2016). Utilization of waste glass in translucent and photocatalytic concrete. Construction and Building Materials. 128. doi: 10.1016/j.conbuildmat.2016.10.063.

[25] Manuello, A., Carpinteri, A., Invernizzi, S., Lacidogna, G., Pagliolico, S., and Torta, A. (2011). Mechanical characterization and AE of translucent self-compacting concrete plates in bending. Conference Proceedings of the Society for Experimental Mechanics Series. 5. doi: 10.1007/978-1-4419-9798-2_15.

[26] Shen, J. and Zhou, Z. (2020). Performance and energy savings of resin translucent concrete products. Journal of Energy Engineering. 146(3). doi: 10.1061/(asce)ey.1943-7897.0000652.

[27] Pilipenko, A., Bazhenova, S., Kryukova, A., and Khapov, M. (2018). Decorative light transmitting concrete based on crushed concrete fines. In IOP Conference Series: Materials Science and Engineering. 365(3). doi: 10.1088/1757-899X/365/3/032046.

[28] Mainini, A. G., Poli, T., Zinzi, M., and Cangiano, S. (2012). Spectral light transmission measure and radiance model validation of an innovative transparent concrete panel for façades. Energy Procedia. 30. doi: 10.1016/j.egypro.2012.11.131.

[29] Han, B., Zhang, L., and Ou, J. (2017). Light-transmitting concrete. Smart and Multifunctional Concrete Toward Sustainable Infrastructures. 273–83.

[30] Zhou, Z., Ou, G., Hang, Y., Chen, G., and Ou, J. (2009). Research and development of plastic optical fiber based smart transparent concrete. Smart Sensor Phenomena, Technology, Networks, and Systems. 7293. doi: 10.1117/12.816638.

[31] Rajak, M. and Rai, B. (2019). Effect of micro polypropylene fibre on the performance of fly ash-based geopolymer concrete. Journal of Engineering and Applied Science. 9(1). doi: 10.2478/jaes-2019-0013.

[32] Rajak, M., Roy, L., and Rai, B. (2017). A systematic review on polypropylene fiber reinforced geopolymer concrete.

[33] Merzbacher, C. I., Kersey, A. D., and Friebele, E. J. (1996). Fiber optic sensors in concrete structures: A review. Smart Materials and Structures. 5(2). doi: 10.1088/0964-1726/5/2/008.

[34] Snoeck, D., Debo, J., and De Belie, N. (2020). Translucent self-healing cementitious materials using glass fibers and superabsorbent polymers. Developments in the Built Environment. 3. doi: 10.1016/j.dibe.2020.100012.

[35] Huang, B. (2020). Light transmission performance of translucent concrete building envelope. Cogent Engineering. 7(1). doi: 10.1080/23311916.2020.1756145.

[36] Mosalam, K. M. and Casquero-Modrego, N. (2018). Sunlight permeability of translucent concrete panels as a building envelope. Journal of Architectural Engineering. 24(3). doi: 10.1061/(asce)ae.1943-5568.0000321.

[37] Dinesh, B. K., Mercy, S. R., and Suji, D. (2019). Effect of fiber pattern in strength of light transmitting concrete. International Journal of Recent Technology and Engineering. 7(5).

[38] Kim, B. (2017). Light transmitting lightweight concrete with transparent plastic bar. Open Civil Engineering Journal. 11(1). doi: 10.2174/1874149501711010615.

[39] Mohan, D. R., Tyagi, P., Sharma, R., and Rajan, H. B. M. (2018). Experimental studies on POF and epoxy-resin based translucent concrete.

[40] Roye, A., Barlé, M., Janetzko, S., and Gries, T. (2009). Faser and textilbasierte lichtleitung in betonbauteilen - lichtleitender beton. Beton- und Stahlbetonbau. 104(2). doi: 10.1002/best.200808235.

[41] Henriques, T. D. S., Molin, D. C. D., and Masuero, Â. B. (2018). Study of the influence of sorted polymeric optical fibers (POFs) in samples of a light-transmitting cement-based material (LTCM). Construction and Building Materials. 161. doi: 10.1016/j.conbuildmat.2017.11.137.

[42] Torres de Rosso, L. and Victor Staub de Melo, J. (2020). Impact of incorporating recycled glass on the photocatalytic capacity of paving concrete blocks. Construction and Building Materials. 259. doi: 10.1016/j.conbuildmat.2020.119778.

[43] Sawant, S. G. S. A. B. and Jugdar, R. V. (2014). Light transmitting concrete by using optical fiber. International Journal of Science and Engineering. 1.

[44] Zhandarov, S. and Mäder, E. (2005). Characterization of fiber/matrix interface strength: Applicability of different tests, approaches and parameters. Composites Science and Technology. 65(1). doi: 10.1016/j.compscitech.2004.07.003.

[45] dos S. Henriques, T., Molin, D. C. D., and Masuero, Â. B. (2020). Optical fibers in cementitious composites (LTCM): Analysis and discussion of their influence when randomly arranged. Construction and Building Materials. 244. doi: 10.1016/j.conbuildmat.2020.118406.

[46] Rai, B., Roy, L. B., and Rajjak, M. (2018). A statistical investigation of different parameters influencing compressive strength of fly ash induced geopolymer concrete. Structural Concrete. 19(5). doi: 10.1002/suco.201700193.

[47] Altlomate, A., Jadan, M., Alatshan, F., and Mashiri, F. (2016). Experiment study of light transmitting concrete. doi: 10.37376/1571-000-020-008.

[48] Pradheepa R. and Krishnamoorthi, S. (2015). Light Transmission of transparent concrete. International Journal of Engineering Science. 3.

[49] Jacek Halbiniak and Paulina Sroka. "Translucent concrete as the building material of the 21st centuary. Cathedral of the Building and Material Processing Technologies, Czestochowa University of Technology, Faculty of Civil Engineering ul. Akademicka. 3, 42-201 Częstochowa, halbiniak@wp.pl

[50] Salih, S. A., Joni, H. H., and Mohamed, S. A. (2020). Effect of plastic optical fiber on some properties of translucent concrete. Engineering And Technology Journal. 32(12).

[51] Robles, A., Arenas, G. F., and Stefani, P. F. (2020). Light transmitting cement-based material (LTCM) as a green material for building. Journal of Applied Research and Technology. 1(1). doi: 10.4995/jarte.2020.13832.

[52] Bashbash, B. F., Hajrus, R. M., Wafi, D. F., and Alqedra, M. A. (2013). Basics of light transmitting concrete. Global Advanced Research Journal Of Engineering, Technology And Innovation. 2(3).

[53] Momin, A. A., Kadiranaikar, R. B., Jagirdar, S., Arshad, M., and Inamdar, A. (2014). Study on light transmittance of concrete using optical fibers and glass rods. IOSR Journal of Mechanical and Civil Engineering. 67.

[54] Tahwia, A. M., Abdel-Raheem, A., Abdel-Aziz, N., and Amin, M. (2021). Light transmittance performance of sustainable translucent self-compacting concrete. Journal of Building Engineering. 38. doi: 10.1016/j.jobe.2021.102178.

[55] Tuaum, A., Shitote, S. M., and Oyawa, W. (2018). Experimental evaluation on light transmittance performance of translucent concrete. International Journal of Applied Engineering Research. 13.

[56] Mehta, P. D. P. K. and Monteiro, P. D. P. J. M. (2014). Concrete: microstructure, properties, and materials, fourth edition. Concrete Microstructure, Properties, and Materials. Fourth Edition.

[57] Naik, R. R. and Prakash, K. B. (2016). An experimental investigation on fibre reinforced transparent concrete. International Research Journal of Engineering and Technology. 1038–44.

[58] He, J., Zhou, Z., and Ou, J. (2011). Study on smart transparent concrete product and its performances. First National Civil Engineering Conference.

[59] Ugale, A. B., Badnakhe, R. R., and Nanhe, P. P. (2019). Light transmitting concrete—litracon. doi: 10.1007/978-981-13-6148-7_25.

[60] Ahuja, A, Mosalam, K. M., and Zohdi, T. I. (2015). Computational modeling of translucent concrete panels. Journal of Architectural Engineering. 21(2). doi: 10.1061/(asce)ae.1943-5568.0000167.

[61] Ahuja, A, Mosalam, K. M., and Zohdi, T. I. (2015). An illumination model for translucent concrete using radiance. 14th International Conference of the International Building Performance Simulation Association (IBPSA 2015).

[62] The Economic Times. (2010). Transparent Cement for the Italian Pavilion at Expo 2010 in Shanghai.

[63] Nachrichten. (2017). Award-winning translucent concrete technology for smart cities.

[64] de Castro, L. (2015). Translucent concrete in prison architecture.

[65] Zhu, B., Guo, Z., and Song, C. (2019). Fiber-optic parameters of light emitting diode active-luminous traffic markings based on light-transmitting concrete. Tongji Daxue Xuebao/Journal Tongji Univ. 47(6). doi: 10.11908/j.issn.0253-374x.2019.06.009.

[66] Fastag, A. (2011). Design and manufacture of translucent architectural precast panels. In fib Symposium PRAGUE 2011: Concrete Engineering for Excellence and Efficiency, Proceedings. 2.

[67] Qin, X., Zhang, X., Qi, S., and Han, H. (2015). Design of solar optical fiber lighting system for enhanced lighting in highway tunnel threshold zone: A case study of huashuyan tunnel in China. International Journal of Photoenergy. doi: 10.1155/2015/471364.

Chapter 28

Identification of service quality parameters for the Indian automobile service industry

Mangesh D. Jadhao[1,a], Arun P. Kedar[2,b] and Ramesh R. Lakhe[3,c]

[1]Department of mechanical engineering, G.H. Raisoni Institute of Engineriring and Technology, Nagpur, India

[2]Department of Mechanical Engineering, Yashwantrao Chauhan College of engineering, Nagpur, India

[3]SQMS Nagpur, India

Abstract

People in the service sector are constantly under pressure to improve service quality. They focus mainly on customer expectations and their requirements. It is necessary to identify the service quality parameters that help the organisation improve service quality in the view of customers. In this research SERVQUAL (service quality) methodology is used to identify the most important service quality parameters. The result of the factor analysis indicates that a minimum of 16 service items under four parameters can measure the service quality of the automobile sector. Except for the tangible parameter, other parameters were different from service quality parameters suggested by other researchers. The study has shown that the most significant factor is customer perception of how to provide services, as a result of which they will be key factors in assessing service quality. service quality parameters

Keywords: Automobile service industry, service quality, SERVQUAL, service quality parameters

Introduction

The contribution of the Indian service industry to state and national income as well as employment makes it the most significant sector. Customer retention is a challenging undertaking for every business because the service sector is so competitive nowadays and clients have many options. Customers rarely express their dissatisfaction with services; instead, they simply stop using them.

Consumers determine the quality of services, and services meet client expectations [6]. Customer satisfaction and service quality are strongly correlated, according to a large number of studies [10, 29]. It has been discovered that service quality affects revenue and expenses [7]. Customer retention [29], repurchase willingness [10, 19], and loyalty [14] have all grown.

The purpose(s) for which a client chooses a specific service centre must be known to the manager of the center. The various service quality factors that have an impact on the current level of customer service quality should be known to the work manager of the service centre. Therefore, every firm is particularly interested in measuring and improving service quality. Therefore, a trustworthy tool for gauging passenger automobile servicing quality is needed by the service centre. The purpose of this study is to determine the most significant service quality indicators and various service components for evaluating passenger automobile service quality.

Though the objective is stated above, this research comprises of:

(1) A literature review that summarises the car service quality parameters used in previous research and identified the research gap.
(2) The research design is then used to determine the service quality parameters that will be employed in this study.
(3) To identify the most critical factors, factor analysis is used.
(4) Finally, the findings are assessed, and the research is completed.

Literature Review

It is possible to define service quality in a variety of ways. One such, often used, defines it as 'the extent to which a service meets customers' wishes or expectations.' The 3D model of service quality [12] includes three distinct facets of service quality: technical, functional, and image quality. The most popular instrument for evaluating service quality was developed by Parasuraman [1], and its name is SERVQUAL. Behra et al. [28] built a neural model of service excellence using the Dutch auto dealer network. A structured diagram, like the reverse SERVQUAL model, was used to illustrate the hierarchical aims and techniques that made up this representation of service quality.

One study investigated used-car owners' preferences for independent garages or reputable car dealers for maintenance services [7]. Berndt [2] used PZB's instrument to assess the quality of car servicing in South

[a]mangesh.jadhav@raisoni.net, [b]arunkedar64@gmail.com, [c]rameshlakhe786@gmail.com

DOI: 10.1201/9781003450252-28

Africa. Katarne et al. [17] assessed the customer service at a garage for cars in a city in India. In order to assess the effectiveness of service quality methods implemented in the same service organisation in response to earlier recommendations, they conducted an additional study [30]. Sharma and Negi, [31] identified the most crucial elements for an auto repair shop by placing the customer's voice first. The SERVQUAL model gauges Karnataka's present level of vehicle service[35]. Jadhao et al. [15], assessed the level of service provided to passenger cars at a service facility in Nagpur. Utilising the Auto SERVPERF model, the study established the automobile after sales service quality dimensions to provide customer satisfaction [32]. Zygiaris et al. [36] studied effect of service excellence on client satisfaction in Saudi auto industry.

Many studies used five SERVQUAL dimensions [3, 4, 7, 20, 22, 28, 31, and 33]. There is a lot of overlap between the service quality parameters that have been discovered and those that are clear from the literature. Table 28.1 lists the parameters established by various studies and authors.

Several studies on automobile service quality measurement were reviewed, and the gap in the previous research discovered that: i) there is no written evidence for the use of the number and types of parameters used in passenger car service quality measurement. ii) In most of the studies, some common parameters and items are reported. iii) It is observed in the studies that major research work is carried out mainly on functional quality. All the studies reviewed in this paper reported that the number of parameters ranged from five to eight in number, and service quality items ranged from 10-39 in number. Five SERVQUAL parameters—reliability, empathy, responsiveness, assurance, and tangibles—appeared in most of the studies.

Depending on the nature of the service industry, different countries have different scale factors and item counts. As a result, there are different types and numbers of service quality parameters based on the country, the type of automotive service, and the respondent. It is vital to investigate service quality in specific contexts or industries [34]. To set the parameters for this scenario, this research was conducted. The

Table 28.1 Service quality parameters

Sr. No.	Service quality dimensions	Authors																Total score
		[36]	[11]	Anantha Raj et al (2014)	Minwir et al (2014)	[31]	[31]	Suhas S. Ambekar (2013)	Wu Shuqin, Liu Gang (2012)	[1]	[2]	S. Keshavraz et. Al (2009)	[7]	[28]	[9]	[6]	[21]	
1	Tangibles	√	√	√	√	√	√	√			√	√	√	√	√	√		12
2	Reliability	√	√	√	√	√	√	√	√		√	√	√	√	√	√	√	14
3	Responsiveness	√	√	√	√	√	√	√	√		√	√	√	√	√	√		13
4	Assurance	√		√	√	√	√	√			√	√	√	√		√		10
5	Empathy	√	√	√	√	√	√	√	√		√	√	√	√		√		12
6	Competence															√			1
7	Courtesy															√		√	2
8	Credibility															√			1
9	Security															√			1
10	Accessibility															√			1
11	Communication															√			1
12	Understanding															√			1
13	Fairness								√										1
14	Trust								√	√									2
15	Commitment				√				√										2
16	Relationship value								√										1
14	Convenience								√										1

Source: Author

significance of this study is further highlighted by the dearth of published research on the topic of service quality in India. This brings up the following research concerns: What does excellent service mean in the automotive industry? In this case, do the parameters mentioned by [27] exist? The objective of this study is to determine the most significant factors influencing service quality in the automotive industry.

Research Design

Identification of service parameters

Customers' criteria for judging service quality suit six parameters that have been validated by prior studies on measuring passenger vehicle service quality, according to exploratory research on passenger car service quality parameters conducted by previous researchers. Empathy, reliability, tangibles, responsiveness, communication, and assurance make up these six criteria. The following description of these six parameters will be useful in determining the service quality item:

- Reliability: This aspect relates to the commitment made by the service centre to its clients. Assurance that the vehicle will be delivered at the specified time.
- Responsiveness: The willingness of a service provider to offer consumers prompt service.
- Assurance: Through customer interactions, the service centre can gain the trust and confidence of its clients.
- Empathy: How the support centre handles clients customers receive the kind, individualised care they deserve from the professionals at the service centre.
- The term 'tangibility' alludes to things like the outward appearance of buildings, contemporary technology, parking, and dress code. Tidiness of the service area.
- Communication: Informing clients about the services rendered and offered.

The six parameters listed above were selected as theoretical parameters for this research study as they were specifically developed for passenger car services. Furthermore, identification of service items under these six parameters through a literature survey and actual discussion with the work manager of the service centre were done.

Identification of service items

For measuring passenger car service quality, a service quality item embodying the various service quality parameters needs to be found [6]. The Dutch automotive service business used 42 characteristics of service excellence [28]. Selected 36 service quality metrics to evaluate the level of customer service at an automobile dealership [2]. Used 39 service quality metrics from earlier studies [22]. Used 22 SERVQUAL measures to measure the customer voice based on a survey of the literature. A total of 32 services were initially listed by Cronin and Taylor [11] in their research study. In their studies, many researchers used recognised service quality measures. Through a literature review, a feedback form, several seminars, and discussions with the job manager, 32 service items for our research study were chosen. The choice of the items was based on their significance to the research topic. The following Table 28.2 displays the specifics of the chosen items.

   The 32 service items were chosen after this research study's thorough literature analysis. Upon assessment by two academicians, three automotive aftermarket services specialists, and two supervisors, all the words used in the questions provided suggestions for the changes, which were then put into practice. Less significant things were removed, and others were consolidated. Due to their lower value in our research study, five items that were chosen from earlier research were eliminated. A total of 27 service quality elements and six service quality metrics were chosen after consultation with important stakeholders and an expert. Table 28.3 lists the six service quality parameters alongside the 27 service quality components.

Research Methodology

Data collection

Data collection for the current study involved the use of a survey instrument. A five-point Likert scale is used for the survey's presentation, with 5 representing a strong agreement and 1 a strong disagreement. There are three components to the instrument. The first part of the survey includes general information on the respondents and the vehicles, while the second half includes 27 questions about perception following service consumption. One question is included in the third section to get feedback on the general level of service quality.

Table 28.2 Service quality items

S. N	Passenger car service items	Source/Researcher
1	Got the appointment as per desired date and time.	[7, 29]
2	Operating hours were convenient.	[2, 7, 21]
3	He had a mobile/door service facility.	[31]
4	It was necessary to pick up and drop off the facility for service/repair.	[31]
5	Received prompt attention on arrival.	[6, 2, 21, 28]
6	The staff were wearing proper uniforms with visible identity cards.	[2, 7]
7	The service advisor understood the problem with the vehicle.	[28]
8	The staff were knowledgeable and skillful at performing their tasks.	[6, 7, 21]
9	Staff had given personal attention.	[2, 11, 28]
10	The friendliness of the service advisor	[2, 6, 21]
11	The staff is concerned about your well-being.	[2, 6]
12	Staff always help willingly.	[21, 28]
13	The service advisor provided a checklist of service and repairs before work had begun.	[2, 6]
14	The service advisor explained why repairs are carried out.	[2, 6]
15	The detailed estimated work cost was provided in advance. (Before the work started)	[6]
16	I was informed of the expected time required to complete the work.	[2, 6, 28]
17	The workshop and reception area were clean.	[2, 6, 7, 28]
18	Modern technical equipment was available.	[7, 28]
19	Contacted when additional repairs were to be done.	[2]
20	Contacted when estimated repairs were more expensive.	[2, 6]
21	Spare parts were immediately available.	[28]
22	Informed about the delay of service in advance.	[2, 6, 28]
23	I completed all the requested work.	[2, 9]
24	Mechanics are trustworthy.	[1, 7]
25	Repairs carried out were error free/work done the first time was right.	[2, 6, 7, 21]
26	The bill provided was clear and correct.	[7, 28]
27	The service advisor explained the actual work performed and it was clearly indicated on the bill.	[2, 6, 28]
28	The cost charged for servicing the vehicle was reasonable.	[21]
29	The vehicle was clean after servicing.	[2, 6, 28]
30	The vehicle was ready when promised.	[2, 6, 7, 28]
31	Focuses on solving customer complaints.	[2, 6]
32	Informed when to schedule next service visit.	[11]

Source: Author

Three of the top auto manufacturers in Nagpur that offer after-sales services have been chosen for our study. Because it can quickly pool vast amounts of data, a handy sampling strategy was adopted in this work [13]. For this study, only those respondents were taken into account who were able to respond to the questionnaire. the car owners who had used a company's official service centre for free service, paid service, or repair and maintenance. To expand the variety of the data, responses to the survey are also gathered using Google survey forms. Three of the biggest auto manufacturers' customers each received 328 questionnaires. A total of 258 survey questionnaires were returned out of 328 surveys. 23 surveys were eliminated from consideration because they were missing general information and scale components. Finally, 235 surveys with a 71.64% response rate—higher than in the previous study [11]—were taken into account for data analysis.

Table 28.4 depicts that in the age group of 31-40 and 41 to above 50 years had relatively equal participation with 43.40% and 44.68% in the research. It can also see that male respondents had more proportion with 67.23% than female respondents 32.76%. Furthermore majority of respondents are service person with 47.65%, business respondents 28.93%, student and hose wife relatively equal with 12.76% and 10.63%.

Table 28.3 Items association with parameters

Parameters	Service quality items
Tangible	• Modern technical equipment was available. • A checklist of work was provided. • The staff wore proper uniforms with visible identity cards. • Reception and workshop areas were clean.
Reliability	• The vehicle was ready when promised. • Got the appointment as per the desired date and time. • Repairs carried out were error free. • What Bill provided was clear and correct. • Focuses on solving customer complaints.
Responsiveness	• Received instant attention on arrival. • Staff always help willingly. • Spare parts are immediately available.
Assurance	• Staff members were knowledgeable and skillful at performing their tasks. • Reasonable service charge. • Mechanics trustworthiness • The service advisor explained the actual work performed and it was clearly indicated on the bill.
Empathy	• Operating hours were convenient. • The service advisor's friendliness • Staff understood the problem with vehicles. • The staff is concerned about your well-being. • It was necessary to pick up and drop off the facility for service/repair. • Clean car after service
Communication	• The service advisor makes clear what has been repaired. • Contacted when estimated repairs were more expensive. • The detailed estimated cost of work was provided in advance (Before the work started). • I was informed of the expected time required to complete the work. • Contacted when additional repairs were to be done.

Source: Author

Table 28.4 Respondent profile

Age	Frequency	Percent (%)	Occupation	Frequency	Percent	Gender	Frequency	Percent (%)
Up to 30 Years	28	11.91	Service	112	47.65	Male	158	67.23
31 to 40 years	102	43.40	Business	68	28.93	Female	77	32.76
41 to above 50 years	105	44.68	Student	30	12.76			
Total	235		Housewife	25	10.63			

Source: Author

Table 28.5 Reliability test of scale

Cronbach's Alpha	Number of Items
0.979	27

Source: Author

Validity and Reliability

For the authors study, content validity is supported by an extensive literature review and a review of the constructs by subject-matter experts. Using factor loadings of measuring items and the corresponding t-values, convergent validity was confirmed. Because all factor loadings of indicators are more than 0.50, the instrument satisfies convergent validity. Using IMB SPSS 20.00 software, a reliability test of the instrument was carried out after the useful responses had been gathered. Table 28.5 indicate the instrument's Cronbach's alpha score was 0.979, it means the questionnaire is quite reliable for researchers to continue their investigations [24].

Sample Adequacy

It is important to confirm that the amount of the collected sample is adequate for factor analysis before doing the analysis. Kaiser-Meyer-Olkin (KMO) measures sample efficiency, and the Bartlett's test of sphericity is one of two tests to be performed to confirm that the data gathered is sufficient for factor analysis [25]. Table 28.6 displays the results of the sampling adequacy test we conducted for our study using the SPSS 20.00 programme.

Table 28.6 KMO and Bartlett's test

KMO Adequacy.		0.928
Bartlett's Test of Sphericity	Approx. Chi-square	2990.011
	df	351
	Sig.	0.000

Source: Author

Table 28.7 Factor analysis of passenger car service items

SERVQUAL Parameters	Description	Factor 1	Factor 2	Factor 3	Factor 4
RS3	Spare parts are immediately available.	.664			
T1	Modern technical equipment was available.	.658			
T4	Reception and workshop areas were clean.	.628			
T2	A checklist of work was provided.	.584			
E4	The staff is concerned about your well-being.	.492			
C4	I was informed of the expected time required to complete the work.	.483			
C1	The service advisor makes clear what has been repaired.	.458			
C3	The detailed estimated cost of work was provided in advance (before the work started).	.432			
C2	Contacted when estimated repairs were more expensive.	.419			
RS1	Received instant attention on arrival.		.664		
RL3	Repairs carried out were error free.		.622		
A2	Reasonable service charge.		.601		
RL2	Got the appointment as per the desired date and time.		.578		
E5	It was necessary to pick up and drop off the facility for service/repair.		.542		
A1	Staff members were knowledgeable and skillful at performing their tasks.		.527		
E2	The service advisor's friendliness		.452		
E1	Operating hours were convenient.		.419		
RL1	The vehicle was ready when promised.			.722	
E3	Staff understood the problem of vehicles.			.716	
RS2	Staff always help willingly.			.652	
A4	The service advisor explained the actual work performed and it was clearly indicated on the bill.			.487	
RL5	Focuses on solving customer complaints.			.452	
E6	Clean car after service				.709
RL4	What Bill provided was clear and correct.				.698
A3	Mechanics trustworthiness				.604
C5	Contacted when extra repairs needed to be made.				.477
T3	The staff were wearing proper uniforms with visible identity cards.				.423

Source: Author

According to Walvekar et al. [16], the index value was 0.928, which is fantastic according to KMO statistics. It suggests that the sample that was gathered is appropriate for factor analysis. The correlation matrix's claim to be a singular matrix is put to the test using Bartlett's sphericity test. The Bartlett's test for the collected data is very significant (about chi-square = 2990.011, p = 0.001), indicating that the data are suitable for factor analysis.

Factor Analysis

An additional analysis of the responses was performed to pinpoint the presence of certain components that influence service quality. In this instance, the principal component-extraction method analysis was used. Acceptable Cronbach's alpha coefficients were those with factor loading of 0.5 or greater [23]. Alpha coefficients of 0.979, which indicate great internal consistencies and reliability, were obtained by these factors. That's why items with factor loadings greater than 0.5 are included in this study. The results of the factor analysis are shown in Table 28.7.

Findings and discussion

The factor loading conveys how each service item relates to the underlying factor. The factor loading displays the connection between a service item and a specific factor. A higher relationship with that factor is indicated by a higher factor loading. There are 27 services that appear to be loaded under four criteria. The eleven service quality items that did not have factor loading above 0.5 were eliminated due to their non-importance, but the remaining items were kept in mind because all of the parameters had factor loading above 0.50. Table 28.8 displays the common minimum, 16 service items that are loaded greater than 0.5, linked to four characteristics taken from factor analysis.

Table 28.9 displays the dependability for the indicated criteria. This agrees with the reliability ratings discovered in other SERVQUAL investigations [5, 6]

Figure 28.1 shows that total four factors were extracted from factor loading due to Eigen value greater than one. A total of 16 service quality items with a factor loading greater than 0.50 were used to extract four factors, which are detailed below.

Factor 1-Tangibility: "tangibility" refers to the tangible attributes of the service centre, such as its comfort and cleanliness, as well as the physical attributes of its employees, materials, and equipment. This factor includes the following: spare parts availability, modern equipment, reception and workshop area cleanliness, and checklist provided.

Factor 2-Kindness: Take care of the customer by performing the service dependably and accurately. This dimension contains items such as: received instant attention on arrival; repairs carried out error free;

Table 28.8 Common service items

S. No	Service items	Factor loading	Factor
1	Spare parts are immediately available.	.664	Tangible
2	Modern technical equipment was available.	.658	
3	Reception and workshop areas were clean.	.628	
4	A checklist of work was provided.	.584	
5	Received instant attention on arrival.	.664	Kindness
6	Repairs carried out were error free	.622	
7	Reasonable service charge.	.601	
8	Got the appointment as per the desired date and time.	.578	
9	It was necessary to pick up and drop off the facility for service/repair.	.542	
10	Staff members were knowledgeable and skillful at performing their tasks.	.527	
11	The vehicle was ready when promised.	.722	Commitment
12	Staff understood the problem of vehicles.	.716	
13	Staff always help willingly	.652	
14	Clean car after service	.709	Trust
15	What Bill provided was clear and correct.	.698	
16	Mechanics trustworthiness	.604	

Source: Author

Table 28.9 Factor's reliability

Factor	Cronbach's alpha	Eigen value
Factor 1- Tangibility:	0.846	10.52
Factor 2- Kindness	0.783	3.88
Factor 3-Commitment	0.872	2.68
Factor 4-Trust	0.893	1.12

Source: Author

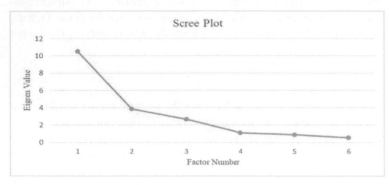

Figure 28.1 Scree plot
Source: Author

reasonable service charged; got the appointment as per desired date and time; had pick up and drop facility for service/repair; staff were knowledgeable and skillful to perform their tasks;

Factor 3-Commitment: Commitment to give services on time. This dimension contains the service items, such as the vehicle being ready at the promised time. Staff understood the problem of vehicles and always helped willingly.

Factor 4-Trust: The service center's employees' courtesy and their capacity to win over customers' confidence and faith. This dimension contains items such as cleaning of the car after servicing, clear and correct bill provided, and mechanics' trustworthiness.

Managerial Implications and Recommendations

The main goal of this study was to identify the crucial service quality indicators for automobile servicing in India. In order to measure the service quality of passenger cars, this study identified 16 service elements under four criteria. While providing service to the clients, work managers can determine the organization's strengths and weaknesses by using these characteristics. The findings demonstrate that the suggested parameters influence long-term client happiness, and management must give this issue more consideration.

Limitations And Research Directions

The findings of the study cannot be generalised to all automakers because some are more represented than others. The sample is also educated, which could not be representative of the market as a whole. Four variables have been identified, but the factor analysis reveals that they are not pure factors, and further investigation into these factors and how they apply to the Indian motor industry is necessary. Because the components uncovered using factor analysis in this paper are different from those found in the earlier study, more research in this area is required.

References

[1] Minwir Al-Shammari, Ahmad Samer Kanina, Perceived Customer Service Quality in a Saudi Automotive Company, International Journal of Managerial Studies and Research (IJMSR), 2(10), November 2014, 173–182
[2] Berndt, A. (2009). Investigating service quality dimensions in South African motor vehicle servicing. African Journal of Marketing Management.1(1), 1–9.
[3] Suhas. S. Ambekar (2013). Service quality gap analysis of automobile service centers. Indian Journal of Research in Management, Business and Social Sciences, 1(1), 38–41.
[4] Anantha Raj A. Arokiasamy and Huam Hon Tat (2014), Assessing the Relationship Between Service Quality and Customer Satisfaction in the Malaysian Automotive Insurance Industry, Middle-East Journal of Scientific Research 20 (9): 1023–1030.

[5] Wu Shuqin, and Liu Gang. An empirical study of after-sales service relationship in China's auto industry. In 1st International Conference on Mechanical Engineering and Material Science (MEMS 2012), 467–470. Atlantis Press, 2012.

[6] Bouman, M. and Van der Wiele, T. (1992). Measuring service quality in the car service industry: building and testing an instrument. International Journal of Service Industry Management. 3(4).

[7] Brito, E. P., Aguilar, R. L., and Brito, L. A. (2007). Customer choice of a car maintenance service provider. International Journal of Operations & Production Management, 27.

[8] Buttle, F. (1996). SERVQUAL: review, critique, research agenda. European Journal of Marketing. 30(1), 8–32.

[9] Chen, C. N. and Ting, S. C. (2002). A study using the grey system theory to evaluate the importance of various service quality factors. International Journal of Quality & Reliability Management. 19(7), 24.

[10] Cronin, J. J. and Taylor, S. A.(1994). SERVPERF versus SERVQUAL: reconciling performance-based and perceptions-minus-expectations measurement of service quality. Journal of Marketing. 58(1), 125–31.

[11] Izogo, E. E. and Ogba, I. -E. (2015). Service quality, customer satisfaction and loyalty inautomobile repair services sector. International Journal of Quality & Reliability Management. 32(3), 250–26.

[12] Gronroos, C. (1984). A service quality model and its marketing implications. European Journal of Marketing. 18(4), 36–44.

[13] Hair et. al (2010). Multivariate data analysis. 7th edition. Pearson New International.

[14] Heskett, J. L. (2002). Beyond customer loyalty. Measuring Service Quality, 12(6), 355–57.

[15] Jadhao, M. D., Kedar A. P., and Lakhe R. R. (2018). Evaluation of passenger car service quality through Fuzzy AHP. International Journal of Engineering Science Invention. 7(4), 30–35.

[16] Kaiser, H. F. (1974). An index of factorial simplicity. Psychometrika. 39(1), 31–36.

[17] Katarne, R., Sharma, S., and Negi, J. (2010). Measurement of service quality of an automobile service centre. In Proceedings of the 2010 International Conference on Industrial Engineering and Operations Management, Dhaka, Bangladesh, pp. 286–91.

[18] S. Keshavarz, , Yazdi, S. M., Hashemian, K., & Meimandipour, A. 2009. Measuring service quality in the car service agencies. Journal of Applied Sciences, 9(24), 4258–4262.

[19] Lee, G. G. and Lin, H. F. (2005). Customer perceptions of e-service quality in online shopping. International Journal of Retail & Distribution Management.

[20] Mason A. N., Narcum J., and Mason K. (2021). Social media marketing gains importance after Covid-19. *Cog. Bus. Manag.* 8, 797.

[21] Mersha, T. and Adlakha, V. (1992). Attributes of service quality: the consumers' perspective. International Journal of Service Industry Management. 3(3).

[22] Al-Shammari, M. and Kanina, A. S. (2014). Customer service quality in a Saudi automotive company. International Journal of Managerial Studies and Research. 2(10). 173–82

[23] Mokhlis, S. and Yaakop, A. Y. (2012). Consumer choice criteria in mobile phone selection: An investigation of malaysian university students. International Review of Social Sciences and Humanities. 2(2), 203–12.

[24] Nunnally, J. C. (1978). Psychometric Theory, 2d Edition. New York, NY: McGraw-Hill.

[25] Pallant, J. (2007). SPSS survival manual: A step by step guide to data analysis using SPSS version 15. New York, NY: McGraw-Hill.

[26] Parameshwaran et al. (2009). Integrating fuzzy analytical hierarchy process and data envelopment analysis for performance management in automobile repair shops. European Journal of Industrial Engineering. 3(4), 450–46.

[27] Parasuraman, A., Zeithaml, V. A., and Berry, L. L. (1988). SERVQUAL: a multiple-item scale for measuring consumer perceptions. Journal of Retailing. 64(1), 12–40.

[28] Behra, R. S., Fisher, W. W., and Lemmink, J. G. M. A. (2002). Modelling and evaluating service quality measurement using neural networks. International Journal of Operations & Production Management. 22(10), 1162–85.

[29] Roland, R., Zahorik, A., and Keiningham, T. (1995). Return on quality (ROQ): Making service quality financially accountable. Marketing. 59, 58–70.

[30] Sharma, S., Katarne, R. and Negi, J. (2011). Impact assessment of service quality strategies in an automobile service. Eighth Aimsinternational Conference on Management. Ahmedabad, India.

[31] Sharma, S. and Negi, J. (2013). Prioritization of voice of customers by using kano questionnaire and data envelopment analysis. International Journal of Industrial Engineering Research and Development. 4(1), 1–9.

[32] Sheriff, N., Roslan, N., and Yusoff, Y. (2020). Determinants of satisfaction for automotive after sales service quality: a preliminary application of autoservperf model. Jurnal Intelek. 15, 197–208.

[33] Srivastava A., Kumar V. (2021). Hotel attributes and overall customer satisfaction: What did COVID-19 change? *Tour. Manag. Persp.* 40:100867.10.1016/j.tmp.2021.100867

[34] Svensson, G. (2006b). New aspects of research into service encounters and service quality. International Journal of Service Industry Management. 17(3), 245–257.

[35] Vijaykanth Urs, M. C., Harirao, A. N., and Kumar, A. N. S.(2014). Service quality gap analysis between personal and fleet users in four wheeler car service centre across karnataka automotive industries. International Journal of Emerging Research in Management & Technology. 4–12.

[36] Zygiaris, S., Hameed, Z., Ayidh, A. M and Ur Rehman, S. (2022). Service quality and customer satisfaction in the post pandemic world: a study of saudi auto care industryFrontiers in Psychology. 13, 842141. doi: 10.3389/fpsyg.2022.842141

Chapter 29

Tribological investigations on Senegalia catechu based green composites

Vinayagamoorthy R[a] and Venkatakoteswararao G[b]

Sri Chandrasekharendra Saraswathi Viswa Mahavidyalaya, Kancheepuram, India

Abstract

This research is focussed to examine the tribological properties of *Senegalia catechu* resin based green composites. Functional bio-resin has been developed by processing the natural resin with polyester through a series of treatments. Composites have been fabricated by varying the compositions of jute reinforcements (20%, 30%, 40%, and 50%) and resin (80%, 70%, 60%, and 50%). The fabricated composites have been tested for its friction and wear characters by changing the various input parameters such as load (5 N, 10 N and 15 N), speed (1 m/s, 2 m/s and 3 m/s), sliding distance (1000 m, 1500 m and 1500 m) on four different categorical samples (I, II, III and IV). The sample which gives the optimum tribological performance has been identified by conducting a list of trial runs and optimising it based on response surface methodology. Optimisation has been done aiming to minimise the output parameters namely specific wear rate and friction co-efficient. Optimium conditions have been arrived at a load of 5 N, speed of 3 m/s, a sliding distance of 1000 m with sample IV giving a specific wear rate of 2.572 mm³/Nm and a friction co-efficient of 0.345. Confirmatory runs have been conducted for the optimum condition and it has been validated that the prediction was accurate.

Keywords: Friction, wear, catechu, bio-fiber, bio-resin

Introduction

Natural composites become the trending scenario in the field of polymer composites due to their custom-made characters that suits the specific application. At the inception, bio-based reinforcements have been introduced as the substitute for glass and carbon and they behave on par with the synthetic ones [1]. Bio-reinforcements would partially increase the biodegradability of the composite. Due to this reason, bio-based resins have been focused to produce composites with full degradability [2]. The synthesis of functional resin from the virgin bio-resin needs a series of chemical and physical techniques and they are not common for all bio-resins [3]. Bio-resins are primarily used for food processing and medicinal uses. Few bio-resins such as banana sap, soya-bean oil and drumstick sap have been processed in the form of matrix resin for composites of structural applications [4–7]. Mechanical characters of such composites are predominantly decided by the interfacial bond between the matrix and reinforcement. A high degree of bonding has been achieved by using soy protein as the matrix material and thus enhances the load bearing capacity of composites [8].

Rubber, a primary user of automotive industry has been in use over the decades. Friction and wear are the two main parameters that decide upon the design of automotive tires. Composites made of grapheme oxide and styrene butadiene rubber are proved to be a high-performance green alternative for present day tires and has wide opportunities for future engine ering applications [9]. The type of matrix and fiber utilised have a dominant contribution on the tribological performance of the composites. A comparison among polytetrafluroethylene (PTFE), polyethylene (PE), polyethylene terephthalate (PET), and polyoxymethylene (POM) showed that, PTFE and PE composites have lowest co-efficient of friction whereas POM and PET composites has high resistance to wear rate [10]. Many research works have been done to examine the friction characters of composites made of artificial resins which includes thermoset and thermoplastics. Characterisation of bio-resin based composites are at the inception stage and very few resins have been utilized as a structural composite material. Hence, there is a research gap on the study of tribological parameters on several bio-resin based composites. The present research is focused on developing functional resin from *Senegalia catechu* tree. Bio-resin is obtained while destruction of the tree either fully or partially and the resin thus obtained has several medicinal values in Indian traditional medicines and also used in food processing industries. The bio-resin is subjected to a series of treatments, followed by development of composites by varying the composition of fiber content. The composites have been tested for its tribological characters for different input conditions and an optimum condition has been arrived.

Experimental Details

Catechu gum, supplied by a local dealer is finely powdered to a size of 0.5 mm by grinding. The finely powdered gum is then mixed with distilled water and soaked for 24 hours. The uniform solution thus formed is

[a]vin802002@gmail.com, [b]venkatakoteswararao02@gmail.com

DOI: 10.1201/9781003450252-29

treated with general purpose polyester solution in 1:1 ratio by continuous heating at 80°C and stirring for about 6 hours. After the preparation of the functional resin, four different composites have been fabricated by varying compositions of woven jute as reinforcements. Methyl ethyl ketone peroxide (MEKP) is used as a catalyst and cobalt napthenate is used as an accelerator during the composite fabrication [11]. The composition of composite samples is listed in Table 29.1. Composites have been developed by using compression molding machine by appropriately placing the fiber and resin layer for settling time of 24 hours. A pin-on-disc testing setup has been used to study the tribological characters of the fabricated composites by using ASTM G99 standard. The samples have been prepared by cutting the composites into the form of pins of size 6 mm² cross-section and 50 mm in height. The disc is made of stainless steel 165 mm in diameter and 8 mm thick as given in Figure 29.1.

Design of Experiments

Response surface method (RSM) is a tool for quantifying relationship between measured responses or output factors and the important input factors. It also allows to find out the trial experimental list, search the significance of input parameters on the responses and to optimise the output parameters under a specified objectives on input parameters [12]. A D-optimal design is one among the several optimal designs that uses iteration-based algorithm and is aimed to reduce the covariance of the estimated parameters. D-optimal design alleviates the experimentation cost by minimizing the number of trial runs as in factorial design and other non-optimal designs. Also, when the design space is constrained and when the process involves both numerical and categorical factors, D-optimal design gives accurate predictions [13]. This design gives two dimensional plots and three-dimensional surface plots using which the behavior of responses on specific input parameters could be easily analyzed. In addition, it gives response equations through which the responses could be accurately predicted for a given set of input parameters. The response y may be approximated in terms of input parameters in the form of a second order mathematical model as given in Equation 1.

$$y = \beta + \beta_1 x_1 + \beta_2 x_2 + \beta_{12} x_1 x_2 + \beta_{11} x_{12} + \beta_{22} x_{22} + \beta \tag{1}$$

Where, y denotes the output response, x_1, x_2, x_{11} etc denotes the input parameters and the co-efficient β, β_1, β_2 etc are to be determined by using the least square method.

The input parameters involved in this design are listed in Table 29.2. Load, speed and sliding distance have been assigned with three numerical levels whereas the sample is assigned with four categorical levels. A list of 32 trial runs have been conducted and during each run, output parameters namely specific wear rate (SWR) and co-efficient of friction (COF) have been noted for analysis. The trial runs and the measured output parameters have been listed in Table 29.3.

Table 29.1 Compositions of composite samples

S. No.	Name	CP resin (%)	Jute (%)
1.	CP1	80	20
2.	CP2	70	30
3.	CP3	60	40
4.	CP4	50	50

Source: Author

Figure 29.1 Disc and pins for tribology test
Source: Author

Table 29.2 Input factor and levels

Input factors.	Type	Levels
Load (N)	Numeric	5, 10, 15
Speed (m/s)	Numeric	1, 2, 3
Sliding distance SD (m)	Numeric	1000, 1500, 2000
Sample	Categorical	I, II, III, IV

Source: Author

Table 29.3 Trial runs and responses

S. No.	Load (N)	Speed (m/s)	SD (m)	Sample	COF	SWR (x 10^{-6}) mm³/Nm
1	5	3	1500	II	0.41	4.68
2	10	3	1000	IV	0.4	4.03
3	10	3	2000	I	0.5	4.67
4	15	2	1000	IV	0.42	3.94
5	15	3	1500	III	0.4	4.69
6	15	1	2000	I	0.52	3.93
7	15	3	2000	IV	0.46	2.40
8	5	2	1000	III	0.35	3.94
9	15	3	1000	II	0.45	2.94
10	10	1	1000	I	0.46	6.40
11	5	2	2000	IV	0.41	3.40
12	5	3	1000	I	0.42	5.40
13	5	1	1000	IV	0.38	4.54
14	5	1	2000	II	0.45	5.42
15	10	1	2000	IV	0.43	3.59
16	15	1	1000	III	0.37	5.43
17	5	3	2000	III	0.39	3.45
18	15	1	1500	II	0.44	3.41
19	15	2	2000	III	0.41	5.34
20	5	1	1500	III	0.37	4.14
21	15	3	1000	I	0.48	2.34
22	5	1	1000	II	0.39	6.34
23	5	2	1500	I	0.46	7.40
24	10	2	1000	II	0.42	3.59
25	15	2	1500	I	0.5	3.68
26	10	2	2000	II	0.48	3.69
27	10	2	1500	III	0.39	3.67
28	15	3	2000	IV	0.46	2.68
29	5	3	1500	II	0.41	4.58
30	15	1	1000	III	0.37	4.59
31	5	1	1000	IV	0.38	4.57
32	10	3	2000	I	0.5	4.55

Source: Author

Results and Discussions

Specific wear rate:

Specific wear rate (SWR) denotes the volume of material removed for unit normal load and sliding distance during the wear test of a pin-on-disc apparatus. As far as the tribological significance is concerned, the SWR plays a vital role in deciding the capacity of materials to suit for applications involving more wear and tear.

A material must possess low wear rate when subjected to worst applications involving rubbing actions. The significance of input factors on SWR is measured by the help of ANOVA analysis as displayed in Table 29.4. The domination of input factors is decided by a p-value not more than 0.05. Here, the input factors namely load, speed and sample majorly affect the SWR. The model appears to be compelling whereas the lack of fit is not compelling. The model has an adequate precision of 11.696, greater than four gives an adequate signal for predicting the responses. R^2 value of 0.94 and Adjusted R^2 value of 0.81 are close to 1 and hence, model correctly fits the data.

Three dimensional plots, interaction plot and normal plot for SWR are given in Figure 29.2. SWR decreases with a hike in the load and speed. This happens due to the formation of transfer film on the surface of the steel disk. As the load goes up, there would be an elevated abrasion on the pin surface leading to enhancement of frictional force. At this instant elevation of interface temperature happens to soften the composite surface. With an elevated load, the pin surface would be modified easily resulting in quicker formation of transfer film. The formed film acts as a shielding for the pulpy pin surface and therefore, wear rate is minimized at an elevated load [14]. Sliding distance does not affect the SWR to a considerable extent. As the fiber content is increased, SWR decreases until 30-40% and for further increase, SWR increases. This

Table 29.4 ANOVA analysis for specific wear rate

Source	Sum of squares	Degrees of freedom	Mean squares	F-value	p-value
Model	38.29	21	1.82	7.15	0.0014
Load	5.43	1	5.43	21.31	0.001
Speed	5.86	1	5.86	22.99	0.0007
SD	0.42	1	0.42	1.63	0.23
Sample	6.50	3	2.17	8.50	0.0042
Error	0.44	5	0.088	---	---

Source: Author

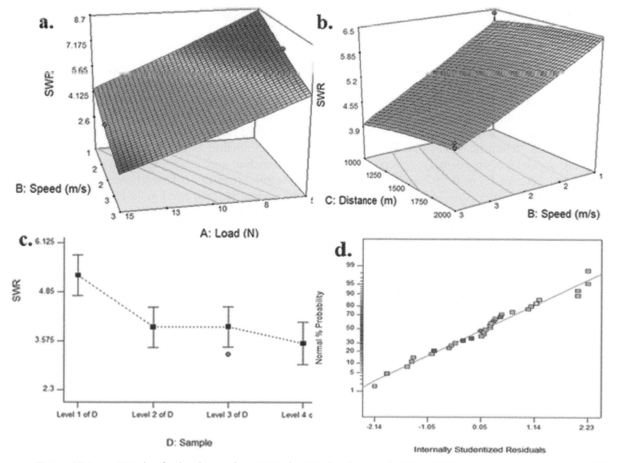

Figure 29.2 (a) 3D plot for load, speed vs SWR (b) 3D plot for speed, SD vs SWR (c) interaction plot for SWR vs sample

Source: Author

shows that, in order to get an optimum wear resistance, the fiber content must be varied between 30% and 40%. This also denotes that, the bonding between the fiber and matrix is maximum in this range [15]. The distribution of points in the normal plot nearly forms a straight line and hence the experimental model is found to be compelling. Mathematical models have been developed for each sample and this helps to predict the output responses under any combination of the input values. The mathematical models for specific wear rate are presented from Equation 2-5.

Sample I

$$SWR = 11.04847 -0.46880 \times load -1.84069 \times speed +1.46296E\text{-}003 \times distance +$$
$$0.036579 \times load \times speed -4.36072 E\text{-}005 \times load \times distance +1.22693 E\text{-}004 \times$$
$$speed \times distance + 3.67368 E\text{-}003 \times load^2 +0.062631 \times speed^2 -2.89763 E\text{-}007 \times distance^2 \quad (2)$$

Sample II

$$SWR = 8.37040-0.30967 \times load -1.29581 \times speed +5.63031E\text{-}004 \times distance + 0.036579 \times$$
$$load \times speed -4.36072E\text{-}005 \times load \times distance + 1.22693E\text{-}004 \times speed \times distance +$$
$$3.67368E\text{-}003 \times load^2 + 0.062631 \times spee^2 d-2.89763E\text{-}007 \times distanc^2 \quad (3)$$

Sample III

$$SWR = 3.37236 + 0.028933 \times load -1.15612 \times speed + 1.45844E\text{-}003 \times distance +$$
$$0.036579 \times load \times speed -4.36072E\text{-}005 \times load \times distance + 1.22693E\text{-}004 \times speed \times$$
$$distance +3.67368E\text{-}003 \times load^2 + 0.062631 \times speed^2-2.89763E\text{-}007 \times distance^2 \quad (4)$$

Sample IV

$$SWR =6.44644-0.12705 \times load -1.07952 \times speed +6.16426E\text{-}005 \times distance +$$
$$0.036579 \times load \times speed -4.36072E\text{-}005 \times load \times distance +1.22693E\text{-}004 \times speed \times$$
$$distance + 3.67368E\text{-}003 \times load^2 + 0.062631 \times speed^2-2.89763E\text{-}007 \times distance^2 \quad (5)$$

Co-efficient of friction

Friction co-efficient denotes the ratio between frictional force resisting the motion of two surfaces in contact and the normal force tightening the two surfaces together. This is again an important parameter to be focused on materials subjected to high wear. In general, a material must have a low co-efficient of friction as it elevates the temperature of rubbing surfaces leading to change in the character of the material. From the ANOVA analysis as given in Table 29.5, it has been observed that, Load, sliding distance and sample majorly affect the friction co-efficient. The model seems to be compelling whereas the lack of fit is not compelling. The model has an adequate precision of 46.88, greater than 4 gives an adequate signal for predicting the responses. An R^2 of 0.997 and an Adjusted R^2 of 0.991 are close to 1 and hence, model correctly fits the data. Three dimensional plots, interaction plot and normal plot for friction co-efficient are given in Figure 29.3. It has been found that, a hike in the load and sliding distance results in the enhancement of the friction coefficient. If the normal load applied is hiked, then the exposure area would also elevate with the tangential force resulting in elevation of friction co-efficient at that instant [16]. In the same way, as the sliding distance is elevated, the pin travels a long distance leading to an increase in friction between the pin and the disc surface [17]. Change in speed does not show a major influence on the friction co-efficient. As the fiber composition is hiked, friction co-efficient decreases until 50%, thereafter it elevates for further

Table 29.5 ANOVA analysis for co-efficient of friction

Source	Sum of squares	Degrees of freedom	Mean squares	F-value	p-value
Model	0.063	21	0.0029	168.52	<0.0001
Load	0.0066	1	0.0066	376.06	<0.0001
Speed	2.9E-5	1	2.9E-5	1.62	0.2324
SD	0.0089	1	0.0089	504.18	<0.0001
Sample	0.033	3	0.011	630.21	<0.0001
Error	0.00	5	0.00	---	---

Source: Author

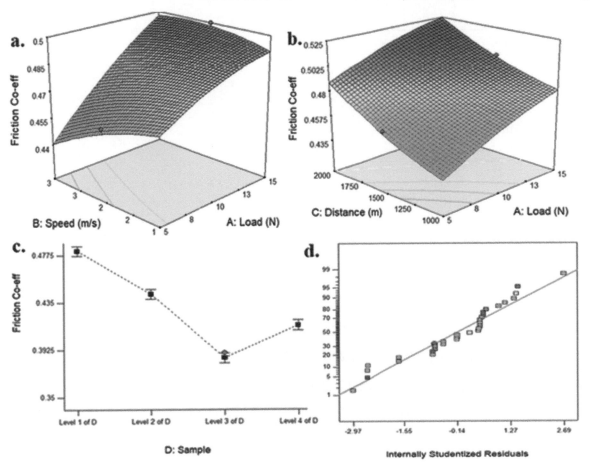

Figure 29.3 (a) 3D plot for load, speed Vs SWR (b) 3D plot for speed, SD Vs SWR (c) interaction plot for SWR vs sample

Source: Author

increase. This shows that 50% of jute would be the optimum condition for a reduced friction. The distribution of points in the normal plot strictly follows a straight line notifying that the experimental model is compelling. The mathematical models for friction co-efficient are presented from Equation 6-9.

Sample I

Friction co-eff = $0.41470 + 5.19232E\text{-}003 \times$ load $-9.23458E\text{-}003 \times$ speed $\times -1.54032E\text{-}006 \times$ distance $+ 1.63152E\text{-}003 \times$ load $\times$ speed $-8.60691E\text{-}007 \times$ load $\times$ distance $+ 1.23272E\text{-}006 \times$ speed $\times$ distance $-1.54155E\text{-}004 \times$ load^2 $-3.54610E\text{-}003 \times$ speed^2 $+ 1.86513E\text{-}008 \times$ distance^2 (7)

Sample II

Friction co-eff = $0.33421 + 5.54958E\text{-}003 \times$ load $+ 2.43967E\text{-}003 \times$ speed $+ 8.33681E\text{-}006 \times$ distance $+ 1.63152E\text{-}003 \times$ load $\times$ speed $-8.60691E\text{-}007 \times$ load$\times$ distance $x+1.23272E\text{-}006 \times$ speed $\times$ distance $-1.54155E\text{-}004 \times$ load^2 $-3.54610E\text{-}003 \times$ speed^2 $+ 1.86513E\text{-}008 \times$ distance^2 (8)

Sample III

Friction co-eff = $0.33722 + 3.50084E\text{-}003 \times$ load $-9.28626E\text{-}004 \times$ speed $-1.37100E\text{-}005 \times$ distance $+ 1.63152E\text{-}003 \times$ load $\times$ speed$-8.60691E\text{-}007 \times$ load $\times$ distance $+1.23272E\text{-}006 \times$ speed $\times$ distance $-1.54155E\text{-}004 \times$ load^2 $-3.54610E\text{-}003 \times$ speed^2 $+1.86513E\text{-}008$ x Distance^2 (9)

Sample IV

Friction co-eff = $0.35381 + 5.44442E\text{-}003 \times$ load $-2.31322E\text{-}003 \times$ speed $-1.64792E\text{-}005 \times$ distance $+ 1.63152E\text{-}003 \times$ load $\times$ speed $-8.60691E\text{-}007 \times$ load $\times$ distance $+ 1.23272E\text{-}006 \times$ speed $\times$ distance $-1.54155E\text{-}004 \times$ load^2$-3.54610E\text{-}003 \times$ speed^2 $+ 1.86513E\text{-}008 \times$ distance^2 (10)

Table 29.6 Optimisation

Parameters	Goals	Lower limit	Upper limit
Load	In range	5	15
Speed	In range	1	3
SD	In range	1000	2000
SWR	Minimise	2.34	7.4
FC	Minimise	0.35	0.52

Source: Author

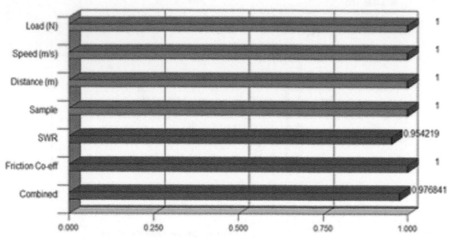

Figure 29.4 Desirability bar chart
Source: Author

Optimisation and confirmation

The model is optimised by setting the goals as given in Table 29.6. The input parameters have been set to the range between minimum and maximum values whereas the responses have been set to minimise in between the minimum and maximum readings. Optimisation has been done by using desirability approach in which among the list of optimum conditions, the condition with desirability close to 1 would be selected. It is found that, a 5 N load, a 3 m/s speed, a 1000 m sliding distance and selection of sample 3 are the ideal values of inputs with a maximum desirability of 0.976. The ideal output values are found to be 2.572 mm³/Nm for SWR and 0.345 for friction co-efficient. Confirmatory trials have been conducted and repeated three times for the optimum condition. During each repeatation, the output values are noted. The mean of three values is calculated for comparison.

The calculated readings are found to be 2.51 mm³/Nm for SWR and 0.339 for friction co-efficient which are very close to the predicted values. Thus, the optimisation carried out by using D-optimal method is fulfilled. The disability chart as given in Figure 29.4 clearly denotes the desirability values of 0.954 for SWR, 1 for friction co-efficient and 0.976 for overall desirability.

Conclusions

Novel catechu resin-based composites have been developed and tested for its friction and wear characteristics. Specific wear rate alleviates during elevation of load and speed because of the development of transfer film on the pin surface. As the fiber content is increased, specific wear rate decreases in the range 30-40 % and for further hike, it increases. This shows that, in order to get an optimum wear resistance, the fiber content must be varied between 30% and 40%. Elevation of load and sliding distance results in the elevation of friction coefficient. Optimization resulted in selected of sample 4, a 5 kN load, a 3 m/s speed, a 1000 m sliding distance as the ideal condition. As sample 4 is found to be optimum, it is culminated that a 50% of jute fiber and 50 % of catechu resin is more suitable for a reduced tribological behavior among the fabricated composites. Confirmatory runs closely match the predicted values showing that the D-optimal optimization is compelling.

References

[1] Cordeiro, N., Belgacem, M., Torres, I., and Moura, J. (2004). Chemical composition and pulping of banana pseudo-stems. Industrial Crops and Products. 19, 147–54.

[2] Vinayagamoorthy, R. and Rajeswari, N. (2014). Mechanical performance studies on Vetiveria zizanioides/jute/glass fiber-reinforced hybrid polymeric composites. Journal of Reinforced Plastics and Composites. 33, 81–92.

[3] Zhang, W., Zhang, X., Wu, Z., Abdurahman, K., Cao, Y., Duan, H., and Jia, D. (2020). Mechanical, electromagnetic shielding and gas sensing properties of flexible cotton fiber/polyaniline composites. Composites Science and Technology. 188, 1-12.

[4] Vinayagamoorthy, R. and Venkatakoteswararao, G. (2021). Development and characterization of bio-composites using *Senegalia catechu* resin. Polymer Composites. 29, S1268-79.

[5] Fei, M., Liu, W., and Jia, A. (2018). Bamboo fibers composites based on styrene-free soybean-oil thermosets using methacrylates as reactive diluents. Composites Part A: Applied Science Manufaturing. 114, 40–48.

[6] Vinayagamoorthy, R., Subrahmanyam, K. S. Murthy, K. S. M. K., Prajwar, K. A., Gopinath, P., Lahari, M. S., and Rangan, K. P. (2020). Influence of nanoparticles on the characters of polymeric composites. IOP Conference Series: Materials Science and Engineering. 954, 012026.

[7] Silva-Guzman, J. A., Anda, R. R., and Fuentes-Talavera, F. J. (2018). Properties of thermoplastic corn starch based green composites reinforced with ´ barley (Hordeum vulgare L.) straw particles obtained by thermal compression. Fibers and Polymers. 19, 1970–79.

[8] Lodha, P. and Netravali, A. N. (2005). Thermal and mechanical properties of environment-friendly 'green' plastics from stearic acid modified-soy protein isolate. Industrial Crops and Products. 21, 49–64.

[9] Mao, Y., Wen, S., Chen, Y., Zhang, Y., Panine, P., Chan, T. W., Zhang, L., Liang, Y., andLiu, L. (2013). High performance graphene oxide based rubber composites. Science Reports. 2, 1-14.

[10] Golchin, A., Simmons, G. F., and Glavatskih, S. (2013). Tribological behaviour of polymeric materials in water-lubricated contacts. Proceedings of the Institution of Mechanical Engineers. 27, 811–25.

[11] Vinayagamoorthy, R. (2022). Bio-fibre reinforced polymeric composites for industrial, medicine and domestic applications. Bio-fiber Reinforced Composite Materials, Composite Science and Technology, Malaysia: Springer Nature, pp. 31-49.

[12] Desirability bar chart Rios, N., Winker, P., and Lin, K. J. D. (2022). TA algorithms for D-optimal of a mixture designs. Computational Statistics & Data Analysi. 168, 1-10.

[13] Vinayagamoorthy, R., Rajeswari, N., and Karuppiah, B. (2014). Optimization studies on thrust force and torque during drilling of natural fiber reinforced sandwich composites. Jordan Journal of Mechanical and Industrial Engineering. 8, 385-92.

[14] Rajmohan, T., Palanikumar, K., Davim, J. P., and Premnath, A. A. (2014). Modeling and optimization in tribological parameters of polyether ether ketone matrix composites using D-optimal design. Journal of Thermoplastic Composite Materials. 29, 161-88.

[15] Vinayagamoorthy, R. (2018). Friction and wear characteristics of fibre reinforced plastic composites. Journal of Thermoplastic Composite Materials. 33, 828-50.

[16] Panin, V. S., Alexcnk, O. V., and Buslovich, G. D. (2022). High performance polymer composites: a role of transfer films in ensuring tribological properties—a review. Polymer. 14, 1-43.

[17] Vinayagamoorthy, R. (2017). A review on the polymeric laminates reinforced with natural fibers. Journal of Reinforced Plastics and Composites. 36, 1577-89.

Chapter 30

Simulation and experimental analysis of lithium-ion battery and supercapacitor hybridisation for EV

Ravikant Nanwatkar[1,a], Deepak Watvisave[2,b] and Aparna Bagde[3,c]

[1]Ph. D. Scholer, Department of Mechanical Engineering, Sinhgad College of Engineering, Vadgaon, Pune, SPPU, India

[2]Ph. D. Guide, Department of Mechanical Engineering, MKSSS's Cummins College of Engineering for Women, Pune, India

[3]Assistant Professor, Department of Computer Engineering, NBN Sinhgad Technical Institutes Campus, Ambegaon, Pune, SPPU, India

Abstract

Hybridisation of energy storage systems in the automobile sector is a prior need of the current situation. The increasing prices and dependency on petroleum fuels will tend to end the storage of petroleum fuels. The extensive use of petroleum fuels raises many environmental issues which will worsen in the upcoming time. To meet the current and future demands of energy in the automobile sector and to replace petroleum fuels it needs energy storage that can match the energy and power density of internal combustion engine (ICE). As per Ragone's plot and ample availability of other non-conventional energy sources, this can be made possible with the hybridisation of different energy storage systems, as the use of anyone can't satisfy the energy and power demands. Many different hybrid combinations like a battery with an ICE, battery with a capacitor, battery with solar, battery with CNG, etc. are possible. The issue is to use the proper combination of both energy systems to avoid further issues in their use for energy sectors, automobiles, and other day-to-day applications. This can be made possible by the selection of appropriate battery chemistries and their structural and thermal behavior for suitable applications. The proposed work focuses on the structural investigation of the hybridisation of lithium-ion batteries and supercapacitors as a single unit and the evaluation of its working parameters. The work starts with an experimental analysis of the lithium-ion battery pack and EDLC supercapacitor pack as a single unit and an evaluation of its current-voltage variation for the applied load. Further the results are compared with simulation analysis in MATLAB using circuit modeling in Simulink for a lightweight electric vehicle. This work will evaluate the current, and voltage variation of HESS along with its change in other working parameters like state of charge, and depth of discharge to make the proposed HESS a novel solution for an electric vehicle.

Keywords: HESS, lithium-ion battery, supercapacitor, Simulink, Simscape, Ragone plot

Introduction

Electric mobility is the field of electric motor-powered on-road vehicles for which the power grid is a primary energy source. These types include pure battery-based, plugin-charged, and hybrid vehicles of different energy storage systems. The basic source of fuel supply to electric vehicles is electrical energy which significantly reduces CO^2 emission. This electricity can be generated from many non-conventional energy resources which helps to decrease the use of fossil fuels. Various rechargeable batteries can be an efficient option for replacing conventional energy sources and reducing their unwanted effects. Comparing various energy storage systems/batteries, lithium-ion battery was found to be the most promising option used for vehicles. These are rechargeable batteries having metallic lithium as an anode with efficiency parameters like elevated energy density, insignificant memory effect, and stubby self-discharge. This li-ion battery has certain limitations of flammable electrolytes and the structural issue of explosions and fires if damaged or charged incorrectly. Various hybrid combinations of energy storage systems are used in electric vehicles like a combination of battery and IC engine, battery with fuel cells, or supercapacitor. Table 30.1 shows the comparative analysis of various power and energy densities of various energy storage systems. The efficient system design of energy storage and its thermal management is a critical issue while designing any of the above hybridisations. In the battery-based system of energy storage, a battery with elevated power density is required to meet the power demands. As a solution for this, the size of the battery can be increased but it will raise the issue of an increase in cost and thermal management in high power load as well as cold temperature conditions. In addition to that there are certain issues related to the cell balancing of the battery as without it, the individual cell voltages tend to vary over a period. This will lead to a reduction in the capacity of the battery pack during operation, resulting in failure of the total battery system during peak rate of charge and discharge conditions. In addition to these issues, the application where instantaneous power input and output are required i.e., where batteries suffer from repeated charge and discharge operations adversely affect the life of the battery. All of the above problems can be resolved by a hybrid energy storage system in which supercapacitor is combined with lithium-ion batteries to achieve better overall

[a]aravikant.nanwatkar@sinhgad.edu, [b]deepak.watvisave@cumminscollege.in, [c]aparnabagade.nbnssoe@sinhgad.edu

DOI: 10.1201/9781003450252-30

Table 30.1 Comparison of Li-ion Supercapacitor with other Energy storage system [5]

Energy storage device	Energy density (Wh/kg)	Power density (W/kg)	Cycle life
EDLC (Supercapacitor)	2-8	500-5000	>100000
Li-ion &Supercapacitor	10-20	900-9000	>100000
Li-acid battery	30-50	100-200	200-300
Li-ion battery	100-265	100-265	300-500
Ni-MH battery	60-120	250-1000	300-500
Zinc-bromide battery	85-90	300-600	2000

Source: Author

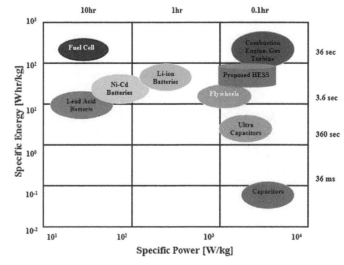

Figure 30.1 Ragone plot of different energy storage system [18, 19]
Source: Author

performance. This is because supercapacitor has more power density, but lower energy density compared to batteries. Figure 30.1 shows the proposed HESS of lithium-ion battery and EDLC Supercapacitor which can meet the energy and power demands to replace conventional IC engines.

Figure 30.2 shows, during higher power requirements i.e., start and peak power phase, supercapacitors will work and during smooth running conditions batteries will work for supplying energy requirements which will surely decrease the battery stresses and increase the life of the battery. The basic idea, in this work, is to use the best features of both devices. The high life cycle and power density of the supercapacitor will be used to improve the battery life and optimise the use of energy density so that there will be a significant reduction in the cost of the battery as well as CO_2 emission. Figure 30.5 shows the block diagram of the experimental setup of the work. This hybridisation should be such that it should be equipped with high power and energy density. Basically, there are three ways for this hybridisation as shown in Figure 30.3. First is the passive type in which the battery and supercapacitor are connected directly to the DC bus. It has advantages like high peak power capability, higher efficiency, and longer battery life cycle, etc. but has a drawback of unachievable system optimisation as there due to the absence of a mechanism for power management that will control the power-sharing between both energy storage systems i.e., Battery and the supercapacitor. Whereas in the active type, the battery and the supercapacitor are connected to the DC bus via the DC-DC converter. It has the advantages of design flexibility, acquiring higher power capacity, less voltage variation, and reduction in weight but with the disadvantage of an increase in the cost of the DC-DC converter. Other possible hybridisation can be other types of batteries as mentioned above with either a supercapacitor or an internal combustion engine. Other types of energy storage can be fuel cells which can be afuture aspect in the automobile field.

The proposed work is based on analysing different hybrid combinations specifically for light weight applications like two-wheelers. Various types of earlier work related to this are explained in the literature survey.

Objectives

• To understand the earlier work done on different types of hybridisations of lithium-ion batteries and supercapacitors for electric vehicles through literature surveys and to know the further scope of work.

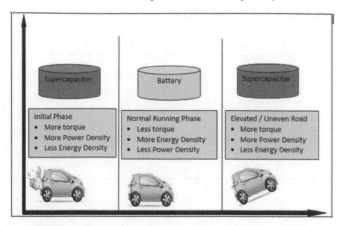

Figure 30.2 Energy and power requirements of proposed hybrid energy storage system for electric vehicle
Source: Author

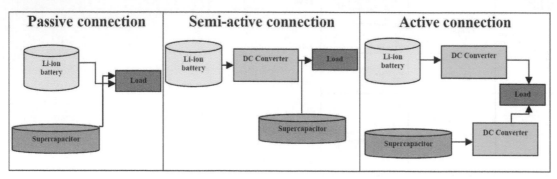

Figure 30.3 Types of hybridisation for battery and supercapaciter [16]
Source: Author

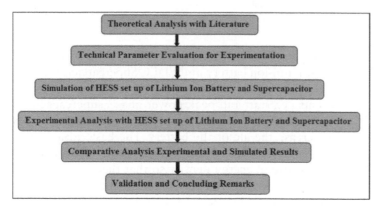

Figure 30.4 Methodology of working
Source: Author

- To study the different charging protocols along with its standards for the proposed hybridised energy storage system.
- To study and understand the standards to be used for components and parts selection for the experimental setup.
- To develop and perform the simulation analysis with EV circuit modeling in Simulink to the performance behavior of HESS as per the applied load for further application.
- To study the different chemistries for lithium-ion batteries and supercapacitors.
- To design the HESS for fast charging by following all safety rules and regulations.
- To compare the simulation results with experimental analysis for HESS of li-ion battery and supercapacitor.

Methodology

As shown in Figure 30.4, the work starts with the identification of the research gap through various literature related to the HESS of batteries and supercapacitors. A further simulation study is performed for HESS

Proposed set up

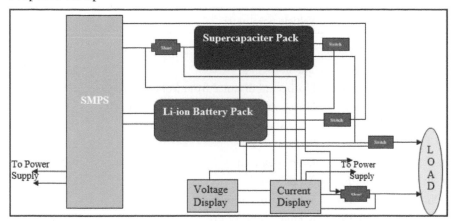

Figure 30.5 Block diagram of the experimental setup of the work
Source: Author

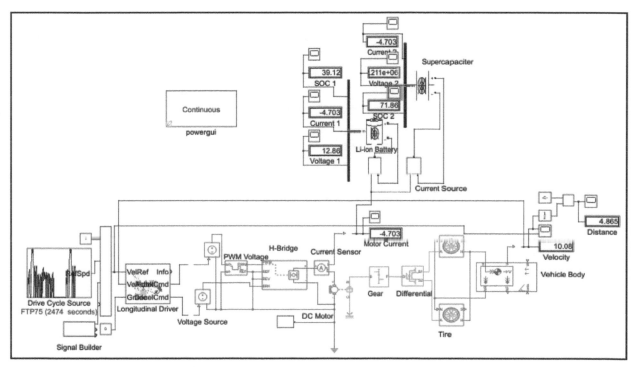

Figure 30.6 Simulink model of electric vehicle
Source: Author

using Matlab/Simulink to evaluate the variation of performance parameters i.e., state of charge, voltage, current, etc. w.r.to time for the FTP 75 drive cycle as shown in Figure 30.6. Next in the experimental set up as shown in Figure 30.18, the battery pack is fabricated using a lithium polymer battery, and a supercapacitor pack is formed using EDLC type. The HESS of both energy storage systems is formed considering a load of 220 watts for approximately 13.5 V and 20 A current capacity. Three different switches are incorporated to operate the battery and supercapacitor individually as well as in hybrid mode. An SMPS is connected to both energy storage devices which work as a charger for both. Display devices are connected through a shunt resistor of mating capacity for evaluating voltage and current variation for the experimental setup. Table 30.2 shows the component specification for the experimental setup.

Literature Survey

In 2012 has worked on Energy recovery through regenerative braking and the results are simulated using Matlab/Simulink PLECS toolbox [1]. In 2012 experimented on HESS with power system analysis toolbox (PSAT) with a compact DC/DC converter that acts as a restrained energy pump for maintaining an inflated amount of ultracapacitor compared to the battery at driving conditions [2]. Worked on novel HESS of lead

Table 30.2 Component specifications for simulation work

Battery parameters		Vehicle parameters	
Nominal voltage of lithium ion battery	12 V	Mass of vehicle	100 kg
Rated capacity	20Ah	The horizontal	2
Initial state of charge	100%	Horizontal distance from front axle to CG	1.4 m
Battery response time	10sec	Horizontal distance from rear axle to CG	1.6 m
No. of batteries in series (each of 3.7V)	3	CG distance above ground level	0.5 m
No. of batteries in parallel (each 2.5A)	8	Frontal area	2m2
Supercapacitor parameters		Drag coefficient	0.25
Rated capacitance	500F	Air density	1.18kg/m3
Equivalent DC series resistance (Ohms)	470kΩ	Vehicle tire parameters	
Rated voltage	13.5V	Rolling radius of tire	0.3 m
Number of series capacitors	5	Rated vertical load	3000 N
Number of parallel capacitors	1	Peak longitudinal force at rated load	3000 N
Initial voltage	0	Slip at peak force at rated load	10%
Operating temperature	25deg	DC motor parameters	
Gear and differential parameters		Rated speed (at rated load)	5000 RPM
Carrier (C) to driveshaft (D) teeth ratio	4	Rated DC supply voltage	50V
Follower (F) to base (B) teeth ratio (NF/NB)	2	No-load speed	7500 RPM
		Rated load (mechanical power)	200W

Source: Author

Table 30.3 Component specifications for experimental work

Particulars	Specifications
A switched-mode power supply (SMPS)	12V, 10 A
Li-ion battery pack	11.1V, 20 A
Supercapacitor pack	13.5V, 500 F
Voltage display unit	0-200 V
Current display unit	0-75 A
Resistor for supercapacitor connections	470 KΩ, 2W
Shunt resistor (2)	75 mV
Switches	3
Load	200 W
Other accessories i.e., solder gun, wire	-

Source: Author

acid battery and supercapaciter using regenerative braking [3]. It worked on supercapacitor characteristics and energy recovery through regenerative braking. Modifications can be done by increasing the supercapacitor operating voltages to enable energy content to be maintained while reducing equivalent series resistance. Ostadi et al. [4] worked on various literature related to HESS of batteries and supercapacitors by connecting them to DC sources to meet energy and power demands of the vehicle including energy management issues. Experimentation showed that the dissociate arrangement with the supercapacitor cell connected to the DC bus and battery cell connected via a bidirectional DC-DC converter is an efficient associating organised in EV/HEV applications. In 2015 worked on lithium-ion batteries and ultracapacitors for network applications with variant materials for cathode, anode, and a lithium-ion battery that results in a variety of output performance characteristics along with equivalent electrical circuits [5]. Their work relates lithium-ion and ultracapacitors for high power density with extensive discharge demand application to improve issues encountered in lithium-ion batteries like high production cost, and high sensitivity for thermal runaway. Capasso et al. [6] worked on HESS with Na-Cl batteries and EDLC using a controlled DC/DC bi-directional power Converter. Further work is proposed on Simulation and experimental study with HESS of lithium-ion and ultracapacitor. Zuo et al. [7] worked on various combinations of HESS with a high capacitive battery and rated capacitive electrode. Further works remained on BSH with fluent high voltage window and integrated

3D electrodes set up. Herath et al. [8] worked on the charging and discharging algorithm of batteries and supercapacitors as per their acceleration and deceleration conditions. The work focused on reducing the strain on the batteries while extending the range of the vehicle compared to the traditional pure battery-based electric vehicle. Soltani et al. [9] worked on lithium ion capacitors that are used as a high-power storage unit for MLTB driving cycle. Further work remained to optimise the li-ion battery and capacitor unit for optimised cost, and size with a higher energy and power density. Sawa et al. [10] worked on HESS with lithium-ion batteries and an ultracapacitor model to evaluate the thermal and electrical performance parameters for different driving cycles. The simulation results in improved dynamic stress, better thermal performance for peak power demand better life span of the battery, and reliability of HESS. The remained work is set up formation for an electric propulsion system test bench to validate the simulation results and to incorporate the intelligent energy management system in the model. Arefin et al. [11] worked on simulations of HESS with battery and supercapacitor new and partially used battery cells. The results showed inverse proportionality between the temperature and the hybrid system efficiency. This hybridisation with increases the battery life span, and the efficiency of the energy storage system and power train. This HESS gives advantages of reduction in battery aging, peak battery current, and a greater number of executed cycles with an increase in power preserving capacity of the system that increases the battery maintenance interval. Kouchachvili et al. [12] worked on battery and supercapacitor HESS by coupling the battery with a super-capacitor, which is basically an electrochemical cell with a similar architecture, but with a better capability rate and cyclability. The Basic principle was a supply of excess energy by the supercapacitor when the battery won't be able to do so. Configurations, design, and performance of HESS had been discussed with active, semi-active, and passive types of HESS. Various applications area of HESS like mobile charging stations, and racing cars, have been discussed with different batteries and supercapacitor combinations, related issues, and future aspects. Immanuel et al. [13] proposed a well-organised hybridisation of battery, supercapacitor, and hybrid capacitor for efficient energy consumption in electric vehicles. The work remits the issue of deficiency in autonomy between two recharge points for the supercapacitor. Experimentation involved analysis of multiple inputs for DC-DC convertor and obtaining electric vehicle profiles for proposed HESS. This work can be further extended for various load profiles with peak crest factors. Vidhya et al. [14] worked on the simulation, design, and power arrangement of the hybrid energy storage system of li-ion battery and supercapacitor which combined a bi directional convertor for a light electric vehicle under Indian driving conditions to get optimized working parameters to improve the life of both energy storage systems. Simulation and experimental analyses were carried out to verify the efficacy of the proposed system with modelled prototype system components of a light electric vehicle. Sankar et al. [15] worked on a smart power converter for an electric bicycle, powered by hybridisation of lead acid battery and supercapacitor. The supercapacitor was connected in parallel to the battery pack via Arduino controller-based power converter that adjudges power between both energy storage systems. Experimental results showed an enhancement in the ascending speed w.r.to time of the bicycle as an undeviating result of the power converter delicate to reaping the remaining current from the high-power adjustable supercapacitor neglecting extensive discharges from the battery to improve its life without a change in maximum speed. The main battery pack was protected from high discharge currents to improve its life cycle. Walvekar et al. [16] worked on the hybridisation of Li-ion batteries and supercapacitors for lightweight electric vehicles. In this paper, the result of various combinations of hybrid energy storage systems and the effect of hybridisation is analysed w.r.t. current, voltage, and state of charge (SOC). Results showed that the use of HESS for pure battery-based electric two-wheelers decreases the higher value of the current of the battery with the corresponding improvement in battery life.

Research Gaps

1) Literature survey showed much work needs to be performed on the implementation of the electric vehicle and their hybridisation related to their structural and thermal investigations as per road and transportation conditions.
2) Mechanism needs to analyse for activation of supercapacitor during high power density requirements and battery during high energy density requirements.
3) HESS can be improvised with a regenerative braking system to charge the supercapacitor after it gets discharged.
4) Systematic experimental study to analyse the effects of noise and vibrations on HESS connections as well as its effects on various performance parameters like state of charge, voltage, and current variations.

The novelty of work:

1) Simulation of electric vehicle for battery and HESS operated conditions for various drive cycles using Matlab/Simulink.

2) Effective hybridization of lithium-ion battery and supercapacitor to compare the results of battery and HESS for given load and working conditions to evaluate the variation of performance parameters like state of charge, current, and voltage through experimental setup.

3) Design of smart control system for activation of battery and supercapacitor in HESS as per the energy and power density requirements for the applied load.

4) Model-based design of HESS of lithium-ion battery and supercapacitor as a whole component to use it for application.

Structural Analysis

List of components required for Simulation and Experimental Analysis

A *Calculations of the battery pack for generating approx.100-watt energy for approx. 2 hours.*
   - Single battery with 3.7 V and 2500 mAh capacity.
   - Using formulae (assuming 80% efficiency of the battery)
   - Therefore 24 cells of batteries with a pack of three $(3.7 \times 3 = 11.1)$ pairs in parallel and eight $(2500$ mAh $\times 8 = 20$ Ah$)$ in series combination.
   - Total energy generated: VA/100 = $(11.1 \times 20)/1000 = 0.222$ Kwh = 222 w/h
   - Total power generated: VA = $11.1 \times 20 = 222$ watt.

B *Calculations for Supercapacitor pack,*
   - We have taken five green cap EDLC (DB) supercapacitor with 2.7 V and 500 faradays, having size 35 mm × 60 mm and connected in series for experimental analysis.
   - Input current limit –1 mΩ, Discharge limit – 470 kΩ,
   - Total voltage $2.7 \times 5 = 13.5$ V
   - Total capacitance (CT) = $\frac{1}{C_T} = \frac{1}{C_1} + \frac{1}{C_2} + \frac{1}{C_3} + \frac{1}{C_4} + \frac{1}{C_5} = \frac{1}{500} + \frac{1}{500} + \frac{1}{500} + \frac{1}{500} + \frac{1}{500} = \frac{5}{500}$
     $C_T = 500/5 = 100$ Faraday
   - Energy calculation (E) = $\frac{1}{2} C_T V^2 = \frac{1}{2} \times 100 \times 13.5^2 = 9112.5$ joules = 2.53125 w/h
   - Power generated = $E/(t_2 - t_1) = (2.53125 / 3) = 843.75$ watt.

Simulation for electric vehicles to evaluate the battery parameters is proposed here. The simulation has been carried out using MATLAB/Simulink considering the hybridisation of a lithium-ion battery which is initially charged 100% and supercapacitor as an energy source and following input parameters,Initially, the battery was charged 100%. Comparing FTP 75 cycle of 2474 seconds, the following results were obtained,

Results and Discussion:

Simulation Analysis:

Simulation of HESS shows using graphical representation for variation in different parameters, and it seems that,

1) Battery state of charge decreases linearly w.r.to. Time with some interruption is shown in Figure 30.7, whereas that of the supercapacitor decreases up to 1400 second and suddenly got rise linearly till 2000 seconds with some interrupted variation for further time cycle as shown in figure 30.8.

2) Battery current and supercapacitor current variation is nearly the same with approximately step variation as shown in Figures 30.9 and 30.10.

3) Voltage variation for the battery is suddenly decreased for the first 100 seconds with interrupted variation for further time cycle, as shown in Figures 30.11 and 30.12.

4) Distance covered by vehicle shows increasing linear relation for the first 1300 seconds, for the next 700 seconds, it is constant, whereas again increases linearly further till the completion of the cycle as shown in Figure 30.13.

5) Figure 30.14 shows the deflection of the stated HESS driving cycle w.r.to standard FTP cycle. It seems that it needs further variation and modification scope to match the FTP cycle.

6) The performance parameters achieved after testing the HESS simulation for 2474 seconds as below, vehicle velocity – 10.08 Km/hr and distance travelled – 4.865 Km.
   The battery state of charge is 39.12, whereas the supercapacitor state of charge is 71.86.

7) The Simscape results reveal the results for other parts of HESS i.e., Figure 30.15 shows step-wise variation except from 1400 seconds to 2000 seconds as shown in Figure 30.15. A similar result was found for the DC motor and other vehicle body parameters such as gear and differential system as

Simulink Results:

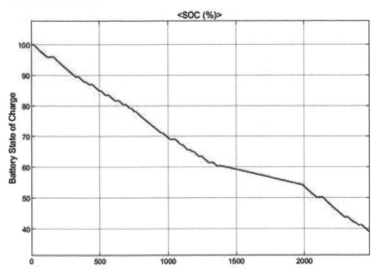

Figure 30.7 Battery state of charge vs time
Source: Author

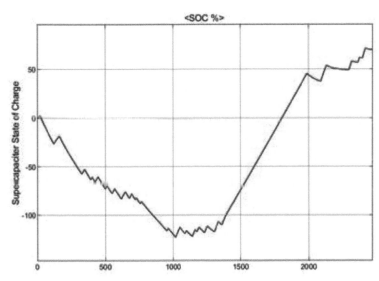

Figure 30.8 Supercapaciter state of charge vs time
Source: Author

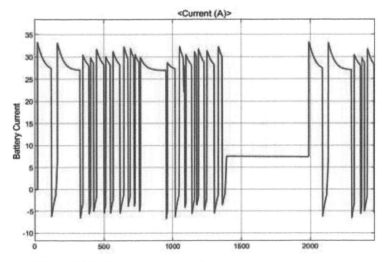

Figure 30.9 Battery current vs time
Source: Author

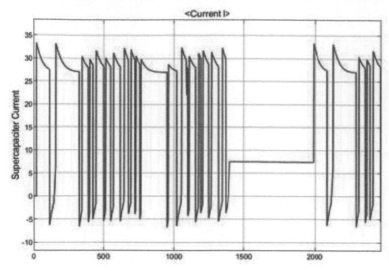

Figure 30.10 Supercapaciter current vs time
Source: Author

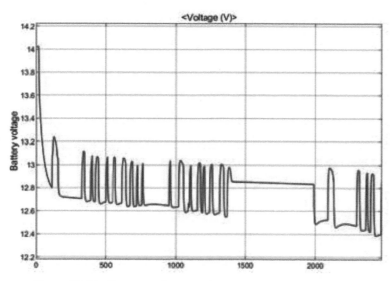

Figure 30.11 Battery voltage vs time
Source: Author

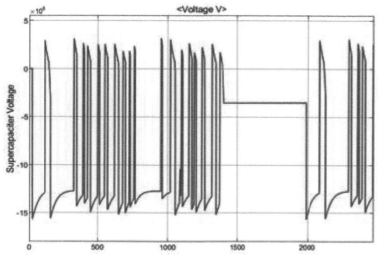

Figure 30.12 Supercapaciter voltage vs time
Source: Author

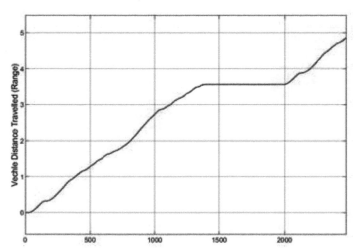

Figure 30.13 Vehicle distance covered vs time
Source: Author

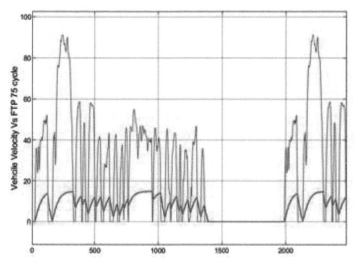

Figure 30.14 ehicle velocity covered vs time
Source: Author

Simscape Results

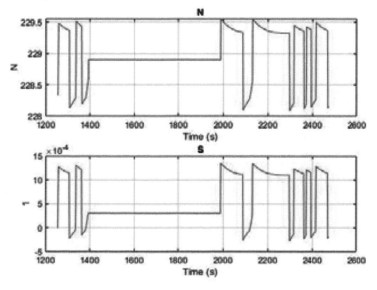

Figure 30.15 Results for load on tire
Source: Author

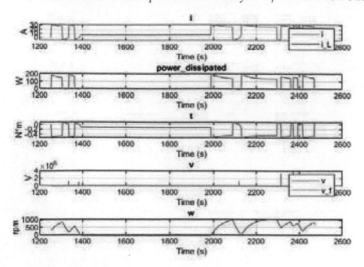

Figure 30.16 Results for DC motor
Source: Author

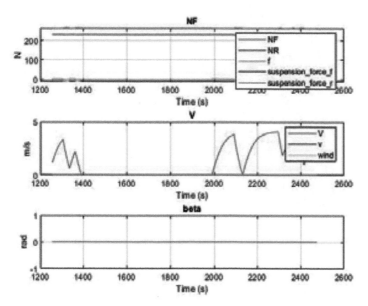

Figure 30.17 Results for effects of vehicle body parameters
Source: Author

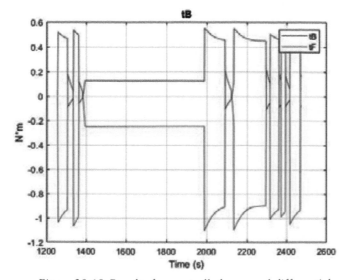

Figure 30.18 Results for controlled gear and differential
Source: Author

Set up for experimentation:

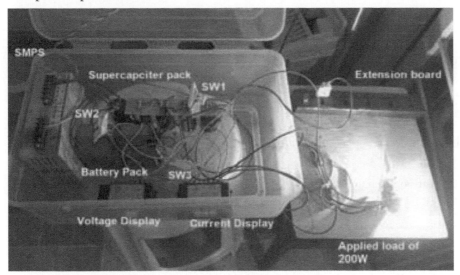

Figure 30.19 BExperimental setup for HESS of Lithium-ion and supercapacitor
Source: Author

Experimental Results (time for experimentation if approx. 3600 seconds):

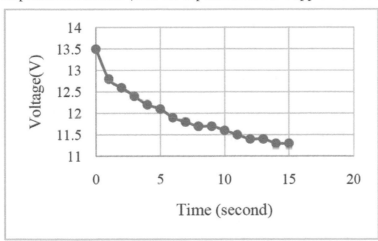

Figure 30.20 HESS voltage variation w.r.to time at no load condition
Source: Author

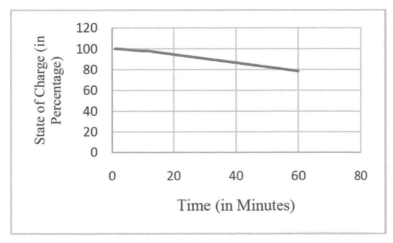

Figure 30.21 HESS SOC variation w.r.to time at load of 220 W
Source: Author

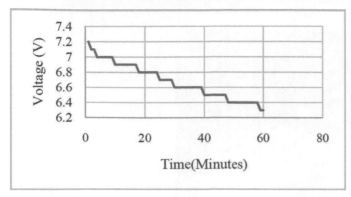

Figure 30.22 HESS voltage variation w.r.to time at load of 220 W
Source: Author

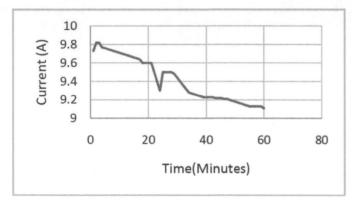

Figure 30.23 HESS voltage variation w.r.to time at a load of 200 W
Source: Author

shown in Figures 30.16, 30.17, and 30.18. This shows the smooth running of the vehicle due to low stresses on the battery.

Experimental Analysis:

1) Initially, the HESS is tested without any load to check the effects on the supercapacitor, it observed that a very negligible effect was observed on the lithium-ion battery pack, but considerable change has been observed on the supercapacitor pack with a change in voltage capacity at constant as shown in Figure 30.19.
2) Further the HESS is tested for a load of 200 watts (four-wheeler headlights) for approximately 1 hour.
3) At loading conditions voltage of HESS decreased stepwise with the initial jerk for the first 2 minutes.
4) The current variation is discrete with an initial jerk for the first 2 minutes with linear variation till 18 minutes, later it is constant for the next 2 minutes. For the next 2-3 minutes it decreases drastically to 9 amperes. The current increases suddenly to 9.6 amperes to decrease in further operation. This is the point where the supercapacitor comes into working as shown in Figure 30.21.
5) HESS state of charge varies linearly w.r.to time as shown in Figure 30.20.
6) It is observed that the supercapacitor charge and discharge time is approximately 3 minutes for the given setup and that for the battery it is approximately 2.5 hrs.

Conclusion

After simulation and experimental analysis of the proposed hybridisation of lithium-ion battery and supercapacitor, we can find that when the HESS is worked for given load considerable results show that the supercapacitor acts only when there is more power requirement else is ideal. As soon as the supercapacitor is discharged it gets charged from the battery when the battery is not working for heavy loads. Further, as the load decrease battery acts for the next working and keeps the supercapacitor ideal. This effectively decreases the load on the battery for high power requirements. This shows a significant solution for automobiles for effective hybridisation to meet current demands of energy and power in the energy and automobile sector. Comparative analysis of simulated and experimental results showed acceptable deflections in current, voltage, and state of charge w.r.to time. In this analysis FTP 75 cycle is considered for comparative analysis with the proposed electric vehicle the graphical representation of which shows the necessity of modification in input parameters to match with the required cycle.

- Regenerative breaking mechanism with a smart control system will be implemented to charge the supercapacitor so that it can charge independently without dependency on the battery pack.
- The mechanical testing for the given experimental setup is proposed further to evaluate the effects of vibration and noise on various joints as well as performance parameters.

References

[1] Soltani, M., Jaguemont, J., den Bossche, P. V., van Mierlo, J., and Omar, N. (2012). Hybrid battery/lithium-ion capacitor energy storage system for a pure electric bus for an urban transportation application. Applied Science Article. 1-19.

[2] Cao, J. and Emadi, A. (2012). A new battery/ultracapacitors hybrid energy storage system for electric, hybrid, and plug-in hybrid electric vehicles. IEEE Transactions on Power Electronics. 27(1). 122-32.

[3] Carter, R. and Cruden, A. (2012). Optimizing for efficiency or battery life in a battery/supercapacitor electric vehicle. IEEE Transactions on Vehicular Technology. 61(4), 1526-33.

[4] Ostadi, A., Kazerani, M. et. al. (2013). Hybrid energy storage system (HESS) in vehicular applications: a review on interfacing battery and ultra-capacitor units. IEEE Transactions IEEE Transportation Electrification Conference and Expo (ITEC). 275-81.

[5] Hamidi, S. A., Manla, E., and Nasiri, A. (2015). Li-ion batteries and Li-ion ultra capacitors: characteristics, modelling and grid applications. IEEE Transactions. 4973-79.

[6] Capasso, C. and Veneri, O. (2017). Integration between supercapacitor and ZEBRA batteries as the high-performance hybrid storage system for electric vehicles. Science Directory, pp. 2539-44.

[7] Zuo, W., Li, R., Zhou, C., Li, Y., Xia, J., and Liu, J. (2017). Battery-supercapacitor hybrid devices: recent progress and future prospects. Advanced Science News publication, pp. 1-21.

[8] Gunawardena, H. P. (2018). Conversion of a conventional vehicle into a battery-supercapacitor hybrid vehicle. American Journal of Engineering and Applied Sciences. 1178-87.

[9] Soltani, M., Ronsmans, J. et. al. (2018). Hybrid battery/lithium-ion capacitor energy storage system for a pure electric bus for an urban transportation application. Applied Sciences. 8(7), 1-19.

[10] Sawa, L. and Poona, H. (2018). Numerical modelling of hybrid supercapacitor battery energy storage system for electric vehicles. Science Directory. 2751-55.

[11] Arefin, M. A. and Mal, A. (2018). Hybridization of battery and ultracapacitor for low weight electric vehicle. Journal of Mechanical and Energy Engineering 2(42), 43-50.

[12] Kouchachvili, L. and Yaïci, W. (2018). Hybrid battery/supercapaciter energy storage system for the electric vehicles. Journal of Power Sources. 374, 237-48.

[13] Jiya, I. N., Gurusinghe, N. et. al. (2019). Hybridization of battery, supercapacitor and hybrid capacitor for load applications with high crest factors: a case study of electric vehicles. Indonesian Journal of Electrical Engineering and Computer Science. 16(2), 614-22.

[14] Vidhya, S. D. and Balaji M. (2019). Modelling, design and control of a light electric vehicle with hybrid energy storage system for Indian driving cycle. Measurement and Control. 52(9-10), 1420–33.

[15] Bharathi, A. and Seyezhai, S. R. (2019). Super capacitor/battery based hybrid powered electric bicycle. WSEAS Transactions on Power Systems. 14, 156-62.

[16] Walvekar, A., Bhateshvar, Y. et al. (2020). Active hybrid energy storage system for electric Two wheeler. In SAE International, pp. 1-6.

[17] Cossalter, V. (2006). Motorcycle Dynamics. Second Edition. Lulu.Com.

[18] Reddy, T. (2010). Handbook of Batteries. McGraw-Hill Professional, 4th Edition.

[19] Moura, S. J., Siegel, J. B., Siegel, D. J., Fathy, H. K., and Stefanopoulou, A. G. (2010). Education on vehicle electrification: battery systems, fuel cells, and hydrogen. IEEE Vehicle Power and Propulsion Conference.

Chapter 31

Experimental investigations of remote street light monitoring using wireless sensor network for charging of electric vehicle

Ravikant Nanwatkar[1,a], Deepak Watvisave[2,b] and Aparna Bagde[3,c]

Ph. D. Scholer, Department of Mechanical Engineering, Sinhgad College of Engineering, Vadgaon, Pune, SPPU, India

Ph. D. Guide, Department of Mechanical Engineering, Sinhgad College of Engineering, Vadgaon, Pune, SPPU, India

Assistant Professor, Department of Computer Engineering, NBN Sinhgad Technical Institutes Campus, Ambegaon, Pune, SPPU, India

Abstract

Many different mechanisms are used for electric vehicle (EV) charging like regenerative braking, battery swapping, etc. But until the increase in the quantity charging stations and their ease in availability, it is difficult to boost the EV market. One of the effective solutions for this is lower the utilisation of energy by streetlights when it is not required and uses the energy to charge the ongoing vehicles. As many streetlights are ON, in daylight, or when there is no need for them. The present work focused on the design of smart solar streetlights using a wireless sensor network so that they will work only when the vehicle is under the street lamp or in the proximity of street lights. So that extra energy which would have been wasted can be used for charging the vehicle. For this, we need to modify the road design to make space available for charging in wireless mode using electromagnetic induction as well as by making a smart street light monitoring system. This paper introduces a wireless sensor network for street light monitoring to overcome the drawbacks of the traditional systems and increase reasonable adjustment, and seasonal variation, enhances quality in the service of human beings, and consumption of electricity using sensors. Work attempts to reduce the required period for which the lamp should turn ON and OFF to automatically detect the vehicle and remove nodes in the network. In this, we intend to control the central monitoring system and all streetlights in real-time as well as incorporate wireless sensors and an IoT-based charging system for the vehicles. Validation of the concept is done by designing hardware prototypes of the sensor node and remote terminal unit and HESS-operated vehicle.

Keywords: Hybrid electric vehicle, IoT system, wireless sensor network, street light monitoring system, frequency band, sensors, microprocessor, server

Introduction

The work starts with the energy principle i.e., the energy cannot be created nor destroyed, but only changes from one form to another. The current automobile industry is moving toward smart electric vehicles (EV) where most of the manual operation is going to be controlled by wireless sensor networks and internet connectivity. These features include identifying the nearest charging station, alarming about any particular part failure of the vehicle, information or signal communication in case of an accident due to foggy situations or urgent unavoidable situations, and many more. As far as Indian road conditions need proper hybridisation of energy sources as one can't rely on pure battery-based vehicles due to a lack of lithium stores. Streetlight monitoring not only controls the excess amount of energy wasted but also that saved energy can be used for charging stations which further can be used for charging the vehicle or other necessary applications. Using the mechanisms proposed in this work we can form a mechanism of variable light intensity of vehicle so that the street light system will give an indication to driver/vehicle about upcoming road conditions, foggy situations, light intensity, etc. and in only necessary conditions the light intensity of vehicle will increase else it will be run in a normal state or OFF. This will conserve the energy which would have been wasted and can be restored in energy storage devices kept at charging stations for further use. This leads to reduced impact on the battery and tends to increase its life. Further solar energy can be stored in an energy storage system as streetlights are replaced by solar panels. The sensory smart vehicle detection systems will indicate the presence of a vehicle and indication to a streetlight to turn ON/OFF. Further, ongoing light-capacity vehicles like two-wheelers or four-wheelers can be charged either in plug in mode through socket or using wireless sensor network.

Intelligent streetlights shown in Figure 31.2, improve the energy network communication by combining related areas into one control system. The smart street light monitoring will perform the lighting operation as per the vehicle detection by virtue of which the amount of energy that would have been wasted

---

[a]ravikant.nanwatkar@sinhgad.edu, [b]deepak.watvisave@cumminscollege.in, [c]aparnabagade.nbnssoe@sinhgad.edu

DOI: 10.1201/9781003450252-31

Figure 31.1 The integrated charging module
Source: https://www.infineon.com/cms/en/discoveries/eluminocity-street-light-makes-cities-smart/

unnecessarily when there is no vehicle on road, can be restored in the energy unit as shown in Figure 31.3. This energy unit can be further utilized for EV battery charging for vehicles. And for other necessary applications in nearby areas as well. The energy unit controller monitors the electricity, which runs by LEDs or sodium vapor lamps. Different sensory mechanisms along with an energy controller unit and street light monitoring system will detect the existence of the ongoing vehicles or persons in the region of the streetlight, identification of whether or any affecting environmental conditions, foggy or dust situation, etc. vary the light intensity as per the conditions. As shown in Figure 31.1, these smart streetlights are equipped with energy storage units with a wired mechanism for charging electric vehicles. Poor visibility at nighttime on highways and regular roads is a major issue for the transportation of vehicles. As driving is generally a visual task, therefore low light intensity, and foggy/smoky environment makes driving very difficult and many times lead to accidents. In the present situation of energy demands, constant electricity supply to the consumers is not feasible as the rate of electricity production is less than its utilization. Therefore, a better solution is to save electrical energy rather than its consumption and production. The street light monitoring found a novel solution for aforesaid issues. The objective of street light monitoring systems in electric vehicles using wireless sensor networks is to control the use of excessive electricity via remote ON/OFF/DIM of lights that reduce costs and amount of required energy, maintenance, and escalate the life of lamps which directly affects the efficiency of e-vehicle. Compared to conventional modern urbanisation and development in automobile industries certain parameters like the safety of drivers, safe driving in foggy situations, and reducing night crime rates need to be considered for the design of an efficient street lighting system. This system can also be used for road accidents to get prompt help for injured people to save their life. Conventionally streetlights were monitored manually, by using high sodium light lamps using optical control circuits. The difficult part for the city is to illuminate the lights on the roads as per the requirements, as it totally depends upon the climatic conditions like in summer day started at 5 AM but in the winter the day started after 7 AM, and in the rainy season lights depend upon circumstances. Therefore, seasonal changes are very important for controlling road illuminations. This system also has a feature to help the victims of road accidents using vision sensors and IoT networks. Many times, on the streets there are no vehicles at the time of night, but the lights are ON, so we can use the vehicle detection sensor for sensing the objects which are passing through the roads. It gives insurance to people who travel mostly at nighttime. In the beginning, road lights are switched ON in the morning time and switched OFF at nighttime by manual process, and our proposed system performs this work automatically on the basis of day and night light conditions. Basically, an automatic controlling system control and monitors energy recovery opportunities on the roads that's riches that area range of power stations. In this effort, wireless sensor networks were developed as a technical method to study the viability of monitoring the streetlight control system and charging the EV batteries for lightweight vehicles. IoT system is implemented to get the information and availability of charging point locations, and their capacity in terms of energy and power. The system contains, a modified road with space available for EV charging with the electromagnetic induction mechanism for wireless charging of electric vehicles i.e., the control center, streetlight equipped with solar panel, energy storage devices, street light monitoring system, Node sensor, and the remote fetal unit. The node sensor is installed at every lamp pole which is used to detect and control the lamp. The microprocessor-controlled electronic device acts as a booster station between the node sensor and the control panel center. The control panel supervises all streetlights in real time. The software is advanced for sensor nodes, servers, and hardware. Figure 31.2 shows the street light monitoring system. The systems applications can increase the

Figure 31.2 Smart city streetlight monitoring with solar panel and EV charging stations
Source: https://www.infineon.com/cms/en/discoveries/eluminocity-street-light-makes-cities-smart/

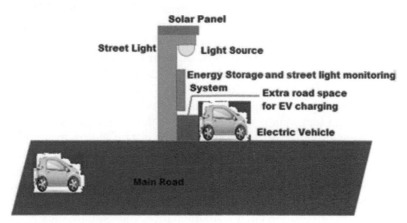

Figure 31.3 Working methodology
Source: Author

scope of controlling the road lights; reduce in electricity required for streetlights with a maintenance cost, which tends to increase the availability of streetlights and energy recovery. Entering wireless remote control for the street, roadway, and area lighting makes financial sense, whether switching to LED or retro-fitting fixtures. With the proposed system, one can able to conserve electricity by only using the accurate amount of light you require and by accurately measuring every watt used. Here we can make maintenance costs by monitoring real-time faults and by using entire operational intelligence to organize day-to-day planning and effectiveness.

In order to replace the conventional combustion engine-based vehicle with an electric vehicle we need to identify its affecting parameters like pollutant emission for variable speed limits, temperature monitoring, and structural issues as well. The smart streetlight equipped with a solar system measures the quality of air under different environmental conditions, detection of nearby vehicles to signal the start and stop of light emissions, and most importantly saves energy. Any unacceptable variation in the aforesaid parameters can be informed to the concerned authority for necessary actions for further improvement. Wireless charging will also reduce the chances of accidents and unnecessary traffic at charging stations. The same system can be further used for identifying the availability of parking spaces in heavy traffic zones by use of cloud connection and proper application by connective vehicle system with that of street light monitoring system using IoT.

Objectives:

1) Identification and selection of sensor and other components.
2) Experimental analysis of the setup for street light monitoring using a wireless sensor network to evaluate technical parameters.

3) Implementation of energy recovery and vehicle battery charging mechanism by identifying its working parameters like state of charge, current, and voltage variation.
4) Validation of the results with battery and supercapacitor-equipped hybrid electric vehicle.

B. *Methodology:*

As shown in Figure 31.3, the work starts with the identification of a research gap through the literature survey on energy generation through streetlights and its monitoring using wireless sensor networks. Here the basic idea is to replace the conventional system with a solar panel-based street light system, - so that in morning solar panel recovers energy and stores it in suitable energy storage devices. Also, smart street light monitoring will glow the lights only when the vehicle is in its radar/proximity. It also saves the waste energy which would be used for charging the energy system equipped on the pole of the streetlight. As an electric vehicle's battery state of charge reaches a minimum level, the driver will get informed about it on the display screen and search for a nearby charging station having sufficient energy level for charging the vehicle battery. For this purpose, during wired charging mode the driver needs to manually scan the printed QR code on the charging station. Then, the defensive shelter of the socket unlocks spontaneously, and the driver can attachment a charging cable so as to rejuvenate the battery of his vehicle. The payment is completed cash-free and appropriately via the smartphone app. Further using the wireless mode, wireless sensor network, and electromagnetic induction working principle the EV gets charged, so, in this case, the driver does need not to get out of the vehicle rather using suitable payment application software payment of battery charging will be done once it gets fully or sufficiently charged. This will also lead to minimising the traffic and crowd situation at charging stations.

Literature Survey

Liu, et. al. [1] worked on detection techniques of long-lasting motion for the dissimilarity of illustration which utilises histogram of direct inclination using a textured-based context model and Gaussian mixture model algorithm. Outputs of the experiment show the robustness and effectiveness of the commencement in recognising objects which may be an animal, vehicles, human beings, and whatever in different illustration states. Zhang, et.al. [2] worked on a WSN based street parking system for controlling the parking space condition by implementing a magnetic sensor node using a vehicle diagnosis and adaptive sampling method for accurately detecting a parking car and an for energy conservation. Tuna et.al. [3] worked on space and design provocations of WSNs considering current network applications with standard and communication protocols based on field tests in electric power system environments. Elejoste et. al. [4] presented an intelligent streetlight management system based on LED lamps, designed to facilitate its deployment in existing facilities using wireless sensor network to minimise the drawbacks of conventional system. Harri, et. al. [5] worked on a framework based on vehicular mobility for automobile vehicles. Further model evaluation of vehicular mobility and its communication with network simulation was carried out. The experiential analysis was carried out for mobility models for ad hoc networks of vehicles aiming for guidance for supply followers to understand effortless streetlights situation compared to the conventional system. Leccese, et. al. [6] presented automatic process systems that can increase the efficiency and management of road light systems. It uses devices like the ZigBee wireless private network that can be much more robust for the management of road lamp systems, the main part is their moderated infrastructure and controlling system. It utilized a number of sensors to control and to give assurance for the wanted parameters of networks; the data is exchanged from one end to another end using transmitters like ZigBee, and recipients are moved to control room used to clear the conditions of the street light lamps and for getting accurate output in the state of failing the system. Peng et. al. [7] proposed an efficient ZigBee-based energy conservation and control sysi.e.,i.e. power sensor nodes which is combination of a gateway, a base station, and sensors to developed and perform both local/remote power parameter measurement and on/off switching for electric appliances. Guo, et.al. [8] worked on a remote sensing system, used for controlling adaptive purpose of metropolitan air quality by deploying various vehicle, as sensor used to measure data including meteorological, traffic status, and environmental data in the city. Kaleem, et. al, [9] proposed electricity-saving outdoor illustration monitoring and control system using ZigBee private wireless network which is able to control and handle outdoor illustrations more effectively than the existing systems. In this system, using ZigBee wireless private network one can control the number of roads with the help of a single model. For proper working of the proposed system, the number of sensors is deployed in the required areas, and after deployment of all sensors. This system can save more than 70% of electricity because electricity saving is more efficient than production. Rashid et. al. [10] worked on each application of WSNs in urban areas in detail with all the problems and technical solution related to it. Zahurul et. al. [11] surveyed advanced agreement for various wireless sensor networks considering the realising communication infrastructure for DRG in Malaysia using IEEE802.15.4 with ZigBee PRO protocol, sensor and embedded system. Banerji et. al. [12] reviewed sources of energy conservation and as long as their working methodologies, detailed concept about relevant research and a present progression of their use in monitoring of structural health for civil engineering structures, considering solar and mechanical energy cultivator for monitoring structures. Nellore et. al. [13] presented a survey of current urban traffic management schemes for priority-based signaling and reducing congestion and the AWT of vehicles to provide a taxonomy of different traffic management schemes used for avoiding congestion. The urban traffic management schemes for the avoidance of congestion and providing priority to emergency vehicles are considered and set the foundation for further research. Tang et. al. [14] worked on comparative analysis of number of energy harvesting technologies applicable to industrial machines by investigating the power consumption of WSNs and the potential energy sources in mechanical systems. Toh et.al. [15] worked on the current state, developments, and emerging scope in area of transportation and mechanisms to improve smart streetlights that will make the future smart cities. Khalifeh et. al. [16] Proposed impotent aeriform vehicles to work as a data carrier for sensor output and transfer the same monitored data cautiously to the center of remote control for further analysis. Next work carried on implementing the issues in the realization of the framework with an experimental evaluation of the design in outdoorsy environments, in various types of environmental obstacles. Zhang et.al. [17] demonstrated the design of a smart street lighting system supported by the combination of NB-IoT and LoRa communication technology by adopting an optimized streetlamp control algorithm, to work as the automatic control of streetlights according to the realtime traffic flow information. Bhaskar et. al. [18] worked on cloud-based EV charging framework to overcome issues of high demand for EV charging stations using various IOT based methodologies like blockchain, behavioral science and economics, artificial and computational intelligence, IoV based digital twins and software, intelligent EV charging with information-centric networking, parking lot microgrids and EV-based virtual storage etc. Omar et.al. [19] made comparative analysis of many exploration studies related to smart street light monitoring systems, providing a comparative results between various systems that highlights the limitations of every of their current and future trends.

Research gap:

- Smart streetlight control system for vehicle detection, light intensity, and accident environmental conditions.
- Effective utilisation of conserved energy for further applications like EV charging and related novelty.

Novelty of work:

- Implementation of solar streets light and conservation of energy to charge the EV battery installed on the poles of the streetlights.
- Energy preservation by reducing the wastage of energy when it is not required on streetlights.
- Accident prevention system by identifying environmental affecting factors like Fogg, dusts etc.
- Wireless and wired charging of EVs as per the requirement using IoT-based applications and the principle of electromagnetic induction.

Experimental Setup

Proposed set up for street light monitoring

In this system, the role of ZigBee (a wireless private network) is to control the number of roads with the help of a single model. We used a PIC and Raspberry Pi microcontroller for controlling and monitoring all the sensors which were used. Also, we form a network so that we can monitor streets with the help of a single setup of hardware which is more cost-efficient than the existing system. Further for accident detection, we will use sensors like smoke sensors and microphones. After a sufficient literature survey we develop and propose street light monitoring using WSN as shown in the Figure 31.3, for monitoring the temperature, day/night environmental conditions, pollution, accident detection, and rescue system for smart e-vehicles, etc. The existing work consists of split roadside units that will be executed on streetlights and a principal server computer. It consists of a microcontroller and XBee network for communication with each system. Android devices are implemented at roadside units preferably for displaying a graphic user interface, application evolution, and interface with servers via the Xbee grid. Also, after recognition of road accidents, the system will spontaneously propel messages to the salvage system.

List of components required for experimental analysis

During the experimental set up all related sensors, microcontrollers, and other devices as mentioned in Table 31.1 are connected with each other as shown in Figure 31.4. And lithium-ion battery-operated

Table 31.1 component specifications of the experimental work

Sr. No.	Name of component	Specifications
1	Energy storage unit	Equipped with lithium-ion battery pack of 20A 12 V.
2	Solar panel	With 5A, 12V, 30W capacity.
3	Microcontroller	Peripheral interface controller: The range of PIC gadgets from 6-pin SMD, 8-pin DIP chips upto 144-pin SMD chips of hardware capabilities, with ADC, discrete I/O pin and DAC modules, and communications ports likeI2C, UART, CAN and USB.
4	Raspberry Pi boards	Clock frequency 1.2GHZ, SOC of chipset: Broadcom BCM 2837, Processor:32bit, Memory:2GB LPDDR2, no of USB 2.0 ports: 2, port extension:40pin GPIO, Video output: HDMI, data storage: micro SD card, network connection:10/100 ethernet, 802.11N Wi-Fi and bluetooth 4.1, peripherals 17GPIO, Supply:5V, 2.5A via micro USB, dimensions:85.60 mm × 53.98 mm × 17 mm. weight 45 g.
5	IR sensors	2.8 V at 15 cm to 0.4 V at 50 cm, use for obstacle detection, shaft encoder, fixed frequency detection
6	Light dependent resistor	Streetlight control, night light control, 100 LUX automatic headlight dimmer.
7	Gas/smoke sensor	1000 ppm domestic gas leakage detector, industrial combustible gas detector, portable Gas detector etc.
8	Temperature sensor	-30-180 degree centigrade with power supplies, battery management, HVAC, Appliances etc.
9	ZigBee	Standard IEEE: 802.15.4, frequency 2.4Ghz, range: 10-100 light of span, data rate: 20kbps to 250kbps, mesh network and device to device mode.
10	LCD 16 × 2	As per requirement
11	Vehicle model	Small toy car with hybrid energy storage of li-ion battery (3.2V and 2.5A) and supercapacitor (100 Faraday and 1.5).

Source: Author

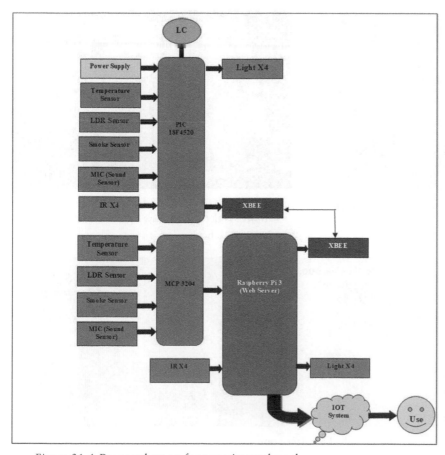

Figure 31.4 Proposed set up for experimental work
Source: Author

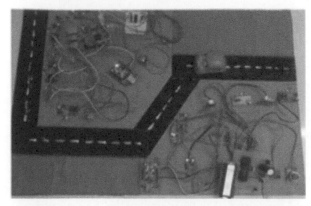

Figure 31.5 Experiential set up
Source: Author

Figure 31.6 Vehicle/object detection at high light intensity for node 1
Source: Author

Figure 31.7 Vehicle/object detection at lowlight intensity for node 1
Source: Author

Figure 31.8 Turn on the LED when a vehicle or object is detected at low light intensity for node 2
Source: Author

Figure 31.9 Turn off the LED at absence of vehicle or object at low light intensity for node 2
Source: Author

Figure 31.10 Turn off the LED when a vehicle or object is detected at low light intensity for node 2
Source: Author

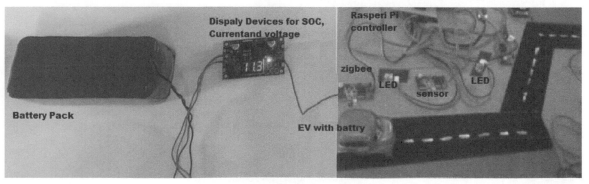

Figure 31.11 Street light monitoring system with the battery unit
Source: Author

Table 31.2 Different sensor data of experiential set up

For node 1

Time	Sound-1	Smoke-1	Light-1	Temp-1	I1-1	I2-1	I3-1
6AM	30.2	22	100.2	23.3	DIM	DIM	DIM
8AM	33.2	130	120.2	24.3	OFF	OFF	OFF
2PM	16.2	115.5	239.6	32.2	OFF	OFF	OFF
4PM	115	115.5	239.6	33.2	OFF	OFF	OFF
6PM	200	133	145.5	27.3	OFF	OFF	OFF
7PM	300	200	146.7	27.3	DIM	DIM	OFF
9PM	123	145	120.5	23.3	ON	DIM	OFF
11PM	234	145.22	120.8	23.3	OFF	ON	ON

For node 2

Time	Sound-1	Smoke-1	Light-1	Temp-1	I1-1	I2-1	I3-1
6AM	30.2	22	100.2	23.3	DIM	DIM	DIM
8AM	33.2	130	120.2	24.3	OFF	OFF	OFF
2PM	1022	288	185	32.2	OFF	OFF	OFF
4PM	234.5	234	185	33.2	OFF	OFF	OFF
6PM	200	133	145.5	27.3	OFF	OFF	OFF
7PM	300	200	146.7	27.3	DIM	DIM	OFF
9PM	123	145	120.5	23.3	ON	DIM	OFF
11PM	234	145.22	120.8	23.3	OFF	ON	ON

vehicle is used for practical performance of the charging of the vehicle. Energy storage units consist of one another battery pack which will be charged by the street light monitoring system and solar energy. Display devices are used to indicate the variation of performance parameters like state of charge, current and voltages. The smart system will indicate the working performance parameters of all connected sensors. Figures 31.5-31.10 show the experimental setup with vehicle detection at different light intensity mode. Figure 31.11 indicated the connections of the street light monitoring system with energy storage units (lithium-ion battery back) to analysis the battery parameters.

Result Analysis:

Table 31.2 shows experimental results for all sensor data recorded for nodes 1 and 2. In this table, we consider all sensor readings after every 2 hours these readings are for node1 and 2 for these all sensors are connected to the PIC microcontroller and the PIC microcontroller is sending it to the raspberry pi through the ZigBee network. Results were taken for sound level, smoke/foggy situation, different light intensities, and nodes. Lithium-ion battery pack with was initially fully discharged, indicates approx. 11% rise to get the state of charge as 11.3% without any measurable variation in current and voltage. This saved energy can be further used for charging the EV batteries using wired and wireless sensor modes.

Graphical representation:

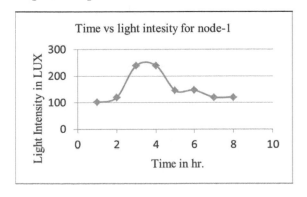

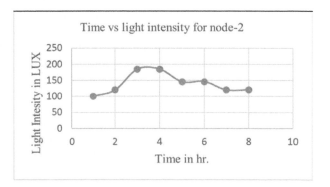

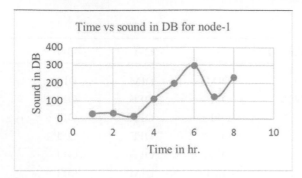

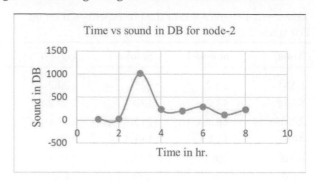

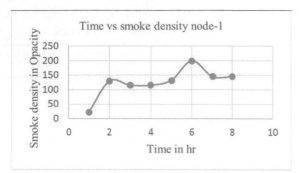

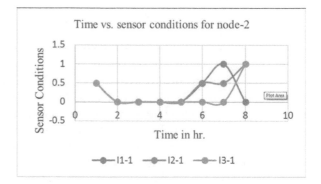

For sensor conditions ON = 1, Off = 0 and for dim = 0.5 use for graphical representations.

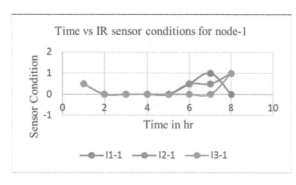

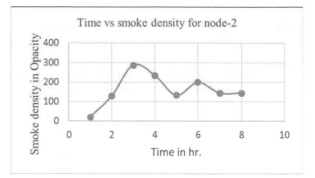

Conclusions

After experimental analysis of proposed system following conclusions are put forward:

- Implementation of a wireless sensor network in the street light monitoring model able to work automatically without using any physical resources, it can monitor the road lights according to the seasonal variation, reasonable changes, and passing vehicles through the roads. The graphical representations show the expected output from the proposed system. The X-axis shows the time every two hours, and the Y-axis shows all required sensors. It shows actual values of sensor changes based on climatic conditions, with the help of a proposed system we can easily monitor the streetlights. After the deployment of all sensors as proposed, it can be said that this system can save more than 70% of electricity. The saved electrical energy can be further used for the charging of electrical vehicle (EV) batteries and other related applications. Wireless mode of EV charging can reduce the physical efforts to move out from the vehicle to recharge the vehicle, crowd situation at the charging point, and most importantly significant reduction in energy and time as well. This type of application for street light monitoring is safe, durable, and with less complexity.

References

[1] Liu, W. et. al. (2010). Robust motion detection using histogram of oriented gradients for illumination variations. The 2nd International Conference on Industrial Mechatronics and Automation. pp. 1-10.

[2] Zhang, Z. et. Al. (2013). A street parking system using wireless sensor networks. International Journal of Distributed Sensor Networks. 15-23.

[3] Tuna, G. et. al. (2013). Wireless Sensor networks for smart grid applications: a case study on link reliability and node lifetime evaluations in power distribution systems. International Journal of Distributed Sensor Networks. 1-10.

[4] Elejoste, P. et. al. (2013). An Easy to Deploy Street Light Control System Based on Wireless Communication and LED Technology. Sensors. *13*(5), 6492-23.

[5] Hrri, J. et. al. (2013). Understanding vehicular mobility in network simulation. IEEE. 1-15.

[6] Leccese, F. (2013). Remote-control system of high efficiency and intelligent street lighting using a ZigBee network of devices and sensors. IEEE. 1-15.

[7] Peng, C. et. al. (2014). Development and application of a ZigBee-based building energy monitoring and control system. Scientific World Journal. 1-15.

[8] Guo, G. et. al. (2015). A mobile sensing system for urban PM2.5 monitoring with adaptive resolution. Journal of Sensors.

[9] Kaleem, Z. et. al. (2015). Energy efficient outdoor light monitoring and control architecture using embedded system. IEEE. 1-14.

[10] Rashid, R. et. al. (2015). Applications of wireless sensor networks for urban areas: a survey. Journal of Network and Computer Applications. 1-28.

[11] Zahurul, S. et. al. (2016). Future strategic plan analysis for integrating distributed renewable generation to smart grid through wireless sensor network: Malaysia prospect. Renewable and Sustainable Energy Reviews. 978-92.

[12] Banerji, S. et. al. (2016). Energy harvesting methods for structural health monitoring using wireless sensors: a review. Resilient Infrastructure. 1-10.

[13] Nellore, K. et. al. (2016). A survey on urban traffic management system using wireless sensor networks. Sensors. 16(2), 1-25.

[14] Tang, X. et. al. (2018). Energy harvesting technologies for achieving self-powered wireless sensor networks in machine condition monitoring: a review. Sensors. 1-39.

[15] Toh, C. K. et. al. (2020). Advances in smart roads for future smart cities. In Proceeding of Ryal Society Publishing. 1-24.

[16] Khalifeh, A. et. al. (2021). Wireless sensor networks for smart cities: network design, implementation and performance evaluation. Electronics. 1-28.

[17] Zhang, J. et. al. A low-power and low cost smart streetlight system based on internet of things technology.

[18] Bhaskar, P. et. al. (2022). Smart electric vehicle charging in the era of internet of vehicles. Energies. 1-24.

[19] Omar, A. et. al. (2022). Smart city: recent advances in intelligent street lighting systems based on IoT. Journal of Sensors. 1-10.

Chapter 32

Comparative analysis of NACA 0015 airfoil with bump using CFD

Aditya Solanki[a], Shashank Jibhakate[b], Vivek Mahadule[c], Yash Belekar[d] and Prasad Hatwalne[e]

Yeshwantrao Chawan College of Engineering, Nagpur, India

Abstract

Airfoils are the bodies used to produce aerodynamic forces when they are placed in an airstream. When the airfoils are subjected to different angles of attack (AOA) in an airstream it produces lift and drag forces. In the below presented work comparative analysis using CFD tool ANSYS Fluent 22 of National Advisory Committee for Aeronautics 0015 (NACA 0015) airfoil with its standard geometry and seven modified geometries, that is, engraving a bump at various location on the upper surface of airfoil along the chord were conducted. The simulation was conducted at different AOA from 0-16° (at an interval of 2°) and coefficient of lift and drag were obtained. In addition to this, pressure and velocity fields obtained from the simulation results were analyzed to understand the effect of bump.

Keywords: NACA 0015, lift and drag force, Airfoil, separation bubble

Introduction

Airfoil is the cross section of the body, which when placed in the fluid flow, experiences lift and drag forces. These forces are influenced by geometry, flow characteristics (which can be represented by Reynolds number) and angle of attack. Symmetric airfoil such as National Advisory Committee for Aeronautics 0015 (NACA 0015) has many applications such as wind turbine blades, sailplanes, aircraft tail, submarine fins, rotary and fixed wings etc. These systems can benefit from improved aerodynamic performance at different angles of attack. Motivated by this need, a few strategies have been proposed to improve this aerodynamic performance, which is, using different active flow control techniques on airfoil [1-3]. But these active flow techniques require external power and sometimes complex sensors. Which is why they aren't widely used. It has also been shown that geometric modification based passive flow control techniques can also achieve this goal. This includes, using winglets [4-5], vortex generator [6-7], Gurney flaps [8-9], shark denticle [10-12], etc. There have been lot of studies happening recently which use shark denticle arranged on the suction side of the airfoil to improve lift and reduce drag. Studies have shown that how shark denticles can be used to reduce drag [13]. Few studies showed that shark denticle can also be used to increase lift of an airfoil [14]. But the problem with denticle is, that they have complex 3D structure which makes it difficult to manufacture these wings using traditional manufacturing process. Domel et al. [14] performed analysis by placing shark denticle inspired bump on NACA 0012 airfoil and it produced improved lift at lower angles of attack. Adding to this, the airfoil with bump is easy to manufacture even using traditional manufacturing process. Villalpando et al. used NACA 63-415 model for prediction of 2D flows. They assessed different turbulence models at high angles of attack (AOA) using FLUENT tool. They observed that SA turbulence model predicts stable recirculation zones, and additionally, while SST k-omega turbulence model is best suited for clean wind turbine blades, SA turbulence model gives better lift near maximum lift and lower AOA [15]. Ravi et al. used NACA4412 airfoil geometry at low Reynolds number. They essentially tested the turbulence models, as the ones installed in commercial CFD software's will assume that the region of boundary layer around the surface of the airfoil is completely turbulent and produced significantly different results as compared to the experimental ones. The authors compared numerical results with experimental results and observed that two turbulence models produced near same values as the actual ones [16]. Srinivosan et al., studied the evaluation of turbulence models for unsteady flow on an oscillating airfoil. The authors used NACA 0015 airfoil to use five different turbulence models on. They observed that although Spalart Allmaras turbulence model overpredicts reattachment, and underpredicts separation, it had good similarity with results of lift and drag coefficient [17]. In this study we performed CFD analysis on NACA 0015 airfoil with shark denticle inspired bump placed on its suction side (upper side) at various location along chord length and calculated the values of lift and drag coefficient at different angle of attack from 0-16° (at an interval of 2°). For this study we are working with Reynolds number of $Re = 2 \times 10^5$.

[a]0712aditya@gmail.com, [b]shashankjibhakate88@gmail.com, [c]vivekmahadule01@gmail.com, [d]belekaryash136@gmail.com, [e]hatwalneprasad1@gmail.com

DOI: 10.1201/9781003450252-32

Theoretical Background

Lift and Drag Forces

When the airfoil is placed in a fluid flow, it experiences various forces across its surface. These forces are caused by pressure gradient and shear stress between the fluid and the surface, all these individual forces produce a resultant force. The component of the resultant force which is perpendicular to the flow is called Lift, and the component along the flow is called drag. These forces depend on various factors like density and velocity of fluid, area of the contact surface, angle of attack and the geometry of the airfoil. Both of these forces are expressed in dimensionless terms called coefficient of lift and coefficient of drag. The lift (C_l) and drag (C_d) coefficient is calculated by the equation [18]

$$C_l = \frac{2F_l}{\rho V^2 A} \tag{1}$$

$$C_d = \frac{2F_d}{\rho V^2 A} \tag{2}$$

Where, F_l = lift produced, F_d = drag produced, ρ = density of fluid, V = Velocity of fluid, A = (Chord × Span) area of the airfoil. Lift and drag coefficient depend on the angle of attack and geometry of the airfoil.

Reynolds Number

Reynold number is a dimensionless number that helps predict the flow pattern. It is the ratio of inertial and viscous forces and is defined as follows:

$$Re = \frac{\rho V L}{\mu} \tag{3}$$

Where, L = characteristic length (chord of airfoil), μ = viscosity of fluid. While performing analysis, the Reynolds number is maintained to ensure similar pressure and friction, flow separation and laminar and turbulence characteristics. In this study, we are working with Reynolds number of $Re = 2 \times 10^5$.

Mach Number

Mach number denoted by 'M' is the ratio of speed of fluid to speed of sound in that medium. In our study the entire flow regime is subsonic, and the flow is assumed incompressible since M < 0.3. While conducting analysis we preserved Mach number as well. This ensures same compressible characteristics and same flow regime.

Proposed Methodology

The proposed investigation is conducted for the following two cases. (At different angles of attack from 0-16° at an interval of 2°).

Case I: CFD analysis of standard NACA0015 airfoil was done and corresponding coefficient of lift and drag are calculated.

Case II: In this case CFD analysis is carried out on modified NACA 0015 airfoil geometry and compared with the standard NACA 0015.

For both the cases lift and drag force are calculated, and following the results, velocity and pressure contours are plotted. Finally, performance from both cases are analysed and compared to examine the effect of change in geometry on aerodynamic performance of said airfoil.

Geometry of proposed airfoils

We considered the chord length (L) of symmetric NACA 0015 airfoil, L = 0.15 m. A shark denticle inspired bump, the front portion of the bump resembles shark denticle and the other half is a streamlined design following the downstream flow to reduce the generated pressure drag. In the first modified geometry, the stream wise location of the bump is at 26% of the chord length, since previous study has shown the optimum position for denticle is at 26% of chord [14].

Referring to the studies, the bump should lie inside of the boundary layer to give the best performance (similar to micro vortex generators) [19]. By the following equation, we can calculate the boundary layer thickness [20]:

$$\delta = \frac{5x}{\sqrt{Re}} \tag{4}$$

where, x = distance from the start of the boundary layer, R_e = local Reynolds number. Based on the thickness of boundary layer, height of bump was taken to be H = 0.7 mm and size of bump in streamwise direction, B = 4.2 mm as it's shown in Figure 32.1.

Geometric Modelling

The coordinates of smooth, standard airfoil were downloaded from the source [21]. 200 points were plotted for maximum smoothness and to minimise the edges. These coordinates were taken in 2D and were then imported to computer aided design (CAD) tool, SolidWorks. This model was then used for further analysis of the standard model, and another new design was created for creating a bumped/modified airfoil.

Figure 32.2 shows the location of bump which is situated at distance(d) from leading edge. The bumped airfoils were then exported to ANSYS Fluent for computational analysis.

CFD Analysis

A dome shaped structure, more specifically, a C shaped mesh was constructed around all the geometries of airfoil. Its length was taken 20 times the chord length, and the width was taken to be 10 times of the same. These dimensions were taken higher than usual to avoid its edges affecting the flow and results.

The mesh as can be seen in Figure 32.3, was then generated and refined, all of the geometries have elements ranging between 100,000 and 150,000 to ensure procurement of refined results.

Turbulent Modelling

The model used for analysis was Spalart Allmaras Turbulence model (SA model) [22] and its equation is as follows:

$$\frac{\partial}{\partial t}(\text{粭}\tilde{}) + \frac{\partial}{\partial_x}(\text{粭}\tilde{}\,u_i) = G_v + \frac{1}{\text{噕}\tilde{}}\left[\frac{\partial}{\partial x_j}\left\{(\hat{1} + \text{粭}\tilde{})\frac{\partial\tilde{}}{\partial\hat{i}x_j}\right\} + C_{b2}\tilde{n}\left(\frac{\partial\tilde{}}{\partial\hat{i}x_j}\right)^2\right] - Y_v \qquad (5)$$

The boundary conditions, have been set as follows:
Velocity was set to 19.71 m/s. The viscosity was taken constant and 1.81 × 10-5 kg/(m s) and lastly, density was taken 1.225 kg/m³. For analysis, 2000 iterations are conducted.

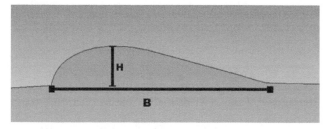

Figure 32.1 Side view of the bump
Source: Author

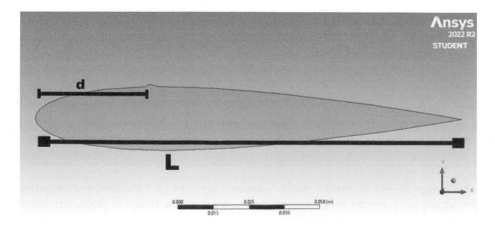

Figure 32.2 Side view of the NACA 0015 with bump
Source: Author

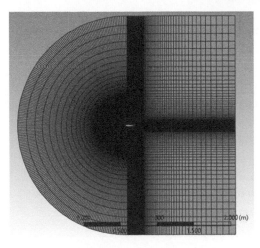

Figure 32.3 Generated mesh
Source: Author

Results and Discussion

Seven different modified geometries were analysed in this study with bump located at 13%, 15%, 17%, 19%, 23%, and 26% of chord length respectively. The results obtained from analysis on all the modifications are as follows:

- The first modification, i.e., bump located at 13% of chord length produced negative lift (Cl = -0.00510) at 0° AOA. The stall was experienced at much lower AOA and lift coefficient was also observed to be slightly lower than the standard airfoil.
- Bump at 15% of chord length is one of the two designs that produced better lift than the rest, including standard geometry. It produces nonzero lift (Cl = 0.00538) at 0° AOA, and overall lift coefficient is observed to be higher than almost all the other designs before stall.
- Bump situated at 17% of chord length also has better lift than standard geometry, however, the resultant curve of 15% and 21% were found to be greater than this geometry. The stall angle was observed to be around 13°. The lift of this geometry is higher (Cl = 0.011) at 0°, however it does not produce better results at higher AOA.
- Bump located at 19% of chord length produces negative lift at 0°, and although it does have stall angle at a higher value when compared to all the other modified geometries, the overall performance isn't as significant and satisfactory.
- Modification at 21% of chord length is the second of the two best geometries. It produces nonzero lift (Cl = 0.00927) at 0 ° AOA and a better lift than standard geometry at other angles, but it experiences stall at a much lower angle, i.e. around 12 ° AOA.
- Bump at 23% performs quite similar to 17%, significant lift at 0° AOA, and noticeable changes in drag coefficients, with higher stall angle than other modified geometry.
- Lastly, 26% of chord length produces negative lift (Cl = -0.01178) at 0° AOA, and the overall performance isn't desirable, though the pressure contour do serve a good subject for significant observation of separation bubble and vaguely, its effect.

Out of all seven designs, two designs were selected. 15% and 21% chord length that produced better lift coefficients.

Figures 32.4 and 32.5 show the coefficient of lift and coefficient of drag of all the geometries with their respective angle of attack. The values of lift and drag coefficient of all geometries at various angles of attack is given in appendix.

Figures 32.6 and 32.7 compare the coefficient of lift and coefficient of drag of two modification that performed best i.e., airfoil with bump at 15% and 21% of chord length with the standard geometry. It can be observed that lift coefficient of both 15 and 21% were higher than the standard geometry. However, the stall angle of these two geometries also arrived at a much lower angle. And drag was slightly more than the standard aerofoil.

From the pressure contour in Figure 32.8(a), it can be observed that the low-pressure zone in 15% modification forms at the top of bump and pressure in that region drops to around -3.78×10^2 pascal whereas in pressure contour of the Figure 32.8(b) the low pressure zone gets dispersed to the back of bump and the pressure drop is less than Figure 32.8(a). Based on this observation we can conclude that the location of the low-pressure zone affects the lift produced by the airfoil.

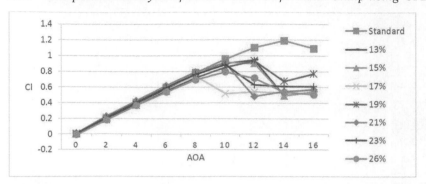

Figure 32.4 The graph showing coefficient of lift of all eight geometries with their respective angle of attack (°)
Source: Author

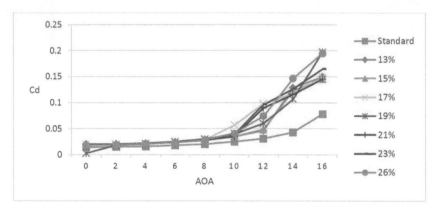

Figure 32.5 The graph showing coefficient of drag of all eight geometries with their respective angle of attack (°)
Source: Author

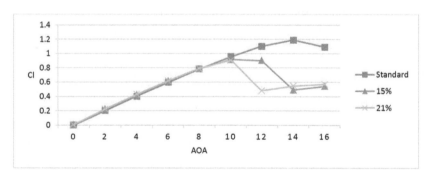

Figure 32.6 The graph showing coefficient of lift of standard airfoil and airfoils with bump at 15% and 21% of chord length
Source: Author

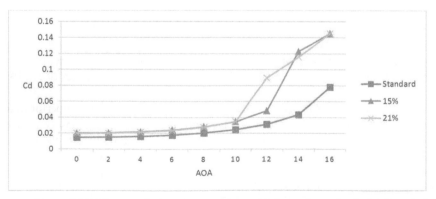

Figure 32.7 The graph showing coefficient of drag of standard airfoil and airfoils with bump at 15% and 21% of chord length
Source: Author

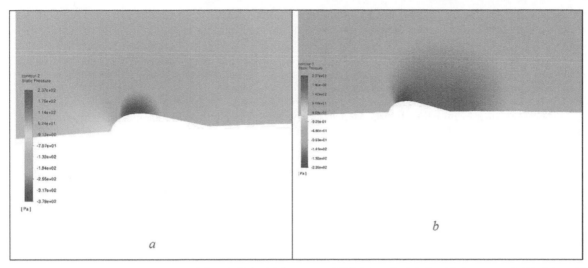

Figure 32.8 Pressure contour of the airfoil with bump at 15% (a) and airfoil with bump at 26% (b) at 0° angle of attack

Source: Author

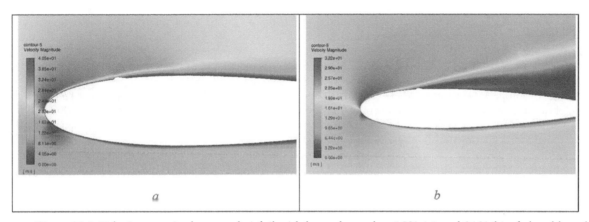

Figure 32.9 Velocity magnitude around airfoil with bump located at 15% (a) and 21%(b) of chord length at 12° angle of attack

Source: Author

The stall of the airfoil with bump at 21% of chord length happens before the airfoil with bump located at 15%. In Figure 32.9 we can observe the pressure contour of both the airfoils at 12° angle of attack. In Figure 32.9(b) we notice that the flow starts to separate because of the bump and is not able to reattach to the surface, this causes stall. Whereas for the airfoil with bump at 15% in Figure 32.9(a) the flow stays attached to the surface.

Conclusion

This research focuses on the effect of bump on symmetric airfoil National Advisory Committee for Aeronautics 0015 (NACA 0015). The modification was made in seven different locations, 13%, 15%, 17%, 19%, 21%, 23%, and 26% of chord length. Computational analysis was conducted on all the models and the results at different angles of attack (AOA) were compared accordingly. The effect of the bump on standard airfoil was found to positively affect the overall lift coefficient. There were certain instances where the coefficient of lift was observed negative, however, the best results were found on 15% and 21% of chord length. These airfoils produced lift even at 0° AOA and the lift coefficient graph was observed to be higher than standard geometry in most AOA. Future studies may include implementation of multiple bumps on a single airfoil or using chambered airfoil for implementation of the bump.

References

[1] Gul, M., O. Ğ. U. Z. Uzol, and I. S. Akmandor. (2014). An experimental study on active flow control using synthetic jet actuators over S809 airfoil. Journal of Physics: Conference Series. 524(1), 012101.

[2] Kazanskiy, P. N., Moralev, I. A., Bityurin, V. A., and Efimov, A. V. (2016). Active flow control on a NACA 23012 airfoil model by means of magnetohydrodynamic plasma actuator. Journal of Physics: Conference Series. 774(1), 012153.

[3] Mostafa, W., Abdelsamie, A., Mohamed, M., Thévenin, D., and Sedrak. M. (2020). Aerodynamic performance improvement using a micro-cylinder as a passive flow control around the S809 airfoil. IOP Conference Series: Materials Science and Engineering. 973(1), 012040.

[4] Putro, S. H. S., Pitoyo, B. J., Pambudiyatno, N., and Widodo, W. A. (2021). Comparison of the winglet aerodynamic performance in unmanned aerial vehicle at low Reynolds number. IOP Conference Series: Materials Science and Engineering. 1173(1), 012002.

[5] SP Setyo, H., Widodo, W. A., and Sonhaji, I. (2021). Numerical study of secondary flow characteristics on the use of the winglets. Journal of Physics: Conference Series. 1726(1), 012012.

[6] Christian, B., Skrzypiński, W., Fischer, A., Gaunaa, M., Brønnum, N. F., and Kruse, E. K. (2018). Wind tunnel tests of an airfoil with 18% relative thickness equipped with vortex generators. Journal of Physics: Conference Series. 1037(2), 022044.

[7] Chengyong, Z., Wang, T., Chen, J., and Zhong, W. (2020). Flow analysis of the deep dynamic stall of wind turbine airfoil with single-row and double-row passive vortex generators. IOP Conference Series: Earth and Environmental Science. 463(1), 012118.

[8] Chng, L., J. Ntouras, A. d., Papadakis, G., Kaufmann, N., Ouro, P., and M. Manolesos. (2022). On the combined use of vortex generators and gurney flaps for turbine airfoils. Journal of Physics: Conference Series. 2265(3), 032040.

[9] Francesco, P., Melani, P. F., Alber, J., Balduzzi, F., Ferrara, G., Nayeri, C. N., and Bianchini, A. (2022). Potential of mini gurney flaps as a retrofit to mitigate the performance °radation of wind turbine blades induced by erosion. Journal of Physics: Conference Series. 2265(3), 032046.

[10] Yuji, Y., Zhang, K., Sasaki, O., Tomita, M., Rival, D., and Galipon, J. (2019). Manufacturing of biomimetic silicone rubber films for experimental fluid mechanics: 3D printed shark skin molds. Journal of The Electrochemical Society. 166(9), B3302.

[11] Joshua, O., Lazalde, M., and Gu, G. X. (2020). Algorithmic-driven design of shark denticle bioinspired structures for superior aerodynamic properties. Bioinspiration & Biomimetics. 15(2), 26001.

[12] Farhana, A., Lang, A., Habegger, M. L., Motta, P., and Hueter, R. (2016). Experimental study of laminar and turbulent boundary layer separation control of shark skin. Bioinspiration & Biomimetics. 12(1), 016009.

[13] Johannes, O. and Lauder, G. V (2012). The hydrodynamic function of shark skin and two biomimetic applications. Journal of Experimental Biology. 215(5), 785-95.

[14] Domel, A. G., Mehdi, S. J. C. Weaver, H. H. -H., Bertoldi, K., and Lauder, G. V. (2018). Shark skin-inspired designs that improve aerodynamic performance. Journal of the Royal Society Interface. 15(139), 20170828.

[15] Fernando, V., Reggio, M., and Ilinca, A. (2021. Assessment of turbulence models for flow simulation around a wind turbine airfoil. Modelling and Simulation in Engineering.

[16] Ravi, H. C., Madhukeshwara, N., and Kumarappa, S. (2013). Numerical investigation of flow transition for NACA-4412 airfoil using computational fluid dynamics. International Journal of Innovative Research in Science Engineering and Technology. 2(7), 2778-85.

[17] Srinivasan, G. R., J. A. Ekaterinaris, and W. J. McCroskey. (1995). Evaluation of turbulence models for unsteady flows of an oscillating airfoil. Computers & Fluids. 24(7), 833-61.

[18] Barnard, R. H. and Philpott, D. R. (2010). Aircraft flight: a description of the physical principles of aircraft flight. Pearson Education.

[19] Fouatih, O. M., Imine, B., and Medale, M. (2019). Numerical/experimental investigations on reducing drag penalty of passive vortex generators on a NACA 4415 airfoil. Wind Energy. 22(7), 1003-17.

[20] Sakiadis, B. C. (1961). Boundary - layer behavior on continuous solid surfaces: II. The boundary layer on a continuous flat surface. AiChE Journal. 7(2), 221-25.

[21] http://airfoiltools.com/airfoil/details?airfoilnaca0015-il

[22] Philippe, S. and Allmaras, S. (1992). A one-equation turbulence model for aerodynamic flows. In 30th Aerospace Sciences Meeting and Exhibit, pp. 439.

APPENDIX

Table 32.1 The values of coefficient of lift (Cl) and coefficient of drag (Cd) of all geometries at different angles of attack

Alpha (°)		0	2	4	6	8	10	12	14	16
Standard	Cl	5.1E-06	0.202168	0.402218	0.59747	0.784234	0.957121	1.10341	1.191499	1.091596
	Cd	0.014679	0.015011	0.016035	0.017833	0.020596	0.024747	0.031241	0.043448	0.077955
13%	Cl	-0.00511	0.198401	0.395116	0.583169	0.757547	0.900863	0.948296	0.494357	0.544535
	Cd	0.020352	0.020586	0.021889	0.024363	0.028385	0.034938	0.046382	0.128711	0.149458
15%	Cl	0.005377	0.221596	0.423542	0.616036	0.787266	0.914932	0.903388	0.491467	0.541854
	Cd	0.020208	0.02049	0.021823	0.024396	0.0284	0.034649	0.048253	0.122659	0.145108
17%	Cl	0.01065	0.216828	0.413028	0.594307	0.741948	0.516137	0.542955	0.527813	0.583108
	Cd	0.019912	0.020292	0.021767	0.024454	0.028715	0.056489	0.097652	0.116826	0.148683
19%	Cl	-0.00146	0.181416	0.368917	0.550333	0.71814	0.854297	0.936942	0.677578	0.773333
	Cd	0.002604	0.018306	0.020531	0.0245	0.029956	0.039697	0.059853	0.106885	0.196839
21%	Cl	0.009277	0.224902	0.428402	0.620544	0.791626	0.903576	0.482098	0.547032	0.569047
	Cd	0.019828	0.020011	0.021308	0.023814	0.02789	0.035079	0.089722	0.115541	0.145005
23%	Cl	0.007432	0.212521	0.41054	0.59659	0.763709	0.888927	0.629254	0.613079	0.61007
	Cd	0.019493	0.020058	0.021737	0.024988	0.030145	0.038668	0.09609	0.126674	0.164308
26%	Cl	-0.01178	0.18035	0.369803	0.538365	0.692982	0.794474	0.716733	0.536589	0.505594
	Cd	0.018336	0.019023	0.020943	0.02314	0.028608	0.041029	0.074058	0.146463	0.194556

Source: Author

Chapter 33

Topology optimisation of aerospace bracket

Syed Abdul Malik Rizwan[a]*, Mohd Abdul Wahed*[b]*, Mohammed Suleman*[c] *and Mohd Ibrahim Ahmed*[d]

Deccan College of Engineering and Technology, India

Abstract

Aerospace and automotive industries are one of the leading application sectors with a wider share of the market than any other sector. They can quickly adapt according to the increasing demand of the product as well as competitive environment. However, as the demand for the product increases simultaneously product cost increases. Also, regulations aiming to reduce fuel consumption and carbon emission require an optimal design of the product that satisfies design requirements. Thus, it leads to the design of lightweight as well as high-performance products in a shorter period of time. In this context, this research work aims to optimize an aerospace bracket using topology optimisation. The metal alloys considered are Ti-6Al- 4V alloy, Al 7075-T6 alloy and stainless steel 304. Based on the obtained results, it was observed that a good amount of weight is reduced for the aerospace bracket of different materials by applying topology optimisation. Firstly, the brackets are designed in SolidWorks and analysis is carried out in ANSYS by applying boundary conditions and then analysing stress, strain and deformation diagrams. Further, topology optimisation is applied on the component and redesigned as per the requirement. Finally, the results revealed a good amount of weight reduction of the component.

Keywords: Aerospace Bracket, ANSYS, Ti-6Al-4V Alloy, Al 7075-T6 Alloy, Stainless Steel 304

Introduction

Topology optimisation (TO) is a process that optimises material and structure within a given geometrical design for a defined set of rules. The goal is to maximise part performance by optimising factors such as external forces, load conditions, boundary conditions, constraints, and material properties [1]. Conventional topology optimisation uses finite element analysis to evaluate the design performance and produce structures to satisfy objective topology such as follows: reduced stiffness to weight ratio, better strain energy to weight ratio, reduced material volume to safety factor ratio and natural frequency to weight ratio.

Stiffness to weight ratio, also known as Specific modulus or specific stiffness is a materials property consisting of the elastic modulus per mass density of a material. High specific modulus materials find wide application in aerospace applications where minimum structural weight is required. The dimensional analysis yields units of distance squared per time. The ratio of a structure's absolute strength to actual applied load is called as factor of safety or safety factor. This is a measure of the reliability of a particular design. With the reduction in mass after optimisation and the change in material, it becomes important to test it for safety [2]. TO has a broad range of applications across industries, in engineering, it is used at the design stage of a new product to optimise and increase stiffness to weight ratio. These designs are often difficult to manufacture using traditional methods. But due to growth and technological advancement in additive manufacturing or so-called 3D printing, the design output by TO can be fed directly into a 3D printer. The TO concept has been commonly used in CAD software's such as SolidWorks, ANSYS etc. Different mechanical properties such as elastic modulus and strength of a truss component can be optimised as a function of aspect ratio. The design, production as well as performance of selective laser melting lattice structures were studied as well as the quality of data reported was investigated to know the best-practice for better prospects. A comprehensive review was provided for topology optimisation of engineering structure design. An effective method was proposed to systematically evaluate the TPU powders and optimise the fabrication process, in order to overcome the challenges of sintering the duel-segment TPU [3]. In this research work, the bracket optimized is used in Kit fox aircrafts manufactured by Denney Aero Craft Company. Light weight components are highly recommended, and hence topology optimisation of bracket can be very useful in making it more efficient [4]. The bracket bears the mechanical forces and thermal stresses. The mechanical forces are applied at the bearing and the bracket is fixed with rigid supports.

Literature Review

According to earlier research, optimisation approaches have recently grown significantly in sophistication and are being used in an expanding range of industries. To clarify the scope of this effort and better

[a]sdrizwan889@gmail.com, [b]wahedmohdabdul@gmail.com, [c]mohammedsuleman016@gmail.com, [d]mohdibrahimahmed72@gmail.com

DOI: 10.1201/9781003450252-33

understand the subject and potential benefits of the results, a thorough analysis of prior literature has been conducted. The overview focuses on engineering optimisation, from the earliest ideas through contemporary approaches used all around the world. The analysis also focuses on optimisation in the chosen baseline program ANSYS and what milestones have already been attained in terms of the functionality and potential of the applied methods [5].

Methodology

ANSYS mechanical's topology optimisation technology gives designers the resources they need to create lightweight, effective components for a variety of applications.

Modern optimisation algorithms and complicated numerical optimisation methods are used in a number of the optimisation strategies that ANSYS employs. The type of problem and the specified parameters determine the optimisation approach to be used. Depending on the model and output specifications, ANSYS parametric and TO solvers employ a number of methodologies and particular algorithms.

In the Workbench interface of ANSYS, there is a direct topological optimisation module. It makes the processes needed to do an analysis much simpler. As part of the normal topology optimisation process, the model is defined, a mesh is created, optimised and non-optimised regions are specified, load scenarios are defined, and optimisation parameters are defined. The study is then conducted, and the outcomes can be examined and analysed afterward [6].

The proper named selections must be assigned to the component before static structural or optimisation investigations may be completed. Named selections can then be used to designate particular model portions with loads, constraints, or conditions. The optimisation region, which is the aerospace bracket, is the primary named selection that needs to be made. Choose a geometry type (body, face, edge, or point), then choose the required geometry, and finally right-click to create a named selection. The named selection in this instance is known as the optimisation region [7].

The implementation of a structural analysis under the previously specified loads and constraints is the following stage after the geometry has been successfully built and the named selections have been defined. The outcomes of the optimised model will be contrasted with the outcomes of this analysis. Two sizing functions are developed in order to provide a relatively fine mesh with fewer elements when more precision is desired. Mesh, right-click, insert, and sizing. The optimisation region will receive its initial sizing function. Select the named selection option under the scope drop-down menu, and then select the "optimisation region" section. Choose a type of element size and enter a value for the element size. As long as the optimization zone lacks unique characteristics like curves or holes, the size function can be configured to be uniform [8].

Choose bearing load from the top ribbon's loads section, and then choose the previously constructed surface at the bracket's bearing end. The maximum load for this illustration is 5560 N. The various solution types that could be used appear when you click the solution tab on the top ribbon. The outcomes that are most relevant to this investigation are total deformation, and stress equivalent (Von-Mises), strain and factor of safety [9].

The model can be solved (solution right click solve) if the loads/constraints have been established. Depending on the mesh quality and the magnitude of the problem, the solution procedure takes roughly 30 minutes. To properly compare the outcomes of the optimised model to those of the preliminary model, the model can be transferred back to ANSYS after being processed with smoothing features and symmetry tools. This procedure serves as a way to validate the optimisation process [10].

Steps followed for topology optimisation

The following simple steps show how topology optimisation is used in typical CAD software to create a simple bracket.

Initial part geometry design - creates a part frame or design space.

1. **Define conserved geometry areas** - defines some of the critical areas to maintain such attachment bolt headspace and material, around the bearing bore diameter.
2. **Define loads, restraints and material** - apply loads and restraints at this stage. A material and its properties are also assigned to the part. As this is one of the most critical step in topology optimisation.
3. **Choose optimisation goal** - decide and choose the purpose of the optimization.
4. **Run topology optimisation** - the software takes over and analyzes the part to evaluate results.
5. **Evaluate the results** - evaluate the results but change levels to identify key FEA areas flagged as critical.

The bracket to be optimised was first designed with the required dimensions using SolidWorks software version 2011. The material used in design is Ti-6Al-4V alloy. The 3D model is then loaded in ANSYS for analysis and optimisation.

Optimization Process:

The bracket shown in Figure 33.1 is designed in SolidWorks 2011 with the referred geometry. Under static structural, loads are applied. One of the ends is fixed; hence fixed rolls are used, while another end consists of a bearing.

After performing analysis with necessary parameters on the CAD model of aerospace bracket, the generated stresses and total deformation are presented in Figures 33.2 and 33.3 respectively.

During topology optimisation of aerospace bracket, 40% of the mass from the bracket is retained and where there are no or minimal loads acting on the bracket, those areas have been excluded. Result can be seen in Figure 33.4.

Analysis on redesigned component

The bracket is redesigned by considering the results obtained without affecting its performance. This stage consists of the bracket taken in design software called space claim, where the bracket is redesigned by fixing the curves and the geometry. Figure 33.5 shows redesigned model of the aerospace bracket.

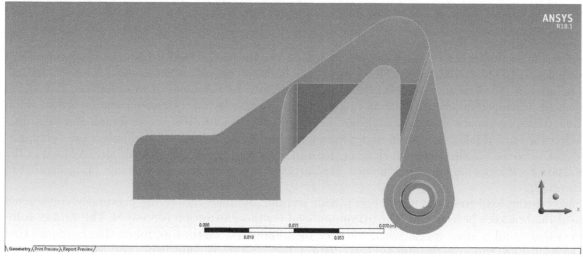

Figure 33.1 CAD model of the bracket
Source: Author

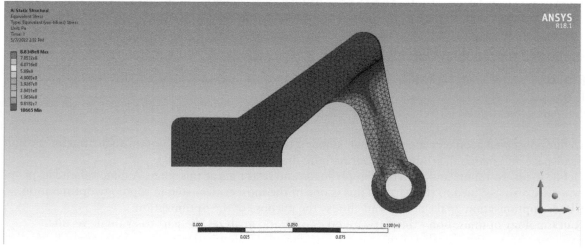

Figure 33.2 Equivalent Von-Misses stresses
Source: Author

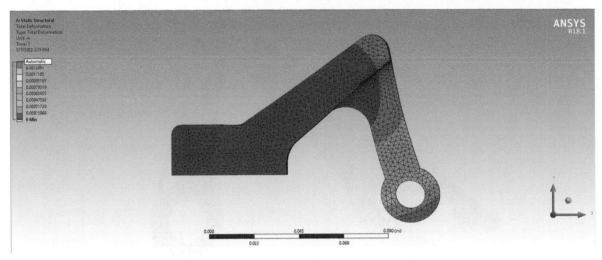

Figure 33.3 Total deformation
Source: Author

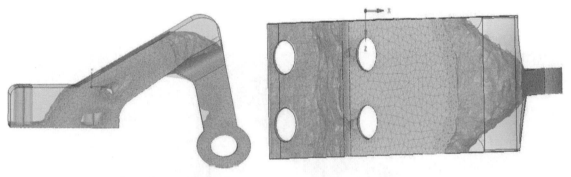

Figure 33.4 Brackets after retaining the mass
Source: Author

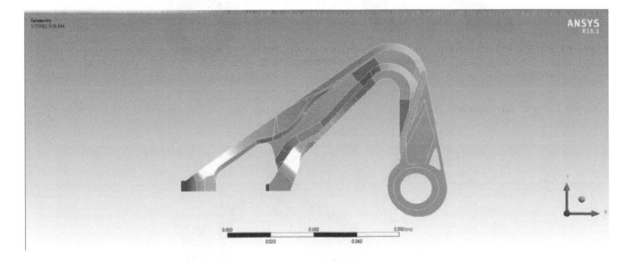

Figure 33.5 Optimised model of aerospace bracket
Source: Author

Analysis of aerospace bracket using Al 7075-T6 alloy:

The optimised bracket is again analysed for the evaluation of results and validation of the optimised model. The results obtained after performing analysis with necessary parameters on the optimised bracket for which the generated stress and total deformation can be seen from Figures 33.6 and 33.7, which can be compared with the pre-optimised bracket results.

It is observed that the bracket made of Ti-6Al- 4V alloy has desired factor of safety for the industrial applications. Figure 33.8 shows the bracket with no regions of minimal factor of safety.

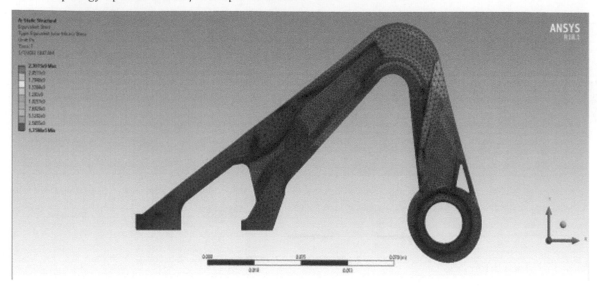

Figure 33.6 Equivalent Von-Mises stresses
Source: Author

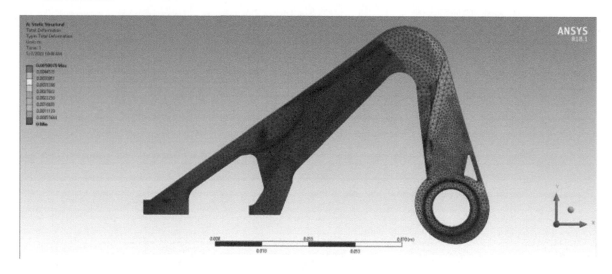

Figure 33.7 Total deformation
Source: Author

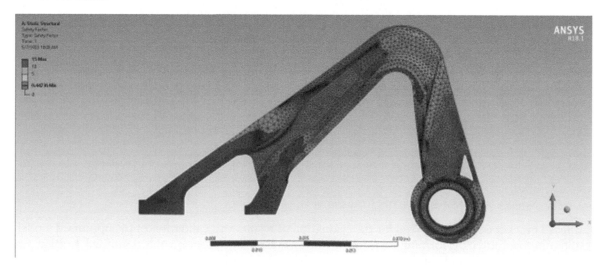

Figure 33.8 Factor of safety
Source: Author

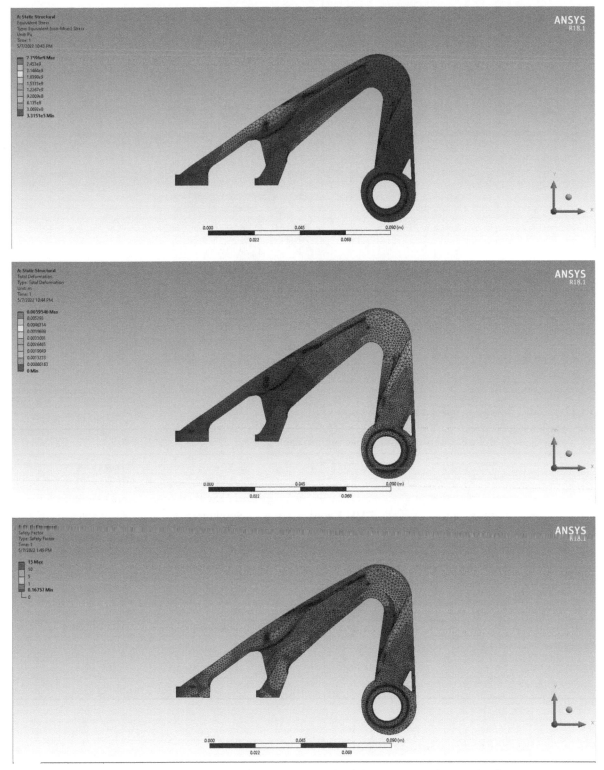

Figure 33.9 Analysis of Al 7075-T6 bracket (a) stresses, (b) deformation (c) factor of safety
Source: Author

Analysis of Aerospace Bracket Using Al 7075-T6 Alloy:

The analysis is repeated with the same load parameters for Al 7075-T6 alloy and then the results are generated.

The above Figure 33.9 (a), (b) and (c) shows the stress, deformation, and factor of safety of Al 7075-T6 bracket. Here, it is observed that the bracket has regions where the factor of safety is undesirable. Though Al 7075-T6 bracket is lighter when compared to Ti-6Al-4V bracket, however due to safety concerns it is not suggested for the industrial application.

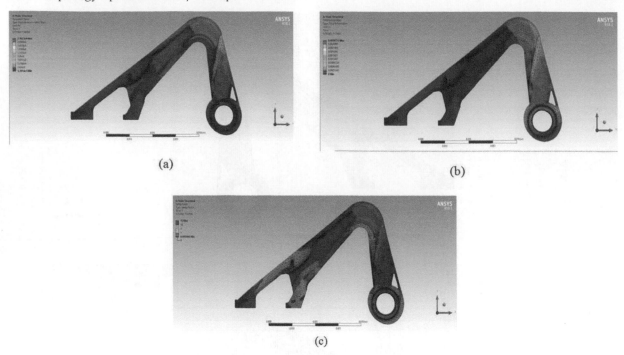

Figure 33.10 Analysis of SS 304 bracket (a) stresses, (b) deformation (c) factor of safety
Source: Author

Table 33.1 Topology optimisation results of aerospace bracket for different alloys

	Before optimisation	After optimisation		
Property	Ti-6Al-4V	Al 7075-T6	SS 304	Ti-6Al-4V
Volume	1.1086e-004 m^2	3.2235e-005 m^2	3.2235e-005m^2	3.2235e-005m^2
Density	4419 kg/m^3	2804 Kg/m^3	7900 kg/m^3	4419 kg/m^3
Mass	0.48990 kg	0.09038 kg	0.25466kg	0.14245kg
Temperature	Room Temperature	Room Temperature	Room Temperature	Room Temperature
Min. eq. stress	5.0393e-6 MPa	0.33151 MPa	0.16615 MPa	0.30523 MPa
Max. eq. stress	576.26 MPa	2759.6 MPa	2823 MPa	2784 MPa
Min. deformation	0 m	0 m	0 m	0 m
Max. deformation	9.4602e-004 m	5.9546e-003 m	2.1867e-003 m	3.8795e-003 m
Safety factor	0.5	0.5	0.5	0.5
Min. safety factor	-	0.16757	7.1528e-002	0.36634
Max. safety factor	15	15	15	15

Source: Author

Analysis of Aerospace Bracket Using SS 304 Alloy:

The analysis is repeated with the same load parameters by using SS 304 alloy and then the results are generated.

The above Figure 33.10 (a), (b) and (c) depicts the stress, deformation and FoS of SS 304 bracket. It is observed that the factor of safety of SS 304 alloy is significantly lower than that of other two alloys. As the factor of safety is very low the bracket made of stainless steel 304 alloy is not applicable for the application. As it may results in component failure. Table 33.1 presents topology optimization results of aerospace bracket for different alloys.

Results and Discussion

- For SS 304 alloy, topology i mprovement resulted in a weight reduction from 0.724-0.254 kg. Although it is heavier than brackets made of other alloys, it does not have a good trustworthy safety factor and could lead to component failure.

- After topology optimisation, the weight of the Al 7075-T6 alloy was lowered from 0.289-0.090 kg. Due to its relatively low factor of safety and potential component failure, this substantial weight reduction came at a cost.
- After topology optimisation, the weight of an aerospace bracket constructed of the Ti-6Al-4V alloy was lowered from 0.4899-0.1424 kg. The weight has been reduced by 71%. This alloy exhibits good strength even after significant mass and weight reduction due to its remarkable strength and thermal characteristics. This alloy usually works well for aircraft applications since the safety factor is adequate.

Conclusions

- The aerospace bracket made of Ti-6Al-4V alloy gives 71% reduction in weight after optimisation and has desired safety factor, also it gives enough strength to the component for safe use. Ti-6Al-4V alloy is highly recommended for topology optimisation (TO) of an aerospace bracket. In Al 7075-T6 alloy, the factory of safety is less and the component tends to fail after optimisation, so this alloy is not desired. For SS 304 alloy, the bracket after optimisation has higher weight than other two alloys. Though having a relatively higher weight, it fails to give satisfactory safety factor that may result in component failure. Reduction in weight at the expense of safety of component is not desired.

References

[1] ZHU, J., ZHOU, H., WANG, C., ZHOU, L., YUAN, S., and ZHANG, W. (2021). A Review of topology optimization for additive manufacturing: Status and challenges. Chinese Journal of Aeronautics. 32(1), 91-110.

[2] Yan, X., Bao, D., Zhou, Y., Xie, Y., and Cui, T. (2022). Detail control strategies for topology optimization in architectural design and development. Frontiers of Architectural Research. 11, 340e356.

[3] Jankovics, D. and Barari, A. (2019). Customization of automotive structural components using additive manufacturing and topology optimization. IFAC-PapersOnLine. 52(10), 212-17.

[4] Topology Optimization in lightweight design of a 3D-printed flapping-wing micro aerial vehicle; Long CHEN, Yanlai ZHANG, Zuyong CHEN, Jun XU, Jianghao Wu; Chinese Journal of Aeronautics, (2020), 33(12): 3206-3219.

[5] Munoz, J. S. Q. (2017). Engineering Optimization Showcase. Thesis.

[6] Bankoti, S., Dhiman, A., and Misra, A. (2015). Comparative analysis of different topological optimization methods with ANSYS. International Journal for Research in Emerging Science and Technology. 2(4).

[7] Esfarjani, S. M., Dadashi, A., and Azadi, M. (2022). Forces in mechanics. Forces in Mechanics. 7, 100100.

[8] Helfesriedera, N., Lechlera, A., and Verla, A. (2020). Method for generating manufacturable, topology-optimized parts for Laminated Layer Manufacturing. Procedia CIRP. 93, 38-43

[9] Tekea, I. T., Akbulutb, M., and Ertasa, A. H. (2021). Topology optimization and fatigue analysis of a lifting hook. Procedia Structural Integrity. 33, 75-83.

[10] Armentania, E., Giannellab, V., Parentec, A., and Pirellic, M. (2020). Design for NVH: topology optimization of an engine bracket support. Procedia Structural Integrity. 26, 211-18.

Chapter 34

Toxicity: sustainable control over non-exhaust pollution

Sachin Jadhav[1,a] and Sanjay Sawant[2,b]

[1]DOT, Shivaji University Kolhapur and SETI, Panhala, Maharashtra, India

[2]Sant Gajanan Maharaj College of Engineering, Mahagaon, Maharashtra, India

Abstract

Non-exhaust pollution is majorly due to rubbing operation between two surfaces in contact. In the operation of brake and clutch, friction material rubs against disc and flywheel and emits wear debris in the atmosphere. The friction material composition is complex and comprises of many ingredients such as binder, filler, reinforcement, and friction modifier. To enhance the properties of friction material, manufacturers added more metallic ingredients. The terrible effects of some of the metallic elements are found on the environment and human health. Some of the toxic elements recorded by the World Health Organization (WHO) are copper, mercury, lead, cadmium, asbestos and chromium etc. These elements invite diseases such as lung cancer, kidney failure, and respiratory track damage. In the present work, two compositions are developed with 10 ingredients and consist of more natural biodegradable and low metallic elements. The compression molding method is used to manufacture friction material and physical, mechanical and tribological characterization is done as per ASTM standards. The results of the developed friction material are compared with commercial friction material. Sample 1 and sample 2 show significant coefficient of friction and very low wear rate. The morphological and thermal behavior of developed friction material is studied. The developed friction material fully fills the requirements needed for brake friction material application.

Keywords: Toxicity, non exhaust pollution, friction material, SEM, TGA

Introduction

Source of wear debris of non-exhaust pollution in a brake is from the clutch, tire and rubbing of metals or nonmetal surfaces in contact. These exhale very tiny particles in the atmosphere called particulate matter (PM) that have a size less than 10 microns. Many countries in the world pass bill in their senate that ban or minimises the percentage of toxicity-produced elements [1, 2]. In the United States, the environmental protection agency (EPA) made agreement with brake friction material manufacturers to minimise the percentage of copper, mercury, lead, cadmium, asbestos and chromium. Some of the toxic elements and their maximum percentage are shown in Table 34.1 [3, 4].

Brake operation is a very complex process. Due to high temperature, various wear mechanisms are observed in between the tribo-pair. The metal oxides listed are Fe_3O_4, MgO, Al_2O_3, CuO, SO_3, CaO, K_2O, SiO_2, MnO_2, Cr_2O_3, $ZrSiO_4$, Ca_2SiO_4, TiO_2, ZnO, Fe, and Fe_2O_3 bad effect on plants and human health [5]. Particle size less than 2.5 microns (PM2.5) remains in the respiratory tract and lower lung region and comes into pulmonary interstitial and vascular space. Due to this reason, WHO recommends $PM_{2.5}$ as an air quality indicator in non-exhaust emission. Urban areas are highly affected by the toxic elements in a rush or heavy traffic conditions. Driving conditions such as full stop or deceleration are also reason for the emission of highly concentrated wear debris into the atmosphere. The metro cities vehicles reduce the quality of air and increase the air quality index (AQI). In the winter season, the combined effect of exhaust pollution from internal combustion engines and non-exhaust particulate matter (PM) increases AQI at peak levels and harms human health such as asthmatic attacks. Sb_2O_3 is oxidising at high temperatures and forms a carcinogenic compound. The copper in the brake friction material composition has a bad effect on human health. Some of the diseases recorded due to copper are lung and kidney spoil, inhalation or reparatory problem, vomiting, headache, etc [6,7]. Several factors influence wear debris formation these include temperature, environment, friction partner, the velocity of brake disc or drum, brake energy, airflow and wheel housing shown in Figure 34.1. The process parameters such as load, speed and sliding distance also play an important role in the formation of wear debris. The hardness of the material also influences the formation of wear debris, as the hardness of friction material increases the wear debris formation decreases [8]. The brake friction material is classified as metallic, semi-metallic and non-asbestos organic. The metallic friction material contains more than 50% metallic elements whereas the semi-metallic range is 15-50% and non-asbestos organic (NAO) friction material has a metallic range below 15%. The metallic friction

[a]spjadhav2112@gmail.com, [b]sanjaysawant2010@gmail.com

DOI: 10.1201/9781003450252-34

Table 34.1 Non-exhaust pollutant concentration and their effective dates [3]

Elements	Concentration (Wt %)	Effective date
Asbestiform fiber	< 0.1	January1, 2014
Cadmium	< 0.01	January1, 2014
Chromium	< 0.1	January1, 2014
Lead	< 0.1	January1, 2014
Mercury	< 0.1	January1, 2014
Copper	< 5	January1, 2021
Copper	< 0.5	January1, 2023

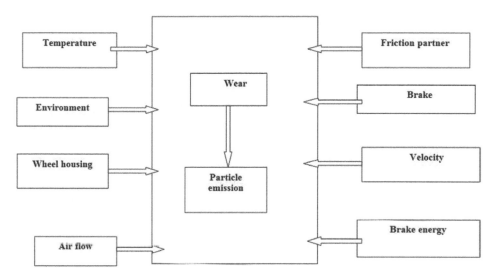

Figure 34.1 Factors influencing particle formation [8]

material consists of iron-based or copper-based powders used to increase strength and withstand higher temperature operating conditions that also improves the heat dissipation rate. For heavy-duty applications, metallic brake friction materials are used. Semi-metallic brake friction materials are used in light motor vehicles and motorcycles. Ceramic fiber and iron or copper powders are used in these friction materials to improve heat resistant properties. In non-metallic brake friction material, organic fibers like (Kevlar, carbon) and inorganic mineral fibers viz. (Wollastonite, glass) is bonded with modified resin and rubber in addition to these the abrasives, lubricant and filler materials are also added to increase tribological, hardness and wear resistance properties [9–11].

In the present work, low metallic two samples are prepared with the help of 10 ingredients. Epoxy resin modified with sodium silicate is used as a binder whereas steel wool and glass fiber are used as reinforced members. The filler materials are vermiculate, cashew shell powder human nails and the abrasive is boron carbide dust. The compression molding method is used to manufacture samples. The physical, mechanical and tribological behavior of the prepared samples is studied by ASTM standards. The morphological and thermal behavior of the samples is studied by SEM and TGA methods respectively and obtained results are compared with commercial brake friction material.

Material and Method

The composition of friction material comprises 10 ingredients classified into a binder, filler, reinforcement, abrasive, friction moodier and elastomer shown in table 34.2. Epoxy resin is modified with sodium silicate to improve wear resistance and hardness properties. Glass fiber and steel wool strengthen the friction material. Cashew shell powder, vermiculate and human nails are used as organic filler materials. Boron carbide has high wear resistance and stands at high temperatures (Table 34.2).

Manufacturing of Friction Material

The sample of friction material is prepared by raw material preparation. In this step, the various ingredient powders are prepared by hammer mill or grinder and sieves of different sizes are used to obtain micro-sized

Table 34.2 Friction material sample 1 and 2 compositions

Material function	Material ingredients	Sample 1 volume %	Sample 2 volume %
Binder	Epoxy resin	32	33
	Sodium silicate	7	7
Filler	Cashew shell powder	31	26
	Vermiculate	12	12
	Human nail powder	3	2
Reinforcement	Steel wool	3	3
	Glass fiber	1	1
Abrasive	Boron carbide	3	8
Friction modifier	Zirconium oxide	5	5
Elanstomer	Rubber crumb	3	3
		100	100

Source: Author

powders. In the next step, homogeneous mixture of all powders is prepared in ball mill machines. The homogeneous mixture is poured into the mold and compressed under a pressure of 17-20 MPa at room temperature. The compressed mixture mold was put into the oven for 6 hours at a temperature of 150-160°C. After 6 hours mold is kept for curing for 24 hours. The samples are removed from the mold and clean to remove the bur with polished paper and keep it ready for different characterization [12-14].

Characterisation of Samples

ASTM standards methods are used to characterise physical, mechanical and tribological properties. The morphological and thermal behavior of the sample is studied with help of a SEM and thermo-gravimetric analysis TGA.

Physical Characterization

The density of the sample is found by ASTM D792 and water and oil absorption tests are conducted by ASTM D570-98. The density of the sample is calculated by Archimedes's principle. The volume displacement method is used to calculate density. The sample is immersed in oil or water for 24 hours in the water and oil absorption test, the percentage of oil or water absorbed by the sample is calculated by the differences in their weights at dry and wet conditions.

Mechanical Characterisation

The hardness of samples is found by Rockwell hardness tester on S scale by ASTM D785 standard. The major and minor loads are applied with 1/16th inch ball indenter. The reading at three spots is noted and the average hardness value is calculated.

Tribological Characterisation

Pin-on disc apparatus is used to evaluate tribological performance of the sample according to ASTM G-99 standard. Pin of 12 mm diameter is prepared. One end of the pin is kept flat to rest on the disc plane surface. The disc is made of cast iron has 165 mm in diameter and 8 mm in thickness. The pin is allowed to press against the disc at various controlling factors shown in the Table 34.3. The wear in micron and friction force is recorded by the pin on the disc apparatus. Coefficient of friction and wear rate is calculated for different load, speed and sliding distance. Wear performance design parameters are shown in Table 34.4.

Result and Discussion

The physical, mechanical and tribological properties results of the friction material samples 1 and 2 are shown in the result Tables 34.5 and 34.6. In physical characterization, density of sample 1 and 2 is lower and higher than commercial friction material (1.5666 and 2.2465 g/cm³ respectively). Water and oil absorption results showed that samples 1 and 2 possess higher absorption rates as compared to commercial friction material samples. Mechanical characterisation result reveals that hardness value of the commercial friction

Table 34.3 Control factors and levels

Factors	Load (N)	Speed (rpm)	Sliding distance (m)
Levels			
1	9.81	900	1000
2	19.62	1100	2000
3	29.43	1300	3000

Source: Author

Table 34.4 Pin on disc wear performance test design parameters

Sample No.	Levels	Load F (N)	Frictional force F_f (N)	Track radius r (mm)	Speed (rpm) N	Sliding distance m	Velocity m/s
1	1	9.81	8.84	15	900	1000	1.4139
	2	19.62	9.36	15	1100	2000	1.7281
	3	29.43	9.48	15	1300	3000	2.0423
2	1	9.81	9.2	75	900	1000	7.0695
	2	19.62	9.26	75	1100	2000	8.6405
	3	29.43	9.38	75	1300	3000	10.2115

Source: Author

Table 34.5 Results table of wear rate and coefficient

Sample No.	Levels	Load F (N)	Frictional force F_f (N)	Wear rate mg/m	Coefficient of friction (μ)	Specific wear rate m^3/N.m 10^{-12}
1	1	9.81	8.84	0.003	0.9011	0.19520643
	2	19.62	9.36	0.003	0.477	0.09760322
	3	29.43	9.48	0.00067	0.3221	0.01445974
2	1	9.81	9.2	0.002	0.9378	0.04537583
	2	19.62	9.26	0.0005	0.4719	0.01134396
	3	29.43	9.38	0.00033	0.3187	0.00504176

Source: Author

Table 34.6 Comparative result table[12–14]

Name of composition	Physical characterisation			Mechanical characterisation	Tribological characterisation		TGA at Temp. 200^0C weight loss in %
	Density	% Water absorption	% Oil absorption	Rockwell Harness Test (HRS)	Wear mg/m	Coefficient of friction (μ)	
Sample1 friction material	1.5666	5.7022	6.4023	61	0.003-0.00067	0.32-0.47	9
Sample 2 friction material	2.2465	4.071	5.1262	76	0.002-0.00033	0.31-0.47	8.76
Commercial friction material	1.89	1.01	0.3	90-95	3.8	0.3-0.4	NA

material is more as compared to samples 1 and 2 shown in Table 34.6. Tribological performance results of a sample 1 and 2 friction materials are calculated at different levels of controlling factors are shown in Table 34.6. With concern to tribological performance, the parameters such as wear rate and coefficient of friction of samples 1 and 2 showed better performance as compared to commercial brake friction material.

Figure 34.2 SEM images of sample 1 a) Before wear test b) After wear test
Source: Author

Figure 34.3 SEM images of sample 2 a) Before wear test b) After wear test
Source: Author

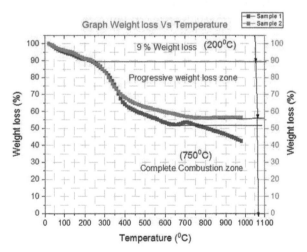

Figure 34.4 TGA graph of sample 1 and 2
Source: Author

SEM Analysis

The scanning electron microscope is used to study surface behavior of the sample before and after wear test. The results showed that on both samples, cracks and wear debris are formed and sample 1 has smaller cracks sizes as compared to sample 2 are shown by yellow marks in Figures 34.2 and 34.3.

Thermo Gravimetric Analysis

The thermo gravimetric analysis (TGA) of two samples are completed under a Nitrogen atmosphere by keeping the total temperature range 1000°C. The rate of isothermal heating was kept 10°C per minute. Weight loss per 10°C is recorded by the TGA machine. The graph of the percentage of weight loss against temperature is plotted. The graph showed that till 200°C the average weight loss in sample 1 is 9% whereas 8.76% is recorded in sample 2. After 200°C progressive weight loss is observed in both the samples and complete combustion temperature is noted beyond the 750°C is shown in Figure 34.4.

Conclusion

Toxicity due to brake friction material is a major challenge and can be solved by using selecting such friction material composition having low toxic and high organic elements. Avoid toxic elements such as copper, mercury, lead, cadmium, asbestos and chromium etc, in the composition of brake friction material. Legislations related to non-exhaust emission should be strictly implemented by the government and obeyed by brake friction material manufacturers of the different countries in the world. The awareness related to non-toxic emission should be created in all the people of world. The drivers should avoid harsh deceleration to minimise wear of brake friction material. Following are the concluding remarks on present study,

- Two samples are prepared with low metallic and more natural elements. The results of samples 1 and 2 are compared with commercial friction material; both the samples have better results and obtain 0.32 to 0.47 coefficient of friction and wear rate 0.006 to 0.00033 mg/m respectively. The controlling factors speed and load has major effects on the wear rate and coefficient of friction. As speed and wear rate are inversely proportional to each other. The SEM analysis showed that sample 2 has more pit marks, wear debris and large cracks sizes as compared to sample 1 after the wear test. The thermal behavior of samples 1 and 2 showed an average weight loss of 9% at 200°C. Samples 1 and 2 are may an alternative to brake friction material for light motor vehicles and motorcycles that helps in reduction in non-exhaust pollution.

References

[1] Confiengo, G. G. D. and Faga, M. G. (2022). Ecological transition in the field of brake pad manufacturing: an overview of the potential green constituents. Sustainability. 14(5). doi: 10.3390/su14052508.

[2] Jadhav S. P. and Sawant, S. H. (2019). A review paper: Development of novel friction material for vehicle brake pad application to minimize environmental and health issues. Materials Today: Proceedings. 19(40), 209–12. doi: 10.1016/j.matpr.2019.06.703.

[3] Lee, P. W., Lee, L., and Filip, P. (2012). Development of cu-free brake materials. SAE Technical Papers. 7. doi: 10.4271/2012-01-1787.

[4] Ciudin, R., Verma, P. C., Gialanella, S., and Straffelini, G. (2014). Wear debris materials from brake systems : environmental and health issues. The Sustanible City. 2. 1423–34. doi: 10.2495/SC141202.

[5] GÜNEY, B. and ÖZ, A. (2020). Microstructure and Chemical Analysis of Vehicle Brake Wear Particle Emissions. European Journal of Science and Technology. 19, 633–42. doi: 10.31590/ejosat.744098.

[6] Kelly, F. J. and Fussell, J. C. (2015). Air pollution and public health: emerging hazards and improved understanding of risk. Environmental Geochemistry and Health. 37(4), 631–49. doi: 10.1007/s10653-015-9720-1.

[7] Tsai, W. T. and Lin, Y. Q. (2021). Trend analysis of air quality index (Aqi) and greenhouse gas (ghg) emissions in taiwan and their regulatory countermeasures. Environments - MDPI. 8(4). doi: 10.3390/environments8040029.

[8] Vasiljević, S., Vasiljević, S., Glišović, J., Stojanović, B., Stojanović, N., and Grujić, I. (2021). Analysis of influencing factors on brake wear and non-exhaust emission with reference to applied materials in brake pads. Mobility and Vehicle Mechanics. 47(2), 45–59. doi: 10.24874/mvm.2021.47.02.04.

[9] Xiao, X., Yin, Y., Bao, J., Lu, L., and Feng, X. (2016). Review on the friction and wear of brake materials. Advances in Mechanical Engineering. 8(5), 1–10. doi: 10.1177/1687814016647300.

[10] Kumar, K. N. and Suman, K. N. (2017). Review of brake friction materials for future development. Journal of Mechanical and Mechanics Engineering. 3(2), 1–29.

[11] bin Selamat, M. S. (2005). Friction materials for brake applications. Journal of Industrial Technology. 14(2), 9–25.

[12] Aranganathan, N. and Bijwe, J. (2016). Development of copper-free eco-friendly brake-friction material using novel ingredients. Wear. 352–53, 79–91. doi: 10.1016/j.wear.2016.01.023.

[13] Asif, M., Chandra, K., and Misra, P. S. (2011). Development of Iron based brake friction material by hot powder preform forging technique used for medium to heavy duty applications. Journal of Minerals and Materials Characterization and Engineering. 10(3), 231–44. doi: 10.4236/jmmce.2011.103015.

[14] Bijwe, J., Kumar, M., Gurunath, P. V., Desplanques, V., and Degallaix, G. (2008). Optimization of brass contents for best combination of tribo-performance and thermal conductivity of non-asbestos organic (NAO) friction composites. Wear. 265(5–6), 699–712. doi: 10.1016/j.wear.2007.12.016.

Chapter 35

Synthesis and study of tin chloride pentahydrate doped polyaniline for carbon dioxide gas sensing

Pranav Awandkar[a] and Shrikrishna Yawale[b]

Government Vidarbha Institute of Science and Humanities, Amravati, India

Abstract

A highly conducting films of polyaniline, tin chloride pentahydrate, PSN ($SnCL_4$. $5H_2O$), is fabricated for CO_2 gas detection and their corresponding observation has been noted. Polyaniline, PSN, film was synthesised through a method by using oxidant to initiate polymerisation. Aniline hydrochloride was used as a monomer and ammonium peroxidisulphate as an oxidant. Polyaniline, PSN, composite has shown low sensitivity for CO_2 gas. The detection of gas is mainly based on change in the resistance of the film on exposure to CO_2 gas at room temperature. Polyaniline/$SnCL_4$. $5H_2O$ composite film was coated on a glass substrate through the screen-printing method. Ethyl cellulose and n-butyl carbitol were used as binder and solvent. Characterisation of the composite was done by X-ray diffraction (XRD) and UV – VIS. At room temperature, gas responses were detected. The advantages of such sensors are that they are easy to fabricate have, high electrical conductivity, and are thermally stable and highly sensitive.

Keywords: Polyaniline, tin (IV) chloride pentahydrate, in situ polymerisation, sensor

Introduction

Conducting polymer-based gas sensors have been widely developed and studied for CO_2 gas detection. They have been preferred due to their properties of sensitivity and good response time. Polyaniline, when exposed to the analyte gas leads to a certain change over its surface electron mobility and polymer matrix which can be detected and can be converted to electrical signal [1]. Polyaniline based gas sensors have been of subject of interest due to their various properties such as high selectivity, fast response and recovery time, low cost, they have high electrical conductivity, thermally stable, and can easily synthesised [2]. Moreover, other oxidative states it may appear are in leucoemeraldine (yellow in appearance), fully oxidised pernigraniline (violet in appearance), and emeraldine (green in appearance). Both leucoemeraldine and pernigraniline are non-conducting in nature whereas emeraldine which is partially doped is conductive [3, 4]. Gas sensors based on polyaniline in combination with different metal oxide has been fabricated and have been used to develop as CO_2 gas sensors. Polyaniline/SnO_2 based gas sensors have been developed and tested for various gases. Bai et. al. [5] developed gas sensors based on PANI/SnO_2 for NH_3 detection at room temperature. Betty et al. [6] developed a sensor mainly made of PANI/SnO_2 for toxic gas sensing. SnO_2 is a n – type of semiconducting metal oxide having various properties and better electron transport ability [7-9]. Various researchers have studied the ability of SnO_2 for CO_2 gas detection and found that SnO_2 has low gas sensitivity and higher response and recovery time at room temperature [7-11]. Therefore, gas sensor based on SnO_2 with combination with different dopant need to be investigated to achieved better sensitivity for CO_2 gas. Whereas tin (IV) chloride pentahydrate is the basic element to synthesis SnO_2 so it becomes necessary to know its ability to sense CO_2 gas.

Experimentation

Material

Aniline hydrochloride, ammonium peroxidisulphate, and tin (IV) chloride pentahydrate ($SnCL_4$. $5H_2O$) were obtained from SDFL India.

Polyaniline, tin (IV) Chloride Pentahydrate, Composite Synthesis

0.2 M of aniline hydrochloride was dissolved in 50 mL of distilled water. $SnCL_4$. $5H_2O$ 5% weight of 0.2 M of aniline hydrochloride was stirred in 50 mL of distilled water in another beaker and was ultrasonicated up to 5 hr. Ammonium peroxidisulphate (APS) 5.71 g was dissolved in 50 mL of distilled water [12]. The above solution of 0.2 M of aniline hydrochloride is briefly stirred for 1 hr and in it, 5% weight of $SnCL_4$: $5H_2O$ which is 5% weight of 0.2 M of aniline hydrochloride is added in the solution is briefly stirred. Then ammonium peroxidisulphate (APS) 5.71 g is added to it drop by drop to polymerise. Polyaniline, $SnCL_4$.

[a]pranavawandkar@gmail.com; [b]spyawale@rediffmail.com

DOI: 10.1201/9781003450252-35

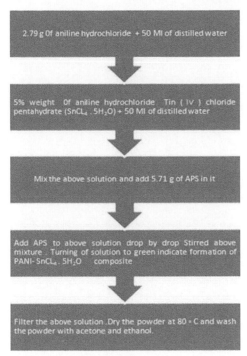

Figure 35.1 Preparation of PSN 5%
Source: Author

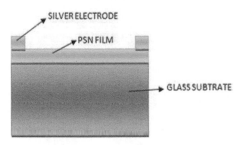

Figure 35.2 Schematic view of PSN sensor
Source: Author

$5H_2O$ (PSN) formed was filtered and cleaned with acetone and ethanol. The powder obtained was dry for 5 hrs at 80°C. Similarly other samples were prepared for 10%, 15%, 20%, and 25% of $SnCL_4 . 5H_2O$. A schematic chart for the preparation of PSN 5% is shown in Figure 35.1.

Sensor Fabrication

Polyaniline, $SnCL_4. 5H_2O$ (PSN), composite was coated on a glass substrate through the screen printing method shown in Figure 35.2. Glass substrate was washed with acetone and heated at 100°C for 1 hr. Ethyl cellulose and butyl carbitol were used as binder and solvent. Polyaniline, $SnCL_4. 5H_2O$ (PSN), composite was coated on a glass substrate with silver electrode.

Characterisation

X – ray Diffraction

XRD of the PSN composites were taken within a range from 10-80° for 2θ shown in Figure 35.3. The XRD pattern for polyaniline, $SnCL_4. 5H_2O$ (PSN), 5% shows a peak at 2θ = 18.10°. For PSN 10% the peak was observed at 2θ = 18.38° and 26.84° broad peak appeared at 2θ = 52.5°. In PSN 15% peak appeared at 2θ = 25.98°, 33.87°, 65.6°, and broad peak appeared at 2θ = 52.54°. In PSN 20% peak were observed at 2θ = 15.93°, 25.97°, 36.21°, 51.90°. In PSN 25% peaks were observed at 2θ = 18.36°, 26.70°, 53.0° 64.86°, 30.72°, 34.12°, 40.53°. The XRD pattern of pure polyaniline shows a peak at 2θ = 20.59° and at 25.05° which indicates its amorphous nature. Whereas several sharp peaks appeared in PSN composite which indicate semi crystallinity. The crystallinity increase with the increase in $SnCL_4.5H_2O$ concentration as more peaks are observed to increase with $SnCL_4.5H_2O$ concentration. The crystalline size calculated from XRD

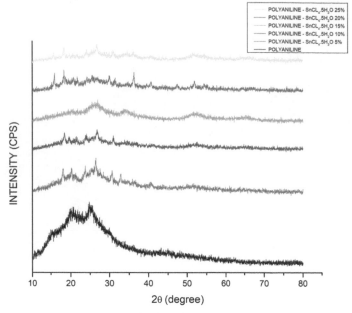

Figure 35.3 XRD of polyaniline – SnCL$_4$5H$_2$O samples
Source: Author

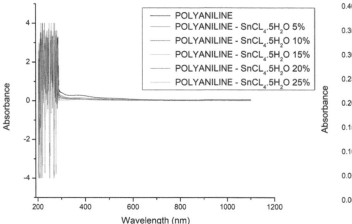

Figure 35.4(a) UV–VIS of PSNs composite
Source: Author

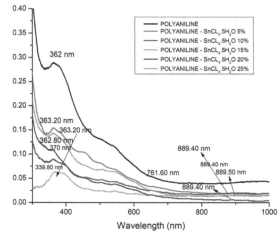

Figure 35.4(b) UV – VIS of PSNs composite within range 300–1100 nm
Source: Author

spectra was found to be 27 nm, 6 nm, 3.3 nm, 26 nm, and 3.8 nm. The degree of crystallinity was found to be 30%, 33%, 35%, 37%, and 38% for PSN 5%, PSN 10%, PSN 15%, PSN 20%, and PSN 25%.

UV – Visible Spectroscopy

The absorbance spectra of the PSN samples were shown in Figure 35.4(a).The pure polyaniline sample shows the absorption band at 761.60, 362 nm which are related to π – polaron transition and π- π* transition in the benzoid ring [13, 14]. In the PSN samples shown in Figure 35.4(b) the same bands appear which are shifted at a different positions such absorption band of 362 nm of polyaniline appears at 363.20 nm, 362.80 nm, 370 nm, 339.80 nm, 363.20 nm in PSN 5%, PSN 10%, PSN 15%, PSN 20%, PSN 25%. More ever other absorption bands were detected at 889.40 nm ,889.50 nm, 889.40 nm, and 889.40 nm for PSN 5%, PSN 15%, PSN 20%, and PSN 25% shown in Figure 35.4(b). Other absorption bands were observed in PSN 5% (254 nm) PSN 10% (270.80 nm) PSN 15% (258.50 nm) PSN 20% (276.20 nm) PSN 25% (279.80 nm) shown in Figure 35.4(c). Very much research has been conducted to synthesize SnO$_2$ through various method by using SnCL$_4$.xH$_2$O and has observed absorption bands that lies within a range of 270-300 nm range and around 355 nm and 380 nm [15-17]. The UV spectra of all PSN samples show similar absorption band near the same wavelength as observed by others which confirm the formation of SnO$_2$ particles in PSN composites. The shift in the band of PSN samples can be due to the presence of SnO$_2$ particles in the polyaniline matrix.

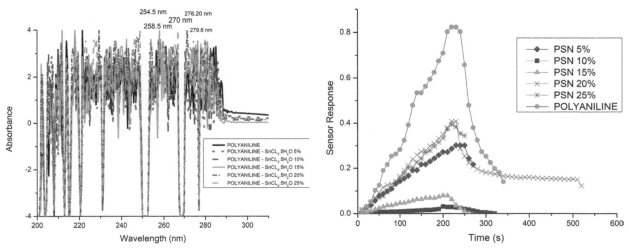

Figure 35.4(c) UV spectra of PSNs composite within the range 200-300 nm
Source: Author

Figure 35.5 Gas sensor response at 150 ppm
Source: Author

Gas Sensing Measurement

The behavior of polyaniline, $SnCL_4. 5H_2O$ (PSN), composites was studied when the CO_2 gas is introduced to it. Dynamic response of sensors was noted. The sensor sensitivity (%) is given by

$$S = ([R_g - R_o]/R_o) \times 100$$

Gas sensor response is given by

$$\Delta R_g/R_o = [R_g - R_o]/R_o$$

R_g and R_o indicate the sensor resistance in gas and air [18]. For all the PSN composite samples when exposed to CO_2 gas for 150 ppm at room temperature, there is a decrease in their electrical conductivity. The observation shows in Figure 35.5 that all of the sensors do not attend their base value upon exposure to the air. For PSN 10% and PSN 15% the response and recovery time are approximately 190 and 80 sec and 180 sec and 40 sec. Whereas PSN 20% shows better sensing ability. The observations are found to be relevant to the observation of the researcher [7].

Conclusions

In this paper, the composite of polyaniline, $SnCL_4. 5H_2O$ (PSN), were successfully synthesised through in situ polymerisation for various percentage of $SnCL_4. 5H_2O$. The XRD spectra confirm semi crystalline behaviors of composites and UV – visible spectra confirm the formation of SnO_2 particles in composites. Composite of polyaniline, $SnCL_4. 5H_2O$ (PSN), shows low sensitivity toward CO_2 gas as compared to pure polyaniline so it can be concluded that SnO_2 based gas sensors have low sensitivity toward CO_2 gas.

References

[1] Steffens, C., Brezolin, A. N., and Steffens, J. (2018). Conducting polymer-based cantilever sensors for detection humidity. Scanning. 2018, Article ID 4782685, 1–6. doi:10.1155/2018/4782685.

[2] Fratoddi, I., Venditti, I., Cametti, C., and Russo, M. V. (2015). Chemiresistive polyaniline-based gas sensors: a mini review. Sensors and Actuators B: Chemical. 220, 534-48. https://doi.org/10.1016/j.snb.2015.05.107.

[3] Bhandari, S. (2018). Polyaniline: structure and properties relationship. Polyaniline Blends, Composites, and Nanocomposites. 23-60.

[4] Feast, W. J. and Friend, R. H. (1990). Synthesis and material and electronic properties of conjugated polymers. Journal of Materials Science. 25, 3796–3805. https://doi.org/10.1007/BF00582445

[5] Bai, S., Tian, Y., Cui, M., Sun, J., Tian, Y., Luo, R., and Li, A. C. D. (2016). Polyaniline@SnO2 heterojunction loading on flexible PET thin film for detection of NH3 at room temperature. Sensors and Actuators B: Chemical. 226, 540-47. https://doi.org/10.1016/j.snb.2015.12.007.

[6] Betty, C. A., Choudhury, S., and Arora, S. (2015). Tin oxide – polyaniline heterostructure sensors for highly sensitive and selective detection of toxic gases at room temperature. Sensors and Actuators B: Chemical. 220, 288-94. https://doi.org/10.1016/j.snb.2015.05.074.

[7] Wang, D., Chen, Y., Liu, Z., Li, L., Shi, C., Qin, H., and Hu, J. (2016). CO2-sensing properties and mechanism of nano-SnO2 thick-film sensor, Sensors and Actuators B: Chemical. 227, 73-84. https://doi.org/10.1016/j.snb.2015.12.025.

[8] Zhang, W., Xie, C., Zhang, G., Zhang, J., Zhang, S., and Zeng, D. (2017). Porous $LaFeO_3/SnO_2$ nanocomposite film for CO2 detection with high sensitivity, Materials Chemistry and Physics. 186, 228-36. https://doi.org/10.1016/j.matchemphys.2016.10.048.

[9] Xiong, Y., Xue, Q., Ling, C., Lu, W., Ding, D., Zhu, L., and Li, X. (2017). Effective CO_2 detection based on LaOCl-doped SnO_2 nanofibers: Insight into the role of oxygen in carrier gas. Sensors and Actuators B: Chemical. 241, 725-34. https://doi.org/10.1016/j.snb.2016.10.143.

[10] Hoefer, U., Kühner, G., Schweizer, W., Sulz, G., and Steiner, K. (1994). CO and CO_2 thin-film SnO2 gas sensors on Si substrates. Sensors and Actuators B: Chemical. 22(2), 115-19. https://doi.org/10.1016/0925-4005(94)87009-8.

[11] Juang, F. -R., Chern, W. -C., and Chen, B. -Y. (2018). Carbon dioxide gas sensing properties of ZnSn(OH)6-ZnO nanocomposites with ZnO nanorod structures. Thin Solid Films. 660, 771-76. https://doi.org/10.1016/j.tsf.2018.03.069.

[12] Sapurina, I., Stejskal, J., Hüsing, U., Laine, N., and Richard, M. (2008). Polyaniline — a conducting polymer. Materials Syntheses, pp. 199-207. Vienna: Springer. https://doi.org/10.1007/978-3-211-75125-1_26 199-207

[13] Nasirian, S. and Moghaddam, H. M. (2015). Polyaniline assisted by $TiO_2{:}SnO_2$ nanoparticles as a hydrogen gas sensor at environmental conditions. Applied Surface Science. 328, 395-404. https://doi.org/10.1016/j.apsusc.2014.12.051.

[14] Nasirian, S. and Moghaddam, H. M. (2014). Hydrogen gas sensing based on polyaniline/anatase titania nanocomposite. International Journal of Hydrogen Energy. 39(1), 630-42. https://doi.org/10.1016/j.ijhydene.2013.09.152.

[15] Matysiak, Wiktor, Tomasz, Tański, Smok, Weronika and Polishchuk, Oleh. (2020). Synthesis of hybrid amorphous/crystalline SnO2 1D nanostructures: investigation of morphology, structure and optical properties. Scientific reports. 10. 14802. 10.1038/s41598-020-71383-2.

[16] Prakash, K., Kumar, P. S., Pandiaraj, S., Saravanakumar, K., and Karuthapandian, S. (2016). Controllable synthesis of SnO_2 photocatalyst with superior photocatalytic activity for the degradation of methylene blue dye solution. Journal of Experimental Nanoscience. 11, 14, 1138-55. doi:10.1080/17458080.2016.1188222

[17] Zhang, Li, Yu, Wei, Han, Cui, Guo, Jiang, Zhang, Qinghong, Xie, Hongyong, Shao, Qian, Sun, Zhiguo and Guo, Zhanhu. (2017). Large Scaled Synthesis of Heterostructured Electrospun TiO_2/SnO_2 Nanofibers with an Enhanced Photocatalytic Activity. Journal of The Electrochemical Society. 164. H651-H656. 10.1149/2.1531709jes. 2018, Article ID 4782685, 1–6.

[18] Bhadra, J., Al-Thani, N. J., Madi, N. K., and Al-Maadeed, M. A. (2013). Preparation and characterization of chemically synthesized polyaniline–polystyrene blends as a carbon dioxide gas sensor. Synthetic Metals. 181, 27-36. https://doi.org/10.1016/j.synthmet. 2013.07.026.

Chapter 36

Experimental research on the influence of FDM parameters on impact strength, tensile strength, and the hardness of parts made of polylactic acid for optimisation

Javed Dhalait[a] and Vishvajeet Potdar[b]

A G Patil Institute of Technology, Solapur,n India

Abstract

In Industry 4.0 additive manufacturing (AM) has become significant in designing and manufacturing components, which are impossible to manufacture with traditional methods. AM allows not only complex part manufacturing but also is helpful in achieving maximum strength with a lower weight. This research paper focuses on the effect of process parameters on the specimen prepared using the fused deposition modeling (FDM) method. Here are three parameters viz. infill pattern, infill percentage, and shell thickness with three levels is investigated. L27 orthogonal array is used for investigation. Polylactic acid (PLA), plastic filament is used in the FDM process. Samples were tested for impact strength, tensile strength, and hardness; results are optimised using Taguchi optimisation in Minitab Software. Results shows infill percentage significantly affects Tensile strength and hardness of components and impact strength is influenced by shell thickness.

Keywords: Additive manufacturing, fused deposition modeling, PLA filament, Taguchi method, optimisation

Introduction

Modernisation and revolution in imaging, architecture, communications, and engineering has provided additive manufacturing (AM) great flexibility and efficiency in manufacturing operations. AM means adding material layer by layer to form an object. It uses CAD model from any CAD software or even from 3D object scanners which is processed in 3D slicer to create a program file. Such program is later fed to 3D printer, and it deposit selected material in layer-by-layer manner. Fused deposition modeling (FDM) is one of the methods in AM process. The materials which are used are thermoplastic polymers in a filament form, fed to nozzle through heated extruder head. The print head is moved under computer control to form the printed shape. In FDM, variety of thermoplastics can be used such as polylactic acid (PLA),, ABS, PET, Nylon, TPU, PC, carbon fiber, PEEK etc. However PLA is most commonly used material in desktop 3D printing as it provides ease of printing, good mechanical properties, low cost and most vitally it is bio-degradable. Many types of research are conducted to reduce the effect of various processing parameters for better performance of printed products. Increase in ultimate tensile strength is observed with increase in infill percentage and no significant effect of layer thickness is observed on ultimate tensile strength. No correlation found between the infill percentage and the hardness [2].

Numerous methodologies and techniques optimally change 3D printer's process and control parameters developed by researchers with advantages and disadvantages [3]. Infill percentage is the major process parameter on all the material's mechanical properties [4]. Layer thickness, infill density, print speed significantly affect energy consumption on the other side these shows less effect on hardness of the components manufactured [5]. Effect of infill percentage, orientation, layer thickness and nozzle temperature parameters values are studied for optimisation [6]. Raster direction angle has less effect on hardness, but layer thickness shows significant influence on UTS and hardness [7]. PLA-CF exhibits better flexural strength and stiffness in comparison with neat PLA. No significant effect is seen in comparison with neat PLA on dimensional accuracy and surface roughness [8]. The material performance is influenced by infill percentage and nozzle temperature and has lowest influence on dynamic properties of the material [9]. 45°/45° raster angle shows better tensile strength compared to other combinations. Higher the layer height higher the tensile strength [11]. Increasing the layer thickness for the same infill pattern improves the hardness [12]. For lightweight and durable components honeycomb structure with 23 shell layers is suggested. To increase the strength, shell thickness should be increased [13]. For ultimate tensile strength layer thickness is most significant parameters and dimensional accuracy. A thinner layer exhibits better bonding capability and improves mechanical properties of nylon material [14]. The tensile strength of the specimen increases with the increase in the raster angle [15]. Maximum tensile strength in components manufactured using ABS filament obtained at 0° raster orientation, better tensile strength is obtained as compared to 90° and 45° raster orientations [16].

[a]dhalaitj@gmail.com, [b]vishwa.potdar@agpit.edu.in

DOI: 10.1201/9781003450252-36

Design of Experimental Setup

Design of Experiment

It is a commanding data collection and analysis tool, used in a variety of experimental conditions, helps in identifying the effect of many input factors and their effect on a desired output (response). Taguchi standardised form of DOE has appealed more researchers for identifying such effects. Taguchi method not only allows reduction in variance but also helps in finding out the best suitable set of parameter. Hence Taguchi method is applied to collect optimised value under varying two parameters with levels are layer thickness and printing pattern. The selected values are mentioned in the table below.

L27 Orthogonal array is prepared as mentioned in Table 36.2 and sample specimens are built and tested.

3D Printer and Test Specimen

Test specimen 3D Drawing is prepared in Autocad software and file is converted in .STL format and later processed in flashprint 3D slicer.

The program file is then fed to Flashforge guider IIs FDM 3D printer. It has layer resolution of ± 0.20mm and positioning precision 11 Microns in XY and 2.5 Microns in Z direction [17]. A WOL 3D flamingo pink PLA filament having 1.75 mm diameter is used for the same.

Table 36.1 Parameter and levels

Level	Parameters		
	Shell thickness	Infill pattern	Infill density
Level I (Low)	0.8	Line	20%
Level II (Medium)	1.6	Hexagon	40%
Level III (High)	2.4	Triangle	60%

Source: Self DOE

Table 36.2 Parameter and levels

S. No.	Infil pattern	Infill percentage	Shell thickness	Sr. no.	Infil pattern	Infill percentage	Shell thickness	Sr. no.	Infil pattern	Infill percentage	Shell thickness
1	Line	20	0.8	10	Hexagon	20	0.8	19	Triangle	20	0.8
2	Line	20	1.6	11	Hexagon	20	1.6	20	Triangle	20	1.6
3	Line	20	2.4	12	Hexagon	20	2.4	21	Triangle	20	2.4
4	Line	40	0.8	13	Hexagon	40	0.8	22	Triangle	40	0.8
5	Line	40	1.6	14	Hexagon	40	1.6	23	Triangle	40	1.6
6	Line	40	2.4	15	Hexagon	40	2.4	24	Triangle	40	2.4
7	Line	60	0.8	16	Hexagon	60	0.8	25	Triangle	60	0.8
8	Line	60	1.6	17	Hexagon	60	1.6	26	Triangle	60	1.6
9	Line	60	2.4	18	Hexagon	60	2.4	27	Triangle	60	2.4

Source: Self DOE

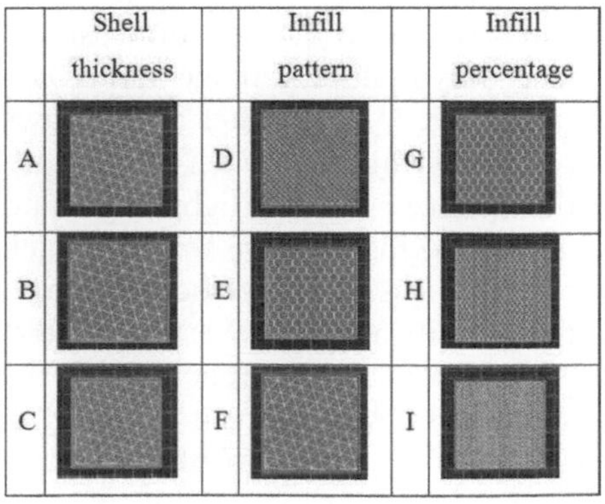

Figure 36.1 Shell thickness, infill pattern, infill percentage

Source: Self (Flash Print -5 software)

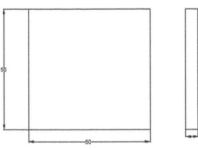

Figure 36.2 Hardness test specimen ASTM D 2240 standard
Source: ASTM D2240 Standard

Figure 36.3 Hardness test specimen
Source: Self (AGPPI Solapur 3D Printer Lab)

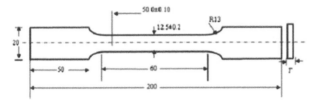

Figure 36.4 Tensile strength sample ASTM D 638 standard
Source: ASTM D638 Standard

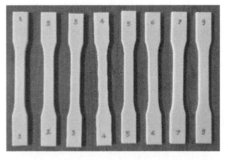

Figure 36.5 Tensile strength specimen
Source: Self (AGPPI 3D Solapur Printer Lab)

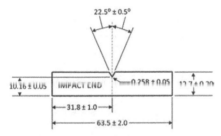

Figure 36.6 Izod impact test specimen (Notched) ASTM D256 standard
Source: ASTM D 256 Standard

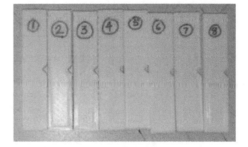

Figure 36.7 Izod impact test specimen
Source: Self (AGPPI Solapur 3D Printer Lab)

Figure 36.8 Flashforge Guider IIs
Source: Self (AGPPI 3D Solapur Printer Lab)

Other parameters kept constant in manufacturing test specimens are extruder temperature -210°, bed temperature -30°, nozzle diameter -0.4mm, print speed -80 mm/s.

Test setup

Rockwell Hardness Tester

As per standard method for plastics load of 60 Kgf is applied using 1/2" ball indentor.

Tensile Strength Tester

The grips of the universal tester hold the specimen at a specified gap and pull until failure. Speed: 5mm/min.

Impact Strength Tester

Izod/charpy impact tester. Maximum load 25J.

Results

Result of Hardness Testing

Following result were obtained after testing Hardness sample specimens.

Result of Tensile Testing

Following result were obtained after testing tensile sample specimens.

Result of Impact Testing

Following result were obtained after testing tensile sample specimens.

Figure 36.9 Rockwell hardness tester
Source: Self (AGPPI Solapur SOM Lab)

Figure 36.10 Indentation on specimen
Source: Self (AGPPI Solapur 3D Printer Lab)

Figure 36.11 Universal tester
Source: Self (Praj metallurgical lab Pune)

Figure 36.12 Izod impact tester
Source: Self (Praj metallurgical lab Pune)

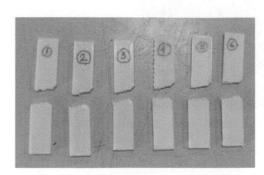

Figure 36.13 Indentation on specimen
Source: Self (AGPPI Solapur 3D Printer Lab)

Table 36.3 Results of hardness testing

Sr. no.	Hardness value (G)	Sr. no.	Hardness value (G)	Sr. no.	Hardness value (G)
1	27	10	28	19	32
2	27	11	28	20	38
3	28	12	28	21	40
4	32	13	82	22	96
5	31	14	87	23	95
6	32	15	86	24	98
7	78	16	52	25	62
8	76	17	48	26	58
9	82	18	55	27	55

Source: Self Result after testing

Table 36.4 Results of tensile testing

Sr. no.	Tensile strength (MPa)	Sr. No.	Tensile strength (MPa)	Sr. no.	Tensile strength (MPa)
1	20.91	10	19.91	19	19.28
2	18.79	11	24.10	20	26.79
3	22.44	12	29.14	21	28.87
4	23.55	13	24.85	22	21.46
5	24.89	14	26.00	23	0.12
6	27.59	15	29.53	24	29.81
7	27.18	16	26.44	25	24.28
8	31.75	17	30.65	26	30.28
9	31.88	18	30.77	27	34.47

Source: Self Result after testing

Table 36.5 Results of impact testing

Sr. no.	Impact value (J)	Sr. No.	Impact value (J)	Sr. no.	Impact value (J)
1	0.10	10	0.10	19	0.08
2	0.13	11	0.12	20	0.12
3	0.13	12	0.13	21	0.16
4	0.12	13	0.12	22	0.14
5	0.10	14	0.12	23	0.12
6	0.10	15	0.12	24	0.14
7	0.10	16	0.13	25	0.13
8	0.14	17	0.13	26	0.12
9	0.13	18	0.14	27	0.12

Source: Self Result after testing

Optimisation

Hardness

Process parameters optimization is found using Taguchi method. The response table for SN ratio and means along with delta and rank value are found using Minitab software shown in Table 36.6 and 3.67.

From Table 36.6 and 36.7 the highest SN ratio of layer thickness is 33.40 at 3rd level. The highest SN ratio of infill pattern is 29.17 at 3rd level. The delta value for layer thickness is 10.50 and the delta value for Infill pattern 1.34. Optimal factor levels of parameter are determined according to the response of each level and setting target value larger-the-better. Here, the parameters are optimised to maximise the hardness of specimens were obtained from the response table. Layer thickness ranks 1 in Tables 36.6 and 36.7 indicating its significance in 3D printing. The maximum hardness is obtained at the layer thickness of 2.4 and 3D infill pattern. Figures 36.14 and 36.15 reveal the main effect plots for SN ratio and means. It also shows that the layer thickness is certainly affecting the hardness.

Table 36.6 SN ratio response table

Level	Infill pattern	Infill percentage	Shell thickness
1	32.21	29.61	634.21
2	33.92	36.09	31.41
3	35.38	35.80	35.88
Delta	3.16	6.47	4.47
Rank	3	1	2

Source: Self (Minitab Software)

Table 36.7 Means response table

Level	Infill pattern	Infill percentage	Shell thickness
1	45.89	30.67	58.44
2	54.89	71.00	39.33
3	63.78	62.89	66.78
Delta	17.89	40.33	27.44
Rank	3	1	2

Source: Self (Minitab Software)

Figure 36.14 SN ratios for hardness
Source: Self (Minitab Software)

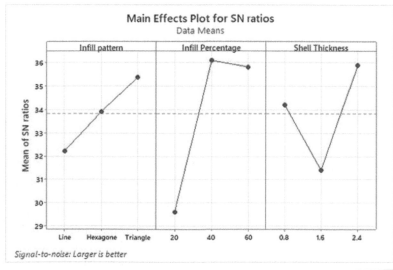

Figure 36.15 Means for hardness
Source: Self (Minitab Software)

Tensile Strength

From the Tables 36.8 and 36.9, it is observed that maximum value for means of means is infill percentage 29.33 at 3rd level. The maximum SN ratio for infill percentage is 29.74 at 3rd level. The delta value for infill percentage is 15.4 and the delta value for shell thickness is 14.57. Optimal factor levels of parameter are determined according to the response of each level and setting target value larger-the-better. Here, the parameters are optimised to maximise the tensile strength of specimens were obtained from the response table. Infill percentage ranks 1 in Tables 36.8 and 36.9 indicating its significance in 3D printing. The maximum tensile strength is obtained at the Infill percentage of 60% and triangle pattern. Figures 36.16 and 36.17 reveal the main effect plots for SN ratio and means. It also shows that the Infill percentage is certainly affecting the tensile strength.

Impact Strength

From Tables 36.10 and 36.11, it is observed that maximum value for shell thickness of means of means is 33.40 at 3rd level. The maximum SN ratio for infill percentage is -17.99 at 3rd level. The delta value for shell thickness is 1.25 and the delta value for infill percentage 0.66. Optimal factor levels of parameter are determined according to the response of each level and setting target value larger-the-better. Here, the parameters are optimised to maximise the impact strength of specimens were obtained from the response table. Shell thickness ranks 1 in Tables 36.10 and 36.11 indicating its significance in 3D printing. The

Table 36.8 Means response table

Level	Infill pattern	Infill percentage	Shell thickness
1	27.94	27.08	13.96
2	28.40	14.29	28.21
3	14.36	29.33	28.53
Delta	14.04	15.04	14.57
Rank	3	1	2

Source: Self (Minitab Software)

Table 36.9 SN ratio response table

Level	Infill pattern	Infill percentage	Shell thickness
1	25.44	23.36	22.38
2	26.82	23.09	26.47
3	23.93	29.74	27.35
Delta	2.89	6.66	4.97
Rank	3	1	2

Source: Self (Minitab Software)

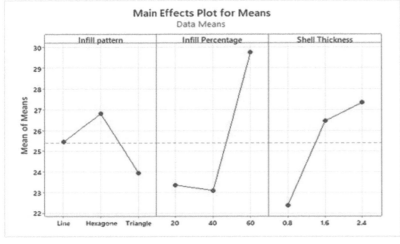

Figure 36.16 Means for tensile test
Source: Self (Minitab Software)

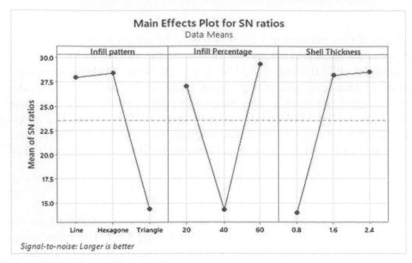

Figure 36.17 Signal to noise ratios for tensile test
Source: Self (Minitab Software)

Table 36.10 Means response table

Level	Infill pattern	Infill percentage	Shell thickness
1	0.1167	0.1189	0.1133
2	0.1233	0.1200	0.1222
3	0.1256	0.1267	0.1300
Delta	0.0089	0.0078	0.0167
Rank	2	3	1

Source: Self (Minitab Software)

Table 36.11 SN ratio response table

Level	Infill pattern	Infill percentage	Shell thickness
1	18.74	18.65	19.03
2	18.21	18.47	18.29
3	18.16	17.99	17.79
Delta	0.58	0.66	1.25
Rank	3	2	1

Source: Self (Minitab Software)

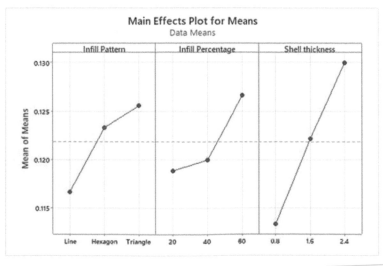

Figure 36.18 Means for impact test
Source: Self (Minitab Software)

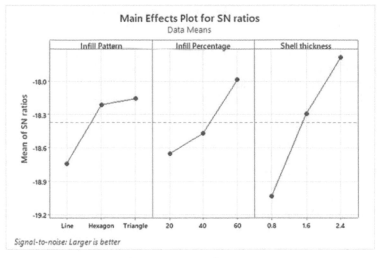

Figure 36.19 Signal to noise ratios for impact test
Source: Self (Minitab Software)

maximum impact strength is influenced by shell thickness. Figures 36.18 and 36.19 reveal the main effect plots for SN ratio and means. It also shows that the shell thickness is certainly affecting the impact strength compared to other parameters.

Conclusion

This research paper aims to investigate the effect of processing parameters of fused deposition modeling (FDM) 3D printing for optimisation. The parameters infill pattern, infill percentage and shell thickness were selected. Sample specimens are manufactured by using polylactic acid (PLA) materials are tested for hardness, impact and tensile strength and results are analyzed by using Taguchi method. Following are some important outcomes of the paper.

Impact Strength: The range of analysis shown that the optimized combination of maximum shell thickness i.e. 2.4 and infill percentage gives a better impact strength. The above combination set has given Impact strength of 0.16 J which is better than other set of combinations. Shell thickness is the most affecting parameter on Impact strength than infill pattern. Hence it is suggested to keep added shell thickness for better impact strength.

Hardness: Line and hexagon infill pattern has not shown significant effect on surface hardness. Compared to all pattern triangle infill patter shows better results. The triangle pattern with infill percentage 40% gives optimum value. The hardness test also indicates that up to 40% infill percent the hardness increases and after that it decreases this may be due to increase in density makes the component soft. So, it is suggested to keep the infill percentage up to 40% for better hardness.

Tensile Strength: Tensile strength is influenced more by the infill percentage than shell thickness. Higher the infill percentage higher the tensile strength. Infill pattern does not show a large variation so it can be used as per availability. The infill percentage and Hexagon pattern shows better optimisation. So, it is suggested that infill percentage 60% for better tensile strength.

The effects obtained directs the need for additional research regarding synchronised changes in many parameters and expected strength or physical properties required in components to be manufactured by using FDM.

References

[1] Peng, G. Ji, Z., Gao, Z., Wu, W., Ye, W., Li, G., and Qu, H. (2021). Effects of printing parameters on the mechanical properties of high-performance polyphenylene sulfide three-dimensional printing. 3D Printing and Additive Manufacturing. 8. doi: 10.1089/3dp.2020.0052

[2] Zisopol, D. G., Nae, I., Portoaca, A. -I., amd Ramadan, I. (2021). A theoretical and experimental research on the influence of fdm parameters on tensile strength and hardness of parts made of polylactic acid. Engineering, Technology & Applied Science Research. 11(4), 745863.

[3] Zohdi, N. and Yang, R. (2021). Material anisotropy in additively manufactured polymers and polymer composites: a review. Polymers. 13, 3368. https://doi.org/10.3390/polym13193368

[4] Algarni, M. and Ghazali, S. (2021). Comparative study of the sensitivity of PLA, ABS, PEEK, and PETG's mechanical properties to FDM printing process parameters. Crystals. 11, 995. https://doi.org/10.3390/cryst11080995

[5] Emmanuel, U., Enemuoh, S. D., Connor, F., and Menta, V. G. (2021). Effect of process parameters on energy consumption, physical, and mechanical properties of fused deposition modeling. Polymers. 13, 2406. https://doi.org/10.3390/polym13152406

[6] Vates, U. M., Kanu, N. J., Gupta, E., Singh, G. K., Daniel, N. A., and Sharma, B. P. (2021). Optimization of FDM 3D printing process parameters on abs based bone hammer using RSM technique. ICRAMEN 2021, IOP Conference Series: Materials Science and Engineering. 1206, 012001. doi:10.1088/1757-899X/1206/1/012001

[7] Hanon, M. M., Dobosa, J., and Zsidaia, L. (2021). The influence of 3D printing process parameters on the mechanical performance of PLA polymer and its correlation with hardness. Procedia Manufacturing. 54, 244–49.

[8] Reverte, J. M., Caminero, M. A., Chacón, J. M., García-Plaza, E., Núñez, P. J., and Becar, J. P. (1010). Mechanical and geometric performance of PLA-based polymer composites processed by the fused filament fabrication additive manufacturing technique. Materials. 13, 1924. doi:10.3390/ma13081924

[9] Dey, A. and Yodo, N. (2019). A systematic survey of FDM process parameter optimization and their influence on part characteristics. Journal of Manufacturing and Materials Processing. 3, 64. doi:10.3390/jmmp3030064.

[10] Shubham, P., Aggarwal, C., and Joshi, S. (2018). optimization of process parameter to improve dynamic mechanical properties of 3D printed abs polymer using taguchi method. International Journal of Mechanical and Production Engineering. 6(6).

[11] Rajpurohit, S. R. and Dave, H. R. (2018). Impact of process parameters on tensile strength of fused deposition modeling printed crisscross poylactic acid. World Academy of Science, Engineering and Technology International Journal of Materials and Metallurgical Engineering. 12(2).

[12] Mahendra, A. and Baroor, A. (2017). Effect of process parameters on the mechanical properties of the components made from acrylonitrile butadiene styrene (ABS) using 3D printing technology. International Journal of Engineering Research in Mechanical and Civil Engineering. 2(11).

[13] Ćwikła, G. Grabowik, C., Kalinowski, K., Paprocka, I., and Ociepka, P. (2017). The influence of printing parameters on selected mechanical properties of FDM/FFF 3D-printed parts. IOP Conference Series: Materials Science and Engineering. 227, 012033. doi:10.1088/1757-899X/227/1/012033.

[14] Basavaraj, C. K. and Vishwas, M. (2016). Studies on effect of fused deposition modelling process parameters on ultimate tensile strength and dimensional accuracy of nylon. IOP Conference Series: Materials Science and Engineering. 149, 012035. doi:10.1088/1757-899X/149/1/012035.

[15] Ezoji, M., Razavi-Nouri, M., and Rezadoust, A. M. (2016). Effect of raster angle on the mechanical properties of PLA 3D printed articles. 12th International Seminar on Polymer Science and Technology. 25, Islamic Azad University, Tehran, Iran.

[16] Letcher, T., Rankouhi, B., and Javadpour, S. (2015). Experimental study of mechanical properties of additively manufactured abs plastic as a function of layer parameters. Proceedings of the ASME 2015 International Mechanical Engineering Congress and Exposition. 1319. Houston, Texas.

[17] https://thinkfab.in/3d-printers/flashforge-fdm/flashforge-guider-iis-2/ (Accessed Feb 12, 2022).

Chapter 37

Competency model for the strategic, tactical, and operational level employees for Industry 4.0

Subbulakshmi Somu[1,a] and Roopashree Rao[2,b]

[1]Dayananda Sagar College of Arts Science and Commerce, India

[2]Somlalit Institute of Management Studies, India

Abstract

Industry 4.0 is marked by the implementation of the internet of things, strong autonomy of machines, and artificial intelligence connected and controlled by the internet interacting in real-time, with minimum error rates and increased efficiency. These changes in the working conditions demand the competencies required by the employees to be aligned with the industrial changes, enabling rapid response to the changes in the external environment. Industry 4.0 workspace could demand upgraded personal and technical skills backed by organizational culture capable of adopting continuous learning and the skills essential to navigate emerging business models and changing transformation lines. This paper is an attempt to identify and organise the competencies required by the people working in Industry 4.0 at different levels of the organisation and conceptually derive a comprehensive model with the help of secondary data. Competency models like the MuShCo approach primarily focusing on the operational level and holistic approach are adapted and equated to frame this exhaustive model which gives an inclusive view of the competencies required by the employees working at the strategic level, tactical level, and the operational level in the organization because Industry 4.0 would affect all industries, albeit at different intensities which would eventually volunteer and demand all the levels of the management to upgrade and be ready. The inclusive model which identifies the competencies required by the employees at all the three levels in the organisation is a novel approach which is not established by the previous studies.

Keywords: Industry 4.0, holistic approach, MuShCo model, operational level, tactical level, strategic level

Introduction

The capacity to innovate and the shortest cycle time puts an organization in the best competitive position compared to its rivals. On the other hand, the demand from customers for customized products creates an unpredictable and heterogeneous market [1]. All these factors drive Industry 4.0, where the human and machine interfaces are integrated to meet all the challenging needs of the market. Thus industry 4.0, demands employee competencies to be aligned with the need of the market, incorporating creativity, team building, problem-solving and specialized knowledge about the work and its processes.

Every industrial revolution has brought with it a change in the physical working conditions of various industrial corporations. The first industrial revolution denotes the later part of the 18th century, which started due to the prevalence of mining of coal, which introduced steam machinery, and the period which saw the introduction of the first machinery into the production process. The next industrial revolution, which is called the second industrial revolution, was witnessed during the former period of the 20th century which was initiated by the introduction of electricity which witnessed the mass production trend in organisations. The 1970s saw the third industrial revolution which witnessed the automation of production systems and also initiated the first information systems in organisations. The fourth industrial revolution, which has entered the industrial scenario is marked by the use and implementation of the internet of things, the strong autonomy of machines, and is marked by artificial intelligence.

Industry 4.0 includes cyber-physical systems, which consist of and employs various soft wares and machines which are controlled and connected by the internet which interact in real-time and minimizes error rates and thereby achieve increased efficiency. This concept also highlights a production chain that is decentralised and emphasises the cooperation of autonomous control units capable of decision-making. Every industrial revolution has brought with it a change in the physical working conditions of various industrial corporations. This fourth industrial revolution would bring about various changes for Indian companies, that will bring in new dimensions which would initiate the industrial players to reinvent the way they are operating. Industry 4.0 deals with the whole value chain being digitalised which leads to people, objects, and systems being connected by actual data exchange. This enables the whole value chain to be equipped with artificial intelligence which aids it to respond to the changes in the external environment rapidly.

Industry 4.0 requires employees to possess a complex set of competencies irrespective of the level of the organization they belong to. This paper is an attempt to identify and classify the competencies required by the employees working at various levels in the organisation. Competency models like the MuShCo

[a]subasuhan@gmail.com; [b]rooparaopeacock@gmail.com

DOI: 10.1201/9781003450252-37

approach primarily focusing on the operational level and holistic approach are adapted and equated to frame this exhaustive model which gives an inclusive view of the competencies required by the employees working at the strategic level, tactical level, and the operational level in the organization because Industry 4.0 would affect all industries, albeit at different intensities which would eventually volunteer and demand all the levels of the management to upgrade and be ready.

Review of Literature

The research conducted by Agolla [2], traces the different industrial revolutions and emphasises the importance of human resources in the present Industry 4.0 scenario which is a result of the innovative solutions provided. This research also observes the characteristics of the first revolution as characterised by labor-intensive production systems, the second industrial revolution as characterised by production processes facilitated by electricity, and the third industrial revolution governed by automation because of information technology and electronics. The research also observes the importance and creativity of human capital which is revered as superior to all other resources. The study on identifying the relevant gap in identifying the required skills required for Industry 4.0 in the cross-discipline arena, attempts to visualise the skills, competencies, and literacies required for the Industry 4.0 environment by reviewing 64 journals that are peer-reviewed across diverse disciplines. This study advocates the necessity of the skills of the 21st century like communication with others and other relevant hard skills also [3].

Concentrating on the necessity of the development of new and unique talents with lasting competitiveness and ideas to match the changing trend in the policy concerning the industry, the research conducted by [4] advocates the construction of trinity application modes engulfing the learning using and creating models which are designed from the top-level design. The domain of total quality management in totality is viewed through the lens of the effect of competencies required for the suitability of Industry 4.0. A cross-sectional survey conducted on 20 engineering professionals who are in the early stage of their professions during the period of 2014, 2016, and 2018 revealed the competencies for sustainable Industry 4.0 [5]. On mapping the competencies and the implications associated with the top-down order, a significant difference was observed, in the aspects of total quality management. The result would aid in the development of the total quality management framework to be suitable for the Industry 4.0 scenario [5].

The essential competencies needed to be able to work in the scenario of Industry 4.0 is outlined [6]. It is also observed that industry 4.0 envelops various technical scenarios which demand extensive development of competencies that match Industry 4.0. This study by evaluating the adequate competencies advocated a model consisting of the competencies required which relates to the descriptors of each competency which would eventually help the various stakeholders like people in the academics, people involved in the formulation of policy, and also the people involved in the development of the Industry 4.0 ecosystem [6]. The subject and the delivery of the concept of upcoming engineering education in the undergraduate area highlights the various competencies which are very crucial for education in the above context, and the judgment in the various dimensions highlights the relevance of the five types of core qualities which are essential are identified by [7]. The research based on the literature review attempts to throw light on the context of the human resources involved in Industry 4.0 and highlights associated three aspects in the context of human resources in the Industry 4.0 scenario, the workforce architecture accompanied by the new interactions to adopt the changes of the Industry 4.0 which would initiate different and innovative interpersonal relations among the employees which would volunteer added attributes to imbibe the required competencies [8].

The importance of a qualified workforce essential for the successful implementation of Industry 4.0 is highlighted. The shift of tasks towards demanding innovative competencies to perform value addition is also highlighted. The demand for innovation and other smarter options to cater to the demand of the ever-changing industrial environment is stressed which highlights the necessity of supporting competencies that calls for the education system to develop necessary process [9]. The prominent application of the competency model in the implementation of an organisation's strategy to enhance the behavior of the employees is observed. Posited on the studies of strategic management, it is observed that adequate classification of the involved knowledge, the interest involved, and the behavior associated results in the cooperation of the individuals associated across domains and specialisations. It is well observed that the weightage allotted to different competencies evolves with the shifting of organisational strategy and the associated shift in the information processing and the understanding of the different roles and drive the competency models to pursue the associated logic throughout the organisational levels [10]. A study was conducted to explore the competencies essential to be Industry 4.0 ready as identified in the literature and scientific mapping by searching the Scopus, Web of Science, and Science Direct databases by the 2010- and 2018-time line and by employing the SciMAT software pointed the competencies that are essential about the professional education and the necessity of the handholding of the organizations, universities and the governments associated [11]. Though previous studies identifies and highlights the competencies required by the employees working in the factories, a comprehensive study combines the MuShCo approach and holistic approach, which can be projected to employees working at all the levels of organisation is not done before and thus the gap is identified and justified for this study.

Industry 4.0

A gamut of technological improvements of the entire transformation line is envisaged during the fourth industrial revolution. The conventional production processes are transformed by technologies of Industry 4.0 like automation, robotics, the internet of things, and artificial intelligence, and by implementing advanced digital technologies, there is an increasing amalgamation of the physical and digital world resulting in cyber-physical production systems. The present industrial players will be affected by different intensities and different industries will be affected by different levels. (i) This industrial revolution will enable organizations to tailor/customise their offerings to individual segments/customers. Due to this the supply chain will become more efficient as the transfer of data will be faster and production carried out at the local sites. (ii) Due to Industry 4.0, there will be a smooth amalgamation between industrial and non-industrial applications, which will eventually result in the mass production of services. There will be a collaboration between IT/telecommunication firms. Conservative manufacturing firms join hands with social media players like, Facebook joining hands with Ascenta which is into the drone business. (iii) Industry 4.0 will make organisations rethink and rework their value chain which would result in the revamping of their traditional supply chain and the traditional organisations have to re-strategize to face the newer challenges. (iv) The various technologies that will dominate will be IT, electronics, robotics, biotech and nanotech, etc. This new industrial scenario will need upgraded social and technical skills. The organisations will be required to develop design orientation instead of production orientation. Their cultures will have to adopt continuous learning and multicultural skills, skills to navigate the different business models, transformation lines, and various technologies. (v) Industry 4.0 would result in at least about 9% of the present jobs being automated according to OECD reports. (vi) If we concentrate on the future demand for skills, it is envisaged that a perfect combination of core professional skills and adequate time management, team leadership, critical thinking, and also employee-oriented culture will be the characteristics of future workplaces.

Industry 4.0 will have both benefits and challenges. The benefits are: optimisation of the cost will be achieved due to reduced human resource expenses and accelerated production effectiveness. Owing to greater flexibility of the value chain of the organisations, the organizations will be able to adjust themselves to the ever-changing demands of the customers and will be able to customise the entire service, and will be able to identify prospective services which can generate value. Due to the increased efficiency of the processes, there will be a marked improvement in the quality of the products, and monitoring and maintenance will be feasible by interconnected systems. With increased optimisation of cost, improved operational efficiency, and a competitive edge to tap newer markets, the organisation will be having a competitive advantage to compete with the other players in the industry.

The challenges are: it will result in the horizontal amalgamation of supply chain and production systems which would increase the complexity of the systems, which would require the workforce involved, to identify and analyse the operation, dependency of the processes, the framework for assimilation and utilisation of data to support the complex working of Industry 4.0. To be able to intelligently assist the production system toward Industry 4.0, the workforce should be more adaptable to analyse complex issues which would volunteer assistance and training procedures using augmented reality. Due to the result of Industry 4.0, industries will be operating with upgraded machinery and flexible production systems, and the need for upgraded skills among the workforce will be volunteered, which would match the ever-changing, upgraded job requirement of workplaces. As the jobs that are essential to operate the workplaces in the Industry 4.0 scenario will demand interdisciplinary skills, the organizations should revamp their training and appraisal procedures which would enable the workforce to be suitable for the changed work scenario. Industry 4.0 in conjunction with benefits like cost optimisation, readying the industries to tap into newer opportunities, enabling industries to achieve better operational efficiencies, and facilitating the industries to take advantage of various external factors poses various challenges like economic, social, technical, and environmental which demands up-skilling of various presentation skills and also demands acquiring of newer skills to be ready for the fourth industrial revolution.

Based on the above reviews the significant gap is identified, as a lack of a comprehensive model which denotes the set of competencies required by the employees working at various levels in the organization such as tactical level, operational level, and strategic level. This view is also supported by [1], who argued that the competency model for each job profile has to be created to identify the competencies required by individuals at different levels in the organization. Continuing with the challenges, the fourth industrial revolution would bring in the industrial working environment, and concerning the factory of the future, our research tries on identifying the skills and qualifications the workers should possess to be ready to contribute to the factory in the future. The skills and qualifications essential for the workers to be ready and to be able to work in the factory of the future and to be ready for the fourth industrial revolution are grounded on the three-tier approach where each tier forms a platform for the other tier.

Competency and Competency Model

The workforce of Industry 4.0 should possess certain competencies required to meet the changing demands of the organisation. Competencies are defined as the set of skills, abilities, knowledge, attitudes, and motivations an individual needs to cope with job-related tasks and challenges effectively [1]. Also from the workplace perspective, it is the fitness of an individual concerning the skills and abilities he/she possesses to perform a particular job or task [12]. The five basic terms associated with competency are (i) Knowledge: the learning and the outcome of learning an individual is possessing, (ii) Skills: an individual's capability to execute a certain job (iii) Self-concepts and values: an individual's self-concept, image, and morals, (iv) Traits: the characteristics of an individual and their behavior to the situation, and (v) Motives- the psychological needs or desires which leads to a particular behavior [12]. The knowledge and skills are termed job-oriented competencies and are easy to develop through training and development in the organisational environment. Personal competencies i.e., self-concepts and values, traits and motives depend on individual capabilities to develop. The four important generalised competencies identified for the employees to adapt to Industry 4.0 are technical competencies which are related to the skills necessary to perform the job effectively, the skills necessary for understanding the workflow, and related to decision-making and problem-solving termed methodological skills, the communication and co-operation skills bundled as social skills and last the individual's attitude, motivation, values, and learning capabilities together contributing for personal skills [13].

According to Shipman, (2000), the seven to nine competencies grouped to suit the needs of the organisation and its environment is a competency model. An explanatory tool that depicts the competencies required by an individual to perform his role within a job, organisation, and industry is termed a competency model. A group of competencies required to perform a particular job inside the organisational setup depending on the demands and environment of the organisation constitutes the competency model. The competency model helps organisations to tailor human resource activities like job design, performance management, career planning, and employment development.

Theoretical Model

The work nature across the levels of organisation differs in the aspects of complexity, functional activity, and the scope of responsibility and hence the competency requirements also. Accordingly, the employees working at the various levels inside the organisation should possess the required competency to meet the challenges of the work. The competencies of the employees in the organisation working across various levels can be studied using two different contrasting theoretical perspectives- the continuity and discontinuity perspectives [12]. The continuity perspective advocates that the competency required at the higher level requires all the lower-level competencies as well as the competencies necessary for the higher levels. As per this view, the employees at the strategic level should also have the competencies of the tactical and operational level in addition to strategic level competencies. The discontinuity approach advocates that to perform successfully at the higher level, the employee should relinquish some of the lower-level skills. This study is approached from the continuity perspective that the employees should continue having the lower lever skills in addition to strategic level skills in the organisation. Industry 4.0, with its radical changes in the technology and the process, demands that employees learn and update themselves constantly irrespective of the level of the organisation hence continuity perspective suits this scenario.

Competencies for Operational level

Employees at the operational level should be equipped with the expertise to work with new technology, production process, and especially with information to cope with the changing landscape of the high-tech working environment [14]. Industry 4.0 demands that employees can analyse the data from consumers, machines, and other resources. It results in workers having cross-functional capabilities to cope with new information systems and manufacturing processes. Automation is ubiquitous and workers need to be flexible for the changes. The automated process which can be handled by the robots should be kept separately on the shop floor from the humans and the interface to interact with these machines should be created and communicated to the employees to work in these interfaces effectively. The "MuShCo" technique is used in this study to derive the competencies required at the operational level. The Mush-co stands for the competencies required by the employees classified into must, should, and could have, the priority of possessing the competency differs from high for must too low for could.

The must-have competencies for the employees working in Industry 4.0 are IT knowledge and abilities, statistical knowledge, and the ability to interact with a modern interface. Computer numerical-control machines (CNC) are commonly used on the shop floor for manufacturing and hence understanding the technology and the process has become a necessity for the workers working at the operational level resulting in the employees possessing IT knowledge and abilities. Industry 4.0 demands that the employees are capable of analysing the data from consumers, machines, and other resources, hence statistical knowledge is essential for the employees [15]. The automation of the process and robots working in the workplace account for the creation of an interface where human and machine efforts could be integrated. This leads to the must-have competency of the ability to interact with modern interfaces needed for the workers working at the operational level.

The employees should have competencies such as specialized knowledge of the manufacturing processes, knowledge about recent technologies, and knowledge management. Process automation makes it necessary for the employees to know the entire process flow on the shop floor so that if any issue arises, it can be resolved easily. Also, the worker should know about the new technology introduced in the market, the benefits of using the technology in the work process as well as the disadvantages if any due to the new technology. The knowledge obtained from learning from new technologies needs to be transferred to other employees for the creation of learning organisations are all part of knowledge management competency, an employee should posse the to work effectively in Industry 4.0.

They could have competencies for the employees which add additional advantages to work in Industry 4.0 are knowledge in computer coding and programming, awareness of ergonomics, and understanding of legal affairs. These are termed as could have competencies because these competencies are not directly related to the work the employees are performing whereas it contributes indirectly to the work efficiency. For example, knowing the ergonomics makes it easy for the employee to understand and work in the human-machine interfaces [15]. This is achieved by positioning the automated process and human-operated machines in the proper places so that the smooth functioning of the production process is ensured.

Competencies for Tactical Level

The tactical level in an organisation is the confluence of operational and strategic levels and hence the knowledge and skills for this level need to be a mix of both levels [16]. The employees at the tactical level should have complete knowledge about the operational process and the new technologies, data analysis, and information processing capabilities as well as interpersonal skills such as communication, problem-solving, and change management. The competencies at this level in this model are derived from the other two levels.

Technical skills

Complete knowledge about the operational processes and workflow is essential for the employees at this level as this helps in the allocation of work and solves problems if any happen during the workflow. To monitor the work that happens on the shop floor, control the workflow, and ensure that employees are

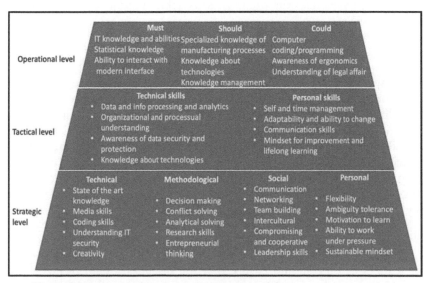

Figure 37.1 Comprehensive competency model
Source: Author

working properly, organisational and processual understanding competency is needed. At the tactical level, the quantity of data to be processed is higher, as the data flow will come from the shop floor, outside customers as well as from the top level. They need to take decisions like workforce scheduling, production planning, demand forecasting, etc which contributes to having data analysis and information processing as essential skills at this level. The employees also should possess knowledge about data integrity and security as they deal with a lot of data at this level. The knowledge about new technologies introduced in the market helps the organisation achieve a competitive advantage as it is easy for the organisation to adapt to new technologies.

Personal Skills

The managers and supervisors at the tactical level need interpersonal skills so that they can manage the employees and ensure that the work is done by others at the operational level. Self-management is an important competence for managers so that they can become role models for others. The employees at this level should play the role of the change agent, to bring about the change in the organization and ensure that the employees are adapting to change, hence should possess change management skills. Since the employees at this level are in the middle of the strategic level and the tactical level, they are playing the role of the bridge in passing the communication between those two levels. They play the channels of communication in passing the instructions from the top level to the lower level as well as the feedback from the lower level to the strategic level, hence they should possess communication skills. Another important competence the employees at this level should possess is aptitude and mindset for learning.

Competencies for Strategic Level

The top level in the organisation includes a lot of strategic decision making which calls for understanding and accounting for the overall operations of the organisation [17]. The employees at this level should have proficiency in understanding their internal environments like work processes, employee needs, value chain of the organisation, as well as their external environment like market conditions, government policies, and competitor strategies. Hence, they should possess more competencies and capabilities compared to the other two levels in the organization. The model used in this study which consists of competencies such as technical, methodological, social, and personal skills.

Technical Competency

The employees at the strategic level should possess updated technical skills regarding new technology introduced in the market, and the advantages of adapting those technologies inside the organisation. They should also be able to predict what could be possible new innovative ideas which could be introduced for that they should possess state of the art knowledge about the technology. In the digital era, media skills are required by senior managers to monitor the external environment, understand customer needs, and wants as well as analyse competitors. The skills related to IT infrastructure such as coding skills, and IT security is essential for the employees at this level to make informed decisions about the organizational strategy and growth. The creativity skills of the employees help the organizations achieve competitive advantage through creating non-imitative core competence.

Methodological Competency

The methodological competencies deal with the capability of an individual to overcome the difficulties faced during the process of achieving organisational objectives. The top-level senior employees should take strategic decisions to ensure that long-term objectives are achieved by the organisation, hence they should possess decision-making skills as their core competency. At the leadership level, conflict-solving skills are essential so that organisational objectives are achieved cohesively and coherently. The data inflow is higher at this level; hence they should possess data handling competencies including analytical capabilities. Strategic decision-making accounts for research skills regarding market conditions, customer preferences, competitor requirements, and government policies. Entrepreneurial thinking is essential as they must be accountable for the decisions they are taking inside the organization.

Social Competency

Interpersonal skills are essential for leaders as they should ensure that the organisational objectives are achieved along with others cohesively. Hence social competencies like communication skills and managing an intercultural team are essential for them. Networking skills help the senior employees at this level to understand the other leader's perspectives about a particular situation, their organisation's current practices, and new trends in the market. Team building and compromising and cooperative skills are essential for top-level employees to achieve the long-term strategy of the organisation through the contribution of everyone in the organisation. Finally, a good leader influences and motivates others by being a good role model demanding leadership skills.

Personal Competency

The intra-personal skills such as being flexible to the situation as well as the view and opinions of the others in the organisation. The skill to understand and analyse ambiguous situations and decision-making during ambiguity are essential for leaders at the strategic level. Tolerance and learning aptitude are essential for employees at this level to work coherently with others and to keep them updated about current trends respectively. The capability to work under pressure helps them to handle difficult situations easily and find a solution in turbulent times. As organisations are focussing on achieving sustainable competitive objectives, the leader should possess this quality as the basic mindset to achieve sustainability.

Contributions to Theory and Practice

It is observed that with industries embracing Industry 4.0, the competencies which will aid in combining the mechanical and the interface of the computers will be looked upon, which will enable the human capital to work following the shift in the structural changes [18]. The necessity of the perfect amalgamation of human-related skills, interpersonal skills in addition to hard-core technical skills is observed. It is also observed that the demand for the competencies is in tandem with the strategic dimensions of the organisations, where organizations will be looking for innovative, service-oriented competencies. As the requirement of competencies is industry specific it is evident that certain industries demand regular upgradation of competencies whereas certain industries demand sustained development of competencies. In this context, organizations are expected to develop competencies of their human capital in tandem with the demands of future workplaces. Development of adequate competencies in the employees would eventually result in the favorable acceptance of the employees in the organisation and the alignment of the employees to the strategy of the organisation which would eventually result in the strategy being achieved which calls for the perfect match between the competencies of the employees with the changing demands of the external industrial environment.

Continuing with the changing external environment, Industry 4.0 affecting all industries at all levels it becomes inevitable for the upgradation of the competencies at every level encompassing the technical, psychological, and the social level which also volunteers the employees to be able to adjust, be competent and have competencies which are interdisciplinary to be able to relate with the different stakeholders. It is also to be noted that the various strata of human capital should develop competencies involving interpersonal relations, adaptability, technical skills, self-reliance, and being able to work in an environment that is controlled digitally.

Conclusion

This paper is an attempt to identify and classify the competencies required by the employees working at various levels in the organisation. Competency models like the MuShCo approach primarily focusing on the operational level and holistic approach are adapted and equated to frame this exhaustive model which gives an inclusive view of the competencies required by the employees working at the strategic level, tactical level, and the operational level in the organisation. The work nature across the levels of organisation differs in the aspects of complexity, functional activity, and the scope of responsibility and hence the competency requirements also. Accordingly, the employees working at the various levels inside the organisation should possess the required competency to meet the challenges of the work. Industry 4.0, with its radical changes in the technology and the process, demands that employees learn and update themselves constantly irrespective of the level of the organization hence continuity perspective suits this scenario. This research in its attempt on adapting and equating the MuShCo to the exhaustive model as explained in the diagram outlines and identifies the competencies required at all levels to be able to embrace and adapt to the changes in the working environment due to industry 4.0. Enumerating the competencies at the operational level, the competencies like IT knowledge and abilities, statistical knowledge, and ability to interact with the modern interface are looked upon as the competencies the employees must possess, the competencies encompassing the special knowledge of manufacturing processes, the latest and management, and technological know-how are looked upon in the should category and the coding and the programming knowledge the awareness of ergonomics and legal know-how could be appreciated.

Coming to the tactical level the technical skills in the various operational processes, workforce scheduling, production planning, forecasting of demand, and interpretation of various data are highly looked upon in tandem with various personal skills like interpersonal skills and the adequate mindset for learning. At the strategic level, the updated technical skills owing to the new technology invasions, the skills to think out of the box and to forecast the future scenario in the technological as well as the customer preferences, the skills enabling them to tap the media and the digital platforms to be innovative and develop core competencies. This level would also demand various methodological skills like conflict solving among others in addition to various data handling, interpersonal, and leadership skills in tandem with leadership skills. In a nutshell, our research advocates that as industry 4.0 is affecting all industries albeit at different intensities, it is evident that to be able to tap the benefits and to minimize the challenges to be able to fit in the new technologically advanced digital workplaces the human resources all levels should develop and upgrade the competencies to be able to think, analyse the future uncertain situations, develop learning aptitude and be able to achieve sustainability, which join hands with the continuity perspective suits this scenario.

References

[1] Hecklau, F., Galeitzke, M., Flachs, S., and H. Kohl, (2016). Holistic approach for human resource management in Industry 4.0. Procedia CIRP, 6th CIRP Conference on Learning Factories. 54, 1–6.

[2] Agolla, Joseph Evans. (2018). Human capital in the smart manufacturing and industry 4.0 revolution. Digital transformation in smart manufacturing. 41–58.

[3] Chaka, C. (2020). Skills, competencies and literacies attributed to 4IR/Industry 4.0: Scoping review. IFLA Journal. 46(4), 369–99.

[4] Zhang, M., Hu, X., Xie, B., and Li, H. (2020). Research on the core capabilities cultivation mode of software engineering talents for new engineering. Proceedings International Conference on Big Data and Informatization Education, pp. 225–28.

[5] Babatunde, O. K. (2020). Mapping the implications and competencies for Industry 4.0 to hard and soft total quality management. Total Quality Management. 33(4), 896–914.

[6] Shet, S. V. and Pereira, V. (2021). Proposed managerial competencies for Industry 4.0 – Implications for social sustainability. Technological Forecasting and Social Change. 173, 121080.

[7] Chen, Z. and Shi, J. (2018). Exploration of talents training on safety engineering major of applied university under emerging engineering education. Proceedings of the 2018 2nd International Conference on Management, Education and Social Science (ICMESS 2018). 176, 540–45.

[8] Flores, E., Xu, X., and Lu, Y. (2020). Human Capital 4.0: a workforce competence typology for Industry 4.0. Journal of Manufacturing Technology Management. 31(4), 687–703.

[9] Horňáková, N., Cagáňová, D., Štofková, J., and Jurenka, R. (2020). Industry 4.0 future prospects and its impact on competencies. Lecture Notes in Mechanical Engineering. 1, 73–84.

[10] Campion, M. C., Schepker, D. J., Campion, M. A., and Sanchez, J. I. (2020). Competency modeling: a theoretical and empirical examination of the strategy dissemination process. Human Resource Management. 59(3), 291–306.

[11] Kipper, L. M. et al. (2021). Scientific mapping to identify competencies required by industry 4.0. Technology in Society. 64.

[12] Vazirani, N. (2010). Competencies and competency model - a brief overview of its development and application. SIES Journal of Management. 7(1), 121–31.

[13] Grzybowska, K. and Lupicka, A. (2017). Economics & Management Innovations (ICEMI) : Key competencies for Industry 4.0. Volkson Press. 1(1), 250–53.

[14] Gehrke, L., Kühn, A. T., Rule, D., Moore, P., Bellmann, C., Siemes, S., ... and Standley, M. (2015). A discussion of qualifications and skills in the factory of the future: A German and American perspective. VDI/ASME Industry, 4(1), 1–28.

[15] Aulbur, W., Arvind, C. J., and Bigghe, R. (2016). Whitepapersummary: skill development for Industry 4.0. Rol. Berger GMBH. BRICS Ski. Dev. Work. Group, India Sect, pp. 1–50.

[16] De Meuse, K. P., Dai, G., and Wu, J. (2011). Leadership skills across organizational levels: A closer examination. Journal of Psychology. 14(2), 120–39.

[17] Mohelska, H. and Sokolova, M. (2018). Management approaches for industry 4.0 – The organizational culture perspective. Technological and Economic Development of Economy. 24(6), 2225–40.

[18] Srivastava, Y., Ganguli, S., Suman Rajest, S, and Regin, R. (2022). Smart HR competencies and their applications in Industry 4.0. A Fusion of Artificial Intelligence and Internet of Things for Emerging Cyber Systems, pp. 293-315. Springer.

Chapter 38

Modeling and optimisation of lifting structure for large corrective actions of wind turbine components

Refugine Nirmal Ignacy Muthu[a], Gurubaran Kanthasamy[b], Anthonyraj Premkumar S[c], Mohammed Farook S[d] and Thamodharan Krishnaraj[e]

Windcare India Pvt Ltd, Gudimangalam, India

Abstract

In this work, we examine two lifting structure's static structures with a moving trolley (SSMT) and moving structure with a hydraulic system (MSHS), which are mounted on the nacelle frame of the wind turbines for service and components replacement activities. We present a novel design that relates to the large corrective action of wind turbines component without using heavy-duty cranes. Using finite element analysis, we calculate the stress, displacement, and bolt strength of both structures which shows MSHS can carry two times more load than SSMT. The working mechanisms of both structures are explained using block diagrams. The interchangeable base frame of MSHS makes the design tailorable to fit any wind turbine. By reducing the number of moving parts, the MSHS is flexible and reduces manual handling during assembly processes. Using Hazard identification and Risk assessment (HIRA) we demonstrated MSHS is safer to work than SSMT. Further, we talk about the problems faced during operations like pulley routing, efficiency losses, the load acting on the hydraulic cylinders, and the methods we applied to optimise the design. The proposed MSHS design is more robust, has a better load-carrying capacity, and improves workplace safety.

Keywords: Design, optimisation, FEA, safety analysis

Introduction

Wind turbine components are replaced using large conventional cranes. In recent times, wind turbines have become larger, taller, and bigger making the replacement work using the classical method a tedious process. Thus, a novel approach is considered for the replacement of wind turbine components. The small portable lifting structures are designed and mounted on the wind turbine's nacelle for replacement work. These portable structures are compact and can be fitted in containers. This helps in better relocation and mobilization to remote regions. Portable cranes are important equipment used for loading and unloading wind turbine components. In this paper, we propose two lifting structures: static structures with a moving trolley (SSMT) and moving structures with a hydraulic system (MSHS) for replacement work. The static structure has a projection outside the nacelle making it risky in terms of assembly, lifting, and lowering. To overcome this drawback movable arm structure is designed with an inbuilt hydraulic system. The movable arm structure can be assembled inside the nacelle and can be extended using a hydraulic arm which makes it safer to work. The movable structure has a hydraulic power unit which is coupled with a structure that generally includes a motor, a fluid reservoir, and a pump. It works to apply the hydraulic pressure needed to drive motors, cylinders, and other complementary parts of a given hydraulic system. When it comes to operation, the structure, and the hydraulic cylinder function uniformly for lowering and replacing the wind turbine components most safely and securely.

In this work, we focus on structural analysis of both structures using finite element analysis and hazard identification and risk assessment (HIRA) to demonstrate that movable structure is safer. We have addressed the problem faced during our replacement work and showed an effective method to overcome those problems.

Working Mechanism

Static structures with a moving trolley (SSMT) [1] and moving structures with a hydraulic system [2] are lifting structures, mounted on the nacelle of the wind turbine. These structures are used for the replacement of wind turbine components like the gearbox, generator, and rotor structure. For this illustration, we chose rotor replacement work.

Static structures with a moving trolley

1. SSMT consists of a lifting structure, block and tackle, wire rope, guide pulleys, tower bottom jig, a chain block, and a winch as shown in Figure 38.1.

[a]design10@windcareindia.com, [b]design2@windcareindia.com, [c]antony@windcareindia.com, [d]mohammed.farook@windcareindia.com, [e]quality@windcareindia.com

DOI: 10.1201/9781003450252-38

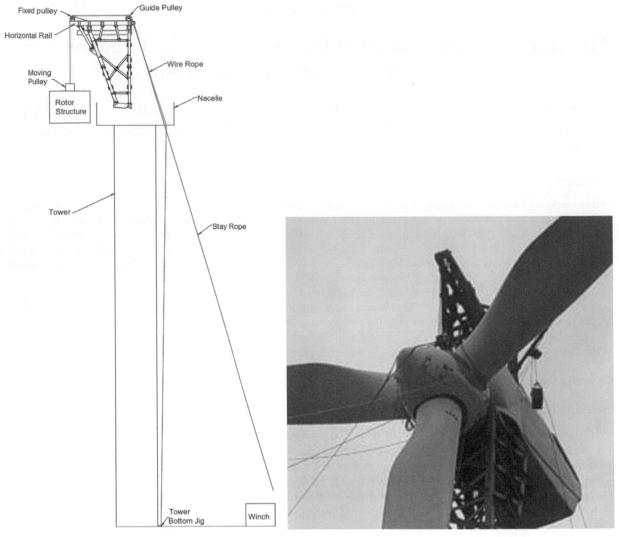

Figure 38.1 SSMT model (a) Schematic (b) Site picture
Source: WindIndia Pvt Ltd

2. Lifting structure will be mounted on the wind turbine's nacelle and projects outwards facing the rotor and away from the CG of the rotor.
3. The rotor structure will be loaded onto the lifting structure and to a chain block.
4. Initially chain block will be loaded to move the rotor from its position.
5. Once the chain block is released the load gets transferred to the lifting structure.
6. Block and tackle with 10 falls will be mounted on the horizontal rail of the lifting structure.
7. The moving pulley of the block and tackle will be attached to the rotor via wire rope.
8. Wire rope starts from the winch and then routed to the pulley attached to the tower bottom jig, then through the guide pulley, and then to the block and tackle setup.
9. Winch is kept on the ground helps to haul the wind turbine components to the desired height.

Moving structure with a hydraulic system

1. MSHS consists of a curved lifting arm, a base structure, hydraulic cylinders, block and tackle, guide pulleys, a tower bottom jig, and a winch as shown in Figure 38.2.
2. The base structure will be mounted on the wind turbine's nacelle. The base frame can be designed to match different nacelle designs making the MSHS tailorable to fit any wind turbine.
3. Hydraulic cylinders will connect the curved arm and the base structure.
4. Block and tackle with 10 falls will be connected at the end of the curved arm.
5. Initially the hydraulic cylinder will be held at a maximum angle such that the arm will be along the CG of the rotor. Once loaded, the angle will be reduced so that the arm will extend outside the nacelle.
6. Once the clearance position is reached the hydraulic cylinder will stop pushing and the winch will release the rope for hoisting.

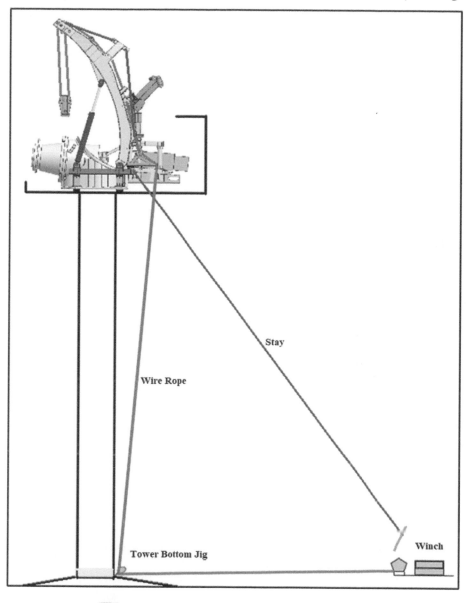

Stay

Wire Rope

Tower Bottom Jig

Winch

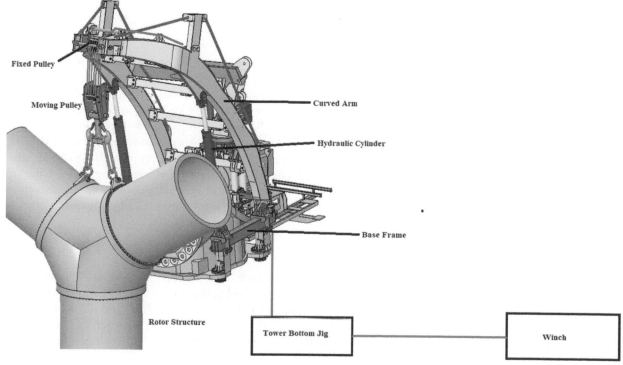

Fixed Pulley

Moving Pulley

Curved Arm

Hydraulic Cylinder

Base Frame

Rotor Structure

Tower Bottom Jig

Winch

Figure 38.2 MSHS model (a) Schematic (b) Schematic (c) Site picture
Source: WindIndia Pvt Ltd

7. The moving pulley of the block and tackle will be attached to the wind turbine component via wire rope.
8. Wire rope starts from the winch then routed to the pulley attached to the tower bottom jig, then goes through a series of guide pulleys and then to the block and tackle setup.
9. Winch is kept on the ground helps to haul the wind turbine components to the desired height.

Methodology

Static structures with a moving trolley

SSMT is designed to be mounted on the nacelle of a wind turbine with a maximum length of 4 m and height of 5.8 m. The model has two pulleys, a front and a back pulley and has provisions for stay support at the rear end. For this analysis, the weight of the rotor is considered as 53 tons. The front pulleys have 10 falls; thus the line pull on the wire rope after the front pulley will be 3500 kg. Also, at the back pulley, the wire rope bends at an angle of 45 degrees thus the load of 3500 kg at an angle of 45-degree load acts on it.

SSMT model is analysed using finite element analysis. The numerical simulations are performed using Solid works [3]. The SSMT is set up with 3D, linear static, and isotropic assumption with ASTM 36 steel as material. Welded regions are modeled using bonded contact and plated touching is modeled using no penetration contact with a frictional coefficient of 0.2. The loading and boundary conditions are shown in Figure 38.3(a). The structure is modeled with 3D solid elements with a total number of nodes and elements as 1637587 and 785253 respectively.

Moving structures with a hydraulic system

A linear static analysis was performed using an Ansys workbench [4] with S355 as material. The structure is designed to lift 90T with a 20-degree angle. The bottom of the leg support is fixed as shown in Figure 38.4(a). The arm is kept at an angle of 12 degrees, standard earth gravity is applied, and a load of 90T is applied to the top of the pulley pin as shown in Figure 38.4(b) and the bolt torque is applied as per the size and bolt grade. The parts, bolts, and stays are modeled using shell, solid, and link elements respectively with a number of nodes and elements are 530182 and 604083.

Fixing & Loading Conditions:

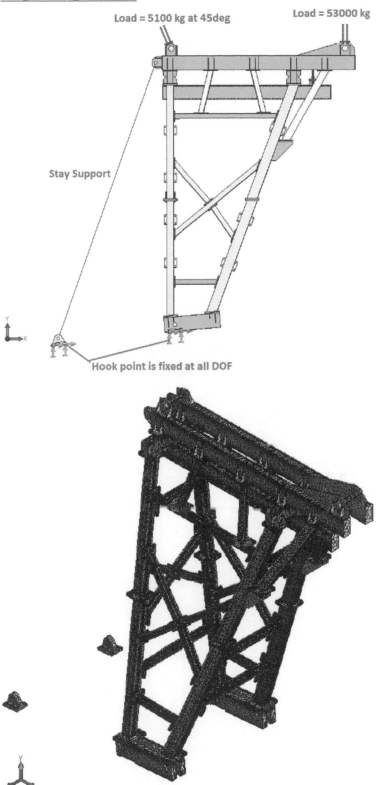

Figure 38.3 SSMT model (a) Loading and boundary condition (b) Mesh details
Source: WindIndia Pvt Ltd

Hazard identification and risk assessment

The HIRA study offers a method to assess hazards and their associated risks [5]. Risk is evaluated based on the severity and the occurrence level and it helps to give recommendations to reduce those risks.

The steps involved in HIRA are as follows,

Table 38.1 Material properties

Material	Young's modulus	Density	Poisson's ratio	Yield strength	Ultimate strength
	Gpa	$\frac{Kg}{m^3}$	v	Mpa	Mpa
ASTM A36 Steel	210	7850	0.3	250	400
S355	210	7850	0.3	355	550
Bolts	210	7850	0.3	660	830
Pins	210	7850	0.3	940	1040

Source: WindIndia Pvt Ltd

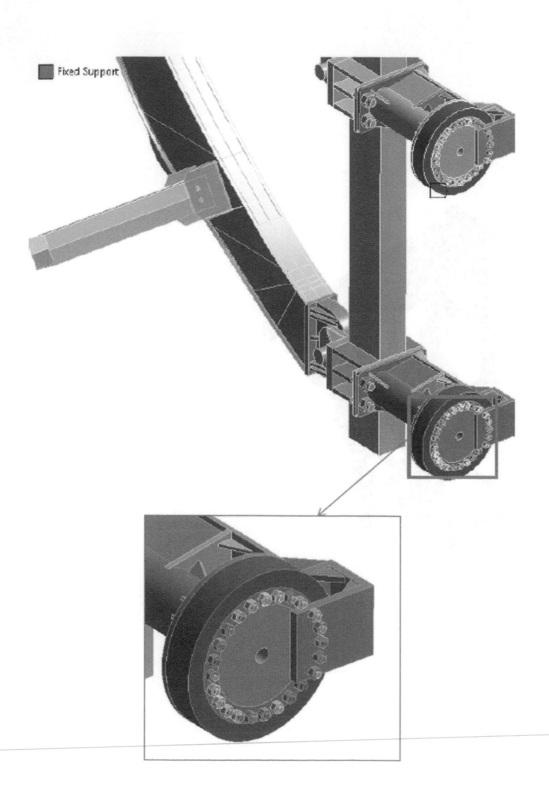

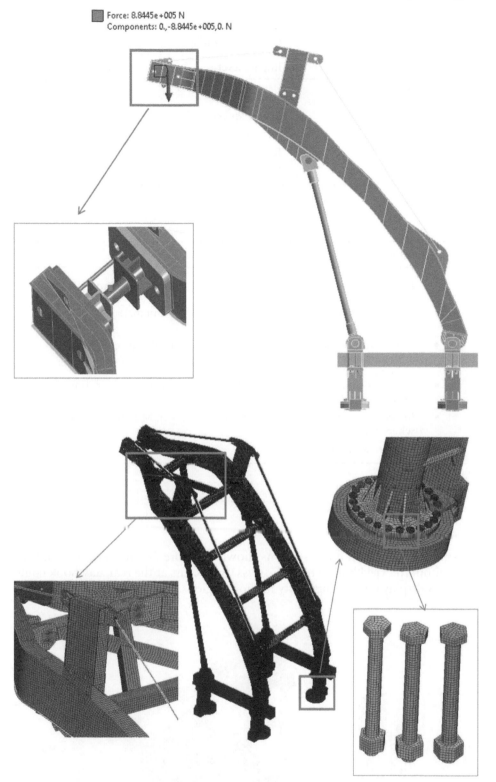

Force: 8.8445e+005 N
Components: 0.,-8.8445e+005,0. N

Figure 38.4 MSHS model (a) Boundary condition (b) Loading condition (c) Mesh details
Source: WindIndia Pvt Ltd

1. Identification of hazard.
2. Assessment of risk.
3. Elimination or reduction of the risk.

Risk can be evaluated mathematically as,
    Risk priority number (RPN) = Severity × Occurrence.
    With RPN being calculated and required actions are taken based on results shown in Table 38.3

Table 38.2 Severity and occurrence rating table

Severity rating		Occurrence rating	
Impact Level	Value	Impact level	Value
Catastrophic or Permanent disability	5	Frequent	5
Fatal	4	Less frequent	4
Serious or major Injury	3	Occasional	3
Minor injury or illness	2	Rare chances	2
No injury or illness	1	Improbable	1

Source: WindIndia Pvt Ltd

Table 38.3 RPN table

RPN number	Risk level	Actions required
> 20	High risk	• Immediate actions to prevent the hazard or eliminate the impact are required. • Activities should be suspended till the hazard is eliminated or control measures are implemented.
>15-20	Medium risk	• Job hazards are unacceptable and must be controlled by engineering, administrative, or personal protective equipment methods as soon as possible. • Activities should be suspended till the hazard is eliminated or control measures implemented.
>10-15	Low risk	• Improve awareness of the potential of the hazard. Monitor and control where needed.
<10	No risk	• No actions are needed.

Source: WindIndia Pvt Ltd

Result

Static structures with a moving trolley

The finite element analysis result shows the von mises stress as shown in Figure 38.5(a). The stress on the lifting structure is found to be within the yield limit. The high stress that occurs at the rear stay hook could not occur in a real system, as FEA displays stress singularity in those regions. The stresses in those regions

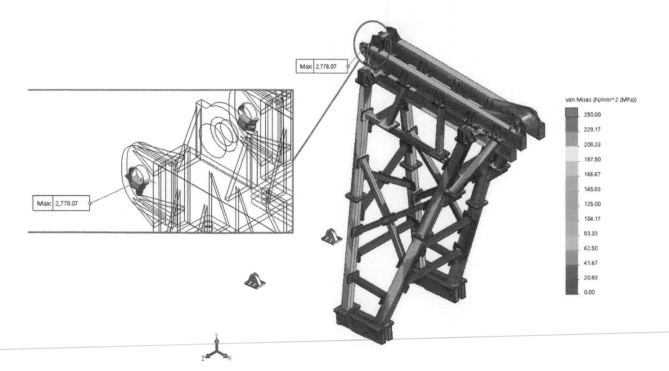

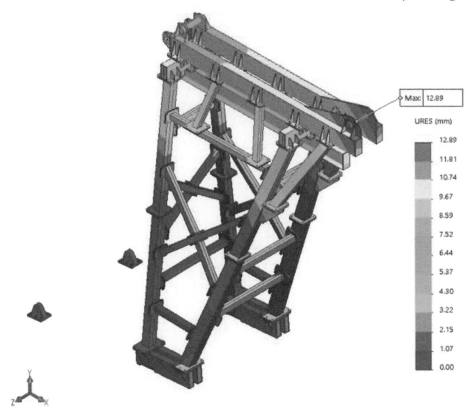

Max: 12.89

URES (mm)

| 12.89 |
| 11.81 |
| 10.74 |
| 9.67 |
| 8.59 |
| 7.52 |
| 6.44 |
| 5.37 |
| 4.30 |
| 3.22 |
| 2.15 |
| 1.07 |
| 0.00 |

Figure 38.5 SSMT Model (a) Von mises stress (b) Total displacement
Source: WindIndia Pvt Ltd

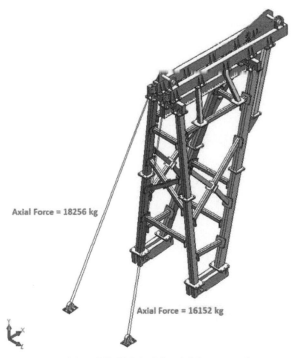

Axial Force = 18256 kg

Axial Force = 16152 kg

Figure 38.6 SSMT Model axial force on the rope
Source: WindIndia Pvt Ltd

are ignored and average stress is considered around that region. The maximum displacement is 12.89 mm as shown in Figure 38.5(b) The axial load on the support stay is 16152 kg as shown in Figure 38.6.

Moving structures with a hydraulic system

The stress results are shown in Figure 38.7(a). High stress displayed in the FEA is due to stress singularity and we ignored those regions. But for a load of 90 tons, the stress on the pulley pin is at 894Mpa which

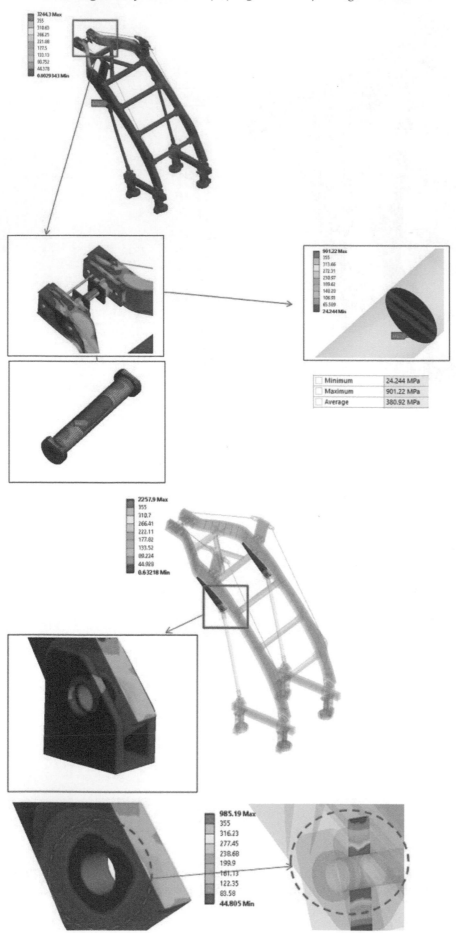

Figure 38.7 MSHS model (a) Von mises stress on the pulley pin, (b) Von mises stress plot
Source: WindIndia Pvt Ltd

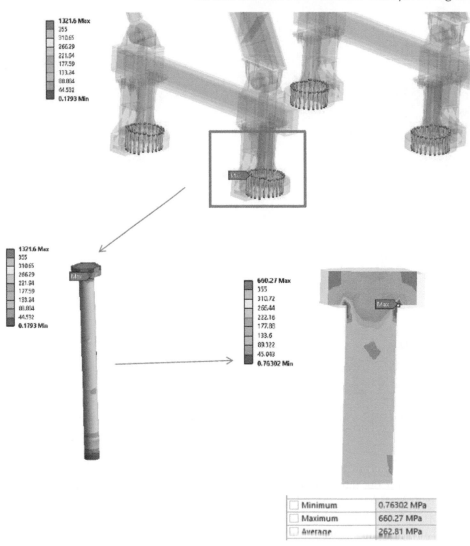

Minimum	0.76302 MPa
Maximum	660.27 MPa
Average	262.81 MPa

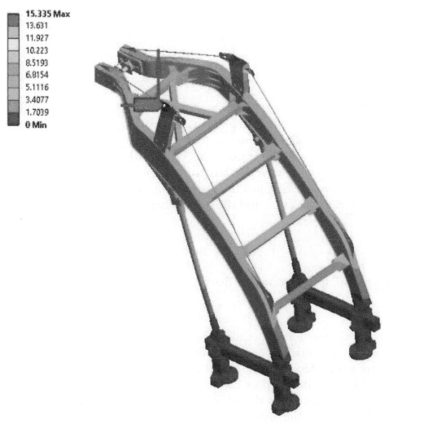

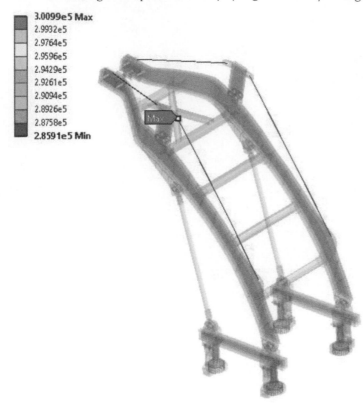

Figure 38.8 MSHS model (a) Von mises stress on the stud, (b) Total displacement, (c) Axial force on the rope
Source: WindIndia Pvt Ltd

is greater than the stress allowable limit but the average stress shown in the region is 373.59Mpa is below the yield limit of 660 Mpa. In the arm and hydraulic cylinder mounting region, as shown in Figure 38.7(b), we observe a high stress of 947.02 Mpa through the thickness but the average stress in the cross-section is 240Mpa and is within the yield limit. From Figure 38.8(a), the maximum stress on the foundation bolt is observed to 1392.8 Mpa but the average stress on the cross-section is 263.8Mpa which is within the stress allowable. We observe a total displacement of 15.35 mm as shown in Figure 38.8(b). From Figure 38.8(c), we observe the maximum axial load acting on the rope is 300 kN thus a rope of 26 mm diameter with a break strength of 372 kN can width stand the load.

Hazard identification and risk assessment

HIRA is calculated and is shown in Table 38.4. From the result, we observe RPN is reduced for the MSHS structure. SSMT will be projected 1.5 m outside the nacelle frame means the worker must work outside in an open space, whereas MSHS can be moved inside and outside the nacelle frame using hydraulic cylinders and the parts can be assembled within the nacelle frame. This reduces the chance of workers working in open spaces. The number of moving is lesser in MSHS structure means the number of bolt or pin connections is lesser. This reduces the workers working height for torquing the bolts. Also, with the use of hydraulic cylinders, horizontal movement is achieved, and the use of chain blocks and trolleys can be eliminated.

Problems and design modifications

The structure experiences major problems during the erection and de-erection process.

Hydraulic Cylinder Load Estimation

The load details are calculated based on simple hand calculations using the formulas shown in Figure 38.9(a) and Table 38.5 we have calculated a load of 75T for individual hydraulic cylinders. From site observation, the hydraulic cylinders lifts are under-designed and could not lift the structure. Thus, additional parameters are included to estimate the hydraulic cylinder capacity. Self-weight, moment, and loads acting on the guide pulley are included in the calculation as shown in Figure 38.9(b). The load acting on the sheaves varies with the angle between the load and the line. As the angle between the lines increases the

Table 38.4 HIRA analysis table

Si No	Activity	Potential hazards of SSMT	Severity	Occurrence	RPN	Control measures of MSHS	Severity	Occurrence	RPN
1	Structure -lowering and Raising	Shifting, moving the structure inside the nacelle was done manually	3	4	12	Shifting, moving the structure inside the nacelle will be done by the nacelle crane hydraulic system	2	3	6
2		Low-capacity lifting frame was used to lower and raise the derrick structure	4	3	12	High-capacity nacelle crane was used to lower and raise the derrick structure.	3	3	9
3	Structure assembly and dismantling	More torquing is required as we used a high number of bolts and nuts	3	4	12	Bolts and nuts were mostly replaced with pin connections which required less torquing than a conventional type	2	2	4
4		While assembling the structure, the technician is required to access outside the nacelle more than 1.5 meter	5	3	15	Accessing outside the nacelle may not be required while assembling or dismantling the structure	2	2	4
5	Rotor, gearbox, and main bearing - lowering and raising	As the structure was fixed type, hence the movement of replacing the components is tough using chain block and trolley	4	4	16	Using the hydraulic system, the movement of components is simple and safe than the conventional type	2	2	4

Source: WindIndia Pvt Ltd

load on the sheave decreases based on the angle factor as shown in Figure 38.10(a). From Figure 38.10(b) we observe six guide pulleys but for our calculation, we consider the first five sheaves as they are directly mounted on the arm. The angle factor of 0.35 is taken as the angle between the ropes is 160 degrees. The compound loading method is chosen for this analysis, with points loads shown in Table 38.6 acting at their respective distance. The calculation gives 120T for individual cylinders.

Pulley Overheating

Many sets of sheaves are attached to gain any mechanical advantage in a moving load. Increasing the number of sheaves on blocks does not mean an improved lifting capacity, because there is a limit beyond which the effort required to overcome friction becomes greater than that necessary to lift the load. Frictional losses also mean there is an incremental increase in friction, which needs extra force to lift the load for adding a further sheave. Too much friction may cause the tackle not to allow the load to be released. For blocks with

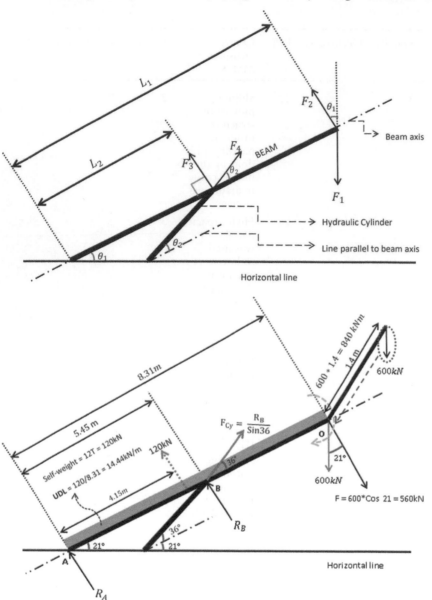

Figure 38.9 1D schematic for hydraulic cylinder load calculation (a) regular calculation, (b) with additional parameters
Source: WindIndia Pvt Ltd

Table 38.5 Hand calculation method

Load acting perpendicularly at the tip of the beam $F_2 = F_1 \times Cos\theta_1$	F_1 = Applied load. F_2 = Load on the arm. θ_1 = Angle between horizontal & beam.
Mechanical advantage M.A = L_1/L_2	L_1 = Distance to pivot. L_2 = Distance to effort.
Load acting perpendicularly at the point hydraulic connects the beam F_3 = MA. F_2	F_3 = Perpendicular force at the point hydraulic connects the beam.
Load supported by the hydraulic cylinder original formula Fcyl = Fp / Sinθ	F_p = Perpendicular force acting at the point hydraulic connects the beam. F_{cyl} = Force acting on the cylinder. θ = angle between beam axis and hydraulic axis.
Load supported by the hydraulic cylinder formula used in this method F4 = F3 / Sin θ_2	F_3 = Perpendicular force acting at the point hydraulic connects the beam. F_4 = Force acting on the cylinder. θ_2= angle between beam axis and hydraulic axis.

Source: WindIndia Pvt Ltd

L x A = S
L = Load in lbs.
A = Angle factor
S = Stress in lbs.

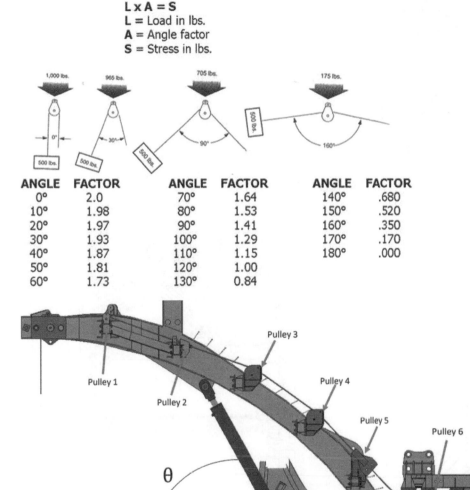

ANGLE	FACTOR	ANGLE	FACTOR	ANGLE	FACTOR
0°	2.0	70°	1.64	140°	.680
10°	1.98	80°	1.53	150°	.520
20°	1.97	90°	1.41	160°	.350
30°	1.93	100°	1.29	170°	.170
40°	1.87	110°	1.15	180°	.000
50°	1.81	120°	1.00		
60°	1.73	130°	0.84		

Figure 38.10 (a) Pulley angle factor, (b) Schematic on guide pulleys

Source: https://www.blockdivision.com/Work-Load-Stress-Formula-Calculation-Pulley-Block-Chain-Rope-Systems-Force-Calculator.html

Table 38.6 Load details on the guide pulley

Pulley number	Load on the rope	Angle	Angle factor	Load acting on the pulley	Distance from point A
	kN	degree		kN	m
Pulley-1	79	160	0.35	27.65	7.3
Pulley-2	83	160	0.35	29.05	5.6
Pulley-3	87	160	0.35	30.45	4.4
Pulley-4	91.5	160	0.35	32.025	2.9
Pulley-5	96	160	0.35	33.6	1.4

Source: WindIndia Pvt Ltd

plain-bore sheaves, reeving beyond 6 or 7 parts of the line provides little advantage; reeving beyond 10 or 11 parts decreases the mechanical advantage because of the cumulative effect of sheave friction. Likewise, with blocks having bronze bushings, reeving beyond 9 or 10 parts has the advantage; and sheaves with roller bearings reach their practical advantage with 15 or 16 parts of the line. On well-maintained sheaves, friction losses are assumed to be about 10% per sheave, having plain bores, used with Manila-rope blocks; 5% per sheave having bronze bushings, and 3% per sheave having roller bearings [6]. Increasing the number of sheaves increases the mechanical advantage but increases the friction loss; thus, the efficiency of the

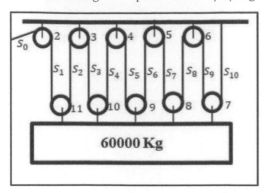

Figure 38.11 Schematic of fixed and moving pulley with 10 falls
Source: WindIndia Pvt Ltd

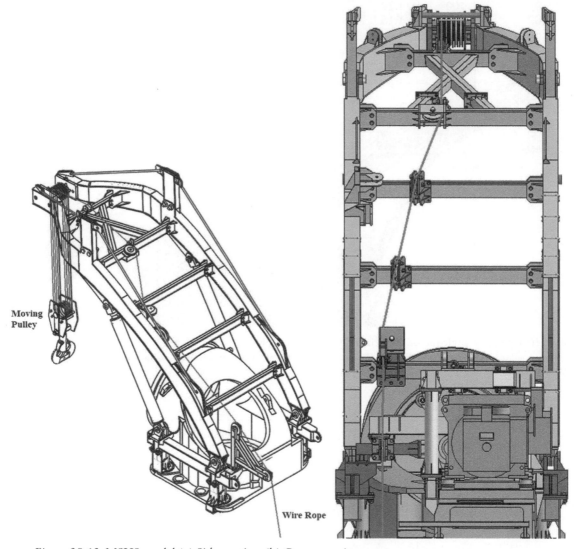

Figure 38.12 MSHS model (a) Side routing, (b) Center routing
Source: WindIndia Pvt Ltd

system will be reduced. By reducing the number of sheaves, friction losses can be reduced—and the system's efficiency is increased.

Observation from early experiments that a sheave made of the gunmetal bush (moderate friction) experiences 5% frictional loss and heating, thus, reducing the load-pulling capacity. Earlier, the frictional losses are considered on a total scale rather than at individual sheaves. From Table 38.7 we observe, the gun metals bushing value of S_0 is 6300 kg. Thus, pulleys are under-constructed. Block and tackle with 10 falls and roller bearing bushing (frictional loss of 3%) is considered a design change for calculating the load-carrying capacity as shown in Figure 38.11. The method shown in [7] is used to calculate the load on each sheave.

Table 38.7 Load acting on S_0 pulley

Initial design calculation	Final calculation
No. of lines supporting the load N = 10 Load = 60000 kg Frictional loss = 5% (Gunmetal Bushing) M.A = N = 10 $S_{10} = \dfrac{60000}{10} = 6000$ Kg	No. of lines supporting the load N = 10 Load = 60000 Kg Frictional loss = 3% (Bearing) M.A = N = 10 $S_{10} = \dfrac{60000}{10} = 6000$ Kg $S_9 = 6000 + (6000*0.03) = 6180$ Kg $S_8 = 6180 + (6180*0.03) = 6365.4$ Kg $S_7 = 6365.4 + 6365.4*0.03) = 6556.362$ Kg $S_6 = 6556.362 + (6556.362*0.03) = 6753.053$ Kg $S_5 = 6753.053 + (6753.053*0.03) = 6955.644$ Kg $S_4 = 6955.644 + (6955.644*0.03) = 7164.314$ Kg $S_3 = 7164.314 + (7164.314*0.03) = 7379.243$ Kg $S_2 = 7379.243 + (7379.243*0.03) = 7600.62$ Kg $S_1 = 7600.62 + (7600.62*0.03) = 7828.639$ Kg $S_0 = 7828.639 + (7828.639*0.03) = 8063.498$ Kg

Source: WindIndia Pvt Ltd

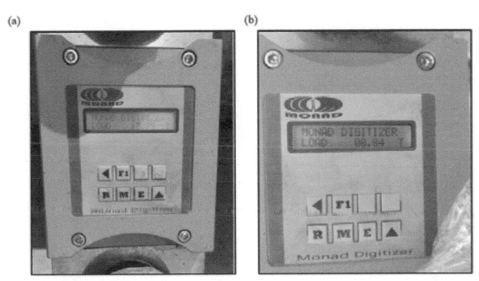

Figure 38.13 Loadcell measurement on tower bottom jig (a) before the design change, (b) after the design change
Source: WindIndia Pvt Ltd

From the calculation, we observe a 2000 kg increase in load on the S_0 sheave for 3% frictional loss. This helps with the better design of the sheaves.

Pulley Routing

From the fixed pulley the wire rope must be routed to the winch through a series of guide pulleys. As we know from Table 38.6, a small amount of load will act on the guide pulley mounting points. These small loads directly act on the structure. From Figure 38.12(a), we observe guide pulleys are mounted on one arm. Thus, one arm is loaded higher than the other arm. From the site observation, we found that for 200 bar pressure, the cylinder on the heavier arm didn't push the cylinder while the other cylinder pushes the arm. Hence, pulley routing has been changed to distribute the load equally. From Figure 38.12(b), we observe the guide pulleys are mounted on the cross bars. This change in the design helps to push the overloaded cylinder.

Conclusion

The structural analysis of both structures is performed using finite element analysis. From this, we work we have demonstrated that the curved arm moving structures with a hydraulic system (MSHS)has a better load-carrying capacity. Using hazard identification and risk assessment (HIRA), we have demonstrated that the MSHS is a lot safer to work than static structures with a moving trolley (SSMT). Further, a hydraulic cylinder of capacity 120T is recommended for the lifting purpose. Gun metal busing in the sheaves is replaced by a roller bearing to improve efficiency and overheating. Alternative methods are discussed to better calculate the load acting on the sheaves. The routing of the guide pulleys is changed to evenly distribute the loading on the cylinders. The loadcell mounted on the tower bottom jig to measure the line pull on the wire rope displays a load of 13.79T before the design change. But after the design change, we observed a reduction in load with 8.84T as shown in Figure 38.13.

Reference

[1] Premkumar, S. A. (2015). System and method for replacing rotor and main shaft assembly of wind turbine. Windcare Technologies.
[2] Premkumar, S. A. (2015). A portable structure for removal and installation of components within a nacelle of the wind turbine. Windcare India Pvt Ltd.
[3] Solidworks Standard Document for FEA modeling.
[4] Ansys Standard Document for FEA modeling.
[5] Occupational health and safety management systems – Requirements with guidelines for use. BS ISO 45001:2018
[6] MacDonald, J. A., Rossnagel, W. E., an Higgins, L. R. (2009). Handbook of Rigging Lifting, Hoisting, and Scaffolding for Construction and Industrial Operations. New York, NY: McGraw-Hill Education.
[7] Glerum, J. O. (2007). Stage Rigging Handbook. Southern Illinois University Press.

Chapter 39

Numerical study on performance of Raft Foundation resting on stone column reinforced soft clay subjected to eccentric load

Asheequl Irshad[a] and Siddhartha Mukherjee[b]

Techno International New Town, Chakpachuria, India

Abstract

Installing stone column is a feasible and favored soil stabilization method for enhancing bearing capacity of weak soil. Since stone columns are stiffer than the materials they replace, they improve the load-bearing capacity of the soil. Additionally, the stone column enhances the foundations' load-settlement characteristics, and it helps to consolidate the surrounding soil at a faster rate. Geogrid encasements are installed to increase load-carrying capacity of these columns in clay soil by restricting lateral displacement. Behavior of the raft foundation, when subjected to eccentric loading, is studied and compared to that of the concentrically applied load. A parametric study is done by varying the loading parameters as well as shear parameters of stone column. The effect due to different arrangements of stone columns under raft foundation is also studied. The lateral displacement of stone column under the stated loading conditions is observed. For eccentric loading, the failure wedge diminishes and shifts towards the direction of the eccentricity. The load-bearing capacity decreases considerably with an increase of eccentricity.

Keywords: Stone column, bearing capacity, eccentric loading

Introduction

The necessity for the improvement of unsuitable and problematic soil is increasing every day as a result of the growing shortage of adequate land space for the development activities of the necessary infrastructure to meet the demand of the ever-growing population. Stone columns have become popular as a soil stabilisation method for speedy construction on weak soil. Stone column installation is one of most efficient and economical ground modification procedures especially suitable for soft clay or loose silty sand. By partially replacing unsuitable underlying soil with a compacted column of stronger materials made of large particle-size aggregates, stone column installation increases load carrying capacity of foundations constructed on weak soil. The lateral support provided by the surrounding soil has significant impact on stone column's performance. Additionally, the possibility of finer particles migrating into the larger size aggregates of the stone column could cause clogging. To overcome these practical issues related to ordinary stone column (OSC), geotextile/geogrid can be used to encase the granular filler materials to prevent lateral bulging under vertical loads and also act as a filter layer. Laboratory testing, analytical solutions, numerical simulations, and field evaluations have been used in previous investigations to assess efficacy of geogrid encased stone columns (GESC) [1-3]. Both bearing capacity evaluation and settlement prediction must be done and reviewed in order to fully understand behavior of foundation installed on the soil bed reinforced with stone columns. Stone columns are typically utilized in groups under varied loading situations. An essential design consideration for a small group of columns is safety against shear failure under loading. Several previous investigators have adopted the concept of unit cell model (UCM) for their analytical studies for prediction of load carrying capacity and settlement of stone column reinforced foundation [4, 5] in which each column in the group is assumed to have tributary cylindrical domain of surrounding soil. In this concept, interactions between individual column domains are not considered and the total capacity of a set of columns arranged in a group is summation of capacity of all unit cell in that group. Stone column also helps to decrease the total and differential settlement, accelerates consolidation process and prevent liquefaction risk [6]. However, UCM is not always applicable to spread footing on small column group to predict bearing capacity or settlement. Numerical analysis has recently attracted a lot of interest for the behavioral study of a foundation built on stone columns reinforced soil. In the numerical modelling, stone column-soil system can be modelled as a single composite continuum where the mass response of the soil composite subjected to foundation loads can be studied conveniently. A review paper with methodological solution to numerical analysis of stone column for ground improvement with both two-dimensional and three-dimensional approaches with the aid of various finite element software was studied [7]). The deformation nature of stone column supported foundations was investigated numerically in terms of settlement of certain nodes within the reinforced soil with respect to total foundation settlement [8]. Composite stone column has a major effect on the strengthening of weak clay soil and enhance the bearing capacity of soil

---

[a]asheequlirshad@gmail.com, [b]siddhartha.mukherjee@tict.edu.in

DOI: 10.1201/9781003450252-39

[9]. The effect of concentric inclined loading on undrained ultimate bearing capacity of stone column group was studied numerically by Ng. et. al. [10]. An FEM analysis on performance of encased stone column constructed on a very soft soil using different types of encasements was done by Alkhorshid et. al. [11].

Although, response of foundation placed on stone column reinforced ground subjected to the concentric load have been well documented, behavioral study of eccentric loaded foundation resting on stone column reinforced softer soil having low bearing capacity has not been investigated so far. Objective of present study is to investigate performance of small raft foundation placed on a group of floating stone column reinforced soft ground under the application of eccentric load emphasising on the estimation of bearing capacity enhancement. With help of finite element programme Plaxis 3D modelling and analysis is performed. For comparison purpose, concentrically loaded foundations were analysed with different shear parameters for both uncased and geotextile encased column with geotextile to observe the load settlement response of overall foundation system at initial stage of investigation. In the main part of investigation, the effect of eccentricity and area replacement ratio were studied and the failure mechanism is compared with the un-reinforced foundation.

Numerical Modelling

General

Mathematical formulation using Plaxis 3D V21 is performed to investigate load response of stone column reinforced foundations (SCRFs). Load response behavior of a group of OSC and GESC are studied under both concentric and eccentric loading conditions and compared. The soil profile considered in this study is relatively soft soil since stone columns are usually installed in areas with soft clay soil and ground improvement effects of stone column is much more notable in those regions.

Numerical Modelling Specification

Finite element analysis of models of both geogrid-encased and un-encased, 0.8 m diameter (D) columns arranged in square and triangular pattern supporting a concrete rigid raft resting on the ground are analysed and studied. For each type of group arrangement, three different values of area replacement ratios (ARR) are considered by varying the spacing of the columns (1.5 D, 2 D and 2.5 D) in the group. ARR is ratio of total area of stone columns to overall area of foundation. To maintain a constant value of ARR, size of the footing is varied keeping same number of columns in the group for respective type of arrangement (Square and triangular). The shape of the footing is considered as square (for square pattern) or rectangular (for triangular pattern). The column length is considered as 8 m and are simulated as floating column. The column length is selected considering critical length consideration ($\geq$ 1.2-2.2 B, where B stands for the width of the raft footing, Ng and Tan [12]). The floating column consideration is achieved by choosing depth of soil layer underneath the base of the stone column sufficiently large enough ($\geq$ 6D)[13] to avoid any interaction of induced stress due to external load. Therefore, the total depth of the soil layer is set as 15 m. Similarly, lateral dimension of the soil contour 20 m × 20 m is considered sufficient as the stress contour results clearly illustrates no influence zone overlapping with the boundary. To maximise the encasement effect, geogrid encasement length should be about 4D from the top [3]. In line with this consideration geogrid encasement length is taken as 3 m ($\approx$ 5D).

Staged Constructions

The analysis consists of the following steps: (1) Stone column installation, (2) Installation of geogrid encasement, (3) Construction of raft foundation, (4) Application of load cases. 1 and 2 are applicable for the models where geogrid-reinforced stone columns are analysed.

Elements, mesh discretisation and boundary conditions

Volume elements are generated as 10-nodded tetrahedral iso-parametric elements for displacement and four Gaussian points for stresses. Medium mesh generated by Plaxis 3D is used in the finite element formulation. Previous study suggests that the 'medium' mesh provides accurate results in considerably low execution time. However, the raft foundation area was refined with local mesh factor 0.5 to enhance the accuracy of settlement response. Bottom boundary is fixed, i.e., both horizontal and vertical displacement are considered zero. Standard fixities are used at vertical boundaries.

Material properties

Both clay soil and column material are modelled by elastic-perfectly plastic Mohr–Coulomb behavior, whose parameter values are shown in Table 39.1. Clay soil is considered as undrained type material

Table 39.1 Material properties

	Layer	Drainage condition	Υ_{sat} [kPa]	ν	Cohesion, C_u [kPa]	Friction angle, $\emptyset$ [°]	E[kPa]
A	Clay soil	Undrained B	16.00	0.30	25.00	-	3.00E3
B	Stone column	Drained	18.00	0.30	0.10	35,40,45	30.00E3

Source: Author

Table 39.2 Material properties of Yoo (2010)

Parameter	Fill	Clayey sand	Silty clay	Gravelly sand	Decomposed granite	Stone column	Geogrid
Depth (m)	0.0-0.7	0.7-2.5	2.5-3.6	3.6-6.2	6.2-10.0	5.4	2.4
$\Upsilon(kN/m^3)$	18.00	19.00	20.00	20.00	21.00	23.00	-
E^{ref}_{50}(MPa)	12	15	0.85	25	45	45	-
E^{ref}_{ur} / E^{ref}_{50}	3	3	3	3	3	3	-
m (-)	0.50	0.5	1.0	0.5	0.5	0.3	-
ν_{ur}	0.20	0.2	0.2	0.2	0.2	0.2	-
Friction angle, $\emptyset$(deg.)	25	28	-	35	40	45	-
Cohesion, c'(KPa)	4	4	-	2	2	5	-
Ψ(deg.)	5	5	-	5	5	10	-
S_u (KPa)	-	-	24	-	-	-	-
Stiffness, J(kN/m)							2500

Source: Author

adopting the "Undrained B" approach of Plaxis software whereas column is taken as a 'Drained' type. To examine impact of frictional angle of column material on bearing capacity, a pure frictional material with different internal friction angles was used for column material. Angle of dilatancy (Ψ) of backfill column material is estimated according to the well-known empirical relation, $\Psi = \varphi - 30°$. Geogrid is considered as linear elastic flexible membrane and its stiffness (J) value in Plaxis is used as the stiffness per unit width. J value of geogrid in this study is taken as 2500 kN/m. Initial stress condition due to gravity is taken into consideration by k_0 procedure considering k = 1 conforming whished-in place condition of stone column. The ratio between Youngs' modulus of column material and surrounding soft soil is taken as 10 which is within the usual range of 10 – 40. The ultimate load bearing capacity for SCRFs is considered as pressure at which the foundation attains maximum settlement of 10% of the foundation width [14].

Modelling Methodology

Stone column groups of variable spacings i.e., 2.5 D, 2 D and 1.5 D in order to vary ARR, were modelled in two different types of configurations (square and triangular). All the models were analysed under the action of concentric and eccentric loading conditions. Adding to this, a detailed numerical study of both uncased and cased stone columns were done. The shear parameters of the stone columns were also varied i.e., $\emptyset$ = 35°, 40° and 45°. For eccentric loading, the load is considered at a distance of B/6, B/8, B/12, B/16 and B/20, where B denotes the least lateral dimension of the foundation. To evaluate the enhancement of load bearing capacity of soil, the same loadings are used to model and analysed unreinforced raft foundations. For the square arrangements nine stone columns with the stated spacings are considered in the group and the raft footing sizes are found to be 5 m × 5 m, 4.25 m × 4.25 m and 3.6 m × 3.6 m corresponding to the ARR 0.18, 0.25 and 0.35 respectively. Similarly, for the triangular configurations, eight stone columns were modelled in symmetrical manner and the raft footing sizes are found to be 5 m × 4.5 m, 4.25 m × 3.75 m and 3.6 m × 3.2 m corresponding to the same variation of ARR.

Validation of FE Mode

Validation of the modelling methodology, carried out in present numerical investigation is adopted from the field test data at Gimhae site in South Korea conducted by Yoo and Lee [3]. The test was performed on a partly encased single stone column of 0.8 m diameter. The soil characteristics employed by Yoo [15] in their numerical and analytical modelling are shown in Table 39.2. Figure 39.1(a) depicts geometry of model utilised in the current formulation.

The settlement response under axial load and lateral displacement obtained from the present numerical formulation are shown with that obtained from the data of Yoo (2010) in Figure 39.1(b) and 39.1(c). With respect to the settlement response, there is a satisfactory agreement between the two data it can be stated that the methodology of the stated numerical model is correct. Same material properties were considered by Imam [8] for the validation of numerical formulation.

Results and Discussion

Raft footing subjected to concentric vertical loading

First part of this research work involves the study of the settlement response of the spread footing subjected to concentric vertical load. Settlement of foundation with bearing pressure curve for a concentrically loaded foundation is shown in Figure 39.2 for two different soil conditions: unreinforced and reinforced with stone columns. In this figure, settlement has been represented as the percentage of the width of foundation for easy comparison of settlement behavior of foundation of different sizes. It is clearly indicated that load-bearing capacity is noticeably increased with installation of stone columns. Also, the use of geogrid as encasement of column material helps to further increase bearing capacity of soil. Ultimate bearing capacity (UBC) is estimated from bearing pressure vs normalized settlement graph. It is observed that bearing capacity values diminishes marginally with increase of foundation sizes though difference is not large (only

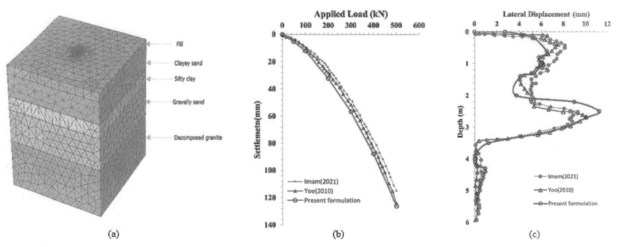

Figure 39.1 FE model validation; (a) Geometry of the model; (b) Axial load vs settlement curve; (c) Lateral displacement vs depth
Source: Author

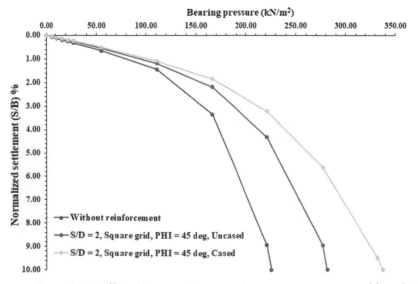

Figure 39.2 Effect of stone column reinforcement on response of foundation
Source: Author

about 10% between smallest and largest size). The average UBC for square footing resting on unreinforced ground surface using 0.1B settlement criteria is found to be 224.4 kPa which is about 21% higher than the UBC calculated by Terzaghi's theory. This discrepancy is probably due to the different UBC criteria between Terzaghi's theory and this study. Figure 39.3 illustrates influence of column material property on bearing capacity improvement. Data shown in Figure 39.3 are corresponding the result of analysis for 3.6 m × 3.6 m size foundation. The improvement is determined by the term BCIR (bearing capacity improvement ratio) which is defined as ratio between bearing capacity of reinforced ground to unreinforced ground. From this figure it is revealed that contribution in bearing capacity improvement of soft ground is greater for larger φ value. It is to be mentioned that column material with φ value of 35° has very nominal effect in bearing capacity enhancement. The maximum enhancement is observed with φ = 45° (30% for OSC and 50% for GESC). It is further observed that the use of geogrid as encasing material has helped in increasing UBC by about 20% over uncased stone column. From this observation, column material with φ = 45° with geogrid encasement is adopted in the main part of this study.

Figure 39.4 presents the deformation mode of unreinforced and stone column reinforced foundation under concentric vertical load. The incremental displacement shades on the surface show quadruple symmetry for unreinforced foundation but not for stone column reinforced foundation. (Figure 39.3a). Failure wages are clearly visible at the bottom of the footing for both the type of foundation (Figure 39.3b) indicating general shear failure. For unreinforced foundation, clear distinct heaving of the soil around the foundation is observed. However, for stone column reinforced foundation the failure wages are restricted to the area of the foundation. Similar observations for unreinforced foundation were also made by previous researchers in their study [16].

Raft Footing subjected to Eccentric vertical loading

Eccentric load is applied on footing for both reinforced and unreinforced cases of foundation soil. Only one-way eccentricity is considered in the present study. Eccentricity value varies from 0 to B/6. Figure 39.5(a)

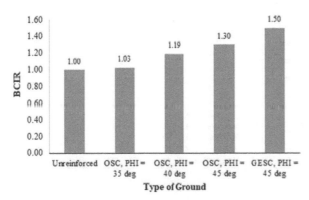

Figure 39.3 Effect of column material property and geogrid encasement on UBC of soil
Source: Author

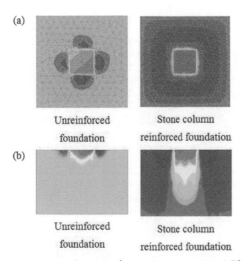

Figure 39.4 Deformation pattern (a) Plan view of displacement shading (b) Incremental displacement
Source: Author

and 39.5(b) presents the displacement effect of eccentricity of applied load on settlement response of surface foundation resting on unreinforced and reinforced foundation respectively. Figure 39.5 shows the result of analysis of foundation of size 5 m × 5 m with ARR = 0.18 only. Similar responses are also observed for the foundation of other sizes considered in this study. As expected, with the increase in eccentricity, ultimate load bearing capacity of soil decreases for both unreinforced and reinforced ground.

From the bearing pressure vs. settlement curve UBC has been obtained for each type of footing at every settlement value. BCIRs of foundation are calculated from the UBC values which are presented in Figure 39.6. In this figure, BCIR values are derived with respect to the UBC of the unreinforced foundation under concentric loading. It is clearly evident that BCIR value decreases with the increase in eccentricity. BCIR value is even found less than unity at eccentricity = B/6, indicating the UBC value goes below the original UBC value for unreinforced foundation under concentric load. It is also noticed that BCIR value increases considerably with the increase in ARR. Maximum rise (about 160%) in UBC is found for ARR value of 0.35.

Furthermore, it is interesting to note that the change in BCIR in both the figures (Figure 39.6a and 39.6b) are almost identical for every ARR value. Therefore, having same ARR value, the grid arrangement pattern of have negligible effect on the UBC of foundation. This argument is clearly visible in Figure 39.7 where the UBC values have been plotted against different eccentricity values. Whether the column pattern is square or triangular, a linear declining trend of ultimate load-bearing capacity with eccentricity values is recorded in all types of foundations. Trend lines of each data sets have been drawn to visualize the actual trend of declination in bearing capacity values. The trend lines are practically overlapping for the data set of same ARR values of different configuration of column with slight variation in slopes for ARR values of 0.35 in square arrangement. Also, the trend lines are found almost parallel to each other with slight deviation (little flatter) for ARR = 0.35 with square grid. Load-bearing capacity improvement has been represented in different manner in Figure 39.8 in which eccentricity of unreinforced foundation

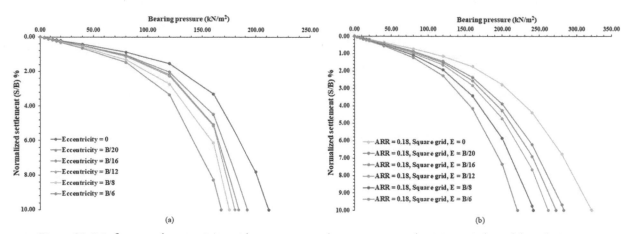

Figure 39.5 Influence of eccentricity with respect to settlement response for (a) unreinforced foundation (b) stone column reinforced foundation

Source: Author

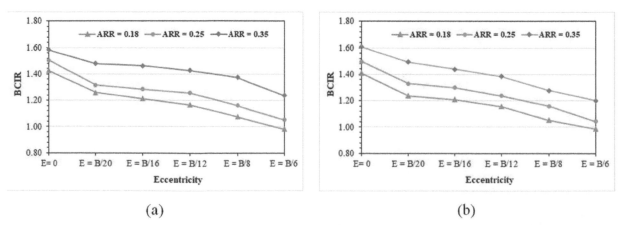

Figure 39.6 Variation of BCIR with load eccentricity for (a) square grid arrangement, (b) triangular grid arrangement

Source: Author

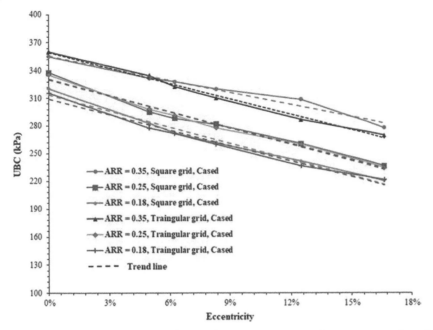

Figure 39.7 Variation of ultimate bearing capacity with eccentricity of applied load
Source: Author

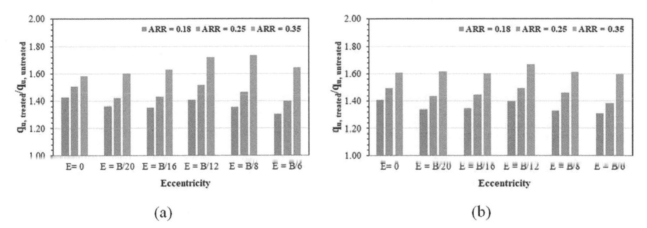

(a)   (b)

Figure 39.8 Variation of bearing capacity improvement under eccentric loading taking eccentricity into consideration for (a) square grid, (b) triangular grid
Source: Author

is taken into consideration. In this figure, bearing capacity improvements are estimated as the ratio of $q_{u,trated}$ to $q_{u,untreated}$, where ultimate bearing capacity (q_u) for both reinforced (treated) and unreinforced (untreated) cases have been taken corresponding to the same eccentricity value. It is seen that the bearing capacity ratios are maintaining approximately same values for every eccentricity value with the exception at B/12 where bearing capacity improvements are found to be maximum. The approximate range of bearing capacity improvements are 1.3-1.4, 1.4-1.5 and, 1.6-1.64 for ARR values of 0.18, 0.25 and 0.35 respectively. It reflects the beneficial effect of stone column reinforcement over unreinforced ordinary soil with the increase in eccentricity value.

Eccentric application of load reduces the bearing capacity primarily attributable to the reduction of failure surface illustrated in Figure 39.9. The failure region and failure wedge under the foundation reduces considerably with the increase in eccentricity and the failure wedges are shifting towards the direction of the eccentricity. It is also to be noted that the unlike the unreinforced foundation, failure wedge is confined within the zone of stone column.

Comparison of lateral displacement

Lateral deflection of columns of all stated model conditions were derived and compared. Figure 39.10(a) and (b) illustrate the deformed pattern of uncased and geogrid cased stone columns. The lateral displacement curve with respect to depth was plotted and illustrated in Figure 39.10(c).

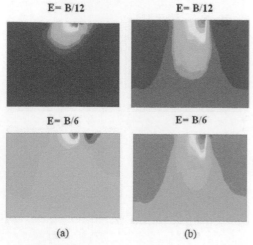

E= B/12   E= B/12

E= B/6   E= B/6

(a)   (b)

Figure 39.9 Failure pattern under the effect of eccentric loading for ARR = 0.18 and E = B/6 and B/12 (a) unreinforced foundation, (b) GESC reinforced foundation
Source: Author

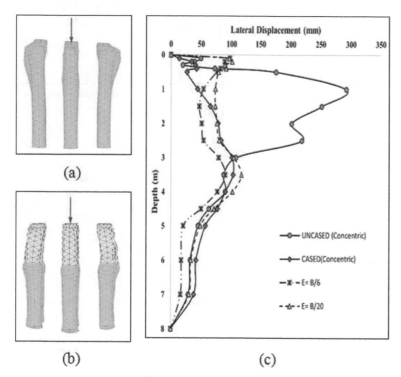

(a)

(b)   (c)

Figure 39.10 Lateral displacement; (a) deformation of uncased column; (b) deformation of cased column; (c) lateral deflection vs depth curve
Source: Author

Conclusion

Finite element study was conducted to examine response of undrained load carrying capacity of unreinforced and stone column reinforced foundations subjected to eccentric loading. Evidently, ultimate bearing capacity of geogrid encased stone column reinforced raft foundation is greater than that of unreinforced foundations regardless of the eccentricity. The following interpretations are derived from the present numerical investigation:

i.   Shear failure mechanism of foundation under concentric loading for unreinforced and reinforced foundation are different. The failure surface in case of unreinforced foundation seems to extend out of the boundary of the foundation, whereas in case of reinforced foundation it is restricted within the boundary of the foundation. For eccentric loading the failure wedge diminishes and shifts towards the direction of the eccentricity. The ultimate load carrying capacity of raft footing diminishes considerably with increase in eccentricity.

ii. Bearing capacity improvement ratio (BCIR) increases with the increase in area replacement ratio (ARR). For constant area replacement ratio (ARR), the bearing capacity enhancement factor is independent of the configuration of stone column arrangement. With increase in eccentricity, ultimate load bearing capacity decreases linearly. Bearing capacity improvement ratio for reinforced foundation with respect to unreinforced foundation, considering the same eccentricity, is nearly equal. Encasement of stone columns helps in minimising the bulging effect and therefore enhances load bearing capacity of the foundation.

References

[1] Murugesan, S. and Rajagopal, K. (2007). Model tests on geosynthetic-encased stone columns. Geosynthetics International. 14(6), 346–54.

[2] Fattah, M. Y. and Majeed Q. G. (2012). Finite element analysis of geogrid encased stone columns. Geotechnical and Geological Engineering. 30(4), 713–26.

[3] Yoo, C. and Lee, D. (2012). Performance of geogrid-encased stone columns in soft ground: full-scale load tests. Geosynthetics International. 19(6), 480–90.

[4] Sexton, B. G. and McCabe B. A. (2013). Numerical modeling of the improvements to primary and creep settlements offered by granular columns. Acta Geotechnica. 8(4), 447–64.

[5] Etezad, M., Hanna, A. M., and Ayadat, T. (2014). Bearing capacity of a group of stone columns in soft soil. International Journal of Geomechanics. 15(2).

[6] Bouassida, M. (2016). Design of column-reinforced foundations. United States: J. Ross Publishing. pp. 224.

[7] Zukri, A. and Nazir, R. (2018). Numerical modelling techniques of soft soil improvement via stone columns: A brief rview. IOP Confefence Series: Materials Science and Engineering. 342, 012002.

[8] Imam et. al. (2021). Relative contribution of various deformation mechanisms in the settlement of floating stone column-supported foundations. Computers and Geotechnics. 134, 104109.

[9] Salam, M. and Wang, Q. (2020). Numerical study on bearing capacity and bulging of the composite stone column. The Open Civil Engineering Journal. 15, 13-28

[10] Ng, K., Idrus, J., and Chew, M. Y. (2021). Bearing capacity of Stone column Reinforced Foundation Subjected to Inclined Loadings. International Journal of Geosynthetics and Ground Engineering. 7, 62

[11] Alkhorshid, N., Araujo, G., Palmeira, E., and Zorrnerg, J. (2019). Large-scale load capacity tests on a geosynthetic encased column. Geotextile and Geomembranes. https://doi.org/10.1016/j.geotexmem.2019.103458

[12] Ng, K. S. and Tan, S. A. (2015). Settlement prediction of stone column group. International Journal of Geosynthetics and Ground Engineering. 1(4), 1–13.

[13] Majeed, Q. G. (2008). Assessment of load capacity of reinforced stone column embedded in soft clay. Building and Construction Engineering Department. M. Sc. Thesis. University of Technology, Baghdad, Iraq.

[14] Lutenegger, A. J. and Adams, M. T. (1998). Bearing capacity of footings on compacted sand. In. International Conference on Case Histories Ii Geotechnical Engineering. 1216–24.

[15] Yoo, C. (2010). Performance of geosynthetic-encased stone columns in embankment construction: numerical investigation. Journal of Geotechnical and Geoenvironmental Engineering. 136, 1148–60.

[16] Zhu, M. and Michalowski, R. L. (2005). Shape factors for limit loads on square and rectangular footings. Journal of Geotechnical and Geoenvironmental Engineering. 131(2), 223–31.

Chapter 40

Implementation of digital twin in supply chain and logistics

Raja V[a], Muralidhar D[b], Mythrayan B[c], Prathiksha K G[d], Venkateshwaran S P[e] and Vikram Sivakumar[f]

PSG College of Technology, India

Abstract

Diversified production sites, increasing world population, natural disasters, crises like COVID-19, and their resulting financial crisis are placing a strain on organisations worldwide. Major problems are mainly because there is no real-time data collection and increasing latency. Real-time monitoring of the factors like demand, credit transactions, and logistics, will enhance the responsiveness of the supply chain (SC) network. There are multiple approaches that can address existing problems and most of them revolve around dynamic response and real-time solutions. Analysing all the existing technologies and methods, it is found that digital twin (DT) can be more productive. Integrating DDT with SC and the logistics network will overrule most of the crises faced in the industries. Creating a DT of SC network will incorporate all the parameters involved in a network with the cloud system. This enables monitoring of all the elements of the supply chain and logistics in real-time many medium-scale and even developed industries are actually facing hardship in their routing and transportation system. Transportation of goods and the routing network is a crucial part of the industry which directly has an impact on the revenue. DT helps distribution centers to develop metrics for predicting the demand, vehicle utilisation, maintenance of temperature, optimisation of routes, transactions and diagnostics of possible outcomes. So, this project is mainly focusing on optimizing the transportation and routing network. It also includes real-time tracking of the transport vehicles. The main reason is that in most well-organised industries there is a major problem in tracking the supply of raw materials. Initially, the network-level analysis is done manually and using various data analytics software. The data collected will be used to provide solutions to real-time problems. This project focuses on building a DT of the logistics network using Microsoft Azure and IoT sensors. Through implementation, it is estimated that even small to medium-scale distribution centers can attain transparency in the SC network. In this study, the impact of real-time data of the transportation network on the overall SC Network is quantified through various analyses.

Keywords: Supply chain, digital twin, logistics, transportation, cloud, real-time

Introduction

The technology of digital twin (DT) began during 2000s. Industries found it as a major hurdle to adopt this technology, but as years passed, it became more accessible and affordable. Though there are many advantages, this technology remains to be underutilised for two major reasons, one being the lack of awareness and the other the complexity of supply chains (SC). However, if adopted efficiently, DT can redefine the existing supply chains. A physical entity can be virtually represented using a DT and is clearly defined as a major development of the physical internet.

Real-time data can be used to create simulations so that easy predictions can be made to determine the way that a product or process would result. Integration of technologies like IoT, software analytics, and artificial intelligence can be used to improve output. With the development of machine learning in areas like big data, these models have become a key factor in modern engineering to improve performance. The virtual digital twin models aid in revealing the information from the past, optimizing the information of the present, and predicting the performance of the future. DT provides a huge variety of options that will facilitate collaborative environments and decision-making processes based on the data from the system. With the help of real-time observation and evaluation of large-scale systems, much research has been done to utilise the values of the digital twin in SCM. SC network is an inherent part of any industry, it can be the food industry, textile industry, manufacturing industry, and so on. In SC, planning is an essential base for procuring, manufacturing, and even logistics networks [1]. In the current era, all the fields are being automated and digitalised to increase productivity, and enhance traceability and transparency. But SC is facing lots of disruptions due to the complexity and range of resources it covers, which also makes it a challenge to digitalise SC. There are various problems like increased lead time, variation between forecast and original demand, and simulations in planning being made naive by limited data. For instance, the Ukraine war and the recent COVID-19 pandemic had and still have caused the SC network to crash and planning has become more complex [2]. These disruptions can be overcome by shifting the physical SC to a more logical and digital manner. This can be implemented with the help of newly emerging technologies like DT and IoT sensors [1]. DT will overcome these problems by planning which is based on data taken from

DOI: 10.1201/9781003450252-40

historical analysis of each individual unit in the SC. As the world is revolutionising toward Industry 4.0, DT will make the SC to be suitable and hassle-free [9]. Initially, DT coordinates physical products (PP) and virtual products (VP) and then improves the virtual model. With the integration of PP and VP, information will be allowed to be shared among individual units and engineers in real-time, thus reducing the lead time. The feedback of customers on various products can be retrieved in real-time so that eagerness to purchase the product can be evaluated. This collected data can be processed to help in making forecasts. The inventory and production levels would be recorded continuously in real-time when it is applied to industries, distribution centers, and warehouses. At last, DT automatically gathers data from various sources [1]. To implement this technology a framework is made using supply chain operations reference model (SCOR). SCOR provides the methodology and tools for benchmarking that helps to make dramatic improvements in the processes of SC. It describes the activities which satisfy the customer demand including plan, source, make, deliver and return. Firstly, the SC is separated into various blocks using SCOR and each block is digitalized. After digitizing each element in the SC, all the blocks are integrated using idea of systems of systems [3]. Considering the various activities of SC such as the production, distribution, and material handling, it is difficult to control the visibility and transparency of the process. Traditionally, enterprise resource planning (ERP) is taking care of all the information sharing within all the blocks of SC in an organization. The main goal of ERP is to coordinate all processes in the SC network. But there are many limitations in ERP system like lack of real-time dynamic data, security risks and so on [4]. DT software can overcome all these limitations. Another important factor is stock missing and theft, which can be tracked using radio frequency identification (RFID) technology. For example, RFID tag is used in supermarket where the RFID tags are stuck on each product and a scanner in each cart. So, no one can leave the market without billing and removing the tag [5].

Apart from digitalised records and IoT-based solutions, digital twin acts as a live ecosystem as it integrates the sensors, trackers, locations, and many such parameters. Thus, if a particular domain gets affected, then it indicates the resulting changes throughout the system. The digital twin is entirely based on the connectivity and responsiveness of its physical counterpart. Through this we can simulate or predict a sequence that may occur in the SC [12].

In this project the SC network is divided into different blocks, in which transportation is the most important segment in the whole network. So, initially the logistics network is analyzed and digitized with the help of GPS tracker and digital cloud like Microsoft Azure. Then the stocks which are delivered to the customers are tracked with RFID technology that was initially done by manual confirmation [13]. Then manual confirmation makes it complex for the industry and also increases the time. Finally, on the whole the virtual system of the logistics will be created and synchronised with the existing physical system. This gives the real-time data of where the delivery truck is there and also the confirmation of the delivery [14].

Need for digital twin and its advantages over conventional methods

Due to the capabilities of real-time monitoring and evaluation of large-scale complex systems, significant research efforts have been made to exploit the values of the digital twin in SCM [15]. As the COVID-19 pandemic has been accelerating the transformation of the digital economy and digital society, DT has become one of the key technological enablers in the new era. DT is often integrated with advanced technologies such as IoT, big data analytics, and blockchain, to develop smart, interconnected SC and logistics systems [16]. Some benefits of DT in the SC are to remotely and instantly monitor the operations.

In this case of the DC centre the data is available in huge quantities and the order confirmation, order generation and data analysis is done manually. When DT is implemented all the parameters can be integrated into a single dashboard thus increasing the efficiency.

Table 40.1 Comparison between digital twin and conventional methods

Digital twin	Conventional methods
Dynamic process- Data obtained in real-time that constantly changes and progresses	Static process- Data obtained at a specific interval of time.
Has the ability to manage the entire SC network using a single dashboard or software.	Requires separate insights on the individual domains of the SCM network
Less manual intervention	More manual intervention
The initial investment is high then further down the line just maintenance	Expensive in the long run

Source: Author

Methodology

The above-stated problems can be rectified by creating some mathematical models and some operations management techniques like PERT, and transportation and then solving comparing these two results. Then combining the growing technologies like DT and RFID in the supply chain to prevent disruptions. Figure 40.1 illustrates the framework of how the project will proceed. The first step is to collect the data from the distribution center. Data like route chart, supplier list, products list and price list is obtained. The samples of these data are given in the following tables. As mentioned in the framework these data can be analysed and DT can be created using relevant software such as Anylogistix, Microsoft Azure or AWS twin maker. The real-time data will be obtained through GPS tracker and RFID tag placed in vehicles used for transportation and on the stocks respectively. A dashboard will be created using the above-mentioned software that will display real-time information. This data can be used by the industry to get an efficient route and predict any disruptions beforehand.

This is the sample data collected from DC which includes the various brands and their products that are being supplied to the customers. The entire data can be utilized when DT is implemented in the DC.

The following table represents the route followed by DC for delivering the products to the respective customers.

Discussion

The RFID tag can be used in the packages in which stocks are being transported to the retailers. This avoids the missing stocks and gives real-time confirmation to the industry. So, it helps to monitor the flow of goods from the industry to retailers and wholesalers.

Figure 40.4 is an example of an RFID tag. These tags will be stuck on the stock boxes before being dispatched for delivery. And, when the stocks reach their respective customers, these tags will be scanned for confirmation as shown in Figure 40.5. This automatically transmits real-time information to the industry.

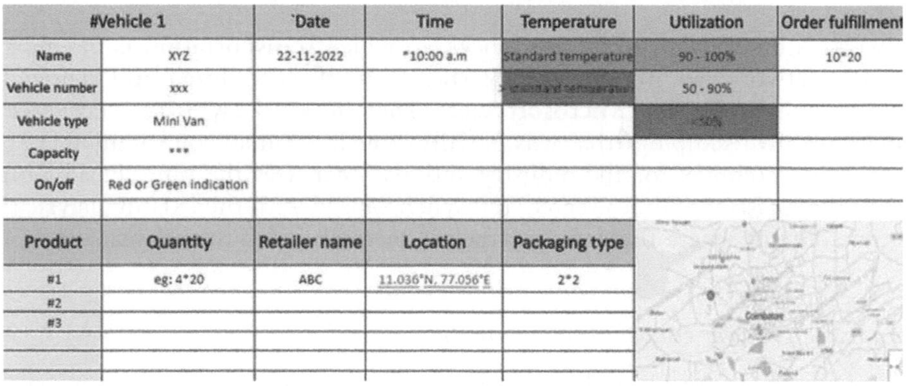

#Vehicle 1		Date	Time	Temperature	Utilization	Order fulfillment
Name	XYZ	22-11-2022	*10:00 a.m	Standard temperature	90 - 100%	10*20
Vehicle number	xxx			Substandard temperature	50 - 90%	
Vehicle type	Mini Van				<50%	
Capacity	***					
On/off	Red or Green indication					

Product	Quantity	Retailer name	Location	Packaging type	
#1	eg: 4*20	ABC	11.036°N, 77.056°E	2*2	
#2					
#3					

Figure 40.1 Dashboard
Source: Author

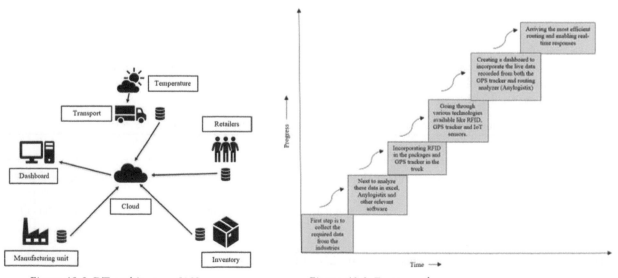

Figure 40.2 DT architecture [10]   *Figure 40.3* Framework
Source: Author   *Source:* Author

As the next step, GPS tracker is to be installed in the vehicle used by the distribution centre. Various GPS tracker devices were checked for their specifications. The purpose of using the tracker is to get dynamic data about the following parameters, current location, destination, schedules, and parking once the vehicle has arrived at the destination. Managing vehicles is more challenging because of the simultaneous balancing of factors like network connectivity issues, increasing costs in maintenance, and concerns regarding safety. Since vehicle tracking using a GPS tracker is a viable option, their vehicles will be tracked using realistic maps and views generated from GPS data. The various GPS tracker devices were checked for their specifications.

Among these, Hirparag GPS Tracker Waterproof Device shown in Figure 40.6 is selected as it was comparatively efficient to serve our purpose. It has data on Live tracking and driving history with which the vehicles' live movement can be tracked using smartphones and sensitive GPS chips and the cloud. Statistics can be obtained on a daily basis. The data of total distance, run time, idle time, stoppage time, maximum speed, and average speed for everyday travel of the vehicle can be obtained. And the daily performance can be compared on graphs. If there are multiple vehicles, with SeTrack all vehicles can be seen on one map. If there is more than one vehicle, all the vehicles can be checked on one map. This GPS tracker is essential to obtain real-time data from the vehicles which is used for delivering the stocks to their respective locations. This is done so that the data can be incorporated to the digital twin.

Table 40.2 Item details collected from distribution centre

Item Name	Purc Rate	Cost	Sale Rate
VVD AYUSH JAGGERY POWDER MRP.45/-	29.25	29.25	33.64
EVE BLACK PEPPER 50GM RS.76	58.14	58.14	62.79
AVT PREMIUM 15 GM	3.58	3.72	3.87
AVT PREMIUM 1 KG	250.90	260.81	268.46
AVT PREMIUM 250 GM	79.97	74.85	85.57
AVT PREMIUM 100 GM	31.99	33.25	34.24
AVT PREMIUM 500 GM	139.94	142.75	149.74
AVT PREMIUM RS.10	7.19	7.47	7.69
AVT COFFEE 500GM	155.93	160.66	166.85
AVT GOLD CUP 50GM	30.91	32.45	33.06
AVT GOLD CUP 4GM	1.29	1.36	1.37
AVT GOLD CUP 500GM	291.11	305.66	311.49

Source: Author

Table 40.3 Customer name and area name

Name	Cell No	Area Name
AYYAN BAKERY -THADAGAM ROAD 1		THADAGAM ROAD 1
BADRI AMMAN STORE - SB COLONY		SAIBABA COLONY
BALAN STORE - PERUR		Perur
CITY CAKE SHOP 3 -THUDIYALUR		Thudhiyalur
ESWARI MINI MART -ERUCOMPANY 2		ERUCOMPANY 2
K P STORE - KOVAPUDUR 2		KOVAIPUDUR 2
K S BAKERY 2- CHETTY VEETHI		Chetty Veethi
KEERTHI TRADING COMPANY		Lotte Products
KUMAR STORE 1 - KARUNYA		KARUNYA
M R S Store - Sukrawarpet		Sukrawarpet
NAVAL PHARMACY - SELVAPURAM 2		Selvapuram 2
S N MEDICAL -RS PURAM		RS PURAM
SAI SNACKS - SELVAPURAM 2		Selvapuram 2
SAKTHI MALIGAI - POOMARKET		POO MARKET
SIVA GANESH STORE		SAIBABA COLONY
SRI BALAJI BAKERY -THADAGAM 3		Thadagam Road 3

Source: Author

Table 40.4 Route assigned for drivers

	MONDAY	TUESDAY	WEDNESDAY	THURSDAY	FRIDAY	SATURDAY
AZAM	SARAMEDU	PN PUDUR	SELVAPURAM 2	THONDAMUTHUR 2	NGGO COLONY	THADAGAM 2
AASHIF	SUKRAWARPET	-	KOVAIPUDUR	VADAVALLI	THUDIYALUR	THADAGAM 2
THANGARAJ	CHETTY VEETHI	RS PURAM	SELVAPURAM 1	KARUNYA	NSN PALAYAM	THADAGAM 1
RAVI	SAIBABA COLONY	POOMARKET	PERUR	LINGANUR	ERUCOMPANY	ERUCOMPNAY 2

Source: Author

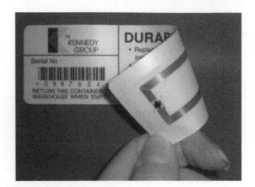

Figure 40.4 RFID tag
Source: Author

Figure 40.5 Scanner
Source: Author

Figure 40.6 GPS tracker
Source: Author

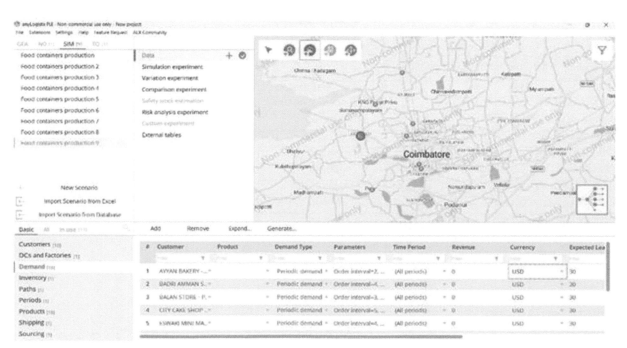

Figure 40.7 Anylogistx sample
Source: Author

The next step is to create a dashboard. For doing this, the data obtained is to be analysed. With this data, a dashboard is created to check the working and feasibility. The software that we are using to create a dashboard at the initial stage is the Anylogistix software. The optimisation will be done for the transportation network. For this, the data of products, prices, retailers, and their corresponding locations were collected from the industry. This data is analysed to find out the products that are highly in demand. The potential retailers were identified and the excel sheet is optimised according to the data required by the software. To do so, an excel is created in the necessary format as required by the software.

Figure 40.7 shows the sample of routing analysis which is done using Anylogistx software. When the coordinates and the data are stored in the software, it shows the customer locations accurately. We can transfer these data to software like google maps using programming language to retrieve the live information.

Conclusion

The project as a whole describes the viability of data twin (DT) in supply chain (SC) management. Starting with analysing the SC framework for the development of DT in association with IoT sensors, it briefly explains the scope, feasibility, and processes involved. The outcome of this project would benefit the organisation to enhance its existing supply chain process and improve its financial outcomes. The overall time required for transportation would be reduced as efficient routing will be updated frequently. This enables the driver to prioritise the deliveries according to the prevailing situation. The stock missing cases will be greatly reduced as there will be a continuous track of the incoming and outgoing stock. Thus the efficiency of the distribution center will be increased.

References

[1] Wang, Y., Wang, X., and Liu, A. (2020). Digital twin-driven supply chain planning. Procedia CIRP 93 (53rd CIRP Conference on Manufacturing Systems 2020). pp. 198–203.

[2] Bygballe, L. E., Dubois, A., and Jahre, M. (2023). The importance of resource interaction in strategies for managing supply chain disruptions. Journal of Business Research. 154. https://doi.org/10.1016/j.jbusres.2022.113333

[3] Zhang, J., Brintrup, A., Calinescu, A., Kosasih, E., and Sharma, A. (2021). Supply chain digital twin framework design: an approach of supply chain operations reference model and system of systems. arXivLabs. 1(1), 1–10.

[4] Marmolejo-Saucedo, J. A. (2020). Design and development of digital twins: A case study in supply chains. Mobile Networks and Applications. 25(6), 2141–60.

[5] Yewatkar, A., Inamdar, F., Singh, R., and Bandal, A. (2016). Smart cart with automatic billing, product information, product recommendation using rfid and zigbee with anti-theft. Procedia Computer Science, 79, 793–800.

[6] Vrinda, N. (2014). Novel model for automating purchases using intelligent cart. IOSR Journal of Computer Engineering. 1(16), 23–30.

[7] Nguyen, T., Duong, Q. H., Nguyen, T. V., Zhu, Y., and Zhou, L. (2022). Knowledge mapping of digital twin and physical internet in supply chain management: a systematic literature review. International Journal of Production Economics. 244, 108381.

[8] Loaiza, J. H. and Cloutier, R. J. (2022). Analyzing the implementation of a digital twin manufacturing system: using a systems thinking approach. Systems. 10(2), 22.

[9] Frederico, G. F., Garza-Reyes, J. A., Anosike, A., and Kumar, V. (2019). Supply Chain 4.0: concepts, maturity and research agenda. Supply Chain Management. 25(2), 262–82.

[10] Park, K. T., Son, Y. H., and Noh, S. D. (2021). The architectural framework of a cyber physical logistics system for digital-twin-based supply chain control. International Journal of Production Research. 59(19), 5721–42.

[11] Alberto, M. -G., Díez-González, J., Ferrero-Guillén R., Verde, P., Álvarez, R., and Perez, H. (2021). Digital twin for automatic transportation in industry 4.0. Sensors. 21(10), 3344.

[12] Barykin, S. Y., Bochkarev, A. A., Dobronravin, E., and Sergeev, S. M. (2021). The place and role of digital twin in supply chain management. Academy of Strategic Management Journal. 20, 1–19.

[13] Voipio, V., Elfvengren, K., Korpela, J., and Vilko, J. (2022). Driving competitiveness with RFID-enabled digital twin: case study from a global manufacturing firm's supply chain. Measuring Business Excellence. 27(1), 40–53. https://doi.org/10.1108/MBE-06-2021-0084

[14] Abideen, A. Z., Sundram, V. P. K., Pyeman, J., Othman, A. K., and Sorooshian, S. (2021). Digital twin integrated reinforced learning in supply chain and logistics. Logistics. 5(4), 84.

[15] Hendrik, H., Li, B., Weißenberg, N., Cirullies, J., and Otto, B. (2019). Digital twin for real-time data processing in logistics. In Artificial Intelligence and Digital Transformation in Supply Chain Management: Innovative Approaches for Supply Chains. Proceedings of the Hamburg International Conference of Logistics (HICL). 27, 4–28.

[16] Schislyaeva, E. R. and Kovalenko, E. A. (2021). Innovations in logistics networks on the basis of the digital twin. Academy of Strategic Management Journal. 20, 1–17.

Chapter 41

Review on parameters affecting chatter during milling in thin plates

Dheeraj Lengare[a] and Dilip Pangavhane[b]

Amrutvahini College of Engineering, Sangamner, India

Abstract

Regenerative vibrations occurring due to the machining process is known as chatter. It leads in wavy surface finish, reduced machining rate, and more uneven distribution of energy. Thin plates are always susceptible for chatter due to their low stiffness. This occurrence of chatter during milling in thin plates is resulting in more energy losses. Chatter has been continuously a part of study from many years. Researchers are looking for more sustainable ways for material removal during milling in thin plates. Thin plates are having applications like in aeronautical, structural, power plants etc. Chatter during milling in thin plate if properly analysed with the different process parameters may provide us with more productive and energy efficient solutions. Chatter, as one of the undesirable phenomenon for machinist, depends on different parameters. This paper reviews the work done on various parameters like tool depth, speed, work piece geometry, fixture, tool, machining parameters and there effects on chatter the most common thin plate materials used in structural and aeronautical applications like titanium alloy, aluminum alloy, steel alloy and CRFP the optimised parameters has been studied in detail. This reviewed work will be useful for researcher to minimise the chatter by optimising the relevant milling process parameters according to the materials of thin plate.

Keywords: Chatter, frequency domain analysis, in-process work piece dynamics, stiffness, time domain analysis

Introduction

The milling process of thin plate work piece is gradually developing towards high cutting removal rate with the developing the numerical control technology. However, the undesirable chatter is always the biggest obstacle to obtain the required surface finish quality as shown in Figure 41.1 and 41.2. Due to time varying characteristics and complicated machining, chatter remains the main obstacle in the machining process. The available studies were mostly focused on stability predictions of chatter. Researchers [1,2] have proposed many typical methods such as frequency and time domain methods. They have provided the dynamic model of cutting system. The model was made by solving the kinematic equation which involved the relations between the depth of cut and the spindle speed. These relations are properly are plotted on the graph called stability lobe diagrams (SLDs). The SLDs provided the bridge between the critical depths of cut and spindle speed [3]. Figure 41.3 is a typical stability lobe diagram, in which the abscissa represents the spindle speed and the ordinate represents the depth of cut. In this Figure 41.3, when a line Parallel to the ordinate axis is drawn, there will be intersection on the curve. The region above the curve is the unstable domain, while the region below the intersection is the stable domain. It means that to keep the machining process stable, the cutting parameters in the stable domain can be selected. Thin plates are plates having width to thickness ratio less than 8. There are different parameters that govern the effect of chatter in thin plates [3]. The parameters if analysed properly will help to reduce the chatter in thin plates.

Researchers had studied the chatter prediction of curved surface thin walled for peripheral milling [5, 6]. They have analysed the dynamic model and 3D SLD for curved plate by considering the variation for work piece dynamics caused by material removal, tool position and height along the tool axis as shown in Figure 41.8. Mathematical and numerical models for chatter prediction and suppression in thin plate milling is verified with the experimental results taken through dynamometer and various signals acquisition devices [5]. The various signal acquisition methods are: force signal, sound signal, current signal, acceleration signal and image signal.

Figure 41.4 shows a generalised process adopted by researchers for chatter prediction in thin plate milling. Experimental work consisting signal acquisition devices like accelerometer, plate and rotating dynamometer are considered in the paper. The data acquisition done by the accelerometer are processed with the help FFT analyser for frequency domain analysis while some researcher [8] has analysed the time domain analysis for chatter prediction.

Hence, extensive work is done on chatter with different advanced technologies, but there is still a huge scope for summarised analysis of chatter occurrence in thin plate while milling operation based on the parameters selected for different plate materials.

[a]lengare27@gmail.com, [b]pangavhanedr@gmail.com

DOI: 10.1201/9781003450252-41

Process Parameters Under Consideration

The stability lobe considers the factors like tool spindle speed and chip width for chatter, but there are some other parameters that also affect the chatter in thin plates. This parameter indirectly affects the chatter in the work piece. There is lot of extensive study done on the effects of these parameters on the chatter. Figure 41.5 shows the various process parameters that are involved during the milling of thin plate. These process parameters affect the milling by contributing to the increase in chatter.

Tool Feed:

Tool feed, also known as feed rate is the travel of tool for one rotation as shown in Figure 41.6. The tool feed position affect the chatter in thin plates as the dynamics of plates get affected [9]. The time

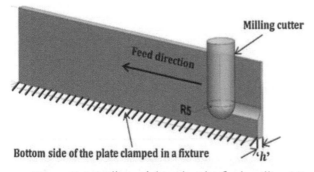

Figure 41.1 Milling of thin plate by flank milling [4]

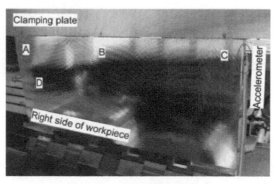

Figure 41.2 Thin plate work piece with chatter [7]

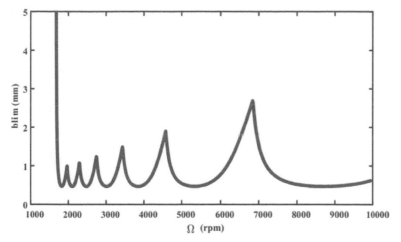

Figure 41.3 Stability lobe diagram [5]

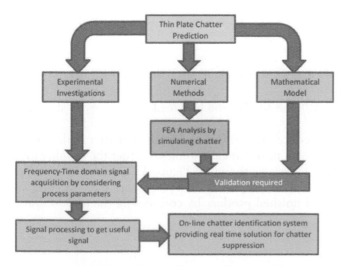

Figure 41.4 Thin plate work-piece chatter analysis
Source: Author

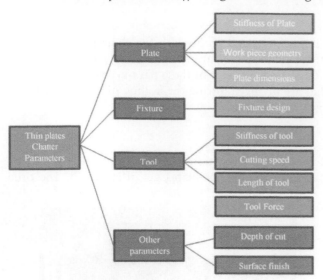

Figure 41.5 Different process parameters under consideration depending on elements in milling of thin plates
Source: Author

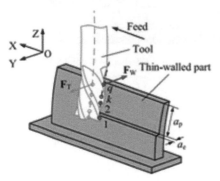

Figure 41.6 Multi point contact milling of thin plate [10]

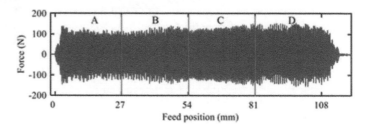

Figure 41.7 Tool force in signal domain [10]

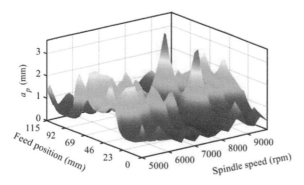

Figure 41.8 3-D Stability lobe diagram considering feed position and spindle speed domain [10]

domain dynamic analysis of thin plate was studied [10] considering the tool feed parameters as shown in Figure 41.7.

The milling chatter prediction was done with DOF reduced model for prediction of chatter in thin plate. Time domain forces are determined for same feed rate with different case studies. Further the stability lobe diagram can be drawn by studying the time domain analysis as shown in Figure 41.8. FEA methods like structural dynamic modification and free interface methods are applied to thin plate work piece by comparing two substructures i.e., material to be removed and finished product by considering the tool feed and spindle speed as domain.

Depth of Cut

The increasing demand for high machining rate is always an area of concern for machinist. The machining rate increases the material removal rate and hence reduces the operation time. Carbide tools now day used are capable of machining four times faster than HSS tools. The machining rate can increased with the increase in the depth of cut. The depth of cut increase may lead to uneven surface finish. Generally, milling of thin plate is regarded with quality surface finish [11]. In order to attain a good surface finish the cutting is done in layers. If during cutting in layers, there remains rough or uneven surface than the next consecutive cut will have to take the error of previous layer and hence this compensated error goes on increasing. Hence researchers had presented mathematical models to predict the deformation during the multilayer machining [12]. Figure 41.9 shows that if the model is implemented the overall error in machining is reduced.

Cutting Speed

Cutting speed is the relative speed between the milling tool and the work piece. In thin plate milling case mostly the work piece is stationary and the milling tool moves along the surface of work piece. The milling tool simultaneously rotates during the process. In some cases researchers had simultaneously milled both sides of plate with different cutting speed of tools so as to suppress the chatter by phasing out the resultant chatter effect on work-piece generated by both tools [7]. Optimisation techniques like multi-objective particle swarm optimisation (MOPSO) are used to optimise values of tool feed, cutting speed and feed rate to get minimum chatter and maximum material removal rate [13].

Work Piece Geometry

Work piece geometry plays a major role in the generation of chatter. As the chatter are regenerative vibrations and one of the parameters for cause of vibration is stiffness. The stiffness parameter gets influenced as the geometry of plate is varied. The reduction in thickness results in the decrease in stiffness as it is observed in the most of experimental verifications conducted by different researchers [4,6,7,11,12,14–19]. The reasons are also shown with some mathematical models presented by the researchers [4,14,20]. The chatter is mainly caused at the edges of thin work piece as there is less clamping at the end. Hence in order to reduce the vibrations at the edges some researchers have optimised the tool orientation at the edge locations so as to reduce the tool forces at edges [17,21]. This change in orientation of the tool can be done during in process milling with 5-axis milling tool [22,23]. So the 5-axis milling tool is capable of milling thin work piece with varying thickness like impellor blade. Sometimes the plate work piece provided may be not exactly flat as mostly the plates are manufactured by rolling process. In such cases double sided milling process may be used with simultaneous milling on both sides [7]. Composites are better alternatives to provide variety of shapes to the work piece [24].

Chatter analysis becomes more complicated when we consider the tri axial components of tool force acting on the work piece. For simplifications, most of the researchers [14] are comfortable to consider only one component of force in their mathematical model as shown in Figure 10 the mathematical model is verified by experimental verification. However, some researchers have also studied two components of tool force F_x and F_y on composite work piece and had plotted SLD for same [24] as shown in Figure 41.6. In the milling process of thin plate the chatter signal i.e. Time-Frequency signals are usually dominated with forced vibrations generated by tool. Researchers [25,26] have used signal processing techniques i.e. Time-Frequency filters which usually eliminate the unwanted forced vibration signals and useful chatter signal study can be more accurately done. Researchers had also presented an online identification system for the regenerative

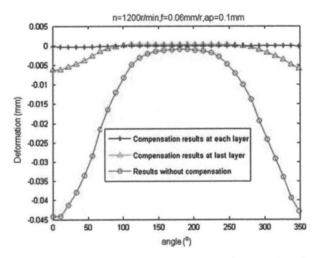

Figure 41.9 Comparing the errors in layer with and without application of model [12]

milling process, which consists of a subspace-based adaptive signal decomposition method, two characterisation indicators, and a lightweight decision predictor [27]. This online chatter identification system will focus on in process chatter generation with machine learning [28] and will provide with real time solution.

Fixture Design

The fixture holding or clamping the work piece on the machine also plays a vital role in the suppression of chatter during milling operation [2,29,30]. Fixture if properly designed is capable of suppressing the vibrations coming during the in process milling. This approach is more useful during the suppression of vibrations generated while milling of less rigid work pieces. Dynamic models of tool-work piece-fixture are capable to understand the vibrations generated [31]. The mathematical model is disturbed due to cutting force of tools. The work piece and tool stiffness are considered for vibrations. Similarly, the fixture in the model can be considered as the damping element. A similar model is analysed in the Figure 41.10 by applying the fixture plate along the plate which suppresses the vibrations [20] as shown in Figure 41.11. Some researchers had even used some external damping for suppressions of vibrations. As shown in Figure 41.12

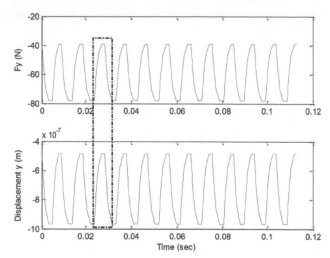

Figure 41.10 Variation of cutting force and displacement with angular tool rotation (in three revolutions) (cutting conditions; speed: 50 m/min, feed: 0.04 mm/tooth, width: 0.50 mm, depth:10 mm) (Tool: 4 flute carbide end mill, I10 mm, rake:0°, helix: 50°, material: AISI 1020, down milling) [14]

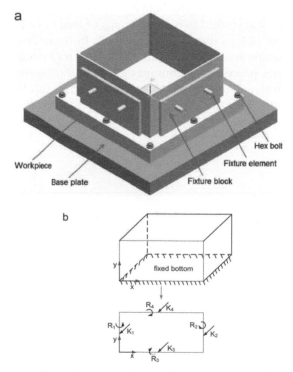

Figure 41.11 (a) A typical tool-work piece-fixture model for thin plate, (b) Equivalent mathematical model [20]

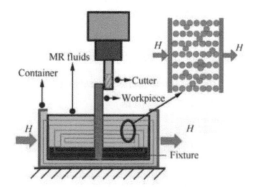

Figure 41.12 Variation of cutting force and displacement with angular tool rotation (in three revolutions) (cutting conditions; speed: 50 m/min, feed: 0.04 mm/tooth, width: 0.50 mm, depth:10 mm) (Tool: 4 flute carbide end mill, I10 mm, rake: 0°, helix: 50°, material: AISI 1020, down milling) [14]

dampers with magneto-rheological fluids are used and the mathematical models are analysed and verified with experimental tests [32]. Some innovative methods uses tuned mass damper (TMD) that are attached to work piece and vibrations are suppressed [18]. Active milling vibration control system that can suppress the relative vibrations between work piece and tool are also presented with specially designed 2-Degree of Freedom active work piece holding stage [19]. Vibration control systems are capable of suppressing considerable amount of chatter vibrations. Sometimes work piece is kept horizontally on the machining bed with screw clamping at the edges and then milling is done to analyse the chatter [33].

Work Piece Dynamics

The in-process work piece dynamics is one of the important parameter that is usually ignored while implementing the milling of thin plates. The in-process dynamics is generated due to varying dynamic displacement, material removal and tool position along the axis of tool. Research has done on this parameters by developing mathematical model by considering dynamic model of plate and tool [6] as shown in Figure 41.13. The modal analysis of tool and work piece is done in FEM to understand the effect material removal during the milling process. The natural frequencies were determined at each tool position.

Stiffness of Plate and Tool

The total stiffness can be computed as given in Equation 1, where E is the Elastic Modulus of the work piece, TH is the wall thickness, H is the wall height and Wt is the wall width [14]. As stiffness depends upon the material of plate. The plate material used during milling operation has a significant effect. Table 41.2 shows the common work piece materials used by different researchers in chatter analysis of thin plates during milling operation.

$$K_y = \frac{E(W_t^3)T_H}{4H^3}$$
(1)

It can be seen by analyzing the different research work the materials used for thin plate milling chatter analysis as shown in Table 41.1. Steel or aluminum materials, the cutting speed can be broadly varied without a detrimental effect for the tool life, when heat-resistant alloys, such as titanium, Inconel, or stainless steel, are machined, the variation range is very narrow [36]. Due to the low stiffness and high material removal rate in titanium alloys, chattering occurs very easily in the machining process, which reduces the surface precision of the work piece and the service life of cutters and machine tools, limiting production efficiency [37]. Composite material may be also used as work piece for chatter analysis due to its high specific strength, stiffness at low weight and a cost effective alternative for aeronautical industry [38]. In the case of structural elements made out of composite materials, delamination, fiber breaking, and matrix cracking are especially dangerous failure modes [39]. Hence it will be interesting to analyse the composite work piece for chatter in milling.

The stiffness effect is also observed when the work piece is thin beam [40]. The mathematical models generally simplify the geometry and basic principles are applied on the simplified geometry. Linear edge model [4] presented simplified the thin plate in simple cantilever plate with Clamp Free Free Free (CFFF) boundary conditions for analysis. Mathematical models based on the stiffness of removed chip elements

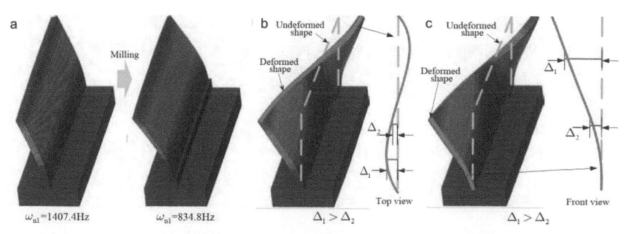

Figure 41.13 Workpiece dynamics variation: a) initial and in-process work-piece mode shapes, b) second mode shape of work-piece, c) mode shape of work-piece along tool axis [6]

Table 41.1 Details of tool used in milling of thin plate (Workpiece) material

Workpiece material	Tool speed (rpm)	Depth of cut (mm)	Radial depth of cut (mm)	feed rate per tooth	Tool	Reference
Ti-6Al-4V	3500-5500	20	1-1.75	350-550 mm/min	A four fluted solid-carbide cylindrical ball end milling tool diameter of 10 mm, helix angle of 30°, primary cutting angle of 10.5° and total cutting flute length 22 mm.	[4]
	200	2	---	400 mm/min	Sintered carbide (Tungaloy UX30), Number of teeth = 11, Tool diameter = 348 mm, Nose radius of teeth = 0.8mm.	[7]
	2123	27	0.6	--	Standard end mills, tool diameter = 12 mm, number of teeth = 4, helix angle = 38°.	[34]
Al 7075-T6 alloy	350	10	0.50	0.06 mm/tooth	Solid carbide end mills. The milling cutter used is 10 mm in diameter, four fluted with a 30° helix angle and 50° helix angle.	[14]
	11200	0.2-1.45	--	---	5-axis ball-end milling.	[16]
	6000-9600	0.4-2	0.2-0.4	600-960 mm/min	Two-flute carbide end mill with 16 mm diameter.	[25]
ZL114A	300	0.3		0.05 mm/rot.	Carbide cutter with tool orthogonal rake of 101, tool clearance of 151 and diameter of 24.1 mm.	[12]
Al alloy 6061-T6	13000	1.6 -12	0.5-0.3	---	2-flute long length carbide end mill with 15.875 mm diameter, 30 deg helix angle, and 78 mm overhang length.	[6]
	18000	0.5–2	0.5 – 2	0.1 mm/tooth	20 mm dia., 4-flute milling cutter.	[11]
	3000	1-2	---	50-100 mm/min	Four toothed HSS (High Speed steel) milling cutter.	[13]
	5000-10000	0.5-9	0.1 -0.7	0.0125 mm/tooth	Sintered carbide end milling cutter, 4 teeth, 10 mm diameter, length 89 mm, edge length 60 mm, overhang 45 mm.	[27]
EN-AW7075	18000	---	---	2880 mm/min	Ball end milling tool, tool diameter = 6mm, number of teeth =2.	[17]
Steel Alloy	600-1800	6	0.3	240-780 mm/min	12 mm diameter, 30 degree helix angle, four teeth tool.	[19]
	5000	2	0.3	500 mm/min	Cylindrical helical milling cutter, diameter 10, teeth 2.	[32]
	3379-9600	6.56-10.8	5.69-9.8	0.2 -0.1 mm/tooth	---	[35]
TC4 titanium alloy	1500	0.54	0.1	60mm / min	flat end milling cutter, 4 teeth, 8 mm diameter, length 60 mm, edge length 25 mm, overhang 40mm.	[26]
aluminum alloy UNS 2024-T3	4000	0.2-1	5	0.1 mm	2 flute bull nose end mill Kendu 4400, 10 mm diameter, 30 degree helix angle and 2.5 mm edge radius.	[33]
CRFP	2000-8000	0.5-2	12	0.05 mm	Diamond-coated cutting steel, having a diameter of 12 mm and two flute.	[24]

Source: Author

Table 41.2 Details of workpiece and machine tool used in chatter analysis

Workpiece material	Parameters studied	Position of workpiece	Workpiece	Machine tool	Reference
Ti-6Al-4V	Forces, stiffness of plate, depth of cut, deflections of plate	Vertical	cantilever plate workpiece of size $121 \times 26 \times 2.5$ mm^3	vertical machining centre (HARDINGE VMC 600 II)	[4]
		Vertical	$450 \times 200 \times 17.5$	Conventionalmilling machine with electro-magnetic chuck	[7]
		---	$40 \times 65 \times 3$ mm^3	Conventional milling machine	[34]
Al 7075-T6 alloy	Cutting force, stiffness of plate, dynamic chip thickness variation, depth of cut, IPW dynamics of plate	Vertical	$100 \times 60 \times 60$ mm^3, 8 ribs with the gap 10.1 mm, thickness of each rib is 3 mm.	Conventional milling machine	[14]
		Vertical	Turbine blade	---	[16]
		Vertical	$85 \times 40 \times 3$ mm^3	five-axis DMU60 CNC machining center	[25]
ZL114A	Deformations in plate	Vertical	Circular ring of diameter 70 mm	MikronUCP-710 machining center	[12]
Al alloy 6061-T6	natural frequencies and mode shapes, effect of material removal, tool position and height along the tool axis, tool speed	Vertical	$115 \times 36 \times 3.5$ mm^3	five-axis machining center DMU 80P	[6]
		Dead center of headstock	Overhanging workpiece of length 25 mm	Makino A55E horizontal milling machine	[11]
		Horizontal		Vertical milling machine	[13]
		Horizontal	Step-like workpiece with equally distributed step heights, i.e., 3 mm, 2 mm and 1 mm.	DMU 50 machining center	[27]
EN-AW7075	cutting forces	Vertical	$50 \times 30 \times 80$ mm^3 turbine blade	HSC five-axis milling machine Deckel Maho DMU 50 evolution	[17]
Steel Alloy	Forces, stiffness of plate, depth of cut, deflections of plate	Horizontal	block	Vertical milling machine (DMU 70V)	[19]
		Vertical	plate $130 \times 60 \times 5$ mm^3	three-axis milling center (YHVT8507	[32]
		Horizontal	workpiece with ribs	CNC Machine tool	[35]
TC4 titanium alloy	IPW dynamics of plate	Horizontal	$100 \times 120 \times 5$ mm^3	5 axis milling NC	[26]
aluminum alloy UNS 2024-T3	Surface roughness, tool speed, mode shapes	Horizontal	50×50 mm^3	5 axis milling NC	[33]
CRFP	Tool Force, modal analysis, depth of cut, spindle speed	Vertical	---	BlueBird MG6037PKK	[24]

Source: Author

were also proposed by considering the Oxley's flow stress model and Johnson cook's material model [14] as shown in Figure 41.14. The stiffness of the removed chips was subtracted and the stiffness of plate is proposed. This model works on certain assumptions and hence has certain limitations too. Limitations like the damping factor and the natural frequency of the material may not be considered in the model. Hence the parameter like damping of plate also plays major role in the prediction of chatter. From available literature, it is observed that for specific kind of workpiece material, a particular types of process parameters were studied. Various researchers had studied different workpiece materials chatter at different workpiece position and machine tools. However, the positions considered are as per the requirement to observe more vibrations to ease the study. The aforementioned data is shown in Table 41.2.

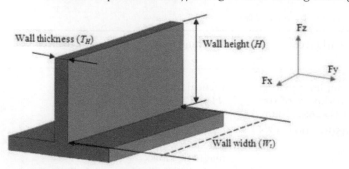

Figure 41.14 Geometrical parameters of thin plate work piece [14]

Tool Type

Some researchers used special end mills having varying pitch angles or helix angles to suppress chatter by studying stability diagrams [34]. According to the generated stability maps, crest-cut tools show outstanding performance in milling of thin-walled parts when compared to the variable pitch and standard end mills [34]. Unlike standard end mills, special end mills have varying pitch or helix angles, or both, which can introduce higher stability limits than standard end mills. Modification of pitch angles between the adjacent teeth's of tool may be made such that the phase difference between the wavy surfaces left on work piece after milling will phase out. Hence modification of tool pitch angle between adjacent teeth's will result in less chatter [36,41]. Hence a novel spindle speed may be obtained. From available literature, it is observed that for specific kind of workpiece material, a particular type of tool was suitable. Various researchers had studied different workpiece materials chatter by using milling tools with mentioned parameters of cutting tool like speed, depth of cut, radial depth of cut and feed rate are listed in Table 41.1.

Conclusion

Chatter during milling in thin plates can be analysed in two aspects a) chatter prediction b) chatter suppression. Basically, both works are related to each other and hence to suppress the chatter, prediction is an important step. Hence optimisation techniques may be used for optimizing the process parameter by considering the data presented in the paper by the specified input parameters selected by statistical analysis and hence chatter suppression process parameters may be obtained [35, 42, 43]. Researchers had developed different mathematical models for chatter predictions and had validated with experimental and numerical methods. The mathematical models consider different process parameters. The more the process parameters considered; more is the accuracy of prediction. Chatters are also predicted by numerical methods like modal analysis [44] of plate. Most of the researchers prefer the aluminum, titanium and steel alloy plates as the work piece. Mostly the material used are applications oriented like aluminum alloy being light in weight is used in aeronautical industry. Steel alloy plates having good strength are used mostly in structural applications. Similarly, titanium alloys are also used by many researchers. However more cost-effective alternatives like Composites CRFP work piece are also used but non-linear properties of composite material should be properly analysed. The tool parameters are different for different material plates, as it depends on the stiffness of materials.

A. The parameters like IPW dynamics, modal analysis are generally analysed on the aluminum and titanium alloy work pieces due to its ease to analyse above parameters. The parameters like depth of cut, surface roughness, spindle speed are analysed on the steel alloy plates. However, these are observations mostly seen, but either cases are also observed.
B. This paper summarise the suitable tools, machine tool, workpiece material, position of workpiece etc. parameters so as to ease the further study of chatter analysis.

References

[1] Altintaş, Y., and Budak, E. (1995). Analytical prediction of stability lobes in milling. CIRP Annals Manufacturing Technology. 44, 357–62. https://doi.org/10.1016/S0007-8506(07)62342-7.
[2] Munoa, J., Beudaert, X., Dombovari, Z., Altintas, Y., Budak, E., Brecher, C., and Stepan, G. (2016). Chatter suppression techniques in metal cutting. CIRP Annals - Manufacturing Technology. 65, 785–808. https://doi.org/10.1016/j.cirp.2016.06.004.
[3] Yue, C., Gao, H., Liu, X., Liang, S. Y., and Wang, L. (2019). A review of chatter vibration research in milling. Chinese Journal of Aeronautics. 32. 215–42. https://doi.org/10.1016/j.cja.2018.11.007.

[4] Khandagale, P., Bhakar, G., Kartik, V., and Joshi, S. S. (2018). Modelling time-domain vibratory deflection response of thin-walled cantilever workpieces during flank milling. Journal of Manufacturing Processes. 33, 278–90. https://doi.org/10.1016/j.jmapro.2018.05.011.

[5] Wang, W. K., Wan, M., Zhang, W. H., and Yang, Y. (2022). Chatter detection methods in the machining processes: A review. Journal of Manufacturing Processes. 77, 240–59. https://doi.org/10.1016/j.jmapro.2022.03.018.

[6] Yang, Y., Zhang, W. H., Ma, Y. C., and Wan, M. (2016). Chatter prediction for the peripheral milling of thin-walled workpieces with curved surfaces. International Journal of Machine Tools and Manufacture. 109, 36–48. https://doi.org/10.1016/j.ijmachtools.2016.07.002.

[7] Shamoto, E., Mori, T., Nishimura, K., Hiramatsu, T., and Kurata, Y. (2010). Suppression of regenerative chatter vibration in simultaneous double-sided milling of flexible plates by speed difference. CIRP Annals - Manufacturing Technology. 59, 387–90. https://doi.org/10.1016/j.cirp.2010.03.028.

[8] Atlar, S., Budak, E., and Özgüven, H. N. (2008). Modeling part dynamics and chatter stability in machining considering material removal. In 1st International Conference on Process Machine Interactions, Hann. pp. 61–72. https://www.researchgate.net/profile/H_Oezgueven/publication/228413807_Modeling_part_dynamics_and_chatter_stability_in_machining_consi ering_material_removal/links/02bfe5132f8b6cdc65000000.pdf.

[9] Brecher, C., Fey, M., and Daniels, M. (2016). Modeling of position-, tool- and workpiece-dependent Milling machine dynamics. High Speed Machine. 2, 15–25. https://doi.org/10.1515/hsm-2016-0002.

[10] Li, W., Wang, L., Yu, G., and Wang, D. (2021). Time-varying dynamics updating method for chatter prediction in thin-walled part milling process. Mechanical Systems and Signal Processing. 159, 107840. https://doi.org/10.1016/j.ymssp.2021.107840.

[11] Kolluru, K., and Axinte, D. (2013). Coupled interaction of dynamic responses of tool and workpiece in thin wall milling. Journal of Materials Processing Technology. 213, 1565–74. https://doi.org/10.1016/j.jmatprotec.2013.03.018.

[12] Chen, W., Xue, J., Tang, D., Chen, H., and Qu, S. (2009). Deformation prediction and error compensation in multilayer milling processes for thin-walled parts. International Journal of Machine Tools and Manufacture. 49, 859–64. https://doi.org/10.1016/j.ijmachtools.2009.05.006.

[13] Mishra, R., and Singh, B. (2022). An ensemble approach to maximise metal removal rate for chatter free milling. Journal of Computational Science. 59, 101567. https://doi.org/10.1016/j.jocs.2022.101567.

[14] Masmali, M., and Mathew, P. (2017). An analytical approach for machining thin-walled workpieces. Procedia CIRP. 58, 187–92. https://doi.org/10.1016/j.procir.2017.03.186.

[15] Sanjay D. Patil, Dheeraj S. Lengare, Arvind J. Bhosale, Kiran B. Bansode and Rashtrapal B. Teltumade (2020). Parameters Affecting the Specific Energy Absorption of Circular Side Impact Beam. International Journal of Engineering and Advanced Technology (IJEAT). ISSN: 2249-8958 (Online), 9(3), February 2020. Retrieval Number: C6259029320 /2020©BEIESP DOI: 10.35940/ijeat.C6259.029320. 3399–3405.

[16] Budak, E., Tunç, L. T., Alan, S., and Özgüven, H. N. (2012). Prediction of workpiece dynamics and its effects on chatter stability in milling. CIRP Annals - Manufacturing Technology. 61, 339–42. https://doi.org/10.1016/j.cirp.2012.03.144.

[17] Biermann, D., Kersting, P., and Surmann, T. (2010). A general approach to simulating workpiece vibrations during five-axis milling of turbine blades. CIRP Annals - Manufacturing Technology. 59, 125–8. https://doi.org/10.1016/j.cirp.2010.03.057.

[18] Wang, M. (2011). Feasibility study of nonlinear tuned mass damper for machining chatter suppression. Journal of Sound and Vibration. 330, 1917–30. https://doi.org/10.1016/j.jsv.2010.10.043.

[19] Long, X., Jiang, H., and Meng, G. (2013). Active vibration control for peripheral milling processes. Journal of Materials Processing Technology. 213, 660–70. https://doi.org/10.1016/j.jmatprotec.2012.11.025.

[20] Zeng, S., Wan, X., Li, W., Yin, Z., and Xiong, Y. (2012). A novel approach to fixture design on suppressing machining vibration of flexible workpiece. International Journal of Machine Tools and Manufacture. 58, 29–43. https://doi.org/10.1016/j.ijmachtools.2012.02.008.

[21] Huang, T., Zhang, X. M., and Ding, H. (2017). Tool orientation optimisation for reduction of vibration and deformation in ball-end milling of thin-walled impeller blades. Procedia CIRP. 58, 210–5. https://doi.org/10.1016/j.procir.2017.03.211.

[22] Budak, E., Ozturk, E., and Tunc, L. T. (2009). Modeling and simulation of 5-axis milling processes. CIRP Annals - Manufacturing Technology. 58, 347–50. https://doi.org/10.1016/j.cirp.2009.03.044.

[23] Ozturk, E., and Budak, E. (2010). Dynamics and stability of five-axis ball-end milling. Journal of Manufacturing Science and Engineering. 132, 0210031–02100313. https://doi.org/10.1115/1.4001038.

[24] Rusinek, R., and Lajmert, P. (2020). Chatter Detection in Milling of Carbon Fiber-Reinforced Composites by Improved Hilbert–Huang Transform and Recurrence Quantification Analysis. Materials. 13, 4105. https://doi.org/10.3390/ma13184105.

[25] Yan, S., and Sun, Y. (2022). Early chatter detection in thin-walled workpiece milling process based on multi-synchrosqueezing transform and feature selection. Mechanical Systems and Signal Processing. 169, 108622. https://doi.org/10.1016/j.ymssp.2021.108622.

[26] Hao, Y., Zhu, L., Yan, B., Qin, S., Cui, D., and Lu, H. (2022). Milling chatter detection with WPD and power entropy for Ti-6Al-4V thin-walled parts based on multi-source signals fusion. Mechanical Systems and Signal Processing. 177, 109225. https://doi.org/10.1016/j.ymssp.2022.109225.

[27] Ren, Y., and Ding, Y. (2022). Online milling chatter identification using adaptive Hankel low-rank decomposition. Mechanical Systems and Signal Processing. 169, 108758. https://doi.org/10.1016/j.ymssp.2021.108758.

[28] Yesilli, M. C., Khasawneh, F. A., and Otto, A. (2022). Chatter detection in turning using machine learning and similarity measures of time series via dynamic time warping. Journal of Manufacturing Processes. 77, 190–206. https://doi.org/10.1016/j.jmapro.2022.03.009.

[29] Ma, H., Guo, J., Wu, J., Xiong, Z., and Lee, K. M. (2020). An active control method for chatter suppression in thin plate Turning, IEEE Transactions on Industrial Informatics. 16, 1742–53. https://doi.org/10.1109/TII.2019.2924829.

[30] Beudaert, X., Erkorkmaz, K., and Munoa, J. (2019). Portable damping system for chatter suppression on flexible workpieces. CIRP Annals. 68, 423–6. https://doi.org/10.1016/j.cirp.2019.04.010.

[31] Deiab, I. M., and Elbestawi, M. A. (2004). Effect of workpiece/fixture dynamics on the machining process output. Proceedings of the Institution of Mechanical Engineers, Part B: Journal of Engineering Manufacture. 218, 1541–53. https://doi.org/10.1243/0954405042418455.

[32] Ma, J., Zhang, D., Wu, B., Luo, M., and Chen, B. (2016). Vibration suppression of thin-walled workpiece machining considering external damping properties based on magnetorheological fluids flexible fixture. Chinese Journal of Aeronautics. 29, 1074–83. https://doi.org/10.1016/j.cja.2016.04.017.

[33] Casuso, M., Rubio-Mateos, A., Veiga, F., and Lamikiz, A. (2022). Influence of Axial Depth of Cut and Tool Position on Surface Quality and Chatter Appearance in Locally Supported Thin Floor Milling. Materials. 15, 731. https://doi.org/10.3390/ma15030731.

[34] Tehranizadeh, F., Berenji, K. R., Yıldız, S., and Budak, E. (2022). Chatter stability of thin-walled part machining using special end mills. CIRP Annals. 71, 365–68. https://doi.org/10.1016/j.cirp.2022.04.057.

[35] Deng, C., Feng, Y., Miao, J., Ma, Y., and Wei, B. (2019). Multi-objective machining parameters optimisation for chatter-free milling process considering material removal rate and surface location error. IEEE Access. 7, 183823–37. https://doi.org/10.1109/ACCESS.2019.2949423.

[36] Iglesias, A., Dombovari, Z., Gonzalez, G., Munoa, J., and Stepan, G. (2018). Optimum selection of variable pitch for chatter suppression in face milling operations. Materials (Basel). 12, 1–21. https://doi.org/10.3390/ma12010112.

[37] Gao, H., and Liu, X. (2019). Stability Research Considering Non-Linear Change in the Machining of Titanium Thin-Walled Parts. Materials, 12, 2083. https://doi.org/10.3390/ma12132083.

[38] Chauhan, M., Mishra, P., Dwivedi, S., Ragulskis, M., Burdzik, R., and Ranjan, V. (2022). Development of the Dynamic Stiffness Method for the Out-of-Plane Natural Vibration of an Orthotropic Plate. Appl. Sci. 12, 5733. https://doi.org/10.3390/app12115733.

[39] Kudela, P., Zak, A., Krawczuk, M., and Ostachowicz, W. (2007). Modelling of wave propagation in composite plates using the time domain spectral element method. Journal of Sound and Vibration. 302, 728–45. https://doi.org/10.1016/j.jsv.2006.12.016.

[40] Schmitz, T. L., and Honeycutt, A. (2017). Analytical solutions for fixed-free beam dynamics in thin rib machining. Journal of Manufacturing Processes. 30, 41–50. https://doi.org/10.1016/j.jmapro.2017.09.002.

[41] Mei, Y., Mo, R., Sun, H., He, B., and Bu, K. (2020). Stability Analysis of Milling Process with Multiple Delays. Appl. Sci. 10, 3646. https://doi.org/10.3390/app10103646.

[42] Barnkob, L., Argyraki, A., and Jakobsen, J. (2020). An Active Control Method for Chatter Suppression in Thin Plate Turning, in IEEE Transactions on Industrial Informatics, 16(3), 1742–1753, March 2020, doi: 10.1109/TII.2019.2924829.

[43] Qu, S., Zhao, J., and Wang, T. (2017). Experimental study and machining parameter optimisation in milling thin-walled plates based on NSGA-II. International Journal of Advanced Manufacturing Technology. 89, 2399–2409. https://doi.org/10.1007/s00170-016-9265-1.

[44] Khandagale, P., Kartik, V., and Joshi, S. (2017). Forced vibration response of a micro-cantilever beam with moving loads In 2017 International Conference on Advances in Mechanical, Industrial, Automation and Management Systems, AMIAMS. 2017 - Proceedings. pp. 220–224. https://doi.org/10.1109/AMIAMS.2017.8069215.

Chapter 42

Experimental studies on pond ash in the proportion with blast furnace slag

Harshal Nikhade[a], Utkarsh Bobde[b], Khalid Ansari[c] and Sneha Hirekhan[d]

Yeshvantrao chavan college of engineering, Nagpur, India

Abstract

This article describes the result of residual ash bottom ash (BA) from laboratory experiments. The material was prepared with a mixture (0.4-1.2%), and alkaline strength building fibers as reinforcement materials, and cement (20%, 25%, and 30%) to bind the materials. The mixing ratio was found to have a significant roll in density, compressive strength, and initial tangent module. The density of the materials due to the fiber mixture for 20% of the mixture percentages varies between 450 kg/m^3 and 170 kg/m^3. Material density variation for 30% of blending percentages was 53.34%. The compression strength of newly prepared materials increases as the cement ratio increases from 20-30%, also strength ranging from 25 kPa to 239.5kPa. As the mix ratio meanwhile, compressive strength diminished due to the addition of EPS beads. Compression resistance varies depending on the mixture ratio percentages from 0.4-1.2%. The pattern of stress and strain curve brittle in behavior. Pond Ash creates environmental pollution and at same time required huge amount of land for deposition of waste. So, utilization of this waste is without harming to environment solve the disposal problem, that material also may provide cheap building materials in the side of sustainable growth of the country.

Keywords: Bottom ash, granular slag, mix proportion, BF slag, density, compressive strength

Introduction

The coal ash which is accumulated in ash pond is generally seen as a pond ash. Pond Ash creates environmental pollution and at same time required huge amount of land for deposition of waste. Pond Ash also pollute soil and water to do heavy metal leaching. It is necessary to deposit this waste in the environmental friendly manner is major concern. Conventionally, granulated fills are extensively consumed for filled applications together with backfill at the back of retaining walls, reform of inferior lying zones and below the ground line pipe. In usages like fill below the ground pipe traces and mine shafts, so low control strength material low energy materials low strength is considered as an operative material in that area. ACI 229R-99 [1] and ASTM D 5971-07 [2, 3]. The slag produced after iron ore is liquefied and reduced among melted iron at furnaces. The quantity of slag generated is almost 0.3 tons iron produced, and the per annum manufacture of slag (BF slag) in Japan more than 2.4 core [4]. According to Indian Mineral Yearbook 2014, the slag moving over the melted pig iron is blushing out in slag container after that sent to slag devastating bush or to preservation pits. In cooling method which is done in the blast furnace, the slag is classified into three types; that is, air-cooled, granulated and expanded slag. The generation of air-cooled slag by allow the liquid slag to cool in a hole in atmospheric situations. Granulated slag is generated after suppression the liquid slag through method of high-pressure water jets. Expanded slag is made from command quenching of melted slag in to water or addition of steam with water and pressurised air. Dubey et al. [5] said that in the manufacture of concrete OPC is one of the key ingredients used. Concrete characteristics is to be preserve with advanced mineral blends like BF-slag dust as fractional substitution of cement up to 5-30%. The compressive strength of BF-slag concrete at dissimilar proportion of slag was calculated as a fractional substitution of cement. Author observed, as the rates of BF slag dust rise, the strength observed to reduced [5]. Hiraskar and Patil found that the BF-slag is a minor product and usages it as aggregates into concrete may will provide effective and environmentally amaible solution in local region. The requirement for aggregates is rise rapidly and so as the need of concrete. so, it was seemly more crucial to search alternatives for aggregates in the future. The outcomes showed that it has characteristics identical to natural aggregates and not affect or harm the character of concrete [6]. The efficiency of GGBS containing Portland cements gave 10-80% replacements and strength efficiencies at 28 days. The strength of concrete varies from 20 MPa with 10% of GGBS to 100 MPa with 80% of GGBS [7]. Song and Saraswath reviewed and discussed the impact of use of GGBS in term of durability reinforced concrete structures like chloride entry and resistance to corrosion, durability in long period, microstructure and porousness of GGBS concrete. The authors observed that partial substitution OPC by GGBS increase the durability [8]. Dubey et al. [5] said that in the manufacture of concrete OPC is one of the key ingredients used. Concrete characteristics

[a]harshal.nikhade@rediffmail.com, [b]utkarshbobdenyss@gmail.com, [c]khalidshamim86@rediffmail.com, [d]sneha_kits@rediffmail.com

DOI: 10.1201/9781003450252-42

is to be preserve with advanced mineral blends like BF-slag dust as fractional substitution of cement up to 5-30%. Compressive strength of BF-slag concrete at dissimilar proportion of slag was calculated as a fractional substitution of cement. Author observed, as the rates of BF slag dust rise, the strength observed to reduced [9]. From the literature review it was observed very few studies carried out pond ash in proportion with blast furnace slag but no studies are carried out to know the behavior of pond ash under compressive loading with a constant deformation rate at 1.0 mm/minute.

Characterization of Materials

To produce the new material, the pond ash was mix with EPS beads, slag, and cement. Ash used in the work had a gravity (G) of 2.10, and hydrometer analysis showed a percentage of very fine sand content of 58, the silt of 38, and clay of 4. According to the standard proctor test, the PA has a dry density (γd) was 1.2 g/cc and the optimum moisture content (OMC) was 17%. The slag had a gravity of 2.10 and a fineness modulus of 2.69 (average sand according to IS 383-1970). From XRF tests determine the chemical properties of the materials BF Slag contains the highest percentage of calcium oxide (36.5). Figure 42.1 shows the photo SEM of pond ash. Chemical composition of bottom ash is shown in Table 42.1.

Test procedure

The quantities of sundry essentials of the amalgamation were calculated on the substructure of method utilised by Lal et. al.[10, 11]. The amalgamation ratios percentages utilised in the current study were 0.4, 0.8 and 1.2%.

Result and Discussion

Density of material

The material density is prime functions of organised materials which notably motivated by the mix of materials and the addition of fiber. The effect of blend ratio on character i.e., density at different percentage of cement to pond ash (PA) ratio is proven in Figure 42.2 (a and b). The material density of tested mix became detected in the variety 70-450 Kg/m³. For individual mix the density of organized substances changed and decrease constantly as the percent of proportion rise of EPS beads to pond ash. For individuals of the blend ratio and curing time the C/ash ratio 30% having maximum density of substances than cement to ash fraction 20% and 25%. The density of developed materials using ash and EPS bead smaller than conventional fill substances that is in variety 2100-1700 kg/m³ [12-17].

Compressive Strength

Compressive strength of EPS bead reinforced PA and slag based materials for considerably influenced by means of the curing period, blend ratio and probabilities of cement to PA ratio. the peak compressive stress

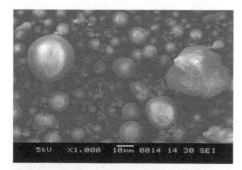

Figure 42.1 Scanning electron microscopic of pond ash
Source: Author

Table 42.1 Chemical composition of bottom ash

Ingredients	Natural structure (%)								
	SiO$_2$	Al$_2$O$_3$	Fe$_2$O$_3$	CaO	K$_2$O	MgO	TiO	Other	LOI
PA	59.21	31.12	3.11	1.01	0.83	0.32	2.01	1.73	0.66

Source: Author

at instance is considered as compressive strength. Figure 42.3 (a and b) shows the pattern of compressive strength of EPS reinforced BA in proportion with slag and cement to PA 20%, 25%, and 25% of different curing durations. Strength ranging from 25-234.9 kPa for 7-28 days period. It is discovered that the addition of EPS to PA decrease the compressive strength of materials continuously from 0.4-1.2%. The improvement within the compressive power additionally relies upon cement to PA percentages and curing duration of the specimen.

Stress strain curve

The strength outcome turned into cast-off to invention the strain- strain capabilities and toughness of PA-based substances. The impact of combination developed mix on compressive strength and axial strength for EPS to ash for 7 and 28 days curing interval for blend value 0.8% is shown in Figure 42.4. The compressive strength consequence was cast-off to invention the stress- pressure capabilities and toughness of pond ash-primarily based substance. The significance of blend ratio on compressive strength and axial

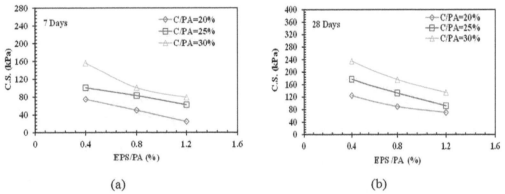

(a) (b)

Figure 42.2 Density of pond ash for (a) 7 days and (b) 28 days
Source: Author

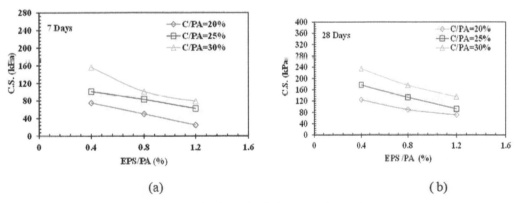

(a) (b)

Figure 42.3 Compressive strength of pond ash for (a) 7 (b) 28 days
Source: Author

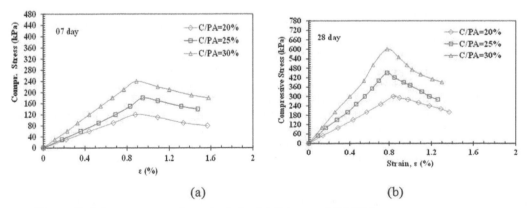

(a) (b)

Figure 42.4 Stress pattern of pond ash for (a) 7 days and (b) 28 days
Source: Author

strength for fiber to ash for 28 days period of curing interval for mix cost 1.2% shown in Figure 42.4. The nonlinear conduct have been determined from stress pattern. The compressive strength decreases from peak point of failure as shown in Figure 42.4 (a and b).

Conclusion

Within the experimental have a look at, the consequence of the accumulation of fiber on density, compressive behavior and stress pattern of the recently prepared EPS containing PA, slag and cement had been investigated. The density of developed material became lower with growing percentages of EPS. With the addition of EPS within the ratio percentages 0.4-1.2%, the density of developed substances were diminished from 70-450 kg/m³. The compressive strength of substances for every mixed percentage and curing time have been within the range of 25-234.9 kPa. The materials the usage of cement in proportion bottom ash ratio of 30% has compressive strength better than cement to pond ash ratio 20% and 25% for mix ratio. The stress performance is found to be nonlinear for each mix ratio for every curing duration.

Acknowledgment

We are grateful to the structural research unit personnel, who helped guide numerous factors till these studies changed into carried out efficiently.

References

[1] Alengaram, U. J. (2022). Valorization of industrial byproducts and wastes as sustainable construction materials. In Handbook of Sustainable Concrete and Industrial Waste Management. pp. 23-43.
[2] Miyamoto, T., Torii, K., Akahane, K., and Hayashiguchi, S. (2015). Production and use of blast furnace slag aggregate for concrete. Nippon Steel & Sumitomo Metal Technical Report, 109, 102.
[3] Lal, K. and Professor, J. A. (2016). To effect on strength properties of concrete of by using GGBS by Partial Replacing cement and addition of GGBS without replacing cement. SSRG International Journal of Civil Engineering. 3(5).
[4] Song, H. -W. and Saraswathy V. (2006). Studies on the corrosion resistance of reinforced steel in concrete with ground granulated blast-furnace slag-an overview. Journal of Hazardous Materials. B138, 226-33.
[5] Dubey A., Chandak, R., and Yadav R. (2012). Effect of blast furnace slag powder on compressive strength of concrete. International Journal of scientific & Engineering Research. 3.
[6] Nadeem M. and Pofale A. (2012). Replacement of natural fine aggregate with granular slag-a waste industrial by-product in cement mortar applications as an alternative construction material. International Journal of Engineering Research and Applications.
[7] Autade P. B. and Wakankar A. B. (2016). Effects of GGBFS as pozzolanic material with glass fiber on mechanical properties of concrete. International Journal of Scientific Engineering and Applied Science. 2(4), 331-37.
[8] ACI Committee 544. (1996). State-of-the-art report on fibre reinforced concrete, ACI.1R-96. USA, Detroit: American Concrete Institute.
[9] Ghugal, Y. M. and Deshmukh, S. B. (2006). Performance of alkali-resistant glass fiber reinforced concrete. Journal of Reinforced Plastics and Composites. 25(6), 617-30.
[10] Lal, B. R. and Nawkhare, S. S. (2016). Experimental study on plastic strips and eps beads reinforced bottom ash based material. International Journal of Geosynthetics and Ground Engineering. 2(3), 2–13. doi: 10.1007/s40891-016-0066-2
[11] Lal, B. R. R. and Badwaik, V. N. (2016). Experimental studies on bottom ash and expanded polystyrene beads–based geomaterial. Journal of Hazardous, Toxic, and Radioactive Waste. 20(2), 04015020. doi: 10.1061/(asce)hz.2153-5515.0000305
[12] Lyons, A. (2010). Materials for architects and builders. 4ᵗʰ Ed., Elsevier, pp. 420, Hong Kong, China.
[13] Saeed, A., Hadee. M. N., Hassan, A., Sabri, M. M. S., Qaidi, S., Mashaan, N. S., and Ansari, K. (2022). Properties and applications of geopolymer composites: a review study of mechanical and microstructural properties. Materials. 15(22), 8250. https://doi.org/10.3390/ma15228250
[14] Polichetty, R. K. C., Thaniarasu, I., Elkotb, M. A., Ansari, K., and Saleel, C. A. (2021). Shrinkage study and strength aspects of concrete with foundry sand and coconut shell as a partial replacement for coarse and fine aggregate. Materials. 149(23), 7420. https://doi.org/10.3390/ma14237420
[15] Khalid, A., Shrikhande, A., Malik, M. A., Alahmadi, A. A., Alwetaishi, M., Alzaed, A. N., and Elbeltagi, A. (2022). Optimization and operational analysis of domestic greywater treatment by electrocoagulation filtration using response surface methodology. Sustainability. 14(22), 15230. https://doi.org/10.3390/su142215230
[16] Nikhade, H. R., and Lal, B. (2023). Studies on Sugar Cane Bagasse Ash and Blast Furnace Slag-Based Geomaterial. In Indian Geotechnical and Geoenvironmental Engineering Conference. pp. 25-32. Singapore: Springer. https://doi.org/10.1007/978-981-19-4739-1_3
[17] Nikhade, H. R. and Lal, B. R. R. (2022). Effect of addition of glass fibre on sugar cane bagasse ash under compressive loading. Materials Today: Proceedings. 61, 1109-14. https://doi.org/10.1016/j.matpr.2021.11.063

Chapter 43

Ex vivo experimental needle insertion force model for minimally invasive surgery

Ranjit Barua[1,a], Surajit Das[2,b], Amit Roychowdhury[1,c] and Pallab Datta[3,d]

[1]Iiest-Shibpur, India

[2]R. G. Kar Medical College and Hospital, India

[3]Niper Kolkata, India

Abstract

The fracture strength of the soft biological tissue aimed by the surgical needles acts as a serious character in clinical processes, for example, robotic-controlled needle steering, catheter insert, suturing, biopsy, etc. Despite the several investigational mechanisms on the fracture toughness of hard tissues, for example, dental tissue, bone, etc., only a very restricted number of investigations have concentrated on soft biological tissues, wherever the effects do not display any constancy mostly because of the casualness of the inserting/puncturing needle geometry. Tissue deformation during needle insertion is one potential source of inaccuracy. In modern medical science, the application of minimally invasive (MI) interventional tools for soft tissue surgery potentially reduced trauma, less blood loss, and less recovery times for the patient. One of the biggest challenges is tissue deformation affected by needle penetrating force, leading the undesired target engagements. The flexibility of soft tissue can create inaccurate needle insertions. With the purpose of the report on this matter, we experimented on the surgical needle insertion investigations on *ex vivo* duck livers with different diameters of the needle. A distinctive value for fracture toughness was attained for the *ex vivo* duck liver tissue by appropriately a link to the stiffness values assessed from the set of insertion experimentations. To confirm the investigational outcomes, a finite element model (FEM) of the *ex vivo* liver was established, and its hyper-viscoelastic properties were projected through an energy-based fracture mechanics method. The perforation forces assessed from the FEM simulations display an outstanding agreement with those developed from the physical investigations for all surgical needle geometries.

Keywords: Needle, microsurgery, stiffness, soft tissue, ex vivo, bevel, fracture mechanics

Introduction

In current medical and healthcare expertise, minimally invasive surgery (MIS) has a significant part, because it has the benefits of being a lesser amount of trauma and having minimum recovery time [1-3]. A surgical needle is the key tool of a tissue biopsy for medical practice. To create a precise clinical diagnosis of biological tissue, it is important to acquire large and thorough biopsy samples. To care for healthy soft tissue, clinicians have complete control over the speed and exact of the surgical needle [4]. Thus, several previous investigators have considered computational approaches (simulation) concerning perforation biopsies processes, which can make available references for the real insertion procedure [5-7]. The Ansys-based simulation was used to investigate the deformation of soft tissue throughout the needle insertion procedure [8]. Another simulation study was done for analysing the optimum needle force [9]. A 2D insertion model was established to simulate the procedure of surgical needle puncturing into soft tissue by using ABAQUS [10], and a PVA gel was used in place of soft tissue and established a finite element model (FEM) to simulate the surgical needle and soft tissue interface during the insertion process. As complex elements in soft tissue result in the difficulty and variety of its biomechanical characteristics. In recent times, investigators have established methods for needle navigation and identified simulations for needle piloting [11-14]. Many needle-directing methods have been executed in tissue mimic hydrogels, e.g., [15-17], with a small number of investigations in actual biological tissues.

In this study, we have described the effects of insertion speed and needle tip irregularity in *ex vivo* duck liver tissue, and analysed the possible errors in the insertion-based model. We use a needle direction-finding technique that integrates with surgical hypodermic needle shafts with bevel and blunt tips. This study offers the significant factors of surgical needle insertion strategy adaptable to soft biological *ex vivo* liver tissue. Fresh *ex vivo* duck liver was selected as the soft tissue for these investigations. We have made a portable robotically controlled surgical needle piloting device at IIEST Shibpur, CHST-Heaton Hall Lab, for the control and performance of needle insertion in *ex vivo* biological tissue (Figure 43.1), this portable needle insertion device can steadily or rotating needle insertion and assisting needle direction-finding. This arrangement was aimed at portability and correction and being made of effortlessly attachable subparts.

[a]ranjitjgec007@gmail.com, [b]drmedico49@gmail.com, [c]mechanicaladditiveengg@gmail.com, [d]pablomech07@gmail.com

DOI: 10.1201/9781003450252-43

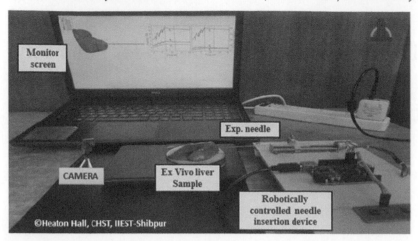

Figure 43.1 Ex vivo needle insertion experimental setup
Source: Author

The insertion speed of the surgical needle is governed by the linear movable (forward and backward) stage and the six DOF F/T sensor was attached with this device.

Materials and Methods

Force Modelling of Ex Vivo Insertion

The *ex vivo* surgical needle insertion study offered in this investigation estimates the insert force with depth of insertion. Total puncturing force (F_x) is the summation of cutting force (f_c) stiffness force (f_s) and friction force (f_f) [10].

$$F_x(y) = \Sigma_{i=f_s+f_f+f_c} f_i \tag{1}$$

At the start, as the needle try to insets the external area of the *ex vivo* tissue, an opposite force arisen due to the tissue's stiffness properties, and it will be dependent on penetration length of the needle insertion (l_i), radius of deformaed tissue (t_d), and reduced modulus ($\grave{E}_r$) [2].

Therefore,

$$f_s(y) = 2t_d\grave{E}_r l_i \tag{2}$$

Reduced modulus ($\grave{E}_r$) can be obtained from the material properties of the surgical needle, and experimental *ex vivo* tissue [14].

$$\grave{E}_r = \frac{1-v_n^2}{\epsilon_n} + \frac{1-v_t^2}{\epsilon_t} \tag{3}$$

[v_n, v_t = poission ratio of exp.needle and exp.tissue;
$\in_n, \in_t$ = Young's modulus of exp.needle and exp.tissue

During insertion, the friction force will be [10],

$$f_f = \mu F_N \tag{4}$$

In this study, each individual experiments was done with constant cutting speed.
Therefore, total puncturing force (F_x) will be [14]:

$$F_x = \begin{cases} f_s & y_I \leq y \leq y_{II} \\ f_f + f_c = \mu F_N + C & y_{II} \leq y \leq y_{III} \\ f_f = \mu F_N & y_{III} \leq y \leq y_{IV} \end{cases} \tag{5}$$

Simulation Model

The hyper-viscoelastic model applied by FEM operated well for the *ex vivo* biological tissue insertion and contact point analysis (Figure 2a, b). Additional intricate physical simulations may be replaced so as to interpretation for these enormous strains and possibly nonlinear hyper-viscoelastic characteristics, devoid of distressing the insertion force dissemination modeling approach offered at this point. Confined *ex vivo* biological tissue stresses as a result of the experimental needle insertion force involvement are impartially small, therefore, the axial strains are minimum. The expected insertion force transfers are responsible for the boundary conditions of insertion force. For the FEM simulation model, it is not essential to analyse the motion of nodules that are not observable or the applied forces on nodules that are not in continual interaction with the needle [5, 12]. Needle force dissemination is used irrespective of linear insertion velocity.

The FEM simulation is executed concerning a vol. of mesh (tetrahedron) comprised of 3250 elements using (Young modulus) E = 450kPa and Poisson's ratio ν = 0.397 to simulate the hyper-viscoelastic model (tissue mimic). The length of the inflexible needle is 10 cm elongated of five beam components, r = 0.2 cm. Let, the puncturing force is $F_x(z_x)$ (nonlinear function); the nodal point of the finite element meshes are (z_x). The insertion period starts as soon as the mediocre adjacent force influences a puncturing force (F_x) [17]:

$$\dot{C}_i > N_i F_X \tag{6}$$

$$\dot{C}_i = \Sigma \dot{N}_i^\theta \psi_i^\theta \tag{7}$$

The number of puncturing constants is N_i and the preliminary path of the needle contact point is $\dot{C}_i$.

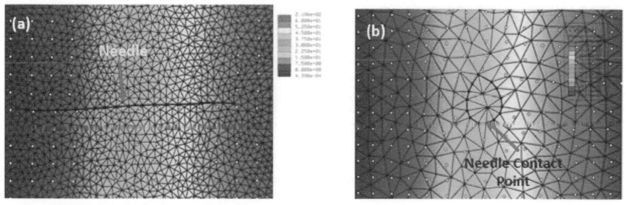

Figure 43.2 FEM simulation model of (a) needle insertion process; and (b) needle- hyper-viscoelastic material model contact point
Source: Author

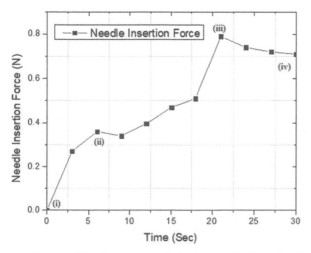

Figure 43.3 Experimental *ex vivo* needle insertion force profile
Source: Author

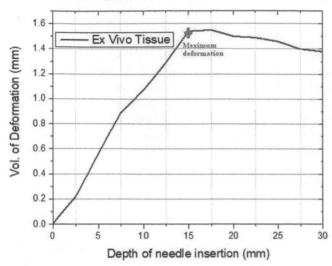

Figure 43.4 issue deformation profile during ex vivo needle insertion
Source: Author

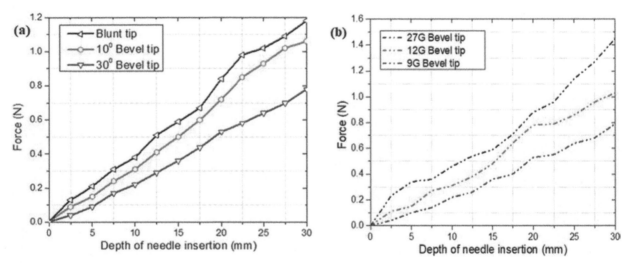

Figure 43.5 Insertion force profile with (a) different needle tip; and (b) different diameter bevel needle during *ex vivo* needle insertion
Source: Author

Ressult and Discussion

Three different segments occurs (Figure 43.3) of needle *ex vivo* biological tissue interaction for the period of *ex vivo* needle insertion process: (A) Tissue deformation period (i-ii): In this period the experimental needle move towards to *ex vivo* biological tissue; after certain time, the applied needle force hit the highest point value and perforate the tissue (ii). (B) Needle insertion period (ii-iii): In this period, the actual cutting force is sum total of insertion and friction force. (C) Withdrawal period (iii-iv): In this period, the surgical needle is taken out from the *ex vivo* tissue. The removal force is caused by the friction force as there is no insertion in this period. It was also seen that, deformation of the Ex Vivo liver tissue gradually increases till the needle punctures, and after that it start decreasing (Figure 43.4).

To study the influence of geometries of surgical needle on the needle insertion process, especially insertion force estimate, we select surgical hypodermic needles with different shapes of tip, different diameters of bevel needle, and different tip angles. The conduct test for every issue is finished five times and every trial is performed in dissimilar spots of the same *ex vivo* biological tissue sample at 5 mm/s constant speed with the purpose of steer clear of the effect of other issues. Figure 43.5a, b shows the insertion force report, where the insertion force increases with needle tip angle, and bevel shaft diameter. In case of blunt needle, force requirement is maximum rather than other needle tip for insertion. We investigated that how needle speed effect the insertion, therefore we experimented three different speeds (2 mm/s, 5 mm/s, and 7 mm/s); and maximum insertion speed was observed at 7 mm/s (Figure 43.6).

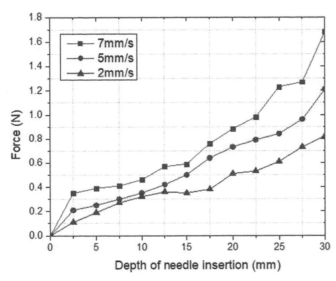

Figure 43.6 Insertion force profile with different insertion speed during *ex vivo* needle insertion
Source: Author

Conclusion

Here, we have done experimentations to examine the properties of the velocity of *ex vivo* needle insertion, and also analysed the effect of asymmetry of the needle tip and diameter of the needle shaft. This investigation provides proof of the idea that steerable surgical needles can be applied to complete valuable clinical and medical tasks, even though much work remains to improve methods that can be applied to patients. These investigations could be enhanced by more precise needle steering methods, superior experimental data, and developed approaches for modern medical imaging technology. This study is the first footstep in emerging that understanding and, by tallying clinical significance, it simplifies the entire needle navigating method that incorporates insertion mechanism, path monitoring, and geometry of needle tip. The forthcoming work comprises describing needle navigation in numerous kinds of clinically appropriate biological tissue, for example, kidney, prostate, etc.

Acknowledgment

The authors would like to thank the Centre for Healthcare Science and Technology, Heaton Hall Lab, IIEST-Shibpur, India.

References

[1] Abolhassani, N., Patel, R. V., and Ayazi, F. (2007). Minimization of needle deflection in robot-assisted percutaneous therapy. The International Journal of Medical Robotics + Computer Assisted Surgery: MRCAS. 3(2), 140–48. https://doi.org/10.1002/rcs.136

[2] Barua, R., Datta, S., RoyChowdhury, A., and Datta, P. (2022). Study of the surgical needle and biological soft tissue interaction phenomenon during insertion process for medical application: A Survey. Proceedings of the Institution of Mechanical Engineers. Part H, Journal of Engineering in Medicine. 236(10), 1465–77. https://doi.org/10.1177/09544119221122024

[3] Barua, R., Das, S., Datta, P., and Roy Chowdhury, A. (2022). Computational FEM Application on Percutaneous Nephrolithotomy (PCNL) Minimum Invasive Surgery Through Needle Insertion Process. In P. Pain, S. Banerjee, & G. Bose (Ed.), Advances in Computational Approaches in Biomechanics (pp. 210-222). IGI Global. https://doi.org/10.4018/978-1-7998-9078-2.ch013

[4] Sadjadi, H., Hashtrudi-Zaad, K., and Fichtinger, G. (2014). Needle deflection estimation: prostate brachytherapy phantom experiments. International Journal of Computer Assisted Radiology and Surgery. 9(6), 921–29. https://doi.org/10.1007/s11548-014-0985-0

[5] Barua, R., Das, S., Datta, S., Datta, P., and Roy Chowdhury, A. (2023). Study and experimental investigation of insertion force modeling and tissue deformation phenomenon during surgical needle-soft tissue interaction. Proceedings of the Institution of Mechanical Engineers, Part C: Journal of Mechanical Engineering Science. 237(5), 1007-14. doi:10.1177/09544062221126628

[6] van Veen, Y. R., Jahya, A., and Misra, S. (2012). Macroscopic and microscopic observations of needle insertion into gels. Proceedings of the Institution of Mechanical Engineers. Part H, Journal of Engineering in Medicine. 226(6), 441–49. https://doi.org/10.1177/0954411912443207

[7] Barua, R., Das, S., Roy Chowdhury, A., and Datta, P. (2023). Experimental and simulation investigation of surgical needle insertion into soft tissue mimic biomaterial for minimally invasive surgery (MIS). Proceedings of the Institution of Mechanical Engineers. Part H, Journal of Engineering in Medicine. 237(2), 254–64. https://doi.org/10.1177/09544119221143860

[8] Chebolu, A. and Mallimoggala A. (2014). Modelling of cutting force and deflection of medical needles with different tip geometries. Procedia Materials Science. 5, 2023-31.

[9] Rossa, C., Sloboda, R., Usmani, N., and Tavakoli, M. (2016). Estimating needle tip deflection in biological tissue from a single transverse ultrasound image: application to brachytherapy. International Journal of Computer Assisted Radiology and Surgery. 11(7), 1347–59. https://doi.org/10.1007/s11548-015-1329-4

[10] Barua, R., Giria, H., Datta, S., Roy Chowdhury, A., and Datta, P. (2020). Force modeling to develop a novel method for fabrication of hollow channels inside a gel structure. Proceedings of the Institution of Mechanical Engineers. Part H, Journal of Engineering in Medicine. 234(2), 223–31. https://doi.org/10.1177/0954411919891654

[11] Zhao C, Zeng Q, and Liu H. (2021). Soft tissue deformation ANSYS simulation of robot-assisted percutaneous surgery. InProceedings of the 10th World Congress on Intelligent Control and Automation 2012 Jul 6 (pp. 3561-3566). IEEE.

[12] Barua, R., Das, S., RoyChowdhury, A., and Datta, P. (2023). Simulation and experimental investigation of the surgical needle deflection model during the rotational and steady insertion process. The International Journal of Artificial Organs. 46(1), 40–51. https://doi.org/10.1177/03913988221136154

[13] Bunni, S. and Nieminen, H. J. (2022). Needle bevel geometry influences the flexural deflection magnitude in ultrasound-enhanced fine-needle biopsy. Scientific Reports. 12(1), 17096. https://doi.org/10.1038/s41598-022-20161-3

[14] Barua, R., Das, S., Datta, S., Datta, P., and Chowdhury, A. R. (2022). Analysis of surgical needle insertion modeling and viscoelastic tissue material interaction for minimally invasive surgery (MIS). Materials Today: Proceedings. 57, 259-64.

[15] Li, A. D. R., Plott, J., Chen, L., Montgomery, J. S., and Shih, A. (2020). Needle deflection and tissue sampling length in needle biopsy. Journal of the Mechanical Behavior of Biomedical Materials. 104, 103632. https://doi.org/10.1016/j.jmbbm.2020.103632

[16] Zhang, L., Li, C., Fan, Y., Zhang, X., and Zhao, J. (2021). Physician-Friendly Tool Center Point Calibration Method for Robot-Assisted Puncture Surgery. Sensors. 21(2), 366. https://doi.org/10.3390/s21020366

[17] Barua, R., Das, S., Datta, S., Roy Chowdhury, A., and Datta, P. (2022). Experimental Study of the Robotically Controlled Surgical Needle Insertion for Analysis of the Minimum Invasive Process. In Emergent Converging Technologies and Biomedical Systems, pp. 473-82. Singapore, Springer.

Chapter 44

A compressive approach in design and analysis of GO-KART

Akshay Anjikar[a] and Shubhangi Gurway[b]

Priyadarshini Bhagwati College of Engineering, Nagpur, India

Abstract

Formula one racing cars like Go-Kart are gaining popularity day by day attracting the users and researchers in its designing and performance-based experiments. Present paper showcases the designing and calculation of all the important components of the generalized Go-Kart system taking utmost care of driver's safety and vehicle durability concern. The major emphasis has been given to the design of safe and functional vehicle based on rigid and torsion free frame. Similarly, the driver's comfort and safety with increased manoeuvrability of the vehicle is again one of the prime motives of the designed model. The designing of various component of Go-Kart right from chassis to steering system to braking system to complete transmission have been covered by considering higher speed with constant fluctuation. The common problem faced by the drivers have also been taken into consideration during design calculation. Electronic failure indication system has been applied in the system which is the new updation to present Go-Kart in the market. The designed components are drafted using CREO is a 3D modelled software package which help us to apply model based approach from designing of different parts of the cart to manufacturing of it and analysed using ANSYS software.

Keywords: GO-KART, design, vehicle durability, driver safety

Introduction

A GO-KART is a no suspension and no differential racing vehicle available in wide range from motor less models to high-power racing machines. The engine as a heart of this system is of type internal combustion, solar, hybrid electric powered. Similarly, due to extensive pollution growth concept of electric powered increasing day by day [1]. On the other side of the design driver's safety and vehicle's durability is again one of the important concerns for the designers.so our main objective of this work is to accommodate the designing and fabrication of Go-Kart and analyses the deformation if any during the designing process.

Review of Literature and Gap Analysis

Significant amount of work has been done in designing and analysis of GO-KART in last few years most of which include the material selection, designing of chassis and other components of the system. Table 44.1 represents the work that have been perform by different authors till 2022 [2-9].

Through literature review it is observed that none of the author have used any electronic component which automatically look out for driver's safety. In our design we have implemented electronic failure indication system for validation of right working of the components which not only increase the system performance but improve driver's safety too.

Design and Calculation

This is the crucial part of the work where selection of materials, components and methods emphasis on major part of the final performance of the kart to be fabricated. In this part of work designing of the different part of the system has been done considering all the standard procedures.

Chassis

The prior step of making the chassis is to find the positions of the components to be fitted on the kart Figure 44.1. After taking all the positions the layout of the chassis should be formed considering the following points:-

- The positioning of the linkages according to transfer of the forces while the vehicle is in motion.
- Providing necessary trusses to make the chassis robust at the where possibility of occurring the failure due to stresses is more.
- The chassis should made compact as much as possible considering the driver comfort.

[a]akshayanjikar@gmail.com, [b]shubhangi.pbcoe@gmail.com

DOI: 10.1201/9781003450252-44

Table 44.1 Detail review of the work done on go-kart design and analysis till present

Sr. No	Paper title	Name of author	Year of publication	Remark	Software used
1	Design and Development of Foldable Kart Chassis	Akash Chaudhary Raghuvanshi et al.	2015	Foldable joint using male and female assembly have been made	SOLIDWORKS 2010 have been used for designing and analysis of the foldable joint.
2	Design, analysis and fabrication of go-kart	Kiral Lal, Abhishek O S	2016	Major emphasis were given on the material selection and designing of frame.	INVENTER software have been used for conducting Frame analysis
3	Design and Analysis of Go-Kart using Finite Element Method	Harshal D. Patil et al.	2016	The paper gives major emphasis on vehicle compactness and high fuel economy.	The design has been modelled in CATIA V5R21, the analysis was done in ANSYS 14.5 and simulation in ADAMS14.
4	Design & Analysis of GOKART chassis	Sanket Nawade et al.	2018	Chassis have been design considering driver's safety including the over weighted diver too considering over influenced position.	CATIA5R18 software have been used for designing the chassis and ANSYSY 14.5 software have been used for analyzing the Optimized strength condition.
5	*Impact behaviour analysis of a newly designed go-kart chassis*	Prateek Mahapatra et al.	2020	This work design Go-kart chassis from three different materials i.e., AISI-4130, Al-6061 and CFRP composite. These Go-kart chassis were tested under different loading conditions.	AutoDesk Fusion 360 software were used in the finite element analysis of three-dimensional model of go-kart
6	Numerical study on strength optimization of Go-Kart roll-cage using different materials and pipe thickness.	Jay Prakash Srivastava	2021	Major emphasis has been given on Roll-cage considering different combination of material grades and pipe thickness.	CAD software SOLIDWORKS version 2018 have been used in chassis design and ANSYS 16.0 have been used or analysis.
7	Comparison of chassis frame design of Go-Kart vehicle powered by internal combustion engine and electric motor.	Naveen Kumar Chandramohan et al.	2021	This paper has convert an IC engine Go-kart into an Electric driven kart where combination of Battery, Motor and Controller and hence it requires a large number of mounts and cross members to support the components	SOLIDWORKS have been used for designing mounts for converted electric driven cast and ANSYS software have been used for analysis.
8	Design and Analysis of Electric Go-kart Suspension	Abhishek Virupaksh Khalipe et al.	2022	Electric Go-Kart have been designed to offers better foothold with street surface and steadier on or rough terrain	CatiaV5 R20 have been used for designing and ANSYS for analysis.

Source: Author

We have made the conceptual design of the chassis according to the points mentioned above. Further the chassis is analysed in the ANSYS software. We have considered both primary as well as conceptual design principles for our project and we preferred to select some of the principals of primary design.

The dimension of chassis according to primary design is:

1. Maximum longitudinal dimension= 56 inches.
2. Maximum transverse dimension = 29 inches.
3. Diameter of chassis material = 1 inch.
4. Thickness of chassis material = 2 mm.

Calculation for forces which are impacted on chassis in various testing.

The chassis is to be analysed with three different impact testing which are as follows.

- Front impact testing
- Rear impact testing
- Side impact testing

The forces are calculated by considering the kinetic energy of the vehicle in movement that is also the work done and we know that work done is the product of force and the displacement and displacement is the product of velocity and the time. Mathematically,

$$WD = \tfrac{1}{2}\, mv^2$$

- Mass of the vehicle including driver = 170 kg
- Vehicle velocity =16.66 m/s
- Impact time (assumed) = 0.13sec

$$WD = 1/2 * 170 * 16.66^2 = \mathbf{23592.22\ J}$$

- Front impact force: The front impact testing is the testing of chassis with forces occurred when the chassis is impacted at front section keeping the rear link fixed. The reaction time considered on the front impact is 0.13 by trial and error basis.

$$WD = F * (t*v)$$
$$34000 = F*0.13*16.66$$
$$\boxed{F = 10922.32\ N}$$

- Side impact test: The side impact testing is the testing of chassis with forces occurred when the chassis is impacted at side section keeping the link of other side fixed. The reaction time considered on the front impact is 0.28 by trial-and-error basis.

$$WD = (F* \text{ displacement})$$
$$23592.22 = F*0.28*16.6 = 5000\ N$$

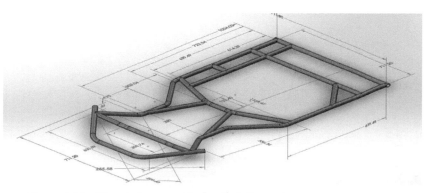

Figure 44.1 Chassis layout with detailed dimensions
Source: Author

- Rear impact test: The rear impact testing is the testing of chassis with forces occurred when the chassis is impacted at rear section keeping the front link fixed. The reaction time considered on the front impact is 0.13 by trial-and-error basis.

$$= F * (t*v)$$
$$23592.22 = F*0.13*16.66 = 10922.32N$$

Further the chassis is tested using above forces in ANSYS software to find out stress incorporated in the chassis. Following are the results of stresses occurred and factor of safety achieved in the chassis after testing in ANSYS. Similarly the detailed analytical result of stress generation is represented in Table 44.2.

Steering System

The steering system is basically used for negotiating a turn of a vehicle. In Go-Kart, a simple mechanical linkage type steering system is used with inversion of four bar chain mechanisms. The mechanism we used is an Ackermann steering mechanism, shown in Figures 44.2 and 44.3. Because of its lesser linkages and simple mechanism. The steering calculations was done in two parts first is the aligning torque calculation which is required for how much efforts do a driver required to apply for the turning and other is steering geometry calculations which is the calculation for the turning radius and the Ackermann percentage.

L = 119.05 (calculated)
M = 1.3576
Mechanical Trail: m = Ml= L*m = 119.05*1.3576*25.4 = 4105.206
Mt = Ml + Mr cos $\sqrt{\lambda^2 + v^2}$ = 8058.174
Aligning torque is same on both the wheels. Therefore Ml = Mr
Therefore,

Table 44.2 Analytical Result of Stress generated in chassis

Tests	Ultimate stress	Developed stress	Deformation	Factor of safety
Front impact test	664N/mm²	464.3	5.72mm	1.43
Side impact test	664N/mm²	316.19	1mm	2.1
Rear impact test	664N/mm²	539.83	15.85mm	1.23

Source: Author

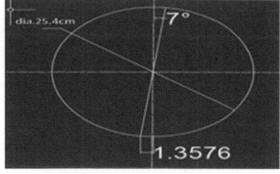

Figure 44.2 Align steering system
Source: Author

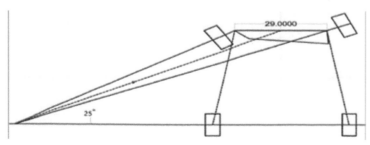

Figure 44.3 Steering geometry
Source: Author

Mt = Total aligning force = 8058.174Nmm

Turning angle diagram

- Turning radius:
- Ackermann angle (α) = $\tan^{-1}\left(\frac{TrackWidth}{2*WheelBase}\right)$ = 25°

- Inner angle (i) = $\tan^{-1}\left[\frac{H}{R-\left(\frac{W}{2}\right)}\right]$ i = 33.89°

- Outer Angle (j) = $\tan^{-1}\left[\frac{H}{R+\left(\frac{W}{2}\right)}\right]$ j = 22.29°

Where,
Wb = Wheelbase = 1041.3 mm.
Tw = Track width = 965.1 mm
R= Radius of curvature.

- Turning radius (R$_t$) = $\frac{Tw}{2} + \frac{Wb}{\sin(\gamma)}$

Where,

γ = Average steer angle

$\gamma = \frac{i+j}{2}$

γ = 28.090°

Turning radius (R$_t$) = 2.69

- Ackerman% = 46.4%

*Steering column length:-*The length of the steering column does not make any impact on the steering geometry and aligning but it is important considering the ergonomics of the driver. And driver comfort. To calculate the steering column length, we have set the inclination of the steering column as 45° for easy handling of the steering column. The height of the steering wheel from the floor plan is found by considering the drivers elbow angle position of the driver. The height of steering wheel which comes 13.66 inches. The steering column length comes as 19.32 inches by taking sine angle of the steering column angle which represented in Figure 44.4 of the work.

For calculation we have assumes Caster angle as 7°, Chamber Angle 0°, king pin inclination 14°, combined angle 14

Braking System Design

The braking system should provide enough braking force.

In our work we have used disc brake having some advantage over drum brake. The braking system of a go-kart is designed in three steps which are as follows:

- Finding the required clamping force which will retard the speed of the vehicle to zero m/s
- Selecting the components to be used according to the clamping force required the components are tandum master cylinder, braking calliper and the brake disc.

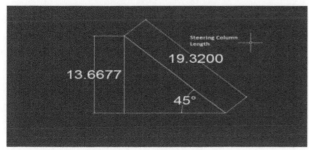

Figure 44.4 Steering column length
Source: Author

- Calculating the actual clamping force and braking torque produced with respect to the components selected.
- Static weight distribution:
  Rear: Front = 59:41
  Weight at front end = 71.4
  Weight at rear end = 98.6 kg
- Weight transferred to front end:
  = C_f * vehicle mass * height of centre of gravity/Wb = 57.84 kg
  Dynamic weight distribution:
- Rear: Front = 40:60
  Kart weight at front axle =102N
  Kart weight at rear axle = 68N
- Braking force = 170*6.25 = **1063N**
- Locking force = 68*0.9* 9.8=**600.372N**
- Torque=locking force* radius of tire = 600.372* 0.139 = **83.42Nm**
- Clamping force = torque/(number of friction surfaces*Cf*radius of disc)=87.9/(2*0.9*0.09)
  Required clamping force = 542.593N
- Selection of components: The components which are selected on the basis of forces required, cost and compatibility including braking specifications are represented in Table 44.3.

Engine

The engine of a Go-Kart can be a four stroke SI engine with a constant mesh gear box or a lawn mower engine using CVT. The power is transmitted from engine to live axle through a chain drive. The engine we deployed in our Kart is 124.7 CC Honda Stunner engine, with constant meshing gear box it has 28.5 kg weight with single cylinder. As the Go-Kart is the racing vehicle and hence the engine and transmission of the vehicle should be producing more power and torque with In the given limit of the CC of the engine and hence we have used the engine of HONDA Stunner.

Transmission

The transmission system comprises of three different parts namely gearbox, chain drive and rear driving shaft.

- Gearbox
  Sequential gears are generally preferable for racing car. As we have employed Honda Stunner engine for our work, following is the gear ratio of the gear box at different gear stages shown in Table 4.
- Chain drive
  Chain drive is basically a way through which the mechanical power is being transmitted. In Go-kart we have used roller bush type chain drive to transmit the power from engine to the rear axle by checking and considering its capacity to transmit power.
- Driven sprocket calculation:

Table 44.3 Braking specification

Type	Specification
Rear disc OD	180 mm
Effective disc thickness	3 mm
Master cylinder dia.	10 mm
Calliper piston diameter	25.4 mm
Brake pedal: Lever	4:1
Pedal force	100 N
Stopping time	2 sec
Stopping distance	12.5 m/s
Deceleration	6.25 m/s2
Braking force	2322.57 N
Braking torque	208.98 Nm

Source: Author

Table 44.4 Gear ratio

Gear no	I st	II nd	III rd	IV th	V th
Ratio	3.1	2	1.5	1.2	1.1

Source: Author

The teeth on the driver sprocket are selected as 14 for smooth movement of chain. The velocity ratio of the sprocket is 1made considering the weight of the vehicle and the maximum speed that can be achieved accordingly.

So, the optimum velocity ratio is selected as 1.57.

We know that velocity ratio is the ratio of teeth on the driven sprocket to the teeth on the driver sprocket. Mathematically:

$$\text{Velocity ratio} = \frac{\text{Teeth on driven sprocket}}{\text{Teeth on driver sprocket}}$$

$$1.5 = \frac{\text{Teeth on driven sprocket}}{14}$$

Teeth on driven sprocket = 21

But no of teeth on sprocket should always be in even numbers therefore the no of teeth is rounded up as 22.

Teeth on driven sprocket = 22

- Load on chain
  Calculated pitch line velocity = 9.078 m/sec
  Load on chain = Rated power/calculated pitch line velocity = $6.54*10^3/9.078$ = **720.423 N.**
  The rear axle
  The rear axle of a go kart is a hollow composite shaft made up of a mild steel material. The shaft is having hubs fitted at both the ends for attachment of the wheels.

- Design of rear axle: Considering the action of twisting moment and bending moment on the axle by assembly of power train and disc brake. The load diagram representing the load of various design component hence proved to be helpful for designing of rear axle which is represented in Figures 44.5 and 44.6. Bending moment on the shaft

- Bearing reactions
  R_a+R_b = 720.423 + 2322.576 + 500.31 + 500.31
  R_a+R_b = 4043.619 N........(1)
  Taking moment at point A=0
  720.423*203.2 + 2322.576*660.4 + 500.31*154.2 + 500.31*787.4 = R_b*939.8
  R_b = **2289.116 N**
  Therefore,
  R_a =4045.81- R_b
  R_a = **1756.694 N**

- Calculating bending moment
  M_a = 0
  M_c = 1756.694*152.4 = $267.72*10^3$ Nmm
  M_d = $267.72*10^3$+(1255.96*50.8) = $331.52*10^3$ Nmm
  M_e = $331.52*10^3$+(533.34*457.2) = $575.36*10^3$ Nmm
  M_f = $575.36*10^3$+(-1789.23*127) = $348.13*10^3$
  $M_{b=0}$
  Considering highest bending moment =M
  M=$575.36*10^3$Nmm

- key and keyway design
  By considering twisting moment
  $T = \frac{\pi}{16} \times \iota \times d_o^3$
  Where,
  T = Twisting moment

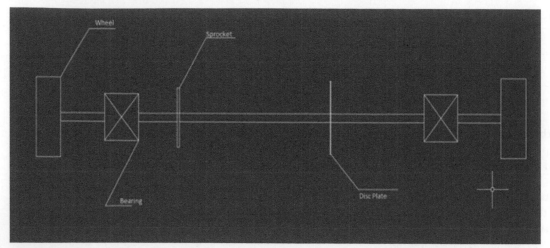

Figure 44.5 Front view of shaft
Source: Author

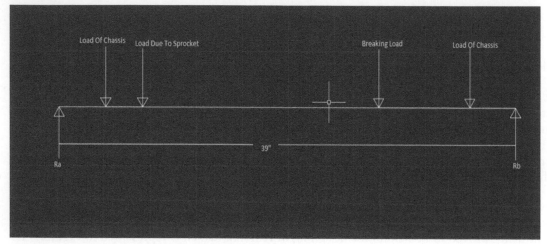

Figure 44.6 Load diagram of shaft
Source: Author

ι = Shear stress.
= 129.99 N/mm2
d_o = outer diameter.
T= $\frac{\pi}{16}$ × 129.99 × 30
T = 765.763 × 10³

By empirical relations
W=width of key = d_o/5
              = 30/5
              = 6 mm
t=thickness of key = d_o/6
              = 30/6
              = 5 mm

- Effect of keyway on strength of shaft
  We have,
  e = 1-0.2(w/d)-1.1(h/d)
  Where,
  W = keyway width, d_o = shaft diameter
  h = keyway depth = 1/2 × t/2
  h = 1/2 × 5/2 = 1.25mm
  e = 1-0.2(6/30)-1.1(1.25/30)
  e = 0.916

Electrical System

The electrical system in our kart was designed to support the mandatory safety equipment's and for instrumentation in particular self-start system. The electrical system we designed for our work mainly comprises of 12 v battery, alternator, rectifier, relay, CDI, ignition coil and spark plug. The main source of power supply is from 12 V battery and alternator. We have divided the wiring in three main parts i.e., the two main power source of the wiring and the junction i.e., relay. Similarly for safety purpose we have used kill switches and self-start system. Figure 44.7 will represent the electrical circuit of the Go Kart.

Electronic Failure Indication System

EFIS is very small device. As we connect EFIS to three components of our electrical circuit it consists of Relays 12VDC (3), LED Red 3V, 30 mA (3), LED Green 3V, 30 mA (3), Resistor 330 ohm (3) and external battery 9V (1)

The main part of EFIS is the relay which operates the two circuit. Relay has five connection points. Both the LED's (i.e., red and green) positive terminal attach to the two points of relay (NC &NO) and negative terminal of LED attach to the battery of 9V supply. The positive terminal of battery is attaching to the relay via resistor. As our LED is of 3 V and 30 mA, we attach resistor of 330 ohm to light up the LED on 9 V supply. The remaining terminal or points of relay are attached to the component of any electrical circuit of vehicle (i.e., Alternator, relay and Kill switch). Until the supply from this component is ON our relay

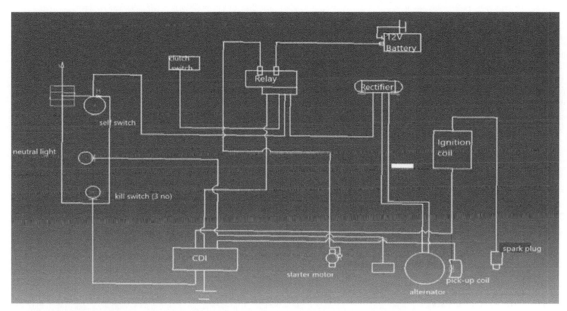

Figure 44.7 Electrical system of Go-Kart
Source: Author

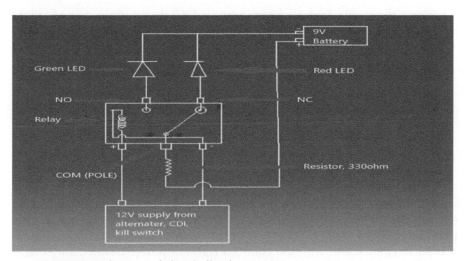

Figure 44.8 Electronic failure indication system
Source: Author

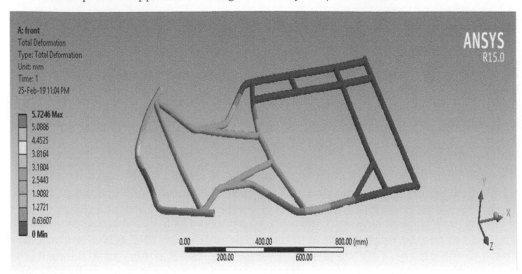

Figure 44.9 Chassis analysis
Source: Author

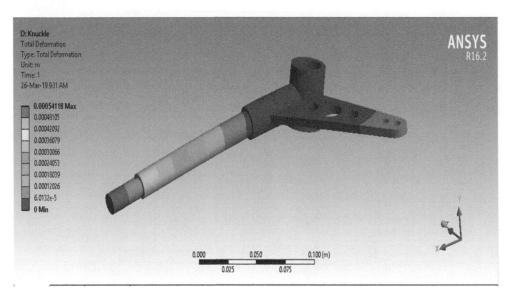

Figure 44.10 Stub axle analysis
Source: Author

light up the Green LED and whenever the supply from the component is OFF that is in the case of failure of this component our relay light up the Red LED. Figure 8 represent the circuit diagram of designed EFIS in our work.

Analysis

The analysis of following parts were done after drafting of it. The software use for drafting was CREO and for analysis was ANSYS.

- Chassis: The analysis of chassis was done in three different steps which are front impact testing, side impact testing and the rear impact testing. Which was done by applying forces calculated earlier. The detail validation is shown in Figure 44.9.
- Stub axle: The stub axels are the front axels on which the front wheels are fitted the analysis of stub axle was done by applying the aligning force of the wheels on its one end calculated in earlier. The detail validation is shown in Figure 44.10.
- Sprocket: The sprocket is used in conjunction with the chain to transmit power from engine to rear axle the analysis of the sprocket was done by applying the tooth load on sprocket calculated earlier. The detail validation is shown in Figure 44.11.

Figure 44.11 Sprocket analysis
Source: Author

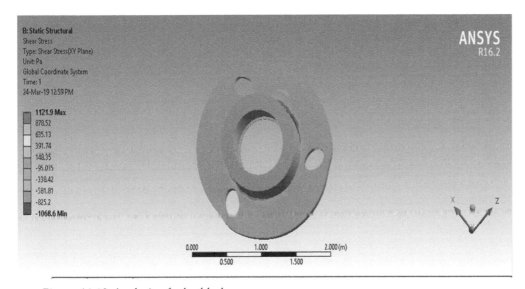

Figure 44.12 Analysis of wheel hub
Source: Author

• Wheel hubs: The wheel hubs are used for fitting wheels on the rear driving axle the analysis of wheel hubs was done by application of the torque on the shaft on its one end the calculation of torque is mentioned in design and calculation part. The detail validation is shown in Figure 44.12.

Conclusion

As Go-Kart is tremendously growing its population day by day, attracting youngsters and sportsman towards it, need of its development with at most care should be taken considering driver's safety and vehicle durability. So, we have designed a Go-Kart considering all the design calculations, CREO software was used to draft the whole system. For drivers' safety concern we have implemented electronic failure indication system which improve the safety performance of the vehicle. All the components of the designed and drafted system were then analysed using ANSYS software where it was observed that all the designed part work well in future.

Reference

[1] Shaisundaram, V. S., Chandrasekaran, M., Shanmugam, M., Padmanabhan, S., Muraliraja, R., and Karikalan, L. (2019). Investigation of Momordica Charania seed biodiesel with cerium oxide nanoparticle on CI engine. International Journal of Ambient Energy. 1–5.

[2] R aghuvanshi, A. C., Srivastav, T., and Mishra, R. K. (2015). Design and development of foldable Kart chassis. Materials Today: Proceedings. 21707–1. doi: 10.1016/j.matpr.2015.07.004.

[3] Lal, K. and Abhishek, O. S. (2016). Design, analysis and fabrication of Go-Kart. International Journal of Scientific Engineering and Research. 7(4).

[4] Harshal, D., Patil, S. S. Bhange, R., an d Deshmukh, A. S. (2016). Design and analysis of Go-Kart using finite element method. International Journal of Innovation in Engineering Research. 3(1).

[5] Sanket, N. and Pathan, D. (2018). Design & Analysis of GOKART chassis. Internaional Journal of Innovations in Engineering & Science. 3(7).

[6] Mahapatra, P., Arora, G., Aggarwal, M., Singh, S., and Manocha, R. (2020). Impact behaviour analysis of a newly designed go-kart chassis, Indian Journal of Science and Technology. 13(23), 2336-44. doi: 10.17485/IJST/ v13i23.502

[7] Srivastava, J. P., Krishna Chaithanya, B., Sai Teja, K., Venugopal, B., Vineeth, S., Rajkumar, M., and Khan, H. (2020). Numerical study on strength optimization of Go-Kart roll-cage using different materials and pipe thickness. Materials Today: Proceedings. doi: 10.1016/j.matpr.2020.08.217

[8] Chandramohan, N. K., Shanmugam, M., Sathiyamurthy, S., Tamil Prabakaran, S., Saravanakumar, S., and Shaisundaram, V. S. (2020). Comparison of chassis frame design of Go-Kart vehicle powered by internal combustion engine and electric motor. Materials Today: Proceedings. doi:10.1016/j.matpr.2020.07.504

[9] Khalipe, A. V., Shedage, A. S., Pawar, M. K., Patil, S. J., and Edake, V. V. (2022). Design and analysis of electric go-kart suspension. International Journal of Advanced Research in Science, Communication and Technology. 2(8).

Chapter 45

Developing the pervious concrete using industrial waste and filter media setup for better refinement of storm water

Sanket Kalamkar[a] and Sanjay Raut[b]

Yeshvantrao chavan college of engineering, Nagpur, India

Abstract

Water problems have been one of the significant concerns. The scarcity of water is mainly due to rapid industrialisation. Increasing urbanisation and diminishing groundwater table we need to focus on sustainability groundwater level, and technologies such as pervious concrete to avoid drainage and evaporation of water. According to the study, runoff water can be treated by charcoal at a particular level while still being used for domestic tasks like washing or irrigation. The main aim of this research is to evaluate the pervious concrete using industrial waste. Treatment of previous concrete with 20% iron slag is studied and compared. Parameters like compressive strength, porosity, turbidity, pH, and hardness were measured with respect to pervious concrete. Pervious concrete is useful for maintaining hydrological stability naturally. It captures the stormwater instead of allowing it to flow in the drainage system. This water slowly percolates into the groundwater. Pervious concrete becomes an aid to the natural infiltration water. The experimental analysis in this research shows that before and after treatment of runoff water has given very satisfactory results. It shows that charcoal is very much effective in decreasing the impurity of runoff water. Switching to pervious concrete with iron slag as admixture allows water to get a sip from porous material directly along with water and allowing the groundwater recharge.

Introduction

A human life requires clean freshwater for daily work and normal living, but on the contrary people lack access to water, and on an average lot of people experience lack of water for one month throughout the year. Further water shortage could affect more and more people. People required to travel a distance for a basic need like drinking and washing. Ultimately this could affect the economic condition of any country. Pervious concrete allows adjacent trees to receive more air and water while still permitting full use of the pavement. Pervious concrete provides a solution for landscapers and architects who wish to use greenery in parking lots and paved urban areas [1]. All the water bodies are under threat from altering climatic patterns due to pollution. Additionally, industrialisation operations have a negative impact on biodiversity and natural resources, which contributes to their decline. Up until the 1990s, urban flooding was only thought to be a problem for local and municipal government, but today disaster and environmental experts are interested in it. Due to its extreme sensitivity and hazards, urban flooding has been classified as a disaster. Extreme mortality and significant economic losses result in every nation.

Urban flooding happens when storm water enters a city at a rate that is faster than it can be absorbed by the soil, transported to a body of water (lake, river, etc.), or collected in a reservoir.

pervious concrete pavement possesses many advantages that improve a city environment such as water treatment by pollutant removal [4]. Pervious concrete can be utilized to solve these problems by allowing some rainfall to seep into the ground, reducing the pollution brought on by the runoff water that is washed off covered surfaces in modern cities [2]. Portland cement pervious concrete is widely used and keeps attracting a lot of attention in the building sector. The void content of pervious concrete is typically high (15–25%). Cement, coarse aggregate, water and very few fine aggregates are used to create pervious concrete, which is distinguished by proper infiltration. Making efficient and productive use of industrial or agricultural waste can decrease potential environmental harm while increasing resource and energy efficiency. The layer of activated charcoal and iron slag below concrete furthermore is a powerful filtering material for elimination of impurities from storm runoff in cities. The development of fully permeable pavement designs as a potential best management practice for stormwater management in areas that carry heavy truck traffic is of significant interest [5]. Parking lots, low-traffic areas, pedestrian walkways, and greenhouses are typical locations for the use of pervious concrete, which helps to build structures that are more environmentally friendly.

Proposed Methodology

Formation of pervious concrete using mix design mix design is made using the reference of IS Code 10262-2019, IS 456 as shown in Table 45.1. The pervious concreter block was prepared with 20% of iron slag and

[a]sanketgk1@gmail.com; [b]sprce22@gmail.com

DOI: 10.1201/9781003450252-45

Table 45.1 Mix design of concrete [3]

Concrete quantity - 0.003365 m3 , Grade M25			
Sr. No.	Item	Amount	Unit
1	Cement	1.94	kg
2	Coarse aggregate (10 mm)	5.67	kg
3	Coarse aggregate (20 mm)	0.00	kg
4	Fine aggregate	0.00	kg
5	Slump	75 to 100	mm
6	Water	0.69	ltr
7	W/C ratio	0.36	

Figure 45.1 Concrete mix
Source: Author

Figure 45.2 Cube casting
Source: Author

Figure 45.3 Cube casting
Source: Author

Figure 45.4 Pervious concrete cubes
Source: Author

results are compared with the controlled specimens (Figures 45.1-45.4). Previous concrete usings iron slag and sand filters can effectively remove heavy metals from bodies of urban storm runoff when used independently [6]. Economical methods for enhancement of qualities of water and proper management of urban storm runoff are still uninvestigated [10]. A mechanical concrete mixer was used to make control samples of pervious concrete and samples of pervious concrete containing iron slag, which were then placed to the experimental molds (15 × 15 × 15 cm).

The developed concrete shown in figure was further cured for 28 days and various tests such as compressive strength, porosity test, have been performed on the developed concrete (Figures 45.5 and 45.6). Compressive strength is the ability of a material or structure to bear loads that tend to reduce size. In short compressive strength resist the compression and tensile strength resist tension. The greatest compressive load that a material can support before breaking is measured through compressive strength tests.

Figure 45.5 Compression test
Source: Author

Figure 45.6 Breaking of concrete cube
Source: Author

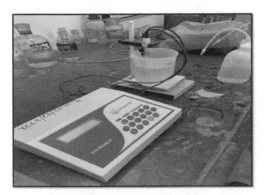

Figure 45.7 pH test apparatus
Source: Author

Figure 45.8 Hardness test
Source: Author

The size of the coarse aggregate, the void ratio, and the binding between the mortar and the coarse aggregate all affect compressive strength. Permeable concrete cubes gain 30% of their strength in 7 days, 70% of their strength in 21 days, and 95% of their strength in 28 days. Compressive strength test is done cubes on 7th day, 14th day and 28th day of curing period.

Pervious concrete pavement has been demonstrated as effective in managing runoff from paved surfaces [7]. Porosity is a measurement of the gaps in a material and a percentage of the total volume of empty space. The test entails applying a known hydrostatic pressure from one side to a concrete specimen of known size that is housed in a specially designed cell, measuring the amount of water that percolates through it over the course of a specified amount of time, and calculating the coefficient of permeability. The test enables measurement of both the water entering and exiting the sample.

Long term health issue is started due to some pollutant that is present in water. This type of pollutant takes a long time for its detection. Investigated were the contaminants and water quality that leached from pervious concrete pavement. The primary goal of this project is to examine how pollutants affect pervious concrete pavement. Water contaminants and water purification system is working or not is confirmed by regular testing. The pH test, hardness test and Turbidity test are conducted on three different pre and post purified water samples to find their pH, hardness, and turbidity.

The pH of a solution is a major indicator of its chemical composition. The pH can regulate nutritional availability, biological processes, microbial activity, and chemical behaviour. Water is acidic or basic is determined by pH scale. Groundwater flows through rocks and soil that can affect the water's pH level (Figure 45.7).

The proportion of calcium and magnesium salts in water is measured by its "hardness." Water absorbs calcium and magnesium primarily as a result of rock weathering. Water becomes harder the more calcium and magnesium are present. In milligram per litre (mg/l) of dissolved calcium and magnesium carbonate, water hardness is typically measured.

Turbidity in a solution is caused by presence of suspended impurity, especially of non-settleable nature. The measurement of turbidity is necessary to decide the dose of coagulant needed to remove the suspended impurity. Presence of turbidity in water causes the incident light to be scattered and absorbed and only part of light gets transmitted in straight line. The same principle is used to determine turbidity.

Figure 45.9 Incubator
Source: Author

Figure 45.10 Nephelo-turbidity meter
Source: Author

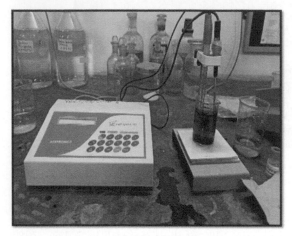

Figure 45.11 pH Test on charcoal
Source: Author

Figure 45.12 Ash content test on charcoal
Source: Author

To determine the credibility of charcoal various tests are done on it such as pH test, moisture content test and ash content test. The evaluation and testing of activated carbon, which is predominantly utilised in industrial applications, is greatly helped by the carbon standards. The forms of the activated carbons include solid, granular, palletised, and impregnated. The numerous chemical and physical characteristics of activated carbon, such as reaction pH, particle size, absorption, and the concentrations and activities of certain chemical compounds, can be determined with the aid of these carbon standards. The aqueous phase isotherm technique, air jet sieving, micro-isotherm technique, ashing, and speedy small-scale column testing are all methods that can be used to assess the attributes described above. To ensure the safe use and application of activated carbons, these standards enable laboratories and other chemical facilities to inspect and evaluate the material.

The pH Value of charcoal is a measure either it is acidic or basic. charcoal have some activity in absorbing Carbon Dioxide and in large quantities may raise the pH however if used in filters only, it has minimal impact on pH. The pH of liquids may change when charcoal is introduced if it has inorganic materials and chemically active groups on its surface. The activated carbon used, and the chemical composition of the treated water determine the pH increase and duration. For 200 to 500 bed volumes, the pH of the effluent can increase to a level over 9 or 10, which can cause the leaching of manganese and other transition metals from reactivated carbon as well as the aluminium from the charcoal in drinking water applications. Charcoal pH testing methods are sensitive to the type of salt solution or water employed during the measurement. Comparing test procedures between the client and the supply is crucial when pH is crucial to an application.

Ash analyses are also a reliable predictor of the quality of wasted carbon utilised in drinking water or groundwater sanitation applications. A high ash concentration could mean that sand is present or that calcium, aluminum, manganese, or iron have been deposited on the activated carbon. At 750°C, the mineral

Figure 45.13 Moisture content test
Source: Author

Figure 45.14 Glass and metal frame
Source: Author

Figure 45.15 Experimental setup
Source: Author

components are transformed into their corresponding oxides to determine the value. The amount of ash depends on the primary raw materials used to make the product, which are mostly silica and aluminum. This method covers the determination of moisture content of charcoal. When water is the sole volatile substance present in charcoal, oven drying is performed.

Experimental Setup

Glass columns with a metal frame measuring 152 × 152 × 700 mm were built to carry out the experiment [8]. The experimental column is fabricated. This column is filled with the samples of pervious concrete and the sample of pervious concrete containing iron slag. The experimental setup consists of glass compartments. Each compartment containing layer of adsorbent such as charcoal and iron slag and last compartment is made to collect filtered water.

Results and Discussion

Concrete Test

The concrete cubes different in proportion is tested to analyse different parameters. Compressive test and porosity test are performed on concrete cubes without iron slag, with 20% iron slag in concrete lab.

Compressive test

Table 45.2 shows the compressive strength of developed concrete. After testing of pervious concrete (without iron slag) we found that compressive strength in 7 days as 8.133 MPa (N/mm²), 14 days as 16 N/mm², 28 days as 27 N/mm². After testing of pervious concrete containing 20% iron slag, we found that compressive strength of cube in 7 days as 20.356 N/mm², 14 days as 27.822 N/mm², 28 days as 37.556 N/mm².

Porosity test

Permeable pavement systems have large infiltration rates except when sediment has accumulated on the surface [11]. The development of fully permeable pavement designs as a potential best management practice for stormwater management in areas that carry heavy truck traffic is of significant interest [12]. The above research gives porosity value in between the range of 15-35%.

Charcoal test

For charcoal the pH test, Ash Content Test and Moisture content test results are as follows:

pH test

pH is usually close to the neutral zone of 8 in collected samples of urban storm runoff waters [9]. The pH of charcoal found to be 7.28 which indicates the alkaline nature of charcoal.

Ash content test

The ash content of charcoal after the test is found to be 8.6%. It is seen that the content of ash content varies according to type of wood, the amount of bark added to the wood in the kiln, and the degree of contamination from sand and earth, the ash percentage of charcoal differ from around 0.5% or more. Ash percentage in good quality lump charcoal is normally around 3%. Fine charcoal may contain a lot of ash, The plus 4 mm residue may only contain 5-10% ash, if material smaller than 4 mm is removed.

Moisture content test

The moisture content test performed on charcoal shows the value of 5.968%. The moisture level of charcoal is often limited by quality criteria to 5–15% of the charcoal's total weight.

Table 45.2 Compressive strength of concrete

Days	Compressive strength (N/mm²)	
	Without iron slag	With 20% iron slag
7	8.133	20.356
14	16	27.82
28	27.556	37.556

Source: Author

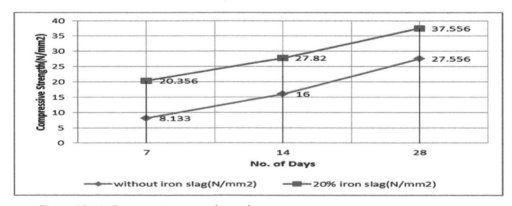

Figure 45.16 Compressive strength graph
Source: Author

Table 45.3 pH test of water

Sample	pH of pre-purified water	pH of post-purified water
1	6.005	6.578
2	6.076	6.620
3	6.159	6.774

Source: Author

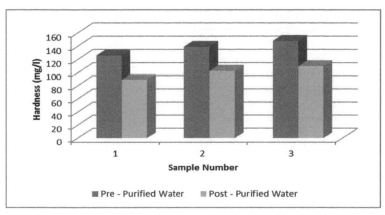

Figure 45.17 Hardness graph
Source: Author

Water test

The relevant water tests were done on three distinct water samples to analyse the effectiveness of each treatment approach and for each parameter. Urban storm runoff water samples were taken at three separate sites. pH, turbidity, and hardness were used as measurement criteria to evaluate the water's quality.

pH test

The pH test result for both the samples before and after passing through the experimental setup is within the neutral range.

Hardness test

The results of hardness test on water are presented in Table 3. Hardness amount is decreased in the water sample.

Turbidity test

According to the results of turbidity test. Suspended solids are decreased considerably after passing sample water through the experimental setup.

Conclusion

Pervious concrete is a reasonably priced and environmentally responsible way to assist sustainable building. Pervious concrete can play a big part because of its capacity to collect storm water, replenish ground water, and lessen storm water flow. Iron slag replacement of 20% in pervious concrete results in a higher improvement in compressive strength and a decrease in coefficent of permeability. As a result, this material can be used for pavement on roads where there is less traffic, such as in parking lots, playgrounds, etc., which aids in recharging. Various characteristics of urban storm runoff water observed to be changed after the treatment with charcoal. The pH of runoff water reached to safe limit. The filtered water will be used for domestic purpose as washing, gardening, cleaning etc. By adding permeable walls, pervious concrete can be used in buildings for cooling as well as rainwater collection purposes. Although pervious concrete has a wide range of uses, the pavement sector is its main application. It is particularly helpful for rural pavements and has a lot of room for additional research, making it a viable road material for ground water recharge in the future.

References

[1] Tennis, P., Leming. M. L., and Akers, D. J. (2004). Pervious Concrete Pavements. pp. 1-25. United States: Portland Cement Association (PCA).
[2] Lee1, M. J. (2013). Water Purification of Pervious Concrete Pavement, Taiwan: The American Society of Civil Engineers.
[3] Ho J.C.M., Liang, Y., Wang, Y. H., Lai, M. H., Huang, Z. C., Yang, D., and Zhang, Q. L. (2022). Residual properties of steel slag coarse aggregate concrete after exposure to elevated temperatures. Construction and Building Materials. 316, 125751.
[4] Yang, J. and Jiang, G. (2003). Experimental study on properties of pervious concrete pavement materials. Cement and Concrete Research. 33(3), 381–86.

[5]   Li, H., Harvey, J., and Jones, D. (2012). Developing a mechanistic empirical design procedure for fully permeable pavement under heavy traffic. Transportation Research Record 2305. pp., 83–94. Washington, DC: Transportation Research Board.

[6]   Hatt, B. E., Deletic, A., and Fletcher, T. D. (2007). Treatment performance of gravel filter media: Implications for design and application of stormwater infiltration systemsJournal of Water Resources. 41(12), 2513–24.

[7]   Brattebo, B. O., and Booth, D. B. (2003). Long-term stormwater quantity and quality performance of permeable pavement systems. Journal of Water Resources. 37(18), 4369–76.

[8]   Koupai, Jahangir Abedi, Soheila Saghaian Nejad, Saman Mostafazadeh-Fard, and Kiachehr Behfarnia. (2016) Reduction of urban storm-runoff pollution using porous concrete containing iron slag adsorbent. Journal of Environmental Engineering 142(2), 04015072.

[9]   Kuang, X. and Sansalone, J. (2011). Cementitious porous pavement in stormwater quality control: pH and alkalinity elevation. Water Science and Technology. 63(12), 2992–98.

[10] Lloyd, S. D., Wong, T., and Porter, B. (2002). The planning and construction of an urban stormwater management scheme. Water Science and Technology. 45(7), 1–10.

[11] Bean, E. Z., Hunt, W. F., and Bidelspach, D. A. (2007). Field survey of permeable pavement surface infiltration rates. Journal of Irrigation and Drainage Engineering. 133(3), 249–255.

[12] Li, H., Harvey, J., and Jones, D. (2012). Developing a mechanisticempirical design procedure for fully permeable pavement under heavy traffic. Transportation Research Record 2305. pp. 83-94. Washington, DC: Transportation Research Board.

[13] Park, S., Ju, S., Kim, H. -K., Seo, Y. -S., and Pyo, S. (2022). Effect of the rheological properties of fresh binder on the compressive strength of pervious concrete. Journal of Materials Research and Technology. 17, 636-48.

Chapter 46

The effect of MoS$_2$ interlayered silica-polyaniline hybrid surfaces for the photomineralisation of organic dyes

Sirajudheen Palliyalil[1,a], Sivakumar Vigneswharan[2,b] and Sankaran Meenakshi[3,c]

[1]Department of Chemistry, Pocker Sahib Memorial Orphanage College, Tirurangadi, Malappuram, Kerala, India

[2]Environmental System Laboratory, Department of Civil Engineering, Kyung Hee University, Republic of Korea

[3]Department of Chemistry, The Gandhigram Rural Institute (Deemed to be University), India

Abstract

Molybdenum sulphide (MoS$_2$), a n-type and 2D semiconductor with a band gap of ~1.6-1.90 eV, is one paradigm of a metal sulphide that exhibits outstanding photocatalytic activity in the visible light spectrum. On the other hand, the very high rate of electron-hole recombination lowers the catalytic effectiveness. MoS$_2$ would, however, lessen the recombination of electron and hole pairs when combined with organic semiconductors and silica. The matrix of silica and polyaniline (PANI) is ideal for incorporating MoS$_2$ catalyst. In this article, the production of a polyaniline/silica/MoS$_2$ (PANI-Si@MoS$_2$) hybrid composite is described, and its ability to degrade dye molecules like Malachite Green and Rhodamine B under visible region is examined. For the MG as well as RhB dyes under visible irradiation, the highest colour removal was 94.6% and 90.3%, respectively. At a starting dye concentration of 30 ppm, a catalyst loading of 0.1 g/L, pH ranges of 8-11, and a photon irradiation time of 60 minutes, the maximum amount of dye is removed. After the dye degradation experiment, the bands of UV of absorption of the molecules of dye vanished from the area where they had been visible in wavelength, emphasising the breaking of bonds in the polyaromatic dye molecules rings. According to the kinetic Langmuir-Hinshelwood investigation, which highlighted the importance of surface contact and revealed that the rate-regulating path in the dye oxidation mechanism is adsorption. The catalyst exhibits good reuse after six successions. It could be determined that the PANI-Si@MoS$_2$ composite has potential and is an ideal material for the cleansing of dye-contaminated wastewater. PANI-Si@MoS$_2$ composite has potential and is an ideal material for the cleansing of dye-contaminated water.

Keywords: Chitosan matrix, photooxidation, metal sulphide, silica, dye mineralisation

Introduction

Any chemical, physical, or biological modifications have a detrimental impact on the quality of water, which renders it unsafe for intake [1]. Due to the fact that several factories release their unprocessed effluents into neighboring waterways, dyes constitute one of the main contaminants in these effluents [2]. Dyes have been regarded as an unpleasant type of pollution as of the destructive consequences that frequently result from inhalation and oral intake, including skin and eye irritation. Dyes are more susceptible to auto-immune illnesses, leukemia, and other diseases when exposed to colored dye effluents in high doses [1, 3]. Since dyes may be found in water in quantities as low as 0.005 mg/L [4], color is often the first pollutant to be noticed, and they inhibit water from re-oxygenating [5]. Due to their intricate structure and unnatural source, dyes are challenging to remove from water [3]. So, as water plays a significant part in preserving the ecosystem of the earth and the quality of life, it is vital to fix it prior to it being spilled into waterways. To treat these sewages, numerous physical, biological, chemical, and techniques are employed [6]. In view of the fact that it hasn't generated secondary pollutants by any means, photocatalysis is one of these techniques that is regarded as being particularly significant [7].

A variety of material photocatalysts are created by several organic, inorganic, and synthetic polymer constituents. Within these substances, polyaniline (PANI) is a synthetic polymer with the ability to conduct electricity since it contains-NH functionality [8]. The surface of polyaniline has active imine and amine groups, which makes it a useful component for treating effluent water [9]. The irradiation would cause valence band (VB) electrons (e$^-$) in the composite to depart and progress to the conduction band (CB), producing holes (h$^+$) in the VB at the same pace. These e$^-$s spawned at the time of illumination of the current

[a]sirajpalliyalil@gmail.com, [b]vigneshwarangri@gmail.com, [c]sankaranmeenakshi@gmail.com

DOI: 10.1201/9781003450252-46

carrying polymer of the hybrid catalyst, are holding on to the semiconductor to avoid the recombination of e⁻—h⁺ couples [10]. To preserve the interface control of the composite, SiO_2 particles might operate as a superstructure guiding component [11, 12].

Metal sulfides, predominantly molybdenum sulfide (MoS_2), are n-type, 2D semiconductors having a 1.9 eV band gap energy [13, 14], demonstrates great photocatalytic performance in the observable region of light. The surprising pace of e⁻—h⁺ coupling, in contrast, decreases its catalytic effectiveness [15]. When SiO_2 is glued to the interface of PANI, it inhibits the composite from aggregating. Furthermore, the silanol groups of silicon dioxide link efficiently with the surfaces of PANI and MoS_2, resulting in the development of composites with diverse functional groups [12]. Furthermore, combining MoS_2 with PANI and SiO_2 improves chemical stability while also reducing both hole and electron coupling rates [16]. The production of polyaniline/silica/MoS_2 (PANI-Si@MoS_2) blended hybrid is investigated, along with its degrading efficacy upon noticeable exposure to light via malachite green (MG) and Rhodamine B (RhB) pigments. FTIR, SEM, XRD, UV-DRS, and TGA-DTA, methods were used to characterise the fabricated materials. The L-H kinetics is employed to scrutinize the interactive relationship between the catalyst and dye toxins. TOC analysis is utilised to resolve the extent of contaminant decomposition.

Materials and Methods

S. D. Fine-Chem Ltd, India, supplied GLR TLC quality Silica gel. From Mumbai CDH Chemicals, India, ammonium molybdate, acetic acid (glacial), sodium hydroxide, ammonium peroxybisulphate (APS), hydrochloric acid, and thiourea are purchased. Aniline ($C_6H_5NH_2$) is purchased from Sigma-Aldrich Chemicals India. MG and RhB, two of the contaminants under research, are purchased from M/s Sree Chemidyes, India, in Bangalore. The chemicals utilized in the experiment have undergone no further modifications.

Fabrication of PANI-Si@MoS₂ Matrix

Initially, silica gel (60–120 mesh) is heated to 110°C for 3 hours to activate the surface. Utilising the Fedorova and Stejskal approach, *in situ* polymerisation is used to create PANI coated silica gel [17]. Typically, SiO_2 (15 g) is dispersed in of double-distilled water (300 mL) for the SiO_2-PANI hybrid production, which is then stirred at 303 K for 2.5 hours. 160 mL HCl (0.5 M) and aniline (0.02 M) are added to the previously mentioned mixture and constantly swirled at 0–4°C. After 2 hours, 160 mL APS (0.025 M) is added at 0-4°C, and the resultant hybrid being maintained at 0-4°C for 24 hours. The aforesaid SiO_2-PANI is subsequently mixed by a certain quantity of MoS_2 particles, which were formerly made by soaking ammonium molybdate (1 g) with of thiourea (2 g), and of water (40 mL) (Milli-Q) and heat for about 6 hours at 180°C in a Teflon-covered autoclave. MoS_2 scattered SiO_2-PANI ((PANI-Si@MoS_2) would then be crystallized in NH_4OH, separated, and properly washed repeatedly with Milli-Q water. Eventually, the hybrid composite is dehydrated in an oven at 80°C after being processed with acetone. The proposed manufacturing of PANI-Si@MoS_2 hetero-junction composite hybrids is depicted in Scheme 46.1.

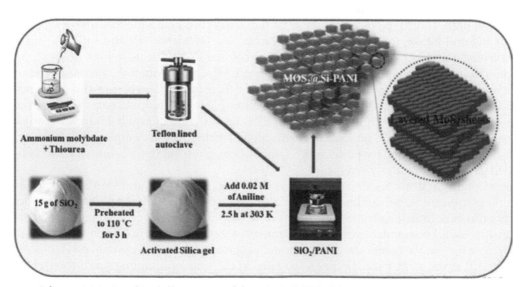

Scheme 46.1 Graphical illustration of the PANI-Si@MoS_2 matrix fabrication
Source: Author

Analytical Methods

The patterns of the X-ray diffractograms of raw PANI, CS, MoS_2, and Si-PANI@ MoS_2 blends were recorded using a diffractometer, the PAN Analytical X-Pert PRO. A scanning electron microscope (TESCAN VEGA3, SEM, Germany) was applied to look into the morphology of the produced composite structure. The chemical interactions have been investigated through spectroscopy with the Fourier transform at a frequency of 400–400 cm^{-1} (FTIR spectrometer, 460 Plus JASCO, Japan). The UV-visible spectrometer (Pharo 300 form, Merck, Germany) was used to investigate dye mineralisation. The bandgap energy in the manufactured matrix is measured with UV-diffuse reflectance spectroscopy (UV-DRS) by the Varian Cary-500 model DRS spectrophotometer. To measure the flat band potential of the hybrid utilizing Na_2SO_4 standard solution (0.5 M), electrochemical impedance spectroscopy (EIS) and Mott-Schottky (M-S) diagrams have been used. Lastly, the pH evaluation was carried out using an expanded ion analyzer as well as a pH electrode (EA940, USA) model.

Investigation of Dye Degradation

PANI-Si@MoS_2 hetero-structure for the decontamination process was used in tests with varied experimental setups in the white light spectrum. PANI-Si@MoS_2 was used to remediate the MG and RhB pigment toxins independently in both darkened and white light conditions. To ensure continuous transmission of light waves, a xenon lamp (250 W) with a 400–800 nm wavelength cutoff has been utilised as the source. The batch analysis was carried out to determine the effects of critical factors such as preparatory dye solution quantities, catalyst dose, irradiance interval, and pH on photon exposure. After being exposed to radiation for a set amount of time, three mL of the reaction solution were collected and filtered. The amount of dye was then calculated using a UV-vis spectrophotometer operating at 617 and 556 nm wavelengths for MG and RhB dyes [18, 19], respectively. The following Equation (1) could be used to calculate and quantify the percentage degradation as the amount of dye decolorisation caused by the photocatalyst:

$$\text{Percentage Degradation} = (\frac{D_i - D_e}{D_i})100\,\% \tag{1}$$

Where D_i represents the original dye intensity in (mg/L) and D_t represents the concentration of dye after a time 't' in (mg/L). The Equation (2) represented the extent of oxidation and the products that resulted in terms of TOC:

$$TOC(\%) = (TOC_i - TOC_f) \div TOC_i \times 100 \tag{2}$$

The terms TOC_i and TOC_f refer to the entire organic carbon contents of the MG and RhB pigment sorbents by the beginning and the completion of reaction, correspondingly, of the mineralisation reaction.

Results and Discussion

XRD and FTIR

XRD analysis is used to demonstrate both the amorphous and crystalline nature of MoS_2, PANI, SiO_2, and the PANI-Si@MoS_2 composite. The XRD analysis findings are displayed in Figure 46.1(a). According to the graph, the SiO_2 had conventional peak maxima at 2θ of 11.68, 25.78, 31.78, and 49.28 [20]. The spectra of PANI could be detected at 2θ of 10 and 30°. The occurrence of benzenoid and quinoid rings in the PANI framework can be seen in the X-ray of conductive PANI, with spikes at angles 2θ of 25.71 and 19.32. The existence of an XRD of MoS_2 with unique archetypal peaks was justified by the card number (JCPDS No. 21-1272). Furthermore, the broadness of the peak position of the PANI-Si@MoS_2 framework supported the quasi-crystalline structure of the crafted hybrid composite [21]. Due to the discrete peak matrices of SiO_2 and PANI with reduced intensity, the strong crystalline phase of the MoS_2 is diminished. Using the 2θ = 14.4° reflectance peak of the X-ray spectra and the Debye-Scherer equation, the size average of the PANI-Si@MoS_2 crystallite of the composite has been calculated and is found to be ~36.94 nm.

The FT-IR spectra are used to elucidate the configurational characteristics of PANI, MoS_2, SiO_2, and the PANI-Si@MoS_2 hybrid, which is shown in Figure 46.1(b). The N-H groups exhibit a broad 3415 cm^{-1} band owing to the stretching mode, while the C-H pair exhibits a delicate 2923 cm^{-1} mode of stretching, both of which are seen in the PANI framework. At 1243 and 1295 cm^{-1}, respectively, the C-N positions of the stretched benzene and quinone rings can be detected [22]. Silica often exhibits distinctive 803, 3393, and 1130 cm^{-1} patterns that match up with the symmetrical and asymmetrical stretching modes of Si-O-Si and OH [20]. Aside from the appearance of the usual peak about 598 cm^{-1} that is distinctive of the MoS_2 in

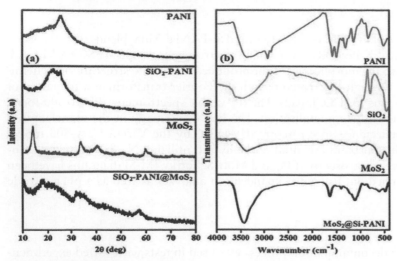

Figure 46.1 (a) X-ray diffraction and (b) FT-IR spectral analysis
Source: Author

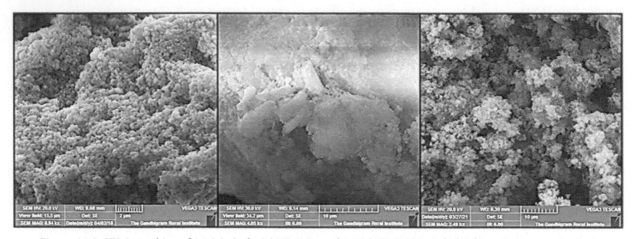

Figure 46.2 SEM graphics of (a) MoS_2 (b) PANI-SiO_2 and (c) PANI-Si@MoS_2 hybrid
Source: Author

the manufactured hybrid, the maxima in the FTIR spectra of PANI-Si@MoS_2 matrix material are identical to those of pure polyaniline as well as SiO_2-PANI [23]. Furthermore, the MoS_2 intertwined with Si-PANI exhibits significant spectral characteristics, indicating that the MoS_2 is possibly constructively entwined with both Si-PANI lattices.

Analysis of Surface Morphology

By using the SEM method, the morphology of the surface and structure of the MoS_2, Si-PANI and PANI-Si@MoS_2 composites were comprehensively investigated and are shown in Figure 46.2(a–c). The SEM investigation reveals that the generated MoS_2 particles are spherical in shape and have a substantial porous nature (Figure 46.2a). The porous template of the clustered nanocrystals aids in the efficient decomposition of the organic pigments. When contrasting with the structure of the PANI-Si@MoS_2 hybrid (Figure 46.2c), which has an imperfect, scratchy, and irregular form with something like a sheet-like layout, the SEM pictures of Si-PANI reveal a much smoother surface (Figure 46.2b). This allowed researchers to distinguish the infiltration of MoS_2 from interfaces within the Si-PANI framework.

UV-DRS Investigation

As shown in Figure 46.3 (a), the UV-DRS are used to validate the perceptible photon-assisted features of the synthesised PANI-Si@MoS_2 hybrid. PANI has a broad frequency spectrum in both the UV and observable light regions [24]. Figure 3(a) shows the nearer rim absorption of PANI-Si@MoS_2 at 420 nm. The graph of $(\alpha h\upsilon)^2$ vs $h\upsilon$ (Tauc Plot) (Figure 3(b)) reveals the indirect bandgap of PANI-Si@MoS_2, which is approximately 2.78 eV and is greater than just the raw MoS_2 (1.85 eV) absorption edge [25–27]. The following explains the photon occurrences on the catalyst, viz., PANI-Si@MoS_2 hybrid composite [26, 28, 29]. The electrons from the π orbit of PANI are moved further into orbital π* at the point of absorption of the photons. Because its energetic levels coincide with the charge transfer as well, they could quickly deploy into the CB of MoS_2. Consequently, it's possible that the partitioning of charges will get faster. Thus, it makes it possible for the e⁻s in the MoS_2 CB to react with oxygen and water to produce reactive oxygen species (ROS). Therefore, these ROS trigger a series of reactions that lead to dye decomposition [30].

Determination of Decomposition Matrices

The photo-decomposition capability of the PANI-Si@MoS_2 hybrid is investigated in the visual range with MG and RhB pigments as representative toxins. Graphs in Figure 46.4(a-d) show the effectiveness of photo-decomposition by the produced material. The organic molecules are progressively broken by the PANI-Si@MoS_2 hybrid by illumination, with a maximal removal percentage of 91.9 and 94.6% for RhB and MG dye pigments, correspondingly, in 50 minutes Figure 46.4(a). The increased pigment depletion is due to an increase in the E_g of the hybrid PANI-Si@MoS_2 blend, which results in an increase in photon absorption at detectable wavelengths by the hybrid catalyst and a concomitant increase in deterioration intensity. The impact of the baseline amount of dye on photodecomposition is investigated by dispersing colored pigment solutions in the range of 10-60 mg/L. The dye removal rate is subsequently assessed by dispersing the initial dye quantities, as shown in Figure 46.4 (b). This report demonstrates that the rise in initial levels of dye from 10-60 mg/L, as well as the fraction of coloring pigment eliminated by the reinforced composites,

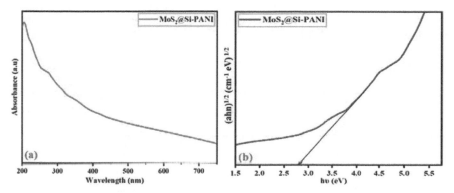

Figure 46.3 (a) DRS and (b) Tauc plot for determining Eg of PANI-Si@MoS_2 hybrid composite
Source: Author

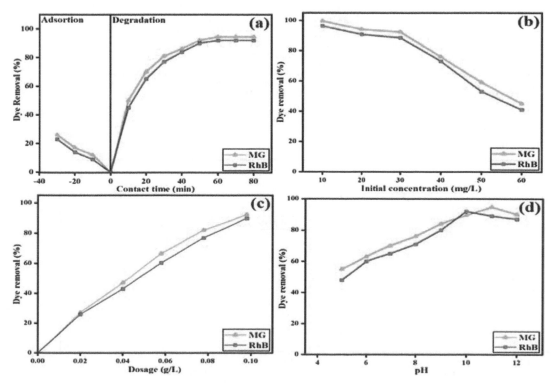

Figure 46.4 Effect of (a) absorption time (b) concentration of dyes (c) composite dosage and (d) pH of solution
Source: Author

has decreased because of the deprivation of vivid site interactions on the hybrid surface. Furthermore, at high primary concentrations of MG and RhB dye solutions, photon entrance into the solution is hindered, lowering the performance of the composite.

The amount of photocatalyst hybrid and its effect on organic dye decomposition are scrutinised with different PANI-Si@MoS$_2$ hybrid doses ranging from 20-100 mg, as shown here at Figure 46.4(c). Extent of decomposition is shown to grow with increasing hybrid catalyst quantity, and significant decomposition has been achieved at 100 mg. This could be attributed to the existence of extraneous façade coordination sites in the PANI-Si@MoS$_2$ composite matrix when dose is increased. Furthermore, the robustness of the PANI-Si@MoS$_2$ catalyst in the illumination might increase the probability of photon absorption, resulting in a greater formation of productive reactive species for the depletion of MG and RhB dye pigments. As a consequence, a 100 mg dose is being employed in future photocatalytic investigations.

By altering the pH level of the solution across 2 and 12 ranges, the solution pH and its effect on the percentage mineralisation of the coloring dye pigments in the PANI-Si@MoS$_2$ hybrid blend have been studied. The PANI-Si@MoS$_2$ catalyst demonstrated superior depleting performance for the removal of RhB and MG colored matter at higher pH ranges, as shown in Figure 46.4(d). At elevated pH levels, the positively charged pigments form bonds by means of catalytic substrates that have a reduced solidity. As a result, the breakdown of MG and RhB pigments takes place very readily between pH 10 and 11, in that order. The composite pHzpc value is found to be (6.2), The layers of MoS$_2$ possess negative charges due to hanging sulfur bonds, while PANI development occurs by aniline monomer coulombic adsorption of its positively charged atoms over the layer of SiO$_2$ through polymerisation. The SiO$_2$ also possesses a negative charge due to its oxygen moiety. The reported pHzpc value is a result of entangling MoS$_2$ over the Si-PANI blend, which also verifies dye decomposition at enhanced pH levels.

Determination of Degradation by UV Spectra

The UV absorbance spectra of RhB and MG dyes are illustrated in Figure 5(a and b). All these MB and RhB had adsorption alone extraction performances of 35.9 and 30%, respectively, in darkened circumstances.

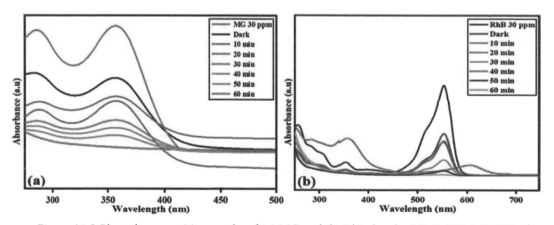

Figure 46.5 Photodecomposition results of (a) MG and (b) RhB dyes by PANI-Si@MoS$_2$ (UV- absorption spectra)
Source: Author

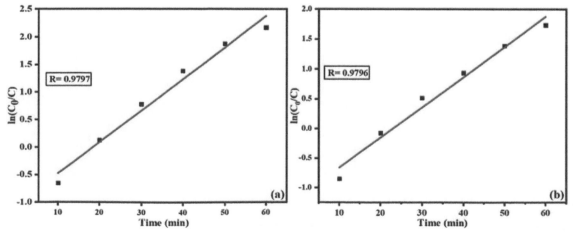

Figure 46.6 L-H kinetics plot for the decomposition of (a) MG and (b) RhB dyes by PANI-Si@MoS$_2$ matrix
Source: Author

While both RhB and MG dyes photodegrad in the perceptible illumination range of optical absorption. When the interaction develops, the intensity ratio of the absorption peaks of both MG and RhB dyes quickly declines. A few of the points in the UV band are either shifted to the left or removed from their reported range of wavelengths, indicating where both RhB and MG chromophores are methodically cracked, and the resulting byproducts are generated further.

Kinetics of Dye Degradation

The L-H kinetic model is utilised to explore the direct proportionality seen between the rate equation and the kinetics of the process for such depletion of MG and RhB pigments by PANI-Si@MoS$_2$ hybrid, as shown in Figure 46.6 (a) and (b), correspondingly. Equation (3) is used to get the constant rate from the inclination of the L-H graph.

$$\ln(C_0/C) = k_{app}.t \tag{3}$$

Where, C_0 denotes the number of organics in the starting position, C is the proportion of pigments at period interval t of the photodecomposition process, and the apparent constant k_{app} is the rate constant of the reaction. The coefficients for L-H correlation (R^2) for RhB and MG dye riddance are 0.9796 and 0.9797, in that order. These findings demonstrated that the degradation of RhB and MG dyes in detectable photoexcitation comprises interfacial sorption, with dyes being removed from the solution by the PANI-Si@MoS$_2$ due to synergistic effects between the catalyst's hydrophilic surface and the contaminants.

Probable RhB and MG Decomposition by-products

The projected oxidation of MG dye pigments by the Vis/PANI-Si@MoS$_2$ matrix is explained below. The framework of MG pigment comprises three main ring structures; upon illumination, the formed attacker entities assault the benzene ring system and break down the number of MG benzene rings, eventually tending to reorganize their appearance as diverse types of interim. The ring opens as a response to the subsequent onslaught. It is also demonstrated that de-methylation and de-chlorination produce molecules with a 274 m/z value. De-methylation of transition products forms again, leading to the decomposition of the chemical with m/z values of 94 and 212 once again. The subsequent intermediary is further oxidised and demethylated, yielding substitutes with m/z values of 94, 109, 110, and 122. Ring-opening is the outcome of continued oxidation that leads to the creation of lactic acid with a m/z value of 90 and reuterin with a 74. Subsidiary decomposition of these intermediates yields CO$_2$ and H$_2$O with low molar masses. N-demethylation is critical in the MG dye breakdown process [31].

The composition of the RhB molecule has three benzene rings in its structure. Irradiating the organic dye in the vicinity of a photosensitiser produces active species that approach the benzene rings of RhB and form various intermediates, causing the ring to collapse. It is also taken into account because illumination produces demethylation and the formation of molecules with a m/z value of 415. The intermediate demethylation byproducts are then separated into m/z values of 357 and 293; following that, the molecule undergoes chromophore breakage and oxidation repeatedly, yielding new chemicals with m/z values of 195, 190, 181, and 149. The expansion of the ring causes ancillary depravation, resulting in the formation of the compounds with m/z values of 90, 79, and 56. Further structural deprivation results the formation of tiny molecules of molecular CO$_2$ and H$_2$O particles.

Quantum Efficiency and Synergetic Action of the Catalyst

The compatible energies of MoS$_2$ and PANI result in a synergetic effect in the PANI-Si@MoS$_2$ heterostructure, which improves the catalytic performance of the composite hybrid. The energy levels are followed by MoS$_2$ and PANI: E(LUMO)>E(CB)>E(HOMO)>E(VB) [25]. The e$^-$s is advanced towards the CB through HOMO to LUMO due to the collision of photons. Following that, the movement of h$^+$s commencing VB to HOMO occurs as polyanilne acts as a transmitter of the h$^+$. As a result, the h$^+$ and e$^-$ congregate in the HOMO of PANI in addition to the MoS$_2$ CB. The aforementioned techniques be very effective for boosting quantum proficiency as the ability to separate h$^+$ and e$^-$s is strengthened [32].

Mott–Schottky Evaluation

This work uses flat band potential (Vfb) data derived from the Mott-Schottky (M-S) curve, applying the EIS method to estimate the band positions of MoS$_2$ and PANI in the blended PANI-Si@MoS$_2$ composite [33]. In order to clarify the band energy potentials, the M-S graphs were employed. The MS graphs are explored in this paper to validate the configuration of the band across MoS$_2$ and PANI in order to determine the

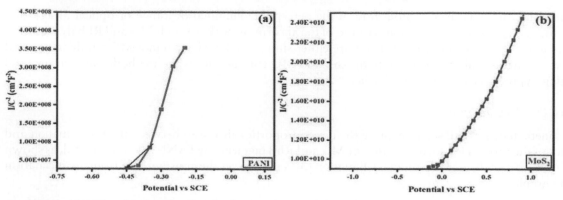

Figure 46.7 Plots Mott–Schottky plot for (a) PANI and (b) MoS$_2$
Source: Author

segregation and conveyance channel of the charge transporters in the trio PANI-Si@MoS$_2$ hybrid hetero-structure. The potential outcomes of the raw PANI and molybdenum sulfide versus SCE were 0.45 and 0.1 V, accordingly, as shown in Figure 46.7. The foregoing outcome confirmed that the PANI has a greater CB frame than MoS$_2$, demonstrating the resulting role in the propagation of photo-excited charge transport. As a result, the photon irradiation caused by the energized e- from the interface of the PANI-Si@MoS$_2$ might enable the CB of conducting polymers PANI to transfer to MoS$_2$. Due to the low HER overcharge potential of MoS$_2$, the excited e's may be utilised efficiently. PANI might aid in the transition from e- to MoS$_2$ by acting as a capillary for the charge convey [34].

In contrast, PANI has quite a high VB level because of the cavities that were formed as a result of irradiating the PANI-Si@MoS$_2$ heterostructure that accumulated there to carry out the oxidative degradation. Since the structural component SiO$_2$ continuously adsorbs the dye molecules and transfers them toward the photocatalyst, the formation of the ternary PANI-Si@MoS$_2$ ternary composite may cause the oxidation of dyes. Owing to its adequate flat band, which exceeds potential and leads to the decay of contaminants, the combined action of PANI and MoS$_2$ may significantly improve the ability to transfer information.

Reasonable Ineralisation Mechanism

The predicted process for the degradation of MG and RhB organic pigments through light-activated oxidation by PANI-Si@MoS$_2$ photocatalyst is shown in Scheme. 46.2. Photooxidation proceeds via two-stage processes. At first, MG and RhB pigment molecules could be tethered to the PANI-Si@MoS$_2$ matrix through adsorption.

The Vis/ PANI-Si@MoS$_2$ composite is then exposed to radiation, which causes the valence band electrons € to progress into the compost's CB. As a consequence, an analogous quantity of h+ has been formed in the VB of the Vis/PANI-Si@MoS$_2$ system. The illuminated Vis/PANI-Si@MoS$_2$ blend generates electrons in PANI, which slide crosswise to MoS$_2$ and prevent the h+ and e- couples from reconnecting. The photo-lytically induced generation of •OH radicals by the H$_2$O molecules constrained in the holes might serve primarily as an oxidising agent more towards the partial or widespread mineralization of dye pigments [35, 36]. The organic dyes can also be degraded by the VB that contains h+. The e that was energised in the CB and produced the radical anion of O$_2^-$ was subdued by the O$_2$ molecules that were dissolved. A number of radical species, including •OH, O$_2^-$, and h+ are created by the whole swing of the h+-e- pair by the photoexcitation. This movement would indeed trigger the extensive dye depletion process [35, 37]. The most plausible process for dye oxidation is depicted below:

$$\text{PANI - Si@MoS}_2 + h\nu \rightarrow \text{PANI - Si@MoS}_2 \, (\text{e}^- + \text{h}^+) \tag{4}$$

$$\text{h}^+ + \text{H}_2\text{O} \rightarrow \text{H}^+ \, \text{OH}^- \tag{5}$$

$$\text{h}^+ + \text{OH}^- \rightarrow \text{OH}^- \tag{6}$$

$$\text{O}_2 \, \text{e}^- \rightarrow \text{O}_2^- \tag{7}$$

$$\text{O}_2 + \text{H}_2\text{O} + 2\text{e}^- \rightarrow \text{H}_2\text{O}_2 \tag{8}$$

$$\text{H}_2 + \text{O}_2 + \text{e}^- \rightarrow \text{OH}^- + \text{OH}^- \tag{9}$$

$$\text{OH}^- + \text{O}_2 + \text{Dyes} \rightarrow \text{H}_2\text{O} + \text{CO}_2 + \text{other products} \tag{10}$$

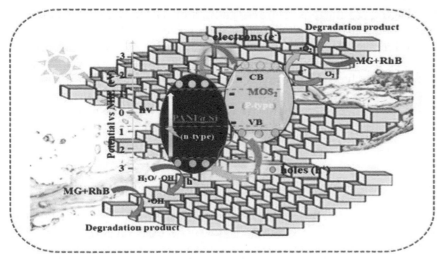

Scheme 46.2 Illustration of the e--h+ pathway Vis/PANI-CS@MoS$_2$ matrix for the detoxification of MG and RhB dye molecules
Source: Author

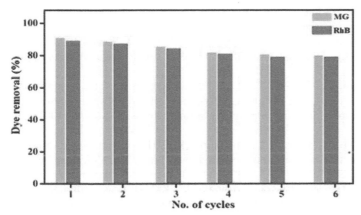

Figure 46.8 Post-use efficiency of the fabricated PANI-Si@MoS$_2$ matrix
Source: Author

Seeing as the delay in rate of e-h+ recombination after their separation speeds up the creation of reactive groups, the e—h+ deferral recombination plays a crucial part in the efficiency of the photocatalyst. Following the photocatalytic process, the TOC assessment showed that the levels of organic substances in the solutions of the RhB and MG dye molecules were reduced to 3.21 and 3.75 mg/L, respectively, from 30 mg/L. The study's conclusion demonstrates that the hybrid hetero-structure of PANI-Si@MoS$_2$ permits greater capability in the mineralization of selected toxic wastes by light-assisted catalytic activity. The analysis reveals that the PANI-Si@MoS$_2$ composite hetero-structure has greater capacity for light-induced catalytic activity in the oxidation of selected poisonous wastes.

Stability and Durability Analysis

The practical implementation is determined by the composite's stability and post-use utilisation. In order to leap down the operating cost of the contaminants' degradation course, commercial exploitation necessitates that the PANI-Si@MoS$_2$ hybrid's firmness be sufficient. The results of five successive cycle degradation trials for the oxidation of targeted pollutants by the catalyst PANI-Si@MoS$_2$ are shown in Figure 46.7, ensuring the stability and strength of the built-in composite. The findings of the reliability assessment suggest that after five iterations of repetitive deprivation phases of MG and RhB organic pigments, the degradation rate merely nominally decreases by 10%. Due to the greater capacity for the MG and RhB coloring pigments' destruction, PANI-Si@MoS$_2$ heterostructure composites might be used as a useful catalyst.

Conclusions

In situ polymerisation is used to construct an efficient PANI-Si@MoS$_2$ hybrid composite heterostructure, which is then analysed using a variety of experimental approaches. In comparison to MoS$_2$, the DRS examination shows that the PANI-Si@MoS$_2$ bandgap is improved to 2.78 eV. (1.80 eV). Under a discernible light source, the created composite demonstrated excellent depletion power for MG and RhB organic dyes. For MG and RhB, the highest mineralisation was 94.6 and 91.9% at primary dye concentrations of 30 mg/L and 0.1 g of composite dose, respectively. Around 60 minutes after illumination, the whole course of reaction reaches equilibrium. The L-H mechanism that suited the data the best focused on adsorption-assisted photocatalytic dye degradation. Up to 6 rounds of reusability are demonstrated by the photocatalyst. As a result, the obtained PANI-Si@MoS$_2$ hybrid composite ternary structure would be used to improve the efficiency of water effluent management.

References

[1] Bhattacharya, S. S. A. (2017). Drinking water contamination and treatment techniques. Applied Water Science. 7, 1043–67. https://doi.org/10.1007/s13201-016-0455-7.

[2] Natarajan, S., Bajaj, H. c., and Tayade, R. J. 92018). Recent advances based on the synergetic effect of adsorption for removal of dyes from waste water using photocatalytic process. Journal of Environment Science. 65, 201–22. https://doi.org/10.1016/j.jes.2017.03.011.

[3] Sirajudheen, P., Poovathumkuzhi, N. C., Vigneshwaran, S., Chelaveettil, M. K., and Meenakshi, S. (2021). Applications of chitin and chitosan based biomaterials for the adsorptive removal of textile dyes from water: a comprehensive review. Carbohydrate Polymers. 273, 118604. https://doi.org/10.1016/j.carbpol.2021.118604.

[4] Singh, R. L., Singh, P. K., and Singh, R. P.(2015). Enzymatic decolorization and degradation of azo dyes: a review. International Biodeterioration and Biodegradation. 104, 21–31. https://doi.org/10.1016/j.ibiod.2015.04.027.

[5] Moyo, S., Makhanya, B. P., and Zwane, P. E. (2022). Use of bacterial isolates in the treatment of textile dye wastewater: a review. Heliyon. 8, e09632. https://doi.org/10.1016/j.heliyon.2022.e09632.

[6] AJ, E. J., Krenkel, P., and Shamas, J. (2015). Wastewater Treatment &water reclamation. Reference Module in Earth Systems and Environmental Sciences (2015). B978-0-12-409548-9.09508–7. https://doi.org/https://doi.org/10.1016%2FB978-0-12-409548-9.09508-7.

[7] Sirajudheen, P., Vigneshwaran, S., Karthikeyan, P., Nabeena, C. P., and Meenakshi, S. (2021). Technological advancement in photocatalytic degradation of dyes using metal-doped biopolymeric composites—present and future perspectives. Energy, Environment, and Sustainability. Singapore: Springer. https://doi.org/10.1007/978-981-16-3256-3_9.

[8] Saravanan, R., Sacari, E., Gracia, F., Khan, M. M., Mosquera, E. E., and Gupta, K. (2016). Conducting PANI stimulated ZnO system for visible light photocatalytic degradation of coloured dyes. Journal of Molecular Liquids. 221, 1029–33. https://doi.org/10.1016/j.molliq.2016.06.074.

[9] Sirajudheen, P., Kasim, V.C.R., Nabeena, C. P., Basheer, M. C., and Meenakshi, S. (2021). Tunable photocatalytic oxidation response of ZnS tethered chitosan-polyaniline composite for the removal of organic pollutants: A mechanistic perspective. Materials Today: Proceedings. 47, 2553–59. https://doi.org/10.1016/j.matpr.2021.05.054.

[10] Sirajudheen, P., Resha Kasim, V. C., Sivakumar, V., Nabeena, C. P., Basheer, M. C., and Meenakshi, S. (2022). Ternary system of TiO2 confined chitosan-polyaniline heterostructure photocatalyst for the degradation of anionic and cationic dyes. Environmental Technology & Innovation. 28, 102586. https://doi.org/10.1016/j.eti.2022.102586.

[11] Jiao, M., Liu, K., Shi, Z., and Wang, C. (2017). SiO2/carbon composite microspheres with hollow core–shell structure as a high-stability electrode for lithium-ion batteries. ChemElectroChem. 4, 542–49. https://doi.org/10.1002/celc.201600658.

[12] Abu Taleb, M., Kumar, R., Al-Rashdi, A. A., Seliem, M. K., and Barakat, M. A. (2020). Fabrication of SiO2/CuFe2O4/polyaniline composite: a highly efficient adsorbent for heavy metals removal from aquatic environment. Arabian Journal of Chemistry. 13 (2020) 7533–7543. https://doi.org/10.1016/j.arabjc.2020.08.028.

[13] Thomas, N., Mathew, S., Nair, K. M., O'Dowd, K., Forouzandeh, P., Goswami, A., McGranaghan, G., and Pillai, S. C. (2021). 2D MoS2: structure, mechanisms, and photocatalytic applications. Materials Today Sustainability. 13. https://doi.org/10.1016/j.mtsust.2021.100073.

[14] Li, Z., Meng, X., Zhang, Z. (2018). Photochemistry Reviews Recent development on MoS2-based photocatalysis : A review. Journal of Photochemistry and Photobiology C: Photochemistry. 35, 39–55. https://doi.org/10.1016/j.jphotochemrev.2017.12.002.

[15] Ghasemipour, P., Fattahi, M., Rasekh, B., and Yazdian, F. (2020). Developing the Ternary ZnO Doped MoS2 Nanostructures Grafted on CNT and Reduced Graphene Oxide (RGO) for Photocatalytic Degradation of Aniline, Sci. Rep. (2020) 1–16. https://doi.org/10.1038/s41598-020-61367-7.

[16] Saha, S., Chaudhary, N., Mittal, H., Gupta, G., and Khanuja, M. (2019). Inorganic–organic nanohybrid of MoS2-PANI for advanced photocatalytic application. International Nano Letters. 9, 127–39. https://doi.org/10.1007/s40089-019-0267-5.

[17] Fedorova, S. and Stejskal, J. (2002). Surface and precipitation polymerization of aniline, Langmuir. 18, 5630–32. https://doi.org/10.1021/la025665o.

[18] Zhang, X., Deng, J., Yang, C., Wang, Z., and Liu,Y. (2022). Selective reduction of nitrite to nitrogen by polyaniline-carbon nanotubes composite at neutral pH. Environmental Research. 14, 114203. https://doi.org/10.1016/j.envres.2022.114203.

[19] Xue, H., Yu, M., He, K. Liu, Y., Cao, Y., Shui, Y., Li, J., Farooq, M., and Wang, L. (2020). A novel colorimetric and fl uorometric probe for biothiols based on MnO2 NFs-Rhodamine B system. Analytica Chimica Acta. 1127, 39–48. https://doi.org/10.1016/j.aca.2020.06.039.

[20] Karthik, R. and Meenakshi, S. (2014). Removal of hexavalent chromium ions using polyaniline/silica gel composite. Journal of Water Process Engineering. 1 (2014) 37–45.

[21] Mohamed, M. A., Salleh, W.N.W., Jaafar, J., Ismail, A. F., Abd Mutalib, M., and Jamil, S. M. (2015). Incorporation of N-doped TiO2 nanorods in regenerated cellulose thin films fabricated from recycled newspaper as a green portable photocatalyst. Carbohydrate Polymers. 133, 429–37. https://doi.org/10.1016/j.carbpol.2015.07.057.

[22] Allahveran, S. and Mehrizad, M. (2017). Polyaniline/ZnS nanocomposite as a novel photocatalyst for removal of Rhodamine 6G from aqueous media : Optimization of in fl uential parameters by response surface methodology and kinetic modeling. Journal of Molecular Liquids. 225, 339–46. https://doi.org/10.1016/j.molliq.2016.11.051.

[23] Thakur, A. K., Deshmukh, A. B., Bilash, R., Karbhal, I., Majumder, M., and Shelke, M. V. (2017). Facile synthesis and electrochemical evaluation of PANI/ CNT/ MoS2 ternary composite as an electrode material for high performance supercapacitor. Materials Science and Engineering B. 223, 24–34. https://doi.org/10.1016/j.mseb.2017.05.001.

[24] Guo, N., Liang, Y., Lan, S., Liu, L., Zhang, J., Ji, G., and Gan, S. (2014). Microscale hierarchical three-dimensional flowerlike TiO2/PANI composite: Synthesis, characterization, and its remarkable photocatalytic activity on organic dyes under UV-light and sunlight irradiation. Journal of Physical Chemistry C. 118, 18343–55. https://doi.org/10.1021/jp5044927.

[25] Yang, C., Dong, W., Cui, G., Zhao, Y., Shi, X., Xia, X., Tang, B., and Wang, W. (2017). Enhanced photocatalytic activity of PANI/TiO2 due to their photosensitization-synergetic effect. Electrochimica Acta. 247, 486–95. https://doi.org/10.1016/j.electacta.2017.07.037.

[26] Gu, L., Wang, J., Qi, R., Wang, X., Xu, P., and Han, X. (2012). A novel incorporating style of polyaniline/TiO2 composites as effective visible photocatalysts. Journal of Molecular Catalysis A: Chemical. 357, 19–25. https://doi.org/10.1016/j.molcata.2012.01.012.

[27] Hashemi Monfared, A. and Jamshidi, M. (2019). Synthesis of polyaniline/titanium dioxide nanocomposite (PAni/TiO2) and its application as photocatalyst in acrylic pseudo paint for benzene removal under UV/VIS lights. Progress in Organic Coatings. 136, 105257. https://doi.org/10.1016/j.porgcoat.2019.105257.

[28] Li, X., Wang, D., Cheng, G., Luo, Q., An, L., and Wang, Y. (2008). Preparation of polyaniline-modified TiO2 nanoparticles and their photocatalytic activity under visible light illumination. Applied Catalysis B: Environmental. 81, 267–73. https://doi.org/10.1016/j.apcatb.2007.12.022.

[29] Zhang, H., Zong, R., Zhao, J., and Zhu, Y. (2008). Dramatic visible photocatalytic degradation performances due to synergetic effect of TiO2 with PANI. Environmental Science & Technology. 42, 3803–07. https://doi.org/10.1021/es703037x.

[30] Gu, L., Wang, J., Qi, R., Wang, X., Xu, P., and Han, X. (2012). A novel incorporating style of polyaniline/TiO2 composites as effective visible photocatalysts. Journal of Molecular Catalysis A: Chemical. 357, 19–25. https://doi.org/10.1016/j.molcata.2012.01.012.

[31] Lu, J., Zhou, Y., Lei, J., Ao, Z., and Zhou, Y. (2020). Fe3O4/graphene aerogels: A stable and efficient persulfate activator for the rapid degradation of malachite green. Chemosphere. 251, 126402. https://doi.org/10.1016/j.chemosphere.2020.126402.

[32] Salem, M. A., Al-Ghonemiy, A. F., and Zaki, A. B. (2009). Photocatalytic degradation of Allura red and Quinoline yellow with Polyaniline/TiO2 nanocomposite. Applied Catalysis B: Environmental. 91, 59–66. https://doi.org/10.1016/j.apcatb.2009.05.027.

[33] Resasco, J., Zhang, H., Kornienko, N., Becknell, N., Lee, H., Guo, J., Briseno, A. L., and Yang, P. (2016). TiO2/BiVO4 nanowire heterostructure photoanodes based on type II band alignment. ACS Central Science. 2, 80–88. https://doi.org/10.1021/acscentsci.5b00402.

[34] Li, T., Cui, J., Gao, L., Lin, Y., Li, R., Xie, H., Zhang, Y., and Li, K. (2020). Competitive Self-Assembly of PANI Con fi ned MoS2 Boosting the Photocatalytic Activity of the Graphitic Carbon Nitride, ACS Sustain. Chemical Engineering Journal. 8, 13352–61. https://doi.org/10.1021/acssuschemeng.0c04089.

[35] Jawad, A. H., Shazwani, N., Mubarak, A., Azlan, M., Ishak, M., Ismail, K., and Nawawi, W. I. (2016). Kinetics of photocatalytic decolourization of cationic dye using porous TiO2 film. Integrative Medicine Research. 10, 352–62. https://doi.org/10.1016/j.jtusci.2015.03.007.

[36] Saravanan, R., Karthikeyan, S., Gupta, V. K., Sekaran, G., Narayanan, V., and Stephen, A. (2023)0. Enhanced photocatalytic activity of ZnO/CuO nanocomposite for the degradation of textile dye on visible light illumination. Material Science Engineering. C. 33, 91–98. https://doi.org/10.1016/j.msec.2012.08.011.

[37] Golshan, M., Zare, M., Goudarzi, G., Abtahi, M., and Babaei, A. A. (2017). Fe3O4@HAP-enhanced photocatalytic degradation of Acid Red73 in aqueous suspension: Optimization, kinetic, and mechanism studies. Materials Research Bulletin. 91, 59–67. https://doi.org/10.1016/j.materresbull.2017.03.006.

Chapter 47

A review on effects of cryotreatment on tungsten carbide tools and its expected outcomes in micromachining

Narendra Bhople[1,a], Narayan Sapkal[1,b], Prateek Malwe[1,c], Dhiraj Deshmukh[2,d], Manoj Nikam[3,e] and Srihari Notla[4,f]

[1]Dr. D.Y. Patil Institute of Technology, Pimpri India

[2]MET Bhujbal Knowledge City, Nasik, India

[3]Bharati Vidyapeeth College of Engineering, Navi Mumbai, India

[4]JSPM RSCOE, Pune, India

Abstract

The trend of micromanufacturing has been increasing from last decade. Researcher realised the need of non-MEMS (micro electro-mechanical system) techniques due to the limitation MEMS such as, types of material processing and manufacturing of complex three dimensional components. Now days micromachining is the preferred manufacturing process to produce microcomponents. This process can be used to process wide variety of materials and able to produce three-dimensional complex geometry. Mechanical microcutting processes is the down scaling of conventional machining process, this considerably affected the material removal mechanism of micromachining. It is found tool life in micromachining significantly reduced due to material heterogeneity, size effect and minimum chip thickness. The probability of catastrophic failure of tool is also high while machining hard metals. Unpredictable tool life is the main hurdle for the development of micromachining. The conventional tool life improvement approach has some constraint in micromachining. It has been found the cryotreatment (CT) of tool is body phenomenon. The metallurgical changes after cryotreatment of tool give longer tool life and improved performance. This paper mainly focus on identifying the key phenomenon of micromilling and how cryotreated tungsten carbide (WC) end mill can contribute to improve material removal mechanism and tool life in micromilling.

Keywords: Cryotreatment, micromachining, size effect, tungsten carbide

Introduction

The trend of miniaturisation started in the beginning of third millennium in biomedical, automobile, electronics, aviation and others sectors of the industry [1]. Initially the micro electro-mechanical system (MEMS) or lithography based processes had wide use to produce semiconductors and micro electric components like sensors, actuators etc. from silicon and limited range of material. In last two decade researchers have been invented non-MEMS techniques which can be use to process almost all type of material and to produce complex three dimensional components. Micromachining is one of the important type of non-MEMS techniques [2–4]. Micromachining or micro cutting is similar to conventional machining, it is the geometrical reduction of all process parameters [3]. Similar to macro machining turning has been using to machine micro cylindrical components while drilling to produce micro holes in nozzle like structure. Micro milling is the demanding fabrication technique, this process has extensive use in the development of micro tooling such as hot embossing, micro injection moulding etc. It has been found all the above micromachining processes has wide use from jewelry manufacturing to aviation industry [5]. Cutting tool is one of the key parameters of micromachining. The quality of final product is mainly depends on the performance of cutting tool.

Many author defined micromachining on the basis of tool diameter or burr size. Tool diameter falls in range of 1 to 999 μm or if undeformed chip thickness is comparable to tool edge radius or material grain size' [6]. Reduced stiffness of cutting tool is the main constrain for miniaturisation of the cutting process. The whole process of machining is significantly affected by size effect, material inhomogeneity and minimum chip thickness. Generally, the failure of the tool occurred by generation of excessive stress, increase in specific cutting energy and fatigue related breakage [5,7]. Machining of harden material is challenging in micromachining due to lower stiffness of cutting tool. Tool radius suddenly chip-off while machining hard metals [8]. The value of cutting forces significantly increases due to increase in radius and this promotes the ploughing mechanism. [9,10]. The existence of residual stresses while sintering and presence of microcracks are the main source for tool failure. The cost of microtools is comparatively higher than that of conventional tools [11]. Micro dies manufactured by harden steel, to machine micro injection dies needs cutting tool with higher strength [8]. Extreme finished surface required for the easy removal of component

[a]nrbhople11@gmail.com, [b]narayan.sapkal@dypvp.edu.in, [c]prateek.malwe@dypvp.edu.in, [d]dhirgajanan@gmail.com, [e]manoj.nikam133@gmail.com, [f]snotla_mech@jspmrscoe.edu.in

DOI: 10.1201/9781003450252-47

from the mold [7]. To achieve the surfaces with desire finish, it is compulsory to machine steel after heat treatment [12]. In micromachining due to reduced stiffness of the tool and while machining hard metals the probability of catastrophic failure of the tool increases [1,7].

Tool life improvement techniques such as application of hard coating, use of cutting fluid and hybrid machining have regular use in conventional machining. These techniques have some constrain in micromachining. Transportation of cutting fluid in such narrow zone is difficult and cutting fluid may affect the provided feed and depth of [8]. Minimum quantity lubrication (MQL) techniques does not give noticeable improvement in machining, droplets do not adhered adequately to the tool due to higher spindle speed [9]. Application of coating increases the cutting edge radius and it will be useless once it get removed from the surface. Laser assisted machining gives higher value of roughness at higher cutting speed [13]. Cryogenic treatment of cutting tool can be a good alternative to improve the life of the micro tools. Cryogenic treatment also called as subsero treatment. This process can be used to treat various metals, composites at low processing cost. Cryogenic treatment can improve the properties of tungsten carbide tools by producing additional eta (η) carbide in matrix along with some metallurgical changes. The desired properties can be achieved by selecting proper cryo-cycle [14]. CT gives refined, uniform and dense microstructure this leads to improvement in physical and mechanical properties of carbide tools.

In conventional machining cryotreated tools performed better than untreated and some coated tools. The application of cryotreated tools while turning improve wear resistance of WC tools up to 29% in flank wear, 67% in crater wear and 81% in notch wear compared to untreated tools [45]. The cryotreated micromilling cutters performed better in wet as well as dry condition. Coated tools after cryotreatment performed better than that of only coated tools [46]. The desired properties can be imparted to cutting tools by altering the cryotreatment parameters for the given cutting tool material [47].

Bhople et.al. found significant improvement in hardness of micro end mill by using shallow and deep cryogenic treatment. The hardness of deep cryotreated tool for 24-hour tool improved by 9%. They also observed variation in the average grain size and formation of additional η carbide. Deeply cryotreated micro end mill performed better in machining test. Bhople et. al. analyses the uncoated, TiN coated and cryotreated micro tungsten carbide tools while micromilling of Ti-6Al-4V. It is found that the TiN coated tool performed better in terms of surface roughness. Application of coating contributed to dissipate heat and to reduce coefficient of friction. surface finish for cryotreated tool affected due to BUE. It is found, 24-hour soaking period at DCT is not sufficient to improve thermal conductivity of WC tool [48-50]. Notable literature is not available related to application of cryotreated tool in micromachining. The main objective of this paper is to study the contribution of cryotreatment to improve the mechanical and physical properties of tungsten carbide tool in macromachining. Moreover, to predict the performance of micro cryotreated tools, especially milling in terms of size effect, minimum chip thickness and material heterogeneity.

Micromachining Process Physics

Though micromachining seems similar to conventional machining, in fact there is difference in actual material removal mechanism. The assumptions made in merchant's theory are not applicable in micromachining. A small vibration of machine tool may affect the accuracy of final product. The occurrence of size effect, experience of material inhomogeneity and minimum chip thickness by cutting tools affect cutting force, surface roughness and tool life significantly. Size effect is the deviation from generalised values of process characteristics, when there is change in dimensions of process parameter(s) such as in machine tools and cutting tools etc. [15]. According to Backer et.al. [16] size effect is the considerable rise in the shear flow stress of the workpiece due to decrease in the uncut chip thickness. It can also explain as, a remarkable and nonlinear increase in the specific cutting energy as the undeformed chip decreases. It is found, the undeformed chip thickness and tool edge radius has strong relation with size effect. When undeformed chip thickness reduces below the cutting edge radius or when it is equivalent to grain size of the workpiece material, it promotes the size effect. Moreover, this influences the deformation mechanism, chip formation and flow [17, 18]. In micro milling feed (radial depth of cut) decides the undeformed chip thickness, therefore feed is important input parameter of micro milling.

Milling needs higher swept angle to plastically deform the material at low undeformed chip thickness or small feed/tooth, this also contribute to increase the specific cutting energy [7]. While machining at submicron level, movable dislocations experience more obstacles hence additional energy required to overcome large atomic bonding within crystal [16]. Size effect also accelerated due to inhomogeneities and tiny defects present in all engineering metals and alloys [19]. It is found geometry of cutting tool also contribute for size effect. Negative rake angle of the tool promotes indenting instead of cutting this also causes increase in specific cutting energy [20]. There is no clear agreement with the factors causing the size effect. However, generation of higher shear strain, reduction of uncut chip thickness below cutting edge radius or when comparable to grain size of the workpiece and work hardening at small undeformed chip thickness these are the major factors which causes size effect.

The assumption of isotropy and material homogeneity cannot be considered for micromachining due to considerable reduction in process parameters [21]. Workpiece has several characteristics which decide its machinability. The phases present in microstructure, crystallographic orientation, grain size, presence of defects and impurities must be consider while machining. The change in crystallography leads to variation in cutting forces this causes deflection while machining. Poly-crystalline material must be considered heterogeneous and discrete. The machining issue belongs to workpiece material, presence of defects and impurities in engineering material are difficult to eliminate [3].When cutting edge radius approaches the grain size of the workpiece, machining occurs by fracturing single grain. For the deformation of single grain need more amount of energy, this leads to high fluctuation of cutting forces [7, 8]. Multiphase material consists different phases, each phase having different hardness. In such case cutting edge face the individual phase this leads to variation in material removal, inconsistency in machining and accelerated tool wear [22]. When grain size larger than uncut chip thickness, spring back may occur while machining [8].

Minimum chip thickness suggest below a certain uncut chip thickness, chip will not form [3] Push-deform (elastic deformation) phenomenon occurs instead of cutting, if the provided feed and depth of cut remain below the critical chip thickness. It is found feed plays the significant role for commencement of chip formation. Whenever the ratio of uncut chip thickness equivalent or less than tool edge radius effective rake angle become negative. Negative rake angle of the tool causes ploughing and reduce the probability of chip formation [3, 23]. To achieve surfaces with better finish, the estimation of proper ratio of uncut chip thickness to cutting edge radius with respect to material is important [24, 25]. It is observed, sharpness of cutting edge radius has strong relationship with minimum chip thickness [25]. If the cutting tool retained sharp cutting edge the elastic deformation can be eliminate.

Cutting force is the key response of machining process, the value of surface roughness and tool life associated with the generating cutting forces. The lower stiffness of the microtools put the limitations to exceed the axial depth of cut, it is the main source for the catastrophic failure of tool [26]. Increased cutting force may cause tool run out this leads to reduce preciseness and deterioration of the surface [24]. Worn cutting edge also increases the value of cutting forces significantly. [27]. Grain size equals to cutting edge radius, negative rake angle and multiphase material are the potential source for higher value of cutting forces [28]. While micro dies manufacturing it must satisfy the attributes like, high accuracy, nano range surface finish and minimal burr size [7]. According to Weule et al. microinjection mold with surface finish Rz equl to or less than 1.00 µm makes easy removal of the product from dies [25]. Feed is the significant parameter for surface roughness hence selection of proper feed is important to achieve desire finish. At lower feed (below minimum chip thickness) plastic deformation or ploughing may occur which increases the roughness while at higher feed roughness increases due to increase in chip load (similar to conventional machining) [7]. Surface roughness also has strong relationship with sharpness of tool. Sharp cutting edge supports chip formation while worn cutting edge promote ploughing or elastic deformation [18, 20, 25].

Fundamentals of Cryotreatment

In the starting of sixteenth century cryogenic treatment began to be used to improve the properties of material. Basically, the word "cryogenic" is Greek word which means "krys" means cold and "genic" treatment [14]. It has been found the properties of the material can be improved by subjecting the material in particular cryocycle.

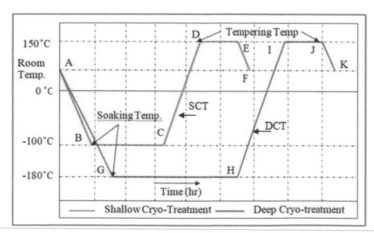

Figure 47.1 Typical cryogenic treatment process [29]

Figure 47.1 shows typical plot of cryogenic cycle. Cryogenic treatment broadly classified in two types. Shallow cryogenic treatment (SCT) and deep cryogenic treatment (DCT). The soaking temperature range for SCT is -80–140°C, while for DCT is -140–196°C [30–34]. The stages involve in cryogenic process are: (a) cooling of component from room temperature to desire negative temperature with certain cooling rate (0.5-1.8°C/min). (b) After achieving required negative temperature holding component for const temperature (between –80–196°C) this is called soaking temperature. (c) Holding of that component for certain duration (8-72 hour) at constant temperature called soaking period. (d) Bring the component to room temperature to tempering temperature (100-200°C). (e) Tempering 1-2 hour and again bring it to room temperature [30–34]. Liquid nitrogen is the preferred gas to conduct CT, it is abbreviated as LN2 [35].

To achieve desire properties, right selection of the cryo parameters is important. The end results also depend on the phases, crystal structure and chemical composition of workpiece before treatment. Soaking temperature is the most significant parameter, it contributes 72% to improve the properties of material. Rate of cooling has 10%, soaking period 24%, and tempering temperature 4%. Tempering period is insignificant for cryo-process [36, 37]. However, few authors obtained significant effect with change in tempering process parameters [38].

Metallurgical Changes in WC-Co after Cryotreatment

Tungsten carbide tools have wide use in manufacturing industry, as diamond has limitation to process ferrous base material [38, 39]. Powder metallurgy is one of the popular carbide tool manufacturing processes. Table 47.1 shows the different phases present in tungsten carbide tool.

Hard phase tungsten carbide and soft phase cobalt binder combine together to form tungsten carbide tool. Cobalt binder holds the hard tungsten particles together and form continuous network structure which gives additional strength to carbide tool [29]. Alpha (α) particles have high hardness and β provide toughness. Figure 47.2 shows the tungsten carbide particles holds by cobalt matrix.

Many researchers obtained significant improvement in tool life due to the application of cryotreated tool in conventional machining. By selecting the appropriate cryo-cycle the strength, hardness, thermal conductivity, toughness and electrical conductivity of the material can be improved. Figure 47.3 (a) and (b) shows the microstructure of tungsten carbide tool before and after CT respectively. The mechanism behind the improvement of tungsten carbide tools due to CT summaries in the below section [29, 38, 40–43].

1) Precipitation of η carbide has major role to improve the properties of carbide tool. CT forms additional fine η carbide along with already present larger size carbides. The voids in the matrix are eliminated by combination of fine and larger η carbides and results in a denser, tougher and more coherent matrix.

2) CT can produce two type of ternary η carbide, constant composition $M_{12}C$ and M_6C in which composition may vary in the range of $Co_{3.2}W_{2.8}C$ to Co_2W_4C. M_6C carbide increases the brittleness and

Table 47.1 Phases present in WC-Co material [29]

Phase symbol	Meaning of symbol
α	Tungsten carbide (WC)
β	Cobalt metal binder (Co)
γ	Carbide of a cubic lattice (TiC, TaC, NbC)
η	Multiple carbides of tungsten and at least one metal of the binder

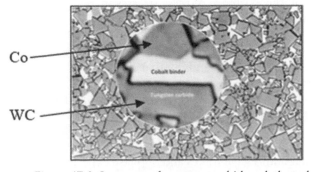

Figure 47.2 Structure of tungsten carbide-cobalt tool [29]

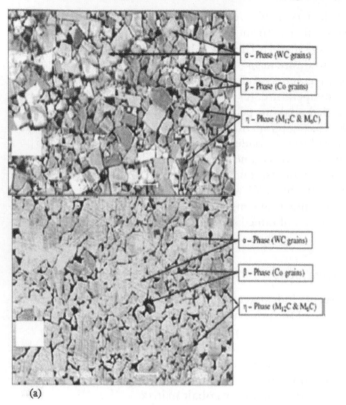

Figure 47.3 Microstructure of tungsten carbide tool (a) Untreated (b) Cryotreated [29]

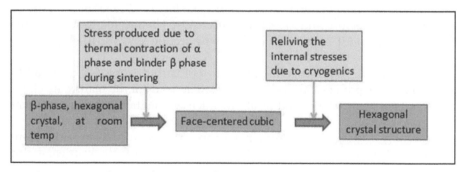

Figure 47.4 The transformation of b-phase crystal structure with temperature
Source: Author

affects the strength of matrix by reducing effectiveness of η phase. The M_6C type η-phase carbide is stable in liquid phase it nucleates and grows while sintering process.

3) $M_{12}C$ (Co_6W_6C) type η carbide precipitated in the solid state and small grains distributed in entire matrix. It is hard and has no any detrimental effect. Co_6W_6C exist only at the surface / coating interface.

4)   Cryo-process enlarges the α-phase particles, due to densification of cobalt phase (β). α-phase particles arrange in such manner which significantly increase thermal conductivity. The increased thermal conductivity is attributed to the decrease in distance between the α- particles.

5)   While sintering micro stresses induce in cutting tool, the existence of these stresses increase the probability of tool brakeage. CT relives that all type of internal residual stresses.

6)   The densification of β-phase occurs while CT, this increases the bonding between carbide particles and cobalt phase.

7)   CT refines and align the tungsten carbide particles into stress free and durable crystallographic orientation this also reduces the risk of stress induce fracture.

8)   Figure 47.4 shows the allotropy of β-phase. It shows room temperature β-phase consist hexagonal crystal structure as the temperature increases during sintering, it transforms to face centered cubic. CT relieves the internal residual stresses this leads to transformation of β- phase to its previous hexagonal crystal structure.

9)   During CT of tungsten carbide, α-phase experience compressive forces and β- phase experience tensile forces. This phenomenon decrease ductility of cobalt phase and significantly increase its hardness.

Expected Outcome of Cryotreated Micro Tools While Machining

While going through the literature of micromanufacturing, it is found the invention of non-MEMS techniques especially micromachining overcomes the many limitations of MEMS based processes. The performance of micromachining majorly depends upon the characteristic of cutting tool. It is found shorter or unpredictable tool life is the main hurdle for the development micromachining processes. If it got possible to retain the sharpness of the tool for longer time, then along with longevity of the tool life other responses can also be improved. In the previous section the fundamentals of cryogenics presented. As discussed in previous section, end results of cryotreatment depend on cryocycle parameters, section size, initial chemical composition and structure of component. More research required to identify the suitable parameters of CT for microtool, as microtool has very small geometrical features.

Size effect can be reduce by using cryotreated tool, CT significantly improve the hardness of tool due to the precipitation η carbide. Whenever the uncut chip thickness will reduce below cutting edge radius or comparable to grain size the probability and amount of elastic deformation may reduce due to the sharp cutting edge obtained by cryotreatment. Further, it can cut the single grains without significant increase in specific cutting energy due to increase in strength. Due to retainment of sharp cutting-edge material may deform at lower swept angle, this can reduce the value of shear stress considerably. Cryotreated tool can overcome the resistance offered by large atomic bonding within crystal while machining at submicron level. In work hardening, tool continuously face the harden part of the workpiece. By selecting suitable cryocycle, hardness of cutting tool can be increase than that of the harden part of workpiece this will ease the machining.

While machinery, multiphase material tool travel through different phases which has different hardness and ductility. As tool enter from soft phase to hard phase it experiences shocks. CT eliminates the voids in the matrix due to combination of large and fine η carbide results in significant improvement in toughness. The cryotreated tool with improved toughness can reduce the probability of catastrophic failure while machining multiphase material. Similarly polycrystalline material, presence of inhomogeneities and impurities in commercial metals produce higher deflection, in such case cryotreated carbide tool can perform better.

Minimum chip thickness has strong relationship with provided feed and depth of cut. If the cutting edge of tool wear immediately, it will promotes elastic deformation and higher value of feed and depth of cut will required for further chip formation. Sharp cutting edge can contribute to commence chip formation at lower feed and depth of cut by avoiding elastic deformation. Finally minimum chip thickness can be reduce or due to the application cryotreated tool chip formation can be start at lower value of provided feed.

Generally worn cutting edge produces higher value of cutting force, this affects the surface roughness and too life. If cutting forces exceeds the particular value, it may cause tool breakage. The existence of residual stresses (while sintering) in the tool is one of the reason for tool breakage in micromachining. These stresses can be relived in cryogenic treatment and probability of tool breakage can be reduces. Formation of ternary η carbide in matrix can retain sharpness of the tool for longer time, this can be reducing the cutting forces.

The interaction of low value of feed and high cutting speed gives better surface finish. In micromachining lower uncut chip thickness promotes size effect and elastic deformation. The densification of cobalt phase while CT increase the bonding between cobalt phase and carbide particles significantly. This bonding holds the carbide particles strongly and offer higher resistance for wear. CT scan permit tool to machine at lower feed (lower undeformed chip thickness) to achieve desire surface finish. Higher cutting speed may produce built up edge (BUE) due to thermal softening of workpiece. The probability of BUE is higher in low thermal conductivity material, in case titanium based material formation of BUE deteriorate the surface [44]. CT enlarge the α-phase particles, this causes reduction in distance between the α-phase particles. Enlarged α-phase particles significantly increase the thermal conductivity of carbide tool. Increased thermal conductivity in cutting zone can reduce the chances of BUE. The application of hard coating on substrate increases the cutting-edge radius this can produce the surfaces with higher roughness value. CT is body phenomenon; the tool properties improvement is the result of metallurgical changes. CT does not affect the cutting edge radius geometrically. The cost of CT can be reduced by processing multiple tool at a time.

Conclusions

Based on the reviewed literature of micromachining and cryogenic treatment of tungsten carbide tool following conclusion can be drawn: Micromachining strongly depends on characteristic of cutting tool. Lower uncut chip thickness, material heterogeneity and worn cutting edge are the main cause of size. Retainment of cutting edge for longer time and right combination of hardness and toughness can improve the life of microtools and overall machining process. Cryotreatment has potential to increase the hardness by precipitation of additional η carbides. Formation of additional η carbides also contributes to improve density and toughness of the matrix. $M_{12}C$ (Co_6W_6C) ternary η carbide is the hard and exist at the surface, which can improve the life of microtools. Densification of β-phase increase the bonding between carbide particles and cobalt phase this imparts the enhanced strength to carbide tool. Cryotreatment relieves internal stresses which produced during sintering, this avoids the catastrophic failure of tool. Enlargement α- phase particles increase the thermal conductivity this can be reduces the probability of built up edge in case of microtools.

Future Scope

Selection of cryo parameter plays the important role to decide the properties of cutting tool. Further research work can be extended to study the effect of different cryogenic parameters (soaking temperature, time etc) on the properties microtools and its application in micromachining. The effect of cryo chamber size and pressure on characteristics of tool can also be studied in further research work.

References

[1] Camara, M. A., Campos Rubio, J. C., Abrao, A. M., and Davim, J. P. (2012). State of the art of micromilling of material, A review. Journal of Materials Science and Technology. 28(8), 673–685.

[2] Takacs, M., Vero, B., and Meszaros, I. (2012). Micromilling of metallic material. Journal of Material Processing Technology. 138, 152–5.

[3] Piljek, P., Keran, Z., and Math, M. (2014). Micromachining – review of literature from 1980 to 2010. Interdisciplinary Description of Complex Systems. 12(1), 1–27.

[4] Chae, J., Park, S. S., and Freiheit, T. (2005). Investigation of micro-cutting operation. International Journal of Machine Tools & Manufacture. 46, 313–32.

[5] Gandarias, E. (2009). Micromilling technology; Global Review. Courtesy of global thesis.

[6] Aramcharoen, A., Mativenga, P. T., Yang, S., Cooke, K. E., and Teer, D. G. (2008). Evaluation and selection of hard coatings for micro milling of hardened tool steel. International Journal of Machine Tools and Manufacture. 48(14), 1578–84.

[7] Aramcharoen, A., and Mativenga, P. T. (2008). Size effect and tool geometry in micromilling of tool steel. Precision Engineering. 33, 402–7.

[8] Bissacco, G., Hansen, H. N., and De Chiffre, L. (2005). Micromilling of hardened tool steel for mould making applications. Journal of Materials Processing Technology. 167, 201–7.

[9] Cardoso, P., and Davim, J. P. (2012). A brief review on micromachining of material. Reviews on Advanced Materials Science. 30, 98–102.

[10] Ajish, T. N., and Dr. Govindan, P. (2014). Effects and challenges of tool wears in micro milling by using different tool coated materials. International Journal of Research in Mechanical Engineering & Technology (IJRMET). 4(2), 2249–5762.

[11] Carou, D., Rubio, E. M., Herrera, J., Lauro, C. H., and Davim, J. P. (2017). Latest advances in the micro-milling of titanium alloys a review. In Manufacturing Engineering Society International Conference 2017, MESIC 2017, 28-30 June 2017, Vigo (Pontevedra), Spain.

[12] Vázquez, E., Rodríguez, C. A., Elías-Zúñiga, A., and Ciurana, J. (2010). An experimental analysis of process parameters to manufacture metallic micro-channels by micro-milling. International Journal of Advanced Manufacturing Technology. 51, 945–955.

[13] Melkote, S., Kumar, M., Hashimoto, F., and Lahoti, G. (2009). Laser assisted micro-milling of hard-to-machine materials. CIRP Annals - Manufacturing Technology. 58, 45–48.

[14] Akincioglu, S., Gokkaya, H., and Uygur, I. (2015). A review of cryogenic treatment on cutting tools. International Journal of Manufacturing Technology. 78, 1609–27.

[15] Vollertsen, F., Biermann, D., Hansen, H. N., Jawahir, I. S., and Kuzman, K. (2009). Size effects in manufacturing of metallic components. CIRP Annals - Manufacturing Technology. 58(2) 566–87.

[16] Backer, W. R., Marshall, E. R., and Shaw, M. C. (1952). Size effect in metal cutting. Transactions of the ASME. 74(1), 61–72.

[17] Mian, A. J., Driver, N., and Mativenga, P. T. (2011). Identification of factors that dominate size effect in micro-machining. International Journal of Machine Tools & Manufacture. 51, 383–94.

[18] Ng, C. K., Melkote, S. N., Rahman, M., and Kumar, A. S. (2006). Experimental study of micro and nano-scale cutting of aluminum 7075-T6. International Journal of Machine Tools and Manufacture. 46, 929–36.

[19] Shaw, M. C. (2003). The size effect in metal cutting. Sadhana, 28(5), 875–96.

[20] Lui, K., and Melkote, S. N. (2007). Finite element analysis of the influence of tool edge radius on size effect in orthogonal micro-cutting process. International Journal of Mechanical Sciences. 49, 650–60.

[21] Elkaseera, A., Popova, K., Dimova, S., Phama, D., Olejnikb, L., and Rosochowskic, A. (2009). Micro milling of coarse grained and ultrafine grained Cu99. 9E: Effects of material microstructure on machining conditions and surface quality. In Procceding of the 6th International Conference on Multi-Material Micro-Manufacture, Germany, pp. 241–244.

[22] Lee, W. B., Cheung, C. F., and To, S. (1999). Material induced vibration in ultra-precision machining. Journal of Material Processing Technology. 89-90, 318–25.

[23] Ducobu, F., Filippi, E., and Riviere-Lorphevre, E. (2009). Chip formation and minimum chip thickness in micro-milling. In Proceedings of the 12th CIRP Conference on Modeling of Machining Operations, pp. 339–46.

[24] Vogler, M. P., DeVor, R. E., and Kapoor, S. G. (2004). On the modeling and analysis of machining performance in micro-endmilling, part I: surface generation. Journal of Manufacturing Science and Engineering, Transactions of the ASME. 126(4), 685–94.

[25] Weule, H., Huntrup, V., and Tritschler, H. (2001). Micro-cutting of steel to meet new requirements in miniaturisation. CIRP Annals - Manufacturing Technology. 50(1), 61–64.

[26] Pratap, T., and Patra, K. (2017). Finite element method based modeling for prediction of cutting forces in micro-end milling. Journal of The Institution of Engineers (India): Series C. 98, 17–27.

[27] Alting, L., Kimura, F., Hansen, H. N., and Bissaco, G. (2003). Micro engineering. CIRP Annals 52(2), 635–657.

[28] Özel, T., Lui, X., and Dhanorker, A. (2007). Modelling and simulation of micro-milling process. In: Proceedings of the 4th International Conference and Exhibition on Designand Production of Machines and Dies/Molds. 2007.

[29] Gill, S., Singh, J., Singh, H., and Singh, R. (2012). Metallurgical and mechanical characteristics of cryogenically treated tungsten carbide (WC-Co). International Journal of Advanced Manufacturing Technology. 58, 119–31.

[30] Thakur, D., Ramamoorthy, B., and Vijayaraghavan, L. (2008). Influence of different post treatment on tungsten carbide-cobalt inserts. Materials Letters. 62, 4403–6.

[31] Reddy, T. V. S., Sornakumar, T., Reddy, M. V., and Venkatram, R. (2009). Machinability of C45 steel with deep cryogenic treated tungsten carbide cutting tool inserts. International Journal of Refractory Metals and Hard Materials. 27, 181–85.

[32] Bensely, A., Venkatesh, S., Lal, D. M., Nagarajan, G., Rajadurai, A., and Junik, K. (2008). Effect of cryogenic treatment on distribution of residual stress in case carburised En 353 steel. Materials Science and Engineering: A. 479, 229–35.

[33] Bensely, A., Prabhakaran, A., Mohan Lal, D., and Nagarajan, G. (2006). Enhancing the wear resistance of case carburised steel (En 353) by cryogenic treatment. Cryogenics. 45, 747–54.

[34] Xuan, F. Z., Huang, X., and Tu, S. T. (2008). Comparisons of 30Cr2Ni4MoV rotor steel with different treatments on corrosion resistance in high temperature water. Mater Design. 29, 1533–39.

[35] Susheel, K. (2010). Cryogenic processing: a study of materials at low temperatures. Journal of Low Temperature Physics. 158, 934–45.

[36] Darwin, J. D., Mohan Lal, D., and Nagarajan, G. (2007). Optimisation of cryogenic treatment to maximise the wear resistance of chrome silicon spring steel by taguchi method. International journal of Materials Sciences. 2(1), 17–28.

[37] Darwin, J. D., Mohan Lal, D., and Nagarajan, G. (2008). Optimisation of cryogenic treatment to maximise the wear resistance of 18% Cr martensitic stainless steel by Taguchi method. Journal of Materials Processing Technology. 195(1–3), 241–7.

[38] Kalsi, N., Sehgal, R., and Sharma, V. (2014). Effect of tempering after cryogenic treatment of tungsten carbide–cobalt bounded inserts. Materials Science. 37(2), 327–35.

[39] Kalpakjian, S., and Schmid, S. R. (2002). Manufacturing Processes for Engineering Materials. New Jersey: Prentice-Hall.

[40] Shabouk, S., and Nakamoto, T. (2003). Micro machining of single crystal diamond by utilisation of tool wear during cutting process of ferrous material. Journal of Micromechatronics. 2(1), 13–26.

[41] Reddy, T. V. S., Ajaykumar, B. S., Reddy, M. V., and Venkataram, R. (2007). Improvement of tool life of cryogenically treated P-30 tools. In Proceedings of International Conference on Advanced Materials and Composites (ICAMC-2007) at the National Institute for Interdisciplinary Science and Technology, CSIR, Trivandrum, India, pp. 457–460.

[42] Seah, K. H. W., Rahman, M., and Yong, K. H. (2003). Performance evaluation of cryogenically treated tungsten carbide cutting tool inserts. Proceedings of the Institution of Mechanical Engineers - Part B: Journal of Engineering Manufacture. 217, 29–43.

[43] Bryson, W. E. (1999). Cryogenics. Cincinnati: Hanser Gardner, pp. 81–10.

[44] Mittal, R., Kulkarni, S., and Singh, R. (2017). Effect of lubrication on machining response and dynamic instability in high speed-micromilling of Ti-6Al-4V. Journal of Manufacturing Process. 28(3), 413–21.

[45] Ozbek, N. A., Cicek, A., Gulesin, M., and Ozbek, O. (2014). Investigation of the effect of cryogenic treatment applied at different holding times to cemented carbide inserts on tool wear. International Journal of Machine Tools and Manufacture. 86, 34–43.

[46] Thamizhmanii, S., Nagib, M., and Sulaiman, H. (2011). Performance of deep cryogenically treated and non-treated PVD inserts in milling. Journal of Achievement in Materials and Manufacturing Engineering. 49(2), 460–466.

[47] Gill, S. S., Singh, H., Singh, R., and Singh, J. (2010). Cryoprocessing of cutting tool material-a review. International Journal of Advanced Manufacturing Technology. 48, 175–92.

[48] Bhople, N., Mastud, S., and Mittal, R. (2023). Metallurgical and machining performance aspects of cryotreated tungsten carbide micro-end mill cutters. Proceedings of the Institution of Mechanical Engineers, Part B: Journal of Engineering Manufacture. 237(3), 492–502.

[49] Bhople, N., Mastud, S., and Mittal, R. (2022). Performance analysis of uncoated, TiN coated and cryotreated micro tungsten carbide tools while micromilling of Ti-6Al-4V. International Journal of Machining and Machinability of Materials. 24(1-2), 110–131.

[50] Bhople, N., Mastud, S., and Satpal, S. (2021). Modelling and analysis of cutting forces while micro end milling of Ti-alloy using finite element method. International Journal of Simulation Multidisciplinary Design and Optimisation. 12(26), 1–10.

Chapter 48

Application of ANN to predict surface roughness and tool wear under MQL turning of AISI 1040

Dr Prashant Prakash Powar[a], Shivraj Kadam[b] and Nitin Desai[c]

Kolhapur Institute of Technology, College of Engineering (Autonomous), Kolhapur, India

Abstract

Minimum quantity lubrication has proven to be more effective and efficient than dry and wet lubrication during machining of materials. However, components surface roughness and tool wear are still of the major concerns affecting the implementation of the said technology in industry. Prediction of these parameters helps to optimise cutting conditions and/or MQL specific parameters. Although various methods are used in the prediction of surface roughness and tool wear such as dimensional analysis, ANOVA etc., present work aims to develop the model for surface roughness and tool wear using artificial neural network. The proposed estimator is based on a neural network with cutting conditions and MQL specific parameters as the input parameters and surface roughness and tool wear as the predicted parameters. The feed forward network that provides ten hidden neurons has been selected. The coefficient of correlation shows good fit with the training, testing, focused and confirmation data sets. The results give encouraging results while using neural network model for the specified conditions used in an experimentation.

Keywords: ANN, surface roughness, tool wear

Introduction

Machining operations in all the manufacturing industries are playing major roles in deciding product quality. Reduction in overall cost with the specified quality of component is a major task in all manufacturing industries. Usage of cutting tools, cutting tools, cutting conditions, specific requirements etc. may help to reduce the problems faced by manufacturers. Many studies have been seen to predict performance of MQL on surface roughness and tool wear on the individual basis. Attempts has also been observed where both the parameters are combined while analyzing the performance of a system. In addition to this, techniques like ANOVA, GRA, Dimensional analysis etc are used in many cases to investigate the effect on surface roughness, cutting forces, tool wear, chip analysis, tool life etc.

The widely used quality index of a product is surface roughness which is used in working on hard turning. Study discussed on MQL which resulted in temperature and good change in chip tool and work interaction [1]. Researchers studied performance of coated tools. MQL performance was found to be greater over dry and that of conventional wet turning [2]. More et al. [3] studied tool wear and performance of machining of cBN–TiN inserts and PCBN inserts. The cost effectiveness point of view cBN–TiN tools shows more capable and can be a complement to PCBN tools.

Artificial neural networks (ANNs) are used extensively in the manufacturing process to model the process viz. turning, milling and drilling [4]. ANN has got the learning and generalisation capabilities. It accommodates nonlinear variables as well as it adapts to changing environment and missing data. Extensive work has been carried out on forces modeling. This work reported that the ANN approach is faster and accurate over many other modeling methods e.g., numerical techniques, analytical modeling. The work has also been seen for cutting forces modeling, using feed-forward multi-layer neural networks with BP algorithm [5].

Researchers have drawn satisfactory results with ANNs for machining related output parameters and some have also drawn focus towards drawbacks such as optimisation of network parameters, the definition of number of layers and neural nodes in an application etc. [6]. Modeling and machining process monitoring is the area wherein ANN is finding its way [7, 8]. The numerous applications of ANN in this manufacturing sector are controlling of cutting processes, predicting the responses such as tool wear, surface roughness, forces [9, 10]. Another study proposed use of radial basis function to predict the parameters and its effect on forces [11]. Some researchers proposed ANN in turning for tool wear classification and monitoring [12]. The ANN was supplied with various inputs to calculate wear observed on flank [13]. The minimum un-deformed chip thickness used as input parameter to predict surface roughness [14]. Hybrid machining modeling was developed by [15] to predict the machining characteristics. ANN model have developed to predict the surface roughness and tool wear and validated with data sets which are not used during the training [17-19].

Based on literature review and preliminary experimentation, the experimentation has been conducted with the MQL specific parameters reported in the paper and cutting conditions on surface roughness and

[a]powar.prashant@kitcoek.in, [b]kadam.shivraj@kitcoek.in, [c]desai.nitin@kitcoek.in

DOI: 10.1201/9781003450252-48

Table 48.1 Parameters with their levels and steps

Input parameters	Min range	Max range	Step
Discharge	100	500	100
Distance of nozzle	2.5	6.5	1
Cutting Speed	100	140	10
Feed	0.1	0.3	0.15
Depth of cut	0.1	0.5	0.1

Source: Author

tool flank wear. Furthermore, paper aims to focus use of artificial neural networks in the field of wear and manufacturing processes is presented in this paper.

Experimental Details

Input Parameter Selection

This work is based on a dataset developed using experimental turning of AISI 1040 using carbide cutting tools. With MQL set up and using carbide insert the experiment is performed. The typical composition is C: 0.35-0.45, Si: 0.1-0.35, Mn: 0.4–0.7. Machining is performed using carbide insert CNMG 12 04 08. The main objective is then to develop an ANN model to predict output parameters viz surface roughness (Ra) and tool wear during turning. The cutting speed, feed, depth of cut along with, MQL set up parameters are considered as two more input parameters. Five levels of each input parameters are considered for each process parameters [16]. Even with the discussions and recommendations of some of the industry experts and through manufacturers catalogue, the values of cutting speed, feed and depth of cut were selected. MQL range referred from the literature is used to select the cutting fluid mixture rate [1]. Input parameters used during experimentation are given in Table 48.1.

Selection of Orthogonal Array

With the referred literature it is found that many a researcher have used Taguchi design of experiments (DOE) to plan the experiment [16]. So, in this study planning of experiment is done using Taguchi DOE approach.

Procedure used to Develop ANN using Neural Network Toolbox

ANN through manual way of implementation may be simple for smaller data sets of instructions. Even for a small set of neurons the use of ANN may help to predict the results. There are many software's available which will develop the ANN to predict the results. One such toolbox which is used in this study is "neural network toolbox" available in MATLAB.

The back propagation algorithm is used in many cases and widely implemented in all ANN's. Choosing and selecting the suitable architecture and parameters for the same is a very important task in ANN. Further there is no prescribed solution or theoretical set of instructions and or guidance to set up the ANN model. Therefore, the main focus is to customise the parameters used in the study and select the architecture to find out the suitable network for a given problem. As reported in literature, the parameters that require attention while developing a good ANN model are, the input and output data sets, the ANN model, the number of repetitions, rate of training, number of hidden nodes and the training function.

Artificial neural network uses a training function which is used to update the ANN parameters by reducing the error of the entire system. It updates every time the network is trained to get a good result and reaches a steady state. In this research, the data has been divided into two groups randomly training data, accounting for 75-80%, validation data and testing data, making up 20% of the total dataset. Most widely used algorithm for Back propagation is the Levenberg-Marquardt (L-M) algorithm. This training algorithm adjusts them iteratively to reduce error between experimental and predicted output values. After the training, testing is needed to evaluate the performance. Testing the ANN is carried out by applying a new input data set, which was not included in the training process. The adequacy of the ANN is evaluated by considering the coefficient of correlation (R). This measures the direct or inverse relationship between the inputs and outputs considered in the experimentation. The strong relationship is when the value is close to 1, no relationship is the value close to 0 and -1 is the inverse strong relationship.

The ANN network with five (5) input nodes viz. quantity of mixture, distance of nozzle block, cutting speed, feed, depth of cut, is used in this study and the same are used to develop the ANN architecture. The network shown Figure 48.1 is used to analyse the output parameters.

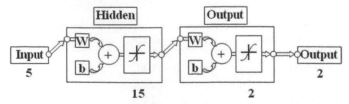

Figure 48.1 ANN network used to analyse surface roughness (Ra) and tool flank wear
Source: Author

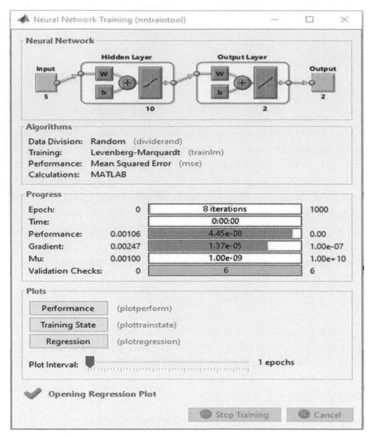

Figure 48.2 ANN network for AISI 1040
Source: Author

In present study, randomly 80% of experimental readings are used to train the network and remaining 20% are used to test the developed network.

The Assessment of the Model

The well-known statistical tools coefficient of correlation is used for this benchmarking. The adequacy of the ANN is evaluated by considering the coefficient of correlation (R).

Training and Testing of Models Developed for AISI 1040

The artificial neural network training with the 25 sets of reading and output patterns has been carried out using neural network toolbox available in 'MATLAB' software. The use of software procedure (training simulation) provides improved performance by allowing change in the learning rate. As discussed in earlier section, the training of data is performed using the ANN network shown in Figure 48.2.

The regression plot generated by an ANN model is shown in Figure 48.3 shows the closeness of predicted results. The plots represent the training, validation and testing of all data. The results indicate a good fit.

The network is trained with 80% of readings available from the experimental data set. The Table 48.2 shows the results of responses obtained from the developed ANN model for the training data set. The trained ANN for each of the input pattern, the predicted values of surface roughness and tool wear are computed. The graphical output of the computed and experimental values for the training data is presented in the figures. The predicted values of output parameters from the ANN network shown in Figure 48.4 are compared with the experimental {E} result. It shows the experimental and predicted {P} values. Acceptable results can be observed in Figure 48.5. Further the comparison of the 20% of the data set considered for testing is shown in a Figure 48.6

The testing data set for validation, the surface roughness and tool wear values are predicted using ANN model used which are compared with the experimental values.

Table 48.3 shows the data obtained from the testing data set. The comparison of the predicted and experimental data shows closeness of the values obtained.

After performing the NNA, the results obtained (predicted) are compared with the experimental values to check the deviation. The R for the training sets for the 20 readings (80%) of the total 25 readings is = 0.97 (97%) and testing sets for the remaining five readings (20%) is = 0.96 (96%) is shown in Table 48.4 which shows a closeness of readings obtained. Further the study of ANN model is carried out with focused experiment data set and confirmation experiment trials data set. In focused experimentation, since the cutting conditions (cutting speed, feed and depth of cut) are the commonly used input parameters and studied

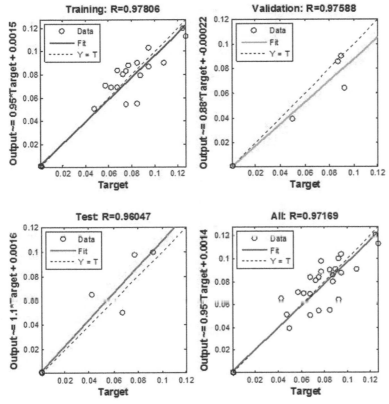

Figure 48.3 Regression plot for AISI 1040
Source: Author

Table 48.2 Experimental and predicted values of training network

Trial	E (Ra) (µm)	P (Ra) (µm)	E (Tw) (mm)	P (Tw) (mm)	Trial	E (Ra) (µm)	P (Ra) (µm)	E (Tw) (mm)	P (Tw) (mm)
2	0.80	0.733	0.057	0.067	15	0.828	0.907	0.092	0.083
4	1.029	0.96	0.127	0.106	16	1.057	1.295	0.09	0.090
5	1.018	1.073	0.125	0.118	17	1.099	0.978	0.087	0.090
6	1.393	1.681	0.075	0.058	18	0.883	0.904	0.067	0.068
7	1.485	1.37	0.067	0.077	19	1.364	1.311	0.077	0.080
8	0.954	0.968	0.075	0.089	20	0.81	0.84	0.085	0.070
9	0.807	0.713	0.095	0.103	21	0.823	0.93	0.085	0.088
11	1.264	1.061	0.107	0.101	22	1.388	1.395	0.087	0.082
12	0.857	0.813	0.095	0.091	24	0.861	0.812	0.047	0.067
14	1.367	1.377	0.072	0.074	25	0.783	0.802	0.062	0.073

Source: Author

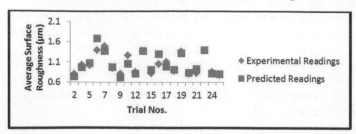

Figure 48.4 Comparison of 80% of experimental values of surface roughness considered for training purpose
Source: Author

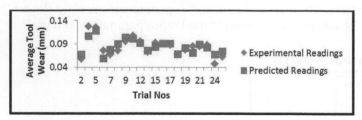

Figure 48.5 Comparison of 80% of experimental values of tool flank wear considered for training purpose
Source: Author

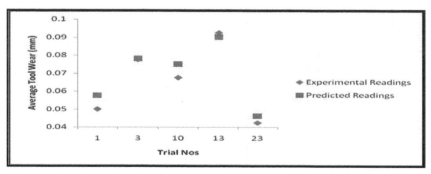

Figure 48.6 Comparison of 20% experimental values of tool flank wear considered for testing purpose
Source: Author

Table 48.3 Experimental and predicted values of testing network

Trial No	E (Ra) (µm)	P (Ra) (µm)	E (Tw) (mm)	P (Tw) (mm)
1	0.828	0.856	0.05	0.057
3	0.933	0.877	0.077	0.078
10	0.984	0.933	0.067	0.074
13	0.760	0.751	0.092	0.090
23	0.805	0.829	0.042	0.046

Source: Author

Table 48.4 Adequacy check for the ANN model

Adequacy check	Training (80%)	Testing (20%)	Focused experiment	Confirmation experiment
R	0.97	0.96	0.94	0.99

Source: Author

by many a researchers, it is decided to keep constant the earlier mentioned input parameters. The major parameters from the point of experimentation are quantity of mixture and distance, which are decided to vary with the constant cutting condition parameters and trials are conducted. Since the results obtained from this trials are not considered in development of ANN model, they are in later stage used for adequacy check. The results obtained are presented in the following table. Similarly the confirmation trial results are used in adequacy check.

Conclusion

In this work, the effect of MQL on turning of AISI 1040 is studied with surface roughness and tool wear as output parameters. The present MQL system has shown effectiveness in overall working of machines. It is learnt that the small reduction with the use of MQL has significant improvement in machinability indices.

Taguchi orthogonal array used in the experiment proved efficient for conducting experiment and the same is further used in neural network development for prediction of parameters. To predict both quality characteristics simultaneously artificial neural network (ANN) approach is used to develop the model. Based on the R value, different networks are found to be optimum amongst other. The results predicted for the 20% results, not used to develop the model, are found in acceptable agreement with the experimental values. The model has been given satisfactory results with experimental results. A satisfactory result of the neural network has been achieved with coefficient of correlation (R) between the model predicted and experimental values are 0.97 for training, 0.96 for testing, 0.94 for focused and 0.99 for confirmation. It must be learnt that results observed cannot be generalised to other ANN structures, other manufacturing processes, other component materials or cutting tools.

References

[1] Dhar, N. R., Kamruzzaman, M., and Mahiuddin, A. (2006). Effect of minimum quantity lubrication (MQL) on tool wear and surface roughness in turning AISI 4340 steel. Journal of Material Processing Technology. 172, 299–304.

[2] Kumar, C. H. R. V., and Ramamoorthy, B. (2007). Performance of coated tools during hard turning under minimum fluid application. Journal of Materials Processing Technology. 185, 210–6.

[3] More, A. S., Jiang, W., Brown, W. D., and Malshe, A. P. (2006). Tool wear and machining performance of cBN–TiN coated carbide inserts and PCBN compact inserts in turning AISI 4340 hardened steel. Journal of Materials Processing Technology. 180, 253–62.

[4] Ezugwua, E. O., Fadera, D. A., Bonneya, J., Da Silvaa, R. B., and Salesa, W. F. (2005). Modeling the correlation between cutting and process parameters in high speed machining of inconel 718 alloy using an artificial neural network. International. Journal of Machine Tools and Manufacture. 45, 1375–85.

[5] Makhfi, S., Velasco, R., Habak, M., Haddouche, K., and Vantomme, P. (2013). An optimised ANN approach for cutting forces prediction in AISI 52100 bearing steel hard turning. Science and Technology. 3(1), 24–32. DOI: 10.5923/j.scit.20130301.03

[6] Pontes, F. J., Silva, M. B., Ferreira, J. R., de Paiva, A. P., Balestrassi, P. P., and Schönhorst, G. B. (2010). A DOE based approach for the design of RBF artificial neural networks applied to prediction of surface roughness in AISI 52100 hardened steel turning. Journal of the Brazilian Society of Mechanical Sciences and Engineering. 32 (no. spe Rio de Janeiro)

[7] Dimla, D. E., Lister, P. M., and Leightont, N. J. (1997). Neural network solutions to the tool condition monitoring problem in metal cutting- A critical review of methods. International Journal of Machine Tools and Manufacture. 39, 1219–41.

[8] Sick, B. (2002). On-line and indirect tool wear monitoring in turning with artificial neural networks: a review of more than a decade of research. Mechanical Systems and Signal Processing. 16, 487–546.

[9] Karri, V. (1999). Performance in oblique cutting using conventional methods and neural networks. Neural Computing & Applications, 8, 196–205.

[10] Liu, Y., and Wang, C., (1999). Neural network based adaptive control and optimisation in the milling process. International Journal of Advanced Manufacturing and Technology. 15, 791–95.

[11] Elanayar, S., and Shin, Y. C. (1995). Robust tool wear estimation with radial basis function neural networks. ASME Journal of Dynamic Systems, Measurement and Control. 117, 459–67.

[12] Ghasempoor, A., Jeswiet, J., and Moore T. N. (1999). Real time implementation of on-line tool condition monitoring in turning. International Journal of Machine Tools and Manufacture. 39, 1883–1902.

[13] Liu, Q., and Altintas, Y. (1999). On-line monitoring of flank wear in turning with multilayered feed-forward neural network. International Journal of Machine Tools and Manufacture. 39, 1945–59.

[14] Grzesik, W. (1996). A revised model for predicting surface roughness in turning. Wear. 194, 143–8.

[15] Li, X., Dong, S., and Venuvinod, P. K. (2000). Hybrid learning for tool wear monitoring. International Journal of Advanced Manufacturing and Technology. 16, 303–7.

[16] Dave, H. K., Desai, K. P., and Raval, H. K. (2013). A taguchi approach-based study on effect of process parameters in electro discharge machining using orbital tool movement. International Journal of Machining and Machinability of Materials. 13(1), 52–66.

[17] Powar, P. P. (2022). Investigations into effect of cutting conditions on surface roughness under MQL turning of AISI 4340 by ANN models. Journal of Mines, Metals and Fuels. 70(8A), 1–479. DOI: 10.18311/jmmf/2022/32017.

[18] Mikołajczyk, T., Nowicki, K., and Bustillo, A. (2018). Predicting tool life in turning operations using neural networks and image processing. Mechanical Systems and Signal Processing. 104, 503–13.

[19] Baig, R. U., Javed, S., Khaisar, M., Shakoor, M., and Raja, P. (2021). Development of an ANN model for prediction of tool wear in turning EN9 and EN24 steel alloy. Advances in Mechanical Engineering. 13(6), 1–14.

Chapter 49

Application of VIKOR analysis in optimising PAW welding parameters of titanium alloy

Kondapalli Siva Prasad[a] and M Sailaja[b]

Anil Neerukonda Institute of Technology & Sciences, India

Abstract

Selection of welding process and choosing accurate welding parameters as per the weld metal thickness is an important task, in order to obtain the desired weld quality. Micro Plasma Arc welding (MPAW) plays an important role in joining thin sheets, as it uses low current. In the present work 0.5 mm thick titanium (Ti-6Al-4V) alloy sheets are welded by considering peak current, base current, pulse rate, and pulse width as input parameters and fusion zone grain size, hardness, and ultimate tensile strength are considered as output responses. Welding is carried out at welding speed of 260 mm/min and plasma gas flow rate of 6 lt/min. For four factor- five level Taguchi L_{25} orthogonal matrix is adopted by considering 95% confidence level in analysis of variance (ANOVA). Empirical mathematical models for the output responses are developed. In order to correlate the results used for maximizing hardness and ultimate tensile strength and minimizing grain size to find the optimal weld input parameters, this work focuses on the application of multi attribute decision making (MADM) methods like VIKOR (VIseKriterijumska Optimizacija Kompromisno Resenje) by considering equal weights. At a peak current of 34 amperes, back current of 11 amperes, pulse rate of 50 pulses/second, and pulse width of 50% optimal output is obtained.

Keywords: Micro plasma arc welding, multi attribute decision making, titanium, VIKOR analysis

Introduction

Ti-6Al-4V grade of Titanium was used in the aerospace industry. It makes up greater than 80% of the titanium-based alloys applied in aircraft sector and around 45% of overall weight of titanium based alloys manufactured. Additionally, it is employed in the chemical, automotive, maritime, and implantable medical device industries. It is a + alloy that offers the best weldability of all + alloys, good workability, and appealing mechanical qualities. Ti-6Al-4V is widely utilised in manufacturing rotating components of cold jet engines [1]. Some elements must be taken into account for titanium welding to be successful. Titanium reacts violently at temperatures above 500–650°C. It reacts with airborne contaminants such as C, O, N, and H. Although modest levels can reduce the ductility and hardness of titanium joints, these metals strengthen titanium. It is also necessary to take into account how welding methods heating as well as cooling related cycles influence composition of alloy and also alloys mechanical features [1]. Karimzadeh et al. [2, 3] evaluated the impact of grain development and porosity distribution on MPAW process parameters by using artificial neural networks(ANN) and observed epitaxial development on the microstructure. From the articles published by researchers on PAW process, it was observed that most of the researchers focused on welding higher thickness of titanium plates. Few works are carried out on joining thin sheets. Also, the works are carried out using continues mode of current. In this paper, pulsed current mode is adopted for thin sheets of titanium.

In the present work, 0.5 mm thick sheets of Ti-6Al-4V are welded using MPAW and the input parameters like peak current, base current, pulse rate, and pulse width are optimised by considering smaller grain size and larger hardness and tensile strength.

Experimentation

Titanium sheets of 0.5mm thick are welded using MPAW by considering input welding parameters like peak current, base current, pulse rate and pulse width (Figure 49.1). Tables 49.1 and 49.2 depicts the properties of Ti-6Al-4V. Table 49.3 represents the welding parameters and their limits. The above input welding parameters are considered from the earlier work [4–6].

Analysis of Fusion zone Grain Size

Each connection has three metallurgical samples removed, leaving only the borders of the faulty part of the welded length. Visual inspection, dye penetrant, X-ray, and Bakelite mounting are used to determine the defective length of the weld. As required by ASTM E 3-1, samples are prepared and mounted. The specimens which were polished are macro etched in a medium of Kroll's reagent (100 ml water, 1-3 ml hydrofluoric acid, and 2-6 ml nitric acid).

[a]kspanits@gmail.com, [b]sailajam073@gmail.com

DOI: 10.1201/9781003450252-49

Figure 49.1 Plasma welding setup
Source: Author

Table 49.1 Percentage of weight of Ti-6Al-4V

O	Fe	H	V	Al	N	C	Ti
0.19	0.20	0.011	5.8	4	0.03	0.05	89.71

Source: Author

Table 49.2 Mechanical properties of titanium (Ti-6Al-4V)

Elongation (%)	Yield strength (MPa)	Ultimate tensile strength (MPa)	VHN
14	880	950	349

Source: Author

Table 49.3 Welding parameters with limits

Limits	Peak current (Amperes)	Base current (Amperes)	Pulse rate (Pulses/second)	Pulse width (%)
1	26	11	10	20
2	28	13	20	30
3	30	15	30	40
4	32	17	40	50
5	34	19	50	60

Source: Author

Figure 49.2 Microstructure of weldment
Source: Author

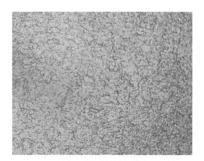

Figure 49.3 Grains of weldment
Source: Author

The weld zone's grain size together with microstructure are made visible by adjusting the etching time. Figure 49.2 displays a micrograph of the source metal, while Figure 49.3 at 100X magnification displays the weld fusion zone and placed in Table 49.4.

Analysis of Hardness

Vicker's micro hardness testing was carried out by applying 500 grams of load as per ASTM E384. The table below displays the average values of each sample's three readings. Figure 49.5 illustrates the hardness variation over the weld joint. T is observed from Figure 49.4, that toughness of the weldment decreases as it moves toward the heat affected zone (HAZ). This is because the pulsed current used in welding causes grain refining to occur in the weld fusion zone where 1, 2, 7, 8 are on HAZ and 3, 4, 5, 6 are on fusion zone (FZ).

Table 49.4 Experimental results

Exp No.	Peak current (PC) (Amperes)	Base current (BC) (Amperes)	Pulse Rate (PS) (Pulses/second)	Pulse width (PW) (%)	Grain Size (Microns)	Hardness (VHN)	Ultimate tensile strength (UTS) (MPa)
1	1	1	1	1	48	310.67	1110.67
2	1	2	2	2	46.4	352	1132
3	1	3	3	3	40	360	1170
4	1	4	4	4	42.67	368.67	1188.67
5	1	5	5	5	44.33	364.33	1164.33
6	2	1	2	3	46.38	351.33	1171.33
7	2	2	3	4	42.33	352.33	1162.33
8	2	3	4	5	42.33	358.67	1174.67
9	2	4	5	1	44.67	364.67	1164.67
10	2	5	1	2	42.71	359	1149
11	3	1	3	5	44.33	360.33	1160.33
12	3	2	4	1	43.98	362	1162
13	3	3	5	2	46.67	354.67	1174.67
14	3	4	1	3	42.67	366.67	1166.67
15	3	5	2	4	38	358	1178
16	4	1	4	2	42.26	364.67	1164.67
17	4	2	5	3	42.67	356.67	1176.67
18	4	3	1	4	43	363	1163
19	4	4	2	5	42.1	359	1179
20	4	5	3	1	46	381	1181
21	5	1	5	4	38.67	368.67	1178.67
22	5	2	1	5	43.33	363.33	1163.33
23	5	3	2	1	45.78	368.33	1162.33
24	5	4	3	2	44.67	364.67	1164.67
25	5	5	4	3	42	368	1172

Source: Author

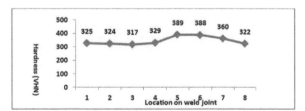

Figure 49.4 Analysis of hardness along weldement
Source: Author

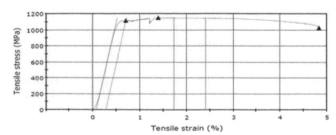

Figure 49.5 Stress strain curve (experiment 16)
Source: Author

Analysis of Tensile Strength

Specimens are prepared and tested on 100 KN computer-controlled universal testing machine. The UTS of weld joints is calculated from the Stress-Strain diagram (Figures 49.5). Figures 49.6 and 49.7 exhibit the

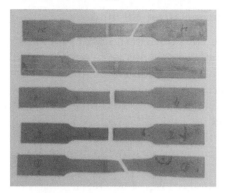

Figure 49.6 Specimens for tensile test
Source: Author

Figure 49.7 Failed tensile specimens
Source: Author

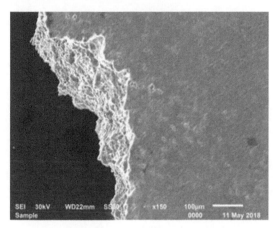

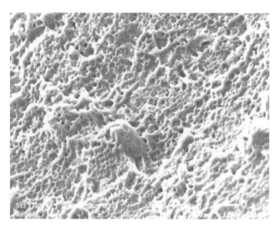

Figure 49.8 Broken section of tensile specimen
Source: Author

Figure 49.9 Dimples in broken zone
Source: Author

Table 49.5 Best and worst values

	Criteria	Grain size (Microns)	Hardness (VHN)	Ultimate tensile strength (MPa)
1	X_a^+	38	381	1188.67
2	X_a^-	48	310.67	1110.67

Source: Author

tensile specimens before testing together with after testing. The weakest zone, HAZ, saw the majority of the samples fail. Nevertheless, a small number of samples had issues at the FZ/HAZ interface. This can be the result of the brittle, hard martensite crystallising.

SEM pictures are taken to better understand how tensile specimens fail. It is clear from Figures 49.8 and 49.9 that the presence of dimples at the failure zone implies a ductile fracture.

To prevent the chance of systematic mistakes creeping into the system, tests were conducted conferring to design matrix at random.

Mathematical Modelling

Considering $Y = b_o + \sum b_i x_i + ?$ linear model empirical models are developed.
Grain size $= 55.06 - 0.1461\ X_1 - 0.164364\ X_2 - 0.0358929\ X_3 - 0.0831188\ X_4$
Hardness $= 306.772 + 1.3193\ X_1 + 0.799007\ X_2 + 0.138561\ X_3 - 0.0210227\ X_4$
UTS $= 1063.89 + 1.6165\ X_1 + 1.80922\ X_2 + 0.448043\ X_3 + 0.331278\ X_4$
coded values of welding input parameters are represented by X_1, X_2, X_3, X_4

Analysis of Variance (ANOVA)

The ANOVA is applied to evaluate how well-developed prototypes work. In view of this technique, model is taken into consideration acceptable within the confidence and the F ratio of the produced model is within the maximum limit of 2.56 at a confidence level of 95%. and the coefficient of determination, is found to be 0.93.

VIKOR Optimisation Method

A multiobjective VIseKriterijumska Optimizacija Kompromisno Resenje (VIKOR) method was developed to solve complex problems and compromisation of ranking list. The solution is obtained by gaining with the initial (given) weights. This technique targets selecting an alternative group and ranking to resolve conflicting criteria [7–10].

Step-(1): Finding the maximum and minimum values according to beneficiary and non-beneficiary output presented in Table 49.5 (X_i^+, X_i^-).

Step-(2): Find S_y (Utility measure) as per Table 49.6 by Equation (1) and R_y (Regret measure) as per Table 49.7 for output responses by Equation (2).

y= 1,2,3,.....J and a = 1,2,3,.....n.

Equal weightage (W_a) is taken

$$S_y = SUM[W_{a*}(X_a^+ - X_{ay})/ (X_a^+ - X_a^-)] .... \tag{1}$$

$$R_y = MAX[W_{a*}(X_a^+ - X_{ay})/ (X_a^+ - X_a^-)]........... \tag{2}$$

Step-(3): Finding VIKOR index (Q)

VIKOR index (Q) is computed as per Equation (3)

$$Q_a = \{T^*[(S_y - S^*)/(S^- - S^*)]\} + \{(1-T)^*[(R_y - R^*)/(R^- - R^*)]\}......... \tag{3}$$

Where $S^* = \min (S_y)$ (y=1,..., J) $S^- = \max (S_y)$ (y=1,..., J)
$R^* = \min (R_y)$ (y=1,..., J) $R^- = \max (R_y)$ (j=1,..., J).

Table 49.6 Utility measure values

S. No.	Grain Size (Microns)	Hardness (VHN)	Ultimate tensile stength (MPa)	S
1	48.00	310.67	1110.67	1.000
2	46.40	352.00	1132.00	0.660
3	40.00	360.00	1170.00	0.246
4	42.67	368.67	1188.67	0.214
5	44.33	364.33	1164.33	0.394
6	46.38	351.33	1171.33	0.494
7	42.33	352.33	1162.33	0.393
8	42.33	358.67	1174.67	0.310
9	44.67	364.67	1164.67	0.402
10	42.71	359.00	1149.00	0.431
11	44.33	360.33	1160.33	0.430
12	43.98	362.00	1162.00	0.403
13	46.67	354.67	1174.67	0.474
14	42.67	366.67	1166.67	0.318
15	38.00	358.00	1178.00	0.155
16	42.26	364.67	1164.67	0.322
17	42.67	356.67	1176.67	0.322
18	43.00	363.00	1163.00	0.362
19	42.10	359.00	1179.00	0.282
20	46.00	381.00	1181.00	0.299
21	38.67	368.67	1178.67	0.123
22	43.33	363.33	1163.33	0.370
23	45.78	368.33	1162.33	0.432
24	44.67	364.67	1164.67	0.402
25	42.00	368.00	1172.00	0.266

Source: Author

Table 49.7 Regret measure values

S. No.	Grain size (Microns)	Hardness (VHN)	Ultimate tensile strength (MPa)	R
1	48.00	310.67	1110.67	0.333
2	46.40	352.00	1132.00	0.280
3	40.00	360.00	1170.00	0.100
4	42.67	368.67	1188.67	0.156
5	44.33	364.33	1164.33	0.211
6	46.38	351.33	1171.33	0.279
7	42.33	352.33	1162.33	0.144
8	42.33	358.67	1174.67	0.144
9	44.67	364.67	1164.67	0.222
10	42.71	359.00	1149.00	0.170
11	44.33	360.33	1160.33	0.211
12	43.98	362.00	1162.00	0.199
13	46.67	354.67	1174.67	0.289
14	42.67	366.67	1166.67	0.156
15	38.00	358.00	1178.00	0.109
16	42.26	364.67	1164.67	0.142
17	42.67	356.67	1176.67	0.156
18	43.00	363.00	1163.00	0.167
19	42.10	359.00	1179.00	0.137
20	46.00	381.00	1181.00	0.267
21	38.67	368.67	1178.67	0.058
22	43.33	363.33	1163.33	0.178
23	45.78	368.33	1162.33	0.259
24	44.67	364.67	1164.67	0.222
25	42.00	368.00	1172.00	0.133

Source: Author

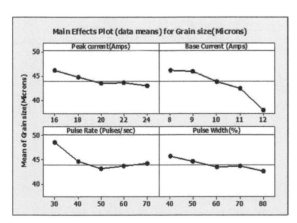

Figure 49.10 Main effects of grain size
Source: Author

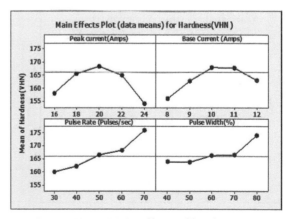

Figure 49.11 Main effects of hardness
Source: Author

Step-(4): Find out the Rank

Table 49.8, the least VIKOR index value was assigned highest rank.

It is understood from Table 49.8, 21st experiment was the optimal combination as it is having highest rank.

Results and Discussions

Main effects of grain size, hardness and ultimate tensile strength (UTS) are depicted in Figure 49.10-49.12.

Table 49.8 VIKOR ranks

S. No.	Utility measure (U)	Regret measure (R)	VIKOR index (I)	Rank (R)
1	1.000	0.333	1.000	25
2	0.660	0.280	0.709	24
3	0.246	0.100	0.145	3
4	0.214	0.156	0.229	5
5	0.394	0.211	0.432	16
6	0.494	0.279	0.613	22
7	0.393	0.144	0.310	11
8	0.310	0.144	0.263	7
9	0.402	0.222	0.457	19
10	0.431	0.170	0.377	14
11	0.430	0.211	0.452	17
12	0.403	0.199	0.416	15
13	0.474	0.289	0.619	23
14	0.318	0.156	0.288	9
15	0.155	0.109	0.110	2
16	0.322	0.142	0.265	8
17	0.322	0.156	0.290	10
18	0.362	0.167	0.333	12
19	0.282	0.137	0.233	6
20	0.299	0.267	0.479	20
21	0.123	0.058	0.000	1
22	0.370	0.178	0.357	13
23	0.432	0.259	0.541	21
24	0.402	0.222	0.457	18
25	0.266	0.133	0.218	4

Source: Author

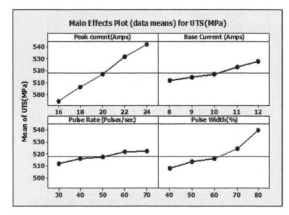

Figure 49.12 Main effects of UTS
Source: Author

With an increase in peak current is increased from 16 to 24 amperes, grain size decreases because of higher heat input. Base current follows the same pattern as that of peak current. When the pulse rate increases from 30-70 pulses/second, the grain size is enlarged. Coarse grain size of fusion zone leads to decrease in strength of the welded joints. Higher pulse rate values lead to agitation in the weld molten pool rigorously, resulting in refinement of grains.

At peak current of 24 amperes, the hardness is high. Equiaxed grains in fusion zone leads to improvement in hardness of the welded joints. Larger current leads to high heat and longer cooling time, there by finer grains are obtained leading to higher hardness. At a pulse rate is 30 pulses/second, the hardness is high. When the pulse rate is increased to 70 pulses/second, the hardness is decreased. At high pulse rate values, the molten bead is agitated rigorously, resulting in refinement of grains.

At a peak current of 24 amperes, the ultimate tensile strength is larger. Equiaxed grains in fusion zone improves hardness, leading to improvement in ultimate tensile strength of the welded joints. At a pulse rate is 30 pulses/second, the ultimate tensile strength is high. When the pulse rate is increased to 70 pulses/second, the ultimate tensile strength is decreased.

Conclusions

Empirical mathematical models are developed for the chosen output responses by considering 955 confidence level and the coefficient of determination of 0.93 is obtained which indicates good coincidence with experimental values. From VIKOR study, highest rank was obtained for 21st combination i.e., peak current of 34 amperes, back current of 11 amperes, pulse rate of 50 pulses/second, and pulse width of 50% are the best combinations for producing materials with smaller grains and greater hardness and tensile strength. Peak current is the most influencing welding parameter followed by pulse rate, pulse width and base current.

References

[1] Donachie, M. J. (2000). Titanium A Technical Guide. (2nd ed). ASM International.

[2] Karimzadeh, F., Salehi, M., Saatchi, A., and Meratian, M. (2005). Effect of microplasma arc welding process parameters on grain growth and porosity distribution of thin sheet Ti6Al4V alloy weldment. Materials and Manufacturing Processes. 20(2), 205–19.

[3] Karimzadeh, F., Ebnonnasir, A., and Foroughi, A. (2006). Artificial neural network modeling for evaluating of epitaxial growth of Ti6Al4V weldment. Materials Science and Engineering A. 432, 184–90.

[4] Prasad, K. S., Rao, C. S., and Rao, D. N. (2016). Optimisation of fusion zone grain size, hardness and ultimate tensile strength of pulsed current micro plasma arc welded Inconel 625 sheets using genetic algorithm. International Journal of Advanced Manufacturing Technology. 85(8-12), 2287–95.

[5] Prasad, K. S., Rao, C. S., and Rao, D. N. (2012) Optimizing fusion zone grain size and ultimate tensile strength of pulsed current micro plasma arc welded inconel 625 alloy sheets using Hooke & Jeeves method. International Transaction Journal of Engineering, Management and Applied Sciences and Technologics. 3(1), 87–100.

[6] Prasad, K. S., Rao, C. S., and Rao, D. N. (2012). Effect of pulsed current micro plasma arc welding process parameters on fusion zone grain size and ultimate tensile strength of inconel 625 sheets. Acta Metallurgica Sinica (English letters). 25(3), 179–89.

[7] Aravind, A. P., Suryaprakash, S., Vishal, S., Sethuraman, M., Deepan Bharathi Kannan, T., Umar, M., and Rajak, S. (2020). Optimisation of welding parameters in CMT welding of Al 5083 alloys using VIKOR optimisation method. In 3rd International Conference on Advances in Mech Engg (ICAME 2020).

[8] Bhadauria, A. S., and Kushwah, A. S. (2018). Analysis of MIG welding by using parametric optimisation techniques taguchi method, VIKOR method and ANOVA. Journal of Recent Activities in Production. 4(3), 12–20. (MAT Journals).

[9] Parmar, S. (2022). Multi-objective parametric optimisation of weld strength of metal inert gas (MIG) welding by using analysis of variance, Taguchi, and VIKOR techniques. International Journal for Research in Applied Science and Engineering Technology. 10(12), 463–71.

[10] Biswas, S. A., Datta, S., Bhaumik, S., and Majumdar, G. (2009). Application of VIKOR based Taguchi method for multi-response optimisation: a case study in submerged arc welding (SAW). In Proceedings of the International Conference on Mechanical Engineering (ICME2009), 26- 28, December 2009, Dhaka, Bangladesh.

Chapter 50

Fusion of human arm synergy and artificial neural network for programming of a robot by demonstration

Priya Khandekar[a] and Pranjali Deole[b]

Shri Ramdeobaba College of Engineering and Management, India

Abstract

In this paper, human arm synergy and artificial neural network are combined to acquire and transfer skill from human demonstration to robot. The structure of a robot manipulator is dissimilar to human hand. This doesn't permit simply copying human hand motions by robot. In this approach, the artificial neural network is utilised to map joint space and Cartesian space configurations of robot and human hand. Whereas, the human and robot hand synergy is utilised to map the human workspace to robot workspace. The experiments were carried out by considering various cases like learning from demonstration and workspace generalisation . Experimentations revealed that the approach generalises well in the tested scenario.

Keywords: Robot programming by demonstration, synergy, artificial neural network, human hand

Introduction

Robotics has rapidly moved from theory to applications and from research lab to industries. This technology is advancing rapidly. In the analysis of such a spatial mechanism, the location of links, joints and end-effector in 3D space is continuously required. To specify a robot's pose, frames must be connected to its joints, linkages, and end-effector. The end-effectuate must follow the planned trajectory in order to manipulate objects or complete certain tasks in the workspace. For this, it is required to regulate the location of each link and joint on the robot. This necessitates research on the robot's kinematic modeling challenge. The two sub problems of the kinematics modeling problem are forward kinematics and inverse kinematics. A forward kinematics problem is concerned with the relationship between the individual joints of the robot manipulator and the position and orientation of the tool. For n-degrees of freedom manipulators, forward kinematics deals with the challenge of determining the position and orientation of the end-effector with respect to a known reference frame, as appeared differently in relation to inverse kinematics, which entails determining a set of joint angles that would bring the end-effector into the desired configuration. Although objects to be manipulated are often stated in Cartesian space because it is easier to see the exact location of the end-effector there, robots commonly operate in joint space. One of the trickiest problems in robotics is the inverse kinematics challenge. Primarily, it is finding a set of all feasible joint variables of a manipulator to arrive at a preferredarrangement of the end- effector frame. Because of the complexity of the non-linear equations that are created during the transition between joint and Cartesian space, solving inverse kinematics is a challenging undertaking. Artificial neural networks (ANN) has been broadly applied in robotics for their flexibility, learning ability and function approximation capacity in nonlinear systems. Inverse kinematics problems are solved by usual techniques including geometric, algebraic, and iterative approaches. These methods give inadequate solution for the redundant robotic manipulator. It takes a lot of time and effort to solve the inverse kinematics problem for redundant manipulators. Because of its inherent learning potential, neural networks have been widely utilised in numerous robotic control applications. Because of their capacity for learning and adaptability to ambiguous and unstructured working settings, they are viewed as intelligent control systems. This is carried out by first acting as a transformation function and then training using data gathered from the robotic system. The network architecture, the learning strategy, and the training data are only a few of the variables that affect network performance.

To solve the inverse kinematic problem of robots, Chiddarwar et al. (2010) proposed an ANN-based fusion technique that merged forward kinematics interactions with the neural network. Using this method, incremental joint angles were produced. This method can be applied to forecast inverse kinematic answers for any kind of robot, regardless of its shape or degree of freedom.

Jha et al. (2014) obtained the inverse kinematics of the 6R manipulator using the methods of multi-layered perceptron back propagation (MLPBP) and multilayered perceptron particle swarm optimisation (MLPPSO). The data set for training MLP was created using the forward and inverse kinematics of a robot.

Kumar et al. (2015) used the geometrical and Denaviet-Hartenberg methods of forward kinematics for producing data and training the neural network. A method called Resilient Back Propogation (RBP) is used to test the network.

[a]khandekarpm@rknec.edu, [b]deoleps@rknec.edu

DOI: 10.1201/9781003450252-50

In order to address the inverse kinematics problem, Feng et al. (2012) integrated an electromagnetism-like technique with a neural network. The training data used in this suggested method was gathered from a subset of the joint space, where the training set was limited to one solution set in order to accomplish one-to-one mapping. They trained the neural network using an effective learning technique. When conventional methods are insufficient, an ANN is employed to solve the inverse kinematics of redundant robotic manipulators. The training algorithm and network topology have an impact on a neural network's performance. The Levenberg-Marquardt (LM) algorithm performs with greater accuracy and has the fastest convergence time, Sari (2014).

Robots have been employed in science and industry for many years, but programming robotic systems is a complex and specialized operation that can only be completed by a few numbers of highly qualified individuals. In current competitive scenario every system must be installed and adjusted separately from other setups and separately programmed to crack the given task. Another strategy is to successfully programme the robot task by a person who possesses the necessary learning, perception, and cognition skills. The fundamental idea behind this approach is to manually direct the robot while intuitively "demonstrating" the lacking data. Robot programming by demonstration refers to transfer skills to robots by providing examples of required behaviour through demonstration Vakanski, et al (2012).

The key benefit of this strategy is the introduction of the need for wide technical knowledge in order to program robots. A thorough overview of the numerous methods created by researchers around the world for robot learning by demonstration was presented by Argall et al. in 2009. Numerous robotics applications revolve around the challenge of understanding a relationship between world state and actions.

Ekvall and Kragic, (2008) developed a robot learning by demonstration technique based on task level planning approach. Their approach divided the entire job to be performed by human operator into various tasks and sub-tasks and then they are co-related to bring out the trajectory for the robot. By using human demonstrations, Maeda et al. (2002) created an easy training process for robots. The two parts of the approach are known as the Teaching Phase and the Planning Phase.Kosmopoulos, (2010) has presented a combination of location based and image based visual servoing technique by demonstration. Kramar, et al (2022) constructed NN based transformation matrix which showed the best accuracy.

Acquiring skills at a trajectory level is considered here because of the rich information contained in the human trajectories. The industrial robot draws inspiration of human hands capabilities. Many robotic tools are built to interact with humans and in particular, to interact with human hands. To communicate the information about the human hand trajectory to robots, the data should be converted to joint angles using inverse kinematics.

The human arm was modeled as 7 degrees of freedom structure in literature by Badler and Tolani in 1996. This mechanism included spherical joints for the wrist, elbow, and shoulder. For finding homogeneous coordinate transformation a fixed coordinate frame attached at the shoulder joint and moving coordinate system at each joint using D-H convention. They proposed an analytical approach for computing the IK of the human arm. For the purpose of computing the inverse kinematic model of the human arm, Mihelj (2006) proposed the method, which predicated on measurement of the hand position, orientation, and radial acceleration of the upper arm. The technique provides a human arm analysis estimate that is sufficient for trajectory planning and rehabilitation purposes.

By using analytical solutions, Al-faiz and Al-Mashhadany (2009) suggested an approach to solve the inverse kinematics for anthropometric limbs. The D-H parameters can be altered to allow the human arm to perform varied motions. Patients' disabilities can be determined via a kinematics examination of limb movement. It requires measuring the end point of the arm as well as respective joint angles. An inverse kinematics tool kit meeting the unique requirements of computer graphics was proposed by Tolani et al. in 2000. Toolkit generalises inverse kinematics constraints and focus on analytical methods. With pose similarity metric learning, Song et al. (2014) established a paradigm for humanoid robot imitation. On the basis of given human pose they adopted related angles as a target pose of the robot. Kinematic mapping algorithm was used by Ficuciello et al. (2014) to imitate human arm movements on anthropomorphic robot arm with seven degrees of freedom.

Proposed Approach

In order to utilise the ability of human arm to carry out certain tasks or trajectories to be followed to manipulate object, the human hand is modeled as kinematics structure as discussed in above section. In view of demonstrating the human arm trajectories in front of robot, a novel framework has been proposed shown in Figure 50.1. The basis of the proposed framework is to convert human arm trajectory to robot trajectory using PbD and soft computing techniques.

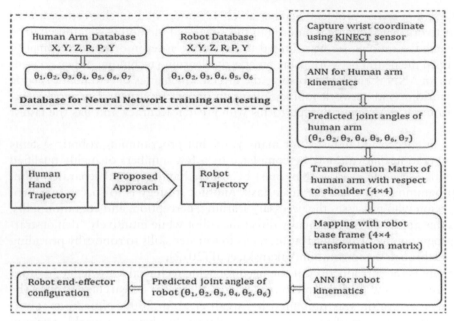

Figure 50.1 Outline of frame transformations and artificial neural network based trajectory
Source: Author

Methodology

1. In order to convert human hand trajectory to robot trajectory, the different trajectories are demonstrated ahead of KINECT sensor. It captures 3D (X, Y, Z) coordinates of human wrist by using software development kinect (SDK).
2. Inverse kinematics of human arm is computed using kinematic structure of human hand so as to develop database of human arm and to train data using neural network toolbox in MATLAB 2014a.
3. The extracted position and orientation of human wrist by Kinect sensor is tested in trained NN in order to predict respective joint angles. Similar method is followed for robot to predict joint angles from the robot dataset.
4. The configuration of a human wrist with regard to a shoulder is determined by predicted joint angles. The human arm configuration must be mapped with the robot workspace in order to transform human arm trajectory to robot trajectory.
5. After mapping, the position of robot base computed from transformation of human shoulder to robot base is tested in robot database. It predicts respective joint angles using NN toolbox.
6. Predicted joint angles put into forward kinematics equations which will determine robot end-effector configuration.

Learning From Human Demonstration

A correspondence problem between the human hand and the robot is shown to be resolved through the use of synergistic interactions. The synergy matrix transforms the human hand's goal-directed trajectory without losing crucial execution data. As a result, a particular demonstration can be programmed into a robot's software, and the skills required to perform it can be acquired through practice. In order to control and improve the performance of new talents, the knowledge obtained from several demonstrations is used. Therefore, the robot should move its joint so that it follows the demonstrated trajectory while attempting to reproduce an observed trajectory. For encrypting human motions, various methods like splines, hidden markov model, hidden markov model combined with NURBS and GMM are developed by various researchers.

The objective of this work is to find out a method to map a set of synergies defined on a reference human hand onto a generic robotic hand. The structure of the human hand is thus defined as classic hand. The classic hand is a kinematic and dynamic model inspired by the human hand that does not closely copy the kinematical and dynamical properties of the human hand. It essentially provides a compromise between the complexity of the human hand model, which accounts for the synergistic arrangement of the sensorimotor system, and the accessibility and simplicity. Suppose the classic hand can be depicted by the joint variable vector $Q_h \in \Re^{n_{qh}}$. Using the notion that input vector $z \in \Re^{n_z}$ (with $n_z \leq n_{qh}$) can represent the subspace of all configurations. This parameterizes the motion of the joint variables along the synergies Qh = Shz being *Sh* ∈ ℜ nqh × n the synergy matrix. For this synergy matrix, a reference position for robot as well as human hand was fixed. A scaling factor is used to account for the differences in the dimensions of the workspace for humans and robots. For the computation of this scaling factor, 2 virtual spheres were fitted on the human and the robotic hand as the minimum volume sphere containing reference points. The ratio between the two spheres' radii is then used to define the virtual object scaling factor: *ksc* =rr/rh. Using synergy matrix and scaling factor, the synergy relationship between robot and human hand can be given as Robot XYZ = S * human hand XYZ. To transfer human hand configurations to robot configurations (Table 50.1), this connection was applied. This method projects human hand synergies to robot hand in the Cartesian task space. The advantage of this paradigm may be used to robot hands with a wide range of kinematics. IRB 120 robot (Figure 50.2) is used to implement the concept of a robot replicating human arm motion using projected synergies.

Robot - ABB IRB 120

ABB IRB 120 robot is used for the validation of the proposed approach.

Artificial Neural Network

A feedforward backpropagation network is used to train the dataset. Network consists of input, output, and hidden layer. The architecture of the network is shown in Figures 50.3 and 50.4. This multilayer neural network has six inputs and seven outputs for human arm and six outputs for robot. As a learning algorithm, the backpropagation algorithm utilised, with the sigmoid transfer function serving as the activation function and the linear function for predicting the output data. The LM algorithm is used for training the network. Offline training is implemented during the study. The neural network training parameters like learning rate, no. of neurons in hidden layer and termination conditions were determined iteratively.

Table 50.1 D-H convention

Link	a_i	α_i	d_i	Θ_i
1	0	-90^0	290	Θ_1
2	270	0^0	0	$\Theta_2 - 90^0$
3	70	-90^0	131	Θ_3
4	0	$+90^0$	240	Θ_4
5	0	-90^0	0	Θ_5
6	0	0^0	72	Θ_6

Source: Author

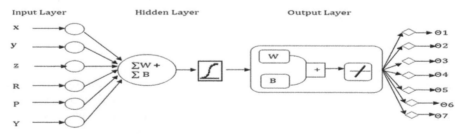

Figure 50.3 Architecture of network for human arm
Source: Author

Figure 50.2 ABB IRB 120 robot
Source: Author

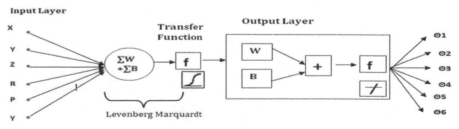

Figure 50.4 Architecture of network for robot
Source: Author

Table 50.2 Selection of trajectory

Trajectory no	Procrustes Index	Variance
1	0.041	0.25
2	0.025	0.16
3	0.0155	0.11
4	0.019	0.124
5	0.0224	0.18
6	0.0184	0.14

Source: Author

Results and Discussion

The methodology presented above was implemented to case problems. In the first case, a task is considered to teach a skill to the robot by human demonstration. In the second case, the learned trajectories are generalised with respect to different workspaces of robot and human operator.

Case 1: Learning from Demonstration

Twenty task demonstrations of a single hand movement were performed for this experiment. To make the situation more realistic, the hand movement suitable for the human operator and feasible for robot was selected. After critical analysis of all 20 task demonstrations, six demonstrations were discarded for reasons such as discontinuity in capturing due to transmission issues or insufficient data points due to occlusion. All the 14 demonstrations were used for trajectory generation. Using proposed artificial NN based approach, the joint angle coordinated corresponding to these 14 Cartesian paths were determined. Additionally, variance was calculated, how far the robot can depart from the intended trajectory is determined by this variance. These joint angle trajectories were given input to the robot controller. Out of 14 joint space trajectories, eight trajectories could not be executed by robot due to unreachable configurations or singularities. Robots could be placed in various, more accessible locations to verify it. Furthermore, offering more demonstrations with greater path variations would result in fewer limitations. For six trajectories successfully executed by robot, the deviation in location of gripper was measured. It was in the millimeter range. In other words, placement is precise enough to allow for successful hand movements. Then these six different trajectories were not bundled together which might result into incoherent trajectory. For selecting a single trajectory from these six, the dissimilarity index computed using Procrustes analysis and variance was used. The joint angle trajectory having low dissimilarity index and low variance was selected. The low value of dissimilarity index indicates the geometrical similarity between demonstrated and followed trajectory whereas variance indicates divergence of robot end-effector from the preferred trajectory. Table 50.2 shows that trajectory three has the lowest Procrustes index and variance. Moreover, it was observed that Procrustes index and variance follow similar trends. Hence, a trajectory having minimum value for both the indices can be selected.

Case 2: Workspace Generalisation

In this case, the generalisation characteristic of developed approach is examined. The robot was given various tasks to complete for this purpose, and its starting configuration was altered from that used for the demonstration. This makes it easier to assess how well the suggested method will work given the structural differences between human and robot hand. In this case, 10 different trajectories were demonstrated to the robot. For these 10 trajectories, three situations were tested.

1. The start point of the human hand trajectory was considered different by keeping the goal point fixed. The robot end-effector initial pose was kept same in all reaching tasks. For this experiment, demonstration 2 was repeated 10 times. The robot could execute seven joint angle trajectories obtained from developed approach. It failed to execute 3 trajectories due to repeated coordinates and very little difference between consecutive trajectory points.
2. The initial pose of the robot end-effector was selected randomly for a task demonstration. For this purpose, demonstration no. 3 was considered. The developed approach was implemented to get robot configurations. Out of three randomly selected poses, robot end-effector could execute desired trajectory with two configurations but failed for third.
3. The robot end-effector was kept in its original position. The start and endpoint of all 10 trajectories was different. These ten trajectories captured by camera were given input to the developed ANN based

algorithm and corresponding joint space configurations were obtained. These configurations were given input to the robot controller. It was observed that robots could execute all ten trajectories. The problem of unavailable configuration or singularity did not arise in this case due to the use of joint space planner. However, out of 10 trajectories, only 6 trajectories were executed smoothly, whereas 4 trajectories were followed with jerk. This can be sorted out by further smoothing trajectories using splines or NURBS or weighted average smoothing.

The experiments mentioned above show that the created system generalises sound under the investigated conditions, satisfactorily deals with the correspondence problem. However, problems like repeated configurations, singularity conditions, collision avoidance and optimal path estimate for the robot end-effector need to be addressed further.

Conclusion

In this paper, a fusion of human hand synergy matrix and ANN method is done to program robot using demonstration. The approach was tested over certain scenarios to check its ability to workspace generalisation and learning from demonstration. From experiments, it was observed that the proposed approach generated executable trajectories for the robot most of the time. However, issues like singularity and reachability were taken care of due joint space motion planning. As a future work, this approach will be tested for pick and place and assembly operation.

References

[1] Chiddarwar, S. S. and Babu, N. R. (2010). Comparison of RBF and MLP neural networks to solve inverse kinematic problem for 6r serial robot by a fusion approach. International Journal of Engineering Applications of Artificial Intelligence. 23, 1083–92.

[2] Jha, P. and Biswal B. B. (2014). Hybrid neural network based prediction of inverse kinematics of robot manipulator, 5th International & 26th All India Manufacturing Technology, Design and Research Conference (AIMTDR) 12th–14th, IIT Guwahati, Assam, India

[3] Kramar, V., Kramar, O., and Kabnov A. (2022). An artificial neural network approach for solving inverse kinematics. Problem for an Anthropomorphic Manipulator of Robot SAR-401. Machines. 10, 241. https://doi.org/10.3390/machines10040241.

[4] Kumar, R. R. and Chand, P. (2015). Inverse kinematics solution for trajectory tracking using artificial neural networks for SCORBOT ER-4u. Proceedings of the 6th International Conference on Automation, Robotics and Applications, Feb 17-19. Queenstown, New Zealand.

[5] Feng, Y., Yao-nan, W., and Yi-min, Y. (2012). Inverse kinematics solution for robot manipulator based on neural network under joint subspace. International Journal of Computer Communication. 3, 459-72.

[6] Sari, Y. (2014). Performance evaluation of the various training algorithms and network topologies in a neural-network-based inverse kinematics solution for robots. International Journal of Advanced Robotic Systems. doi: 10.5772/58562.

[7] Ekvall S. and Kragic D. (2008). Robot learning from demonstration: a task-level planning approach. International Journal of Advanced Robotic Systems,5,(3), 729-8806, 223-234,.

[8] Argall, B., Chernova, S., Velso, M., and Browning, B. (2009). A survey of robot learning from demonstration, Robotics and Autonomous System.

[9] Kosmopoulos D. (2010). Robust Jacobian matrix estimation for image-based visual serving. Pre-print submitted to Elsevier.

[10] Matjaz, M. (2006). Human arm kinematics for robot-based rehabilitation. Robotica. 24(3), 377-83.

[11] Badler, N. and Tolani, D. (1996). Real time inverse kinematics of the human arm. 5(4), 393-401.

[12] Faiz, A. L. and Mashhadany, A. L. (2009). Analytical solution for anthropomorphic limbs model. IEEE Symposium on Industrial Electronics and Applications. pp. 4-6. Kuala Lumpur, Malaysia

Chapter 51

Law relating to health and health care: A systematic mechanism and critical study with reference to Bangladesh

Md. Kamruzzaman^a and Inderpreet Kaur^b
Chandigarh University, Chandigarh, India

Abstract

As a developing country, Bangladesh has seen an increase in total GDP in recent years. But it can be further improved by developing "health-care" (HC) services because it has enormous infrastructure problems all over the country. Bangladesh's HC system is now clearly poised to undergo reform at any process level, including prevention, diagnosis, and treatment. Although the Bangladeshi government is trying to develop the HC sector, due to health corruption in this sector, the improvement has not accelerated yet. For this reason, lots of Bangladeshi people are facing acute diseases. Regarding the prevention, diagnosis, and treatment of disease, this research will illustrate the law relating to health and HC to ensure excellent health and well-being. Firstly, this paper investigates health under Bangladeshi law from different perspectives related to the HC system. A massive gap has been investigated in this research after comparing Bangladeshi and international health law (HL). Secondly, a practical scenario is investigated and compared with international HC law. It is evident that the Bangladeshi HC system did not achieve the satisfactory standard level concerning international law. A staggering 70% of Bangladesh's population lives in rural areas, with no restrictions on access to hospitals and clinics. But proper HC infrastructure and some new medical practices are urgently needed to ensure HC quality. Finally, this research provides suggestions for developing a HC system to ensure the health of all Bangladeshi people that need to be immediately implemented by the Bangladeshi government. This research has practical implications in the HC system for any developing country to maintain their citizens' safety.

Keywords: Bangladeshi HL, HC system, human HC suggestions, international HL, law relating

Introduction

Every person has the right to a life of grandeur, which can only be achieved via excellent health and a sense of well-being. Because people's health and morals are essential to their well-being as well as the state's peace and stability, the legislature must protect and promote these critical interests. It is impossible to build a powerful nation on a foundation of straws. Healthy populations live longer and are more productive. Therefore, national HC mostly contributes to economic success. As a result, a nation's growth is dependent not only on its industrial, agricultural, and service balance sheets but also on its success on the human development index [1].

The state of a nation's HC is one of the evaluative markers of that nation's performance. That is far more significant for a nation like Bangladesh. In order to address health-related issues at a cost that is acceptable and affordable to the general public as well as the nation, as well as in a manner that is responsive to their social, cultural, and economic well-being, the government, as well as health professionals, are responsible for developing and implementing applicable human HC policies and plans. They should be carried out so that all citizens benefit equitably. The author of this study attempted to shed light on various aspects of human HC on a national and international level, including the meaning, history, and HC issues in Bangladesh, government efforts in the health sector through various constitutional legislative provisions, the current state of the HC sector and HC policies and programs; and suggestive reforms in the HC sector that could eliminate health hazards [2, 3].

HC in Bangladesh requires revolutionary transformation since the health and well-being of its people are possibly more important than anything else. Bangladesh's HC system is now clearly poised to undergo reform at any process level, including prevention, diagnosis, and treatment. A healthy population may undoubtedly help a country's economic growth and development. HC services aim to protect and improve the public's health by nourishing the HC system, growing human resources, and strengthening public-health capability and norms. These are essential aspects of the human HC sector [4].

The Bangladeshi government has enacted and implemented significant legal provisions in the HC sector at both the national and international levels to protect humans from HC-related risks, which has resulted in significant improvements in health indicators such as life expectancy, infant mortality rate, and maternal mortality rate as a result of the increased thrust of HC services across the country, as well as an increase in the number of government and private HC providers [5, 6].

^aassociates.shadhin@gmail.com, ^binderpreet.e9827@cumail.in

DOI: 10.1201/9781003450252-51

Despite the remarkable achievements, the Bangladeshi HC system confronts many deficits which need to be amended if a sound and adequate HC infrastructure are to be attained for a healthy Bangladesh. Hence, the author has laid out some gaps to be implemented successfully to move the nation toward a healthy nation in this research. The research mainly focused on the below-mentioned problem of HC in Bangladesh.

- The persisting infrastructure is not being utilised to the extent intended because of the scarcity of doctors, nurses, and other health personnel. Hence, there is a need to enhance the number of medical personnel [7].
- Lack of proper review of health policies at regular intervals [8].
- Lack of an exhaustive, coordinated, and intermingled approach among different performers to produce more fruitful consequences [9].
- Inadequate infrastructure, investments in the medical field, government oversight of how the funds are being used, and digital health technologies [10].
- Lack of improvement in public hospitals, living circumstances for people, and rural HC institutions [11].
- Lack of insurance to comprehend the current environment of the Bangladesh HC system [12].

Health Under Bangladeshi Laws

HC is one of the essential determinants of a nation's development, as healthy citizens promote a healthy nation. Because there is a strong link between a person's health and their standard of living, the government is obligated to ensure the well-being of the people. This can be accomplished by announcing a comprehensive legislative structure, providing favorable conditions for restoring equilibrium in the HC sector. Countries that have developed a universal approach to HC have accepted comprehensive legislation that has organised the human HC system under a single canopy and combined steps to provide structured and synchronised HC services to their population. Access to HC, provision of proper infrastructure, unfairness, carelessness, unfair practices, dishonesty in HC systems, standards in HC systems, occupational and environmental-related health issues, health issues relating to reproduction, infringement of rights, allotment of resources, professional behaviour, and other issues are all addressed by legislation. Also, the legislature should take steps to make the right to HC a fundamental right so that it may be enforced under the constitution [13–15].

Human health is one of our most basic needs. Both the law and health have implications for society and social welfare. Because human health is deteriorating day by day despite various health schemes and policies, the framers of the Bangladeshi Constitution inserted numerous provisions concerning human HC, and the Bangladeshi Supreme Court has also been instrumental in safeguarding people's health through various landmark decisions. Furthermore, the HC issue will be controlled through the proper implementation of laws created based on constitutional provisions [16].

The Bangladesh Constitution, which went into effect on November 4, 1972, sheds light on several aspects of people's lives, including socioeconomic, political, and other issues. Bangladesh's constitution focuses on ensuring social, economic, and political justice among the masses, protecting individual rights and promoting national health. The purpose of the constitutional substructure is to achieve the goals outlined in the preamble, which establishes rights and imposes duties on citizens, as well as commands the state to protect the various rights of its citizens, one of which is the right to health, as the nation's prosperity is dependent on a healthy population [17].

The Bangladeshi constitution declares health to be a key indicator of human progress. Bangladesh's constitution has clauses about the right to health. Constitutional directions impose on the state the responsibility of determining the conditions conducive to well-being. Since independence, Bangladesh has viewed the people as owners of rights and the state as obligated to act as the primary provider of HC. Not only that, but Bangladesh is a member of the United Nations and has accepted several international accords that pledge to protect individuals' human HC rights [18].

Although HC does not appear as a sector in many places in the Bangladeshi Constitution, there are indirect and implied allusions to people's HC and the state's role in their development. In 2011, the National Health Bill was introduced, which views human HC as a public good and HC as a human right [19].

The Constitution's Preamble addresses social, economic, and political justice, with social justice referring to the general public's equal access to HC facilities. The concept of democratic socialism attempts to improve the people's HC situation. Also included in the preamble is equality of position and opportunity, which refers to equality in all domains, including medical practice, admission to medical, educational institutions, and so on, to improve citizens" health. The preamble aims to provide for a welfare state with socialistic social patterns that relate to the right to HC for the public [20].

Article 15(a) imposes a legal obligation on the state to maintain social order to advance the people's welfare. Still, this welfare cannot be achieved without human HC, and the state is obligated to implement particular policy ideas to expand people's HC [21].

One of the fundamental obligations of the state, according to Article 18 (1), is to keep track of the increase in the level of nutrition and living standards of its citizens and the improvement in human health. As a result, the state is required to take steps to prohibit the consumption of intoxicating drinks as well as excessive doses of drugs that are harmful to one's health, except for medical reasons, following the Bangladeshi Constitution, which expresses the state's brotherly role in safeguarding citizens" health. When it comes to intoxicating drinks, narcotics, smoking, and other substances that have such negative consequences, the government's responsibility is to intervene effectively to reduce their consumption [22].

HC Under International Laws

International law, refers to the norms, agreements, and treaties that bind nations together. International law applies to the government just as domestic law applies to citizens and others. Every state government has a responsibility to enforce and comply with international law [23].

In international law, HL is an emerging field. It has been argued that health protection is a pressing social need that should now be addressed by international law. Health is a fundamental requirement that requires strong protection under international law. Therefore, it is crucial to focus on the ability of individuals to function correctly in society, pursue personal life services, and create conditions for health, including access to safe drinking water and sanitation, access to health-related information and education, and access to safe, healthy food opportunities. The right to the highest attainable standard of health is a key human rights norm in international HL, as it underscores the need to protect individual health and strive for equity in health on a global scale. It is a human rights norm linked to the protection of health. Globalisation continues to affect the world economy and the HC of populations and individuals worldwide. It poses several significant challenges to the health sector that require national and international responses. Plans that also show that not only a secure approach to HC should be prioritised [24].

Health research shows that health inequalities within and between countries are increasing. Many countries continue to grapple with the negative health impacts of environmental degradation, climate change, urbanisation and other factors. Due to the growing influence of international trade and multinational corporations, changes in disease patterns are also being screened, some of which are related to lifestyle. Because of these developments, attention to international health should focus not only on the spread of infectious diseases but also on domestic and international health inequalities and issues such as HC and medical methods and human health. in international and national emergencies and armed conflicts [25].

Some human rights documents proclaim the right to the best possible quality HC, sometimes referred to as the right to health. The term "human rights" first appeared in the Charter of the United Nations, ratified in San Francisco on June 25, 1945, and includes civil, cultural, economic, political, and social rights. Human rights are promoted in international forums, recognised in international conventions, and voluntarily recognised by states [26].

They have universally recognised principles that protect individuals and groups from activities that interfere with human dignity, fundamental freedoms, and principles of rights to ensure individual well-being and growth. Their powers are based on the shared commitment of all ratifying governments to respect, safeguard, and honour them. Governments have a responsibility to achieve these goals. The importance of international cooperation in exercising rights for states to perform their duties was emphasised.

Everyone has the right to a minimum standard of living, which includes food, clothing, housing, medical care, and essential services, as stated in Article 25 of the Universal Declaration of Human Rights (UDHR). This article also mentions disability, widowhood, old age, or lack of livelihood in other situations outside the control of the right to security.

The International Covenant on Economic, Social, and Cultural Rights (ICESCR), commonly known as the Economic Covenant, has come into play in human health. The Economic Covenant recognises the right of everyone to the best possible physical and mental health. It sets out the steps states must take to realise the rights set out in Article 12, including the rights to maternal, child, and reproductive health, the rights to a healthy natural and working environment, the rights to disease prevention, treatment, and control, and access to the right to sanitation, goods, and services.

In order to increase agricultural output and help combat famine and malnutrition, the United Nations established the FAO in 1945. It provides funds to states to help them improve their food production and guidance to states on how to improve their land's productivity, help alleviate the strain of starvation in their country, and contribute to improving the health of the masses.

The UN established the Global Fund to Fight AIDS, tuberculosis, and malaria in 2001, contributing to improving health by providing cash to statewide programs in countries with high disease burdens. Public, private, and governmental organisations were given the authority to apply for funds to promote their country's health infrastructure development and improvement. All applications had to be aimed at improving AIDS, tuberculosis, and malaria treatment and prevention on a national level.

UNICEF was formed by a UN official decision in 1946 to improve the welfare of children around the world. UNICEF supports children's educational efforts and works to reduce child and infant mortality by working directly in countries where children are affected by disease, starvation, and war. UNICEF works with governments and non-profit organisations to adopt national policies to support children's welfare and health.

The UNHCR was established in 1950 by UN official order to function as the UN's vehicle for ensuring refugee well-being and providing access to life-saving cardinal HC. Because they are not citizens of the country they are compelled to live in, refugees' HC is jeopardised. As a result, it is the responsibility of the UNHCR to ensure that the needs of displaced individuals are met regardless of their country of origin or status within their country of residence.

Present Scenario of Human HC

Despite years of national plans to control most of these diseases, infectious diseases remain a serious public health problem. Infectious diseases such as hepatitis, tuberculosis, dengue fever, malaria, and pneumonia continue to plague Bangladesh due to a growing reluctance to receive medical treatment. The WHO recommends introducing dengue vaccines in high-burden countries such as Bangladesh. Dengue cases doubled from 12,317 in 2006 to 27,247 in 2009.

Additionally, the leading cause of death worldwide is non-communicable diseases. Poor diet, physical inactivity, exposure to tobacco smoke, and harmful alcohol use put children, adults, and seniors at risk for NCDs. Poor eating habits and a lack of exercise may cause people's blood pressure, blood sugar, blood lipids, and obesity, which can result in cardiovascular disease, long-term respiratory diseases, cancer, and diabetes. Human health is significantly impacted by subpar sewage infrastructure, wastewater treatment facilities, and sanitation.

A key aspect of the world people wants to live in is for everyone to have access to clean, accessible water, and there is enough fresh water on Earth to make that happen. Every year, millions of people, including children, die from inadequate water, sanitation, and hygiene due to poor economic conditions or poor infrastructure. More than 2 billion people are currently at risk of limited freshwater resources. At least one in four people will live in a country with chronic or recurring freshwater shortages by 2050.

According to the minimum standard of the WHO, one doctor is required for one thousand patients. According to the Medical Council of Bangladesh, Bangladesh has 0.389 physicians per 1,000 inhabitants, which is fewer than what the WHO requires. The doctor-to-population ratios in various nations are as follows:

- Doctor and population ratio in Australia is 3.374:1000.
- Doctor and population ratio in Brazil is 1.852:1000.
- Doctor and population ratio in China is 1.49:1000.
- Doctor and population ratio in France is 3.227:1000.
- Doctor and population ratio in Germany is 4.125:1000.
- Doctor and population ratio in Russia is 3.306:1000.
- Doctor and population ratio in USA is 2.554:1000.

According to the UN, most under-five deaths are due to avoidable or curable conditions such as congenital disorders, pneumonia, diarrhea, malaria, and other illnesses. Generally speaking, generally speaking. The under-five mortality rate for rural children is 50% higher than for children in large cities. According to a paper published in the international journal The Lancet in 2017, adopting preventive approaches, such as enabling diarrhea and pneumonia treatments, and providing measles and tetanus vaccinations, helped increase hospital birth rates and reduce child deaths. In 2017, Bangladesh's infant mortality rate was 32 per 1,000 live births, compared with the global average of 12. There were similar differences in neonatal mortality in Bangladesh, with 24 infant mortality per 1,000 live births, compared with the global average of 18.

Although the Bangladeshi government has made significant efforts in recent years to address the problem of malnutrition, it persists in the country and remains a significant public health concern. Nutrition is inextricably linked to a country's human resource development, efficiency, and wealth. Improving people's nutritional status has been a global effort, particularly in underdeveloped countries.

The general population's mental health is one of our country's most neglected issues. A national mental health survey was recently conducted, which found that almost 150 million people in Bangladesh require mental HC, with 70% to 92% of these patients failing to receive the necessary treatment. According to the WHO, Bangladesh has the highest rate of teen suicide globally.

Despite the government's efforts, the Bangladeshi public HC system faces significant obstacles in providing care, including primary, secondary, and tertiary care institutions. As a result, it would be appropriate to assess Bangladesh's existing HC system compared to those of other industrialised countries. The government's HC system is divided into three tiers: primary, secondary, and tertiary care facilities, with public and private providers providing services ranging from single doctors to super-specialist tertiary care corporate hospitals.

Bangladesh initially accepted a public-sector-led HC delivery model, with most services being provided for free to all citizens. Teaching hospitals, secondary level hospitals, community health centres or rural hospitals, dispensaries, primary health centres, sub-centres, and other facilities are included. Primary health facilities were linked to district hospitals, which provided secondary care, and government-run medical college hospitals, which provided tertiary care.

Over 80% of HC requirements are met by private HC facilities, which account for most of the HC in Bangladesh. Private hospitals charge substantially more for treatment than United, IBN Sina Hospital, or other public hospitals. As a result of private hospitals overhauling state facilities, HC costs have increased, making private-sector HC expenses for most Bangladeshis.

People generally prefer the private sector to the public sector because most public HC services are provided in rural areas. Poor service quality results from experienced HC professionals' unwillingness to visit rural areas, so the majority of the public HC system delivering services in rural and remote areas relies on inexperienced and unmotivated interns who are directed to spend time in public HC. Other key factors that entice people to the private sector include the distance from public sector facilities, excessive wait times, and inconvenient operating hours.

The most significant problem the Bangladeshi government has faced since independence has been providing HC to people. Bangladesh's population is more susceptible to diseases due to rising poverty, population pressure, and climatic variables. Infrastructure has been identified as a critical component in delivering public health services. The primary criterion for assessing a country's HC policy and welfare system is its health infrastructure. It expresses a preference for HC facility development as an investment.

Health infrastructure focuses on tangible potential building in public HC delivery mechanisms rather than solely the country's health policy (HP) results. The HC system necessitates solid planning and management abilities and policies that are well-executed and administered by government agencies in collaboration with private HC providers.

Challenges to Growth of Human HC Sector

Today, Bangladesh faces a double burden of infectious and chronic diseases. History shows that treatment alone does not reduce the burden of infectious diseases; prevention also requires government action. However, the government has made various efforts through Upazilla Health Complexes (UHC), Primary HC in Bangladesh (PHCB), and others to address the double burden of disease effectively. Bangladesh's National Health Policy (BNHP) has several components that primarily focus on infectious diseases but rarely adhere to the prevention of chronic diseases. PHCB is another initiative that supports the treatment of certain chronic diseases, but not prevention. The government has given up prevention and public health functions to a certain extent, negatively affecting human HC.

Another factor affecting the HC landscape in Bangladesh is the high out-of-pocket costs of the population. Generally, most Bangladeshi patients pay for their hospital visits and doctor appointments through direct cash aftercare without payment arrangements. Furthermore, the World Bank and the National Committee revealed that health insurance policies cover only 3% of Bangladeshis. This low number comes in a budding health insurance market that only applies to urban, middle- and high-income people. Medicare is primarily limited to urban areas, while in other areas, mainly rural ones, people continue to pay out of pocket.

Infrastructure is another pain point for the Bangladeshi HC industry. The country faces severe resource shortages in both human and capital goods. HC infrastructure requires human resources, i.e., highly skilled human resources ranging from doctors to other HC support staff (such as nurses, laboratory technicians, pharmacists, etc.) to provide adequate HC to the population. Still, Bangladesh has a doctor-to-population ratio of 0.34 per 1,000 people, compared with 1.9 and 3.2 in China and the OECD. There is a shortage of health professionals, but most of these professionals happen to be concentrated in urban areas where consumers are more affordable, leaving rural areas unattended. There is an 81% shortage of specialists in rural community health centres in Bangladesh. There are fewer than 3,000 doctors in rural Bangladesh's 25,308 primary HC centres.

Public sector HC in Bangladesh has received a lot of criticism due to a lack of focus on promoting public HC delivery. In Bangladesh, the public versus private HC debate is often highly vindictive, and policymakers need to take practical rather than ideological stances in both areas. The country will need to intervene adequately in every sector to improve the quality and affordability of public and private HC services.

The medical device industry is one of the fastest-growing industries in the country and is very similar to the health insurance market but is the smallest piece of the HC puzzle in Bangladesh. The HC industry faces many regulatory challenges that hinder the growth and development of the medical device industry. Recently, the government has been optimistic about removing regulatory barriers to importing and exporting medical devices. Foreign and regional investors regard the medical device industry as the most promising field in the future, which is lucrative and has been insisted on in other countries.

A staggering 70% of Bangladesh's population lives in rural areas, with no restrictions on access to hospitals and clinics. Given this impression of the human HC system, it is clear that proper HC infrastructure and some new medical practices are urgently needed to ensure HC quality even in the most affected parts of the country.

HC Policy in Bangladesh

The right to health does not automatically entail that one has that right, according to Mary Robinson, the High Commissioner for Human Rights. It does not imply that a resource-constrained government must offer prohibitively expensive HC services. But in order to provide HC for everyone in the shortest amount of time, governments and public bodies must put policies and action plans into place. It is difficult for the human rights community and public health experts to ensure this occurs.

The birth of a powerful country cannot be achieved by relying on straws. National health contributes to economic progress because healthy people live longer and are more skilled. Governments and HC providers must develop and implement practical human HC policies. They plan to manage health-related matters at a cost that is generally acceptable and affordable and in a way that is responsive to their social, cultural, and economic well-being. Their implementation should enable all citizens to benefit equally from HC services. HC has increased significantly for the development of individual health, thereby affecting the health of families and nations. These government actions will primarily ensure HC for every citizen of the country. Thus, Bangladesh seems to be able to tackle the various health problems plaguing the country today confidently.

HC is an essential part of the HC system. It is often one of the broadest areas of global government and personal spending. Health promoters are keen to recognise and encourage the adoption of policies that will encourage the health and wellbeing of communities. Because of established accountability, policies and programs are developed in response to the population's needs. A human rights-based approach recognises relationships, enables people to assert their rights, and motivates policymakers and service professionals to deliver on their commitment to creating a more inclusive HC system. HP symbolises the decisions, plans, and actions taken to achieve society's specific HC goals.

The Ottawa Charter recognises a healthy public policy that enhances the health and well-being of the individuals and communities it affects. In addition, the World Health Organization explained that a clear HP could define a vision of the future, set priorities and the expected roles of different groups, build consensus and inform people.

There are many categories of health policies (including "personal HC policies", and "medicine policies") related to public health such as "vaccination policies", "tobacco control policies", "breastfeeding promotion policies", etc., and may cover topics such as financing and delivery of HC, the right approach, HC quality, and health equity.

Suggestions for Better Human HC Infrastructure

Some recommendations must be successfully followed because the Bangladeshi HC system has several flaws that must be addressed. Some recommendations must be successfully followed if a solid and adequate HC infrastructure for a healthy Bangladesh is to be achieved.

- There is a need to increase the number of medical staff in the national HC system, such as doctors, nurses, and pharmacists, to address the shortage of medical personnel, which might significantly improve the general public's health. The existing infrastructure is not being exploited to its full potential because of the scarcity of doctors, nurses, and other health professionals.
- The government must regularly review health policy to assess the impact of various plans and programs to identify regions where HC services lag, and special attention must be given to these areas. Bangladesh's health policies have been developed and changed regularly. Although these policies have good programs to reach the poor and those living in rural regions, their execution is still weak, owing

to corruption and poor governance at the local level. As a result, proper implementation of health policies is required.

- There is also a need for better coordination among the various actors involved in HC, whether directly or indirectly, such as the federal government, state governments, and civil society. A more comprehensive, coordinated, and intertwined strategy would provide more fruitful results and significantly improve our HC system.

- Furthermore, developing digital health technology is critical for improving the HC system's productivity and results and delivering better HC results in approach, quality, affordability, disease burden minimization, and effective monitoring of citisens' health entitlements.

- In addition, further investments in the HC sector are required. Inequalities in access to quality and affordable health services are primarily due to insufficient financial allocations to health. In recent years, government spending on health has stabilised at around 1% of GDP, well behind the international average. Despite smaller grants, these funds are inefficient due to fragmented planning and vertical disease programs, and there is an urgent need to increase the government's share of public health spending.

- Though investment in the HC sector of 2-3% of GDP is assured, it is still insufficient to bring about significant changes, as the government is required to monitor the use of these allotted funds, as there is a lack of monitoring on the part of the government of the funds and resources dedicated to the improvement of the HC sector.

- Government hospitals are running out of beds, rooms, and medicine, resulting in insufficient resources and infrastructure. Therefore, infrastructure upgrades are required. In addition, infrastructure needs to be improved in terms of equipment. Most public health agencies do not have sufficient infrastructure to conduct medical tests such as X-rays, blood tests and other complex tests. These devices should be made available to the public through public-private partnerships or by encouraging public and private entities to produce such devices locally at competitive prices.

- Because there are insufficient beds, rooms, and medicines, government hospitals face a lack of resources and infrastructure. As a result, adjustments to the infrastructure are required. In addition, there is a need for infrastructure improvements in terms of equipment, as most public health institutions lack the necessary equipment for medical examinations like X-rays, blood testing, and other complex tests. The equipment should be made available to the general public through a public-private partnership or by supporting localised production of such equipment at competitive prices by both public and private entities.

- The government should make more efforts to improve HC facilities in rural areas because few government hospitals exist in these areas. Even those that do exist lack the majority of medical facilities. Furthermore, rural communities have a scarcity of skilled medical doctors in comparison to urban areas. Most HC professionals are concentrated in urban areas, where consumers have more disposable income, leaving rural areas untreated. Then there is an urgent need to correct this alarming scenario by expanding medical infrastructure and expertise.

- There is also a need for an international surveillance network because different diseases have erupted in one country. There is a risk of disease spreading to other countries globally; as a result, WHO must declare that states should be held accountable for sharing information about disease outbreaks to protect the human community. So, necessary efforts can be made to prevent disease transmission from spreading further.

- Health surveys are also required to monitor the impact of public health and disease interventions, among other things, using digital techniques.

- There is also a pressing need for HC regulation to be resurrected. An excellent regulatory structure is required to achieve significant public health results. There are numerous enforcement, monitoring, and assessment gaps, resulting in a shaky human HC system. As a result, public health regulation must be revitalised through government collaboration, which can be accomplished by revising and implementing public health regulations and increasing public awareness of current rules and enforcement procedures.

- Another key step is to strengthen and increase the effectiveness of health promotion activities by extending them to rural areas and holding diabetes days, heart days, and other similar activities in villages, which will help to raise awareness at the grassroots level, as well as help young people to understand the harms of smoking and promote physical activity. These events focus on encouraging people to make healthy decisions.

- As clean drinking water and sanitation are significant determinants of health, which would directly contribute to a 70-80% reduction in the burden of infectious illnesses, it is necessary to concentrate on inadequate sanitation, lack of access to fresh water, poor hygiene, and incorrect waste disposal methods.

- Another significant component is population stabilization, which is necessary to ensure the quality of life for all inhabitants.
- Furthermore, the effects of climate change and calamities on HC must be minimised. The public health concerns associated with climate change include temperature extremes and weather disasters, vector dissemination, food, or water illnesses, etc. The public HC system must be appropriately trained and equipped at predetermined levels to respond effectively to catastrophes and emergencies.
- The need of the hour is for mass awareness of HC issues, cooperation, and focused efforts to implement government policies and programmes to accomplish the goal of a healthy nation. The lack of public knowledge is a challenge that must be overcome while developing a HC strategy. Even if the therapy is free, there is no doubt that people will take advantage of it until and unless the masses are educated and informed about the signs of the diseases, their repercussions and complications, and the treatment options available.
- The current chaos in public HC has prompted people to seek treatment in private hospitals. Suppose the government truly cares about the health of its citizens. In that case, it must improve public hospitals so that they can compete with private competitors and provide affordable, high-quality, universal HC. The role of private firms in the HC sector is growing, which means that HC facilities are becoming more expensive and out of reach for the poor.
- All three pillars of the public HC system are dysfunctional. Primary, secondary, and tertiary HC systems are critical components of the public HC system because they provide emergency treatment to patients in an emergency. Yet, they are insufficient to service such a large population. Primary, secondary, tertiary, and outpatient care are all undeveloped, and there is a need for improvement in all three sectors of HC.
- In addition, the lack of health insurance complicates the approach to HC, as about 75% of the Bangladeshi population pays for HC services out of pocket, putting an immense financial strain on the country. Bangladesh's inpatient surgical operations have been limited to critical sickness, and they have frequently been limited to one-time lump-sum payouts. As a result, insurance is required to comprehend the existing state of Bangladesh's HC system, aid a large number of people, and make a significant improvement in the HC sector by eliminating its flaws.
- Bangladesh's constitution needs to include HC as a fundamental right. While different regulations, national health policies, and programs have contributed significantly to the development of the HC industry, much remains to be done to improve the industry. Existing HC systems need to be restructured to achieve equity and social justice, which can be achieved by enacting a comprehensive legislative framework. Therefore, legislation should be supplemented by making the right to HC a fundamental right that can be enforced.

Conclusion

The government of Bangladesh has developed and implemented essential legal provisions in the HC sector at the national and international levels to protect people from HC-related dangers. Significant improvements in health indicators include life expectancy, infant mortality, and maternal mortality. Other programs launched by the Bangladeshi government to protect public health include the National Health Mission, National Tobacco Control Program, and National Elderly Care Program, which provide different goals and strategies for protecting human health and strengthening the HC system.

A universal HC strategy is an explicit goal of international organisations and national governments. Bangladesh's National Health Policy 2011 aims to achieve the highest level of health and well-being for all ages by introducing preventive and primary HC and a universal approach to high-quality HC services that do not put citizens in financial hardship. Bangladesh's progress on several health-related indicators has driven the country's progress toward the Millennium Development Goals. Nonetheless, much remains to be done to achieve the goal of universal health coverage.

Despite the extraordinary achievements, Bangladesh's HC system still has some deficiencies that must be addressed if the national health policy goals are to be achieved. Low health budget allocations are an important cause of inequalities in access to quality and affordable health services. Government spending on health has remained around 1% of GDP for many years, putting Bangladesh well behind the international average. These funds are not used effectively due to fragmented planning and vertical disease programs. Therefore, the government must increase its contribution to public health spending. In addition, the country has a severe shortage of labor and infrastructure. Despite Bangladesh's role as a global pharmacy, medical supplies are scarce, and access to life-saving medicines remains a problem. On the other hand, the private sector, which provides care to about 70% of the population, is unregulated and prohibitively expensive for the public. Bangladesh also faces significant problems in providing safe drinking water, sanitation, solid waste management and drainage. Access to health services is disparate, with more than 90%

of urban residents having access to health services, compared to 39% in rural Bangladesh. Rural and urban disparities are also stark in access to health services.

References

[1] Shrotryia, V. K. and Mazumdar, K. (2017). The history of well-being in South Asia. The Pursuit of Human Well-Being. Cham: Springer. pp. 349–80.

[2] Kruk, M. E., and Freedman, L. P. (2008). Assessing health system performance in developing countries: a review of the literature. Health Policy. 85(3), 263–76.

[3] Sun, D., et al. (2017). Evaluation of the performance of national health systems in 2004-2011: an analysis of 173 countries. PloS One. 12(3), e0173346.

[4] Joarder, T., Chaudhury, T. Z., and Mannan, I. (2019). Universal health coverage in Bangladesh: activities, challenges, and suggestions. Advances in Public Health. Volume 2019, 1–12.

[5] Tsai, M.-F., et al. (2019). Understanding physicians' adoption of electronic medical records: health care technology self-efficacy, service level and risk perspectives. Computer Standards & Interfaces. 66, 103342.

[6] Granja, C., Janssen, W., and Johansen, M. A. (2018). Factors determining the success and failure of eHealth interventions: systematic review of the literature. Journal of Medical Internet Research. 20(5), e10235.

[7] Dussault, G., and Franceschini, M. C. (2006). Not enough there, too many here: understanding geographical imbalances in the distribution of the health workforce. Human Resources for Health. 4(1), 1–16.

[8] Angeles, G., Ahsan, K.Z., Curtis, S.L., Spencer, J., Streatfield, P.K., Chakraborty, N. and Brodish, P. (2022). Measurement Challenges in Designing and Conducting Surveys on Urban Population: Experience from Bangladesh Urban Health Surveys. Survey Methods: Insights from the Field. Retrieved from https://surveyinsights.org/?p=16730. DOI:10.13094/SMIF-2022-00001

[9] Saha, K. K., et al. (2015). Bangladesh National Nutrition Services: assessment of implementation status. World Bank Publications.

[10] Mavalankar, D., et al. (2005). Building the infrastructure to reach and care for the poor: trends, obstacles and strategies to overcome them. https://www.iima.ac.in/sites/default/files/rnpfiles/2005-03-01mavalankar.pdf

[11] Darkwa, E.K., Newman, M.S., Kawkab, M. et al. (2015). A qualitative study of factors influencing retention of doctors and nurses at rural healthcare facilities in Bangladesh. BMC Health Serv Res 15, 344. https://doi.org/10.1186/s12913-015-1012-z

[12] Islam, A., and Biswas, T. (2014). Health system in Bangladesh: challenges and opportunities. American Journal of Health Research. 2(6), 366–74.

[13] Andersen, R. M. (2008). National health surveys and the behavioral model of health services use. Medical Care. 46(7), 647–53.

[14] Petersen, P. E., and Yamamoto, T. (2005). Improving the oral health of older people: the approach of the WHO Global Oral Health Programme. Community Dentistry and Oral Epidemiology. 33(2), 81–92.

[15] Viner RM, Ozer EM, Denny S, Marmot M, Resnick M, Fatusi A, Currie C. (2012). Adolescence and the social determinants of health. Lancet. 2012 Apr 28;379(9826):1641-52. doi: 10.1016/S0140-6736(12)60149-4. Epub 2012 Apr 25. PMID: 22538179.

[16] Ali M, Emch M, Tofail F, Baqui AH. (2001). Implications of health care provision on acute lower respiratory infection mortality in Bangladeshi children. Soc Sci Med. 2001 Jan;52(2):267–77. doi: 10.1016/s0277-9536(00)00120-9. PMID: 11144783.

[17] Saifuddin, S. M., Chhina, H., and Zaman, L. (2022). Perspectives on diversity and equality in Bangladesh. In Research Handbook on New Frontiers of Equality and Diversity at Work. Edward Elgar Publishing.

[18] Rahman, R. M. (2006). Human rights, health and the state in Bangladesh. BMC International Health and Human Rights. 6(1), 1–12.

[19] Agyemang, K. K., Adu-Gyamfi, A. B., and Afrakoma, M. (2013). Prospects and challenges of implementing a sustainable national health insurance scheme: the case of the Cape Coast metropolis, Ghana. Prospects. 3(12), 140–148.

[20] Mia, B. (2008). Human rights education and realization in Bangladesh: Implementation of right to education. Human Rights. 28(3), 583–603.

[21] Nahar, N., Blomstedt, Y., Wu, B. et al. (2014). Increasing the provision of mental health care for vulnerable, disaster-affected people in Bangladesh. BMC Public Health 14, 708. https://doi.org/10.1186/1471-2458-14-708

[22] Rayhan, A., and Khan, T. I. (2020). The constitution of people's republic of Bangladesh-solemn expression of people's will; a legal basis of social changes. Indian Journal of Law and Justice. 11, 234.

[23] Glahn, G. and Taulbee, J. L. (2015). Law among nations: an introduction to public international law. Routledge.

[24] Labonté, R. and Gagnon, M. L. (2010). Framing health and foreign policy: lessons for global health diplomacy. Globalization and Health. 6(1), 1–19.

[25] Currie, C., et al. (2008). Researching health inequalities in adolescents: the development of the Health Behaviour in School-Aged Children (HBSC) family affluence scale. Social Science & Medicine. 66(6), 1429–36.

[26] Meier, B. M. (2010). Global health governance and the contentious politics of human rights: Mainstreaming the right to health for public health advancement. The Stanford Journal of International Law. 46, 1.

DOI: 10.1201/9781003450252-

Chapter 52

Energy and cost saving potentials of agricultural crop residues as alternative fuel: an Indian cement plant case study

Prashant Sharma[1,a], Arun Kumar Sharma[2,b], Vivek Gedam[2,c], Murali Krishna M.[2,d] and Dr. S.V.H. Nagendra[3,e]

[1]Shri Ram Institute of Science & Technology, Jabalpur, India

[2]Gyan Ganga Institute of Technology & Sciences, Jabalpur, India

Abstract

As an agro nation, the strength of India's biofuels projects is concentrated in the rural sector based on agro based waste biomasses. Agriculturally based economy is deemed as the base of Indian economy. Crop residue are the left-over parts of crop production system including the parts which are left in the fields after harvesting. The bio energy generation from the residues are always benefitted the energy demands cost effectively. The aim of the research is to find out the capacity and cost saving opportunity in a cement plant, while using the crop residues left over on the farms. Gasification routes have been optimised for biofuel generation and using as the basic source of energy for the cement plant. There is total 10 major crops residues including the Indian crops of both Rabi (summer) and Kharif (winter) seasons are considered as the biomass having gasification routes. A simulation study has been carried out.

Keywords: Alternative fuel, biomass, crop residue, gasification, simulation

Introduction

There were 159 nations and regions that produced cement, either in integrated cement facilities or via crushing trade in clinker, in last decade. India is 2nd biggest manufacturer of cement with capacity of 545.0 Mt/yr having both integrated and grinding plants. India is the third largest emitter of CO_2 in the year 2022 followed to China and USA, having the share about 7.18%. The production of cement is currently the third most fuel user and second highest CO_2 producer in India's industrial sectors (IEA, 2021), accounting for 9% of the entire 11 national green-house gas inventories [1].

India approved the second commitment phase of the Kyoto Protocol, which binds nations to cutting emissions, reiterating its commitments to environmental change. India aims to reduce the carbon emission by 24% by 2020 in comparison with 2005 level. In which all the sectors need to find out the different sources and different methods for heat generation.

It has been estimated by various researchers that about 30-40% of cost of energy are accounted in total cost for production of cement. The combustion of traditional fuels like coal generates a significant amount of greenhouse gases (GHG) which accounts 5% of global human caused, non-biogenic carbon dioxide productions. As a result, there is a need for replacing and improving the energy supply route that generates fewer emissions in the quick and moderate phrase, such as: substituting presently utilised raw resources with components with a lower energy dense to yield or emit less GHG emission or create more efficient via process improvement or coal and oil substitute.

To manufacture one ton of cement, 110–130 kg of coal based on their calorific value and 110 kWh of current is essential [1]. Because of the substantial use of coal in kilns, cement plants are vulnerable to growing oil and gas rates, hence an alternate fuel is required to lower energy bills.

In cement industries the various waste derived fuels have been utilised called Refused Derived Fuel (RDF) refers to fuel formed from ignition of solid wastes that can be obtained energy from waste materials. The example for RDF are municipal solid waste (MSW), reprocessing, waste, manure and biomass waste etc. [2] Various researchers have been carried out there study for different alternative fuels are given in Table 52.1.

Since coal and oil are supplanted with substances that would normally be diminished or burned with correlating carbon output and finished by-products, utilising alternative fuels (AFs) has the advantage of reducing the amount of non-renewable coal and oil and lowering carbon dioxide pollution in the cement production. A cement manufacturing facility in Alcanar, South Catalonia, Spain, began co-processing a unique sort of refuse-derived fuel (RDF) known as EnerFuel in 2009 [16]. Some researchers have been carried out the LCA for the different fuel [17].

[a]pra.sharma30683@gmail.com, [b]arunksharma232@gmail.com, [c]vgedam@ggits.org, [d]muralikvamsi@gmail.com, [e]svhnagendra@gmail.com

DOI: 10.1201/9781003450252-52

Table 52.1 Previous researches for different fuels in cement plant

Fuel	Replacement	Findings	Authors
Rubber wastes and plastic wastes	Partial	Plastic wastes: ash <2%, CV: 9230 cal/g). Rubber wastes: ash >21% CV:8600 cal/gm.	[2, 3]
Previously dried sewage sludge	Complete/viable supplementary fuel	CV: 8300 J/g, sewage sludge might substitute up to 14% of the raw resources required to make cement while affecting crucial raw feed properties. May cut the use of carbon fuels by almost 70% in a contemporary cement kiln. A modest rise in the production of belite	[4]
municipal solid wastes	20% RDF and 80% Coke	Fewer harmful ecological effects than if all fuels were solely petroleum coke. Significant net economic advantage to community, individuals, and localities.	[5-8]
Spent activated carbon	Partial	Whereas the energy content is 18% less, it remains adequate for the intended application. Burning biochar produces 10 times greater ash as burning oil coke.	[9]
shredded non-recycled plastics and paper residue (NRPP) called "engineered fuel"	20% fossil-based carbon and 80% biogenic	The lower heating value (LHV) of that kind of substance is approximately 17 MJ/kg. Because EF is primarily made of wood fiber, a biogenic resource, using it reduces the CO_2 emissions caused by making cement. when utilised instead of high-quality coals, decreases carbon dioxide emissions by approximately three tons of CO_2 per ton of EF. No negative impact on the characteristics of generated cement or the stacks pollution of cement factory	[10]
Wood derived fuel (WDF)	20% WDF with 80% coal.	Co-combustion resulted in 15% less greenhouse gases and 14% less overall fuel usage as compared to coal use alone.	[11, 12]
Spent pot lining (SPL)	Complete replacement	According to the findings, coal fuel may be replaced with the ultimate processed fuels in the cement sector to decrease pollutions out from boiler. Reduce NO and CO_2. High energy content and enough carbon for co-firing with coke or burning as feedstock in the cement production.	[13]
Meat and bone meal (MBM)	A Comparative analysis	Comparing MBM burning to burning coal, the exhaust temperature is less.	[14]
Carbon black	Carbon black with coal at 0, 1.5, 3 and 5% by mass	The findings imply that carbon black may be employed in order to reduce energy bills by 1-3%. The significant Sulphur and ash concentration of carbon black, though, may have an impact on the properties of clinker and kiln cement.	[15]

Solid biofuel, derived from household food waste, Wood based fuel are also used as a potential energy sources in cement industry [18]. From the above discussion it has been understand that no more studies have been carried out in Indian cement industry for having alternative fuel in Indian cement plant.

India is facing another problem i.e., pollution control. Among all the pollution air pollution are the major threat for the humans. The foremost reason of air-pollution in Indian cities are burning of agriculture residues after harvesting called Parali. The idea for this research is to bring this agriculture residue to the cement plant and use it as the alternative fuel for cement production. The aim of the current research is to find out the possibility of using the agriculture waste as the alternative fuels in the cement manufacturing having route of gasification process [19].

Top accessible path for having solid wastes in the cement industry

Basically, there are two different ways to utilise alternative fuels in cement manufacturing i.e., direct combustion in cement kilns and gasified or incinerated beforehand. The byproduct of the second ones are utilised as raw materials as required.

Direct combustion

Direct combustion of alternating fuel is one of the traditional processes but having some limitations. It requires the preconditioning process for having uniform heating and temperature throughout and to achieve it uniform preprocessing unit must be applied.

Gasification

The gasification process has been carried out for biomass fuels as alternative fuels produced syn gas (mixture of hydrogen, carbon mono oxides, carbon dioxide, and methane). This gas can be used as fuel and can co-fired in furnaces. Various type of biomass wastes can be used as the product gas has homogeneity. For industrial scale application the fluidised bed gasification are preferable as having good temperature control, high rate of reactions, higher carbon conversion ratio and finally the high quality of syn gas. There can be another option like plasma gasification. During the gasification process there are high temperatures, thus the carbon-based composites are detached into identical modest particles and still dangerous wastes are converted into a valuable product gas. The slag generated after gasification could be directly utilised as raw substances in cement process. On the other hand if the temperature is lower during the gasification process due to the incomplete reactions the amount of slag and ash are higher than the syn gas.

Major Crops Production in Vicinity of the Cement Plant

Basically, there are two districts i.e., Satna and Katni placed in the vicinity of the plant. There are two seasons for crop productions are Kharif (monsoon crops or autumn crops) and Rabi (grown in winter and harvested in spring). The major crops for Kharif are arhar/tur, jowar, maize, rice, soyabean and urad and for Rabi season are gram, masoor, rapeseed and mustard and wheat as per the data available by the government of India, Ministry of agriculture. (Table 52.2 (a) and (b))

Table 52.2(a) Total production in the district (Rabi season) [ministry of agriculture, India]

	Gram	Masoor	Rapeseed and mustard	Wheat
Total area (Hectare)	134969	40692	5633	245491
Total production (Tonnes)	88692	22305	3315	694072

Table 52.2(b) Total production in the district (Kharif season) (ministry of agriculture, India)

	Arhar/Tur	Jowar	Rice	Urad	Soyabean	Maize
Total area (Hectare)	32565	5645	189207	17885	82513	4127
Total production (Tonnes)	12348	5078	476306	5707	18757	8104

Whenever an agriculture product is consumed, leftovers are produced. The stalk, vacant berry bunches, leaflets, and shoots, besides other things, are examples of agricultural portions that are frequently useless. A particular crop's projected annual waste production is primarily influenced by three key variables the number of crops grown, the proportion of dried residual to yield, and the breed or type. As increased output amounts of harvests are required to generate considerable portions of leftovers, the proportion of the yield has the biggest impact on the overall volume of by-products amongst three components.

Regarding the dried residual to crop proportion, this element could only have a noticeable impact when considering several crops that are generated in the similar amounts. But irrespective of type or variant, this percentage frequently falls inside a short amount for a particular crop. The agricultural type and/or variation can be more easily controlled since they have already been chosen to fit the environment and the needs of the current companies that processed those harvests. The most common agricultural crop wastes generated in India are sugarcane leaves, paddy straw, rice husks, and maize stover, that are frequently burnt to prepare the soil for the following cropping season [20].

Analysis of biomass (crop residue)

The ultimate and proximate analysis has been done by applying ASTM standard to obtain the characteristics of different biomass. The sampling of crop residue has been carried out from the field following the ASTM D5231 standard. 10 samples of each crop residues have been collected and the results have been estimated using the average sampling method. Table 52.3 shows the ASTM standards applied for the analysis.

The following results have been obtained as proximate and ultimate analysis

Simulation Study

A simulation study has been carried out for crop residues as the fuel for calcination process considering Aspen Plus as the simulation tool [20]. Aspen Plus is a programme that analyses the unit and overall process balance using rigorous and analytical techniques cantered on the system's principles of resource and energy sustainability. It is employed to simulate thermal, biological, or conventional transformation processes. Table 52.4 gives the reactions of biomass gasification takes place.

Table 52.3(a) Standards for ultimate and proximate analysis of crop residue

Analysis	Components	ASTM standards
Proximate analysis (Dry basis)	Volatile matter	D 3175
	Fixed carbon	Calculated
	Ash	D 3174
	Moisture (moisture-included basis)	D 3173
Ultimate Analysis (Dry basis)	Carbon	D 3178
	Hydrogen	D 3178
	Oxygen	Calculated
	Nitrogen	D 3179
	Sulfur	D 3177
	Ash	D 3174
	HHV	D 5865

Table 52.3(b) Ultimate and proximate analysis of crop residue

Analysis	Components	Gram	Masoor	Rapeseed and mustard	Wheat	Arhar/ Tuwar	Jowar	Rice	Urad	Soyabean	Maize
Proximate analysis (Dry basis)	Volatile Matter	21.0	63.0	59.0	65.3	67.2	55.1	50.5	70.0	66.0	70.3
	Fixed Carbon	35.0	18.0	21.0	17.7	23.0	23.0	15.9	17.5	16.5	16.0
	Ash	38.0	5.5	4.0	7.0	4.0	11.7	18.7	6.1	6.6	5.3
	Moisture (moisture-included basis)	6.0	13.5	16.0	10.0	5.9	10.3	15.0	6.4	11.0	8.4
Ultimate analysis (Dry basis)	Carbon	41.0	40.0	37.5	44.9	44.5	41.6	34.2	45.0	40.8	44.2
	Hydrogen	3.4	6.3	5.6	5.5	5.5	5.2	5.2	6.4	3.5	5.8
	Oxygen	10.0	34.0	35.0	32.0	39.5	30.1	30.0	35.3	37.8	35.1
	Nitrogen	1.3	0.6	0.6	0.4	0.6	0.7	0.9	0.7	0.2	1.3
	Sulfur	0.4	0.1	0.0	0.2	0.1	0.1	0.2	0.1	0.1	0.0
	Ash	38.0	5.5	5.3	7.0	4.0	12.0	15.0	6.1	6.6	5.3
GCV		4057	3943.1	3502.1	4201	3924	3925	3290	4350	3111	4124

The simulation has been carrying out for generation of energy that will be used for the calcination process. The following assumptions has been made for the modelling in Aspen Plus

• The process is considered as steady state and no heat loss have been considered.
• There is no air leakage
• No pressure loss
• Motion is turbulent
• The oxides of different components like Fe, Si, Al, and other metals have been neglected.
• Calciner function in adiabatic condition.
• Process is steady state and isothermal.
• Pressure drops, Heat loss and Tar formation are neglected.

The chemical conversion reactions that has been taken place during the gasification process are as:

The Peng Robinson equation of state is used for the analysis. The biomass obtained from the farms are in wet condition. Thus the biomass is considered as the wet biomass i.e. it contains considerable amount of moisture. For the analysis MCINCPSD stream class has been considered which is generally used for conventional and non conventional both type of materials. The supply temperature and pressure of raw material i.e. biomass is about atmospheric pressure and the temperature i.e 300K and 1 bar respectively. (Figure 52.1)

Table 52.4 Reactions taking place during the gasification process [21]

Reactions	Chemical equation
Char gasification	$C + H_2O \leftrightarrow CO + H2 + 131$ KJ/mol
Boundouard	$C + CO_2 \leftrightarrow 2CO + 172$ KJ/mol
Methane decomposition	$\frac{1}{2} CH_4 \leftrightarrow \frac{1}{2} C + H_2 + 74.8$ KJ/mol
Water gas shift	$CO + H_2O \leftrightarrow CO_2 + H_2 - 41.2$ KJ/mol
Steam reforming	$CH_4 + H_2O \leftrightarrow CO + 3H_2 + 206$ KJ/mol

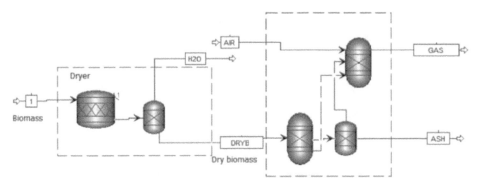

Figure 52.1 Aspen plus model for gasification unit
Source: Author

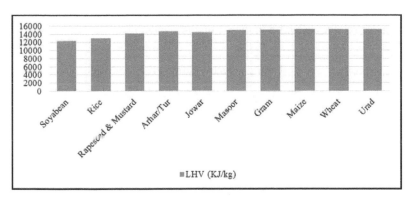

Figure 52.2 LHV obtained from the gasification simulation process for each crop residue
Source: Author

The biomass characterisation has been obtained by the its proximate and ultimate analyis given in previous section. For heat of combustion method the HCOALGEN enthalpy and DCOALIGHT density model has been applied for the analysis.

Initially a dryer (RSTOIC reactor) is applied for drying the biomass to remove additional surface moisture as possible. High temperature nitrogen is utilised as moisture removing fluid. The dry biomass is then suuplied to the seperator (i.e. flash seperator) which seperates the nitrogen gas contained water particles from dry biomass. The dry biomass then supplied for gasification process in a reactor (i.e. RYIELD reactor). The property of RYIELD reactor is best suitable here as it does not depends on kinetics and stoichiometry. The ouput from the yielding is send to the another reactor (i.e. RGIBBS reactor). The RGIBBS reactor works on the principle of GIBBS free energy without the specification of chemical equillibrium.

In RGibbs reactor combustion of SYN gas taken place. The afterburn gases supplied for further seperation process to remove the polluting particles. A RStoic reactor is utilised as calciner for cement manufacturing. Lime stone is added to the reactor for the calcination process. During the calcination the carbonates decomposition is the main process. The gaseous and solid products are seperated out using seperator block.

Simulation Results

Energy yields

To find out the effect of crop residues gasification, the LHV of the product gas (syn gas) has been examined. The energy yield are the sum of the mass yield of every constituent in product gas (i.e. CO_2, CH_4, CO, H_2

and N_2) multiplied by their respective LHV. The Figure 52.2 shows the LHV generation potentials per kg for each of the residue.

The generation of crops residues i.e., stalks, cobs etc per kg of grain generation and syn gas potential for every crop residue along with total LHV per kg of grains are given in Table 52.5. Table 52.6 gives the total production and total cultivated area for each crop in the vicinity of cement plant and *Table 52.7 gives the* total energy that can be produced through gasification

Table 52.5 Residue available and total heating capacity for 1 kg of each grain

Residue available for 1 kg of grains

Crop Name	Residue name	Amount in Kg	LHV (KJ/kg)	Total LHV per kg of grain
Gram	Stalks	1.3	15085.9	19611.6
Masoor	Stalks	1.8	15003.6	27006.6
Rapeseed and mustard	Stalks	1.8	14162.3	25492.1
Wheat	Stalks	1.5	15159.2	22738.8
Arhar/Tur	Stalks	2.5	14730.5	36826.3
Jowar	Cobs, Stalks	2.2	14489.2	31876.2
Rice	Stalks	2	13066.9	26133.8
Urad	Stalks	1.1	15223.3	16745.7
Soyabean	Stalks	1.7	12302.3	20913.8
Maize	Stalks	2	15128.8	30257.7

Table 52.6 Total production and total cultivated area for each crop in the vicinity of cement plant (i.e., Satna and Katni, M.P.) [ministry of Agriculture, India]

Crop	Total area (Hectare)	Total production (Tonnes)
Gram	134969	88692
Masoor	40692	22305
Rapeseed and mustard	5633	3315
Wheat	245491	694072
Arhar/Tur	32565	12348
Jowar	5645	5078
Rice	189207	476306
Urad	17885	5707
Soyabean	82513	18757
Maize	4127	8104

Table 52.7 Total energy potential through gasification

Crop Name	LHV (KJ/kg)	Total LHV per kg of grain	Total production (Tonnes)	Total energy yields (KJ)
Gram	15085.9	19611.6	88692	1.74×10^{12}
Masoor	15003.6	27006.6	22305	0.60×10^{12}
Rapeseed and mustard	14162.3	25492.1	3315	0.08×10^{12}
Wheat	15159.2	22738.8	694072	15.78×10^{12}
Arhar/Tur	14730.5	36826.3	12348	0.45×10^{12}
Jowar	14489.2	31876.2	5078	0.16×10^{12}
Rice	13066.9	26133.8	476306	12.45×10^{12}
Urad	15223.3	16745.7	5707	0.10×10^{12}
Soyabean	12302.3	20913.8	18757	0.39×10^{12}
Maize	15128.8	30257.7	8104	0.25×10^{12}

The plant daily production capacity is varying in between 2100-2200 tonnes/day, then for the analysis the average value of about 2200 tonnes/day is considered. From the plant visit it has been observed that the heat required during the calcination process is about 2170 KJ/kg of limestone. As coal is used as fuel for energy requirements for calcination process, the total amount of 1.17 metric-tonn of coal required per day to satisfy the energy demand and in summary about 1 kg of coal is used for clinkerization of 1.869 kg of clinker at present condition as the coal is supplied from eastern coal fields. Thus for 300 days of working about 1.432×10^{12} KJ energy is required while the capacity of generation of energy through biomass is about 32.01×10^{12} KJ which is about 4.47% of total requirement of the plant.

Saving of coal

In working of the plant the total coal required as fuel is about 1.17 Kilotonn for 24 hours working and the energy generated from 1 kg of coal combustion can clinkerizd about 1.87 kg of clinker and totally depends upon the quality of coal i.e. calorific value of coal. This estimation has been obtained from the plant survey. The aim of the work is to determine the replacement of traditional fuel concerning the cost and environment. The crushing capacity of the plant is about 500 tonnes of clinkers per day and the average clinkerization achived varies from 470-480 tonnes per day and the total coal requirement is about 256.6 tonne of coal per day working. The rate of coal in India about 400 indian rupess per tonne. The coal freight rates per tonne for the plant as the coal is supplying from eastern and south-central India is about 480-500 INR per tonne. Thus the final cost of the fuel is approximately about 880-900 INR per tonne of coal. This plant can save approximately 70 billion INR per year.

Conclusion

The crop rsidues are the biomass which are not required by the farmers and they usually burn them in the farms. Thus these can be collected from the farms easily and also buy from the farmers and utilised in plants. The crop rsidues are the biomass which are not required by the farmers and they usually burn them in the farms. Thus these can be collected from the farms easily and also buy from the farmers and utilised in plants. A detailed analysis for utilising the crop residues from the farms using biomass gasification routes for fulfilling the energy demand of the cement industry have been carried out considering a cement plant in central India i.e. Satna district. About 10 types of cross residues are examnied and analysed using proximate and ultimate analysis. The syn gas generated after biomass gasification process have been utilised as basic fuel for clinkerization process. The total energy yields are found out after gasification process for all the crop residues. A lot of energy can be generated. It has been estimated that for 300 days of plant working, about 1.432×10^{12} KJ energy is required while the capacity of generation of energy through biomass is about 32.01×10^{12} KJ which is about 4.47% of total requirement of the plant. This work can provide the solution for problems facing by both the farmers and cement plants. As the farmers will find out the cost saving solution of mitigating the crop residues called parali and cement plants getting fuel in very low cost compare then traditional fuel.

References

[1] Lamas, W. D. Q., Palau, J. C. F., and De Camargo, J. R. (2013). Waste materials co-processing in cement industry: Ecological efficiency of waste reuse. Renewable and Sustainable Energy Reviews. 19, 200–7. doi: 10.1016/j.rser.2012.11.015.

[2] Hashem, F. S., Razek, T. A., and Mashout, H. A. (2019). Rubber and plastic wastes as alternative refused fuel in cement industry. Construction and Building Materials. 212, 275–82. doi: 10.1016/j.conbuildmat.2019.03.316.

[3] Asamany, E. A., Gibson, M. D., and Pegg, M. J. (2017). Evaluating the potential of waste plastics as fuel in cement kilns using bench-scale emissions analysis. Fuel. 193, 178–86. doi: 10.1016/j.fuel.2016.12.054.

[4] Rodríguez N. H., Ramírez S. M., Blanco-Varela M. T., Donatello S., Guillem M., Puig J., Fos C., Larrotcha E., Flores J. (2013). The effect of using thermally dried sewage sludge as an alternative fuel on Portland cement clinker production. The Journal of Cleaner Production. 52, 94–102. doi: 10.1016/j.jclepro.2013.02.026.

[5] Güereca, L. P., Torres, N., and Juárez-López, C. R. (2015). The co-processing of municipal waste in a cement kiln in Mexico. A life-cycle assessment approach. The Journal of Cleaner Production. 107, 741–8. doi: 10.1016/j.jclepro.2015.05.085.

[6] Reza, B., Soltani, A., Ruparathna, R., Sadiq, R., and Hewage, K. (2013). "Environmental and economic aspects of production and utilization of RDF as alternative fuel in cement plants: A case study of metro vancouver waste management. Resources, Conservation & Recycling. 81, 105–14. doi: 10.1016/j.resconrec.2013.10.009.

[7] Zhu, N., Hu, P., Xu, L., Jiang, Z., and Lei, F. (2014). Recent research and applications of ground source heat pump integrated with thermal energy storage systems: A review. Applied Thermal Engineering. 71(1), 142–51. doi: 10.1016/j.applthermaleng.2014.06.040.

[8] Fyffe, J. R., Breckel, A. C., Townsend, A. K., and Webber, M. E. (2016). Use of MRF residue as alternative fuel in cement production. Waste Management. 47, 276–84. doi: 10.1016/j.wasman.2015.05.038.

[9] Rodríguez, N. H., Martínez-Ramírez, S., and Blanco-Varela, M. T. (2016). Activated carbon as an alternative fuel. Effect of carbon ash on cement clinkerization. The Journal of Cleaner Production. 119, 50–58. doi: 10.1016/j.jclepro.2016.01.093.

[10] Bourtsalas, A. C. T., Zhang, J., Castaldi, M. J., and Themelis, N. J. (2018). Use of non-recycled plastics and paper as alternative fuel in cement production. The Journal of Cleaner Production. 181, 8–16. doi: 10.1016/j.jclepro.2018.01.214.

[11] Saidur, R., Abdelaziz, E. A., Demirbas, A., Hossain, M. S., and Mekhilef, S. (2011). A review on biomass as a fuel for boilers. Renewable and Sustainable Energy Reviews. 15(5), 2262–89. doi: 10.1016/j.rser.2011.02.015.

[12] Hossain, M. U., Poon, C. S., Lo, I. M. C., and Cheng, J. C. P. (2017). Comparative LCA on using waste materials in the cement industry: A Hong Kong case study. Resources, Conservation & Recycling. 120, 199–208. doi: 10.1016/j.resconrec.2016.12.012.

[13] Ghenai, C., Inayat, A., Shanableh, A., Al-Sarairah, E., and Janajreh, I. (2019). Combustion and emissions analysis of spent pot lining (SPL) as alternative fuel in cement industry. Science of the Total Environment. 684, 519–26. doi: 10.1016/j.scitotenv.2019.05.157.

[14] Ariyaratne, W. K. H., Malagalage, A., Melaaen, M. C., and Tokheim, L. A. (2015). CFD modelling of meat and bone meal combustion in a cement rotary kiln - Investigation of fuel particle size and fuel feeding position impacts. Chemical Engineering Science. 123, 596–608. doi: 10.1016/j.ces.2014.10.048.

[15] Sarawan, S., and Wongwuttanasatian, T. (2013). A feasibility study of using carbon black as a substitute to coal in cement industry. Energy for Sustainable Development. 17(3), 257–60. doi: 10.1016/j.esd.2012.11.006.

[16] Mari, M., Rovira, J., Sánchez-Soberón, F., Nadal, M., Schuhmacher, M., and Domingo, J. L. (2018). Partial replacement of fossil fuels in a cement plant: Assessment of human health risks by metals, metalloids and PCDD/Fs. Environmental Research. 167, 191–7. doi: 10.1016/j.envres.2018.07.014.

[17] Georgiopoulou, M., and Lyberatos, G. (2018). Life cycle assessment of the use of alternative fuels in cement kilns: A case study. The Journal of Environmental Management. 216, 224–34. doi: 10.1016/j.jenvman.2017.07.017.

[18] Tsiligiannis, A., and Tsiliyannis, C. (2019). Renewable energy in cement manufacturing: A quantitative assessment of energy and environmental efficiency of food residue biofuels. Renewable and Sustainable Energy Reviews. 107(August 2018), 568–86. doi: 10.1016/j.rser.2019.03.009.

[19] Sharma, P., Gupta, B., Pandey, M., Sharma, A. K., and Mishra, R. N. (2020). Recent advancements in optimization methods for wind turbine airfoil design: A review. Materials Today: Proceedings. 47(xxxx), 6556–63. doi: 10.1016/j.matpr.2021.02.231.

[20] Sharma, P., Gupta, B., Sharma, A. K., Parmar, H., and Baredar, P. (2019). Municipal solid waste in combination with coal as alternative fuel for cement plant: Energy feasibility analysis. SSRN Electronic Journal. 1–8. doi: 10.2139/ssrn.3417104 .

[21] Sharma, P., Gupta, B., and Pandey, M. (2022). Hydrogen rich product gas from air–steam gasification of Indian biomasses with waste engine oil as binder. Waste and Biomass Valorization. 13(6), 3043–60. doi: 10.1007/s12649-022-01690-4.

Chapter 53

A review on the application of twisted tube heat exchanger with nano-fluids as a coolant in automobile radiator

Md NaushadAlam[a] and Dr. Akash Langde[b]

Anjuman college of Engineering and Technology, Nagpur, India

Abstract

In almost every industrial operation, a shell and tube heat exchanger (STHX) is used. The system, however, has drawbacks, such as pressure drop, ineffective utilization of the shell side, and low flow zones nearby the baffles, which can lead to fouling, corrosion, and induced vibration, which can lead to equipment failure. Nanofluids are heat transmission fluids that have better heat transfer (HT) performance and thermos physical characteristics can be utilised to improve the performance of several devices. Cooling fluids containing ultrafine nanoparticles have demonstrated exceptional behavior in trials as compared to pure fluids, including higher heat transfer coefficient (HTC) and better thermal conductivity. According to recent research, nanofluid look to be a promising replacement for conventional coolants in automobile radiators. Nanofluid has the potential to increase cooling rates in automobiles and heavy-duty engines by improving efficiency, lowering weight, and simplifying thermal management. This paper presents a review of the twisted tube heat exchangers and the application of nanofluid in the radiator of the automobile.

Keywords: Heat exchanger, nanofluid, radiators, twisted tube

Introduction

A heat exchanger (HX) shown in Figure 53.1 works efficiently to transfer heat from one fluid to another. Shell and tube heat exchanger (STHX) is a common type of heat exchanger. It is made up of a succession of finned tubes in which one fluid flows through the circular or any cross-section tube and the other fluid passes over it to heat or cool it. Inside the tube or plate system, high-pressure, high-temperature steam or water is moving at a high-speed during heat exchanger operation. A heat exchanger takes advantage of the fact that energy flows where there is a temperature difference. As a result, heat will be transmitted from the heat reservoir at high temperature to the heat reservoir at lower temperature. The temperature difference created by the circulating fluids forces the energy to move between them. The heat energy passing through a HX could be latent heat or sensible heat from the circulating fluids. The fluid that provides energy is referred to as hot fluid, whereas the fluid that accepts energy is referred to the as cold fluid. In a heat exchanger, it is self-evident that as cold fluid temperature increases results in the decrease in hot fluid temperature. The objective of a HX is to either cool or heat the fluid that is being cooled or heated.

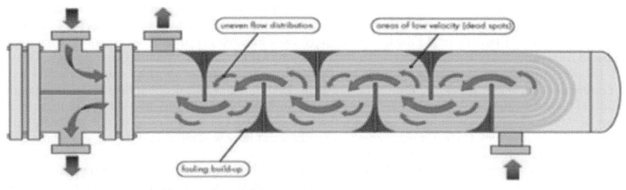

Figure 53.1 Conventional heat exchanger [1]

STHX, condensers, automobile radiators, evaporators, cooling towers and air pre-heaters are some of the common examples of heat exchangers. One of the fastest developing fields of heat transmission technology is heat transfer improvement. Twisted-tube heat exchangers are practiced to improving the HTC on the tube side of STHX, which helps to reduce heat exchanger size (Figure 53.2) [1, 2]. Depending on the geometry of twisted tube, swirl is created, which helps to rise in a temperature gradient. The mean velocity of fluid and effective Reynolds number also depends on twisted tube geometry (Figure 53.3a and b) [3-5].

[a]naushad.ngp@gmail.com, [b]akashlangde@gmail.com

DOI: 10.1201/9781003450252-53

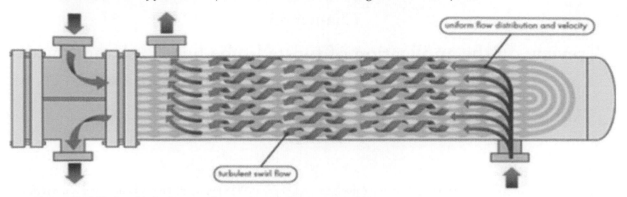

Figure 53.2 Twisted tube heat exchanger bundle [1]

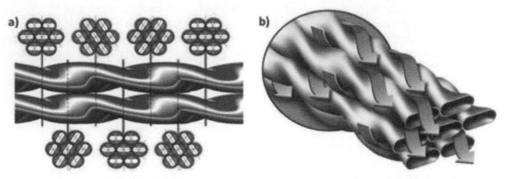

Figure 53.3 (a) Tube alignment and support (b) Shell-side interrupted swirl flow [1]

It was found that the very scanty literature is available on twisted tube heat exchangers that can be used in the radiators. This paper discusses the recent advancement of twisted tube concept, which has ability to overcome drawbacks of traditional concept and how it can be suitable for the radiator of automobile. This paper also discusses the usage of nanofluids as a coolant in vehicle radiators and the issues that come with it.

Twisted Tube Heat Exchanger

Tan et al. [6] looked at how a twisted oval tube with dissimilar twist ratio and axis ratio affected HT and pressure loss. When making an analogy between smooth circular tubes, the twisted oval tubes demonstrated a greater HT and a larger pressure drop. Their findings also revealed that when the axis ratio increases, heat transfer, and pressure loss rise, whereas both decrease as the twist ratio increases. Tang et al. [7] looked at how a twisted oval tube having tri lobed twist affected HT and friction factor. The tri-lobed twisted tube provided a thermal boost better than the oval twisted tube, according to their findings. Sun et al. [8] measured the temperature rise in built in twist belt on outer side of threaded tubes using a variety of nanofluids at varying mass fractions as the test fluid. Their findings revealed that the best friction factor and HTR were achieved by applying an upgraded tube in which Cu and water nanofluid with mass fraction of 0.05% flowed. Pozrikidis [9] examined flow characteristics of square cross-section channels with a twisted tube inserted. Their findings revealed that the axial flow was influenced by the secondary flow which occurred across cross section of tube.

In laminar flow area, Cheng et al. [10] investigated HT and behavior of fluid in twisted tube having varied cross section forms. Their findings revealed that cross-section shape of twisted tube did not affect the HTR. The performance of thermal factor (maximum) of 2.69 was achieved with a pentagon shaped twisted tube having 0.17 twist pitch ratio. The heat transfers and flow properties of nanofluid inside twisted square channels were investigated by Khoshvaght-Aliabadi et al. [11]. As a reference model, they conducted studies on water run through twisted square channels. Their findings revealed that twist pitch modifications had a substantial impact on thermo hydraulic performance of square channels twist. Yang et al. [12] examined consequence of water flow in elliptical shaped twisted tubes in an experimental setting. Various geometrical characteristics (twist pitch and aspect ratio) were examined for their impacts. When compared to a straight tube, it was discovered that the assumed tubes improve both pressure drop and HT. Additionally, tubes with smaller twist pitches and greater tube aspect ratios had higher HTR and friction factor values. The pressure loss and HT in heat exchangers utilizing testing fluid i.e., natural gas was reported by Li et al. [13]. Their findings revealed that using a twisted tube reduces the length of the gas coil by up to 25%

when compared to a straight tube. Alempour et al. [14] practiced finite volume method to model heat transfer and turbulent flow inside 3D tube with cross sections of circular and elliptic shape. According to the findings, heat transfer increased by 5% and 20% in the elliptical tube with twist pitches of 0.4, 0.2 and aspect ratio of 1.75, respectively, when related to smooth tube with related cross section and heat transfer enhanced by up to 30% when the twist pitch was suddenly reduced from 0.2-0.1 along the path. In three turbulent, transitional, and laminar flow regimes, According to Thantarate et al. [15], the twisted tube aspect ratio, should be high and pitch should be low for good heat transmission performance. In the event of a multipass heat exchanger, following this criterion to construct connecting ends will be extremely difficult, as the surface of the highly twisted metallic pipe will be overlapped at some point, clogging the channel. As a result, the end connections should be circular.

Danielsen [16] observed the vibration, fouling, heat transfer rate and pressure drop of twisted tube heat exchangers. Fouling accumulations were reduced on the tube side, where the flow was dispersed evenly and at a virtually constant speed along the tubes. There was also a decrease in vibration, which improved system steadiness and reduced the likelihood of failure. Morgan et al. [1] evaluated the usage of a twisted tube exchanger in the laboratory and compared it to a typical rod baffles exchanger, concluding that the heat transfer coefficient increased averagely by 1.7 times. Sun et al. [8] studied pressure drop effect and HTC on twisted tubes numerically and discovered a 3.89 decrease in pressure drop and a 4.01% rise in HTC. Cheng et al. [10] investigated different twisted tube cross section configurations with Reynolds numbers between 50-2000. It was determined that the HTR of twisted shape square tube is superior to square shape smooth tube, despite the increased pressure drop. It was also revealed that twist pitch has a significant impact on twisted tube HT performance, with the smallest twist pitch ratio delivering the greatest results.

Using computational fluid dynamics (CFD), a STHX with a single twisted tube pack in five dissimilar twist angles is studied and compared to a standard STHX with one segmental baffles [17]. When HTR per unit pressure drop was compared to shell side mass flow rate, it was clear that the HX having twisted-tube bundle outperforms the segmental baffled HX in both perpendicular and tangential shell-side nozzle situations. According to the authors, for shell side flow rate, optimal bundle twist angles for such a HXs are 65 and 55 degrees.

Thantharate and Zodape [15] determined the feasibility of the use of twisted tube heat exchangers in automobile radiators. The authors compared a twisted tube with a plain tube in multiple tube pass (four passes) of 0.3 m length each pass for four flow rates of 1.5 lpm, 1.37 lpm, 0.5 lpm and 0.24 lpm resulting in Reynolds number of Re 62-7000 covering turbulent and laminar range. For a high Reynolds number range, the performance of the twisted tube is better, the reason is attached flow through tubes. The authors also discovered that plain tubes perform 8.71% better than twisted tubes at low flow rates, whereas twisted tubes perform 45% better at high flow rates.

In summary, twisted tubes should be employed in applications like radiators where the flow rate is large. The size of the heat exchanger must be reduced for these applications. When it comes to car radiators, the extra room might be used to expand the passenger area. Twisted tubes and rubber pipes can be connected in radiator applications when there is a high flow rate and low temperature range. Rubber pipes are simple to twist. However, it is important to make sure that the flow is kept continuous while utilising such twisted rubber pipes for the end connections. Through the connections, the flow shouldn't be interrupted, and a velocity profile should be established.

Application of nanofluid as a coolant in radiators

Radiators are nothing but air-cooled HXs that transport heat from the IC engine to the surrounding air. During combustion the air and fuel mix to produce power in an automotive engine. A part of the total generated power is used to provide power to the vehicle; the remainder is squandered as heat and exhaust. If this extra heat is not taken out, temperature of an engine increases, causing viscosity breakdown and overheating of the lubricating oil, stress between engine parts and metal weakening of overheated engine parts, resulting in faster wear of the corresponding moving components [18, 19].

As a result, we'll need to install a cooling system to eliminate the engine's waste heat. The pressure cap, reserve tank, heater hoses, lower hose, heater core, thermostat, water pump, transmission cooler, fan, upper house, and radiator, are the main components of a cooling system (Figure 53.4). A vehicle's cooling arrangement works by moving liquid coolant through channels in the engine heads and block. The heat from the engine is then absorbed by the liquid coolant that runs via channels. The fluid then travels to the radiator via a rubber pipe. The heated liquid is cooled at this point by an air stream that enters the engine part through grills. As soon as it has cooled, it is returned to the engine and the procedure is repeated.

Generally, water and ethylene glycol (EG) mixture are utilised in radiator of vehicle engines as an automotive coolant. Because of their weaker thermal conductivity, these fluids perform poorly in HT as compared to water [20]. The weight and design of a vehicle's front-end module are influenced by air cooled heat

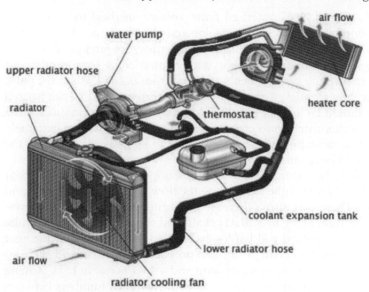

Figure 53.4 Automotive cooling system components [20]

Table 53.1 Study of nanofluid in automobile radiator

Sr. No.	Author	Nanofluid	Levels
1	[23]	Al_2O_3/EG, Al_2O_3/Water and Al_2O_3/Water + EG	Al_2O_3 significantly improves rate of heat transmission. The major elements impacting the HTR are volume concentration and flow conditions. The temperature of the inlet has less impact. The HT enhancement of nanofluids is because of Brownian motion.
2	[24]	Al_2O_3/mono EG + water (50:50)	Al_2O_3 nanoparticles significantly improve heat transmission. The Nusselt number rises from 3.89-28.7% at the minimum flow rate of coolant. In the laminar regime, an empirical relation for Nusselt number is also constructed.
3	[25]	ZnO/Water	By raising the volume percent, heat transmission improves. Heat transmission reduced after 0.2 vol. percent. The HTR was only slightly affected by the inlet temperature.
4	[26]	Al_2O_3/Water	They discovered that the HTR of nanofluid was greatly influenced by the volume percent and Reynolds number, but that the pumping power also rose when these two parameters were raised.
5	[27]	TiO_2/water	When compared to the base fluid, nanofluid significantly improves HTR. The Reynolds number and volume concentration have a considerable influence on the heat transfer coefficient.
6	[28]	TiO_2/EG + water	By raising the volume percent, heat transmission improves. Heat transmission reduced after 0.2 vol. percent. The heat transfer rate was only slightly affected by the inlet temperature.
7	[29]	Al_2O_3/(Water+ EG) CuO/(Water+EG)	The HTC is substantially influenced by Reynolds' number and volume percent, i.e., at a constant volume percent of 6 percent. In comparison to the base fluid, the convective HTC rises by 61% for Al_2O_3 and 92.5% for CuO at a constant Reynolds number of 5500.
8	[30]	TiO_2/water	In the presence of nanoparticles, heat transmission increased considerably. In comparison to water, an 18% rise in Nusselt number was attained. Numerical investigations appear to be in good accord with experimental findings.
9	[31-33]	CuO/water	OHTC increases as the flow rate and volume percent increase but reduces as the input temperature rises. The use of nanofluids resulted in a significant increase in HT.

Source: Author

exchanger (AC condenser, radiator, evaporator, etc.). An optimisation procedure is required to address these difficulties. This optimisation goal necessitates sophisticated design tools that can reveal the best solution. At high speeds, overcoming aerodynamic drag consumes around 65% of a truck's overall energy output. Different types of nanofluids have been studied for pressure drop and thermal performance to ensure its suitability for such application. Table 53.1 lists some of most widely utilised nanoparticle in radiators. The usage of nanofluids as a coolants will allow radiators to be better and smaller positioned. Engine of truck can be operated at upper temperatures and, coolant pumps may be shrunk, letting for increase in horsepower while still fulfilling severe pollution rules, because there can be less fluid due to the greater efficiency.

In comparison to conventional coolants, these unique and advanced coolant concepts offer intriguing heat transfer characteristics. There is a lot of work on nanofluids' improved HT capabilities, particularly convective heat transfer and thermal conductivity. Eastman et. al. [21] found that nanofluids have a significantly greater thermal conductivity than traditional coolants. He also found improved convective heat transport. With these features, nanofluids look to be promising in industries such as heat exchangers. However, several factors such as longstanding stability, increased pumping power and pressure drop, nanofluids' performance in fully developed regions and turbulent flow, higher production costs and lower specific heat of nanofluids, may obstruct development and applications of nanofluids.

By boosting efficiency, lowering weight, and simplifying thermal management systems, nanofluids have potential to increase cooling rates in automotive and heavy-duty engines by up to 30%. With the same sized cooling system, better cooling rates for car and engine of truck may be employed to extract more heat from engines with high horsepower. Alternatively, a more compacted cooling system with lighter and smaller radiators is advantageous. As a result, the fuel economy and high performance of cars and trucks give benefits. The high boiling point of nanofluids allows them to be employed to raise the regular coolant working temperature and then throwaway extra heat over the current coolant system. Because of the less- pressure compared to 50-50 blend of water and ethylene glycol, they have gotten a lot of interest in the application as coolant of an engine [22].

Ali et al. [34] validated the performance of a copper nanoparticles-based nanofluid in car radiator. When compared to ethylene glycol-based fluid, nanofluid improved HTR and HTC. Using alumina-based nanofluid, Chidambaram et al. [35] investigated radiator performance. The HTC improved as the rate of flow of pure water or nanofluid was improved. The temperature of the incoming nanofluid did not affect radiator efficiency, according to their findings. In an automobile radiator, the authors examined the overall HTC of oxides of copper and iron-based nanofluids. The coefficient of heat transfer improved as the fluid flow rate was increased, but declined when the inlet fluid temperature was raised. In addition, when the percentage of nanoparticles increased, the coefficient of heat transfer increased dramatically, particularly for ferric oxide nanofluid.

By using oxide nanofluids made of tungsten and alumina, Nieh et al. [36] were able to determine the heat dissipation capability of automobile's radiator. With equal volume ethylene glycol and water base fluids, as well as 0.2 weight percent chitosan dispersant, six distinct volume ratios of nano-coolant were generated. For nanofluids varying concentrations and dissimilar rates of flow and nano-coolant intake temperatures, different thermo-physical characteristics, pumping power, pressure loss and heat dissipation capacity were examined. The outcomes demonstrated that nano-coolants outperformed base fluids in terms of heat dissipation. Furthermore, tungsten nanoparticle-based nano-coolants had a higher heat dissipation capability than alumina-based nano-coolants. Ali et al. [25] used a water-based zinc oxide nanofluid to test HTC enhancement in a car radiator. Nanofluids were shown to have a higher HTR than water. The highest HTR reached up to 0.2 volume percent nanofluid, after which it began to decrease. The application of TiO_2 based nanofluid improved radiator performance [28]. HTC improved greatly with increased circulation rate and somewhat with increased nanofluid inlet temperature, according to the findings. The HTC improvement in a cooling system of automobile using TiO_2 nanofluid reported by several authors [37-42]. By using nanofluid, there was a considerable saving in the front part (area) of the radiator for the same rate of heat transfer. Additionally, the aerodynamic drag reduction resulted in a saving in fuel consumption. Wu et al. [43] discovered that graphene nanoplatelets based nanofluid improved radiator performance. They discovered that increasing graphene nanoplatelets concentration, temperature of inlet fluid, and mass flow rate improved coefficient of heat transfer. In laminar flow zones, nanofluid was found to be effective and suited for compact thermal systems. In an automobile radiator, Subhedar et.al. [24] calculated the HT potential of coolant (Al_2O_3/ethylene glycol). According to the findings, increasing inlet fluid temperature and increasing nanoparticle volume fraction has a minimal impact on HTC enhancement. Furthermore, for the same HTC, the front-part area of the radiator was lowered, making the cooling system lighter, producing a lesser amount of drag, and saving fuel costs. Said et al. [40] reported on the use of TiO_2, and Al_2O_3 nanoparticle spread in distilled ethylene glycol and water to enhance performance of automotive radiator. Because of their anti-corrosive qualities, they picked these nanofluids. According to corrosion tests, nanofluids based on Al_2O_3 nanoparticles corroded at a lower rate than nanofluids based on TiO_2 nanoparticles. In addition, with larger concentrations of nanoparticles in nanofluids, the radiator performed better in terms of thermal performance.

According to Kumar et al. [41], the front of an automobile radiator can be downsised by 10% when high conductivity nanofluids are used. This decrease in aerodynamic drag can result in up to a 5% reduction in fuel consumption. The usage of nanofluid helped to reduce wear and friction, as well as parasitic losses and the operation of machinery like compressors and pumps, resulting in a fuel savings of more than 6%. In the future, it is possible that savings could be improved even more. The authors created and standardised an instrument that can simulate the flow of coolant in a radiator to evaluate whether nanofluids degrade radiator material. Impact angle and weight loss measurements are a function of fluid velocity and are used to detect radiator material erosion. The authors also reported that nanofluids created with trichloroethylene glycols and base fluid ethylene at velocities of 9 m/s and impact angles of 900–300 had no erosion. With nanofluid (copper-based) at an impact angle of 900 and a speed of 9.6 m/s, erosion was detected. The comparable decline rate for operating vehicles was intended to be 0.065 mils per year. According to early research, water-based copper nanofluid has a higher rate of wear than the base fluid, which could be attributed to copper nanoparticle oxidation. Alumina nanofluids had a lower rate of wear and friction than the base fluid.

Conclusion

Twisted tube type heat exchangers have been examined in terms of their construction, thermal properties, performance, and application. This form of heat exchanger has been found to have several advantages over the traditional STHX with segmental baffles. When looked at the substitute of shell and tube type traditional equipment, heat exchangers with twisted tube offer higher economic performance as determined by the cost per unit heat load.

It has been discovered that nanofluids could be a viable candidate for use in automobiles. Because nanofluids improve heat transfer, radiators of automobiles can be manufactured more compact and energy-efficient. A reduced or compact design can minimise drag, improve fuel economy, and lighten the vehicle's weight. Many researchers have stated that the exact mechanism of improved HT for nanofluids is still unknown. Nanofluids pose several obstacles that must be recognised and overcome before they can be used in automobile radiators. The stability of nanofluids and the high cost of their manufacture are two major obstacles to their widespread use. If these issues can be fixed, nanofluids have the potential to become widely used as a coolant in heat exchange systems.

References

[1] Morgan, R. (2019). Twisted tube heat exchanger technology. Journal of Chemical Information and Modeling. 53(9), 1689–99.

[2] Kumar, A., Singh, S., Chamoli, S., and Kumar, M. (2019). Experimental investigation on thermo-hydraulic performance of heat exchanger tube with solid and perforated circular disk along with twisted tape insert. Heat Transfer Engineering. 40(8), 616–26. https://doi.org/10.1080/01457632.2018.1436618.

[3] Gorjaei, A. R., and Shahidian, A. (2019). Heat transfer enhancement in a curved tube by using twisted tape insert and turbulent nanofluid flow. Journal of Thermal Analysis and Calorimetry. 137(3), 1059–68. https://doi.org/10.1007/s10973-019-08013-1.

[4] Eiamsa-ard, S., Nivesrangsan, P., Chokphoemphun, S., and Promvonge, P. (2010). Influence of combined non-uniform wire coil and twisted tape inserts on thermal performance characteristics. The International Communications in Heat and Mass Transfer. 37(7), 850–56. https://doi.org/https://doi.org/10.1016/j.icheatmasstransfer.2010.05.012.

[5] Blazo Ljubicic. (1996). Twisted tube heat exchangers technology and application, 2(3), 1–12.

[6] Tan, X., Zhu, D., Zhou, G., and Zeng, L. (2012). Experimental and numerical study of convective heat transfer and fluid flow in twisted oval tubes. International Journal of Heat and Mass Transfer. 55(17), 4701–10. https://doi.org/https://doi.org/10.1016/j.ijheatmasstransfer.2012.04.030.

[7] Tang, X., Dai, X., and Zhu, D. (2015). Experimental and numerical investigation of convective heat transfer and fluid flow in twisted spiral tube. International Journal of Heat and Mass Transfer. 90, 523–41. https://doi.org/https://doi.org/10.1016/j.ijheatmasstransfer.2015.06.068.

[8] Sun, B., Yang, A., and Yang, D. (2017). Experimental study on the heat transfer and flow characteristics of nanofluids in the built-in twisted belt external thread tubes. International Journal of Heat and Mass Transfer. 107, 712–22. https://doi.org/https://doi.org/10.1016/j.ijheatmasstransfer.2016.11.084.

[9] Pozrikidis, C. (2015). Stokes flow through a twisted tube with square cross-section. The European Journal of Mechanics - B/Fluids. 51, 37–43. https://doi.org/https://doi.org/10.1016/j.euromechflu.2014.12.005.

[10] Cheng, J., Qian, Z., Wang, Q., Fei, C., and Huang, W. (2019). Numerical study of heat transfer and flow characteristic of twisted tube with different cross section shapes. Heat and Mass Transfer. 55(3), 823–44. https://doi.org/10.1007/s00231-018-2471-7.

[11] Khoshvaght-Aliabadi, M., and Arani-Lahtari, Z. (2016). Proposing new configurations for twisted square channel (TSC): Nanofluid as working fluid. Applied Thermal Engineering. 108, 709–19. https://doi.org/10.1016/j.applthermaleng.2016.07.173.

[12] Yang, S., Zhang, L., and Xu, H. (2011). Experimental study on convective heat transfer and flow resistance characteristics of water flow in twisted elliptical tubes. Applied Thermal Engineering. 31(14–15), 2981–91. https://doi.org/10.1016/j.applthermaleng.2011.05.030.

[13] Li, X., Zhu, D., Yin, Y., Tu, A., and Liu, S. (2019). Parametric study on heat transfer and pressure drop of twisted oval tube bundle with in line layout. International Journal of Heat and Mass Transfer. 135, 860–72. https://doi.org/https://doi.org/10.1016/j.ijheatmasstransfer.2019.02.031.

[14] Alempour, S. M., Abbasian Arani, A. A., and Najafizadeh, M. M. (2020). Numerical investigation of nanofluid flow characteristics and heat transfer inside a twisted tube with elliptic cross section. Journal of Thermal Analysis and Calorimetry. 140(3), 1237–57. https://doi.org/10.1007/s10973-020-09337-z.

[15] Thantharate, M. (2013). Experimental and numerical comparison of heat transfer performance of twisted tube and plain tube heat exchangers. International Journal of Scientific Engineering and Research. 4(7), 1107–13.

[16] Danielsen, Sven Olaf. (2009). Investigation of a twisted-tube type shell-and-tube heat exchanger, Research Journal of Engineering Sciences, 10(3), Sept. 2021, 36–38.

[17] Jahanmir, G. S., and Farhadi, F. (2012). Twisted bundle heat exchangers performance evaluation by CFD (CJ12/5054). The International Communications in Heat and Mass Transfer. 39(10), 1654–60. https://doi.org/https://doi.org/10.1016/j.icheatmasstransfer.2012.09.004.

[18] Bhogare, R. A., and Kothawale, B. S. (2014). Performance investigation of automobile radiator operated with Al2O3 based nanofluid. IOSR Journal of Mechanical and Civil Engineering. 11(3), 23–30. https://doi.org/10.9790/1684-11352330.

[19] Web Page https://www.carparts.com/blog/main-components-of-your-cooling-system/. accessed on 10/03/2022.

[20] Chougule, S. S., and Sahu, S. K. (2014). Comparative study of cooling performance of automobile radiator using Al2O3-water and carbon nanotube-water nanofluid. Journal of Nanotechnology in Engineering and Medicine. 5(1), 1–6. https://doi.org/10.1115/1.4026971.

[21] Eastman, J. A., Choi, U. S., Li, S., Thompson, L. J., and Lee, S. (1996). Enhanced thermal conductivity through the development of nanofluids. MRS Online Proceedings Library. 457(1), 3–11. https://doi.org/10.1557/PROC-457-3.

[22] Zeinali Heris, S., Nasr Esfahany, M., and Etemad, S. G. (2007). Experimental investigation of convective heat transfer of Al2O3/water nanofluid in circular tube. International Journal of Heat and Fluid Flow. 28(2), 203–10. https://doi.org/https://doi.org/10.1016/j.ijheatfluidflow.2006.05.001.

[23] Peyghambarzadeh, S. M., Hashemabadi, S. H., Jamnani, M. S., and Hoseini, S. M. (2011). Improving the cooling performance of automobile radiator with Al 2O3/water nanofluid. Applied Thermal Engineering. 31(10), 1833–38. https://doi.org/10.1016/j.applthermaleng.2011.02.029.

[24] Subhedar, D. G., Ramani, B. M., and Gupta, A. (2018). Experimental investigation of heat transfer potential of Al2O3/water-mono ethylene glycol nanofluids as a car radiator coolant. Case Studies in Thermal Engineering. 11, 26–34. https://doi.org/10.1016/j.csite.2017.11.009.

[25] Ali, H. M., Ali, H., Liaquat, H., Bin Maqsood, H. T., and Nadir, M. A. (2015). Experimental investigation of convective heat transfer augmentation for car radiator using ZnO–water nanofluids. Energy. 84, 317–324. https://doi.org/https://doi.org/10.1016/j.energy.2015.02.103.

[26] Gokhal, G. S. (2021). Heat transfer performance of water based nanofluids: A review. Materials Today: Proceedings. 37, 3652–55. https://doi.org/10.1016/j.matpr.2020.09.787

[27] Siraj Ali Ahmed, Mehmet Ozkaymak, Adnan Sözen, Tayfun Menlik and Abdulkarim Fahed. (2018). Improving car radiator performance by using TiO2-water nanofluid. Engineering Science and Technology, an International Journal. 21(5), 996-1005. https://doi.org/10.1016/j.jestch.2018.07.008

[28] Madderla Sandhya, D. Ramasamy, K. Sudhakar, K. Kadirgama and W.S.W. Harun. (2020). Hybrid nano-coolants in automotive heat transfer–an updated report. Maejo International Journal of Energy and Environmental Communication. 2(3), 43–57. https://doi.org/10.54279/mijeec.v2i3.245040

[29] Vajjha, R. S., Das, D. K., and Namburu, P. K. (2010). Numerical study of fluid dynamic and heat transfer performance of Al2O3 and CuO nanofluids in the flat tubes of a radiator. International Journal of Heat and Fluid Flow. 31(4), 613–21. https://doi.org/10.1016/j.ijheatfluidflow.2010.02.016.

[30] Hussein, A. M., Bakar, R. A., Kadirgama, K., and Sharma, K. V. (2014). Heat transfer enhancement using nanofluids in an automotive cooling system. The International Communications in Heat and Mass Transfer. 53(November 2017), 195–202. https://doi.org/10.1016/j.icheatmasstransfer.2014.01.003.

[31] M. Naraki, S.M. Peyghambarzadeh, S.H. Hashemabadi and Y. Vermahmoudi. (2013). Parametric study of overall heat transfer coefficient of CuO/water nanofluids in a car radiator. International Journal of Thermal Sciences. 66, 82–90. https://doi.org/10.1016/j.ijthermalsci.2012.11.013.

[32] Teng, T.-P., and Yu, C.-C. (2013). Heat dissipation performance of MWCNTs nano-coolant for vehicle. Experimental Thermal and Fluid Science. 49, 22–30. https://doi.org/https://doi.org/10.1016/j.expthermflusci.2013.03.007.

[33] Bhimani, V. L., Rathod, P. P., and Sorathiya, A. S. (2013). Experimental study of heat transfer enhancement using water based nanofluids as a new coolant for car radiators. International Journal of Emerging Technology and Advanced Engineering. 3(6), 295–302.

[34] Ali, M., El-Leathy, A. M., and Al-Sofyany, Z. (2014). The effect of nanofluid concentration on the cooling system of vehicles radiator. Advances in Mechanical Engineering. 2014, 1–13. https://doi.org/10.1155/2014/962510.

[35] Chidambaram, K., Jayasingh, T. R., Vinoth, M., and Thulasi, V. (2014). An experimental study on the influence of operating parameters on the heat transfer characteristics of an automotive radiator with nano fluids. International Journal of Recent Trends in Mechanical Engineering. 2(May), 7–11.

[36] Nieh, H.-M., Teng, T.-P., and Yu, C.-C. (2014). Enhanced heat dissipation of a radiator using oxide nano-coolant. International Journal of Thermal Sciences. 77, 252–61. https://doi.org/https://doi.org/10.1016/j.ijthermalsci.2013.11.008.

[37] Devireddy, S., Mekala, C. S. R., and Veeredhi, V. R. (2016). Improving the cooling performance of automobile radiator with ethylene glycol water based TiO2 nanofluids. The International Communications in Heat and Mass Transfer. 78, 121–6. https://doi.org/10.1016/j.icheatmasstransfer.2016.09.002.

[38] Selvam, C., Mohan Lal, D., and Harish, S. (2017). Enhanced heat transfer performance of an automobile radiator with graphene based suspensions. Applied Thermal Engineering. 123(October), 50–60. https://doi.org/10.1016/j.applthermaleng.2017.05.076.

[39] Leong, K. Y., Saidur, R., Kazi, S. N., and Mamun, A. H. (2010). Performance investigation of an automotive car radiator operated with nanofluid-based coolants (Nanofluid as a Coolant in a Radiator). Applied Thermal Engineering. 30(17–18), 2685–92. https://doi.org/10.1016/j.applthermaleng.2010.07.019.

[40] Said, Z., El Haj Assad, M., Hachicha, A. A., Bellos, E., Abdelkareem, M. A., Alazaizeh, D. Z., and Yousef, B. A. A. (2019). Enhancing the performance of automotive radiators using nanofluids. Renewable and Sustainable Energy Reviews. 112, 183–94. https://doi.org/https://doi.org/10.1016/j.rser.2019.05.052.

[41] Kumar, A., and Subudhi, S. (2019). Preparation, characterization and heat transfer analysis of nanofluids used for engine cooling. Applied Thermal Engineering. 160, 114092. https://doi.org/https://doi.org/10.1016/j.applthermaleng.2019.114092.

[42] Bhadouriya, R., Agrawal, A., and Prabhu, S. V. (2015). Experimental and numerical study of fluid flow and heat transfer in an annulus of inner twisted square duct and outer circular pipe. International Journal of Thermal Sciences. 94, 96–109. https://doi.org/https://doi.org/10.1016/j.ijthermalsci.2015.02.019.

[43] Wu, J. M., and Zhao, J. (2013). A review of nanofluid heat transfer and critical heat flux enhancement—research gap to engineering application. Progress in Nuclear Energy. 66, 13–24. https://doi.org/https://doi.org/10.1016/j.pnucene.2013.03.009.

Chapter 54

Vibration analysis of advanced healthy and damaged rail – a step towards fault detection

Vyankatesh P. Bhaurkar[a], Laxmikant S Dhamande[b] and Harshal P. Varade[c]

Sanjivani College of Engineering, Kopargaon, India

Abstract

The current work is the part of project of crack detection in recently introduced rails in Indian railway. The focus is given on modal analysis of 60 kg (UIC 60) type rail. The previous countrywide used rail is of 55 kg type and is now about to replace by 60 kg type. So, in current work, latest applied rail is used for numerical and experimental analysis. The existence of crack can be easily studied from the analysis of healthy rail. So initially, modal analysis of healthy rail as a fixed beam is carried out from finite element analysis (FEA) and vibration analysis approach. The health of non-cracked rail is also confirmed. In the next step, the modal analysis is done on single cracked rail with similar approaches. It is observed that, as compared to light-weight structures, the very heavy rail is difficult for acquiring the data into frequency domain. The comparison of healthy and cracked rail has a sharp distinguishing line from frequency point of view which is crucial factor in crack detection in rails.

Keywords: Crack, finite element analysis, fixed-fixed beam, modal analysis, non-destructive testing, rail, vibration

Introduction

Indian rail is one of the wide networks, spread over the entire nation. Huge number of employees are the part of this network. Daily, large number of passengers are using this mode of transfer. Many advanced technologies have been implemented in rail systems from last few decades and this improvement is continuously going on. A rail system is one of the good examples of mechanical engineering. Variety of mechanical components and processes are included in the construction of these systems such as suspension, wheels, rails, slippers, riveting, welding and many more. The quality and performance of these all components should be up-to the mark, as the safety of many lives depends on this factor. Hence it is the prime responsibility of design engineers in rail systems to ensure the long life of all components.

Rail and sleepers are main components of the railway track. These components are continuously subjected to complex force systems. It includes, impact load of rail wagon, contact forces of wheel. It also experiences the harsh changes in atmospheric humidity, temperature. Any small fault in these components causes huge loss in the form of human lives. So, it is one of the prime duties to identify faults or damages in such components. The current work is the part of the same study. Currently in the rail system, many types of rails with different cross-sections are used. These rails also differ in the materials and their mechanical properties [1]. Few of them are, UIC 54, UIC60. In current work, UIC 60 is taken for the study.

On the rail surface, four types of loads are acting, they are vertical, lateral, longitudinal and torsional. To sustain all these types of forces, the cross-section of rail and its material is decided very carefully. Thus, the rail cross-section is exposed to different types of stresses. Over a period of time, hairline cracks are generated in the rail. These cracks start to propagate due to continuous use. These cracks can remove material layers from the top surface of rail as shown in Figure 54.1. This causes imbalance in the force distribution system and consequently causes derailment of railway wagon. To prevent such accidents, it is important to detect the damage at its early stage, when the cracks are very thin and not deep. To detect very thin hairline crack, vibration testing is a proven technique.

The rails are modelled as simply supported beams on elastic foundation by Yang et al [1]. The study of foundation of rail tracks is also important. Here, authors point out the importance of two-axle study compared to that of single axle. The vital point to note here is, the boundary conditions used for the rail. In existing literature, it is observed that, the rail is modelled as a type of beam than fixed-fixed beam, for the simplicity in calculations. There are so many techniques available such as dye penetrant testing, ultrasonic testing, which help to locate the faults [2]. But there are some limitations in every technique. The phased array ultrasonic testing system is applied in rail crack detection by Kim [3]. It includes ultrasonic transducer and other hardware along with control software. The moving mass technique is also employed for detecting the cracks in the beams, by research community [4]. It has been proved that the displacement analysis can fulfil the requirement of comparison of healthy and damaged beam. The regular I section is used for

[a]bhaurkarvyankateshmech@sanjivani.org.in, [b]dhamandelaxmikantmech@sanjivani.org.in, [c]varadeharshalmech@sanjivani.org.in

DOI: 10.1201/9781003450252-54

Figure 54.1 Actual photograph showing cracks and material de-layering from top rail surface
Source: Author

rails, but it is difficult to find out exact cross-sectional area and its other features. These features are needed to construct CAD model. The vibration is also used as a tool for crack detection by many researchers along with moving mass approach [5]. The material of component carries lot of importance in the analysis process. Even the vibration signal analysis in steel and functionally graded material is also different [6]. The research community is continuously working on the use of advanced techniques based on vibration and statistical concepts. The acoustic emission (AE) found application in the damage detection of rails [7]. In this work, the technique of improved Least-Square Generative Adversarial Networks for vibration signal monitoring is reported. This indicates that the task of crack detection in rails using vibration as a tool, is not so simple. At a time, researcher must consider many factors for the study. Along with AE technique, the lamb wave analysis is also employed for analysis of cracks in switch rails [8]. Here, the continuous wavelet transform (CWT) is used to analyse the lamb wave dispersion. The wavelet transform is recently used by many researchers because of its accuracy. The crack detection signals are improved by many techniques such as Shannon entropy-improved adaptive line enhancer [9].

The research community also worked on the factors affecting effective implementation of AE techniques. These factors are propagation distances, types, and depths of AE sources. They proved to be important in the phenomenon of rail crack detection [10]. Hence, such literature is helpful to the railway department for the smooth working of rail system and to prevent harsh situations. Huge literature is available where the crack in any structure is modelled as breathing crack. The moving thermal source in thermography is employed by Boue et al [11] for breathing crack detection. The fracture growth in the rails is also one important area for the scope of the research. This is because many latest techniques help to identify the pattern of the crack growth and thereby preventing any accidental situations [12]. Open or breathing crack means the crack which is moving through alternate closing-opening cycle when it is subjected to vibrations. There so many attempts made by scientists, for single and multiple crack detection in structures, and this led us to many new techniques. The hybrid Cuckoo-Nelder-Mead Optimization Algorithm is used for multiple crack detection in cantilever beams by Moezi et al [13]. The importance of vibration phenomenon studied by numerical and experimental approach is reported by many scientists [14, 15]. The Euler-Bernoulli beam theory is employed for the study of multi-step beam. The breathing crack can be analysed using natural frequency study. Some techniques can be easily adopted for multiple crack detection also [16, 17]. The use of finite element analysis (FEA) in crack detection has been stated many decades back. But still, it is promising technique in the numerical study for complex structures. The accuracy in crack detection is found improved after the use of FEA approach. It also helps to validate the results from mathematical and experimental analysis. In the literature, the transverse fatigue crack is focused. The concept of anti-resonance is proved as useful tool by many researchers [18-20]. Few authors have expressed exact mathematical model for deriving anti-resonant frequencies. While few of them have directly implemented it in experimental analysis.

In current work, single cracked rail is analysed. An attempt is made of vibration analysis of UIC 60 kg type rail used in India. Same rail used throughout the study. It is a step towards the project of unknown fine cracks detection in the rail. The cross-section of rail is built-up in CAD model. FEA approach is used for numerical study. Modal analysis is performed in commercial FEA software for healthy as well as cracked rail. Dye penetrant testing is employed to assure the crack-free surface of rail. The dedicated test set-up is used for vibration analysis of rail. The 4-channel FFT analyser is used to extract vibration signatures. A single saw crack is generated in the rail. The comparison study of numerical and experimental analysis for both, healthy and cracked rail, is performed. The study of cracked rail gives an idea about change in natural

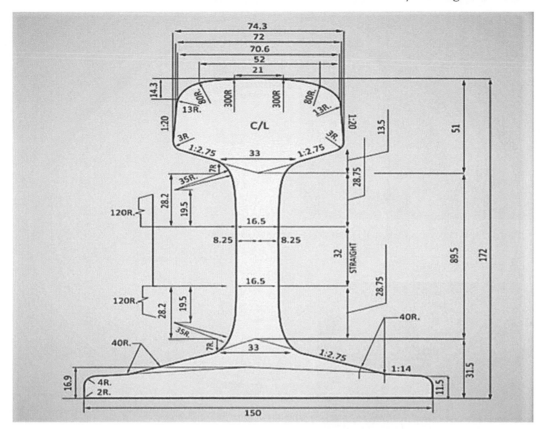

Figure 54.2 Cross-section and dimension details of UIC 60 kg type rail [1]

frequencies between healthy and cracked conditions. This study is useful for the further work of unknown crack detection in rails. If the crack detection is addressed properly, then inspection and rectification decisions can reduce potential risk of rail breaks and derailments. Also, this step leads us further to reduce the cost in inspection and allied remedies. The aim is to detect the crack or any damage in the rails that can lead to the safety of the passengers and off course to the railway department and to the government.

Actual rail details

In the previous section, we come across the fact, that many types of rails are used currently in railway track systems. In the current study, UIC 60 kg type rail is used. Its weight is 60 kg for 1 meter length. It is of I section. Its cross-section and material details are provided in Figure 54.2.

Experiment set-up and FFT analyser

To perform the experimental testing of healthy as well as cracked rail, dedicated vibration test set-up is used throughout the work [14]. The rail is mounted with specially designed rail clamps. The rail is clamped with fixed-fixed boundary condition on the test set-up with the clamps. The 4-channel Adash V4PRO FFT analyser is used for extracting the vibration signal of the rail. The contour plot of the FFT analyser exhibits all the peaks of resonance which are used to find all the natural frequencies of the rail. Figure 54.3 shows the FFT analyser used in current project.

Non-destructive testing of rail

Before proceeding to the experimental testing of actual rail, it is passed through one NDT testing. The purpose of this testing is to confirm that the rail surface is free from any crack. Simply it means that, to assure that the rail is really healthy. In current work, dye penetrant testing is performed on the rail. And it is observed that, only few dents are available on the surface but no deep crack. This confirms that, this rail now can be tested under healthy rail analysis. Figure 54.4 shows the testing details of rail.

Healthy rail –numerical and experimental study

FEA is the process of simulating the behaviour of a part or assembly under given conditions, so that it can be assessed using the finite element method (FEM). FEA is used by engineers to help simulate the physical

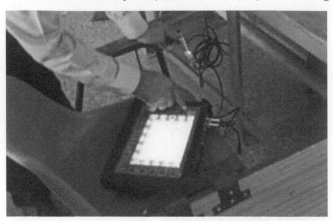

Figure 54.3 A photograph showing FFT analyser used for experimentation
Source: Author

Figure 54.4 A photograph showing dye penetrant testing on the rail
Source: Author

phenomena and thereby reduce the need for physical prototypes. FEA uses mathematical models to understand and quantify the effects of real-world conditions on a part or assembly. These simulations, which are conducted via specialised software, allow engineers to locate potential problems in a design, including areas of tension and weak spots.

The simulations used in FEA are created using a mesh of millions of smaller elements that combine to create the shape of the structure that is being assessed. Each of these small elements is subjected to calculations, with these mesh refinements combining to produce the final result of the whole structure. The points where the values can be determined are called nodal points.

The 60 kg rail which is used in current study, is already shown in Figures 54.2 and 54.4. In the next step, the CAD model of the same cross section is built-up using commercial CAD software package [18]. This CAD model is shown in Figure 54.5.

Once the model is created, the next step is to export it to commercial FEA software for numerical analysis. The purpose of this numerical study is to find out the natural frequencies of healthy as well as cracked rail. Thus, in this step, the modal analysis is carried out using FEA software package. Figure 54.6 shows the meshing of rail developed in the software. The element used for meshing is tetrahedron 10 node element, due to its fine meshing and improved accuracy. The configuration details for this development are given in Table 54.1 [18]. The rail is treated as fixed-fixed beam, which is its real-time position on the railway track.

After performing modal analysis of healthy rail in software, we can extract infinite number of natural frequencies. But in current work, first six frequencies are considered. This is because, the difference between consecutive frequencies is becoming small, as we go for the higher natural frequencies. Also, from available literature, it is evident that generally first few natural frequencies only are taken for analysis purpose [14]. These six natural frequencies are expressed in graphical and tabular manner as shown in Figure 54.7 and Table 54.2.

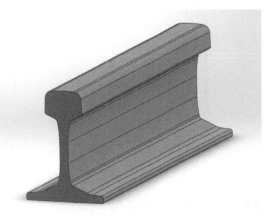

Figure 54.5 A CAD model of rail developed with commercial CAD software
Source: Author

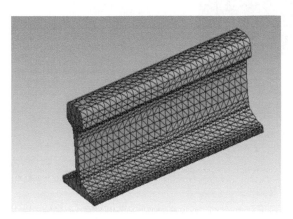

Figure 54.6 A meshing refinement of rail cross-section with FEA software
Source: Author

Table 54.1 Configuration details of healthy rail used for numerical analysis

Sr. No	Parameter	Detail
1	Density of material (ρ)	7850 kg/m³
2	Modulus of elasticity (E)	210 GPa
3	Cross-sectional area of rail	76.72 * 10-4 m²
4	Length of rail	640 mm
5	Poisson's ratio (μ)	0.265

Source: Author

Table 54.2 First 6 natural frequencies of healthy rail taken from FEA software

Sr. No.	Mode	Frequency (Hz)
1	1st	1216.9
2	2nd	1729.7
3	3rd	1890.8
4	4th	2120.5
5	5th	2662.9
6	6th	3240.3

Source: Author

The experimental testing of healthy rail is carried out to verify the natural frequencies obtained from numerical analysis. For this the dedicated test set-up is used. The 4-channel FFT analyser is used to extract natural frequencies of rail. One important observation needs to be mentioned here that, the usual FFT

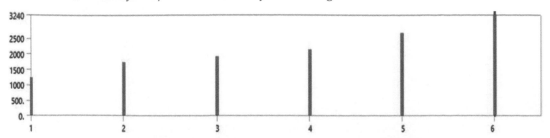

Figure 54.7 Graph of first six natural frequencies taken from FEA software
Source: Author

Figure 54.8 Testing of rail on vibration test set-up using FFT analyser
Source: Author

Table 54.3 Natural frequencies of healthy rail from experimental testing

Sr. No.	Natural frequency no.	Natural frequency from FFT (Hz)
1	1st Natural frequency	1164
2	2nd Natural frequency	1694
3	3rd Natural frequency	1872

Source: Author

hammer has not given satisfactory result as it is not impacting with higher strength. So heavy construction hammer is used in current experimentation.

Before experimental study, the non-destructive testing of rail is accomplished, which is of the type of dye penetrant testing. It helps to assure that the beam is healthy and not having any deep crack.

The Figure 54.8 shows testing of rail on vibration test set-up using FFT analyser. The results obtained from this experimental testing are shown in Table 54.3.

Cracked rail – numerical and experimental study

In this part of work, single cracked rail is analysed. Same rail used for healthy analysis, is used for cracked rail analysis also. The study of cracked rail gives an idea about change in natural frequencies between healthy and cracked conditions. This study is useful for the further work of unknown crack detection in rails. If the crack detection is addressed properly, then inspection and rectification decisions can reduce potential risk of rail breaks and derailments. Also, this step leads us further to reduce the cost in inspection and allied remedies. From the literature survey, it is interpreted that there is a need for better prediction of rail defects over a period based on operating conditions and maintenance strategies. The issues and challenges related to rail maintenance are outlined for further research in this area. The aim is to detect the crack or any damage in the rails that can lead to the safety of the passengers and off course to the railway department and to the government.

The study of single cracked rail is carried out in a similar way as that of healthy rail. The execution of procedural steps are same, it means initially numerical and then experimental analysis. The single crack in the rail is developed by a saw of width 0.4 mm. Hence the actual width of the crack is found 0.450 mm. This is the least width saw available for laboratory testing. The location of the crack is at 200 mm from the left fixed end. This location of crack is decided for further study of multiple cracks detection in the same rail. This study is already undertaken by the authors. The crack is generated for the depth of 4 mm. The reason behind this depth, is the intention of locating unknown fine crack. Also, very low depth crack is not giving a significant change in natural frequency compared to healthy beam. So, to have this significant change, the crack depth started at 1 mm and moved to 4 mm.

The CAD model of cracked rail, generated in the software, for numerical study is shown in Figure 54.9. The model of this rail after meshing is shown in the Figure 54.10. Similar to the healthy rail case, the natural frequencies are obtained for cracked rail also using numerical study. The variation in these frequencies are directly obtained from software and are shown in Figure 54.11. Also they are expressed in tabular way in Table 54.4. It is clear from these figures and tables, that the first 6 frequencies are in increasing manner.

Also, the subsequent experimental analysis is performed in a similar way of healthy rail. The results of this experimentation are reported in the Table 54.5.

Results and Discussion

The rail type used in current study is recently implemented. It is one of the heaviest rail types of 60 kg weight for 1 meter length. The important observation made during experimentation is that, it is difficult

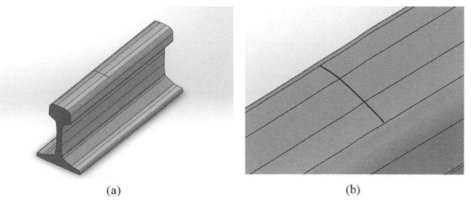

(a)   (b)

Figure 54.9 A CAD model of single cracked rail [a) whole cracked rail, b) zoom view of crack]
Source: Author

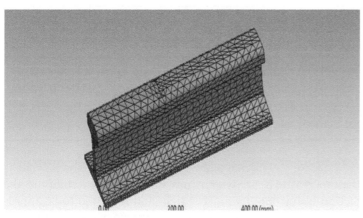

Figure 54.10 A meshing refinement of cracked rail cross-section with FEA software
Source: Author

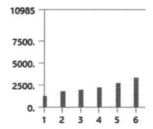

Figure 54.11 Graph of first 6 natural frequencies of cracked rail taken from FEA software
Source: Author

Table 54.4 First 6 natural frequencies of single cracked rail taken from FEA software

Sr. No.	Mode	Frequency (Hz)
1	1st	1215.9
2	2nd	1730.1
3	3rd	1889.4
4	4th	2121.3
5	5th	2663.3
6	6th	3239.6

Source: Author

Table 54.5 Natural frequencies of single cracked rail from experimental testing

Sr. No.	Natural frequency no.	Natural frequency from FFT (Hz)
1.	1st Natural Frequency	1160
2.	2nd Natural Frequency	1692
3.	3rd Natural Frequency	1876

Source: Author

Table 54.6 Comparison of natural frequencies of healthy rail

Sr. No.	Natural frequency no.	Natural frequency (Hz)		% Error in FEA and FFT reading
		Numerical (FEA)	Experimental (FFT)	
1	1st Natural Frequency	1216.9	1164	4.5 %
2	2nd Natural Frequency	1729.7	1694	2.1 %
3	3rd Natural Frequency	1890.8	1872	1.0 %

Source: Author

Table 54.7 Comparison of natural frequencies of single cracked rail

Sr. No.	Natural frequency no.	Natural frequency (Hz)		% Error in FEA and FFT reading
		Numerical (FEA)	Experimental (FFT)	
1.	1st Natural Frequency	1215.9	1160	4.8 %
2.	2nd Natural Frequency	1730.1	1692	2.2 %
3.	3rd Natural Frequency	1889.4	1876	%

Source: Author

task to vibrate such a heavy structure. The hammer of FFT analyser is not showing repeatability in the readings. Hence the decision of use of construction heavy duty hammer is right and it proved to be appropriate for the repeatability of readings of natural frequency. The readings of experimentation shown in previous sections are the average values.

Because of this step, the good agreement is found between the results of numerical and experimental study. It has been observed that when larger harmonics i.e., higher natural frequencies are extracted, then this difference between readings increased. Hence, we consider initial few natural frequencies only.

It is good practice to compare the results of numerical analysis with experiment readings. Tables (54.6 and 54.7) gives this comparison for healthy as well as cracked rail.

It is observed that the natural frequencies obtained from numerical and experimental study are in good agreement. The percentage error is within 10% which is generally taken as final limit. Also, it is again proved here, that the existence of crack reduces the natural frequency. From the comparison tables, it is evident that the difference of natural frequency of healthy and cracked rail, is quite tough to identify. Hence for heavy structures, the vibration as NDT tool may not be used effectively. This is because, it may create confusion in the study of unknown crack detection for the industries. This obviously is not a favourable condition.

Conclusion

Damage detection in mechanical components is necessary in every field of industries. This need gives rise to numerous branches of fracture mechanics. Till date, many advanced techniques are available for fault detection. The current work supports that the agreement between numerical and experimental study always assures compatibility of proposed technique.

The heavy structure is not responding to light impact of hammers. To achieve the significant vibration signals of heavy rails, heavy hammer is used for experimentation. The step of dye penetrant testing on healthy rail again proved to be helpful and it is reflected in the repeatability of readings. Because of this, the good agreement is found between numerical and experimental natural frequencies of healthy rail.

It is again observed that, the crack reduces the natural frequency of component. But the difference is very small. This indicates that, there is lot of scope of research of the use of vibration as NDT tool for heavy structure like rail.

References

[1] Yang, Y. B., Wang, Z. L., Shi, K., Xu, H., Mo, X. Q., and Wu, Y. T. (2020). Two-axle test vehicle for damage detection for railway tracks modeled as simply supported beams with elastic foundation. Engineering Structures. 219, 1–13.

[2] Xiaohui, C., Wen, X., Siddiqui, M. A., and Li, C. R. (2020). Defect detection method for rail surface based on line-structured light. Measurements. 159, 1–17.

[3] Kim, G., Mu-Kyung, S., Kimb, Y., Segon, K., and Ki-Bok, K. (2020). Development of phased array ultrasonic system for detecting rail cracks. Sensors and Actuators A: Physical. 311, 1–11.

[4] Chatterjee, A., and Vaidya, T. (2019). Experimental investigations: dynamic analysis of a beam under the moving mass to characterise the crack presence. Journal of Vibration Engineering and Technologies. 7, 217–26.

[5] Ariaei, A., Ziaei-Rad, S., and Ghayour, M. (2009). Vibration analysis of beams with open and breathing cracks subjected to moving masses. Journal of Sound and Vibration. 326, 709–24.

[6] Changjian, J., Linquan, Y., and Cheng, L. (2019). Transverse vibration and wave propagation of functionally graded nanobeams with axial motion. Journal of Vibration Engineering and Technologies . 1, 257–266, https://doi.org/10.1007/s42417-019-00130-3

[7] Kangwei, W., Xin, Z., Qiushi, H., Yan, W., and Yi, S. (2019). Application of improved least-square generative adversarial networks for rail crack detection by AE technique. Neurocomputing. 332, 236–48.

[8] Jinrui, Z., Hongyan, M., Wangji, Y., and Zongjin, L. (2016). Defect detection and location in switch rails by acoustic emission and Lamb wave analysis: A feasibility study. Applied Acoustics. 105, 67–74.

[9] Qiushi, H , Xin, Z., Yan, W., Yi, S., and Viliam, M. (2018). A novel rail defect detection method based on undecimated lifting wavelet packet transform and Shannon entropy-improved adaptive line enhancer. Journal of Sound and Vibration. 425, 208–20.

[10] Xin, Z., Naizhang, F., Yan, W., and Yi, S. (2014). An analysis of the simulated acoustic emission sources with different propagation distances, types and depths for rail defect detection. Applied Acoustics. 86, 80–88.

[11] Boué, C., and Holé, S. (2020). Comparison between multi-frequency and multi-speed laser lock-in thermography methods for the evaluation of crack depths in metal. Quantitative Infra-Red Thermography Journal. 17(4), 223–234. doi: 10.1080/17686733.2019.1635351.

[12] Orringer, O., Morris, J. M., and Jeong, D. Y. (1986). Detail fracture growth in rails: Test results. Theoretical and Applied Fracture Mechanics. 5, 63–95.

[13] Moezi, S. A., Zakeri, E., and Zare, A. (2018). Structural single and multiple crack detection in cantilever beams using a hybrid cuckoo-nelder-mead optimization method. Mechanical Systems and Signal Processing. 99, 805–31.

[14] Montiel-Varela, G., Domínguez-Vazquez, A., Gallardo-Hernández, E., Luigi, B., and García-Illescas, R. (2016). Experimental and numerical study for detection of rail defect. Engineering Failure Analysis.7, 1–14. doi: 10.1016/j.engfailanal.2017.07.024.

[15] Zerbst, U., Lundén, R., Edel, K., and Smith, R. (2009). Introduction to the damage tolerance behaviour of railway rails – a review. Engineering Fracture Mechanics. 76, 2563–601.

[16] Greco, A., Pluchino, A., Cannizzaro, F., Caddemi, S., and Caliò, I. (2018). Closed-form solution based genetic algorithm software: Application to multiple cracks detection on beam structures by static tests. Applied Soft Computing. 64, 35–48.

[17] Zumpano, G., and Meo, M., (2006). A new damage detection technique based on wave propagation for rails. International Journal of Solids and Structures. 43, 1023–46.

[18] Bhaurkar, V. P., and Thakur, A. G. (2019). Investigation of crack in beams using anti-resonance technique and FEA approach. Journal of Engineering Design and Technology. 17(6), 1266–84. https://doi.org/10.1108/JEDT-10-2018-0179.

[19] Hai, T. T., Nguyen, K., and Lien, P. T. B. (2019). Characteristic equation for antiresonant frequencies of multiple cracked bars and application for crack detection. Nondestructive Testing and Evaluation. 34(1), 1–25. DOI: 10.1080/10589759.2019.1605604

[20] Elshamy, M., Crosby, W. A., and Elhadary, M. (2018). Crack detection of cantilever beam by natural frequency tracking using experimental and finite element analysis. Alexandria Engineering Journal. 57, 3755–66.

Chapter 55

Study on Six Sigma methodology to improve the brand value of beverage industry

Pratheesh Kumar S[a], Naveen Anthuvan, R.[b], Anand K.[c], Mohanraj R.[d], Arunkoushik R.[e] and Saran V.[f]

PSG College of Technology, Coimbatore, India

Abstract

A brand is described as a set of tangible and intangible characteristics that are used to develop an organization's reputation. It also indicates a company's worth in the perspective of a customer. A company's market value is determined by its brand image. A company's positive brand image can impact the purchase behavior of its target audience. This study investigates the brand value of the beverage industry and the different aspects that influence it. To address this, the DMAIC technique was used to analyse the impact of the components that determine brand value. Six Sigma tools, such as voice of customer (VOC), quality function deployment (QFD), pareto analysis, cause and effect diagram, five whys, and brainstorming, demonstrated that marketing strategies and beverage flavor were the most influential variables impacting the beverage industry's brand value. As a result, the current study gives the necessary inputs for a company that needs to develop its brand value, which will result in greater revenue and market share.

Keywords: Beverage industry, brand value, DMAIC, six sigma

Introduction

The beverage industry is crucial since one of the most popular ways to preserve food is by using beverages. Interest in the food and beverage business is growing along with worldwide consumption. Building a name or image for something – a good, service, person, place, or even an entire company in the minds of customers is known as branding. Calculating a brand's equity will help you assess its value. In an effort to categorise how much consumers identify with distinct brands; brand equity was first utilised in marketing literature [1]. The effect of brand perception on a company's growth line is shown in Figure 55.1. Together, marketing initiatives that create and uphold a unified brand identity are more likely to succeed. Increasing a company's brand recognition is an effective way to increase its market share. The power of a company's brand image directly correlates to its capacity to influence its target market to make purchases. The best strategy for a business to increase its market share is to create an integrated plan for brand development and marketing, as well as to give customers one simple way to interact with the brand [2]. Successful businesses understand the importance of branding and view it as a potential area for investment. As seen by vast market recognition and the steadfast support of devoted clientele, businesses have achieved great success. Profits rise as a result of effective branding. Return on investment has increased by 31% for brands with a good reputation for quality. Deep brand loyalty among employees makes them more willing to go above and beyond in their efforts to boost sales. Six Sigma is based on a set of fundamental ideas that prioritise ongoing internal growth [3]. Six Sigma is a tactic for increasing brand recognition and patronage. Even if a firm is profitable, its reputation is still vital. The six-sigma method can be used to identify the numerous obstacles to creating a powerful brand. There are steps that can be taken to keep these issues under control, as well as adequate ways to evaluate them and put in place crucial actions to reform activities.

The most effective of these approaches are examined in this article in light of Six Sigma implementation methods; several authors have addressed the same intelligent strategy for enhancing an industry's brand value. Hung and Sung [4] carried out numerous tests to reduce the high failure rate brought on by process volatility. When compared to the baseline, the DMAIC approaches significantly reduced the miniaturised custard bun fault rate by 70%. The fault rate decreased from 0.45% (the project baseline) to 0.141% after six months of improvement operations. Eng and Keh [5] focused on the relationship between advertising and brand value and how it affects a company's long-term success. They developed two models and carried out empirical testing using information and samples from the existing literature. The first model ignores the synergistic effects of advertising and brand value on firm performance, in contrast to the second model, which investigates the impact of digital marketing and brand value on brand performance. According to the findings, brand value and marketing expenses have a favourable impact on future return on assets (ROA) for the same time frame (four years). The company-level findings and the brand-level observations are in agreement. The quality of the secondary data used in the study has a significant impact on the results.

[a]spratheeshkumarth@gmail.com, [b]rna.prod@psgtech.ac.in, [c]mechanand@gmail.com, [d]mohanraj839@gmail.com, [e]21p118@psgtech.ac.in, [f]21p121@psgtech.ac.in

DOI: 10.1201/9781003450252-55

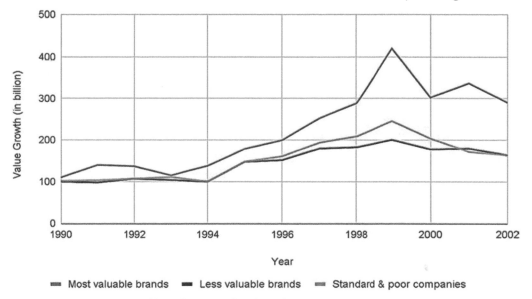

Figure 55.1 Impact of brand in growth value of companies [1]

Maheshwar [6] used the DMAIC method to reduce the production company's error rates. The use of the DMAIC stages directly led to changes in the assembly process. Company was able to reduce production difficulties, accelerate the manufacturing process, standardise procedures, and increase operator enthusiasm by abiding by the guidelines. The number of errors was reduced by the application of statistical techniques.

Desai [7] provided an example of the application of Six Sigma in a small engineering consulting firm. It demonstrated how Six Sigma might boost output and revenue by utilising all of the DMAIC techniques. When rejection rates were lowered in half, the business's bottom line increased by 40%. Radpour and Honarvar's [8] found that social media marketing can result in more contented and devoted customers. To do this, survey on social networking and brand equity was performed. In order to test the study's hypotheses, the gathered data was subsequently analysed using a structural equation model. According to the research, utilising all facets of social network marketing can aid in developing a customer-focused brand value. Isoraite [9] examined and compared the views expressed by many authors on the subject of brand awareness. The study assessed how effectively the author used internet marketing techniques to gauge brand familiarity. Research questions produce research issues. The best way to increase consumer awareness of a brand is to promote its name, goods, or services online through teaching, reteaching, and persuasion techniques. Increase the visibility of your business by luring in more clients, assisting them in developing positive opinions of your brand, and convincing them to purchase more of your products. The company's usage of social media aided its efforts to market its name and sell its goods.

According to review of literatures, a product's brand value has an impact on its commercial viability. This study uses six sigma to enhance the reputation of the beverages sector, since it has been demonstrated to be the most effective quality control method and is differentiated by its data-driven solutions. This article makes a unique contribution to the discipline by viewing six sigma as a useful tool for addressing the importance of increasing a product's brand value. In order to increase customer satisfaction, the study also takes into account the research perspective of incorporating six sigma into the implementation approach. Additionally, a detailed plan for integrating six sigma practises, methodologies, and performance improvement into day-to-day operations of the food industry is put out, which should boost the success rate of the sector. This study also examines an online survey to shed light on specific issues in the beverage sector. The study's findings show promising features in using six sigma as a strategy to raise the brand value of the beverage industry.

Industrial Practices

One example of the many products produced by the beverage industry is syrup. Other products include soft drinks, fruit juices, and caffeinated drinks like coffee and tea. Additionally, alcoholic drinks like wine, beer, whisky, and distilled spirits are produced. These beverages come in a wide variety of flavours. Even though the ingredients and manufacturing procedures utilised in the beverage sector differ, many of them follow similar challenges in marketing sector. A product in the form of a concentrate has majority of its original constituents removed. Fruit juice, for instance, becomes a classic illustration of a meal or beverage that has been dried by taking out all the water. Water treatment, raw material collection, concentrate production, automated packaging and export are the five primary tasks of a concentrate manufacturing plant.

Mango, apple, guava, lemon, orange, and other fruit flavours are available in a variety of container types. Additionally, they provide a huge variety of flavours in carry-on tetra packs. There are several beverage firms who manufacture their own tetra packs. After being assessed on a number of criteria, such as price, cleanliness, and customer happiness, products are given star rating in commercial sector.

Process flow

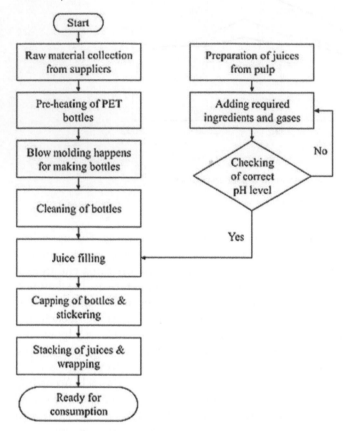

Figure 55.2 Flow chart of process involved in beverage production
Source: Author

The process flow of the beverage sector initially involves the various raw materials obtained from various sources (Figure 55.2). Then on the basis of inventory level, the period wise production is determined. Then bottles are produced by blow molding after they are preheated. Simultaneously the pulp content is prepared and processed. After the juice content is prepared, it is filled in molded bottles. Then bottles are capped and stickered. Then the bottles are wrapped, in respective batches and then they are packed and dispatched for consumer consumption.

Supply chain and logistics

Supply chain management largely allows the flow of resources, information and the transportation of finished goods from plants to local markets. The below mentioned SIPOC diagram explains the procurement of various raw materials and the distribution of the finished goods (Figure 55.3). The raw materials are obtained from both internal and external sources. The beverage industries use groundwater supply for juice preparation. They also consider buying water from other local sources, in case of a deficit. The company gets the pulp prior to the requirement. For transport purposes, they use contract-based transport services. Generally, the beverages are transported to the distributors, who distribute them to the local dealers.

Implementation of DMAIC

Define phase

The project goals and client deliverables will be defined during this define phase [4]. In this phase, the tools are utilised to find the reasons for the low brand value.

Voice of Customer: We utilised an online survey method for obtaining the voice of the customer. The survey presented a series of questions to its respondents. These questions helped in determining the reason

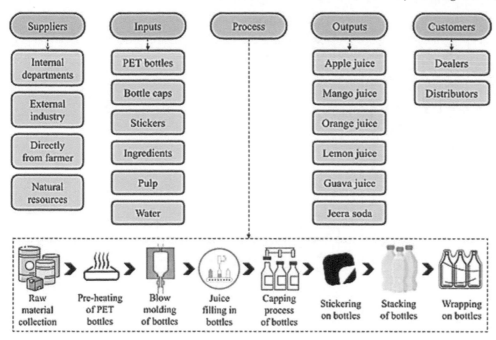

Figure 55.3 SIPOC diagram of the distribution of goods in a beverage industry
Source: Author

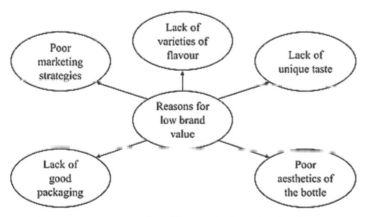

Figure 55.4 Reasons for affecting the brand value from VOC
Source: Author

for the low brand value of the particular beverage industry. The main reasons identified with the help of respondents are the beverages lacked a unique taste, the containment design demonstrated a lack of aesthetic presentation, lack of flavour variety, poor marketing and advertising strategies and lack of good packaging techniques (see Figure 55.4).

Quality Function Deployment: By utilising this tool, the technical requirements that are needed to fulfil the customer requirements are determined. In this tool a house of quality diagram is used in which the customer requirements were considered along the row and the technical were considered along the column and carbonated soft drinks are considered as the benchmark (Figure 55.5). On the basis of respondents given from VOC, the weightage is given. The relationship between technical and customer requirements are given using three different notations with varying weight such as strong influence (9), moderate influence (3), and weak influence (1). From this tool we could find that marketing strategies utilised for various beverages should be targeted for the achievement of a higher brand value.

Measure phase

The DMAIC's second phase involves measuring the process' existing performance and identifying the elements that influence its behaviour [10]. In this phase, the tools are utilised to find the impact of factors for the low brand value.

Pareto Chart: Using this tool, the main reasons for the poor brand value of the beverage industry are identified. The causes for the low brand value are taken along the x-axis. The frequencies of occurrences

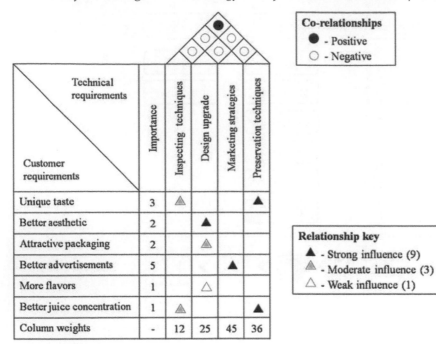

Figure 55.5 QFD diagram
Source: Author

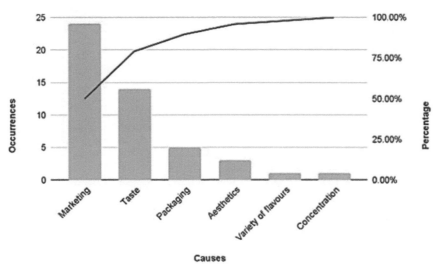

Figure 55.6 Pareto analysis of causes for low brand value
Source: Author

are taken along the left y-axis. The trend line depicts the relationship between the cumulative percentage of occurrences, which is taken along the right y-axis and the causes for low brand value. It is an evident from the below diagram that the marketing strategies and the beverage taste have the maximum influence on the brand value of the beverage industry (Figure 55.6).

Relationship Matrix: The relationship matrix is utilised to determine the most influencing factors that serve as the reason for problems arising in the industry. The factors are considered along the row and the problems faced are considered along the column. In this tool the relation between factors and issues are given using three denotations of varying strength such as strong, moderate and weak. It is found that of all these factors, the product taste and better advertisements had the most influence on the various problems mentioned (Figure 55.7).

Analyse phase

During this phase, the important product or process performance parameters are analysed and benchmarked [11]. In this phase, the tools are utilised to find the root cause of the problem faced to achieve the high brand value.

Problems / Factors	Quality of the product	Marketing	Brand value
Taste of the product	▲		▲
Aesthetics of the bottle			▲
Attractive packaging	△	△	▲
Better advertisements		▲	▲
Variety of flavors			△

▲ - Strong   ▲ - Moderate   △ - Weak

Figure 55.7 Relationship matrix diagram
Source: Author

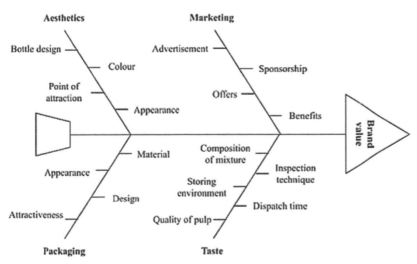

Figure 55.8 Fishbone diagram for low brand value
Source: Author

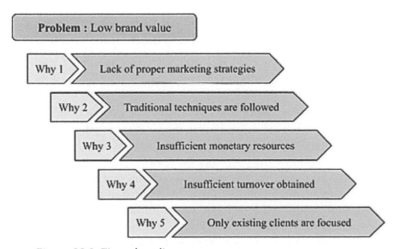

Figure 55.9 Five whys diagram
Source: Author

Cause and Effect Diagram: In this study, the main causes for the low brand awareness were identified such as marketing, taste, aesthetics and packaging from various tools. The factors that influence those causes are identified using fishbone diagram. Factors such as lack of advertisements, lack of sponsorships, lack of offers for the customers, influence the brand value in the field of marketing. Similarly, the factors influencing other main causes are also identified and grouped (Figure 55.8).

Five Whys: Using this tool, the root cause for the low brand value of the beverage industry is determined. Low brand value is considered as the problem here. For this problem the five whys technique is implemented and the answers obtained for each "Why?" are further questioned to find the root cause (Figure 55.9).

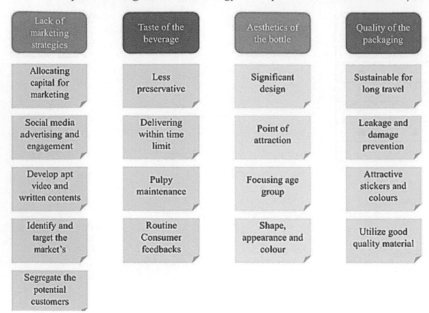

Figure 55.10 Affinity diagram of solution for root causes
Source: Author

We conclude that the root cause of the problem is that the company has not reached out to any new clients as they focused on their existing clients only.

Improve Phase

We can enhance the process at this stage to eliminate the sources of faults [12]. In this phase, the tool is utilised to find the solution for the root cause for low brand value.

Brainstorming: From the previously utilised tools, the various causes for the low brand value of the beverage industry are identified and the possible solutions for these causes can be found by using this technique. Four main causes are considered for our discussion. The possible solutions for these causes are obtained and grouped with the help of an affinity diagram (Figure 55.10). These solutions can be implemented for improving the brand value.

Control Phase

The control phase is the final step, and it guarantees that the gains are maintained and the continuing performance is monitored [13]. In this phase, documentation is utilised to document each and every measure taken to improve the brand value.

Documentation: The tool that focuses on the future is documentation. Teams should devote time to write detailed reports on the obstacles they encountered and how they overcame them. These reports also serve as a helpful guide for management in terms of understanding how the new procedures work [14]. We used this tool to record and store the data such as the opinion of the customer on reasons for low brand value, relation of the customer requirements to the technical requirements, the main reasons that influence brand value of the beverage industry, the strength of the relationship between problems faced in the industry and the influencing factors, the main causes for the low brand value and the factors influencing them, the root cause for low brand value and the possible solutions for those various causes, obtained from various six sigma tools. These data are stored in the form of bar graphs, tables, figures, and written content. By the use of this tool, the various implemented solutions are monitored and are used to control the brand value. These records can also be used in handling problems which arise due to similar circumstances, in the future.

Conclusion

In this study, a beverage industry that produces beverages in various flavours is considered. The factors that affect the brand value of the beverage industry are identified and improved by using DMAIC methodology. In this DMAIC methodology various six sigma tools are used for the identification of the factors that majorly influence the brand value. The factors were found to be marketing strategies, aesthetics of the bottles, delivery time and taste of the beverage. Improvement in these areas such as better advertising strategies, allocation of a significant amount to marketing, delivery of the beverage within the time limit, proper pulpy maintenance, attractive design of the bottle and so on can improve the brand value of the company which can lead to better market share and revenue.

References

[1] Wood, L. (2000). Brands and brand equity: definition and management. Management Decision. 38(9), 662–9.

[2] Deanna Wang, H.-M. (2010). Corporate social performance and financial-based brand equity. Journal of Product and Brand Management. 19(5), 335–45.

[3] Gupta, V., Acharya, P., and Patwardhan, M. (2012). Monitoring quality goals through lean Six-Sigma insures competitiveness. International Journal of Productivity and Performance Management. 61(2), 194–203.

[4] Hung, H.-C., and Sung, M.-H. (2011). Applying six sigma to manufacturing processes in the food industry to reduce quality cost. Site Reliability Engineering (SRE). 6(3), 580–91.

[5] Eng, L. L., and Keh, H. T. (2007). The effects of advertising and brand value on future operating and market performance. Journal of Advertising. 36(4), 91–100.

[6] Maheshwar, G. (2012). Application of six sigma in a small food production plant of India: a case study. International Journal of Six Sigma and Competitive Advantage. 7(2/3/4), 168.

[7] Desai, D. A. (2008). Improving productivity and profitability through Six Sigma: experience of a small-scale jobbing industry. International Journal of Productivity and Quality Management. 3(3), 290.

[8] Radpour, R., and Honarvar, A. R. (2018). Impact of social networks on brand value based on customer behavior using structural equations. International Journal of Customer Relationship Marketing and Management. 9(3), 50–67.

[9] Išoraitė, M. (2016). Raising brand awarenees through internet marketing tools. Independent Journal of Management and Production. 7(2), 320–39.

[10] Kaid, H., Noman, M. A., Nasr, E. A., and Alkahtani, M. (2016). Six sigma DMAIC phases application in Y company: a case study. The International Journal of Intelligent Collaborative Enterprise. 5(3/4), 181.

[11] Maminiaina, A. R. (2019). Business management, master of management program, and parahyangan catholic university. A Thorough literature review of customer satisfaction definition, factors affecting customer satisfaction and measuring customer satisfaction. International Journal of Advanced Research. 7(9), 828–43.

[12] Abdur Rahman, Salaha Uddin Chowdhury Shaju, Sharan Kumar Sarkar, Mohammad Zahed Hashem, Kamrul Hasan S M, Ranzit Mandal and Umainul Islam. (2017). A case study of six sigma define-measure-analyse-improve-control (DMAIC) methodology in garment sector. Independent Journal of Management and Production. 8(4), 1309.

[13] Smętkowska, M., and Mrugalska, B. (2018). Using six sigma DMAIC to improve the quality of the production process: A case study. Procedia - Social and Behavioral Sciences. 238, 590–6.

[14] Ramu, G. (2016). Seattle, WA: Quality Press: The Certified Six Sigma Yellow Belt Handbook.

Chapter 56

Implementation of Six Sigma to improve service quality and customer satisfaction in e-commerce industry

Pratheesh Kumar S[a], Rajamani R[b], Maharasi P B[c], Mohanraj R[d], Morsshini K[e], Nitis Prabhu M[f] and Pachaiyappan V[g]

PSG College of Technology, Coimbatore, India

Abstract

The internet has changed the language of e-commerce transactions for clients towards the purchase and selling of products and services during last decade. Sales in the e-commerce industry have been drastically increased as a result of internet's rapid growth and these figures continue to rise year by year. As sales expands, the issues associated with the e-commerce industry also increases. The goal of this review paper is to identify the key challenges in the e-commerce industry and give appropriate solutions using various six sigma methodologies. This study presents a six-sigma improvement model based on DMAIC cycle that incorporates a set of statistical tools for quality improvement in e-commerce industry. Six major challenges in e-commerce industry such as delivery defects, delivery delays, customer complaints, cancellation, return, website and server errors were recognised and addressed. In each phase of the DMAIC cycle, appropriate tools such as project charter, pairwise ranking, cause and effect diagram, standard operating procedure, and reaction plan have been selected and used to overcome the aforementioned challenges. On comparing the various challenges with the help of pairwise ranking in measure phase, delivery defects which was ranked 1 with a maximum occurrence of five times turned out to be a primary concern in the e-commerce industry. The DMAIC tool implementation described in this paper can be applied to e-commerce as well as other service industries for product and service quality improvement, process improvement, cost reduction, capital profitability, and customer satisfaction.

Keywords: Affinity diagram, cause and effect diagram, DMAIC, E-commerce

Introduction

E-commerce or electronic-commerce refers to any form of business where users buy and sell goods and services over an electronic network, primarily the internet using their own devices [1]. The core objectives of e-commerce are to satisfy the customer's requirements, assist in corporate decision making, implement high speed transactions, and impose lower cost [2]. The significance of this study is to identify key challenges existing in the e-commerce industry and to overcome those challenges by using appropriate DMAIC tools which enhances service quality and customer satisfaction. The E-commerce business is thriving as a result of smartphone usage, easy and affordable access to technology, and the convenience of shopping from anywhere at any time. Because the primary goal of industries is to enhance sales and profitability, e-commerce demands improvement. The most common stumbling blocks in e-commerce are delivery defects, cancellation, delivery delays, customer complaints, return, website, and server. Six Sigma, a quality control methodology, was created to address these issues. According to survey on existing research works, no significant research has been conducted to address the challenges in the e-commerce industry using Six Sigma technique. This study, which is unique in its nature in defining the specific perception on the deployment of Six Sigma techniques, was conducted in the e-commerce industry to solve this research gap. To better understand the concepts of Six Sigma and e-commerce, pertinent literature was reviewed and presented below.

According to Reddy et al. [3] inadequate infrastructure, a confusing tax structure, the fact that the vast majority of the population does not have internet connection, and a general lack of understanding are just a few of the significant challenges that E-Commerce sellers must overcome. Despite the challenges, various companies, such as Flipkart, Myntra, Jabong, and countless more, have grown to become industry leaders. The e-commerce sector's cash on delivery (COD) and instalment payment plan (EMI) options are only two instances of the inventive and customer-friendly strategies used to drive sales. The survey method is used in this study to identify the challenges that such organisations face in India, as well as the solutions that must be implemented. In India, the rising cost of fuel is a persistent factor that determines how much profit firms may generate on each transaction.

Haji [4] discusses how the advent of e-commerce and the digital economy influences all aspect of human existence, from business and education to government administration, economic and cultural progress, political mobilization, and interpersonal touch. According to Maiki and Taivolo [5], the rise of online commerce has both benefits and drawbacks for individuals everywhere, particularly those in rural and isolated areas. Furthermore, e-commerce is regarded as a powerful tool capable of addressing a wide range

[a]spratheeshkumarth@gmail.com, [b]rrm.prod@psgtech.ac.in, [c]21p112@psgtech.ac.in, [d]mohanraj839@gmail.com, [e]21p114@psgtech.ac.in, [f]21p115@psgtech.ac.in, [g]21p116@psgtech.ac.in

DOI: 10.1201/9781003450252-56

of societal and economic concerns. E-contribution commerce amounted for 12.2% of total retail sales in 2018, 14.1% in 2019, and is predicted to reach 22% by 2023. The IBEF is expected to take place in 2019. According to analysts, India will overtake the United States as the world's second largest e-commerce industry by 2034. According to Clemes et al. [6], the internet is increasingly being used for global communication and trade. According to research on the factors that influence web consumer buying behavior in China, the seven essential decision factors that influence consumers in China to engage in online shopping activities are perceived risk, consumer resources, service quality, subjective norms, product variation, convenience, and website factors. According to Hsu et al. [7], in order to compete, e-commerce enterprises must ensure that consumers feel safe making purchases on their sites and that favorable evaluations are easily accessible. Many things have become easier for consumers, particularly business owners, thanks to technological advancements. People may now shop online from the comfort of their own homes, all they need is a mobile device and an internet connection. According to Soegoto et al. [8], customers feel at ease making online purchases because of the vendor earned consumers may consider a ranking system relevant to the quality of service of online business sellers and consumers are not concerned about making transfer payments. According to Yoon and Occea [9], the use of the web and e-commerce has resulted in the creation of a new online ecosystem that is both efficient and effective. E-capabilities commerce's have swiftly transformed how firms communicate with their customers, the public, and government authorities. E-commerce has had a significant influence on small enterprises and the service industry.

The growth of e-commerce has been aided by the spread of mobile phone and internet connectivity, government efforts to promote digitalisation, higher levels of public awareness, a broader range of payment and delivery options, and the entry of foreign e-business firms [10]. The government should take steps to provide a favorable legislative framework for internet commerce and reduce the number of impediments that businesses face. A lack of trustworthy Internet connections, consumer concerns about online security, the high cost of COD, strong rivalry, the current tax system, and tough cyber regulations are all impediments to the spread of e-commerce in India. The majority of Americans still prefer to shop in person rather than online. Six Sigma is heavily dependant on two quality management approaches known as DMAIC and DMADV. DMAIC stands for define, measure, analyse, improve, and control and is a four-step approach. This method is based on the deming cycle, a continuous process improvement paradigm. It is a means of enhancing organizational procedures. The DMAIC cycle consists of five linked steps: Establishing the tone by establishing the target goal and prerequisites right now, we're measuring the process. Currently measurement result interpretation, root cause analysis, and remedy planning: To address shortcomings, the technique must be changed and improved. Monitoring and sustaining the improved process with precision and consistency. It is prudent to first identify and then eliminate external variables that contribute to organisational expenses before addressing internal cost concerns. To accomplish Six Sigma, businesses must investigate the factors that lead to process variation, estimate the impact of those factors, and assess the corresponding financial losses. DMAIC integration into the Six Sigma framework has the potential to increase efficiency while also allowing for faster and more suitable responses to new challenges and problems. Six Sigma is defined by Takao et al. [11] as a quality technique with the goals of minimising operational instability, saving money, and increasing customer satisfaction. As a result, this method has risen in popularity in recent years and will continue to help organisations globally, particularly SMEs, optimise their processes. Estimates indicate that firms will struggle to remain competitive in the next years unless they develop and achieve a Six Sigma level, which improves quality and productivity metrics. This study lends credence to the concept that corporate acceptance of improvement activities is increasing around the world. Six Sigma activities, in contrast to other quality improvement techniques, are based on meticulously monitored data and occurrences. Making judgments based on intuition or responding based on personal experience in the absence of relevant evidence is unacceptable for increasing a company's efficiency and productivity and instead cures the symptoms.

According to Kaushik [12], the Six Sigma technique is undergoing significant revisions and enhancements in the SME sector as a result of the adoption of the DMAIC cycle, which was previously employed by larger-scale organisations. According to Jacobs et al. [13], even a high-quality level-3 business loses 15-20% of sales owing to rework, inspection, testing, and other waste causes. In comparison to the Six Sigma standard, which allows for no more than a 0.00034 percent margin of error, this quality level is extremely low (3.4 ppm). Six Sigma breakthrough techniques, according to Utomo [14], focus on reducing cycle time and increasing customer satisfaction while outlining the parameters of high-quality service and their associated costs. Six Sigma's implementation in industry is newer than its application in economics. Six Sigma helps to reduce process variability in the manufacturing sector, resulting in a more consistent business and higher quality product for clients.

Pareto diagrams, cause and effect diagrams, histograms, control charts, scatter diagrams, graphs, and check sheets are among the seven quality control tools used to tackle quality concerns, according to Magar and Shinde [15]. These devices are commonly employed in the industrial business to control general operations and optimise processes. In order to improve production, these devices are utilised to discover and eliminate the origins of difficulties. Hristoski et al. [16] discussed some of the variables that contribute to the decline in the competitiveness of e-Commerce firms. Using brainstorming and cause-and-effect diagrams, the study introduces the notion of causality and attempts to apply it to competition challenges. The use of visualisation improves research quality and provides a suitable instrument. In this case, the problem is a decline in corporate competitiveness. Using the proposed diagram-based technique, researchers will have a solid basis in data, allowing them to conduct case studies, comparative overviews, and more research on this and related subjects. Poka Yoke, according to Thareja [17], is a paradigm that advises the operator or designer to minimise inadvertent system errors. According to Philip Crosby's concept of quality, adjustments are made to the process or items to ensure compliance on the first try. The provision of such a service is crucial to encouraging long-term improvements and achieving really world-class operations. This paper provides a clear demonstration of the practical application of Poka Yokes by using examples drawn from actual practice.

Prabowo [18] conducted a thorough review of the international literature on Poka-Yoke implementation. The analysis of this study is divided into two sections: First, mistakes must be avoided; second, those that have already occurred must be identified. Prevent error is an approach that can assist you in preventing quality concerns. The prevent mistakes strategy employs both the control method and the warning method. When a mistake or a quality issue develops, a technique called as "detection error" may be used. the detect mistakes technique employs the contact method, the fixed value method, and the motion step method. Zandi [19] provided four categories in his article: This is accomplished with four actions: 1) developing a fishbone diagram for a fuzzy bi-level environment, 2) presenting a deciding model for a country's e-commerce maturity level that includes local, regional, and international e-markets, 3) accounting for the interdependence of these e-markets, and 4) synthesising qualitative and quantitative data of maturity criteria and sub-criteria. This self-evaluation can reveal a country's e-commerce strengths and weaknesses, as well as its readiness to move on to the next stage of e-commerce development. Furthermore, it is frequently used to transmit information to developing countries. Shinde et al. [20] intended to use this method to discover faults in India's engineering curriculum, with both students and teachers serving as stakeholders. Fishbone diagrams are used to investigate the root causes of problems using a number of components such as individuals, academics, resources, and educational institutions. Reasonable solutions are given after the root reasons have been discovered. At engineering schools, a case study is utilised to thoroughly investigate difficulties between students and teachers.

According to Pramono et al. [21] the new seven tools' implementation, use, and application are dependent on in-depth knowledge of the process, formal training in problem-solving techniques, the appropriateness of tools chosen for use, and the use of simple models at all levels of the organisation to aid communication and learning. The right set of quality tools and techniques can engage everyone in the process of continuous improvement and problem solving, monitoring and evaluating processes, developing a continuous improvement mindset, applying quality improvement lessons to day-to-day business operations, and strengthening teamwork through problem solving. Dahlgaard et al. [22] discovered that conventional quality tools are quantitative procedures, but new quality tools are qualitative approaches. Among the new tools are an affinity diagram, a matrix diagram, a tree diagram, an arrow diagram, a relationship diagram, a matrix data analysis, and a process decision program chart (PDPC). Even though the most basic quality tools can resolve up to 95% of quality issues in some cases. The seven quality control instruments, according to Sokovic et al. [23], constitute the backbone of any endeavor. These seven quality tools were first emphasised by Ishikawa, a well-known management expert. His first seven tools are regarded as "primitive" or "ancient" tools. Following it, new instruments with various purposes were developed. These approaches paved the way for Kaizen and Juan's quality control systems.

By reviewing the work of numerous researchers and planning to identify technological solutions to the various problems in e-commerce industry. This study raises the topic of how to improve process output efficiency and timeliness, as well as accuracy, controls, policy compliance, and customer service. Over the years, businesses have utilised a variety of methods to improve, control, and manage the quality of their products and services, but no quality control approach has been as widely employed as Six Sigma. It outperforms all other methods that are gaining popularity due to its concentration on statistical methodologies.

Challenges in E-Commerce Industry

Affinity Diagram

Affinity Diagram is an analytical tool used to organise many ideas into subgroups with common themes or common relationships after brainstorming [6]. Figure 56.1 shows the affinity diagram which groups various issues in e-commerce into six major categories were obtained by referring literatures such as books, journal papers and finally consolidated along with a practical perspective after brainstorming.

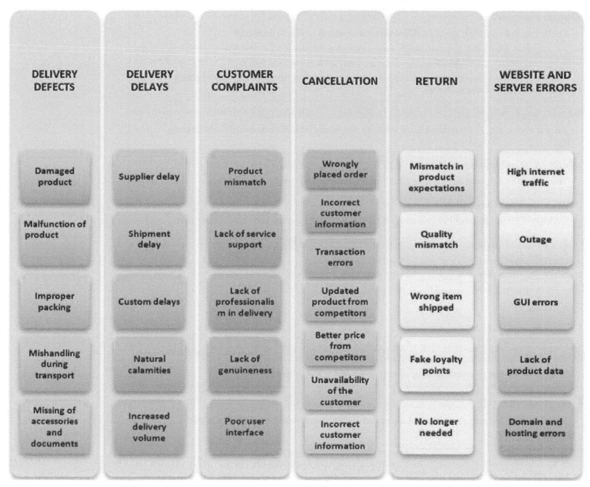

Figure 56.1 Affinity diagram
Source: Author

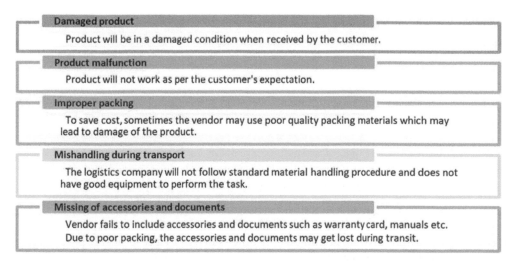

Damaged product
Product will be in a damaged condition when received by the customer.

Product malfunction
Product will not work as per the customer's expectation.

Improper packing
To save cost, sometimes the vendor may use poor quality packing materials which may lead to damage of the product.

Mishandling during transport
The logistics company will not follow standard material handling procedure and does not have good equipment to perform the task.

Missing of accessories and documents
Vendor fails to include accessories and documents such as warranty card, manuals etc. Due to poor packing, the accessories and documents may get lost during transit.

Figure 56.2 Reasons for delivery defect
Source: Author

Delivery Defect

Delivery defects occur during the delivery of the product due to improper packaging by the vendor and improper material handling by the logistics partner. Figure 56.2 shows various reasons for delivery defects.

Cancellation

Cancellation is the process of deciding to refrain from the purchase of a product from an e-commerce website due to certain reasons. Figure 56.3 shows various reasons for cancellation.

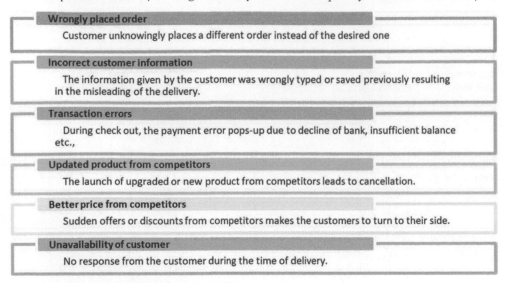

Figure 56.3 Reasons for cancellation
Source: Author

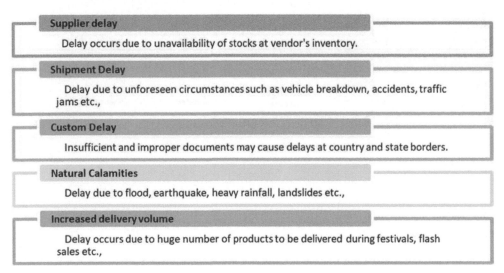

Figure 56.4 Reasons for delivery delays
Source: Author

Delivery Delays

Delivery delays happen when unforeseen circumstances of a logistic nature have occurred and the delivery date gets probably postponed. Figure 56.4 shows various reasons for delivery delays.

Customer Complaints

A customer complaint is an expression of dissatisfaction on consumer's behalf to a responsible party. Figure 56.5 shows various reasons for customer complaints.

Return

A product return in e-commerce is the process where a customer places a return order back to the retailer, and in turn will request for a refund in the original form of payment or replacement of the item. Figure 56.1 shows various reasons for the return of the product.

Website and Server Error

It is a message from the web server that something went wrong. In some cases, it could be a mistake made by the customer end, but often, it is the site's fault. Each type of error has an HTTP error code dedicated to it [8]. Figure 56.1 shows various reasons for website and server error.

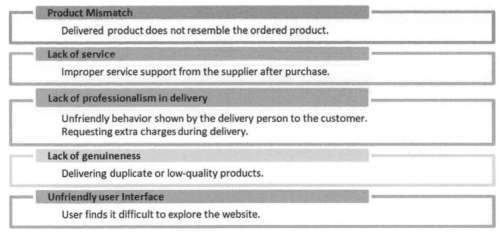

Product Mismatch
Delivered product does not resemble the ordered product.

Lack of service
Improper service support from the supplier after purchase.

Lack of professionalism in delivery
Unfriendly behavior shown by the delivery person to the customer.
Requesting extra charges during delivery.

Lack of genuineness
Delivering duplicate or low-quality products.

Unfriendly user Interface
User finds it difficult to explore the website.

Figure 56.5 Reasons for customer complaints
Source: Author

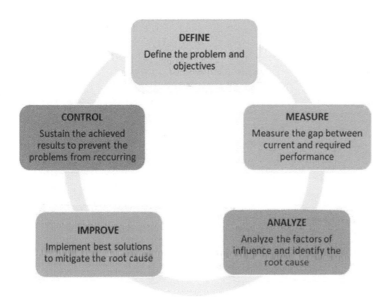

Figure 56.6 DMAIC cycle
Source: Author

Six Sigma

Six Sigma is a quality control philosophy which was developed by Motorola, Inc in 1986 [8]. Basically, Six Sigma is a problem-solving technique that uses human assets, data, measurement and statistics to identify the vital few factors to decrease waste and defects while increasing customer satisfaction, profit and shareholder value. Six Sigma decreases the process variations while increasing the performance and employee morale. Six Sigma helps in reducing the cycle time and reducing defects in the manufacturing of not more than 3.4 defects per million opportunities [9]. It uses certain tools and techniques which define and evaluate each step of the process thereby improving efficiencies in business structure, improving the quality of the process.

DMAIC Implementation

DMAIC is an acronym which stands for define, measure, analyse, improve, control. This methodology aims not only to improve the quality of product but also to improve the quality of the process which produces output. Figure 56.6 gives a brief description about the various phases involved in DMAIC implementation cycle.

Define

Project charter. It is a formal, typically short document that contains information of the project such as reasons for project, objectives and constraints of the project, main stakeholders, risks identified, benefits of

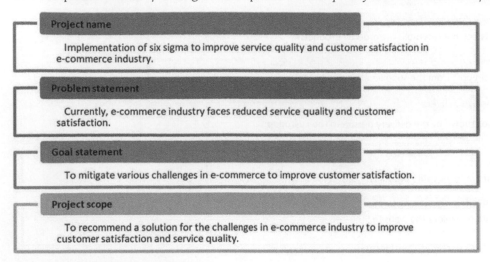

Figure 56.7 Project charter
Source: Author

Table 56.1 Pairwise ranking

Challenges	A	B	C	D	E	F	No of times occurred	Rank
A	-	A	A	A	A	A	5	1
B	-	-	C	B	E	B	2	3
C	-	-	-	C	C	C	4	2
D	-	-	-	-	D	D	2	3
E	-	-	-	-	-	E	2	3
F	-	-	-	-	-	-	0	6

Source: Author

the project and common overview of the budget [8]. Figure 56.7 shows the project charter. Here it formally authorises the existence of the project and provides a reference source for the future.

Measure

Pairwise ranking. It is a participatory technique that permits analyzing, identifying and prioritising problems to implement adequate improvements and solutions. There are many variations of the system, but it assists to rank all items against one another [8]. Here this pairwise ranking helps to prioritise the problems and needs of the customer. Table 56.1 shows the pairwise ranking. The challenges considered for pairwise ranking were delivery defect (A), delivery delay (B), customer complaint (C), cancellation (D,) returns (E), website and server errors (F).

Analyse

Cause and effect diagram. It is a visual quality tool that helps to identify, sort, and display possible causes of a specific problem or quality characteristics. Having categorised the e-commerce issues into six major challenges, the next step is to find the root causes behind each challenge using cause and effect diagram which is shown in Figure 56.8.

Improve

Standard Operating Procedure (SOP). It is a set of written instructions that documents a routine activity that is to be followed by members of a corporation [9]. SOPs are essential part of excellent quality systems which can enhance consistency and reduce human error. Figure 56.9 shows SOP which is used here to standardise the process.

Control

Reaction plan. It specifies a course of action that needs to be taken when process control parameters deviate from the desired level and it includes both immediate and long-term actions. Figure 56.10 shows the

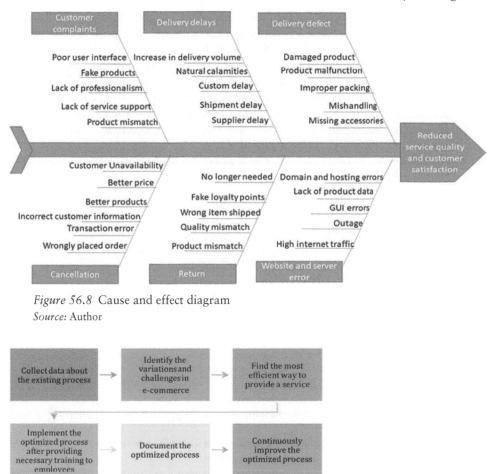

Figure 56.8 Cause and effect diagram
Source: Author

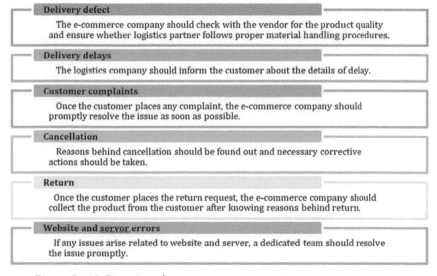

Figure 56.9 Standard operating procedure
Source: Author

Delivery defect

The e-commerce company should check with the vendor for the product quality and ensure whether logistics partner follows proper material handling procedures.

Delivery delays

The logistics company should inform the customer about the details of delay.

Customer complaints

Once the customer places any complaint, the e-commerce company should promptly resolve the issue as soon as possible.

Cancellation

Reasons behind cancellation should be found out and necessary corrective actions should be taken.

Return

Once the customer places the return request, the e-commerce company should collect the product from the customer after knowing reasons behind return.

Website and server errors

If any issues arise related to website and server, a dedicated team should resolve the issue promptly.

Figure 56.10 Reaction plan
Source: Author

reaction plan which is used here to instruct people who are executing the process thereby properly responding to failures.

Conclusion

The reasons for various issues in the e-commerce industry, such as delivery defects, delivery delays, customer complaints, cancellation, return, and website and server errors, were discussed in this paper, and appropriate solutions to overcome these challenges were proposed using DMAIC methodology, resulting in increased service quality, customer satisfaction, reduced processing time and cost. The project charter was used to provide a formal introduction to the project during the defined phase. Throughout the measuring phase, pairwise ranking was used to establish a link between different activities. During the analyse phase, a cause-and-effect diagram was used to determine the root causes of each problem. Standard operating procedure (SOP) standards were presented during the improve phase to standardise the procedure in the e-commerce industry. Finally, a reaction plan was used in the control phase to prescribe reactive measures when something negative happened. Six sigma implementations have lagged in e-commerce and other service companies. Better productivity and efficiency can be attained if a greater number of service organizations implement the six sigma methodologies stated above, as well as continuous improvement, which will benefit both customers and service providers.

References

[1] Zhao, W., Hu, F., Wang, J., Shu, T., and Xu, Y. (2023). A systematic literature review on social commerce: Assessing the past and guiding the future. Electronic Commerce Research and Applications. 57(101219), 101219.

[2] Reinartz, W., Wiegand, N., and Imschloss, M. (2019). The impact of digital transformation on the retailing value chain. Int. Journal of Marketing Research. 36(3), 350–66.

[3] Reddy, N. A., and Divekar, B. R. (2014). A study of challenges faced by E-commerce companies in India and methods employed to overcome them. Procedia Economics and Finance. 11, 553–60.

[4] Haji, K. (2021). E-commerce development in rural and remote areas of BRICS countries. Journal of Integrative Agriculture. 20(4), 979–97.

[5] Mäki, M., and Toivola, T. (2019). Global ecommerce development – joining the resources of universities and enterprises for new markets. In INTED2019 Proceedings, Valencia, Spain, 2019.

[6] Clemes, M. D., Gan, C., and Zhang, J. (2014). An empirical analysis of online shopping adoption in Beijing, China. Journal of Retailing and Consumer Services. 21(3), 364–75.

[7] Hsu, M.-H., Chuang, L.-W., and Hsu, C.-S. (2014). Understanding online shopping intention: the roles of four types of trust and their antecedents. Internet Research. 24(3), 332–52.

[8] Soegoto, E. S., Christiani, A., and Oktafiani, D. (2018). Development of E-commerce technology in world of online business. IOP Conference Series: Materials Science and Engineering. 407, 012031.

[9] Yoon, H. S., and Occeña, L. G. (2015). Influencing factors of trust in consumer-to-consumer electronic commerce with gender and age. International Journal of Information Management. 35(3), 352–63.

[10] Kumar, N. (2018). E–Commerce in India: An Analysis of Present Status, Challenges and Opportunities. International Journal of Management Studies, 5(2), 3.

[11] Vendrame Takao, M. R., Woldt, J., and Da Silva, I. B. (2017). Six Sigma methodology advantages for small-and medium-sized enterprises: A case study in the plumbing industry in the United States. Advances in Mechanical Engineering, 9(10), 1687814017733248.

[12] Kaushik, P. (2011). Relevance of six sigma line of attack in SMEs: A case study of a die casting manufacturing unit. Journal of Engineering and Technology. 1(2), 107.

[13] Jacobs, B. W., Swink, M., and Linderman, K. (2015). Performance effects of early and late six sigma adoptions. Journal of Operations Management. 36(1), 244–57.

[14] Utomo, U. (2020). A systematic literature review of six sigma implementation in services industries. IJIEM-Indonesian Journal of Industrial Engineering and Management. 1(1), 45.

[15] Magar, V. M., and Shinde, V. B. (2014). Application of 7 quality control (7 QC) tools for continuous improvement of manufacturing processes. International Journal of engineering research and general science, 2(4), 364–371.

[16] Hristoski, I., Kostoska, O., Kotevski, Z., and Dimovski, T. (2017). Causality of factors reducing competitiveness of e-Commerce firms. Balkan and Near Eastern Journal of Social Sciences. 3(2), 109–27.

[17] Thareja, P. (2019). Poka Yoke: Poking into Mistakes for Total Quality!. SSRN, 1(1), 5–26.

[18] Prabowo, R. F., and Aisyah, S. (2020). Poka-Yoke method implementation in industries: A systematic literature review. IJIEM - Indonesian Journal of Industrial Engineering and Management. 1(1), 12.

[19] Zandi, F. (2013). A country-level decision support framework for self-assessment of E-commerce maturity. iBusiness. 5(1), 67–78.

[20] Shinde, D. D., Ahirrao, S., and Prasad, R. (2018). Correction to: Fishbone diagram: Application to identify the root causes of student–staff problems in technical education. Wireless Personal Communications. 100(2), 665–5.

[21] Pramono, S. N., Ulkhaq, M. M., Rachmadina, D. P., Trianto, R., Rachmadani, A. P., Wijayanti, W. R., and Dewi, W. R. (2018). The use of quality management techniques: The application of the new seven tools. International Journal of Applied Science and Engineering, 15(2), 105–112.

[22] Dahlgaard, J. J., Khanji, G. K., and Kristensen, K. (2008). Fundamentals of total quality management. Routledge, 5(3), 85–103.

[23] Soković, M., Jovanović, J., Krivokapić, Z., and Vujović, A. (2009). Basic quality tools in continuous improvement process. Journal of Mechanical Engineering, 55(5), 1–9.

Chapter 57

A review on the hydro dynamic analysis of fluidisation of nano particles using vibrations

Khwaja Izhar Ahmad[a] and Dr. Akash M Langde[b]

Anjuman College of Engineering, Nagpur, India

Abstract

In various industrial processes tiny particles are required to be dispersed thoroughly. In order to achieve uniform distribution of such particles, fluidisation is a popular technique. Fluidisation is being widely used in industries due to its efficiency in powder handling and due to its good heat and mass transfer. There are many ways and methods to achieve fluidisation of powders/nano particles such as fluidisation by aeration but it is not enough to fluidise very small and cohesive particles thoroughly, whereas some assisted techniques are quite efficient in fluidising such powders thus making these techniques more popular in today's industrial scenario. The industrial application of fluidised and semi-fluidised bed systems is highly promising due to the efficiency of these methods. Despite many studies that have been conducted to better comprehend fluidisation technology, there are still many unknown factors such as the effect of cohesiveness of irregular and regular shaped particles, bed material, liquid viscosity and surface tension, etc. on a system intended for industrial application. To find an improved system for fluidisation process which require high rates of fluidability of particles numerous combinations of acoustic, aeration and vibratory techniques have been used. A series of analysis and tests, as Arthur P. Fraas correctly pointed out, have revealed heat transmission and high rates of mixing with a wide range of flow circulation patterns. Further experiments on above stated parameters will help in establishing effects of combined methods of acoustic and vibrations on minimum fluidisation velocity. It will show how mechanical and acoustic vibrations combined can noticeably affect the U_{mf}. Since vibration is an effective means to overcome the interparticle van der Waals forces of fine powders in reducing U_{mf}. Further experimentation on above stated method will help in finding out a critical value i.e., minimum vibration velocity which will help in improving U_{mf} as excessive vibration may affect the desired results as far as U_{mf} is concerned. This paper deals with combined acoustic, aeration, and vibrating technique to improve fluidability of nano particles.

Keywords: Acoustic and vibration techniques, aeration, fluidisation, nano particles

Introduction

By passing a fluid (liquid or gas) through a bed of particle solids, the material is "fluidised," or given the properties of a fluid. Fluidised beds are reactors that facilitate the fluidisation of particles. Fluidised beds play a crucial role in a variety of industrial processes due to their multiple benefits, which include a rapid rate of low-pressure dips, consistent temperature distribution and heat and mass transfer [1].

The graphic representation below gives an idea of various techniques which can be used for fluidisation of various powders to find out minimum fluidisation velocity and other parameters related to fluidisation.

Methods of fluidisation

With aeration only i.e., without assistance	Acoustic field assisted various combination of SPL and frequency	Mechanical vibrating techniques	Combined effect of acoustic field, mechanical and other assisted

Fluidised bed hydrodynamics are quite complicated and must be comprehend in order to enhance fluidised bed operations. To comprehend the behavior of a fluidised material, a number of parameters are employed. Therefore, examining a variety of materials with distinct properties will aid in identifying common and distinct behavior. Glass beads, for instance, offer a consistent fluidizing bed. On the other hand, because of their variable size, shape, and density, it is difficult to fluidise biomass particles. The majority of published research focuses on the impacts of various materials on the flow structure of a fluidised bed but does not address the implications of bed height or the employment of various approaches that can be coupled to facilitate the fluidisation process. It is essential to comprehend the effect of specific features and qualities on the hydrodynamics of a fluidised bed [2,3].

[a]kiahmad@anjumanengg.edu.in, [b]kashlangde@gmail.com

DOI: 10.1201/9781003450252-57

The minimum fluidisation velocity *(Umf)*, which calculates the drag force required to obtain solid suspension in the gas phase, is a crucial quantity for describing fluidised bed conditions. Hamaker conducted a critical analysis of the attractive force between tiny particles of any substance, as well as between a particle and a surface. Van der Waals forces, which he attributed in large part, were the impetus for his development of universal formulations for calculating the interactions between two spherical particles. Fluidisation of nano particles by aeration only and assisted with supportive techniques, such as mechanical vibration, is the purview of current work. Extensive research survey will help to design and formulate the research plan to achieve the goal [2].

Wide-range of studies carried out on the fluidisation behavior of number of powders/particles having different material, size, and shape. Based on the study and empirical observations of the particles he categorised them into four distinct categories viz. A for aerated, B for bubbling, C for cohesive and D for large particles. This categorisation is very popular and adopted by worldwide researchers for performing experiments to find out fluidisation behavior of particles. It is a well-known fact that the Geldart C particles are sometimes tough to fluidise due to the large cohesive forces which lead to channeling in the bed and crack formations. The Figure 57.1 gives an idea about the classification and behavior pattern of particles.

At least six distinct types of fluidisation systems for gas-solid fluidised beds were taken into consideration in a study. These fluidisation systems were as follows: bubbling fluidisation, fixed bed, slugging fluidisation, rapid fluidisation, turbulent fluidisation, and pneumatic conveying. A graphical illustration of the fluidisation systems that are now in use in fluidised beds can be shown in Figure 57.2. In the immovable bed regime, the air that flows over the particles does not have a velocity that is sufficient to transfer the particles. When the surface velocity of gas (Ug) hits a certain threshold, the system enters a regime known as bubbling fluidisation. The lowest bubbling velocity is the speed at which bubbles first appear, merge, and cause solid mixing to occur in this regime [4].

Improving fluidisation quality by preventing channeling, slugging, and breaking out agglomeration with the help of an acoustic field is a widely used and effective technique. The formation of large dense agglomerates was seen to allow for the fluidization of C particles. The granular temperature of A and B particles was measured, and it was found that the granular temperature of A particles is significantly higher than

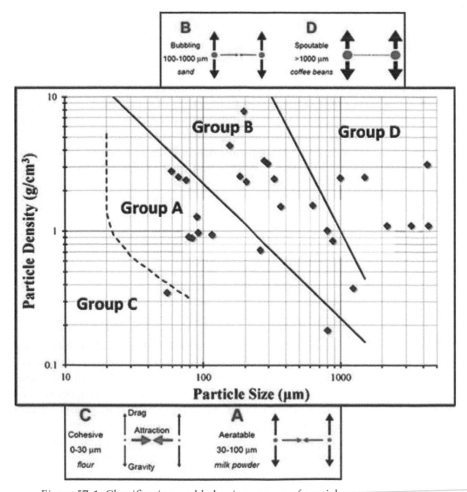

Figure 57.1 Classification and behavior pattern of particles
Source: Author

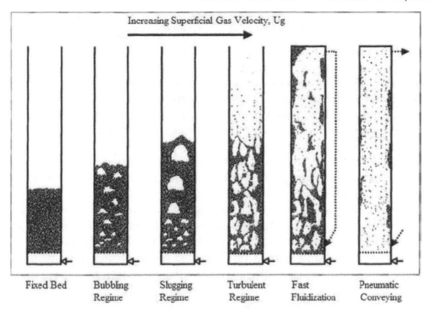

Figure 57.2 Regimes of fluidisation in a gas-solid fluidised bed

Source: Escudero, David Roberto. (2014). Characterization of the hydrodynamic structure of a 3D acoustic fluidized bed. Graduate Theses and Dissertations. Paper 13928.

that of the larger B particles. It is notoriously challenging to fluidise cohesive powders using only aeration, so assisting techniques are commonly used to achieve much higher fluidisation quality. Researchers have reported using a wide variety of aiding techniques, such as the application of a magnetic field or acoustic field, mechanical vibration, mechanical stirring, micro jets, surface modification, the addition of another particle, and so on, in order to enhance fluidisation quality. Some experimental results show that when the vibration is added to the particle bed, the minimum fluidisation velocity becomes smaller and experiments by some researchers has shown that fluidisation with the help of an acoustic field is both straightforward and effective.

Maclaren (2021) has shown via experiments that vibration can reduce both *Umf* and the minimum bubbling velocity (*Umb*) in Geldart Group B and D particles. Zhao [13] experimented with mechanical vibration and stirring method because "the idea of combing the vibration and stirring is that the vibration mainly introduces vertical force while the stirring mainly introduces horizontal force, which can be used to disturb the bed materials in two directions." And latter concludes that "both vibration and stirring can improve the fluidisation quality of nano particle agglomerates. However, the fluidisation quality is no longer improved, or even reduced when the vibration or stirring parameter increases to a critical value" [6].

Hashemnia [9] carried out experimentation to investigate the effect of vibration frequency and amplitude on the quality of fluidization and flow mode of a vertically vibrated granular flow and concludes that "increasing the vibration amplitude resulted in the particles average velocity increase due to the larger kinetic energy of the fluidised bed. Therefore, it was concluded that increasing vibration amplitude in processes such as mixing, segregation and vibratory finishing would be more efficient than increasing vibration frequency for fluidisation" [9].

The effect of variable acoustic field and frequency on fine powder gas solid suspension was examined by the researchers. Another study demonstrated the use of microjets to improve the fluidisation of agglomerates of metal oxide nano powders. Some other researchers investigated the significance of inter-particle forces and the effect of temperature on the fluidisation of Geldart's Group C and A powders.

Experimentation is carried out to investigate the thermal effect on fluidisation at temperatures ranging from 20-800°C and acoustic parameters ranging from SPL 130_150 dB and frequency 50-200 Hz. Furthermore, the effect of nano particle density, mixture composition and sound intensity on nano powder mixing quality was also investigated. The effects of pressure, temperature, and inter particle forces on the hydrodynamics of a gas-solid fluidised bed were critically reviewed.

The effects of combined acoustic vibrations and mechanical on the fluidisation of cohesive powders was also investigated by the researchers. The experimental setup includes a Plexiglas column, a porous distributor plate with a plenum underneath, manometers, rotameters, and a microphone embedded in the bed. The minimum fluidisation velocity is calculated by measuring the pressure drop. An acoustic field of 148 dB at 80 Hz was used to improve fluidisation. Aside from the above, mechanical vibration with a frequency of up to 9.5 Hz and an amplitude of 1 to 10 mm is applied in the horizontal direction to improve fluidisation quality.

The frequency beyond resonance frequency is not useful to improve the quality. Many authors studied the effect of frequency and its optimum value. Some other studies showed that the *Umf* initially decreased upto a certain value i.e. resonance frequency and then increased on further increase in the sound frequency and fluidisation of SiO_2 nanoparticles assisted with acoustic field was also carried out. The main components of the setup are glass column, porous distributor plate, manometer and rotameters. The main components of the setup are glass column, porous distributor plate, manometer and rotameters. Maximum SPL applied was 130 dB and max. frequency upto 3 MHz. Apart from above, various waveforms are applied including sinusoidal, triangular, square, rampup, sin(x)/x etc. A SEM analysis is carried out to inspect the agglomerates.

Literature review reveals that a particular frequency improves the quality of fluidisation which is termed as resonance frequency for the bed. The connection between resonance frequency and agglomerate clusters was theoretically derived. It was found that the bed's resonance frequency is sensitive to the particles' properties. For a given sound pressure level (SPL) and bed weight, channeling free fluidisation only occurs within a specific frequency range. Another study investigated the acoustic field assisted fluidisation of nanoparticle and studied the agglomerates. Minimum fluidisation velocity and maximum bed expansion noted at frequency of 50 and 100 Hz. Large ellipsoid shaped bubbles also noted at higher frequencies around 200 Hz and above, that wasn't noted with aeration alone. Further, the validation of surface coating on nanoparticles to achieve better flowability of fine cohesive particles was also investigated. Surface coatings improve the quality of fluidisation by weakening the cohesiveness and consequently reduction in minimum fluidisation velocity. Figure 57.3 is shows effect of variable sound frequency (at constant SPL 140 dB) on *Umf* on the fluidisation behaviour of three cohesive samples adapted.

The increase in SPL reduces the *Umf* continuously but frequency of particular range improves the fluidisation. Sinusoidal and triangular waveforms were reportedly shows better results than that of the square and ramp-up waves [10, 11].

Mechanical assisted techniques influidisation are technique in which various mechanical methods, such as use of stirrer, vibrations etc. are used to prevent channeling, improving mixing and quality of fluidisation. Mechanical agitation is a popular technique, where mechanical action is imparted to the bed, so that agitation can be created at bed. Vibrating fluidised beds have been used in many different chemical and physical procedures. Increased solid-gas contact efficiency, fluidisation pressure drops and, lower minimum fluidization velocity and more uniform and stable fluidised bed layers are just some of the advantages of the vibrating fluidised bed over the traditional fluidised bed. These benefits can be used in a wide variety of contexts, including chemical reactions, industrial processes, and home projects. Vibrational forces break up cohesive forces, allowing Group C particles in a vertically vibrating gas-fluidised bed to be fluidised [Chevilank, et al.]. Vibration has also been shown to decrease *Umf* in Group B and D particles [Gupta, R., and Mujumdar] and Group C and A particles [Erdesz, K. and others]. Bed void fraction at *Umf* decreases with increasing vibration strength, which was used in a previous study to account for the decrease in *Umf* in Group C and A particles (which allows for bed expansion) [12,13].

The feasibility of using vibration for different sized powders is suggested by the results of a small number of studies examining the effect of vibration on bed void age for large particles.

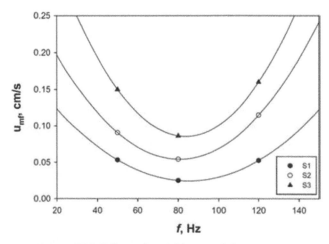

Figure 57.3 Effect of variable sound frequency (constant SPL 140 dB) on U*mf* on the fluidisation behavior of three cohesive samples (Chirone et al., 2018)

Source: Roberto Chirone, Federica Raganati, Paola Ammendola, Diego Barletta, Paola Lettieri, Massimo Poletto. (2018). Powder Technology 323 A comparison between interparticle forces estimated with direct powder shear testing and with sound assisted fluidization. pp. 1–7

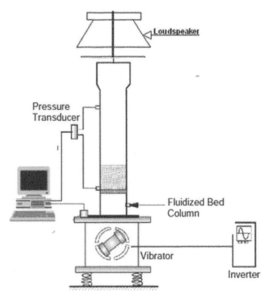

Figure 57.4 Experimental setup of fluidisation of nano particle
Source: Author

Nano particle fluidisation: Over the past decade, nano particles and nano composite materials have been the subject of extensive study due to the unique properties of nano structured materials that make them appealing for a wide range of industrial applications. Nanoparticle construction significantly alters and, in many cases, improves the catalytic, mechanical, electronic, optical, and other physical and chemical properties of a material. Mechanical vibration aids in reducing channeling and spouting in a bed of nano sized powders, as demonstrated by experiments involving mechanical methods, acoustic systems, and aeration. Example of a standard experimental setup shown in Figure 57.4.

The above diagram depicts a possible fluidisation system that could be used to carry out the experiment. This setup includes a fluidised bed of nano particle agglomerates, a vibrating system, and acoustic and aeration components. The fluidised bed consists of a vertical column of clear acrylic with openings at the base and a sound wave generator at the top. The column is an established length of acrylic pipe of a specified diameter and height. It has been demonstrated experimentally that nano powders can be easily and smoothly fluidised into the form of stable, very porous agglomerates with negligible elutriation, and that Umf decreases in a vibrated bed as compared to a non-vibrated bed.

J. Ruud van Ommen has written that currently, fluidisation of nanopowders is only applied in a limited number of commercial processes. The two most important large-scale processes involving fluidisation of nano powders are the production of fumed metal oxides and carbon black.

Methodology

Zhao [13] has pointed out that to expand the use of nano particles, different methods can be proposed to improve their fluidisation. The assisted methods are divided into two groups. One group is through introduction of external fields into the fluidised bed to break the agglomerates, like electric field, acoustic field, centrifugal fields, vibration, stirring, pulsed gas flow, micro jets, and magnetic field. Another method is adding coarser particles or surface modification [12].

The extant studies illustrate that the assisted methods can reduce the minimum fluidisation velocity and increase the bed expansion to a certain degree. However, in most cases, it was found that a single field has some limitations on enhancing fluidisation. Like nano particle agglomerates tend to adhere to the wall of fluidised bed at static electric fields because of electrophoretic deposition. As seen some studies had illustrated that the fluidisation index, bed expansion ratio may decrease and the minimum fluidisation velocity may increase with the increase of amplitude or frequency. Zhao [13] further adds that investigation on the fluidisation quality of Al_2O_3 and Fe_2O_3 nano particles by applying a combination of vibration and magnetic field showed that large bubbles and agglomerates disappear in the bed.

To enhance fluidisation various combined techniques such as acoustic and mechanical vibrations can be used to comprehend the process profoundly and to find out effect on various parameters such *Umf and Umb*.

Experimentation will be carried out on the set up as shown in the Figure 4 for different nano particles to find out the behaviour of particles and to find out minimum fluidisation velocity *Umf* and the minimum bubbling velocity *Umb* [14].

Conclusions

Particle fluidisation can be improved by vibration, as demonstrated by experimental study of the effect of vibration on the void age of a fluidised bed. Effects of changing vibration amplitude on bedtime urination. As an added bonus, you can use it to calculate the height of the particle bed and the layers of particles, as well as the effect that vibration energy has on those levels.

Through experimentation in fluidisation using assisted techniques such as mechanical vibration it can be investigated to find out the effect of vibration frequency and amplitude on the quality of fluidisation and flow mode of a vertically vibrated granular flow. This research will aid in determining the minimum fluidisation velocity for different nano powders, which will have applications in industry and contribute to the achievement of sustainable development goals, which are currently the top priority for countries all over the world.

Reference

[1] Gupta , R., and Mujumdar, A. (1980). Aerodynamics of a vibrated fluid bed. The Canadian Journal of Chemical Engineering. 58, 332–8.

[2] Jin, H., Tong, Z., Zhang, J., and Zhang, B. (2004). Homogeneous fluidization characteristics of vibrating fluidised beds. The Canadian Journal of Chemical Engineering. 82, 1048–53.

[3] Ruud van Ommen, J., Manuel Valverde, J., and   Robert, P. (2012). Fluidization of nanopowders: a review. Journal of Nanoparticle Research, 14(3), 737.

[4] Chirone, R., Massililla, L., and Russo, S. (1993). Bubble free fluidization of a cohesive powder in an acoustic field. Chemical Engineering Science. 48(I), 41–52.

[5] Tatemoto, Y., Mawatari, Y., and Noda, K. (2005). Numerical simulation of cohesive particle motion in vibrated fluidised bed. Chemical Engineering Science. 60, 5010–21.

[6] Jin, H., Zhang, J., and Zhang, B. (2007). The effect of vibration on bed voidage behaviors in fluidised beds with large particles. Brazilian Journal of Chemical Engineering. 24(03), 389–97.

[7] Xuliang-Yang, Yadong Zhang, Yu-Yang and Enhui-Zhou (2016). Fluidization of geldart D type particles in a shallow vibrated gas-fluidised bed. Powder Technology. 305, 333–9.

[8] Zhou, E., Zhang, Y., Zhao, Y., Luo, Z., He, J., and Duan, C. (2018). Characteristic gas velocity and fluidization quality evaluation of vibrated dense medium fluidised bed for fine coal separation. Advanced Powder Technology. 29, 985–95.

[9] Raganati, F., Chirone, R., and Ammendola, P. (2018). Gas-solid fluidization of cohesive powders. Chemical Engineering Research and Design. 133, 347–87.

[10] Hashemnia, K., and Pourandi, S. (2018). Study the effect of vibration frequency and amplitude on the quality of fluidization of a vibrated granular flow using discrete element method. Powder Technology. 335–45.

[11] Lehmann, S., Hartge, E.-U., Jongsma, A., de Leeuw, I.-M., Innings, F., and Heinrich, S. (2019). Fluidization characteristics of cohesive powders in vibrated fluidised bed drying at lowvibration frequencies. Powder Technology. 357, 54–63.

[12] Lehmann, S., Hartge, E.-U., Jongsma, A., de Leeuw, I.-M., Innings, F., Heinrich, S. (2019). Fluidization characteristics of cohesive powders in vibrated fluidised bed drying at low vibration frequencies. Powder Technology. 357, 54–63.

[13] Zhou, Y., and Zhu, J. (2020) . Group C+ particles: Extraordinary dense phase expansion during fluidization through nano-modulation. Chemical Engineering Science. 214. 1–10.

[14] Zhao, Z., Liu, D., Ma, J., and Che, X. (2020) . Fluidization of nanoparticle agglomerates assisted by combining vibration and stirring methods. Chemical Engineering Journal. 388. 1–10.

[15] McLaren, C. P., Metzger, J. P., Boyce, C. M., and Müller, C. R. (2021). Reduction in minimum fluidization velocity and minimum bubbling velocity in gas-solid fluidised beds due to vibration. Powder Technology. 382, 566–572.

Chapter 58

Behavior analysis and optimisation of process parameters in petrochemical industry using PSO: a case study

Janender Kumar[1,a], Virat Khanna[1,b], Munish Mehta[2,c] and Sandeep Chauhan[3,d]

[1]Maharaja Agrasen University, Solan, India

[2]Lovely Professional University, Jalandhar, India

[3]State Institute of Engineering & Technology, Nilokheri, India

Abstract

The availability analysis and enhancement of system parameters in polypropylene unit of a polymer section at refinery using particle swarm optimisation (PSO) algorithm has been performed in the present paper. The polypropylene system consists of four important subsystems which are arranged in series and parallel configuration: blender, extruder, classifier, and rotary feeders. Markov approach is utilised to know availability of each subsystem and parameters are optimized by varying the particle size and iteration size. Maximum availability of 97.608% was attained at particle sizes of 45 and 97.603% at the 70th iteration, according to the successful findings. The crucial evaluation has been discussed with plant personnel for maintenance strategies.

Keywords: Maintenance planning, availability, optimisation, polymer section

Introduction

The random failure of working equipment in process industries reduces the productivity and increases the burden of production targets. The improper maintenance strategies added other losses in it. The actual availability analysis of an operating equipment of a system becomes necessary in the plant to increase production and minimize the wastages. The polypropylene system is intricate in design, and timely repairs can keep the equipment functional for an extended period of time. The study has been performed in plastic granules (pellets) producing section i.e., polypropylene unit. The purpose behind study was to recognise the operating equipment which have more contribution in reducing availability and optimisation of process parameters for best combination of failure and repair rate. Further, to set maintenance priority accordingly. The review of literature shows that system performance evaluation techniques like failure mode and effect analysis, Lambda-Tau methodology, Markov approach etc. The researchers have also optimized process parameters using genetic algorithm, nature inspired algorithm etc. In process industry like thermal power plant, performance of equipment was analysed. Steam and power generation units are considered for reliability analysis. Markov birth-death process is utilised to know the performances of both systems [1]. Ammonia synthesis unit comprises five subunits in a fertiliser plant. It was facing the issue of lesser availability. Due to the complex and repairable system of the unit, a reliability tool has been utilised for performance evaluation of its subunits via mathematical modeling [2]. The study has been executed for enhancing the productivity and examine the variations of subsystems performance with respect to time in cattle feed plant. C-program and matrix method used for purpose of solving the differential equations developed with the help of transition diagram [3]. In the paper, reliability assessment has been done by comparing two states; continuous and discrete for the components of a system. Monte Carlo algorithm utilised to examine the reliability of a multistate system [4]. The warm standby repairable system of one unit and transfer switch subsystem of second unit are considered for reliability analysis [5]. In the fertiliser plant, reliability, availability and maintainability examination was carried out for suggesting maintenance strategies to maintenance staff, on behalf of the calculated results in urea decomposition system. Lamda-tau methodology and petri net technique adopted for study [6]. The thermal power plant having capacity of 210 mw, an analysis conducted for availability of its components. Reliability, availability, maintainability technique implemented to identify critical components and also suggested the importance of preventive maintenance program [7]. In the competitive environment, it becomes necessary to improve the quality of product to satisfy customer needs. The reliability analysis of automatic gun i.e., weapon system is studied with the new proposed intuitionistic fuzzy set approach and find out the problematic component of the system due to which it was not working properly i.e., magazine spring [8]. The system behavior analysed in paper mill which consists of filter, screener and decker subsystems of a washing unit in it. Mean time between failure

---

[a]dhingra76@gmail.com, [b]khanna.virat@gmail.com, [c]munishmehta1@rediffmail.com, [d]sandeep2140@gmail.com

DOI: 10.1201/9781003450252-58

and mean time to repair was discussed properly to reduce the operational as well as maintenance cost of system [9]. A steady state availability of a pump system carried out with semi-Markov approach and further Monte Carlo simulation is used for validation of results [10]. The reliability and availability analysis has been performed in urea synthesis system which consists of CO_2 booster, compressor, NH_3 preheater, liquid ammonia feed pump and recycle solution feed pump subsystems in a fertilizer plant. Markov birth-death process implemented to know availability of each subsystem [11]. The analysis performed at port for ships to reduce their waiting time and handling time. so that it could leave as soon as possible. The berth allocation problem identified and sort out with nature inspired algorithm which is influenced from group of birds flying in the sky [12]. A thermal cycling test and peel strength analysis performed on anisotropic conductive adhesive flex on board assembly of an electronic device under various parameters to know the reliability of assembly. The results indicates that no failure observed [13]. The reliability analysis performed for subsea oil production system which comprises three subsystems. The piecewise deterministic Markov process, a reliability tool is utilised to know the availability of components and output compared with PN- Monte Carlo simulation. The result of both approaches was found close to each other. The validation of results was done with petri net approach [14]. The availability analysis for carbonated soft drink glass bottle has been performed in a beverage plant. Mathematical model with help of Markov process is derived to know the availability of each subsystem. The process parameters were further optimised with algorithm based on swarm intelligence. The findings were discussed with plant engineers for planning maintenance strategies [15]. The reliability of machines could be increased by reducing the breakdown and proper maintenance. An approach of using multiple maintenance modes was suggested to incorporate. The integrated model combines preventive and corrective maintenance modes for minimising maintenance cost with the assistance of particle swarm optimisation (PSO) algorithm [16]. Behavior analysis of coal handling system in a power generation plant studied to find out the availability of its subsystems. The purpose behind this was to increase the productivity via planned maintenance strategies. Markov approach or chain has been used for developing availability equations and PSO algorithm for optimising the process parameters to get maximum output from different subsystems. The outcomes were shared with plant personnel [17]. As per the available literature, the researchers had only worked in limited process industries like sugar, power plant, milk and cement etc. however, work related to petrochemical industry was observed very less. The efforts have been made in polypropylene unit to optimize the performance.

System Configuration

It consists of eight subsystems as also shown in Figure 58.1.

1. Blender subsystem (E): Powder and liquid additives are mixed in blender for homogenisation and further fed to next process. It consists of parallel system. If one fails system capacity reduced. But when both fails, system failed.
2. Extruder subsystem (F): Powder and mixed additive is melted in extruder due to external heating and shearing took place (250oc). This melt is passed through screen pack (mesh/jaali) to separate foreign particles. This melt is now passed through die plate where a cutter continuously cut the melt into pellets in water. Water is used for cooling of pellets and conveying to downstream. Extruder is critical equipment in polypropylene unit at Naphtha cracker unit. Extrusion process is designed to mix flux and the combined flow of resins, additives charged from material feed equipment to give final product i.e., pellets. Melt is extruded through die plate nozzles in form of hot melt strands which are cut into pellets. Die plate is made of special grade forged stainless steel with a 3 mm tile of titanium carbide on the surface having holes to extrude the polymer. The number of holes of die plate are 2640. During normal extruder operation the Tic surface of die plate get damaged. The Tic layer can be repaired in the form of replating. It just has one unit whose failure causes the system to completely fail.
3. Pre-dewatering screen: Dewatering screen is there to separate water and pellets. Here, water is collected at bottom in tank and then again fed into pellet cutting area. This system never fails.
4. Centrifugal dryer: Moist pellets go to centrifugal dryer where moisture and pellets separates. Moisture is sucked by wet air fan and went to atmosphere. This system rarely fails.
5. Classifier (G): Pellets are sent to classifier for screening. Three-streams comes out from classifier. Oversize and undersise are collected on ground and normal pellets are transfer to storage silos with help of rotary feeders. It is having single unit. Failure of which results in complete failure of system.
6. Rotary feeders (H): These are used to carry or transfer pellets after separation of different sizes towards storage area. This subsystem has 3-units out of which one in standby and two are in parallel. The failure of a single unit has no impact on the system's regular operation. as standby unit comes into play. Capacity reduced when two units fails, failure of all three units results system shutdown.

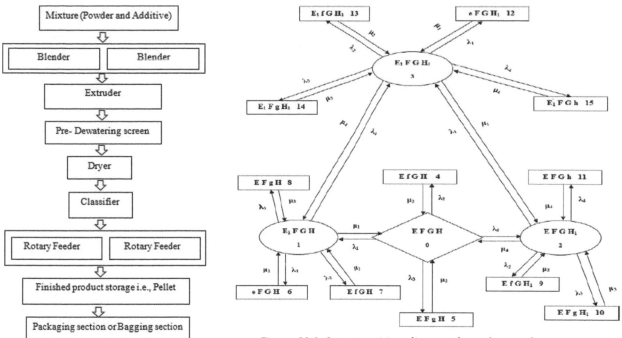

Figure 58.1 Block diagram for polypropylene system
Source: Author

Figure 58.2 State transition diagram for polypropylene system
Source: Author

7. Final product storage: The pellets are stored in storage area. This subsystem never fails.
8. Bagging or packaging section: Pellets are further sent to bagging area for packaging where bags are filled with 25 kg of pellets. This subsystem rarely fails.

Assumptions and Notations

For the purpose of creating transition diagrams and notations, the following assumptions are made.

1. Process parameters like failure and repair rate remains constant.
2. All the subsystems are in working mode and perfect initially.
3. After getting repaired, subsystem behaves new one.
4. At its early stages, maintenance is provided without delay.
5. The system covers three distinct states like working, reduced & failed.

Notations

In Table 58.1, the notations are listed.

Table 58.1 The notations related with transition diagram

E, F, G & H	Describe the operating states of blender, extruder, classifier and rotary feeder sub-system.
E_1 & H_1	Describe the reduced states of blender and rotary feeder subsystem.
e, f, g & h	Describe the failed states of blender, extruder, classifier & rotary feeder sub-system.
E_0 (b)	Probability of full capacity working states at time 'b'.
E_1 (b) – E_3 (b)	Probability of reduced capacity state.
E_4 (b) – E_{15}(b)	Probability of the unit in failed states.
$\lambda_1 \lambda_2 \lambda_3 \lambda_4$	Failure rates
$\mu_1 \mu_2 \mu_3 \mu_4$	Repair rates
d/db	Derivative w.r.t. time (b).
	Working state (full capacity)
	Reduced capacity state
	Failed state

Source: Author

Performance Evaluation

At the facility, the performance of the polypropylene unit was assessed using a probabilistic method based on the Markov birth-death process [1, 15]. In order to calculate the steady state availability of the afore-mentioned unit, mathematical equations related to the transition diagram are produced via this procedure with the use of data on failure and repair rates.

$$\left(\frac{d}{db} + \sum_{i=1}^{4} \lambda_i\right) E_0(b)$$
$$= \mu_1 E_1(b) + \mu_2 E_4(b) + \mu_3 E_5(b)$$
$$+ \mu_4 E_2(b) \tag{1}$$

$$\left(\frac{d}{db} + \sum_{i=1}^{4} \lambda_i + \mu_1\right) E_1(b)$$
$$= \mu_1 E_6(b) + \mu_2 E_7(b) + \mu_3 E_8(b) + \mu_4 E_3(b)$$
$$+ \lambda_1 E_0(b) \tag{2}$$

$$\left(\frac{d}{db} + \sum_{i=1}^{4} \lambda_i + \mu_4\right) E_2(b)$$
$$= \mu_1 E_3(b) + \mu_2 E_9(b) + \mu_3 E_{10}(b) + \mu_4 E_{11}(b)$$
$$+ \lambda_4 E_0(b) \tag{3}$$

$$\left(\frac{d}{db} + \sum_{i=1}^{4} \lambda_i + \mu_1 + \mu_4\right) E_3(b)$$
$$= \mu_1 E_{12}(b) + \mu_2 E_{13}(b) + \mu_3 E_{14}(b) + \mu_4 E_{15}(b) + \lambda_4 E_1(b) + \lambda_1 E_2(b) \tag{4}$$

$$\left(\frac{d}{db} + \mu_m E_i(b)\right)$$
$$= \lambda_m E_j(b) \tag{5}$$

For Equation (5) we have:

m = 1, i = 6, j = 1; i = 12, j = 3   m = 2, i = 4, j = 0; i = 7, j = 1; i = 9, j = 2; i = 13, j = 3

m = 3, i = 5, j = 0; i = 8, j = 1; i = 10, j = 2   m = 4, i = 11, j = 2; i = 14, j = 3; i = 15, j = 3

Initial conditions at time b = 0 are:

$E_i(b) = 1$ if i = 0 $E_i(b) = 0$ if i ≠ 0

Put d/db = 0 at t → ∞ in Equations (1) to (5) and simplify it recursively to attain the long-term availability of polypropylene unit and we get:

$$E_4 = \left(\frac{\lambda_2}{\mu_2}\right) E_0 ; \quad E_5 = \left(\frac{\lambda_3}{\mu_3}\right) E_0 ; \quad E_6 = \left(\frac{\lambda_1}{\mu_1}\right) E_1$$

$$E_7 = \left(\frac{\lambda_2}{\mu_2}\right) E_1 ; \quad E_8 = \left(\frac{\lambda_3}{\mu_3}\right) E_1 ; \quad E_9 = \left(\frac{\lambda_2}{\mu_2}\right) E_2$$

$$E_{10} = \left(\frac{\lambda_3}{\mu_3}\right) E_2 ; \quad E_{11} = \left(\frac{\lambda_4}{\mu_4}\right) E_2 ; \quad E_{12} = \left(\frac{\lambda_1}{\mu_1}\right) E_3$$

$$E_{13} = \left(\frac{\lambda_2}{\mu_2}\right) E_3 ; \quad E_{14} = \left(\frac{\lambda_4}{\mu_4}\right) E_3 ; \quad E_{15} = \left(\frac{\lambda_4}{\mu_4}\right) E_3$$

$$(\lambda_1 + \lambda_4) E_0 = \mu_1 E_1$$
$$+ \mu_4 E_2 \tag{6}$$

$$(\mu_1 + \lambda_4) E_1 = \mu_4 E_3$$
$$+ \lambda_1 E_0 \tag{7}$$

$$(\mu_4 + \lambda_1)E_2 = \mu_1 E_3$$
$$+ \lambda_4 E_0$$

(8)

$$(\mu_1 + \mu_4)E_3 = \lambda_4 E_1$$
$$+ \lambda_1 E_2$$

(9)

Solving above Equations 6-9 and assuming, we get

$$E_1 = ME_0 \qquad\qquad E_2 = NE_0 \qquad\qquad E_3 = RE_0$$
$$S_1 = \lambda_1 + \lambda_4 \qquad S_2 = \mu_1 + \lambda_4 \qquad S_3 = \mu_4 + \lambda_1 \qquad\qquad S_4 = \mu_1 + \mu_4$$

Now putting values of S_1, S_2, S_3 and S_4 and E_1, E_2 & E_3 in Equations 6-9: we have

$$S_1 E_0 = \mu_1(ME_0) + \mu_4(NE_0) = S_1$$
$$= \mu_1 M + \mu_4 N$$

(10)

$$S_2(ME_0) = \mu_4(RE_0) + \lambda_1 E_0 = S_2 M$$
$$= \mu_4 R + \lambda_1$$

(11)

$$S_3(NE_0) = \mu_1(RE_0) + \lambda_4 E_0 = S_3 N$$
$$= \mu_1 R + \lambda_4$$

(12)

$$S_4(RE_0) = \lambda_4(ME_0) + \lambda_1(NE_0) = S_4 R$$
$$= \lambda_4 M + \lambda_1 N$$

(13)

Further solving, we get values of M, N & R

$$M = (S_1 S_3 + \mu_1 \lambda_1 - \mu_4 \lambda_4)/\mu_1 S_3$$
$$+ S_2 \mu_1$$

(14)

$$N = (S_1 S_2 + \mu_4 \lambda_4 - \mu_1 \lambda_1)/\mu_4 S_3$$
$$+ S_2 \mu_4$$

(15)

$$R = \lambda_1 \lambda_4 (S_2 + S_3)/S_2 S_3 S_4 - S_3 \lambda_4 \mu_4$$
$$- S_2 \lambda_1 \mu_1$$

(16)

Now, using normalising conditions i.e., sum of all state probabilities is equal to one, we get

$$\Sigma_{i=0}^{15} Ei = 1, \text{ Hence } E_0 + E_1 + \cdots E_{15} = 1$$

Now putting the values of E_1, E_2, E_3 ... E_{15} in above equation we get, value of E_0.

$$(E_0 + ME_0 + NE_0 + RE_0 + \lambda_2/\mu_2 E_0 + \lambda_3/\mu_3 E_0 + \lambda_1/\mu_1 ME_0 + \lambda_2/\mu_2 ME_0 + \lambda_3/\mu_3 ME_0 + \lambda_2/\mu_2 NE_0$$
$$+ \lambda_3/\mu_3 NE_0 + \lambda_4/\mu_4 NE_0 + \lambda_1/\mu_1 RE_0 + \lambda_2/\mu_2 RE_0 + \lambda_3/\mu_3 RE_0 + \lambda_4/\mu_4 RE_0 = 1)$$

$$E_0 = 1/[(1 + M + N + R) + \lambda_2/\mu_2 + \lambda_3/\mu_3 + (\lambda_1/\mu_1 + \lambda_2/\mu_2 + \lambda_3/\mu_3)M + (\lambda_2/\mu_2 + \lambda_3/\mu_3 + \lambda_4/\mu_4)N$$
$$+ (\lambda_1/\mu_1 + \lambda_2/\mu_2 + \lambda_3/\mu_3 + \lambda_4/\mu_4)R]$$

To get the polypropylene unit's maximum steady state availability (AV) which is equal to the sum of all functioning and decreased capacity states, add the probabilities of each state.

$$Av = E_0 + E_1 + E_2 + E_3 \qquad\qquad Av = [1 + M + N + R]E_0 \qquad\qquad Av = 97.27\ \%$$

Having; $\lambda_1 = 0.000132$, $\mu_1 = 0.0039$; $\lambda_2 = 0.000123$, $\mu_2 = 0.0135$; $\lambda_3 = 0.000233$, $\mu_3 = 0.0104$; $\lambda_4 = 0.00036$, $\mu_4 = 0.0185$

Behavior Analysis

The system behavior analysed for each sub-systems by varying the available data. The information gathered from the maintenance department's history charts is utilised in availability equation for creating number of tables. Tables 58.2-58.5 depict the trend of availability on basis of failure and repair rates parameters and graphs from Figure 58.3-58.6 shows the behavior pattern for all subsystem in context of process parameters variation. In Table 6 optimal values for different subsystems are represented.

　　Due to availability of numerous combinations between failure and repair rate parameters in Tables 58.2-58.5, it observed that which subsystem's affect the availability more than any other on basis of available data. So, maintenance priorities are decided for each as shown in Table 58.7.

Optimisation Modeling

The process parameters are optimised with PSO algorithm for maximum availability of operating equipment. This algorithm is inspired by a flock of birds who are flying in multidimensional space in search of

Table 58.2 The system trend against variation in blender subsystem (E) parameters

λ_1/μ_1	0.0039	0.0055	0.0094	0.020325	0.03125	Constant values
0.000132	0.936268	0.93674	0.937061	0.937195	0.937217	$\lambda_2 = 0.00012$, $\mu_2 = 0.00278$
0.00018	0.935466	0.936329	0.936917	0.937163	0.937203	$\lambda_3 = 0.000233$, $\mu_3 = 0.0104$
0.000286	0.932911	0.935006	0.936449	0.93706	0.937159	$\lambda_4 = 0.00036$, $\mu_4 = 0.0185$
0.000488	0.925354	0.931014	0.935013	0.93674	0.937021	
0.00069	0.91481	0.925296	0.932903	0.936262	0.936815	

Source: Author

Table 58.3 The system trend against variation in extruder subsystem (F) parameters

λ_2/μ_2	0.00278	0.00379	0.00592	0.00971	0.0135	Constant values
0.000123	0.945472	0.954357	0.963513	0.970572	0.972715	$\lambda_1 = 0.000132$, $\mu_1 = 0.0039$
0.000168	0.934597	0.946474	0.958796	0.968352	0.971264	$\lambda_3 = 0.000233$, $\mu_3 = 0.0104$
0.000269	0.911445	0.929518	0.94854	0.963486	0.968075	$\lambda_4 = 0.00036$, $\mu_4 = 0.0185$
0.00047	0.870357	0.898833	0.929592	0.954348	0.962055	
0.000667	0.832813	0.87011	0.911386	0.945381	0.956109	

Source: Author

Table 58.4 The system trend against variation in classifier subsystem (G) parameters

λ_3/μ_3	0.0104	0.0147	0.0244	0.047915	0.07143	Constant values
0.0002325	0.936268	0.942039	0.947656	0.951877	0.953327	$\lambda_1 = 0.000132$, $\mu_1 = 0.0039$
0.0003134	0.929494	0.937177	0.944686	0.950348	0.952298	$\lambda_2 = 0.00012$, $\mu_2 = 0.00278$
0.00048	0.915849	0.927322	0.938628	0.947216	0.950186	$\lambda_4 = 0.00036$, $\mu_4 = 0.0185$
0.000753	0.894334	0.911612	0.928867	0.942129	0.946746	
0.001026	0.873808	0.896427	0.919307	0.937095	0.943331	

Source: Author

Table 58.5 The system trend against variation in rotary feeder subsystem (H) parameters

λ_4/μ_4	0.0185	0.024	0.033	0.044	0.055	Constant values
0.000359	0.936268	0.936397	0.936486	0.93653	0.93655	$\lambda_1 = 0.000132$, $\mu_1 = 0.0039$
0.000482	0.936014	0.936245	0.936406	0.936485	0.936521	$\lambda_2 = 0.00012$, $\mu_2 = 0.00278$
0.000733	0.935275	0.935802	0.936169	0.936351	0.936436	$\lambda_3 = 0.000233$, $\mu_3 = 0.0104$
0.00113	0.93353	0.934747	0.935603	0.93603	0.936229	
0.00153	0.931101	0.933266	0.934801	0.935572	0.935934	

Source: Author

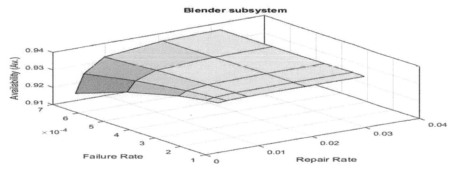

Figure 58.3 The system trend against variation in blender subsystem (E) parameters
Source: Author

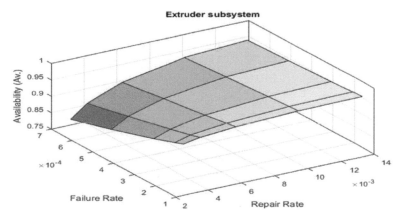

Figure 58.4 The system trend against variation in extruder subsystem (F) parameters
Source: Author

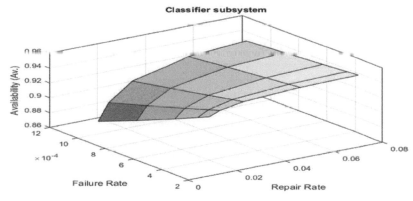

Figure 58.5 The system trend against variation in classifier subsystem (G) parameters
Source: Author

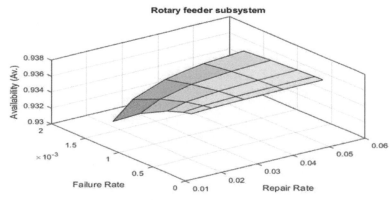

Figure 58.6 The system trend against variation in blender subsystem (H) parameters
Source: Author

Table 58.6 Optimal values for different subsystems

Subsystem	Failure rate(λi)	Repair rate(μi)	Maximum availability level
E	$\lambda_1 = 0.000132$	$\mu_1 = 0.03125$	93.72
F	$\lambda_2 = 0.000123$	$\mu_2 = 0.0135$	97.27
G	$\lambda_3 = 0.0002325$	$\mu_3 = 0.07143$	95.33
H	$\lambda_4 = 0.000359$	$\mu_4 = 0.055$	93.66

Source: Author

Table 58.7 Maintenance priority for Pellet manufacturing subsystems [15]

Subsystem	Increase in failure rates	Decrease in availability	Increase in repair rates	Increase in availability	Maintenance priority
Blender (E)	0.000132-0.00069	2.15	0.0039-0.03125	2.20	4th
Extruder (F)	0.000123-0.000667	11.26	0.00278-0.0135	12.33	1st
Classifier (G)	0.0002325-0.001026	6.23	0.0104-0.07143	6.95	2nd
Rotary feeder (H)	0.000359-0.00153	0.516	0.0185-0.055	0.488	3rd

Source: Author

food. They have social behavior, self-motivation as well as group motivation. This algorithm works upon position and velocity of flying birds. The individual bird is a particle and has personal best value (p-best). The number of birds in group are considered as population size. Each bird has fitness value and objective function for calculating this value. The position and velocity continuously updated of each particle (bird). That's why it works on iteration. The particle who has personal best value in a group would be considered global best (g-best) and will be the optimum solution for the problem.

With $N_i = (N_{i,1} + N_{i,2} + \ldots + N_{i,n})$ and $F_i = (F_{i,1} + F_{i,2} + \ldots + F_{i,n})$, the initial velocity and location for the ith particle are represented where n is considered as population size and p shows iteration number. The following PSO equation for updating the velocity and position of particle in space used:

$$N_i^{p+1} = w^*N_i^{p+1} + k_1^* l_1^* (\text{p-best}_i - F_i^p) + k_2^* l_2^* (\text{g-best}_i - F_i^p)$$

PSO algorithm is executed with certain parameters which are also mentioned in above equation. In this w is inertia weight, k_1 and k_2 are cognitive and social parameters, l_1 and l_2 numbers are randomly selected. The iterations stops when same values of availability start to come continuously or instantly trend of coming value changes. The block diagram of PSO algorithm is represented in Figure 58.7 and variables are indicated in Table 58.8. In study failure and repair rate parameters are selected (8). The following parameters with ranges are considered and given below:

λ_1: [0.000132-0.00069] λ_2: [0.000123-0.000667] λ_3: [0.0002325-0.001026] λ_4: [0.000359-0.00153]
μ_1: [0.0039-0.03125] μ_2: [0.00278-0.0135] μ_3: [0.0104-0.07143] μ_4: [0.0185-0.055]

The flow chart for PSO algorithm is mentioned below:

Result and Discussion

The availability of polypropylene system is computed as 97.27 % through mathematical modeling with the aid of Markov process. The following leading process parameters such as: $\lambda_1 = 0.000132$, $\mu_1 = 0.0039$; $\lambda_2 = 0.000123$, $\mu_2 = 0.0135$; $\lambda_3 = 0.000233$, $\mu_3 = 0.0104$; $\lambda_4 = 0.00036$, $\mu_4 = 0.0185$ are utilized for calculation of availability value. Further, the process parameters are optimized to get maximum availability of the system using MATLAB R2017a tool. The optimized optimised value of availability is obtained 97.608 % when particle size of 45 and iteration size was 100 as described in Table 58.9. At this level, the following process parameters provide the optimal combination between failure and repair rates are:

$\lambda_1 = 0.0002$, $\mu_1 = 0.0126$; $\lambda_2 = 0.0002$, $\mu_2 = 0.0135$; $\lambda_3 = 0.0004$, $\mu_3 = 0.0367$; $\lambda_4 = 0.0009$, $\mu_4 = 0.0249$.

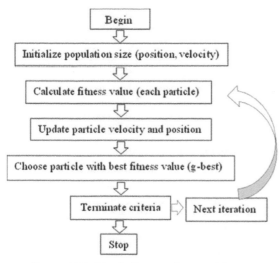

Figure 58.7 Block diagram for particle swarm optimisation algorithm
Source: Author

Table 58.8 Particle swarm optimisation variables [17]

S. No.	PSO (variables)	Value
i.	Cognitive factor (k1)	1.5
ii.	Cognitive factor (k2)	1.5
iii.	Random Number (l1)	0.9706
iv.	Random Number (l2)	0.0318
v.	Particle size (P.S)	05-50 (Step size '5')
vi.	Iteration size (I.S)	10-100 (Step size '10')
vii.	Inertia weight (w)	0.7299

Source: Author

Table 58.9 Highlights the impact of particle size on availability of pellet manufacturing system through PSO

Particle size (PS)	Failure rate				Repair rate				Av.
	λ_1	λ_2	λ_3	λ_4	μ_1	μ_2	μ_3	μ_4	
5	0.0004	0.0003	0.0009	0.0013	0.0081	0.0084	0.0384	0.043	0.94172
10	0.0 004	0.0004	0.0002	0.0006	0.0128	0.0119	0.0443	0.0319	0.96338
15	0.0004	0.0004	0.0002	0.0006	0.013	0.0119	0.0455	0.032	0.96346
20	0.0004	0.0004	0.0002	0.0006	0.0131	0.0119	0.0458	0.0321	0.96348
25	0.0002	0.0002	0.0004	0.0008	0.0127	0.0133	0.0374	0.0254	0.97602
30	0.0002	0.0002	0.0004	0.0008	0.0127	0.0134	0.0374	0.0254	0.97603
35	0.0002	0.0002	0.0004	0.0008	0.0127	0.0134	0.0374	0.0254	0.97603
40	0.0002	0.0002	0.0004	0.0008	0.0127	0.0134	0.0374	0.0254	0.97603
45	0.0002	0.0002	0.0004	0.0009	0.0126	0.0135	0.0367	0.0249	0.97608
50	0.0002	0.0002	0.0004	0.0009	0.0126	0.0135	0.0366	0.0249	0.97608

Source: Author

The reflection of varying particle size is indicated on Figure 58.8. The optimized value of availability is obtained 97.603 % when iteration size of 70 and constant particle size was 45 as described in Table 58.10. At this stage the optimal combination of failure and repair rate process parameters are $\lambda_1 = 0.0002$, $\mu_1 = 0.0130$; $\lambda_2 = 0.0002$, $\mu_2 = 0.0134$; $\lambda_3 = 0.0004$, $\mu_3 = 0.038$; $\lambda_4 = 0.0008$, $\mu_4 = 0.0254$. The reflecti-on of varying iteration size is indicated on Figure 58.9. The result was also shared with maintenance authorities for consideration.

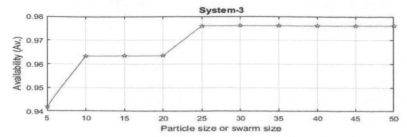

Figure 58.8 Highlights the impact of particle size on availability of pellet production system
Source: Author

Table 58.10 Highlights the impact of iteration size on availability of pellet manufacturing system through PSO

Iteration size (IS)	Failure rate				Repair rate				Av.
	λ_1	λ_2	λ_3	λ_4	μ_1	μ_2	μ_3	μ_4	
10	0.0003	0.0003	0.0005	0.0009	0.0142	0.012	0.0385	0.0253	0.97101
20	0.0002	0.0002	0.0004	0.0008	0.0118	0.0134	0.0362	0.0252	0.97515
30	0.0002	0.0002	0.0004	0.0008	0.0127	0.0134	0.0371	0.0252	0.97573
40	0.0002	0.0002	0.0004	0.0008	0.0127	0.0134	0.0373	0.0252	0.97599
50	0.0002	0.0002	0.0004	0.0008	0.0127	0.0134	0.0374	0.0254	0.97601
60	0.0002	0.0002	0.0004	0.0008	0.0127	0.0134	0.0375	0.0255	0.97602
70	0.0002	0.0002	0.0004	0.0008	0.013	0.0134	0.038	0.0254	0.97603
80	0.0002	0.0002	0.0004	0.0008	0.013	0.0134	0.038	0.0254	0.97603
90	0.0002	0.0002	0.0004	0.0008	0.013	0.0134	0.038	0.0254	0.97603
100	0.0002	0.0002	0.0004	0.0008	0.013	0.0134	0.038	0.0254	0.97603

Source: Author

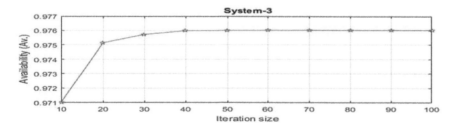

Figure 58.9 Highlights the impact of iteration size on availability of pellet production system
Source: Author

Conclusion

The critical points of analysis are concluded in this section and are mentioned below:

- In behavior analysis of the four sub-systems, it has been noticed that extruder (F) subsystem is contributing more in context of availability as compared to another subsystems. When failure rates rise, availability falls by 11.26%, and when repair rates rise, availability rises to 12.33%. That's why first ranking shall be provided to it during maintenance planning as also highlighted in Table 7. After mathematical modeling of the system, the availability is computed as 97.27% whereas optimal value through optimisation algorithm is observed as 97.608%. Results are compared and an increase of 0.34% in availability has been noticed with the use of PSO algorithm. The findings are discussed with maintenance staff of the plant for enhancing the life of assets through perfect strategy of maintenance programs.

References

[1] Arora, N. and Kumar, D. (1997). Availability analysis of steam and power generation systems in the thermal power plant. Microelectronics Reliability. 37(5), 795–99.

[2] Kumar, S., Tewari, P. C., and Kumar, S. (2009). Performance evaluation and availability analysis of ammonia synthesis unit in a fertilizer plant. Journal of Industrial Engineering International. 5(9), 17–26.

[3] Garg,D., Kumar, K., and Singh, J. (2010). Availability analysis of a cattle feed plant using matrix method. International Journal of Early Childhood Special Education. 3(2), 201–19.

[4] Zhang, C. and Mostashari A. (2011). Influence of component uncertainty on reliability assessment of systems with continuous states. International Journal of Industrial and Systems Engineering. 7(4), 542–52.

[5] Yuan, L. and Meng, X. (2011). Reliability analysis of a warm standby repairable system with priority in use. Applied Mathematical Modelling. 35(9), 4295–4303. doi: 10.1016/j.apm.2011.03.002.

[6] Sharma, S. P. and Garg, H. (2011). Behavioural analysis of urea decomposition system in a fertiliser plant. International Journal of Industrial and Systems Engineering. 8(3), 271–97.

[7] Adhikary, D. D., Bose, G. K., Chattopadhyay, S., Bose, D., and Mitra, S. (2012). RAM investigation of coal - fired thermal power plants: a case study. International Journal of Industrial Engineering Computations. 3, 423–34.

[8] Kumar, S. P. and Yadav, M. (2012). The weakest t - norm based intuitionistic fuzzy fault- tree analysis to evaluate system reliability. ISA Transactions. 51, 531–38. doi: doi:10.1016/j.isatra.2012.01.004.

[9] Garg, H. (2013). Reliability analysis of repairable systems using Petri nets and vague Lambda-Tau methodology. ISA Transactions. 52(1), 6–18. doi: 10.1016/j.isatra.2012.06.009.

[10] Kumar, G., Jain, V., and Gandhi, O. P. (2013). Availability analysis of repairable mechanical systems using analytical semi-Markov approach. Qual Eng. 25(2), 97–107. doi: 10.1080/08982112.2012.751606.

[11] Aggarwal, A. K., Kumar, S., Singh, V., and Garg, T. K. (2015). Markov modeling and reliability analysis of urea synthesis system of a fertilizer plant. Journal of Industrial Engineering International. 11(1), 1–14. doi: 10.1007/s40092-014-0091-5.

[12] Ting, C., Wu, K., and Chou, H. (2014). Expert systems with applications particle swarm optimization algorithm for the berth allocation problem. Expert Systems with Applications. 41(4), 1543–50. doi: 10.1016/j.eswa.2013.08.051.

[13] Kiilunen, J. and Frisk, L. (2015). Reliability analysis of an ACA attached flex-on-board assembly for industrial application. Soldering & Surface Mount Technology. 26(2), 62–70. doi: 10.1108/SSMT-03-2013-0007.

[14] Zhang, H., Innal, F., Dufour, F., and Dutuit, Y. (2014). Piecewise deterministic Markov processes based approach applied to an offshore oil production system. Reliability Engineering and System Safety. 126, 126–34.

[15] Kumar, P. and Tewari, P. C. (2017). Performance analysis and optimization for CSDGB filling system of a beverage plant using particle swarm optimization. International Journal of Industrial Engineering Computations. 8, 303–14. doi: 10.5267/j.ijiec.2017.1.002.

[16] Lin, D., Jin, D., and Chang, D. (2020). A PSO approach for the integrated maintenance model. Reliability Engineering and System Safety. 193, 106625. doi: 10.1016/j.ress.2019.106625.

[17] Malik, S. and Tewari, P. C. (2020). Optimization of coal handling system performability for a thermal power plant using PSO algorithm,. Grey System: Theory and Application. 10(3), 359–76. doi: 10.1108/GS-01-2020-0002.

Chapter 59

Machining of titanium alloys by using micro abrasive jet machine: an experimental investigation

Vinod V. Vanmore[1] and Uday A.Dabade[2]

[1]Assistant Professor, Sanjeevan Engineering & Technology Institute Panhala, Kolhapur, Maharashtra, India

[2]Professor, Walchand College of Engineering, Sangli, Maharashtra, India

Abstract

Ceramics, silicon, glass, titanium and nickel alloys, and other difficult-to-cut materials are now widely used in the MEMS, electronic device, and aerospace industries. The increased cost is due to the machining of these materials. One of these materials' most convenient micromachining technologies is micro abrasive jet machining (MAJM). This method has several distinct advantages, including a small heat-affected zone, low cutting forces, high machining versatility, and high flexibility. Fine abrasive particles (aluminum oxide or silicon carbide) and highly compressed air or gas (helium, nitrogen, or air) are directed on the target surface via a fine nozzle in this machining process. The abrasives exiting the nozzle at high speeds impinge on the target surface, causing material removal due to erosive action. This method had a very high etching rate compared to other micro-fabrication techniques. Furthermore, it does not require a clean room environment, making it particularly appealing for low-cost industrial practices for machining difficult-to-cut materials. This research aimed to create MAJM for difficult-to-machine materials like the titanium alloy (Ti-6Al-4V) plate. The new design and fabrication of the Laval nozzle were first reported in order to increase the machining productivity of micromachining. The circular cross-sectional nozzle was designed for high-speed, precise etching and patterning on difficult-to-machine materials. Using Taguchi's design of experiment methodology, this study investigates the effect of various parameters such as air pressure, abrasive size, and standoff distance on machining performance. The analysis of variance (ANOVA) method was used to determine the significance of each factor. The developed MAJM experimental setup investigates whether the Laval nozzle reduces the dimensional variation of the machined hole.

Keywords: ANOVA, MAJM, material removal rate, Taguchi method, Ti-6Al-4V

Introduction

Titanium and nickel-based ferrous and super alloys, ceramic materials, composite materials, and cobalt-chromium alloys have all been developed in recent decades for high-strength, heat-resistant applications in the automobile, aviation, nuclear, healthcare, and electronic sectors. All these materials are stronger and harder than typical engineering materials. However, these materials' applications are currently limited because converting the final component costs half the total cost of the product. This is because of the decreased cutting speed and shallower depth of cut caused by excessive tool wear. As a result, these materials are categorised as difficult to cut. For these materials, traditional machining processes are inefficient. Many attempts have been made in recent years to improve material machinability through the use of external energy-assisted machining. Among the numerous external energy-assisted machining processes, micro abrasive jet machining (MAJM) has piqued the interest of metal-cutting researchers. Much research has been conducted in recent years. Machining superalloys and refractory metals has become critical in order to meet the demands of ever-increasing jet engine technology [1]. The complex designs of jet engine parts present machining challenges that are beyond the capabilities of traditional machining procedures. One such issue is drilling small deep holes in superalloys. Non-traditional (or advanced) machining techniques are well suited to creating cost-effective holes in such cases. Other conditions driving these advanced machining technologies (AMPS) for drilling include micro-drilling holes.

Metals are particularly interesting because they are relatively low-cost substrates, are widely available, recyclable, disposable, and have good structural strength. Titanium alloys are especially intriguing due to their current biomedical applications. As a result, the goal is to create and test a system capable of predicting the geometric evolution of micro holes in titanium alloy (Ti-6Al-4V) . For drilling purposes, "micro" is typically considered as large as 2 mm (0.078 in.), and here selected, micro holes on difficult-to-machine materials from various machining processes for experimentation. Aerospace, automotive, medical, electronic, optics, jewellery, printed circuit boards, semiconductors, and mechanical watches all use microholes.

Workpiece Materials and Experimental Procedure

Titanium alloy

MAJM was used to machine difficult-to-machine materials such as Ti-6Al-4V. Titanium and its alloys are appealing materials due to their high strength-to-weight ratio at high temperatures and excellent corrosion

DOI: 10.1201/9781003450252-59

Table 59.1 Chemical composition of the titanium alloy

Element	Al	V	C	Mn	Si	Cr	Fe	Sn	Ti
Content (wt.%)	5.98	3.91	0.006	0.0021	0.008	0.010	0.04	0.002	Remaining

Source: Author

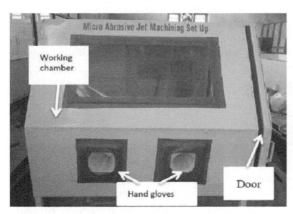
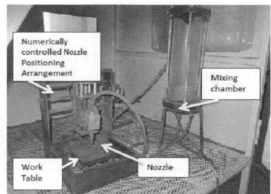

Figure 59.1 Developed MAJM setup
Source: Author

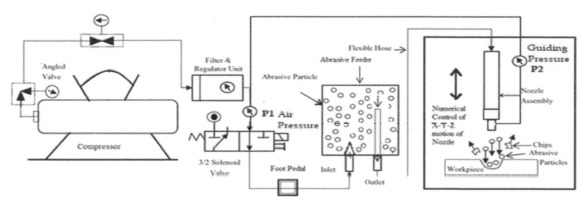

Figure 59.2 Shows the schematic of MAJM
Source: Author

resistance. Titanium's primary application has been in the aerospace industry. However, market trends have documented a shift from military to commercial and aerospace to industries. Titanium and its alloys, on the other hand, have poor thermal properties and are classified as difficult-to-machine materials [2]. For the purposes of experimentation, the plate thickness was set at 1.5 mm. Table 59.1 shows the chemical composition. The Ti-6Al-4V sheet size is 50 mm 25 mm 1.5 mm with a better surface finish.

Experimental Procedure

The basic mechanical structure, which measured 3 × 3 × 5 feet, was built. The MS sheet was used to machine and weld the fluidised mixing chamber (abrasive feeder) and working chamber. Abrasive particle openings in an abrasive feeder are tight and leak tight. Figure 59.1 depicts the MAJM experiment setup that was created. The Laval nozzle [3], which is made of 750-micron tungsten carbide, can withstand a variety of operations. Better machining results are provided by the concentricity of the abrasive jet and the uniform mixing of the particle flow perpendicular to the processing direction.

Figure 59.2 displays the MAJM schematic. The compressed air gas cylinder in the circuit is activated by an angled valve. The straight valve was used to set the desired pressure. The pressure gauges P1 and P2 in the pneumatic lines display the cylinder pressure and the guiding pressure for the nozzle stream, respectively. The compressed air is delivered via air conditioning equipment, such as a filter-regulator unit, which eliminates moisture and provides dry air to the abrasive feeder via a solenoid valve that switches the operation on and off instantly [4]. The abrasive particles in the abrasive feeder benefit from the dry compressed air. The kinetic energy of the air-abrasive mixture is sufficient for it to exit the cylinder. The nozzle is injected with a high-energy air-abrasive mixture, resulting in a high-velocity stream of the air-abrasive mixture. An air-abrasive combination's high-velocity jet strikes a work surface and removes material by erosion. Silicon carbide (SiC) particles with sizes of 50 and 100 microns are used. The nozzle is made of tungsten carbide and has a diameter of 750 microns. Air pressure (P1) and guiding air pressure (P2), stand-off distance, and abrasive size are critical micro-AJM process parameters.

Calculation for Response Variables

Table 59.2 shows the calculations and their evaluation procedure [18] in this work. The machining of Ti_6Al_4V is considered in the current study. The workpiece sample for the hole-cutting operation was a plate with dimensions of 50 mm 25 mm 1.5 mm. Following the test, the samples were cleaned with high-pressure air. Micro holes are measured using a Dino-Lite Premier Digital Microscope (AM3713TB). Display the formulas and methodology used to calculate experimental values.

Taguchi Method-Based Design of Experiments

Design of experiment (DoE), a Taguchi system component, is a useful tool for designing experiments. Dr. Genechi Taguchi was a researcher in Japan's Electronic Control Laboratory in the late 1940s, where he conducted extensive research using DoE methods. He worked hard to make this experimental technique more user-friendly before applying it to improve the quality of manufactured items (simple to apply). The Taguchi technique, also known as the Taguchi approach, is Dr. Taguchi's standardised version of DoE, first introduced in the United States in the early 1980s [16]. Today, engineers in a wide range of industrial processes use it as one of the most effective tools for improving quality. Taguchi's Taguchi approach may be used to meet budget-friendly problem-solving and product/process design optimization projects. Engineers, scientists, and researchers who are aware of and use this approach can significantly reduce the time required for experimental studies. As a result, many industrial firms and researchers employ Taguchi-based DoE in their research. The Taguchi methodology [5] can also be used to optimise process parameters. Nozzle pressure, stand-off distance, volume flow rate, mixture, abrasive grain size, abrasive geometry, abrasive particle type, and nozzle diameter are all factors to consider for previously investigated abrasive jet process parameters for machining of brittle materials [6–8]. However, air pressure, stand-off distance, and abrasive particle size were determined to be the most important characteristics [9–13]. The abrasive particles infringing the specimen resulted in a rough workpiece surface once the pressure exceeded 0.9 bar. When the SOD is less than 1 mm, many abrasive particles enter the nozzle tip, making it difficult to effectively cut the work surface; when the SOD exceeds 3 mm, the jet divergence increases, resulting in poor machining. When the abrasive particle size is less than 50 microns, the material removal rate is reduced, and the nozzle is blocked when the particle size is greater than 100 microns. As a result, certain uncut portions of the workpiece's top and bottom surfaces were visible. Table 59.3 shows the range of parameters for testing that were chosen based on previous research findings.

The Taguchi L9 orthogonal array was used in the experiments. This experiment employed three variables and three levels. Nine experiments with nine levels were carried out Table 59.4. The L9 orthogonal array for Ti-6Al-4V material is shown, along with the experimental results. Figure 59.3 depicts micrographs of Ti-6Al-4V with micro holes. A review of the literature and laboratory trials aided in identifying the most important abrasive jet parameters that influence machining time, material removal rate (MRR), hole characteristics (top hole, bottom hole), and hole taper angle. The most important aspects are highlighted below. Air pressure (bar), standoff distance (mm), and abrasive size (m) are all measured. When the pressure was

Table 59.2 Response variables measurement

Sr. No	Response variable	Formula	Elements
1	Material removal rate (MRR) (mg/min)	$MRR = \dfrac{(Wb - Wa)}{Machining\,Time(\,T)}$	Where, Wb = Weight before machining and Wa = Weight after machining, T = Machining time in minutes
2	Taper angle (θ), (degree)	$\theta^0 = Tan\text{-}1(DT\text{-}DB/2h)$	where, DT = Top hole diameter, DB = Bottom hole diameter h = thickness of plate = 1.5 mm
3	Radial over cut (ROC) (µm)	$ROC = \dfrac{((DT + DB)/2) - DN]}{2}$	where, DT = Top hole diameter, DB = bottom hole diameter and DN = nozzle diameter 750 µm
4	Micro hole diameter and machining time	Dino-lite premier digital microscope (AM3713TB) and STOP Watch respectively	

Source: Author

Table 59.3 Process parameters and their levels

Sr. No	Parameters	Level 1	Level 2	Level 3
1	Air pressure (bar)	7	8	9
2	Standoff distance (mm)	1	2	3
3	Abrasive size (SiC) µm	50	100	Mix (50% of 50 + 50% of 100)

Source: Author

Table 59.4 L9 Orthogonal array and experimental results for titanium alloy

Sr. No.	Process (input) parameters			Response (independent) variables					
	Air pressure (bar)	SOD (mm)	Abrasive size (µm)	MRR (mg/min)	Hole diameter (µm)		ROC (µm)	Time (min)	Taper angle (θ^0)
					Top	Bottom			
1	7	1	50	2.33	1122	936	165.5	6.45	7.9
2	7	2	100	3.48	1150	926	202.5	3.45	9.5
3	7	3	MIX	4.92	1032	852	146	3.25	7.7
4	8	1	100	6.76	903	701	76	3.4	8.6
5	8	2	MIX	5.41	998	824	130.5	2.59	7.4
6	8	3	50	4.24	888	732	112	4.25	6.7
7	9	1	MIX	2.61	1032	852	146	6.14	7.7
8	9	2	50	4.24	999	832	132.75	4.25	7.1
9	9	3	100	3.69	1024	796	130	5.15	9.7

Source: Author

less than two bar, the particle velocity was sufficient to reach the bottom surface of the workpiece, resulting in an uncut region.

Experimental Results Analysis and Discussion

For response variable analysis, the Minitab 16 is used. To investigate how different responses to process parameters are affected, ANOVA (at a 95% confidence level) and Main Effect Plots are used. To confirm the significance of process parameters, the probability value p is used. For a process parameter to significantly affect the response variable at the 95% confidence level, the p-value should be less than 0.05. The effect of process parameters on various reactions is discussed in detail in the sections that follow. According to the ANOVA results, the three parameters, air pressure, SOD, and abrasive size, all have a significant impact on the micro-hole characteristics.

Effect of process parameter on MRR in Titanium alloy

Figure 59.4 shows the effect of pressure on titanium material removal rate during machining (Ti-6Al-4V). The rate of material removal increases initially with increasing pressure and then decreases, yielding an optimum value. The energy of impact increases as the pressure increases, resulting in a greater depth of penetration in the workpiece and a faster rate of material removal. This increase in material removal rate with pressure should be maintained. The abrasive flow rate, on the other hand, decreases with increasing pressure, and thus the material removal rate. Furthermore, titanium has been shown to absorb oxygen, hydrogen, and nitrogen, which may cause embrittlement [14]. When compared to steel, titanium is also known to be notch sensitive [15]. All of these variables could have influenced the increased erosion rate. According to the ANOVA results (Table 59.5), air pressure, which reached 48.35%, had a significant impact on overall performance. The contribution percentages for stand of distance and abrasive size were lower, at 2.50% and 11.05%, respectively. A stand of distance has the lowest contribution percentage, most likely 2.50%.

Effect of process parameter on ROC in Titanium

Figure 59.5 depicts the effect of pressure on radial overcut during Ti-6Al-4V machining. As air pressure rises, ROC tends to decrease, which is influenced by abrasive particle size. However, once the optimum value is reached, the ROC increases as air pressure increases [17]. SOD is a more important factor for radial

Experiment no. 1	Experiment no. 2	Experiment no. 3
Top dia. – 1122 μm	Top dia. – 1150 μm	Top dia. –1032 μm
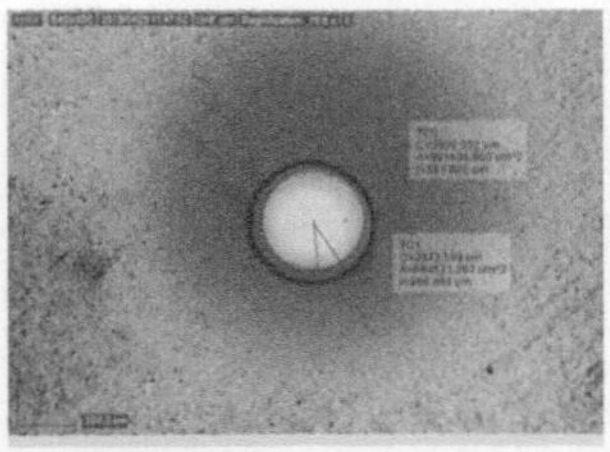	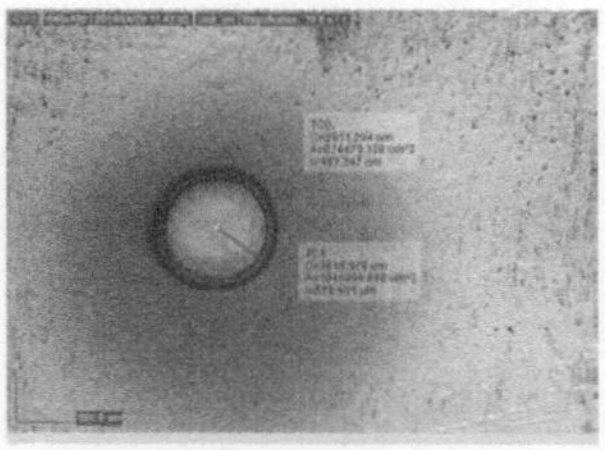	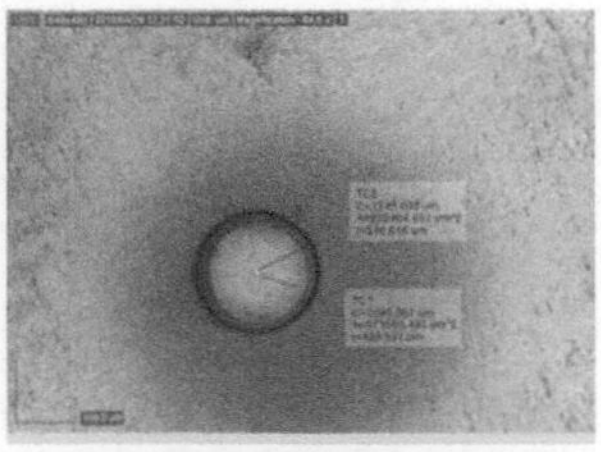
Bottom dia.– 936 μm	Bottom dia.– 926 μm	Bottom dia.– 852 μm
Experiment no. 4	Experiment no. 5	Experiment no. 6
Top dia. – 903 μm	Top dia. – 998 μm	Top dia. – 888 μm
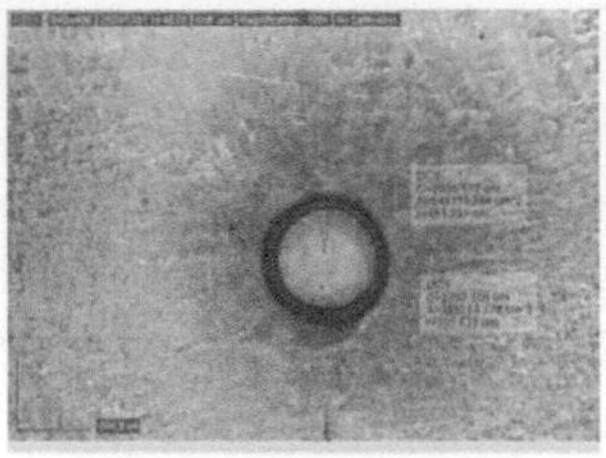	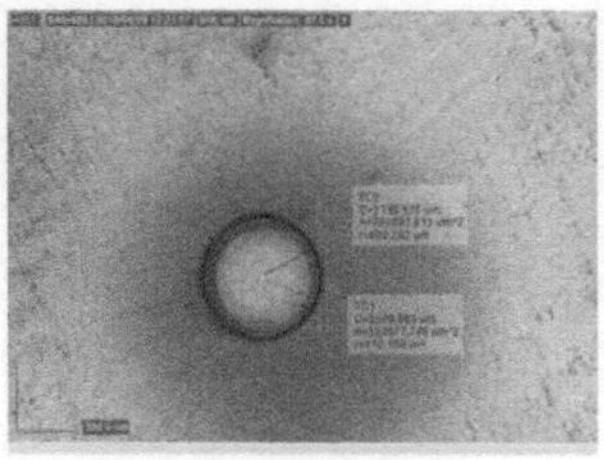	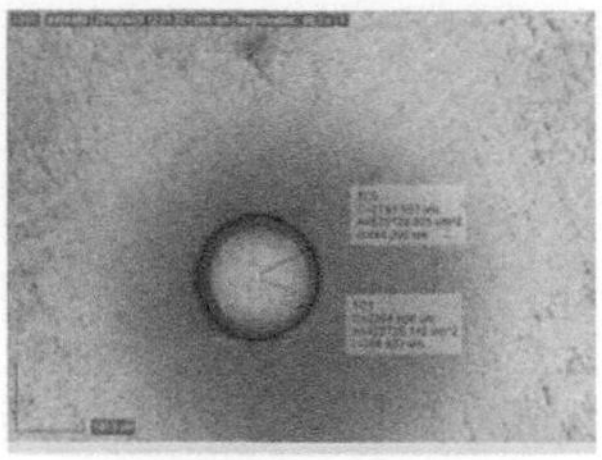
Bottom dia.– 701 μm	Bottom dia.– 824 μm	Bottom dia.– 732 μm
Experiment no. 7	Experiment no. 8	Experiment no. 9
Top dia. – 1032 μm	Top dia. – 999 μm	Top dia. – 1024 μm
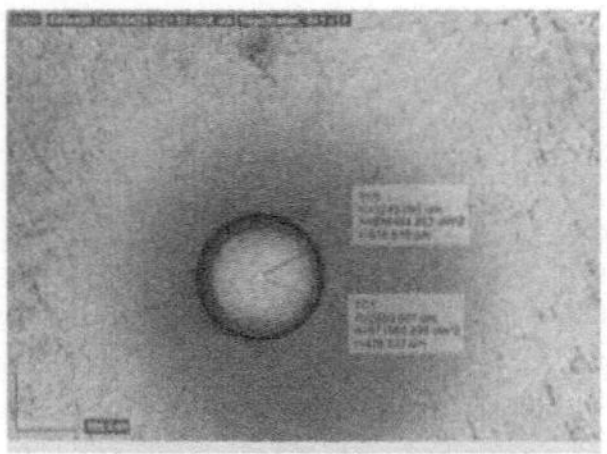	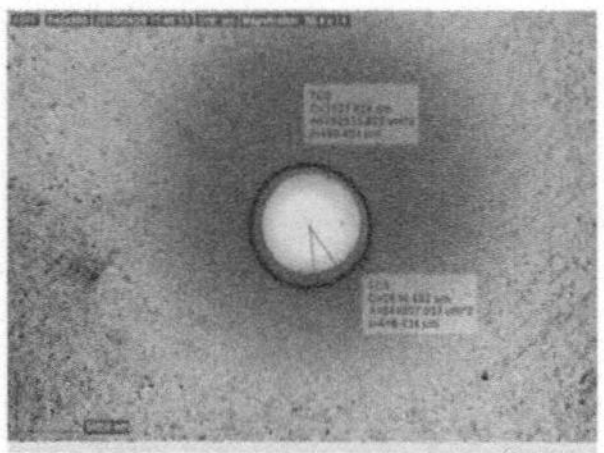	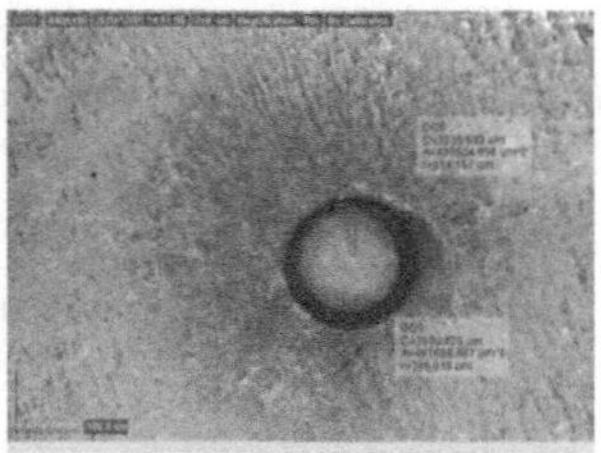
Bottom dia.– 852 μm	Bottom dia.– 832 μm	Bottom dia.–796 μm

Figure 59.3 Micrographs of micro holes on titanium
Source: Author

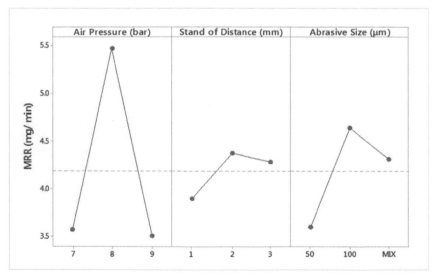

Figure 59.4 Main effect plot for MRR on titanium alloy
Source: Author

Table 59.5 ANOVA for MRR on titanium alloy

Source	DF	Seq SS	Contribution	Adj SS	Adj MS	F-value	P-value
Air pressure (bar)	2	7.4173	48.35%	7.4173	3.7086	1.27	0.441
Stand of distance(mm)	2	0.3829	2.50%	0.3829	0.1914	0.07	0.939
Abrasive size(µm)	2	1.6946	11.05%	1.6946	0.8473	0.29	0.775
Error	2	5.8465	38.11%	5.8465	2.9232		
Total	8	15.3412	100.00%				
		R-sq=85.83%	R-sq(adj)=78.68%				

Source: Author

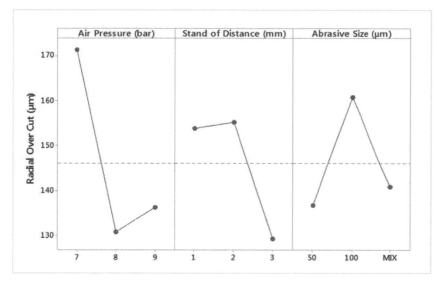

Figure 59.5 Main effect plot for ROC on titanium
Source: Author

Table 59.6 ANOVA for ROC on titanium

Source	DF	Seq SS	Contribution	Adj SS	Adj MS	F-Value	P-Value
Air pressure (bar)	2	2900.4	53.55%	2900.4	1450.2	11.83	0.078
Stand of distance(mm)	2	1273.9	23.52%	1273.9	637.0	5.20	0.161
Abrasive size(µm)	2	996.7	18.40%	996.7	498.3	4.07	0.197
Error	2	245.1	4.53%	245.1	122.5		
Total	8	5416.1	100.00%				
		R-sq=84.80%	R-sq(adj)=76.77%				

Source: Author

overcut because increasing SOD increases ROC and MRR due to an increase in particle kinetic energy. At the nozzle exit, only the carrier gas has the highest kinetic energy. As abrasive particles strike the workpiece with a lot of kinetic energy, a lot of material is removed. When SOD increases, the jet flares, resulting in an increase in ROC and taper angle. The ANOVA results for ROC in Ti-6Al-4V (Table 59.6) show that air pressure had the greatest impact on overall performance, reaching 53.55%. Air pressure and abrasive size contribute less, at 23.52% and 18.40%, respectively.

Effect of process parameter on taper angle in titanium alloy

Figure 59.6 show the effect of air pressure and abrasive size on taper angle during Ti-6Al-4V machining. The taper angle decreases as air pressure increases, which is influenced by the size of the abrasive particles. However, once the ideal value is reached, the taper angle increases with increasing air pressure. SOD influences the taper angle because as SOD increases, so does the taper angle and ROC due to an increase in particle kinetic energy. At the nozzle exit, only the carrier gas has the highest kinetic energy. The results of the ANOVA for taper angle in Ti-6Al-4V (Table 59.7) show that abrasive size, accounting for 35.87% of the

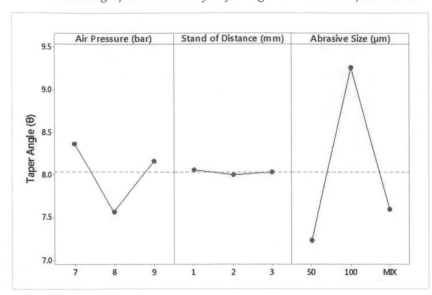

Figure 59.6 Main effect plot taper angle on titanium alloy
Source: Author

Table 59.7 ANOVA for taper angle on titanium

Source	DF	Seq SS	Contribution	Adj SS	Adj MS	F-value	P-value
Air pressure (bar)	2	1.4156	32.50%	1.4156	0.7078	4.79	0.173
Stand of distance(mm)	2	1.0822	24.85%	1.0822	0.5411	3.66	0.215
Abrasive size(μm)	2	1.5622	35.87%	1.5622	0.7811	5.29	0.159
Error	2	0.2956	6.79%	0.2956	0.1478		
Total	8	4.3556	100.00%				
		R-sq=75.80%	R-sq(adj)=70.73%				

Source: Author

overall performance. Air pressure and standoff distance have lower contribution percentages, at 32.50% and 24.85%, respectively.

Conclusions

According to the literature gap and research objectives, a micro abrasive jet machining setup with a unique Laval nozzle was developed, and experimental work was performed on difficult-to-cut materials such as titanium alloy (Ti-6Al-4V). To understand the quality and characteristics of the machined micro hole, macroscopic examinations of machined micro holes were performed. Material removal rate machined micro hole characteristics (top and bottom micro hole diameter, Taper angle, and so on), and machining time were all investigated. The following are the key research findings and contributions of this investigation:

- The developed MAJM setup is unique for difficult-to-machine materials due to the new Laval nozzle's extremely high jet velocity and inherent concentricity. For the analysis of process parameters and response variables, micro-hole abrasive jet machining of Titanium plates with SiC abrasive particles of 50 μm, 100 μm, and a mixture of both with equal volume was used. The maximum percentage variation of micro-hole diameter reported for Ti_6Al_4V at different levels of process parameters is 25.13%. The variation is due to flaring of the air abrasive jet as well as material properties. The new Laval nozzle significantly reduces machining time. Within the scope of the experimental work performed, the investigation reveals that the minimum machining time for titanium is 2.59 minutes at a thickness of 1.5 mm with various levels of process parameters. The taper angle observed of the machined micro holes for Ti-6Al-4V seems to be lower with the novel Laval nozzle set-up developed for MAJM. A higher level of pressure and lower level of stand-off distance increase the material removal rate and reduce variation in machined micro-hole diameter within the scope of experimental work.

References

[1] Jain, V. K. (Ed), (2013). Micromanufacturing processes. CRC Press.

[2] Zhu, D., Zhang, X., and Ding, H. (2013). Tool wear characteristics in machining of nickel-based superalloys. International Journal of Machine Tools and Manufacture. 64, 60–77. doi: 10.1016/j.ijmachtools.2012.08.001.

[3] Vanmore, V. V., and Dabade, U. A. (2018). Development of laval nozzle for micro abrasive jet machining [MAJM] processes. Procedia Manufacturing. 20, 181–6. doi: 10.1016/j.promfg.2018.02.026.

[4] Verma, A. P., and Lal, G. K. (1984). An experimental study of abrasive jet machining. International Journal of Machine Tools and Manufacture. 24(1), 19–29. doi: 10.1016/0020-7357(84)90043-X.

[5] Cavazzuti, M. (2013). Design of experiments, optimisation methods: from Theory to Design. Heidelberg: Springer Berlin, pp. 13–14.

[6] Balasubramaniam, R., Krishnan, J., and Ramakrishnan, N. (2000). Empirical study on the generation of an edge radius in abrasive jet external deburring (AJED). The Journal of Materials Processing Technology. 99(1), 49–53. doi: 10.1016/S0924-0136(99)00350-7.

[7] Nouhi, A., Spelt, J. K., and Papini, M. (2018). Abrasive jet turning of glass and PMMA rods and the micro-machining of helical channels. Precision Engineering. 53, 151–62. doi: 10.1016/j.precisioneng.2018.03.010.

[8] Nouraei, H., Wodoslawsky, A., Papini, M., and Spelt, J. K. (2013). Characteristics of abrasive slurry jet micro-machining: A comparison with abrasive air jet micro-machining. The Journal of Materials Processing Technology. 213(10), 1711–24. doi: 10.1016/j.jmatprotec.2013.03.024.

[9] Unde, P. D., Gayakwad, M. D., Patil, N. G., Pawade, R. S., Thakur, D. G., and Brahmankar, P. K. (2015). Experimental investigations into abrasive waterjet machining of carbon fiber reinforced plastic. Journal of Composites. 2015, 1–9. doi: 10.1155/2015/971596.

[10] Dhanawade, A., and Kumar, S. (2017). Experimental study of delamination and kerf geometry of carbon epoxy composite machined by abrasive water jet. Journal of Composite Materials. 51(24), 3373–90. doi: 10.1177/0021998316688950.

[11] Nazari, M., Rashidi, S., and Esfahani, J. A. (2019). Mixing process and mass transfer in a novel design of induced-charge electrokinetic micromixer with a conductive mixing-chamber. The International Communications in Heat and Mass Transfer. 108, 104293. doi: 10.1016/j.icheatmasstransfer.2019.104293.

[12] Dehnadfar, D., Friedman, J., and Papini, M. (2012). Laser shadowgraphy measurements of abrasive particle spatial, size and velocity distributions through micro-masks used in abrasive jet micro-machining. The Journal of Materials Processing Technology. 212, 137–1491. doi.org/10.1016/j.jmatprotec.2011.08.016

[13] Zhang, D., and Liu, L. (2011). A study on speed of fluid in swirling abrasive jet nozzle and drilling hole performance. 294, 3434–39. doi: 10.4028/www.scientific.net/AMR.291-294.3434.

[14] Hong, H., Riga, A. T., Gahoon, J. M., and Scott, C. G. (1993). Machinability of steels and Titanium alloys under lubrication. Wear. 162–164(PART A), 34–39. doi: 10.1016/0043-1648(93)90481-Z.

[15] Haşçalik, A., and Çaydaş, U. (2007). Electrical discharge machining of Titanium alloy (Ti-6Al-4V). Applied Surface Science. 253(22), 9007–16. doi: 10.1016/j.apsusc.2007.05.031.

[16] Roy, R. K. (2001). Design of experiments using the Taguchi approach: 16 step to product and process improvement. (4th Edition), John Wiley & Sons, pp. 120–25.

[17] Narutaki, N., Murakoshi, A., Motonishi, S., and Takeyama, H. (1983). Study on machining of Titanium alloys. CIRP Annals Technology. 32(1), 65–69.

[18] Tomy, A., and Hiremath, S. S. (2019). Machinability and characterisation of machined hole on quartz using developed μ-AJM set-up. Advances in Materials and Processing Technologies. 5(2), 242–57. doi: 10.1080/2374068X.2018.1564868.

Chapter 60

Experimental investigation of straightness error in 3 axis CNC machines using neural network

Jamuna Ravichandran[1,a], Mohanraj Manoharan[2,b] and Thilagham K T[2,c]

[1]Associate professor, government College of Engineering, Burugur, India

[2]Government College of Engineering, Salem, India

Abstract

Precision engineering is a key area of importance in the modern manufacturing industry 4.0 enhancement of accuracy and quality at low cost is the need of the hour. Artificial neural network (ANN) has been applied in calculating error compensation values for axis positioning in CNC machines by offline technique. The ANN and regression models were generated, and results are compared. Among the models developed, ANN showed better results with a mean square error of 0.12 and a mean absolute percentage error (MAPE) of 3.53. In this research, a laser Measurement system had been used to compute the straightness error in real-time 3- axis VMC as per VDI 3441 Germany standard. In order to reduce the straightness error, the back propagation neural network of the ANN tool was developed and applied. It was inferred from this work that, ANN could be more suitable for the prediction and determination of positioning and straightness errors.

Keywords: Accuracy, artificial neural network, laser interferometer, position error, precision, straightness error

Introduction

Accuracy at low cost is the prime focus. There is a linear relation between precision and cost. Engineers in the past decade have contributed much in balancing precision, cost, and quality. The three major contributors of volumetric errors are straightness errors, Backlash errors and squareness errors. There has been much work done in backlash errors and neural networks. Error elimination offline is time-consuming and increases the machine downtime. In reality, it is infeasible to eliminate the error totally. This paper provides a methodology to predict error.

Literature Review

As per Feng et al [27], the two straightness measurement methods are displacement method (±2.5%±16) micro meter and measurement method by angle (±0.2% μm). The method comprising of measuring aangle has a much higher effectiveness. Machine straightness error is deviating away from work piece straightness error.

Straightness error can be shown as

$$X = x[y] + x[z]$$

$$Y = y[x] + y[z]$$

$$Z = z[x] + z[y]$$

Ocafor et al [1] derived kinematic errors models for geometric and thermal error in VMC using homogeneous transformations mix. Ramesh et al [2, 13] compared and analysed different techniques of geometric error measurements and error origins. Neural network (NN) is the key source in developing compensation module [3]. Geometric accuracy tests had been done on VMC using dial gauge. Circularity error of spindle was measured by Renishaw Ball bar. The same software also measured vibration. [4]. VM101 linear encored measurement has been used to measure linear positioning error. Computation and prediction of backlash error had been done by artificial neural network (ANN) and compensation of error was done to improve accuracy of machine [5]. Angular errors [6] had been measured using laserautocollimator. Three sets of experimental data had been used to train ANN in MATLAB. Angular errors were lesser at the home position and higher at the end of the stroke [8, 9]. Laser interferometer had been used to measure individual error components. Online, error estimation had been done by integrating ANN and sensor. The kinematic model had been developed and error compensation had been done [10]. Error data sets had been measured using HP5010 Laser interferometer and geometric error modeling has been done using NN [11]

[a]jamuna.ravichandran2010@gmail.com, [b]mohanraj.cad@gmail.com, [c]thilagham.met@gmail.com

DOI: 10.1201/9781003450252-60

compensation method has been developed. Fines [12][14] implemented the compensation values in a real time machine compensation values in a real time machine and NN had proved a better result. Comparative study of squareness estimation has been done using laser interferometer and double ball bar [15]. By applying pitch error compensation method [16] on CNC driven axes, the positioning accuracy of machine tool had been improved. Using Renishaw laser measurement, positioning error of CNC Lathe, dynamic tests of CNC VMC, straightness error in a deep drilling machine and milling machine had been measured. According to two different methods, laser interferometry and CCD method have been computed [17]. The package comprising error compensation was inserted into the machine controller to improve the precision after measuring errors of CNC machine tool by laser interferometer [18]. After measuring the geometric error components using QC-20 ball bar, GA-PSO ie Genetic Algorithm was implemented to compute error values [19]. Parametrised error models of 14 errors using body diagonal method had been used to compute errors. OMM method can be employed to save time [20]. Magnetic ball bar had been used to measure 3D accuracy of a machine tool [21], [22]. Accuracy, the precision required for the error, the purpose of a measured machine tool, time and cost factors affect the choice of measuring system [23]. After measurement of errors in machine tools, diagnosis of error origins has been done. Extra NC program processor (ENPP) had been implemented on machine tool [24]. After measuring the positioning error by laser measuring system, Matlab data analysis functions had been used to compute positioning error of machine tool [25]. Condition monitoring methods using laser calibration and Ballbar test had increased the quality, accuracy, and productivity of machine tools [26]. A tuning method measures tracking error was introduced to match arial dynamics [27]. The health status of CNC Machines have been monitored by NN model [28]. Experiments have been done to improve contouring accuracy [29, 30] executing curved path with strong curvature.

According to Fines et al., positioning error was evaluated by NN. The results obtained by applying NN had been incorporated to correct the linear positioning error and backlash error. Measurement was carried on the machine (OMM) and it was used to inspect machined part instead of CMM.

Features of CNC Vertical Machining Center

This HMT VTC 800 Model was used to investigate the performance of CNC machine is shown Table 60.1.

Measuring tool: laser interferometer and its specifications

Renishaw laser interferometer system had been used to find linear error and error due to straightness. Renishaw laser measurement system comprises of XL-80 laser system, measurement optics, XC-80 compensator, and other metrology accessories. It is easy to carry and use. Tripod base has been used to mount XL-80. Laser system consists of two mirrors. The first retroreflector has been joined to beam splitter to get reference arm. The second retroreflector comprises the variable length measurement arm. Environmental conditions have been automatically compensated by XC-80compensator. After the alignment of the laser beam, it enters into optics used for measurement. Measurement optics has been fixed on three axis CNC vertical milling machine spindle by involving magnetic base. XC-80 compensator is automatically used to compensate the environmental conditions. In XL-80 laser system, the laser beam passes into the beam steerer.

After the aligning the laser beam and it enters into linear measurement optics. This linear measurement optics is fixed on the three axes CNC vertical milling center machine spindle using the magnetic base. The laser measuring system with Table 60.2 specifications has been used for experiments.

Table 60.1 Specifications of vertical milling machine

Particular	Characteristic
X-Axis	710 mm (max travel)
Y-Axis	500 mm (max travel)
Z-Axis	500 mm (max travel)
Main Spindle Power	5 kW
Spindle speed	50-400 rpm
Spindle taper	BT-40
Control System	SINUMERIK 840D
Number of Tools	24
X, Y, Z positional tolerance	0.01 mm

Source: Author

Table 60.2 Specification of laser interferometer system, Renishaw xl-80

Element	Characteristic
Linear resolution	1 nm
Linear measurement accuracy	±0.5 ppm
Maximum travel velocity	4m/s
Environmental range	0-40°C
Laser	Helium-neon
Dynamic capture rate	50 kHz
Linear range as standard	80 mm

Source: Author

Calibration process

The machining center has a working volume of 710 × 500 × 500 mm. Accuracy ,error and repeatability have been computed in three axis CNC machine has been in use for two to three hours with an exercise prior to measuring linear positioning error. It was done to calibrate the machine to decrease thermal error. According to Standards, the machine was subjected to transition state of thermal to reach steady state equilibrium. Vertical machining centre was made to go with a low feed rate of 50 mmpm. After the warm-up cycle, the displacement measurement was taken in both direction (forward and reverse.). It is said to be bidirectional measurements. As per VDI3441 Germany standard, each run consists of one forward stroke and one reverse stroke. Five runs have been measured as per VDI3441 for each axis. Linear positioning data has been recorded at equal and every 20mm intervals. The part program comprises of a sequence of moves which begins at X-axis in the forward direction (f) and returning to the starting position in opposite direction (r).

Straightness Error

Surface Straightness

It is a 2D tolerance of a cylinder along the axial direction.

Axis Straightness

It is 3D tolerance that controls the central axis of a workpiece preventing it from twisting. Straightness is a 2D version of flatness. Straightness error is the translational error of the machine elements in the two perpendicular directions other than its own axis. Straightness is a slight movement from the true line of travel orthogonal to the direction of travel in the horizontal plane.

Two types of straightness measurement

There are two straightness error namely horizontal straightness error (deviates in horizontal plane) and vertical straightness error (deviates in vertical plane). Axis straightness is closely linked to axis parallelism and axis perpendicularity because both are controlling a central axis with a cylindrical tolerance zone. Squareness is also expressed as biaxial straightness.

Symbol of Straightness

Surface straightness is found out on the surface of the workpiece. Axis straightness is the size dimension of the axis. Straightness errors can also be got by integration of the angular errors at certain points.

Methodology

Linear positioning error has been measured first shown in Figure 60.1. By handling laser interferometer on vertical machining machine Straightness error has been computed for each axis the form of horizontal and vertical straightness separately. The polynomial models of straightness error have been computed. Prediction of straightness error was done by ANN. The comparison has been done between ANN and experimental values. A systematic method was developed to measure the errors by incorporating the machine tool metrology principles.

Experimental procedure

The flowchart to measure errors are shown in Figure 60.1. The reference beam and measurement beam are reflected back by passing via the beam splitter forming interference fringes at ML10 laser head. The

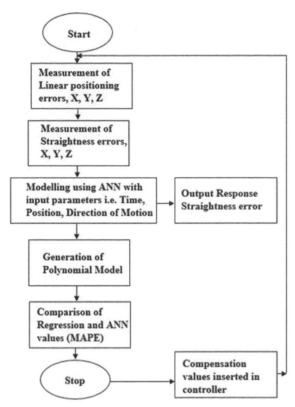

Figure 60.1 Systematic approach developed to measure the errors
Source: Author

path of one reference beam is fixed and second retroreflector changes position which is measured by fringe count. By changing the second retro reflector counts the number of fringes. Half the wavelength of laser beam multiplied by fringe number gives the length measured. At first linear positioning error is measured for X-axis ,the setting up procedure for Y axis is similar to the X-axis. In case of Z-axis, laser interferometer has been mounted on the table and angular retroreflector was kept With a feed rate of 300 rpm, CNC Machine has been made to wander in three axes simultaneously without rotation of spindle for the warm-up period of one hour. During the course of the warm-up cycle, three sensors measure the temperature of CNC machines. After each warm-up cycle, the machine has been intended to move in equal intervals (20 mm) in preplanned axis and dwell for four seconds to calculate linear positioning error.

Experimental Results

Axis positioning error

The linear positioning error of x-axis has been done. The graph has been plotted between X-axis nominal position in mm and error in µm. Error 1-f indicates the measurement of errors while the machine has been moving in the forward direction. Similarly, 1-r indicates the measurement of errors while the machine has been moving at the same point in the reverse direction. The linear positioning error of Z -axis and Y -axis are computed. The maximum displacement error in X-axis, Z-axis, and Y-axis is 7µm, 22 µm, and 9 µm respectively. Backlash error is the key reason.

Artificial neural network models of straightness error

By using feed forward back propagation method, straightness models have been created. Total positioning error at any point was measured by laser interferometer in workspace volume of predefined volume. Straightness errors measured by Renishaw laser measurement systems which was predicted by neural network. The MATLAB 8.1.0 version had been used for creating the ANN model. Fines et al [7] had discussed that compensation of the positioning error by using the neural network. This network comprises of three layers- one inner layer, one output layer, and one hidden layer. The input layer depicts the nominal position of the axis in mm, time in seconds and direction of motion of table with respect to the reference axis. The hidden layers of the network was used to know the features of the relationship between the nominal position in mm, time in seconds and direction of motion of table with respect to reference axis and straightness error of machine tool. The output layer represents the straightness error of the respective axis.

Levenberg-Marquardt propagation was selected as training function and Mean square error (MSE) was kept as performance function, This method has been adapted to minimise the backlash error [8].

Results and Discussion

Accuracy after Straightness Error Compensated

The source values for straightness errors in X, Zand Yaxis at varying the nominal position of the X, Zand Yaxis. The feed forward network architecture was used to model for a time in seconds, position in mm and direction of motion of table with respect to the reference axis, straightness errors in the axis is shown in Figures 60.2, 60.3, 60.4 and 60.5. The models had been trained using back propagation method of gradient

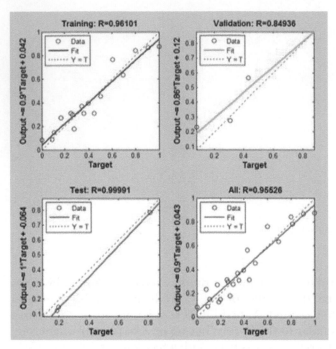

Figure 60.2 Fitness plots for y-axis vertical (forward)
Source: Author

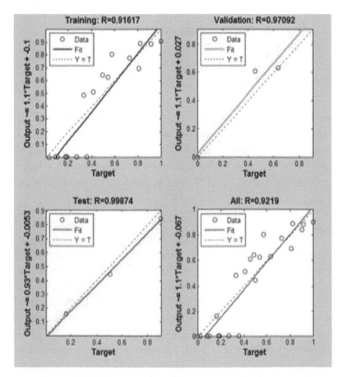

Figure 60.3 Fitness plots for y-axis vertical (reverse)
Source: Author

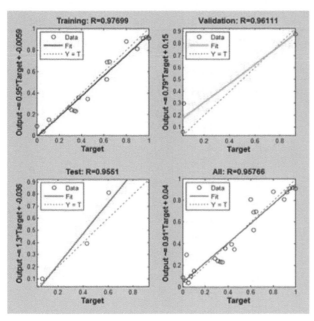

Figure 60.4 Fitness plots for y-axis horizontal (forward)
Source: Author

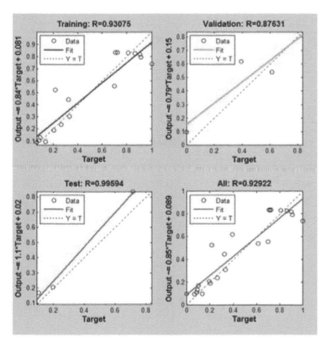

Figure 60.5 Fitness plots for y-axis horizontal (reverse)
Source: Author

descent method (GDE). The 1-2-1 back propagation algorithm has been used for training of straightness error in the axis. The training, validation and test data has been observed closely to fit line in fitness plot.

It was planned to use a learning process of 50 epochs and goal of 0. The ANN models have been trained and completion of learning processes has been reported at the epochs of 32, 30 and 26 in the axis respectively.

The polynomial models of straightness error

Polynomial functions were created for straightness errors (sex, sey, sez) with the position along each axis. Two straightness models were created for each individual axis and the equations used to create polynomial functions are expressed below:

X axis horizontal and vertical straightness are:

Table 60.3 Error percentage of polynomial and ANN model

Axis	Direction	Position (mm)	Exp Value	Regression Value	ANN	% of Error (REG)	% of Error (ANN)
X	Horizontal Forward	0	-3.5	3	-3.512	14.28	0.28
		580	6	6.634	5.8	10.56	3.3
	Horizontal Reverse	0	3.5	3.58	3.55	2.28	1.4
		580	5.6	6.2355	5.54	11.34	1.07
	Vertical Forward	0	0.5	0.52	0.49	4	2
		440	-6.4	5.3878	-6.43	15.9	0.46
	Vertical Reverse	0	-0.5	-0.54	-0.51	8	2
		440	-7.6	-6.298	-7.428	17.13	2.2
Y	Horizontal Forward	-355	-0.22	-0.19	-0.21	13.63	4.5
		45	-1.5	-1.5658	-1.49	4.38	0.6
	Horizontal Reverse	-355	-0.22	-0.190	-0.218	13.63	4.5
		45	1.5	1.32	1.52	12	1.3
	Vertical Forward	-395	1.5	1.397	1.47	7.3	2
		5	1.86	1.63	1.86	12.36	0
	Vertical Reverse	-395	-1.5	-1.55	-1.41	3.33	4
		5	0.86	0.725	0.85	15.6	1.1
Z	Horizontal Forward	-392	1	-0.86	0.98	14	2
		-92	1	0.864	0.98	13.6	2
	Horizontal Reverse	-392	2	1.8895	1.99	11.05	0.5
		-92	2	1.65	1.96	17.5	2
	Vertical Forward	-332	-1	0.9	-0.98	10	2
		28	1	0.75	0.98	25	2
	Vertical Reverse	332	1	0.78	0.84	22	16
		28	-2	-1.803	1.98	9.85	1

Source: Author

$$Sexh = 3.5 + 0.15X_1 - 0.026X_2 + 0.37X_3 \tag{1}$$

$$Sexv = 0.5 - 0.013X_1 + 0.063X_2 - 4.79X_3 \tag{2}$$

Y axis horizontal and vertical straightness are:

$$Seyh = -12.06 - 0.01X_1 + 0.04X_2 - 0.47X_3 \tag{3}$$

$$Seyv = 1.979 + 0.06X_1 + 0.06X_1 - 0.96X_3 \tag{4}$$

Z axis horizontal and vertical straightness are:

$$Sezh = 0.402 + 0.01X_1 + 0.001X_2 + 0.001X_3 \tag{5}$$

$$Sezv = 0.6877 + 0.001X_1 + 0.001X_2 + 0.001X_3 \tag{6}$$

The error has been analysed and data obtained was very close to the fitted regression line. The coefficient of determination (R2) values for each axis is given in Table 60.3. The calculation process of the straightness error value was very quick. The disadvantage of the polynomial model is its inferior accuracy.

Validation of ANN and Polynomial Model

The experimental values have been obtained from laser interferometer, the predicted values have been obtained from ANN models. The error percentage for ANN model and Polynomial model were calculated using the Equation 7 and 8

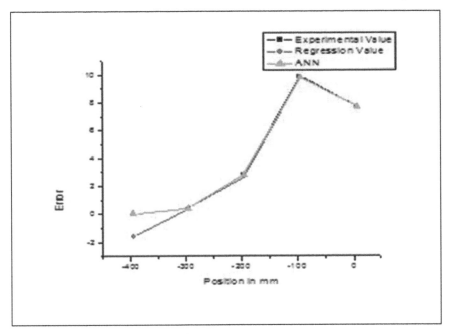

Figure 60.6 Y-axis horizontal (reverse)
Source: Author

$$\% \ of \ error \ for \ ANN = \frac{Exp.Value - Pred.Value}{Exp.Value} \times 100 \qquad (7)$$

$$\% \ of \ error \ for \ Polynomial = \frac{Exp.Value - Reg.Value}{Exp.Value} \times 100 \qquad (8)$$

The average absolute percentage of error in Polynomial model for straightness error in X, Y and Z axis was computed. The comparison plot for experimental, predicted and regression values for straightness errors in the axis are shown Figure 60.6.

The ANN models had low values of error compared to polynomial models. Lastly, the predicted error values have been inserted to the CNC controller to change the origin of the slides axes to improve the accuracy of the machine by error compensation.

Conclusion

The application of artificial intelligence in laser interferometer measurements was experimented using artificial neural network (ANN) and the results are compared with the conventional experimental models . The results are indicating that the neural network (NN) could act as a powerful tool for computing compensation values for linear errors in. Training and designing of a network had positively corrected major portion of straightness errors in CNC machines . Integration of compensation unit with real time application was fruitful.

A novel methodology has been developed to compute errors. Three different methods were studied and it is proved that three methods were made to achieve the final results successfully to satisfy the machine requirement. It is expected that some analysts had to be done previously to see which method could be well tailored to the application. Straightness error which is the major influential parameter in linear positioning error. The ANN model had predicted the straightness error in the respective X, Z and Y axis more accurately. Also, it showed minimum error percentage of straightness error in X, Y, and Z-axis. The average absolute percentage error of the ANN model was found to be lesser than 10 %. In the end, the predicted error values were sent to the CNC controller to change the origin of the slides axes to improve the accuracy of the machine by error compensation Hence, the accuracy ,reliability and performance of the machine are improved after the prediction and compensation of errors.

Acknowledgment

The authors would like to be thankful to Shri. V. A. P. Sharma, , Shri. S. S. Avadhani, Mrs. S. Sudha, Mr. Ramamohan and Mr. C. T. Shivaraj of Central Manufacturing Technology Institute, Bangalore for their support in error measurements using laser measuring system.

References

[1] Okafor, A. C., and Ertekin, Y. M. (2000). Vertical machining center accuracy characterization using laser interferometer, Part 1 linear positional errors. Journal of Material Processing and Technology. 105(3), 394–406.

[2] Ramesh, R., Mannan, M. A., and Poo, A. N. (2000). Error compensation in machine tools- a review Part-1: geometric, cutting-force induced and fixture dependent error. International Journal of Machine Tools and Manufacture. 40, 1235–56.

[3] Prakash, V., Narendra Reddy, T., Sajin, S., Shashi Kumar, P. V., and Narendranath, S. (2014). Real-time positioning error compensation for a turning machine using neural network. Procedia Material Science. 5, 2293–300.

[4] Anjali, N., Khandare, S. S., and Mehta, J. P. (2013). Accuracy enhancement of 3 axis vertical milling machine center. Proceedings of the Ist international and 16th national conference on Machine and mechanisms. (pp. 953–956).

[5] Liu, H., Xue, X., and Tan, G. (2010). Backlash error measurement and compensation on the vertical machining center. Engineering. 2, 403–407.

[6] Lavrov, E. A., Epikhin, V. M., Mazur, M. M., Suddenok, Y. A., and Shorin, V. N. (2015). Development of methods precision length measurement using transported laser interferometer. Physics Procedia. 72, 222–6.

[7] Fines, J. M., and Agah, A. (2008). Machine tool positioning error compensation using artificial neural networks. Engineering Application of Artificial Intelligence. 21, 1013–26.

[8] Demuth, H., and Beale, M. (2000). Neural network toolbox user's guide. The Mathworks Inc.

[9] Mahdavinejad, R. A. (2011). Prediction of angular errors on a vertical CNC Milling machine. Transactions of Mechanical Engineering. 35(M2), 181–95.

[10] Quafi, A. E., Guillot, M., and Bedrouni, A. (2000). Accuracy enhancement of multi-axis CNC machines through online neuro compensation. Journal of Intelligent Manufacturing. 11, 535–45.

[11] Tan, K. K., Huang, S. N., Lim, S. Y., Leow, Y. P., and Liaw, H. C. (2006). Geometrical error modeling and compensation using neural networks. IEEE Transactions on systems, Man and Cybernetics. 36(6), 797–809.

[12] Veldhulis, S. C., and Elbestawi, M. A. (1995). A strategy for the compensation of errors in five –axis machining. Annals of the CIRP 44(1), 373–7.

[13] Ramesh, R., Mannan, M. A., and Poo, A. N. (2000). Error compensation in machine tools- a review Part-II: Thermal error. International Journal of Machine Tools and Manufacture.40, 1257–84.

[14] Fines, J. M. (1995). Machine Tool Positioning Error Compensation using ANN. Ph.D. Thesis

[15] Lee, H. H., Lee, D. M., and Yang, S. H. (2014). A technique for accuracy improvement of squareness estimation using a double ball bar. Measurement Science and Technology. 25(9), 941–953.

[16] Kopac, J., Pusavec, F., and Kramar, D. (2013). How to improve the positional accuracy of HSC Machine tools. In 7th Brazilian Congress on Manufacturing Engineering.

[17] Begovic, E., Plancic, I., Ekinovic, S., and Ekinovic, E. (2014). Laser interferometry- measurement and calibration method for machine tools. In 3rd Conference. Maintenance, pp. 19–25.

[18] Barman, S., and Sen, R. (2010). Enhancement of accuracy of multi-axis machine tools through error measurement and compensation of errors using laser Interferometry Technique. Mapan. 25(2), 79–87.

[19] Linares, J. M., Chaves–Jacob, J., Schwenke, H., Longstaff, A., Fletcher, S., Flore, J., Uhlmann, E., and Wintering, J. (2014). Impact of measurement procedure when error mapping and compensating a small CNC machine using a multilateration Laser interferometer. Fraunhofer –Publica.

[20] Wang, H., Wang, J., Chen, B., Xiao, P., Chen, X., Cai, N., and Ling, B. W. K. (2015). Absolute optical imaging position encoder. Measurement. 67, 42–50.

[21] Bryan, J. B. (1982). A simple method for testing measuring machines and machine tools, part 1: Principle and Application. Precision Engineering. 4(3), 61–69.

[22] Bryan, J. B. (1982). A simple method for testing measuring machines and machine tools, part 2: Principle and Application. Precision Engineering. 4(3), 125–38.

[23] Ekinovic, S., Prcanovic, H., Begovic, E. (2013). Calibration of machine tools by means of laser measuring systems. Asian Transitions on Engineering. 2(06), 17–21. (ATE ISSN: 2221-4267).

[24] Rahman, M. (2004). Modeling and measurement of multi-axis machine tools to improve positioning accuracy in a software way. Oulu, Finland, Ph.D. Thesis.

[25] Zhang, Y., Chu, X., and Yang, S. (2015). Research of error detection and compensation of CNC machine tools based on laser Interferometer. In 2nd International Conference on Machining Materials Engineering.

[26] Naveen Kumar, B. K., Darshan Kumar, C. D., and Rachappa, C. T. (2013). Condition monitoring of CNC machine tool accuracy with renishaw equipment. International Journal of Innovation Research in science, Engineering and Technology. 2(12), 7861–66.

[27] Feng, W. L., Yao, X. D., Azamat, A., and Yang, J. G. (2015). Straightness error compensation for large CNC gantry type milling centers based on B- spline curves modeling. International Journal of Machine Tools and Manufacture. 88, 165–174.

[28] Bhatt, H. H., and Saradara, K. D. (2016). Controlling accuracy in CNC machine. International Journal of Engineering Development and Research. 4(1), 500–3.

[29] Jiang, W. (2020). Research on error monitoring model of CNC machine tool based on Artificial Intelligence. Journal of Physics: Conference Series. 1574, 012125.

[30] Convey, J. R., Darlong, A. L., Erresto, C. A., Farouki, R.T., and Palomann, C. A. (2013). Experimental study of contouring accuracy for CNC machines executing curved paths with constant and curvature-dependent feedrates. *Robotics and Computer Integrated Manufacturing.* 29, 357.

Chapter 61

Characteristics study on dissimilar friction stir welded AA6082 and AA2014 weldments

Thilagham K T[a], Jamuna Ravichandran[b] and Mohanraj Manoharan[c]

Government College of Engineering, Salem, India

Abstract

The purpose of this research is to investigate the properties of annealed aluminum alloys AA2014 and AA6082 that were joined using friction stir welding. The weld was characterised using tensile and metallurgical studies. The butt joined dissimilar alloy is welded with a high speed steel threaded pin at varying rotational and feeding rates. Thus, the high strength was achieved with a tool revs of 900 rpm and 40 mm/min traverse. The macrostructures clearly show weld nuggets, thermo mechanically affected zones, and heat affected zones. The weld nugget was revealed by an AA6082 alloy α-aluminum solid solution. The observations include narrow HAZ near AA6082 and also significant variation in precipitate size across the weld.

Keywords: Feeding rate, friction stir welding, tool rotational speed

Introduction

Friction stir welding creates a variety of high-strength joints for structural, automotive, and aerospace applications [1]. Because of potential flaws in the weld zone, such as solidification micro-structural and porosity, these aluminum alloys AA2014 and AA6082 cannot be welded using the fusion method. Tensile behavior also was significantly lower than the base material, while using fusion method. Friction stir welding (FSW) has a number of advantages over fusion welding, particularly in terms of strength.

Friction stir welding was used by Sameer M. D. et al. to join the AA6082-T6 aluminum alloy and the AZ91 magnesium alloy. By adjusting the different materials mostly on advance and retreating sides, the microstructure and mechanical characteristics of various joints were compared. The joint macro-structure also revealed the material mixing between joints that were similar and those that were dissimilar. The highest UTS, 172.3 MPa, was attained when Mg had been placed on AS, and the lowest UTS, 156.25 MPa, was attained when Mg had been placed on RS [2].

Anganan et al. [3] used FSW and conventional metal inert gas welding to examine the mechanical characteristics of AA-6082-T6 joints. Tests for tensile strength, metallography, and hardness were performed on the joints. When compared to MIG joints, the Friction stir welding joints had the best tensile strength. Additionally, it was noted that the HAZ of Weld joints was smaller in width compared to MIG joints. The outcomes demonstrated that FSW joints possess superior qualities to MIG joints [3].

Scialpi et al. [4] looked into the effect of different shoulder geometric shapes on the mechanical and microstructure properties of friction stir-welded joints. The base metal, AA6082, had a thickness of 1.5 mm, and welding had been performed using two separate pins with three different shoulder geometries at 1810 rpm of tool rotation. 460 mm/min of feed rate, and a plunge depth of 0.1 mm. Because it enhances the joint's transverse and longitudinal strength as well as the crown surface, the tool with both the fillet and cavity produced a clean crown with the least amount of flash in the experiment compared to the two other tools [4].

Minton et al. used a Parkson type vertical mill machine to use the weld on AA6082-T6 Aluminum plates that were 4.6 mm and 6.3 mm thick. Similar to this, Mynors (2006) looked into the viability of performing friction stir welding on a conventional milling machine. For the experiment, a shortened tool for 4.6 mm sheet and a single universal tool for 6.3 mm sheet with 19 mm radius of silver steel were created. The samples that need to be welded have been kept on the feed table with a baking plate forming a butt joint. The aluminum sheets' chosen t thicknesses are 4.6 mm and 6.3 mm. In FSW four levels, the spindle speeds were set to 1550 rpm. Following the peak feed speed of 3.175 mm/sec, the feed speed gradually decreased. In order to determine the minimum spindle speed, the feed was set to 0.2646 mm/s and the weld speed was decreased to 620 rpm [5].

Magdy et al. [6] conducted research on AA6082 FSW welds to characterise multiple passes. The 120 × 100 × 6 mm³ AA6082 plates were used. The tool used in FSW rotated at 850 rpm and it was performed in three passes that frequently overlapped. A tool with a 5 mm length and a 15 mm flat shoulder was used for this. The FSP runs were completed in a direction that was different to the rolling direction. The grain sizes increased together with the number of passes made on the joint, thus the tool's transverse speed along the

[a]thilagham@gmail.com, [b]jamuna.ravichandran2010@gmail.com, [c]mohanraj.cad@gmail.com

DOI: 10.1201/9781003450252-61

joint remained constant. It was found that the material loses strength properties and softens as the total number of passes is increased one at a time. On the other hand, as transverse speed increased, strength and hardness increased [6].

Cavaliere et al. [7] used a conical-shaped tool made of C40 steel for the experiment to assess the joints made between the dissimilar joints on AA6082 and AA2024. The tool advances at a rate of 80 to 115 mm/min while the rotational speed stays constant at 1600 rpm. The Vickers indenter with such a 5N head was used for 15 seconds to do micro hardness testing on the whole welded zone to identify its mechanical qualities. According to the report, advancing at a rate of 115 mm/min is sufficient for the dissimilar joints on AA6082 and AA2024 to reach their maximum tensile strength. It was concluded that by retaining 6082 on the forward-facing side, a uniform hardness profile with higher welding speed and increased tensile strength was obtained [7].

These studies imply the characterization on FSW of AA6082 at higher welding rpm with feed rate. Here AA 6082 plate form in T651 was heat treated by stretching and then artificially aged with 370MPa was chosen. T4 heat treatment was used on the AA2014. Additionally, it was found that the chosen profiles were broken by high rotating speeds under load, as indicated in Table 61.3. To withstand the load, the pin profile has been modified. The plates were then annealed to reduce hardness.

The AA6082 alloy was treated for 3 hours at 413°C (775°F), followed by controlled cooling at 10-26°C (50-500°F), followed by air cooling. Similarly, AA2014 alloy was also treated for 3 hours at 413°C (775°F). The alloy is then cooled every hour between 10 and 26°C (50 and 500 °F), followed by air cooling. Both plates were annealed prior to welding to reduce hardness to 50HV. The objective is to illustrate the metallurgical characteristics of an annealed weld.

This current work focuses on determining the FSW characterisation properties of annealed AA2014 and AA6082 plates. So, the AA2014 and AA6082 specimens were configured for the dissimilar joint. Further characterisation studies were performed on the FSW welded samples, including tensile tests, microstructure, macrostructure, and hardness surveys.

Materials and Methods

Material and sample preparation

The aluminum plates were cut into 25 × 25 mm^2 segments for chemical analysis. Table 61.1 shows the elemental analyses of the selected AA2014 and AA6082 as evaluated using spectroscopy.

After that, the AA2014 as well as AA6082 materials were machined to the precise specifications of 100 × 50 × 6 mm^3 in order to be fixed as in FSW milling machine. They were welded together with cylindrical taper pin profiles. Friction stir welding was used to create weldments with dimensions of 100 × 100 × 6 mm^3 using the welding parameters shown in Table 61.2.

Welding parameters

It was chosen so that AA6082 is on the advancing side and AA2014 is on the receding side to enhance strength properties. The plates also were friction stir welded using the above parameters. The penetration depth of 1mm was given to join the weld. The welding was carried out using the FSW-3T-Hydraulic milling machine. Figure 61.1 displays the configuration geometry for the FSW of the AA2014 and AA6082 weld. Further, the welds were prepared for mechanical testing.

Table 61.1 Chemical composition (Wt %)

	Si%	Fe%	Cu%	Mn%	Mg%	Zn%	Ti%	Al%
AA6082	0.063	0.286	4.155	0.58	0.521	0.019	0.104	Bal.
AA2014	0.77	0.236	4.49	0.65	0.67	0.67	0.103	Bal

Source: Author

Table 61.2 Welding parameters used for dissimilar FSW aluminum plates

Tool	Welding speed (mm/sec)	Rotational speed (rpm)
HSS (Threaded)	30	700
	40	900

Source: Author

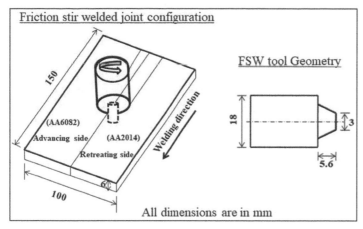

Figure 61.1 Configuration geometry of FSW of AA2014 and AA6082 weld
Source: Author

Table 61.3 Showing weld trial parameters

S. No	Tool rotational speed (rpm)	Traverse speed(mm/min)	samples
1.1	700	30	
1.2	900	40	
2.1	700	40	
2.2	900	30	

Source: Author

Mechanical studies

The tensile properties of the AA2014 and AA6082 welds were determined using ASTM A370, and they were sized in the as-welded condition at room temperature using AWS B4.0. The specimen dimensions have been cut in accordance with standards, and the testing was carried out on a universal testing unit with a capacity of 60 tonnes.

Microstructure studies

The specimen was polished in 220, 400, 600, and 800 emery sheets, and then cloth polished. The specimen was then etched with freshly made Keller's reagent, which contained 190 ml water, 5 ml HCl, 3 ml HF, and 2 ml HNO_3. The microstructures were photographed using a Metzer microscope. After completing the microstructure analysis, the specimen was etched with Tucker's reagent to reveal macrostructure details with the naked eye. The photograph of the macro structure was then taken. Micro hardness testing was performed across the welded joint in polished and etched sections in accordance with ASTM A370. The Zwick Microhardness device with 0.3 kg of pure aluminum was used to conduct the hardness test.

Results and Discussion

Sidhu et al. [8] used the threaded cylindrical profile pin to weld the plates. The optimum weld formed very little flush on the surface. When welds were welded at 30 mm/sec with a rotational speed of 700 rpm, the flush formed was minimal; however, when welds were welded at 40 mm/sec with a rotational speed of 900 rpm, the flush formed increased by 5%.

Tables 61.3 and 61.4 are illustrating the tensile test results of the AA2014 and AA6082 welds. The optimum value of 150 MPa was revealed for 40 mm/min as well as 900 rpm combination of weld parameters. Furthermore, the weld had exceptional weld strength properties. Layer by layer onion rings were also observed to form. The UTM failure mode was found in the area of the AA 6082 HAZ.

Table 61.4 Showing tensile test results

S. No	Ultimate tensile strength	Breaking load	Comments
1.1	136.66 MPa	4.1 kN	Fracture occurred near HAZ of the weld zone
1.2	143.33 MPa	4.3 kN	Fracture occurred at the middle of the weld zone
2.1	150.00 MPa	4.5 kN	Fracture occurred at the zone near AA6082
2.2	150.00 MPa	4.5 kN	Fracture occurred at the zone near AA6082

Source: Author

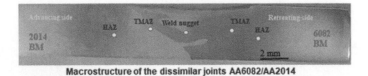

Macrostructure of the dissimilar joints AA6082/AA2014

Figure 61.2 Showing macrostructure of dissimilar joints on AA6082 and AA2014
Source: Author

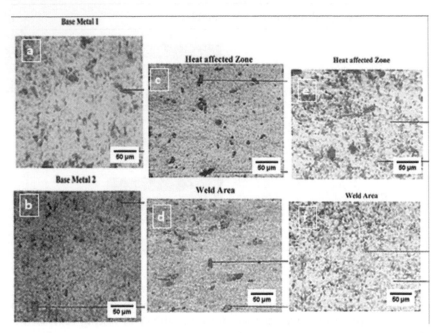

Figure 61.3 Showing microstructure of varies zones such as (a) Base metal AA2014, (b) Base metal AA6082, (c) HAZ zone near AA6082, (d) weld nugget AA6082, and (e) HAZ zone near AA2014, (f) weld nugget AA2014
Source: Author

The macrostructure of the AA2014 and AA6082 welds is depicted in Figure 61.2. All of the zones are clearly visible in the macrostructure, including the weld nugget (WN), the thermo-mechanically affected zone (TMAZ), and the heat affected zone (HAZ) (HAZ). This same solid solution of AA6082 dominates weld nugget. The TMAZ becomes extremely narrow near AA6082. It can be seen from both sides.

The microstructure of the AA2014 and AA6082 welds is depicted in Figure 61.3. Several zones were clearly defined. The weld nugget was mostly revealed with an α-aluminum solid solution of AA6082 alloy. The HAZ was narrow near AA6082. There were no differences in precipitate size revealed by the weld cross section [9, 10]. However, the precipitate sizes varied significantly across the weld.

Table 61.5 Hardness survey of dissimilar joints on AA6082 and AA2014 revealed the hardness results. The micro hardness in the weld nugget region was 48.2 HV due to stirring. Sadeesh et al., [11] found that the fine grains increased the strength of the weldment. Additionally, refined grains were discovered in this study at the nugget region.

Conclusions

The Friction stir welding (FSW) properties of aluminum alloy AA2014-AA6082 plates annealed to a lower 50HV are determined as follows. Both plates were successfully welded using friction stir welding. The weld strength in the given optimised condition was 150 MPa. The microstructure clearly showed all of the zones. The analysis reveals that precipitates are uniformly distributed in the weld center. Because of the stirring action, the hardness value in the weld region was relatively low. It reveals 48.2Hv near AA6082 in the weld center.

References

[1] Mishra, R. S. (2005). Mazy friction stir welding and processing. Material Science and Engineering R. 50, 1–78.

[2] Sameer, M. D., and Birru, A. K. (2019). Mechanical and metallurgical properties of friction stir welded dissimilar joints of AZ91 magnesium alloy and AA 6082-T6 aluminum alloy. Journal of Magnesium and Alloys. 7(2), 264–71.

[3] Anganan, K., Murali, J. G., Krishnan, M. M., and Marimuthu, K. (2014). Study of mechanical properties and experimental comparison of Mig and Friction stir welding processes for aa6082-t6 aluminum alloy. 2014 IEEE 8th International Conference on Intelligent Systems and Control (ISCO), pp. 74–78.

[4] Scialpi, A., De Fillippis, L. A. C., and Cavaliere, P. (2006). Influence of shoulder geometry on microstructure and mechanical properties of friction stir welded 6082 aluminum alloy. Materials and Design. 28, 1124–29.

[5] Minton, T., and Mynors, D. J. (2006). Utilisation of engineering workshop equipment for friction stir welding. Journal of Material Processing Technology. 177, 336–9.

[6] El-RRayes, M. M., and El-Danaf, E. A. (2012). The influence of multipass friction stir processing on the micro structural and mechanical properties of aluminum alloy 6082. Journal of Materials Processing Technology. 212, 1157–68.

[7] Cavaliere, P., De Santis, A., Panella, F., and Squillace, A. (2008). Effect of welding parameters on mechanical and micro structural properties of dissimilar AA6082-AA2024 joints produced by friction stir welding. Materials and Design. 30, 609–16.

[8] Sidhu, M. S., and Chatha, S. S. (2012). Friction stir welding–process and its variables: A review. *International Journal of Emerging Technology and Advanced Engineering*, 2(12), 275-279.

[9] Jones, M. J., Heurtier, P., Desrayaud, C., Montheillet, F., Alléhaux, D., and Driver, J. H. (2005). Correlation between microstructure and microhardness in a friction stir welded 2024 aluminium alloy. Scripta materialia, 52(8), 693–697.

[10] Hassan, K. A., Norman, A. F., Price, D. A., and Prangnell, P. B. (2003). Stability of nugget zone grain structures in high strength Al-alloy friction stir welds during solution treatment. Acta Materialia, 51(7), 1923–1936.

[11] Sadeesh, P., Kannan, M. V., Rajkumar, V., Avinash, P., Arivazhagan, N., Ramkumar, K. D., and Narayanan, S. (2014). Studies on friction stir welding of AA 2024 and AA 6061 dissimilar metals. Procedia Engineering, 75, 145–149.

Chapter 62

Effects of nanoparticles on emission control of IC engine – a review

Govind Murari[a], Binayaka Nahak[b] and Tej Pratap[c]

Motilal Nehru National Institute of Technology Allahabad, Prayagraj, India

Abstract

A growing population causes the expansion of the automotive industry, resulting in negative effects on human health, fuel economy, conventional energy sources, and air quality due to toxic emissions and greenhouse gases. Due to their high energy density, nanoparticles may result in slight improvements in these causes, since they boost catalytic activity in fluids, oxygen liberat ion ability for combustion and oxidation, and ease of slipping during tribropair to reduce friction. Meanwhile, nanoparticles allow easy atomisation, reducing viscosity, and shortening ignition delay; all of these aspects contribute to emissions control and improved performance of IC engines. Thus, the current study discusses the various forms of nanoparticles, including metallic, metallic oxide, and carbon nanoparticles, as well as their synthesis processes, including coprecipitation, hydrothermal, microemulsions etc., which have a profound impact on engine performance. This paper examines the effects of various nanoparticles in fluids and during lubrication in order to convert poisonous gases into nonpoisonous ones such as CO_2, H_2O, and N_2. A further brief investigation of the effect of these nanoparticles on engine performance variables (BSFC, BTE, NOx, HC, CO) is conducted based on their dispersion into lubricants and fuel. The characterisation of nanoparticles in liquid and solid media are examined in the interim to investigate the influence on stability and other physiochemical parameters.

Keywords: Characterisation, emissions control, lubrication, nanoparticles types, synthesis, tribological performance

Introduction

Nanotechnology revolutionised the 21st century due to its exceptional ability to transform the characteristics of lubricants, fuels, and coatings through the use of nanoparticles of different sizes and shapes. It usually deals with Nanoparticles, which are the tiny particles ranging from 1 to 100 nm. Back in 1970 and 1980, these particles were called ultrafine particles by Granqvist and Buhrman. Around 1990, the name nanoparticles were proposed by national nanotechnology (United States) [1, 2]. Globally, transportation consumes more than 20% of the world's energy and is estimated to increase by 30% by 2040, producing about 23% of carbon dioxide emissions and about 14% of greenhouse gas emissions [3, 4]. Meanwhile, traditional fuels such as petrol, diesel, and kerosene oil degrade the environment and shows various health risks. Additionally, diesel-biodiesel fuel blends exhibit several adverse effects such as low cloud point and pour point, poor fuel atomisation, low calorific value, and more NOx emissions in engines [5]. It is therefore very important to reduce poisonous emissions and improve engine performance from an energy saving, environmental, and human health protection point of view. Thus, research has shifted to using renewable sources of energy like biodiesel and nanoparticles in diesel or biodiesel (animal fats, vegetable oils) to reduce adverse environmental and health impacts. To reduce these detrimental effects, a variety of approaches are currently being employed. These include catalytic reduction, particulate filters, engine design modifications, and fuel additives. Nanoparticles are also used in engineering domains such as tribology, surface design, and micro-nano-manufacturing. Among various application these days, nanoparticles in the form of fuel additives are being focused in IC Engine because they enhance engine performance and reduce harmful emissions. Nanoparticles, in particular, improve combustion properties, protect engine components from corrosion, provide engine cleanliness, and enhance combustion properties [3–5]. As a result of nanoparticles' outstanding properties, they provide a larger surface area to volume ratio, allow for a faster catalytic reduction, and enhance catalytic activity [5]. The catalytic application of nanoparticles reduces poisonous emissions via thermocatalysis and photocatalysis. It is also important to note that various schemes for the synthesis of nanomaterials provide varying morphological properties of nanoparticles, such as their size and shape, which play a crucial role in the preparation of nano fluids, otherwise they can affect nano fluid stability, including thermophysical properties [1, 2]. Carbon nanoparticles in all forms (0D, 1D, 2D and 3D) are equally important because they improve tribological performance by promoting self-lubrication ability. Additionally, the forms such as SWCNT and MWCNT drastically enhance ignition rates, ignition delays, and combustion durations, thereby reducing emissions [5]. With the addition of alumina to B20 (20% biodiesel and 80% diesel), combustion time and ignition delay are significantly reduced, and peak pressure, cylinder pressure, and maximum heat release rate are improved, and HC and CO might be reduced by 26.72 and 48.43%, while NOx may be reduced by 11.27%. It was found that the

[a]govind.2020rme03@mnnit.ac.in, [b]binayaka@mnnit.ac.in, [c]tpratap@mnnit.ac.in

DOI: 10.1201/9781003450252-62

amount of carbon nanotube in diesel and biodiesel at different proportions enhanced the power by 3.67%, 8.12%, and 7.12%. However, the amount of NOx emissions was increased by 27.49% [5]. This work aims to determine whether different types of nanoparticles (metallic, metallic oxide, and carbon nanoparticles) improve the properties of dispersion media, such as fuel, and how they are synthesised and, in addition, whether they affect engine performance as a means of saving energy, protecting the environment, and protecting the human health.

Catalytic application of nanoparticles on emissions control

Metallic nanoparticles

There is special attention being given to these nanoparticles (NPs) for their improvement of combustion, tribological properties, and the reduction of emissions. These nanoparticles also have the capability of enhancing contact surface area and rapid oxidation, and are capable of producing zero-carbon energy. Because of their high energy density, nanoparticles such as Al, B, Mg, and Zr could be used in fuel or fuel additives to enhance the energy density of propellants and condensing explosives. Since metallic nanoparticles in the porous matrix have high surface areas, they are considered catalysts and their dispersion abilities help promote stability against migration and coalescence of nanoparticles [4]. A rapid development of nanoscience demands the encapsulation of metallic nanoparticles in nanoshells or nanopores to overcome stability issues of metallic NPs. Despite its high energy density, Beryllium is extremely hazardous and expensive, which limits its use [2]. When fuel is ignited, iron nanoparticles release rapid heat, indicating that they are oxidising rapidly. Due to these iron particles, the heat transfer rate is enhanced, resulting in faster fuel evaporation and shorter ignition delays, thus enhancing the probability of the fuel igniting. Due to mixed nanoparticles absorbing heat energy in the combustion chamber at a faster rate than diesel engines, this is quite possible. When 2 wt% of Al is added to ethanol, the ignition delay is reduced by 32%, mainly due to a greater thermal conductivity so that ignition takes place earlier. Adding Iron NPs increases viscosity, density, and calorific value, resulting in an improvement in BTE by 2%, a reduction in BSFC by 3.07%, and a reduction in HC, CO, and NOx of approximately 6, 12, and 11% [2, 4].

Metal oxide nanoparticles

In this category of nanoparticles, oxygen is liberated and extra oxygen is introduced into the combustion chamber. This results in a significant reduction of harmful exhaust emissions, which in turn has improved combustion and reduced ignition delay, resulting in an improved BTE and a decrease in BSFC. Although metals and metal oxides are harder, at the nanoscale they act as lubricants by providing rolling action and easily filling wear scars and preventing further damage. NPs CeO_2 as fuel additive exhibit redox properties of storing and releasing oxygen, i.e. oxygen deficient (CeO_2 into CeO_2-x) and reoxidised (CeO_2- x oxides into CeO2), i.e. a combination of these properties reduces NOx and provides oxygen for CO and HC [2, 4].

Carbon nanoparticles

Due to their superior thermal and chemical properties, CNTs, MWCNTs, graphene oxide, and graphene Nano platelets are used to enhance performance and reduce emissions in diesel or biodiesel blended fuels. The CNTs occupy the intermolecular space, which facilitates easier slipping as well as a reduction in viscosity, which enhances fuel atomization. They may also act as anti-knock agents by trapping free radicals [4].

Thermocatalysis

It usually focuses on the oxidation of poisonous emissions like in automobiles where a 3-way catalytic converter converts automobile exhaust into nontoxic compounds. $Pd@Ce_{0.5}Zr_{0.5}O_2/Al_2O_3$ nanoparticles can easily generate the hydroxyl group (OH) during high-temperature oxidation of CO, HC and NOx, which in turn promotes oxidation and three-way catalytic reactions, resulting in high catalytic activity and hydrothermal stability [2].

Photo catalysis

Usually used to remove organic pollutants from the environment by facilitating surface redox reactions. Photocatalytic activity is increased when metallic nanoparticles are used, such as silver nanoparticles and gold nanoparticles that increase the absorption of sunlight. Platinum nanoparticles accelerate surface redox reactions. By combining metal nanoparticles like Ag and Au with TiO_2, sunlight is absorbed more efficiently, and CO_2 is reduced, and organic pollutants are photodegraded.

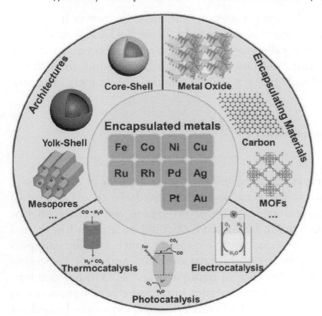

Figure 62.1 Schematic illustration showing encapsulated metal nanoparticles for catalytic applications [2]

Electrocatalytic reduction

CO_2 is converted into C_2+ through electrocatalytic reduction with H_2 that has an important demand in the chemical and energy industries, thus reducing the depletion of conventional sources of energy and reducing emissions [2]. A visualisation of the different forms of metallic nanoparticles as well as their action on emission control can be seen in Figure 62.1.

Schemes of nanoparticles synthesis

Wet chemistry routes, also known as bottom-up approaches, produce NPs in batches of self-assembling atoms, molecules, or clusters. Although this approach has a limitation in terms of the surface structure of the generated NPs, which might change their physiochemical properties. Furthermore, this method also consumes large amounts of energy to maintain high temperatures and pressures, which increases costs [6]. In top-down approaches, the bulk materials are converted into micro-nano ranges through the use of various mechanical, physical, and chemical methods to produce wide ranges of NPs. This is an environmentally friendly technique known as green. As a result, this approach is simpler, cheaper, and more readily available to production [6, 7]. Bottom up approaches: As a result of co-precipitation, water-soluble salts are mixed with water-insoluble salts in liquid medium and are capable of processing nanoparticles such as ZnO, NdF_3, Au, and hybrid nanomaterials such as $Au-Fe_2O_3$, Au-NiO, and $Au-CO_3O_4$. By using a chemical reaction between precursors and appropriate agents, the hydrothermal process is able to produce nano-TiO_2, $Au-TiO_2$ with a flower shape, $Pt-TiO_2$ with Ag-ZnO core-shell, nanotubes, nanorods, and graphene nanosheets with manageable morphological parameters [1, 6–8]. Specifically, sonochemical processes combine hydrolysis, thermolysis, and ultrasonic irradiation, thereby accelerating the reaction due to their combined non- invasive energy sources. Nano sized hollow iron oxide, ferromagnetic magnetite nanocubes can be prepared using this approach [7]. It is widely used to process nanocrystalline metals and alloys using inert gas condensation. The present approach relies on the condensation action of evaporating metal atoms to process nanocrystalline metals and alloys. Through collisions between metallic atoms with gaseous atoms, it loses energy, making it capable of processing nanomaterials such as gold-palladium [9]. Micro-emulsions are formed when three or more components are mixed and then react to form an emulsion, generally, the micro-emulsion contains polar and no polar phases, usually water and oil. Using the given method, metallic nanoparticles (Pt, Pd, Rh, Ir, Ag, and Au), metallic oxides (ZrO2, TiO2, SiO2, Fe_2O_3), and nano hybrid materials ($CdS-TiO_2$, $CdS-Ag_2S$, CdSe- ZnS, and CdSe-ZnSe, Au-Ag, Au-Pd, and Pd-Pt) can be processed. In the presence of surfactant and oxidant, a thermal decomposition method is used to prepare highly crystalline and monodisperse maghemite nanoparticles. In microwave assisted processes, electromagnetic waves with wavelengths between 1 mm and 1 m are used to heat the materials, which offers an advantage in terms of rapid processing, reduced reaction times, and high energy efficiency, and is particularly useful for iron oxide, magnetite, and hematite nanoparticles. The reduction of metal ions in solution, in particular, makes use of a variety of chemical reducing agents while also using a stabilising agent to produce nanoparticles of varied sizes and shapes [6, 10, 11]. Top down approaches: Through bombarding action, sputtering forms

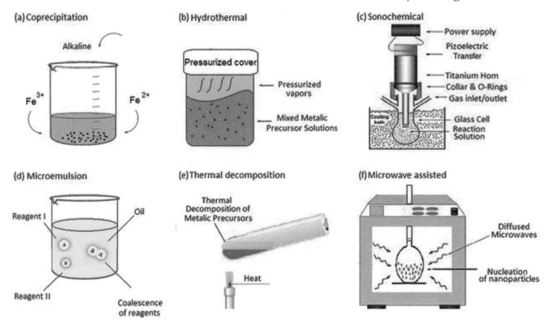

Figure 62.2 Various schemes used in synthesizing different forms of nanoparticles [7]

nanocrystalline thin films of different metals and their oxides (Cu, Ag, TiO_2, Al_2O_3) by transferring the momentum of the targeted atom [6, 10]. An ultrasound process involves heating or cooling a dispersion liquid medium by using the physical and chemical effects of ultrasonic irradiation. The method can process nanoparticles derived from carbon, titanium, alumina, and ZrO_2, whose morphological properties are easily controlled [12, 13]. By utilising the kinetic energy of rollers or balls, the milling process directly transforms the bulk material into micro-nano structural shapes. The intensity, milling time, and temperature all affect the size and shape of the produced NPs. Due to its quick processing and improved control over size and shape, laser ablation has become a popular alternative to conventional chemical methods. Pyrolysis is a type of thermal decomposition that focuses on using heat to dissolve chemical bonds. Figure 62.2 shows the different schemes used to synthesise nanoparticles [6]. In chemical vapor deposition, gaseous reactant settles as a thin film, and when it combines with other gaseous molecules, the substrate is heated to induce the reduction of ions, which leads to uniform, nonporous NPs with good purity [6].

Various types of emissions and its controlling measures

Carbon monoxide: It occurs when there is insufficient oxygen and low combustion temperatures resulting in incomplete combustion. CO reduction is a crucial part of human health because it has a 240 times higher affinity for reacting with hemoglobin than oxygen. Therefore, it reduces oxygen carrying capacity of blood at tissue level and interrupts cellular respiration, resulting in tissue hypoxia. Therefore, control over CO is an important factor. In the presence of metal oxide nanoparticles, CO emissions are reduced because the oxygen contents in metal-based oxides are increased, resulting in a high catalytic activity of NPs because their surface area to volume ratio is high [14–16].

Nitrogen oxide emission (NOx): The term is used to describe N_2O_2, NO, and NO_2 that may adversely affect human health and the environment. When oxygen is present and post combustion temperatures are high, NOx is formed more quickly. In particular, it depends on the residence time, the oxygen atom availability in the test fuel, and the temperature of the cylinder. As a result of oxygen donating catalyst in the combustion chamber, metal oxide containing oxygen content results in complete combustion. Among the few oxides, SiO2 has a higher specific heat than Al_2O_3 and TiO_2, while oxides with a low specific heat have higher average kinetic energy and also higher rise in temperatures, so low specific heat results in higher NOx.

Hydrocarbon emissions: The formation of this substance usually results from test fuel not participating in the combustion and evaporation of the fuel. Due to the addition of nanoparticles, the surface-to-volume ratio is increased, and oxygen is readily available, facilitating a more complete combustion, which reduces HC emissions [14–16].

Smoke: When hydrocarbon fuel is burned improperly, this undesirable by-product is formed. It depends on the fuel type and operation conditions. As a result of reducing ignition delay and vaporization rate as well as enhancing ignition properties, smoke pollutant is reduced and proper combustion occurs [14–16].

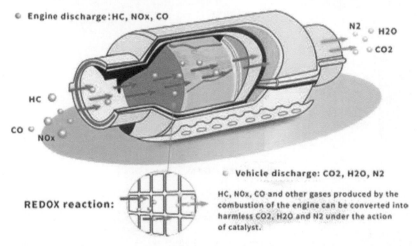

Figure 62.3 Polluting emissions from automobile engines and its control strategies [15]

Brake thermal efficiency: A ratio can be defined as the result of efficiently converting the heat from the engine into work, i.e., how much power is obtained from the engine in relation to how much energy is supplied to it. Neochloris oleoabundans algae methyl ester containing RuO_2 nanoparticles improves BTE by improving atomisation and fuel evaporation.

BSFC: A ratio is used to define the quantity of fuel consumed to the amount of effective power produced. This is a measure of the efficiency of a combustible fuel air mixture that rotates a crankshaft [14–16].

EGT: It usually indicates the temperature of exhaust gases, which is an indication of the temperature of the fuel during combustion. The presence of oxide nanoparticles improves oxygen content, resulting in improved EGT values. As nanoparticles alter the quality of fuel atomisation, oxygen content, viscosity, and fuel evaporation, cetane number determines ignition delay. As the cetane number increases, the delay of ignition is reduced and therefore fuel combustion is improved, which reduces emissions [14–16]. Figure 62.3 illustrates how poisonous gases are converted into nonpoisonous gases.

Nanoparticles as lubricant for emission control

0 D-nanoadditives include ceramic nanoparticles and metal nanoparticles that form solid matrices that significantly improve tribological performance. It is particularly responsible for achieving uniform dispersion of 20 g/L-Al_2O_3 in Ni-Cu alloy, thereby creating uniform crystal growth. As a result, morphologies of prepared samples change in the form of crystalline coating sizes (reduced from 91 nm to 16 nm), and a high performance tribofilm is produced with superior microhardness and attractive ductility. In addition to 0 D additives, 1D-nano additives also act as a solid lubricant. Copper nanowires are a good approach for tribofilm formation. Additionally, the carbon rich film protects the composite coating from direct interaction with its counterpart. This reduces surface adhesion and atom transfer among tribopairs, thus exhibiting low COF and high wear resistance [17, 18]. Due to their unique flake structure, high thermal conductivity, low surface energy, and smoother surface on an atomic level, 2 D- Nano additives drew considerable interest. A tribofilm formed by graphene, MoS_2, carbides and nitrides of metals shows excellent friction and wear characteristics. Tribological performance is enhanced by the GO/MOS_2 composite coating due to its excellent lubrication. 3 D-nanoadditives are designed to create synergistic effects through the agglomeration of 0 D-nanoadditives and 1 D with 2 D-nanoadditives [18]. The goal 3 D-nano additive is to solve problems related to poor dispersion and insufficient mechanical strength to achieve high performance tribofilms [18]. The presence of the following forms of nanoparticles in lubricants improves the rheological properties, roughness of contact surfaces, and density of asperities, which are important for controlling friction force through rolling effects of nanoparticles, thus reducing fuel consumption and reducing toxic emissions. By adding 2% TIO_2 nanoparticles to engine oil, friction losses are reduced by 77.7%, while emissions are reduced by 1719% for CO, 89% for CO_2, 1317% for NOx, and about 1216% for HC [17]. Figure 62.4 illustrates the different forms of nanoparticle lubrication. Table 62.1 summarises current studies of different nanoparticle concentrations in a particular dispersion medium and their effects on emission control.

Stability phenomena of Nano fluid and their Characterisation Aspects

It is crucial to prepare stable nano fluids because the Van der Walls interaction between nanoparticles cause strong aggregation, clogging, and sedimentation between them. Due to which a declination in thermal conductivity, viscosity, and specific heat would have occurred. The formation of a stable fluid requires the use

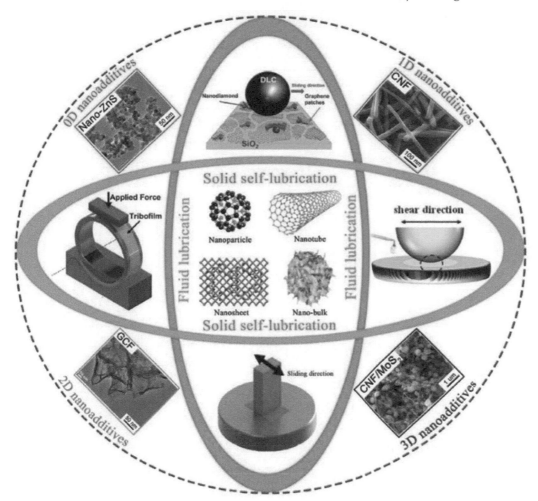

Figure 62.4 Application of micro-nano structural morphologies to enhance tribological performance under self-lubrication [18]

of surfactants or strong forces on clustered nanoparticles, as well as the spreading of surface active agents for dispersion in aqueous solutions. While the colloids designation is determined by its dispersion stability, defined as the ratio of quality of nano fluid preparation. There are three ways to define stability of nano fluid stability: chemical stability, kinematic stability, and dispersion stability. There are particle- particle interactions as well as particle-fluid interactions that determine nano fluid chemical stability. Kinematic stability as well as dispersion stability are based on the physical interaction and movement of nanoparticles [18, 29, 30]. The following instruments are available for measuring the stability of nano fluids: Nanoparticles can be distinguished by SEM and TEM by their shape, size, and distribution, whether they are in a nano fluid or as dried samples. There are cryo-SEM and cryo-TEM instruments readily available when we feel that we are having difficulty distinguishing between nano fluids or that the microstructure has not changed. With sedimentation balance methods, the tray of the balance is immersed in a fresh suspension of nanoparticles where we measure the weight of sedimented nanoparticles. As a result of nanoparticle sedimentation, the three omega method particularly detects thermal conductivity growth. The electrophoretic behavior, zeta potential defines the stability, as the electrostatic potential in between the nanoparticles increases, which results in a good suspension stability. Furthermore, XRD, EDX, XPS, IR, and Raman can be used to study NP structural properties. In the case of XRD, it can provide information about crystallinity and phase, as well as a rough idea of particle size, whereas EDX can determine the elemental composition of ultrasonochemically synthesised NPs. Figure 62.5 illustrates the factors that affect the stability of nano fluids [29–31].

Conclusions and Future Scope

In this article, a brief review of catalytic applications of nanoparticles and their synthesis process for carbon-based nanoparticles is presented. In order to reduce the adverse effects of polluting emissions on the environment, emission control variables are discussed. Moreover the nanoparticles as a lubricant was reviewed for emissions control. Further the comparative analysis was done over different fractional presences of

Table 62.1 Effects of nanoparticle in dispersion medium on performance of IC Engine [19, 28]

Types of nanoparticles	Name of dispersion medium	Optimum quantity of nanoparticles	Effects of NPs on engine performance		Effects of NPs on emissions control		
			BSFC	BTE	NOx	HC (Reduction)	CO (Reduction)
Alumina (Al_2O_3) nanoparticles [19]	Cashew nut shell biodiesel. (B100A)	10 wt %	Increased by 3.8%	Decreases by 1.1%	Decreased by 10.23%	7.4%	5.3%
Al_2O_3 [20]	MSME20	26.2 nm	Decrement in BSFC	Improve by 1.39%	Relatively similar to initial	35.48%	13%
Al_2O_3 and Fe_2O_3 [21]	Biodiesel (8 0% diesel and 20% MME)	40, 80, and 120 ppm	Reduction in 7.66%	Improved by 8.8%	Effectively reduces	6.39%)	10.24%
Al_2O_3 and TiO_2 [22]	B100: 100% rubber seed oil methyl ester	25 ppm and 50 ppm	Decreases by 10.56%	Increase by 5.2%	Increased by 21%	28%,	44%,
CNT, TiO_2 and Al_2O_3 [23]	B30 blend	100 ppm	Reduced by 5.98%	Increase by 9.83%	Reduction in 10.37%	30.68 %	27.89%
Al_2O_3 [24]	castor oil biodiesel	25 ppm	14.86%	Improved by 23.1%	Reduction of 10.92%	4.82%	30.5%
$H2O_2$+ CeO_2 [25]	B_2O	H_2O_2: (0.5%, 1%, and 1.5%) CeO_2: (40, 80 ppm)	1.88%	Improved 1.2%	Reduction in 22.7%	41–48%	56–60%
CuO_2 [26]	rapeseed methyl ester(RME)	1–100 nm	12.73%	Improve by 12.43%	Decreases by 12.74%	8%	5%
Al_2O_3+MWCNT [27]	tamarind seed methylester	60 ppm	Relatively less than convention	Improve by 20%	Reduces by 7–9%	24–68%	15–51%
CuO [28]	Convention diesel fuel	<77nm	8%	Increase by 14.6%	Reduction of 4.7%	13.4%	20.8%

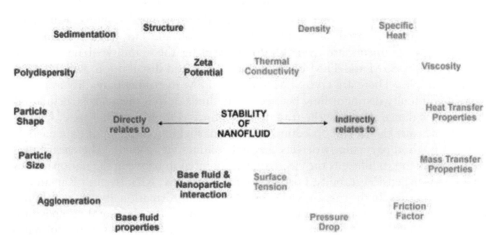

Figure 62.5 Schematic illustration of factors that relate stability of nano fluid [29]

nanoparticle in suitable medium to discuss their impact on emission control. Finally the stability phenomena of nano fluid and their characterization was discussed. The following points can be summarises as a concluding remark with future scope of this study.

In terms of emissions control, various catalytic nanoparticles improved performance due to their high energy density, oxygen liberating capabilities, and heat-absorbing abilities. Nanoparticle synthesis is crucial for choosing the most appropriate method for preparing a particular type of nanomaterial. The use of nanoparticles as three-way catalytic converters is crucial since they convert toxic gases into nontoxic ones (CO_2, N2, H_2O). As a result, nano additives which contain (0D, 1D, 2D and 3D) work as a solid lubricant to enhance friction and wear resisting behavior, and they also serve as an emission control additive. Inappropriate stability of nano fluids due to dispersion of nanoparticles adversely affects thermophysical properties such as viscosity, density, and thermal conductivity. A cryo-SEM and cryo-TEM are capable of distinguishing microstructural variations during their use. The development of a nanoparticle impact assessment system is necessary due to be created owing to the toxicity and exposure to nanotechnology issues in the field of real world application such as the environment, health, and the transportation vehicle industry (coatings, nano additives, paints, engines, mirrors, tires etc.) An experimental based statistical regression (modelling method) model must be developed in order to make easy predictions of fuel consumption, NOx, CO, SO_2, opacity of nano fluid fuel at various loads and to help make vehicles smarter, more efficient, stronger, and more durable. It is necessary to conduct in-depth and exhaustive research on surface reactive nanoparticles, organic nanoparticles as additives, and the behavioral aspects of these nanoparticles, such as shelf life, stability, and thermal conductivity, in order to comprehend the impact that friction-wear behavior (piston ring cylinder liner) has on emission levels. Study may concentrate on the nanoparticle's interaction between static and flow conditions, the impacts of interfacial contact, the mobility of nanoparticles, nano induced surface structures and in-depth their impact on emission level.

References

[1] Rane, A. V., Kann, K., Abitha, V. K., and Thomas, S. (2018). Methods for synthesis of nanoparticles and fabrication of nanocomposites. In Synthesis of Inorganic Nanomaterials. Bhagyaraj, S. M., Oluwafemi, O. S., Kalarikkal, N., Thomas, S., eds. Duxford United Kingdom: Woodhead Publishing, pp. 121–139.

[2] Gao, C., Lyu, F., and Yin, Y. (2020). Encapsulated metal nanoparticles for catalysis. Chemical Reviews, American Chemical Society. 121(2), 834–81.

[3] Ağbulut, Ü., Karagöz, M., Sarıdemir, S., and Öztürk, A. (2020). Impact of various metal-oxide based nanoparticles and biodiesel blends on the combustion, performance, emission, vibration and noise characteristics of a CI engine. Fuel. 270, 117521.

[4] Khan, S., Dewang, Y., Raghuwanshi, J., Shrivastava, A., and Sharma, V. (2020). Nanoparticles as fuel additive for improving performance and reducing exhaust emissions of internal combustion engines. International Journal of Environmental Analytical Chemistry. 102(2), 319–41.

[5] Lv, J., Wang, S., and Meng, B. (2022). The effects of nano-additives added to diesel-niodiesel fuel blends on combustion and emission characteristics of diesel engine: A review. Energies. 15(3), 1032.

[6] Habibullah, G., Viktorová, J., and Ruml, T. (2021). Current strategies for noble metal nanoparticle synthesis. Nanoscale Research Letters. 16(1), 1–12.

[7] Mosayebi, J., Kiyasatfar, M., and Laurent, S. (2017). Synthesis, functionalization, and design of magnetic nanoparticles for theranostic applications. Advanced Healthcare Materials. 6(23), 1700306.

[8] Tri, P. N., Ouellet-Plamondon, C., Rtimi, S., Assadi, A. A., and Nguyen, T. A. (2019). Methods for synthesis of hybrid nanoparticles. In Noble Metal-Metal Oxide Hybrid Nanoparticles. Mohapatra, S., Nguyen, T. A., Nguyen-Tri, P., eds. Duxford United Kingdom: Woodhead Publishing. pp. 51–63.

[9] Pérez-Tijerina, E., Gracia Pinilla, M., Mejía-Rosales, S., Ortiz-Méndez, U., Torres, A., and José-Yacamán, M. (2008). Highly size-controlled synthesis of Au/Pd nanoparticles by inert-gas condensation. Faraday Discuss. 138, 353–362.

[10] Ayyub, P., Chandra, R., Taneja, P., Sharma, A., and Pinto, R. (2001). Synthesis of nanocrystalline material by sputtering and laser ablation at low temperatures. Applied Physics A Materials Science and Processing. 73(1), 67–73.

[11] Malik, M., Wani, M., and Hashim, M. (2012). Microemulsion method: A novel route to synthesise organic and inorganic nanomaterials. Arabian Journal of Chemistry. 5(4), 397–417.

[12] Yan, Q., Qiu, M., Chen, X., and Fan, Y. (2019). Ultrasound assisted synthesis of size-controlled aqueous colloids for the fabrication of nanoporous zirconia membrane. Frontiers in Chemistry. 7, 337.

[13] Vaitsis, C., Mechili, M., Argirusis, N., Kanellou, E., Pandis, P. K., Sourkouni, G., Zorpas, A., and Argirusis C. (2020). Ultrasound-assisted preparation methods of nanoparticles for energy-related applications. Nanotechnology and the Environment. 77–103.

[14] Ashok, B., and Nanthagopal, K. (2019). Eco friendly biofuels for CI engine applications. In Advances in Eco-fuels for a Sustainable Environment. Azad, K., eds. Duxford United Kingdom: Woodhead Publishing, pp. 407–40.

[15] Dey, S., and Dhal, G. (2019). Materials progress in the control of CO and CO2 emission at ambient conditions: An overview. Materials Science for Energy Technologies. 2(3), 607–23.

[16] Kalaimurugan, K., Karthikeyan, S., Periyasamy, M., Mahendran, G., and Dharmaprabhakaran, T. (2019). Performance, emission and combustion characteristics of RuO2 nanoparticles addition with neochloris oleoabundans algae biodiesel on CI engine. Energy Sources, Part A: Recovery, Utilization, and Environmental Effects. 1–15.

[17] Wozniak, M., Batory, D., Siczek, K., and Ozuna, G. (2020). Changes in total friction in the engine, friction in timing chain transmissions and engine emissions due to adding TiO2 nanoparticles to engine oil. Emission Control Science and Technology. 6(3), 358–79.

[18] Chen, Y., Yang, K., Lin, H., Zhang, F., Xiong, B., Zhang, H., and Zhang, C. (2022). Important contributions of multidimensional nanoadditives on the tribofilms: From formation mechanism to tribological behaviors. Composites Part B: Engineering. 234, 109732.

[19] Radhakrishnan, S., Munuswamy, D., Devarajan, Y., T Arunkumar., and Mahalingam, A. (2018). Effect of nanoparticle on emission and performance characteristics of a diesel engine fueled with cashew nut shell biodiesel. Energy Sources, Part A: Recovery, Utilization, and Environmental Effects. 40(20), 2485–93.

[20] Raju, V. D., Reddy, S. R., Venu, H., Subramani, L., and Soudagar, M. E. (2021). Effect of nanoparticles in bio-oil on the performance, combustion and emission characteristics of a diesel engine. Liquid Biofuels: Fundamentals, Characterization, and Applications. 613–37.

[21] Nutakki, P., and Gugulothu, S. (2021). Influence of the effect of nanoparticle additives blended with mahua methyl ester on performance, combustion, and emission characteristics of CRDI diesel engine. Environmental Science and Pollution Research. 29(1), 70–81.

[22] Srinivasan, S., Kuppusamy, R., and Krishnan, P. (2021). Effect of nanoparticle-blended biodiesel mixtures on diesel engine performance, emission, and combustion characteristics. Environmental Science and Pollution Research. 28(29), 39210–26.

[23] Fayaz, H., Mujtaba, M., Soudagar, M., Razzaq, L., Nawaz, S., Nawaz, M., Farooq, M,. Afzal, A,. Ahmed, W,. Khan, T,. Bashir, S,.Yaqoob, H,. Seesy, A,. Wagesh, S,. Ghamdi, A,. Elfasakhany, A,. (2021). Collective effect of ternary nano fuel blends on the diesel engine performance and emissions characteristics. Fuel. 293, 120420.

[24] Al-Dawody, M., and Edam, M. (2022). Experimental and numerical investigation of adding castor methyl ester and alumina nanoparticles on performance and emissions of a diesel engine. Fuel. 307, 121784.

[25] Mohan, S., and Dinesha, P. (2022). Performance and emissions of biodiesel engine with hydrogen peroxide emulsification and cerium oxide (CeO2) nanoparticle additives. Fuel. 319, 123872.

[26] Fayad, M., Ibrahim, S., Omran, S., Martos, F., Badawy, T., Jubori, A,. Dhahad, H,. Chaichan, M,. (2022). Experimental effect of CuO_2 nanoparticles into the RME and EGR rates on NOX and morphological characteristics of soot nanoparticles. Fuel. 331, 125549.

[27] Dhana Raju, V., Kishore, P., Nanthagopal, K., and Ashok, B. (2018). An experimental study on the effect of nanoparticles with novel tamarind seed methyl ester for diesel engine applications. Energy Conversion and Management. 164, 655–66.

[28] Ağbulut, Ü., Sarıdemir, S., Rajak, U., Polat, F., Afzal, A., and Verma, T. (2021). Effects of high-dosage copper oxide nanoparticles addition in diesel fuel on engine characteristics. Energy. 229, 120611.

[29] Bhanvase, B. A., and Barai, D. (2021). Nanofluids for heat and mass transfer: Fundamentals, sustainable manufacturing and applications. Londond, United Kingdom: Academic Press. pp. 69–97.

[30] Ghadimi, A., Saidur, R., and Metselaar, H. (2011). A review of nanofluid stability properties and characterization in stationary conditions. International Journal of Heat and Mass Transfer. 54(17-18), 4051–68.

[31] Khan, I., Saeed, K., and Khan, I. (2019). Nanoparticles: Properties, applications and toxicities. Arabian Journal of Chemistry. 12(7), 908–931.

Chapter 63

Numerical analysis of a blunt body at very high reynolds number for the application on ejection seat system

Md. Mahbubur Rahman[a], Vaibhav Anant Bandal[b], Sunil Chandel[c] and D. G. Thakur[d]

Defence Institute of Advanced Technology (DIAT), Pune, India

Abstract

The ejection seat system is an essential component of an aircraft during an emergency. As the ejection seat with the occupant is a complex blunt body, the aerodynamic analysis of a blunt body will be helpful in analyzing the ejection seat system. For this purpose, a blunt benchmark body like a sphere is considered to look at the effects of flow at very high Reynolds numbers. A 3-dimensional Reynolds Averaged Navier-Stokes code with turbulence modeling is considered for the computational fluid dynamics (CFD) analysis using an unstructured grid. The results show that the Reynolds Averaged Navier-Stokes code with the Realizable k-ε turbulence model can successfully predict the aerodynamic coefficients of blunt bodies at very high Reynolds numbers. In the end, the computed drag, pressure, and skin friction coefficients at Reynolds numbers ranging from 4000000-6000000 have been validated with the experimental results and discussed in detail.

Keywords: Blunt body, sphere, ejection seat, CFD, Reynolds number

Introduction

Flow over a sphere has been studied rigorously as the flow dynamics are pretty complex. The wake at a high Reynolds number (Re) for flow over a sphere is quite large, and the pressure drag contribution is the highest, which makes the sphere a blunt body. Hence, the analysis of flow over a sphere can help us understand the physics of flow for different types of blunt bodies like the ejection seat system, missiles, torpedoes, etc.

Many authors have carried out experiments, and computational fluid dynamics (CFD) analysis on a sphere using Large Eddy Simulation (LES) and its formulations and even Direct Numerical Simulation (DNS). Sakamoto and Haniu [1] experimentally investigated the vortex shedding from the sphere at Re from 300-400000 using hot wire techniques. Vortex shedding frequency and its variation with Re were studied in a water channel. Taneda [2] used different methods to analyse the flow behavior for Re 10000-1000000 on a sphere. Achenbach [3] experimentally investigated flow past a smooth sphere at Re 50000 to 6000000. At 0.45% of the turbulence level, different parameters like drag coefficient, static pressure, and total skin friction were calculated.

The effects of small scales are essential in wake formation. Hence, capturing the flow details of such a complex flow at very high Re requires proper resolution of small eddies as done in DNS, detached eddy simulation (DES), and LES. Nagata et al. [4] investigated flow over a sphere at high Mach numbers and low Re using DNS for predicting the aerodynamic coefficients. They measured the aerodynamic coefficient considering a rigid sphere at Re 50 to 300 and Mach numbers 0.3 and 2. Rodriguez et al. [5] studied the aerodynamic characteristics of a sphere at Re = 3700 by DNS. The flow parameters, like wall pressure, skin friction, and drag coefficient, were measured using an unstructured grid. LES using a dynamic Smagorinsky-type subgrid stress model and DES have been carried out by a few authors like Constantinescu and Squires [6] studied the instabilities during the flow transitions at a sub-critical regime where the Re was 10000 first and the sub and supercritical regimes both at Re 10000 to 1000000 using DES [7]. Jindal et al. [8] studied LES with a classical Smagorinsky subgrid-scale model using an unstructured grid at Re = 1140000 (supercritical regime) for a sphere. The drag coefficient, pressure, and skin friction were compared with experimental data. Wang et al. [9] investigated the capability of ANSYS Fluent, a commercial CFD software using the Splart-Allmaras model with LES, to study the forced convection of airflow around a sphere at Re 1000 to 150000 with a Prandtl number of 0.71. They found that Fluent can provide good results in the case of complicated structures having flow separation. Hassanzadeh et al. [10] predicted instantaneous and time-averaged characteristics of flow using LES on the multi-block grid system at Re = 5000. Yen et al. [11] measured the drag coefficient and flow separation angle on a sphere using unsteady simulation at Re 300000-600000. The turbulent wake structure downstream of a sphere was analysed using a hybrid spatially evolving DNS model at Re 3700 by VanDine et al. [12]. A detailed experimental and numerical technique had been discussed on a stationary sphere by Tiwari et al. [13]. Rashid and Faruque [14] studied

---

[a]mahbub.rizvee@gmail.com, [b]vaibhavbandal98@gmail.com, [c]sunildiat2010@gmail.com, [d]thakur@diat.ac.in

DOI: 10.1201/9781003450252-63

flow over a sphere using standard k-ε and standard k-ε turbulent model at Re = 30000. Cocetta et al. [15] investigated stratified flow past a sphere at Re 200 and 300. The wake structure and flow separation on the sphere was measured. The computational requirement for DNS and LES is high; hence, 3-D RANS has been solved using a Realizable k-ε turbulence model for flow over a sphere to check the applicability of this numerical approach for an ejection seat system at Re 4000000, 4500000, 5000000, 5500000 and 6000000.

Governing Equations

The well-known Navier-Stokes equations are the equations for momentum conservation for Newtonian fluids. Five equations in total govern the motions of a Newtonian fluid, of which (1) is the conservation of mass (Continuity); the following (2) is the conservation of momentum in three directions of the Cartesian coordinate system; and (3) is the conservation of energy.

$$\frac{\partial \rho}{\partial t} + \nabla . \rho U = 0 \tag{1}$$

$$\frac{\partial \rho U}{\partial t} + \nabla . \rho UU = -\nabla p + \nabla .(\mu \nabla U) + F \tag{2}$$

$$\frac{\partial \rho e}{\partial t} + \nabla .(\rho Ue) = -\nabla (pU) + \nabla .(\mu U \nabla U) + \rho F.U - \nabla .q \tag{3}$$

Where, p is the pressure, r is the density, and e is the total specific energy. U, F, and q are the velocity field, force field, and heat flux vector, respectively. At Mach numbers less than 0.3, the density fluctuations are less than five percent, so the assumption of incompressible flow, in this case, is good enough.

Numerical Methodology

Computational Domain and Meshing

For a blunt body analysis, a sphere of diameter 1.3 m is chosen. The computational domain for the current study is shown in Figure 63.1. It is extended in the upstream direction by 5 m and downstream direction by 20 m. The lateral sides are at a distance of 1.5 m. The top and bottom faces are at a distance of 2.275 m.

An unstructured grid is generated using tetrahedral and prism layer cells in ICEM-CFD. The mesh details of the Mesh-3 configuration have been presented in Table 63.1. For all the simulations of Re 4000000, 4500000, 5000000, 5500000, and 6000000, a Y+. value of one was maintained to resolve the turbulent boundary layer.

The grid size near the sphere is kept fine to adequately capture the flow details, while the remaining domain includes a coarser mesh, as shown in Figure 63.2, where the flow features are insignificant. The overall mesh quality is kept above the acceptable limit. A mesh convergence study is conducted on four mesh configurations for Re = 4000000. The drag coefficient obtained from all mesh configurations is presented in Table 63.2, which is compared with the experimental drag coefficient at Re = 4000000 [3]. The results show that the Mesh-3 configuration provides better results when compared with the experimental results; hence, it is selected for further analysis.

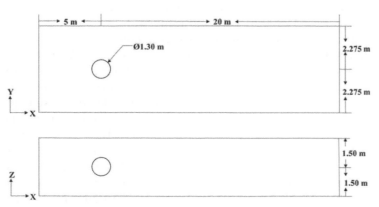

Figure 63.1 Dimensions of the fluid domain

Source: Author

Table 63.1 Details of Mesh-3 Configuration

Parameter	Value
Global mesh size	500 mm
First prism cell height	0.0087 mm
Prism layer count	35
Growth rate	1.2
Total prism layer thickness	25.73 mm

Source: Author

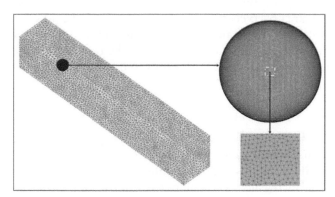

Figure 63.2 3-D volume grid (left) and surface grid on the sphere (right)
Source: Author

Table 63.2 Details of the mesh convergence study

Mesh configuration	No of elements	Computed drag	Experimental drag
Mesh-1	5.2 million	0.147	0.184
Mesh-2	8.4 million	0.172	
Mesh 3	9.4 million	0.183	
Mesh-4	11.1 million	0.1832	

Source: Author

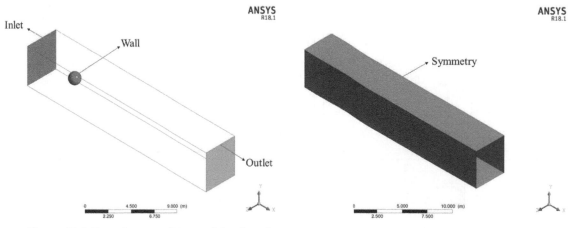

Figure 63.3 Boundary conditions of the domain
Source: Author

Boundary Conditions

Air with constant viscosity and density is the working fluid. Uniform velocity is specified at the inlet, the outlet with a "pressure outlet," the sphere wall with "no-slip condition," and the remaining faces with "symmetry," as shown in Figure 63.3. The temperature of the air is taken as 288.16 K, and operating pressure is 1 atm.

Numerical Approach

The 3-D RANS equations are solved with a pressure-based solver and the SIMPLE algorithm. The Realizable k-ε turbulence model is used to predict the flow separations accurately. The simulation data is obtained for residuals below 10^{-5}.

Results and Discussions

Flow Field Details

The velocity contours are shown in Figure 63.4 for Re ranging from 4000000 to 6000000. The highest attained velocities at the top and bottom sides of the sphere increase with the Re. Flow separates far behind the sphere, thereby reducing the Wakefield. The flow separation region behind the sphere displays low velocities in the wake region where low-pressure values are observed, which are responsible for the pressure drag. No significant differences can be seen in the wake formed in all five cases, as the physical nature of the flow has no considerable changes. Figure 63.5 shows the pressure contours where the highest pressure is seen to be developed on the front of the sphere surface. The pressure is also seen to be increasing on the front face of the sphere with the Re.

Force and Pressure Coefficients

- The drag coefficient is the combination of the pressure drag and viscous drag. The sphere is a blunt body; hence, the pressure drag contribution is significant, which is seen from the wake formed. Drag increases with Re; however, at Re 4500000 and 5000000, a small discrepancy is seen, but the over-all behavior of the drag coefficient is seen to be increasing. The computed drag values shown in Figure 63.6 match well with the experimental values of Achenbach [3], and the percentage of error for all five cases is below 5%. Figure 63.7 displays the predicted skin friction coefficient with the angle where the maximum value is at angles ranging from 700-900. It happens because of the high velocities attained in that range. Similar behavior is seen at all five Re. The pressure coefficient distribution is

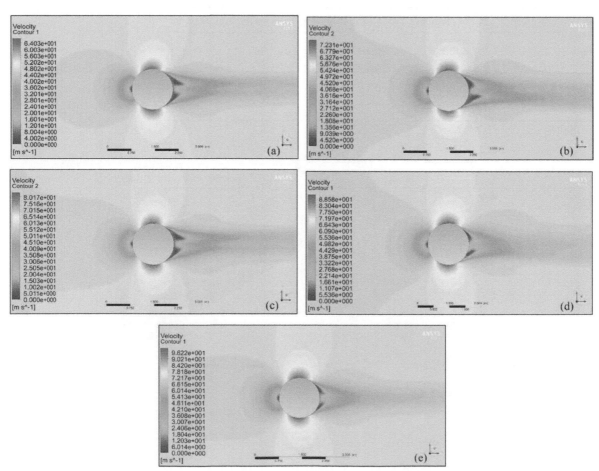

Figure 63.4 Velocity contour on the symmetric plane; (a) Re = 4000000 (b) Re = 4500000 (c) Re = 5000000 (d) Re = 5500000 (e) Re = 6000000

Source: Author

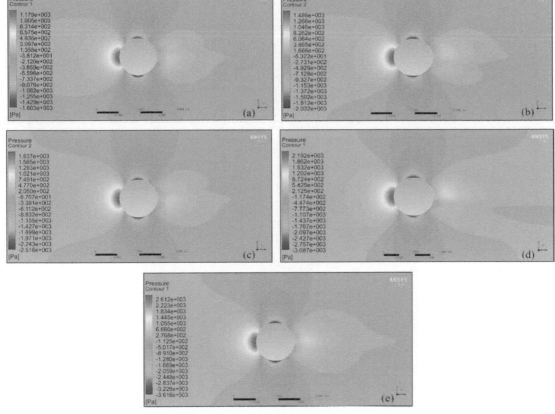

Figure 63.5 Pressure contour on the symmetric plane; (a) Re = 4000000 (b) Re = 4500000 (c) Re = 5000000 (d) Re = 5500000 (e) Re = 6000000
Source: Author

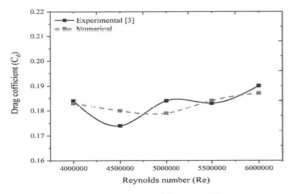

Figure 63.6 Variation of drag coefficient
Source: Author

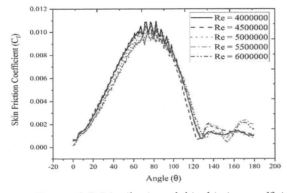

Figure 63.7 Distribution of skin friction coefficient over the sphere
Source: Author

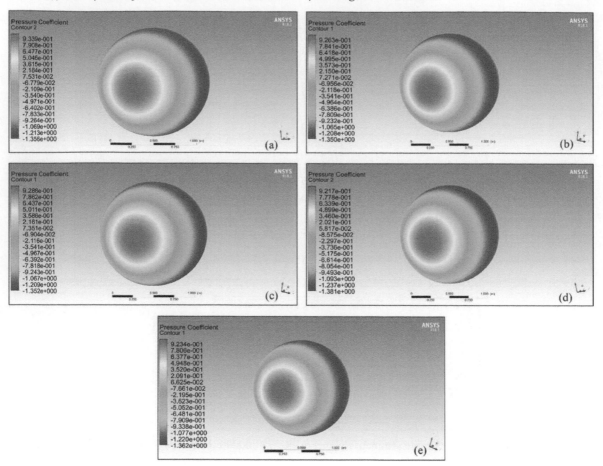

Figure 63.8 The pressure coefficient contour on the spheres; (a) Re = 4000000 (b) Re = 4500000 (c) Re = 5000000 (d) Re = 5500000 (e) Re = 6000000
Source: Author

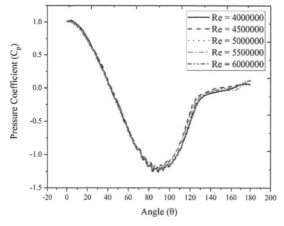

Figure 63.9 Variation of pressure coefficient over the sphere
Source: Author

displayed in Figure 63.8. The maximum pressure coefficient is at the front face, as expected, where the pressure is high due to the flow impact. Figure 63.9 displays the pressure coefficient distribution. The behavior is similar to that observed at Re = 1140000 in Achenbach [3]. The lowest value of the pressure coefficient is seen at an angle of around 900.

Conclusions

The use of unstructured grids has increased the number of cell counts compared to the cases of structured grids, but the drag, pressure, and skin friction coefficient predictions are well. The predicted drag values show that the turbulent boundary layers have been adequately captured. At very high Re, the behavior of the pressure coefficient and skin friction coefficient is quite similar. The computed results have matched the experimental data with a good agreement suggesting that steady-state RANS equations can sufficiently predict the aerodynamic characteristics of blunt bodies at very high Re. Similar flow characteristics are observed in complex blunt bodies like the ejection seat system, for which unstructured grids are preferred. Hence, a similar approach can be applied to an ejection seat system to obtain the flow details and the aerodynamic coefficients.

References

[1] Sakamoto, H. and Haniu, H. (1990). A study on vortex shedding from spheres in a uniform flow Journal of Fluids Engineering, Transactions of the ASME. 112, 4.

[2] Taneda, S. (1978). Visual observations of the flow past a sphere at Reynolds numbers between 104 and 106. Journal of Fluid Mechanics. 85, 1.

[3] Achenbach, E. (1972). Experiments on the flow past spheres at very high Reynolds numbers. Journal of Fluid Mechanics. 54, 3.

[4] Nagata, T., Nonomura, T., Takahashi, S. ,Mizuno, Y., and Fukuda, K. (2016). Investigation on subsonic to supersonic flow around a sphere at low Reynolds number of between 50 and 300 by direct numerical simulation. Journal of Fluid Mechanics. 28, 5.

[5] Rodriguez, I., Borell, R., Lehmkuhl, O., Perez Segarra, C. D., and Oliva, A. (2011). Direct numerical simulation of the flow over a sphere at Re = 3700. Journal of Fluid Mechanics. 679.

[6] Constantinescu, G. S. and Squires, K. D. (2003). LES and DES investigations of turbulent flow over a sphere at Re = 10,000. Flow, Turbulence and Combustion. 70, 1–4.

[7] Constantinescu, G. and Squires, K. (2004). Numerical investigations of flow over a sphere in the subcritical and supercritical regimes. Physics of Fluids. 16, 5.

[8] Jindal, S., Long, L. N., Plassmann, P. E., and Sezer-Uzol, N. (2004). Large eddy simulations around a sphere using unstructured grids. 34th AIAA Fluid Dynamics Conference and Exhibit.

[9] Wang, Y. Q., Jackson, P. L., and Ackerman, J. D. (2006). Numerical investigation of flow over a sphere using LES and the Spalart-Allmaras turbulence model. Fluid Dynamics Journal. 15(1).

[10] R. Hassanzadeh, B. Sahin, and M. Ozgoren. (2011) Numerical investigation of flow structures around a sphere. International Journal of Computational Fluid Dynamics. 25(10).

[11] Yen, C. H., Hui, U. J., We, Y. Y., Sadikin, A., Nordin, N., and Taib, I. et al. (2017). Numerical study of flow past a solid sphere at high Reynolds number. IOP Conference Series: Materials Science and Engineering. 243(1), 012042.

[12] VanDine, A., Chongsiripinyo, K., and Sarkar, S. (2018). Hybrid spatially-evolving DNS model of flow past a sphere. Computers & Fluids. 171, 41-52.

[13] Tiwari, S. S., Pal, E., Bale, S., Minocha, N., Patwardhan, A. W., Nandakumar, K. et al. (2020). Flow past a single stationary sphere, 1. Experimental and numerical techniques. Powder Technology. 365, 115–48.

[14] Rashid, M. M. and M.G.M. Al Faruque. (2020). Numerical investigation of flow over a spherical body at high Reynold's number. International Journal of Science and Engineering Research. 11, (3).

[15] Cocetta, F., Gillard, M., Szmelter, J., and Smolarkiewicz, P. K. (2021). Stratified flow past a sphere at moderate Reynolds numbers. Computers & Fluids. 226.

Chapter 64

Spinning effects on projectile undergoing normal impact using finite element analysis

Pranay Vaggu[a] and S K Panigrahi[b]

Defence Institute of Advanced Technology (DU), Pune, India

Abstract

The main objective of the projectile is to damage the target by concentrating the projectile energy on the concentric area of the target. The introduction of the rifled barrel in small arms for stabilization purposes made the projectile spin. The present study deals with the effect of spin on 7.62 mm projectiles by using LS-DYNA based on numerical method using finite element analysis (FEA). For the current study, 7.62 mm projectiles of steel core, lead core, and target of Al 7075-T651 are considered. Materials are defined by using the Johnson cook (JC) Strength model, and the failure model. A numerical model is prepared in LS-DYNA software. The numerical model results are validated with experimental results available in the literature and the difference between the experiment and numerical results is 0.7%. The validated model is used to analyse the effect of spinning. In the validated model along with impact velocity, spin is given to the projectile to study the effect of spinning. Lead material is also considered for the projectile. By observing the results from numerical simulations, it is concluded that for both projectile materials, the variation pattern of residual velocity, and residual mass parameters is similar, they are varying with only magnitude. From residual energy analysis, it is observed that the spin effect is advantageous in steel projectile, while lead projectile doesn't have much significant advantage with spin.

Keywords: Impact velocity, lead, residual energy, residual mass, residual velocity, spin, steel

Introduction

When the projectile is fired from the firearm, the projectile is pushed by high-pressure gases and the barrel guides the direction to the projectile, it travels through the air and reaches the target. At the target, the projectile impacts on target and creates damage to the target. Many researchers have studied the various influential parameters in penetration for both armors and projectiles and developed different armor materials, armor material arrangement configurations have been developed against projectiles for defense applications at sub-ordinance velocity range. The current study is focused on defeating the target.

These studies are conducted using experimental, analytical, and numerical analysis methods. Forrestal et al. [1] proposed penetration theory by conducting experiments on Al 7075-T651 targets impacted by ogive-nosed rods. Gupta et al. [2] conducted a study on thin aluminum plates of various thicknesses and studied the parameters related to the mass of the projectile, projectile diameter, residual velocity, length-to-diameter ratio, and deformation mechanism of aluminum plates. Børvik et al. [3,4] has conducted an experimental and numerical study related to armor grade steels ballistic perforation resistance. Radin et al. [5] is one of the earliest experimental studies for multi-layered targets by using multi-layered plates of soft aluminum, and polycarbonate, in the sub-ordinance velocity regime. Senthil et al. [6,7] also studied the layered configurations of armors using armor steel, mild steel, Al 7075-T651.

In literature [8], a simple law of conservation of energy through internal ballistics is given. In this, it is mentioned that some of the energy liberated by propellant is utilised for engraving the driving band and overcoming friction in the bore. Because of using the driving band, the projectile spins, which makes the projectile stabilise while traveling through the air. When the projectile reaches the target, it consists of to types of velocities i.e., translational velocity which is generally referred to as impact velocity, and rotational velocity which is generally referred to as the spin of the projectile. It is observed that in most of the studies related to the projectile, the rotational velocity is neglected because the rotational energy of the projectile is very less compared to the translational velocity of the projectile. Some literature has studied the effects related to the rotational velocity of the projectile, which is available in the reference [9–12]. In the earlier studies, the spinning effect analysis is conducted on projectiles of L/D ≈ 11. There are very few studies are conducted on the spin effect study for working small arms caliber projectiles. The present study is conducted on the effects of spinning i.e., rotational energy in small arms projectile. 7.62 mm caliber projectiles are one of the well-known projectiles in the research field. In the study of Senthil et al. [7], only the core of the projectile is considered for analysis, and equivalent dimensions of the projectile core are proposed.

[a]vaggupranaybaa@gmail.com, [b]panigrahi.sk@gmail.com

DOI: 10.1201/9781003450252-64

Research Methodology

The present study is conducted numerically i.e., Finite element analysis (FEA) by using LS-DYNA software. Initially, the projectile is given only translational velocity, and the residual velocity is calculated from numerical simulation. The calculated residual velocity from the LS-DYNA model is compared with the results available in the literature for model validation. The validated model conditions are used to study the effect of spinning. Steel and lead materials are considered for the projectile. Al 7075-T651 material is considered for the target. Materials are defined using Johnson-cook (JC) strength, JC failure model, and parameters are given in Table 64.1.

For the present study, 7.62 mm projectiles are considered. In reference [7], the equivalent projectile dimensions are given for the 7.62 AP projectile shown in Figure 64.1. The target diameter is 152 mm and the thickness is 20 mm.

Both projectile and target have meshed with hexahedron elements. The target is spitted into different regions to reduce the computational time. The inner square region is of 10 mm side, then the second region is of 15 mm radius and the third region is of 25 mm radius, and the outer region is from 25 mm radius to 76 mm radius. The mid-region is the main impact region, so this region of target and projectile have the same element size and the outer regions of targets have meshed with different element sizes. An inner square region and projectile element size are selected as 0.25 mm, and up to 15 mm radius zone it is 0.5 mm element size, up to 25 mm radius 1 mm element size, from 25 mm radius to 76 mm radius (till edge) is 2 mm element size is considered, 50 elements through-thickness direction is considered. The discretised models are shown in Figures 64.2 and 64.3.

Initially, the projectile is assigned with steel material, and the impact velocity is 867.8m/s given, the target is assigned with Al 7075-T651 material, and target edges are assigned with fixed edges boundary conditions. The friction coefficient considered is 0.2. By using LS-DYNA explicit solver, numerical simulation is

Table 64.1 Material model parameters for target and projectile metals [7,13–15]

Parameter	Units	Al 7075-T651	Projectile steel	Lead
ρ	kg/m^3	2810	7850	11340
E	GPa	71.7	202	15.008
υ		0.33	0.32	0.34
Johnson-Cook strength parameters				
A	MPa	520	2700	240
B	MPa	477	211	300
n	-	0.52	0.065	1.0
c	-	0.0025	0.005	0.1
ε	1/s	5e-4	1e-4	5e-4
T_0	K	293	293	293
T_m	K	893	1800	327.5
m	-	1.61	1.17	1
C_p	J/kg.K	910	452	129
Johnson-Cook failure model				
D_1	-	0.096	0.4	0.3
D_2	-	0.049	0	0
D_3	-	3.465	0	0
D_4	-	0.016	0	0
D	-	1.099	0	0

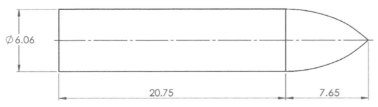

Figure 64.1 Schematic of the projectile (equivalent core only) [7]

Figure 64.2 Projectile mesh model
Source: Author

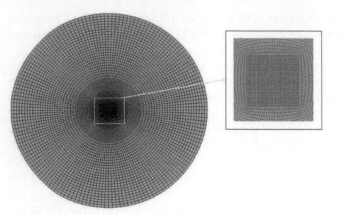

Figure 64.3 Target mesh model
Source: Author

Table 64.2 Model validation-comparison of simulation results with literature result

Impact velocity(m/s)	Residual velocity(m/s)		Error(%)
	Numerical Simulation	Experiment Result [16]	
867.8	583.9026	579.7	0.7

Source: Author

conducted and residual velocity is computed. The obtained residual velocity is validated with the literature [16] experimental results. The results are presented in Table 64.2.

From the results, it can be observed that the error percentage is 0.7%, which is considered to be acceptable limit. So, this model is extended to analyse, the effects of spinning for steel core. The spinning velocities are considered from reference [9]. For simple representation, 13000rad/s is represented as '1x', 26000rad/s is '2x', 52000rad/s is '4x', and no spin condition is represented as '0x'. Lead material is also one of the core materials for the projectile. In this study additionally, the projectile is assigned with lead material and the spin, no spin conditions are studied.

Effects of spinning on projectile penetration

From the literature [8], it is known that at the time of impact, the projectile has both translational and rotational energies. In general consideration, translational energy will make the projectile move forward inside the target, by overcoming the target resistance. In this process, the projectile will lose its energy and get slow down. If the target is harder than the projectile, the projectile will get deformed. It is proposed that, if the projectile has rotational energy along with its translational energy, it will have some additional energy to penetrate the target. The rotational energy of the projectile will exert some force on the target surface in the penetration path, which makes the target gets damaged, as well as the projectile will lose less translational energy. From the literature [10] studies, if the rotational velocity is not sufficiently high, then the projectile will lose more translational energy, and as it makes more contact with the target, it will experience more damage.

Results and Discussions

Initially, the projectile is assigned with steel material. Along with translational velocity (impact velocity) of 867.8 m/s, rotational velocity (spin) is given to the projectile. The considered velocity is well above the ballistic limit velocity [16], the projectile perforates the target, and residual velocity is obtained. It is observed that the projectile has experienced less damage, and ductile hole failure has been observed in the target. The results are tabulated in Table 64.3. The same results are shown in graphical format in Figure 64.4 and simulation images are shown in Figures 64.5 and 64.6.

It is observed that initially giving the spin of 1x to the projectile will provide additional energy in penetration, and residual velocity is increased, but because of spin, the projectile is making more contact with the target, which caused the projectile residual mass to decrease. At the spin of 2x, it is observed that the residual velocity of the projectile is increased, but the residual mass of the projectile is less compared to 0x condition, and still residual mass is greater than 1x condition. At 4x spin, it can be observed that both residual mass and residual velocity have decreased compared to 1x, and 2x conditions, but residual velocity is more compared to 0x condition, while residual mass is less compared to 0x condition.

Table 64.3 Simulation results of Steel projectile impacted on al 7075-t651 target

Spin condition (rad/s)	Impact velocity (m/s)	Residual velocity (m/s)	Initial mass (kg)	Residual mass (kg)
0x	867.8	583.9026	1.1309e-2	1.1243e-2
1x		592.7130		1.1238e-2
2x		593.1727		1.1242e-2
4x		590.2109		1.1225e-2

Source: Author

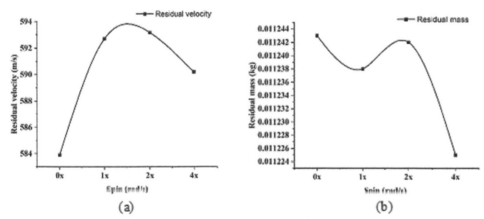

Figure 64.4 Graphical representation of results for steel projectile (a) Residual velocity(m/s) vs Spin (rad/s), (b) Residual mass(kg) vs Spin (rad/s)

Source: Author

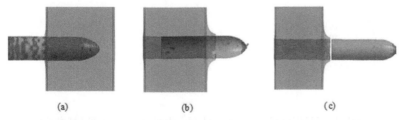

Figure 64.5 Simulation images of steel projectile penetrating Al 7075-T651 target with 0x spin at (a) 21e-6sec (b)47e-6sec (c)78e-6sec

Source: Author

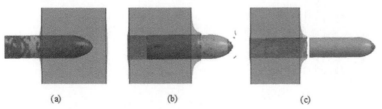

Figure 64.6 Simulation images of steel projectile penetrating Al 7075-T651 target with 2x spin at (a) 21e-6sec (b)47e-6sec (c)78e-6sec

Source: Author

By changing the projectile material to lead, under the same validated conditions, the simulations are conducted. It is observed that the lead projectile has experienced severe damage, and only a small portion of the fragment is left after perforation. Observed residual velocity and residual mass results are tabulated in Table 64.4, and a graphical representation is given in Figure 64.7. and simulation images are shown in Figures 64.8 and 64.9.

A similar kind of steel projectile behavior is observed in the case of the lead projectile also. Initially, providing the spin of '1x' to the projectile leads to a very small increase in residual velocity and a reduction in residual mass. At the spin of 2x, the residual velocity of the projectile is increased, but the residual mass of the projectile is decreased compared to 0x condition, and still residual mass is greater than 1x condition. At the 4x spin condition, both residual mass and velocity decreased and lowest of all impact conditions.

Table 64.4 Simulation results of lead projectile impacted on al 7075-t651 target

Spin condition (rad/s)	Impact velocity (m/s)	Residual velocity (m/s)	Initial mass (kg)	Residual mass(kg)
0x	867.8	750.3009	1.634e-2	4.6568e-4
1x		750.7700		4.4292e-4
2x		757.2600		4.6081e-4
4x		746.5402		2.9802e-4

Source: Author

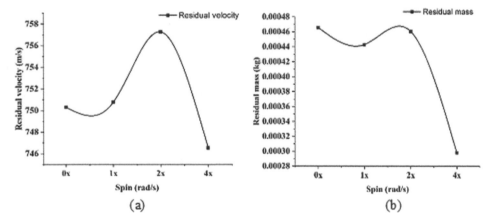

Figure 64.7 Graphical representation of results for lead projectile (a) Residual velocity(m/s) vs Spin (rad/s), (b) Residual mass(kg) vs Spin (rad/s)

Source: Author

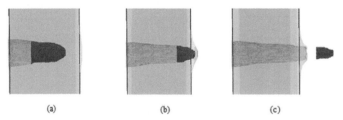

Figure 64.8 Simulation images of lead projectile penetrating Al 7075-T651 target with 0x spin at (a) 4e-5sec (b) 52e-5sec (c) 64e-5sec

Source: Author

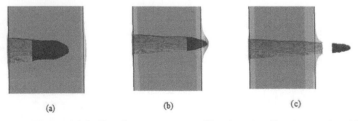

Figure 64.9 Simulation images of lead projectile penetrating Al 7075-T651 target with 2x spin at (a) 4e-5sec (b) 52e-5sec (c) 64e-5sec

Source: Author

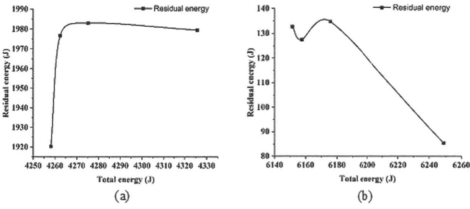

Figure 64.10 Total energy vs residual energy plots for (a) steel projectile, (b) lead projectile
Source: Author

The main motive of the projectile is to create damage to the target. In considered cases, the projectile has successfully perforated the target. The residual projectile part still had some energy, which needed to be observed to get a basic understanding of the spin effect. The total energy vs residual energy plots have been plotted for both steel and lead projectiles shown in Figure 64.10.

One of the observations from the study is that steel projectile has suffered less damage compared to lead projectile because of the high strength of steel. In general conditions, the energy of the projectile is the function of the mass of the projectile, velocity. With spinning, the spin velocity of the projectile, and the inertia of the projectile also influence the total energy of the projectile. From Figure 64.10 (a) for the steel projectile, it can be observed that with the spin of 1x residual energy is increased and reached its maximum at 2x, as spin increased to 4x and thereby decreasing the residual velocity. From Figure 64.10 (b), in the case of the lead projectile, it can be observed that with 1x spin there is a decrease in residual energy, for 2x spin residual energy has increased, with further increase of spin to 4x, there is a drop of residual energy. As the rotation of the projectile makes more contact with the target, if there is high spin velocity then the projectile with less strength will suffer more damage. By comparing projectile behavior for understanding the approximate influence of spin in penetration, it can be observed that at the initial 1x spin, both projectiles have different behavior. In steel projectile, residual energy is more compared to 0x condition, while lead projectile has less residual energy compared to 0x condition. At the 2x spin condition, both steel and lead have the highest residual energies compared to their respective 0x condition. At 4x spin conditions, the residual energy is decreased. In the case of the projectile of steel material, residual energy is greater than 0x, while in lead projectile residual energy is less than 0x and minimum of all its impact conditions.

Conclusions

From the study, it is concluded that both residual velocity and residual mass follow similar behavior with spin. The only difference observed is the magnitude of the change of parameters. By observing the residual velocity, and residual mass results, the change of pattern of these parameters with spin is observed. It is evident that providing spin to the projectile made the projectile have additional energy. The projectile with steel material has experienced less damage and an advantageous spin effect has been observed, while in the case of the lead projectile, the spin advantage is less. In both projectile cases, the 2x spin condition is the most favorable. Steel projectiles have residual energy of 1983.1 J and lead projectiles have 134.71 J of residual energy at 2x spin condition.

References

[1] Forrestal, M. J., Luk, V. K., Rosenberg, Z., and Brar, N. S. (1992). Penetration of 7075-T651 aluminum targets with ogival-nose rods. International Journal of Solids and Structures. 29(14-15), 1729–1736. Doi: https://doi.org/10.1016/0020-7683(92)90166-Q.

[2] Gupta, N. K., Ansari, R., and Gupta, S. K. (2001). Normal impact of ogive nosed projectiles on thin plates. International Jurnal of Impact Engineering. 25(7), 641–60. Doi: https://doi.org/10.1016/S0734-743X(01)00003-3.

[3] Børvik, T., Hopperstad, O. S., Berstad, T., and Langseth, M. (2001). Numerical simulation of plugging failure in ballistic penetration. International Journal of Solids and Structures. 38(34-35), 6241–64. Doi: https://doi.org/10.1016/S0020-7683(00)00343-7.

[4] Børvik, T., Dey, S., and Clausen, A. H. (2009). Perforation resistance of five different high-strength steel plates subjected to small-arms projectiles. International Journal of Impact Engineering. 36(7), 948–64. Doi: https://doi.org/10.1016/j.ijimpeng.2008.12.003.

[5] Radin, J., and Goldsmith, W. (1988). Normal projectile penetration and perforation of layered targets. International Journal of Impact Engineering. 7(2), 229–59. Doi: https://doi.org/10.1016/0734-743X(88)90028-0.

[6] Senthil, K., Iqbal, M. A., and Gupta, N. K. (2017). Ballistic resistance of mild steel plates of various thicknesses against 7.62 AP projectiles. International Journal of Protective Structures. 8(2), 177–198. doi: https://doi.org/10.1177/2041419617700007.

[7] Senthil, K., and Iqbal, M. A. (2021). Prediction of superior target layer configuration of armour steel, mild steel and aluminium 7075-T651 alloy against 7.62 AP projectile. Structures. 29, 2106–19. doi: https://doi.org/10.1016/j.istruc.2020.06.010.

[8] Farrar, C. L., and Leeming, D. W. (1982). Internal Ballistics part II, Military Ballistics, A Basic Manual. Brassesy's Publisher's Limited. pp. 39–50.

[9] Gálvez, F., Chocron, S., Cendón, D., and Sánchez-Gálvez, V. (2005). Numerical simulation of the tumbling of kinetic energy projectiles after impact on ceramic/metal armours. WIT transactions on modelling and simulation. 40.

[10] Pranay, V., and Panigrahi, S. K. (2022). Effects of spinning on residual velocity of ogive-nosed projectile undergoing ordnance velocity impact. Proceedings of the Institution of Mechanical Engineers, Part C: Journal of Mechanical Engineering Science. 236(3), 1685–97. doi: https://doi.org/10.1177/09544062211020030.

[11] Pranay, V., and Panigrahi, S. K. (2022). Design and development of new spiral head projectiles undergoing ballistics impact. International Journal of Structural Integrity. 13(3), 490–510. doi: https://doi.org/10.1108/IJSI-01-2022-0008.

[12] Pranay, V., and Panigrahi, S. K. (2022). Effecys of spinning on residual velocity of projectile for normal and oblique impact. 13th International High Enegry Materials Conference and Exhibits, TBRL, Chandigarh, India, 26–28.

[13] Jørgensen, K. C., and Swan, V. G. (2014). Modeling of armour-piercing projectile perforation of thick aluminium plates. In Proceedings of the 13th International LS-DYNA Users Conference. 2014.

[14] Flores-Johnson, E. A., Saleh, M., and Edwards, L. (2011). Ballistic performance of multi-layered metallic plates impacted by a 7.62-mm APM2 projectile. International Journal of Impact Engineering. 38(12), 1022–32. doi: https://doi.org/10.1016/j.ijimpeng.2011.08.005.

[15] Becker, M., Seidl, M., Legendre, J. F., Mehl, M., and Souli, M. (2018). Numerical ricochet model of a 7.62mm projectile penetrating an Armor steel plate. In Proceedings of the 15th International LS-DYNA Users Conference. 2018.

[16] Forrestal, M. J., Børvik, T., and Warren, T. L. (2010). Perforation of 7075-T651 aluminum armor plates with 7.62 mm APM2 bullets. Experimental Mechanics. 50(8), 1245–51. doi: https://doi.org/10.1007/s11340-009-9328-4.

Chapter 65

Wire arc additive manufacturing process parameters optimisation using Rao algorithms

Pooja Patel[a], Miit Pabari[b] and R Venkata Rao[c]

Sardar Vallabhbhai National Institute of Technology, Surat, India

Abstract

Additive manufacturing (AM), as opposed to subtractive manufacturing procedures, involves combining materials to produce products from 3D model data, often layer by layer. The Wire Arc Additive Manufacturing (WAAM) procedure involves the deposition of weld metal to produce a whole component. The WAAM process has several advantages viz. better Buy-to-Fly (BTF), and a significantly high deposition rate when compared with feed or powder bead systems. Furthermore, this wire-based method is also cost-effective to a greater extent than powder-based techniques due to the high price of the raw material. Additionally, metal wire poses less of a threat to the environment and operators' health than metal powder. WAAM is used for the production of a wide range of parts because of its inexpensive equipment costs, high precision deposition rate, and outstanding structural integrity. In this study, we investigated how modifying certain input parameters like travel speed, welding current, and voltage, can significantly affect the geometry of single weld beads in the WAAM process using Wire Arc Additive Manufacturing (WAAM) The considered output parameters are bead height and bead width. The Rao optimization algorithms are used to solve this bi-objective optimization problem. Rao Algorithms are a recently developed set of three metaphor-less and parameter-free algorithms for solving optimization problems. Rao algorithms are employed in the present work as they are easier to use and convenient to apply. Using the Rao algorithms, the optimal process parameters (v = 300 mm/min, U = 23 V, and I = 140 A) are found out as the optimal input parameters for achieving the required profile of weld beads in WAAM of 308L Stainless Steel. The proposed Rao algorithms can include any number of objectives and may be extended to process parameters optimization of other additive manufacturing processes also.

Keywords: Bead geometry, optimization, process parameters, Rao algorithms, wire arc additive manufacturing

Introduction

Additive manufacturing (AM), as opposed to subtractive manufacturing procedures, involves combining materials to produce products from 3D model data, often layer by layer. The Wire Arc Additive Manufacturing (WAAM) procedure involves the deposition of weld metal to produce a whole component. The WAAM process has several advantages viz. better Buy-to-Fly (BTF), and a significantly high deposition rate when compared with feed or powder bead systems [1]. Furthermore, this wire-based method is also cost-effective to a greater extent than powder-based techniques due to the high price of the raw material. Additionally, metal wire poses less of a threat to environment and operators' health than metal powder [2]. Many types of arcs can be used for welding in WAAM, although Wire Arc Additive Manufacturing (WAAM) is preferable for manufacturing of large-scale components since its deposition rate is twice as high as either plasma arc or gas tungsten welding.

A wide variety of published studies investigate the mechanical and microstructural characteristics of parts produced from Mg-, Al-, Ni-, and Ti-based alloys using WAAM. Previous studies used steels such as 316 L, 304, and 304 L, which are austenite stainless steels, or martensitic steels like 15-5 PH or 17-4 PH, as the feedstock material [3,4]. However, only a few studies have examined the optimum bead geometry of WAAM 308L components. An austenite stainless steel, 308L steel has good general corrosion resistance along with a low amount of carbon and strong mechanical qualities, and thus, it finds applications in numerous industries including mining, manufacturing, automobile, gas, and oil industries [5].

In this work the impact of critical process factors such as travel speed, voltage, and welding current on the geometry of weld bead is explored. Additionally, the optimal values for each process parameter are determined using Rao algorithms, and the findings have produced significant proof to support the utilization of 308L components in a variety of industrial applications.

It has been observed that the research works on WAAM process and the aspects of its parameters optimization are few in number. So far everything that has been done on WAAM is experimental and little effort was put into developing the mathematical models and applying optimization techniques to find the true optimal parameters. Le et al. [1] developed the mathematical models but only response surface methodology (RSM) has been used by the researchers to optimize the parameters and applications of advanced optimization techniques have not been utilized. RSM only studies the effect of some input parameters over the output parameters but does not give the optimum process parameters of WAAM. Many advanced

[a]poojapatel15702@gmail.com, [b]miitcp10@gmail.com, [c]rvr@med.svnit.ac.in

DOI: 10.1201/9781003450252-65

optimization techniques such as Genetic Algorithm, Particle Swarm optimization, Artificial Bee Colony, etc. are available in the literature. However, they require tuning of their algorithm-specific parameters, which increases the burden on the designer or the process planner. There might be a chance of incorrect tuning leading to inferior results. Therefore, the optimum process parameters suggested by those techniques may not be optimal. Hence, in the present study, recently developed Rao algorithms that do not require any algorithm-specific parameters have been applied to see if there can be any improvement in the optimized results.

Procedure and Objective Function

Le et al. [1] determined a range of appropriate processing parameters which would generate weld beads with specified form and quality, using experimental data by using a 308L wire of diameter 1mm.

The purpose of the study is to use the Rao optimization algorithm to consider the influence of important process parameters, like travel speed, welding current, and voltage, on weld bead geometry and find the optimal combination of welding process variables for creating a weld bead with the quality and shape that meets the requirements, with the height and width of the bead being functions of the processing parameters.

We are using the following process optimization parameters in the given range for optimization purposes: Welding current I(A) levels of 100, 120, and 140; voltage U(V) levels of 17, 20, and 23; and travel speed v(mm/min) levels of 300, 400, and 500.

The second-order regression equation is utilized in this study to create the mathematical models for prediction of bead height and width, as given in Eq. (1):

$$z = a_0 + \sum_{i=1}^{3} a_i a_i + \sum_{i=0}^{3} a_{ii} x_i^2 + \sum_{i,j=0,i\neq j}^{3} a_{ij} x_i x_j \tag{1}$$

where z is the output parameter – i.e., BW (mm) or BH (mm); x_i and x_j are the process optimization parameters; a_0, a_i, a_{ii}, *and* a_{ij} are the coefficients. Following are the comprehensive models for predicting both the bead width (BW) (Eq. (2)) and the bead height (BH) (Eq. (3)):

$$BW \text{ (mm)} = -16.190 + 0.143 \times I + 1.027 \times U + 0.508 \times 10^{-3} \times v + 0.375 \times 10^{-3} \times I \times$$
$$U - 0.016 \times 10^{-3} \times I \times v - 0.263 \times 10^{-3} \times U \times v - 0.567 \times 10^{-3} \times I^2 - 0.018 \times U^2 + 3.67 \times 10^{-6} \times v^2 \tag{2}$$

$$BH \text{ (mm)} = 7.575 + 0.085 \times I - 0.623 \times U - 0.014 \times v + 8.33 \times 10^{-6} \times I \times U + 0.024 \times 10^{-3} \times$$
$$I \times v + 0.048 \times 10^{-3} \times U \times v - 0.292 \times 10^{-3} \times I^2 + 0.013 \times U^2 + 5.223 \times 10^{-6} \times v^2 \tag{3}$$

Process Parameter Optimisation and Optimisation Algorithm

The bead geometry is highly significant in the GMAWAM process. Weld beads having large width and little fluctuation in height are normally preferred because they provide deposit stability. Thus, the issue of optimizing the processing parameters is stated as given:

Find: Travel Speed (v (mm/min)), Voltage (U (V)), and Welding Current (I (A)) and to maximize BH and BW

Constrained to: $300 \leq v \leq 500$ (mm/min); $17 \leq U \leq 23$ (V); $100 \leq I \leq 140$ (A).

The Rao optimization algorithms [6] are used to resolve this multi-objective optimization problem.

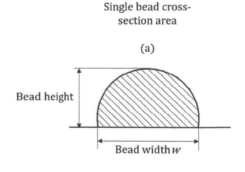

Figure 65.1 (a) Weld bead geometry parameters to be optimised, (b) Workpiece formation from continuous layers of welding beads

Rao Algorithms

The Rao Algorithms are a set of three metaphor-less and parameter-free algorithms for solving optimization problems. Accurately configuring algorithm-specific control parameters is a challenging task requiring increased calculations. Thus, the Rao algorithms, similar to the TLBO and Jaya Algorithms, were employed to solve this objective function as they are easier to use and are accessible to a wider range of audience.

In these algorithms, *f(x)* is regarded as the desired function to be maximized (or minimized). Assume that for every iteration *i*, there are 'p' candidate solutions and 'q' design variables. The worst candidate obtains the worst value of *f(x)* and the best candidate attains the best value of *f(x)*. The value of $Y_{v,c,i}$ is then altered as per the equation of particular Rao algorithm used. Also the vth variable's value for the cth candidate on the ith iteration is denoted by $Y_{v,c,i}$.

Rao-1 Algorithm

To demonstrate how the proposed Rao-1 algorithm works, we consider that the benchmark function taken into account is to be maximized. A population size of ten has been assumed alongside three design variables — v(mm/min), U(V), I(A), and with thousand iterations as the termination criterion.

$$\text{Rao-1 equation: } Y'_{v,c,i} = Y_{v,c,i} + r_{1,v,i}\left(Y_{v,best,i} - Y_{v,worst,i}\right) \quad (4)$$

$$\text{Objective function: maximise } f(x) = BW/BW_{max} + BH/BH_{max} \quad (5)$$

A flowchart depicting the process flow for the Rao algorithms has been illustrated in Figure 65.2.

Since *f(x)* is a maximization function, its highest value is ideal and its lowest value is undesirable. BW_{max} is the maximum value of the bead width when it is solved individually using the Rao algorithm with the same ranges of variables. BH_{max} is the maximum value of the bead height when it is solved individually using the Rao algorithm with the same ranges of variables. Eq. (5) indicates the combined objective function with the same ranges of the variables and giving equal weightages to the two objective functions.

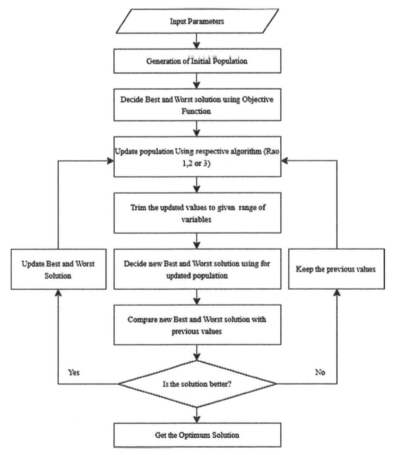

Figure 65.2 Process flow of Rao algorithms
Source: Author

Table 65.1 Calculation of benchmark function using Rao-1 algorithm.

Candidate	I(A)	U(V)	v(mm/min)	BW	BH	f(x)
1	125.3187	18.7932	363.1555	4.7978	3.5686	0.7944
2	127.7249	20.9974	300.0000	5.6718	3.8326	0.8959
3	113.5926	21.9329	492.9511	4.8457	2.3178	0.6644
4	118.8112	19.2639	394.5640	4.6096	3.0610	0.7488
5	112.6802	19.5028	300.0000	5.1342	3.6203	0.8279
6	126.3762	17.0000	306.8641	4.3979	4.2142	0.8339
7	111.1274	18.7069	300.0000	4.8650	3.6666	0.8110
8	100.0000	17.0000	300.0000	3.9959	3.5700	0.7287
9	117.0322	21.8790	327.0966	5.5138	3.1825	0.8440
10	105.1690	18.0139	319.4883	4.4332	3.4393	0.7500

Source: Author

Table 65.2 Results obtained after using Rao 1,2 and 3 algorithms.

Objectives	I(A)	U(V)	v(mm/min)	BW	BH	f(x)
Max *BW*	129.4753	23	300	6.0821	3.7917	0.925
Max *BH*	140	17	300	4.3357	4.4604954	0.8564
Max *f(x)*	140	23	300	6.0193	3.9359	0.936031413

Source: Author

Optimizing the combined objective function will give the optimum values of process parameters considering both the objectives simultaneously. The values of the process parameters, *BW* and *BH*, as well as the sample calculation for combined objective function, at the end of first iteration, are shown in Table 65.1.

Rao-2 Algorithm

The combined objective function remains the same, but the algorithm equation changes.

$$Y'_{v,c,I} = Y_{v,c,i} + r_{1,j,i} \left(Y_{v,best,i} - Y_{v,worst,i} \right) + r_{2,v,i} \left(|Y_{v,c,i} \text{ or } Y_{v,l,i}| - |Y_{v,l,i} \text{ or } Y_{v,c,i}| \right) \tag{6}$$

Rao-3 Algorithm

The combined objective function remains the same, but the algorithm equation changes.

$${''}_{v,c,i} = Y_{v,c,i} + r_{1,v,i} \left(Y_{v,best,i} - |Y_{v,worst,i}| \right) + r_{2,v,i} \left(|Y_{v,c,i} \text{ or } Y_{v,l,i}| - (Y_{v,l,i} \text{ or } Y_{v,c,i}) \right) \tag{7}$$

For more details regarding Eq. (4), Eq. (6), Eq. (7) reader may refer here .

Results and Discussions

Using the Rao algorithms, the optimum parameters for the manufacturing process are determined so that the final beads have the greatest possible height and width. We then determine the optimum parameters for the travel speed, welding current, and voltage by optimizing the combined objective function, *f(x)*. It is found that all three algorithms produced the same results which are tabulated as follows.

In the case of bead width, it is observed that travel speed (*v*(mm/min)) and voltage (*U*(V)) have opposing impacts. Rise in voltage from 17 V to 23 V causes increase in width of bead, however, it decreases when travel speed is increased from 300 to 500 mm/min. In addition, width of the weld bead rises proportionally with welding current up to about 120 A, after which it remains roughly constant. Meanwhile, as the welding current rises from 100 to 140 A the height of the weld bead increases, but decreases with both increased velocity and voltage. Table 63.2 shows the results obtained after using Rao algorithms.

The results obtained after using Rao algorithms (*I* = 140 A, *U* = 23 V, *v* = 300 mm/min) were superior as compared to the RSM algorithm used by Le et al [1] which gave the optimized processing parameters as *I* = 122 A, *U* = 20 V, *v* = 368 mm/min.

Table 65.3 Variation of objective function according to weightage of BW and BH.

Percentage BW	f(x)	BW	BH
90%	0.978946515973717	6.0797	3.8253
80%	0.971624869216491	6.0717	3.8583
70%	0.957488695554704	6.0571	3.8903
60%	0.946760054418324	6.0345	3.9209
50%	0.936031413281944	6.0193	3.9359
40%	0.925302772145564	6.0193	3.9359
35%	0.919938451577374	6.0193	3.9359
30%	0.913857978227859	6.0193	3.9359
29.9%	0.914466844597821	6.0193	3.9359
29.8%	0.914432251497844	4.3357	4.4605
29%	0.91672934457778	4.3357	4.4605
28%	0.919600710927701	4.3357	4.4605
25%	0.928214809977463	4.3357	4.4605
20%	0.942571641727066	4.3357	4.4605
10%	0.971285305226272	4.3357	4.4605

Source: Author

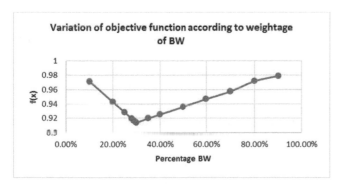

Figure 65.3 Variation of combined objective function according to weightage of *BW*
Source: Author

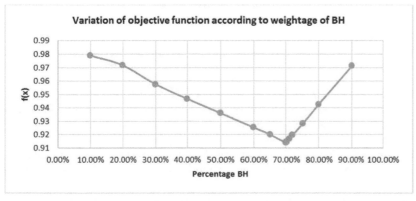

Figure 65.4 Variation of combined objective function according to weightage of *BH*
Source: Author

In addition, we also explored other possibilities by varying the weightage of bead height and width in the combined objective function. This is done to demonstrate how the optimal values for percentage change in weightage in bead width and bead height would be obtained, according to different specifications. The weights assigned to *BW* and *BH* are shown in Table 65.3 which depicts the resultant variation to the objective function. The summation of weightages given to *BW* and *BH* is 100%. Depending upon his or her

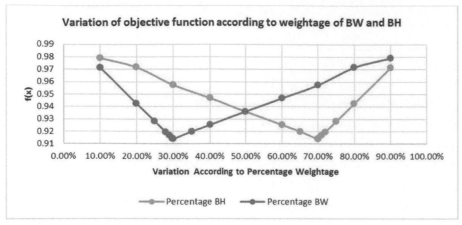

Figure 65.5 Variation of combined objective function according to weightage of *BW* and *BH*
Source: Author

preferences (i.e., weightages assigned to the two objectives), the process planner may choose any one of these solutions.

Figure 65.3 we can see how the combined objective function shifts depending on the weightage of *BW*. Figure 65.4 illustrates how the combined objective function shifts depending on the weightage of *BH*. And Figure 65.5 shows the variation of combined objective function according to the weightage of *BW* and *BH*.

Conclusions

- The voltage, travel speed, and welding current have a significant impact on the form and size of individual weld beads produced by WAAM using GMAW method. The width of the bead is significantly affected by the Voltage and Travel Speed, and the height of the bead by the speed and welding current.
- Using the Rao algorithms, the optimal process parameters (v = 300 mm/min, U = 23 V, and I = 140 A) are found out. With these ideal characteristics, we can produce single-weld beads of the appropriate height and width with a high degree of uniformity.
- The proposed multi-objective approach can include any number of objectives and the approach may be extended to process parameters optimization of other additive manufacturing processes also.

References

[1] Le, V. T., Mai, D. S., Doan, T. K., and Paris, H. (2021). Wire and arc additive manufacturing of 308L stainless steel components: Optimisation of processing parameters and material properties. Engineering Science and Technology, an International Journal. 24,1015–26. https://www.sciencedirect.com/science/article/pii/S2215098621000094

[2] Wu, B., Pan, Z., Ding, D., Cuiuri, D., Li, H., Xu, J., and Norrish, J. (2018). A review of the wire arc additive manufacturing of metals: properties, defects and quality improvement. Journal of Manufacturing Processes. 35, 127–39. https://doi.org/10.1016/j.jmapro.2018.08.001

[3] Wu, W., Xue, J., Zhang, Z., and Yao, P. (2019). Comparative study of 316L depositions by two welding current processes. Materials and Manufacturing Processes. 34, 1502–8. https://doi.org/10.1080/10426914.2019.1643473

[4] Wang, D., Chi, C. T., Wang, W. Q., Li, Y. L., Wang, M. S., Chen, X. G., Chen, Z. H., Cheng, X. P., and Xie, Y. J. (2019). The effects of fabrication atmosphere condition on the microstructural and mechanical properties of laser direct manufactured stainless steel 17–4 PH. Journal of Materials Science & Technology. 35, 1315–22. https://doi.org/10.1016/j.jmst.2019.03.009

[5] Bajaj, P., Hariharan, A., Kini, A., Kürnsteiner, P., Raabe, D., and Jägle, E. A. (2020). Steels in additive manufacturing: a review of their microstructure and properties. Materials Science and Engineering A. 772. 138633. https://doi.org/10.1016/j.msea.2019.138633

[6] Rao, R. V. (2020). Rao algorithms: Three metaphor-less simple algorithms for solving optimisation problems. International Journal of Industrial Engineering Computations. 11, 2–3. http://dx.doi.org/10.5267/j.ijiec.2019.6.002

Chapter 66

A case study to improve the productivity of ladle gearbox manufacturing

Pratheesh Kumar S[a], Rajesh R[b], Raashika R[c], Mohanraj R[d], Varunkumar T C[e], Kelvin Mark V[f] and Lakshmanan A[g]

PSG College of Technology,Coimbatore, India

Abstract

Gearbox manufacturing company was confronted with the problem of only being able to produce seven gear boxes per month, rather than the projected twelve gear boxes per month. A process-oriented approach was used by the organisation. Further research revealed that the procedure requires a substantial amount of time that is not used to create value (i.e., transport time, waiting time, quality rework time, poor work environment and poor inventory management). Company decided to employ lean tools to boost the product's production rate by eliminating as many wastes as possible during the production process. The layout was altered from a process to a cellular pattern as part of the project. 5S techniques were then implemented, which enhanced the working atmosphere while also giving employees a greater sense of control over the machine they were using. At long last, the company put in place an inventory-reduction Kanban system, allowing it to go from batch production to JIT manufacturing.

Keywords: 5S, gearbox manufacturing, kanban, lean manufacturing, productivity

Introduction

Company considered for case study was developed with the goal of marketing coir fibre extraction machinery both locally and globally. As part of its strategy for growth and diversification, the company started making foundry equipment such ladles and ladle handlers. Project orders with a focus on heavy, non-standard activities started coming in as well. With its excellent infrastructure and internationally competitive products given at a cheap price it began exporting machines to various other nations. Because of the company's low-maintenance and tough goods, which are backed by teams of highly qualified and experienced personnel and highest quality standards, absolute on-time deliveries, and rapid after-sales services, they offer amazing value in the market.

Value Stream Mapping

One of the methods used to enhance flow of information and materials in a lean manufacturing or lean company is known as value stream mapping. In addition, this tool aids in the identification of waste and the formulation of a strategy for its reduction [1, 2]. A product's two basic flows are the production flow from raw materials into the hands of customers and the design flow from the design to launch. This includes both value and non-value added operations. To summarise, value streaming is critical tool in any organisation for the following eight reasons.

i. Assembling, welding, and so on are only a few examples of how it helps to see the complete production process.
ii. You may eliminate waste by using mapping to identify where it happens in your value chain.
iii. As an added benefit, it provides a consistent vocabulary for discussing industrial processes.
iv. Bringing up the decisions that were made helps to keep the narrative on track. If you don't, a lot of things will happen by chance on your shop floor.
v. To put it another way, it shows the relationship between data flows and material flows. The market is devoid of anything similar.
vi. It lays the groundwork for a plan to put the strategy into action. Planning the entire door-to-door flow is something it helps with, filling a critical void in many lean programs. Lean manufacturing can be implemented using value stream maps as a guide.
vii. The lead time, the distance travelled, and the stock are far more crucial than quantitative methods and layout concepts when it comes to calculating the non-value-added stages. The value stream mapping allows you to explain in great detail how your facility should work in order to produce flow, however this is a qualitative technique. When expressing a sense of urgency or serving as a reference point, numbers are a powerful tool. Value stream mapping is an effective technique for describing how you want to influence those figures.

DOI: 10.1201/9781003450252-66

Cellular Manufacturing

Component families are collections of related items that require similar processing, and cellular manufacturing is an architecture in which machines are arranged in groups based on process requirements [3, 4]. Using cellular manufacturing in medical device production is common. These clusters of humans are referred to as cells. Thus, a cellular layout refers to a piece of equipment architecture created specifically for use in cell production. Group technology is a method for organising processes into cells (GT). Products are sorted into groups based on their similarity in size, shape, and function as well as their shared production characteristics (type of processing required, available machinery that performs this type of process, processing sequence). In order to run all of the equipment in a cell and take responsibility for its output, those who work in cellular layouts undergo cross-training. In other circumstances, the cells will be fed to an assembly line, where they will be used to make the final product. If you want a cell, you can get one by dedicating certain equipment to making a group of parts rather than transferring the machineries into its own cell. As a result, the organisation avoids the hassle of having to change the current structure of its layout [5, 6]. The cellular production process has been automated in the flexible manufacturing system (FMS). Computer-assisted flow management systems (FAMs) allow producers to reap the features of mass production without giving up the adaptability afforded by low-volume production [7].

The following are the three advantages of cellular manufacturing,

i.   By cutting down on things like processing time, material handling, WIP inventory, and setup time, cellular manufacturing helps keep costs down.
ii. Small-batch production offers for greater versatility in the manufacturing process.
iii. Workers are less likely to get bored on the job when they are cross-trained to use all of the plant's machinery. Because workers are in charge of their own cell manufacturing, they feel more in control and ownership of their work.

As an illustration, consider a project containing four machines and six parts.

The data from Tables 66.1 and 66.2 can be used to determine whether it is better to keep machines 1 and 2 together in one cell while moving machines 3 and 4 to another cell in order to have faster processing time.

5S Method

Clean, tidy, efficient, and able to consistently produce high-quality output are all hallmarks of a 5S workspace. This method is fast since it is methodical, terms-based, and easy to learn and execute [8–10]. As a result of how it accomplishes this, the 5S method is termed as such because each stage begins with the letter "s". Individuals who work in the region implement 5S as a process of continuous improvement rather than having it imposed from above. To them, all the improvements and changes they bring about are their own doing and their own doing alone.

1) Benefits of 5s

The number of benefits driven by the implementation of 5S

i.   You save a lot of time by not having to squander it looking for stuff or moving around tonnes of debris in the first step.
ii. Goods (such parts, equipment and personnel) are therefore guaranteed to be in the most ergonomic and efficient locations. As a result, several of the seven types of production waste are no longer present. Each of these objects is placed in the most ergonomic location and at the lowest height and orientation possible in order to minimise handling.
iii. If your working environment is clutter-free and any malfunctioning signs are evident, you can take action to avoid more serious failures and other delays.

Table 66.1 Manufacturing cell layout.

Machines	1	2	3	4	5	6
M1		1		1		1
M2		1		1		1
M3	1		1		1	
M4	1		1		1	

Source: Author

Table 66.2 Modified cellular layout.

Machines	2	4	6	1	3	5
M1	1	1	1			
M2	1	1	1			
M3				1	1	1
M4				1	1	1

Source: Author

iv. It's critical to check that everyone is following the same practises before moving on to step v. As a result, the most efficient working method is adopted, and standard operating procedures are formed. The result is a decrease in the number of issues including mistakes, delays, and other inefficiencies. A simple way for increasing productivity and reducing waste in your company is to use relevant tools and methods in the proper locations. As long as you have a dedicated team behind you, you can achieve your goal!

v. There is no slipping and your activities are constantly challenged by including your employees on a regular basis, which results in modifications. Last but not least is

vi. Sustain, which is responsible for making certain that everyone's duty is upheld on an ongoing basis.

Kanban System

Everything you need to know about a product's journey from inception to completion is on this card. It even includes which elements you'll need at later stages. Work in progress (WIP), production, and inventory movement are all handled via the card system [11, 12]. Manufacturing processes such as just-in-time (JIT) can be implemented in a Kanban-based organisation. Just-in-time manufacturing, also known as "pull" type production, uses real orders as signals for when a product should be produced based on those orders. Only when and in the exact quantity needed can a company manufacture with demand-pull manufacturing [13, 14]. It's possible to keep raw materials and components at a minimum while also reducing inventory levels. This requirement necessitates meticulous planning of production schedules and resource flows. The two most common types of kanban are, production Kanban and withdrawal kanban.

Venkataraman [15] and his colleagues completed the study. The decision-making procedure in a manufacturing system can be examined using a multi-criteria decision-making model and an analytical hierarchy approach. The company behind the lawsuit wanted to boost exports with these initiatives. The organization has decided that the lean manufacturing system is the most suitable method for achieving its quality, cost, and timeliness objectives. This single-piece flow crankshaft passed all testing and validation requirements, and the customer approved its use in any variant that they wished to make in their firm because it was manufactured using low-cost machinery created in-house. A 40% reduction in production lead time was accomplished through the implementation of lean manufacturing systems. Errors were also lowered as a result of the implementation, and the overall process capability was improved. At a Malaysian auto company in 2013, Ahmad Naufal Bin Adnan and colleagues [16] conducted an investigation into the pre-requisite activities for implementing Kanban, starting with the design of Kanban flow, gathering manufacturing data, calculating the optimal Kanbans in the systems, and establishing the pull mechanism and rule. The researchers then evaluated Kanban performance using lean parameters. The implementation was restricted to the BLM Cylinder Head Cover assembly process. the scope. They also claim that the Kanban approach has decreased lead times, reduced inventory on the floor, and simplified storage space. Researchers want to show that using the Kanban approach improves a manufacturing system while still maintaining the Just in Time principle. Using the 5S concept, J. Michalska and colleagues [17] presented a strategy for implementing it in their organization in 2007, which was eventually accepted. According to our findings, implementing the 5S rules leads to significant changes in the organization, such as process improvement through cost reduction, increased effectiveness and efficiency in the processes, maintenance and improvement of machine efficiency, increased safety and reduction of industry pollution, and the implementation of decisions, as well as our findings suggest. According to our findings, training staff on the 5S criteria is crucial to their long-term performance. The most important thing is to break activities down into smaller, more manageable components and to keep working on them. In a corporation, every program for improvement begins with the 5S approach. Any organization can use this approach. Organizing and streamlining your office will have a positive impact on your productivity. Lean manufacturing background, an overview of manufacturing wastes and an introduction to the tools and methods required to transform a company into high-performance lean enterprises are all addressed in detail, according to the work done in 2008. Most companies use a technique known as value stream mapping to find areas for implementing various lean techniques. Non-Value-added operations are eliminated in order to reduce costs in lean manufacturing. The automobile, electronics, and consumer products industries have all profited from this new technology. The value stream mapping (VSM) technique is used in this study to map the current working conditions of a production line. Waste sources are highlighted on this map, as well as possible lean methods for reducing waste. A Kanban system is proposed for the pre-machining segment to eliminate the clutter, while a single-piece flow concept is proposed for the machining region. Both should be considered. After that, a system for future state map was created and lean technologies were used to improve it. As early as 2011, Sree and colleagues [18] published a paper outlining their efforts to create an efficient and performant document clustering algorithm. In developing cluster merging criteria, it is largely concerned with investigating and using cluster overlapping phenomena. The majority of the effort is going towards developing a new way of determining the overlap rate that will improve both time efficiency and "the veracity." The Hierarchical clustering method is used to describe the implementation of the Expectation-Maximisation (EM) method in the Gaussian Mixture Model for counting parameters and joining two sub-clusters when their overlap is highest. Experiments show that this method can improve the effectiveness of clustering while also reducing computation time in both public and document clustering data. Depending on how much overlap there is between components of a data set that satisfy the Gaussian distribution, a human operator or clustering computer may be able to identify more clusters in the data. When comparing the two ways, there may be a significant difference between intuitively defined clusters and real clusters that match the mixture's components. EMS plant case studies were used to apply Nittaya Ngampak et al [19] analysis of systematic layout planning (SLP) for cellular manufacturing layout design and analytic hierarchy process (AHP) for facility layout selection. The current layout of this factory needs to be altered to meet the many types of medical crises. Being flexible in the face of change is also important. The data from this analysis was used to create a cellular manufacturing strategy for the items chosen. Manufacturing cellular layouts were investigated, planned, and implemented utilizing SLP. As a result of this evaluation, the selection criteria for plants were defined. Each of these metrics has a monetary value ascribed to it by AHP. Next, the best cell arrangement design was selected. As a result of this case study, a realistic guideline for EMS layout design and selection has been established. On the other hand, Maria-Florina Balcan and colleagues [20] provide a new comprehensive paradigm for studying clustering using similarity information that directly addresses the question of what qualities of a resemblance measure are adequate to accurately cluster and what techniques. For instance, we may use learning-theoretic and game-theoretic concepts from mathematical biology to design unique, extremely effective algorithms for data clustering. Two primary clustering goals are taken into account, namely: In contrast to the goal of a list clustering algorithm, which is to generate a small set of clusterings, of which at least one is roughly appropriate, the objective of a hierarchy clustering method is to generate a hierarchy such that the desired clustering is a trimming of the branches of this tree. Using learning theory, a concept of clustering complexity for a certain characteristic is constructed and explored across a large range of attributes, with strict upper and lower bounds on both. They also show that our methods apply to the situation when a consistent sample size is employed, such as in property testing, and how this can be done. To demonstrate the accuracy of these incredibly efficient algorithms, sophisticated research based on regularity-type results is required.

Need for Study of Production Time

The ladle gear box is now being manufactured by the company according to the process layout. As a result, extra non-value-added time is required, such as transportation time and quality rework time. Inadequate working conditions result in low tool life, and sometimes incorrect tool placement results in a lengthy search process that takes much of one's time. In addition, they have a big amount of inventory in the process. In the current phase of the investigation, the following objectives are being pursued:

i.   Keeping the current situation in mind, the work must be completed for their new plant, which is 120 by 120 feet in size, and the layout must be changed from a process layout to a cellular pattern in order to eliminate non-value-added times.
ii. 5S strategies should be implemented in order to improve the nature of the workplace and the sense of ownership felt by employees.
iii. The development of a kanban system is required to improve inventory management.

Methodology

The methodology used in this study is illustrated as a flow chart in Figure 66.1.

Current State Mapping

Current Condition

The company has been employing a department arrangement in which the lathe, drilling, milling, and gear hobbing machines are all located in distinct departments since it began operations. Once a part has completed one of its process sequences, it must be transported to another department for additional processing, resulting in some non-value time that the client is unwilling to pay for.

i.   When a part is reworked, it must be returned to its department for quality assurance purposes. Because the item is too heavy to be moved manually, a crane must be utilized for the lifting, which adds non-value-added time to the process.

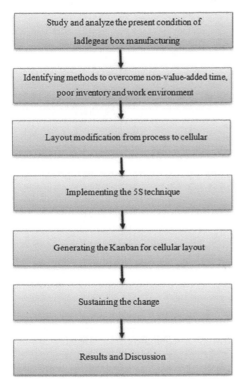

Figure 66.1 Methodology flow chart
Source: Author

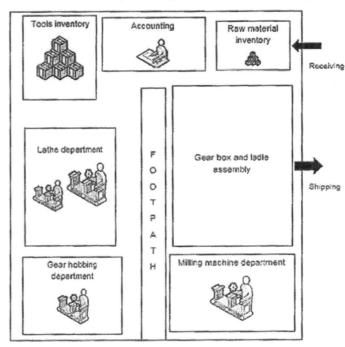

Figure 66.2 Present plant layout
Source: Author

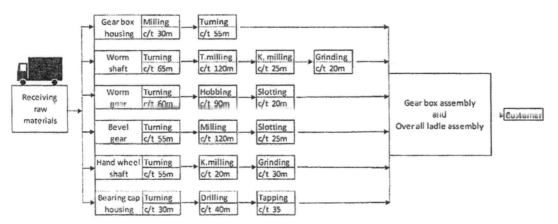

Figure 66.3 Current state mapping
Source: Author

ii. As a result of the worker spending time searching for the tool itself because it is not in the order and the worker practicing in a terrible working environment, the quality of the work is compromised.

iii. In addition, they do not have effective inventory management, which results in an excess of inventory and a shortage of inventory of raw materials on hand.

Present Layout of the Plant

The size of the layout is 60 × 120 square feet. The present layout of the plant is shown in Figure 66.2

Current State Mapping

The current state of the company illustrates in Figure 66.3. where the milling denotes thread milling and k.milling denotes keyway milling.

Time for a Gear Box

At the moment, the percentage of value-added time in the factory will be 0.9218 percent of the overall time spent in the plant. In addition, seven gear boxes will be manufactured each month. Figure 66.4 shows the time required for the fabrication of each component of a gear box assembly.

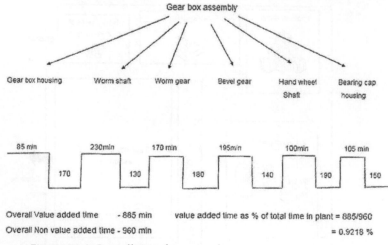

Overall Value added time   - 885 min   value added time as % of total time in plant = 885/960

Overall Non value added time - 960 min   = 0.9218 %

Figure 66.4 Overall time for a gear box
Source: Author

Table 66.3 Step 1 ROC.

	Worm wheel shaft (1)	Hand wheel shaft (2)	Bevel gear (3)	Worm gear (4)	Gear Box housing (5)	Bearing cap housing (6)
Lathe 1 (1)	1					
Thread milling (2)	1					
Keyway milling (3)	1	1				
Grinding (4)	1	1				
Lathe 2 (5)		1				
Drilling (6)		1				
Lathe 3 (7)			1	1		
Milling (8)			1			
Hobbing (9)				1		
Slotting (10)			1	1		
Vertical milling (11)					1	
Lathe (12)					1	1
Drilling (13)					1	1

Source: Author

Overall value-added time = 885 min
Overall non-value-added time = 960 min
Value added time as % of total time in plant = 885/960
= 0.9216 %

Layout Modification

From process layout to cellular layout, the design has been altered. As a result, the parts that have a similar process sequence are grouped together and organized in the form of cells. Because there are two types of algorithms that can be used to sort out the machines, it is possible to use both. The rank order clustering algorithm is employed in this piece of work.

Step 1: The part and the machines are tabled, and if the component makes use of the machine, it is indicated as 1, and if it does not, it should be left blank or zero as shown in the Table 66.3.

Step 2: The next step is weight has to be given to position from right to left by giving $2^0, 2^1, 2^2, 2^3, 2^4, 2^5$ and ranking it based on the value it gets from the weight as shown in Table 66.4.

Step 3: The machines have to be arranged based on the rank and they tabled as shown in Table 66.5. And now the weight has to been given to column in the descending order.

Step 4: Since there is no change in order due to the same rank they gets and parts have been highlighted based on the process as shown in Table 66.6.

Step 5: By concluding

Table 66.4 Step 2 ROC.

Weight	2^5	2^4	2^3	2^2	2^1	2^0	Total	Rank
Part	1	2	3	4	5	6		
M1	1						32	3
M2	1						32	4
M3	1	1					48	1
M4	1	1					48	2
M5		1					16	5
M6		1					16	6
M7			1	1			12	7
M8			1				8	9
M9				1			4	10
M10			1	1			12	8
M11					1		2	13
M12					1	1	3	11
M13					1	1	3	12

Source: Author

Table 66.5 Step 3 ROC.

Weight	Rank	P1	P2	P3	P4	P5	P6
2^12	3	1	1				
2^11	4	1	1				
2^10	1	1					
2^9	2	1					
2^8	5		1				
2^7	6		1				
2^6	7			1	1		
2^5	10			1	1		
2^4	8			1			
2^3	9				1		
2^2	12					1	1
2^1	13					1	1
2^0	11					1	
Total		7680	6528	112	104	7	6
Rank		1	2	3	4	5	6

Source: Author

i.   In cell 1, the machines 1,2,3,4,5,6
ii. In cell 2, the machines 7,8,9,10
iii. In cell 3, the machines 11,12,13

The Figure 66.5 is the model layout formed from the above result of rank order clustering algorithm.

It was created using the CATIA program, and the new plant dimensions were taken into consideration when creating the layout shown in Figure 66.6. Also, the dimensions are 120 feet by 120 feet, and the machines have been carefully organized in accordance with their sizes and available walk-through area.

5S Method

As a result of their poor working environment and the continual damage to equipment in their shop, the gear manufacturing business has implemented 5S. When compared to milling, grinding, and lathe work, thread milling, gear hobbing, and drilling machines have terrible work environments, hence attempt to implement 5S is made in these machines.

Table 66.6 Step 4 ROC.

Rank	P1	P2	P3	P4	P5	P6
3	1	1				
4	1	1				
1	1					
2	1					
5		1				
6		1				
7			1	1		
10			1	1		
8			1			
9				1		
12					1	1
13					1	1
11					1	

Source: Author

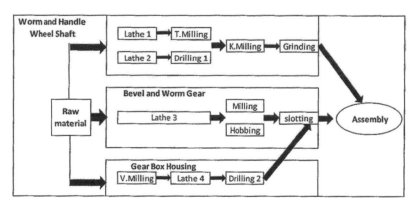

Figure 66.5 Layout model
Source: Author

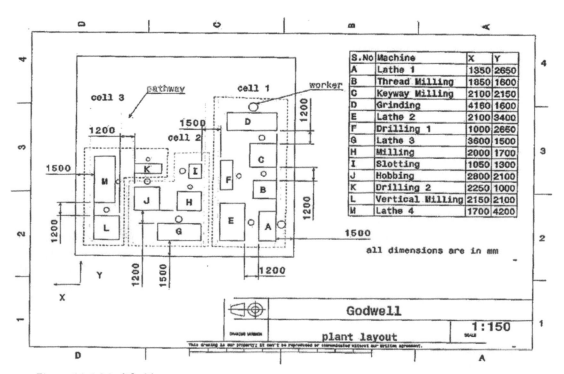

Figure 66.6 Modified layout
Source: Author

Kanban System

The kanban system is being developed for their new facility in response to issues such as excess inventory, a lack of inventory, and a higher in-process inventory that they were experiencing at their old plant. So the above problem could be resolved by installing the kanban system, and the continual improvement could be achieved by doing adequate maintenance on a regular basis.

Card Generation

Production Kanban Card

The production kanban card is shown in Figure 66.7.

Withdrawal Kanban Card

The withdrawal kanban card is shown in Figure 66.8.
    The Figure 66.9 is the flow diagram of the kanban cards.

i.   Within the cell, it was a single kanban system, where the worker from the succeeding station gets the job from the previous station by using the withdrawal card. And there are no external persons involved for the movement of jobs.
ii. Overall it's a dual kanban system, where there is an external member who uses the withdrawal or transportation card for the movement purpose.

Overall Flow in the Industry

i.   Once an order has been placed, the supplier must deliver the materials within the specified time frame, after which they will be received by the storage department.
ii. When the part is finished in the cell, the external member takes the parts and gives the production card to the concerned worker, who then returns the parts to the external member.
iii. The worker then uses his withdrawal card to retrieve the jobs from storage and begins the machining operation as soon as he receives the production card.

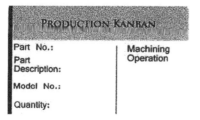

Figure 66.7 Production kanban card
Source: Author

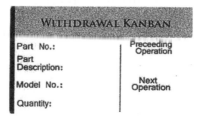

Figure 66.8 Withdrawal kanban card
Source: Author

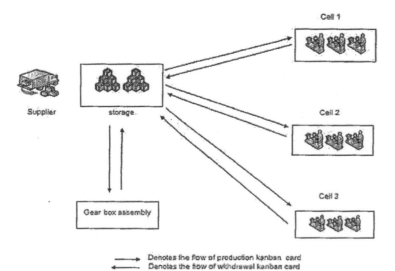

Figure 66.9 Kanban card flow
Source: Author

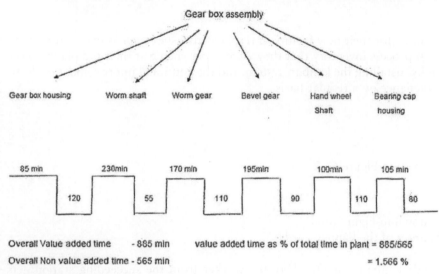

Overall Value added time   - 885 min   value added time as % of total time in plant = 885/565

Overall Non value added time - 565 min   = 1.566 %

Figure 66.10 Future state mapping
Source: Author

iv.   In the gear box assembly area, when a worker receives a production card, he sends a signal to the storage department, which then retrieves the necessary parts for the assembly process.

v.   It is recommended that workers begin operations immediately after receiving a production card; otherwise, the worker and the machine will be idle.

Flow within the Cell

i.   When the external member hands over the production card to the last member in the cell, the worker retrieves the part from the previous station using his withdrawal card and hands over the production card, which has been placed on the material to be handed over to the previous station's worker, thereby kicking off the manufacturing process.

ii.   After the worker receives the production card, he moves to the previous station, retrieves the part using his withdrawal card, and passes the production card to the worker who is in charge of the project.

Overall value-added time = 885 min
Overall non-value-added time = 565 min
Value added time as % of total time in plant = 885/565
= 1.565%

And after the layout modifications, 5S, kanban system the value-added time as percentage of total time in plant will be 1.566% as shown in Figure 10, which is increase of 0.6442% when compared with current state mapping. Now they can produce nine gear boxes per month.

Sustaining Change

i.   A regular maintenance schedule for cellular layout should be followed in order to ensure that the machines are operating properly.

ii.   The employee should be compensated for his or her ownership of the machines.

iii. In addition, the inventory must be checked on a regular basis by the management.

Conclusion

i.   By switching from a process-oriented to a cellular-oriented arrangement, the overall transport time can be reduced by up to 40%.

ii.   Rework of work will be easier as compared to process layout since, anytime there is a quality issue, the work may be sent back to the prior operations in a much shorter amount of time than with process layout.

iii. Cellular manufacturing reduces the amount of time required for manufacturing. Furthermore, it will provide considerably greater flexibility, allowing production to be carried out wherever demand changes, whether it is higher or lower.

iv. In addition to creating a positive work atmosphere, the use of 5s encourages employees to assume greater responsibility and ownership for the equipment.

v. The kanban system decreases in-process inventory to a greater extent, which encourages the company to adopt just-in-time (JIT) manufacturing, which minimizes the total amount of raw materials on hand.

References

[1] Forno, A. J. D., Pereira, F. A., Forcellini, F. A., and Kipper, L. M. (2014). Value Stream Mapping: a study about the problems and challenges found in the literature from the past 15 years about application of Lean tools. The International Journal of Advanced Manufacturing Technology, 72, 779–790.

[2] Bhim, S., Garg, S. K., Sharma, S. K., and Grewal, C. (2010). Lean implementation and its benefits to production industry. International Journal of Lean Six Sigma. 1(2), 157–68.

[3] Michalska, J., and Szewieczek, D. (2007). The 5S methodology as a tool for improving the organisation. Journal of Achievements in Materials and Manufacturing Engineering. 24(2), 211–14.

[4] Kumar, V. (2010). JIT based quality management: concepts and implications. International Journal of Engineering Science and Technology. 2, 40–50.

[5] Singh, B., Garg, S. K., Sharma, S. K., and Grewal, C. (2010). Lean implementation and its benefits to production industry. International journal of lean six sigma, 1(2), 157–168.

[6] Frein, Y., Di Mascolo, M., and Dallery, Y. (1995). On the design of generalized kanban control systems. International Journal of Operations & Production Management, 15(9), 158–184.

[7] Antony, J. (2011). Six Sigma vs Lean: Some perspectives from leading academics and practitioners. International Journal of Productivity and Performance Management, 60(2), 185–190.

[8] Gomes, D. F., Lopes, M. P., and de Carvalho, C. V. (2013). Serious games for lean manufacturing: the 5S game. IEEE Revista Iberoamericana de Tecnologias del Aprendizaje. 8(4), 191–6.

[9] Veres, C., Marian, L., Moica, S., and Al-Akel, K. (2018). Case study concerning 5S method impact in an automotive company. Procedia Manufacturing. 22, 900–905.

[10] Korenko, M., Földešiová, D., Máchal, P., Kročko, V., and Beloev, C. (2014). 5s method as a set of measures to improve quality of production in the organisation. Agricultural, Forest and Transport Machinery and Technologies. 1(1), 27–32.

[11] Junior, M. L., and Godinho Filho, M. (2010). Variations of the kanban system: Literature review and classification. International Journal of Production Economics. 125(1), 13–21.

[12] Rahman, N. A. A., Sharif, S. M., and Esa, M. M. (2013). Lean manufacturing case study with Kanban system implementation. Procedia Economics and Finance. 7, 174–180.

[13] Huang, Chun-Che, and Andrew Kusiak. (1996). Overview of Kanban systems. 169–189.

[14] Monden, Y. (1994). Adaptable kanban system maintains just-in-time production. In Toyota Production System Boston, MA: Springer,. pp. 15–35.

[15] Gupta, S. M., Al-Turki, Y. A., and Perry, R. F. (1999). Flexible kanban system. International Journal of Operations and Production Management, 19(10), 1065–1093.

[16] Pattanaik, L. N., and Sharma, B. P. (2009). Implementing lean manufacturing with cellular layout: a case study. The International Journal of Advanced Manufacturing Technology. 42(7), 772–9.

[17] Green, J. C., Lee, J., and Kozman, T. A. (2010). Managing lean manufacturing in material handling operations. International Journal of Production Research. 48(10), 2975–93.

[18] Anand, G., and Kodali, R. (2009). Simulation model for the design of lean manufacturing systems–a case study. International Journal of Productivity and Quality Management. 4(5-6), 691–714.

[19] Sivakumar, R., Boobal, A., Gowtham, M., and Perumal, P. S. (2021). To reduce the setting piece rejection rate in gear hobbing process by advanced product quality planning. In Advances in Materials Research. Singapore: Springer. pp. 375–383.

[20] Nallusamy, S. (2016). Lean manufacturing implementation in a gear shaft manufacturing company using value stream mapping. International Journal of Engineering Research in Africa.

Chapter 67

The effect of downstroke angle of attack on the aerodynamic performance of dragonfly during take-off

Shubham Tiwari[a] and Sunil Chandel[b]

DIAT(DU) Pune, India

Abstract

A numerical analysis is performed to study the effect of the angle of attack during mid-downstroke (α_D) on the aero-dynamic performance of dragonfly flight during the take-off procedure. A commercial software ANSYS Fluent is used to perform the two-dimensional simulation of tandem foils oscillation along an inclined stroke plane with asymmetric upstroke and downstroke. The downstroke angle of attack is varied between $45° \leq \alpha_D \leq 90°$ whereas pitch amplitude for upstroke is the same for all cases. The results show that increasing the value of increases the vertical force as well as generates higher drag. This is due to stronger LEV created for higher and higher pressure region at the lower surface of the foils due to added mass effect. It is also observed that the presence of the forefoil reduced the performance of the hindfoil. The results obtained can help in the optimization of flapping wing-based MAVs.

Keywords: Dipole jet, dragonfly, forewing, hindwing, inclined stroke plane, take-off

Introduction

Micro air vehicles are inspired by flying insects, showing efficient flying ability at low Reynolds number. There has been extensive research on dragonfly flight compared to other flying insects due to their remarkable agility and rapid maneuvering abilities. Dragonfly flaps their wing along an inclined stroke plane with asymmetric upstroke and downstroke. According to steady-state aerodynamics, the lift generated by dragonfly wings can support only 40% of their weight. Therefore, most of their weight is supported by the unsteady flow phenomena [1]. The unsteady flow phenomenon includes delayed stall, wake capture, rotational lift and added mass. Wang [2] numerically studied dragonfly hovering flight by varying its stroke plane angle and mean pitch angle. It was found that dragonfly uses drag to support 75% of their weight and a vortex pair is generated in the wake, creating a downward dipole jet that becomes faster and narrower with increased stroke plane angle. A dragonfly's two-wing configuration helps them change its flight modes by varying the phase difference between its forewing and hindwing. There have been various studies on the different flight modes of dragonflies, such as gliding, hovering, turning, forward and backward flight [3–14]. The highest vertical force coefficient is generated by the in-phase oscillation of the dragonfly wings and the highest thrust is obtained from phase difference [15].

Dragonfly flight generates a highly complex 3D flow structure, and the aerodynamic forces significantly vary along the wingspan. The two-dimensional numerical analysis helps us to understand the significance of the unsteady flow phenomenon involved in dragonfly flight. There are different techniques by which an insect generates a large amount of force required to become airborne such as flapping with leg jump, clap and fling, and slow take-off. The experimental study by Li et al. [16] shows that the dragonfly has the ability to take off in one wingbeat cycle. They also observed that the higher angle of attack during downstroke generates a larger vertical force coefficient, which helps faster take-off.

The take-off flight of dragonflies was further studied by Shanmugam and Sohn [17]. They performed numerical simulation of a dragonfly wing flapping along an inclined stroke plane and studied the effect of phase difference, wing spacing, and downstroke angle of attack. They observed that the aerodynamic forces depend significantly on the kinematic parameters of flapping foils. We note that there have been very limited studies on the effect of the downstroke angle of attack on the aerodynamic performance of dragonflies during take-off. Hence, in this study, we have analyzed the effect of the angle of attack during mid-downstroke on the vertical force generation of dragonflies during take-off flight. We have performed a numerical simulation of two elliptical foils flapping along an inclined stroke plane of at Re = 160.

Numerical Methodology

Two elliptical airfoils with tandem arrangements are considered for the present study. The airfoil has a thickness of 0.1c and the pitch center is located at its centroid. The foil spacing is 1.3c. The foil kinematics is based on the experimental observation of Li et al. [16] and it is shown in Figure 67.1.

---

[a]shubh6tiwari@gmail.com, [b]sunildiat2010@gmail.com

DOI: 10.1201/9781003450252-67

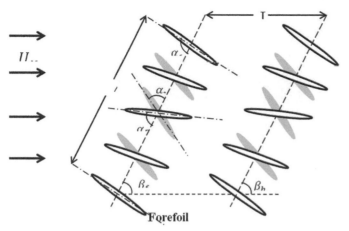

Figure 67.1 The schematic diagram for forefoil and hindfoil kinematics
Source: Author

The downstroke is represented by a white ellipse, whereas the grey ellipse denotes the upstroke. The forefoil and hindfoil undergo plunging as well as pitching motion along an inclined stroke plane based on the following equation of motion:

For forefoil

$$x_f(t) = \frac{A_o}{2}\cos(\beta_f) + \frac{A_o}{2}\cos(2\pi ft)\cos(\beta_f) \tag{1}$$

$$y_f(t) = \frac{A_o}{2}\sin(\beta_f) + \frac{A_o}{2}\cos(2\pi ft)\sin(\beta_f) \tag{2}$$

$$\alpha_f(t) = \alpha_o + \alpha_m\cos(2\pi ft + \pi/2) \tag{3}$$

For hindfoil

$$x_h(t) = \frac{A_o}{2}\cos(\beta_h) + \frac{A_o}{2}\cos(2\pi ft)\cos(\beta_h) \tag{4}$$

$$y_h(t) = \frac{A_o}{2}\sin(\beta_h) + \frac{A_o}{2}\cos(2\pi ft)\sin(\beta_h) \tag{5}$$

$$\alpha_h(t) = \alpha_o + \alpha_m\cos(2\pi ft + \pi/2) \tag{6}$$

Here, subscript f and h represent forefoil and hindfoil, respectively. The instantaneous airfoil displacement along the x- and y- direction is represented by $x(t)$ and $y(t)$ whereas the instantaneous pitch angle is represented by $\alpha(t)$. β is the stroke plane angle with respect to the horizontal. f and A_0 represents the flapping frequency and stroke amplitude, respectively. The initial pitch angle and pitch amplitude are represented by α_o and α_m.

A two-dimensional unsteady and incompressible flow is considered in this study. The model used to resolve the flow structure is the one-equation Spalart-Allmaras (S-A) model. The governing equation for the present study is equations [7] and [8].

$$\nabla \cdot u = 0 \tag{7}$$

$$\frac{\partial u}{\partial t} + (u \cdot \nabla)u + \nabla p - \frac{1}{Re}\nabla^2 u = 0 \tag{8}$$

Here u denotes fluid velocity and is pressure. The commercial software ANSYS Fluent 17.2 is used for solving the governing equations. PISO scheme is employed for pressure-velocity coupling and second order upwind scheme is used for spatial discretization of pressure and momentum.

The computational domain consists of two different types of region as shown in Figure 67.2. Two circular moving regions enclosing two airfoil and an outer stationary square domain of size $60c \times 60c$. The airfoils and their respective circular domain have the same motion, which is provided by a user-defined

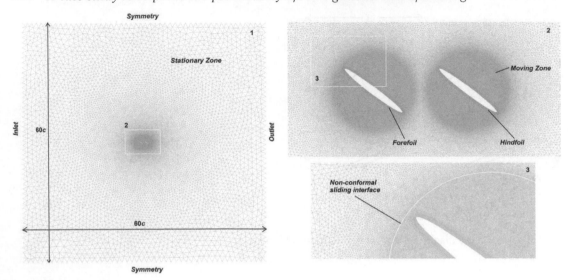

Figure 67.2 Computational model with different mesh zones
Source: Author

Table 67.1 Mesh and time-step convergence study.

Case	Number of mesh elements	Time Step size	(Total)
1	170000	T/300	0.7208
2	210000	T/600	0.7317
3	290000	T/900	0.7280

Source: Author

function (UDF). Based on this approach, there is no mesh deformation in the circular domain resulting in better results. The diffusion-based smoothing method and remeshing is used inside the stationary zone to update the mesh. The interface between the moving and stationary zones is a non-conformal sliding interface. The surface of the airfoil has no-slip condition. The inlet is defined as velocity inlet, outlet is defined as pressure outlet, and side boundaries are considered as symmetrical.

The chord length c is 0.01m, and the Reynolds number based on freestream velocity is 160. The advanced ratio (J) is 0.1 based on the take-off flight of a dragonfly [16]. U_R ($= U_\infty + U_F$) represents the reference velocity. U_F represents the flapping velocity of the foil given by $U_F = \pi A_o f$

The coefficient of pressure (C_P), as well as aerodynamic forces, are given by

$$C_P = \frac{P - P_\infty}{0.5\rho U_R{}^2} \tag{9}$$

$$C_V = \frac{F_V}{0.5\rho U_R{}^2 c} \tag{10}$$

$$C_H = \frac{F_H}{0.5\rho U_R{}^2 c} \tag{11}$$

Here, the cycle-averaged vertical and horizontal force coefficient is denoted by C_V and C_H whereas instantaneous vertical and horizontal force coefficient is denoted by C_v and C_H. The instantaneous vertical and horizontal force is denoted by F_V and F_H respectively and ρ denotes density.

We have performed grid convergence and time step convergence study by considering three different cases of mesh and time step. The case is considered as in-phase oscillation of tandem foils with following parameters: $A_o/c = 2.5$, $\beta = 80°$, $L = 1.3c$, $\alpha_0 = 105°$, $\alpha_m = 19°$ (during downstroke), $\alpha_m = 25°$ (during upstroke), $f = 30$ Hz, $J = 0.1$ and $Re = 160$. The result in Table 67.1 shows that the difference in the value of C_V (total) for cases 2 and 3 is approximately 0.5%. Hence, we considered the mesh and time step size of case 2 for further simulations.

The accuracy of the present computational model is examined by validating the results of the experiment and numerical study performed by Lua et al. [18]. Two elliptic airfoils oscillating along a vertical stroke plane in a tandem configuration are simulated with the parameters: $A_o/c = 1.5$, $\alpha_m = 30°$, $\alpha_0 = 90°$, $\beta = 90°$, $f = 0.667\ Hz$, $Re = 5000$ and $c = L$ (=0.04m). The numerical results obtained by the present computational model are shown in the figure, along with the experimental and numerical results of Lua et al. [18]. It can be observed from Figure 67.3 that our results agree well with the experimental results.

Result and Discussion

The role of pitch amplitude during downstroke on the aerodynamic force generation during the take-off flight of a dragonfly is studied in this paper. The downstroke and upstroke of a dragonfly flapping is asymmetric in nature with different values of pitch amplitude. It has been observed that the angle of attack during mid-downstroke during take-off flight is larger compared to forward flight [16]. In this study, the angle of attack during mid-downstroke is varied from 45° to 90°. It is achieved by changing the downstroke pitch amplitude α_m from 15° to 60°. The value of during upstroke is same for all cases to be 25°. The value of α_0 is 105° relative to the stroke plane. Based on the experimental observations and previous numerical studies, the value of stroke amplitude A_0 and flapping frequency (f) is 2.5_C and 30 Hz, respectively.

The effect of the mid-downstroke angle of attack on the instantaneous vertical force coefficient is shown in Figure 67.4. It can be observed that C_V of forefoil and hindfoil show similar trends relative to α_D. The downstroke generates positive vertical force; hence, the wing kinematics during downstroke plays a major role in lift production for dragonfly flight. The peak in C_V is obtained during mid-downstroke, and it can be observed that the increase in α_D leads to higher C_V. During the upstroke, the forefoil and hindfoil are subjected to wake capture which results in higher negative with increase in α_D.

The instantaneous horizontal force coefficient C_H variation with respect to α_D is similar to C_V and is shown in Figure 67.5. It is shown that a higher α_D produces higher drag during downstroke, whereas thrust is produced during the upstroke. The forefoil and hindfoil show similar behavior with varying α_D.

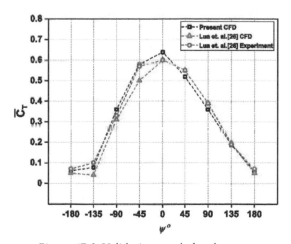

Figure 67.3 Validation result for the present computational model
Source: Author

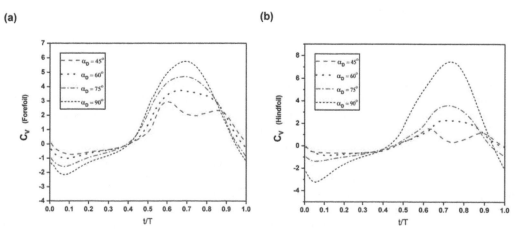

Figure 67.4 Instantaneous vertical force coefficient during one oscillation cycle for different. (a) forefoil (b) hindfoil
Source: Author

(a)   **(b)**

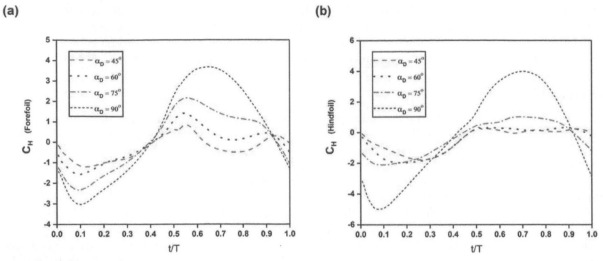

Figure 67.5 Instantaneous horizontal force coefficient during one oscillation cycle for different α_D. (a) forefoil (b) hindfoil
Source: Author

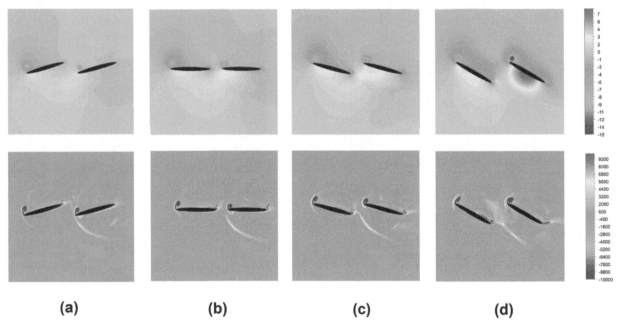

(a)   **(b)**   **(c)**   **(d)**

Figure 67.6 Pressure coefficient (top row) and vorticity (bottom row) contour during mid-downstroke for four different α_D (a) $\alpha_D = 45°$ (b) $\alpha_D = 60°$ (c) $\alpha_D = 75°$ (d) $\alpha_D = 90°$. Vorticity is in s^{-1}
Source: Author

The change in aerodynamic force coefficient with respect to α_D as shown in Figures 67.4 and 67.5 can be explained with the help of vorticity and pressure coefficient contour. Figure 67.6 shows the vorticity and pressure contour for four cases of α_D during mid-downstroke. It can be observed from the pressure contour that the strength of LEV on the upper surface of the airfoils increases with . This leads to an increase in the value of the negative pressure coefficient on the upper surface of the airfoil as shown in figure, resulting in higher C_V. Additionally, the added mass effect creates a higher pressure region on the lower surface of the airfoils. It is observed that increasing increases the added mass effect because it becomes hard for the fluid to flow through the gap between the forefoil and the hindfoil. The velocity at the gap increases, creating a stronger hindfoil LEV resulting in higher C_V than forefoil for $\alpha_D = 90$.

Figure 67.7 shows the cycle-averaged aerodynamic force coefficients with respect to α_D. It is shown in the Figure 67.7a that the maximum total C_V is obtained at $\alpha_D = 90°$ and the value of C_V increases with an increase in α_D for both the foils. The forefoil C_V is greater than hindfoil C_V for all value of α_D except for $\alpha_D = 90°$ which shows that the presence of the forefoil reduces the performance of the hindfoil. As shown

(a)   **(b)**

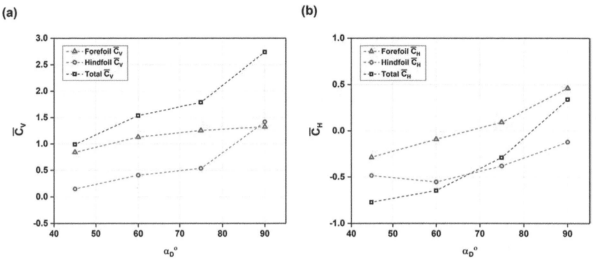

Figure 67.7 Cycle-averaged vertical ( ) and horizontal force () coefficient for different
Source: Author

in Figure 67.7b, the cycle average horizontal force coefficient increases with an increase in α_D resulting in higher drag. It can be observed that the hindfoil generated thrust for all cases of α_D whereas forefoil produces thrust only for $\alpha_D \leq 60°$.

Conclusion

In this study, a two-dimensional numerical analysis of tandem flapping foils is performed to study the effect of the downstroke angle of attack α_D on the dragonfly take-off flight. Two elliptical airfoils in tandem arrangement oscillating in-phase along an inclined stroke plane with asymmetric upstroke and downstroke is considered in this analysis. It is observed that the increase in α_D results in higher C_V and greater drag. The downstroke is mainly responsible for generating the vertical force required for the take-off flight. The highest C_V and C_H is obtained at $\alpha_D = 90°$. The higher α_D results in stronger LEV which creates a negative pressure region on the upper surface of the airfoils producing higher C_V. It is also observed that at higher α_D. The added mass effect increases and a higher-pressure region is created on the lower surface of the airfoil, resulting in lift enhancement.

Reference

[1] Norberg, R. Å. (1975). Hovering flight of the dragonfly aeschna juncea L., kinematics and aerodynamics. In: Swimming and Flying in Nature. Boston: Springer. pp. 763–81.

[2] Wang, Z. J. (2004). The role of drag in insect hovering. Journal of Experimental Biology. 207, 4147–55.

[3] Bomphrey, R. J., Nakata, T., Henningsson, P., and Lin, H. T. (2016). Flight of the dragonflies and damselflies. Philosophical Transactions of the Royal Society B: Biological Sciences, 371(1704), 20150389.

[4] Zhang, J., and Lu, X. Y. (2009). Aerodynamic performance due to forewing and hindwing interaction in gliding dragonfly flight. Physical Review. E.80, 017302.

[5] Levy, D. E., and Seifert, A. (2009). Simplified dragonfly airfoil aerodynamics at Reynolds numbers below 8000. Physics of Fluids. 21, 071901.

[6] Mou, X., and Sun, M. (2012). Dynamic flight stability of a model hoverfly in inclined-stroke-plane hovering. Journal of Bionic Engineering. 9, 294–303.

[7] Xu, N., and Sun, M. (2014). Lateral dynamic flight stability of a model hoverfly in normal and inclined stroke-plane hovering. Bioinspiration & Biomimetics. 9, 036019.

[8] Lai, G., and Shen, G. (2012). Experimental investigation on the wing-wake interaction at the mid stroke in hovering flight of dragonfly. Science China Physics, Mechanics & Astronomy. 55, 2167–78.

[9] Shumway, N. M., Gabryszuk, M., and Laurence, S. J. (2018). Flapping tandem-wing aerodynamics: dragonflies in steady forward flight. In: 2018 AIAA Aerospace Sciences Meeting. Kissimmee, FL, USA, paper no. 1290. Reston, VA: AIAA.

[10] Chen, Y. H., and Skote, M. (2015). Study of lift enhancing mechanisms via comparison of two distinct flapping patterns in the dragonfly Sympetrum flaveolum. Physics of Fluids. 27, 033604.

[11] Hefler, C., Qiu, H., and Shyy, W. (2018). Aerodynamic characteristics along the wing span of a dragonfly Pantala flavescens. Journal of Experimental Biology. 221, jeb171199.

[12] Sun, X., Gong, X., and Huang, D. (2017). A review on studies of the aerodynamics of different types of maneuvers in dragonflies. Archive of Applied Mechanics. 87, 521–54.

[13] Bode-Oke, A. T., Zeyghami, S., and Dong, H. (2018). Flying in reverse: Kinematics and aerodynamics of a dragonfly in backward free flight. Journal of The Royal Society Interface. 15, 20180102.

[14] Li, C., and Dong, H. (2017). Wing kinematics measurement and aerodynamics of a dragonfly in turning flight. Bioinspiration & Biomimetics. 12, 026001.

[15] Lan, S., and Sun, M. (2001). Aerodynamic force and flow structures of two airfoils in flapping motions. Acta Mechanica Sinica -PRC. 17, 310–31.

[16] Li, Q., Zheng, M., Pan, T., and Su, G. (2018). Experimental and numerical investigation on dragonfly wing and body motion during voluntary take-off. Scientific Reports, 8(1), 1011.

[17] Shanmugam, A. R., and Sohn, C. H. (2019). Numerical investigation of the aerodynamic performance of dragonfly-like flapping foil in take-off flight. Proceedings of the Institution of Mechanical Engineers, Part G: Journal of Aerospace Engineering. 233, 5801–15.

[18] Lua, K. B., Lu, H., Zhang, X. H., Lim, T. T., and Yeo, K. S. (2016). Aerodynamics of two-dimensional flapping wings in tandem configuration. Physics of Fluids, 28(12).

Chapter 68

Occurrence of coronary heart disease: analysis and classification using machine learning

Bharati Karare[1,a], Lalit Damahe[1,b], Sunny Gurmeet Singh Gandhi[2,c] and Seema Mendhekar[3,d]

[1]YCCE, NAGPUR, India

[2]FEAT, DMIHER, Wardha, India

[3]Nagpur College of Pharmacy, Nagpur, India

Abstract

It is observed, over the last few years that heart disease has been on the rise in the world and it is becoming one of the reasons for death. It is therefore important to go for early diagnosis and first aid. The universally available Framingham dataset was used for this research. Machine learning techniques were used to predict the risk of coronary heart disease (CHD) in later life. The study analysed more than 4,000 records and 15 attributes, as well as potential risk factors. The results of the proposed machine learning model show that a person's age and systolic blood pressure are important factors for developing coronary heart disease at 10 years of age. If we know in advance the potential risk, the risk of life-threatening coronary heart disease can be avoided and both the mortality and long-term costs associated with disease management can be reduced.

Keywords: Cardiovascular disease, coronary heart disease, machine learning

Introduction

Today, in the 21st century, with the rapid development and digitisation of technology, data has proved to be important for industries in every field like healthcare, education, etc. Digitisation allows all hospitals and medical institutions to collect their patient data electronically and analyse it and use it for further research. Such data includes the patient's medical history, symptoms displayed, diagnosis, duration of illness, recurrence, and any deaths. Therefore, the amount of such medical data is constantly increasing every day. To fully evaluate the influence of each element or sub-division, an analysis was undertaken in the study discussed here by reading multiple previous studies from well-known medical web pages or papers to compile the numerous risk factor classes linked to coronary heart disease (CHD). Cohort studies are the most popular approach for identifying risk factor categories, but alternative methods include systematic reviews, qualitative research filters, and so on. In many cases, a combination of research approaches has been employed to improve findings. The numerous risk factors are described in detail, including how HDL cholesterol levels were measured, how diabetes affects CHD, and how smoking, age, and gender affect CHD. There were four categories of risk factors: demographic, behavioral, and medical history. There were two or more sub-divisions in each factor. The goal is to assess both traditional and potential risk factors in order to accurately forecast the development of CHD in people who have never been diagnosed. Machine learning (ML) models can be created using artificial intelligence that can make decisions and predictions using self-learning and past experiences. According to this model in Figure 68.1, training and testing are conducted using the data set according to the application. It first extracts all the features, then selects the key features from the data set and uses those features to train the classification algorithm, then trains the model to be able to identify patterns from the features, then validates the model. Predicts heart disease by classifying it into general and cardiac categories.

Related Work

Krishnani et al. [1] proposed a machine-learning model for predicting CHD. In this study, K-fold cross validation was used for generating the random data. The final result predicts the risk of heart disease efficiently. Gonsalves et al. [2], proposed a model for analysing and predicting CHD based on the three supervised ML algorithms. This model used medical historical data to predict CHD. South African Heart Disease dataset was used. Akella et al. [3] build a machine-learning model for predicting coronary artery disease. The Cleveland dataset was used to predict and analyse patient health. The UCI machine learning heart disease dataset is used to accurately predict heart disease based on machine-learning techniques and in-depth learning. Heart disease has been analysed using 14 key features in this dataset. The good results were achieved and validated using an accuracy and confusion matrix [4]. Chowdary et al. [5], proposed a new ensemble

[a]kararebharati@gmail.com, [b]damahe_l@rediffmail.com, [c]sunnyg.feat@dmimsu.edu.in, [d]mendhekarseema41@gmail.com

DOI: 10.1201/9781003450252-68

approach for predicting cardiovascular diseases based on ML techniques like logistic regression, random forest [8], K-nearest neighbors (KNN) [6], etc. Kwon et al. [7] proposed a hybrid approach for the prediction of coronary heart disease based on neural network and feature correlation analysis. This approach used two phases, in the first phase, select an important feature that is useful for CHD risk and assign a rank for every feature. In the second phase, measure the correlation between the selected feature and the data of the neural network. A research based on deep learning approach was used for predicting heart disease over the 900 samples of the dataset which is obtained from medical Norte hospital in Mexico [10]. In this study ensemble machine learning, Bidirectional LSTM, and CNN approaches were used. The proposed neural model gives 91% accuracy. Based on studied literature, it is found that most ML techniques, such as random forest, AdaBoost, KNN, MLP, decision trees, and other bagging algorithms, are reused extensively to predict coronary heart disease.

Proposed Methodology

The hospitals encounter several cases of CHD every day. The various hospitals, all over the world, keep the data of all the patients, in and out, every day. These records include everything from patients' medical history to various parameters of the patient's health. The data from all the hospitals combine is humungous and has very prolific potential. But all this data is lying unused in various parts of the world. This raw data could be used to extract, study, and analyse the patterns and make predictions to reach some important breakthroughs in the health industry. So, have picked up one such dataset (Framingham dataset from Kaggle Website) and used it to come to an important conclusion regarding CHD. Our objective is studying about the pattern in the particular dataset with respect to the risk associated with the patient based on several health factors which include age, blood pressure, ECG, etc. After training each model using the same summarised data and calculating various factors like the accuracy of the model, AUC, ROC, and F1 score (which measures the model's accuracy by combining precision and recall). Finally compared all four models based on the results of these factors and the predictions made by these models and came to a conclusion as to which model is more suitable and should be used for further implementation abbreviations and acronyms

Data Set and Analysis

In this research, we have utilised the cardiovascular patient data from Framingham which is publically available and collected from Kaggle. The motive of this research is to predict coronary heart disease in patients has an issue of heart disease for 19 years. This dataset contains up to 4000 patient records. The dataset is explored, and extended as per requirements and is used to analyse various attributes. The data that is used to analyse and make predictions should be complete and must have 0 ambiguities. Given dataset has some missing values in some of the parameters, the analysis and predictions based on this dataset may lead to inaccurate results. Since the accuracy of predictions is of higher importance here, it is required to find the missing variables and proceed accordingly. It is observed that 12.7% of the entire dataset is missing data. Figure 68.1 shows the total missing data in the dataset.

The blood glucose component has the largest amount of missing data, at 9.15%, whereas the others have very little. There is a need to exclude these records because they count just 12% of the overall data.

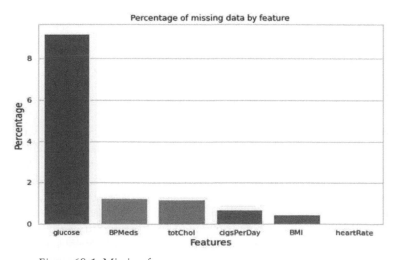

Figure 68.1 Missing features

Source: Author

Data Distribution

The data on the prevalent stroke, diabetes, and blood pressure meds are poorly balanced. The plot for the occurrence of CHD with respect to the age of the patients is presented. This is necessary to find because it is important to find which age groups are more prone to the risk.

Figure 68.2 describes the people with the highest risk of developing CHD between the ages of 51 and 63 i.e., the blue bars. To analyse and describe the data, stacked bar charts are useful and converting the raw values to percentages, and plotting and creating bars using various python libraries. Some important and decisive attributes with respect to the disease outcomes as shown in Figure 68.3.

It's tough to draw inferences based on what's noticed because the datasets is unbalanced. It has been observed that men may be somewhat more likely to develop heart disease problems than women. The odds of developing CHD are higher in hypertensive and diabetic patients and are almost similar between smokers and non-smokers.

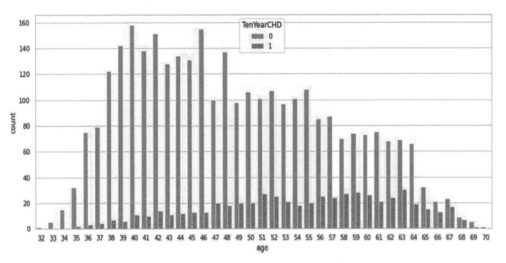

Figure 68.2 Count vs age
Source: Author

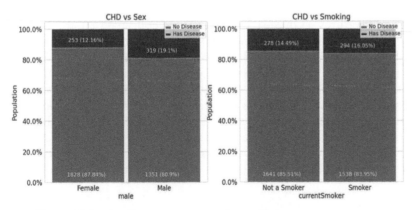

Figure 68.3a Attributes vs coronary heart disease
Source: Author

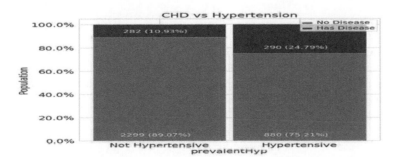

Figure 68.3b Attributes vs coronary heart disease
Source: Author

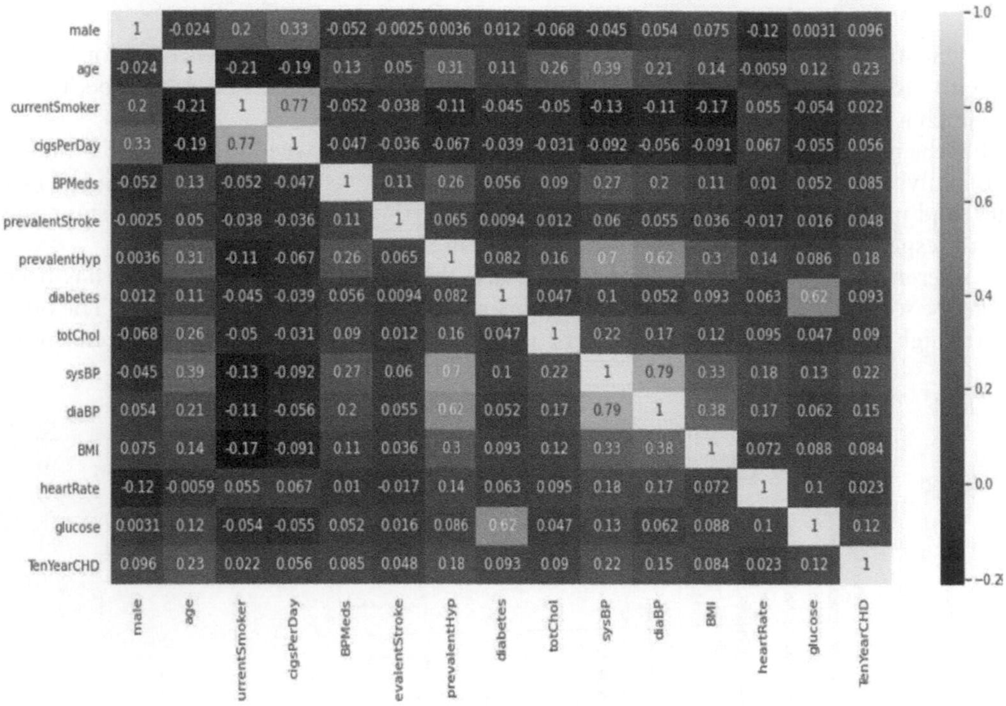

Figure 68.4 Heat map
Source: Author

Correlation Heat Map

A correlation heat map is presented in Figure 68.5. There have been no features that have a greater than 0.5 association with the 10-year chance of getting CHD, indicating that they are significantly predictive. As a result, it is mandatory to do feature selection in order to identify the greatest features.

Feature Selection

Let's start with creating replicates of overall features set in the requested dataset collected that are jumbled and randomised. It builds a classifier with a large dataset and measures the significance of every feature set using a significant parameter with high values denoting huge meaning. Sometimes, the feature set that is deemed to be very inconsequential is eliminated, and at the end of this process, the Z score is used to determine if an actual feature has greater relevance is the best shadow feature. Because all features' sets are accepted or denied, or the amount of classifiers operations reaches a predefined limit, the process stops. The most crucial characteristics for forecasting the likelihood of developing cardiovascular illness within the next ten years are determined. It can be shown that blood pressure and aging are now two major factors particularly important for predicting a person's 10-year risk for developing cardiovascular heart disease. To create neural network models for selecting the relevant features for predicting cardiovascular problems.

Statistical Analysis of Top Feature

By analyzing the dataset some top features are selected and statistical analysis is performed on these top features. These selected features play a significant role in model creation. Tables 68.1 and 68.2 summarised the details of data and attributes.

Classification Using Models

Since the dataset is imbalanced i.e. for every positive case there are about six negative cases. There is a need of a classifier and to address this objective a classifier that mostly predicts negative classes must be selected; thus, have high accuracy but poor specificity or sensitivity. To address this, there is a need to balance the dataset using the SMOTE.

SMOTE

The disadvantage of unbalanced classification is that there are few samples of the minor category for models to remember the selection boundary effectively. Another method for addressing such an issue is to

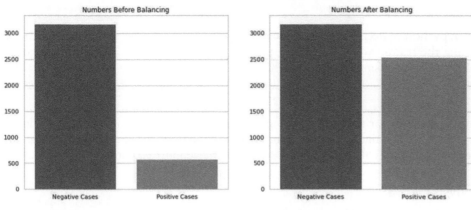

Figure 68.5 Balancing data
Source: Author

Table 68.1 Data summary

Dep. variable	Ten Year CHD	No. observation	3751
Model	Logit	Df Residuals	3744
Method	MLE	Df Model	6
Date	Sat, 25 Apr 2021	Pseudo R-squ	0.02354
Time	22:16:47	likelihood	-1564.0
Converged	True	LL-Null	-1601.7
Covariance type	Non-robust	LLR p-value	3.174e-14

Source: Author

Table 68.2 Attribute summary

Attributes	coef	std err	z	P>|z|	[0.025	0.975]
age	0.0223	0.006	3.982	0	0.011	0.033
totChol	-0.0029	0.001	-2.72	0.01	-0.01	≈0
sysBP	0.0245	0.003	7.436	0	0.018	0.031
diaBP	-0.027	0.006	-4.6	0	-0.04	-0.02
BMI	-0.0499	0.012	-4.19	0	-0.07	-0.03
heartRate	-0.03	0.004	-8.06	0	-0.04	-0.02
glucose	0.0043	0.002	2.616	0.01	0.001	0.007

Source: Author

oversample data in minor categories. Whenever training a network, just duplicate minority class values in the training dataset. It can assist balance the classification performance but does not provide any further prediction model. It is preferable to synthesise fresh samples from the minor category as opposed to just copying existing ones.

The given complete dataset is divided into training and testing sets, and then feature scaling is applied.

Logistic Regression

Using the suitable libraries in python of logistic regression, we search for optimum parameters using grid search. The classifier is then trained using the training dataset, and the predictions are made. The accuracy score of this classifier is 67.6% and F1 score is 62.41%.

For plotting the ROC Curve and AUC, keeping the probabilities for the positive outcome only we first calculated the AUC, then calculated the ROC curve respectively.

K-Nearest Neighbors

Using the suitable libraries in python for KNN, we search for optimum parameters using grid search. The classifier is then trained using the training dataset, and the predictions are made. Accuracy of KNN of 82.53%. The F1 score is 82.27%.

Decision Trees

Using the suitable libraries in python of decision trees, we search for optimum parameters using grid search. Classifier is then trained using the training dataset, and the predictions are made. The accuracy of the decision tree is 72.4%. The F1 score for decision tree is 67.62%

Support Vector Machine

Using the suitable libraries in python for support vector machines, we search for optimum parameters using grid search. The classifier is then trained using the training dataset, and the predictions are made. The accuracy of the support vector machine is 86.46%. The F1 score for the support vector machine is 85.31%

Comparative Result Analysis

We have analysed the data using different ML models using Framingham data. Accuracy, AUC, and F1 score for each of these models are calculated and compared with each other.

Comparative analysis of various classifiers is presented in Table 68.3. The obtained accuracy (86.55%), AUC (93.39%), and F1 score (84.93) scores for the SVM model are higher compared to other models. Support vector machine classification outperforms and gives better accuracy, AUC, and F1 score. This shows that the SVM model predicts coronary heart disease more precisely.

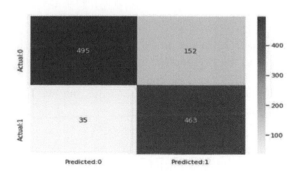

Figure 68.6 Prediction using logistic regression
Source: Author

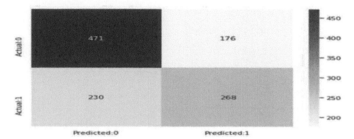

Figure 68.7 Positive rate comparison using LR
Source: Author

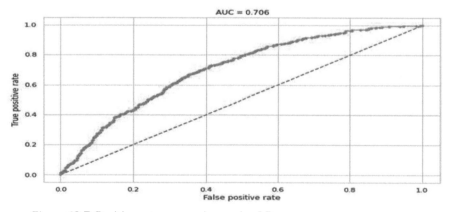

Figure 68.8 Prediction using KNN
Source: Author

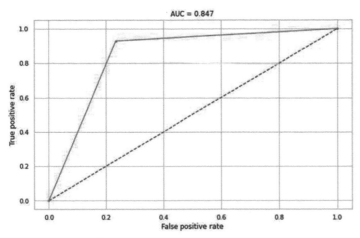

Figure 68.9 Positive rate comparison using KNN
Source: Author

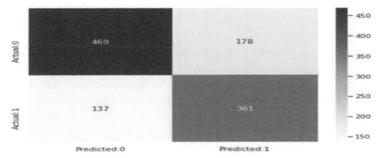

Figure 68.10 Prediction using decision trees
Source: Author

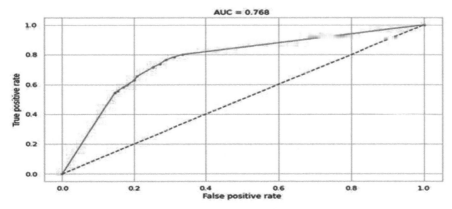

Figure 68.11 Positive rate comparison using decision trees
Source: Author

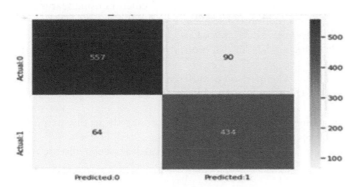

Figure 68.12 Prediction using SVM
Source: Author

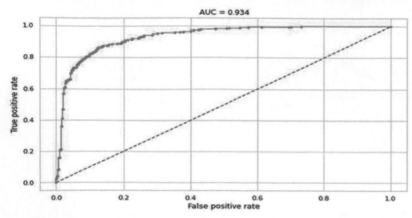

Figure 68.13 Positive rate comparison using SVM
Source: Author

Conclusion and Discussion

Our findings suggest that, given sufficient data and carefully selected important variables in the CHD dataset to the model, ML algorithms can accurately predict the likelihood of CHD problems. In this study, we created four machine-learning models using the Kaggle dataset and achieved about 86% accuracy. While this shows significant performance in classification results over earlier studies on the same data. The goal of this research is to fix the dataset's imbalanced classification problems to achieve even higher AUC and F1 score. In addition to enhancing overall prediction accuracy, SMOTE oversampling techniques were used to solve this problem in the dataset, and the performance evaluation of all prediction models improved dramatically. The oversampled dataset was subjected to four different prediction algorithms. In future work, it is possible to upgrade the model to predict the 10-year risk of CHD of every patient that would be personally calculated using his/her medical history. Such systems can be modified to monitor the health of every patient and warn of hasardous situations due to changes in the human lifestyle.

References

[1] Krishnani, D., Kumari, A., Dewangan, A., Singh, A., and Naik, N. S. (2019). Prediction of coronary heart disease using supervised machine learning algorithms. In TENCON 2019 - 2019 IEEE Region 10 Conference (TENCON), pp. 367–72. doi: 10.1109/TENCON.2019.8929434.

[2] Gonsalves, A., Thabtah, F., Mohammad, R., and Singh, G. (2019). Prediction of coronary heart disease using machine learning: An experimental analysis. CDLT '19: Proceedings of the 2019 3rd International Conference on Deep Learning Technologies. 51–56. ISBN:9781450371605. 10.1145/3342999.3343015.

[3] Akella, A., and Akella, S. (2021). Machine learning algorithms for predicting coronary artery disease: efforts toward an open source solution. Future Science OA. 7(6), FSO698. https://doi.org/10.2144/fsoa-2020-0206.

[4] Bharti, R., Khamparia, A., Shabaz, M., Dhiman, G., Pande, S., and Singh, P. (2021). Prediction of heart disease using a combination of machine learning and deep learning. Computational Intelligence and Neuroscience. 11. https://doi.org/10.1155/2021/8387680.

[5] Chowdary, G. J., Suganya, G., and Premalatha, M. (2020). Effective prediction of cardiovascular disease using cluster of machine learning algorithms. Journal of Critical Reviews. 7(19).

[6] Almustafa, K. M. (2020). Prediction of heart disease and classifiers' sensitivity analysis. BMC Bioinformatics. 21, 278. https://doi.org/10.1186/s12859-020-03626-y

[7] Kim, J. K., and Kang, S. (2017). Neural network-based coronary heart disease risk prediction using feature correlation analysis. Journal of Healthcare Engineering. https://doi.org/10.1155/2017/2780501

[8] Padmajaa, B., Srinidhib, C., Sindhuc, K., Vanajad, K., Deepika, N. M., Rao, E. K., and Patro, F. (2021). Early and accurate prediction of heart disease using machine learning model. Turkish Journal of Computer and Mathematics Education. 12(6), 4516–28.

[9] Dipto, I., Islam, T., Rahman, H., and Rahman, M. (2020). Comparison of different machine learning algorithms for the prediction of coronary artery disease. Journal of Data Analysis and Information Processing. 8, 41–68. doi: 10.4236/jdaip.2020.82003.

[10] Khdair, H., and Dasari, N. M. (2021). Exploring machine learning techniques for coronary heart disease prediction. International Journal of Advanced Computer Science and Applications. 12(5).

[11] Sajja, T. K., and Kalluri, H. K. (2020). A deep learning method for prediction of cardiovascular disease using convolutional neural net

Chapter 69

Design and development of waste plastic recycling unit

Abhishek Dixit[a], Sumit Dhage[b], Mahesh Kavre[c] and Ajit Khushwah[d]

Terna Engineering College, India

Abstract

This study was intended to design and develop a waste plastic recycling unit, which will help to recycle different types of used plastics and would thereby help in waste management and disposal. This project consists of design of plastic shredding unit which crushes the plastic bottles into fine pieces and a compression molding unit which will mold the pieces into bricks and blocks of the bottles while considering the required forces for crushing of plastic bottles as well as the ergonomics and aesthetics of the machine. The optimum load to crush bottles is designed to ensure that there is no strain on the user of operator. Once the bottles are crushed, it will be further transferred in a compression molding machine to create plastic blocks for various purposes which will help in recycling of the waste. After researching various research papers, we understood the need for a simplified unit for recycling the waste plastic. Most of the researchers focused on either one of the following solutions. Currently, there are not many optimal recycling units available in the market that shreds the bottles as well as molds it into the bricks. The aim is to combine both process and thus simplifying the entire process. Thereafter, the study highlights the future work that could be done to improve the capabilities of the project.

Keywords: Analysis, disposal, plastics, recycling, shredder, thermoplastics

Introduction

Background

From ancient age, humans are known to produce waste and have been disposing it wasteful matters, which is not a new issue. But the types of waste generated, its methods of disposal systems and perceptions of proper methods of disposal have varied as per time. The applications of plastics have been increasing rapidly due to its low manufacturing costs and ease of manufacture. Therefore, such huge chunks of waste create a huge opportunities for challenges in disposal. Wide variety of applications of plastics have forced organisations that are face with problem of proposing new methods for waste disposal [7, 8]. Disposal of plastic waste is considered to be a problem due to low biodegradability.

In this day and age sustainability has become a significant factor in the field of construction. Plastics have become a viable option for course aggregate this providing a suitable option to tackle plastic waste. Hence, recycling of waste has become a necessity for lowering the damage done to the environment and wastage of precious resources. The municipal waste generated has become an interesting field of research for many scholars. The scope of converting such waste to useful recycled products is indeed tempting [10]. The vast sources of concrete waste generated replaced by recycled plastic pavement blocks. This move would prove to be more beneficial in terms of economics as well as environment.

Plastics

Plastic is derived from a Greek 'plastikos' which translates to "fit for molding". This perfectly sums up the properties of plastics. The term literally means plasticity and flexibility of the material to be molded into any shape and form. Plastics is a combination of various materials in which such substances are mixed in a fixed proportion to form a compound as per the requirements. Plastics have widespread applications in field of engineering and technology. Electronics, construction, product packaging are a few of the vast fields under which plastic could be used [11]. However, the use of plastics is increasing exponentially with the expected growth of plastic waste as shown in Figure 69.1 exceeding all expected rates in 2015. For (Placeholder1)m 1950 to 2015, the plastic has been produced at a rate of approximately 8% and almost half of the plastics waste has been generated in the previous decades. Such waste has impacted the environment wherein a small island has formed consisting of only plastic waste which was dropped into the oceans. This has an adverse effect on the local biodiversity of the oceanic ecosystem.

Effect of plastic on environment

Central board of Pollution control predicted that there would be approximate 15,342 tons of plastic waste in the country, of which around 9205 tons was reported to be recycled and about 6000 tons was

[a]abhishekdixit268@gmail.com, [b]sumitdhage2001@gmail.com, [c]maheshkavre@ternaengg.ac.in,
[d]ajitkhushwah@ternaengg.ac.in

DOI: 10.1201/9781003450252-69

Figure 69.1 Plastics
Source: Author

Table 69.1 Properties of plastics [10]

Properties	PLA	PS	PP	PE	ABS	PET
Melting temp. (°C)	160	140	170	120	140	240

left uncollected and littered [9]. The metropolitan areas of Mumbai, Delhi, Chennai, Kolkata has had an estimated total of 3500 tons of plastic waste per day, which has an adverse effect of the lifestyle of humans and leads to an unhygienic outlook towards the nation.

Properties of plastics

Plastics are used in various industrial applications due to the aesthetic appearances of plastics that appeal to the consumers and is less costly and lighter than other materials. The properties like toughness, abrasive resistance, appearance, insulation formability, machinability are much better in plastics.

Following Table 69.1 shows the melting points for different types of plastic.

Literature review

One of the non-biodegradable materials used widely is plastic. Plastic can be recycled with ease and at a low cost. The plastics polymers are separated and then distinguished for effective recycling. Plastic must be sorted immediately for various economic reasons. In this work a new method of spectroscopy (NIRS) has been used for the instant identification of plastics used by consumers. The main function of the NIRS is to quickly monitor and identify structural of molecular properties of the plastic waste. The paper demonstrates an economic machine able to separate plastics. The system is controlled using a wireless controller. Also, a controller is used to control the spectroscopy device from a specified distance has been used to prevent nugatory environment in plastic recycling plants [1].

It illustrates one of the critical environmental problems in India – industrial waste [2]. There is an increase in the demand of energy due to rapid hike in population, migration of majority of population into cites, industrialisation and a hike in fuel prices. Thus, creating a necessity for dependence on renewable energy. The extracted oil from the E-waste is used for inputs in electricity generation.

Waste recycling of plastics is a universal answer to the global issue of disposal of waste and is becoming increasingly common in industries. However, there is little research in such aspects, the plastic recycling has recently even expanded into automobile industry. This paper illustrates the research done over the assessment of impacts on life cycle of recycling in a Chinese company [3].

In their proposed paper, they have made a machine which eliminate the hydraulically operated machine by pneumatic controlled system for injection of molten plastic by using appropriate design procedure. As the needs of peoples for various product has changed for reliability of a product as well as its price. This gives rise to the plastic made product which have a better life. But its drawback is that it has no end of his life rather than breaking or any fault arise which has a very severe effect on the environment [4].

Ghodrati et al. [5] proposed that the research shows that there are several dimensions taken into consideration and then only factor of safety is decided. The implementation of 5S Practices can be seen by doing their research methodology and then they found the proper data which need further analysis and then they can summarise it. How the 5S would benefit them. Ad focusing on the company activities. Performance of company can be studied by the previous steps by extracting its sub factors. The design of questionnaire can we studied by using given literature review. Improvement and weak points are found after the first pilot run test of questionnaire. Expert's opinion taken into consideration as performing indicator as well as performance factor.

Waleola et al. [6] proposed that due to use of plastic largely it has effects of environment pollution and thrown on a landfill which also consume space. There is a process in which the plastic is recycle and new plastic products are formed such as particle board, containers, etc. This plastic can be converted into a small piece by shredding it for efficient transport of it from one place to further process. This process of shredding of plastic was found in a University from Akure. In the state of ONDO, Surrounded by Nigerian environment. It has feeding unit, shredding unit, machine strengthen frame and at last the most important power unit in their shredding model system. By seeing the performance of this machine which can be further evaluated that it can be correlated between its speed with regression < 1 the various parameters of it have a linear relationship (recovery efficiency (RE), throughput (TP), specific mechanical energy (SME) and shredding time (ST)) and variable operation speeds.

Research Gap

The research conducted in the field of waste plastic recycling is usually focused on two major bifurcations i.e., Segregation and Recycling. Segregation consists of differentiating the plastic waste on the basis of its material. Some materials are easily recycled whereas few require much more energy. The Recycling is done by melting the waste into ingots or reels of plastic. We have identified that there is no such paper that focuses on the segregation and recycling of the waste plastic simultaneously.

Thus, we hope to demonstrate one such method through our paper which will segregate and recycle waste plastic based on compression molding and segregate based on the melting points of the plastic waste.

Methodology

Methodology is the first step in planning of the project. It states the complete procedure of the project which starts at problem identification and ends with testing of the prototypes and review of the results from these tests. The detail of methodology steps is shown in Figure 69.2 in a type of flow chart.

Design and Development

Selection of Motor

The three phase induction motor having 3 HP power capacity with rotation speed of 1440 rpm used for the shredder. It is mounted on the frame and connected to the shaft of the reduction unit. This motor is used to rotate the blades of shredder to shred the plastic material.

The torque required to cut a plastic container of 5 mm thickness is 140 Nm

We have formula HP = (Torque X rpm)/5252

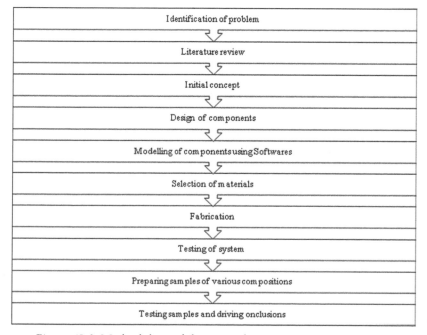

Figure 69.2 Methodology of the research
Source: Author

So, if we select 3 HP motor with 140 rpm then the torque will be 152 Nm which will be sufficient to shred a plastic material having thickness of 5 mm.

So, we selected 3HP motor with reduction unit to reduce the rpm to 140 with the reduction ratio of 10.

Selecting a Reduction Unit

For the shredder unit standard motor of 3HP operating at 1440 rpm was selected and it is easily available in the market. But our required rpm is in the range of 100-150 rpm as per our calculations. So, we have decided to go for a speed reduction unit. For that purpose, we made a survey to purchase the speed reduction unit from market. Finally, we got our reduction unit from steel-o-fab industries in Andheri. As per our need we purchased a speed reduction unit having speed reduction ratio of 10:1 to get 140 rpm which is required for our machine. As the reduction ratio is 10, so we purchased a reduction unit having worm and worm type of gears as shown in Figure 69.3.

Design Of Cutter Blade

Cutting forces required to shred plastic:-

F = (Breaking strength) × (Cross-sectional area)

Breaking strength= (Ultimate strength) × (F.O.S)

C.S. Area= (Thickness of blade) × (Width of plastic)

Cutting forces of blade

Diameter of blade: 120 mm

Width of blade: 5 mm

Mass: 0.2 kg

Area of blade: 4200 mm^2

Momentum (P) = Mass × Velocity

P = (Force/Area)

Force (f) = 1153 N

f > F

1153 N >1080N

Therefore, 5 mm of blade was selected as it produces required cutting forces for shredding plastic.

Structural Analysis of Blades

The structural analysis was performed on the models of Blades using the calculated forces as given in Table 69.2.

The above Figures 69.4 and 69.5 illustrate the structural stress on the blades of the shredder due to the forces acting on it. The mesh profile of the blades is also showcased in the figure along with a similar illustration of the total deformation of the blades.

Design of Shredder Shaft

The shaft is subjected to torsional forces as well as bending. When the shaft is subjected to combined bending and torsion, shaft is designed on the basis of both these forces.

Figure 69.3 Reduction unit

Source: Author

Table 69.2 Cutting forces required for types of plastic [11]

TYPE	THICKNESS OF PLASTIC			
	0.25mm	0.5mm	1mm	2mm
PET	135N	270N	540N	1080N
HDPE	43N	85N	170N	337N
PP	60N	118N	236N	472N
PS	81N	162N	324N	647N

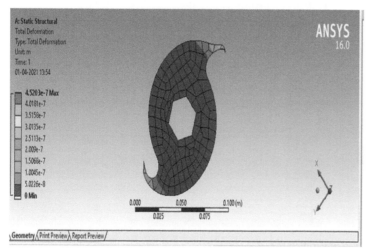

Figure 69.4 Equivalent stress in blades
Source: Author

(2) According to maximum principal stress theory:
We have

$$M_e = \frac{\Pi}{32}\, \sigma_b d^3$$

Where, Me = Equivalent bending moment
       T = Torque
       M= Maximum moment
Total load acting on the shaft = 3 kg
According to maximum principal stress theory: d = 19 mm
Shaft of 20 mm was used for the Shredder according to the analysis shown in Figures 69.6 and 69.7.

Manufacturing and Assembly

Plastic recycling machine having a hopper bolted on the top of the cutter box by using the nut and bolts as shown in Figure 69.8. The cutter box is mounted on the frame made of square pipe of steel material. The shaft of the cutter box is coupled to the reduction unit by using the Jaw coupling. The other shaft of reduction unit is coupled to the motor by using the same type of jaw coupling. The Power is transmitted from motor to the reduction unit where the speed of motor is reduced to 140 rpm with the reduction ratio of 10 and having worm and worm wheels. In the reduction unit the speed of motor is reduced with increase in the torque by keeping the HP rating of the motor constant. The cutters are rotated by the shaft and the plastic material is fed to the blades to get shredded.

Compression Moulding of Shredded Plastic

The Plastic which is shredded in the shredder is now recycled by melting it in an electric oven as shown in Figure 69.9 and then compressing it in a mold. The shredded plastic is dumped in a mold and pressed using an external force or a plug. The applied pressure forces the material to be in contact with the mold areas, the pressure and temperatures are sustained until the material is cured. The process also incorporates the

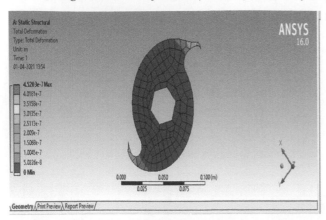

Figure 69.5 Total deformation in blades
Source: Author

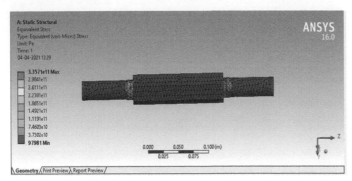

Figure 69.6 Equivalent stress of shaft
Source: Author

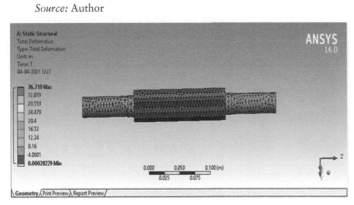

Figure 69.7 Total deformation of shaft
Source: Author

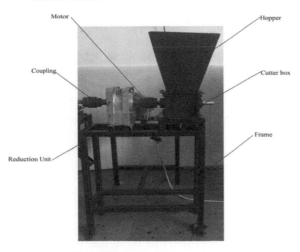

Figure 69.8 Shredder machine
Source: Author

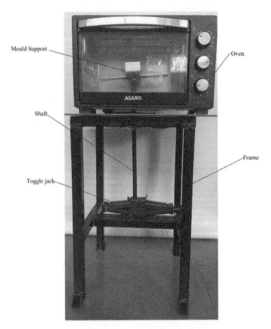

Figure 69.9 Compression moulding machine
Source: Author

Figure 69.10 Plastic flakes from shredder
Source: Author

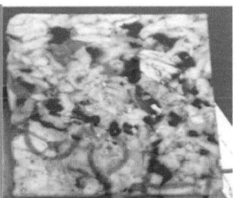

Figure 69.11 Compressed plastic pad
Source: Author

use of thermosetting plastics in a partially cured form, as granules or a semi solid paste. It is a very common and most popular method of molding as compared to complex processes like injection molding. There is little to no wastage of material, making it pretty economical and profitable.

Results and Discussion

The main objective of the project is to reutilise the used plastic waste. If the plastic is not recycled, it will clearly have a detrimental effect on the environment. We have been successful in making such a prototype that can recycle the waste plastic and compress it together to form a block of plastic. This compression moulding method could also be utilised to make use of moulds of pavement blocks and a lot more useful applications. The only perquisite required is a mould of the required application. In our model, we have demonstrated the shredded plastic from the shredder and have used it in compression moulding to form small rectangular pads as shown in Figures 69.10 and 69.11, that can be used as a writing pad after certain finishing operations like sanding and polishing.

Conclusion

Firstly, research and study were conducted on various issues related to plastic waste, their recycling systems, and applications of recycled plastic and then the problem statement was made. In the view of the problem statement a system was designed for turning waste plastic into small pieces called flakes i.e. The Shredder. Firstly 2D and 3D modelling of the system was done using software's like AutoCAD and PTC

Creo. According to the speed and power required by the shredder motor and reduction unit was selected. After designing, the material selected for shredder parts was stainless steel. The fabrication of the shredder cutter blades, spacer was done on milling machines, the shaft was machined on the lathe machine. The frame was made of mild steel L-columns and was welded together. All the parts of the shredder were assembled and tested. Trials of cutting plastic bottles and cans were done and small flakes of required size were obtained. Further in this project, the flakes of plastic produced from shredder will be melted and then along with other materials and additives like glass powder, cement, ceramics etc. various compositions will be made and compressed into small blocks or plates. Then these samples will be tested for strength under tension and compression test. After the test, the results will be analysed and then the various samples will be assigned for particular applications.

References

[1] Prasad, D. B. R., and R. S. (2017). Electrical distributed generation by industrial waste plastics and muncipal waste using pyrolysis process. In 2017 International Conference on Current Trends in Computer, Electrical, Electronics and Communication (CTCEEC). doi:10.1109/ctceec.2017.8455006.

[2] Gu, F., Guo, J., Zhang, W., Summers, P. A., and Hall, P. (2017). From waste plastics to industrial raw materials: A life cycle assessment of mechanical plastic recycling practice based on a real-world case study. Science of the Total Environment. 601–2.

[3] Shukla, P. G., and Shukla, G. P. (2013). Design and fabrication of pneumatically operated plastic injection molding machine. International Journal of Engineering and Innovative Technology. 2(7), 98.

[4] Ghodrati, A., and Zulkifli, N. (2013). (Mechanical and manufacturing department, engineering faculty/university Putra Malaysia), The impact of 5S implementation on industrial organisations' performance. International Journal of Business and Management Invention. 2(3), 43–49.

[5] Ayo, A. W. Olukunle, O. J., and Adelabu, D. J. (2017). Department of agricultural and bio- environmental engineering technology, rufus giwa polytechnic, Owo, Ondo State, Nigeria, development of a waste plastic shredding machine. International Journal of Waste Resources. 7, 2. doi: 10.4172/2252-5211.1000281.

[6] Ragaerta, K., Huysveld, S., Vyncke, G., Hubo, S., Veelaert, L., Dewulf, J., and Du Boisc, E. (2020). Design from recycling: A complex mixed plastic waste case study. Resources, Conservation and Recycling. 155, 104646.

[7] Rickert, J., Cerdas, F., and Herrmann, C. (2020). Exploring the environmental performance of emerging (chemical) recycling technologies for post-consumer plastic waste. Procedia CIRP. 90, 426–31.

[8] Khedkar, S. B., Thakre, R. D., Mahantare, Y. V., and Gondne, R. (2012). Study of implementing 5S techniques in plastic molding. International Journal of Modern Engineering Research. 2(5), 3653–6.

[9] Al-Salem, S. M., Letter, P., and Baleens, J. (2009). Centre for CO2 Technology, Department of Chemical Engineering, School of Process Engineering. Torrington Place, London WC1E 7JE, UK: University College London (UCL).

[10] Tominaga, A., Sekiguchi, H., Nakano, R., Yao, S., and Takatori, E. (2019). Advanced recycling process for waste plastics based on physical degradation theory and its stability. Journal of Material Cycles and Waste Management. 21, 116–24.

[11] Sahajwalla, V., and Gaikwad, V. (2018). The present and future of e-waste plastics recycling. Current Opinion in Green and Sustainable Chemistry. 13, 102–107.

Chapter 70

Initiatives for solar powered agriculture by Indian Government

Nivedita S. Padole[1,a], R. M. Moharil[1,b], Ankita Wasankar[1,c] and Vignesh Pillai[2,d]

[1]Department of Electrical Engineering, Yeshwantrao Chavan College of Engineering Nagpur, Maharashtra, India

[2]Department of Panchakarma Datta Meghe Ayurvedic Medical College Hospital & Research Center Nagpur, Maharashtra, India

Abstract

The sustainable economic development of the country is possible only if all the sectors including agriculture get developed. Development of rural area and agricultural sector has one of the limitation due to low voltage distribution feeder and longer distances of distribution lines of electric power. Decentralized renewable energy will be the key component to overcome these limitations. This will also help to boost the food productivity in the Asian countries. So as to support the objectives of sustainable growth, agricultural policies must be linked with economic, environmental, and social policies. The Government of India (GoI) has started many initiatives for agricultural growth and farmers' welfare. This paper discusses several initiatives of the Central and various State government of India, using solar and the impact analysis of these initiatives on the farmers and the rural development.

Keywords: Solar photovoltaic, Agricultural feeder, Deendayal Upadhyaya Gram Jyoti Yojana, Ministry of New and Renewable Energy, Pradhan Mantri Kisan Urja Suraksha evam Utthan Mahabhiyam

Introduction

India's primary economic activity is agriculture, and there is a considerable amount of fertile land available. Additionally, due to its high population, it requires a large-scale production of food grains. For over 300 days each year, India receives clear sunshine. As a result, farming and solar energy is a successful blend. The solar photovoltaic(SPV) systems are favoured for providing power agricultural loads [1]. SPV have been a key factor in the 'Green Revolution', which has increased agricultural production. Remote rural areas can use solar energy to fulfil their own energy needs of heat and electricity. Usage of solar energy helped to reduce the import of petroleum products from other countries. Economic growth can be sparked by having access to affordable, sustainable electricity.

Raising agricultural productivity per unit of land, reducing rural poverty through a socially inclusive strategy that includes both agricultural and non-agricultural employment, and ensuring that agricultural growth corresponds to needs for food security were previously the main challenges faced by Indian farmers. Therefore, a strong foundation for a far more productive, competitive globally, and diversified agricultural sector will need to be established.

The Indian economy is directly impacted by agricultural policies within the country. Irrigation policies in particular have a broader impact on agricultural output because they have a direct impact on the economy, food security, and environment. India is at the top in usage of groundwater for irrigation in the world. Majority of Indian crops its crops are irrigated through groundwater [2-4]. Such extensive groundwater use for irrigation has negative effects on the environment and the economy. Groundwater levels are inevitably stressed, and pumping up groundwater demands a tremendous amount of energy. A problem even gets worse as water levels drop. Either diesel- or electricity-powered pumps provides this energy.

Free or cheaper electricity to the farmers results into the over-extraction of ground water. This cheaper electricity tempts farmer to use low-cost poor-quality pumps which are inefficient, making the distribution of power to agriculture pumps difficult. The proper set of policies must be developed and implemented for farmers. Their influence should be assessed in order to assure positive growth in agriculture. It is crucial to implement efficient agricultural methods in order to address difficulties with water and energy conservation. Water can be saved in irrigation by using contemporary methods like drip and sprinkler irrigation systems. Community-level incentives will also affect ground water conservation.

This paper presents an overview of the GoI's initiatives for empowering the agriculture and rural sector. Different policies are introduced in Section II. Role of SPV in development of agricultural segment in India is explained in section III. Section IV provides the brief introduction to various schemes and its impact analysis for solar powered agriculture in India.

---

[a]niveditapande1@gmail.com, [b]rmm_ycce_ep@yahoo.com,, [c]ankitawasankar21@gmail.com, [d]vigfrndz@gmail.com

DOI: 10.1201/9781003450252-70

Policies Initiated by GOI for Agriculture Sector

More than half of India's population relies on agriculture as a source of income, making it one of the country's major economic sectors. Agriculture supports the nation's rural economy and ensures food security. Agriculture was the only industry to emerge as a bright spot for India's economic recovery in the first quarter of FY 2020–21, when the country's GDP had a 23.9 percent negative growth rate. As a result, agriculture becomes the nation's most important sector. Therefore, it requires substantial support from the public sector to achieve sustainable growth.

The policies initiated by GoI in agriculture [5] are mentioned below.

Gramin Bhandaran Yojana

The Government of India has a scheme named the Warehouse Subsidy Scheme or the Gramin Bhandaran Yojana. Grants are offered under this programme for the development, remodelling, or renovation of food godowns. However, this scheme is only relevant to the food godowns in India's rural regions.

Year of launch: 2001

Major objectives of this scheme are mentioned below:

Vision

1. Construct contemporary storage facilities in remote places.
2. Motivate farmers to produce as much as possible.

Mission

1. Farm products, processed farm product, and agricultural inputs can all be stored by farmers.
2. Enhance the marketability of agricultural products.
3. To avoid suffering, prevent the selling of produce right away following harvest.

National Mission for Sustainable Agriculture (NMSA)

The National Mission for Sustainable Agriculture (NMSA), which focuses on unified agriculture, water usage efficacy, managing soil strength, and synchronising supply preservation, has been developed to rise agricultural manufacture, particularly in rainfed zones.

Year of launch: 2014

Vision:

1. To increase the productivity of agriculture.
2. To shield natural resources by captivating appropriate soil and moisture conservation measures.

Mission:

1. Make Farmers use new and modern techniques in farming.
2. Encourage farmers to boost their yield.
3. Motivate farmers to grow various crop varieties.

Pradhan Mantri Krishi Sinchai Yojana (PMKSY)

In India, 65 million hectare (45%) of net sown part is sheltered by various irrigation schemes. For 76 million hectare (55%) of net sown area is an unirrigated land significantly rely on rainfall. Farming on unirrigated land is a risky and under productive activity.

The goal of this scheme [6] is to guarantee that all agricultural fields in the nation have admittance to some type of protective irrigation system so as to produce "per drop more crop" and make the long-awaited rural prosperity a reality.

Year of launch: 2015

The scheme1 provides the following outcomes:

1. The farm has easier access to water and also increase the area that can be farmed with reliable irrigation (Har Khet ko pani)

2. The incorporation of water source, distribution, and effectual usage, to maximise the water consumption by adopting the right practises and technologies.
3. Promote the use of water-saving technology like precision irrigation (More crop per drop).
4. Increase the effectiveness of on-farm water use to cut waste and extend and lengthen water availability.

Paramparagat Krishi Vikas Yojana (PKVY)

PKVY is a inclusive part of the National Mission of Sustainable Agriculture's (NMSA), Soil Health Management (SHM) programme. Under PKVY, organic agriculture is encouraged over the implementation of the organic village by cluster concept and Participatory Guarantee Systems (PGS) certification.

Year of launch: 2015

The scheme provides the following outcomes.

1. Supporting certified organic farming by encouraging commercial organic output.
2. The produce won't have any pesticide residue and will help to improve consumer health.
3. Farmers' incomes will increase, and it will open up prospective markets for traders.
4. It will encourage farmers to mobilise natural resources for the production of inputs.

Each farmer in the cluster will have 50 acres available for organic farming as part of the scheme, and there will be at least 50 farmers in total. In this manner, 10,000 clusters will develop over a three-year period, spanning 5.0 lakh acres for organic farming. The farmers won't be held responsible for the cost of certification. Rs. 20,000 per acre will be received by each farmer for planting seeds, growing crops, harvesting them, and getting the produce to market, over three years. By integrating farmers, it will boost domestic output and organic produce certification.

Pradhan Mantri Fasal Bima Yojana (PMFBY)

From the Kharif 2016 season onward, the Pradhan Mantri Fasal Bima Yojana (PMFBY) programme was introduced in India by the Ministry of Agriculture & Farmers Welfare, New Delhi.

Year of launch: 2016

The scheme provides the following outcomes:

1. Offering financial assistance to farmers that experience crop loss or damage as a result of unforeseeable circumstances.
2. Stabilizing farmers' income to support their continued involvement in agriculture.
3. Inspiring farmers to embrace progressive and contemporary agricultural practises.
4. Finance flow is to be ensured for the agricultural sector, to increase crop diversity, food safety, and the industry's growth and competitiveness while also shielding farmers from production risk.

Micro Irrigation Fund scheme

To attain the goal of "per drop more crop," the National Bank for Agriculture and Rural Development (NABARD) has established a micro irrigation fund plan with an amount of Rs. 5000 crores. The nodal Ministry of Agriculture and Farmers Welfare (MoA&FW), GoI, supervises this scheme.

The plan aims to make irrigation less difficult. Majority of the farmers rely on rain to produce their crops. The government wants to give farmers irrigation infrastructure through this scheme.

Year of launch: 2019

Vision:

1. Farmers are able to start up brand-new irrigation projects.
2. To enable the farmer to fulfil his own irrigation requirements.
3. Farmers are no longer only reliant on the rain.

Mission:

1. Extend the reach of irrigation systems.

2.   The farmers should start micro-irrigation programmes.
3.   Consolidate the nationwide installation of Micro-irrigation systems.

The government has a five-year target of micro-irrigating 100 lakh hectares of land. The Micro Irrigation Fund has approved projects totalling Rs. 3805.67 crore to irrigate 12.53 lakh hectares as of September 2020.

Apart from the above said schemes, e-National Agriculture Market (eNAM), Kisan Credit Card (KCC), Soil Health Card (SHC), PM Kisan Samman Nidhi Yojana have been launched by GoI for farmers welfare and empowerment.

PV's Contribution to India's Agricultural Sector

Need of Solar Powered Pumps

The use of water pumps for irrigation has been around for a while. In essence, water pumps run on fuel or electricity, and traditional fuels are used to generate energy. Burning fossil fuels has a bad effect on the environment. There are several geographical areas where there is a lack of energy and a greater need for water, yet these areas also have significant solar power intensities. The country ranks eighth in terms of greenhouse gas emissions due to its heavy reliance on fossil fuels. The primary issue in rural areas is load shedding, which has an adverse effect on crop productivity. The agricultural sector must switch to solar-powered pumps in order to reduce output losses and boost productivity. For irrigation purposes, solar-powered pumps are a tried-and-true technique for farmers. The price of solar modules decreased due to the massive growth of the photovoltaic industry, which also resulted in lower pump prices.

Switching to solar water pumps in place of conventional water pumps will help India to achieve the goal of 100 GW of solar energy by 2022. Grid-connected, net metered solar pumps are crucial for giving farmers access to reliable power for irrigation, which in turn helps them generate a secondary income. There have been many solar-powered system success stories in India including solar-powered drip irrigation system. India's Sunderbans, specifically the Gosaba Island, has solar-powered drip irrigation equipment installed under the "Cropping System Intensification in the Salt-affected Coastal Zones" (CSISCZ) project [7, 8]. Farmers benefited more from this drip irrigation system that was driven by solar energy. In comparison to the previous methods, there was a 20–30% increase in yield, a 40–60% reduction in labour costs, and a growth in cropping intensity of up to 300%.

On March 8th, 2019, the PM-KUSUM policy was introduced [9]. By 2022, it intends to have 25,750 MW of solar capacity installed, with financial backing from the federal government amounting Rs. 34,422 crores. The Ministry of Finance has authorized the installation of 20 lakh freestanding solar pumps and the solarization of 15 lakh grid-connected agricultural pumps for the 2020–2021 fiscal year. If this plan is effective, there will likely be a 10-15 GW installed capacity boost. The 25-year proposal calls for farmers to exploit dry or unusable land. By putting the plant above the required height, a developer can use cultivable land while ensuring that farmers continue to cultivate their crops. The midday load might be met by these installations. Farmers and the Distribution Company benefit equally as a result of the reduction in the price of fuel for them (DISCOM). The DISCOM benefits from a decrease in the load in rural areas and from hitting RPO targets.

Necessity for Feeder Separation

Feeder separation is one of the favorite methods for distributing energy to farm and non-farm users distinctly over specified feeders. In rural areas, feeder separation enables continuous delivery to non-agricultural customers and power supply to farms in the planned manner. Both domestic and non-domestic consumers benefit from enhanced load management and expanded power supply. According to 2014 World Bank research, before feeder separation [10-12]. In Rajasthan and Gujarat 80% of the consumers has experienced low voltage issues before feeder separation, but after feeder separation, this number has dropped down to 6%. Increased employment and a better quality of life have resulted from the increase in energy supply brought out by feeder separation.

Deendayal Upadhyaya Gram Jyoti Yojana (DDUGJY)

Deendayal *Upadhyaya Gram Jyoti Yojana* (DDUGJY) [13], with a total cost of Rs. 43033 crore, was approved by the Indian government in December 2014 which enables the separation of agricultural and non-agricultural lines, facilitating well-planned supply to both farm and non-farm consumers in rural regions, as well as strengthening and allowance of rural sub-transmission and distribution arrangement, with distribution metering.

Objectives

1. To make power available in every village.
2. Feeder separation to ensure that farmers have access for enough amount of power and the villagers should receive the power without interruption.
3. Enhanced sub-transmission and distribution networks to boost the supply's dependability and quality.
4. Metering to cut down on losses.

Benefits

1. Distinguishing agricultural from non-agricultural feeders.
2. Improving the sub-transmission and distribution network in rural areas, including metering distribution transformer feeders and customers.
3. Electrification of rural areas in accordance with Cabinet Committee on Economic Affairs (CCEA) approval in order to meet the Rajiv Gandhi Grameen Vidyutikaran Yojana (RGGVY) targets.
4. To make sure that the infrastructure for rural distribution is strengthened, feeder separation, and quick electrification.
5. To ensure a smooth and quick deployment of electricity must be closely monitored.

Policies initiated by GoI for Solar Powered Agriculture

Photovoltaics (solar panels) can supply electricity for electric fences, distant water pumps, and farm operations. Renovations can be made to buildings and barns to use natural light instead of electric lighting. Extending power lines is frequently more expensive than using solar energy.

PM – KUSUM Policy in India

The PM-KUSUM scheme was launched in India by the MNRE Sources to support the development of farmers and to strengthen the rural supply structure. The programme intends to endorse grid-connected solar energy and the use of solar pumps in the agricultural sector.

Year of launch: 2019

The scheme has four main objectives:

1. To increase agricultural income for farmers
2. To decrease the farm sector's reliance on fossil fuels and to increase access to dependable power.
3. To cut agriculture's power subsidy (MNRE 2019).

The three components of the scheme are as follows:

Component A:

10,000 MW of distributed, grid-connected, ground-mounted renewable power plants with individual capacity up to 2 MW.

Component B:

Installation of 17.50 lakh standalone solar-powered agriculture pumps, each with a 7.5 HP pump capacity.

Component C:

Solarization of 10 lakh grid-connected agriculture pumps with a maximum individual capacity of 7.5 HP.

Table 70.2 presents the top 10 state wise details of the sanctions as on 30.11.2022 under PM-KUSUM scheme.

As per MNRE (05.09.2022), in PM-KUSUM scheme, 4.9 GW Capacity allotted under Component A with 63.95 MW installed Capacity. 8.07 lakhs pumps are sanctioned under Component B with 1.37 lakh installed numbers. 25.43 lakhs pumps are sanctioned under Component C (IPS + FLS) with 1056 solarized numbers under IPS. Scheme is extended till 31.03.2026.

The amendments state that under Component-B and Component-C of the Scheme, individual farmers in the North Eastern States, the UTs of Jammu & Kashmir and Ladakh, and the States of Uttarakhand and Himachal Pradesh, will be eligible for Central Financial Assistance (CFA) for pump capacity up to 15 HP

Table 70.2 Top 10 State's details of sanctions under PM-KUSUM Scheme

S. No.	Component A		Component B	Component C					
	State	Sanctioned solar capacity (MW)	State	Sanctioned quantity (Nos)	State	Total sanctioned individual pump solar (IPS) (Nos)	State	Total sanctioned feeder level solar (FLS) (Nos)
1	Rajasthan	1200	Maharashtra	200000	Kerala	45100	Uttar Pradesh	400000
2	Maharashtra	500	Haryana	197655	Odisha	40000	Karnataka	337000
3	Madhya Pradesh	500	Rajasthan	158884	West Bengal	23700	Chhattisgarh	330000
4	Gujrat	500	Punjab	63000	Tamil Nadu	20000	Gujrat	300500
5	Odisha	500	Madhya Pradesh	57000	Rajasthan	10764	Madhya Pradesh	270000
6	Telangana	500	Uttar Pradesh	36842	Tripura	2600	Maharashtra	250000
7	Tamil Nadu	424	Chhattisgarh	25000	Jharkhand	1000	Bihar	160000
8	Uttar Pradesh	225	Jharkhand	16717	Uttarakhand	200	Punjab	125000
9	Punjab	220	Karnataka	10314	Punjab	186	Rajasthan	100000
10	Haryana	65	Gujrat	8082	Madhya Pradesh	0	Meghalaya	10000

Table 70.3 Scheme Beneficiaries with 7.5 hp pump capacity from Nagpur, Maharashtra

Sr. No.	No. of beneficiaries	Pump capacity (hp)	Year of installation (Nagpur District)
1	145	3	2022
2	45	5	2022
3	14	7.5	2022

during the scheme's extended tenure beyond 2022 [15] and until March 2026. Though, CFA will be limited to 10% of all installations for pumps under 5 HP.

The State Government will offer a 30% subsidy for components B and C, with the farmer contributing the remaining 40%. For the farmer's share, bank funding may be provided, requiring the farmer to finance only 10% of the cost upfront and the remaining 70% as a loan.

The eligible categories for PM-KUSUM Scheme are: 1. An individual farmer, 2. A group of farmers, 3. FPO or Farmer producer organisation, 4. Panchayat, 5. Co-operatives, 6. Water user associations.

The following documents are required to apply for the PM-KUSUM scheme: 1. Aadhar card, 2. A land document, 3. A bank account passbook, 4. A declaration form, 5. Mobile number, 6. Passport size photo.

After a successful online application for KUSUM Scheme, farmers must deposit 10% of the total cost to set up a solar pump to the supplier sent by the department. After the subsidy amount is approved, which typically takes 90 to 10 days, the solar pump set will be activated.

Punjab Solar Water Pumping Scheme

With a goal of encouraging solar energy utilisation for irrigation in the agricultural segment, the Indian government's MNRE created the Solar Water Pumping Scheme in FY 1999-2000 by offering an 80% subsidy on the construction of solar pumping systems.

Punjab's government also offered a 10% subsidy, and from 2000–2001 to 2003–2004, 1850 solar pumps with a 2 HP capacity were installed there, setting the standard for the rest of the country. In Punjab, Punjab Energy Development Agency (PEDA) erected 105 solar pumps with a 2 HP capacity as part of the National Solar Mission between 2013 and 2014. For these solar pumping systems, the GoI and Government of Punjab (GoP) had supplied 30% and 40% of the subsidy, respectively. During the years 2018–19 and 2019–20, 2970 solar pumps with 3 and 5 HP capacities were installed in Punjab with subsidies from the GoI and GoP of 30% and 50%, respectively.

The details of year-by-year [16] accomplishments are as follows:

Table 70.4 Scheme Beneficiaries from Nagpur, Maharashtra

Farmer name	District	Taluka	Village	Year of installation	Pump capacity	Pump type	Pump category	Pump sub type
Gulab Shamrao Bonde	Nagpur	Umred	Katara	2022	7.5	DC	Water Filled	Submersible
Chetan Ramesh Malewar	Nagpur	Mauda	Tarsa	2022	7.5	DC	Water Filled	Submersible
Mahadeo Maroti Zade	Nagpur	Umred	Pimpalkhut (ri)	2022	7.5	DC	Water Filled	Surface
Krunal Shriram Kathane	Nagpur	Kalameshwar	Budhla	2022	7.5	DC	Water Filled	Submersible
Sachin Vishvanath Ikhar	Nagpur	Ramtek	Dudhala (R)	2022	7.5	DC	Water Filled	Submersible
Pranay Shankarrao Kathane	Nagpur	Kalameshwar	Budhla	2022	7.5	DC	Water Filled	Submersible
Viththal Gopalrao Giradkar	Nagpur	Umred	Umred (Rural)	2022	7.5	DC	Water Filled	Submersible
Satish Shankar Khandale	Nagpur	Kamptee	Pandherkawada	2022	7.5	DC	Water Filled	Submersible
Gautam Ganpat Kumbhare	Nagpur	Umred	Shedeshwar	2022	7.5	DC	Water Filled	Submersible
Pralhada Shambhu Bavankule	Nagpur	Ramtek	Masala	2022	7.5	DC	Water Filled	Submersible
Rajnibai Subhashrao Chaure	Nagpur	Katol	Shekapur	2022	7.5	DC	Water Filled	Submersible
Amol Chandrabhan Kirpan	Nagpur	Ramtek	Chokhala	2022	7.5	DC	Water Filled	Submersible
Yashodabai Vithobaji Badave	Nagpur	Katol	Mhaskhapra	2022	7.5	DC	Water Filled	Surface
Alhad Rameshrao Mohite	Nagpur	Katol	Wajbodi	2022	7.5	DC	Water Filled	Submersible

Suryashakti Kisan Yojana (SKY)

The Suryashakti Kisan Yojana (SKY) schemes was introduced by the Gujarat state of India via the PM-KUSUM policy [17, 18]. Under SKY, farmers are permitted to install irrigation pump sets with net-meter1ing SPV power on their land. Utilizing water wisely and increasing agricultural output as a result of daylong access to power, and a new source of income for farmers all contribute to the scheme's positive effects on rural economies.

The basic goal is to capture solar energy and use it to create electricity. Presently, farmers are able to generate their own electricity, utilise it for irrigation, and then sell the surplus power to the grid. Farmers who are already connected to the grid are the target customers for the programme. It would be ideal to have as many farmers as possible on a given AG feeder.

Year of launch: 2018

Funding details are given below: 1. Minimum initial farmer investment of 5%, 2. 35% Loan on the farmer's behalf, 3. 30% Gujarat Government Subsidy (through a loan), 4. GoI subsidy of 30%.

Table 70.5 Year wise pumps installed National Solar Mission

Sr. No.	Year	No. of Pumps
1	2000-01	500
2	2001-02	500
3	2002-03	700
4	2003-04	150
5	2013-14	105
6	2017-18	2605
7	2018-19	365
	Total	4925

Table 70.6 SCHEME BENEFICIARIES FROM Amritsar, Punjab

Beneficiaries under component B of PM-KUSUM scheme (Amritsar)								
Sr No.	Farmer name	Location District	Location Taluka	Location village	Year of installation	Pump capacity (HP)	Pump type	Pump category	Pump sub type
1	Rajindersingh	AMRITSAR	Amritsar -I	Other	2022	5	AC	WaterFilled	Submersible
2	SatnamSingh	AMRITSAR	Amritsar -I	Other	2022	5	AC	WaterFilled	Submersible
3	Gurmejsingh	AMRITSAR	Ajnala	Other	2021	7.5	AC	WaterFilled	Submersible
4	Rattan singh	AMRITSAR	Ajnala	Kot Razada	2022	5	AC	WaterFilled	Submersible

Table 70.7 Suryashakti Kisan Yojana Statistics 2022

Total AG consumers	15 Lakhs
Total number of AG feeders	7060
Total districts covered	33
Total contract load	172 Lakh hp (Avg: 11.43 hp/ farmer)
Solar PV potential	21,000 MW
Total project cost	Rs. 1,05,000 crores
Gov. of India subsidy	30 %
Gov. of Gujrat subsidy	30 %
Farmer's loan	40 %

Eligibility criteria and required documents: 1. The applicant must be a permanent resident of Gujarat, 2. Aadhar Card, 3. Resident certificate, 4. Income certificate, 5. Passport size photograph, 6. Mobile number, 7. Email ID etc

This scheme offers following benefits: 1. A reduction in electricity costs, 2. More money from the sale of extra electricity, 3. 12 hours of grid-quality power during the day, 4. Investment return between 8 to 18 months, 5. PV system ownership following loan payback, 6. A PV system's performance is guaranteed for seven years, 7. State government insurance on PV systems, 8. Crops may be grown on land beneath PV modules, 9. The PV module height-increasing option, 10. The growth of the rural economy.

Mukhyamantri Saur Krishi Vahini Yojana (MSKVY)

Distributed solar power plants are also encouraged through the Mukhyamantri Saur Krushi Vahini Yojana (MSKVY) programme in Maharashtra.

Year of launch: 2018

Decentralised solar project requirements:

1. Required Capacity: 2 MW to 10 MW
2. Required land: Minimum 10 Acres to Maximum 50 Acres
3. Required distance from MSEDCL Substation: Maximum 5 Km
4. Connectivity: 11 kV /22 kV bus bar of MSEDCL 33/22/11 kV substation.

This scheme offers following benefits: 1. Supply of power to AG customers during the day, 2. Increasing farmers' income by installing solar farms on their arid, unusable land, 3. A reduction in the amount of subsidies paid by agriculture customers, 4. Reducing the need for a transmission system, 5. Reducing T&D losses with decentralised solar power plants.

The Ggovernment of Maharashtra developed the MSKVY vide G.R. 14.6.2017 & 17.03.2018 [19] with the intention of providing daytime power to AG consumers. Under this scheme, decentralised solar projects with a capacity ranging from 2 MW to 10 MW will be installed within a 5 Km radius of a distribution sub-station where agriculture load dominates.

In addition to reducing the need for a transmission system, decentralised solar energy facilities can be connected directly to the MSEDCL's existing 33/11 kV sub-stations. Such plants close to these sub-stations might be built, especially by farmers, providing them a chance to boost their income by using their arid, unusable land for solar or other renewable energy-based power plants.

According to MSEDCL's estimation, 8500 MW of total solar capacity must be contracted and put into service during the course of the next three years, from 2020–2021 to 2022–2023, in order to facilitate the delivery of daytime power to AG users.

Karnataka Surya Raitha Scheme

Karnataka Surya Raitha Scheme is a new initiative that the Karnataka government recently unveiled. Under this unique programme, the government will give solar water pump sets to state farmers who are unable to purchase them because of a lack of funds. Farmers benefit from the Surya Raitha Scheme since they do not need to turn on their IP Sets at night for irrigation purposes. The solar water pumps thereby prevent the wastage of both water and power.

Year of launch: 2018

The MNRE subsidies and loans obtained by the DISCOM on behalf of farmers served as the project's funding sources. The feed-in tariff was allocated INR 6 per unit by the DISCOM for debt repayment.

M. AP Grid connected Solar Brushless DC (BLDC) Pump Sets Scheme

The Grid connected Solar Brushless DC (BLDC) Pump set Scheme was introduced by the government of Andhra Pradesh for the benefit of the state's farmers. Under this scheme [20], all new connections will

Table 70.8 List of AG Feeders Solarized in Maharashtra.

Name of District	Capacity of solar plant commissioned (MW)	Substation Name	Substation PT Capacity (MVA)	AG Feeder Name
Amravati	16	33/11 kV Temburkheda	5	11 kV Temburkheda
Amravati	16	33/11 KV Temburkheda	5	11 KV Shahapur
Nanded	10	33/11 kV Bodhadi	5	11 kV Bodhadi
Nanded	10	33/11 KV Bodhadi	5	11 KV Sawari
Ahmednagar	5	33/11 kV Khadakwadi	5	Palashi AG
Ahmednagar	4	33/11 kV Hingni	5	Deodaithan AG
Latur	4	33/11 kV Budhoda	5	11 kV Budhada
Ahmednagar	3.92	33/11 kV Hingni	5	Mekhni
Ahmednagar	3.92	33/11 kV Hingni	5	Kolhewadi
Hingoli	3.12	33/11 kV Potra	5	V Potra

Table 70.9 Top 10 Districts from Maharashtra

Sr. No.	District	Total number of solar plants commissioned	Spare capacity in MW
1	Solapur	215	1237
2	Ahmednagar	186	1163
3	Nasik	165	1006
4	Pune	123	777
5	Aurangabad	113	773
6	Jalgaon	144	675
7	Sangli	117	672
8	Kolhapur	103	595
9	Osmanabad	98	522
10	Beed	129	481

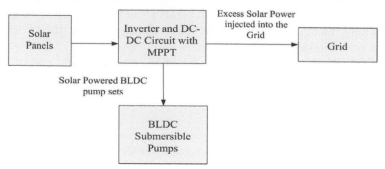

Figure 70.1 Block diagram of Grid connected BLDC Pump Set at the farmer-end
Source: Author

henceforth be powered by solar energy. Solar pumps with a 3 HP or 5 HP capacity will be provided by the state of Andhra Pradesh. This government programme for pump sets aims to convert agricultural connections into solar connections.

Year of launch: 2016

Farmers can use solar energy to power their operations all day long by using solar pump sets that are connected to the grid. Another advantage of dispersed solar power generation is decreased network power losses.

The current pump set at the farmer's end can be swapped with either an energy-competent AC pump set or a brushless DC (BLDC) motor pump set paired with a SPV. BLDC motors have a lesser inertia, improved torque control, and higher efficiency than conventional AC motors, which results in a 20–25% increase in efficiency. Additional vital fact regarding installation of a DC pump set is that the inability to use the grid to power the pump avoids the load from entering the system and compels the pump to be powered entirely by the SPV. Therefore, it is suggested to use solar power and DC pump technology to replace the current pumpset at the farmer's end. States such as Chhattisgarh, Madhya Pradesh, Uttar Pradesh, Rajasthan, and Haryana by now launched proposals for the purchase of Solar BLDC motor pump sets. MNRE has released specifications for these motors.

As seen in Figure 70.1, the BLDC submersible pump is powered by DC electricity from SPV that is routed via a DC-DC circuit and a Maximum Power Point Tracker (MPPT). Using an inverter, extra solar energy is converted to AC power and added to the grid.

Farmers can gain from the following:

1. Obtain free entree to energy-competent solar pumping equipment value a few lakhs.
2. Consistent water release from BLDC pumps for 8 to 9 hours each day.
3. The possibility of the farmer earning additional revenue.

The farmer could operate the pump solely on solar power under this arrangement and sell any excess energy to the grid for INR 1.5/kWh in feed-in tariff. The initiative was totally funded by the DISCOM out of its fund.

The state government's grid-connected solar BLDC pump sets scheme will assist farmers in ensuring irrigation costs are low and a daytime electricity supply. The Andhra Pradesh government would also lessen the cost of subsidies for new agricultural connections.

Conclusion

Various policies initiated by Government of India (GoI) and different states of India are analyzed in this paper. These policies are initiated for the farmers assistance and rural sector. The power to the agricultural pumps is the major concern for distribution company. Irrigation water pumps are the major load in the rural sector. BLDC pump will help for using same solar panels for longer duration, pumping may start early due to the low starting torque. Solar power pumps help to reduce the burden over distribution feeder. If agriculture feeder is supported by the solar power generation, then electricity will be available for farmer at cheaper rate and for larger duration.

References

[1] Rahman, A., Agrawal, S., and Jain, A. (2021). Powering Agriculture in India: Strategies to Boost Components A & C under PM-KUSUM Scheme, Council on Energy, Environment and Water (CEEW), New Delhi, India, 2021.

[2] Patel, R. B. and Patel.R. D. (2019). Smart Energy Management for the Grid Connected Solar Agricultural Prosumers and Consumers. Metering India—Leema, New Delhi, India, 2019.

[3] Padole, N., Moharil, R. M., and Munshi, A. (2022). Performance investigation based on vital factors of agricultural feeder supported by solar photovoltaic power plant. Energies. 15(1), 75. doi: 10.3390/ en15010075.

[4] Badulescu, N. and Tristiu, I. (2017). Integration of photovoltaics in a sustainable irrigation system for agricultural purposes. In Proceedings of the 2017 International Conference on ENERGY and ENVIRONMENT (CIEM), Bucharest, Romania, pp. 36–40.

[5] Programms and Schemes by Ministry of Agriculture and Farmers Welfare, Government of India. https://agricoop.nic.in/sites/default/files/FFH201819_Eng.pdf

[6] Operational Guidelines of Pradhan Mantri Krishi Sinchayee Yojana (Pmksy). Department of agriculture and farmers welfare, Government of India. Pradhan Mantri Krishi Sinchayee Yojana (pmksy.gov.in)

[8] M ahanta, K., Burman1, D., Sarangi1, S., Mandal1, U., Maji1, B., Mandal, S., Digar, S. and Mainuddin, M. (2019). Drip irrigation for reducing soil salinity and increasing cropping intensity: case studies in Indian Sundarbans. Journal of the Indian Society of Coastal Agricultural Research. 37, 64–71.

[9] Akshay Urja, A. (2018). Bi-Monthly Newsletter of the Ministry of New and Renewable Energy, Government of India. 12, 1–52. Available from: 670406a017f54c9386fcde911ee5abe6.pdf (mnre.gov.in)inistry of New and Renewable Energy. 'Vikaspedia' Online Information Guide Launched by the Government of India. PM KUSUM Scheme; MNRE: New Delhi, India, 2019. Available online: PM KUSUM scheme — Vikaspedia PM KUSUM Reforms (26.04.2022). file_s-1650960108170.pdf (mnre.gov.in)   Best Practices on e-Platform of CEA; MSEDCL: Mumbai, India, 2012. Available from: Microsoft Word - Final_Best_Practices_CEA.doc

[10] Central Electricity Authority (Technical Standards for Connectivity to the Grid) Amendment Regulations. 2013. Guidelines Deendayal Upadhyaya Gram Jyoti Yojana (DDUGJY), Scheme of Government of India for rural electrification. https://www.ddugjy.gov.in/assets/uploads/1548234273fykio.pd PM KUSUM (mnre.gov.in)

[11] Extension of Pradhan Mantri Urja Suraksha evam Utthaan Mahabhiyaan (PM-KUSUM) (02-08-2022) https://mnre.gov.in/img/documents/uploads/file_f-1659439385100.pdf

[12] Punjab Energy Development Agency. https://www.peda.gov.in/assets/media/gallery/SWP.pdf

[13] Suryashakti Kisan Yojana (SKY). https://pmmodiyojana.in/suryashakti-kisan-yojana/

[14] Compendium of Central & State Government Policies on Renewable Energy Sources In India, Indian Renewable Energy Development Agency Limited (IREDA). www.cbip.org/Policies2019/policies.aspx#myPage

[15] Mukhyamantri Saur KrIshi Vahini Yojana (MSKVY). https://www.mahadiscom.in/solar-mskvy/index_new.php

[16] Solar P V Water Pumping Programme by New & Renewable Energy Development Corporation of Andhra Pradesh Ltd., A State Gov. Company. NREDCAP | New & Renewable Energy Development Corporation of AP Ltd.

Chapter 71

Analysis and design of precast RCC box structure

Sushama Jibhenkar[a] and Sneha Hirekhan[b]

Department of Civil Engineering, Yeshwantrao Chavan College of Engineering, Nagpur, India

Abstract

This paper is related to the analysis and design of RCC box structure for road under bridge of cross-traffic works in railroads. We utilised precast RCC box structures since we cannot disrupt the movement of the trains. When designing the RUB, several load situations involving the weight of the concrete, the superimposed dead load (SIDL) and the earth pressure were taken into account by using the ultimate limit state and serviceability limit state. In this study, a 3D model was employed to effectively analyse the loading conditions mentioned in the IRS Bridge rule and IRC six regulations. STAAD Pro software was used to analyse the precast box taking into account combinations of loads. The bottom slab, top slab, and side walls of the RCC box structure were designed using the STAAD Pro results. From the results obtained, the structural components of the RCC box structure are built to withstand the highest possible shear force and moment.

Keywords: Cross traffic works, IRC & IRS codes, RCC box structure, STAAD Pro

Introduction

A bridge is a construction assembly which thus permits people to cross a gap in the ground without blocking the entrance below. The box pushing technique is employed in railways whenever an underpass is required, such as for a road under bridge (RUB), the extension of existing railway culverts, canal crossings, and so on. Because the operation must be completed without interfering with rail movement, the box pushing strategy is used over the usual method.

The study and design of a precast RCC box structure are carried out in this article according to Indian standards, particularly Indian railway standards, IRC, IRS, and IS codes. The box structure is designed using STAAD pro software and the behavior of RCC box under different traffic conditions are observed. Box structure is analysed for different classes of IRC loadings. In addition to the component's own weight, the upper slab of a box structure must withstand weight of the structure, living loads from moving vehicles, soil pressure acting on the side panels, and pressure acting on the bottom of the base. In view of the most outrageous bending moment and shear force ends have been made.

Limit State Design

The term "limit state design" (LSD) refers to a structural engineering design process. The method is actually a refinement and simplification of technical knowledge that had been well documented before the implementation of LSD. LSD merely comprises the application of data to assess the amount of protection required by or throughout the design process, beyond the idea of a limit state.

The structure must meet two main requirements when it comes to limit state design:

1. A limit state is a collection of performance criteria that must be met when a structure is subjected to loads (for example, vibration levels, deflection, stiffness, consistency, buckling, torsion, and failure).
2. Assumptions are a part of any design process. The loads that a component will be subjected to must be estimated, member sizes must be determined for inspection and technical requirements must be specified. All engineering design criteria have the same purpose in mind: to ensure that the structure is both safe and functional.

Ultimate Limit State

When exposed to the highest design load for which it was intended, the structure need not crumble to achieve the ultimate limit state. If all figured bending, shearing, tension, or applied load are less than the factored resistance estimated for the segment under consideration, the structure is said to meet the ultimate limit state requirements. Magnification ratio is used for workloads, and reduction factor is used for member resistance. Instead of load, the limit state criteria might be configured in terms of stress. When the factored "maximised" loads are much less than factored "reduced" resistance, the structural member being examined is proved to be safe.

[a]sushamajibhenkar@gmail.com, [b]snehahirekhan01@gmail.com

DOI: 10.1201/9781003450252-71

Serviceability Limit State

A structure must meet the serviceability limit condition criterion if it is to continue operational for its planned use when subjected to ordinary loading, and it must not induce occupant dissatisfaction. When the core characteristics of a construction do not deviate more than predefined limitations set by building regulations, the floors meet established vibration parameters, and other conceivable requirements as specified by the appropriate building regulation, the structure is said to achieve the serviceability limit state. Crack lengths in concrete, which must normally be kept below prescribed dimensions, are an example of further serviceability limit restrictions. If the serviceability standards are not met, for example, if the beams deflect more than the SLS limit, the structure will not certainly fail structurally. The intention of the SLS testing is to make sure that citisens in the building are not disturbed by large deformations of the surface, vibration caused by strolling, or disgusted by extreme swinging of the constructing during gusty winds, or by an overpass swinging from sideways, and to maintain beam deformations negligible enough just to avoid flaky coating on the ceiling above cracking, affecting the structure's appearance and longevity. Several of these restrictions are dependent on the designer's choice of flooring (sheetrock, acoustic tile), hence displacement limits in building regulations are normally explanatory and give the decision to the engineer of reference.

Literature Review

In this paper attempts were made to demonstrate the structural analysis and design of RCC box type minor bridge using manual approach and by computational approach (STAAD-pro) using IRS - CBC codes [1]. The structural elements (top slab, bottom slab, side wall) were designed to withstand Ultimate Load criteria (maximum bending moment and shear force) due to various loads (Dead Load, Live Load, SIDL, LL surcharge, DL surcharge) and serviceability criteria (Crack width) and a comparative study of the results obtained from the above two approaches were carried out to validate the correctness of the results.

This paper illustrated the work to be carried out for the widening of existing roads using box pushing techniques for rail under bridges. The design of pre-cast box was done using STAAD pro. In railways whenever there was a need to make an underpass, either for canal crossing, RUB'S (Rail under bridges), programme of widening existing railway culverts etc, box pushing technique is used. The author concluded from this study that, with the box pushing technique, there was no interruption to the traffic moving around [2].

This paper dealt with box culverts made of RCC. The scope of this paper was further restricted to the structural design of box. The structural design involved consideration of load cases (box empty, full, surcharge loads etc.) and factors like live load, effective width, braking force, dispersal of load through fill, impact factor, co-efficient of earth pressure etc. Relevant IRC Codes were required to be referred. The structural elements were required to be designed to withstand maximum bending moment and shear force [3].

This project provided development of designing box culvert by allowing easy access to a daunting task. The main parameter in this program were concrete and steel property, number of vents, dimension of the box and height of the earth fills. The main objective of the author in this project was to redesign and analyse a box culvert, as the box culvert at the current site had been washed out [4].

In this paper the structure was designed as per standard practice bridges (short span less than 6 m) by IRC: 6 – 2000 and IRC: 21 – 2000 [8]. Both the codes were based upon Working Stress Method. The design based on both the approaches WSM & LSM were also compared and a parametric study was carried out by keeping steel grade as constant with different Grade of Concrete and finding the % reduction in Steel Reinforcement as well as % reduction in section dimensions due to LSM approach over WSM approach [5].

Objectives

1. To plan the construction according to Indian norm and consider the different load combinations on the RCC box and evaluation of different bridge parameters.
2. To take into account a variety of load scenarios, such as effective living load, dead load and earth pressure coefficient.
3. To model and analyse RCC box structure using STAAD Pro software.
4. Also, to determine the total design loads (dead and imposed) acting on the various parts of box structure.
5. IS code guidelines.

Methodology

RCC box of size 5.5 x 4.5 m was considered for analysis and design purpose as per the site requirement in BHOPAL district. The design of the RCC box structure was carried out by considering 1 m unit width of the box and for the analysis purpose the modelling was done for full length on STAAD Pro software. The top slab, walls and base slab was modelled as beam elements. Implementation of loads such as dead load, super imposed dead load (SIDL), live load for maximum moment and shear force and earth pressure for surcharge was applied on the elements as per the IRC and IRS code. After the application of loads for various combination, the results obtained from the analysis for maximum shear force and bending moment were compared between the top, bottom and side wall of the box segment.

Model Description

Mathematical model

The roof slab, side walls and base slab of size 0.55 m and unit width were modeled as beam elements. The soil was modeled as spring having modulus of sub grade reaction as 38400 kN/m³ approximately.

   The element was considered as beam element and the box was modeled as per the parameters given in Table 71.1.

Load cases

DEAD: Self load
LIVE-M: Live load for maximum moment
LIVE-S: Live load for maximum shear force
EP: Earth pressure with dl + ll surcharge
EPDLS: Earth pressure with dl surcharge
EPLLS: Earth pressure with LL surcharge
SIDL: Super imposed dead load

Figure 71.1 shows the beam element numbering in which 1 indicates the top slab, 2 and 3 indicates the side walls and 4 to 15 refers to the bottom slab of RCC box segment. The geometry of the model was created in STAAD Pro software with joint coordinate and member incidence command, thus indicating the element numbering.

   Figure 71.2 shows the model of RCC box segment in 3D view, which indicates 1m unit width strip being modeled for size 5.5 x 4.5 m. The loads were applied on each element in which the top slab comprised of moving train load, bottom slab bearing the load of road traffic and side walls with earth pressure load.

   Following are the types of loads taken into consideration during analysis:

- Dead load: The self-weight of the box was considered by software as self-weight.
- Super imposed dead load: This included dead load of ballast, track, ballast retainer etc, which was calculated below and applied on the top surface as shown in Figure 71.3.

Following shows the calculation for total SIDL considered in this study:

- Dispersion width for SIDL = (Fill width + Ballast cushion + length of sleeper)
$$= (0.30 + 0.35 + 2.745) = 3.395\text{m}$$

Table 71.1 Details of structure

Sr. No.	Details of structure	
	Particulars	Details
1.	Size of the box(m)	5.5 x 4.5
2.	Thickness of roof slab(m)	0.55
3.	Thickness of walls(m)	0.55
4.	Thickness of base slab(m)	0.55
5.	Dead load (KN/m²)	65.00
6.	Unit weight of soil (KN/m²)	18.00
7.	Angle of repose (Φ)	30.00
8.	fck (KN/m²)	35000.00
9.	fy (KN/m²)	500000.00

Source: Author

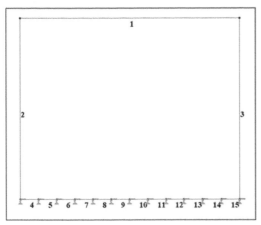

Figure 71.1 Basic frame model showing beam numbers
Source: Author

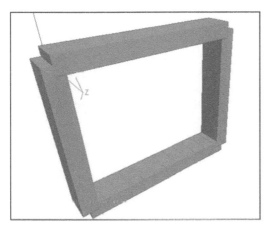

Figure 71.2 3D model of RCC box
Source: Author

Figure 71.3 Frame showing superimposed dead load
Source: Author

- Minimum spacing of track = 5.1 m
- Effective dispersion width = 3.395 m
- UDL due to weight to of track = 65 /3.395 = 19.146 kN
- Weight of earth fill = Density of soil × Depth of fill
  = 18 x (0.3)/3.395 = 1.591kN
- Total SIDL = (19.15+1.59) = 20.736 kN

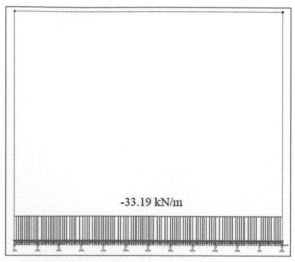

-33.19 kN/m

Figure 71.4 Frame showing live load
Source: Author

Figure 71.4 shows live load on bottom slab which was applied by considering the IRC class 70 R loading for road condition in which impact percentage was taken as 25% of 70R loading.

All earth-retaining structures must be designed to withstand active pressure from the earth fill. The active pressure due to earth fill was calculated by the formula, based on Coulomb's theory for active earth pressure given below:

Pa =1/2 wh²Ka

where:

Pa= Active earth pressure per unit length of wall.

W =Unit weight of soil

h= Height of wall

ø = Angle of internal friction of back fill soil

δ = Angle of friction between wall and earth fill

Where value of δ was assumed.

1.δ = 1/3ø for concrete structures.

2.δ = 2/3ø for masonry structures

I = Angle which the earth surface makes with the horizontal behind the earth retaining structure.

A = Angle which the back surface of earth retaining structure makes with vertical.

Ka = Coefficient of static active earth pressure condition

Ka=

$$\frac{Cos^{2}(\emptyset - \alpha)}{Cos^{2}\alpha \ Cos\ (\alpha + \delta)\left[1 + \sqrt{\frac{\sin(\emptyset+\delta)\sin(\emptyset-i)}{Cos(\alpha+\delta)Cos(\alpha-i)}}\right]^{2}}$$

Lateral earth pressure was applied on the beam elements of RCC box model in three conditions; when Ko = (1-sin ø); Ko = Ka and when Ko = 2 Ka on one side and Ko = Ka on other side of the side walls and vice versa.

(In graph numbers indicate various load combinations such as 1-at top, 2-at bottom, 3-bottom+top, 4-top(ka), 5-bottom (ka), 6-bottom + top(ka), 7-top (2Ka/Ka), 8-bottom (2Ka/Ka), 9-Top + bottom (2Ka/Ka), 10-Top (Ka/2Ka), 11-bottom (Ka/2Ka), 12- top + bottom (Ka/2Ka).

Results

The bending moment of top slab, vertical wall and bottom slab obtained from combination of loads in serviceability limit state condition 1, serviceability limit state condition 2, ultimate limit state condition 1 and ultimate limit state condition 2 (Table 12, IRS concrete bridge code: 1997) [7] for maximum criteria is compared.

From Figure 71.5, it was observed that the maximum bending moment of 342 kN-m was obtained for load combination 6 for SLS 1 condition, which was on top slab of RCC box segment i.e., when the traffic is on both top and bottom slab.

From Figure 71.6, it was observed that the maximum bending moment for SLS 2 condition was obtained for load combination 6 as 319 kN-m on top slab of RCC box segment.

From Figure 71.7, it was observed that the maximum bending moment for ULS 1 condition was obtained for load combination 1 and 6 which as 610 kN-m on top slab of RCC box segment.

From Figure 71.8, it was observed that the maximum bending moment for ULS 2 condition was obtained for load combination 1 as 551 kN-m on top slab of RCC box segment when the traffic flow was considered only on top slab.

Figure 71.9 shows shear force result for top slab, side wall and bottom slab in SLS-1. From the comparison it was observed that the maximum shear value as 408 kN for load combination 10 and 12 for top slab of RCC box segment i.e., when the active earth pressure was acting on the side walls.

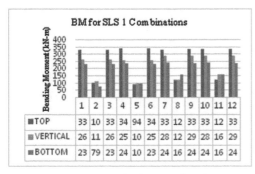

Figure 71.5 Graph showing bending moment result for top slab, side wall and bottom slab in SLS-1
Source: Author

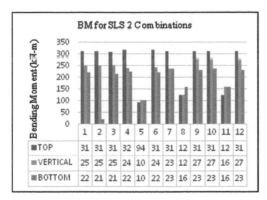

Figure 71.6 Graph showing bending moment result for top slab, side wall and bottom slab in SLS-2
Source: Author

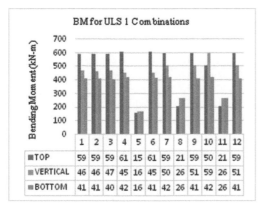

Figure 71.7 Graph showing bending noment result for top slab, side wall and bottom slab in ULS-1
Source: Author

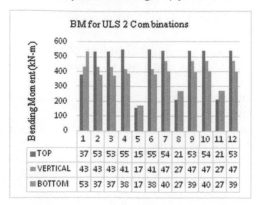

Figure 71.8 Graph showing bending moment result for top slab, side wall and bottom slab in ULS-2
Source: Author

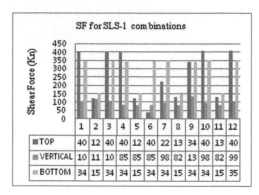

Figure 71.9 Graph showing shear force result for top slab, side wall and bottom slab in SLS-1
Source: Author

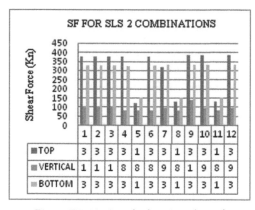

Figure 71.10 Graph showing shear force result for top slab, side wall and bottom slab in SLS-2
Source: Author

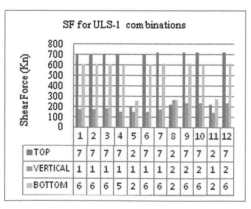

Figure 71.11 Graph showing shear force result for top slab, side wall and bottom slab in ULS-1
Source: Author

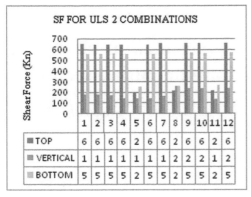

Figure 71.12 Graph showing shear force result for top slab, side wall and bottom slab in ULS-2
Source: Author

From Figure 71.10, it was observed that the maximum shear force for SLS 2 condition was obtained for load combination 9,10 & 12 as 382 kN on top slab of RCC box segment.

From Figure 71.11, it was observed that the maximum shear force for ULS 1 condition was obtained for load combination 9 as 722 kN on top slab of RCC box segment.

From Figure 71.12, it was observed that the maximum shear force for ULS 2 condition was obtained for load combination 10 as 659 kN on top slab of RCC box segment.

Conclusion

Based on the results obtained, the following conclusions were made:

The design and study of the box were completed using manual calculations in accordance with IRC standards. The underlying components top slab, base slab and side walls of pre cast RCC box are intended to endure ultimate standard loads conditions (maximum bending moment and shear force) due to various loads acting on box element. Upon the comparison made between combination of loads in serviceability and ultimate limit state conditions, the maximum value of bending moment and shear force were obtained as 610 kN-m and 722 kN on top slab of RCC box segment. When the traffic flow was on top slab, the critical shear forces were found at the bottom slab and the top slab near the face of side wall and hence the design was given accordingly. Box can undoubtedly be set over delicate establishment by expanding base piece projection to hold base strain inside safe bearing limit of ground soil. The present work was done on precast RCC box which was cost effective whereas modeling and design in STAAD-PRO was easier and gave more economic design.

- **Future scope:** This research can be carried out on a larger bridge, with the cost of the substructure estimated in detail. This study can be compared to the composite bridge.

References

[1] Kazmi, Z. A., Ashad, I., and Shrivastav, V. (2017). Analysis and design of box type minor railway bridge. International Journal of Civil Engineering and Technology (IJCIET).

[2] Rahman, M. A., and Raju, G. (2018). Design and analysis of railway under bridge (Rub) by using box pushing method. International Journal For Technological Research in Engineering. 5(10).

[3] Chanakya, P., and Naidu Damodar, C. H. (2018) Design aids of RCC box culvert by using STAAD pro. International Journal for Research in Applied Science and Engineering Technology (IJRASET).

[4] Vani, D., Kottary, A., and Vishwas, G. (2019) Design of box culvert using visual basic. International Research Journal of Engineering and Technology (IRJET).

[5] Sakalkar, B., Kumar, R., and Singh, P. (2019) Design & analysis of RCC box culvert using WSM & LSM. IJSRD - International Journal for Scientific Research & Development. 6(11).

[6] IRC 6 (1996). Standard specification and code of practice for road bridges section II Loads and stresses & The Indian Road Congress.

[7] IRS (1997). Concrete bridge code.

[8] IRC: 21-2000. Standard specification and code of practice for road bridges. Section III.

[9] AASHTO (2003). Guide Specification for Horizontally Curved Highway Bridges. Washington, DC: American Association of State Highway and Transportation Officials.

[10] Krishnaraju, N. (2019). Design of Bridges. (Third Edition) New Delhi: Oxford and IBH Publishing Co. Pvt. Ltd.

Chapter 72

A review of role of ICT and knowledge management processes to enhance supply chain performance in manufacturing sector

Shivanand Prabhuswamimath[1,a], Mahantesh Halagatti[2,b], Adarsh Patil[1,c], Vinayak Banakar[2,d] and Chetan Hiremath[3,e]

[1]School of Mechanical Engineering, KLE Technological University, India

[2]School of Management Studies, KLE Technological University, India

[3]Kirloskar Institute of Management Studies, Harihar, India

Abstract

The main motive behind this work is to identify the potential barriers that hinders implementation of supply chain management (SCM) practices like customer and supplier relationships, lean practices, quality management practices and analyse the relationship between information and communication technology (ICT), knowledge management (KM) processes and supply chain practices in industries and their relevance in the present era. SCM is important for organisations achieve their mission and vision and ganization top management are faced with many complex situations. Effective supply chain management is the need of the hour for survival of an organisation in multifaceted complex environment and to attain it, organisations need to keep an eye on all the barriers which are hindering supply chain activities. Small and medium enterprises (SME) which are a backbone for developing countries in terms of contributing to GDP and creating job opportunities in local regions, are lagging behind large enterprises in effective management of supply chain activities. SCM blended with ICT usage gives manufacturing sector a gives a good platform to compete in globalised which has positive impact in procurement, enhancing collaboration between suppliers and the organisation and for the proper information flow without any hiccups there by reducing lead time for the products. E-commerce has provided a good platform for manufacturing units to adopt ICT tools to improve the performance of the complete supply chain, by carefully addressing the barriers for ICT implementation. Knowledge is an important asset of every organisation and KM is enabler of SCM, which is gives sustainable direction to industries which adds value to product throughout the complete supply chain which processes data and information required at the right place and right time. Growth of any organisation depends on how valuable knowledge is captured, transformed and disseminated and by using ICT in knowledge management processes can be very crucial in complex environments.

Keywords: Barriers, ICT, knowledge management, manufacturing, organisational performance, small and medium enterprises, supply chain management

Introduction

Supply Chain Management Practices

In 1980, the concept of supply chain came into existence and from their industry and academic professional are using to get competitive edge for the organisations. Supply chain management (SCM) is the blending and fusing of the activities that relate to acquiring materials and services, transform them into intermediary goods and the final product, and channelise them to customers.

Information and Communication Technology and Supply Chain Management

Industry 4.0 has become need of the hour for manufacturing sector which encompasses information and communication technology (ICT) and other advanced technologies like big data analytics, Internet of things (IoT), cyber physical systems. ICT tools are great enablers for accessing timely information for decision making throughout the supply chain areas like transaction execution, coordination and collaboration, and decision support. The ICT tools include software, hardware or IoT devices which is both hardware and software.

Knowledge Management, Role of ICT in Knowledge Management and Importance of Knowledge Management in Supply Chain Management

Many studies in literature highlight the importance of KM practices like developing strategy for Knowledge creation and utilisation, arranging training sessions, guiding and documentation. New knowledge to bring changes to product/services can be acquired internally or externally. Storage/retention of new knowledge is

[a]shivanand@kletech.ac.in, [b]mahantesh.halagatti@kletech.ac.in, [c]adarsh@kletech.ac.in, [d]vinayak.banakarh@kletech.ac.in, [e]hirechetan@gmail.com

DOI: 10.1201/9781003450252-72

very important from future scope as well as the resources that are being invested for creation of knowledge [1–4].

Objectives

The aim of this paper is to study, analyse and structure the available data on supply chain practices, role of ICT and knowledge management in supply chain activities. The review evaluates key findings by the author and aims at similarities and differences in reported findings in literature. The paper focusses on review of literature in the fields of operations, supply chain practices, knowledge management and ICT role along the supply chain manufacturing sector considering the major journals and conferences [5–8].

Supply Chain Management Practices in Manufacturing Sector and its Importance

SCM is one of the main areas in manufacturing sector where focus is shifted from one-dimensional subject dealing with logistics and material flow, into a multidimensional concept including strategic differentiation, value enhancement, value to the customer, supplier collaborative relationship and so on. The SCM practices have an important role in enhancing organisational performance and to compete in globalised world.

Literature Review

Barriers for Supply Chain Management Practices

Table 72.1 gives the number of papers taken for literature survey from Journals, Conferences, Articles and Book chapters including the year of publication. All papers conclude that, barriers for SCM practices need to address to enhance supply chain performance in manufacturing sector [9–14].

From Table 72.2, we can observe that in all research work conceptual models have been developed for research work by formulating hypotheses and testing them. Both survey and case study methods have been used. For analysis of barriers to SCM practices implementation, interpretive structural modeling (ISM) technique is used which gives relationships among barriers, which are the causes of the problem/issue. SCM practices implementation level is related to the intensity of barriers. Barriers identified holds good for all types of organisations with respect to product variety, geographical presence, and manufacturing/service nature.

From Table 72.3, we can observe that, all the research work reviewed highlights the different barriers for effective implementation of SCM practices in manufacturing sector. All the barriers as observed by each of the authors take care of both the internal and external environment of organisation, but each of workers barriers should be combined to get a complete exhaustive list of barriers for SCM practices implementation and need to be reviewed timely to take care of new entrants of barriers [15–23].

Table 72.1 Publication year and number of papers/articles/conferences/books

Year	Number of papers
2000	01
2002	01
2005	01
2006	01
2007	01
2008	02
2019	01
2010	01
2011	02
2012	01
2013	01
2014	05
2015	01
2016	01
2017	02
2019	01

Source: Author

Table 72.2 Methodology, tools used by different authors

Reference	Methodology, tools used, perspective and context
[1]	Case study approach used for 15 organisations covering 90 in depth interviews. Conceptual framework for SCM which includes business processes of SCM, management criteria of SCM, network relationship of SCM.
[2]	Focus of this paper is on the Bullwhip effect. Bullwhip effect in manufacturing units is well analysed through case studies and economic data analysis.
[3]	Case study research highlighting the research process and quality of research with respect to three garment industries.
[4]	Research framework involving SCM practices, organisational performance, and competitive advantage.
[5]	Case studies of 29 organisations involving four different sectors: construction supply chain, assistive technology supply chain, apparels and computer consumables. Survey method used for gathering information.
[6]	An exigency framework developed for barriers affecting the adoption of SCM practices along with advantages. Work includes survey aspects and case study procedure. Around 50 case studies conducted for study, which includes retailers, finished goods manufacturers, first-tier and lower-tier suppliers, service providers.
[7]	Structural equation modelling is used to depict relationship between SCM practices and performance of manufacturing units in Taiwan. The survey procedure included hypothesis statements depicting: Customer relationship management, leadership, human resource management (HRM), total quality management (TQM), SCM, business process management (BPM)
[8]	A framework for supply chain implementation was constructed using driving forces for SCM, challenges, organisational performance, and strategy initiative management constructs. Survey method used to compare Malaysian and Iranian organisations SCM implementation level and the challenges faced by them.
[9]	Survey method used for 100 various multiple manufacturing industries with a total of more 200 respondents to assess green supply chain management activities and comparison highlighted between lean and green manufacturing.
[10]	The importance and adoption of SCM to SMEs is highlighted.
[11]	Key aspects of this paper include barriers for sustainable supply chain management adoption in manufacturing sector through survey methodology and interpretive structural modeling (ISM).
[12]	Conceptual model developed for SCM implementation by considering the different elements like: mass production, organisational culture and departmentalisation, technical issues. Survey method used for analysis.
[13]	Survey method used, conceptual model developed, and hypothesis were framed for different constructs like: Goals of organisations and supply chain.
[14]	Survey method used involving interpretive structural modeling (ISM), digraph, MICMAC analysis to analyse and classify barriers
[15]	ISM approach is used for analyzing barriers for SCM practices adoption.
[16]	This paper focuses on SMEs in India. Elements of survey include stimulators, challenges, financial resources, supply chain practices and organisational performance for which the conceptual model is framed.

Source: Author

Categorisation of Barriers for Supply Chain Management Practices Implementation in Manufacturing Sector

From Table 72.3, we can observe that individual barriers can be categorised depending upon their nature and intensity and this categorisation is highlighted in Table 72.4.

Adoption and implementing of the supply chain methods is of utmost importance to all manufacturing sector units, which cannot be left to chance. For attaining this goal, organisational professionals are striving to understand and analyse SCM practices and thereby to achieve the benefits of SCM. Every manufacturing unit objective is to improve the efficiency of supply chain to meet customers demand at right time. And to achieve this, identification and analysis of barriers which hinder the SCM practices implementation is need of the hour. Technology, information, and measurement related barriers play a vital role and need to be addressed. Human resources barriers involving cultures, skills, attitude towards change also to be viewed seriously as people are going to carry the complete implementation work. Organisational structure parameters like size, structure and investment capacity must be analysed for proper implementation of SCM

Table 72.3 Barriers to supply chain management practices implementation

Reference		Barriers to supply chain management practices implementation
[1]		Operations management, work relationship, authority and responsibility relationship structure, facility layout, management tools, organisational culture.
[2]		Information flow, coordination, vendor management system, inventory system, decentralised and centralised planning flow, decision making.
James H. Foggin et al	2004	Costly information flow channels, manual practices, poor visibility of information flow.
[5]		Organisation size, structure, characteristics, resources, employee involvement, financial constraints, industry type that includes competition, different products/services, local/global, Government rules.
Joanne Meehan et al	2008	Workforce skills, authority, power,top management commitment for training and improvement programmes, workforce interest to participate, trust among members knowledge/skills of electronic trading, location from customers/supplier.
[6]		Changing customer's needs, global competition, lesser lead time for product cycles, acquisition and merger, ICT adoption, top management support, long term and short-term methodology, trust, change management, training and career development programmes
Tumaini Mujuni Katunzi et al	2011	Hiding information, improper supply chain visibility, trust, knowledge, bullwhip effect, demand updating, inventory holding, price changes, scarce goods management
Mohd Nizam Ab. Rahman1 et al	2011	Knowledge, trust, power, infrastructure, ICT, skillset of workers,
[12]		Large scale manufacturing, organisational relationship, culture, structure, technology constraint.
Sergio Palomero et al	2014	Shortage of skilled workforce to manage supply chain, training programmes, technical knowledge, people involvement, handling conflicts, poor responsibility roles, trust among people in organisation, mission and vision aspects, top management structure, flexibility for change in organisations, consultants, ICT infrastructure, performance indicators, financials resources.
Raja Irfan Sabir et al	2014	Information technology facilities, information sharing, trust, management philosophy, incompatibility in the organization.
[12]		Large scale manufacturing, organisational structure, technology adoption challenges.
Arvind Jayanta et al	2014	Financial management, ICT, organisational culture, Top management commitment, technology adoption, government support, knowledge, technical expertise, global competition, awareness of customer needs, environmental awareness, training courses.
Ganeshan Wignaraja	2015	Organisational size, technological capabilities, human resource aspects like skill, age, knowledge, foreign ownership, financial resources availability.
[14]		Owner attitudes, resources, ICT infrastructure, knowledge, and expertise of ICT willingness of members of supply chain.
[20]		Perceived benefits, customer pressure and needs, supplier pressure and coordination, competitor pressure in terms of price, top management support, government support, staff and their skills, security, resistance to change.
Aditya Nugroho et al	2017	Dissemination of information, human resources capability, web-based business, inventory level, customer service and needs.
Noor Hidayah Abu1 et al	2018	Finance availability, organisational structure and culture, resources for ICT, skills, competition, training and mentoring aspects, support from suppliers.
[16]		Top management commitment, resources and funds, transportation facilities, coordination among members, modern technologies, demand forecast system, sharing information with suppliers, Shortage of technical manpower and experts, sophisticated information system, trust among members, location of suppliers and customer.

Source: Author

Table 72.4 Categorisation of barriers to SCM practices implementation

Different ways of categorising barriers	Categorised barriers
A	• Strategic barriers (vision, mission, objective, decision making process) • Individual barriers (education, skills, knowledge, culture) • Cultural barriers (unwillingness to share information, mistrust) • Organisational barriers (finance, structure, measurement system) • Technological barriers (ICT usage)
B	• Information dissemination barriers • Human resource barriers • Customer service barriers • Electronics based business barriers • Storage level barriers
C	• Planning and control barriers • Work structure barriers • Organisational structure barriers • Top management and authority barriers • Organisational culture and workers attitude barriers • Barriers for proper channelising the information path
D	• Internal barriers (organisational policy, employee involvement) • External barriers (customers, competition, marketing, suppliers)
E	• Technology barriers (cost of ICT, technical expertise, ICT vendors) • Financial barriers (availability of finance, access to finance) • Knowledge barriers (logistics skills, customer relationship) • Involvement and support barriers (Mutual cooperation, adoption to change) • Outsourcing barriers (supplier culture, structure)

Source: Author

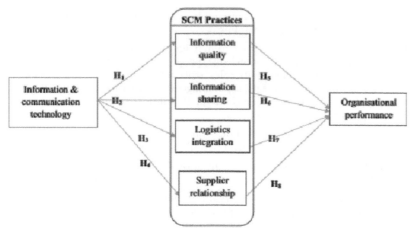

Figure 72.1 Relationship between SCM and ICT
Source: Author

practices to reduce lead time to put the product into market, total cost of product cost and cost of holding inventory and to improve quality of products/services, response to customers and sales forecast [24–29].

Role of Information and Communication Technology in Supply Chain Management

Still there exists a gap between theory and practice in SCM field regarding ICT usage by considering the cost of ICT tools and factors affecting ICT implementation in manufacturing sector [30]. ICT decides the quality of decision making along supply chain members and size and nature of organisation need to be considered for ICT tools selection and implementing them and proper integration of above aspects helps in better decision making as shown in Table 72.5 [31].

Information and Communication Technology and Supply Chain Integration

Conceptual framework has been proposed as shown in Figure 72.1 involving SCM practices like Information, logistics and supplier relationship aspects along with ICT adoption and Organisational performance as shown in Figure 72.1. Public food distribution agencies were selected for study and questionnaire was distributed to 121 officers involved in food industry. For analysis of data from the survey partial least square

Table 72.5 Exploring relationship between ICT And SCM

Author, Year	Research methodology, objectives, ICT tools used, research findings
Prashant R. Nair, 2019	ICT tools used: Enterprise resource planning (ERP), barcoding, web services, electronic data interchange (EDI), radio-frequency identification (RFID), decision support systems (DSS), cloud computing, Industry 4.0, SMAC-social, mobile, analytics and cloud stack
Anil Kumar et al, 2019	Using survey method framework is developed for ICT integration in SCM practices. Partial least square (PLS) and –structured equation modeling (SEM) deployed for analysis of Data. Proposed conceptual framework involving SCM practices like information attribute aspects like quality, ease of sharing, relationship with suppliers, logistics aspects of integration, ICT and organisational performance
Monica Colina et al, 2015	Survey method used: 288 small and medium enterprises (SMEs)
	Conceptual model framed between strategy, ICT and SCM practices.
	To analyse the structure of the conceptual model and to test hypotheses, structural equation modelling was deployed.
	Research Findings: Positive relationship exists between ICT integration in SCM and strategies implemented adopted by manufacturing organisations to implement SCM practices
Kanabar Kavitha et al, 2019	Survey method used: 100 companies comprising 40 small scale, 47 medium scale and 13 large scale from Mumbai.
	ICT tools: Tel/Fax, wireless LAN Internet, LAN, E- mail, CRM, company website, RFID, ERP, EDI, barcode
	Research Findings: large companies spend more on the integration of the ICT tools with the supply chain companies compared to SME's.
Liu, Wen-Liang et al, 2017	Using Survey procedure, around 1000 enterprises in manufacturing arena were interviewed for SCM practices in Taiwan.
	Proposed conceptual model developed involving ICT, dynamic capabilities of suply chain, commitment and trust.
	Structural equation modelling was deployed to validate conceptual model developed and the hypotheses framed for analysis of data.
	Research findings: ICT integration with SCM practices play a crucial role in combating global competition.
Maria Kollberg et al, 2004	Proposed research model encompassing ICT and its effects, level of ICT integration, factors influencing ICT integration in supply chain framework.
	Research findings: The proposed model had a set of control parameters that play key role to study the impact of ICT in supply chain integration.
Tony Cragg et al, 2015	For collection of data on ICT usage in SCM, interviews were taken up at SME's in the agricultural machinery sector using series of semi-structured, face-to-face, in-depth were through case study method.
	Research framework modelled incorporating ICT and barriers to effective SCM practices.
	Research findings: ICT solutions were proposed to help SMEs.
Jaana Auramo et al 2005	Survey and case study methods: Around 50 progressive manufacturing units were surveyed to identify what IT solutions implementation in SCM followed by in-depth case studies reveal the mechanisms for getting advantages of IT in SCM.
	Research findings: Different propositions like focused e-business, improved efficiency, use of e-business solutions highlighted the use of ICT in process re-design.
	ICT Tools: Intranet, extranet, EDI, RFID
Alain Yee-Loong Chong, 2009	Survey method used to investigate ICT adoption in SCM arena for Electrical and Electronic (E&E) enterprises in Malaysia.
	Conceptual model developed and tested encompassing complexity of product design, volume of production, adoption level of trust and e-collaboration tools
	Research findings: The factors affecting the adoption of ICT are trust, product complexity and product volume and frequency.
Ming-Lang Tsenga, 2011	The DEMATEL procedure used for data analysis which is practical for visualizing the structure of complicated relationships with matrices or digraphs.
	Research Findings: ICT helps to gain marketing performance and customer satisfaction in Vietnam industries in textile zone.

Source: Author

and structured equation modeling were deployed. The results indicate that ICT and SCM practices have a positive relationship. And also SCM practices have a significant impact on the performance of the enterprise. This work can be extended to other sectors like manufacturing, pharmaceutical etc. The findings also help in the decision-making process in food distribution sector. The constructs used in the study is of good importance in complex environment of business and the results are in conformance with other studies in this area of research.

Conclusions

Adoption and implementing of the supply chain methods (SCM) is of utmost importance to all manufacturing sector units, which cannot be left to chance. For attaining this goal, organisational professionals are striving to understand and analyse SCM practices and thereby to achieve the benefits of SCM. Every manufacturing unit objective is to improve the efficiency of supply chain to meet customers demand at right time. And to achieve this, identification and analysis of barriers which hinder the SCM practices implementation is need of the hour. Technology, information, and measurement related barriers play a vital role and need to be addressed. Human resources barriers involving cultures, skills, attitude towards change also to be viewed seriously as people are going to carry the complete implementation work. Organisational structure parameters like size, structure and investment capacity have to be analysed for proper implementation of SCM practices to reduce lead time to put the product into market, total cost of product cost and cost of holding inventory and to improve quality of products/services, response to customers and sales forecast. Information and communications technology (ICT) will provide positive impact in a supply chain, because it provides better access to information, making logistics aspects more accurate, faster and cheaper in on both front-end and back-end processing. From literature it is observed that significant association exists between the different SCM practices like information sharing, logistics integration and supplier relationship. ICT helps in communication and data management thereby reducing costs and time in the transmission of information. Finance resource availability for ICT adoption in supply chain path plays a very crucial role reap benefits for manufacturing sector. Organisational issues like top management commitment, culture, employee skill set, and behavior are key to proper integration of ICT into supply chain activities. Knowledge management (KM) processes play a vital role in dynamic and flexible SC and KM blended with ICT will enhance SC performance and better decision making.KM helps SC to evolve for better results and helps in consolidating information required for organisation both from suppliers and customers. KM is a platform that helps organizations in SC integration, inter and intra relations along SC, formulating strategy in complex environment, inventory, reverse logistics and so on [32–34].

References

[1] Douglas M. Lambert, Matias G. Enz, (2017). Issues in Supply Chain Management: Progress and potential, Industrial Marketing Management, 62, 1–16, ISSN 0019-8501

[2] Sahin, Funda and Powell E. Robinson. (2002). Flow Coordination and Information Sharing in Supply Chains: Review, Implications, and Directions for Future Research. Decis. Sci. 33: 505–536.

[3] Seuring, S. (2005). Case Study Research in Supply Chains — An Outline and Three Examples. In: Kotzab, H., Seuring, S., Müller, M., Reiner, G. (eds) Research Methodologies in Supply Chain Management. Physica-Verlag HD.

[4] Suhong Li, Bhanu Ragu-Nathan, T.S. Ragu-Nathan and S. Subba Rao, (2006). The impact of supply chain management practices on competitive advantage and organizational performance, The International journal of Management Science, 34(2), 107–124.

[5] C.M. Harland, N.D. Caldwell, P. Powell, J. Zheng, (2007). Barriers to supply chain information integration: SMEs adrift of elands, Journal of Operations Management, 25(6), 1234–1254.

[6] Fawcett, Stanley and Magnan, Gregory and Mccarter, Matthew. (2008). Benefits, Barriers, and Bridges to Effective Supply Chain Management. Supply Chain Management: An International Journal. 13. 35–48. 10.1108/13598540810850300.

[7] Ou, Chin and Liu, Fang-Chun and Hung, Yu-Chung and Yen, David. (2010). A structural model of supply chain management on firm performance. International Journal of Operations & Production Management - INT J OPER PROD MANAGE. 30. 526–545. 10.1108/01443571011039614.

[8] Manzouri, Malihe and Ab Rahman, Mohd and Arshad, Haslina and Ismail, Ahmad Rasdan. (2010). Barriers of supply chain management implementation in manufacturing companies: A comparison between Iranian and Malaysian companies. Journal of The Chinese Institute of Industrial Engineers. 27. 456–472. 10.1080/10170669.2010.526379.

[9] Dube, Anil and Gawande, Rupesh and Coe, B. (2011). Green Supply Chain management – A literature review.

[10] Chin, Thoo and Abdul Hamid, Abu Bakar and Rasli, Amran and Baharun, Rohaizat. (2012). Adoption of Supply Chain Management in SMEs. Procedia - Social and Behavioral Sciences. 65. 614–619. 10.1016/j.sbspro.2012.11.173.

[11] Zaabi, Shaikha and AL Dhaheri, Noura and Diabat, Ali. (2013). Analysis of interaction between the barriers for the implementation of sustainable supply chain management. The International Journal of Advanced Manufacturing Technology. 68. 10.1007/s00170-013-4951-8.

[12] Dubihlela, Job and Omoruyi, Osayuwamen. (2014). Barriers To Effective Supply Chain Management, Implementation, And Impact On Business Performance Of SMEs In South Africa. Journal of Applied Business Research. 30. 1019–1031. 10.19030/jabr.v30i4.8651.

[13] Ullah, Inayat and Narain, Rakesh and Singh, Amar. (2015). Supply Chain Management Practices in SMEs of India: Some Managerial Lessons from Large Enterprises. International Research Journal of Engineering and Technology. 2. 1176–1196.

[14] Toktaş-Palut, Peral and Baylav, Ecem and Teoman, Seyhan and Altunbey, Mustafa. (2014). The impact of barriers and benefits of e-procurement on its adoption decision: An empirical analysis. International Journal of Production Economics. 158. 77–90. 10.1016/j.ijpe.2014.07.017.

[15] Jayant, Arvind and Azhar, Mohd. (2014). Analysis of the Barriers for Implementing Green Supply Chain Management (GSCM) Practices: An Interpretive Structural Modeling (ISM) Approach. Procedia Engineering. 97. 10.1016/j.proeng.2014.12.459.

[16] Singh, Rajesh and Kumar, Ravinder. (2020). Strategic issues in supply chain management of Indian SMEs due to globalization: an empirical study. Benchmarking: An International Journal. ahead-of-print. 10.1108/BIJ-09-2019-0429.

[17] Parmar, N.S. (2016). A literature review on supply chain management barriers in manufacturing organization. International Journal of Engineering Development and Research, 4, 26–42.

[18] Cooper, Martha and Lambert, Douglas and Pagh, Janus. (1997). Supply Chain Management: More Than a New Name for Logistics. International Journal of Logistics Management, The. 8. 1–14. 10.1108/09574099710805556.

[19] Meehan, Joanne and Muir, Lindsey. (2008). SCM in Merseyside SMEs: benefits and barriers. TQM Journal. 20. 223–232. 10.1108/17542730810867245.

[20] Hamadneh, Samer and Alshurideh, Muhammad and Alzoubi, Haitham and Akour, Iman and Al Kurdi, Barween and Joghee, Shanmugan. (2023). Factors affecting e-supply chain management systems adoption in Jordan: An empirical study. Uncertain Supply Chain Management. 11. 411–422.

[21] More, Dileep and Babu, A. (2011). Supply chain flexibility: A risk management approach. Int. J. of Business Innovation and Research. 5. 255–279. 10.1504/IJBIR.2011.040098.

[22] Al-Shboul, Moh'D and Garza-Reyes, Jose Arturo and Kumar, Vikas. (2018). Best Supply Chain Management Practices and High-Performance Firms: The Case of Gulf Manufacturing Firms. International Journal of Productivity and Performance Management. 67. 10.1108/IJPPM-11-2016-0257.

[23] Acharyulu, G. (2014). Supply Chain Management Practices in Printing Industry. Operations and Supply Chain Management: An International Journal, 7(2), 39–45.

[24] Kumar, Anil and Kushwaha, G. (2017). Supply chain management practices and organisational performance in fair price shops. 85–99. 10.17270/J.LOG.2018.237.

[25] Lambert, Douglas and Cooper, Martha. (2000). Issues in Supply Chain Management. Industrial Marketing Management. 29. 65–83. 10.1016/S0019-8501(99)00113-3.

[26] Choy, K.L. and W.B., Lee and Lo, Victor. (2003). Design of a case based intelligent supplier relationship management system - The integration of supplier rating system and product coding system. Expert Syst. Appl. 25. 10.1016/S0957-4174(03)00009-5.

[27] Zaid, Ahmed and Baig, Javeria. (2020). The Impact of Supply Chain Quality Management Practices, Supply Chain Quality Management Capabilities and Knowledge Transfer on Firm Performance: A Proposed Framework.

[28] Farooque, Muhammad and Zhang, Abraham and Thurer, Matthias and Qu, Ting and Huisingh, Donald. (2019). Circular supply chain management: A definition and structured literature review. Journal of Cleaner Production. 228. 10.1016/j.jclepro.2019.04.303.

[29] Singh, Rajesh and Kumar, Ravinder. (2020). Strategic issues in supply chain management of Indian SMEs due to globalization: an empirical study. Benchmarking: An International Journal. ahead-of-print. 10.1108/BIJ-09-2019-0429.

[30] Walker, Helen and Di Sisto, Lucio and Mcbain, Darian. (2008). Drivers and barriers to environmental supply chain management practices: Lessons from the public and private sectors. Journal of Purchasing and Supply Management. 14. 69–85. 10.1016/j.pursup.2008.01.007.

[31] Olugu, Ezutah and Wong, Kuan Yew and M Shaharoun, Awaluddin. (2011). Development of key performance measures for the automobile green supply chain. Resources Conservation and Recycling - RESOUR CONSERV RECYCL. 55. 567–579. 10.1016/j.resconrec.2010.06.003.

[32] Luthra, Sunil and Qadri, Mohd and Garg, and Haleem, Abid. (2013). Identification of critical success factors to achieve high green supply chain management performances in Indian automobile industry. International Journal of Logistics Systems and Management.

[33] Govindan, Kannan and Mathiyazhagan, K. and Kannan, Devika and Haq, A. (2013). Barriers analysis for Green Supply Chain Management implementation in Indian Industries Using Analytic Hierarchy Process. International Journal of Production Economics. 147. 555–568. 10.1016/j.ijpe.2013.08.01.

[34] Tiwari, Atul and Tiwari, Anunay and Samuel, Cherian. (2015). Supply chain flexibility: A comprehensive review. Management Research Review. 38. 10.1108/MRR-08-2013-0194.

Chapter 73

Effect of compaction pressure and sintering time on characteristics of nickel based superalloy

Geetika Salwan[a], Rayapati Subbarao[b] and Subrata Mondal[c]

NITTTR Kolkata, India

Abstract

Nickel based superalloys have good proficiency of retaining strength and toughness, even at elevated temperatures, which makes them more suitable as gas turbine blades. Turbine blades are mostly prepared by investment casting method, which consumes more time and increases cost. This led the researchers to study powder metallurgy technique for verifying its suitability in making the turbine blades. This process has a unique capability of alloying the elements for better uniformity and the ability to increase temperature bearing capability using refractory elements. In such a scenario, present work uses powder metallurgy technique for making a nickel-based superalloy. By controlling various parameters like time of blending, compaction pressure, temperature of sintering and time of holding, requisite superalloy with desirable properties can be prepared. In this work, nickel-based superalloy with typical composition is prepared and tested for its characteristics upon changing such parameters. Mechanical properties like hardness and physical properties like density are determined with and without heat treatment. Microstructural analysis is carried out to study porosity in the sample. Compaction pressure of 850 MPa is found to be giving density and hardness, comparable to the standard material. Similarly sintering holding time of 4 hours for a constant temperature of 1200°C is found appropriate. For the same pressure and sintering holding time, microscopic behavior is observed, which shows less porosity and proper association of the metallic elements. Thus, the current work gives the hint of appropriate compaction pressure and sintering temperature with holding time while using powder metallurgy in preparing such superalloys for various applications.

Keywords: Compaction pressure, nickel based superalloy, physical and mechanical characteristics, powder metallurgy, sintering time

Introduction

Nowadays nickel based superalloys are extensively used as high temperature structural materials. Particularly, for making the blades and discs of gas turbine engines for aviation jet engines and power plant. Difficulty in deforming and machining restricts its application. Its thermal conductivity is low and work hardening rate is very high, which makes it difficult to machine the material. Also, traditional machining cost of the material is very high due to high tool wear rate, poor surface finish and high consumption of cutting fluid. Investment casting method is normally used for making the turbine blade, but large number of alloying elements in superalloy causes segregation, hence results in crack formation during hot working. Therefore, powder metallurgy technique is checked in order to determine its suitability for preparing the alloy. By this technique, uniform composition and structure with fine grain can be obtained. This process costs less, is versatile and has the ability to form complex shape. Earlier, Tang et al. [1] used powder metallurgy technique to prepare GH4049 super alloy. They successfully prepared the sample with 99% density through super solidus liquid phase sintering in vacuum furnace. Mechanical properties of the prepared sample is found to be same as that of wrought superalloy GH4049. Yang et al. [2] used the powder metallurgy technique to prepare nickel-based superalloys, for making disk of turbine blade. Zielińska [3] studied the mechanical and microscopic properties of Rene 77, manufactured by investment casting method. During casting, surface layer of ceramic mold was coated with cobalt aluminate, which had significant impact on its properties. Sreenu et al. [4] studied about mechanical and microscopic properties of nickel-based superalloys processes through hot isostatic pressing route and found improved stress rupture life. Danninger [5] found the influence of powder metallurgy in manufacturing industry and observed that it can be used for the tailor-made novel materials with less wastage and energy consumption is also reduced by adopting this process. Guo et al. [6] have discussed the effect of heat treatment on Ti-6Al-4V alloy prepared by hot isostatic pressing of powder at different temperatures and compaction pressures. Huda [7] has developed the heat treatment process for IN 792 alloy made by conventional powder metallurgy technique. But, he has not discussed the effect of changing the compaction pressure and sintering temperature. Subbarao and Chakraborty [8] conducted microscopic studies of different superalloys for determining their suitability as gas turbine blade materials. Somani et al. [9] prepared IN100 superalloy by means of powder metallurgy technique and deliberated the hot deformation behavior in the temperature range of 1000–1200°C. Recently, Subbarao [10] has computationally studied the application of nickel-based alloys

[a]geetikasalwan@gmail.com, [b]rsubbarao@nitttrkol.ac.in, [c]subratamondal@nitttrkol.ac.in

DOI: 10.1201/9781003450252-73

as turbine blade materials. Understanding the growing demand of the high strength nickel-based superalloys and the development and processing of alloys by powder metallurgy technique, this study is being carried out in order to achieve the desired combination of mechanical and physical properties. It is clear that, various computational and experimental studies have been done earlier for determining the suitability of superalloys as turbine blade materials. It is also observed that superalloys can be prepared by powder metallurgy techniques but, with limitation on the process parameters. Studies on the effect of compaction pressure, sintering temperature and holding time on physical and mechanical properties of superalloy are yet to be carried out. In this contest, the present work deals in the preparation of a nickel-based superalloy using conventional powder metallurgy technique and also, determination of effect of processing on its hardness and density with and without heat treatment of the prepared sample is investigated. Fractional porosity derived from sintered density is determined and its variation with respect to compaction pressure and sintering holding time is analysed.

Experimental Work

Powders used are of loba chemie and SRL make, which is obtained from designated vendors. Composition of powders used for making the nickel-based superalloy by powder metallurgy technique is shown in Table 73.1. Variation of the size of the metal powder used ranges from 0.01 mm to 0.1 mm. The apparent density of the powder is 4 gm/cm³. Weighing of the individual powders is done using a digital balance. Powders with appropriate weight percentage are put inside a container made of stainless steel and mixed by using ball milling mixing method for the duration of 7 hours. Specimens are prepared by single ended cold compaction process, with suitable die punch set at a compaction pressure of 700 MPa, 800 MPa and 850 MPa in a 10-ton capacity UTM. Cylindrical specimens of 12mm diameter and 10 mm height are obtained as shown in Figure 73.1. Then, sintering is done at 1200°C in a vacuum of 10^{-2} torr using a tubular vacuum furnace for different holding time. Temperature raises in three steps, i.e. from 0-600°C in 100 min, 600-1000°C in 80 min, and 1000-1200°C in 50 min. At 600°C and 1000°C, 10 min. holding time was given. Holding times taken for different samples are 2 hours, 3 hours and 4 hours. Direct aging and solution annealing is done in the same vacuum furnace. Aging of sample is done at 800°C for 4 hours and solution annealing is done at 1000°C for 2 hours. Cooling is done in the furnace itself. Solution annealing is done in order to dissolve the second phases and hence the additional solute is accessible for precipitation hardening. This is used to prepare strengthening phases and to control the carbides and the topologically close packed phases. Hardness of the sintered and heat treated samples are measured by Rockwell hardness measuring machine. Density is determined by Archimedes principle using distilled water. The fractional porosity (γ) in the sintered specimen is considered as the ratio of the density of the sintered component and the density of the solid material. Density of the solid material (ρ_s) is measured based on the weight % of the individual elements present in the alloy.

Results and Discussion

In this section, SEM and EDS analyses of the blended powder is carried out initially. Then, the effect of compaction pressure and sintering holding time on mechanical properties. Later, the effect of heat treatment

Figure 73.1 Cylindrical specimen prepared for the study
Source: Author

Table 73.1 The composition of the powders used for making the super alloy

Elements	Ni	Co	Cr	Al	Ti	Mo	C	B
Weight %	Bal	18.5	15	4.2	3.5	5.2	0.08	0.01

Source: Author

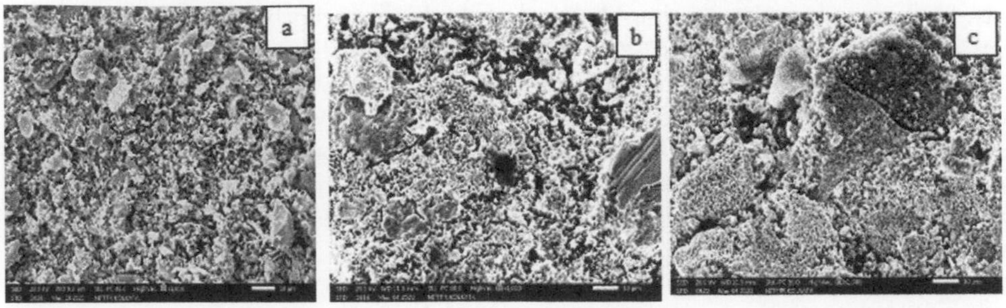

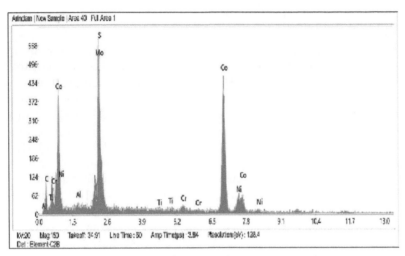

Figure 73.2 SEM topological image of (a) blended powder (b) compacted sample (c) sintered sample
Source: Author

Figure 73.3 EDS outline of the sample
Source: Author

on mechanical properties is discussed. Blended powder analysis is done using scanning electron microscope (SEM) in order to see that powder is properly mixed or not. It is observed that powders are of different sizes and properly entangle with each other as shown in Figure 73.2. Small size particles are uniformly distributed around the bigger size particles. From the EDS analysis, shown in Figure 73.3, it is confirmed that proper mixing of the powder is done. All the alloying elements are present in the selected spot of the blended powder.

Sintered Sample

In this section, the effect of compaction pressure on density, hardness and porosity of the sintered sample is discussed. Along with the compaction pressure, the effect of holding time of sample inside the furnace is also discussed. Holding time has great influence on the diffusion of elements within themselves, which affects the mechanical properties of any material.

Density of the sintered sample

Density of the sample is measured by Archimedes principle. With the increase of compaction pressure, density of the sample is increasing as shown in Figure 73.4. When the compaction pressure increases from 700 to 800 MPa, density is approximately the same. But when compaction pressure is increased to 850 MPa, there is a remarkable escalation in density. Holding time of sample inside the furnace also affects density. It is observed that increase in holding time inside the furnace, increases the density of the material. When holding time raises from 2-3 hours, density increases by about 61%. This is because atoms get sufficient time at high temperature to get diffused with each other due to the activation of liquid phase sintering, by which, porosity of the sample reduces. When the holding time increases from 3-4 hours, rate of increase in density is 7%.

Hardness of the sintered sample

The hardness of the sample increases with the increase of compaction pressure, It, has the reversed effect on increase in sintering holding time. When a material stays for longer time at high temperature, grain size increases, consuming other grains due to substantial level of atomic diffusion in solid state. Neck formation between the particles occurs, which reduces the surface free energy. Hence the hardness reduces. From Figure 73.5, it is observed that, the hardness of the sample increases with increase in compaction pressure but reduces with the increase in sintering holding time. The higher the sintering holding time, lesser is the hardness.

While relating the properties of the prepared sample with the standard sample, it is found that the Rockwell hardness of the sintered sample made by applying the compaction pressure of 850 MPa and 1200°C sintering temperature with 2 hour holding time has almost same hardness as that of the standard sample of same composition. This is due to the retention of the sample at high temperatures for more time, which results in the formation of single phase in the alloy, which makes the material, more ductile. But density of the sintered sample with 4 hours holding time with 850MPa compaction pressure has almost same density as that of the standard sample, as shown in Figures 73.6 and 73.7. This is because, elements within the sample get more time to diffuse with each other. Hence, porosity reduces and density gets improved.

Porosity of the sintered sample

Porosity acts as stress raiser, hence, has great influence on the mechanical properties. In some of the components like bearing, porosity is intentionally tailored into the component to satisfy its purpose. This control of porosity within the object is only possible in powder metallurgy technique. Fractional porosity for all the samples are evaluated and it is observed that porosity reduces with increase in compaction pressure and sintering holding time as shown in Figure 73.8. By microscopic study also, porosity in the samples are studied. Pore morphology of the sintered sample, prepared with compaction pressure of 850MPa and 1200°C sintering temperature with all the three-holding time is studied, using optical microscope at 2.5x magnification and shown in Figure 73.9. From the pore morphology it is observed that porosity has been

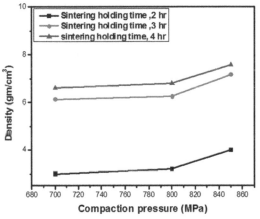

Figure 73.4 Density of the samples at different holding time

Source: Author

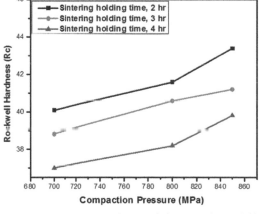

Figure 73.5 Hardness of the sample at different with holding time

Source: Author

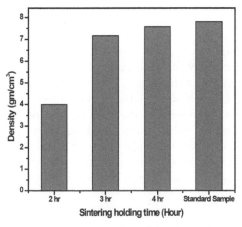

Figure 73.6 Density of the sintered sample compacted at 850 MPa

Source: Author

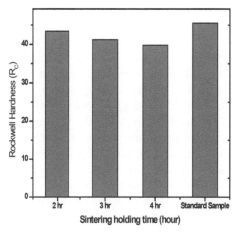

Figure 73.7 Hardness of the sintered sample compacted at 850 MPa

Source: Author

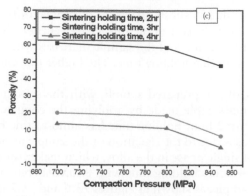

Figure 73.8 Effect of compaction pressure and sintering holding time on porosity
Source: Author

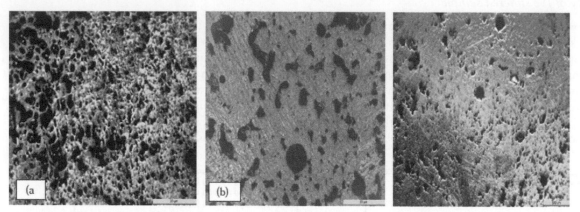

Figure 73.9 Pore morphology of the sample compacted at 850 MPa and sintered at 1200°C with (a) 2 hours (b) 3 hours (c) 4 hours holding time
Source: Author

reduced with increase in sintering holding time. Both number of pore and size of pores is decreases as holding increases. Along with porosity, some surface cracks are also seen. Grid lines, which are seen in the image, are nothing but grinding marks. After taking the sample from the furnace, grinding is done to remove the outer surface layer.

Heat treatment of the sintered sample

In this section, study of heat-treated sample is done. Heat treatment is done in the tubular vacuum furnace and cooling is done within the same. Solution annealing and ageing is done on the sintered sample. Heat treatment has a great influence on the microstructure of the material. Variation in microstructure leads to changes in mechanical properties and hence the working capability of the material changes. This led to the demand of determining the outcome of heat treatment on density and hardness at different holding times.

Effect of heat treatment on density

From the study, it is observed that because of heat treatment, density increases as shown in Figure 73.10. In all the three cases of holding time, heat treated sample has more density than the sintered sample. When sintering holding time is 2 hours, density of sample undergone solution annealing is more than the aged sample as shown in Figure 73.10(a). It is observed from the Figure 73.10(b) and 73.10(c), that, when holding time becomes 3 and 4 hours, density of solutioned annealed sample reduces then the density of the aged sample. To dissolve precipitates, present in the material, solution annealing is done and transform the material into a single phase structure. This leads to the decline in density of the sample. Even it is observed when holding time is 4 hours and compaction pressure is 850 MPa, the density of the solution hardened sample is less than the sintered sample, as shown in Figure 73.10(c). In aged sample, density is maximum in both the samples with 3 and 4 hour holding times.

Effect of heat treatment on hardness

With heat treatment, hardness of the sintered material increases as shown in Figure 73.11. It is observed that hardness of the heat-treated sample is more than sintered sample in different holding time as shown in

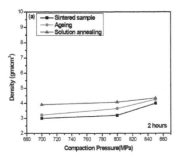
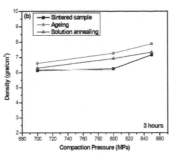
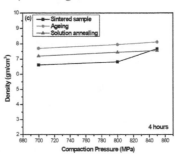

Figure 73.10 Density of sintered and heat treated samples at different sintering holding times (a) 2 hours (b) 3 hours (c) 4 hours.

Source: Author

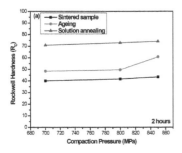
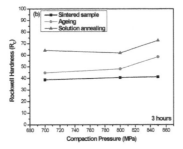
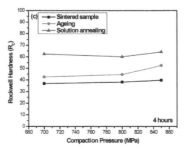

Figure 73.11 Hardness of sintered and heat treated samples at different sintering holding times (a) 2 hours (b) 3 hours (c) 4 hours.

Source: Author

Figure 73.11 (a), 73.11(b) and 73.11(c). With the increase in holding time hardness is reducing. By ageing and solution annealing, hardness of the material increases because of the formation of precipitation phase inside the material. Solutioned annealed sample in all the three sintered conditions has more hardness than the aged sample. There is enhancement of hardness with the increase of compaction pressure after ageing. But the rate of increase is maximum for 3 hours sintering holding time as shown in Figure 73.11 (b). In all the three cases of holding times, rate of increase in hardness is more, when compaction pressure increases from 800 to 850 MPa. But the hardness is lesser as higher sintering holding time. In all three cases, even after heat treatment the same result is observed.

Conclusions

In this study, it is observed that nickel-based superalloy can be successfully prepared by powder metallurgy technique, with controlled properties. Proper control of compaction pressure and sintering holding time can enhance the property of the material as desired. The following conclusions are obtained from the experimental study:

1. Density of the prepared sample is increasing with the increase in sintering holding time and compaction pressure. Heat treatment of the sample also results in increase of density of the sintered sample. It is primarily due to the decrement in total fractional porosity of the sample.
2. Hardness of the sample is increasing when there is increase in compaction pressure. But, with increase of sintering holding time, hardness reduces. This is due to the retention of the sample at high temperatures for more time, which results in the formation of single phase in the alloy, which makes the material, more ductile.
3. It is observed that with proper control of the process parameters used in powder metallurgy technique, specimen with 99% of density can be prepared. Maximum density is found when the sample is compacted at 850 MPa, with 4 hours sintering holding time. Whereas, maximum hardness is found when the sample is compacted at 850 MPa with 2 hours sintering holding time.

References

[1] Tang, C. F., Pan, F., Qu, X. H., Duan, B. H., and He, X. B. (2009). Nickel based super alloy GH4049 prepared by powder metallurgy. Journal of Alloys and Compounds. 474, 201–5.

[2] Yang, L., Ren, X., Ge, C., and Yan, Q. (2019). Status and development of powder metallurgy nickel- based Disk super alloys. International Journal of Material Research. 110, 10.

[3] Zieliñska, M., Sieniawski, J., and Porêba, M. (2007). Microstructure and mechanical properties of high temperature creep resisting superalloy René 77 modified $CoAl_2O_4$. Archives of Materials Science and Engineering. 28(10), 629–32.

[4] Sreenu, B., Sarkar, R., Kumar, S.S. S., Chatterjee, S., and Rao, G. A. (2020). Microstructure and mechanical behaviour of an advanced powder metallurgy nickel base superalloy processed hot isostatic pressing route for aerospace applications. Materials Science and Engineering A. 797, 140254.

[5] Danninger, H. (2018). What will be the future of powder metallurgy?. Powder Metallurgy Progress. 18(2), 070–079. http://dx.doi.org/10.1515/pmp-2018-0008.

[6] Guo, R., Xu, L., Wu, J., Yang, R., and Zong, B. Y. (2015). Microstructural evolution and mechanical properties of powder metallurgy Ti–6Al–4V alloy based on heat response. Materials Science and Engineering A. 629, 327–34.

[7] Huda, Z. (2007). Development of heat-treatment process for a P/M superalloy for turbine blades. Materials and Design. 28, 1664–7.

[8] Subbarao, R., and Chakraborty, S. (2018). Microscopic studies on the characteristics of different alloys suitable for gas turbine components. Material Today Proceedings. 5(5, Part 2), 11576–84.

[9] Somani, M. C., Muraleedharan, K., Prasad, Y. V. R. K., and Singh, V. (1998). Mechanical processing and microstructural control in hot working of hot isostatically pressed P/M IN-100 superalloy. Materials Science and Engineering A. 245, 88–99.

[10] Subbarao, R. (2021). Gas turbine blade failure scenario due to thermal loads in case of Nickel based super alloys. Materials Today: Proceedings. 46(Part 17), 8119–26.

Chapter 74

Analysis of rail joint by finite element method

Arvind Bodhe[1,a], Pragati Dethe[2,b] and Anoop Vishwakarma[3,c]

[1]G.H. Raisoni University, Saikheda, India

[2]Science College Pauni, India

[3]BPCL, Kochi, India

Abstract

In India, Railway plays an important role in the development of the country. The speed of rail is based on the quality of the tract. This paper illustrates a comparative study of various insulated rail joints (IRJs). This work is focused on material strength and failure during high speed and compares the results of different materials. Fiberglass, polyhedral methylene adipamide, and polytetrafluoroethylene (PTFE) are used for providing insulation. A wheel load of 175kN and 133KN is considered for finite element method (FEM) analysis. During FEM analysis joint assembly is modeled in the 3D structure of the rail track, and wheel. Three materials like Nylon 66, fiberglass, and PTFE are used for analysis. It is observed that PTFE is found more suitable for this type of IJRs.

Keywords: FEM, fiberglass, IRJs, Nylon, PTFE, rail joint

Introduction

Indian Railways is working on the modification for a high-speed train [1]. They focused on the development of better performance of the track. The railway track circuit consists of a wheel and axel hunt track, and an insulating system [2, 13]. decrease the consistent idea of the hailing framework, in this manner addressing a serious gamble to the thriving of rail tasks. The main purpose of IRJ is to allow a rail line flagging framework to find trains by maintaining a shorting circuit's framework.

The main course mishap is of Failure of this joint [3, 4]. The life of the IJRs is estimated at about 12.5 years [4]. The failure of IRJs also depends on the load produced during the movement of the wheel on it. The stress resulted in more deformation of insulating materials and stay as permanent deformation on it [5, 6]. This permanent deformation causes disturbing the circuit [8, 7, 14]. The failure of the end post and insulation plate allows the open circuit for the track circuit shown in Figures 74.1 and 74.3 and plays a dangerous role to generate mishaps. A cross sectional view as shown in Figure 74.2 for getting clear view of end post, insulation, fish plate and bushing of IRJ.

Material Properties and Modeling

In this analysis, three different materials are used. The material must possess a low coefficient of erosion, good insulating, and great sturdiness. It must have heartless toward UV Outstanding electrical properties, high protection from fire, high protection from microbiological assault great protection from most synthetic compounds low thermal conductivity, and high stiffness [10, 11].

Geometry Modeling

A three-dimensional model is developed as shown in Figure 74.4 for FEM. A load of 175KN and 133KN is considered for analysis. The various parts like end post, rail, insulation, fish plate, bushing nut & bolt were modeled in the CAD software and made assembly of its analysis as shown in Figures 74.3 and 74.4.

Fiber Glass Fiber glass comparatively called glass-maintained plastic or GRP or glass-fiber-created plastic or GFRP, is a fiber foster polymer made of a plastic organization created by fine strands of glass. The structure of fiberglass is shown in Figure 74.5. The properties of fiber glass are shown in Table 74.1.

The physical properties of fiber glass are as follows:

* Outstanding electrical properties
* High protection from fire
* High protection from microbiological assault
* Great protection from most synthetic compounds low thermal conductivityHigh stiffness

Nylon 66

Nylon is perhaps of the most broadly involved plastic on the planet, particularly as a course and wear material. Nylons are regularly utilized as substitutions for bronze, metal, aluminum, steel and various metals, as

DOI: 10.1201/9781003450252-74

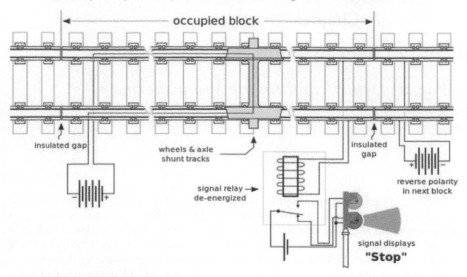

Figure 74.1 Rail track circuit
Source: https://www.railwaysignallingconcepts.in/category/network-rail/

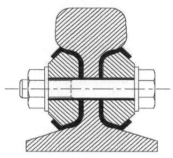

Figure 74.2 Cross-sectional view of IRJ
Source: https://skill-lync.com/student-projects/week-2-railwheel-and-track-160

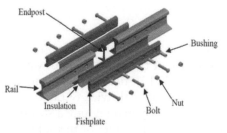

Figure 74.3 Geometry representation of each component involve in rail joints
Source: https://www.semanticscholar.org/paper/A-Review-On-Insulated-Rail-Joints-(IRJ)-Failure-Dangre/039e1bf99046490341d8d85f8628e0d164ef8301

Figure 74.4 CAD model is taken for FEA simulation
Source: https://www.researchgate.net/figure/Exploded-view-of-IJ-assembly_fig1_295833697

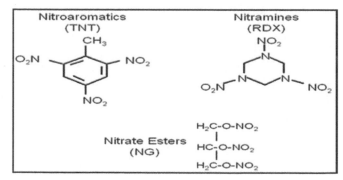

Figure 74.5 Structure of fiber glass
Source: https://www.technicaltextile.net/articles/glass-fibre-classification-production-structure-properties-applications-9418

Table 74.1 General properties of fiber glass

Sr. No	Properties	Value
1	Young's modulus (N/m^2)	4510E6
2	Poisson's ratio	0.2
3	Operating temp (°c)	-35 to 550
4	Melting temp (°c)	1250
5	Thermal conductivity (W/m-k)	0.06
6	Coefficient of warm development (/ °c)	7.58*106

Source: Author

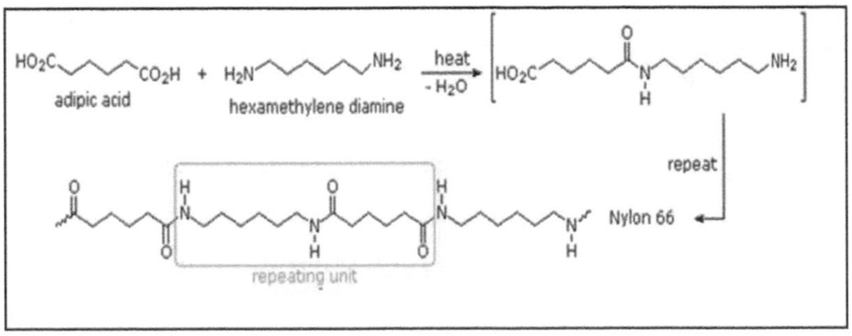

Figure 74.6 Chemical structure of Nylon 66
Source: https://www.researchgate.net/figure/Structure-of-polyamide-66-and-polyamide-6_fig1_233893450

Table 74.2 General properties of Nylon 6

Sr. No	Properties	Value
	Density kg/m^3	1140
1	Young's modulus (N/m2)	1620E6
2	Poisson's ratio	0.41
3	Operating temp (°c)	98
4	Melting temp (°c)	265
5	Thermal conductivity (W/m-k)	0.24
6	Coefficient of warm development (/ °c)	7.1*105

Source: Author

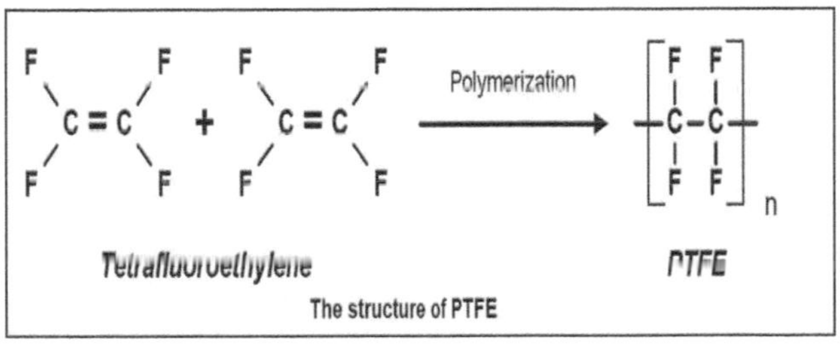

Figure 74.7 Chemical structure of PTFE
Source: https://www.researchgate.net/figure/Chemical-structure-of-the-Teflon-FEP-film_fig1_245072573

well as various plastics, wood, and flexible. Nylon 66 is acquired by the polymerization of adipic corrosive with hex-methylene diamine. The structure of nylon 66 is shown in Figure 74.6. The properties of Nylon 66 are shown in Table 74.2.

The physical properties of Nylon 66 are as follows

- Electrically insulating
- Impervious to many oils, lubes, diesel, petroleum, cleaning liquids
- Solid and extreme
- Effectively machined
- It has great sliding properties
- No mold or bacterial impacts
- Debased by light as normal filaments
- Extremely durable set by intensity and steam

Poly-tetra-fluro-ethylene

It is generally known as PTFE. It is obtained by polymerization of water emulsion of tetrafluoroethylene under pressure in presence of catalysts like benzyl peroxide H_2O_2. The structure of PTFE is shown in Figure 74.7. The properties of PTFE are shown in Table 74.3. The physical properties of PTEF are as follows:

- Low coefficient of erosion
- Great sturdiness however by and large low mechanical strength

Table 74.3 General properties of PTFE

Sr. No	Properties	Value
1	Young's modulus (N/m^2)	410E6
2	Poisson's ratio	0.47
3	Operating temp (°c)	265
4	Melting temp (°c)	339
5	Thermal conductivity (W/m-k)	0.25 to 0.39
6	Coefficient of warm development (/ °c)	13.9*10-5

Source: Author

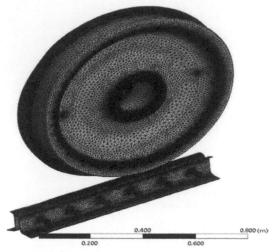

Figure 74.8 Grid formation
Source: Author

- High unambiguous gravity
- High thickness and waxy touch
- Dielectric strength
- Heartless toward UV
- Expansive temperature reach (- 200 to 260°C) and non-combustible water absorption

Methodology

Grid Generation (Meshing)

The mechanical physical preference, mechanical APDL solver preference is selected for study [8, 12]. The fine mesh size and fast transition for mesh development, for the mesh size- for minimum edge size is 0.01mm and maximum edge 0.01m as in Figure 74.8. After GIT we found that the total no of elements in the mesh is 5053661 and the total node present in the model is 1006758. The quality (aspect ratio and skewness) of the mesh is good [10].

Remote Displacement

For the remote displacement we have to select remote point in connection and then it has applied remote displacement and set only rotation and translation in x and y direction as so bellow in the setting. In the above setting as shown in Table 74.4, only y-direction displacement is allowed and rotational motion allows about the x-axis. So, the wheel can move and rotate in a single plane. For supporting the rail track, it applies fixed support [9]. Bolt and sidebar are bonded to the track frame for the analysis. The joint is partially acting like a cantilever and partially overhang so both types of analysis is considered by the software.

Results and Discussions

During the FE analysis the total deformation and directional deformation are noted at the load of 133466N as shown in Figures 74.9 and 74.10 for three materials. The Figures 74.11 and 74.12 show the total deformation and directional deformation at load of 175000N. These two different magnitude loads applied on

Table 74.4 Representation of remote displacement setting

Details of "Remote Displacement"	▾ ⏻ ☐ ✕
⊟ **Scope**	
Scoping Method	Remote Point
Remote Points	Remote Point
Coordinate System	Global Coordinate System
X Coordinate	0.20438 m
Y Coordinate	0.81881 m
Z Coordinate	0.12585 m
Location	Click to Change
⊟ **Definition**	
Type	Remote Displacement
☐ X Component	0. m (ramped)
Y Component	Free
☐ Z Component	0. m (ramped)
Rotation X	Free
☐ Rotation Y	0. ° (ramped)
☐ Rotation Z	0. ° (ramped)
Suppressed	No
Behavior	Deformable
⊞ **Advanced**	

Source: Author

(a) Nylon 66   (b) Fiber glass   (c) PTFE

Figure 74.9 Total deformation shows at the load of 133466N (a) Nylon 66 (b) Fiber glass and (c) PTFE
Source: Author

the rail joint for material and observations are recorded. Nylon has more deformation at both magnitudes of the load.

It is found that nylon has more deformation than composite material. At load 133466 N total deformation in Nylon 66 has 0.15mm and composite material has 0.14mm and PTFT has 0.12mm. as shown in Figures 74.10 and 74.11 and Table 74.5. At load 175000N Nylon has more deformation than PTFE as shown in Table 74.5. And it is revealed from the values obtained that PTFE is found more suitable. The

(a) Nylon 66

(b) Fiber glass

(c) PTFE

Figure 74.10 Directional deformation shows at the load of 133466N (a) Nylon 66 (b) Fiber glass and (c) PTFE
Source: Author

(a) Nylon 66

(b) Fiber glass

(c) PTFE

Figure 74.11 Total deformation shows at the load of 175000N (a) Nylon 66 (b) Fiber glass and (c) PTFE
Source: Author

Figure 74.12 Directional deformation shows at the load of 175000N (a) Nylon 66 (b) Fiberglass and (c) PTFE
Source: Author

Table 74.5 Total deformation in meter

S. No.	Load	Nylon 66	Fiber glass	PTFE
1	133466N	0.00015023	0.0001427	0.0001278
2	175000N	0.0001970	0.0001872	0.0001676

Source: Author

Table 74.6 Direction deformation in meter

S. No.	Load	Nylon 66	Fiber glass.	PTFE
1	133466N	8.0439e-5	6.499e-6	5.334e-6
2	175000N	1.055e-5	8.524e-6	6.996e-6

Source: Author

direction of deformation is within allowable limits as shown in Figures 74.9 and 74.11. Table 74.6 shows the Direction deformation in meter

Conclusion

In this analysis of track and wheel assembly, an examination of weaker links for analysis had been done, and observed limiting force and stress produced on that component. During analysis, the track and insulating materials at loads of 133466N and 175000N found good outcomes. These loads are selected concerning the passenger bogy and cargo body. That shows PTFT material performed well under the high load condition. An analysis of the aspect and possibility of this design was conducted for a better performance of IJRs. Amongst Nylon66, Fiberglass, and PTFE material used in insulation in IJRs through analysis. It is detected that PTFE is found better and more appropriate for this type of insulating rail joints.

Acknowledgment

We thank the Ministry of Railway for their kind support to provide the relevant data. We thank to the Management of G H Raisoni University, Saikheda for providing technical and laboratory support. We also expressed gratitude to the Dean of Engineering and Technology for continuous support.

References

[1] Government of India, Ministry of Indian Railway, RESEARCH DESIGNS AND STANDARDS ORGANISATION LUCKNOW - 226 011 (1998). Manual for glued insulated rail joint., 01–31

[2] Zong, N., and Dhanasekar, M. (2012). Analysis of rail ends under wheel contact loading. International Journal of Mechanical and Aerospace Engineering. 6, 452–460.

[3] Elsayed, H., Lotfy, M., Youssef, H., and Sobhy, H. (2019). Assessment of degradation of railroad rails: Finite element analysis of insulated joints and unsupported sleeper. Journal of Mechanics of Materials and Structures. 14(3), 429–48.

[4] Elshukri, F. A. (2016). An experimental investigation and improvement of insulated rail joints (IRJs) end post performance, A thesis submitted for the degree of Doctor of Philosophy in the Faculty of Engineering of the University of Sheffield, 1–134

[5] Mayers, A. (2017). The effect of heavy haul train speed on insulated rail joint bar strains. Australian Journal of Structural Engineering. 18(3), 148–59.

[6] Gallou, M. (2018). The Assessment of Track Deflection and Rail Joint Performance. Loughborough University.

[7] Yang, Z., Boogaard, A., Chen, R., Dollevoet, R., and Li, Z. (2018). Numerical and experimental study of wheel-rail impact vibration and noise generated at an insulated rail joint. International Journal of Impact Engineering. 113, 29–39.

[8] Soylemez, E., and Ciloglu, K. (2016). Influence of track variables and product design on insulated rail joints. Transportation Research Record. 2545(1), 1–10.

[9] Németh, A., Major, Z., and Fischer, S. (2020). FEM modelling possibilities of glued insulated rail joints for CWR Tracks. Acta Technica Jaurinensis. 13(1), 42–84.

[10] Samantaray, S. K., Mittal, S. K., Mahapatra, P., and Kumar, S. (2019). Assessing the flexion behavior of bolted rail joints using finite element analysis. Engineering Failure Analysis. 104, 1002–13.

[11] Németh, A., and Fischer, S. (2021). Investigation of the glued insulated rail joints applied to CWR tracks. Facta Universitatis. Series: Mechanical Engineering. 19(4), 681–704.

[12] Németh, A., and Fischer, S. (2018). Investigation of glued insulated rail joints with special fiber-glass reinforced synthetic fishplates using in continuously welded tracks. Pollack Periodica. 13(2), 77–86.

[13] Federal Railroad Administration (2011). Rail defect reference manual.

[14] Mandal N. K., and Brendan Peach. (2010). An engineering analysis of insulated rail joints. International Journal of Engineering Science and Technology. 2(8), 3964–3988.

Chapter 75

Experimental and numerical investigation of vertical equi-section water charged gravity-assisted heat pipe for different heat input and fill ratio

Devendra Shaharea[1,a], Saurabh Sarkar[2,b] and Harshu Bhimgade[3,c]

[1]YCCE, Nagpur, India

[2]Hochschule Harz, University of Applied Sciences, Department of Automation and Computer Science, Friedrichstraße 57-59, Germany

[3]IIT Kanpur, India

Abstract

Over the years, rapid advancement in the gravity assisted heat pipe technology has been advanced with various applications in the multiphase fluid flow thermal energy transfer field for enhancing heat transfer rates and compactness of the system. Effective phase change of working fluid in heat pipes (HPs) perform a vital role in many industrial and commercial uses, results into the improved thermal performance and decreasing energy consumption. In this paper, copper heat pipe with water charged of equi-section of evaporator, adiabatic and condenser 150 mm of length each, inside and outside diameter of 35mm and 38mm respectively for different heat input and fill ratio has been investigated experimentally and numerically. ANSYS 18.1 FLUENT code is implemented for the computational fluid dynamics (CFD) analysis of HP. It is revealed from the experiment that for 30W, 40W, 50W and 60W heat input, evaporator and condenser section average surface temperature 99.4°C, 100.7°C, 101.7°C, 103.1°C and 85.1°C, 92.6°C, 97.9°C, 98.8°C were recorded respectively. For different fill ratio 30%, 50% and 100% the evaporator and condenser section average surface temperature 97.76°C, 98.36°C, 99.32°C and 71.4°C, 74.46°C, 77.96°C were recorded respectively. Volume of fluid (VOF) method has been incorporated and it is observed that for constant 50W heat input, numerical error of the evaporator and condenser average temperature was 2.025% and 2.216% respectively.

Keywords: CFD analysis, gravity-assisted heat pipe, heat transfer, multiphase fluid flow, phase change

Introduction

In the field of thermal engineering, enhancement of heat flow is the main topic of several studies [1]. A heat pipe is a heat exchange device which exchanges heat within a sealed container by phase change phenomenon [2]. Mainly HPs consist of three parts, adiabatic, heat source known as condenser and heat sink known as evaporator [3]. Due to gravitational forces, the fluid inside a closed thermosyphon or wickless HP returns back from the condenser to evaporator section by capillary action [4,5]. Now a day's heat pipes are being implemented in much area, prominently in heating and cooling systems, heat pumps, water-heating systems and in electronics [6–9]. HPs can be used at cryogenic conditions with appropriate fluids [10]. HPs offer high thermal efficiency with simple structure and compactness with reversibility [11–14] as shown in Figure 75.1.

Aim of this article is to simulate and demonstrate the accuracy of evaporation and condensation phenomenon in a thermosyphon using volume of fluid (VOF) model. To represent the entire system process, the relevant source terms in the governing flow equations have been considered the primary equations of fluid, which governs the heat and mass transfer between the liquid and vapour phases. The CFD results presented in this study demonstrate that the VOF approach can represent the complex dynamics inside thermosyphons. The CFD simulation was able to capture the phenomenon of the thermosyphon, including the evaporator vapour formation and condenser film condensation, according to the flow representation.

The projected findings and the experimental results corresponded well when the HPs walls average surface temperature was compared to the results of experiment conducted under the identical conditions. Using the thermosyphon's effective thermal resistance, the thermal performance of the device has also been evaluated at various temperatures, and it has been discovered that raising the heating power inputs improves the performance of HP of different material as shown in Figure 75.2.

Experimental Investigation

10The RTDs used in the experimental setup has a measurement error of +/-0.2°C. The electric band heater and adiabatic wall of the HP is insulated with a layer of 128 density ceramic blanket followed by a layer of aluminium foil to prevent any heat loss.

[a]devshahare0501@gmail.com, [b]u37748@hs-harz.de, [c]harshu@iitk.ac.in

DOI: 10.1201/9781003450252-75

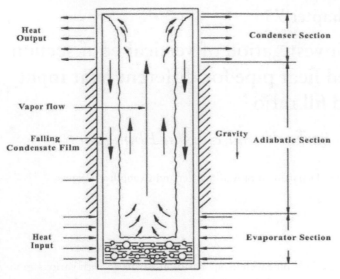

Figure 75.1 Gravity -assisted heat pipe working
Source: Author

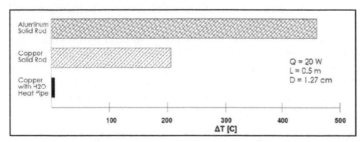

Figure 75.2 Solid aluminium and copper rod temperature compared to a copper heat pipe charged with water
Source: Author

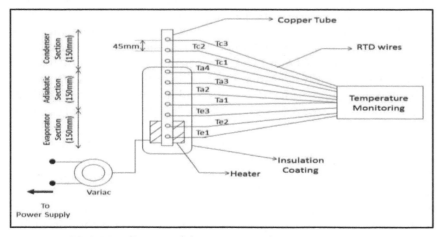

Figure 75.3 Line diagram of the experimental setup
Source: Author

For different power heat input tests were carried out and the temperature monitored along the HP. At the steady state condition all the readings of temperature were recorded. The line sketch and experimental arrangement is shown in Figures 75.3 and 75.4 respectively. The copper rod is hermetically sealed at one end with the help of the brazing process, where the heater is at the bottom of the rod. Then 10 holes are drilled on the copper rod at pre-defined positions on which the temperature sensors are placed. To prevent air leakage, we have placed a small copper tube inside the drill holes on the rod. Then, with the help of the brazing process, it can be sealed and then with proper adhesive applies to prevent the leakages. The electric band heater and adiabatic section are then covered with a layer of 128 density ceramic insulation and a layer of aluminium foil to reduce heat loss to the environment.

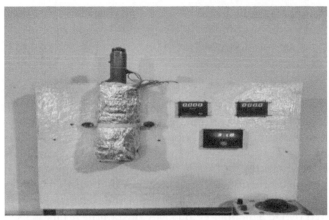

Figure 75.4 Experimental setup
Source: Author

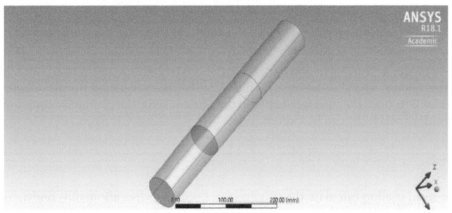

Figure 75.5 Three-dimensional model generated in Space Claim with fluid domain on the wall thickness of 0.003 m
Source: Author

Table 75.1 Constructional details of heat pipe

Sections	Material/type	Dimensions /specifications
Heater (Electrical Input)	Band type heater	38mm (D) x 150mm(L) 300 W
Adiabatic	128 density ceramic blanket	Length-150 mm Inner Dia.-35mm Outer Dia.-38 mm
Condenser (Exposed to air)	Copper Tube	Length-150 mm Inner Dia.-35mm Outer Dia.-38 mm
Temperature Sensor	RTD sensors Accuracy Grade A	Range-10K to 473 K Element Size- 2mm x 2mm Least Count: 0.1 K
Evaporator	Distilled water	144 ml water filled up to length of heater

Source: Author

Numerical Analysis

Model Description

In this study, a closed two-phase thermosyphon 3-D heat pipe (shown in Figure 75.5) simulated using the commercial software ANSYS Fluent 18.1 by using the VOF technique. Both the Lagrange technique and the Euler approach are used in the numerical computations of multiphase flows. The continuum fluid phase is seen by Euler-Lagrange approach and second phase in which droplets or bubbles are seen in which dispersed phase volume percentage cannot be greater than 10%.

The Euler technique was utilised since the current application believes the second phase to have a volume percentage larger than 2 10%. A phase cannot have its volume completely occupied by the phase above it since this method makes use of the idea of a phase volume fraction which is to be considered as a continuous function of space and time.

N-S Equations for Volume of Fluid Model

In this paper, the VOF approach is adopted which simulates two non-mixing fluids i.e. the liquid-vapour phases. In this study, it is considered that each cell in the computational domain is occupied by liquid or vapour phase or a mixture of the two phases. VOF model is able to solve the motion of different fluids phases that can be tracked by solving of Navier-Stokes equations set for the liquid- vapour volume fraction throughout the computational domain [16]. A user defined function (UDF) is incorporated which measures the mass and heat transferred between the phase change processes. Calculations of the mass and energy transfer have been made using source terms suggested by De Schepper et al [17]. Constant heat flux is used for boiling and the latent heat in the UDF code is set to 2455 kJ/kg.

Continuity equation for volume fraction equation model By applying the physical principle of conservation of mass to the fluid, the continuity equation has the following form:

$$\nabla.(\rho\vec{u}) = -\frac{\partial\rho}{\partial t} \tag{1}$$

Where 'ρ' is the density, 'u' is the velocity and 't' is the time. Solution of the above equation for the volume fraction of one of the phases is used to track the interface between the phases. Thus, the continuity equation of the VOF model for the secondary phase (L) can be expressed as:

$$\nabla.(\alpha_L\rho_L\vec{u}) = -\frac{\partial}{\partial t}(\alpha_L\rho_L) + S_m \tag{2}$$

Where 'Sm' is the mass source term used to calculate the mass transfer during evaporation and condensation. The continuity equation shown above can be called the volume fraction equation and this equation will not be solved for the primary phase; the primary-phase volume fraction is computed based on the following constraint:

$$\sum_{L=1}^{n}\alpha_L = 1 \tag{3}$$

When the cell is not fully occupied by the primary phase (V) or the secondary phase (L), a mixture of the phases L and V exist. Thus, the density of the mixture is given as the volume-fraction-averaged density and takes the following form:

$$\rho = \alpha_L\rho_L + (1 - \alpha_L)\rho_V \tag{4}$$

Momentum Equation for VOF Model

The forces acting in the fluid were considered to be gravitational, pressure, friction and surface tension. In order to consider the effect of surface tension along the interface between the two phases, the continuum surface force (CSF) model proposed by Brackbill et al. has been added to the momentum equation

$$F_{CSF} = 2\alpha_{LV}\frac{\alpha_L\rho_L C_V\nabla\alpha_L + \alpha_L\rho_V C_L\nabla\alpha_L}{\rho_L + \rho_V} \tag{1}$$

Where 'σ' is the surface tension coefficient and C is the surface curvature. By taking into account the above forces, the momentum equation for the VOF model takes the following form:

$$\frac{\partial}{\partial t}(\rho\vec{u}) + \nabla.(\rho\vec{u}\vec{u}) = \rho\vec{g} - \nabla p + \nabla.\left[\mu\left(\nabla\vec{u} + \nabla\vec{u^T}\right) - \frac{2}{3}\mu\nabla.u I\right] + F_{CSF} \tag{2}$$

Where 'g' is the acceleration of gravity, 'p' is the pressure, and 'I' is the unit tensor. The momentum equation depends on the volume fraction of all phases through the physical properties of density and viscosity. Thus, the dynamic viscosity 'μ' is given by

$$\mu = \alpha_L \mu_L + (1 - \alpha_L)\mu_V \tag{3}$$

A single momentum equation is solved throughout the computational domain, and the calculated velocity is shared among the phases.

Energy Equation for VOF Model

The energy equation for the VOF model has the following form:

$$\frac{\partial}{\partial t}(\rho e) + \nabla.(\rho e \vec{u}) = \nabla.(k.\nabla T) + \nabla.(\rho \vec{u}) + S_E \tag{1}$$

where 'SE' is the energy source term used to calculate the heat transfer during evaporation and condensation. The VOF model treats the temperature 'T' as a mass-averaged variable and the thermal conductivity is calculated as:

$$k = \alpha_L k_L + (1 - \alpha_L)k_V \tag{2}$$

The VOF model also treats the internal energy 'e' as a mass-averaged variable in the following form:

$$e = \frac{\alpha_L \rho_L e_L + \alpha_V \rho_V e_V}{\alpha_L \rho_L + \alpha_V \rho_V} \tag{3}$$

where 'eL'and 'eV' are based on the specific heat (cp) of the phase and the shared temperature, given by the caloric equation of state:

$$\begin{aligned} e_L &= c_{p.L}(T - T_{sat}) \\ e_V &= c_{p.V}(T - T_{sat}) \end{aligned} \tag{4}$$

A single energy equation is also solved throughout the domain for both phases, and the calculated temperature is shared among the phases.

Meshing and Boundary Conditions

High end meshing domain had to be used in order to properly capture the multi-phase flow inside the thermosyphon. Tetrahedral elements were produced for this purpose using the ICEM ANSYS meshing interface, and the meshing quality and size were determined using the grid independence test. Finally 67,345 elements and 32,789 nodes have been selected in the meshing domain as shown in Figure 75.6.

No-slip boundary condition on the inner walls of HP has been applied. Constant heat flux is considered on the evaporator walls. In the adiabatic section, provided it is insulated, a boundary condition is defined as a zero heat flow. The heat produced by the condensation of vapour is used to cool the condenser portion. The experimental setup presupposes that water is used to cool the condenser. A convective heat transfer coefficient boundary condition given to condenser section. The following formula has been used to get the relevant heat transfer coefficients:

$$h_c = \frac{Q_c}{2\pi r L_c (T_{c,avg} - T_\infty)}$$

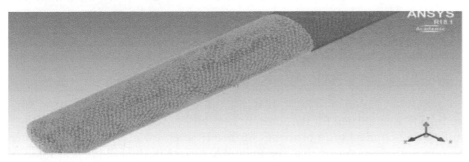

Figure 75.6 Tetrahedral meshing using ICEM ANSYS
Source: Author

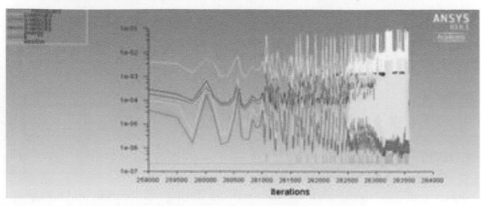

Figure 75.7 Tetrahedral meshing using ICEM ANSYS
Source: Author

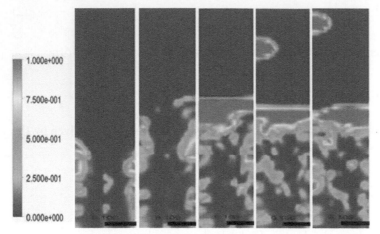

Figure 75.8 Vapour volume fraction produced in multi-phase flow
Source: Author

where,
h_c is the heat transfer coefficient
L_c is the condenser section length
$T_{c, avg}$ is the average temperature of the condenser section
T_φ is the ambient temperature

Solution and Convergence

Figure 75.7 shows the convergence of solution at the end of iteration. As soon as the liquid pool in the evaporator heated to saturation temperature the evaporation process starts and phase change occurs. This saturated vapour is then moves upward to the condenser section, where it condenses along the cold walls forming a thin liquid film.

An unsteady simulation is carried out for two-phase flow with a time step of 0.0005s. The time step is decided on the basis of the Courant number which is kept less than 3, which is the defined as the time step over the time a fluid takes to move across a cell. For VOF models, 250 is the maximum possible allowed Courant number [18]. The solution approaches a steady state condition after around 60s. The solution is converged by using SIMPLE algorithm. Pressure-velocity coupling and momentum and energy equations are solved by using first-order upwind scheme.

Results and Discussion

Proposed heat pipe was investigated for heat input of 30W, 40W, 50W and 60W and fill ratio of 30%, 50% and 100% re1spectively. Also the HP is simulated for the constant heat input of 50W. The solution reached quasi-steady state for the boiling and condensation processes in the HP after around 60s. The simulated and recorded surface temperature with thermocouples along the outer surface of the HP for heat input of 50W is shown in Table 75.2, respectively. Ten thermocouples have been place on the outer wall from top to bottom of HP to monitor the average vapour temperature for the evaporator, adiabatic and condenser

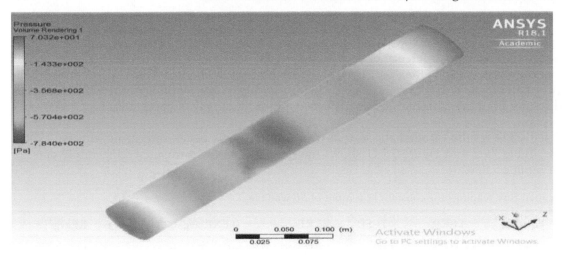

Figure 75.9 Pressure gradient across the thermosyphon obtained through simulation
Source: Author

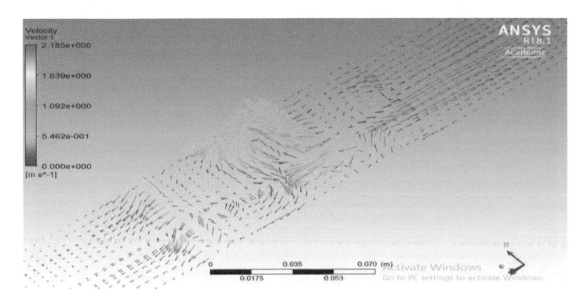

Figure 75.10 Pressure gradient across the thermosyphon obtained through simulation
Source: Author

sections respectively. From the simulation, it has been observed that the results of the VOF model shows less variation with the experimental data. The computational and experimental average relative error of temperatures is 2.025%, 1.79 % and 2.216 % recorded respectively.

Vapour Volume Fraction Contour

Initially the evaporator section is filled with water was heated by supplying constant heat input. Due to constant heat supply to the evaporator section water temperature reaches to boiling temperature and then water starts to evaporate and hence phase change occurs. For constant heat input of 50W, Figure 8 shows the volume fraction contours of surface bubble formation or evaporation and thin liquid film of condensed vapour in the evaporator and condenser section respectively. A red colour indicates the presence of only vapour volume fraction = 1 while a blue colour presents the presence of only liquid vapour volume fraction = 0. It is observed that after the start of evaporation process water vapours or bubbles moves towards the condenser section where it disappears completely (from right to left side of Figure 75.8) and condensed liquid on the inner walls of HP flows back gravitationally to evaporator section.

The simulation was further incorporated to investigate the pressure drop across the thermosyphon that was experimentally not feasible. The results showed a maximum pressure of 60.45 Pa at the each ends of the HP. Low pressure of 50.7 Pa was interpreted in the adiabatic region as shown in Figure 75.9. Vector plot of steam flow inside the thermosyphon was revealed showing the thermosyphoning effect inside the HP as shown in Figure 75.10.

Experimental Results

During the phase change process, the surface temperature along the length of the HP was recorded. It is found that the temperature difference of each section of HP was found to be in the range of 11.5°C. Surface temperature distribution along the length of the HP is nearly isothermal which falls in line with the research paper of S.H. Noie [15]. Hence, we can say that the heat pipe was working isothermally. Deviation of simulated results from experimental results happened due to the supply of continuous heat input to the evaporator section.

The predicted CFD results of adiabatic section temperature shows better agreement with the experimental results. Table 75.2 and Figure 75.11 shows the simulated and experimental temperature data for constant heat input of 50W.

Table 75.2 Temperature readings for constant heat input of 50 wattage

Section	Position	TEXP.	TCFD	RE	TEXP. avg	TCFD avg	REavg
		oC	oC	%	oC	oC	%
Evaporator	Te1	102.4	104.3	1.855			
	Te2	101.6	103.8	2.16	101.7	103.76	2.025
	Te3	101.1	103.2	2.077			
Adiabatic	Ta1	99.1	101.7	2.623			
	Ta2	98.1	100.5	2.44	98.77	100.55	1.79
	Ta3	98.9	100.6	1.71			
	Ta4	99.0	99.4	0.4			
Condenser	Tc1	98.5	96.4	2.131			
	Tc_2	97.4	95.2	2.25	97.9	95.73	2.216
	Tc_3	97.8	95.6	2.25			

Source: Author

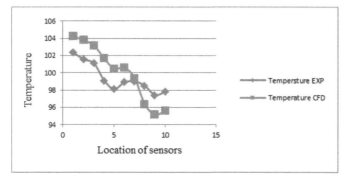

Figure 75.11 Predicted CFD and experimental temperature distribution for heat input of 50 W
Source: Author

Table 75.3 Fill ratio effect on the operating temperature with constant heat input of 30W

Temperature sensors	30% fill ratio	50% fill ratio	100% fill ratio
Te1	99.8	100.1	100.4
Te2	98.2	99.6	99.38
Te3	95.3	95.4	98.2
Ta1	89.9	90.4	94.5
Ta2	88.6	91.1	93.8
Ta3	82.8	85.6	93.2
Ta4	79.5	82.5	90.5
Tc1	76	78.9	82.4
Tc2	68.4	71.8	75.1
Tc3	69.8	72.7	76.4

Source: Author

The effect of different fill ratio on the operating temperature of heat pipe with constant wattage (Table 75.3 and Figure 75.12) was studied. The effect of different fill ratio on the operating temperature of heat pipe was obtained which falls in line with research paper of S.H. Noie [15] and hence we can say that operating temperature increases with increase in input wattage at all fill ratio.

The thermosyphon was operated at different wattages at 100% fill ratio from 30W to 60 W at an interval of 10W, as shown in Table 75.4 and Figure 75.13. Temperatures by the RTDs were measured at a steady-state along the heat pipe. The higher operating temperature was revealed by a subsequent increase in operating heat input.

From the Figure 75.14 it is concluded that the investigated heat pipe in this paper is more efficient than a hollow copper tube of the same diameters and length, as the conductivity of the HP was found to be at least 14 times more than that of a copper tube for the same amount of heat transfer.

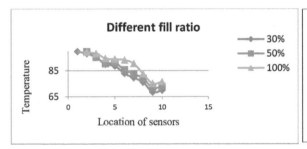

Figure 75.12 Fill ratio effect on the operating temperature with constant heat input of 30W
Source: Author

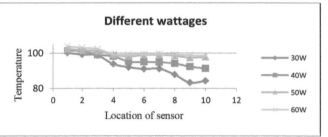

Figure 75.13 Heat input effect on the surface temperature of HP with constant fill ratio of 100%TABLE 3 Fill ratio effect on the operating temperature with constant heat input of 30W
Source: Author

Table 75.4 Heat input effect on the surface temperature of HP with constant fill ratio of 100%

Temperature sensors/Heat input	30	40	50	60
Te1	100.1	101.4	102.4	103.8
Te2	99.2	101.2	101.8	103
Te3	98.7	99.1	101.1	102.3
Ta1	93.8	98.2	99.1	99.3
Ta2	92.0	95.2	98.1	99.1
Ta3	91.0	95.0	98.9	99.2
Ta4	91.1	94.8	99.0	99.5
Tc1	87.8	94.1	98.5	99.1
Tc2	83.2	92.3	97.4	98.7
Tc3	84.3	91.3	97.8	98.5

Source: Author

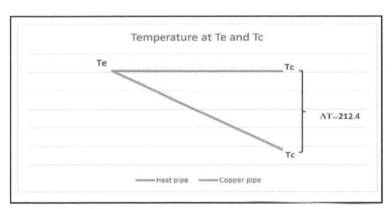

Figure 75.14 Comparison of heat pipe with the hollow copper tube of same dimension
Source: Author

Conclusion

- In this paper, vertical orientation equi-section heat pipe with distilled water charged has been investigated for different heat input and fill ratio experimentally and numerically.
- It is found from the experiment that for the given heat input ranges from 30 to 60W, highest and lowest average surface temperature of evaporator and condenser walls 103.1°C and 85.1°C were recorded respectively.
- For different fill ratio 30%, 50% and 100%, the average surface temperature of evaporator and condenser walls 97.76°C, 98.36°C, 99.32°C and 71.4°C, 74.46°C, 77.96°C were recorded respectively.
- Volume of fluid (VOF) method has been incorporated and it is observed that for constant 50W heat input, numerical error for average temperature was 2.025% and 2.216% respectively shown in Table 75.2.
- The CFD analysis with FLUENT show that the VOF method can be considered and adopted for the complex phenomena inside the thermosyphon. From the flow visualization, it can be concluded that the CFD simulation can execute the operation of the gravity assisted HP or thermosyphon, including the volume fraction of one phase or the mixture of phases in either section.
- The simulated temperature for the proposed HP showed good agreement with the experimental results in the same assumed condition.
- Thermosyphon works nearly isothermally at constant fill ratio. With an increasing fill ratio, the operating temperatures for the respective heat transfer rates were reducing. At the minimum fill ratio, maximum operating temperature is achieved.
- Increased heat transfer rate leads to increased operating temperature of heat pipe for all fill ratios.
- Conductivity of heat pipe was found to be at least 14 times more than copper.
- The constructed heat pipe had negligible temperature gradient and was found to work isothermally as per the heat pipe principle.

References

[1] Malvandi, A., Ganji, D. D., and Kafash, M. H. (2015). Magnetic field effects on nanoparticle migration and heat transfer of alumina/water nanofluid in a parallel-plate channel with asymmetric heating. The European Physical Journal – Plus. 130, 63. https://doi.org/10.1140/EPJP/I2015-15063-Y.

[2] Alizadehdakhel, A., Rahimi, M., and Alsairafi, A. A. (2010). CFD modelling of flow and heat transfer in a thermosyphon. International Communications in Heat and Mass Transfer. 37, 312–8.

[3] Zamani, R., Kalan, K., and Shafi, M. B. (2018). Experimental investigation on thermal performance of closed loop pulsating heat pipes with soluble and insoluble binary working fluids and a proposed correlation. Heat and Mass Transfer. https://doi.org/10.1007/s00231-018-2418-z.

[4] ESDU (1980). Heat pipes - general information on their use, operation and design. ESDU Manual 80013.Faghri, A. (1995). Heat Pipe Science and Technology. Washington, D.C.: Taylor & Francis.

[5] Kerrigan, K., Jouhara, H., O'Donnell, G. E., and Robinson, A. J. (2011). Heat pipe-based radiator for low grade geothermal energy conversion in domestic space heating. Simulation Modelling Practice and Theory. 19, 1154–63.

[6] Jouhara, H., and Meskimmon, R. (2010). Experimental investigation of wraparound loop heat pipe heat exchanger used in energy efficient air handling units. Energy. 35, 4592–99.

[7] Mathioulakis, E., and Belessiotis, V. (2002). A new heat-pipe type solar domestic hot water system. Solar Energy. 72, 13–20.

[8] Weng, Y., Cho, H., Chang, C., and Chen, S. (2011). Heat pipe with PCM for electronic cooling. Applied Energy. 88, 1825–33.

[9] Alhuyi Nazari, M., Ahmadi, M. H., Ghasempour, R., and Shafii, M. B. (2018). How to improve the thermal performance of pulsating heat pipes: a review on working fluid. Renewable and Sustainable Energy Reviews. 91, 630–8.

[10] Jouhara, H. (2009). Economic assessment of the benefits of wraparound heat pipes in ventilation processes for hot and humid climates. International Journal of Low-Carbon Technologies. 4, 52–60.

[11] Parand, R., Rashidian, B., Ataei, A., and Shakiby, K. (2009). Modeling the transient response of the thermosyphon heat pipes. Journal of Applied Sciences. 9, 1531–37.

[12] Ochsner, K. (2008). Carbon dioxide heat pipe in conjunction with a ground source heat pump (GSHP). Applied Thermal Engineering. 28, 2077–82.

[13] Du, J., Bansal, P., and Huang, B. (2012). Simulation model of a greenhouse with a heat-pipe heating system. Applied Energy. 93, 268–76.

[14] Noie, S. H. (2005). Heat transfer characteristics of a two-phase closed thermosyphon, Applied Thermal Engineering. 25(4), 495–506.ANSYS FLUENT (2013). Theory Guide (Release 15.0). Multiphase Flows. ANSYS Inc., pp. 465–600 (Chapter 17).

[15] De Schepper, S. C. K., Heynderickx, G. J., and Marin, G. B. (2009). Modeling the evaporation of a hydrocarbon feedstock in the convection section of a steam cracker. Computers and Chemical Engineering. 33(1), 122–32.

[16] Fadhl, B., Wrobel, L. C., and Jouhara, H. (2013). Numerical modelling of the temperature distribution in a two-phase closed thermosyphon, Applied Thermal Engineering. 60(1-2), 122–31.

Chapter 76

Implementation of origami application in a solar PV system: a review

Ninad Jawajwar[a], Sakshee Tembhurkarb[b], Parimal Pawar[c], Devendra Shahare[d]

Yeshwantrao Chavan College of Engineering, Nagpur, India

Abstract

Fossil fuels cannot be used to produce energy at the current rate of use for a very long time. Additionally, from a green perspective, these exhaustible energy sources are a major contributor to a wide range of environmental degradation and health issues. Renewable energy sources including wind, solar, tidal, hydro, and geothermal energy are the most renowned ecologically friendly alternatives to fossil fuels. Using photovoltaic cells and arrays, which are frequently used to convert solar irradiance into electricity, solar energy is obtained from the sun's rays. Despite the many benefits of solar energy, there are a few downsides that should raise some alarm. Solar panels are substantially less efficient even though they are directly exposed to sunlight. Due to some elements like temperature and energy losses, the conversion efficiency is substantially lower. Second, additional solar arrays are needed to provide more electricity to increase energy extraction. Due to the significant amount of space that these panels occupy, the system is large and sturdy for the same amount of energy output. This review provides the developments in solar energy harvesting technologies, conversion efficiency, and better design for space usage. Researchers identified the solution in origami, a type of craft. The ancient art of origami is folding a single sheet of paper into 2D or 3D structures that employ deformability. In addition, several efficiency optimisation strategies are being investigated.

Keywords: Photovoltaic system, origami, efficiency, reflector, concentrator

Introduction

Solar power can be produced using a variety of solar technologies, such as photovoltaic, concentrated solar power systems (CSP), and solar thermal power plants. Solar energy, which is supplied by the sun in the form of radiant light, heat, solar irradiance, or electromagnetic radiation, can also be converted into heat and light to produce electricity [2]. The process of turning sunlight into power is known as solar energy. Using semiconducting material, photovoltaic (PV) technology turns the quantity of solar irradiance received by a certain solar cell or panel into energy. The semiconductor substance that is utilised to make solar cells demonstrates either a photovoltaic effect or a photoelectric effect. Several solar cells are stacked in an array shape to create a single unit or single unit panel in a PV system. These modules are then put together to form a sizable module of an array that can be installed on a roof and utilised to generate electricity. Fixed-mounted PV systems and mobile (tracker-based) PV systems can be used to categorise these modules positioned on the roof or in any other location. When it comes to concentrated solar power systems (CSP), lenses and mirrors are used, and the structure of the standard module is altered. The CSP system relies on the phenomena of sunlight diverging from a broad area into a narrow beam. A narrow beam then can be directed over solar panels to generate electricity [5]. Parabolic troughs, solar tower collectors, concentrated linear Fresnel reflectors, and stirling dishes are just a few of the concentration techniques that can be used [12]. The solar power system is then powered by a concentrated beam of sunlight. Due to its regenerative nature and environmentally friendly application, solar power is currently capturing the interest of the entire world. India's land receives around 5000 trillion kWh of energy annually, and some areas can even get solar energy at rates of 4–7 kWh per m^2 per day. All of these non-renewable sources of fuel will eventually run out due to the continued use of fossil fuels. In addition, the usage of all these resources results in the release of greenhouse gases, which ultimately cause global warming impact. Renewable energy sources are the best option for obtaining clean and sufficient energy, whereas non-renewable energy sources will put humanity at risk in the future. In terms of the solar energy source, the total efficiency of producing electricity from solar irradiance is only about 15-20%. Thus, a few enhancements can be made to boost the effectiveness of solar systems, such as the installation of solar trackers, the application of solar trackers can increase the efficiency of the solar system as it increases the exposure duration of panels to sunrays [13] and also by the efficient use of maximum power point tracking (MPPT) system [4].

An inventive idea for solar panels comes from the Japanese paper-folding art of origami. Action origami, modular origami, wet folding origami, Pure land origami, tessellation, and kirigami are a few styles of origami that permit the cutting and folding of paper. Depending on the space available, the mechanism, and

[a]ninadjawajwar10@gmail.com, [b]saksheetembhurkar@gmail.com, [c]miparimalpawar@gmail.com, [d]devshahare0501@gmail.com

DOI: 10.1201/9781003450252-76

the operation, these strategies can be used in a variety of ways and applications to harness solar energy. The surface area of conventional solar panels is constrained by their flat shape, which lowers their potential efficiency. Due to the large difference between the folded and unfolded configurations, origami structures have been extensively studied in the context of deployable systems [14]. Rigidly foldable origami is commonly used in deployable systems because its folding kinematics can be predicted with precision [14]. Thus, a deployable, contracting structure that can also follow the sun for solar energy is built using origami and other modifications. Robust solar panels can have their size significantly reduced by adopting origami techniques. These solar panels have improved technology that makes them portable and versatile enough to be used in any setup.

Working of Solar Panels

An assembly of solar photovoltaic cells—commonly referred to as photovoltaic cells—mounted in a rectangular housing-like structure is what is referred to as a solar photovoltaic module and is used to generate electricity. The production of energy from panels is influenced by several elements that must be taken into account. Sunlight is used as a source of energy by solar panels to create electricity. Solar panels are constructed from solar cells, also known as PV (photo-voltaic) cells, which are the basis of a solar array. The term "array" or "solar array" refers to the collection of solar panels [1, 3].

Principal of Solar Cell

The two primary semiconductor types found in solar cells are n- and p-type silicon. P-type silicon is created using substances like boron or gallium, which have one fewer outer electron than silicon. P-type silicon produces electron vacancies or holes because boron or gallium lack the one electron required to bond with neighboring silicon atoms. Additionally, n-type silicon is produced using substances like phosphorus, which has an extra electron in its outer orbit compared to silicon. The outer level orbit of the phosphorus atom contains five electrons. As a result, the junction's two sides have unequal distributions of holes (p-type) and free electrons (n-type). The loosely bound electrons in solar cells are excited by photons from sunshine, which causes them to move in one direction. As a result, electron-hole pairs develop at both junctions, and the solar panel generates energy [13].

Factors influencing the Efficiency of Solar Panels

Solar irradiance

The amount of energy received from the sun as electromagnetic radiation per square meter is known as solar irradiance. Watts per square meter (W/m²), which are SI units, are used to measure it. The performance of the solar array or solar panel system is affected by it, making it one of the most crucial elements to take into account [11]. The solar constant, or W/m², is the amount of energy that the earth receives from the sun per square meter, and it ranges between 1350 and 1370 W/m². The actual amount of solar energy is not able to reach the earth's surface because of the dispersion of sunlight by clouds or dust particles, which results in a large loss of energy. Sun panel systems directly respond to variations in solar irradiation by producing less electricity or energy. Solar energy is captured using concentrated photovoltaic thermal technology,

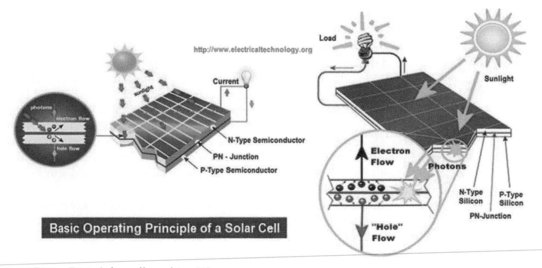

Figure 76.1 Solar cell working [1]

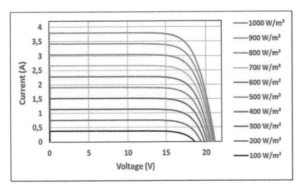
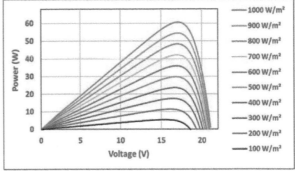

(a). I–V and P–V characteristics with constant temperature (25 °C) and variable irradiance.

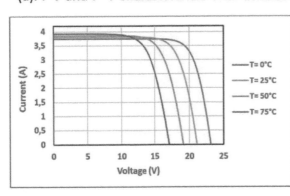
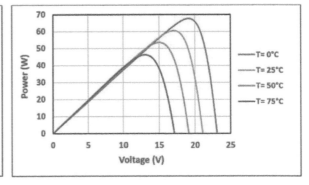

(b). I–V and P–V characteristics with constant irradiance (1000 W/m²) and variable temperat

Figure 76.2 Characteristics changes in PV module due to the effect of solar irradiance [10]

which uses solar concentrators to raise the solar density. Energy production rises as the concentration ratio rises. The temperature has an impact on the performance of primary and secondary enriched materials as well as single-junction and multi-junction semiconductors. Temperature and concentration ratio has an impact on the CPVT system's functionality and dependability. The CPVT system is also impacted by optical composition, reflectivity, and material type [12]. The output of the solar panel system varies with changes in temperature and solar irradiance [10].

Electrical and thermal considerations for CPVT systems

Sun photovoltaic cells use semiconductors to convert solar energy into electricity. In a p-n junction, the photon is absorbed by the electrons in the valence band, which then excites the electron to travel into the conduction band. Free electrons begin to gravitate toward holes as a result of this. The photon hitting the surface must have more energy than the bandgap energy to cause the photo generation of charges, hence bandgap energy is the one determining the energy output [5]. Thermalisation, below bandgap, emission losses, Carnot, and Boltzmann are just a few of the intrinsic losses brought on by photon energy that is incompatible with bandgap energy.

Reduced electrical performance is the result of these losses. Voltage reduction is caused by Carnot and Boltzmann losses, whereas current reduction and thermalisation are caused by below-band gap and emission losses. Temperature and sun irradiation both have an impact on a cell's I-V curve [5]. The short circuit current I_{sc} has a proportionate relationship with solar irradiation while the open circuit voltage V_{oc} has an inverse relationship with temperature. When using multijunction solar cells, a high fill factor (FF) is seen at low temperatures or at relatively high temperatures. This is indicated by the I-V circuit's squareness. High temperatures cause the I-V to flatten and the FF value to drop, which affects the PV cells' electrical output [5]. The electrical characteristics of solar cells V_{oc}, I_{sc}, FF, and hence alter with concentrated solar irradiation, as well as how they react to temperature. A multijunction PV cell enables the sorting of photon energy by including many junctions with various bandgap energies to maximise the PV cell's efficiency and, consequently, power output. Depending on the semiconductor material, bandgap energy varies [5]. The thermalisation is brought down below bandgap losses by increasing the number of connections, which boosts the solar cell's conversion efficiency.

The CPVT system's optical tolerance is important, especially when the concentration ratio is increased and the sun's divergence angle is taken into consideration, which is ±0.265. Sun divergence is the least

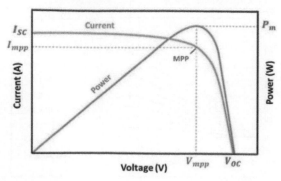

Figure 76.3 Curves showing the PV module's I-V and P-V voltages [9, 10]

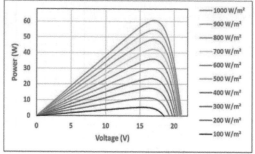

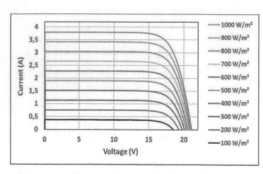

(a). I–V and P–V characteristics with constant temperature (25 °C) and variable irradiance.

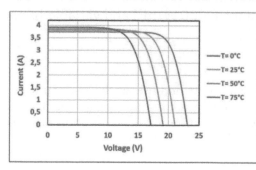

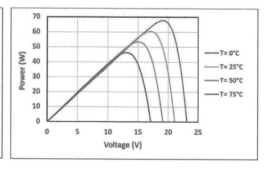

(b). I–V and P–V characteristics with constant irradiance (1000 W/m²) and variable temperature.

Figure 76.4 Characteristics changes in the PV module due to the effect of temperature [10]

acceptance angle that is narrow enough to catch solar radiation. The acceptance angle is the sensitivity of the tracking system to minimize the divergence for a high concentration ratio which can be achieved by primary or secondary optic [5]. A solar concentrator's optical efficiency is dependent on the incidence angle, with minimal scattering and maximal internal absorption. The performance of concentrators is also influenced by their material and optical characteristics.

Temperature

Since the normal temperature for testing solar panels is around 25°C, the working temperature range for solar panels is between 15°C and 35°C. At these temperatures, solar panels or photovoltaic cells operate at their most efficient levels. The efficiency of photovoltaic systems and the amount of energy they produce are both negatively impacted by the solar system's temperature [11]. The optimal working temperature of a solar cell is typically exceeded by solar panels, which inhibits their function and eventually lowers efficiency. The solar panel temperature coefficient is the most crucial factor to take into account in this situation. It is often expressed as the percentage drop in the power production of a solar panel system for every 1°C increase from the standard temperature of 25°C [1]. The majority of solar panels typically have a solar temperature coefficient of between -0.3% and -0.5% per degree celsius. As a result, the output of the solar cell decreases at higher temperatures in comparison to lower temperatures [9].

Temperature effect on PV module

The PV array power output equation can be used to express the PV module's temperature effect [6].

$$Ppv = Y_{pv} \times f_{pv} \times \left(\frac{G_T}{G_{T,STC}}\right) \times [1 + \alpha_p \times (T_c - T_{c,STC})]$$

Where [6] G_T is the solar radiation directing on the solar system array at the current time step (kW/m^2), Y_{pv} is the rated capacity of the PV array, implying that its output power under standard test circumstances is (kW), f_{pv} is the PV derating factor (%), and PV cell temperature at the current time step is T_c, and $T_{c,STC}$ is the PV cell temperature under standard test. $G_{T,STC}$ stands for incoming radiation at standard test conditions (kW/m^2), α_p for temperature coefficient of power (%/°C), and T_c for PV cell temperature. The PV module temperature coefficient of power is essential because it helps determine how far the power output of the PV module deviates from its values at STC [6]. The PV module's efficiency can be expressed as,

$$\eta_{mp,STC} = \frac{Y_{pv}}{(A_{PV} \times G_{T,STC})}$$

The terms Y_{pv} and A_{PV} stand for the rated power output of the PV module under standard test conditions (kW/m2), $G_{T,STC}$ stands for radiation at STC (kW/m^2), and $\eta_{mp,STC}$ stands for the PV module's efficiency under standard test conditions (%) [6].

Soiling and Shading

The traditional stationary panels are often installed on the roof at a specific angle. As a result of the panel's permanent location, dust begins to build up on its surface, preventing sunlight from reaching it and reducing the panel's output [7, 11].

Sun Tracking

A system called sun tracking typically adjusts the angle of a solar panel array assembly with respect to the sun. The entire device is referred to as a solar tracker. To find the ideal angle between sunlight and the panel, a solar tracker used sophisticated algorithms and tools [7].

Cooling Effect

At the peak of the solar day, a PV module's efficiency and power output decrease as a result of energy loss as heat energy, which reduces the module's power output. A multi-concept cooling technique combining air, conductive materials, and water can be used to maintain the temperature of the PV module [6].

PV module efficiency

The voltage and current of the module are impacted by the temperature of the module and its surroundings. The PV module's maximum power is expressed by the following [6],

$$P_{mp} = V_{mp} \times I_{mp}$$

$$P_{mp} = V_{oc} \times I_{sc} \times FF$$

The maximum power of a photovoltaic module is denoted by P_{mp}, while the maximum voltage and current are indicated by V_{mp}, I_{mp}, and FF, respectively. V_{oc} and I_{sc} stand for the maximum open circuit voltage and short circuit current, respectively [6]. I_{sc} rises while FF and V_{oc} fall as the module temperature rises. PV cell efficiency can be written as,

$$\eta = \left(\frac{P_{max}}{E}\right) \times A$$

E represents the solar irradiance under standard test conditions (W/m^2), P_{max} is the maximum power, and A represents the surface area of the module (m). Another formula for calculating efficiency is [6],

$$\eta_{pv} = \eta_{rT} [1 - \beta(T_{pv} - T_{rT})]$$

Where η_{pv} denotes the efficiency of PV cells, η_{rT} denotes the efficiency of the photovoltaic module at the reference temperature, typically 25°C, T_{pv} denotes the temperature of the PV module cell, β denotes the temperature coefficient of power, and T_{rT} denotes the temperature at which the PV module is used as a reference.

Effect of heating and cooling of PV module

Heat generation begins as soon as the PV module is exposed to continuous sunlight. At a certain point when the PV module's output starts to decline, overheating begins [11]. This is a result of both excessive sun exposure and a warm surrounding environment. As the temperature increases, the efficiency and output power both decline significantly; the magnitude of the decline depends on the material used to make the solar cell. PV modules can be cooled to avoid overheating [6].

Implementation of Origami in a Solar Panel System

Solar energy is used more frequently as a solution as a result of rising energy demand. Solar panels, which are a collection of photovoltaic cells, are widely utilised. Despite the benefits, the issues that need to be addressed are low efficiency and space consumption. Transporting and maintaining stationary panels might be challenging. Origami folds are therefore used in panels to provide deployability [8]. Panels with Miura folds are used for this. A tilt and swing mechanism based on a dual-axis tracking system is added to maximise the efficiency of the solar panel system.

Different Types of Origami Folding Techniques

Origami is a traditional Japanese paper folding technique. Folding-based technological developments have been created and used for the deployment of numerous items, including stent implants and automotive airbags, among others.

Miura Folds

The Miura fold is a technique for condensing a flat surface, such as a sheet of paper [3]. On the surface, the Miura folds to create a parallelogram tessellation. A parallelogram forms a mirror neighbor across each crease when the creases are in a straight line, whereas in a different direction, the zigzag creases form a parallelogram that is the translation of its neighbor across the crease [3]. One zigzag design creates mountains and valleys that alternate. A continuous motion can be used to carry out the single stiff form of origami known as the Miura pattern, which flattens the parallelogram in each step. As a result, it can be used with hard materials like solar panels. The Miura folding technique can be used to fold a big, compact structure and then expand it in a single motion. The contraction enhances mobility, while the retraction mechanism facilitates panel opening and closing. A stiff structure can be split into small volumes using Miura folds, which can then be deployed in a single continuous motion [2].

Folding Fan Design

For closing and opening operations, a folding fan constructed of thin material, such as paper or feathers, can revolve around a pivot. Closed fan patterns take significantly less room, making them suitable for usage in panels that can be easily deployed and transported to nearby homes [3].

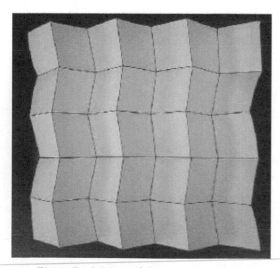

Figure 76.5 Miura folding pattern [15]

Bucky lab's Xcone Structure

This cone-shaped construction, created by Bucky Lab, was intended to provide refuge in off-the-grid locations [3]. The Xcone construction is compact and can be folded and unfolded like a cone. After the panels are mounted, this structure can be utilised as a shelter with electricity. This construction is stable and rigid.

Triangular Tessellation

Similar to the Miura pattern, this design has a triangular tessellation that transforms into a well-triangular grid. The development of a hyperbolic paraboloid with very minimal irregular deformation results from the tension and compression combined canceling each other out. Although smooth curves can be made, the material and mechanism utilised in solar panels determine the actual curves.

Methodology

The solar panel assembly must have an intuitive design and be simple to deploy and retract. The solar panel array assembly model was developed so that it can withstand any climatic and meteorological conditions and perform admirably under static and dynamic loadings. Mirrors are employed as reflectors and space is

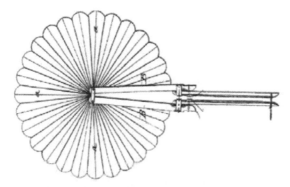

Figure 76.6 Folding fan pattern [3]

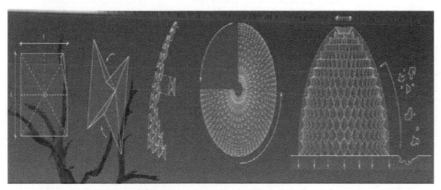

Figure 76.7 Bucky Lab's Xcone structure [3]

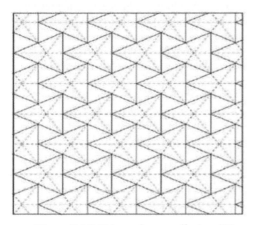

Figure 76.8 Triangular tessellation [3]

utilised for several panels using an appropriate origami mechanism. Each panel is attached exactly taking space consumption into account. The components and their functions are as follows:

- Solar panels and reflector array: Solar panels and reflectors are arranged in an array using a tessellation origami mechanism so that the surface of the panels receives the most sunlight and that other rays can be reflected onto the panels by the reflectors.
- Tracking mechanism: A dual-axis tracking mechanism is used in the system to track the solar irradiance at every point the sun travels, capturing the most solar rays possible. The panel-reflector pairs are positioned to utilise space and enhance the efficiency of the system [13].
- Adjustable base: An adjustable base can be placed anywhere, such as on a rooftop or the ground, and is made for simple expansion and contraction of assembly. The base's height can be adjusted and bear both static and dynamic stresses.
- Cooling system: Passive air cooling is added to the system to control panel temperature, lessen heat loss from usable efficiency loss, and increase PV system output.

Conclusion

Origami implementation in solar panels can be an optimal solution to the problem of space utilisation by employing deploy ability in the photovoltaics (PV) system. Reflectors are the simplest means to improve the efficiency of the panel. Thus, a combination of solar panels with reflectors would be more helpful in achieving the stated problems. The control of temperature and solar irradiance can be done using passive cooling and dual axis tracking system which fits perfectly with the origami PV panel and reflector system. Space utilisation, efficiency enhancement, cooling, and tracking can be achieved by using the stated mechanism. A flexible, compact, and energy-efficient photovoltaic system is altogether possible with origami.

References

[1] Sreega, R. S., Nithiyananthan, K., and Balavelayutham, N. (2017). Design and development of automated solar panel cleaner and cooler. International Journal for Electrical and Electronics Engineering.

[2] Binyamin, J. and Taheri, P. (2018). An origami-based portable solar panel system. IEEE 9th Annual Information Technology, Electronics and Mobile Communication Conference (IEMCON). https://doi.org/10.1109/iemcon.2018.8614997.

[3] Yogesh, S., Yogalakshmi, M., Abishek, R. Prasath, A., and Madhusudanan, G. (2021). Origami based folding techniques for solar panel applications. International Journal of Electrical Engineering and Technology. 12(3), 158-64.

[4] Venkateswari, R. and Sreejith, S. (2019). Factors influencing the efficiency of photovoltaic system. Renewable and Sustainable Energy Reviews. 101, 376-94. https://doi.org/10.1016/j.rser.2018.11.012

[5] Alzahrani, M., Shanks, K., and Mallick, T. K. (2021). Advances and limitations of increasing solar irradiance for concentrating photovoltaics thermal system. Renewable and Sustainable Energy Reviews. 138, 110517. https://doi.org/10.1016/j.rser.2020.110517

[6] Idoko, L., Anaya-Lara, O., and McDonald, A. (2018). Enhancing PV modules efficiency and power output using multi-concept cooling technique. Energy Reports. 4, 357-69. https://doi.org/10.1016/j.egyr.2018.05.004

[7] Maghami, M. R., Hizam, H., Gomes, C., Radzi, M. A., Rezadad, M. I., and Hajighorbani, S. (2016). Power loss due to soiling on solar panel: a review. Renewable and Sustainable Energy Reviews. 59, 1307-16. https://doi.org/10.1016/j.rser.2016.01.044

[8] Lahondes, Q. and Miyashita, S. (2022). Temperature driven soft reversible self-folding origami string. IEEE 5th International Conference on Soft Robotics (RoboSoft). https://doi.org/10.1109/robosoft54090.2022.9762090

[9] Kumar, M. S. et al. (2019). Effect of temperature on solar photovoltaic panel efficiency. International Journal of Engineering and Advanced Technology. 8(6). Blue Eyes Intelligence Engineering and Sciences Engineering and Sciences Publication. 2593-95. https://doi.org/10.35940/ijeat.f8745.088619

[10] Aboubakr, H. E. L. et al. (2022). Solar PV energy: from material to use, and the most commonly used techniques to maximise the power output of PV systems: a focus on solar trackers and floating solar panels. Energy Reports. 8. 11992-2010. https://doi.org/10.1016/j.egyr.2022.09.054

[11] Tyagi, V. V. et al. (2013). Progress in solar pv technology: research and achievement. Renewable and Sustainable Energy Reviews. 20, 443-61. https://doi.org/10.1016/j.rser.2012.09.028

[12] Rosell, J. I., X. Vallverdú, Lechón, M. A., and Ibáñez, M. (2005). Design and simulation of a low concentrating photovoltaic/thermal system. Energy Conversion and Management. 46(18–19), 3034-46. doi: 10.1016/j.enconman.2005.01.012. Available from http://dx.doi.org/10.1016/j.enconman.2005.01.012

[13] Awasthi, A et al. (2020). Review on sun tracking technology in solar PV system. Energy Reports. 6, 392-405. doi: 10.1016/j.egyr.2020.02.004.

[14] Lang, R. J. (Ed.). (2009). Origami 4 (1st ed.). A K Peters/CRC Press. https://doi.org/10.1201/b10653

[15] Nishiyama, Y. (2012). Miura folding: Applying origami to space exploration. International Journal of Pure and Applied Mathematics. 79, 269-79.

Chapter 77

A review of microstructure, mechanical properties, and welding of SS316L

Dipali Bhoyar[1,a] and Dr.Nischal Mungle[2,b]

[1]Laxminarayan Institute of Technology, Nagpur, India

[2]Yeshwantrao Chavan College of Engineering, Nagpur, India

Abstract

In today's industrial sector, productivity and quality are crucial factors. The primary goal of industries is to enhance productivity while providing goods of higher quality at lower costs. The most important and frequent procedure used to connect two similar and different elements is welding. This review article provides a summary of the mechanical behavior, microstructure, and use of welded joints made of austenitic grade steel. Due to its high weight percentages of chromium and nickel, austenitic grade stainless steel has great corrosion resistance. The SS 316 offers great weldability and is less susceptible to the sensitisation issue because to its low carbon content. The purpose of this study is to stimulate future SS 316 welded designs and applications by introducing the potential for more research in this area.

Keywords: Welding, austenitic stainless steel, Taguchi, ferritic stainless steel

Introduction

Austenitic stainless steel plays a significant part in the production and manufacture of goods because of its excellent qualities and simple weldability, especially for all forms of welding. Numerous welding methods, including gas metal arc (MIG), electroslag welding, electron beam welding, immerse arc welding, tungsten inert gas welding (TIG), and thermite welding, are accessible for the fabrication of ASS [1]. The outstanding corrosion resistance, fabricability, and strong mechanical qualities at high temperatures of AISI 316 austenitic stainless steels make them very popular in many industrial applications [2]. Their popularity has been fueled by their accessibility to consumers and lower prices. Chromium, nickel, and molybdenum are added to SS316L, to increase its corrosion resistance in moderately corrosive environments. In order to reduce dangerous carbide precipitation as a result of welding, 316L is an additional low-carbon variant of 316. Molybdenum increases resistance to corrosion resistance and chloride pitting. The carbon content further categorizes austenitic stainless steels into "L" types, "straight types," and "H" types. The C content of the L grades ranges from 0.03-0.08 weight percent, the straight types from 0.03-0.08 weight percent, and the H types from 0.04-0.10 weight percent. Table 77.1 and 77.2 shows chemical composition mechanical properties of SS316 respectively.

The arc of welding is formed by the plasma arc, which is composed of ionised gas, liquid metal, slag, vapor, molecules and gaseous atoms. Arc welding parameters such as the voltage of arc, welding current, speed, protecting gas, and filler material greatly influence the resistance to corrosion, weld bead shape, strength, and cooling rate of stainless steel of weld joints.

Welding current is the greatest significant thing that affects the rate of melting, rate of deposition of metal, amount of molten parent metal, depth of penetration, width of the joint, and the amount of molten parent metal [3]. Arc voltage is the difference in the electrical potential between the tip of the electrode and the surface of the molten weld region. The length of the arc and the kind of electrode affect the welding arc voltage. Arc voltage affects how a weld bead appears. Porosity, welding reinforcement electrode, spatter, and weld width all increase with increasing arc voltage [4, 5]. The speed of welding is defined as the linear velocity at which the welding flame travels in respect to the plate and the weld junction. With a reduction in speed of welding, the cooling rate and heat input (HI) rise [6, 7]. The quantity of energy delivered proportionally per unit of weld length is measured as HI. Its influence on cooling rate, which may have an impact on the characteristics and structure of the weld area and Heat affected zone, makes it a crucial attribute along with preheat and interpass temperature [8, 9]. The heat lost during welding is measured as a "cooling rate," or CR. The parameters of the final solidification microstructure are significantly influenced by Merchant [10] investigated the effects of arc voltage, welding speed, and welding current on the cooling rate, duration of solidification, and mild steel hardness welded by MMAW. When welding samples using different current and voltage values, it had discovered that the rate of cooling reduced as the welding current increased while the solidification duration increased. With faster welding, the cooling rate increased while the solidification time decreased.

[a]npdmungle@gmail.com, [b]bhoyar.dipali@gmail.com, [c]npdmungle@gmail.com

DOI: 10.1201/9781003450252-77

Table 77.1 Chemical composition of SS316

Cr	Ni	Mn	Mo	Si	C	N	P	S	Fe
16.25%	11.5%	1%	2.5%	0.5%	0.04%	0.05%	0.023%	0.015%	68.5%

Source: Author

Table 77.2 Echanical properties of SS316L

UTS (MPa)	YS (MPa)	% elongation	Elongation at break	Rockwell hardness	Brinell hardness
580	290	40%	50%	95	219

Source: Author

Table 77.3 Physical properties of SS316

Density (kg/m^3)	Youngs modulus (GPa)	Mean thermal expansion coefficient (µm/m/°C)	Mean thermal conductivity (W/m*K)	Specific heat capacity (J/kg*K)	Elastic resistivity (nΩ*m)	Calomal potential (mV)
8000	193	16.5	18.9	500	740	-50

Source: Author

In this study, an effort is made to review the mechanical properties, microstructure, and welding of SS316L. The objective of this review is to promote future SS 316 welded designs and applications and to highlight the potential for more research in this area.

Welding Methods

Tungsten inert Gas Welding

A separate filler, a mixture of shielding gases, a non-consumable tungsten electrode, and other components are used in this manual arc process. The majority of thin workpieces are processed using this method, which may be finished at moderate speeds but demands exceptional operator skills. A stable arc and high weld quality are its defining characteristics. With this technique, stainless steel, aluminium, and other light materials may be joined together.

The effects of employing H_2 in argon as a protecting gas for TIG welding of SS316L were investigated by Durgutlu [11]. They observed that, among combinations such as pure argon, 5% H_2-Ar, and 1.5% H_2-Ar, the weld performed with the shielding gas 1.5% H_2-Ar generated the maximum strength. The weld metal's mean grain size increased with an increase in H_2 concentration. The breadth and weld penetration depth were both increased as well. In a combination with 5% H_2-Ar, it is discovered that the grain characteristics are more effective. For all protective gases combinations, hardness of the weld material was less than that of HAZ and parent metal.

Kuuo et al. [12] investigated the angular distortion, morphology of weld, and surface appearance of 316 SS welding with mild steel with and without activated flux. It was discovered that the flux powder that was coated had a density of 5–6 mg/cm^2. Without the usage of the flux, the surface was even and clean, but applying flux such as Fe_2O_3, Cr_2O_3, and CaO caused a significant quantity of slag to build on the surface. However, using SiO_2 resulted in far less slag formation. With the usage of SiO_2 flux, the distortion (angular) of TIG welded part was reduced, and the weld metal was not left with any voids or cracks.

Arivazagan et al. [13] analysed the dissimilar welding between SS304 and low alloy SS316 by using GTAW, EBW, and friction welding (FRW). The findings demonstrated that the welding created by electron beam has a highest (682 MPa) than the joints prepared by TIG (636 MPa) and FRW (495 MPa). Additionally, as compared to EBW and FRW, the impact strength of GTAW weldments is stronger.

Ganesh et al. [14] predicted residual stresses, heat cycle, and distortion in SS316L weld joints produced by TIG welding using a FEM model utilizing SYSWELD. The authors came to the conclusion that numerical modelling in addition with NDT methods offers a powerful technique for estimating the impact of the welding methods on the weld features after observing good agreement between the forecasts of model and the experimental readings of distortion, temperature, and residual stresses (Figure 77.1).

Devaraju [15] employed 316L as a filler metal in an attempt to weld together two dissimilar materials, duplex stainless steel (DSS) and 316L. The author observed in comparison to the 316L welded zone, the DSS welded zone had a better hardness and impact strength. As presented in Table 77.4.

In order to examine the impact of different multi and single component fluxes on geometries of weld bead, Kulkarni et al. [16] used A-TIG welding of SS316L on 8 mm thick plates. The multi-component flux

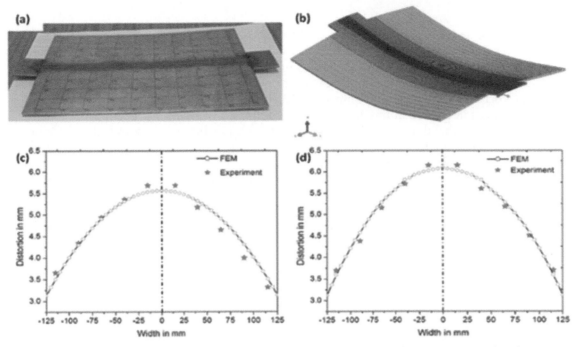

Figure 77.1 Comparison of distortion, (a) welding experiment, (b) welding FEM, c). Right edge transverse distortion, (d). Left edge transverse distortion [14]

Table 77.4 UTM test results [15]

Sample	YS (MPa)	UTS (MPa)	% elongation	Hardness (BHN)
Parent metal (316L)	170	485	40	210
Parent metal (DSS)	450	621	25	293
Welded sample	610	735	17	260

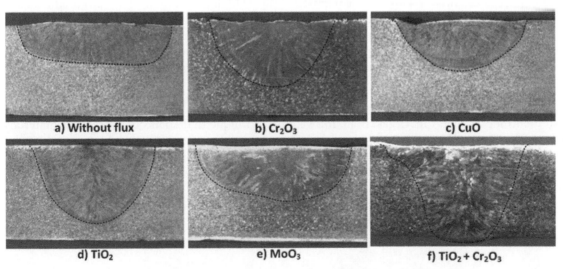

Figure 77.2 Macrographs showing weld bed geometries (a) without flux, (b) Cr2O3, (c) CuO, (d) TiO_2, (e) MoO_3 and (f) TiO_2+ Cr_2O_3 [16]

mixture, as found by the authors, caused through penetration in a single pass and increased the concentration of δ-ferrite in A-TIG weld regions related to parent metal. Figure 77.2 shows Macrographs showing welded bed geometries (a) without flux, (b) Cr2O3, (c) CuO, (d) TiO_2, (e) MoO_3 and (f) TiO_2+ Cr_2O_3.

Fouad [17] investigated the impact of HI, filler metals, and addition of N_2 (2% vol) on the weld metal microstructure characteristics of SS316. With increased weld current, the UTS, YS, elongation percentage, and fatigue life decrease. With rising HI and nitrogen content in the protecting gas, weld metal grain size and crystallisation temperature rise.

Resistance Spot Welding

Bayram et al. [18] examined at how welding time, welding cooling conditions, and various welding environment affected the quality of SS316L resistance spot welds. Due to the growth in nugget size, it was discovered that the tensile shear force bearing capability of welded samples improved with increasing input of heat linked to duration of weld. Additionally, it was discovered that during the whole welding process, welded samples performed in nitrogen environment had a slightly greater tensile shear load bearing capability than those acquired in normal atmosphere.

Tensile shear strength of the aluminum alloy6063 and SS316L junction was researched by Noor et al. [19]. They discovered that the joint strength improved with the rise in current of welding. The increase in nugget diameter was responsible for the joints' increased tensile shear strength (TSS). Additionally, it was discovered that the TSS for joints undergone heat treatment were essentially identical to those of non-heat-treated joints.

To determine the size and form of the nuggets and to forecast the strength of the resistance spot weld, Gopalkrishnan et al. [20] conducted tests by altering the process variable such as tip diameter electrode (6, 8, 9 mm), welding current (7, 9, 10 kA), and heating duration (6, 7, 8 cycles). The highest load supported by the joint, according to the authors, was achieved at an electrode spot size of 8 mm, a weld current of 9 kA, and a heating cycle time of eight cycles.

Friction stir welding (FSW)

In comparison to traditional arc welding methods, FSW offers various advantages, including reduced energy input, process temperatures, residual stresses, welding times, and distortion. Ramirez [21] developed the parameters for FSW of dissimilar cooper/SS316L 2mm thick plates. Investigations into the impact of speed of spindle (1000, 1400 rpm), speed of welding (100, 200, 350 mm/min), and offset of tool (0, 0.6, 1.6 mm) on the microstructure and joint consolidation revealed that increasing the offset reduces the robustness and leads to pores formation, reduced elongation,and lesser tensile strength.

Meshram et al. [22] attempted to weld SS316 that was 4 mm thick using the FSW technique and observed a defect-free weld while using tool rotational speeds of 1200 rpm and torch speeds of 9 mm/min. With the increase in the nugget zone hardness, they were able to produce a weld's ultimate tensile strength that was virtually identical.

Laser Welding

To search the impact of the pulse energy on the features of the weld fillet, Ventrella et al. [23] used a pulsed neodrymium:ytrium aluminium garnet laser-weld. With a 4 ms pulse length, the pulse energy was adjusted in steps of 0.25 J from 1.0 to 2.25 J. They noticed that while the pulse energy increased, the UTS of the welded junctions primarily increased and then declined. Figure 77.3 shows micrograph of the (a) fusion zone and (b) HAZ of an SS316L welded joint.

Gas metal arc welding

GMAW, also identified as metal inert gas, is an arc welding technique that connects two parent materials by providing an electrode in form of wire continuously into the weld pool through a welding gun. In order to shield the weld pool from oxidation, a protecting gas is also provided through the welding gun. In order to weld a 5 mm thick plate of SS316, Madavi [24]investigated the effects of various oxide fluxes (ZnO, Al_2O_3, and SiO_2). The authors noted that due to the development of a well-ordered structure of grain, and structure of the weld formed with flux as compared to those formed without flux, flux welding results in

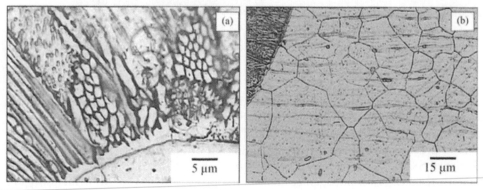

Figure 77.3 Micrograph of the (a) fusion zone and (b) HAZ of an SS316L welded joint [24]

Table 77.5 UTS and penetration depth for without and with flux (experimental results)[24]

Sample	Input parameter			UTS, MPa (without flux)	UTS, MPa (with flux)	Penetration depth, mm (without flux)	Penetration depth, mm (with flux)
	Flux (g/cm^2)	Voltage (V)	Current (A)				
1	SiO$_2$	20	170	268.10	308.29	1.428	1.678
2	ZnO	24	170	258.68	298.23	1.028	1.134
3	Al$_2$O$_3$	28	170	284.13	282.94	1.400	1.450
4	ZnO	20	190	278.48	304.04	1.012	1.587
5	Al$_2$O$_3$	24	190	266.64	268.13	1.176	1.787
6	SiO$_2$	28	190	258.40	312.89	2.276	2.066
7	Al$_2$O$_3$	20	210	289.26	317.56	1.179	1.579
8	SiO$_2$	24	210	278.38	319.86	2.089	2.485
9	ZnO	28	210	292.10	326.58	1.874	2.147

an improvement in ultimate tensile strength of 10.80%. The flux reduces the amount of passes required to complete a weld, which increases output by 15% as shown in Table 77.5.

To inspect effects of HI on joint's impact strength, Dalakoti et al. [25] attempted to weld two different materials, SS316 and mild steel. 120, 125, 140 A of current, 20 to 24 V of voltage, and 26 to 32 cm per minute of welding speed are the welding parameters. The greatest impact strength was found to be 3.2 j/mm^2 at 4.5 kJ/cm of the HI, and the minimum impact strength was found to be 2.24 j/mm^2 at 10 kJ/cm of HI, as observed by the authors. Impact strength declines as HI increases.

Gnanarathinam et al. [26] studied behavior of corrosion of SS316L welded component with different welding processes (TIG, MIG, and arc) under various solutions (Ferric chloride, nitric acid, and oxalic acid). These acids are used for quick corrosive purposes at 1 N concentration. Once every day for 12 days, the weight reduction was measured. Steel that was TIG welded showed more corrosion when exposed to ferric chloride, while steel that was arc welded showed less corrosion. In an oxalic acid solution, MIG welded steel showed greater corrosion whereas steel welded with TIG showed less corrosion. TIG welded steel exhibited greater corrosion in nitric acid solution, but arc welded steel exhibited less corrosion.

Microwave Hybrid Heating

The novel non-traditional technique for joining processes has evolved as the microwave hybrid heating (MHH) method. Internal material heating occurs as a result of the material being subjected to oscillating electromagnetic radiation for the duration of microwave joining. This procedure is special because it uses volumetric heating, uses less energy, and produces high-quality goods while being environmentally friendly. The joining cost of MHH-based joints is significantly increased by the high cost of the powder used at interface in MHH-based joints. Transferring this technique to the welding industry is challenging without a significant decrease in the price of the interface powder. Maintaining a vacuum is necessary to prevent oxide formation at the joints, which has its own costs and complicates the experimental setup.

Kumar et al. [27] joined stainless steel of various grades using the microwave joining technique (SS304 and SS316). In a microwave oven operating at 900 W and 2.55 GHz, experiments were conducted. The produced joints were found to be crack-free and to have homogenous microstructures as a result of the joint's uniform volumetric heating, according to the authors.

Srinath et al. [28] presented an innovative method for connecting bulk austenitic SS316 utilising a microwave source applicator. A square butt junction was created by joining SS316 plates with dimensions of 25 x 12 x 6 mm^3 using an interface material made of epoxy resin and 40 μm fine nickel powder. The authors found that the joint's average hardness was 420±15 HV interface in contrast to the 290±14 HV hardness in the welded region. Because of the development of metal carbides in the fusion region, the hardness of the weld junction was detectedas 30% higher than the parent material and 309 MPa was the average tensile strength with 12% effective elongation. Table 77.6 shows research reported on welding of SS316 and its outcome.

Stainless Steel Joint Mechanical Properties

The mechanical behavior of joint affects the safety of the components that are in use. ASS makes about 2/3rd of all stainless steel production across all categories. Despite this, the poor strength, low impact resistance,

Table 77.6 Research reported on welding of SS316 and its outcome

Material	Type of welding	Parameters	Outcomes	Ref
SS316L	TIG	Current (90,110, 120 A), flow rate of gas (1.1, 0.7,0.9 LPM) and bevel angle (60,65,70)	At 110A current, a gas flow rate of 0.7 LPM, 60° bevel angle, greater tensile strength was attained.	[2]
A387 and SS316	MIG	Voltage (16, 18, 20 V), current (120, 140, 160 A), bevel angle (50, 60, 70)	The maximum impact strength was attained using a 20 V welding voltage, 50° of bevel, and 140 A welding current.	[29]
SS316L	Friction Welding	Frequency (25, 40, 45 HZ), Burn-off distance (1 , 3 mm), Friction pressure (60, 160, 240 MPa) Amplitude (±1 ,2.5, 3 mm)	TS:512-610 MPa, PMR: failure location, the final - ferrite percentage is greatly influenced by the burn-off rate, with the high burn-off rate resulting in the low amounts of -ferrite.	[30]
SS316	FSW	Speed (rotational) : 1500rpm Force applied:30KN Holding Time:5,10 S	Fracture: at the interface A small number of $Cr_{23}C_6$ were formed but no sigma phase.	[31]
SS316, AISI 409	MIG	Current (100, 112, 124 A), Nozzle-plate distance (10, 12, 15 mm), and Gas flow rate (10, 15, 20 LPM)	Welding current at 112 A, the gas flow at 15 l/min, and nozzle-plate distance at 15 mm are the ideal parameters.	[32]
SS316L	SAW	Current (165.91, 200, 250, 300, 335.08), Voltage (18.29, 20, 22.5, 25, 26.70), Travel speed (13.29, 15, 17.5, 20, 21.70)	Maximum rate of dilution and penetration area noted at welding currents of 335.068 A, 26.70 V for the arc, and 13.29 m/s for travel speed.	[33]
SS316L	TIG	Current (60, 80, 100 A), Speed (50, 65, 80 mm/min), Root gap (1, 1.5, 2 mm)	Speed of welding (46.41% influence) has higher impact on bend strength and current (97.65%) has highest contribution on TS	[34]
SS316L	TIG	Current: 80, 90, 100 and 120 A Argon flow: 15 lpm Diameter of filler rod: 2.4 mm	HI increased with an increase in current, width and depth of weld pool increase, and chromium percentage reduced	[35]
SS316 and SS304	RSW	Current (5,6,7 A), Time (5,6,7 cycles), pressure (2, 2.2,2.5 kg/m³)	The primary regulating factor affecting the TSS is weld current, asymmetric fusion zone was obtained	[36]
SS316 and CoCrFeMnNi	Laser welding	Peak pulse of 2.20 kW, duration: 10 ms, total energy: 17.9 J	Defect free joint, there was a rise in hardness at the fusion zone, and microstructures were made up of just one FCC phase.	[37]
SS316H, SS316L-A, SS316L-B	GMAW	Current:95A, number of layers: 3, Voltage: 23 V, HI: 0.7 KJ/min, travel speed: 203 mm/min	The resistance to corrosion of E316H weld deposits is higher than that of E316L.	[38]
SS316 and AZ31B	Laser welding	2.5 kW power, wavelength of 1030 nm, travel speed 1.8 m/min	The maximum tensile strength was greater than 100 MPa, in the weld zone, a thin layer with dimensions on the order of microns formed.	[39]
copper/316 L	TIG	current of 100 A, torch speed of 5 mm·s⁻¹	The weld metal showed a well defined phase separation.	[41]
SS316L	TIG	Current (150.160,170 A), Gas flow rate (10,15,20 l/min), speed (180,190,200 mm/min)	The ideal process parameters are observed to be gas flow rate (15 L/min), torch speed (190 mm/min), and current (150 A).	[42]
SS316 and Monel 400	FSW	rotating speed:450 to 650 rpm torch speed: 50 to 200 mm/min	At a rotating speed of 450 rpm and welding speeds ranging from 50 to 150 mm/min, sound welds were produced.	[43]
SS316	CO₂ laser weld	Power: 3.5 kW, Shilding gas: He, HI: 210J, travel speed: 1 m/min	The UTS point increased on by 5% between the tensile tests conducted after and before vibratory stress reduction treatment.	[44]
SS316, SS308, SS317M	GTAW	Current (75, 85, 86 A), Voltage (14,13, 12 V), Speed (15.78, 16.66, 15.78 mm/min)	The material ER-317M had the greatest yield strength of 330 MPa and microhardness of 188 HV.	[45]
SS304 and SS316	TIG	Current (30, 45, 60 A), voltage (40, 60 and 80 V),	The UTS was found with 60A current, which was 528.36 MPa.	[46]

Material	Type of welding	Parameters	Outcomes	Ref
Inconel82 and SS316	TIG	Current 90A, travel speed 8 cm/min, voltage 14V, interpass temp 50°C	Dendritic structure in Inconel, ductile failure, Precipitations of the secondary phase and substantial segregation are seen.	[47]
SS316	Laser weld	focusing length: 75 mm, spot diameter: 40 μm, , pulse duration: 0.2 and 20 ms	The pulse energy, focal location, and torch speed tuned to 8 J, 2.2 mm, and 4.25 mm/s, to achieve maximum weld penetration and minimal weld breadth.	[48]
Inconel 718 and SS316	EBW	Fabrication temperature is 920°C, the beam current is 12 mA, the beam focus is 22 mA, and the beam speed is 918 mm/s.	The findings from joining point to strategies for producing parts out of many materials.	[49]
Monel 400 and SS316	TIG	Voltage: 12.5 V, Current: 130 A	At the weld zone, migrated grain boundaries were discovered.	[50]

PMR: parent metal region, FSW: friction stir welding, SAW: submerged arc welding, RSW: resistance spot welding, TSS: tensile shear strength.

and low hardness value of austenitic stainless steel after welding limit its use in several applications. As a result, several researchers conducted studies to enhance the properties of SS welded joint.

The mechanical and micro structural characteristics of SS316L were improved by Xie [51] using the ultrasonically wave aided gas tungsten pulse arc technology. The TS, corrosion and wear resistance of A-TIG welded Inconel 600 and SS316L dissimilar weld joint were increased by Chandra shekar et al. [52] using the laser shock peening (LSP) process. The author also noted that LSP approach converts the TS (residual) into compressive residual stress, improved the mechanical performance.

The electromagnetic vibration approach was utilised by Sabzi and Mersagh [53] to increase the SS316L TS, yield strength, toughness, and hardness. They discovered that the quantity of dilution significantly lowers when the voltage of electromagnetic vibration in welding is raised. It is a result of high vibration-induced convective heat transfer and strong turbulence. The weld pool temperature is lowered by the electromagnetic vibration that is being used, which lowers the high CR of the weld metal. This phenomenon causes the grain to become more refined, the quantity of δ-ferrite to decrease, the unmixed zone to extend, and the solidification mode to switch from (A) austenite to (AF)austenitic-ferritic.

Conclusion

As the use of the welded components may be in a sensitive area, it is crucial to inspect the mechanical characteristics of the welded material because its primary function is to firmly join the two metals together. Checking the weld's tensile strength and the variables that affect that strength are crucial. The main issue with SS316 welded joints is the development of inter-metallic compounds at the joint, which have an impact on the weld's performance and characteristics. Filler metals (such Zn) can be used in lap joints without causing IMC to form because they increase strength, although the tool pin depth is also very important.

The superior performance of laser beam welding was attributable to the high class joint, which was characterized by no underfill, full penetration, and freedom from microcracks and porosity. Higher peak outputs and longer pulse durations significantly increase the loss of alloying elements. If the welding parameters are not optimal, problems like gas evolution will occur while welding austenitic steel with a laser. It was discovered that in resistance spot welding, the expansion of nugget size caused the tensile shear load bearing capability of welded specimens to rise with rising HI related to weld duration. TIG welding with different metals is shown to produce satisfactory results. By reducing the grain size of HAZ, activating fluxes enhance the mechanical characteristics of the joint. The TIG-MAG welded samples showed strong weld joints with no fractures, holes, or porosity. Due to an increase in the retained δ-ferrite concentration of SS welds, A-TIG welding can boost the UTS of weldments. FSW welding is the most defect-free type of welding. The weld in FSW had no intermetallic -phase. The major cause of stainless steels' deterioration is the precipitation of the -phase. The findings of tensile test have demonstrated that the mechanical characteristics of the FSW of SS316L are comparable to those of the parent metal.

References

[1] Garg, H., Sehgal, K., Lamba, R., and Kajal, G. (2019). A systematic review: Effect of TIG and A-TIG welding on austenitic stainless steel. Singapore: Springer. https://doi.org/10.1007/978-981-13-6412-9_36.

[2] Balaji, C. (2012). Evaluation of mechanical properties of SS316L weldments using tungsten inert gas welding. International Journal of Engineering Science and Technology.

[3] Dhobale, A. L. and Mishra, P. H. K. (2015). Review on effect of HI on tensile Strength of Butt Weld Joint Using Mig Welding. *International Journal of Innovative Technology and Exploring Engineering.* 2, 1–13.

[4] Kulkarni, S. S. and Halakatti, P. G. (2015). Study of influence of welding parameters on mild steel. International **Journal** of Advance Research in Engineering and Technology. 2, 49–50.

[5] Fande, A. W., Taiwade, R. V., and Raut, L. (2022). Development of activated tungsten inert gas welding and its current status: A review. Materials and Manufacturing Processes. 37, 841–76. https://doi.org/10.1080/10426914.2022.2039695.

[6] Sen, W. H. (2005). Effect of welding variables on cooling rate and pitting corrosion resistance in super duplex stainless weldments. Materials Transactions. 46, 593–601. https://doi.org/10.2320/matertrans.46.593.

[7] Raut, L. P. and Taiwade, R. V. (2021). Wire Arc Additive Manufacturing: A Comprehensive Review and Research Directions. Journal of Materials Engineering and Performance. 30, 4768–91. https://doi.org/10.1007/s11665-021-05871-5.

[8] Kumar, S. D., Sahoo, G., Basu, R., and Sharma, V. (2018). Mohtadi-Bonab MA. Investigation on the microstructure—mechanical property correlation in dissimilar steel welds of stainless steel SS 304 and medium carbon steel EN 8. Journal of Manufacturing Processes. 36, 281–92. https://doi.org/https://doi.org/10.1016/j.jmapro.2018.10.018.

[9] Raut, L. P. and Taiwade, R. V. (2022). Microstructure and Mechanical Properties of Wire Arc Additively Manufactured Bimetallic Structure of Austenitic Stainless Steel and Low Carbon Steel. *Journal of Materials Engineering and Performance.* https://doi.org/10.1007/s11665-022-06856-8.

[10] Samir, M. (2015). Investigation on Effect of HI on Cooling Rate and Mechanical Property (Hardness) Of Mild Steel Weld Joint by MMAW Process. International Journal of Modern Engineering Research. 5, 34–41.

[11] Durgutlu, A. (2004). Experimental investigation of the effect of hydrogen in argon as a shielding gas on TIG welding of austenitic stainless steel. **Materials & Design.** 25, 19–23. https://doi.org/10.1016/j.matdes.2003.07.004.

[12] Kuo, C. H., Tseng, K. H., and Chou, C. P. (2011).. Effect of activated TIG flux on performance of dissimilar welds between mild steel and stainless steel. *Key Engineering Materials.* 479, 74–80.

[13] Arivazhagan, N., Singh, S., Prakash, S., and Reddy, G. M. (2011). Investigation on AISI 304 austenitic stainless steel to AISI 4140 low alloy steel dissimilar joints by gas tungsten arc, electron beam and friction welding. Materials & Design. 32, 3036–50. https://doi.org/https://doi.org/10.1016/j.matdes.2011.01.037.

[14] Ganesh, K. C., Vasudevan, M., Balasubramanian, K. R., Chandrasekhar, N., Mahadevan, S. and Vasantharaja, P. et al. (2014). Modeling, prediction and validation of thermal cycles, residual stresses and distortion in type 316 LN stainless steel WELD joint made by TIG welding process. Procedia Engineering. 86, 767–74. https://doi.org/10.1016/j.proeng.2014.11.096.

[15] Devaraju, A. (2015). An Experimental study on TIG welded joint between Duplex Stainless Steel and 316L Austenitic Stainless Steel. International Journal of Mechanical Engineering. 2, 1–4. https://doi.org/10.14445/23488360/IJME-V2I10P101.

[16] Kulkarni, A., Dwivedi, D. K. and Vasudevan, M. (2019). Effect of oxide fluxes on activated TIG welding of AISI 316L austenitic stainless steel. 18:4695–702. https://doi.org/10.1016/j.matpr.2019.07.455.

[17] Fouad, R. A., Ali, E. A., Shaaban, A. R., and El-Nikhaily A. E. (2022). Effect of welding variables on the quality of weldments. In: Eds. Cooke, K. O., Cozza, R. C. Rijeka: IntechOpen. https://doi.org/10.5772/intechopen.103175.

[18] Kocabekir, B., Kaçar, R., Gündüz, S. and Hayat, F. (2008). An effect of HI, weld atmosphere and weld cooling conditions on the resistance spot weldability of 316L austenitic stainless steel. Journal of Materials Processing Technology. 195, 327–35. https://doi.org/10.1016/j.jmatprotec.2007.05.026.

[19] Noor, M. M., Zhen, A. T. Y., Jamaludin, S. B., Hayazi, N. F. and Shamsudin, S. R. (2013). Joining of dissimilar 6063 aluminium alloy-316L stainless steel by spot welding: Tensile shear strength and heat treatment. Advanced Materials Research. 795, 492–95. https://doi.org/10.4028/www.scientific.net/AMR.795.492.

[20] Gopalakrishnan, K., Gokulakrishnan, S., Gopalakrishnan, K. and Gokulakrishnan, S. (2015). Testing on Resistance Spot Weld Mechanical Testing on Resistance Spot Nuggets Weld of Steel Sheets Nuggets of AISI Stainless Steel Sheets. International Journal of Mechanical, Aerospace, Industrial, Mechatronic and Manufacturing Engineering. 1, 6-12.

[21] Ramirez, A., Benati, D. and Fals, H. (2011). Effect of tool offset on dissimilar Cu-AISI 316 stainless steel friction stir welding. Proceedings of the Annual International Offshore and Polar Engineering Conference. 548–52.

[22] Meshram, M. P, Kodli, B. K., and Dey, S. R. (2014). Mechanical Properties and Microstructural Characterisation of Friction Stir Welded AISI 316 Austenitic Stainless Steel. Procedia Materials Science. 5, 2376–81. https://doi.org/10.1016/j.mspro.2014.07.482.

[23] Ventrella, V. A., Berretta, J. R., De Rossi, W. (2010). Pulsed Nd:YAG laser seam welding of AISI 316L stainless steel thin foils. Journal of Materials Processing Technology. 210, 1838–43. https://doi.org/10.1016/j.jmatprotec.2010.06.015.

[24] Madavi, K. R., Jogi, B. F., and Lohar, G. S. (2021). Investigational study and microstructural comparison of MIG welding process for with and without activated flux. Materials Today: Proceedings. 51, 212–16. https://doi.org/10.1016/j.matpr.2021.05.240.

[25] Dalakoti, M., Kumar, A., and Khanna, P. (2020). Influence of HI on Microstructure and Impact Strength in Dissimilar Joining Using Mig Welding. International Research Journal of Engineering and Technology.

[26] Gnanarathinam. A, Palanisamy, D., Gangaraju, M., Arulkirubakaran, D. and Balachander, E. (2022). Investigation of corrosion behavior of welded area of austenitic stainless steel under different environments. Materials Today: Proceedings. https://doi.org/10.1016/j.matpr.2022.09.351.

[27] Kumar, A., Sehgal, S., Singh, S. and Bagha, A. K. (2019). Joining of SS304-SS316 through novel microwave hybrid heating technique without filler material. Materials Today: Proceedings. 26, 2502-2505. https://doi.org/10.1016/j.matpr.2020.02.532.

[28] Srinath, M. S., Sharma, A. K., and Kumar, P. (2011). A novel route for joining of austenitic stainless steel (SS-316) using microwave energy. Proceedings of the Institution of Mechanical Engineers, Part B: Journal of Engineering Manufacture. 225, 1083–91. https://doi.org/10.1177/2041297510393451.

[29] Arunkumar, S. P., Prabha, C., Saminathan, R., Khamaj, J. A., Viswanath, M., and Paul Ivan, C. K. et al. (2022). Taguchi optimisation of metal inert gas (MIG) welding parameters to withstand high impact load for dissimilar weld joints. Material Today Proceedings. 56, 1411–7. https://doi.org/10.1016/j.matpr.2021.11.619.

[30] Bhamji, I., Preuss, M., Threadgill, P. L., Moat, R. J., Addison, A. C. and Peel, M. J. (2010). Linear friction welding of AISI 316L stainless steel. Materials Science and Engineering A. 528, 680–90. https://doi.org/10.1016/j.msea.2010.09.043.

[31] Zhou, L., Zhou, W. L., Huang, Y. X., and Feng, J. C. (2015). Interface behavior and mechanical properties of 316L stainless steel filling friction stir welded joints. International Journal of Advanced Manufacturing Technology. 81, 577–83. https://doi.org/10.1007/s00170-015-7237-5.

[32] Ghosh, N., Pal, P. K., and Nandi, G. (2017). GMAW dissimilar welding of AISI 409 ferritic stainless steel to AISI 316L austenitic stainless steel by using AISI 308 filler wire. Engineering Science and Technology, an International Journal. 20, 1334–41. https://doi.org/10.1016/j.jestch.2017.08.002.

[33] Kumar, R., Dikshit, I., and Verma, A. (2020). Experimental investigations and statistical modelling of dilution rate and area of penetration in submerged arc welding of SS316-L. Material Today Proceedings. 44, 3997–4003. https://doi.org/10.1016/j.matpr.2020.10.201.

[34] Bharath, P., Sridhar, V. G., and kumar, M. S.(2014). Optimisation of 316 Stainless Steel Weld Joint Characteristics using Taguchi Technique. Procedia Engineering. 97, 881–91. https://doi.org/https://doi.org/10.1016/j.proeng.2014.12.363.

[35] Moslemi, N., Redzuan, N., Ahmad, N., and Hor, T. N. (2015). Effect of current on characteristic for 316 stainless steel welded joint including microstructure and mechanical properties. Procedia CIRP. 26, 560–64. https://doi.org/10.1016/j.procir.2015.01.010.

[36] Verma, A. B., Ghunage, S. U., and Ahuja, B. B. (2014). Resistance Welding of Austenitic Stainless Steels (AISI 304 with AISI 316). 5th Int 26th ALL India: 1–6.

[37] Oliveira, J. P., Shen, J., Zeng, Z., Park, J. M., Choi, Y. T., and Schell, N. et al. (2021). Dissimilar laser welding of a CoCrFeMnNi high entropy alloy to 316 stainless steel. Scripta Materialia. 206, 114219. https://doi.org/10.1016/j.scriptamat.2021.114219.

[38] Cui, Y. and Lundin, C. D.(2007). Austenite preferential corrosion attack in 316 austenitic stainless steel weld metals. Material Design. 28, 324–28. https://doi.org/https://doi.org/10.1016/j.matdes.2005.05.022.

[39] Casalino, G., Guglielmi, P., Lorusso, V. D., Mortello, M., Peyre, P. and Sorgente, D. (2017). Laser offset welding of AZ31B magnesium alloy to 316 stainless steel. Journal of Materials Processing Technology. 242, 49–59. https://doi.org/https://doi.org/10.1016/j.jmatprotec.2016.11.020.

[40] Liu, F., Xu, B., Song, K., Tan, C., Zhao, H., and Wang, G. (2022). Improvement of penetration ability of heat source for 316 stainless steel welds produced by alternating magnetic field assisted laser-MIG hybrid welding. Journal of Materials Processing Technology. 299, 117329. https://doi.org/https://doi.org/10.1016/j.jmatprotec.2021.117329.

[41] Xu, Y., Hou, X., Shi, Y., Zhang, W., Gu, Y., and Feng, C. (2021). Correlation between the microstructure and corrosion behaviour of copper/316 L stainless-steel dissimilar-metal welded joints. Corrosion Science. 191, 109729. https://doi.org/https://doi.org/10.1016/j.corsci.2021.109729.

[42] Vinoth, V., Sudalaimani, R., Veera, A. C., Kumar, S. C., and Prakash, S. K. (2021). Optimisation of mechanical behaviour of TIG welded 316 stainless steel using Taguchi based grey relational analysis method. Materials Today: Proceedings. 45(3), 7986–93. https://doi.org/https://doi.org/10.1016/j.matpr.2020.12.1002.

[43] Aghaei, A. and Dehghani, K. (2015). Characterisations of friction stir welding of dissimilar Monel400 and stainless steel 316. The International Journal of Advanced Manufacturing Technology. 77, 573–79. https://doi.org/10.1007/s00170-014-6467-2.

[44] Mohanty, S., Arivarasu, M., Arivazhagan, N., and Phani Prabhakar, K. V. (2017). The residual stress distribution of CO2 laser beam welded AISI 316 austenitic stainless steel and the effect of vibratory stress relief. Material Science Engineering A. 703, 227–35. https://doi.org/https://doi.org/10.1016/j.msea.2017.07.066.

[45] Ostovan, F., Shafiei, E., Toozandehjani, M., Mohamed, I. F. and Soltani, M. (2021). On the role of molybdenum on the microstructural, mechanical and corrosion properties of the GTAW AISI 316 stainless steel welds. Journal of Materials Research and Technology. 13, 2115–25. https://doi.org/https://doi.org/10.1016/j.jmrt.2021.05.095.

[46] Ramakrishnan, A., Rameshkumar, T., Rajamurugan, G., Sundarraju, G., and Selvamuthukumaran, D. (2021). Experimental investigation on mechanical properties of TIG welded dissimilar AISI 304 and AISI 316 stainless steel using 308 filler rod. Materials Today: Proceedings. 45, 8207–11. https://doi.org/https://doi.org/10.1016/j.matpr.2021.03.502.

[47]. Jang, C., Lee, J., Sung, K. J., and Eun, J. T. (2008). Mechanical property variation within Inconel 82/182 dissimilar metal weld between low alloy steel and 316 stainless steel. International Journal of Pressure Vessels and Piping. 85, 635–46. https://doi.org/https://doi.org/10.1016/j.ijpvp.2007.08.004.

[48] Solati, A., Bani Mostafa Arab N, Mohammadi-Ahmar A, Fazli Shahri HR. Multi-criteria optimisation of weld bead in pulsed Nd:YAG laser welding of stainless steel 316. Journal of Process Mechanical Engineering. 233, 151-164. https://doi.org/10.1177/0954408918756654.

[49] Hinojos, A., Mireles, J., Reichardt, A., Frigola, P., Hosemann, P. and Murr, L. E. et al. (2016). Joining of Inconel 718 and 316 Stainless Steel using electron beam melting additive manufacturing technology. Materials & Design. 94, 17–27. https://doi.org/https://doi.org/10.1016/j.matdes.2016.01.041.

[50] Mishra, D., Vignesh, M. K., Raj, B. G., Srungavarapu, P., Devendranath Ramkumar, K., and Arivazhagan, N. (2014). Mechanical characterization of Monel 400 and 316 stainless steel weldments. Procedia Engineering.75, 24–8. https://doi.org/10.1016/j.proeng.2013.11.005.

[51] Xie, W. and Yang, C. (2020). Microstructure, mechanical properties and corrosion behavior of austenitic stainless steel sheet joints welded by gas tungsten arc (GTA) and ultrasonic–wave–assisted gas tungsten pulsed arc (U–GTPA). Archives of Civil and Mechanical Engineering. 20, 43. https://doi.org/10.1007/s43452-020-00044-y.

[52] Chandrasekar, G., Kailasanathan, C., and Vasundara, M. (2018). Investigation on un-peened and laser shock peened dissimilar weldments of Inconel 600 and AISI 316L fabricated using activated-TIG welding technique. Journal of Manufacturing Processes. 35, 466–78. https://doi.org/10.1016/j.jmapro.2018.09.004.

[53] Sabzi, M. and Dezfuli, S. M. (2018). Drastic improvement in mechanical properties and weldability of 316L stainless steel weld joints by using electromagnetic vibration during GTAW process. Journal of Manufacturing Processes. 33, 74–85. https://doi.org/10.1016/j.jmapro.2018.05.002.

Chapter 78

Effect of vehicle vibrations on the human body

Apurva Weikea[a], Girish Mehta[b] and Sagar Shelare[c]

Priyadarshini College of Engineering, Nagpur, Maharashtra, India

Abstract

Human bodies are impacted by forces produced by mechanical vibrations. The sitting position is one of the most typical positions for human bodies. In healthy individuals, body shaking is a type of workout that enhances neuromuscular function. The overall vibration increases the risk of diseases affecting the spine and surrounding neurological structure. Also, it may result in headaches, dizziness, and gastrointestinal issues. These symptoms occur when a large group of individuals goes on a lengthy vehicle trip. Therefore, the present work aims to understand the properties of full-body vibration in those who have had spinal cord injuries and to assess the most efficient vibration procedures. Examining the past can offer insights into the future in many technological domains. The driver- and passenger-related technologies that evolved throughout time and the research required for such innovations, as they influenced both vehicle design and evaluation, are examined in this study. This paper examines cutting-edge whole-body vibration workout methods, pointing out the potential benefits of vibration as a stimulant for human muscles and laying the groundwork for further investigation.

Keywords: Vehicle vibrations, mechanical vibrations, human body, spinal chord

Introduction

Human vibration is a multi-disciplinary topic that involves several fields. The transportation system is a part of vibration research. Vehicles (air, sea, and land) subject individuals to powered shaking that might be random, transitory, or periodic. Total body vibration is well recognised as a significant occupational hazard worldwide. Much research has found that prolonged exposure to high levels of total body vibration increases the risk of complications in the nervous system and the lumbar spine [1]. Vibration is caused by a variety of methods in construction, mining, industry, agriculture, and community services. Body vibration arises when the humanoid frame stays on a vibrating surface, such as in all means of transportation and when occupied nearby various manufacturing operations. Hand-transmitted vibration occurs when motion enters the body through the hands, such as when revolving or the fingers or hands carry harmful control tools or vibrant workpieces. The frequency, direction, and size of the vibration motion influence the human reaction to vibration [2]. Mechanical vibrations generate forces that have an impact on human bodies. Individuals typically suffer low back discomfort throughout their lives [3]. Low back pain (LBP) is linked to activity restriction, disability, and employee absenteeism. Based on the Worldwide Chronic Disease Burden 2010 Report, LBP causes more suffering than any other condition or disorder, as measured by years lived with disability (YLDs). YLDs from LBP are predicted to have grown from 58.2 million in 1990 to 83.0 million in 2010 [4-5]. LBP incidence and problems rise with age and are larger in men than in women. Our physiques are extremely complex compared to the normal processes that occur from place to place. The essential wonder that constantly impacts the organisms and the entire body is vibration. People are sensitive to low-frequency vibrations, and the condition develops after they are repeatedly exposed to vibration [6]. When the bio-dynamic study of the humanoid frame began in 1918, Hamilton observed the vibration of the finger condition as a result of vibrant hand instruments [7]. The mechanics of the human body differ from person to person and throughout time since it is a complex dynamic system. Many scientific models have been established to describe human biodynamic activities based on various types of data [8].

Other factors thought to be associated with vibration's harmful effects, in addition to its physical properties, include the duration of exposure, the structure of exposure (periods, intermittent, continuous), the type of apparatuses, methods, or automobiles that formed the environmental conditions (noise, airflow, ambient temperature, humidity), vibration, and the dynamic response of the human body. Workplace tremors can result in a variety of complaints and health issues, particularly in the lower back and upper limbs [9]. Body vibration can induce musculoskeletal problems in the spinal system, with the most commonly reported negative consequences being low back discomfort, early spine degeneration, and ruptured intervertebral discs. Through joint laboratory research, the interplay among vibration and other ergonomic, individual, and environmental aspects will be studied. The study will also analyse existing standard procedures for risk observation and assess the safety provided by chairs and gloves to enhance strategies for disorder prevention [10]. Drivers and passengers are subjected to many sorts of vibrations, the intensity and frequency of which impact the traveller's comfort and, indirectly, safety. Human vibrations can produce imbalances, muscular discomfort, nausea, vomiting, and disorientation. They frequently mirror motion sickness symptoms.

[a]apurvaweike12@gmail.com, [b]girishmehta1980@gmail.com [c]sagmech24@gmail.com

DOI: 10.1201/9781003450252-78

Human vibration is a multidisciplinary subject that includes ergonomics, engineering, physics, medicine, and many other fields [11]. It should be mentioned that those who are subjected to lengthy periods of vibration in their everyday jobs, such as mechanism workers, and particularly those who use work vibration (e.g., rollers, drills, hammers, concrete compactors), are at high risk for a variety of disorders. Continuing work experience with high levels of body vibration is linked to a higher risk of problems with the nervous system and the lumbar spine [12].

Mechanical vibrations were classified as two forms of oscillatory motion.

- General vibrations are delivered to the human body via the legs and pelvis (whole-body vibration).
- Hand-transmitted vibrations are micro-vibrations delivered to the human body directly.

Reverberations caused by road irregularities or additional infrastructure elements (e.g., speed bumps) are felt by the motorist and travellers in the automobile, as are vibrations caused by the vehicle's operating state, such as motor oscillations, tyre motion, exhaust system vibrations, and transmission system vibrations [1, 13].

The study of human vibration spans several academic fields, including ergonomics, engineering, physics, medicine, and many more. Instead of great need, there is a scarcity of available literature. Therefore, if the aim is to undertake a complex study of human vibration, it necessitates a wide scope and meticulous preparation that takes numerous factors into account. The purpose of this review paper is to provide information on the effects of vibrations, specifically the effects of vehicle vibrations on the human body. Accordingly, the research papers are reviewed, and based on the review, conclusions are made. These findings will be beneficial to researchers and scientists working on vehicle vibrations.

Vibrations on Human Body

Vibration of the Entire Body

A powered inducement characterised by an oscillatory wave is referred to as "vibration." Through a vast contact area, whole-body vibration transfers natural vibration to the human body. The frequency, amplitude, and magnitude of the oscillations are the biomechanical properties that determine their strength. The degree of oscillatory motion, the vibration's frequency, and the vibration's displacement between peaks (in mm) are what define the shaking's amplitude by the reappearance frequency of oscillatory rotations and the magnitude of the vibration by the acceleration. Low-frequency, low-amplitude, powered stimulation of the humanoid body has been proposed as a harmless and effective method of increasing well-developed assets. Muscular strength and power gains in individuals exercising using specifically built workout equipment have lately been observed [14]. The use of vibrations as a workout intervention is a relatively new concept. The whole-body vibration environment is generally defined in terms of acceleration due to the simplicity of the quantity, its direct relationship with stress or force, and the humanoid sensitivity to shaking. Whole-body vibration has been planned as a preventative and rehabilitative technique for osteoporosis, and it is now time to investigate its usefulness in humans. The amount of the vibration is determined by the extent of this motion. Oscillatory motion can take several forms, including sinusoidal, multi-sinusoidal, random stationary, and transient. Russian scientists were the first to use vibration as a workout interposition, discovering that it remained helpful in increasing strength in well-trained participants [15] and [16]. In contrast, a rise in gravitational load (hyper gravity) will increase force-generating ability and muscle cross-section area. Activities that increase gravitational weight are performed in training plans intended to enhance strength and power. These types of exercises have been found to generate the necessary responses in thin muscles that involve both structural and neurological aspects [17]. Body vibration is an automatic stimulation that causes the physique to tremble. WBVE is frequently performed during a light workout on a vibrant stand, where the sinusoidal vibration produced penetrates the foot and spreads throughout the individual standing on a platform. The WBV signal affects soft, firm, and hard tissues in varying degrees. Its frequency (measured in hertz, Hz), amplitude (measured in metres or millimetres), and direction (x, y, z) distinguish it. Vibration has been shown in animal experiments to have an anabolic impact on bone material. Suggest that vibration-induced low-magnitude, high-frequency stresses can excite bone growth in weight-bearing parts of the skeleton [18]. It is frequently acknowledged that shaking at work, mostly in sectors requiring automobile movement, is enough to cause a bodily reaction in people [19]

Human Body Vibrations During Travel

The human body's vibration is an interesting phenomenon that is described as follows: A rider on a two-wheeler is depicted in Figure 1, given below. While riding, the human body along with the vehicle body are in a dynamic motion, which means they acquire displacement, velocity, acceleration, jerk, jounce, crackle, and pop (refer to Dr. Mehta's lecture on the dynamics of vibrations). This moving weight is constantly applying a rolling load to the road surface.

According to Newton's 3rd law, "the opposite ground reaction will be experienced by tires." The induced force will then be transmitted to the humanoid frame, complete with shock absorbers, a seat, and a chassis. As the transmitted force is periodic and cyclic, the human body turns into a cyclic vibration. The human body comprises various linkages, from the head to the arm and the legs; these linkages are made up of bones with certain joints. Of course, human body vibration certainly affects these joints and bones. As a result, the current work is intended to determine the effect of road conditions on human body vibrations that result in spinal-related problems.

An additional component is rehabilitation and physical therapy, which can substantially influence total healing. Finally, earlier bioinformatics symptoms such as bladder and bowel problems, pain, stiffness, and difficulty with regular breathing play a role [20]. Paediatric spinal illnesses are diagnosed and treated differently from adult spinal problems. If spinal cord diseases are identified and treated, permanent severe neurologic impairment can be reduced. Children can suffer from a wide range of congenital, developmental, traumatic, and neoplastic illnesses. Due to the fast growth of children, issues with the spine may become apparent as a result of the development of scoliosis, discomfort, or neurological abnormalities such as weakness or numbness [21]. Spinal cord injury (SCI) can cause significant impairment. Males and females are equally at risk for traumatic spinal cord injury (T-SCI), which has a male-to-female ratio of around 4:1 during adolescence and early adulthood [22]. The mental health of this group of persons with spinal cord injuries 60% additionally had at least one other emotional illness, showing a significant 56% increase in the chance of mental combination over the general population. Better health and time since injury were linked to a lower probability of psychopathology [23]. Physical ability, health, and financial concerns were commonly mentioned as problems. Stress was linked to depressive symptomatology and nervousness but not to lifetime approval [24]. The association between challenges and physical welfare procedures was aided partly by visible anxiety. According to social support, the link between perceived stress, unhappiness, and worry varied. This finding suggests that those with weak societal support are particularly vulnerable to the harmful effects of anxiety on their physical health. Anxiety-management plans tailored to the needs of bioinformatics experts are required [25]. The incidence of depressive and anxious mood states in individuals with cervical spondylosis myelopathy, a degenerative spine disorder characterised by symptoms such as neck discomfort, numb and clumsy hands, walking problems, sphincter dysfunction, and impotence [26]. Depression and anxiety levels are closely related to impaired mobility in CSM patients [27]. Specifically, 11.4% of those who participated met the criteria for probable major depressive illness [28]. There was a link between a definite major depressive illness and worsening subjective health, decreased life satisfaction, and increased difficulty carrying out everyday tasks. Most economic or injury-related characteristics were unrelated to the likelihood of severe depression. Both physical and psychological signs suggested a significant depressive disease [29]. Preliminary research supports the relevance of psychosocial variables, including perceived social support and campus connectivity, in the development of mental health disorders. The hypothesis that felt social support and connectivity would mitigate the associations between perceived stress, anxiety, and depression was supported in part [30]. There were clinically significant symptoms in 48% of the individuals [31]. Minority individuals, especially women, were far more likely to exhibit depressive symptoms. After controlling for years of schooling and wealth, both of which were significantly associated with depressive symptoms, this risk dropped but did not disappear. Although education and income had little effect on these relationships, ageing factors were kind of positively associated with depression

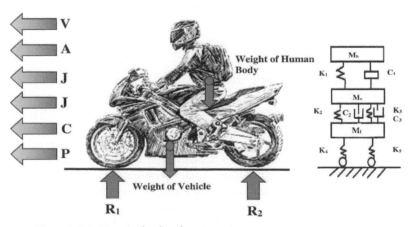

Figure 78.1 Human body vibration
Source: Author

[32]. These vibrations vary in the presence of acute or chronic diseases. Everything in the world is formed of energy that vibrates at different frequencies [33]. At the quantum level, even things that appear solid are made up of vibrating energy fields. From a theoretical and scientific point of view, "beings" are made up of several energy levels: physical, mental, emotional, and spiritual [34]. It reveals that a car structure with quadratic equation-checking hydraulic bases has high accuracy in minimising vibrations and so improves car comfort, but its nonlinear checking feature has little consequence on the basic occurrences of the car arrangement [35]. The dynamic modelling of a human body sitting on a seat with viscoelastic qualities is offered to quantitatively analyse the ride quality of passengers subjected to vertical vibrations. The metrological capabilities of several transducers were evaluated by determining their occurrence of floor noise, linearity, and responsiveness to thermal, response function, and electromagnetic instabilities [36]. With the single-frequency calibration approach, the ordinary instrumental uncertainty was less than 5%, and this number was decreased to 2% [6]. As an approximation of vibration exposure, a time function for distinct frequency bands has been constructed. Even within the investigated bands, the time function reveals the variability of the period and regularity of the vibration. The time function may be calculated using the constructed vibration representation, which reflects the isolated vibration dynamics phenomenon [37].

Measurement of Vibration

While mathematical simulations can specify the intensity and form of shaking as well as its information to some extent, shaking measuring techniques have gained popularity in recent years to fill the gaps left by academic representations [38]. The significance of shaking monitoring of the automobile engine and the humanoid body was investigated [39] based on the ISO 2631-1 (1997) vibration standard, as well as a comparison of observed vibration values taking into consideration resonance regularities and mode figures at diverse locations of the vehicle seat. During the examination of structural dynamic characterization, comparable frequency ranges are observed [40]. The humanoid frame and the automobile system are coupled at frequencies ranging from 10 to 50 hertz [41]. Furthermore, below 80 Hz, similar learning discovered 3 resonance regularities and 3 associated shaking patterns of the automobile worktable [42]. The vibration was assessed using a small measuring device with accelerometers, a procurement unit, and a regular mainframe [41]. The goal of the investigation was to conduct internal vibration dimensions at a low cost without the use of any specialised equipment. The vehicle vibration was subjected to similar tests [43].

Effects of Vibrations on Human Body

Low Back Discomfort and Internal Spinal Load Measurements

Vibration exposure may reflect just a portion of the inner forces operating on the structural components of the lumbar spinal column. Furthermore, it is unclear whether measurements of daily vibration contact are appropriate for assessing the hazard of lasting severe health impacts, such as lumbar back problems. It has been argued that the ISO 2631-1 regular vibration contact estimates are inappropriate because nervous system and lower back problems are dimensions included in driving vocations [44]. The EU Directive does not take into account the effect of other factors on the hazard of small spinal conditions; age, physical features, positions, and generational WBV contact are all hazards for the growth of severe healthiness properties in exposed workers' spines [45-46]. Table 78.1 shows the driver distribution based on the regular contact action levels specified by the EU Order on mechanical shaking. Information is presented as percentages (%) [47]

A VDV is the Shaking Dose Rate, it is the regular shaking contact rate normalized to an 8-hour position period [44] [45] [46].

The Impact on Bone Density

After 6–12 months of WBVE, 4 or 5 of the relatively optimistic trials reported a substantial growth (0.4–10.2%) in bone thickness, whereas the other three found only nonsignificant increases. One study revealed a decrease in calcium phosphate formation in urine samples despite only giving participants WBVE for three weeks. Five studies were carried out on postmenopausal women aged 55–88 years [48]. Another study looked at the impact of WBVE on young (4–19-year-old) girls and boys who have functional limitations such as reduced mobility. WBVE was used in the remaining tests on healthy men and women aged 19 to 38 years [49]. Spirometry measurements were shared among the four initial investigations. All four 1960s investigations included measurements of oxygenation, volume, and respirational frequency. Spirometry measurements increased significantly from baseline or control circumstances in three of the four investigations. The common research limitation was the shaking rate, which extended from 15 to 2 Hz in each sample. Overall features of the trial design were altered just slightly, making analysing their influence on

Table 78.1 Driver distribution based on the regular contact action levels

Daily shaking exposure measurements	Drivers			
	Forklift vehicles (N=67)	Movements of the earth (N=49)	Vehicles of utility (N=86)	Entire sample (N=202)
A (47) maximum (ms^{-2} r m s) < 0.5 > 0.5	54 (80.6) 13 (19.4)	39 (79.6) 10 (20.4)	86 (100) 0 (0)	179 (88.6) 23 (11.4)
VDV maximum (ms$^{-1.75}$) <9.1 >9.1	32 (47.8) 35 (52.2)	19 (38.8) 30 (61.2)	86 (100) 0 (0)	137 (67.8) 65 (32.2)
VDVsum (ms$^{-1.75}$) <9.1 >9.1	31 (46.3) 36 (53.7)	14 (28.6) 35 (71.4)	77 (89.5) 9 (10.5)	122 (60.4) 80 (39.6)
A (47) sum (ms^{-2} rms) < 0.5 > 0.5	48 (71.4) 19 (28.4)	20 (40.8) 29 (59.2)	86 (100) 0 (0)	154 (76.2) 48 (23.8)

Source: Author

spirometry measurements unfeasible. Mikala and Bhambhini identified the most reliable research projects and discovered a substantial rise in heart rate during vibration at frequencies ranging from 3 to 6 Hz in 14 females [50].

Managers in various physically demanding jobs are accountable for their employees' health, care, and optimum bodily functions. Recognising places where bodily presentation might be affected provides a framework for physicians and academics to aim for and ultimately manage workplace components. This study aims to summarise numerous identified bodily reactions to WBV and hypothesise if these may translate into meaningful changes in bodily presentation.

Table 78.2 summarises the findings of a group study on whole-body vibration training and its effects on human bone compactness.

Effects of Spinal Cord Injuries

Cervical: The most common nontraumatic cause of cervical spine myelopathy is cervical spondylosis. Unlike most other spinal diseases, where therapeutic therapy is generally the primary option, early surgery is critical for interfering with the natural course of cervical spondylitis myelopathy (CSM) and improving the neurological prognosis. Indeed, there is a strong indication that surgery within one year of the onset of symptoms improves the prognosis in CSM.

Between C2 and C5, some respiratory muscles and all arm and leg muscles are paralysed.
Between C5 and C6: Leg, trunk, hand, and wrist paralysis. The muscles that move the shoulder and elbow are weak.

Between C6 and C7: Leg, trunk, and the portion of the hand and wrist paralysis shoulder and elbow mobility are normal.

Between C7 and C8: Leg, trunk, and hand paralysis.

C8 to T1: Weakness of the muscles that move the fingers and hands, as well as paralysis of the legs and trunk.

Thoracic: Because the thoracic spine is so stiff and strong, it is the least likely area of your spine to be damaged. The thoracic spine turns from the base of your neck to the bottom of your ribs. The thoracic hole is a deep region within the humanoid frame. It is similarly identified as the rib cage hole. The thoracic hole is threatened through the thoracic divider. The thoracic divider is made up of the muscle, fascia, and rib cage.

T2 to T4: Leg and trunk paralysis, lack of feeling below the nipples, normal shoulder, and elbow motion.
T5 to T8: Leg and lower trunk paralysis; lack of feeling below the rib cage.
T9 to T11: Leg paralysis, lack of feeling below the navel.
Lumbar: Lumbar spondylosis covers all progressive syndromes disturbing the lumbar vertebrae's discs, vertebral frames, and connected intersections. Spondylosis is a descriptive term used to describe spinal abnormalities rather than a clinical diagnosis. Spondylosis most commonly affects people over the age of 40's lumbar spine. Morning stiffness and soreness are frequent complaints.
T11 to L1: Loss of feeling in the hips and legs due to paralysis.

Table 78.2 The whole vibration training and its properties on human bone compactness

Reference	Learning type	Purpose	Physical activity	Subject	Measurement way	Vibration	Result
[51]	RCT 2groups: 1. PPV 2. VPN	WBVE effects on bone density after three weeks	WBVE + high protein intake 15 sessions in total	Ten healthy subjects (22-29 years,6 men)	For three weeks, urine samples were collected every workday.	D: V Acc:3.5g F:30Hz W:9	Calcium and phosphate discharges in urine were significantly decreased (p = 0.006)
[17]	CT 2gruops: 1. Control 2. WBVE	WBVE's 12-month impact on the skeleton and muscular tissue	WBVE 10 min per day	50 early females (15-20 ages old) with low bone compactness and an increased risk of fracture	DXA and CT	D: V Acc:0.3 g F:30Hz W:6	Cortical and trabecular bone have improved significantly (p = 0.025 and 0.001, respectively)
[18]	RCT 1. Gait training 2. WBVE	WBVE for 8 months on balance and bone density	1. Gait exercise three periods per week 55 min each period with 5 min elasticity	28 post-menopausal ladies with no training	DXA	D: L A:3 mm F: 12.6 Hz V: 8	Significant increase in femoral neck bone density (p 5 0.011)
			2. WBVE Three periods/work week 6 16 minutes with 1 minute rest between sets.				
[52]	RCT 2 groups: 1. ALNzWBVE 2. ALAN	WBVE after 12 calendar months and the properties of alendronate on bone compactness in the minor back	1. Only ALN. Regular dose 5 mg 2.ALNzWBVE 4 min once per week.	50 postmenopausal ladies (aged 55-88) with osteoporosis k	DXA	W: 5 A: 0.7-4.2 mm D: V F: 20 Hz	Significant increase in bone density in the minor back in both ALNzHKVT also ALN groups (p, 0.0001)
[53]	RCT 2 groups: 1. Placebo 2. WBVE	WBVE after 12 months and its impact on bone density	1. Placebo 26 Ten min/day 2. WBVE 2 6 10 minutes/day	70 postmenopausal women	DXA	V: 6 D: V F: 30 Hz A: 0.2 g	WBVE prevents bone degeneration. Femur neckline 0.27% vs. 0.69% in comparison to placebo Trochanter: 0.07% as opposed to 0.19%. 0.51% vs. 0.65% for spinal vertebrae

Reference	Learning type	Purpose	Physical activity	Subject	Measurement way	Vibration	Result
[48]	RCT 2 groups: 1. Placebo 2. WBVE	WBVE for 6 months and its properties on bone compactness in the shinbone and back cortical bone	WBVE 5 days/week 10 min/day	20 young people having functional limitations (14 guys), ages 4 to 19	3-D QCT	F: 90 Hz A: 0.3 g D: V V: 4	WBVE improves proximal tibia function significantly (p 5 0.0033) as compared to placebo. Spinal vertebrae improved as well, although not much (p 5 0.14)
[54]	RCT 1. Control 2. WBVE D: V	WBVE for 8 months on bone thickness, muscular performance, and balance	1. Here was no variation in exercise habits in the controller group. 2. WBVE 3–5 /week. 4 min.	56 untrained individuals topics (21 males) (and 21 men), 19–38 years	DXA and pQCT	A: 2–8 g 2 groups: D: V F: 25–45 Hz W: 7	There is no additional bone density derived from WBVE
[55]	RCT Three sets: 1. Control 2. Strong point exercise 3. WBVE	WBVE for six months and its impact on hips bone compactness, muscular strong point, and position controller	1. WBVE and strong point physical activity 2. Controller group did not variation keep fit ways. 3. Strength training performed 72 sessions.	70 postmenopausal women ladies aged 58 to 74	DXA	A: 2.28–5.09 g W: 9 D: V F: 35–40 Hz	WBVE causes a significant increase in the bone frame in the hip (p 5 0.03). There has been no improvement in the other categories.

Source: Author

L2 to S2: Depending on the level of the damage, there are various shapes of limb weakness and numbness. Sacral spondylosis is characterised by degenerative alterations in the vertebral joints and discs. Pain, stiffness, and muscular weakness are common complaints.

S3 to S5: Numbness in the lower stomach.
Symptoms of a spinal cord disorder

- Back pain
- Amputation of limbs
- Uncontrollable muscle spasms
- loss of sensation
- Reflex alterations
- Gastrointestinal management
- Urinary disability
- Weakness

Conclusion

The goal of this article is to investigate recent issues with vehicle vibrations in the literature, as well as the problems that vibration causes for passengers. Additionally, it provides a broad overview of the potential future directions for issues that still need to be resolved in practise. As a result, the presented research frequently reaches different conclusions about the influence of several elements. Efforts to characterise reaction under realistic vibration circumstances in a way that is considerate of the harmful consequences of whole-body vibration Having the right vibrational frequency within our bodies can even help us avoid infections and stay healthy in the face of diseases and illnesses. New cars should be well equipped with essential technology. Vehicle-to-vehicle connections will become increasingly important as cars become more autonomous. Future studies should focus on automated and flying vehicles since they will rule the globe in the near future. In order to enhance the safe and pleasant riding experience, this study summarises the impacts and design parameters on vibration as well as the techniques for limiting the transmission of vibration at certain frequency ranges. More research is needed to corroborate the data provided by this small set of studies in terms of intervention design and group criteria constraints.

References

[1] Živanović, S., Pavic, A., and Reynolds, P. (2005). Vibration serviceability of footbridges under human-induced excitation: A literature review. Journal of Sound and Vibration. 279(1–2), 1–74. doi: 10.1016/j.jsv.2004.01.019.

[2] Griffin, M. J. and Erdreich, J. (1991). Handbook of Human Vibration. Acoustical Society of America. Academic Press.

[3] Directive, E. U. and Provisions, G. E. (2002). Directive 2002/44/EC of the European Parliament and the Council of 25 June 2002 on the minimum health and safety requirements regarding the exposure of workers to the risks arising from physical agents (vibration)(sixteenth individual Directive within the meaning of Article 16 (1) of Directive 89/391/EEC). The Official Journal of the European Union. 117(13), 6–7.

[4] Seidel, H., Hinz, B., Hofmann, J., and Menzel, G. (2008). Intraspinal forces and health risk caused by whole-body vibration—predictions for European drivers and different field conditions. International Journal of Industrial Ergonomics.38(9–10), 856–67.

[5] Dhutekar, P., Mehta, G., Modak, J., Shelare, S., and Belkhode, P. (2021). Establishment of mathematical model for minimization of human energy in a plastic moulding operation. Material Today Proceeding. 47, 4502–07. doi: 10.1016/j.matpr.2021.05.330.

[6] Amirouche, F. M. L. (1987). Modeling of human reactions to whole-body vibration.

[7] Dai, J. S. (2006). An historical review of the theoretical development of rigid body displacements from Rodrigues parameters to the finite twist. Mechanism and Machine Theory. 41(1), 41–52. doi: 10.1016/j.mechmachtheory.2005.04.004.

[8] Qassem, W., Othman, M. O., and Abdul-Majeed, S. (1994). The effects of vertical and horizontal vibrations on the human body. Mechanism and Machine Theory. 16(2), 151–61.

[9] Mehta, G., Deogirkar, S., Borkar, P., Shelare, and S., Sontakke, S. (2019). Estimation of Vibration Response of a Bridge Column. SSRN Electronic Journal. doi: 10.2139/ssrn.3356326.

[10] Griffin, M. J. (1990). Handbook of Human Vibration. London: Ed, Elsevier Academic Press.

[11] Bovenzi, M. (2006). Health risks from occupational exposures to mechanical vibration. La Medicina del Lavorovol. 97(3), 535–41.

[12] Burdzik, R. (2011). The influence of the rotational speed of engine on vibrations transferred on vehicle construction. Scientific Journal of Silesian University of Technology. 72, 13–22.

[13] Warczek, J. (2017). A study on exposing the driver of a commercial vehicle to mechanical vibration. Zesz. Nauk. Transp. Śląska.

[14] Issurin, V. B., Liebermann, D. G., and Tenenbaum, G. (1994). Effect of vibratory stimulation training on maximal force and flexibility. Journal of Sports Science. 12(6), 561–66.

[15] Issurin V. B. and Tenenbaum, G. (1999). Acute and residual effects of vibratory stimulation on explosive strength in elite and amateur athletes. Journal of Sports Science. 17(3), 177–82.

[16] Duchateau J. and Enoka, R. M. (2002). Neural adaptations with chronic activity patterns in able-bodied humans. American Journal of Physical Medicine & Rehabilitatio. 81(11), S17–S27.

[17] Gilsanz, V., Wren, T. A. L., Sanchez, M., Dorey, F., Judex, S., and Rubin, C. (2006). Low-level, high-frequency mechanical signals enhance musculoskeletal development of young women with low BMD. Journal of Bone and Mineral Research. 21(9), 1464–74.

[18] Gusi, N., Raimundo, A., and Leal, A. (2006). Low-frequency vibratory exercise reduces the risk of bone fracture more than walking: a randomized controlled trial. BMC Musculoskeletal Disorders. 7(1), 1–8.

[19] Schust, M., Blüthner, R., and Seidel, H. (2006). Examination of perceptions (intensity, seat comfort, effort) and reaction times (brake and accelerator) during low-frequency vibration in x-or y-direction and biaxial (xy-) vibration of driver seats with activated and deactivated suspension. Journal of Sound and Vibration. 298(3), 606–26.

[20] S. N. Waghmare, S. D. Shelare, C. K. Tembhurkar, and S. B. Jawalekar, "Pyrolysis system for environment-friendly conversion of plastic waste into fuel," in *Lecture Notes in Mechanical Engineering*, Springer, 2020, pp. 131–138. do: 10.1007/978-981-15-4748-5_13.

[21] Migliorini, C. E., New, P. W., and Tonge, B. J. (2009). Comparison of depression, anxiety and stress in persons with traumatic and non-traumatic post-acute spinal cord injury. Spinal Cord. 47(11), 783–88.

[22] Migliorini, C., Tonge, B., and Taleporos, G. (2008). Spinal cord injury and mental health. Australian and New Zealand Journal of Psychiatry. 42(4), 309–14.

[23] Rintala, D. H., Rohinson-Whelen, S., and Matamoros, R. (2005). Subjective stress in male veterans with spinal cord injury. Journal of Rehabilitation Research and Development. 42(3).

[24] Ramteke, H. P. and Mehta, G. D. (2022). Flexible coupling—a research review. Eds. R. Kumar, V. S. Chauhan, M. Talha, and H. Pathak, Singapore: Springer Singapore, pp. 887–92. doi: 10.1007/978-981-16-0550-5_81.

[25] Stoffman, M. R., Roberts, M. S., and King Jr, J. T. 92005). Cervical spondylotic myelopathy, depression, and anxiety: a cohort analysis of 89 patients. Neurosurgery. 57(2), 307–13.

[26] Belkhode, P., Modak, J. P., Vidyasagar, V., and Shelare, S. (2021). Procedure of collecting field data: causes, extraneous variables, and effects. Mathematical Modeling and Simulation, CRC Press, pp. 33–47.

[27] Bombardier, C. H., Richards, J. S., Krause, J. S., Tulsky, D., and Tate, D. G. (2004). Symptoms of major depression in people with spinal cord injury: implications for screening. Archives of Physical Medicine and Rehabilitation. 85(11), 1749–56.

[28] Modak, J. P., Mehta, G. D., and Belkhode, P. N. (2004). Computer aided dynamic analysis of the drive of a chain conveyor. ASME International Mechanical Engineering Congress and Exposition. 47136, 1037–43.

[29] Uhlemann, T. H.-J., Lehmann, C., and Steinhilper. R. (2017). The digital twin: realizing the cyber-physical production system for Industry 4.0. Procedia CIRP. 61, 335–40. doi: https://doi.org/10.1016/j.procir.2016.11.152.

[30] Krause, J. S., Kemp, B., and Coker, J. (2000). Depression after spinal cord injury: relation to gender, ethnicity, aging, and socioeconomic indicators. Archives of Physical Medicine and Rehabilitation. 81(8), 1099–1109.

[31] Aglawe, K. R., Dhande, M., Matey, M., and Shelare, S. (2022). State of the art and materials based characteristics in power converters for electric vehicles. Materials Today: Proceedings. 58, 726–35. doi: 10.1016/j.matpr.2022.02.384.

[32] Pouraminian, M., Amarlou, M., Pourbakhshian, S., and Khodayari, R. (2015). Evaluation of the vibrational properties of three-span continuous concrete bridge by dynamic finite element method. 5(1), 27–30.

[33] Ramteke, A. L., Waghmare, S. N., Shelare, S. D., and Sirsat, P. M. (2022). Development of sheet metal die by using CAD and simulation technology to improvement of quality BT. Proceedings of the International Conference on Industrial and Manufacturing Systems (CIMS-2020). 687–701.

[34] Tarabini, M., Saggin, B., Scaccabarozzi, D., and Moschioni, G. (2012). The potential of micro-electro-mechanical accelerometers in human vibration measurements. Journal of Sound and Vibrations. 331(2), 487–99.

[35] Burdzik, R. (2017). Novel method for research on exposure to nonlinear vibration transferred by suspension of vehicle. International Journal of Non-Linear Mechanics. 91, 170–80.

[36] Gajbhiye, T., Shelare, S., and Aglawe, K. (2022). Current and future challenges of nanomaterials in solar energy desalination systems in last decade. Transdisciplinary Journal of Engineering & Science. 13, 187–201. doi: 10.22545/2022/00217.

[37] Lakušić, S., Brčić, D., and Tkalčević Lakušić, V. (2011). Analysis of vehicle vibrations–new approach to rating pavement condition of urban roads. Promet-Traffic & Transportation. 23(6). 485–94.

[38] Belkhode, P. N., Ganvir, V. N., Shelare, S. D., Shende, A., and Maheshwary, P. (2022). Experimental investigation on treated transformer oil (TTO) and its diesel blends in the diesel engine. Energy Harvest System. 9(1), 75–81. doi: 10.1515/ehs-2021-0032.

[39] Hoy, D., Brooks, P., Blyth, F., and Buchbinder, R. (2010). The epidemiology of low back pain. Best Practice & Research: Clinical Rheumatology. 24(6), 769–81.

[40] Hoy, D. et al. (2014). The global burden of low back pain: estimates from the Global Burden of Disease 2010 study. Annals of the Rheumatic Diseases. 73(6), 968–74.

[41] Mehta, G. D. and Modak, J. P. (2011). An approach to estimate vibration response at all bearings of countershaft due to all machine components on it. 13th World Congress in Mechanism and Machine Science, Mexica. pp. 19–25.

[42] Murray, C. J. L. et al. (2012). Disability-adjusted life years (DALYs) for 291 diseases and injuries in 21 regions, 1990–2010: a systematic analysis for the Global Burden of Disease Study 2010. Lancet. 380(9859), 2197–2223.

[43] Pankoke, S., Buck, B., and Woelfel, H. P. (1998). Dynamic FE model of sitting man adjustable to body height, body mass and posture used for calculating internal forces in the lumbar vertebral disks.Journal of Soundd and Vibrations. 215(4), 827–39.

[44] Bovenzi, M. (2010). A longitudinal study of low back pain and daily vibration exposure in professional drivers. Industrial Health. 48(5), 584–95.

[45] Bovenzi, M. and Hulshof, C. T. J. (1999). An updated review of epidemiologic studies on the relationship between exposure to whole-body vibration and low back pain (1986–1997). International Archives of Occupational and Environmental Health. 72(6), 351–65.

[46] Kraus, J. F., Franti, C. E., Riggins, R. S., Richards, D., and Borhani, N. O. (1975). .Incidence of traumatic spinal cord lesions. Journal of Chronic Diseases. 28(9), 471–92.

[47] Botsis, T and Hartvigsen, G. (2008). Current status and future perspectives in telecare for elderly people suffering from chronic diseases. Journal of Telemedecine Telecare. 14(4), 195–203.

[48] Ward, K., Alsop, C., Caulton, J. Rubin, C. Adams, J., and Mughal, Z. (2004). Low magnitude mechanical loading is osteogenic in children with disabling conditions. Journal of Bone and Mineral Research. 19(3), 360–69.

[49] Lo, L., Fard, M., Subic, A., and Jazar, R. (2013). Structural dynamic characterization of a vehicle seat coupled with human occupant. Journal of Sound Vibration. 332(4), 1141–52.

[50] Stein, G., Chmúrny, R., and Rosík, V. (2011). Compact vibration measuring system for in-vehicle applications. Measurement Science Review. 11(5), 154.

[51] Cardinale, M. Leiper, J. Farajian, P., and Heer, M. (2007). Whole-body vibration can reduce calciuria induced by high protein intakes and may counteract bone resorption: A preliminary study. Journal of Sports Science. 25(1), 111–19.

[52] Iwamoto, J. Takeda, T. Sato, Y., and Uzawa, M. (2005). Retracted article: effect of whole-body vibration exercise on lumbar bone mineral density, bone turnover, and chronic back pain in post-menopausal osteoporotic women treated with alendronate. Aging Clinical and Experimental Research. 17(2), 157–63.

[53] Rubin, C., Recker, R., Cullen, D., Ryaby, J., McCabe, J., and McLeod, K. (2004). Prevention of postmenopausal bone loss by a low-magnitude, high-frequency mechanical stimuli: a clinical trial assessing compliance, efficacy, and safety. Journal of Bone and Mineral Research. 19(3), 343–51.

[54] Torvinen, S. et al. (2003). Effect of 8-month vertical whole body vibration on bone, muscle performance, and body balance: a randomized controlled study. Journal of Bone and Mineral Research. 18(5), 876–84.

[55] Verschueren, S. M. P., Roelants, M., Delecluse, C., Swinnen, S., Vanderschueren, D., and Boonen, S. (2004). Effect of 6-month whole body vibration training on hip density, muscle strength, and postural control in postmenopausal women: a randomized controlled pilot study. Journal of Bone and Mineral Research. 19(3), 352–59.

Chapter 79

Experimental analysis of solar powered proton exchange membrane electrolyser and fuel cell system

Rajesh G. Bodkhe[1,a], Rakesh L. Shrivastava[1,b], Rajkumar B. Chadge[1,c], Prashant D. Kamble[1,d] and Vinodkumar Soni[2,e]

[1]Department of Mechanical Engineering, Yeshwantrao Chavan College of Engineering, Nagpur, India

[2]Department of Indian Railways, Nagpur, India

Abstract

This paper mainly focuses on green hydrogen generation using the water electrolysis process. The PEM electrolyser of MEA size 25 cm² was developed and tested using solar power in the Sainergy Research & Development Laboratory, Fuel Cell India Private Limited, Chennai. The off-grid solar P-V system was developed at the site to operate the PEM electrolyser testing setup for the generation of green hydrogen and to run the PEM Fuel cell system to generate the DC electricity. Hourly solar irradiance from solar P-V panel was recorded for the average global solar radiation on a horizontal surface at the Chennai location. The P-V and I-V curves of solar panels were found using experimental analysis. The effects of temperature on the performance of solar panels were studied. This paper also provided a comprehensive review of the environmental impact of climate change, air quality in the atmosphere, and the electrolysis method for hydrogen fuel production and conversion into electricity. They went over the individual approach in great detail. The work of international researchers has been cited and praised. Still, some research gaps must be investigated, which is the primary goal of the current research work.

Keywords: Electricity, electrolyser, hydrogen fuel cell, proton exchange membrane, renewable energy, solar PV panel

Introduction

In our country, stationary systems, transportation systems, industry operations, and household operations all rely on a combustion power generation system based on fossil fuels. As a result, air quality issues arise in everyday life as shown in Figure 79.1. It has an impact on the human respiratory system, plants, animals, environmental conditions, pollution, and climate change, among other things. We are experiencing all three seasons in a single day or week. Due to high investment, power transmission grid lines are still not reachable in remote and hilly areas. This is an investment that no government can afford. Solar PV panels have been used to directly generate electricity for over two decades. However, there are some limitations to solar panels, such as the fact that their efficiency is only about 15–18%. Whether predictability issues arise when using a solar panel for direct power generation, particularly in remote or hilly areas.

Many researchers are focusing on increasing efficiency. Few researchers investigated alternative methods of harnessing solar energy for the generation of hydrogen or any other type of renewable fuel. This is the superior option. This renewable fuel can also be used to power the fuel cell. The global demand for energy in daily life is rapidly increasing. At the moment, coal, electricity, and gas power are used to charge mobile phones, laptops, transportation, automobiles, electric cars, and machines, and to generate heat. The majority of the time, there is an imbalance between available energy and demand.

Currently, the majority of power is generated using fossil fuels, but these fuels are depleted as a result of their use and are limited in nature. Fossil fuels may not be available anytime soon. Because of combustion product gases such as CO_2, N_2, and sulphur, which contribute to global warming, fossil fuels harm the environment. Hydrogen is regarded as an alternative clean energy source capable of overcoming the depletion of fossil fuels [1]. These fossil fuels are a major source of the GHG effect, which causes global warming. Everyday pollution is rapidly increasing. This is detrimental to everyone. Today's major issue that every human being faces is the inability to obtain clean air in the atmosphere. For the development of sustainable energy, fuel cells are compatible with hydrogen derived from renewable energy sources. If pure H_2 is used as a fuel, the output is pure water and heat. It generates clean energy. The fuel cell is a technology that is increasingly being used for both primary and backup power in places like telecommunication towers, hospitals, airports, banks, and power plants. Fuel cell technology is cutting-edge technology.

Literature Review

This paper provided a thorough comparison of fossil fuel-based combustion fuel generation methods and renewable source fuel generation methods such as solar energy. Globally, now a day's hydrogen and fuel

[a]rgbodkhe@gmail.com, [b]rlshrivastava@gmail.com, [c]rbchadge@rediffmail.com, [d]drpdkamble@gmail.com, [e]veekaysoni@gmail.com

DOI: 10.1201/9781003450252-79

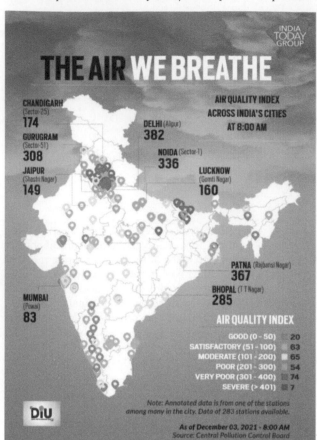

Figure 79.1 Overview of air pollution in India
Source: Central pollution central board

cells are used for electricity generation, transportation, heat, industry, and low-carbon energy storage systems are investigated in this study. This review demonstrates that in today's competitive world, cost, performance, and significant improvements are still required for hydrogen to become a clean energy carrier. However, such competitiveness is required for hydrogen's long-term prospects in the global energy system, which fully justifies global interest in and policy support for these technologies [2].

Hydrogen can be produced using both renewable and non-renewable resources. Researchers all over the world are working hard to produce hydrogen. Fuel cell technology, on the other hand, provides a clean and efficient mechanism for energy conversion with no pollution. Only a few researchers conducted laboratory-scale experimental research on a single fuel cell. Further research and development in the fuel cell stack are needed to reduce costs, increase durability, and optimise and improve performance. Researchers should carefully examine current fuel cell industry trends to overcome the obstacles and challenges to commercialising hydrogen fuel cell technology for large-scale power generation [3]. The author discusses recent advances in PEMFC, advanced materials for fuel cell membranes, physical properties, structure, and performance of different polymer electrolyte membranes, prospects of the metal-organic framework for fuel cells, and key factors for commercial applications of fuel cell technologies. Although more research is needed to scale up the technologies, the use of new materials in fuel cells remains promising [4]. As a renewable energy source, a fuel cell is a promising candidate. In general, fuel cells run on pure hydrogen and air as shown in Figure 79.1. DC electricity requires a constant supply of fuel and oxidants. Hydrogen can be burned in the same way as fossil fuels, but the only by-products are water vapor and heat, making it a zero-emission fuel. This cell is usable during non-sunny hours. It does not rely on crude oil in the same way that an internal combustion engine does. These are modular in design, and their efficiency is proportional to their size. Thermodynamic efficiency is approximately 80%. These are quiet, have no moving parts, have a longer life, do not need to be replaced like batteries, are efficient, and are preferred because they never run out of power or require charging like batteries. It does not affect climate change [5].

In India, conventional thermal power plants currently generate power on a large-scale using fossil fuels such as coal. Heat engines use petrol and diesel to produce power and transport, but they increase greenhouse gas emissions daily. Very few electric cars are on the road because the investment is high, there are few charging stations available, and the car's limitation is that it can only travel within the city after charging. Batteries store electricity but are sensitive to extreme heat and cold, especially in tribal areas, remote areas, or hilly areas where electricity is not readily available. The battery can be used in these areas, but it discharges in 3 to 4 hours. It may become overheated as a result of its continuous use, and as a result, it must be replaced every few years due to declining performance. The battery's limitations are a major source of concern [6]. Hydrogen has been identified as a potential future renewable energy source that will be used on a large scale to generate electricity and transport around the world. Hydrogen is the most environmentally friendly energy carrier [7].

Potential WTER routes that do not have a negative societal impact on human life on this planet for their daily energy needs through the use of solar energy, wind energy, wave energy, and tidal energy for sustainable energy development. It has the potential to reduce global conflict over energy reserves. Through WTER, it will provide opportunities for the development of new technologies [8]. This study presents the experimental results of solar photovoltaic cells coupled with an electrolyser to generate hydrogen gas for residential applications. When compared to traditional energy systems, this system would be more cost-effective. If a photovoltaic system is operated in parallel then the system can fulfil the dynamic load of houses and also supply non-coincidental power demand. If standalone systems are used then there may be the chance to incorporate energy storage devices. This device must provide power at the desired rate and time while also storing excess energy. Very common lead acid batteries are used to store energy in PV systems [9]. Regenerative fuel cells, which are similar to batteries, are one of the promising new energy storage devices as shown in Figure 79.2. It is possible to put such a system design into action by combining regenerative fuel cells and batteries to create a hybrid power source, in recent literature many researchers reviewed various techniques of steam reforming and gasification. Recently, plasma technology has emerged as an important method for producing hydrogen from hydrocarbons or alcohols. Water electrolysis is the procedure used to produce hydrogen to obtain eco-friendly technology, but which renewable energy source was specifically mentioned is not covered here.

More research and development in the field of hydrogen production, according to the researcher, are required to develop the best hydrogen production technologies [10]. The off-grid solar P-V system was developed to operate the PEM electrolyser testing setup for the generation of green hydrogen and to run the PEM Fuel cell system to generate the DC power for stationary applications such as charging mobile phones, batteries, laptops, etc. Solar irradiance was measured using a digital solar power meter at the location as shown in Figure 79.3. The P-V and I-V curves of solar panels were found using MATLAB simulation. The effects of temperature on the performance of solar panels were studied as shown in Figure 79.4.

Description of Experimental Setup and its specifications

A single cell of 25 cm² was designed and developed at the research and development laboratory, Sainergy Fuel Cell India Private Limited, Chennai. The hydrogen gas production capacity was about 50 cc/min. The

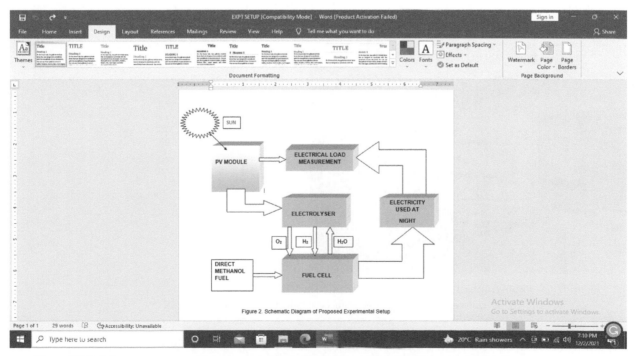

Figure 79.2 Solar power generation through PEM electrolyser and fuel cell system [10]
Source: Hydrogen production technologies overview. Journal of Power and Energy Engineering. Mentioned in references [10]

Figure 79.3 Photograph showing measurement of solar irradiance using solar power meter
Source: Research & development laboratory, sainergy fuel cell India Private Limited, Chennai

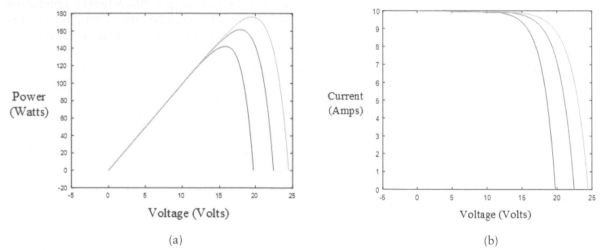

(a)   (b)

Figure 79.4 P-V (a) and I-V curve (b) showing the effect of temperature on the performance of PV panel
Source: P-V (a) and I-V curve (b) showing the effect of temperature on the performance of PV panel based on experimental
Analysis

Figure 79.5 Photograph of water electrolyser setup using solar power
Source: Research & development laboratory, sainergy fuel cell India Private Limited, Chennai

experimental setup as shown in Figure 79.5 was operated through solar power output. The experimental
setup consists of a PEM electrolyser, DC programmable power supply, Temperature controller box, and
high-pressure peristaltic pump. The experimental setup as shown in Figure 79.5 was interfaced with the

Table 79.1 Number of observations taken after testing a water electrolyser based on experimentation

Sr. No.	Voltage (V)	Current (A)	Power (W)	Temperature (°C)	H_2 Flow rate (cc/min)
1.	2.1	7.28	15.29	59.53	50.96
2.	2.1	7.27	15.27	59.6	50.89
3.	2.1	7.29	15.31	59.6	51.03
3.	2.1	7.3	15.33	59.74	51.1
4.	2.1	7.32	15.37	59.84	51.24
5.	2.1	7.35	15.44	60.42	51.45
6.	2.1	7.37	15.5	60.18	51.59

Source: Author

data acquisition software which showed the operating parameters and hydrogen production rate on screen. Specification of the water electrolyser is given in Table 79.1.

Conclusion

The system continuously runs for various water flow rates to get a constant hydrogen production rate. The operating voltage was in the range of 2 V to 2.5 V. The operating current was in the range between 7 amps to 9 amps. The speed of the peristaltic pump varied from 8 rpm to 25 rpm. The cell temperature was kept constant at about 60°C. The deionised pressurised water passed through the anode side and water splits into hydrogen and oxygen after an electrochemical reaction takes place at PEM between the anode and cathode electrode. The purity of generated hydrogen gas was about 99.9%. The generated hydrogen gas can be used to run the Fuel cell system. Table 79.2 shows the number of readings noted during experimentation. Given the number of readings, it is proposed that by keeping the voltage constant, current, and power input increase. Therefore the hydrogen production rate increases. Figure 79.6 shows the effect of current on hydrogen flow rate.

The hydrogen gas generation capacity was increased from 50 cc/min to 500 cc/min after increasing the size of the cell from 25 cm² to 50 cm². The development of catalyst material for the polymer electrolyte membrane is the remaining challenge for the water electrolyser. The main concern was about compressing the hydrogen gas at a pressure of say about 150 to 200 bar and its storage in the cylinder.

In today's scenario, hydrogen's prospects are very challenging. It is critical to reduce the cost of renewable energy production and to develop more efficient coal-fired power generation with carbon capture technology on a global scale.

Electrolysis can help to reduce the intermittent nature of renewable energy production. Integrating electrolysers with a renewable energy system will open up new avenues for power generation in the future.

Acknowledgment

The authors are sincerely thankful to Dr. Srinivas and Mr. Baskar Directors of Sainergy Fuel Cell India Private Limited, Chennai for their guidance, motivation, and kind support to carry out this current research work in their research and development laboratory. The authors gratefully acknowledge the researchers for their efforts taken in the field of hydrogen fuel cell technology. However, a lot of scopes are still available for research and development shortly

References

[1] Hübert, T., Boon-Brett, L., Black, G., and Banach, U. (2011). Hydrogen sensors—A review. Sensors and Actuators B: Chemical. 157, 329–352. https://doi.org/10.1016/j.snb.2011.04.070J.

[2] Iain Staffell, Daniel Scamman, Anthony Velazquez Abad, Paul Balcombe, Paul E. Dodds, Paul Ekins, Nilay Shah and Kate R. Ward. (2019). The role of hydrogen and fuel cells in the global energy system. Energy and Environmental Science. 12, 463–91.

[3] Sharaf, O. Z., and Orhan, M. F. (2014). An overview of fuel cell technology: Fundamentals and applications. Renewable and Sustainable Energy Reviews. 32, 810–853.

[4] Abhishek Dhand, Srinidhi Suresh, Aman Jain, O. Nilesh Varadan, M.A.K. Kerawalla and Prerna Goswami. (2017). Advances in materials for fuel cell technologies-a review. International Journal for Research in Applied Science and Engineering Technology (IJRASET). 9(5), 1672–82.

[5] Carmo, M., and Fritz, D. (2013). A comprehensive review on PEM water electrolysis. International Journal of Hydrogen Energy. 38(12), 4901–34.

[6] Schmidt-Rohr, K. (2018). How batteries store and Release energy: Explaining basic electrochemistry. Journal of Chemical Education. 95, 1801–10.

[7] Dincer, I., and Acar, C. (2018). Smart energy solutions with hydrogen options. International Journal of Hydrogen Energy. 43, 8579–99. https://doi.org/10.1016/j.ijhydene.2018.03.120

[8] Kothari, R. et. al. (2010). Waste to energy: A way from renewable energy sources to sustainable development. Renewable and Sustainable Energy Reviews. 10, 3164–70.

[9] Maclay. J. D., Brouwer, J. et al. (2011). Experimental results for hybrid energy storage systems coupled to photovoltaic generation in residential applications. International Journal of Hydrogen Energy. 36(19), 12130–40.

[10] El-Shafie, M., Kambara, S., and Hayakawa, Y. (2019). Hydrogen production technologies overview. Journal of Power and Energy Engineering. 7, 107–54. http://www.scirp.org/journal/jpee.

Chapter 80

Optimisation of abrasive and corrosion behavior of AA2218 reinforced with Si_3N_4 / Al_2O_3 / ZnO_2 hybrid composites

S. Nanthakumar[1,a], V. M. Madhavan[2,b], Pon. Maheskumar[3,c], R. Jayaraman[4,d], P. Muruganandhan[5,e] and R. Girimurugan[3,f]

[1]PSG Institute of Technology and Applied Research, Coimbatore, India

[2]Sona College of Technology, Salem, India

[3]Nandha College of Technology, Perundurai, India

[4]Vinayaka Mission's Kirupananda Variyar Engineering College, Salem, India

[5]Mahendra Institute of Technology, Namakkal, India

Abstract

Silicon nitride (Si_3N_4), aluminium oxide (Al_2O_3), and zinc oxide (ZnO_2) reinforcement particles were combined with AA 2218 alloy matrix material. Stir casting was used to create AA 2218 composites in the following 0, 3, 6, 9 and 12 weight percent. The AA 2218 composites were tested for a number of mechanical properties, including micro-hardness, TS and compressive strength. Reinforcement percentages were raised to improve mechanical qualities. AA 2218 composites were submitted to a number of wear tests, including abrasive and erosion wearing tests, to measure the wear resistance of composites. Wear testing was used to account for the mass loss. An ANOVA was utilised to study the influence of different weight percent of aluminum alloy 2218 composites on abrasive and erosion wear test outcomes, as well as to determine which input and output process parameters had the greatest impact on wear resistance.

Keywords: Abrasive wear, aluminium oxide, erosion wear tests, mechanical qualities, micro - hardness, silicon nitride, zinc oxide

Introduction

Aluminium Matrix Composites had superior mechanical characteristics as well as highest elastic modulus, TS and highest heat stability over monoliths alloy [1, 2]. AMCs had also been shown in some studies to have excellent wear resistance properties under a variety of circumstances [3, 4]. Sliding and abrasive characteristics of materials, such as cylinder heads, pistons and others, had been observed by several researchers [5]. It was the presence of reinforcing particles such as C or N_3 that caused gas cavities to react with alloying materials. Non-ferrous metals (aluminum, magnesium, copper, titanium or nickel) were frequently utilised as the matrix material, and the dispersion of the particles ensured that the composite was chemically and thermodynamically stable [6–9]. When it comes to making composites, casting was widely seen as the most promising process. In situ interactions between the matrix and reinforcing particles had been successful [10]. Al_2O_3 was suitable dispersive reinforcing particle for AA because of its highest strength, thermal stability, and thermal expansion coefficient [11, 12]. Si_3N_4 and ZnO_2 were both appropriate ceramic particles for the manufacture of Al matrix composites because of their amazing strength, high thermal stability, and ability to withstand corrosion and erosion [13–15]. It is possible to use an optical microscope to study the microstructure of the AA 2218 in situ composites with different weight % of reinforcement in terms of mechanical parameters as well as hardness, UTS and ultimate compressive strength [16, 17]. Taguchi and ANOVA were used to improve the outcomes by evaluating its abrasiveness and eroding behavior in various conditions were mixed in a variety of proportions (0, 3, 6, 9 and 12 wt.%) [18]. It was decided how much reinforcement (Si3N4, Al_2O_3, and ZnO_2) was required based on the weight of the matrix material itself (AA 2218) [19–21]. Reinforced particles were introduced to the matrix at the following percentages: 3, 6, 9 and 12 weight percentage through stir casting; specifics are shown in Tables 80.1 and 80.2. A mold is needed for the abrasive and erosion test.

Materials

Mechanical properties

AA 2218 composites with varying weight percentage were subjected to mechanical testing after their production. ASTM E10-07, hardness ranges were measured using a Micro Vickers Hardness machine for each

[a]kumarmechpsg@gmail.com, [b]madhavanvm@sonatech.ac.in, [c]ponmaheskumarmech@gmail.com,
[d]jayaramansanjay12@gmail.com, [e]mnandhan55@gmail.com, [f]dr.r.girimurugan@gmail.com

DOI: 10.1201/9781003450252-80

Table 80.1 Chemical arrangement of AA2218 in wt.%.

Fundamentals	Copper	Magnesium	Iron	Nickel	Silicon	Zinc	Aluminium
wt.%	4.5	1.8	1.0	2.3	0.9	0.25	Bal

Source: Author

Table 80.2 Matrix and reinforcements weight percentage

S. No	AA2218(g)	Reinforcements			Weight percentage (wt. %)
		Silicon nitride (Si3N4) in g	Aluminium oxide (Al2O3) in g	Zinc Oxide (ZnO2) (K2ZrF6 + KBF4) in g	
1	1	0	0	0	0
2	1	25	25	25	3
3	1	50	50	50	6
4	1	75	75	75	9
5	1	100	100	100	12

Source: Author

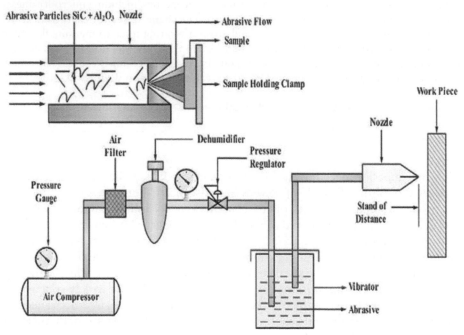

Figure 80.1 Abrasive flow in abrasive jet machine
Source: Author

weight percent of sample. There were 25 second intervals of 0.5 kg indent load with a typical indent load. Each sample's hardness was measured in a variety of places. Depending on the quantity of reinforcement, a total of five samples were employed. The UTS of the AA 2218 composite samples was determined in accordance with ASTM E08-8 by conducting an experiment on universal testing equipment. To minimum scrapes and imperfections, the specimens were cleaned to a grit of 1200 using silicon nitride emery paper prior to the testing. At 2.5 m/min, the cross sectional moved at a speediness of 10 kN while being loaded. The ultimate compression strength of AA 2218 composites with varied reinforcing quantities was determined by ASTM E09-9 standard [22, 23]. Using a universal testing machine, authors conducted the tests. The ultimate compression strength was determined using five specimens.

Abrasive wear test

Researchers in this study examined the weight percent loss of AA 2218 composites using an abrasive jet machine (AJM). As part of this experiment, high-speed stream was used to squirt abrasive particles through the nozzle [24, 25]. Another method of identifying wear resistance had been attempted by using another technique of an AJM to identify the wear resistance of the same AA 2218 samples that were previously used in a POD device with room and higher temperature circumstances with different input circumstances [26, 27]. Figure 80.1 depicts the experiment's setup. Because the settings were precisely controlled, this machining technique differed from other traditional sand blasting procedures. When cutting or machining brittle or hard materials, the AJM was a common tool of choice. These 20–50 mm diameter abrasive particles contain both Si3N4 and Al2O3 [28–30]. The mass loss of AA 2218 composite specimens was calculated via the common parameters. Table 80.3 lists the input parameters.

Erosion wear test

Because of the small particles present in semi-solid circumstances, slurry erosion wear resulted in a gradual loss of material. In a slurry erosion test, AA 2218 composites with varying weight percent of reinforcement (0, 3, 6, 9 and 12) were used. When small hard particles struck the specimens, they damaged the specimens' surfaces in this test and loss of weight was found in the specimens at varying input values [31, 32]. Slurry was created for the erosion test using a variety of mixtures, depending on the input requirements. The slurry and specimen were placed in a pot and circulated using a spinning spindle [33]. Mass loss was determined for each input condition while the specimen was submerged in slurry and circulated at different rpms on a spindle. Slurry erosion testers are depicted in Figure 80.2 and input parameters for slurry erosion tests are listed in Table 80.4.

Optimisation techniques

Three different methods of optimisation were employed in this study in order to get the best possible outcomes. The Taguchi technique and the ANOVA were two of these methods. MINITAB software is responsible for the invention of the Taguchi and ANOVA methodologies [34]. There were many different kinds of process variables used in the testing and the Taguchi technique was used to find the ones that had the most of an impact [35, 36]. A total of five elements were employed in the Taguchi design, which resulted in a total of 25 runs in the testing process. Testing resulted in mass loss as the primary end result. ANOVA approaches were utilised to find the percentage of each process variable's contribution to Abrasive and

Table 80.3 Abrasive wear testing input factors

S. No.	Input factors	values
1	A-composites (wt. %)	0,3,6,9,12
2	B-abrasive grain size (μm)	30,60,90,120,150 (20 – 140 μm)
3	C-abrasive flow rate(g/min)	10,20,30,40,50
4	D-velocity (m/min)	175,225,275,325,375
5	E-time (s)	40,80,120,160,200

Source: Author

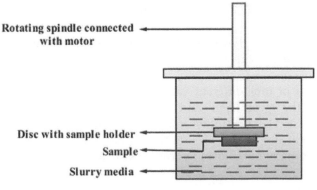

Figure 80.2 Slurry erosion tester
Source: Author

Table 80.4 Erosion wear testing input factors

S. No.	Input factors	Values
1	F-composites (wt. %)	0,3,6,9,12
2	G-adding of sand in slurry (wt. %)	30,60,90,120 and 150
3	H-speed (rpm)	350,700,1050,1400 and 1750
4	I-time (s)	40,80,120,160,200

Source: Author

Erosion tests. This optimisation method was used to analyse a number of initial solutions. In this study, AA 2218 composites were subjected to abrasion and erosion testing. The different weight percentages of reinforcements were used to identify test results like mass loss. For all outcomes, different input conditions and samples were used in the testing process [37, 38]. Traditional approaches were utilised to determine the best input influenced variable, the most contributed variable, the finest and average values of both inlet and outlet outcomes, as well as the finest specimens among the different wt. percent of reinforcements used in the experiment.

Results

Study of mechanical characteristics

Mechanical tests were conducted on AA 2218 with varying percentages of reinforcement (0, 3, 6, 9 and 12). Mechanical testing findings for aluminum alloy 2218 compounds are presented in Table 80.5. Reinforcement particles were incorporated into the matrix material using an on-site technique in this mixture Reinforcements had the greatest impact on the composites' hardness and tensile strength. Composites had an interface because of the strong interfacial connection between the matrix and reinforcements. Dislocations were impeded by the presence of reinforcing particles like Si3N4, Al_2O_3, and ZnO_2. Cracks were prevented in the composite construction because of the strong bonding, and the strength of the structure was increased in diverse conditions due to the plastic deformation. The AA 2218 composites had the necessary ductility to boost strength when plasticity was restricted. The AA 2218 composites' flexural strength was evidently enhanced by the reinforcement particles' ability to prevent crack deformation in the composite structure. The volume percentage of reinforcing particles was banned at the highest wt. percent of composites due to the reduction in matrix fluidity. Grain-to-grain bonding ensures that loads were transferred equally to reinforcements through the matrix [39–42]. Increased reinforcing particles boosted mechanical characteristics.

Investigation of optimising outcomes

- Abrasive wear on Taguchi Analysis
  The Taguchi technique was utilised in this work to determine the wear of numerous input parameters, including abrasion and erosion wear. The L25 orthogonal array with AA2218 compound specimens in varied proportions of compound samples was utilized to carry out this wear test (0, 3, 6, 9 and 12). DOE from MINITAB software was used to test AA 2218 composite samples for abrasive wear. The abrasive wear test used five parameters, each with five runs; a total of 25 values were obtained and mass loss (g) was thought about output variable. This method was used to determine which process parameter had the greatest impact. Mass loss was shown to be reducing as reinforcement was increased. Because of their strong bonding strength and low plastic deformation, aluminum alloy 2218 composites with high weight percentages withstand wear when exposed to abrasive particles in a variety of operating conditions. When using AA 2218 composite, the minimal mass loss occurred in 12 % of the material when working at 90 μm grain size, abrasive flow rate 20 g/min, 175 m/min velocity and 200 s times. Table 80.6 shows the L25 orthogonal array with abrasive wear.
- Taguchi method – erosion wear
  It was determined that mass loss (g) was a result of four different variables: the composite (weight percent), the addition of sand (weight percent), the speed (rpm) and time (s). An orthogonal array known as L25 was also employed in the wear test. There were five runs for each weight percent of composite. During erosive wear, there was no influence of particles on the surface of the composite. High-reinforced composite (12%) had the lowest material removal in erosive wear tests because of its

Table 80.5 The outcome of mechanical characteristics results

Reinforcement	Micro vickers hardness	Compressive strength	Tensile strength
(wt%)	(HV)	(MPa)	(MPa)
0	122	372	442
3	140	385	469
6	154	399	481
9	165	403	491
12	180	415	495

Source: Author

Table 80.6 L25 Orthogonal array with abrasive wear

S.NO	Composite	Abrasive grain size	Abrasive flow rate	Velocity	Time	Mass loss
Units	(Wt. %)	(µm)	(g/min)	(m/min)	(s)	(g)
1	0	30	10	175	40	0.02050
2	0	60	20	225	80	0.02630
3	0	90	30	275	120	0.03310
4	0	120	40	325	160	0.04530
5	0	150	50	375	200	0.04472
6	3	30	10	275	160	0.01781
7	3	60	20	325	200	0.01900
8	3	90	30	375	40	0.01760
9	3	120	40	175	80	0.01610
10	3	150	50	225	120	0.01720
11	6	30	10	375	80	0.01310
12	6	60	20	175	120	0.01030
13	6	90	30	225	160	0.01140
14	6	120	40	275	200	0.01182
15	6	150	50	325	40	0.01040
16	9	30	10	225	200	0.00960
17	9	60	20	275	40	0.00650
18	9	90	30	325	80	0.00392
19	9	120	40	375	120	0.00453
20	9	150	50	175	160	0.00380
21	12	30	10	325	200	0.00342
22	12	60	20	375	160	0.00372
23	12	90	30	175	200	0.00320
24	12	120	40	225	40	0.00281
25	12	150	50	275	80	0.00350

Source: Author

increased strength. During erosion testing, 12% of the AA 2218 composite had reduced material clearance of 0.83502 g, and the controlled input variables were 60% sand, 350 rpm speed, and 200 second's time. The outcomes of the wear tests are shown in Table 80.7.

- Signal to noise ratio – abrasive wear
  The SNR and data mean were measured using the MINITAB program. Tables 80.8 and 80.9 show the mean and SNR, respectively. An abrasive wear test was carried out using AA 2218 composites under varied input conditions, with the statistical software providing the results. The mean values for each of the input parameters were calculated. There is a wide range of variation in the DOE's mean values. Various process parameters were evaluated using an abrasive wear test, which yielded data such as means and S/N ratios. Main effect charts for data mean are shown in Figure 80.3.
- Erosion wear – S/N ratio
  AA 2218 composites had the lowest material removal in 9% composites when tested for erosion wear. Input parameters such as sand particle additions were altered in weight percent and speed and time at various rates. Micro-cutting and ploughing were taking place at various shallow angles on composite surfaces as minute particles made contact. Mean response tables are depicted as main effect plots for data means in Figure 80.4.
- ANOVA – abrasive wear

Table 80.7 L25 Orthogonal arrays with erosion wear

S. No	Composite	Adding of sand	Speed	Time	Mass loss	S. No	Composite	Adding of sand	Speed	Time
Units	(Wt. %)	(Wt. %)	(rpm)	(s)	(g)	Units	(Wt. %)	(Wt.%)	(rpm)	(s)
1	0	30	350	40	1.30764	7	3	60	1050	200
2	0	60	700	80	1.31895	8	3	90	1400	40
3	0	90	1050	120	1.35808	9	3	120	1750	80
4	0	120	1400	160	1.37833	10	3	150	350	120
5	0	150	1750	200	1.40756	11	6	30	700	80
6	3	30	700	160	1.20482	12	6	60	1400	120

S. No	Composite	Adding of sand	Speed	Time	Mass loss	S. No	Composite	Adding of sand	Speed	Time	S. No
Units	(Wt. %)	(Wt. %)	(rpm)	(s)	(g)	Units	(Wt. %)	(Wt. %)	(rpm)	(s)	Units
15	6	150	700	40	1.08631	21	12	30	1750	160	0.93378
16	9	30	1050	200	1.01859	22	12	60	350	200	0.83502
17	9	60	1750	40	1.04782	23	12	90	700	120	0.84361
18	9	90	350	80	0.94205	24	12	120	1050	40	0.87283
19	9	120	700	160	0.97228	25	12	150	1400	80	0.90206
20	9	150	1050	120	0.98785						

Source: Author

Table 80.8 Abrasive wear means response

Level	Composite	Abrasive grain size	Abrasive flow rate	Velocity	Time
Units	(Wt. %)	(µm)	(g/min)	(m/min)	(s)
1	0.041596	0.013420	0.012350	0.010600	0.011530
2	0.017950	0.013200	0.013130	0.013420	0.012150
3	0.012300	0.013570	0.014870	0.016390	0.014340
4	0.006130	0.016300	0.018030	0.018000	0.017200
5	0.003250	0.016297	0.017153	0.018697	0.017381
Delta	0.040362	0.003190	0.006362	0.005397	0.006241
Rank	1	5	2	4	3

Source: Author

Table 80.9 Abrasive wear S/N ratio response

Level	Composite	Abrasive grain size	Abrasive flow rate	Velocity	Time
Units	(Wt. %)	(μm)	(g/min)	(m/min)	(s)
1	31.80	38.35	41.98	42.10	41.90
2	36.45	38.82	42.14	39.65	41.85
3	39.05	41.85	41.60	39.35	41.44
4	46.58	41.09	39.46	39.86	39.81
5	51.24	41.07	39.22	39.14	39.70
Delta	2028	1.60	2.78	2.97	2.30
Rank	1	5	3	2	4

Source: Author

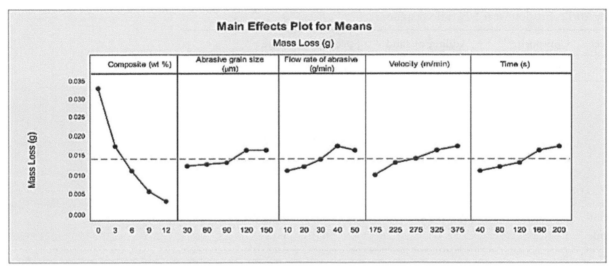

Figure 80.3 Mass loss (means) main effects plots
Source: Author

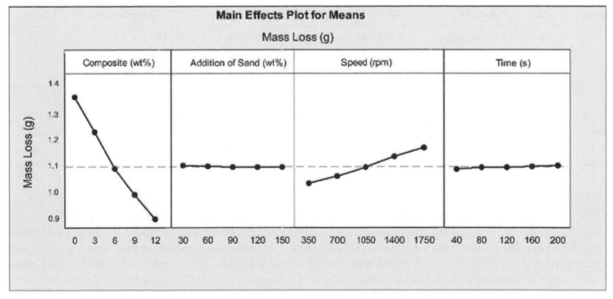

Figure 80.4 Mass loss (means) main effects plot
Source: Author

Each variable's contribution can be determined by using ANOVA technique. The abrasive wear test was evaluated using five factors in this study. According to tables composites had 84.14%, velocity 3.68 mm/min, time 2.65s, flow rate 4.56 g/min, abrasive 3.56% μm.

- ANOVA–erosion wear

Table 80.10 Erosion wear mean response

Level	Composite	Addition of sand	Speed	Time
Units	(Wt. %)	(Wt. %)	(rpm)	(s)
1	1.4356	1.2231	1.0581	1.1072
2	1.2399	1.2231	1.0852	1.1089
3	1.1127	1.2221	1.1121	1.1117
4	0.9934	1.2211	1.1491	1.1145
5	0.8752	1.1207	1.1757	1.1180
Delta	0.4820	0.1030	0.1014	0.1123
Rank	1	3	4	2

Source: Author

Table 80.11 Erosion wear S/N ratio responses

Level	Composite	Adding of sand	Speed	Time
Units	(Wt. %)	(Wt. %)	(rpm)	(s)
1	-2.61031	-0.87904	-0.39052	-0.75621
2	-1.78365	-0.81377	-0.59981	-0.81243
3	-0.91497	-0.78536	-0.82653	-0.84768
4	0.06426	-0.78958	-1.02978	-0.84927
5	1.16868	-0.80929	-1.26089	-0.81057
Delta	3.75897	0.78870	0.06045	0.07622
Rank	1	2	4	3

Source: Author

Table 80.12 Abrasive wear results attained form ANOVA

Source	DF	Adj. SS	Adj.MS	F	P	% of contribution	Rank
A	4	0.002932	0.000742	33.30	0.004	84.14	1
B	4	0.000132	0.000033	1.45	0.367	3.68	3
C	4	0.000137	0.000032	1.39	0.382	2.65	5
D	4	0.000131	0.000032	1.36	0.388	4.56	2
E	4	0.000092	0.000025	0.60	0.742	3.56	4
Error	4	0.000046	0.000012	-	-	1.40	-
Total	24	0.003441	-	-	-	100	-

Source: Author

The percentage of each processing variable examined in erosion wear was calculated using ANOVA. The composite 95.96% achieved the maximum percentage of contribution from the outcomes. Depending on the operating conditions, different input parameters are affected. Erosion testing used a speed (rpm) factor of 3.97% since the disc speed was altered simultaneously. The number of particles that came into touch with samples increased. To make the wear, duration (s) 0.06% and adding of sand 0.0019% had small of an effect. According to the aforementioned findings, just a small amount of material was removed from the 12% composite sample, as demonstrated by the tests. Erosion test results are presented in Tables 80.10-13.

Conclusions

AA 2218 composites were effectively manufactured using the stir casting technique and strengthened particles were effectively disseminated and homogenous dispersion was attained. Since the AA2218 composite

Table 80.13 Erosion wear results attained form ANOVA

Source	DF	Adj. SS	Adj.MS	F	P	% of contribution	Rank
F	4	0.699371	0.174817	-	-	95.96	1
G	4	0.046352	0.009113	-	-	3.97	2
H	4	0.000362	0.000094	-	-	0.06	3
I	4	0.000024	0.000002	-	-	0.0019	4
Error	8	0	0	-	-	0	
Total	24	0.746109	-	-	-	100	

Source: Author

strong bond strength prevented cracks and dislocations, the mechanical characteristics of the materials were evidently high. At the same time, the composites' plastic deformation was decreased due to the good particle bonding produced through the in-situ technique. Increasing the reinforcements improved the mechanical characteristics of the material. AA2218 composites had successfully undergone abrasive wear and slurry erosion wear tests because to the increased reinforcing wear resistance that was achieved by reducing plastic deformation and avoiding composite flaws. AA 2218 composites account for about 13% in terms of material removal under the various conditions of the aforementioned wear tests. The surface roughness was inspected after abrasive and erosion testing. Fine cracks and grooves had been found on the composites' surfaces during abrasive wear tests. Composites with less than 12% reinforcing in AA 2218 composites show no wear at all. As the rate of erosion increased, a layer of work hardened material was formed on the composite's surfaces.

References

[1] Chen, Q., Li, X., Wang, Y., Wang, J., Chang, L., Zhang, Y., Ma, H. and Jia, X., 2021. Optimization of thermo-electric properties of Al-doped Zn1– xAlxO under high pressure and high temperature. Journal of Alloys and Compounds, 886, p.161200.

[2] Adun, H., Adedeji, M., Adebayo, V., Shefik, A., Bamisile, O., Kavaz, D. and Dagbasi, M., 2021. Multi-objective optimization and energy/exergy analysis of a ternary nanofluid based parabolic trough solar collector integrated with kalina cycle. Solar Energy Materials and Solar Cells, 231, p.111322.

[3] Mohanavel, V., Prasath, S., Yoganandam, K., Tesemma, B. G., and Kumar, S. S. (2020). Optimization of wear parameters of aluminium composites (AA7150/10 wt%WC) employing Taguchi approach. Materials Today: Proceedings. 33(Part 7), 4742–45. https://doi.org/10.1016/j.matpr.2020.08.356.

[4] Uddin, M.M., Rahaman, M.H. and Kim, H.C., 2022. Highly stable hydrogen sensing properties of Pt–ZnO nanoparticle layers deposited on an alumina substrate for high-temperature industrialapplications. Sensors and Actuators B: Chemical, 368, p.132088.

[5] Esfe, M.H., Toghraie, D., Esfandeh, S. and Alidoust, S., 2022. Measurement of thermal conductivity of triple hybrid water based nanofluid containing MWCNT (10%)-Al2O3 (60%)-ZnO (30%) nanoparticles. Colloids and Surfaces A: Physicochemical and Engineering Aspects, 647, p.129083.

[6] Prašnikar, A., Jurković, D.L. and Likozar, B., 2021. Reaction path analysis of CO2 reduction to methanol through multisite microkinetic modelling over Cu/ZnO/Al2O3 catalysts. Applied Catalysis B: Environmental, 292, p.120190.

[7] Ha, N.N. and Ha, N.T.T., 2021. New insight into the mechanism of carbon dioxide activation on copper-based catalysts: A theoretical study. Journal of Molecular Graphics and Modelling, 107, p.107979.

[8] Liao, J. and Zhao, B., 2021. Phase equilibria study in the system "Fe2O3"-ZnO-Al$_2$O$_3$-(PbO+ CaO+ SiO$_2$) in air. Calphad, 74, p.102282.

[9] Toropova, A.P., Toropov, A.A., Leszczynski, J. and Sizochenko, N., 2021. Using quasi-SMILES for the predictive modeling of the safety of 574 metal oxide nanoparticles measured in different experimental conditions. Environmental Toxicology and Pharmacology, 86, p.103665.

[10] Ravikumar, M. M., Kumar, S. S., Kumar, R. V., Nandakumar, S., Rahman, J. H., and Raj, J. A. (2022). Evaluation on mechanical behavior of AA2219/SiO2 composites made by stir casting process. AIP Conference Proceedings. 2405, 050010. https://doi.org/10.1063/5.0078029.

[11] Shakhgil'dyan, G.Y., Savinkov, V.I., Shakhgil'dyan, A.Y., Alekseev, R.O., Naumov, A.S., Lopatkina, E.V. and Sigaev, V.N., 2021. Effect of Sitallization Conditions on the Hardness of Transparent Sitalls in the System ZnO–MgO–Al$_2$O$_3$–SiO$_2$. Glass and Ceramics, 77, pp.426-428.

[12] Yuan, H., Yuan, H., Casagrande, T., Shapiro, D., Yu, Y.S., Enders, B., Lee, J.R., van Buuren, A., Biener, M.M., Gammon, S.A. and Baumann, T.F., 2021. 4D Imaging of ZnO-Coated Nanoporous Al2O3 Aerogels by Chemically Sensitive Ptychographic Tomography: Implications for Designer Catalysts. ACS Applied Nano Materials, 4(1), pp.621-632.

[13] Liu, T.-L., and Bent, S. F. (2021). Area-selective atomic layer deposition on chemically similar materials: achieving selectivity on Oxide/Oxide patterns. Chemistry of Materials. 33(2), 513–23. doi: 10.1021/acs.chemmater.0c03227.

[14] Khatibi, M., Nemati-Farouji, R., Taheri, A., Kazemian, A., Ma, T. and Niazmand, H., 2021. Optimization and performance investigation of the solidification behavior of nano-enhanced phase change materials in triplex-tube and shell-and-tube energy storage units. Journal of Energy Storage, 33, p.102055.

[15] Meyer, M., Mehrabi, M., and Meyer, J. P. (2021). Modeling and multi-objective optimization of heat transfer characteristics and pressure drop of nanofluids in microtubes. Heat Transfer Engineering. 42(21), 1811–26. doi: 10.1080/01457632.2020.1826740.

[16] Pavlišič, A., Huš, M., Prašnikar, A. and Likozar, B., 2020. Multiscale modelling of CO2 reduction to methanol over industrial Cu/ZnO/Al2O3 heterogeneous catalyst: Linking ab initio surface reaction kinetics with reactor fluid dynamics. Journal of Cleaner Production, 275, p.122958.

[17] Sineva, S., Shevchenko, M., Shishin, D., Hidayat, T., Chen, J., Hayes, P.C. and Jak, E., 2020. Phase equilibria and minor element distributions in complex copper/slag/matte systems. Jom, 72, pp.3401-3409.

[18] Simya, O.K., Balachander, K., Dhanalakshmi, D. and Ashok, A., 2020. Performance of different anti-reflection coating and TCO layers for kesterite based thin film photovoltaic devices using Essential Macleod simulation program. Superlattices and Microstructures, 145, p.106579.

[19] Sun, L., Chen, E., and Guo, T. (2020). Field emission enhancement of composite structure of ZnO quantum dots and CuO nanowires by Al2O 3 transition layer optimization. Ceramics International. 46(10), 15565–71. doi: 10.1016/j.ceramint.2020.03.103.

[20] Kim, M., Zhao, W., Tsapatsis, M., and Stein, A. (2020). Three-dimensionally ordered macroporous mixed metal oxide as an indicator for monitoring the stability of ZIF-8. Chemistry of Materials. 32(9), 3850–9. doi: 10.1021/acs.chemmater.9b05395.

[21] Delgado Otalvaro, N., Kaiser, M., Herrera Delgado, K., Wild, S., Sauer, J., and Freund, H. (2020). Optimization of the direct synthesis of dimethyl ether from CO2 rich synthesis gas: Closing the loop between experimental investigations and model-based reactor design. Reaction Chemistry & Engineering. 5(5), 949–60. doi: 10.1039/d0re00041h.

[22] Rajbhandari, P.P. and Dhakal, T.P., 2020. Low temperature ALD growth optimization of ZnO, TiO_2, and Al_2O_3 to be used as a buffer layer in perovskite solar cells. Journal of Vacuum Science & Technology A, 38(3), 032406.

[23] Amiri, M., Eskandarian, A. and Ziabari, A.A., 2020. Performance enhancement of ultrathin graded Cu (InGa) Se2 solar cells through modification of the basic structure and adding antireflective layers. Journal of Photonics for Energy, 10(2), pp.024504-024504.

[24] Gohi, B.F.C.A., Zeng, H.Y., Xu, S., Zou, K.M., Liu, B., Huang, X.L. and Cao, X.J., 2019. Optimization of znal/chitosan supra-nano hybrid preparation as efficient antibacterial material. International Journal of Molecular Sciences, 20(22), p.5705.

[25] Hemmat Esfe, M., Amiri, M. K., and Bahiraei, M. (2019). Optimizing thermophysical properties of nanofluids using response surface methodology and particle swarm optimization in a non-dominated sorting genetic algorithm. Journal of the Taiwan Institute of Chemical Engineers. 103, 7–19. doi: 10.1016/j.jtice.2019.07.009.

[26] Chen, Z.-Z., Chen, H.-Q., Huang, L., Zhang, Y.-H., and Hao, N.-J. (2022). Research progress on silica nanofluids for convective heat transfer enhancement. Gongcheng Kexue Xuebao/Chinese Journal of Engineering. 44(4), 812–25. doi: 10.13374/j.issn2095-9389.2022.02.10.002.

[27] Yin, M., Yun, Z., Fan, F., Pillai, S.C., Wu, Z., Zheng, Y., Zhao, L., Wang, H. and Hou, H., 2022. Insights into the mechanism of low-temperature H2S oxidation over Zn–Cu/Al2O3 catalyst. Chemosphere, 291, p.133105.

[28] Shahverdi, N., Yaghoubi, M., Goodarzi, M., and Soleamani, A. (2019). Optimization of anti-reflection layer and back contact of perovskite solar cell. Solar Energy. 189, 111–119. doi: 10.1016/j.solener.2019.07.040.

[29] Dasireddy, V. D. B. C., and Likozar, B. (2019). The role of copper oxidation state in Cu/ZnO/Al_2O_3 catalysts in CO_2 hydrogenation and methanol productivity. Renewable Energy. 140, 452–60. doi: 10.1016/j.renene.2019.03.073.

[30] Pei, P., Huang, S., Chen, D., Li, Y., Wu, Z., Ren, P., Wang, K. and Jia, X., 2019. A high-energy-density and long-stable-performance zinc-air fuel cell system. Applied Energy, 241, pp.124-129.

[31] Yang, J.H., Yun, D.J., Kim, S.M., Kim, D.K., Yoon, M.H., Kim, G.H. and Yoon, S.M., 2018. Introduction of lithography-compatible conducting polymer as flexible electrode for oxide-based charge-trap memory transistors on plastic poly (ethylene naphthalate) substrates. Solid-State Electronics, 150, pp.35-40.

[32] Wu, J., Zhu, X., Shapiro, D.A., Lee, J.R., Van Buuren, T., Biener, M.M., Gammon, S.A., Li, T.T., Baumann, T.F. and Hitchcock, A.P., 2018. Four-dimensional imaging of ZnO-coated alumina aerogels by scanning transmission X-ray microscopy and ptychographic tomography. The Journal of Physical Chemistry C, 122(44), pp.25374-25385.

[33] Norek, M., Zaleszczyk, W., and Łuka, G. (2018). Optimization of UV luminescence from ZnO thin film: A combined effect of Al concave arrays and Al2O3 coating. Materials Letters. 229, 185–8. doi: 10.1016/j.matlet.2018.07.021.

[34] Gieraltowska, S., Wachnicki, L., and Godlewski, M. (2018). ALD oxides-based n-i-p heterostructure light emitting diodes. Acta Physica Polonica A. 134(2), 596–600. doi: 10.12693/APhysPolA.134.596.

[35] Doumit, N. and Poulin-Vittrant, G., 2018. A new simulation approach for performance prediction of vertically integrated nanogenerators. Advanced Theory and Simulations, 1(6), p.1800033.

[36] Seidel, C., Jörke, A., Vollbrecht, B., Seidel-Morgenstern, A., and Kienle, A. (2018). Kinetic modeling of methanol synthesis from renewable resources. Chemical Engineering Science. 175, 130–8. doi: 10.1016/j.ces.2017.09.043.

[37] Sommer, S., Bøjesen, E. D., Blichfeld, A. B., and Iversen, B. B. (2017). Tailoring band gap and thermal diffusivity of nanostructured phase-pure ZnAl2O4 by direct spark plasma sintering synthesis. Journal of Solid State Chemistry. 256, 45–52. doi: 10.1016/j.jssc.2017.08.023.

[38] Zhu, C., Jian, P., and Wang, Y. (2017). Facile synthesis of copper zirconium composite oxide and its application for the catalytic reduction of nitrobenzene methyl ether. Spec. Petrochemicals, 34(6), 6–11.

[39] Yan, Z., Wang, Y., Wang, X., Xu, C., Zhang, W., Ban, H. and Li, C., 2023. Unraveling the Role of H_2O on Cu-Based Catalyst in CO_2 Hydrogenation to Methanol. Catalysis Letters, 153(4), pp.1046-1056.

[40] Dasgupta, K., Mondal, A. and Gangopadhyay, U., 2021. Mathematical Modelling of Bifacial Dual SIS Solar Cell and Optimization of Tilt Angle. Silicon, pp.1-11.

[41] Phimu, L.K., Dhar, R.S. and Singh, K.J., 2022. Design optimization of thickness and material of antireflective layer for solar cell structure. Silicon, 14(13), pp.8119-8128.

[42] Dhahad, H.A., Hasan, A.M., Chaichan, M.T. and Kazem, H.A., 2022. Prognostic of diesel engine emissions and performance based on an intelligent technique for nanoparticle additives. Energy, 238, p.121855.

Chapter 81

Low power analog four-quadrant multiplier based on dynamic threshold MOSFET

Ankita Tijare[1,a], Kuldeep Pande[1,b] and Aniket Pathade[2,c]

[1]Department of Electronics, Yeshwantro Chavan College of Engineering, Nagpur

[2]Jawaharlal Nehru Medical College, Datta Meghe Institute of Medical Sciences, Wardha

Abstract

This research work describes a dynamic threshold CMOS-based low power-low voltage, and with increased gain four-quadrant analogue multipliers (FQM). The basic Gilbert's multiplier block and the cross coupled structure of complementary metal–oxide–semiconductor (CMOS) transistors make up the proposed multiplier. Through the creation of a negative resistance, the structure of cross-coupled arrangement raises the circuit's gain. The designed multiplier's supply voltage is low and uses extremely little power because of the sub-threshold area, dynamic threshold voltage (DTMOS), and a caudate design. The DTMOS transistor's low threshold voltage and strong transconductance, which are caused by the positive source-body voltage, make it appealing. Ten DTMOS transistors were used in the design of the proposed FQM, which is self-biased. The suggested multiplier has a maximum value of total harmonic distortion (THD) of 4.5%, and a supply voltage of 400 mV, a -3 dB bandwidth of 330 kHz, and a 7.5dB voltage gain. The power consumption is 66 nW. Tanner T-SPICE Simulation tool was used to simulate it using Taiwan Semiconductor Manufacturing Company Limited 90 nm CMOS technology specifications.

Keywords: DTMOS, four quadrant analog multiplier, low voltage, ultra-low power, wide bandwidth

Introduction

Typically, analog applications like modulation, frequency doublers, gain amplifiers, neural networks, etc. use four-quadrant analogue multipliers (FQM). Gilbert [1] announced the existence of the first FQM based on Bipolar Junction Transistor. The topologies of FQMs are divided into voltage-mode [2, 11] and current-mode [12–15]. Low-voltage and thus low-power orbit have lately grown more and more common due to the increasing demand for portable devices. Furthermore, the performance of multipliers with small power supplies is constrained by the 0.54V power source to aim for in low-power systems in near future. In order to design a four quadrant multiplier with a low supply and low power consumption, MOS transistors like QFGMOS and FGMOS [11, 12] and bulk driven metal–oxide–semiconductor field-effect transistor (MOSFET) [3, 9] have been used. However, these techniques decrease the transistor's transconductance. Also used for reducing power supply are transistors with a dynamic threshold voltage (DTMOS), studied by Babacan [2].

The first four-quadrant analogue multiplier was created using bipolar transistors, and it is known as the widely used gilbert cell [8]. Since then, a sizable amount of CMOS-related works has been disclosed. The development of CMOS current-mode architectures has been pushed by the rising demand for low voltage/ low power integrated circuits. The majority of current-mode structures operate on the same principles as their voltage-mode counterparts by using MOS drain current in either the strong inversion [9] or weak inversion region [10, 11].

It has proven possible to create current mode FQM circuits using stacked and folded MOS translinear loops (MTL) [11, 12] design approaches. However, these methods are either sensitive to body affect or call for higher supply voltages, which consume more power.

The FQM based on voltage differencing [7] has a supply voltage of 1V and a power consumption of 627W. A high-performance multiplier with a wide bandwidth of 19.3GHz is provided, constructed using a twin X second-generation current, however it consumes too much power [8]. Multiplier in ref [9] is bulk driven and developed in the saturation region. Also the multiplier is built in sub threshold area uses little power and accepts input voltages up-to 240V [10]. uses floating GATE CMOS transistors, which results in tiny power supply and low power consumption, although the FGMOS transistor lowers the transistor's transconductance.

DTMOS transistors significantly increase transconductance in addition to restricting its decrease. In addition, creating circuits in the sub-threshold region has aided in their realisation [4, 10]. X Architecture reported by Babacan [2] was created utilising a DTMOS transistor for the voltage-mode circuits and consumes low power because of low voltage, although it relatively takes up a large chip. The low voltage multiplier was implemented using a bulk drive method [3], but this method lowers the transistor's transconductance. Multiplier designed has low bandwidth withe less power consumption [3]. The multiplier

---

[a]ankita.tijare@gmail.com, [b]ycce.kuldeep@gmail.com, [c]aniketpathade@gmail.com

DOI: 10.1201/9781003450252-81

consumes little energy because it operates below the threshold [4]. Multiplier architectural designed in [5] has a good gain because of the use of inGaZnO thin-film transistor (TFT) at the cost of huge power supply. An operational amplifier-based FQM with a projected supply voltage of 2.4V was presented by Riewruja and Rerkratn [6].

A very small-power consumption, truncated-voltage with high-gain dynamic threshold MOSFET (DTMOS) analog multiplier with better performance than earlier low-voltage devices is reported in this study. The remainder of this essay is structured as follows: The DTMOS transistor is introduced in Section A. The suggested DTMOS based multiplier is shown in Section B. Results of the simulation are displayed in part D. Final thoughts are at the conclusion.

Dynamic Threshold MOSFET

One of the main concerns of designers is lowering the circuits power supply and their power consumption. Various methods that are stated in the previous section have been suggested in the past for this aim. The usage of circuit design using DTMOS transistors is one of these avant-garde concepts. In 1994, Assaderaqhi unveiled DTMOS. The gate and its body are connected, just like a MOSFET transistor. Along with having a lower threshold voltage and more transconductance than a MOSFET, DTMOS (Figure 81.1) provides advantages over that technology. DTMOS is a good choice for low-voltage and low-power circuit design because of these benefits. Floating bulk issues like stability in threshold voltage transient and kick are also resolved by DTMOS. Bulk technology can be used to implement it [16–19].

The threshold voltage is given in equation (1):

$$V_{TH} = V_{T0} - \gamma\left(\sqrt{|2\phi_F|} - \sqrt{|2\phi_F| - V_{BS}}\right) \tag{1}$$

Where V_{BS} is the source-to-body voltage, V_{TH} is the threshold voltage when V_{BS} is not equal to 0, at $V_{BS} = 0$ V_{T0} is the threshold voltage, and the body effect parameter $\gamma = (t_{ox}/\varepsilon_{ox})\sqrt{2q\varepsilon_{Si}N_A}$ $2\Phi_F$ =surface potential, t_{ox} = oxide thickness, ε_{Si} = silicon permittivity, N_A = doping concentration, q =charge.

The MOSFET's current is equal to:

$$I_{DS,MOS} = I_{Do}e^{\frac{-V^{GS}}{\eta V_T}}e^{\frac{-(\eta-1)V_{BS}}{\eta V_T}}\left[1 - e^{\frac{-V_{DS}}{\eta V_T}}\right] \tag{2}$$

$$I_{D0} = 2\eta V_T^2 X \beta e^{\frac{V_{TH}}{\eta V_T}} \tag{3}$$

If the MOS bulk terminal is linked to the appropriate source terminal ($V_{BS} = 0$, $V_{TH} = V_{T0}$), then:

$$I_{DS,MOS} = I_{Do}e^{\frac{-V^{GS}}{\eta V_T}}\left[1 - e^{\frac{-V_{DS}}{\eta V_T}}\right] \tag{4}$$

$$I_{D0} = I_{Do}e^{\frac{-V^{GS}}{\eta V_T}} \tag{5}$$

$$I_{D0} = 2\eta V_T^2 X \beta e^{\frac{V_{TH}}{\eta V_T}} \tag{6}$$

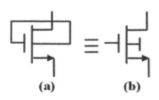

Figure 81.1 DTMOS transistor
Source: Author

$$g_{m,MOS} = \frac{I_{DS,DTMOS}}{\eta V_T} \tag{7}$$

In DTMOS, the MOS bulk terminal is linked to the appropriate gate terminal ($V_{GS} = V_{BS}$).

$$I_{DS,DTMOS} = I_{Do} e^{\frac{-V^{GS}}{\eta V_T}} e^{\frac{-(\eta-1)V_{BS}}{\eta V_T}} \left[1 - e^{\frac{-V_{DS}}{\eta V_T}} \right] \tag{8}$$

$$I_{DS,DTMOS} = I_{Do} e^{\frac{-V^{GS}}{\eta V_T}} \left[1 - e^{\frac{-V_{DS}}{\eta V_T}} \right] \tag{9}$$

$$I_{DS,DTMOS} = I_{Do} e^{\frac{-V^{GS}}{\eta V_T}} \tag{10}$$

$$I_{D0} = 2\eta V_T^2 X \beta e^{\frac{V_{TH}}{\eta V_T}} \tag{11}$$

$$g_{m,DTMOS} = \frac{I_{DS,DTMOS}}{\eta V_T} \tag{12}$$

$$g_{m,DTMOS} = \frac{I_{DS,DTMOS}}{\eta V_T} + \frac{(\eta-1)I_{DS,DTMOS}}{\eta V_T}$$
$$= \frac{I_{DS,DTMOS}}{\eta V_T} \tag{13}$$

Circuit Implementation

Conventional Multiplier

The standard Gilbert cell-based four-quadrant analogue multiplier is depicted (Figure 81.2). Gilbert cells (M1–M6), diode-connected loads (M7 and M8), and a tail current source make up the circuit (I_{SS}) [1]. The standard multiplier's output voltage is given in Equation 14.

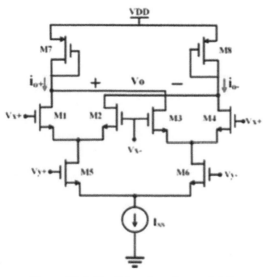

Figure 81.2 Traditional analog multiplier
Source: Author

$$V_{out} = (1/(g_{m7,8} + g_{ds7,8}))(i_{OUT}) \qquad (14)$$

Where g_m and g_{ds} are sub threshold regions small signal parameters.

Proposed Multiplier Based on DTMOS

Figure 81.3 depicts the offered small voltage, extremely very low-power, and increased gain FQM DTMOS based multiplier. Gilbert cells (DTM1–DTM6), diode-connected loads (DTM7 and DTM8), and a cross-coupled structure make up the suggested multiplier (DTM9 and DTM10). Through the creation of a negative resistance, the cross-coupled structure raises the circuit's gain. The suggested multiplier's low supply voltage and extremely low power requirements are made possible by the sub-threshold region and DTMOS. By removing the conventional structure's tail current source, more voltage headroom is created. As a result, the linearity is also improved. By rremoving the tail, we can lowers CMRR, which can be offset by lowering the incongruity using effective designing techniques. Channel length modulation affects the circuit that has $L = L_{min}$. Thus, the impact of channel length modulation is diminished as a result of $L = 2L_{min}$. Ten n-DTMOS transistors were used in the self-biased design of the suggested multiplier. Our suggested multiplier's DTMOS transistor dimensions are (W/L)14 = 4 m/0.2 m and (W/L)5–10 = 1 m/0.2 m. The proposed multiplier's output voltage is stated in Equation (15).

$$V_{out} = (1/(g'_{m7,8} - g'_{m9},10 + g'_{ds7,8} + g'_{ds9,10}))(i_{OUT}) \qquad (15)$$

Simulation Results

The suggested FQM performs amplitude modulation and frequency doubling. The corner and frequency response results, and the DC characteristics are all demonstrated in this section. The TSPICE tool is used to simulate the DTMOS transistor based offered multiplier using the TSMC CMOS 90nm technology characteristics. By connecting the MOSFET transistor's body and gate, DTMOS is produced. 0.4 V supply voltage, and the FQM uses 66 nW of power.

The suggested multiplier's effectiveness as an amplitude modulator is assessed using a carrier signal at 100 kHz (vx) and a signal at 4 kHz (vy) (AM). Additionally, the frequency doubling employs two alike signals with a frequency of 100 kHz and amplitude of 100mVP-P. The output waveforms of the amplitude modulator and frequency doubler, respectively, are shown in Figures 81.4 and 81.5. A modest sinusoidal signal and dc voltage are applied to the suggested multiplier inputs in order to plot the frequency response. The suggested FQM is then compared to earlier efforts employing TSMC 90 nm CMOS technology specifications. The output voltage in the presence of noise is 6.76491n V/Rt (Hz) which is shown in Figure 81.6. Figure 81.7 represents the application of Analog multiplier as Low pass filter by exhibiting the 3dB bandwidth is 330 KHz.

Table 81.1 depicts the four-corner analysis for the designed multiplier where the bandwidth achieved 330 KHz at 7.5 dB gain in typical-typical configuration. Where in fast-fast region the maximum bandwidth is achieved.

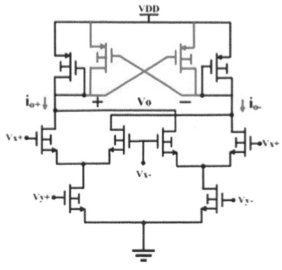

Figure 81.3 Proposed architecture of analog multiplier
Source: Author

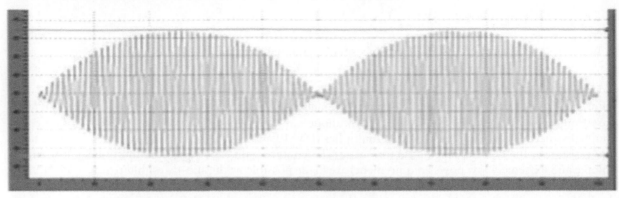

Figure 81.4 Multiplier output as amplitude modulator (AM)
Source: Author

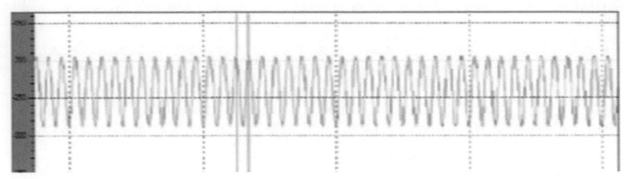

Figure 81.5 Multiplier output as frequency doubler
Source: Author

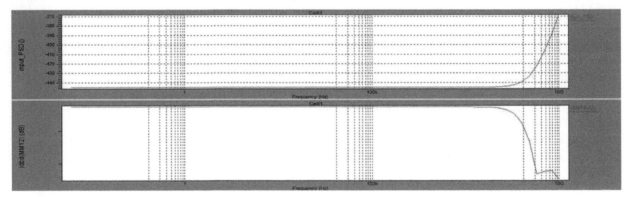

Figure 81.6 Noise analysis
Source: Author

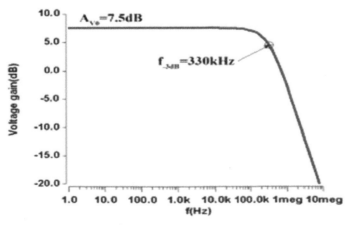

Figure 81.7 AC frequency response
Source: Author

Table 81.1 Four corner result perusals

Specifications	T-T	F-F	F-S	S-F	S-S
Voltage gain	7.5	5.6	4.6	5	8.6
Bandwidth (kHz)	330	1300	766	248	85

Source: Author

Conclusion

A low-voltage, extremely low-power, and high-gain four-quadrant analogue multiplier based on DTMOS is reported in this study. The gain of the suggested multiplier is raised as a result of the cross-coupled structure. The DTMOS-based multiplier created in the sub-threshold area also uses incredibly little power. The outcomes of the technological corners and Monte Carlo are used to confirm the proposed multiplier's performance. The suggested device functions as both a frequency doubler and an amplitude modulator. The suggested multiplier can be used in a variety of analogue signal processing scenarios. Due to its greater dynamic range and better THD, the multiplier circuit exhibits better FOM than previous multipliers. Another benefit of this circuit is that which needs high voltage headroom for adequate dynamic range, it can operate on scaled CMOS technology.

References

[1] Gilbert, B. (1968). A precise four-quadrant multiplier with subnanosecond response. IEEE Journal of Solid-State Circuits. 3, 365–73.

[2] Babacan, Y. (2019). Ultra-low voltage and low-power voltage-mode DTMOS-based four-quadrant analog multiplier. Analog Integrated Circuits and Signal Processing. 99, 39–45.

[3] Panigrahi, A. and Paul, P. K. (2013). A novel bulk-input low voltage and low power four quadrant analog multiplier in weak inversion.. Analog Integrated Circuits and Signal Processing. 75, 237–43.

[4] Xin, X., Cai, J., Xie, R., and Wang, P. (2017). Voltage-mode ultra-low power four quadrant multiplier using sub-threshold PMOS. IEICE Electronics Express, DOI:10.1587/elex.14.20170063

[5] Pydi Ganga Bahubalindruni , P. G., Tavares, V. G., Borme, J., Oliveira, P. G. D., Martins, R., Fortunato, E., et al. (2016). InGaZnO thin-film-transistor-based four-quadrant high-gain analog multiplier on glass. IEEE Electron Device Letters. 37, 419–21.

[6] Riewruja, V. and Rerkratn, A. (2011). Four-quadrant analogue multiplier using operational amplifier. International Journal of Electronics. 98, 459–74.

[7] Gupta, P. and Pandey, R. (2019). Voltage differencing buffered amplifier based voltage mode four quadrant analog multiplier and its applications. International Journal of Engineering. 32, 528–35.

[8] Rajpoot, J. and Maheshwari, S. (2019). High performance four-quadrant analog multiplier using DXCCII. January 2020 Circuits Systems and Signal Processing 39(1):54–64. DOI:10.1007/s00034-019-01179-x

[9] Mejia, G. Z., Armendariz, A. D., Ramirez, H. S., Perez, J. M. R., Marin, C. A. G., and Sanchez, A. D. (2019). Gate and bulk-driven four-quadrant CMOS analog multiplier. Circuits, Systems, and Signal Processing. 38, 1547–60.

[10] Khan, S. I. and Mahmoud, S. A. (2019). Highly linear CMOS subthreshold four-quadrant multiplier for teager energy operator based sleep spindle detectors. November 2019 Microelectronics Journal 94(12):104653 DOI:10.1016/j.mejo.2019.104653

[11] Keleş, S. and Keleş, F., (2019). Ultra low power wide range four quadrant analog multiplier. March 2020 Analog Integrated Circuits and Signal Processing 102(2018). DOI:10.1007/s10470-019-01491-1

[12] Absi, M. A. A. and Sabban, I. A. A. (2015). A new highly accurate CMOS current-mode four-quadrant multiplier. Arabian Journal for Science and Engineering. 40, 551–8.

[13] Maryan, M. M., Azhari, S. J., and Ghanaatian, A. (2018). Low power FGMOS-based four-quadrant current multiplier circuits. Analog Integrated Circuits and Signal Processing. 95, 115–25.

[14] Alikhani, A. and Ahmadi, A. (2012). A novel current-mode four-quadrant CMOS analog multiplier/divider. AEU-International Journal of Electronics and Communications. 66, 581–6.

[15] Popa, C. (2013). Improved accuracy current-mode multiplier circuits with applications in analog signal processing. IEEE Transactions on Very Large Scale Integration (VLSI) Systems. 22, 443–7.

[16] Assaderaghi, F., Sinitsky, D., Parke, S. A., Bokor, J., Ko, P. K., and Hu, C. (1997). Dynamic threshold-voltage MOSFET (DTMOS) for ultra-low voltage VLSI. IEEE Transactions on Electron Devices. 44, 414–22.

[17] Pruthwiraj, B., Mishra, A. K., and Vaithiyanathan, D. (2022). A low power CMOS analog multiplier using regulated cascode current mirror based on translinear principle. 2022 IEEE 3rd Global Conference for Advancement in Technology (GCAT). 07–09.

[18] Sousa, A. J. S. D., Andrade, F. D., Santos, H. D., and Gonçalves, G. (2019). CMOS analog four-quadrant multiplier free of voltage reference generators. 2019 32nd Symposium on Integrated Circuits and Systems Design (SBCCI). 26–30.

[19] Danesh, D., Jayaraj. A., Chandrasekaran, S, T., and Sanyal, A. (2019). Ultra-low power analog multiplier based on translinear principle. 2019 IEEE International Symposium on Circuits and Systems (ISCAS). 26–29.

Chapter 82

A simple approach for camera calibration using the technique for order performance by similarity to ideal solution method

Rohit Zende[a] and Raju Pawade[b]

Dr. Babasaheb Ambedkar Technological University, Lonere, Maharashtra, India

Abstract

This experimental work focuses on the development of a simple method for camera calibration. Camera calibration is a crucial process for machine vision systems. Error in the calibration process reduces the accuracy of the measurement therefore it is vital to design a robust method that will minimize the errors in the measurement. A method based on a technique for order performance by similarity to ideal solution is found to be very effective in wide applications of engineering as well as non-engineering systems where the output responses are chosen based on the rank assigned to it. In this experimental study, slip gauges were used for calibration purposes where camera distance, light source, and a number of measurements were chosen for the analysis. The measurements were done for four different calibration cases. The measured data were further used for the decision-making purpose using the technique for order performance by similarity to ideal solution method where rank was assigned to nine experimental runs. Taguchi design of experimentation was used to design nine experimental runs. The experimental run belonging to the first rank was further used for finding the optimal measuring conditions. Analysis of variance shows the camera distance, and the light source as the most influencing factors with an average contribution of 76.51% and 18.56% respectively. The confirmation experiment shows that the technique for order performance by similarity to ideal solution method was found to be best applied in the process of camera calibration. The average percentage error obtained in optimal conditions is about 0.9798% which shows the suitability of the method.

Keywords: TOPSIS, camera calibration, taguchi method, clip gauges

Introduction

Every object is defined in 3D coordinates whereas the image obtained using the camera is expressed in a 2D coordinate system. This clearly shows that there is a need to define the relationship between the three-dimensional coordinate system and the 2D coordinate system by performing camera calibration [1]

During camera calibration, a geometric model is defined using two basic parameters intrinsic and extrinsic. Intrinsic parameters are those which are related to the camera lens such as optical length, focal length, etc. However, the extrinsic parameters are those which are related to the camera orientation concerning objects which are mainly defined by translational parameters and rotational parameters. Three categories classify the camera calibration algorithm namely traditional camera calibration algorithm, camera self-calibration algorithm, and active vision-based camera calibration algorithm [2]

Traditional camera calibration includes the calibration piece or block on which calibration is performed. Camera self-calibration includes the image sets which are defined by the constraint equations following each image. Active vision-based camera calibration includes the calibration performed using a controlled camera with the advantage of the camera motion information. Being a simple method and widely accepted in industrial applications. The present experimental study is focused on the traditional camera calibration algorithm.

There are studies on camera calibration as reported in the literature. A few of them are discussed here. Kannala and Brandt [3]2 used a fish eye lens camera for calibration where they proposed a generic model for the calibration process. The experimental results showed the accuracy of the proposed method and it found the best one for wide angle view. Sirisantisamrid et al. [4] proposed a new method to find the initial camera parameters using a single image. Authors have developed a coplanar calibration method to determine the initial camera parameters. The performance of the method was tested using a set of synthetic and real images. This method was found to be more effective as it requires less computational time than Zhang's calibration method. Zheng and Peng [5] developed a new practical method to calibrate the roadside camera used in traffic surveillance. The least-square optimization method was used during the calibration process which was demonstrated for the synthetic and real images. The proposed method was found to be more accurate where the objective for calibration was to determine the camera's intrinsic parameters such as camera focal length, three rotation angles, the principal point, and the translation vector. Zhang et al. demonstrated a method based on the camera's intrinsic parameters and distortion coefficients for

[a]zenderohit@gmail.com, [b]rspawade@dbatu.ac.in

DOI: 10.1201/9781003450252-82

camera calibration. Authors have accurately estimated the distortion center from target points and their image which is further decoupled and independently solved by a linear method. By comparing it with the iterative method; this method shows improvement in the computational efficiency more than 20 times without compromising the calibration accuracy. Juarez-Salazar and Diaz-Ramirez [7] calibrated the camera projector using a checkerboard. The physical checkerboard was placed on the reference plane, and this was superimposed by the projected checkerboard. This image was captured using a camera where the checkerboards were recovered using the color information. The experimental results showed the flexibility and usefulness of the developed method for in situ calibration applications. Among the intrinsic and extrinsic parameters that are crucial in camera calibration, it is important to determine their effectiveness in the calibration results. This could be done by using the technique for order of preference by similarity to ideal solution (TOPSIS) method. In the past researchers have used TOPSIS to rank the parameters in order of their similarity in machining applications.

Ramesh et al. [8] optimised the cutting parameters in turning the magnesium alloy AZ91D with the use of polycrystalline diamond inserts under dry conditions. In the investigation of the cutting parameters cutting speed, feed rate, and the depth of cut among the 27 experimental runs they employed grey relational analysis (GRA) for optimisation of process parameters which are further compared by using the TOPSIS method. Gopal and Prakash [9] studied the end milling effect on magnesium (Mg) metal matrix composite (MMC) using a carbide tool. Magnesium MMC was produced by reinforcement of the cathode ray tube (CRT) glass particle size. Reinforcing material particle size and weight, tool diameter, speed, feed, and depth of cut (DoC) was the input factors while feed force, cross-feed force, thrust force, temperature, and surface roughness was the output parameters. Analysis of variance (ANOVA) was performed to identify the significance of the input factors on the output parameters. Optimum parameters were identified by using GRA and TOPSIS methods. Nguyen et al. [10] analysed powder mixed dielectric fluid in electrical discharge machining (PMEDM) with workpiece material, electrode material, polarity, pulse on time, current, pulse off time, and powder concentration as the process parameters. They determined the optimal parameters by using a combined Taguchi-TOPSIS method. Experimental results show the suitability of the TOPSIS method under multi-objective optimization problems. Ananthakumar et al., [11] investigated the effect of plasma arc cutting (PAC) parameters like cutting speed, arc current, gas pressure, and stand-off distance on Monel 400 superalloy with performance measure of heat affected zone (HAZ), kerf taper (KT), and MRR. Experiments were planned to use the response surface methodology (RSM) which was further optimized using the TOPSIS method.

The literature review shows the necessity to work on a robust camera calibration system. Camera calibration methods for industrial applications need to be simple, cost-effective, and robust. Camera self-calibration algorithms and active vision-based camera calibration algorithms are flexible but have the limitation of cost-effectiveness with their robustness. Camera intrinsic parameters are mostly unchanged, but camera extrinsic parameters mostly affect the camera calibration process keeping in this view, in this study camera extrinsic parameters are chosen. TOPSIS method was found to be effective for multi-objective optimization but optimization of the camera parameters in the calibration process using the TOPSIS method is not found in the literature review. Hence this experimental study is intended to focus on the optimization of the camera parameters using the TOPSIS method.

Experimental Setup

The experimental setup includes the selection of appropriate slip gauges, a smartphone camera, and a light source. Slip gauges are the primary gauges having precise dimensions due to which they are used for calibration of the secondary instruments like vernier caliper, micrometer, etc. Thus, for this experiment; slip gauges are chosen as a calibration block. Distance between the camera and the object is one of the most influencing parameters which affects the image quality [12–14]

In image processing, an image is represented by the combination of the Red, Green, and Blue colors (RGB) therefore the effect of these color intensities affects the image quality [15–18]

A number of measurements change the output value of the parameter; hence, the distance between the camera and object, three RGB light sources, and the number of measurements is chosen as input parameters for the experiment. Taguchi design of experiment is found to be one of the popular methods used for an experimental design where it gives the optimised combinations of the input parameters for minimum experimental runs [19, 20].

The Taguchi method also describes the interactions of the input factors while the design of experimental runs using the Taguchi method is called an orthogonal array. For this study L9 orthogonal array is used for the experimentation which has 9 experimental runs having combinations of the input factors with their levels which are shown in Table 82.1. Based on the screening experiments, levels of the input factors are selected. Minitab 19 software is used for Taguchi design of experiment and analysis.

Table 82.1 Input parameters with levels

Levels	Input parameters		
	Camera distance (mm)	Light source	Number of measurements
1	500	Red	3
2	600	Green	5
3	700	Blue	7

Source: Author

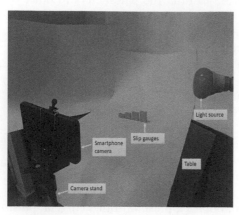

Figure 82.1 Experimental setup
Source: Author

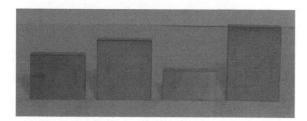

Figure 82.2 Experimental setup
Source: Author

The experimental setup is shown in Figure 82.1. To capture an image, a smartphone camera has a capacity of 48 megapixels is used. Before capturing the image, the smartphone camera plane and table plane are set to a horizontal position using a spirit level. Figure 82.2 shows the image captured in a red light source background.

In image processing noise present in the image needs to be removed to increase the accuracy of the image. However, slip gauge images captured using a smartphone camera has a distinct boundary of slip gauge block due to the red color background which is also verified with green and blue color background therefore most of the image processing steps are neglected. Using Python programming language (version 3.8.12) with Open-CV library (version 4.6.0) is used for image processing where a color image is converted into a grey scale image and this grey scale image is further used in Coslab software to measure the pixel height of the slip gauge. This experimental work comes under camera calibration using a standard calibration workpiece. To calibrate the camera; the pixel value of the captured image is determined, and this pixel value is compared with the actual value of the calibration piece thus value for one pixel is calculated with an equal millimeter length. Figure 82.3 shows the captured image converted into a grey scale image which is measured in a pixel.

In Figure 823, three measurements for each slip gauge are shown and for analysis average pixel value for each slip gauge is considered. Likewise, for each slip gauge, five measurements and seven measurements were also taken for the analysis. Using Taguchi L9 orthogonal array nine experimental runs were designed which is shown in Table 82.2 with input and output factors.

During the experiment, each experimental run was performed three times thus each experimental run has three replicas, and the average values of the replicas are considered for the analysis which is shown in Table 82.2.

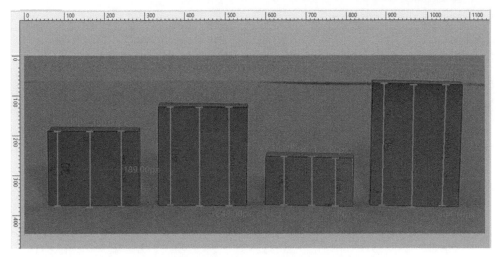

Figure 82.3 Greyscale image measured in pixels using Coslab software
Source: Author

Table 82.2 Taguchi l9 orthogonal array with input and output parameters

	Input parameters			Output parameters (Slip gauges)			
Exp. run	Camera distance in mm	Light source	Number of measurements	20 mm (pixel length)	30 mm (pixel length)	40 mm (pixel length)	50 mm (pixel length)
1	500	Red	3	121.0000	187.4444	246.2222	305.9578
2	500	Green	5	121.0327	185.4907	246.4147	306.1067
3	500	Blue	7	121.4776	185.7361	247.0233	306.9600
4	600	Red	5	99.2667	146.9333	196.5333	245.2667
5	600	Green	7	98.6667	145.9048	197.3333	244.5467
6	600	Blue	3	99.3333	145.8889	197.0000	245.3711
7	700	Red	7	87.6190	131.9048	173.4286	215.6667
8	700	Green	3	87.4444	128.5556	172.7778	214.8889
9	700	Blue	5	87.4667	129.6667	172.2667	214.4667

Source: Author

Methodology

Multi-criteria decision-making (MCDM) methods include the decision-making of the desired parameters with simple mathematical calculations. Due to their simplicity and robustness, these methods are applied to wide application areas including engineering and medical fields. The beauty of these methods is that for conflicting criteria such as one output parameter should be maximum while at the same time, other parameter should be minimum, these methods are found to be the best suited. There are many methods used in MCDM among which the TOPSIS method was found to be popular. There are two criteria used beneficial criteria and non-beneficial criteria. Beneficial criteria are those which are desirable to maximum value for example MRR should be maximum which can be treated as a beneficial criterion. Non-beneficial criteria are those which are not desirable and require the minimum values for example tool wear should be minimum and is considered as non-beneficial criteria. In this experiment, all the slip gauges are measured using an image processing technique where a maximum error that occurred during the calibration needs to be addressed hence all the output measured values are considered as a beneficial criterion to optimize the maximum error. In the TOPSIS method rank is assigned based on a maximum value of the closeness coefficient. The experimental run belongs to the first rank and is considered an optimal input parameter condition. Detailed explanations and steps to calculate the closeness coefficient using the TOPSIS method are explained in [21, 22].

Four slip gauges are measured in pixel height using Coslab software. For each experimental run, there is a different pixel value for each slip gauge. One pixel value for the 20 mm slip gauge is calibrated for the actual value of the slip gauge which is considered as 20 mm and using this value measurement was

Table 82.3 Measured output values of slip gauges

Exp. run	Input parameters			Output parameters (Slip gauges)			
	Camera distance in mm	Light source	Number of measurements	20 (in mm)	30 (in mm)	40 (in mm)	50 (in mm)
1	500	Red	3	19.5989	30.6887	40.1144	49.7479
2	500	Green	5	19.6639	30.3533	40.2740	49.9266
3	500	Blue	7	19.6929	30.3032	40.2686	49.9411
4	600	Red	5	20.2359	29.8208	39.9298	49.8038
5	600	Green	7	20.1535	29.6608	40.3070	49.8076
6	600	Blue	3	20.2790	29.5746	40.1060	49.8941
7	700	Red	7	20.1500	30.3708	39.7462	49.3402
8	700	Green	3	20.3323	29.6923	40.0129	49.6816
9	700	Blue	5	20.3126	29.9959	39.8026	49.4860

Source: Author

Table 82.4 Closeness coefficient and its associated rank calculated using the topsis method

Exp. run	Camera distance in mm	Light source	Number of measurements	Weighted normalised matrix for slip gauges				Euclidean separation		Closeness coefficient	
				20 mm	30 mm	40 mm	50 mm	Si +	Si -	Pi	Rank
1	500	Red	3	0.081464	0.085095	0.083440	0.083352	0.003092	0.003255	0.512876	5
2	500	Green	5	0.081734	0.084165	0.083772	0.083651	0.002931	0.002628	0.472748	8
3	500	Blue	7	0.081855	0.084026	0.083761	0.083675	0.002866	0.002535	0.469408	9
4	600	Red	5	0.084112	0.082688	0.083056	0.083445	0.002573	0.002868	0.527104	3
5	600	Green	7	0.083769	0.082244	0.083841	0.083452	0.002954	0.002710	0.478468	7
6	600	Blue	3	0.084291	0.082006	0.083423	0.083597	0.003126	0.003068	0.495308	6
7	700	Red	7	0.083755	0.084213	0.082675	0.082669	0.001930	0.003181	0.622355	1
8	700	Green	3	0.084513	0.082332	0.083229	0.083241	0.002863	0.003168	0.525264	4
9	700	Blue	5	0.084431	0.083174	0.082792	0.082913	0.002319	0.003200	0.579753	2

Source: Author

performed for 30 mm, 40 mm, and 50 mm slip gauges. This process was applied for each experimental run and after calibration of the 20 mm slip gauge the same process was repeated for 30 mm, 40 mm, and 50 mm slip gauge calibration. While calibrating each slip gauge other slip gauges were considered for the measurement. Thus, each slip gauge is calibrated for a single time and measured three times. The average of this measured value for each slip gauge was considered for analysis. The measured values of slip gauges are shown in Table 82.3. Table 82.4 shows the closeness coefficient and its associated rank calculated using a TOPSIS method. Four slip gauges are equally important and hence weight for each output parameter is assigned as 0.25 which is used to calculate the weighted normalized matrix in the TOPSIS method.

From Table 82.4, it is clear that experimental run 7 is having the highest closeness coefficient 0.622355 with an assigned rank of 1. Experimental run 7 has a camera distance of 700 mm, a red-light source, and a number of measurements are 7. These input parameters are considered initial parameter settings for optimization. The best combination of the experimental runs according to the TOPSIS method is 7-9-4-8-1-6-5-2-3

Results and Discussion

To understand the effect of the input parameters on the measurements of slip gauges; the main effect plot for the slip gauges was plotted which is shown in Figure. 82.4. A camera distance of 500 mm resulted in more error in slip gauge measurements while a camera distance of 700 mm gives the minimum error

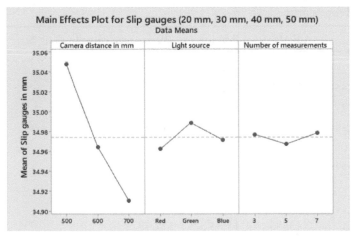

Figure 82.4 Greyscale image measured in pixels using Coslab software
Source: Author

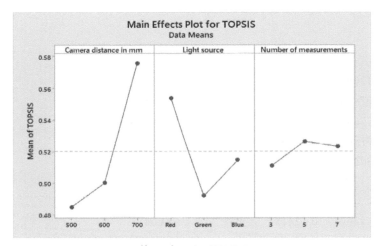

Figure 82.5 Main effect plots for TOPSIS
Source: Author

(Figure 82.4). The reason behind this is that the camera kept at a distance of 700 mm from the target is more prone to accurate image boundary profile. Measurements performed under a red-light source has the minimum errors in slip gauge measurements while a green light source has moderate errors in slip gauge measurements. As it is known the different wavelengths of light sources influence the image quality. For five measurements the minimal error in slip gauge measurements reveals that at least five measurements are needed to minimise the errors in slip gauge measurements.

The main effects plot for the TOPSIS method is shown in Figure 82.5 in which the main effects plot was plotted for the closeness coefficient. The higher value of the TOPSIS for a camera distance of 700 mm and the lower at a 500 mm camera distance is found (Figure 82.5). The red-light source has a high TOPSIS value while for five number of measurements; the TOPSIS value is higher. Therefore, these values are considered the final optimal camera parameters which are considered for the analysis.

Analysis of variance (ANOVA) for the slip gauge measurements and TOPSIS method was performed which is shown in Table 82.5 from which it is cleared that for all the slip gauge measurements; the camera distance has an average contribution of 76.51%, the Light source has an average contribution of 18.56% and the number of measurements has an average contribution of 2.85%. Camera distance produces large errors in slip gauge measurements which supports the significance of the camera distance as obtained by ANOVA. ANOVA for the TOPSIS method also highlights the significance of the camera distance by the contribution of 66.54% which is followed by the light source with a contribution of 27.67% and the number of measurements with a contribution of 1.86%.

As discussed earlier initial conditions for the camera calibration is camera distance – 700 mm, light source – Red, number of measurements – 7, and optimal conditions for the camera calibration are Camera distance – 700 mm, light source – Red, number of measurements – 5. For these optimal conditions confirmation experiment was performed and the measured values of the slip gauges are shown in Table 82.6.

Table 82.5 ANOVA analysis for the slip gauges and topsis method

Source	DF	Adj SS	Adj MS	F-Value	Percentage contribution (%)
Slip gauge of 20 mm					
Camera distance in mm	2	0.703546	0.351773	93.73	95.55
Light source	2	0.015027	0.007514	2	2.04
Number of measurements	2	0.010261	0.005131	1.37	1.39
Error	2	0.007506	0.003753		
Total	8	0.736341			
R-Square = 98.98%					
Slip gauge of 30 mm					
Camera distance in mm	2	0.87771	0.43885	15.34	71.49
Light source	2	0.26881	0.13441	4.7	21.89
Number of measurements	2	0.02407	0.01204	0.42	1.96
Error	2	0.0572	0.0286		
Total	8	1.2278			
R-Square = 95.34%					
Slip gauge of 40 mm					
Camera distance in mm	2	0.212045	0.106023	112.38	62.50
Light source	2	0.107674	0.053837	57.06	31.74
Number of measurements	2	0.01765	0.008825	9.35	5.20
Error	2	0.001887	0.000943		
Total	8	0.339256			
R-Square = 99.44%					
Slip gauge of 50 mm					
Camera distance in mm	2	0.24827	0.124135	9.59	74.04
Light source	2	0.051945	0.025972	2.01	15.49
Number of measurements	2	0.009205	0.004602	0.36	2.75
Error	2	0.025898	0.012949		
Total	8	0.335318			
R-Square = 92.28%					
TOPSIS method					
Camera distance in mm	2	0.014174	0.007087	16.97	66.54
Light source	2	0.005895	0.002948	7.06	27.67
Number of measurements	2	0.000397	0.000198	0.48	1.86
Error	2	0.000835	0.000418		
Total	8	0.021302			
R-square = 96.08%					

Source: Author

The percentage error measured for slip gauge values is shown which are calculated for concerned actual slip gauge values (Table 82.6). From Table 82.6 it is observed slip gauges 20 mm and 40 mm show the minimum percentage error at initial conditions with values -0.7499% and 0.6345 % respectively. Slip gauges 30 mm and 50 mm show the minimum percentage error at optimal conditions with values -1.1586% and 1.0247% respectively. Neglecting the negative values and considering the positive values of percentage error it is found that the average percentage error for initial conditions is 0.9850% while the average percentage error for optimal conditions is 0.9798%. This shows that optimal conditions have a minimum percentage error than the initial conditions thus it supports the selection of the optimal values for the camera calibration.

Table 82.6 confirmation experiment

	Measured values of slip gauges		Percentage error (%)	
	Initial conditions for camera calibration	Optimal conditions for camera calibration from Taguchi analysis	For Initial conditions	For Optimal conditions
Slip gauges	Camera distance - 700 mm, light source - Red, number of measurements - 7	Camera distance - 700 mm, light source - Red, number of measurements - 5		
20 mm	20.1500	20.1631	-0.7499	-0.8156
30 mm	30.3708	30.3476	-1.2359	-1.1586
40 mm	39.7462	39.6319	0.6345	0.9202
50 mm	49.3402	49.4877	1.3195	1.0247

Source: Author

Conclusion

The objective of this experimental study was to develop an effective and simple camera calibration method and to find the optimal conditions for camera calibration. From the experimental study following conclusions are drawn.

- ANOVA shows that the camera distance is the most significant factor for the slip gauge measurement with an average contribution of 76.51% followed by the light source with a contribution of 18.56% and 2.85% for a number of measurements. The initial conditions for the camera calibration obtained using the TOPSIS method are camera distance – 700 mm, light source – Red, number of measurements – 7, and optimal conditions for the camera calibration obtained using the Taguchi method was camera distance – 700 mm, light source – Red, number of measurements – 5. The best combination of the experimental runs obtained using the TOPSIS method is 7-9-4-8-1-6-5-2-3. The confirmation experiment shows that the average percentage error in optimal conditions for camera calibration is 0.9798% which shows the best suitability of the TOPSIS method for camera calibration.

Acknowledgment

The Authors acknowledge that the experiment was performed the Dr. Babasaheb Ambedkar Technological University, Lonere, Maharashtra, India. The experiment was performed with the followed instructions without any conflicts.

References

[1] Wang Qi, Fu Li, and Liu Zhenzhong. (2010). Review on camera calibration. In 2010 Chinese Control and Decision Conference, Xuzhou, China, pp. 3354–3358. doi: 10.1109/CCDC.2010.5498574.

[2] Long, L. and Dongri, S. (2019). Review of camera calibration algorithms. In Advances in Computer Communication and Computational Sciences: Proceedings of IC4S 2018. Springer Singapore, 924, 723–732. doi: 10.1007/978-981-13-6861-5_61.

[3] Kannala, J. and Brandt, S. S. (2006). A generic camera model and calibration method for conventional, wide-angle, and fish-eye lenses. IEEE Transactions on Pattern Analysis and Machine Intelligence. 28(8), 1335–40. doi: 10.1109/TPAMI.2006.153.

[4] Sirisantisamrid, K., Tirasesth, K., and Matsuura, T. (2009). A determination method of initial camera parameters for coplanar calibration. International Journal of Control, Automation and Systems. 7(5), 777–87. doi: 10.1007/s12555-009-0510-3.

[5] Zheng, Y. and Peng, S. (2014). A practical roadside camera calibration method based on least squares optimization. IEEE Transactions on Intelligent Transportation Systems. 15(2), 831–43. doi: 10.1109/TITS.2013.2288353.

[6] Zhang, Z., Zhao, R., Liu, E., Yan, K., and Ma, Y. (2018). A single-image linear calibration method for camera. Measurement. 130, 298–305. doi: 10.1016/j.measurement.2018.07.085.

[7] Juarez-Salazar, R. and Diaz-Ramirez, V. H. (2019). Flexible camera-projector calibration using superposed color checkerboards. Optics and Lasers in Engineering. 120, 59–65. doi: 10.1016/j.optlaseng.2019.02.016.

[8] Ramesh, S. Viswanathan, R., and Ambika, S. (20166). Measurement and optimization of surface roughness and tool wear via grey relational analysis, TOPSIS and RSA techniques. Measurement. 78, 63–72. doi: 10.1016/j.measurement.2015.09.036.

[9] Gopal, P. M. and Soorya Prakash, K. (2018). Minimization of cutting force, temperature and surface roughness through GRA, TOPSIS and Taguchi techniques in end milling of Mg hybrid MMC. Measurement. 116, 178–92. doi: 10.1016/j.measurement.2017.11.011.

[10] Nguyen, H.-P., Pham, V.-D., and Ngo, N.-V. (2018). Application of TOPSIS to Taguchi method for multi-characteristic optimization of electrical discharge machining with titanium powder mixed into dielectric fluid. International Journal of Advanced Manufaturing Technology. 989(5–8), 1179–98. doi: 10.1007/s00170-018-2321-2.

[11] Ananthakumar, K., Rajamani, D., Balasubramanian, and Paulo Davim, E. (2019). Measurement and optimization of multi-response characteristics in plasma arc cutting of Monel 400TM using RSM and TOPSIS. Measurement. 135, 725–37. doi: 10.1016/j.measurement.2018.12.010.

[12] Chenchen, L., Fulin, S., Haitao, W., and Jianjun, G. (2014). A camera calibration method for obstacle distance measurement based on monocular vision. Fourth International Conference on Communication Systems and Network Technologies. pp. 1148–51. doi: 10.1109/CSNT.2014.233.

[13] Zhang, J., Yu, H., Deng, H., Chai, Z., Ma, M., and Zhong, X. (2019). A robust and rapid camera calibration method by one captured image. IEEE Transactions on Instrumentation and Measurement. 68(10), 4112–21. doi: 10.1109/TIM.2018.2884583.

[14] Setyawan, R. A., Soenoko, R., Mudjirahardjo, P., and Choiron, M. A. (2018). Measurement accuracy analysis of distance between cameras in stereo vision. Electrical Power, Electronics, Communications, Controls and Informatics Seminar (EECCIS). pp. 169–172. doi: 10.1109/EECCIS.2018.8692999.

[15] Guo, J.-M., Tian, Y.-C., and Lee, J.-D. (2008). Object matching using hybrid modified RGB color model and HRR-based background detection. 34th Annual Conference of IEEE Industrial Electronics, Orlando. pp. 2992–2997. doi: 10.1109/IECON.2008.4758437.

[16] Lee, H., Kim, H., and Kim, J.-I. (2016). Background subtraction using background sets with image- and color-space reduction. IEEE Transactions on Multimedia. 18(10), 2093–2103. doi: 10.1109/TMM.2016.2595262.

[17] Sun, Y. Liu, M., and Meng, M. Q.-H. M. Q.-H. (2019). Active perception for foreground segmentation: an rgb-d data-based background modeling method. IEEE Transactions on Automation Science and Engineering. 16(4), 1596–1609. doi: 10.1109/TASE.2019.2893414.

[18] Bhatti. U. A. et al. (2021). Advanced Color edge detection using clifford algebra in satellite images. IEEE Photonics Journal. 13(2), 1–20. doi: 10.1109/JPHOT.2021.3059703.

[19] Amanna, A. E., Ali, D., Gadhiok, M., Price, M., and Reed, J. H. (2012). Cognitive radio engine parametric optimization utilizing Taguchi analysis. 1, 5. doi: 10.1186/1687-1499-2012-5.

[20] Awty-Carroll, D., Ravella, S., Clifton-Brown, J., and Robson, P. (2020). Using a Taguchi DOE to investigate factors and interactions affecting germination in Miscanthus sinensis. Scientific Reports. 109(1), 1602. doi: 10.1038/s41598-020-58322-x.

[21] Rao, R. V. (2006). Machinability evaluation of work materials using a combined multiple attribute decision-making method. International Journal of Advanced Manufacturing Technology. 28(3–4), 221–27. doi: 10.1007/s00170-004-2348-4.

[22] Tripathy, S. and Tripathy, D. K. (2016). Multi-attribute optimization of machining process parameters in powder mixed electro-discharge machining using TOPSIS and grey relational analysis. Engineering Science and Technology, an International Journal. 19(1), 62–70. doi: 10.1016/j.jestch.2015.07.010.

Chapter 83

Synthesis, characterisation and optical studies of Cd-doped silver nanoparticles using aqueous extract of Gmelina philippensis Cham

Nida S. Shaikh[a], Mashooq Ahmed Wani[b], Rahimullah S. Shaikh[c] and Zakir S. Khan[d]

Government Vidarbha Institute of Science and Humanities, Amravati, India

Abstract

The green synthesis of silver nanoparticles extracted from Gmelina philippensis Cham. is environmentally friendly and easily affordable. The extracted silver nanoparticles were doped with cadmium nitrate, cherished their optical properties for various applications. Aqueous leaf extracts are helpful for surface stabilising and reducing silver nitrate to silver nanoparticles. Physicochemical techniques like XRD, FT-IR, and UV-Vis were used to observe the synthesis of cadmium-doped nanoparticles From X-ray diffraction analysis the prepared material reveals crystalline nature, and an average particle size of 6 nm was calculated by using FWHM. From surface plasmonic resonance (SPR) study, bands being recorded around 442 nm and 449 nm wavelengths, the prepared cadmium-doped silver nanoparticles exhibit a distinct absorption peak in UV-Visible spectroscopy. The spectral absorption by Fourier-transform infrared spectroscopy (FT-IR) determines the possible involvement of the functional group in the biosynthesis and stabilisation of Cd-doped silver nanoparticles. The peaks at 1454 cm^{-1} and 1400 cm^{-1} have been attributed to the asymmetric bending of methyl groups of flavonoids and lipids. The optical emission of silver nanoparticles due to SPR, which is based on the collective oscillation of conduction electrons brought on by an electromagnetic field, was investigated using Raman scattering and photoluminescence experiments. These minor discoveries will be helpful in experiments testing the use of metal-enhanced fluorescence for a variety of applications, including photonic devices, fluorescence imaging, biosensors, and a light-emitting diode.

Keywords: Cadmium doped silver nanoparticles (Cd-AgNPs), FT-IR and UV-visible study, green synthesis, Raman spectra

Introduction

In today's world nanoscience and nanotechnology assume much more attention and their presence can be seen in all branches of science and technology which includes the field of optics, medicine, catalysis, engineering, agriculture, and electronics [1–4]. Based on their specific characteristics such as morphology, distribution, and size nanoparticles reveal new or some enhanced properties of materials [5, 6]. Nanomaterials possess dynamic optical properties such as absorption, reflection, transmission and light emission which makes them distinguishable from the same bulk material. By modifying its shape, size and surface functionality a variety of optical effects can be produced for a variety of applications. Today silver is the most frequently used nanoparticle in everyday life as well as in research labs [7]. Silver nanoparticle which is a type of noble metal is extensively used in pharmaceuticals and also in optical materials such as surface-enhanced Raman spectrometry sensors [8], calorimetric sensors [9] chemiluminescence sensors [10] and fluorescence sensors [11, 12]. Because of these reasons, researchers are diverting their interests to metallic nanoparticles.

In recent years nanotechnology has achieved a remarkable leap, and with that respect to needs, many new synthetic routes have been developed for the synthesis of nanoparticles with particular shapes and sizes. Plant-mediated nanoparticle synthesis is highly recommended due to its economical, eco-friendly, and suitable for human usage as compared to other chemical techniques which include the use of harmful chemicals. It is pertinent to note that in the biogenesis of silver nanoparticles a large number of studies using a variety of plant extracts have been presented. Resembling this, many plant components such as alkaloids, steroids, flavonoids, tannins, and saponins have distinct stabilizing and reducing agents which in the synthesis of nanoparticles are very useful [13]. The Gmelina philippensis is a plant genus that belongs to the Labiatae family, which was previously known as Verbenaceae [14]. Gmelina is a climbing shrub with pendant branches that reaches up to 3-8 m in height, currently considered a beautiful ornamental plant. The stem is cylindrical, and monopodially branched, twigs carry spines up to 1-2 cm in length, and leaves are elliptic to obovate, smooth, and 5-7.5 cm long, 3-4 cm broad. The inflorescence is terminal with yellow slightly fragrant flowers which emerge from a pendant structure of overlapping bracts.

The cadmium as a dopant is useful for enhancing electrical properties of silver nanoparticles. The cadmium-doped silver nanoparticle from plant extract was reported for the first time. In this present work,

[a]nida.aafy@gmail.com, [b]mashooqwani25@gmail.com, [c]rahimgvish@gmail.com, [d]zakirskhan@gmail.com

DOI: 10.1201/9781003450252-83

we have synthesised cadmium-doped silver nanoparticles using an aqueous extract of Gmelina philippensis Cham. and salt of cadmium nitrate. We strongly aimed to investigate the influence of cadmium doping on the structural and optical properties of silver nanoparticles by XRD, FT-IR, UV-Visible spectroscopy, Raman spectroscopy, and Photoluminescence (PL).

Experimental Methods

Materials and Methods

The fresh and healthy leaves of Gmelina philippensis Cham. were plucked from the botanical garden of Government Vidarbha Institute of Science and Humanities, Amravati (M.S), India. Analytical grade silver nitrate, cadmium nitrate, and deionised water were used to synthesise Cd-doped silver nanoparticles.

Preparation of leaf ectraction

After being cleaned and washed with distilled water, the leaves of Gmelina philippensis, Cham. were then washed with deionised water. The leaves were powdered with an electric blender after being allowed to air dry at room temperature. 10 grams of powdered leaf and 250 ml of water were combined in a Soxhlet apparatus for the extraction illustrated in Figure 83.1. This process takes 48 hours. Finally, the resultant extract was filtered using Whatman filter paper number 1.

Biosynthesis of silver and Cd- doped silver nanoparticles

For the synthesis of Cd-doped silver nanoparticles, 1 mM aqueous solution of silver nitrate and 0.01 ppm solution of cadmium nitrate was used. 45 ml of 1 mM silver nitrate solution was thoroughly mixed with 5 ml of aqueous leaf extract, and 5 ml of 0.01ppm of cadmium nitrate solution while being constantly stirred for 15 minutes at room temperature. More characterization was done on the synthesised doped silver nanoparticles (Figure 83.2).

Characterisation of biosynthesised silver nanoparticles

X-Ray diffraction analysis

The crystallographic structure of Cd-doped AgNPs was investigated using X-ray diffraction. In the sample preparation procedure, a thin film of the sample was applied onto a glass slide by dropping 100 µL of the sample and drying for 30 minutes. The XRD pattern was captured using a Rigaku Miniflex 600 X-ray diffractometer with a 40 kV operating voltage and a current strength of 15 mA.

Figure 83.1 Preparation of plant extract using Soxhlet apparatus

Source: Author

Figure 83.2 Visual observation of Cd-doped silver nanoparticles after addition of leaf extract and cadmium nitrate solution to $AgNO_3$ solution at different time intervals (a) 1 mM $AgNO_3$ solution (b) leaf extract (c) 0.01 ppm solution of cadmium nitrate (d) just after addition (e) after 1hr (f) after 24 hrs
Source: Author

Fourier transform-IR spectra (FT-IR) Investigation

The chemical and molecular elucidation structure were examined by Fourier transform-IR spectra (FT-IR). To ascertain the presence of biomolecules coating the surface of the produced Ag-NPs, they were subjected to FT-IR (Shimadzu IR Prestige-21) spectroscopy analysis. The FT-IR measurement was conducted using KBr discs. The wavelength range of 500–4000 cm^{-1} was used to capture the spectra.

Raman Spectra Analysis

The structural behavior and chemical composition of the synthesised Cd-doped silver nanoparticle was examined from Raman spectroscopy. The Raman spectrum of the Cd-doped silver nanoparticles was recorded at room temperature. A CCD-based monochromator with a spectral range of 150-1700 cm^{-1} was used to collect and detect scattered light. The sample solution was placed in the path of the laser beam in a standard 1 cm × 1 cm cuvette.

UV- Visible Spectroscopy

The UV-Vis spectroscopy technique is a simple, remarkably sensitive method for the analysis of the stability of metal nanoparticles in an aqueous solution [21, 22]. Using a Shimadzu double-beam spectrophotometer, the surface plasmon resonance (SPR) of silver nanoparticles was examined at wavelength intervals of 200 -800 nm, keeping deionised water used as a blank.

Photoluminescence Investigation

The Photo-excitation (PLE) and Photo-emission (PL) of the synthesised compound were carried out by using Hitachi F-7000 Fluorescence Spectrophotometer.

Results and Discussion

X-Ray diffraction analysis

Silver nanoparticles doped with cadmium can be seen in an X-ray diffractogram that supports their crystalline nature (Figure 83.3a). The diffraction peak at a 2θ value of 31.68 broad peaks confirms the Cd-AgNPs formation. We introduced 1 mM aqueous solution of silver nitrate and 0.01 ppm of cadmium nitrate was used for preparation. The concentration content of cadmium ions in the synthesis of silver nitrate nanoparticles does not lead to substantial changes in the position of the profile peak size distribution. Consequently, at the same time with cadmium ion concentrations during the synthesis of silver nanoparticles, we observed optical density growing intensely in the spectra. From, earlier reports reflections corresponding to the (111), (220), and (311) faces of cubic cadmium sulfide match well with our synthesised material [15 16]. Here silver nanoparticles doped with a growing concentration of cadmium ion show no change in spectra. In X-ray diffraction, we have observed one broadband in the region of 2θ = 10-30°, also one intense peak of 31-32°depicated in Figure 83.3b. The presence of two such broad bands in the X-ray analysis of the nanoparticles is due to the small particle size distribution of the cadmium nitrate. The average crystallite size was calculated from the full width at half maximum (FWHM) diffraction peaks by the Scherrer formula depicted below,

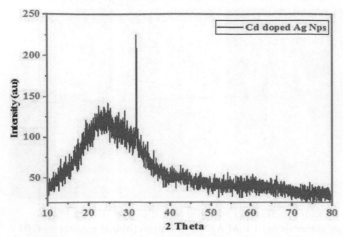

Figure 83.3a X-ray diffractogram patterns of Cd-doped silver nanoparticle
Source: Author

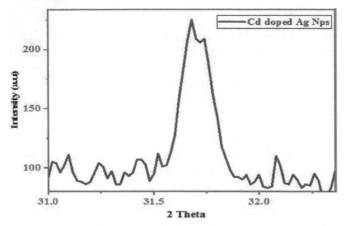

Figure 83.3b Broad peaks of Cd-doped silver nanoparticle recorded from the x-ray diffraction pattern
Source: Author

$$D = \frac{0.9.\ \lambda}{\beta Cos\theta} \qquad\qquad (1)$$

Where β-Width of diffraction peak in radian = 0.28967
λ- wavelength of X- rays (1.54 °A)
θ- diffraction angle
D-Average diameter

The average size of the nanoparticle that we obtained is in the order of 6 nm, leading to the broadening of the Ag ions peaks and consequently, their overlap with the formation of broad bands. The Cd peak presence observed in the X-ray diffractograms is clearly due to its small content insufficient for determination by X-ray diffraction.

Fourier transform-IR spectra (FT-IR) Investigation

The spectral absorption was performed to determine the possible involvement of the functional group in the biosynthesis and stabilization of Cd-doped silver nanoparticles as shown in Figure 83.4. The broad absorption peak at 3319.00 cm^{-1} was due to the O-H stretching of phenolic compounds, while 2395.59 cm^{-1} represents the N-H stretching of secondary amine. A strong band at 1616 cm^{-1} was assigned to the bending vibration of the amide group in proteins [17].

The peaks at 1454 cm^{-1} and 1400 cm^{-1} have been attributed to the asymmetric bending of methyl groups of flavonoids and lipids [18]. The absorption peak at 1114.86 cm^{-1} was caused by C-O-C vibration. Notably, from the examined spectral absorption bands results, it can be concluded that the bioactive components such as flavonoids, alkaloids, saponins, terpenoids, phenols, amino acids, and tannins are responsible for the synthesis and stabilization of silver nanoparticles in an aqueous medium.

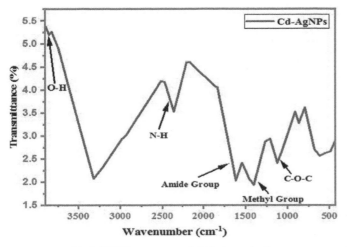

Figure 83.4 FT-IR spectra of Cd-doped silver nanoparticles extracted form plant
Source: Author

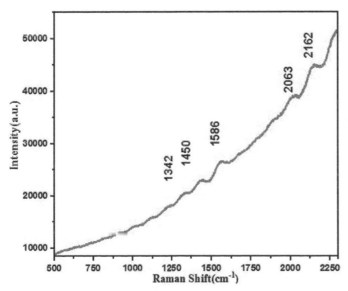

Figure 83.5 Raman spectra of Cd-doped silver nanoparticles
Source: Author

Raman Spectra Analysis

To find out the possible functional groups of capping agents associated with the stabilization of silver nanoparticles we prefer Raman spectroscopy. Surface-enhanced Raman spectroscopy was used to analyse the colloidal dispersion of synthesised co-doped silver nanoparticles extracted from plant, as shown in Figure 83.5. The peak at 1342 cm^{-1} was assigned as symmetric vibrations of the C=O bond. The other peak at 1450 cm^{-1} ascribed due to the O-H bending in the prepared compound. While C-N stretching vibration peak around 1586 cm^{-1} containing elements of the amide groups, showed that nanoparticles were interacting with the peptide linkage's backbone. The peak at 2162 cm^{-1} was caused by the stretching of C-C. [19, 20].

UV- Visible Spectroscopy

The absorbance spectra of aqueous leaf extract, 1 mM aqueous solution of silver nitrate, 0.01 ppm solution of cadmium nitrate, and Cd-doped silver nanoparticles synthesised using aqueous leaf extract of Gmelina philippensis Cham. at different time, intervals (i.e., 1 hour., and 24 hours.) as shown in Figure 83.6. Silver nanoparticle formation was verified by a change in color from colorless to reddish brown, which over time turned dark brown [23]. The SPR peak of synthesised Cd-doped silver nanoparticles was obtained from UV-visible analysis at 442 nm and 419 nm. The various bioactive components present in the plant extract when it is mixed with an aqueous solution of silver nitrate then it is responsible for making a broad plasmon band. By the above observation, it was concluded that the intensity of the peak of SPR

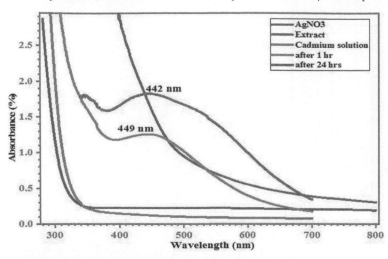

Figure 83.6 UV-Vis. spectra of pure and cadmium doped silver nanoparticles extracted from plant
Source: Author

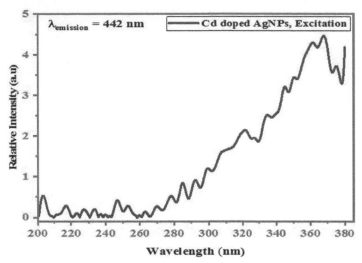

Figure 83.7a Photoluminescence excitation (PL) of spectra of Cd-doped silver nanoparticles extracted from plant at room temperature
Source: Author

increases with an increase in reaction time due to a continuous increase in the number of silver nanoparticles formed because of the reduction of silver ions present in the aqueous of solution of silver nitrate by leaf extract.

The peak of SPR of synthesised Cd-doped silver nanoparticles appears at 449 nm. No other absorption peaks in the red and near-infrared range show throughout the whole synthesis process, suggesting that all the synthesised silver nanoparticles have a spherical configuration. This is because although anisotropic particles, depending on their forms, exhibit two or three SPR bands, small spherical or quasi-spherical nanocrystals only exhibit one. The existence of tiny, spherical silver nanoparticles is shown by the absorption band at 442 and 449 nm.

Photoluminescence Investigation

The photoluminescence excitation (PL) of spectra of Cd-doped silver nanoparticles using the leaves of Gmelina philippensis Cham. fixed at room temperature depicted in Figure 83.7(a). The excitation spectra of Cd-doped silver nanoparticles samples at 442 nm and 449 nm, exhibit excitation bands ranging from 300-380 nm, bearing a strong absorption peak around 362 and 378 nm. Upon 352 nm excitation, a series of emissions are recorded around 400-650 nm illustrated in Figure 83.7(b). The most dominant emission peak of the Cd-doped silver nanoparticles was recorded at 442 nm. It is due to the excitation of electrons from the orbital into the state above the Fermi level and it is responsible for the visible luminescence property of the silver nanoparticles [24]. The cadmium ions incorporated with silver nitrate nanoparticles lead to a substantial change in the emission region. We observed optical density growing

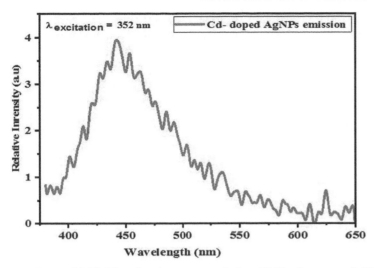

Figure 83.7b Photoluminescence emission (PLE) of spectra of Cd-doped silver nanoparticles extracted from plant excited at 352 nm
Source: Author

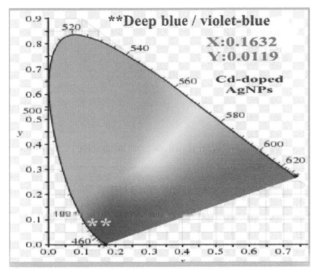

Figure 83.8 CIE diagram, shows color emission of Cd doped Ag NPs
Source: Author

intensely. The characteristic peaks of the said Cd-doped AgNPs were recorded in 420-500 nm regions bearing blue bands. In an earlier report, the photoluminescence characteristics of AgNps, emission peak was observed at 473 nm, upon excitation at 343 nm [25]. This provides evidence of doping affects the photoluminescence property of silver particles. These nanoparticles may be used in various biosensor devices because their optical properties depend on intraband and interband transitions between electronic states.

CIE chromaticity coordinates of Cd-doped AGNPs

The Commission Internationale de l'Eclairage (CIE) coordinates determines color formation of the compounds. The CIE coordinates (x, y) of the particular materials evaluate color bands corresponding to their particular spectra. The chromaticity coordinates of Cd-doped AgNPs have illustrated in Figure 83.8. The calculated CIE coordinates are **(0.1632, 0.119). At 442-449 nm, synthesised materials showed a formation of deep blue/violent-blue color region bands.** Therefore, these nanoparticles find excellent applications in biosensors, as well as in photonic devices.

Conclusion

The synthesis of the cadmium-doped silver nanoparticles was accomplished using an easy, affordable, and environmentally friendly process. This study examined how cadmium nitrate dopant affected silver nanoparticles extracted from the said plant. By using XRD, FT-IR, Raman and UV-Visible spectroscopy, the cd-doped silver nanoparticles have been characterised. The average size of cd-doped silver nanoparticles extracted from the plant leaf was found 6 nm from FWHM. From FT-IR analysis different spectral absorption bands of functional group of various elements were recorded including flavonoids, alkaloids, saponins, terpenoids, phenols, amino acids, and tannins are responsible for the synthesis and stabilization of silver nanoparticles in an aqueous medium. The peak of SPR of synthesised Cd-doped silver nanoparticles appears at 449 nm. No other absorption peaks in the red and near-infrared range were shown throughout the whole synthesis process Cd-doped silver nanoparticles were studied for their photoluminescence for the first time. The confirmation of emission regions was predominantly found in NUV to visible regions. Herein, Cd-doped AgNPs, we found deep blue to violet-blue at 442 nm and 449 nm from CIE chromaticity. Notably, the creation of biosensor devices, photonics devices, as well as other uses may benefit from the use of the Cd-doped silver nanoparticles.

References

[1] Ovais, M., Zia, N., Ahmad, I., Khalil, A. T., and Raza, A. (2018). Phyto-therapeutic and nanomedicinal approach to cure alzheimer disease : phyto-therapeutic and nanomedicinal approaches to cure alzheimer's disease : present status and future opportunities.Frontiers in aging neuroscience, 10.

[2] Khalil, A.T., Raza, A., Ayaz, M., (2018). Multifunctional theranostic applications of biocompatible green-synthesised colloidal nanoparticles. 102, 4393–08.

[3] Adelere, I. A., and Lateef, A. (2016). A novel approach to the green synthesis of metallic nanoparticles : the use of agro-wastes , enzymes and pigments. Nanotechnol Rev. 5(6), 567–87.

[4] Shaikh, N. S., Shaikh, R. S., and Kashid, S. (2020). In vitro biosynthesis of silver nanoparticles using flower extracts of parasitic plant Cascute reflexa and evaluation of its biologocal properties. 3, 121–30.

[5] Murugapandi, M., and Muniyappan, N. (2018). Biogenic synthesis , characterization and pharmacological study of silver nanoparticles using enicostema axillare leaves. International Journal of Pharmacognosy and Chinese Medicine. 2(5), 1–10.

[6] Riyanto, R., Mulwandari, M., Asysyafiiyah, L., Sirajuddin, M. I., and Cahyandaru, N., (2021). Direct synthesis of lemongrass (cymbopogon citratus) essential oil- silver nanoparticle (eo-agnps) as biopesticides and application for lichens inhibition on stones. SSRN Electronic Journal. 8, 9701.

[7] Sivaraj, R., Vanathi, R., (2017). Anticancer potential of green synthesised silver nanoparticles : A Review .International journal of Current Research. 7(10), 21539-21544.

[8] Gomezb, H.E., Lucía Z., Lopeza, F., Barrerac, E.L.S., Rivera, A.N. (2019). Applied surface science study of the green synthesis of silver nanoparticles using a natural extract of dark or white Salvia hispanica L . seeds and their antibacterial application. Applied Surface Science. 489, 952–61.

[9] Ren, Y. Y., Yang, H., Wang, T., and Wang, C., (2016). Bio-synthesis of silver nanoparticles with antibacterial activity. Materials Chemistry and Physics. 235, 121746, 2019.

[10] Jouyban, A., and Rahimpour, E. (2020). Optical sensors based on silver nanoparticles for determination of pharmaceuticals: An overview of advances in the last decade. Talanta. 217, 121–71.

[11] Wu, Z., Song, N., Menz, R., Pingali, B., Yang, Y. W., and Zheng, Y. (2015). Nanoparticles functionalised with supramolecular host-guest systems for nanomedicine and healthcare. Nanomedicine. 10(9), 1493–1514.

[12] Das, P., Dutta, T., Manna, S., Loganathan, S., and Basak, P. (2022). Facile green synthesis of non-genotoxic, non-hemolytic organometallic silver nanoparticles using extract of crushed, wasted, and spent Humulus lupulus (hops): Characterization, anti-bacterial, and anti-cancer studies. Environmental Research. 204(PA), 111–962.

[13] Dawadi, S., Katuwal, S., Gupta, A., Lamichhane, U., (2021). Current Research on Silver Nanoparticles: Synthesis, Characterization, and Applications Sonika. Journal of Nanomaterials. 6687290.

[14] Katta, V. K. M., and Dubey, R. S. (2021). Green synthesis of silver nanoparticles using Tagetes erecta plant and investigation of their structural, optical, chemical and morphological properties. Materials Today: Proceedings. 45, 794–98.

[15] Sayed, H. M., Ahmed, A. S., Khallaf, I. S., Asem, A. Botanical profiling of Gmelina Philippensis Cham. cultivated in Egypt. Bull. Pharm. Sci., Assiut University. 41, 55–80.

[16] Mousavi, S. M., et al. (2018). Green synthesis of silver nanoparticles toward bio and medical applications: review study. Artificial Cells, Nanomedicine, and Biotechnology. 46(sup 3), S855–72.

[17] Desai, R., Mankad, V., Gupta, S. K., and Jha, P. K. (2012). Size distribution of silver nanoparticles: UV-visible spectroscopic assessment. Nanoscience and Nanotechnology Letters. 4(1), 30–34.

[18] Singla, S., Jana, A., Thakur, R., Kumari, C., Goyal, S., and Pradhan, J. (2022). Green synthesis of silver nanoparticles using Oxalis griffithii extract and assessing their antimicrobial activity. OpenNano. 7, 100–47.

[19] Raja, S., Ramesh, V., and Thivaharan, V. (2017). Green biosynthesis of silver nanoparticles using Calliandra haematocephala leaf extract , their antibacterial activity and hydrogen peroxide sensing capability. Arabian Journal of Chemistry. 10(2), 253–61.

[20] Gao, J., Fu, D., Lin, C., Lin, J., Han, Y. Yu, X., (2004). Formation and photoluminescence of silver nanoparticles stabilised by a two-armed polymer with a crown ether core. Langmuir. 20(22), 9775–9.

[21] Prakash, P., Gnanaprakasam, P., Emmanuel, R., Arokiyaraj, S., and Saravanan, M. (2013). Green synthesis of silver nanoparticles from leaf extract of Mimusops elengi, Linn. for enhanced antibacterial activity against multi drug resistant clinical isolates. Colloids Surfaces B Biointerfaces. 108, 255–9.

[22] Oliveira, R.N., Mancini, M.C., Oliveira, F.C.S., Passos, T.M., Quilty, B. FTIR analysis and quantification of phenols and flavonoids of five commercially available plants extracts used in wound healing. Reviews Materials. 21(3), 767–79.

[23] Joshi, N., Jain, N., Pathak, A., Singh, J., Prasad, R., and Prakash, C. (2018). Biosynthesis of silver nanoparticles using Carissa carandas berries and its potential antibacterial activities. Journal of Sol-Gel Science and Technology. 86, 682–89.

[24] Szekeres, G. P., and Kneipp, J. (2019). SERS probing of proteins in gold nanoparticle agglomerates. Frontiers in Chemistry. 7.

[25] Ganesan, V., Deepa, B., Nima, P., and Astalakshmi, A. (2014). Bio-inspired synthesis of silver nanoparticles using leaves of millingtonia hortensis L.F. International Journal of Advanced Biotechnology and Research. 5, 93–100.

Chapter 84

Impact of concentration of polyacrylonitrile on electrochemical performance of carbon nanofibers

Diptee Jamkar[1,a], Hemlata Ganvir[1,b], Renuka Mahajan[2,c], and Pavan Kumar[3,d]

[1]Yeshwantrao Chavan College of Engineering, Nagpur, India

[2]Nagpur College of Pharmacy, Nagpur, India

[3]DMIMSU, India

Abstract

In the article, we describe the synthesis of carbon nanofibers (CNFs) using electrospinning solution polyacrylonitrile (PAN) in N,N dimethyl formamide by varying PAN concentrations and following the steps of stabilisation and carbonisation in a quartz of tube shaped furnace. Using scanning electron microscopy (SEM), the morphology of the CNFs was examined. In the region of 400–500 nanometer, the diameter of CNFs at 6wt%, 8wt%, and 10wt% indicates nanoscale fibers with porosity. Using cyclic voltammetry (CV) and galvanostatic charge discharge (GCD), the electrochemical study of as synthesised CNFs were investigated (GCD). The CV curves of CNFs have distorted rectangular shape, which is indicative of a perfect double layer electric capacitor. The discharge time of 10wt% CNFs is longer than that of 6wt% and 8wt% CNFs. Increasing the PAN concentration to 10 wt% is what makes the CNFs so effective electrochemically.

Keywords: electrochemical performance, electrospinning, PAN nanofibers carbon nanofibers, supercapacitor

Introduction

Graphene and carbon nanofibers (CNFs) have both gained popularity as potential supercapacitors in recent years. Graphene is monolayered or multilayered graphitic nanosheets of carbon. In the domain of nanoscience, CNFs are of tremendous significance [1-3]. CNFs with submicron and nanometer dimensions have gained attention due to their exceptional chemical, electrical, and mechanical capabilities and distinctive 1D nanostructure [4]. Polyacrylonitrile (PAN) is used to make CNFs. Electrospinning PAN, stabilisation, and carbonisation yield hierarchical CNFs. According to Zang et al. catalysis, sensing, adsorption, energy conversion, and storage and biological applications have quickly adopted this novel type of carbon material with 1D nanostructure and large specific surface area [5].

Precursors including rayon, PAN, and mesophase pitch are used in the synthesis of carbon fibers (CFs). The first prototype to carbon fibers was rayon. Using rayon-based CF as a precursor for producing commercial CFs has been hampered by its low productivity. Thus, PAN and mesophase pitch have been favored as precursors for the manufacture of carbon fibers in industrial settings. PAN-based CFs have better mechanical qualities, especially tensile strength, than mesophase pitch-based ones. Because of this, PAN has been widely employed as a precursor for carbon fibers for nearly three decades [6-9]. Electrospun nanofibers were reported by Alarifi1, et al., have been used as a precursor in the development of PAN-derived carbon nanofibers. Electrospun PAN fibers were stabilised and then carbonised to create PAN-based carbon nanofibers [3].

By stabilising and carbonising electrospun precursor polymer nanofibers, Reneker et al. created the first continuous electrospun CNFs [10]. CNFs have demonstrated considerable potential to produce inexpensive, structure-flexible, and property- customisable porous electrodes for use in rechargeable batteries and supercapacitors. CNFs have a considerably bigger surface area which is functionalised than carbon nanotubes (CNTs) is due to distinct carbon orbital stacking on the surface. Recent electrochemical investigations have demonstrated that CNFs can enhance the electron transfer kinetics reactions, reduce surface fouling of electrode, and boost electrocatalytic activity [11]. The preparation of molybdenum carbide (MO_2C) supported CNFs for the liquid-phase hydrodeoxygenation of guaiacol [12] at varying carburisation temperatures and heating rates was fascinating.

Strong temperatures for carburisation and modest rates of heating produce well-defined β-MO_2C crystals with high catalytic activity. Synthesis and use of flexible bifunctional nano electrocatalytic textile materials, Fe3O4 NP-CNFs, for the effective degradation and complete mineralisation of carbamasepine in H2O [13]. CNFs based ruthenium nanoparticles (Ru-CNF) are synthesised by ozone-assisted atomic layer deposition (ALD) of Ru NPs on electrospun CNF. The Ru-CNF snano catalyst system hydrolytically dehydrated methylamine boranc ($CH_3NH_2BII_3$) [14]. Ru-CNF preserved 72% of catalytic activity after five reuse cycles and converted >99% in air at ambient temperature.

[a]jamkardipti13@gmail.com, [b]hrwasnik@gmail.com, [c]renukamahajan@gmail.com, [d]pavank.feat@dmimsu.edu

DOI: 10.1201/9781003450252-84

Recently published uses of electrospun CNFs include the creation of 3D cross-linked nitrogen-enriched porous CNF (CCNF) from a polyacrylonitrile/zinc chloride (PAN/ZnCl2) precursor, which can be used in supercapacitors. In alkaline electrolyte, the CCNFs had a cycling stability nearly of 98% over 60000 cycles, while in acidic electrolyte, they demonstrated a high specific capacitance about 215 F g^{-1} at 1 A g^{-1} [15].

On the contrary, electrospinning as a scalable, low-cost, nonmanufacturing process has been widely studied for creating continuous nanofibers of a wide range of metals, polymers, metal oxides, polymer-derived carbon, ceramics, etc. Electrospinning also enables the fabrication of porous, low-cost, nanofibrous electrodes for rechargeable batteries and supercapacitor. CNFs prepared by carbonizing the electrospun polymer nanofibers exhibit exceptional thermal and electrical conductivities and high connectivity for the transfer of thermal as well as electrical currents, attracting the attention of materials scientists and engineers worldwide.

Due to its high enough capacity, good power density and durable life cycle with low value of internal resistance, supercapacitors (SCs) are a significant energy storage device. When weighed against more conventional energy storage options like secondary batteries, supercapacitors emerge as a promising alternative [16–22]. Supercapacitors performance of is typically determined by the chemical and physical properties of electrode materials [23–25, 33, 35]. The electrode materials have a significant impact on the electrochemical characteristics of supercapacitors; therefore, it is crucial to create new materials with high value capacitance and power density to meet the needs of industry.

Large surface area, moderate diameter, functionalities of surface to ensure a quick oxidation-reduction process, and good electrical conductivity are all important factors in the electrochemical study of electrodes. The development of carbonaceous materials as electrodes have received a lot of attention recently due to its excellent electrical conductivity, non-toxicity, low cost, high chemical stability, and ease of fabrication [26, 27, 35]. The CNFs synthesised by electrospinning, exhibiting high surface area and porosity, stand out as a promising option for electrodes. CNFs are currently only useful for supercapacitors with low power and energy densities, though.

CNFs are modified to enhance their electrochemical characteristics. Nonwoven carbon webs were created by Kim et al. using the electrospinning method [28].

More activation time during carbonization was found to be beneficial to the formation of microporous structural fibers. Thus, fiber surface area was increased, which in turn enhanced specific capacitance. Polymethyl hydrosiloxane (PMHS) added to the PAN precursor solution might also be used to create carbon nanofibers with a microporous structure [29, 35].

With this context, we constructed supercapacitor electrodes employing continuous CNFs of 6wt%, 8wt%, and 10wt%. The study's goal was to highlight the benefits of electrospun CNTs for storing electrical energy, namely their continuity and high electrical conductivity. The CNFs were made by first electrospinning a PAN/N,N-dimethylformamide (DMF) solution of changing PAN concentration, then stabilizing the filaments, and then carbonizing the filaments. The electrochemical performance of novel CNFs-based electrodes were compared.

Experimental Details

Materials and Method

Sigma-Aldrich supplied PAN (98.5% purity, Mw = 150,000) and N,N-DMF with 99% purity. All the compounds were utilised as received, without any additional purification or alteration. By use of SEM, the morphologies and structures of the prepared CNFs were analysed (Zeiss EVO18). On a CH Instruments of Model- CHI6112D electrochemical workstation equipped with a conventional symmetrical three-electrode system, electrochemical study was carried out using cyclic voltammetry (CV) method, galvanostatic charge-discharge (GCD) method. For electrochemical measurement, 1 M KOH was utilised as the electrolyte. From galvanostatic charge-discharge, the electrode's specific capacitance was computed.

Preparation of Carbon Nanofibers (CNFs)

Electrospinning was used to create PAN nanofibers in three distinct weight percent (6%, 8%, and 10%). The following procedure was used to create a 6 wt% PAN nanofiber solution in DMF. First, 6 wt% PAN-DMF solution was prepared by dissolving PAN powder in DMF and stirring the mixture for 1 hour at 60 °C. The produced solution was then transferred to a syringe fitted with a stainless spinneret and linked to a DC power supply positive high voltage for the electrospinning process. The nanofibers were collected using an electrically grounded, 3.3 cm diameter, revolving aluminium disc that was created in the lab. The syringe needle was placed in the collector and a high DC positive voltage of 18 kV was delivered. A constant 22 cm was set as the distance between the syringe's tip and the collector.

With a computer-controlled syringe pump, the rate of flow for the electrospinning solution was maintained at 0.5 ml/hr [30,31]. Following electrospinning the solution, the PAN nanofibers were removed from the foil, and then dried in an oven at 90°C for 9 hours before undergoing stabilization. Following their initial preparation, PAN nanofibers were stabilised and carbonised in a quartz furnace of tube shape. For the objective of oxidative stabilizing the PAN, the electrospun PAN nanofibers were initially heated with rate of 0.8°C min^{-1} before being held at 300°C for an hour in air. As the temperature was increased from 300 to 850°C in a Nitrogen environment at a rate of 10 °Celsius min^{-1}, the fibers were carbonised and then annealed at 850°C for 30 minutes. After cooling to 400°C in N2, the furnace was held at this temperature (400°C) by adding air for 1 hour to activate the carbonised PAN nanofibers. The same processing settings were used to synthesise CNFs containing 8 wt% and 10 wt% PAN-DMF solutions for the sake of comparison [30, 31, 34, 35].

Characterisations

A SEM was used to examine the structure and morphology of 6 wt%, 8 wt%, and 10 wt% CNFs (Zeiss EVO18). The standard symmetrical three-electrode system was used in these studies of electrochemical performance on a CH Instruments electrochemical workstation. In this example, solution of a 1 M KOH was utilised as the electrolyte in electrochemical measurements. Electrochemical characteristics of samples were determined using a CHI 6112D electrochemical workstation, which was equipped to perform cyclic voltammetry (CV) method and Galvanostatic charge-discharge (GCD) method, Galvanostatic charge-discharge was used to determine the electrode's specific capacitance.

Result and Discussion

Scanning Electron Microscopy (SEM)

Standard SEM images of CNFs at 6 wt%, 8 wt%, and 10 wt% s are shown in (a), (b), and (c) of Figure 84.1. The diameter of these carbon fibers is measured to be between 400 and 500 nm, and they are incredibly smooth. The diameter of Increment in the polyacrylonitrile concentrations from 6 wt% to 10 wt% was found to improve the mean diameter.

Electrochemical Properties

In 1 M H2SO4 aqueous solution, the electrochemical characteristics of CNFs of 6wt%, 8wt%, and 10wt% concentrations were examined. Figure 84.2 (a), (b), and (c) illustrate the cyclic voltammetry curves of CNFs 6wt%, 8wt%, and 10 wt% at scan rates ranging from 5 mV/s to 100 mV/. In a cyclic voltammetry (CV) experiment, the potential of working electrode was linearly increased in relation to time. CV experiments differ from linear sweep voltammetry in that, once the potential which is set is attained, potential of the working electrode's is ramped in the opposite direction to return to the initial

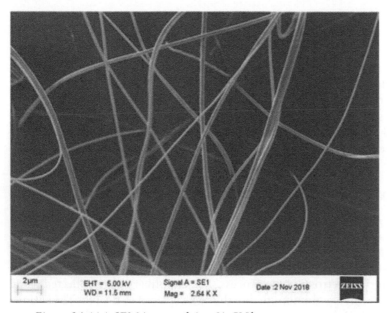

Figure 84.1(a) SEM image of 6 wt% CNfs
Source: Author

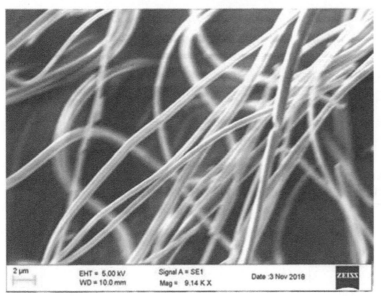

Figure 84.1(b) SEM image of 8 wt% CNfs
Source: Author

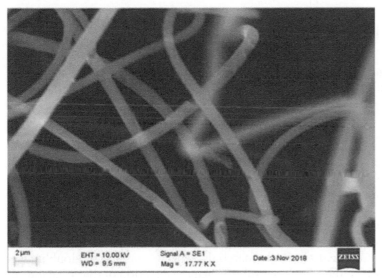

Figure 84.1(c) SEM image of 10 wt% CNfs
Source: Author

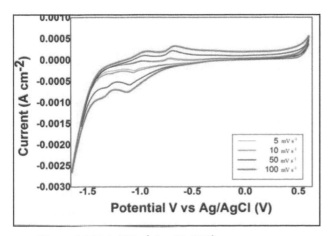

Figure 84.2(a) CV of 6 wt% CNfs
Source: Author

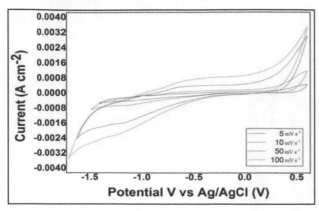

Figure 84.2(b) CV of 8 wt% CNfs
Source: Author

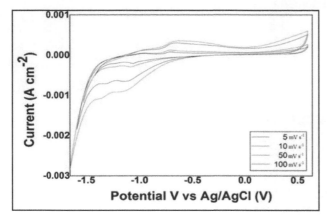

Figure 84.2(c) CV of 10 wt% CNfs
Source: Author

potential. These cycles of potential ramps may be repeated as often as required. The cyclic voltammogram is generated by plotting the current at the working electrode vs the applied voltage to generate the trace.

Generally, CV is utilised to investigate the electrochemical characteristics of materials based on carbon. CV supplied detail regarding the electrodes' reversible nature. The geometry of the curves represents a frame of rectangular type, showing that optimal electrode materials generate an electric double layer quickly and evenly at the electrode/electrolyte interface topography. When the direction of the voltage scanning is reversed, the steady-state current can be reached swiftly. Since this is the case, the CV curve for the perfect electrode material is depicted as nearly having a rectangular shape [32–35]. At a given scan rate, the integral area of the 10wt% CNFs CV curve is greater than that of CNFs. 10wt% carbon nanofibers have greater capacitance than 6wt% and 8wt%.

The rate performance of the CNFs at 6wt%, 8wt%, and 10wt% was measured using galvanostatic charge-discharge (GCD) measurements. High densities of 1.4 A/g, as displayed in Figure 84.3, necessitate the capacity to maintain a high capacitance during quick charge/discharge operations. According to the CV curves, the electric double-layer capacitive behavior is depicted in the picture as a charge/discharge curve in the shape of a triangle.

The specific capacitances (SC) of electrodes were calculated from the discharge time curves with the following equation:

$$Sc = \frac{I.\Delta t}{m.\Delta V} \tag{1}$$

Where I is the desired current, Δt is the time of discharge, V is the potential window, and m is the loaded mass of the sample. SC is specific capacitance. Although SC often decreases with increasing numbers of charge-discharge cycles, the maximum number of cycles that may be obtained through galvanostatic studies is proportional to the amount of the current. Parameters such as current density and discharge time can be adjusted to maximise the number of charge-discharge cycles that can occur without a noticeable drop

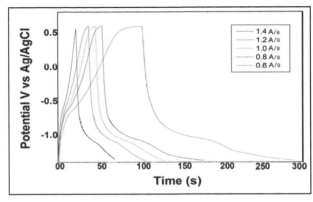

Figure 84.3(a) GCD of 6 wt% CNfs
Source: Author

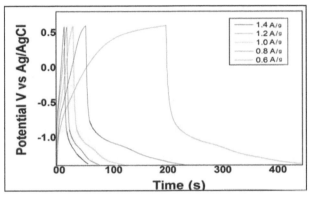

Figure 84.3(b) GCD of 8 wt% CNfs
Source: Author

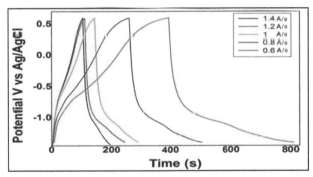

Figure 84.3(c) GCD of 10 wt% CNfs
Source: Author

in specific capacitance. Specific capacitance for three-electrode systems based on conducting polymer (CP) materials such as polypyrrole and polyaniline, with electrodes consisting of nickel, stainless steel, carbon, etc. [32]. From Figure 84.3 (a), (b) and (c) discharging time of CNFs were calculated.

Time taken to discharge 10 wt% CNFs is longer than that required by 6wt% and 10wt% CNFs, respectively. The synergistic effects of carbon nanofibers are responsible for the increased supercapacitance at 10 wt% CNFs.

Conclusion

Electrospinning and carbonisation have been used to successfully create carbon nanofibers (CNFs) with 6wt%, 8wt%, and 10wt% carbon by weight. 6wt%, CNFs, 8wt% CNFs, and 10wt%, CNFs all have diameters between 400 and 500 nm. With increasing polyacrylonitirle (PAN) concentration in dimethylformamide (DMF), the specific capacitance of CNFs was observed to rise. The CNFs' 10wt% discharge time is primarily to blame for their high specific capacitance. The synergistic effects of CNFs account for the increased specific capacitance seen at 10 wt% CNFs.

Acknowledgment

The present work is supported financially by RTM Nagpur University Research Project Support Grant No.Dev./RTMNURP/AH/ 1672 (II).

References

[1] Blake, P., Yang, R., Morozov, S. V., Schedina, F., Ponomarenko, L. A., Zhukov, A. A., Nair, R. R., Grigorieva, I. V., Novoselov, K. S., and Geim, A. K. (2009). Influence of metal contacts and charge inhomogeneity on transport properties of graphene near the neutrality point. Solid State Communications. 149, 1068–71.

[2] Katsnelsona, M. I. and Novoselov, K. S. (2007). Graphene: New bridge between condensed matter physics and quantum electrodynamics. Solid State Communications. 144, 3–13.

[3] Meyer, J. C., Geim, A. K., Katsnelsond, M. I., Novoselo, K. S., Obergfelle, D., Rothe, S., Girita, C., Zettl, A. (2007). On the roughness of single-and bi-layer graphene membranes. Solid State Communications. 143, 101–9.

[4] Dahal, B., Mukhiya, T., Ojha, G. P. Alagan Muthurasu. (2019). In-built fabrication of MOF assimilated B/N co-doped 3D porous carbon nanofiber network as a binder-free electrode for supercapacitors. Electrochimica Acta. 301, 209–219.

[5] Morgan, P. (2005). Carbon Fibers and their Composites. Boca Raton: CRC Press.

[6] Zhang, L., Aboagye, A., Kelkar, A., Lai, C., and Fong, H. (2014). A review: carbon nanofibers from electrospun polyacrylonitrile and their applications. Journal of Material Science. 49, 463–480.

[7] Lee, S., Kim, J., Ku, B. C., Kim, J., and Joh, H. (2012). Structural evolution of polyacrylonitrile fibers in stabilization and Carbonization. Advances in Chemical Engineering and Science. 2, 275–82.

[8] Buckley, J. D., and Edie, D. D. (1993). Carbon-carbon materials and composites. Noyes. 1254, 35012.

[9] Seo, M. K., Park , S. H., Kang, S. J., and Park, S. J. (2009). Carbon fibers (III): recent technical and patent trends. Carbon Letters. 10(1), 43–51.

[10] Alarifi1, I. M., Khan, W. S., and Asmatulu, R. (2018). Synthesis of electrospun polyacrylonitrile derived carbon fibers and comparison of properties with bulk form. PLOS ONE. 13(8), 201345.

[11] Muthirulan, P., and Velmurugan, R. (2011). Direct electrochemistry andelectrocatalysis of reduced glutathione on CNFs–PDDA/PB nanocomposite film modified ITO electrode for biosensors. Colloids and Surfaces B: Biointerfaces. 83, 347–54.

[12] Ochoa, E., Torres, D., Moreira, R., Pinilla, J. L., and Suelves, I. (2018). Carbon nanofiber supported Mo2C catalysts for hydrodeoxygenation of guaiacol: theimportance of the carburization process. Applied Catalysis B: Environmental. 239, 463–74.

[13] Liu, K., Yu, J. C., Dong, H., Wu, J. C. S., and Hoffmann, M. R. (2018). Degradation and mineralization of carbamasepine using an electro-fenton reaction catalysed by magnetite nanoparticles fixed on an electrocatalytic carbon fiber textile cathode. Environmental Science and Technology. 52(21), 12667–74.

[14] Khalily, M. A., Yurderi, M., Haider, A. Bhushan Patil, Ahmet Bulut, Mehmet Zahmakiran and Tamer Uyar. (2018). Atomic layer deposition of ruthenium nanoparticles on electrospun carbon nanofibers: a highly efficient nanocatalyst for the hydrolytic dehydrogenation of methylamine borane. ACS Applied Materials Interfaces. 10(31), 26162–9.

[15] Jiang, Q., Pang, X., Geng, S. S. Yuhui Zhao. (2019). Simultaneous cross-linking and pore-forming electrospun carbon nanofibers towards high capacitive performance. Applied Surface Science. 479, 128–36.

[16] Kim, B. H., Yang, K. S., Kim, Y. A., Kim, Y. J., Bai An, and Oshida, K. (2011). Solvent- induced porosity control of carbon nanofiber webs for supercapacitor. Journal of Power Source. 196, 10496–501.

[17] Niu, Z., Liu, L., Zhang, L., Zhou, W., Chen, X., and Xie, S. (2015). Programmable nanocarbon- based architectures for flexible Supercapacitors. Advanced Energy Materials. 5(23), 1500677.

[18] Zhang, L. L., and Zhao, X. S. (2009). Carbon-based materials as supercapacitor electrodes. Chemical Society Reviews. 38, 2520–31.

[19] Zhang, W., Lin, H., Lin, Z., Yin, J., Lu, H., Liu, D., and Zhao, M. (2015). 3D Hierarchical porous carbon for supercapacitors prepared from lignin through a facile template- free method. ChemSusChe. 8, 2114–22.

[20] Dong, Y., Lin, H., Zhou, D., Niu, H., Jin, Q., and Qu, F. (2015). Synthesis of mesoporousgraphitic carbon fibers with high performance for supercapacitor. Electrochim Acta. 159, 116–23.

[21] Li, L., Wu, Z., Yuan, S., and Zhang, X.-B. (2014). Advances and challenges for flexible energy storage and conversion devices and systems. Energy and Environmental Science. 7, 2101–22.

[22] Simon, P., and Gogotsi, Y. (2008). Materials for electrochemical capacitors. Nature Materials. 7, 845–54.

[23] Ran, F., Zhang, X., Liu, Y., Shen, K., Niu, X., Tan, Y., Kong, L., Kang, L., Xu, C., and Chen, S. (2015). Super long-life supercapacitor electrode materials based on hierarchical porous hollow carbon microcapsules. RSC Advances. 5, 87077–83.

[24] Li, Y. T., Pi, Y.-T., Lu, L.-M., Xu, S.-H., and Ren, T.-Z. (2015). Hierarchical porous active carbon from fallen leaves by synergy of K2CO3 and their supercapacitor performance. Journal of Power Sources. 299, 519–28.

[25] Gu, W., and Yushin, G. (2014). Review of nanostructured carbon materials for electrochemical capacitor applications: advantages and limitations of activated carbon, carbide-derived carbon, zeolite-templated carbon, carbon aerogcls, carbon nanotubes, onion-like carbon, and graphene. Wiley Interdisciplinary Reviews: Energy and Environment. 3, 424–73.

[26] Wang, G., Zhang, L., and Zhang, J. (2012). A review of electrode materials for electrochemical supercapacitors. Chemical Society Reviews. 41, 797–828.

[27] Zhang, Y., Feng, H., Wu, X., Wang, L., Zhang, A., Xia, T., Dong, H., Li, X., and Zhang, L. (2009). Progress of electrochemical capacitor electrode materials: A review. International Journal of Hydrogen Energy. 34, 4889–99.

[28] Kim, C., Yang, K. S., and Lee, W. J. (2004). The use of carbon nanofiber electrodes prepared by electrospinning for electrochemical supercapacitors. Electrochemical and Solid-State Letters. 7, A397–A399.

[29] Kim, B. H., Yang, K. S., Woo, H. G., and Oshida, K. (2011). Supercapacitor performance of porous carbon nanofiber composites prepared by electrospinning polymethylhy drosiloxane (PMHS)/polyacrylonitrile (PAN) blend solutions. Synthetic Metals. 161, 1211–16.

[30] Bhute, M. V., Mahant, Y. P., and Kondawar, S. B. (2017). Titanium dioxide / poly(vinylidene fluoride) hybrid polymer composite nanofibers as potential separator for lithium ion battery. Journal of Materials NanoScience. 4(1), 6–12.

[31] More, A. M., Sharma, H. J., Kondawar, S. B., and Dongre, S. P. (2017). Ag- SnO2/Polyaniline composite nanofibers for low operating temperature hydrogen gas sensor. Journal of Materials NanoScience. 4(1), 13–18.

[32] Dirican, M., Yanilmaz, M. M., Fu, K., Lu, Y., Kizil, H., and Zhang, X. (2014). Carbon-enhanced electrodeposited SnO2/carbon nanofiber composites as anodefor lithium-ion batteries. Journal of Power Sources. 264, 240–7.

[33] Ju, Y. W., Choi, G. R., Jung, H.R., and Lee, W. J. (2008). Electrochemical properties of electrospun PAN/MWCNT carbon nanofibers electrodes coated with polypyrrole. Electrochimica Acta. 53, 5796–803.

[34] Zhao, P., Yao, M., Ren, H., Wang, N., and Komarneni, S. (2019). Nanocomposites of hierarchical ultrathin MnO2 nanosheets/hollow carbon nanofibers for high-performance asymmetric supercapacitors. Applied Surface Science. 463, 931–8.

[35] Jamkar, D. V., Lokhande, B. J., and Kondawar, S. B. (2019). Improved electrochemical performance of free standing electrospungraphene incorporated carbon nanofibers for supercapacitor. Journal Material NanoScience. 6(1), 32–37.

Chapter 85

Study of feed ratio alteration on structural, magnetic and dielectric properties of aniline/pyrrole copolymer

Hemlata Ganvir[1,a], Vikrant Ganvir[1,b], Pavan Kumar[1,c] and Rupesh Gedam[2,d]

[1]Yeshwantrao Chavan College of Engineering, Nagpur, India

[2]Visvesvaraya National Institute of Technology, Nagpur, India

Abstract

At room temperature, $FeCl_3$ was used to chemically synthesise pyrrole and aniline copolymers by varying the feed ratio of monomers. FTIR, XRD and Scanning electron microscopy were employed for structural characterisation of prepared copolymers. The FTIR spectra and X-ray diffractograms of synthesised copolymers reveals alterations in the location of recognizable bands and the formation of new bands. Both the monomers have different reactivity in the feed solution, which results in random linking of two units causing thereby shifting or formation of new bands. In addition, the SEM examination reveals that the produced copolymers have a globular shape that changes with the monomer feed ratio. The samples' dielectric constant was tested at a constant frequency of 1 kHz in the temperature range, 313K–673K, which decreases with increase in temperature. Further magnetic susceptibility measurements were conducted in the temperatures region of 315K–445K using the Guoys method, confirming diamagnetic nature of copolymer materials.

Keywords: Aniline, copolymer, dielectric constant, magnetic susceptibility, pyrrole

Introduction

The synthesis of aniline/pyrrole-based copolymers has received a considerable lot of focus recently. A probable explanation is that it is very challenging to manufacture novel polymers with outstanding electrical properties and durability in comparison with polypyrrole (PPY) and polyaniline (PANI). The copolymerisation process of aniline/pyrrole and other monomers makes it possible to make a new PANI- or PPY sort of copolymer that has both the good characteristics of PANI/PPY and some new ones [1–3]. The primary driver for preparing copolymer focuses on the probability that these substances will prove to be novel conjugated compounds exhibiting p-bond molecules. Copolymerisation of two monomers is expected to enhance the number of conductive polymers that may be produced from a given set of monomers, as shown by the foregoing research. Thus, two entirely different monomers can be subjected to simultaneous polymerization in the common reaction medium and required to introduce chemical heterogeneity. Thus, the term 'copolymerisation' is used to describe research into the effects of complexation on the initiation and spread of copolymer chains.

Over the past two decades, several studies have been performed on electronically conducting polymers. PPY and PANI are the most desirable conducting polymers and due to their excellent electrical properties, environmental stability, economical, and superior physiochemical characteristics [4–6]. These polymers can be made into composites or copolymers for specialised needs [7, 8]. However, by expanding homopolymers to copolymers, functional materials with unique properties and structures may be synthesised. Further copolymerisation adds heterogeneity to the polymer backbone and helps polymerise monomers that are hard to polymerise alone [9].

Considering the aforementioned benefits, these two have been co-polymerised extensively by electrochemical approach (poor yield) and to a lesser amount via chemical method. In this study, pyrrole and aniline were treated to chemical polymerisation at room temperature, and the effect of feed ratio variation on structural properties was investigated and reported. As aniline and pyrrole in doped form exhibits very good dielectric properties depending upon the synthesis methods and dopant employed An effort has been made to study the variation of dielectric constant of their copolymers. Also both of these monomers are diamagnetic in nature so it was interesting to study the magnetic behavior of their copolymers.

The findings of an investigation into the effect of feed ratio on the structural, magnetic and dielectric characteristics of synthesised copolymers are given.

Experimental

Materials and Methods

Both pyrrole (SRL chemicals, Mumbai) and aniline (SRL chemicals, Mumbai) were distilled to remove impurities and make them suitable for use as basic monomers. An acidic medium was provided by H_2SO_4,

[a]hrwasnik@gmail.com, [b]vyganvir@gmail.com, [c]pavank.feat@dmimsu.edu.in, [d]rupeshgedam411@gmail.com

DOI: 10.1201/9781003450252-85

(Qualigen Chemicals, Mumbai), and copolymerisation was initiated using anhydrous $FeCl_3$. After the two monomers were combined in the right feed quantity, the sample was subjected to 15-20 minutes of constant stirring. 30 ml of sulphuric acid (0.5 M) was added to the feed. This clear solution was then agitated continuously with more anhydrous $FeCl_3$ (0.25M) until its color changed to clear solution. After being vigorously agitated for about an hour at room temperature, this copolymer base reaction mixture was kept as it is to polymerise for duration of 24 hours. It was necessary to filter the reaction mixture to get the required copolymer. Impurities were removed by rinsing the resulting precipitate (sample) with acetone and distilled water in turn. Additional drying in the oven at 60°C was followed by weighing to ensure a steady weight for the sample.

To confirm that the aforesaid technique synthesised pyrrole aniline copolymer and not the two monomers, a solubility test was performed. The resulting product/copolymer is soluble in DMF and DMSO, but insoluble are PPY and PANI [10].

Infrared Spectroscopy

Transmission spectra were collected using a Schimadzu IR spectrometer (FTIR-8101A) after an FTIR analysis of the copolymers, performed in the 400-4000 per cm wave number region.

X-Ray Diffraction

Synthesised samples' XRD data were acquired using a PANalytical Xpert Pro XRD machine. Radiation was produced within a vacuum tube with a copper electrode, emitting in the copper k alpha1line. The sample was ground into powder and then mounted in an aluminium holder.

The sample was held in this holder and scanned in the reflection mode throughout a two spectrum while the stage of the diffractometer rotated through 5-40°.

Scanning Electron Microscopy

Scanning electron microscopy was used to examine the texture of the copolymer samples. Metallic platinum was sputtered onto the amorphous powder (sample). Scanning was performed at several magnifications using a JEOL 6380 A scanning electron microscope.

Measurement of Dielectric Constant

A LCR metre (model VLC 2) was utilised to determine the dielectric constant of the test specimen. A frequency of 1 KHz was used throughout the investigation. For the temperature range of 313 K – 673 K, the sample's capacitance was measured. To determine the geometric capacitance, the appropriate formula was applied.

Measurement of Magnetic Susceptibility

The material's magnetic susceptibility was measured and analysed using Guoy's Method. The sample was dried at a temperature of 60 degrees Celsius for this purpose, and then it was gently inserted into a thin glass tube. After determining the sample's weight at room temperature (RT), the sample was heated to the appropriate temperature. At each of the several magnetic fields, the rate of heating was held constant at an increase of 10 degrees Celsius per minute. The cooling procedure resulted in a change in the sample's mass, which was recorded. Magnetic susceptibility was estimated using the formula:

$$\chi = \frac{\beta \Delta m_{sample}}{m_{sample}}$$

where m = mass of sample
β = Guoy's tube constant
Δm_{sample} = change in mass of the sample

Results and Discussion

In the current investigation, chemical copolymerisation was accomplished by using $FeCl_3$ as an initiator, and the feed ratio of monomers was adjusted in a number of different ways. This technique was adapted for 0.25M $FeCl_3$ concentrations (initiator). Table 85.1 displays the proportion of monomers that are converted into copolymers when the synthesis takes place at a temperature of 30 degrees Celsius (303 degrees Kelvin).

The information shown in Table 85.1 makes it abundantly evident that the feed ratio and the concentration of the initiator have a significant effect on the proportion of copolymers that are converted. Because FeCl$_3$ is a better initiator for pyrrole than it is for aniline [10–12], a sample that contains a high proportion of pyrrole is more likely to be polymerised to a significant degree.

Table 85.2 lists the copolymers' FTIR spectra's distinctive bands. Many IR peaks in homopolymers have either moved to higher or lower wavenumbers or are missing in copolymers [13–15]. This moving of the peaks to the low wavenumber region can be explained by the fact that hydrogen bonds and electrostatic interactions change the chains from being coiled to being stretched. Due to this, the degree of conjugation of the copolymer chains is enhanced [15].

These IR bands may shift owing to aniline-pyrrole hetero association. The feed ratio affects the amount of comonomer in the feed, resulting in various additional bands in copolymer samples. New side groups change the bond's chemistry and vibrational frequency. New attachments produce bonds that shift bond lengths. The copolymerisation induced polar effects through interacting permanent and transient charges. The copolymer of aniline and pyrrole has smaller wavenumber peaks, indicating near constitutional units

Table 85.1 Sample designation and percentage conversion of monomers to copolymer samples

Designation	Pyrrole : Aniline	Percentage conversion
X5	1 : 1	36.3988 %
Y5	1 : 2	13.2592%
Z5	2 : 1	54.5236%

Source: Author

Table 85.2 IR band positions of aniline, pyrrole, copolymer samples

PANI	X5	Y5	Z5	PPy
----	416	----	418	-----
618	----	----		610
----	----	700, 750	750	----
----	785	----	----	780
816	----	----	800	----
----	----	----	----	830
----	----	----	----	910
----	935	930	940	----
	1049	1049	1050	1048
1114	----	----	----	----
1160	----	----	----	----
----	----	----	----	1185
----	----	----	----	1250
----	----	----	----	1290
1302	----	----	----	----
1385	----	----	----	----
1493	1400	1400	1400	1451
----	----	----	----	1510
1599	1647	----	1600	----
1706	1718	----	----	----
----	1860	1830	----	----
----	2350	2360	2350	----
----	----	----	----	2843
----	2920	2920	2920	2906
----	----	----	----	2963
-----	----	----	----	3441

Source: Author

[16]. Figure 85.1 shows the copolymers' x-ray diffractograms, and Table 85.3 lists their peaks. All copolymers have distinct peaks compared to homopolymers [17, 18] and a common peak at 24° and 32.8°. Despite feed ratio heterogeneity, hetero linkage yields comparable angular peak positions. The monomers' pyrrole and aniline units bind to this shared link (i.e., C-H or C-N or N-H). All copolymers must preserve the random attachment of two reactant units in the primary chain. Hetero association may form tiny units with various crystal planes, reducing homopolymer unit cell size. Thus, tiny crystallites might have a single lateral plane but different diagonal planes. Thus, feed ratio impacts copolymer geometry.

Figure 85.2 is a scanning electron micrograph of produced copolymers. During aniline and pyrrole copolymerisation, agglomerated particles form spherical geometries, according to SEM images. The spherical constructions are between 10 and 20 μm in diameter [19]. Few samples possess a flaky structure. The feed

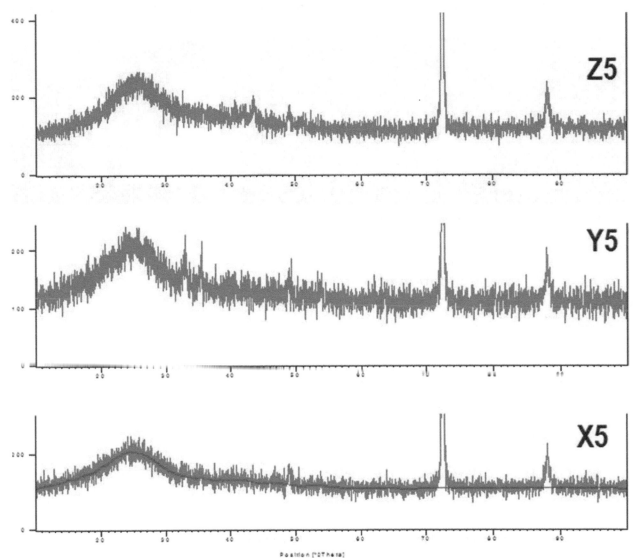

Figure 85.1 X-ray diffractograms of copolymer samples
Source: Author

Table 85.3 2q peak positions for aniline, pyrrole, copolymer samples

PANI	X5	Y5	Z5	PPy
9.5	10	6	12	23.6
14.7	22.5	11	24.5	25.8
20.4	24	24	26	
25.1	28	25.7	28.4	
	32.8	32.8	32.7	
	35.7	35	40.8	
	--	39.5	--	

Source: Author

monomers modify the form of the copolymer. Uneven, disordered (amorphous) spherical agglomerates are seen in the copolymer. As the feed varies, copolymerisation occurs with random consumption of both units, resulting in the formation of novel structures [20].

Figure 85.3 depicts the dielectric curves of the produced copolymer samples. As shown in the graph, the dielectric constant drops as the temperature rises. All the samples display a minor rise between 573K and 623K. This may be due to the collision of monomer units with salt ($FeCl_3$ ions) species in the solution, which disturbs the internal geometry of the monomers. Moreover, because the copolymer chain is heterogeneous, both monomer units respond differently to the applied frequency, resulting in fluctuations [21–23].

Figure 85.4 shows magnetic susceptibility $1/\chi$ values vs temperature. At a temperature range of 315–445K and a magnetic field of 100 Gauss, magnetic susceptibility was measured. Negative susceptibility values were observed for all copolymer samples, indicating that charge carriers are unaffected by magnetic

Figure 85.2 SEM images of copolymer samples
Source: Author

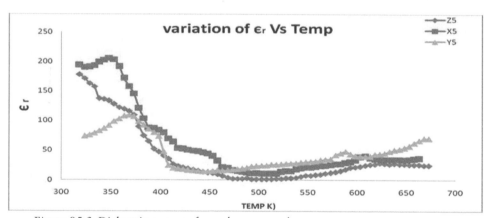

Figure 85.3 Dielectric curves of copolymer samples
Source: Author

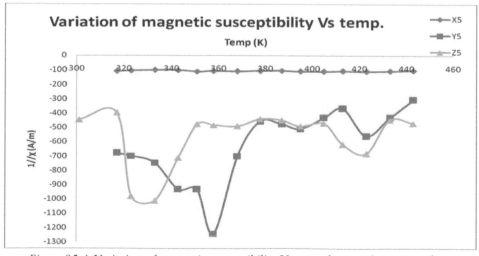

Figure 85.4 Variation of magnetic susceptibility Vs temp for copolymer samples
Source: Author

fields. The samples have been observed to be diamagnetic. Polarons (spin-carrying charge carriers) and bipolarons govern the electrical conductivity phenomena of polypyrrole and polyaniline (charge carriers without spin) [24–28]. Charge carriers (with spin) were found to be aligned in the direction of the applied magnetic field by ESR analyses of PANI and PPy. Polarons are always assigned either a +1 or -1 charge. In terms of the spin dynamics of PANI, it has been proposed that polaron–polaron interaction or polaron ionisation to bipolaron (spinless charge defects) might modify the sample's spin concentration and, therefore, its magnetic susceptibility [29].

Conclusion

By adjusting the feed ratio, pyrrole and aniline were effectively copolymerised chemically at room temperature using $FeCl_3$ as an initiator. Since both the monomers have different monomer reactivity ratio in the feed [12], there is a correlation between the concentration of the initiator and the percentage of conversion of monomers to copolymers. It is observed that sample with higher percentage of pyrrole gives better yield however sample with higher aniline content gets poorly copolymerised. The FTIR and XRD analysis shows several new bands with shifts in the characteristics band positions. Further, all the copolymer samples exhibit some common linkage among aniline and pyrrole units which is reflected in the similar band positions of FTIR and XRD spectra. The scanning electron micrographs reflect prominent globular structures for copolymer with higher pyrrole content rather than copolymer with higher aniline content. As the feed varies, cross linking amongst the monomer units gets disturbed and bonds gets shifted and cause different orientation of polymer chains. This produces agglomerated spheres of different sizes distributed in the matrix.

The dielectric study of the samples indicate a common trend of decrease in the dielectric constant with increase in temperature with small transition peaks. These peaks are also in the same temperature range in which magnetic susceptibility curves show transitions. Magnetic susceptibility measurements reveal the linear behavior of the sample with equimolar concentration of the monomers. However all samples indicate diamagnetic nature revealing bipolarons as the charge carriers.

References

[1] Panero, S., Prosperi, P., Bonino, F., Scrosati, B., Corradini, A., and Mastragostino, M. (1987). Characteristics of electrochemically synthesised polymer electrodes in lithium cells—III. Polypyrrole, Electrochimica Acta. 32, 1007–11. doi.org/10.1016/0013- 4686(87)90025-9.

[2] Nikolaidis, M. G., Ray, S., Bennett, J. R., Fasteal, A. J., and Cooney, R. P. (2010). Electrospun functionalised polyaniline copolymer-based nanofibers with potential application in tissue engineering. Journal Macromolecular bioscience, 10(12), 1424–31. https://doi.org/10.1002/mabi.201000237.

[3] Otero, T. F., and Azcano, C. B. (2000). Optimization of poly-3-methylthiophene to be used in advance polymer batteries. International Journal of Hydrogen Energy. 25, 221–33. doi.org/10.1016/S0360-3199(99)00048-8.

[4] Ansari, R., and Keivani, M. B. (2006). Polyaniline conducting electroactive polymers thermal and environmental stability studies. Journal of Chemistry. 3(4), 202–17. https://doi.org/10.1155/2006/395391.

[5] Ansari, R. (2006). Polypyrrole conducting electroactive polymers: synthesis and stability studies. Journal of Chemistry. 3(4), 186–201. https://doi.org/10.1155/2006/395391.

[6] Chandrakanthi, N., and Careem, M. (2000). Thermal stability of polyaniline. Polymer Bulletin. 44, 101–108. https://doi.org/10.1007/s002890050579

[7] Eftekhari, A., Li, L., and Yang, Y. (2017). Polyaniline supercapacitors. Journal of Power Sources. 347, 86–107. https://doi.org/10.1016/j.jpowsour.2017.02.054.

[8] Yang, L., Wang, S., Mao, J., Deng, J., Gao, Q., Tang, Y., and Schmidt, O. G. (2013). Hierarchical MoS_2/Polyaniline nanowires with excellent electrochemical performance for lithium-ion batteries. Advanced Materials. 25(8), 1180–84. https://doi.org/10.1002/adma.201203999.

[9] Wang, H. L., MacDiarmid, A. G., Wang, Y. Z., and Gebier, D. D. (1996). Application of polyaniline (emeraldine base, EB) in polymer light- emitting devices. Synthetic Metals. 78(1), 33–37. doi.org/10.1016/0379-6779(95) 03569-6.

[10] Moon, D. K., Yun, J. Y., Osakada, K., Kambara, T., and Yamamoto, T. (2010). Synthesis of random copolymers of pyrrole and aniline by chemical oxidative polymerization. Molecular Crystals and Liquid Crystals. 464(1), 177–185. doi.org/10.1080/15421400601030878.

[11] Rabek, J. F., Linden, L. A., Adamczak, E., Sanetra, J., Starzyk, F., and Pielichowski, J. (1995). Polymerization of thin pyrrole films on poly(Ethylene Oxide) - FeCl3 coordination complex. Material Science Forum. 191, 225–34. https:/doi.org/10.4028/www.scientific.net/MSF.191.225.

[12] Wasnik, H. R., Kelkar, D. S., and Ganvir, V. Y. (2014). Yield analysis of copolymers: effect of temperature, feed ratio and initiator concentration on the copolymerization. Journal of Polymer Engineering. 35(2), 99–103. DOI: https://doi.org/10.1515/polyeng-2014-0027.

[13] Nicho, M. E., and Hu, H. (2000). Fourier transform infrared spectroscopy studies of polypyrrole composite coatings. Solar Energy Materials and Solar Cells. 63(4), 423–35. doi.org/10.1016/S0927-0248(00)00061-1.

[14] Furukawa, Y., Tazawa, S., Fujii, Y., and Harada, I. (1988). Raman spectra of polypyrrole and its 2,5-^{13}C-substituted and C-dueterated analogues in doped and undoped states. Synthetic Metals. 24(4), 329–41. doi.org/10.1016/0379-6779(88)90309-8.

[15] Partch, R. E., Gangoli, S. G., Matijevic, E., Cai, W., and Arajs, S. (1991). Conducting polymer composites I: surface-induced polymerization of pyrrole on iron(III) and cerium (IV) oxide particles. Journal of Colloid and Interface Science. 144(1), 27–35. 10.1016/0021-9797(91)90234-Y.

[16] Zhang, X., Zhang, J., and Liu, Z. (2005). Tubular composite of doped polyaniline with multi-walled carbon nanotubes. Applied Physics A. 80, 1813–17. doi.org/10.1007/s00339-003-2491-z.org.

[17] Lu, W., Yin, S., Wu, X., Luo, Q., Wang, E., Cui, L., and Guo, C. Y. (2021). Aniline–pyrrole copolymers formed on single-walled carbon nanotubes with enhanced thermoelectric performance. Journal of Materials Chemistry C. 9, 2898–2903. doi.org/10.1039/DOTCO5757F.

[18] Palaniappan, S., Sydulu, S. B., and Srinivas, P. (2010). Synthesis of copolymer of aniline and pyrrole by inverted emulsion polymerization method for supercapacitor. Applied Polymer Science. 115(3), 1695–701. doi.org/10.1002/app.31028.

[19] Li, N., Shan, D., and Xue, H. (2007). Electrochemical synthesis and characterization of poly (pyrrole-co-tetrahydrofuran) conducting copolymer. European Polymer Journal. 43(6), 2532–539. doi.org/10.1016/j.eurpolymj.2007.01.048.

[20] Xu, P., Han, X. J., Wang, C., Zhang, B., and Wang, H.-L. (2009). Morphology morphology and physico-electrochemical properties of poly (aniline-co-pyrrole). Synthetic Metals. 159, 430–34. doi.org/10.1016/j.synthmet.2008.10.016.

[21] Dey, A., De, S., De, A., and De, S. K. (2004). Characterization and dielectric properties of polyaniline–TiO2 nano-composites. Nanotechnology. 15(9), 1277. **doi** 10.1088/0957-4484/15/9/028.

[22] Upadhyay, J., and Kumar, A. (2014). Investigation of structural, thermal and dielectric properties of polypyrrole nanotubes tailoring with silver nanoparticles. Composites Science and Technology. 97, 55–62.

[23] Bhattacharya, A., and De, A. (1996). Conducting composites of polypyrrole and polyaniline a review. Progress in Solid State Chemistry. 24(3), 141–181. https://doi.org/10.1016/0079-6786(96)00002-7.

[24] Krinichnyi, V. I. (2014). Dynamics of spin charge carriers in polyaniline. Applied Physics Reviews. 1(2), 021305. https://doi.org/10.1063/1.4873329.

[25] Harima, Y., Patil, R., Yamashita, K., Yamamoto, N., Ito, S., and Kitani, A. (2001). Mobilities of charge carriers in polyaniline films. Chemical Physics Letters. 345(3–4), 239–44. https/doi.org/10.1016/S000-2614(01)00877-6.

[26] Toušek, J., Toušková, J., Chomutová, R., Křivka, I., Hajná, M., and Stejskal, J. (2017). Mobility of holes and polarons in polyaniline films assessed by frequency-dependent impedance and charge extraction by linearly increasing voltage. Synthetic Metals. 234, 161–165. https://doi.org/10.1016/j.synthmet.2017.10.015.

[27] Wang, P. C., and Yu, J. Y. (2012). Dopant-dependent variation in the distribution of polarons and bipolarons as charge-carriers in polypyrrole thin films synthesised by oxidative chemical polymerization. Reactive and Functional Polymers. 72(5), 311–6. https://doi.org/10.1016/j.reactfunctpolym.2012.03.005.

[28] Brédas, J. L., Scott, J. C., Yakushi, K., and Street, G. B. (1984). Polarons and bipolarons in polypyrrole: Evolution of the band structure and optical spectrum upon doing. Physical Review B, 30(2), 1023–1025. doi:10.1103/physrevb.30.1023

[29] Jalilian, N., Ebrahimzadeh, H., Asgharinezhad, A. A., and Molaei, K. (2017). Extraction and determination of trace amounts of gold (III), palladium (II), platinum (II) and silver (I) with the aid of a magnetic nanosorbent made from Fe3O4-decorated and silica-coated graphene oxide modified with a polypyrrole-polythiophene copolymer. Microchimica Acta. 184(7), 2191–2200.

Chapter 86

Correlation of modal test and CAE simulation results of car door assembly

Pandurang Maruti Jadhav[1,a], Dr. Kishor Bhaskar Waghulde[1,b] and Dr. Rupesh V. Bhortake[2,c]

[1]Dr. D. Y. Patil Institute of Technology, Pimpri, Pune, India

[2]Marathwada Mitramandal's Institute of Technology, Lohegaon, Pune, India

Abstract

While focusing on customer satisfaction in terms of the NVH of vehicles, overall noise and tactile vibrations are the main concerns. At the same time, car door dynamic behavior, noise, and vibrations are equally important. Car doors are the first interaction part of any vehicle. Customer opinion and decisions about any vehicle will be influenced by car door operations, aesthetics, noise, and vibrations. Finite element analysis (FEA) and modal testing was performed to validate the car door assembly FEM; this will be used in next level noise, vibration, and harshness (NVH) analysis at the vehicle level model. Modal testing of car door assembly has been done using Siemens (Scadas and Test Lab) systems at ARAI NVH Lab. Modal testing was done by exciting at hinge locations, and output responses were taken on the door surface to find out the modal parameters in terms of Eigen frequency with associated mode shapes, and driving point dynamic stiffness (DPDS) analysis was done to capture locale dynamic behaviors. To develop 3D CAD model of car door assembly a reverse engineering technique was used. All parts have been purchased from Maruti Suzuki genuine parts outlet and scanned using CMM machines. All 3D CAD models have been developed from scanned data using Siemens, NX 12.0 software. The FE model was built in HyperMesh and dynamic analysis was carried out in Altair OptiStruct/Nastran solver. Door assembly connections, material properties, and damping were fine-tuned to correlate the Test and FEA results. Car door assembly FEM was validated with door global (modal) and local (DPDS) dynamic behaviors. Both results are well correlated between 0–100 Hz, which indicates that the developed FEM of car door assembly is suitable for next level NVH detailed research study at the vehicle level.

Keywords: CAE, car door, car door NVH, correlation, modal analysis, modal test, NVH

Introduction

Car door is key accessary, when we think about the car aesthetics design, car handling performance, passenger safety, car body reliability, car body vibrations, car overall noise and the first interaction point of all human beings who want to use or access the car. For better alignment of door to the above important performance attributes, it is need cum requirement to study the car door. Car door means, car door assembly which includes door main frame structure, hinges, latch, window glass, window glass regulator mechanism, window glass seals, door primary and secondary seals, door handle and door mounted side mirror.

As an aim of providing customer satisfaction or ride comfort, car door is the one of the aggregates which we must focus and should work on its dynamic behaviors. As a car door assembly is a complex dynamic structure which only works as support aggregate but it is at the top of all such aggregates. To understand the dynamics of door assembly, there are two ways, first once is physical testing of sample door in test lab and second one is the CAE or FE analysis/simulation of door. To do this research study a door of Maruti Suzuki Swift Dzire (front right-hand side) has been selected.

Final aim is to find the effect of car door design and door closing on noise and vibration results. Noise will be studied inside and outside of the car and vibrations on door surface. To meet the final aim finite element analysis (FEA) and modal testing was performed to validate the car door assembly FEM. This FEM will be used in next level noise, vibration, and harshness (NVH) analysis at the vehicle level model. Modal testing of car door assembly has been done using Siemens (Scadas and Test Lab) systems at ARAI NVH Lab. As a first stage of finding door NVH effect study, this correlation of car door FEM is done based on global (modal) and local (DPDS) dynamic behaviors. Generally modal test correlation had been done using modal analysis results only [1-3].

To build the sample car door assembly for modal testing and 3D CAE simulation, all parts of car door have been purchased from Maruti Suzuki genuine parts outlet as spare parts. First these parts were scanned using CMM machine to generate surface data then build the physical sample of car door assembly from same spare parts at Maruti Suzuki service centre (Figure 86.1a). This sample door assembly has been testing at ARAI for modal and DPDS behavior. Both tests represent door assembly global and local dynamic behaviors with natural frequencies with associated mode shape. Siemens, NX 12.0 software was used to

[a]pmjadhav@rediffmail.com, [b]kishor.waghulde@dypvp.edu.in, [c]ruprani_2000@yahoo.com

DOI: 10.1201/9781003450252-86

develop 3D CAD model (Figure 86.1b) from scanned surface data of all parts. Developed 3D CAD model has been used for CAE/FE simulation.

Modal Test Set-Up and Results

Modal test was carried out on passenger vehicle door of selected Maruti Suzuki Swift Dzire car. Door assembly was suspended using bungee cords as shown in below Figure 86.2. Door was spatially divided into 17 response points for measurements and total DOFs were 51 numbers. All points response has been recorded in cartesian coordinate system as shown in Figure 86.2. Door was excited using PCB make 500 lbf impact hammers at hinge locations in all three axes viz. X, Y and Z and vibration response of all points were measured using triaxle accelerometer of PCB and Dytran make. Based on the measured data, the modal parameters viz. natural frequencies with associated mode shapes were extracted. As a second parameter DPDS measurements were carried out at all 17 points using single triaxle accelerometer. Each point was excited at a time by an impact hammer and dynamic stiffness was measured between 0 – 500 Hz frequency range.

Modal Test Results

Recorded modal test data was post processed to find out modal frequencies with associated mode shapes (Figure 86.3). There were clear response peaks between the frequency range of 0 – 100 Hz. It shows that global modes of door assembly will be observed between 0 – 100 Hz so this is the frequency range related to or concerned to the door assembly for NVH assessment. From extracted all modes, first 6 global modes of the door assembly were identified and listed as shown in below Table 86.1. Mode shapes were logical and showing global torsion and bending behaviors.

DPDS (Drive Point Dynamic Stiffness) Results

Under modal testing global behaviors of the door were extracted in term of frequency with associate mode shape. DPDS results will give local behaviors of the door assembly between the required frequency ranges.

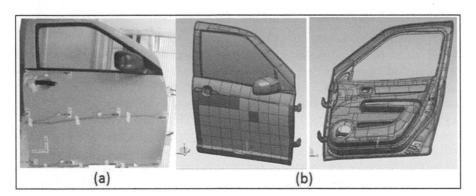

Figure 86.1 Sample car door and 3D CAD model
Source: Author

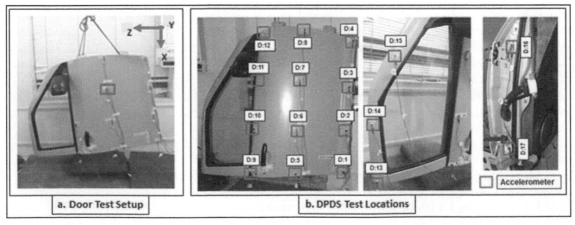

Figure 86.2 Car door types of analysis and types of NVH analysis in last 20 years
Source: Author

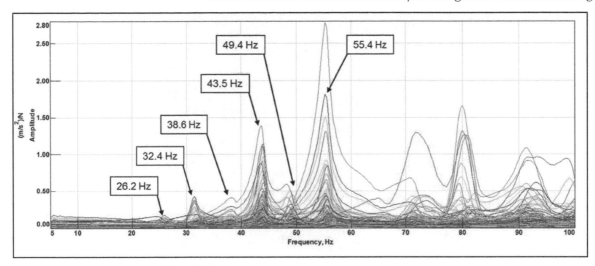

Figure 86.3 Modal test data of all points
Source: Author

Table 86.1 Test modal frequency list

Mode No.	Natural frequencey, Hz	Mode shape description
1.	26.2	1st Lateral bending
2.	32.4	1st Vertical bending
3.	38.6	2nd Vertical bending
4.	43.5	2nd Lateral bending
5.	49.4	1st Torsional bending
6.	55.4	2nd Torsional bending

Source: Author

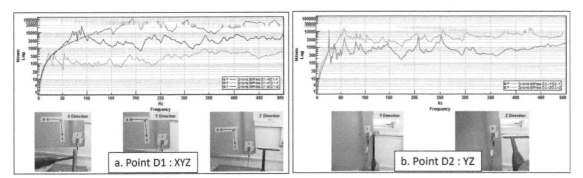

Figure 86.4 DPDS results of point D1 and D2
Source: Author

To get overall local behaviors of the door, it is spatially divided into 17 points and extracted the dynamic responses for the frequency ranges of 0 – 500 Hz in term of dynamic stiffness by exciting each location in respective directions. Sample results of point D1 and D2 are presented here (Figure 86.4)

CAE Simulation and Results

Door assembly finite element model

From scanned surface data of all spare parts of the door were used to develop the 3D CAD models of all parts using Siemens, NX 12.0 software. Same model has been used to build the FE model of the door assembly parts (Figure 86.5). Mid-surfaces of all parts were extracted, and discretisation/meshing were done, except mirror assembly and window glass regulator motor. HyperMesh was used to develop the entire door assembly with 2D shell and 3D Hex.elements. Mirror assembly and regulator motor assembly was modelled with 2D shell elements on outer surfaces as an envelope and adjusted the measured masses

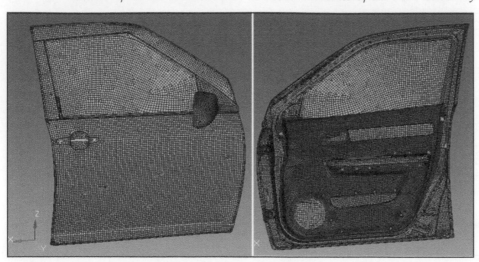

Figure 86.5 Finite element model of car door assembly
Source: Author

Table 86.2 Material properties used

Sr. No.	Material	Modulus of elasticity (N/mm²)	Poisons ratio	Density (Tones/mm³)
1	Steel	$2.1\,e^5$	0.29	$7.85\,e^{-9}$
2	Plastic	$1.1\,e^3$	0.3	$1.0\,e^{-9}$
3	Glass	$6.5\,e^4$	0.24	$2.4\,e^{-9}$
4.	Adhesive	1.9	0.36	$1.2\,e^{-9}$

Source: Author

by adjusting thickness. All parts were meshed with 10 mm average element length and maintained general elemental quality criteria to minimise the calculation deviation.

Steel material was assigned to all structural parts and Plastic for trim parts. Plastic and aluminium were assigned to mirror and regulator motor assembly respectively. Glass material properties were given to side window glass (Table 86.2). Latch assembly, handle, speaker, and switch box were represented with RBE3-Conm2 with respective masses.

Soft connections were modelled with RBE3-Hex-RBE3 and assigned adhesive material. All structural parts connections were done with seam weld, and it was represented with 2D shell elements with minimum thickness of connection two panels. Whether seal was represented with RBE3-CBush-RBE3 with appropriate stiffness properties (k1:15, K2:78 and K3:15 N/mm). Hinge pin and hinge brackets were modelled with 3D hex. Elements and hinge pin representation was done using RBE2 with keeping pin axis rotational degree of freedom free.

Modal Analysis of Car Door Assembly

Complete Finite Element model of the Car door assembly was performed free-free modal analysis in Altair OptiStruct solver for frequency range of 0 – 100 Hz to find global behavior of the door assembly. Most of the global modes were observed between 25 – 60 Hz and having pure global behaviors. Above 60 Hz complex mode shapes were there which may not be described in standard behaviours with visual review. After post processing entire frequency range of 0 – 100 Hz and out of these modes, first 6 flexible global bending frequencies with associated mode shapes were listed in below (Table 86.3). Out of these six frequencies two frequencies of mode no. 2 and 5 with its mode shapes are presented in below Figure 86.6.

Driving Point Dynamic Stiffness Analysis of Car Door Assembly

Similar to the physical door sample test setup for the DPDS analysis. In FE model approximately same locations were considered and done the FRF (Frequency Response Function) analysis (Figure 86.7).

All 17 points FRF analysis in 51 DOF were done but most significant and concerned DOF was the Y direction which was perpendicular to the door outer surface. To assess the DPDS results only Y DOF was evaluated closely between frequency range of 0–100 Hz. These DPDS results of point 1 and 2 in Y direction were shown in Figure 86.8.

Table 86.3 FE based modal frequency list

Mode No.	Frequencey (Hz)	Mode shape description
1	26.16	1st Lateral bending driven by mirror
2	35.02	1st Vertical and lateral bending
3	40.29	2nd Vertical bending
4	42.77	2nd Lateral bending
5	52.59	1st Torsional bending
6	58.89	2nd Torsional bending

Source: Author

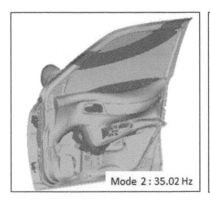 

Figure 86.6 Flexible modes of car door assembly
Source: Author

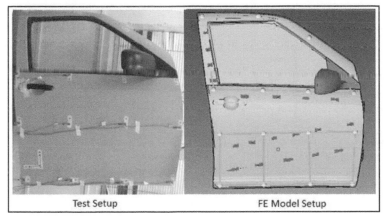

Figure 86.7 DPDS response locations
Source: Author

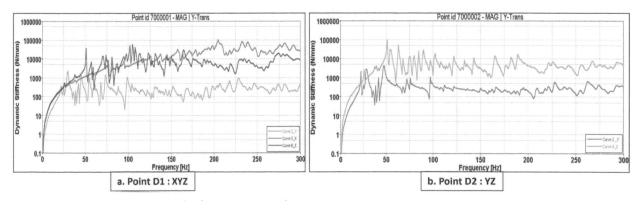

Figure 86.8 DPDS FRF results for point D1 and point D2
Source: Author

Table 86.4 Correlation of modal test and FEA frequency list

Mode No.	Modal test frequencey (Hz)	Modal test - mode shape description	CAE / FE model frequencey (Hz)	FE Modal - mode shape description	Difference in frequencey (Hz)	% Deviation as compared to model test
1	26.2	1st Lateral bending	26.16	1st Lateral bending driven by mirror	-0.04	-0.15
2	32.4	1st Vertical bending	35.02	1st Vertical & lateral bending	2.62	8.09
3	38.6	2nd Vertical bending	40.29	2nd Vertical bending	1.69	4.38
4	43.5	2nd Lateral bending	42.77	2nd Lateral bending	-0.73	-1.68
5	49.4	1st Torsional bending	52.59	1st Torsional bending	3.19	6.46
6	55.4	2nd Torsional bending	58.89	2nd Torsional bending	3.49	6.30

Source: Author

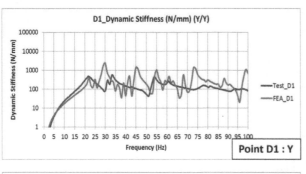
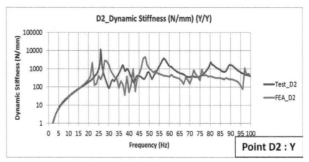
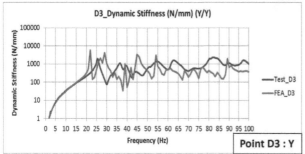
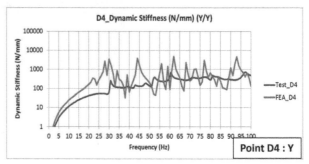

Figure 86.9 DPDS results comparison of test and FEA in Y direction of point D1, D2, D3 and D4
Source: Author

Correlation of Modal Test Results and Cae Simulation Results

As a first part of the correlation of the door assembly dynamic results, modal frequencies with its associate mode shape behavior were compared for first six global flexible modes (Table 86.4). These comparisons of mode shape behaviors were matching with minor deviation in frequency values of corresponding modes. Max. deviation in frequency value was 8.09 % which shows that the modal dynamic behaviors in term of frequency value and its associated mode shape were correlating well.

Out of 17 points, 15 points were on the outer surface of door assembly which has been assessed for Y direction DPDS results. Dynamic stiffness trend or pattern between 0 – 100 Hz is matching with test results for point D1, D2, D3 and D4 (Figure 86.9). Few points which are on the edges of door assembly were having deviation as compared to frequency peaks and in term of amplitude of stiffness, it is mainly due to variation in damping of FE model and damping of physical test sample. Mid-point of door surface, point D6 and D7 are exactly matching in term of pattern and amplitude/ stiffness value. Points related to the Door top/glass window were also having good agreement between test results and FEA results in term of dynamic stiffness trend and amplitudes between 0 – 100 Hz (Figure 86.10).

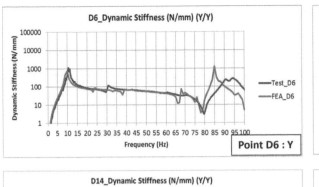 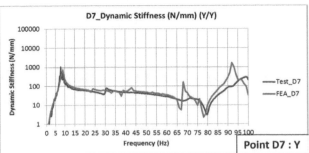

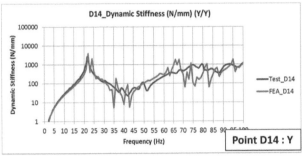 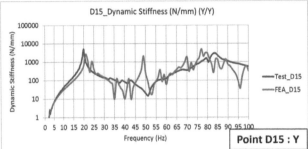

Figure 86.10 DPDS results comparison of test and FEA in Y direction of point D6, D7, D14 and D15
Source: Author

Conclusion

As discussed, global results in term of modal frequency with its associated modes of door assembly and local stiffness results assessed as DPDS. Based on global behaviors of door assembly finite element (FE) model results were well correlated with test results in term of frequencies with its associated mode shapes. Same FE model showing local stiffness/DPDS also better correlated in term of stiffness pattern as well as stiffness values/amplitudes between 0 – 100 Hz, except few points near the door edges.

Based on these correlations aim of validation of the door assembly FE model has gone well. This indicates that the developed FE model of car door assembly was suitable for further detailed research study at the vehicle level.

Acknowledgement

I would like to thank my research mentor cum guide Dr. Kishor B. Waghulde for timely guidance, inputs and in detailed explanation about the research topic. Also, I would like to thanks ARAI Mr. R Ramkumar for proving guidance and testing facility at ARAI. I sincerely thanks to my research centre Dr. D. Y. Patil Institute of Technology, Pimpri, India for providing opportunity and facilitate to work on this highly sensitive and current topic "Design and Development of Methodology to Simulate, Correlate and Improve Car Door NVH Performance".

References

[1] Qi, X., Wang, Y., Hui Guo, and Zhao, L. (2017). Characteristic analysis of passenger car door vibration transfer function. In 2017 2nd International Conference on Artificial Intelligence and Engineering Applications (AIEA 2017).

[2] Wan Iskandar Mirza, W. I. I., Abdul Rani, M. N., Musa, M. F., Yunus, M. A., Peter, C., and Aziz Shah, M. A. S. (2019). Identifying the appropriate frequency response function driving point of a car door using finite element analysis and modal testing. IOP Conference Series: Journal of Physics: Conference. Series. 1–10. doi:10.1088/1742-6596/1262/1/012007 .

[3] Chandru , B. T. and Suresh P. M. (2016). Finite element and experimental modal analysis of car door. Journal of Chemical and Pharmaceutical Sciences. 2, August 2016, 85–88.

Chapter 87

Numerical study on the effects of additional dimethyl ether into intake manifold on exhaust pollutants and performance of marine diesel engine

Doan Cong Nguyen[a]

University of Transport Technology, Hanoi 100000, Vietnam

Abstract

The International Maritime Organization (IMO) has issued regulations on emissions from ships, including the International Convention for the Prevention of Pollution from Ships (MARPOL), which sets limits on SO_x and NO_x emissions. Measures to reduce emissions from ships can include operating ships with alternative fuels. Currently, dimethyl ether (DME) is one of the promising alternative fuels for diesel engines. One effective solution to using DME for diesel engines is adding DME to the intake manifold of the engine. This paper presents the research results of the effects of the addition DME into intake manifold on the performance and exhaust pollutants of marine diesel engine Yanmar S185L-ST using AVL-Boost simulation software. The amount of DME is provided into the intake manifold with replacement rates of 5%, 10%, and 15% of diesel fuel when the engine is operating at the propeller characteristics and load characteristics. The results show that when DME is added to the intake manifold, the engine power is not significantly changed, NO_x emissions decrease by up to 30.48% at low load modes, and soot emissions decrease by up to 68.15% at full load modes. The results of the simulation study are the basis for researching, designing, and manufacturing a system that provides DME into the intake manifold of marine diesel engines to evaluate the ability to reduce harmful emissions.

Keywords: AVL-boost, dimethyl ether, exhaust emissions, marine diesel engine Yanmar S185L-ST

Introduction

The goal of mitigating the climate impact of shipping requires a variety of solutions. To reduce greenhouse gas emissions from ships, the International Maritime Organization (IMO) has prepared a number of proposals to meet and implement relevant requirements. IMO has introduced provisions in Annex VI on the prevention of air pollution from ships under the International Convention for the Prevention of Pollution from Ships (MARPOL), with objectives to reduce future emissions from international sea shipping. To reduce the level of environmental pollution from diesel engine, IMO has also introduced mandatory standards set out in Annex VI of the MARPOL 73/78 regulating the content of toxic substances. harmful in diesel engine exhaust, especially NOx emissions.

In Vietnam, the issue of efficient fuel use and management is of particular concern with one of the prominent goals being the development of a transportation system to meet the increasing demand for quality transportation. high, save fuel, and limit environmental pollution. Promote the application of new technologies, using renewable energy to replace traditional fuels in transportation such as liquefied petroleum gas [1], nature gas [2], and biofuel [3, 4]. These studies produced useful results for developing alternative fuels in Vietnam.

Among alternative fuels for diesel engines, dimethyl ether (DME) is a competitively priced fuel that can completely replace traditional diesel fuel. DME has a rich source of production materials, including recycled sources [5]. Compared with traditional diesel fuel, DME used on engines has basic advantages including a high hydrogen/carbon content ratio, so the combustion process produces less CO_2 emissions, low boiling point helps DME evaporate quickly when burning. injected into the engine cylinder, high oxygen content along with no carbon-carbon linkage in the molecule leads to less soot formation, high cetane number (>55) due to low autoignition temperature and the ability to evaporate quickly, leading to improved combustion quality, does not contain sulfur, so it does not form H_2SO_4 and sulfate particles in the exhaust gas, helping to reduce environmental pollution and increase the timeserving of exhaust gas treatment catalysts [6, 7].

On the other hand, the theoretical air/fuel (A/F) ratio of DME (A/F = 9) is smaller than that of diesel fuel (A/F = 14.6) which reduces the flame zone in the cylinder. On the other hand, the latent heat of vaporization of DME is two times greater than that of diesel fuel, so it is possible to cool the working medium when DME is vaporised and at the same time reduce the temperature of the flame area in the cylinder. Therefore, using DME reduces the NO_x emissions of the engine significantly. DME has a high cetane number which shortens the ignition delay. Accordingly, the pressure increase rate of $dp/d\varphi$ when

---

[a]doannc@utt.edu.vn

DOI: 10.1201/9781003450252-87

burning DME is 1.8 ÷ 1.9 times lower than that of diesel fuel, so it reduces noise and makes the engine operate more smoothly [8, 9]. In last decades, scientists attended in studies on using DME as an alternative fuel in diesel engine to reduce mineral fuel consumption rate as well as exhaust pollutants [10, 11]. Many comparative studies on characteristics of DME and other traditional and alternative fuels had been carried out [12, 13]. Other studies considered about physical and chemical properties, fuel supplying methods, injection strategies and combustion characteristics of DME [14, 15]; implementation of DME in gas engine [16].

As a liquid fuel for internal combustion engine, DME can be aslo used as a fuel to replace a proportion of conventional fuel or a fuel additive [17] for engine performance ehancement and emissions reduction. Studies focused on implementation of DME in internal combustion engines had been carried out as mentioned in [18–24]. Konno conducted a study on using DME in 4-strokes, one cylinder, semi-combusiton chamber conventional diesel engine [19]. In which, DME and diesel fuel were provided by two seperate injection systems. DME was injected into main combustion chamber while diesel fuel was supplied into semi combustion chamber. The study results showed that, NO_x emission of the engine fuelled with duel fuel reduced significantly especially at high engine speed regimes. In the same topic, a research conducted by Fleisch [21], NOx emission of a diesel engine equiped in heavy duty truck reduce considerabley about 35% on average when the test vehicle was evaluated following a driving cycle. In addition, the operation characteristic of exhaust gas recirculation (EGR) system improved as the vehicle fuelled with DME. In the research of Sorenson [22], an evaluation on performance and pollutants of engine fueled with heavy DME fuel compounding of DME (92%), water (4%) and methanol (4%) was conducted in one cylinder engine for fuel cost reduction. In the case of fueled with blend of DME, output power of the engine was similar the case of fuelling with pure DME. However, the study results showed that, at partial and idling conditions, CO and HC emissions of the test engine increased when the engine fueled with DME blend incompare to the case of fueled with pure DME. Blending DME with low cetane number fuel such as bio-diesel will contribute to improvement in chemical and physical properties of fuel blend, thereby a result in fuel economy enhancement and pollutants reduction. A comparison study on emission characteristics of diesel engine fuel with conventional diesel fuel and blend of bio-diesel and DME was conducted [20]. The results show that, the apperance of DME in the fuel blend could result in smoke and NOx reduction, however, CO and HC increased in opposite trend. Because DME owns high cetane number, which contributes to combustion process improvement, so that DME is used as fuel additive to improve combustion process of other low cetane fuels [23, 24]. For example, in the research of Karpuk [24], the engine fueled with methanol as primary fuel while as DME was uesed as additive to improve ignition ability of the engine. In experiment, DME was continuously produced by on board catalytic system.

However, there are still many problems to be solved when supplying DME to the engine. To effectively use this fuel for diesel engines, it is necessary to solve scientific-technical problems such as ensuring durability, reliability, and safety for the engine and service systems. On the other hand, DME has low viscosity and poor lubricity leading to an increased possibility of fuel leakage, causing surface wear of moving parts in the fuel injection system. Therefore, using DME to partially replace traditional diesel fuel by supplying it to the intake manifold is an effective solution. The supply of DME to the intake manifold does not affect the operation of the available fuel system on the engine and it is easy to change the ratio between the two types of fuel (diesel and DME) supplied to the engine. For the above reasons, theoretical and experimental studies on the supply of DME to the intake manifold to evaluate the fuel economy and the possibility of reducing harmful emissions for diesel engines are very urgent and have a high scientific significance and meet practical requirements. In this work, a simulation study on evaluating the pollutants and performance characteristics of a marine diesel engine operating with additional DME into intake manifold was conducted. AVL Boost software had been used for the simulation work. The aim of this work is to evaluate the effects of DME on performance and pollutants characteristics of the test engine. The findings of this study lay the groundwork for expanding the use of DME in practical settings for marine diesel engines in Vietnam, with the goals of reducing emissions and increasing efficiency while using fossil fuels.

Material and Methodology

Test fuels and engine specifications

The selected research object is the Yanmar S185L-ST marine diesel engine. Yanmar S185L-ST engines are used as main engines or auxiliary engines to drive generators on ships. Yanmar S185L-ST engine is a 4-stroke medium-speed diesel engine, turbocharger, exhaust gas turbine, cylinders arranged in a line, direct fuel injection, Bosch high-pressure pump, water-cooled, compressed air starter. Table 87.1 shows the main technical parameters of the test engine. Commercial diesel and DME were used as test fuels in this study, and their properties are shown in Table 87.2.

Table 87.1 Specifications of the test engine

Item	Technical parameters	Unit
Name of engine	Yanmar S185L-ST	-
Type of engine	4-strokes, four cylinder, direct injection, turbo charger, firing order 1-5-3-6-2-4	-
Bore x Stroke	185 × 230	mm
Displacement	37.09	dm^3
Compression ratio	13.6	-
Max. power/engine speed	442/900	kW/rpm
Specific fuel consumption	205	g/kWh

Source: Author

Table 87.2 Properties of diesel fuel and DME (at a pressure of 0.5 MPA)

Property	Diesel	DME
Lower heating value (MJ/kg)	42.76	28.43
Cetane number (-)	49	55
A/F stoichiometric (kg/kg)	14.6	9.1
C/H/O content (%wt)	86/14/0	52.2/13/34.8
Liquid density (kg/m^3) at 20°C	840	660[a]
Latent heat of vaporization (kJ/kg)	220	425
Kinematic viscosity at 20°C (mm^2/s)	3.53	0.15
Flash point (°C)	380	235
Sulfur content (ppm)	428	0

Source: Author

Simulation model building

To study the effect of DME supply into the intake manifold on the performance and environmental characteristics of the marine diesel engine Yanmar S185L-ST, the author used a simulation research method using AVL-Boost software based on actual structure parameters such as the gas exchange system geometry, the volumes of plenums, the properties of the boost system, the engine combustion chamber geometry, the valve timing and lift diagrams and also the fuel system properties, injection rate profile, injection pressure and blow-by [25, 26]. The process of creating a model includes several processes, including the creation of the model, choosing the governing equations, and handling the initial data as well as boundary conditions. According to the selected option, DME will be injected into the intake manifold of the engine, so on the diesel-DME model, in addition to the basic elements such as the original diesel model, there will be an injector element (I1) to simulate the DME provisioning process. The injector element is connected to the intake manifold of the engine, after the compressor. The simulation model of engine Yanmar S185L-ST with DME supply nozzle installed into the intake manifold after construction is shown in Figure 87.1.

To ensure the accuracy of the model, the original engine was tested at the Ship Power Plants Laboratory of the University of Transport Technology to get data to verify and calibrate the model. Simulation and experimental results on power and fuel consumption are compared when the engine operates according to external characteristics. The power deviation is 3.34%, and the fuel consumption rate error is 0.63% (Figure 87.2), showing that the model ensures the reliability to meet the requirements for further simulation studies.

Selection combustion model

Chmela [27] developed the combustion model and modeling approach used in the AVL Boost program. The vaporization rate of the fuel had to be considered because, throughout the simulation procedure, DME fuel was provided as an external fuel into the intake manifold. Direct injection of diesel fuel into the combustion chamber was done in the meanwhile to create the internal mixing process. In order to analyse the combustion process, the Mixture Controlled Combustion (MCC) model was chosen. Pre-mixed combustion (PMC) and diffusion-controlled combustion processes are essentially taken into account by the MCC model, and heat release is calculated in accordance with (1):

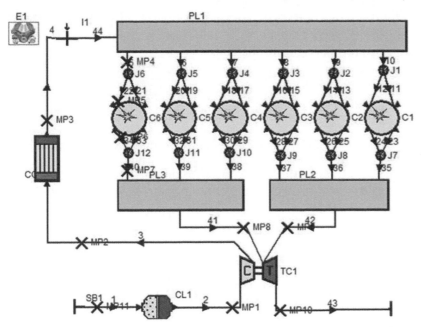

Figure 87.1 Yanmar S185L-ST engine model with DME injector in the intake manifold on AVL-Boost software
Source: Author

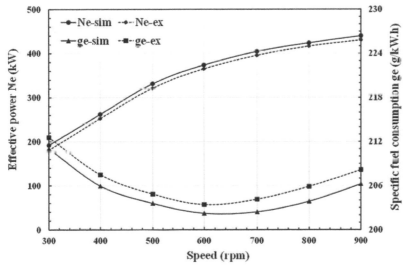

Figure 87.2 Comparison of effective power and specific fuel consumption at full load condition
Source: Author

$$\frac{dQ}{d\alpha} = \frac{dQ_{MCC}}{d\alpha} + \frac{dQ_{PMC}}{d\alpha} \tag{1}$$

The heat release resulting from the pre-mixed combustion is described using the vibration function, as shown in (2) and (3):

$$\frac{\left(\dfrac{dQ_{PMC}}{Q_{PMC}}\right)}{d\alpha} = \frac{a}{\Delta\alpha_c}.(m+1)y^m.e^{-a.y^{(m+1)}} \tag{2}$$

$$y = \frac{\alpha - \alpha_u}{\Delta\alpha_c} \tag{3}$$

where Q_{PMC} is the amount of heat from the fuel that goes into the pre-mixed combustion, Q_{MCC} is the amount of heat from the fuel that goes into the mixing-controlled phase, α is crank angle, $\Delta\alpha_c$ is the duration of pre-mixed combustion., α_{id} is the timing of the ignition delay, m is the shape parameter, and a is the Vibe parameter. According to (4), the fuel quantity available (f_1) and the turbulent kinetic energy density (f_2) determine the amount of energy released for mixing controlled phase:

$$\frac{dQ_{MCC}}{d\alpha} = C_{Comb} \cdot f_1(M_F, Q) \cdot f_2(k, V)$$ (4)

where $f_1(M_F, Q) = M_F - \frac{Q}{LHV}$; $f_2(k, V) = exp(C_{rate} \cdot \frac{\sqrt{k}}{\sqrt[3]{V}})$, C_{Comb} is the constant of combustion, C_{rate} is the constant of mixing rate, k is the local density of turbulent kinetic energy, M_F is the vaporised fuel mass, LHV is the lower heating value, Q is the cumulative heat release for the mixture-controlled combustion and V is the cylinder volume.

Studying procedure

The use of DME fuel to partially replace diesel fuel is made on the basis that the heat value of these two parts is equivalent. That is, the heat of the dual fuel input to the engine does not change compared to the heat of diesel input in the case of single diesel fuel. Thus, the replacement DME fuel ratio is calculated according to (5):

$$E_{DME} = \frac{M_D \cdot 100}{M_{D0}}$$ (5)

In which, E_{DME} – replacement DME fuel ratio, %;
$\qquad M_D$ – amount of diesel fuel to be replaced, g;
$\qquad M_{D0}$ – initial amount of diesel fuel, g.

The amount of alternative DME fuel is determined by (6):

$$M_{DME} = \frac{M_D \cdot LHV_D}{LHV_{DME}}$$ (6)

In which, M_{DME} – amount of replacement DME fuel, g;
$\qquad LHV_D$ – lower heating value of diesel fuel, MJ/kg;
$\qquad LHV_{DME}$ – lower heating value of DME fuel, MJ/kg.

With the principle of replacing DME as mentioned above, proceed to change the input parameters in the declaration of the amount of diesel supplied to the cycle. The amount of diesel fuel adjusted down is 5%, 10%, and 15%, respectively. After reducing the amount of diesel supplied, proceed to add the amount of DME supplied to the cycle according to (6). At that time, the fuel mixture used for the engine when replacing 5%, 10%, and 15% of diesel fuel with DME, respectively, is denoted as DME5, DME10, and DME15.

Yanmar S185L-ST engine can be used as the main engine or auxiliary engine to drive generators on ships, so the working modes of the engine are determined according to the propeller characteristics (100% N_e, 85% N_e, 75% N_e, 50% N_e, and 25% N_e at engine speeds of 900 rpm, 853 rpm, 819 rpm, 720 rpm, and 567 rpm, respectively; and load characteristics (100% N_e, 75% N_e, 50% N_e, and 25% N_e at engine speed of 900 rpm. Performance characteristics (effective power N_e and specific energy consumption SEC), and environment characteristics (NO$_x$ and soot emissions) are determined when changing the DME fuel replacement rate.

Results and Discussion

The effects of DME on performance characteristics

According to propeller characteristics, when DME replacement ratio is increased, the effective power of engine decreases. When using DME5 fuel, the effective power decreased insignificantly (not more than 4.7%). When using DME10 and DME15 fuel, the largest reduction in effective power was 9.43% and 14.2%, respectively (Table 87.3).

The SEC parameter is used to evaluate fuel economy of engine when using DME. In this study, SEC parameter is used instead of specific fuel consumption because there is difference in input fuel properties. DME has a lower LHV than diesel, with 28.43 MJ/kg for DME and 42.76 MJ/kg for diesel, respectively.

Table 87.3 Comparison of engine effective power according to propeller characteristics

Speed (rpm)	Effective power (kW)						
	Diesel	DME 5	Reduction (%)	DME 10	Reduction (%)	DME 15	Reduction (%)
567	110.03	104.86	4.7	99.65	9.43	94.41	14.2
720	221.35	212.37	4.06	203.27	8.17	194.12	12.30
819	331.58	319.22	3.73	306.65	7.52	293.97	11.34
853	375.85	362.15	3.65	348.33	7.32	334.27	11.06
900	442.08	426.45	3.54	410.65	7.11	394.61	10.74

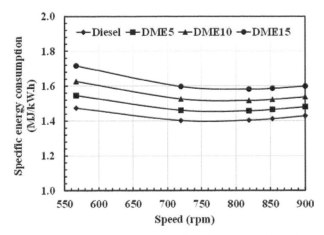

Figure 87.3 Comparison of specific energy consumption according to propeller characteristics
Source: Author

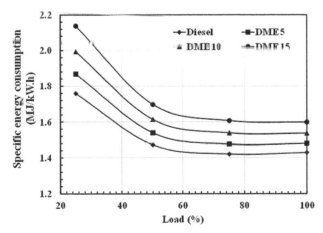

Figure 87.4 Comparison of specific energy consumption according to load characteristics at engine speed of 900 rpm
Source: Author

The absolute mass of DME supplied to intake manifold must be accounted in the total input energy for the test engine. As a result, the determination of SEC is based on the following (7):

$$SEC = \frac{m_D.LHV_D + m_{DME}.LHV_{DME}}{N_e} \tag{7}$$

Where *SEC* is the specific energy consumption (MJ/kWh); m_D and m_{DME} represent the mass flow of diesel fuel and DME (kg/h), respectively; LHV_D and LHV_{DME} are the lower heating value of diesel fuel and DME (MJ/kg), respectively.

Analysis of the graphs in Figures 87.3 and 87.4 shows that when increasing the DME fuel replacement rate, SEC decreases. When using DME5, DME10, and DME15, SEC decreased on average by 4.06%,

8.52%, and 13.44%, respectively, according to propeller characteristics; SEC decreased on average by 4.59%, 9.72%, and 15.47%, respectively, according to load characteristics at engine speed of 900 rpm. The effective power and SEC decrease as the amount of DME replacement increases due to the fact that DME displaces air in the intake manifold, which affects the combustion of diesel fuel.

The effects of DME on exhaust pollutants

The simulation results of environmental characteristics (NO_x and soot emissions) of the engine operating according to propeller characteristics when changing the DME fuel replacement ratio are shown in Figures 87.5 and 87.6. When increasing the DME fuel replacement rate, NO_x emissions decrease at low speed modes, and increase at high speed modes. The largest reduction in NOx emissions was 30.15% at engine speed 567 rpm when using DME15. The largest increase in NO_x emissions was 21.8% at engine speed 900 rpm when using DME10. This can be explained by the fact that the theoretical air/fuel ratio of DME is

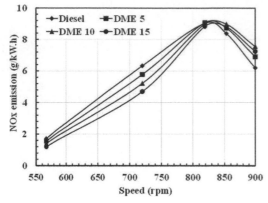

Figure 87.5 Comparison of NOx emission according to propeller characteristics
Source: Author

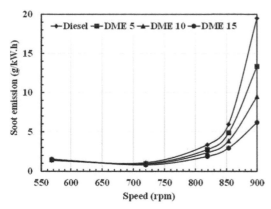

Figure 87.6 Comparison of soot emission according to propeller characteristics
Source: Author

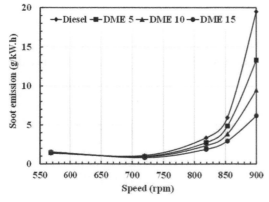

Figure 87.7 Comparison of NOx emission according to load characteristics
Source: Author

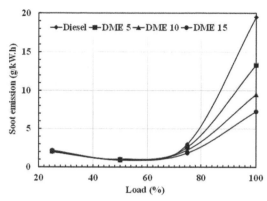

Figure 87.8 Comparison of soot emission according to load characteristics
Source: Author

smaller than that of diesel fuel, which reduces the flame area in the cylinder, leading to a reduction in NOx emissions at low-speed modes. However, at high-speed modes, the additional amount of DME makes the combustion process more optimal, the high combustion chamber temperature is the cause of the increase in NO_x emissions.

When increasing the DME fuel replacement rate, soot emissions decrease at high-speed modes. The largest reduction in PM emissions was 31.58% when using DME5, 51.26% when using DME10, and 68.15% when using DME15 at engine speed 900 rpm. This can be explained by the fact that DME has a high oxygen content, there is no carbon-carbon bond in the molecule, resulting in less soot formation during combustion.

In addition, the simulation results of environmental characteristics of the engine operating according to load characteristics at engine speed 900 rpm when changing the DME fuel replacement ratio are shown in Figures 87.7 and 87.8. Analysis of the graphs in Figures 87.7 and 87.8, shows that:

–   When increasing the DME fuel replacement rate, NO_x emissions decrease at low load modes and increase at high load modes.
–   When increasing the DME fuel replacement rate, soot emissions are significantly reduced at high loads. The largest reduction is 62.71% at 100% load when using DME15.

The raise of NO_x level is the result of higher in-cylinder temperature and the appearance of oxygen atoms in fuel molecular [28]. Furthermore, DME has a high cetane number and extremely good vaporisation ability, resulting in a quick combustion reaction when diesel fuel is injected into the combustion chamber. Moreover, the oxygen content in molecular acts as an oxidiser, promoting the combustion reaction and raising the combustion temperature. It is found that the NO_x emission increases by 11.45%, 21.80%, and 30.93% when using DME5, DME10, and DME15, respectively, compared with that of the original engine at full load condition.

Due to DME's own oxygen and lack of carbon-carbon structure in its molecules, which results in fewer precursors forming during combustion, soot emissions are greatly decreased at high loads. Additionally, there is a homogeneous combination of DME and charged air at the end of the compression stroke that will lessen fuel-rich regimes in the combustion chamber before the injection moment. As a result, when the engine runs on DME5, DME10, and DME15, the combustion process emits less soot.

Conclusion

The results of the simulation study show that the use of dimethyl ether (DME) fuel supplied to the intake manifold to partially replace diesel fuel is an effective solution to reduce NO_x and soot emissions for marine diesel engines. When increasing the DME fuel replacement rate, NO_x emissions decreased by up to 30.48% at low load modes, increasing at high load modes. soot emissions are reduced by up to 68.15% using DME15. The results of the simulation study are the basis for researching, designing, and manufacturing a system that provides DME to the intake manifold of engine to evaluate the ability to reduce harmful emissions for diesel engine.

Acknowledgment

This work was supported by the Ship Power Plants Laboratory, University of Transport Technology (UTT), Hanoi, Vietnam.

References

[1] Tuan, N. T. and Dong, N. P. (2021). Theoretical and experimental study of an injector of LPG liquid phase injection system. Energy for Sustainable Development. 63, 103–12.

[2] Khanh, N. D., Vinh, N. D., Long, H. D., Thanh, N. V., and Tuan, L. A. (2019). Performance and emission characteristics of a port fuel injected, spark ignition engine fueled by compressed natural gas. Sustainable Energy Technologies and Assessments. 31, 383–89.

[3] Vu, N. H., Minh, D. Q., Kien, N. T., Thin, P. V., and Phuong P. X. (2020). An extensive analysis of biodiesel blend combustion characteristics under a wide-range of thermal conditions of a cooperative fuel research engine. Sustainability. 12, 7666. doi:10.3390/su12187666.

[4] Hoang, A. T., Tran, V. D., Dong, V. H., and Le, A. T. (2019). An experimental analysis on physical properties and spray characteristics of an ultrasound-assisted emulsion of ultra-low-sulphur diesel and jatropha-based biodiesel. Journal of Marine Engineering and Technology, 21(2), 73–81.

[5] Lecksiwilai, N., Gheewala, S. H., Sagisaka, M., and Yamaguchi, K. (2016). Net energy ratio and life cycle greenhouse gases (GHG) assessment of bio-dimethyl ether (DME) produced from various agricultural residues in Thailand. Journal of Cleaner Production. 134. 523–31.

[6] Park, S. H. and Lee, C. S. (2014). Applicability of dimethyl ether (DME) in a compression ignition engine as an alternative fuel. Energy Conversion and Management. 86, 848–863.

[7] Fang, Q., Huang, Z., Zhu, L., Zhang, -J.-J., and Xiao, J. (2011). Study on low nitrogen oxide and low smoke emissions in a heavy-duty engine fuelled with dimethyl ether. Proceedings of the Institution of Mechanical Engineers, Part D: Journal of Automobile Engineering. 225(6), 779–86.

[8] Teng, H, McCandless, J. C., and Schneyer, J. B. (2004). Thermodynamic properties of dimethyl ether – an alternative fuel for compression-ignition engines. SAE Tech Paper, SAE 2004-01-0093.

[9] Huang, Z. H., Wang, H. W., Chen, H. Y., Zhou, L. B., and Jiang, D. M. (1999). Study of combustion characteristics of a compression ignition engine fuelled with dimethyl ether. Proceedings of the Institution of Mechanical Engineers, Part D: Journal of Automobile Engineering. 213, 647–52.

[10] Golovitchev, V. I., Nordin, N., and Chomiak, J. (1998). Neat, dimethyl ether : Is it really diesel fuel of promise?. SAE Technical Paper Series. 982537, pp. 255–268.

[11] Sorenson, S. C. (2001). Dimethyl ether in diesel engines: Progress and perspectives. Transactions of the ASME. Journal of Engineering for Gas Turbines and Power, 123(3), 652–8.

[12] Andren, M., Martin, V., and Svedberg, G. (1999). Combined production of power and alternative fuels in connection with pulp mills. SAE Technical Paper Series. 1999-01-2470, pp. 1–8.

[13] Verbeek, R. and Van der Weide J. (1997). Global assessment of dimethyl ether: comparison with other fuels. SAE Technical Paper Series. 971607, pp. 1–12.

[14] Bek, B. H., and Sorenson S. C. (2001). A mixing based model for dimethyl ether combustion in diesel engines. Transactions of the ASME. Journal of Engineering for Gas Turbines and Power. 123(3), 627–32.

[15] Wakai, K., Nishida, K., Yoshizaki, T., and Hiroyasu, H. (1998). Spray and ignition characteristics of dimethyl ether injected by a D.I. diesel injector. The Fourth International Symposium COMODIA 98, 537–542.

[16] Yao, M., Zheng, Z., and Qin J. (2006). Experimental study on homogeneous charge compression ignition combustion with fuel of dimethyl ether and natural gas. Transactions of the ASME. Journal of Engineering for Gas Turbines and Power, 128(2), 414–20.

[17] Cong, D. N., Duc, K. N., and Nguyen, V. (2021). The effects of dimethyl ether enriched air (DMEA) on exhaust pollutants and performance characteristics of an old generation diesel engine. International Journal of Sustainable Engineering. 14(5), 1143–56.

[18] Huang, Z., Qiao, X., Zhang, W., Wu, J., and Zhang, J. (2009). Dimethyl ether as alternative fuel for CI engine and vehice. Frontiers of Energy and Power Engineering in China, 3(1), 99–108.

[19] Konno, M., Kajitani, S., Oguma M., Iwase, T., and Shima, K. (1999). No emission a characteristics of a CI engine fueled with neat dimethyl ether. SAE Technical Paper Series. 1999-01-1116, 1–8.

[20] Hyun, G., Oguma, M., and Goto, S. (2002). Spray and exhaust emission characteristics of biodiesel engine operating with the blend of plant oil and DME. SAE Technical Paper Series. 2002-01-0864, 9.

[21] Fleisch, T., McCarthy, C., Basu, A., Udovich, C., Charbonneau, P., Slodowske, W., Mikkelsen, S., McCandless, J. (1995). A new clean diesel technology: Demonstration of ULEV emissions on a navistar diesel engine fueled with dimethyl ether. SAE Technical Paper Series. 950061, 1–10.

[22] Sorenson, S. C., Mikkelsen, S. E. (1995). Performance and emissions of 0.273 liter direct injection diesel engine fuelled with neat dimethyl ether. SAE Technical Paper Series. 950064, pp. 1–11.

[23] Guo, J., Chikahisa, T., Murayama, T., and Miyano, M. (1994). Low NOx methanol diesel engine with DME torch ignition method. Transactions of the Japan Society of Mechanical Engineers. Part B. 60(577), 3179–3184.

[24] Karpuk, M. E., Wright, J. D., and Dippo, J. L. (1991). Dimethyl ether as an ignition enhancer for methanol-fueled diesel engines. SAE Technical Paper Series. 912420, pp. 119–131.

[25] AVL. Thermodynamic cycle simulation Boost, AVL LIST GmbH Hans-List-Platz 1, A-8020 Graz, Austria, 2011.

[26] AVL. BOOST v2011.1 – Users Guide, AVL LIST GmbH Hans-List-Platz 1, A-8020 Graz, Austria, 2011.

[27] Chmela, F., and Orthaber, G. (1999). Rate of heat release prediction for direct injection diesel engines based on purely mixing controlled combustion. SAE Technical Paper. 1999-01-0186.

[28] Raine, R. R., Stone, C. R., and Gould, J. (1995). Modeling of nitric oxide formation in spark ignition engines with a multizone burned gas. Combustion and Flame. 102(3), 241–255.

Chapter 88

Formulation and optimisation of mathematical model of relationship between organisational performance and TQM practices in Indian industry

Arun Kedar[1,a], Sunil Prayagi[1,b] and Yogesh Deshpande[2,b]

[1]Yeshwantrao Chavan College of Engineering, Nagpur, India

[2]Shri Ramdeobaba College of Engineering and Management, Nagpur, India

Abstract

Due to cutthroat competition in business and to survive in this competitive environment many manufacturing industries across globe are practicing total quality management (TQM) as a strategy to improve organisational performance in terms of quality of products, productivity, and financial indicators. This paper aims to formulate and optimise mathematical relationship in the form of model amongst organisational TQM practices and their performance in Indian context. An exploratory survey instrument has been developed to collect field data from industries those have implemented TQM initiative. Ten TQM implementation factors were derived, and mathematical models were formulated and further optimised. The values for identified organisational performance parameters were estimated through mathematical models. Comparison of these values with values obtained from various respondents indicate that these models can be efficiently used to find performance parameters from available set of TQM implementation factors. Sensitivity analysis highlights customer involvement as the most sensitive while management support as least sensitive factor for TQM initiative. The findings of the study reveal that Indian manufacturing companies can use these derived TQM implementation factors and developed research instrument for obtaining in depth status of quality management strategy being adopted in industry and link it with related benchmarked performance measures of the organisation.

Keywords: Indian industry, mathematical model, organisational performance, survey, total quality management

Introduction

In today's globalise economy, competition is becoming ever more intense. The organisational progress and existence depends on significant approaches introduced and meticulously implemented by the industries. With the process of liberalisation, most of the developing nations including India have initiated broad series of policy drives to offer a proper situation for business investment and globalisation. The recent study by Adem and Virdi [1] illustrates benefits from total quality management (TQM) practices achieved by manufacturing organisations in Ethiopia. Literature review demonstrates role of TQM in improving organisational performance [2, 3, 4]. Kotabe [5] identified need of developing innovations for emerging market. Various organisational performance improvement approaches are practiced by industries. Kedar et al. [6] in their study compared and briefed the characteristics of six organisation performance approaches representing their similarity and difference. Past studies highlight significance of critical factors for effective TQM application in industries [7-10]. In Indian perspective, study by Wali et al. [11], is possibly the first representative study of quality aspects. They have identified twelve critical factors of TQM. Several research studies presenting association between TQM and organisational performance are available. Amongst large number of studies presented on implementation of TQM methodology and its effect on the performance of the organisations, only the studies relevant to theme are cited. In recent studies, Singh et al. [12] confirm positive relationship between TQM and organizational performance pertaining to satisfaction results of various business domain. Modarres and Pezeshk, [13] identified considerable positive relationship of TQM with innovation and learning in the organisation. Pradhan, [14] recognised key factors which lead to performance improvement through TQM implementation.

Most of the research studies presents direct linkage of association amongst TQM practices and performance, Shafiq et al. [15] states organisations have adopted TQM to enhance product and service quality. Modgil and Sharma [16] states TQM practice in industry improves share of company in the market along with productivity and overall performance. The study of Bolatan et al., [17] discloses a very robust link between TQM and quality performance of the manufacturing unit. Kulkarni et al. [18] states how quality circles as an employee base tool helps in improvement of productivity resulting in remarkable savings of the organisation. Saraph et al. [19] developed the research instrument that can be used for assessment of quality management in organisations. Quality of products is used by Ahire [20] to measure operational performance of TQM practice. Productivity as a parameter to evaluate performance of TQM is also focused by Samson and Terziovski, [21]. Hendricks and Singhal [22] explore the hypothesis representing link of TQM

[a]arunkedar64@gmail.com, [b]sunil_prayagi@yahoo.com, [c]deshpandeyv@rknec.edu

DOI: 10.1201/9781003450252-88

with operative performance. In earlier study Brah et al. [23] evaluates operational performance through customer satisfaction, employee satisfaction, suppliers' performance, product quality. TQM strategy effectively and positively links quality of product and innovation in product performance [24]. The effect of TQM on productivity was discussed by Khan, [25]. Link between quality, customer satisfaction, business results, human resource and time with TQM factors is proposed by Shrivastava et al. [26]. A strong relationship is identified amongst implementation of TQM approach and organisation performance for Joiner [27].

This paper recognises most relevant TQM implementation factors and presents linkage between TQM practice and organisational performance. Further section of the paper present scope of present research along with necessity for formulation of field data-based model. Section IV presents methodology, followed by model formulation and model optimisation in section V and VI respectively. Section VII presents case study. Further section VIII deals with results and discussion and finally section IX describes conclusion.

Scope of Present Research

TQM approach represents complex phenomenon involving man, machine, facilities, and sub systems. Approximate generalised models have been developed for complex phenomenon involving man machine system [28, 29]. Based on earlier mathematical models, the present research identifies interacting factors of TQM implementation and organisational performance and proposes unique approximate generalised models for TQM approach. Managers can use the approximate generalised model to evaluate influence of TQM practices on organisational performance of companies.

Necessity for Framing Field Data Based Model

Various factors contribute for successful TQM implementation. Journals associated to quality management approach of assessing the effect of TQM on organisational performance do not portray quantifiable interface of the associated independent parameters with dependent variables in developed generalised field data base models for TQM approach. Industrial production system represents a very intricate structure that is analogous to man machine system. The empirical data-based model for this TQM methodology is framed on parallel lines for studies carried out with experimentation [28, 29] excluding the test planning mechanism. Analytical linking is established with data collected from industry through research instrument and the values of TQM factors (independent variables) and organisational performance factors (dependent variables) were calculated. Established relationship is optimised to find independent variables values for maximisation and minimisation of the associated objective function [30].

Methodology

Research Instrument Development

This study being exploratory in nature, a main tool for data collection is a research instrument in the form of questionnaire. For this study, the unique research instrument was developed and validated based on available literature for similar studies. A major source of identifying various questions for developing the research instrument was [31-34]. The research instrument for data collection was developed after review of related literature from earlier studies, visits in TQM practicing industries followed by discussion and suggestions from practicing managers. National and international quality award models were also well thought out in the course of questionnaire development. Five-point Likert scale was used for recording replies from respondents in a scale of 1 to 5 stating 'No emphasis' to 'Very strong emphasis' respectively. Respondents were provided with 49 statements identified through exhaustive literature review representing factors of TQM. Fifteen statements directing toward impact of TQM program on identified parameters of organisational performance were provided in concluding section of research instrument. Respondents were asked to reply in five-point scale from 1 to 5 stating 'No improvement' to 'Very large improvement' respectively.

Independent Variables

Various elements of TQM approach were illustrated in this part of study. Based on extensive literature review, several dimensions like continuous improvement, strategic planning, customer delight, process management, management leadership, education/training, teamwork, rewards and recognition, suppliers' quality management, information architecture and management of human resource were thought off in developing independent variables as the preliminary contributions to the implementation of TQM practice. In all 49 questions representing these domains were included in the research instrument.

Dependent Variables

Organizational overall performance leads to dependent variables. For most of the manufacturing industries performance parameters like quality of product, productivity indicators and financial measures indicate dependent variables. Through associated literature review, this research assesses organisational performance expressed as dependent variables. These performance parameters were described by five sub parameters. Table 88.1 depicts these performance indicators along with their dependent sub parameter.

Instrument Administration

For collection of field data, the developed and pilot tested research instrument was handed over to various cadres of managers amongst TQM practicing companies across India by post, through e-mail and in some cases in person. The respondents represent diverse functional area from wide variety of manufacturing sectors like automobile, furniture, metal, chemical etc. Respondents were expert in their work domain with lot of experience. The estimated sample size for this study was 225. All the respondents were briefed about objective of the research. Data collection includes obtaining information like respondent designation, name of industry, performance improvement plans followed by the industry, total employees, industry, type, industry location, industry age and yearly sales volume along with various question items pertaining to dependent variables and performance indicators. About 88 % of the respondents were holding managerial and higher position which indicate that enough thought is given to the various elements of questionnaire. 50% of the respondents were highly experienced with average 20 + years of experience. The manufacturing companies represent heterogeneous combination almost from all sectors. The active participation of respondents with constant follow up results in receiving 86% response rate.

Internal consistency and validity

Reliability is assessed from internal consistency analysis. For individual construct Cronbach's alpha value was obtained. The alpha values vary from 0.7473 to 0.9035. No items were dropped from each factor as internal consistency was more than 0.70. Instrument exhibits content validity as items selected were based on literature review followed by experts' opinion of industrial managers and academicians. Multiple correlation coefficients attained above 0.5 ($p < 0.05$) indicating presence of criterion-related validity. Unifactorial test was performed to test construct validity and result indicate instrument display construct validity. The result of data testing indicates the data collected is ready for further analysis.

Factor Analysis

In this study 49 items were identified which represent input to the TQM implementation methodology. These items mentioned in research instrument were subjected to exploratory factor analysis (EFA). EFA

Table 88.1 Performance indicator and dependent sub-parameter

S. N.	Performance indicator	Dependent sub-parameter	Tag
1	Quality of product	Satisfaction of customer	Y1
2		Defects	Y4
3		Rework	Y7
4		Customer claims	Y10
5		Scrap/waste	Y14
6	Productivity indicators	Production capacity	Y3
7		Inventory	Y6
8		Delivery schedule	Y9
9		Cycle time	Y11
10		Net profit as % of sale	Y13
11	Financial measures	Sales growth	Y2
12		Return on asset	Y5
13		Cost of production	Y8
14		Cost of supply chain	Y12
15		Cost of man power	Y15

Source: Author

identified underlying factors which illustrate the correlation pattern within observed items. These items represent implementation factors of TQM. Varimax rotation in EFA yields ten factors which account for 71.5 % cumulative variance. In this analysis both KMO value representing sampling adequacy is estimated as 0.93 while Bartlett's Test was significant with Chi square value of 6001.7 at $p < 0.001$. Table 88.2 highlights extracted factors with cumulative % variance.

Factor 1 is tagged as 'Process Management' (Fact1) and ten items load on it. Seven items load on factor 2, this factor is named as 'Involvement of employees' (Fact2). The third factor denotes one of the important requisites of TQM, 'Customer involvement and satisfaction' (Fact3) includes seven items. Factor 4 captures ten items and is labeled as, 'Education and training' (Fact4). Factor 5 comprises the nine items, and named as, 'Communication and information' (Fact5). Factor 6 loaded with three items depicts reward for quality improvement. It is named as, 'Employee appraisal' (Fact6). Factor 7, loaded with three items represent critical dimension of quality management focusing on communication and involvement with supplier. It is labeled as, 'Supplier involvement' (Fact7). The last three factors, Factor 8, 'Employee satisfaction' (Fact8), Factor 9, 'Employee empowerment (Fact9)' and Factor 10, 'Management support (Fact10)' are loaded with one item each.

Further analysis depicts ranking of the ten extracted factors based on practice thus highlighting the level of implementation for quality management principles. The various means for the perception of practice for extracted factors were analysed. An overall mean for each factor was obtained to look at the level of an importance perceived by the respondents. Mean value for each factor is evaluated. The values range from 3.150 to 4.136, which corresponds to a 'considerable emphasis' to 'high emphasis' level of practice. Communication and information (4.1260), involvement of employees (4.031), and customer involvement and satisfaction (4.008) were the three highest practices in this study, while employee satisfaction (3.66), employee appraisal (3.550) and management support (3.150) were the bottom three. From this result, it can be observed that most of the respondents rated at 'considerable emphasis' to 'high emphasis' for degree of emphasis, organisation currently places on the attributes mentioned in their companies, indicating that companies are stressed to practice the approaches under study successfully.

Model Formulation

For any complex process, structure or engineering phenomenon, it is essential to evaluate quantitative correlation amongst dependent variables (outcome) and independent variables (input). In such case the correlation amongst identified independent variable presents a mathematical model for phenomenon under consideration. These mathematical models act as tool to evaluate the outcome of the program/process. Quantifiable relationship amongst TQM factors (independent variables) and organisational performance parameters (dependent variables) is stated as below:

First dependent variable, i.e. customer satisfaction (Y1):

$$Y1 = f\,[\text{Fact1, Fact2, Fact3, Fact4, Fact5, Fact6, Fact7, Fact8, Fact9, Fact10}] \tag{1}$$

Considering an exponential function for this complex system as depicted by equation 1, this equation can be written as

$$Y1 = K1*[\text{Fact1}]^{a1}[\text{Fact2}]^{b1}[\text{Fact3}]^{c1}[\text{Fact4}]^{d1}[\text{Fact5}]^{e1}[\text{Fact6}]^{f1}[\text{Fact7}]^{g1}[\text{Fact8}]^{h1}[\text{Fact9}]^{i1}[\text{Fact10}]^{j1} \tag{2}$$

Table 88.2 Principle component analysis indicating extracted factors

Factors' Extracted	Eigen value	variance percentage	cumulative variance percentage
1	6.1	11.5	11.5
2	5.8	10.8	22.3
3	5.4	10.0	32.3
4	5.2	9.6	41.9
5	5.0	9.1	51
6	3.7	6.6	57.6
7	3.1	5.4	63
8	2.6	4.4	67.4
9	2.5	4.1	71.5
10	1.4	4.5	76

Source: Author

Further adopting the procedure of multiple regression analysis along with supportive computer program, the values of K1, a1, b1, c1, d1, e1, f1, g1, h1, i1 and j1 are estimated as 1.0000, -0.0218, 0.4000, 0.3807, -0.0164, -0.3439, 0.3112, 0.2159, 0.1475, 0.0095 and -0.082 respectively.

This analysis yields the generalised field data base model for customer satisfaction (Y1) as a dependent variable in TQM environment as,

$$Y1 = 1.0000 * [Fact1]^{-0.0218} [Fact2]^{0.4000} [Fact3]^{0.3807} [Fact4]^{-0.0164} [Fact5]^{-0.3439} [Fact6]^{0.3112}$$
$$[Fact7]^{0.2159} [Fact8]^{0.1475} [Fact9]^{0.0095} [Fact10]^{-0.082} \quad (3)$$

Above Equation (3) symbolise generalised field data-based model for customer satisfaction (Y1) as dependent variable. By using same approach other fourteen mathematical models were formulated for fourteen dependent variables (viz. Y2 to Y15) for TQM practice related to quality of product, productivity indicators and financial measures.

Model Optimisation

Total fifteen models were developed for the TQM practice adopted by Indian industries. Along with development of mathematical models for all fifteen performance parameters, these models were further optimised. The purpose of optimisation was to obtain the best set of TQM factors demonstrating input to the system. Objective function may be in the form of minimisation or maximisation depending of the performance parameter. The fifteen formulated models broadly represent quality of product, productivity indicators and financial measures as organizational performance.

These models present nonlinear arrangement, therefore for optimisation purpose they are converted into linear form.

Model for customer satisfaction (Y1),

$$Y1 = K1 * [Fact1]^{a1} [Fact2]^{b1} [Fact3]^{c1} [Fact4]^{d1} [Fact5]^{e1} [Fact6]^{f1} [Fact7]^{g1} [Fact8]^{h1} [Fact9]^{i1} [Fact10]^{j1} \quad (4)$$

Further solving, taking log of both sides of equations

$$\log Y1 = \log K1 + a1 * \log [Fact1] + b1 * \log [Fact2] + c1 * \log [Fact3] + d1 * \log [Fact4] + e1 *$$
$$\log [Fact5] + f1 * \log [Fact6] + g1 * \log [Fact7] + h1 * \log [Fact8] + i1 *$$
$$\log [Fact9] + j1 * \log [Fact10] \quad (5)$$

Substituting appropriate values in Equation (5), the linear model for this situation in linear polynomial form will be expressed like following equation

$$Z = K1' + a * X1 + b * X2 + c * X3 + d * X4 + e * X5 + f * X6 + g * X7 + h * X8$$
$$+ i * X9 + j * X10 \quad ...(6)$$

Eq^n (6) represents maximization type objective function.

Zmax = 9.68 and antilog X1 to X10 yields value for all 10 independent variables.

Consequently Z value for Fact1, Fact2, Fact3, Fact4, Fact5, Fact6, Fact7, Fact8, Fact9 and Fact10 are 2.42, 7.30, 7.65, 3.13, 2.42, 4.89, 5.17, 2.76, 2.80 and 0.66 respectively.

Similar approach for optimization is followed for remaining models. On the basis of computed scores of independent factors, the optimal values are obtained for different performance indicators. Table 88.3 depicts results of optimisation for other models.

Case Study

The major aim of the case study is to provide a practical example of performance improvement of the Indian manufacturing company that has implemented TQM initiative. One of the objectives of case study was to assess the TQM implementation practices and performance improvement of the organization. The study was conducted in company that has already implemented this initiative. The case study helps in evaluating the company's TQM implementation and overall business performance. To do so, research instrument was administered amongst twenty managers in the company and their responses were analysed. In order to have a better understanding of the impact of TQM implementation, the information about overall business performance was obtained over the last three years. The results of the findings of a case study are

Table 88.3 Optimisation results: models Yi to Y 15

Model Y1 Max Z		Model Y2 Max Z		Model Y3 Max Z		Model Y4 Min Z		Model Y5 Max Z	
Z	9.680	Z	18.293	Z	16.110	Z	0.332	Z	12.274
Fact1	2.419	Fact1	2.419	Fact1	2.419	Fact1	2.419	Fact1	8.679
Fact2	7.299	Fact2	7.299	Fact2	7.299	Fact2	2.239	Fact2	7.299
Fact3	7.650	Fact3	7.650	Fact3	7.650	Fact3	3.419	Fact3	7.650
Fact4	3.151	Fact4	3.151	Fact4	3.151	Fact4	9.009	Fact4	3.129
Fact5	2.419	Fact5	2.419	Fact5	2.419	Fact5	8.509	Fact5	2.419
Fact6	4.889	Fact6	4.889	Fact6	4.889	Fact6	0.979	Fact6	4.889
Fact7	5.170	Fact7	5.170	Fact7	5.170	Fact7	1.339	Fact7	5.170
Fact8	2.759	Fact8	2.759	Fact8	2.759	Fact8	2.759	Fact8	2.759
Fact9	2.800	Fact9	0.560	Fact9	0.560	Fact9	0.560	Fact9	2.800
Fact10	0.659	Fact10	3.290	Fact10	3.290	Fact10	0.659	Fact10	0.659

Model Y6 Min Z		Model Y7 Min Z		Model Y8 Min Z		Model Y9 Max Z		Model Y10 Min Z	
Z	0.611	Z	0.838	Z	1.429	Z	12.150	Z	0.910
Fact1	2.419	Fact1	2.419	Fact1	2.419	Fact1	2.419	Fact1	8.679
Fact2	2.239	Fact2	2.239	Fact2	2.239	Fact2	7.299	Fact2	7.299
Fact3	3.419	Fact3	3.419	Fact3	3.419	Fact3	7.650	Fact3	3.419
Fact4	9.009	Fact4	9.009	Fact4	3.151	Fact4	3.129	Fact4	3.129
Fact5	8.509	Fact5	8.509	Fact5	2.419	Fact5	2.419	Fact5	8.509
Fact6	0.977	Fact6	0.979	Fact6	0.979	Fact6	4.889	Fact6	0.979
Fact7	1.339	Fact7	1.339	Fact7	1.339	Fact7	5.170	Fact7	1.339
Fact8	0.550	Fact8	0.550	Fact8	0.550	Fact8	0.550	Fact8	0.550
Fact9	2.800	Fact9	2.800	Fact9	0.560	Fact9	2.800	Fact9	2.800
Fact10	0.659	Fact10	3.290	Fact10	3.290	Fact10	3.290	Fact10	0.659

Model Y11 Min Z		Model Y12 Min Z		Model Y13 Max Z		Model Y14 Min Z		Model Y15 Max Z	
Z	0.897	Z	0.4684	Z	17.914	Z	0.378	Z	0.349
Fact1	8.679	Fact1	8.679	Fact1	2.419	Fact1	2.419	Fact1	8.679
Fact2	2.239	Fact2	7.299	Fact2	7.299	Fact2	2.239	Fact2	2.239
Fact3	3.419	Fact3	3.419	Fact3	7.650	Fact3	3.419	Fact3	3.419
Fact4	9.009	Fact4	9.009	Fact4	3.151	Fact4	9.009	Fact4	9.009
Fact5	8.509	Fact5	8.509	Fact5	2.419	Fact5	8.509	Fact5	8.509
Fact6	0.9799	Fact6	0.979	Fact6	4.889	Fact6	0.979	Fact6	4.889
Fact7	1.339	Fact7	1.339	Fact7	5.170	Fact7	1.339	Fact7	1.339
Fact8	0.550	Fact8	0.550	Fact8	2.759	Fact8	0.550	Fact8	0.550
Fact9	0.560	Fact9	2.800	Fact9	0.560	Fact9	0.560	Fact9	0.560
Fact10	0.659	Fact10	0.659	Fact10	3.290	Fact10	3.290	Fact10	0.659

Source: Author

discussed in Table 88.4. The observed and predicted values of various performance measures are fairly close to each other.

Results and Discussion

These derived mathematical models provide indices, which are indicating the way in which organisational performance (dependent terms) is affected by interaction of several TQM implementation factors (independent terms). Table 88.5 shows the sequence of indices of TQM implementation factors with organisational performance parameter.

Table 88.4 Observed values and predicted values of performance measures

Performance measures	Observed output	Predicted output
Customer satisfaction	4.35	4.43
Defects	3.95	3.47
Rework	4.15	4.20
Claims from customer	4.00	3.34
Scrap and waste	3.80	4.07
Production capacity	4.65	4.37
Inventory	3.95	3.80
Delivery schedule	4.35	4.02
Cycle time	4.20	4.05
Net profit as % of sale	4.20	4.28
Sales growth	4.60	4.17
Return on asset	4.25	4.82
Cost of production	4.20	4.20
Cost of supply chain	3.95	3.40
Cost of man power	3.80	4.02

Source: Author

Model for customer satisfaction (Y1)
    The model for customer satisfaction (Y1) is as below:

$$Y1 = 1.0000 * [Fact1]^{-0.0218} [Fact2]^{0.4000} [Fact3]^{0.3807} [Fact4]^{-0.0164} [Fact5]^{-0.3439} [Fact6]^{0.3112} [Fact7]^{0.2159} [Fact8]^{0.1475} [Fact9]^{0.0095} [Fact10]^{-0.082}$$

The above stated model reveals main interpretations as stated below:

1. Involvement of employees is having highest absolute index of 0.4. Analysis reveals involvement of employees is highly influencing factor for customer satisfaction model. Positive value of this TQM factor indicates involvement of employees has strong bearing on customer satisfaction.
2. Employee empowerment turns out to be weakest influencing factor for this model with the absolute index as 0.0095. Therefore, employee empowerment needs improvement in the TQM implementation initiative of the organisation.
3. Order of influence of rest of the TQM implementation factors representing Customer satisfaction model are Fact3, Fact5, Fact6, Fact7, Fact8, Fact10, Fact1 and Fact4 having absolute indices 0.3807, 0.3439, 0.3112, 0.2159, 0.1475, 0.082, 0.0218, 0.0164, respectively. The negative value of indices for some TQM implementation factors viz. communication and information, management support, process management and education and training points towards strengthening these domains for performance improvement.
4. The value of constant as presented in customer satisfaction model is unity i.e., 1. Since constant value is unity, it does not have any effect on increase in the value computed from the multiplication of the corresponding factors of TQM implementation of the model.

A summary of interpretation of other models is depicted in Table 88.4.
    Sensitivity analysis shows customer involvement and satisfaction factor is the most sensitive for 85% of the TQM implementation drive whereas with 39% management support turns out to be least sensitive which states that the factor management support needs strong improvement. Also, both supplier involvement and involvement of employees are found to be additional sensitive factors. Further, employee empowerment, communication for information and employee satisfaction domain demand strong improvement for TQM practicing Indian industries. Performance improvement parameters values are estimated for all developed mathematical models in this study. Table 88.6 shows comparison between observed values and computed values from mathematical model along with % error. These % error values suggest both observed values and computed values are in close agreement with each other.

Table 88.5 The order of influence of indices of the tqm implementation factors

Performance parameters	Order of TQM factors according to intensity of influence
Y1	Fact2, Fact3, Fact5, Fact6, Fact7, Fact8, Fact10, Fact1, Fact4 and Fact9
Y2	Fact3, Fact5, Fact7, Fact1, Fact4, Fact2, Fact8, Fact6, Fact9 and Fact10
Y3	Fact2, Fact3, Fact4, Fact1, Fact5, Fact6, Fact7, Fact10, Fact9 and Fact8
Y4	Fact3, Fact5, Fact1, Fact7, Fact4, Fact6, Fact9, Fact8, Fact2 and Fact10
Y5	Fact3, Fact5, Fact7, Fact4, Fact6, Fact9, Fact8, Fact2, Fact10 and Fact1
Y6	Fact3, Fact6, Fact7, Fact5, Fact4, Fact1, Fact2, Fact9, Fact8 and Fact10
Y7	Fact3, Fact7, Fact4, Fact1, Fact6, Fact5, Fact8, Fact2, Fact9 and Fact10
Y8	Fact7, Fact4, Fact8, Fact6, Fact2, Fact9, Fact1, Fact2, Fact5 and Fact10
Y9	Fact3, Fact4, Fact2, Fact1, Fact9, Fact10, Fact7, Fact6, Fact8 and Fact5
Y10	Fact3, Fact7, Fact1, Fact5, Fact6, Fact4, Fact8, Fact10, Fact2 and Fact9
Y11	Fact3, Fact2, Fact7, Fact5, Fact6, Fact1, Fact4, Fact10, Fact9 and Fact8
Y12	Fact3, Fact7, Fact4, Fact5, Fact6, Fact9, Fact8, Fact10, Fact2 and Fact1
Y13	Fact3, Fact5, Fact2, Fact7, Fact6, Fact4, Fact8, Fact1, Fact9 and Fact10
Y14	Fact3, Fact7, Fact4, Fact5, Fact2, Fact1, Fact6, Fact8, Fact10 and Fact9
Y15	Fact3, Fact7, Fact4, Fact2, Fact5, Fact6, Fact9, Fact1, Fact10 and Fact8

Source: Author

Table 88.6 Comparison between observed values and computed values

Organisational performance sub parameters	Mean observed	Mean computed	% error
Satisfaction of customer (Y1)	3.7	3.6	2.2
Sales growth (Y2)	3.9	3.6	7.9
Production capacity (Y3)	4.0	3.7	8.2
Defects (Y4)	3.5	3.4	1.4
Return on asset (Y5)	3.6	3.4	5.8
Inventory (Y6)	3.7	3.5	4.1
Rework (Y7)	3.8	3.8	1.6
Cost of production (Y8)	3.9	3.7	3.6
Delivery schedule (Y9)	3.8	3.6	5.3
Customer Claims (Y10)	3.8	3.7	3.4
Cycle time (Y11)	3.8	3.7	1.3
Cost of supply chain (Y12)	3.8	3.6	5.6
Net profit as % of sale (Y13)	3.6	3.3	7.0
Scrap/wasteY (14)	3.5	3.3	5.9
Cost of man power (Y15)	3.5	3.3	5.7

Source: Author

Conclusion

In today's competitive environment, there is need for enhancing business performance for which industries are targeting from 'customer satisfaction to customer delight' by adopting various strategies. Industries are adopting several initiatives for performance improvement.

Mathematical models were developed for Indian industries those have implemented total quality management (TQM) program for performance improvement. Table 88.4 depicts mean values for all fifteen identified organisational performance parameters. These values are derived from responses received from industries and from developed mathematical models. The estimate of percentage error for various parameters reveals that industries practicing TQM can readily use these models for evaluating performance output for specified set of input parameters in the form of adopted TQM strategy. Also, the optimisation technique using linear programming approach was used to find optimal values of independent variables for minimising/maximising i.e., optimising the performance parameters for TQM practice. On the basis of computed scores of independent factors, the optimal values are obtained for different performance indicators. Top level management can use these results for decision making so as to improve organisational performance.

The data of the company's overall business performance could be used as input for formulating an effective improvement plan. Therefore, evaluating overall business performance was also an important part of TQM implementation. This study can be very useful to the organization attempting to identify those characteristics often mentioned in the TQM literature that may provide an opportunity to increase the level of quality and improve performance. The periodic use of the research instrument will help in the monitoring process in identifying those areas of quality management where improvements should be made, and in prioritizing quality management efforts.

The developed research instrument can be directly used by all levels of management to measure the impact of TQM implementation in various divisions of the industry. Also, industries can identify the weak linkage of TQM implementation factor and organisational performance. Decision makers can effectively use the results obtained through analysis for strengthening the quality management strategy of the organisation.

References

[1] Adem, K., and Virdi, S. (2021). The effect of TQM practices on operational performance: an empirical analysis of ISO 9001: 2008 certified manufacturing organizations in Ethiopia. The TQM Journal. 33(2), 407–40.

[2] Abdi, M., and Singh, A. (2021). Effect of total quality management practices on nonfinancial performance: an empirical analysis of automotive engineering industry in Ethiopia. The TQM Journal. [ahead-of-print, no. ahead-of-print]. 34, 1116–1144.

[3] Bhaskar, H. (2020). Establishing a link among total quality management, market orientation and organizational performance: An empirical investigation. The TQM Journal. 32(6), 1507–24.

[4] Sahoo, S., and Yadav, S. (2020). Influences of TPM and TQM practices on performance of engineering product and component manufacturers. 17th Global Conference on Sustainable Manufacturing, (vol. 43, pp. 728–735).

[5] Kotabe, M., and Kothari, T. (2016). Emerging market multinational companies' evolutionary paths to building a competitive advantage from emerging markets to developed countries. Journal of World Business. 51(5), 729–43.

[6] Kedar, A. P., Lakhe, R. R., Deshpande, V. S., Wakhare, M. V., and Washimkar, P. V. (2008). A comparative review of TQM, TPM and related performance improvement programs. Paper presented at the International conference on Emerging Trends in Engineering and Technology 2008, Nagpur, India. 16-18 July 2008, pp. 725–30.

[7] Lakhe, R. R., and Mohanty, R. P. (1994). Total quality management- concepts, evolution, and acceptability in developing economies. The International Journal of Quality and Reliability. 11(9), 9–25.

[8] Mohanty, R. P., and Lakhe, R. R. (1998). Factors affecting TQM implementation :An empirical study in Indian industry. Production Planning and Control. 9(5), 511–20.

[9] Motwani, J. G., Mahmoud, E., and Rice, G. (1994). Quality practices of Indian organizations: An empirical analysis. International Journal of Quality and Reliability Management. 11, 38–52.

[10] Rao, S. S., Raghu-Nathan, T. S., and Solis, L. E. (1997). A comparative study of quality practices and results in India, China and Mexico. Journal of Quality Management. 2, 235–50.

[11] Wali, A. A., Deshmukh, S. G., and Gupta, A. D. (2003). Critical success factors of TQM: a select study of Indian organizations. Production Planning and Control. 14(1), 3–14.

[12] Singh, V., Kumar, A., and Singh, T. (2018). Impact of TQM on organisational performance: the case of Indian manufacturing and service industry. Operations Research Perspectives. 5, 199–217.

[13] Modarres, M., and Pezeshk, J. (2018). Impact of total quality management on organisational performance: exploring the mediating effects of organisational learning and innovation. International Journal of Business Environment. 9(4), 356–89.

[14] Pradhan, B. L. (2017). Confirmatory factor analysis of TQM implementation constructs: evidence from Nepalese manufacturing industries. Management Review International of Journal. 12(1), 26–56.

[15] Shafiq, M., Lasrado, F., and Hafeez, K. (2019). The effect of TQM on organisational performance: empirical evidence from the textile sector of a developing country using SEM. Total Quality Management and Business Excellence. 30(1–2), 31–52.

[16] Modgil, S., and Sharma, S. (2016). Total productive maintenance; total quality management and operational performance : An empirical study of Indian pharmaceutical industry. Journal of Quality in Maintenance Engineering. 22(4), 353–77.

[17] Bolatan, G. I. S., Gozlu, S., Alpkan, L., and Zaim, S. (2016). The impact of technology transfer performance on total quality management and quality performance. Procedia – Social and Behavioral Sciences. 235, 746–55.

[18] Kulkarni, S., Welekar, S., and Kedar, A. (2017). Quality circle to improve productivity: A case study in a medium scale aluminium coating industry. International Journal of Mechanical Engineering and Technology. 8(12), 793–809.

[19] Saraph, J. V., Benson, P. G., and Schroeder, R. G. (1989). An instrument for measuring the critical factors of quality management. Decision Sciences, 20(4), 810–29.

[20] Ahire, S. L. (1996). An empirical investigation of quality management in small firms. Production and Inventory Management Journal. 2nd quarter, 37, 44–50.

[21] Samson, D., and Terziovski, M. (1999). The relationship between total quality management practices and operational performance. Journal of Operations Management, 17, 393–409.

[22] Hendricks, K. B., and Singhal, V. R. (1997). Does implementing an effective TQM program actually improve operating performance? Empirical evidence from firms that have won quality awards. Management Science. 43, 1258–74. https://doi.org/10.1287/mnsc.43.9.1258

[23] Brah, S. A., Tee, S., and Rao, B. (2002). Relationship between TQM and performance of Singapore companies. International Journal of Quality and Reliability Management. 19(4), 356–79.

[24] Prajogo, D. I., and Sohal, A. S. (2003). The relationship between TQM practices, quality performance, and innovation performance: An empirical examination. International Journal of Quality and Reliability Management. 20(8), 901–918.

[25] Khan, J. (2003). Impact of total quality management on productivity. The TQM Magazine. 15(6), 374–80.

[26] Shrivastava, R. L., Mohanty, R. P., and Lakhe, R. R. (2006). Linkages between total quality management and organisational performance: an empirical study for Indian industry. Production planning and Control. 17, 13–30.

[27] Joiner, T. A. (2007). Total quality management and performance: The role of organization support and co-worker support. International Journal of Quality and Reliability Management. 24(63), 617–27.

[28] Modak, J. P., and Bapat, A. R. (1994). Formulation of generalised experimental model for a manually driven flywheel motor and its optimization. Applied Ergonomics, U.K. 25(2), 119–22.

[29] Bansod, S. V., Patil, S. G., and Modak, J. P. (2003). Design of experimentation for the formulation of an approximate experimental model for the effectiveness of seats in industry. International Journal of AMSE France, Modeling Series D. 24, 55–68.

[30] Rao, S. S. (1984). Optimization Theory & Applications. Wiley Eastern Ltd. (2nd Ed.).

[31] Benson, P. G., Saraph, J. V., and Schroeder, R. G. (1991). The effects of organizational context on quality management: An empirical investigation. Management Science. 39, 1107–24.

[32] Flynn, B. B., Schroeder, R. G., and Sakakibara, S. (1994). A framework for quality management research and an associated measurement instrument. Journal of Operations Management. 11, 339–66.

[33] Black, S., and Porter, L. (1996). Identification of the critical factors of TQM. Decision Sciences. 27(1), 1–21.

[34] Grandzol, J. R., and Gershon, M. (1998). A survey instrument for standardizing TQM modeling research. International Journal of Quality Science. 3(1), 80–105.

Chapter 89

Critical review on weldability of 316 austenitic and 410 martensitic stainless steel

Aditya Thaware[1,a], Sachin Ambade[1,b], Prachi Gawande[1,c], Radhika Katyarmal[1,d], Taha Josh[1,e] and Sagar Shelare[2,f]

[1]Yeshwantrao Chavan College of Engineering, Nagpur, India

[2]Priyadarshini College of Engineering, Nagpur, Maharashtra, India

Abstract

Stainless steel has a varied range of applications. It is used for the smallest thing used in homes like grills, cooker, saucepans to large equipment like tanks, furnace, vessels, kiln. The requirement of stainless steel never stops but no matter which grade of steel used, it will corrode after some time. The need of corrosion resistant stainless steel can be fulfilled by the weldment of 316 ASS and 410 MSS because both these metals are highly corrosion resistant. The use of dissimilar metal weldments is required in numerous industrial circumstances, especially in the petrochemical, aircraft, marine, and other such industries, for financial gain and to improve component performance. The mechanical and anti-corrosive qualities of austenitic and martensitic stainless steel have drawn attention towards it recently. As 316 ASS contains chromium and nickel, it has a high level of corrosion resistance. It is utilised all over the world for a variety of purposes, including tanks, storage containers, aerospace, pharmaceutical, food, and beverage equipment, and mining and chemical equipment. They are employed in sectors including the chemical, power plant, and many others due to their strength, ability to protect from corrosion, formability, and characteristics in high temperatures. But due to a lack of nickel, which is driving up the price of nickel, the price of 316 ASS is rising quickly. 410 MSS is therefore widely used in marine turbine blades, steam turbine blades, and other applications. Due to its low price and ideal mix of hardness, strength, and wear-resistance in different conditions. For welding 316 ASS with 410 MSS, a variety of welding processes including TIG, MIG, EBW, SMAW, and SAW are used. The weldment of 316 ASS and 410 MSS will provide better corrosion resistive properties. This paper aspires to provide a comprehensive review of the weld between 316 austenitic stainless steel and 410 martensitic stainless steel and its mechanical characteristics. It will cover the domain of mechanical properties and CMT welding of 316 ASS and 410 MSS, together.

Keywords: Dissimilar welds, corrosive resistive, welding processes

Introduction

Welding is a simple, quick, and efficient technique that is employed in many different industries, including pipeline construction and shipbuilding. Barrels, cutlery, guns, and other early stainless steel applications were its only uses. The features like corrosion resistance, high strength, and durability were improved by composition improvement, which led to a growth in the use of stainless steel. One form of stainless steel utilised globally in the chemical, textile, agricultural, pharmaceutical, railroad, power, and nuclear industries is Austenitic stainless steel. Different series are used to classify austenitic stainless steel. One of them is the 300 series, which has 8% nickel and 18% chromium. At both high and low temperatures, nickel demonstrates exceptional toughness and strength. Additionally, it has increased resistance to oxidation and corrosion. But because the price of nickel is rising daily, the cost of the 300 series is also rising daily. Various industries that require high strength, hardness, and wear resistance use martensitic stainless steels. In steam and marine turbine blades, generators, pressure vessels, pumps, and diesel engines, 410 MSS is used. Compared to austenitic stainless steel, they are less expensive and more cost efficient. Electron Beam Welding (EBW), and Submerged Arc Welding (SAW), Metal inert gas (MIG), tungsten inert gas (TIG), and shielded metal arc welding (SMAW) methods are used to join austenitic and martensitic stainless steel. Table 89.1 shows the chemical composition of the 316 ASS and 410 MSS.

Since ASS has strong mechanical characteristics, 316 ASS is frequently used in industries that employ a range of temperatures. This metal resists corrosion well and is utilised in a variety of industrial applications. A composite material's strength and resistance to ageing can both be increased by adding nickel to it. Austenitic stainless steel is sensitive to heat cracking. Variable penetration is another issue with 316 ASS. Low sulphur steels typically have wide welds with shallow penetration. Martensitic stainless steel welding issues include cold fracture and embrittlement of the welded joint.

The present work reported the review on weldability of 316 austenitic and 410 martensitic stainless steel in last decades. the various criteria required for the process selections are discussed in details. Also, the various mechanical properties which affect the weldability of 316 austenitic and 410 martensitic stainless steel is highlighted. Finally paper is concluded with the recent trends and future perspectives of weldability of 316 austenitic and 410 martensitic stainless steel in welding phenomena. The work done in this paper is

DOI: 10.1201/9781003450252-89

Table 89.1 Chemical Composition (% wt.)

Grades	C	Cr	Fe	Mn	Mo	Ni	N	P	Si	S
316 ASS	0.08	16-18	62-72	0-2	2-3	10-14	0-0.01	0-0.45	0-0.75	0-1.0
410 MSS	0.08-0.15	11.5-13.5	83.5-88.5	0-1	0	0-0.75	0	0-0.045	0-0.03	0-0.03

Source: Author

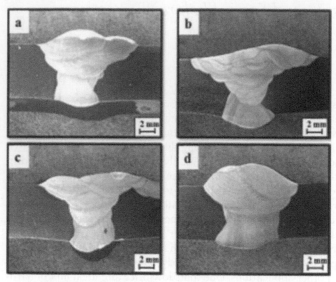

Figure 89.1 Effect of welding current on the micro morphology of 316 stainless steel butt-welding joint with welding current (A) 80A, (B) 90A, (C)100A, (D) 110A [3]

has been done by many researchers before but they no research has been done on 316 ASS and 410 MSS together. The parameters used in this work is different and less used, CMT welding is also employed.

Process Selection

Sharifitbar et al. [1] used resistance upset butt welding to investigate the welded joints of MSS and ASS, and the outcome indicate that the strength of welds improve with higher welding forces. As welding power is raised, the range in which hardness changes is constrained, and the martensitic stainless steel heat affected zone (HAZ) hardness greatly increases. Tamizi et al. [2] studied RSW of MSS and found pull-out failure mode and interfacial failure mode in which crack is propagated. Moslemi et al. [3]. As shown in Figure 89.1 width and depth of fusion weld increases by growing arc current in TIG welding of 316 ASS. Ramkumar et al. [4] found that the weld interface showed signs of the formation of a (unmixed) zone after using pulsed current TIG welding. They also observed low Mo separation in weldments that used nickel-molybdenum rich filler wires using four different filler wires. Mironov et al [5] studied that super austenitic stainless-steel FSW has been investigated, and it was seen that the stir zone's recrystallisation process was incomplete. Zhang et al. [6] determined that the grain transition from columnar to equiaxed during the solidification of the weld pool has increased as a outcome of double-sided arc welding method. Ragavendran et al. [7] suggested metal inert gas and hybrid laser welding is better than hybrid laser and laser TIG welding. Sun et al. [8] results indicated that TIG welding shows satisfactory hot cracking resistance with filler wire. AISI 316 austenitic stainless steel was studied and welded using laser and electron beam technology, and it resulted that the failure of the joint occurred in the HAZ region [9]. Ambade et al. [10] used GTAW with filler and no filler and discovered that autogenous welds had higher mechanical strength and less sensitization than filler welds as the welding current increased dendrite size and interdendritic spacing also increased. 316 ASS and DSS 2205 and steel were both utilised for shielded metal arc welding and full penetration welds were obtained and to avoid welding related defects best range of ferrite content in weld is resulted [11].

Jeraldnavinsavio et al. [12] used 316 L and 430 plates and welded utilising the TIG welding procedure with filler materials of ER310 and ER2594, and it was found that the weldments were sound, lacking in porosity and penetration when welded with ER2594. For 316 austenitic stainless steel ferritic/ martensitic, electron beam welding was utilised for better optimising the mechanical properties by positioning the electron beam [13]. Reddy et al. [14] found that gas tungsten arc welding was used for AISI 4140 and AISI 316 with filler as well as no filler, and martensitic emergence at the HAZ of AISI 4140 was concluded.

Thakre et al. [15] investigated that used gas tungsten arc welding for martensitic and ASS dissimilar welded joints, dissimilar welded joints of P91 and 304L are successfully formed using filler P91. SS304 and EN 8 were studied utilising the tungsten inert gas welding technique, which produced a noticeable difference in the weld and interface microstructure. [16]. Ambade et al. [17] investigated that welding of 409 M FSS with shielded metal arc welding and discovered large dendritic grain development and the structure became finer. It became finer as seen from the microstructure because there was growth seen in the welding heat input width [18]. The effect of delta ferrite on ferritic ASS welds using gas tungsten arc welding was inquired, and was discovered the grain growth was accompanied by martensitic development in the HAZ of FSS as shown in Figure (89.2) [18].

Reddy et al. [19] used two welding technique friction and electron beam welding. Higher ferrite content were observed in electron beam welds of DSS whereas friction welds contain same amount of austenitic as well as ferrite. GTAW was used to complete deposition regarding 21 satellite, preheat and post heat weld overlays in three different thicknesses on the surface of 410 MSS [20]. Bhaduri et al. [21] investigated that the blades of steam and gas turbines are operated under extremely strong centrifugal and bending stresses. The blades in turbine stages sensitive to resonance have high loads; any relaxation of the tight requirements during production, assembly as well as quality assurance result in blade breaking. The blades are frequently made of MSS. The standard corrective method, which entails replacing the broken blades, greatly extends time of the turbine outage as a result, decreases the turbine's performance. The weld repair on the blade, however, can cut down on the amount of downtime required for blade replacement, making it very cost-effective. A low carbon martensitic stainless steel known as "JFE410DB-ER" has been created for motorcycle brake discs and has improved heat resistance and corrosion resistance over ordinary steels [22]. Additionally, corrosion resistance is essential for maintaining braking performance and for aesthetic reasons.

Takano et al. [23] investigated that without drilling a pilot hole, a special screw known as a self-drilling and tapping screw is inserted into the external ornamental steel sheet. The tip alone is made of tool steel or another high strength steel and is linked to the body of the self-drilling and tapping screw. The self-drilling screw is constructed entirely of high carbon steel enhanced by quenching. Self-drilling and tapping screw joints have a tendency to corrode, which has increased demand for high corrosion steel in recent years. Priharni et al [24] investigated that the last piece of extraction of energy machinery which are mostly seen in thermal power plants. AISI Type 403 is commonly used to make the rotors/disks and turbine blades. However, the main mechanisms of corrosion fatigue, pitting corrosion and corrosion cracking are usually found to fail as a result of mechanical corrosion caused by turbine blades operating in hostile environments and at high rotational speeds. MSS are frequently heat treated because it gives good mechanical properties and a low level of corrosion resistance.

A new type of steel with enhanced ductility, superior mechanical strength, and great corrosion resistance is described as super MSS [25]. They have a low carbon content and are made from the composition of martensitic steels that have been refined. They also have 13% chromium, 5% nickel, and 2% molybdenum. Yukio et al. [26] investigated that KL-HP12CR, a seamless pipe made of martensitic stainless steel has weldability, mechanical characteristics, and it is highly corrosion resistant. Therefore, has been developed for line pipe applications. The lowering of both C and N content enhances weldability. C reduction also works well to increase CO_2 corrosion resistance. Avcu et al. [27] investigated that as we reduce the temperature in the formation of martensite there is an is an increase in carbon content which allows the formation of a fully martensitic microstructure.

Zunko et al. [28] found out that when greater hardness and wear resistance are needed, such as during rasping, cutting extruding, crushing, processes, and processing martensitic stainless steels are typically chosen. Due to less contact time, contact area, and processing temperature in such applications than in industrial chemical food processing. For this purpose there is need of good mechanical properties, while those for corrosion resistance decrease. In contrast to ferritic or austenitic grades, these materials have a higher mechanical strength thanks to an increased carbon content of 0.2% to more than 1%. The development of a martensitic microstructure is made possible by this action. A heat treatment that includes hardening and tempering is required to modify the desirable characteristics in these grades. Sashank et al. [29] found that dissimilar welding is generally favored in nuclear reactors where high-temperature implementation is needed. Dissimilar welding between Austenitic and mmartensitic stainless steel is done. There is enormous urge for those materials which can bear corrosive environments with high rupture strength and pressure is needed. Rajasekhar et al. [30] found that steel sheets of this grade are often used in components that require amalgamation of tensile strength, durability and corrosion resistance. The transformation is very slow due to the presence of high alloy, and therefore it needs high air cooling for hardening to achieve maximum hardness. Appropriate heat treatment techniques can vastly change the properties of MSS.

Kaçar et al. [31] investigated that the tensile strength, hardness, durability and corrosion resistance of various welds of X5CrNi18-10 austenitic stainless steel and X20CrMo13 martensitic steel were observed.

The welding tensile strength of the duplex electrode (E2209-17) was found to be considerably observed less than that of the austenite electrode (E308L-16). Wua et al. [32] discovered that martensitic stainless steel hard welds are used to create products that are resistant to corrosion, wear and oxidation. A medium carbon steel or low alloy steel is welded using this technique with multiple layers of martensitic stainless steel. By revealing the surface of a part worn due to friction, corrosion or oxidation, the areas of use of the welded components can be expanded and equipment and maintenance costs can be significantly minimized. Rajasekar et al [33] found that when there is need of high strength, toughness and excellent corrosion resistance, MSS is preferred. Kaladhar et al [34] found that ASS are laborious to machine because of their great hardness, strength, ductility and poor thermal conductivity. There are several factors to consider when grinding austenitic stainless steel.

Davison et al. [35] studied that a representative selection of these new grades is described using the terms 'service proven', 'stress corrosion resistance' and 'crevice corrosion resistance'. Sen et al [36] investigated that due to strong corrosion resistance and excellent mechanical properties ASS is the most popular grade. Compared to conventional cordon steels, ASS have high ductility, low yield point and relatively high tensile strength. They can be used successfully in a variety of environments, including cryogenic temperatures and extremely hot temperatures such as those encountered in furnaces and jet engines. It contains 16-25% chromium and can also contain nitrogen in solution for excellent corrosion resistance. If the nickel used to stabilize the austenitic structure wasn't so expensive, these would be used even more. Tembhurkar et al. [37] found that separate lap joints of austenitic and ferritic stainless steels were spot welded to determine the maximum strength of the joints. Lap joints are based on tensile stress and weld point failure testing is structural. The von Mises stress is shown to be 9% greater than the maximum experimental strength. The data from tensile tests on a universal testing machine and FEM performed in ANSYS are used to confirm the conclusion. Ambade et al. [38] investigated that due to the difference in welding heat input, SMAW has a wider HAZ than GMAW and GTAW. GTAW has a narrower HAZ than GMAW and SMAW. Due to welding, the structure of the fusion zone and HAZ coarsens as the welding heat input changes. As compared to GMAW and GTAW, SMAW increases the corrosion rate of welded joints with changes in weld heat input. Compared to GMAW and SMAW, GTAW has a lower corrosion rate. Figure 1 shows the effect of welding current on the micro morphology of 316 stainless steel butt-welding joint with welding current.

Kuddusa et al. [39] investigated that 316L electrodes can be used to bond Cr-Mn ASS and 30 ASS. Ch-Mn ASS and AISI 30 ASS were also bonded by TIG welding. Ferrite and austenite form weld zones in the weld matrix. Welds were free of precipitates and ferrite was present as vermicular and lathy ferrite. Welds, 30 ASS and Cr-Mn ASS have the highest hardness. Also, the Cr-Mn-ASS base metal had excellent tensile strength. Durugkar et al. [40] found that finite element analysis is the most accurate and adaptable modelling technique for reproducing transient thermal conditions in welds. Using temperature-dependent material properties is essential for estimating the plate temperature distribution. Thermal profiles generated from temperature distributions can be used to predict material residual stress, deformation, and microstructural changes. Further studies are needed to accurately evaluate the effects of weld settings on the thermal and mechanical responses of the weld as well as the metallurgical events in the HAZ during non-welded welding of ferritic stainless steels. Figure 89.2 shows the formation of martensite at the grain boundaries in the HAZ of AISI 430 stainless steel.

Mechanical Properties of Materials

Sharifitbar et al. [1] found that ASS HAZ's hardness did not diminish but rather increases with increasing distance from the contact, while martensitic stainless steel HAZ's hardness grows gradually. A profile of hardness at different welding strengths is shown in Figure 89.3 (a). Moslemi et al. [3] as shown in Figure

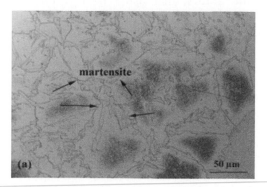

Figure 89.2 Formation of martensite at the grain boundaries in the HAZ of AISI 430 stainless steel [3]

89.3 (b) used Vickers hardness test at different current as Shand resulted that the hardness is different from base metal to HAZ and by varying the carbon content hardness value decrease toward weld metal from base metal. The HLM weld shown greater strength, ductility, and acceptable toughness compared to autogenous welds, as well as increased hardness [7]. Sun et al. [8] concluded that MSS hardness near HAZ is higher than other parts of weldments and ASS hardness change were small. Tjong et al. [9] used Vickers hardness test and stated the HAZ as well as weld metal have high hardness distributions. As the distance to the base metal increases from the center line, the hardness value decreases. Strength, hardness, and degree of sensitisationise higher in the absence of filler welds [10]. Tensile strength and hardness decrease as welding current rises. By using electron beam position in welding mechanical properties were improved at room temperature [13]. Maximum hardness was found at head affected zone of AISI 4140 and in without filler gas tungsten arc wel dments weld zone were found [14]. Thakre et al. [15] noticed uneven distribution along the welded joint in microhardnes. Ambade et al. [17] showed that as there is increase in weld heat input hardness decreases and at high heat input there is lower hardness. Ghasemi et al. [18] claimed that the weld metal's hardness and tensile strength were higher than HAZ and base metal.

The mechanical test performed on the samples will be tensile test, hardness test and impact test. Tensile testing is an engineering test in which a sample is subjected to a controlled tension until failure. The test process involves placing the test specimen in the testing machine and slowly extending it until it fractures. The data is manipulated so that it is not specific to the geometry of the test sample. The most common testing machine used in tensile testing is the universal testing machine. The Vickers test is often easier to use than other hardness tests since the required calculations are independent of the size of the indenter, and the indenter can be used for all materials irrespective of hardness. The basic principle, as with all common measures of hardness, is to observe a material's ability to resist plastic deformation from a standard source. The Vickers test can be used for all metals and has one of the widest scales among hardness tests. The SI unit for hardness is called as Vickers pyramid number (HV). The Charpy impact test measures the energy absorbed by a standard notched specimen while breaking under an impact load. This test continues to be used as an economical quality control method to determine the notch sensitivity and impact toughness of engineering materials such as metals, composites, ceramics, and polymers.

Recent Trends and future perspective

The primary concern for other advanced processes is the corrosion behavior of dissimilar welding. The main problem solver will be that the weldment will be cost effective and highly corrosion resistance. The welding process used here is cold metal transfer. This recent technique is highly beneficial because it minimizes heat affected zone and almost solves the porosity (weld defect) problem.

Future research should focus on developing cutting-edge methods like hybrid welding of dissimilar metals. Industries currently need materials made of 316L ASS that are stronger and more resistant to corrosion. To obtain materials that are as strong and compatible as 316L ASS, production has undergone significant modification. In engineering applications where the use of various materials for welding purpose is necessary, excellent and popular grades call for greater research in this area. Filler materials are still clearly chosen when welding 316L ASS through various alloys. Unwanted precipitates occurred in procedures including GTA, GMA, and SMA welding, and the rate of deformation increased.

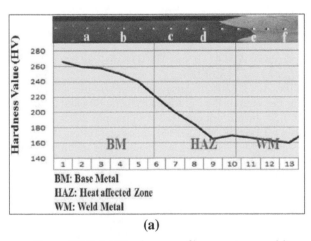
(a)

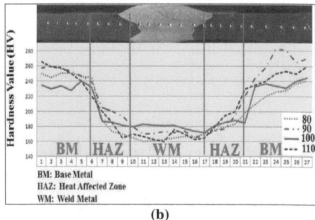
(b)

Figure 89.3 (a) Hardness profile at various welding power; (b) Comparison of the number of points versus hardness (HV) of 316 stainless steel welding joint [3]

Conclusions

Following are the major conclusions drawn from the overview:

- The application of 316 ASS and 410 MSS are wide and used in variety of industries because of their good corrosion and mechanical properties. With increase in welding power the hardness changes and welding strength also rises. After welding, change in microstructure such as growth in grain size, dendritic grain is seen. Varying heat input changes size of heat affected zone. By keeping the power density as high as possible its help in maintaining low heat input, high welding speed, and increase in penetration and weld quality. Change in microstructure such as growth in grain grain-size, dendritic grains are also seen after welding stainless-steel.

References

[1] Sharifitabar, M. and Halvaee, A. (2010). Resistance upset butt welding of austenitic to martensitic stainless steels. Materials & Design. 31, 3044-50. https://doi.org/10.1016/j.matdes.2010.01.026

[2] Tamizi, M., Pouranvari, M., and Movahedi, M. (2016). Welding metallurgy of martensitic advanced high strength steels during resistance spot welding. Science and Technology of Welding and Joining. 22, 327-35. https://doi.org /10.1080/13621718.2016.1240979

[3] Moslemi, N., Redzuan, N., Ahmad, N., and Hor, T. N. (2015). Effect of current on characteristic for 316 stainless steel welded joint including microstructure and mechanical properties. Procedia CIRP. 26, 560-564. https://doi. org/10.1016/j.procir.2015.01.010

[4] K. Ramkumar, K. D., Chandrasekhar, A., Srivastava, A., Preyas, H., Chandra, S., Dev, S., and Arivazhagan, N. (2016). Effects of filler metals on the segregation, mechanical properties and hot corrosion behaviour of pulsed current gas tungsten arc welded super-austenitic stainless steel. Journal of Manufacturing Processes. 24, 46-61. https://doi.org/10.1016/j.jmapro.2016.07.006

[5] Mironov, S., Sato, Y. S., Kokawa , H., Inoue, H., and Tsuge, S. (2011). Structural response superaustenitic stainless steel to friction stir welding welding. Acta Materilia. 59, 5472-81. https://doi.org/10.1016/j.actamat.2011.05.021

[6] Zhang, Y. M., Pan, C., and Male, A. T. (2013). Welding of austenitic stainless steel using double side arc welding process Material science and technology, 17, 1280-1284. 10.1179/026708301101509205

[7] Ragavendran, M. and Vasudevan, M. (2020). Laser and hybrid laser welding of type 316L(N) austenitic stainless steel plates. Materials and Manufacturing Process. 35, 1-13. https://doi.org/10.1080/10426914.2020.1745231

[8] Sun, Z. and Han, H.-Y. (2013). Weldability and properties of martensitic/ austenitic stainless steel joints. Material Science and Technology. 10, 823-29. https://doi.org/10.1179/mst.1994.10.9.823

[9] Tjong, S. C., Zhu, S. M., Ho, N. J., and Ku, J. S. (1995). Microstructural characteristics and creep rupture behavior of electron beam and laser welded AISI 316L stainless steel. Journal of Nuclear Materials. 227(1-2), 24-31. https://doi.org/10.1016/0022-3115(95)00142-5

[10] Ambade, S., Patil, A. P., Tembhurkar, C. K. and Meshram, D. B. (2021).Effect of filler and autogenous welding on microstructure, mechanical and corrosion properties of low nickel Cr-Mn ASS; Journal of Nuclear Material. 227, 24-31. https://doi.org/10.1080/2374068X.2021.1970989

[11] Verma, J., Taiwade, R. V., Khatirkar, R. K., Kumar, A. (2016). A comparative study on the effect of electrode on microstructure and mechanical properties of dissimilar welds of 2205 austenoferritic and 316L austenitic stainless steel. Material. 57, 494-500. https://doi.org/10.2320/matertrans.M2015321

[12] Jeraldnavinsavio, D., Farid, A. M., Ramanamurthy, E. V. V., Porchilamban, S., and Ravikumar, S. (2019). Evaluation of mechanical properties and micro structural charecterization of dissimilar TIG welded AISI 316L and AISI 430 plates using ER310 and ER2594 filer. Materials Today Proceedings. 16, 1212-18. https://doi.org/10.1016/j.matpr.2019.05.216

[13] Hara, N., et al. (2017). Mechanical property changes and irradiation hardening due to dissimilar metal welding with reduced activation ferritic/martensitic steel and 316L stainless steel. Fusion Science and Technology. 56(1), 318-22.

[14] Reddy, M. P. et al. (2014). Assessment of mechanical properties of AISI 4140 and AISI 316 dissimilar weldments. Procedia Engineering. 75, 29-33. https://doi.org/10.1016/j.proeng.2013.11.006

[15] Thakarea, J. G., Pandeyb, G., Mahapatrac, M. M., Mulik, R. H. (2019). An assessment for mechanical and microstructure behavior of dissimilar material welded joint between nuclear grade martensitic P91 and austenitic SS304 L steel. Journal of Manufacturing Processes. 48, 249-59. https://doi.org/10.1016/j.jmapro.2019.10.002

[16] Singh, D. K, Sahoob, G., Basua, R., Sharmac, V., and Mohtadi-Bonab, M. A. (218). Investigation on the microstructure—mechanical property correlation in dissimilar steel welds of stainless steel SS 304 and medium carbon steel EN 8. Journal of Manufacturing Processes. 36, 281-91. https://doi.org/10.1016/j.jmapro.2018.10.018

[17] Ambade, S., Tembhurkar, C., Patil, A. P., Pantawane, P., and Singh, R. P. (2022). Shielded metal arc welding of AISI 409M ferritic stainless steel: study on mechanical, intergranular corrosion properties and microstructure analysis. World Journal of Engineering. 266-73. dpoi:10.1108/WJE-03-2021-0146

[18] Ghasemi, R., Beidokhti, B., Fazel-najafabadi, M. effect of delta ferrite on the mechanical properties of dissimilar ferritic-austenitic stainless steel welds. Archives of Metallurgy and Materials. 63, 437-43. doi: 10.24425/118958

[19] Reddy, G. M. and Srinivasa Rao, K. (2009). Microstructure and mechanical properties of similar and dissimilar stainless steel electron beam and friction welds. International Journal of Advanced Manufacturing. 45, 875-88. https://doi.org/10.1007/s00170-009-2019-6

[20] Laridjani, M. S., Amadeh, A. (). Stellite 21 coatings on AISI 410 martensitic stainless steel by gas tungsten arc welding. Material Science and Technologyl. 26, 1184-90.

[21] Bhaduri, A. K., Gill, T.P.S., Albert, S. K., Shanmugam, K., and Iyer, D. R. (2001). Repair welding of cracked steam turbine blades using austenitic and martensitic stainless-steel consumables. Nuclear Engineering and Design. 206, 249–59.

[22] Katsuhisa, Y. Yoshihiro, O., and Takumi, U. (2008). Martensitic stainless steel "JFE410DB-ER" with excellent heat resistance for motorcycle brake disks. JFE Technical Report.

[23] Takana, K., Sakakibara, M., Murata, W., Matsui, T., and Yoshimura, K. (1996). Development of high-strength martensitic stainless steel YUS 550 for artificial use. Nippon Steel Technical Report.

[24] Prifiharni, S., Sugandi, M. T., Pasaribu, R. R., Sunardi, and Mabruri, E. (2019). Investigation of corrosion rate on the modified 410 martensitic stainless steel in tempered condition IOP Conf. Series: Materials Science and Engineering. 541. doi:10.1088/1757-899X/541/1/012001 (2019)

[25] Oulabbas, A., Tlili1, S., and Meddah, S. (2021). Comparative study of corrosion behaviour of martensitic and supermartensitic stainless steels in two corrosive media. International Journal on Emerging Technologies. 12(2), 269-76.

[26] Yukio, M., Mitsuo, K., and Tomoya, K. (2006). Martensitic stainless steel seamless pipe for linepipe. JFE Technical Report. 7.

[27] Iakovakis, E., Avcu, E., Roy, M. J., Gee, M., and Matthews, A. (2022). Wear resistance of an additively manufactured high-carbon martensitic stainless steel Scientific Reports. 12.

[28] Zunko, H. and Turk, C. (2022). Martensitic stainless steels for food contact applications BHM Berg-und huttenmannische monatshefte. 167, 408–15.

[29] Sashank, S. S., Rajakumar, S., Karthikeyan, R. (2021). Dissimlar welding of austenitic and martensitic stainless steel joints for nuclear applications:a review. 3rd International Conference on Design and Manufacturing Aspects for Sustainable Energy (ICMED-ICMPC 2021).

[30] Rajasekhar, A. (2015). Heat treatment methods applied to AISI 431 Martensitic stainless steels. International Journal of Scientific & Engineering Research. 6(4),547.

[31] Kacar, R. and Baylan, O. (2001). An investigation of microstructure/property relationships in dissimilar welds between martensitic and austenitic stainless steel. Material and Design. 25(4), 317-29.

[32] Wu, W., Hwu, L. Y., Lin, D. Y., and Lee, J. L. (2000). The relationship between alloying elements and retained austenite in martensitic stainless steel welds.Scripta Materialia. 42(11) 1071-76.

[33] Rajasekar, A. (2005). Corrosion behaviour of martensitic stainless steel? Role of Composition and Heat Treatment Procedures. 4(4).

[34] Kaladhar, M., Kambagow111, V., and Rao, C. S (2012). Machining of austenitic stainless steels: a review. International Journal of Machining and Machinability. 12(1/2), 178-92.

[35] Davison, R. M., Laurin, T. R., Redmond, J. D., Watanabe, H., Semchyshen, M. (1986). A review of worldwide developments in stainless steels. 7(3), 111-19

[36] Sen, P. (2014). Review about high performance of austenitic stainless steel. International Journal for Innovative Research in Science & Technology. 1(9/01), 1.

[37] Tembhurkar, C., Ambade, S., Kataria, R., and Tikle, A. (2022). Spot welding analysis of dissimilar joint by finite element analysis. 50(5), 2052-56.

[38] Ambadea, S. P. Sharmab, A., Patilc, A. P., Puri, Y. M. (2021). Effect of welding processes and heat input on corrosion behaviour of ferritic stainless. 41(5), 1018-23.

[39] Abdul Irshad Kuddusa *, Sachin P. Ambade , Ankur V. Bansodb and Awanikumar P. Patil Microstructural, mechanical and corrosion behaviour of dissimilar welding of Cr-Mn ASS and AISI 304 ASS by using 316L electrode 2018 Irshad kuddus 2018 Elesvier[1].pdf

[40] Durugkar, P. S., Ambade, S. P., Patil, A. P., and Puri, Y. M. (2013). Review on Finite Element Analysis of Temperature Distribution in Heat Affected Zone by Different Welding Process. International Journal of Advanced Materials Manufacturing and Characterization. 3(1), 285–90.

Chapter 90

Design and Finite Element Method (FEM) analysis of electric two-wheeler alloy wheel

Sumitkumar Tulshidas Bambole[a], Dr. Imran Ahemad Khan[b] and Himanshu Suresh Dhandre[c]

Priyadarshini College of Engineering, Nagpur, India

Abstract

The wheel is a significant component of the vehicle. This research is related to the analysis of stress and displacement of electric two-wheeler vehicle alloy wheels. However, if the strength of alloy wheel is increase which result in increasing the quantity of alloy wheel material. Due to this the wheel becomes heavier and it will directly affect the mileage of vehicle. During this research designing of alloy wheel of two-wheeler is done by using Automotive Research Association of India (ARAi) and International Organization for Standardization (ISO) standard for maintaining the quality of the product. These two standard norms will be helpful in restructuring the design and optimising the material, so as to obtain reliable, low cost and light weighted alloy wheel. Also, the futuristic point of view adds some extra features so if any accident happens, it will not result in damage to the chassis. By using CAD and FEM Software the designing of two-wheeler wheel is obtained by subjected load and pressure condition. To achieve less weight and increase efficiency, it was updated and reconfigured.

Keywords: Alloy wheel, alloy wheel reverse engineering, CATIA design, impact analysis, stress and force analysis in Ansys

Introduction

The significance of wheel in a vehicle is very important. The vehicle load is carried by wheel and tire which also provide cushioning effect and steering control. The vehicle can be run without an engine but in the absence of wheels it is not possible. Alloy wheels are different from standard steel wheels because they are lighter, which enhances the vehicle's steering control and speed. However, every alloy wheels are not eventually having lighter weight as compare with steel, due to low unsprung mass, this lighter wheel performs quite better handling as well as permitting the suspension to follow the terrain negligible, and thus boosting grip. Reduced vehicle mass generally can aid in lowering energy use.

A wheel is a circular object that can be used to conduct work in machines, facilitate movement or transit while carrying a load (mass), or rotate on its axis. Generally, alloy wheels are manufactured from an alloy of magnesium or aluminium material, or sporadically an admixture of both to increase the alloy wheel's strength. Since alloy wheels typically use an excess amount of alloy (mixture of aluminum and magnesium), their weight directly affects the vehicle's fuel economy. The wheel is a crucial structural component of the suspension system of the vehicle and supports both static and dynamic loads. It is a proven fact that less unstrung weight results in more precise running and uses less fuel.

Literature Review

With increasing competition in the global market for alloy wheels, it need to be able to offer high quality alloy wheels [1]. The wheels are one of the most important components for a vehicle's performance. The total weight of vehicle is minimise by using alloy wheel. In this work, the design is performed in CATIA software. In this work, optimization of wheel hub is obtained by using FEM [2]. Alloy wheels are lighter than regular steel wheels, which increases speed, but some alloy wheels are heavier than steel wheels of the same dimension [3]. Vehicle weight reduction has a direct impact on fuel efficiency. Therefore, in this research work, modelled a full two-wheeler wheel design and analysed it under various loads and pressures condition [4,5]. Aluminum alloys are used in critical vehicle parts because they combine two important properties of the material, strength and light weight [6]. The objective of the work is to guide the wheel design process using an example implementation of an electric racing car [7]. The mesh size affects the simulation of the rim design. Since different mesh sizes are known to give different results, finite element calibration should be performed from the benchmark literature studies [8,9]. Given the equivalent von-mises stress, which is less than the ultimate strength, and hence deflection, magnesium alloys are preferred as the material of choice for wheel rim construction [10]. Current work reduces tire wear, improves fuel consumption and handling, and increases driving safety by eliminating steering and tracking issues. The main advantages of this metal are the low weight of the wheel, high precision and design possibilities.

[a]sumitbambole16@gmail.com, [b]iak20041978@rediffmail.com, [c]himanshu.dhandre7@gmail.com

DOI: 10.1201/9781003450252-90

During literature review it is found that very few literatures are available on two-wheeler alloy wheel of electric vehicle. No one had done any work related to design and analysis of two-wheeler alloy wheel of electric vehicle. In this research work same is done. This will going to help for designing the chassis of two wheeler alloy wheel of electric vehicle.

Methodology

Every research project adheres to a specific preferred technique, which is the foundation for achieving the study goal. Through appropriate design analysis, this automobile-focused study will minimise shape faults seen in the alloy wheels of electric two-wheelers. The objective of this research is to reduce the weight of the wheel which will lead to more precise operation and reduced fuel consumption of the vehicle. The problem will be identified based on the analysis, and the literature pertinent to the topic will be thoroughly reviewed. Using CATIA V5 CAD software, an alloy wheel for an electric two-wheeler will be created in three dimensions. Ansys software will then be used to thoroughly analyse the issue's root cause. The software tool will then be used to assess whether the design update is compatible, and in-line quality enhancement will then be investigated. At the conclusion of the research, the appropriate conclusion will be offered based on the findings.

3D CAD Model of Electric two-wheeler alloy wheel

This electric two-wheeler alloy wheel 3D CAD model was created using CATIA and is based on standard dimensions. Figure 90.1 depicts the outer and hub regions of the wheel, with a total of five spokes, an outer dia of 480 mm, and a hub dia of 100 mm, all in accordance with the CAD design of the wheel rim prior to optimisation.

FEA Modelling

Ansys has received a 3D model of an alloy wheel for an electric two-wheeler for additional analysis. Alloy wheel composed of aluminum alloy Al 6063 T6 with the following specifications are silicon 0.4, titanium 0.15, zinc 0.25, copper 0.1, iron 0.4, magnesium 4.0-4.9, manganese 0.4-1.0. With the default coordinate system, it exhibits difficult behavior. Construction materials made of the aluminum alloy (Al6063 T6) have youthful moduli of 69.89e-05 MPa, Poison ratios of 0.33, 6.7549e-05 MPa for bulk moduli, and 2.5902e-05 MPa for shear modulus as shown in Table 90.1. This model consists of 88283 elements and 151376 nodes as shown in Table 90.2.

Figure 90.1 3D Model of alloy wheel
Source: Author

Table 90.1 Material properties

AL 6063-T6	
Mass density	2.7e-06 kg/mm2
Young's modulus	6.89 x 10-5 MPa
Poisson's ratio	0.33

Source: https://asm.matweb.com/search/SpecificMaterial.asp?bassnum=MA6063T6

Table 90.2 Number of node and element

Statistics	
Nodes	151376
Elements	88283

Source: Author

The different boundary conditions are applied on a wheel as follow. The fixed support is applied at central hole and five peripheral holes of a wheel as shown in the Figure 90.2(a). The force of 1000N is applied to the wheel at a location 2.7823e-015, 0, and -94947mm as shown in Figure 90.2(b). The pressure of 0.248Mpa is applied on the circumference of the wheel rim as shown in Figure 90.2(c).

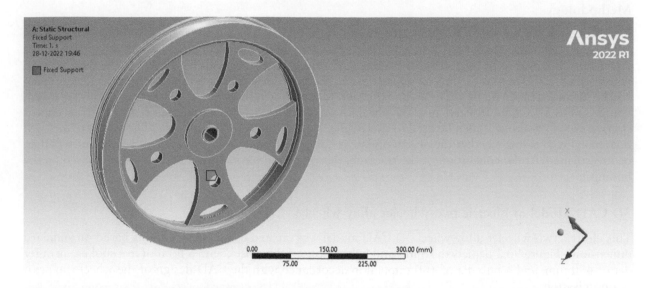

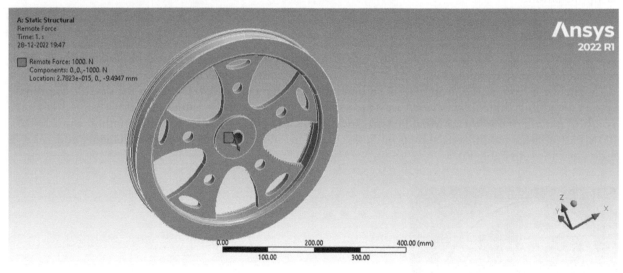

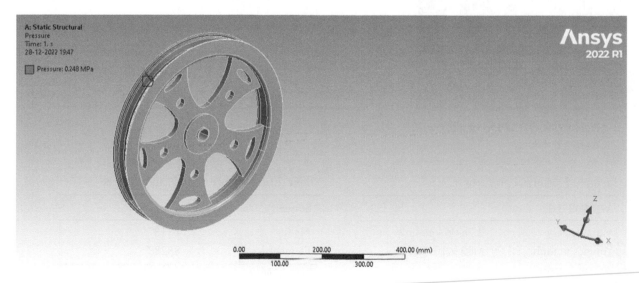

Figure 90.2 Boundary conditions (a) Fixed support, (b) Remote force, (c) Pressure
Source: Author

Analysis

FEA Analysis of Static Structural Total Deformation

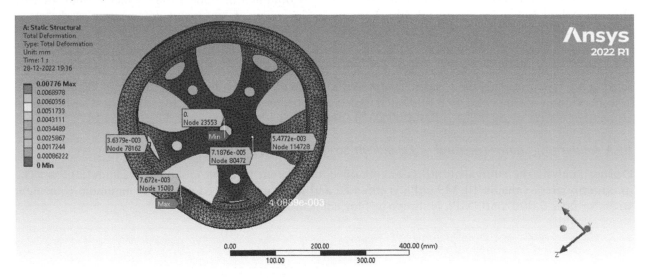

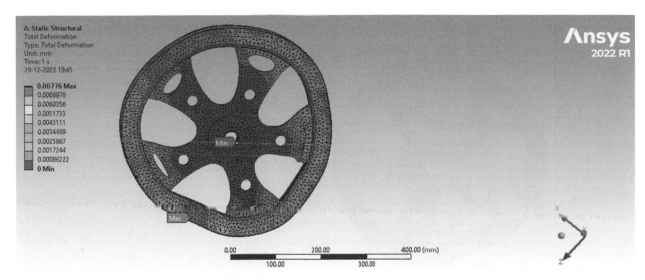

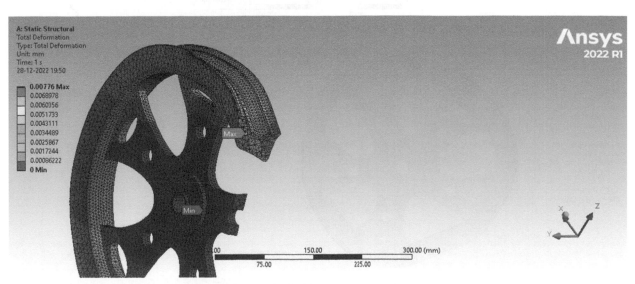

Figure 90.3 Total deformation, (a) Deformation showing at different location, (b) Maximum deformation of a wheel, (c) Cut section view of maximum deformation
Source: Author

After the application of the mentioned force and pressure at specified boundary conditions, the total deformation of a wheel is obtained and different deformation at different points is shown in Figure 90.3(a). The maximum deformation of 0.00776mm is obtained at node 15083 which is indicated by the red color shown in Figure 90.3(b). The cut section view of maximum deformation is shown in Figure 90.3(c).

FEA Analysis of Static Structural Equivalent Elastic Strain

During analysis the equivalent elastic strains are obtained at different location is shown in Figure 90.4(a). The maximum strain value is 0.00015097mm is obtained and it is shown by red color. The minimum equivalent elastic strain value is 0.00000003717mm is shown by blue color in Figure 90.4(b).

FEA Analysis of Static Structural of Equivalent Stress

During analysis the equivalent stress are obtained at different location is shown in Figure 90.5(a). The maximum stress value is 10.349 MPa is obtained and it is shown by red color. The minimum equivalent stress value is 0.00070503MPa is shown by blue color in Figure 90.5(b).

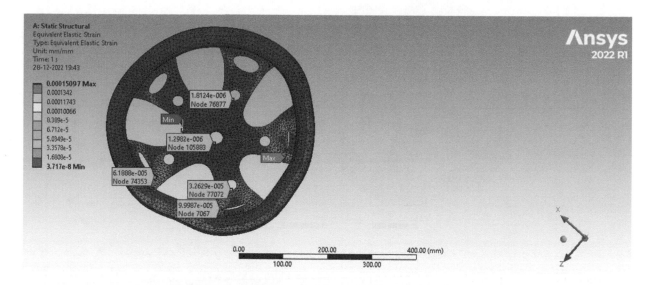

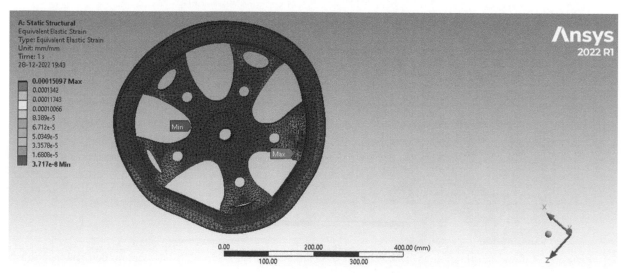

Figure 90.4 Structural equivalent elastic strain, (a) Strain values showing at different location, (b) Minimum and maximum strain value

Source: Author

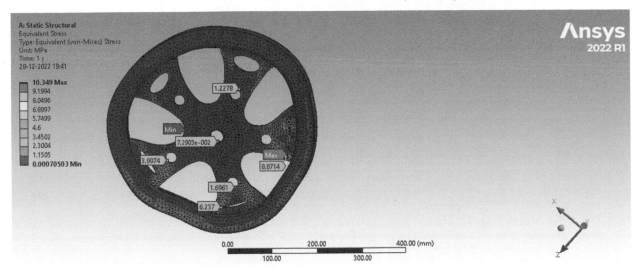

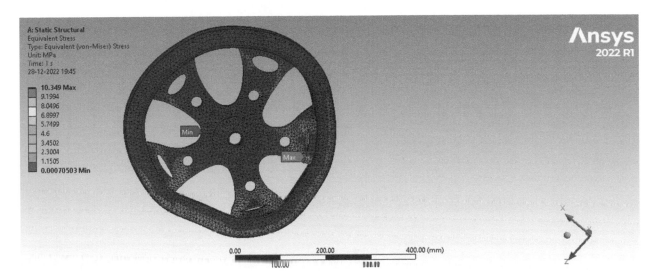

Figure 90.5 Structural equivalent stress, (a) Stress values showing at different location, (b) Minimum and maximum stress value

Source: Author

Conclusion

The FEM analysis of two-wheeler alloy wheel is done by using material Al6063 - T6. The weight of this wheel is 7.5kg which is approximately half weight of the two-wheeler alloy wheel available in the market. During analysis it is obtain that the maximum stress induced in the wheel is quite lesser than the yield stress of the material Al6063 – T6. Therefore, this lighter weighted alloy wheel is best suited for the two wheeler electric vehicle.

References

[1] Maryani, E., Purba, H. H., and Sunadi, S. (2021). Analysis of aluminium alloy wheels product quality improvement through DMAIC method in casting process: A case study of the wheel manufacturing industry in Indonesia. Journal Européen des Systèmes Automatisés. 54(1), 55–62.

[2] Sharma, A., Yadav, R., and Sharma, K. (2021). Optimization and investigation of automotive wheel rim for efficient performance of vehicle. Materials Today Proceedings. 45(Part 2), 3601–4.

[3] G. Srihari, P. Saikiran, K. Durga Prasad, R. Raman Goud, M. Srikanth, Harinadh Vemanaboina; Static structural analysis of two-wheeler rim with different spoke patterns and materials. AIP Conf. Proc. 27 October 2023; 2869 (1): 030002. https://doi.org/10.1063/5.0168322

[4] T.B. Korkut, E. Armakan, O. Ozaydin, K. Ozdemir, A. Goren, Design and comparative strength analysis of wheel rims of a lightweight electric vehicle using Al6063 T6 and Al5083 aluminium alloys, Journal of Achievements in Materials and Manufacturing Engineering 99/2 (2020) 57–63. DOI: https://doi.org/10.5604/01.3001.0014.1776

[5] Sanjayaa, Y., Prabowoa, A. R., Imaduddina, F., and Nordin, N. A. B. (2021). Design and analysis of mesh size subjected to wheel rim convergence using finite element method. Procedia Structural Integrity. 33, 51–58.

[6] Kumar, Mr. R. Naveen and Amarnath, Mr. M.A. and Gowthaman, Mr. M. and Kumar, Mr. S.Ram, Design and Analysis of Alloy Wheel Rim (March 7, 2020). IJRAR - International Journal of Research and Analytical Reviews (IJRAR), E-ISSN 2348-1269, P- ISSN 2349-5138, 7(1), 990–995, March 2020, Available at SSRN: https://ssrn.com/abstract=3677086

[7] Seshagiri Rao, G. V. R., Gupta, A., Prashanth, B., D. V., and Reddy D. V. R. (2017). Impact analysis of aluminium wheel. International Journal of Mechanical Engineering and Technology (IJMET). 8(Issue 7), 889–896.

[8] Theja, M. S., and Krishna, M. V. (2013). Structural and fatigue analysis of two wheeler lighter weight alloy wheel. IOSR Journal of Mechanical and Civil Engineering (IOSR-JMCE). 8(2), 35–45. e-ISSN: 2278-1684, p-ISSN: 2320-334X.

[9] Tebaldinia, M., Petrogalli, C., Donzella, G., and La Vecchia, G. M. (2017). Estimation of fatigue limit of a A356-T6 automotive wheel in presence of defects. 3rd International Symposium on Fatigue Design and Material Defects, FDMD, 2017, Lecco, Italy, pp. 19–22, September.

[10] Deng, Y. J., Zhao, Y. Q., Lin, F., and Zang, L. G. (2022). Influence of structure and material on the vibration modal characteristics of novel combined flexible road wheel. Defence Technology. 18, 1179–89.

Chapter 91

Empirical modeling for solar radiation prediction in f-solve optimisation technique

Veeresh G. Balikai[1,a], M. B. Gorawar[1,b], Rakesh Tapaskar[1,c], P. P. Revankar[1,d], P.G. Tewari[1,e] and Vinayaka H. Khatawate[2,f]

[1]KLE Technological University, Hubballi, India

[2]DJS College of Enggineering, Mumbai, India

Abstract

Solar energy has been explored ever since dawn of civilisation as a source of energy to meet process requirements through myriad routes of conversion. The countries world over has come to a consensus of adopting sustainable development goals. The renewable devices operate with efficiency governed by solar insulation, temperature, wind speed and humidity. Solar energy data helps build devices to harness energy to a larger extent. The solar energy correlations developed exhibited inverse relationships of maxima and minima by two independent models that generated coefficients (a = 0.251, b = 0.52) and (a = 0.406, b = 0.256). Photovoltaic technology has brought in cost competitiveness leading to more installation of solar based power.

Keywords: f-solve method, modeling, solar radiation

Introduction

The models on heat balance, atmospheric parameters, altitude, aerosol scattering and absorption factor, clearness index and cloudiness is based on local climatic and parameters such as declination angle, altitude, and ratio of relative sunshine hour. Levenberg-Marquardt algorithm increased accuracy and convergence speed [1–11]. ANN-SFP variable weather results had accuracy better than ANN-HDS model and sunshine duration influenced performance more compared to other factors [12, 13]. ANN-SVM indicated temperature strongly correlated to insolation, while in support Vector machine regression, support vector machine classification and decision tree improved performance of ANN-SVMR. The evolutionary algorithm and extreme learning machine were robust and reliable [14–16]. The convex-optimisation in SVM increased of train and test speed, while particle swarm -back propagation neural network (BPNN) produced better accuracy than genetic algorithm. Multi-gene genetic programming and genetic algorithm grouped data to optimise [17–20]. The genetic algorithm relates irradiance and site parameters, not available in machine learning [21]. The k-NN decomposition and ANN was used to predict 60-minute data as against bio-mimicry approach [22, 23]. The Bees Algorithm for radiance prediction and models using J48 machine learning along with Gaussian and exponential functions with solar data collected through Meteonorm [24–26]. The long short-term memory (LSTM) algorithm gave better results compared to Weather Research and Forecasting meso-scale model with results comparable to genetic algorithm. The genetic algorithm, particle swarm optimised and hybrid support vector-genetic algorithm were used to find constants to compare with statistical regression [27–30].

Mathematical Modelling for Prediction of Solar Radiation

Solar radiation was predicted using f-solve technique to evaluate Angstrom constants.

a) **Empirical models for solar radiation prediction**: The average global radiation (H) on basis of clear day radiation (H_c) at location and average fraction of possible sunshine hours (S/So) as indicated by Equation (1) [31].

$$\frac{H}{H_c} = a + b(\overline{S}/\overline{S}_o) \tag{1}$$

$$\frac{\overline{H}}{\overline{Hg}} = a + b(\overline{S}/\overline{S}_o) \tag{2}$$

[a]veeresh_gb@kletech.ac.in, [b]mb_gorwar@kletech.ac.in, [c]rptapaskar@kletech.ac.in, [d]pp_revankar@kletech.ac.in, [e]pg_tewari@kletech.ac.in, [f]vinayak.khatawate@djsce.ac.in

DOI: 10.1201/9781003450252-91

The high order accurate estimations on basis of equation (1) lead to Prescott model (Equation 2) [32]. The average daily extra-terrestrial radiation (Ho) is estimated using Equation 3

$$H_o = \frac{24}{\pi} \times I_{sc}\left(1 + 0.033\cos\left(\frac{360}{365}n\right)\right)(\sin\omega_s \cos\phi\cos\delta + \omega_s \sin\phi \sin\delta)$$ (3)

Model S1: Tiwari and Suleja model [33] proposed linear model for regression constants by Equation 4, defined for locations in India [33].

$$a = -0.110 + 0.235 \cos(\delta) + 0.323(\overline{S}/\overline{S}_o)$$ (4a)

$$b = -1.449 - 0.553 \cos(\delta) - 0.694(\overline{S}/\overline{S}_o)$$ (4b)

Model S2: Gopinathan [34] estimated regression co-efficient using Equation 5 for world wide application to estimate regression co-efficient [34].

$$a = -0.309 + 0.539 \cos(\delta) - 0.0693h + 0.29(\overline{S}/\overline{S}_o)$$ (5a)

$$b = 1.527 - 1.027 \cos(\delta) + 0.0926h - 0.359(\overline{S}/\overline{S}_o)$$ (5b)

Model S3: Glover and Mc Colloch [35] used curve fitting method [35].

$$a = 0.26\cos(\phi) \text{ and } b = 0.52$$ (6)

Model S4: Garg and Garg (1985) utilised of monthly mean daily radiation and bright sunshine hours to arrive at 'a' and 'b' indicated in Equation (7) [36]

$$a = 0.3156 \text{ and } b = 0.4520\cos(\phi)$$ (7)

Model S5: The analysis based on TMY data is indicated by Equation (11).

$$a = 0.42247 \cos(\delta) + 0.00077(\overline{S}/\overline{S}_o)$$ (8a)

$$b = 0.2665 \cos(\delta)$$ (8b)

The model S1 and model S2 used average solar radiation respectively.

b) **Simulation models for solar radiation:** is based on simulation models developed on ANN: NARX a time series model and MLPANN model.

Model S7: (MLPANN): Iterative network training constitutes core segment of NARX time series model with a functionality to reduce deviation as indicated in Eq.9.

$$F(W) = \sum_{p=1}^{p}\left[\sum_{k=1}^{k}(d_{kp} - O_{kp})^2\right]$$ (9)

Overall regression of training, validation and test data showed predicted solar global radiation with coefficient of variance of 99.959% in close agreement with target data.

c) **Prediction of Solar radiation using optimisation model (Model S8):** The parameters used in optimisation are interdependent exhibiting nonlinear relationship indicated by formulation in Equations (10a) and (10b)

$$k_t = x(1) + x(2) \times (\overline{S}/\overline{S}_o)$$ (10a)

$$0 \leq (x(1), x(2)) \leq 1$$ (10b)

$$a = 0.28 \cos(\delta) + 0.19 \ (\overline{S}/\overline{S}_o) \tag{11a}$$

$$b = 0.18 \cos(\delta) + 0.13 \ (\overline{S}/\overline{S}_o) \tag{11b}$$

The results analysed for %MAPE, %MBE, %RMSE, t-stat and regression coefficients indicated good agreement up to 99.9% with TMY data. Equation 12 represents the generalised form regression equation to determine regression constants 'a' and 'b',

$$b = C + \cos(\delta) + y \ (\overline{S}/\overline{S}_o) + Z \times h \tag{12}$$

d) **Prediction of hourly solar radiation using decomposition method:** The decomposition model estimated daily average solar global radiation for horizontal or inclined surface as illustrated by equation 13.

$$I_h = \frac{\pi}{24} \times \left(a_h + b_h \cos\omega\right) \left[\frac{\cos\omega - \cos\omega_s}{\sin\omega_s - \dfrac{\pi\omega_s}{180}\cos\omega_s} \right] \overline{H}_g \tag{13}$$

$$a_h = 0.409 + 0.5016\sin(\omega_s - 60) \tag{13a}$$

$$b_h = 0.660 - 0.4767\sin(\omega_s - 60) \tag{13a}$$

ANN based NARX Network building: Network architecture is vital to ensure efficient performance of ANN based time series model to achieve higher accuracy in predictions. The ANN based NARX time series model indicated in Figure 91.1 shows the three-layer model developed to forecast daily average solar global radiation.

Experimental Investigations on Solar Radiation Modelling

The observations of hourly average solar global radiation and bright sunshine hours at test location was done using devices shown in Figure 91.2 depict experimental set up to measure hourly solar insolation

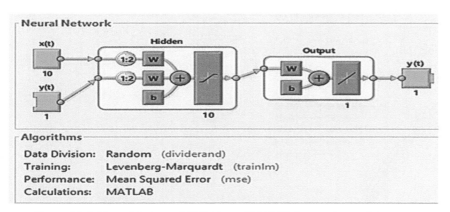

Figure 91.1 Pictorial view of ANN based NARX time series model
Source: Author

Figure 91.2 Solar insolation measurement using –pyranometer and sunshine recorder
Source: Author

and bright sunshine hours of the day. The intensity of the heat flux of sun beam was measured using the pyranometer that presents the global radiation at location.

The potential sunny duration of the day was recorded on basis of burn-mark on graduated paper due to converging sun beam from spherical glass sphere. This data is useful to the non-uniform burn traces were corrected as per the WMO standards for assessment of total recorded bright sunshine hour at the location in relation to predicted day length. The relative sunshine was used to predict hourly average solar global radiation for the test location.

Results and Discussions

The prediction models were assessed based on statistical parameters as presented in subsequent sections along with the economic viability studies. The Figure 91.3 highlights variation of global radiation for representative month of January. Models 3-8 indicated excellent results and models 1 and 2 showed good performance on basis of % RMSE. The lowest and highest % RMSE were reported by models 8 (0.75%) and model 1(5.48%) respectively. The models exhibited similar trend for % MBE and MAPE, the models

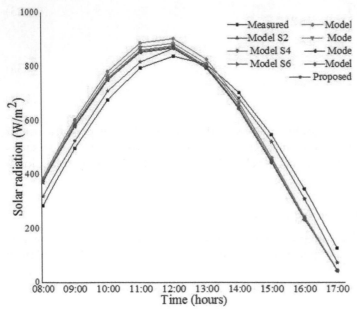

Figure 91.3 Insolation prediction (Jan)
Source: Author

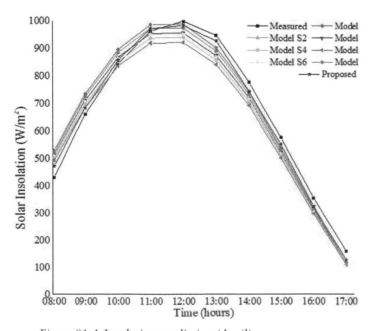

Figure 91.4 Insolation prediction (April)
Source: Author

2-8 exhibited outstanding performance and model 1 had good agreement. The lowest and highest % MBE and MAPE were reported by model 7 (-0.67 & 1.72) and model 1 (5.30 and 5.29) respectively. The model S5 was unique indicating overestimates at 8:00 hour and underestimates during other part of the day. The models S1 to S8 yielded high R^2value above 0.95 establishing the consistency in the high of degree fit with measured data producing % RMSE between 0.04% (S7) and 1.41% (S2). The models exhibited % MBE and MAPE below 5% with S3 (-0.04 and 4.66), S2 (2.08 and 4.66) and S6 (0.05 and 2.08) exhibiting satisfactory results. The figure 91.4 depicts pridicted hourly average solar radiation for month of April representing summer climatic condition. The overall trend of variation of solar insolation are identical with relatively larger magnitudes of insolation against other seasonal values.

The Figures 91.5 and 91.6 exhibited variation in predicted hourly insolation during July and Octber with reference to models S1 to S8 that indicated irregular deviations. The models S2, S3, S4, S5, S6 and S8 estimated higher % deviation during morning as compared to evening that reversed in case of S1 and S7. The S1 and S5 gave lowest (11.20%) and highest (22.72%) average % deviations respectively that indicated best and poor performing models. The t-stat values of all the models were less than t-critical indicated the hypothesis made was true with 95% confidence level. The model S8 provided lowest (-0.05) t-stat value, indicated model S8 had very good agreement with measured data compared to other models with respect

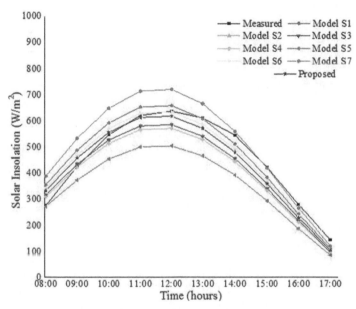

Figure 91.5 Insolation prediction (July)
Source: Author

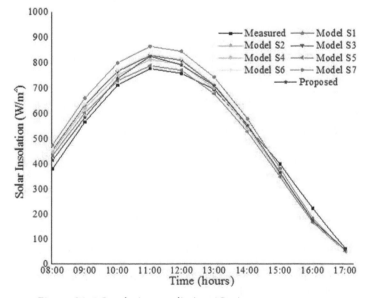

Figure 91.6 Insolation prediction (Oct)
Source: Author

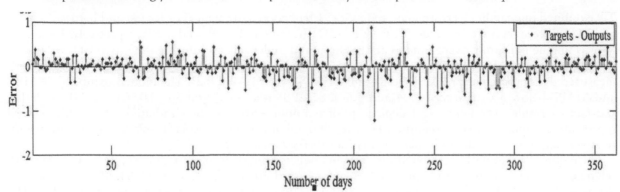

Figure 91.7 Simulated results of ANN based NARX time series model (S6)
Source: Author

to t-stat test. In summary the investigated prediction models lead to high hourly average insolation during April month by S1 (987 W/m^2) and S8 (982 W/m^2) at 12:00 noon compared to highest measured value of 998 W/m^2. The f–solve regression coefficients were better due to higher coefficient of determination and lower standard deviation than other models. The Figure 91.7 depicts revised output of NARX time series model with reference to solar global radiation either overestimated or under-estimated in training, validation, and test data segments. The location had diverse climate of cloud and clear sky (winter), recurrent surface inversion (autumn) and low solar radiation (January and October). The high cloud density significantly reduced atmospheric transmittance that resulted in lower solar radiation.

Conclusions

The conclusions on solar radiation prediction models developed for test locations and their experimental validations are as follows.

- Radiation regression constants were inversely related to maxima and minima yielded by S3 and S5 as (a = 0.251, b = 0.52) and (a = 0.406, b = 0.256) respectively. The 'b' value had a greater degree of variance than 'a' and influenced prediction to a greater extent, hence model S5 was recommended to predict coefficients. The results of S1 to S5 and S8 models indicated good fit with measured data during all months exhibiting acceptable limits of variation in statistical indicators. The S8 was recommended for evaluation of regression constants on account of lowest values of RMSE, MBE and t-stat as compared to all other prediction models. The model S8 also reported highest R^2 for all months and hence gave better estimates than conventional least square approach. The monthly average daily solar global radiation predicted using model S8 captured seasonal variations in a consistent manner that reflected actual observations at location. The peak monthly average global radiation of 6.7 kWh/m^2/day was predicted for month of April as against the lowest value of 4.5 kWh/m^2/day for the month of July.

References

[1] Talabi, H. B., Moradi, M. H., and Ayob, S. B. M. (2014). A review on classification and comparison of different models in solar radiation estimation. International Journal of Energy Research, John Wiley & Sons. 38(6), 689–829.

[2] Sekihara, K., and Kano, M. (1957). On the distribution and variation of solar radiation in Japan. Journal of Metrology. 8, 144–49.

[3] Bird, R. E., and Hulstrom, R. L. (1981). A simplified clear sky model for direct and diffuse Insolation on horizontal surface. SERI. 1, 1–46.

[4] Muneer, T., and Saluja, G. S. (1985). A brief review of models for computing solar radiation on inclined surfaces. Energy Conversion Management. 25, 443–58.

[5] Malik, Q., and Abdullah, R. H. (1996). Estimation of solar radiation in brunei darussalam, RERIC. International Energy Journal. 18, 75–81.

[6] Abdullah, Y. M. (2008). Study the relationships between cloudiness and meteorological factors in Mosul City. Rafidain Journal of Science. 19, 67–76.

[7] Ahmad, M. J., and Tiwari, G. N. (2010). Solar radiation models – review. International Journal of Energy and Environment (IJEE). 1(3), 513–32.

[8] Yadav, A. K., and Chandel, S. S. (2012). Artificial neural network based prediction of solar radiation for Indian stations. International Journal of Computer Applications. 50(9), 1–4.

[9] Hasni, A., Sehli, A., Draoui, B., Bassou, A., and Amieur, B. (2012). Estimating global solar radiation using artificial neural network and climate data in the south-western region of Algeria. Energy Procedia. 18, 531–7.

[10] Premalatha, N., and Valan Arasu, A. (2016). Prediction of solar radiation for solar systems by using ANN models with different back propagation algorithms. Journal of Applied Research and Technology. 14, 206–14.

[11] Mohammed, L. B., Hamdan, M. A., Abdelhafez, E. A., and Shaheen, W. (2013). Hourly solar radiation prediction based on nonlinear autoregressive exogenous (Narx) neural network. Jordan Journal of Mechanical and Industrial Engineering. 7, 11–18.

[12] Wang, F., Mi, Z., Su, S., and Zhao, H. (2012). Short-term solar irradiance forecasting model based on artificial neural network using statistical feature parameters. Energies. 5, 1355–70.

[13] Benkaciali, S., Haddadi, M., Khellaf, A., Gairaa, K., and Guermoui, M. (2016). Evaluation of the global solar irradiation from the artificial neural network technique. Revue des Energies Renouvelables. 19(4), 617–31.

[14] Belaid, S., and Mellit, A. (2016). Prediction of daily and mean monthly global solar radiation using support vector machine in an arid climate. Energy Conversion and Management. 118, 105–18.

[15] Jiménez-Pérez, P. F., and Mora-López, L. (2016). Modeling and forecasting hourly global solar radiation using clustering and classification techniques. Solar Energy. 135, 682–91.

[16] Tao, H., Ebtehaj, I., Bonakdari, H., Heddam, S., Voyant, C., Al-Ansari, N., Deo, R., and Yaseen, Z. M. (2019). Designing a new data intelligence model for global solar radiation prediction: application of multivariate modeling scheme. Energies. 12(1365), 1–24. doi:10.3390/en12071365.

[17] Quej, V. H., Almorox, J., Arnaldo, J. A., and Saito, L. (2017). ANFIS, SVM and ANN soft-computing techniques to estimate daily globalsolar radiation in a warm sub-humid environment. Journal of Atmospheric and Solar–Terrestrial Physics. 155, 62–70.

[18] Xue, X. (2017). Prediction of daily diffuse solar radiation using artificial neural networks. International Journal of Hydrogen Energy. 42(47), 23 November 2017, 28214–28221 https://doi.org/10.1016/j.ijhydene.2017.09.150.

[19] Pan, I., Pandey, D. S., and Das, S. (2013). Global solar irradiation prediction using a multi-gene genetic programming approach. Journal of Renewable and Sustainable Energy. 5(6), 1–32. doi: 10.1063/1.4850495.

[20] Wu, J., Chan, C. K., Zhang, Y., Xiong, B. Y., and Zhang, Q. H. (2014). Prediction of solar radiation with genetic approach combingmulti-model framework. Renewable Energy. 66, 132–9.

[21] Al-Hajj, R., and Assi, A. (2017). Estimating solar irradiance using genetic programming technique and meteorological records. AIMS Energy. 5(5), 798–813. DOI:10.3934/energy.2017.5.798.

[22] Unit Three Kartini and Chen, C. (2017). K-NN Decomposition artificial neural network models for global solar irradiance forecasting based on meteorological data. International Journal of Computer Electrical Engineering. 9(1), 351–9.

[23] Khanna, M., Srinath, N. K., and Mendirattam, J. K. (2017). A novel approach for long term solar radiation prediction. ICTACT Journal on Soft Computing. 08(1), 1574–81.

[24] Bagheri, T. H., Moradib, M. H., and Bagheri, T. F. (2013). New technique for global solar radiation forecast using bees algorithm. IJE Transactions B: Applications. 26(11), 1385–92.

[25] Yadav, A. K., and Chandel, S. S. (2015). Solar energy potential assessment of western Himalayan Indian state of Himachal Pradesh using J48 algorithm of WEKA in ANN based prediction model. Renewable Energy. 75, 675–93.

[26] Kaplanis, S., Kumar, J., and Kaplani, E. (2016). On a universal model for the prediction of the daily global solar radiation. Renewable Energy. 91, 178–88.

[27] Araujo, J. M. S. (2020). Performance comparison of solar radiation forecasting between WRF and LSTM in Gifu, Japan. Environmental Research Communications. 2(045002), 1–12.

[28] Chen, L., and Li, Y. (2020). A state-of-art method for solar irradianceforecast via using fisheye lens. International Journal of Low-Carbon Technologies. 00, 1–15. Oxford University Press.

[29] Ghimire, S., Deo, R. C., Downs, N. J., Raj, N. (2018). Self-adaptive differential evolutionary extreme learning machines for long-term solar radiation prediction with remotely-sensed MODIS satellite and Reanalysis atmospheric products in solar-rich cities. Remote Sensing of Environment. 212, 176–98.

[30] Meenal, R., & Selvakumar, A. I. (2017, February). Temperature based model for predicting global solar radiation using genetic algorithm [GA]. In 2017 International Conference on Innovations in Electrical, Electronics, Instrumentation and Media Technology (ICEEIMT) (pp. 111–116). IEEE.

[31] Srivastava, R. C., and Pandey, H. (2013). Estimating angstrom-prescott coefficients for India and developing a correlation between sunshine hours and global solar radiation for India. Hindawi Publishing Corporation ISRN Renewable Energy. Article ID 403742, 7 p.

[32] Rajesh, K., Aggarwal, R. K., and Sharma, J. D. (2013). New regression model to estimate global solar radiation using artificial neural network. Sage publications Advances in Energy Engineering (AEE). 1(I), 66–73.

[33] Tiwari, G. N., and Suleja, N. (1997). Solar Thermal Engineering System. India: Narosa Publishing House.

[34] Gopinathan, K. K. (1998). A General formula for computing the coefficients of the correlation connecting global solar radiation to sunshine duration. Solar Energy. 41, 499–502.

[35] Glover, J., and Mcculloch, J. S. G. (1958). The empirical relation between solar radiation and hours of sunshine. Quarterly Journal of the Royal Metrological Society. 84, 172–5.

[36] Garg, H. P., and Garg, S. N. (1987). Improved correlation of daily and hourly diffuse radiation with global radiation for Indian stations. Solar & Wind Technology. 4, 113–26.

Chapter 92

Ergonomic design of sewing machine seating structure by measuring anthropometric data

Riddhi Chopde[1],[a] and Dr. Tushar Deshmukh[2],[b]

[1]G H Raisoni university, Amravati, India

[2]Sant Gadge Baba Amravati, University, India

Abstract

In textile industries, there are different types of work that are consecutively repeated like cutting, stitching, weaving, ironing etc. with continuously hand and legs movement. The operator's body moves in the same pattern for long time. Due to seating arrangements, workplace layout, environmental conditions, operators suffering from a number of health issues related to musculoskeletal pain. All the type of given work needs correct seating posture, proper height from floor, distance from operator to machine, appropriate light and thermal conditions. For daily dynamic motions it is difficult to work with the same work layout with different body dimensions. In industries operators have stools or benches for seating, leads to awkward position for working and every operator has different seating posture angle while working on sewing machine as well as the sight angle. It required adjustable seating arrangement also causes neck and vision problems during working. To overcome all these problems of sewing machine operators, need to develop the new seating arrangement and analyse the new design structure for better working and for resolving MSD problems, as the design is based on ergonomic factors it required adjustable seating arrangement by considering the operator's weight also angles of the hands as per the different height of the operator. In this study, by measuring the anthropometric data of operators with workstation, establish the relationship between ergonomical and operational factors. Also, re-design the sewing machine seating arrangement by considering dependent and independent variable parameters of the workstation on the sewing machine operator. It includes a DC motor with drive which adjust the chair with its appropriate height. The design is based on posture of workers and ergonomic considerations. A new adjustable chair increases the comfort for the operators and reduces the body aches to do the work easy and may increase the productivity rate of the manufacturer.

Keywords: MSD problems, sewing machine workstation, sewing machine operators, anthropometric data, ergonomics

Introduction

The purpose of this paper is to discuss the new design of seating system in garment manufacturing industry where the sewing machine operators complain about the musculoskeletal problems in their day to day life due to regular and repetitive work. It is very difficult to do a job continuously either by standing position or by seating. Musculoskeletal disorders of the machinist lead to risk about the body exposure area [1]. From the industrial survey it is found that, due to several awkward working postures, incorrect agronomical gestures, some environmental conditions, and thermal effects, machinists are not comfortable during working. After the habituate for the workplace they are suffering from various unwanted health conditions leads to different disorders, sick leaves, complaints etc.\The workers comfortness play important role in product quality and productivity. To manage health issues and improve the quality of work this study introduces the new design for the working chair of machinist by measuring anthropometric parameters in different clothing units. By establishing the relation between ergonomical and operational factors, it is easy to find the probable solutions considering all human aspects for workers confortness. The design of the chair is done with CAD Tool. With the DC geared motor, the height of the chair will be adjusted. So, operators can work with their convenience as they all have the different height, weight, eyesight angle, different body gestures. The detailed design modifications are explain in design chapter in the paper. The data collection is by considering dependent and independent variables of anthropometry for the workstation related to operators' postures. The survey is done with questionnaire at different places in textile industries at Nangaon Khandeshwar MIDC, Amravati, Maharashtra. Musculoskeletal disorders of the machinist lead to risk about the body exposure area while working. Different levels of body pain risks are found outby questionnaire during the survey.

Methodology

Ergonomic Issues

According to a literature survey it is found that, the Garment manufacturing Factory operators suffering from various musculoskeletal disorders. They are suffer from multiple types of health issues due to

[a]r_chopde02@rediffmail.com, [b]tushar.d69@gmail.com

DOI: 10.1201/9781003450252-92

improper posture and repetitive body motions during working. Due to that there is a work dissatisfaction between the workers. Types of pain and their causes are as follows:

i. Improper work postures - Musculoskeletal disorders like neck and shoulder pain, upper limb disorder, wrist pain, back pain, knee pain, ankle pain, elbow pain, joint swelling.
ii. Long hours sitting or standing in one position, Hip joint pain, backaches and breathing problems.
iii. Difficult workstation - logistical weaknesses
iv. Too short break – More absenteeism
v. Working conditions of shift workers - accident severity
vi. Completion of target within time - Signs of psychological problems
vii. Working in traditional frameworks -Mental trauma, fatigue of workers
viii. Vibrations and noise on the workstation – Severe headeque among the workers [2–4].

Questionnaire

To overcome the above research problems, study need to do the detail survey in various clothing industries for finding out the probable outcome. With the help of this it can do the required modifications in previous structure. The new design will become more reliable for workers point of view. The main aim of all this is to reduce the musculoskeletal problems among the operators having repetitive tasks for all the 8 hrs. working per day. The textile industry operator makes eight shirts per day approximately. There is continuous seating as well as stitching by simultaneously movement of hand and legs. Eyesight must be engaged at needle point in every condition of light. All these aspects should be considered while designing new workstation it may a chair or table. We can meet 150 clothing factory workers during the survey. They answered well to all the questions related to different conditions, problems, effects on work etc.

Table 92.1 shows the asked questions with their answers:

Table 92.1 Questionnaire and answers during survey

S.N.	Questions Related to workstation	Average outcomes	Questions related to operators	Average outcomes	Effect on work
Q1	Workplace is ergonomically perfect or not?	below satisfactory level	What are the ergonomical factors to be considered for man machine relationship?	repetitive work position, awkward posture, stationary/dynamic position, vibration, temperature, noise.	Work stress, increase sick leaves
Q2	Implementation of advanced techniques of working methods are adopted or not?	Moderate	What are the different methods to be identified for comfort of the operators?	Redesign the workstation according to anthropometric data of operators	comfortable workplace
Q3	The workstation design is fully prepared with anthropometric measurement or not?	Moderate	Operators can do their work in given time or not?	Lacuna of completion in given task due to MSD problems	low productivity
Q4	Operators are satisfied with the current working or not?	Below satisfactory level	Given tasks are bounded with target time of not? What happens if fail?	Yes. Over time and stress is there	Work pressure
Q5	Improvement in performance is measured or not?	Not regularly due to repetitive work	What factors should be needed to improve the worker and manager relationship?	Friendly environment, sometimes professional sometimes casual, adjective nature for both	slow or faulty production
Q6	Working environment is friendly or not?	Satisfactory	Which parameters should be considered for improvement?	Chair back, chair cushion, chair height, Table height etc.	Increase quality and productivity

Source: Author

Measurement of Anthropometric data

In some literatures there are descriptions of postures and their measurements which are easy to overview with concerning study. Following are the seating posture and dimensions for collection of anthropometric data. Follows to questionnaire it is easy to take the anthropometric measurements of operators during the same survey. Related to ergonomics of workplace like improper gesture of human body, repetitive motion of hands and legs, vibrations, pressure on elbows and hips, noise, and stress the measurements are done. To design new workstation all details of workplace is needed. Apart from standard anthropometric measurements like weight, height, length, body circumference, the dependent and independent variables parameters are required. Based on work task variables, health related variables and background variables all dependent, independent parameters are measured. The anthropometric data related to workplace collected by questionnaire are as shown in Table 92.2.

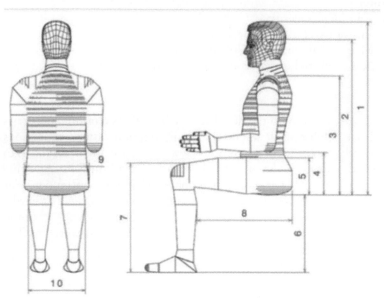

Figure 92.1 Anthropometric measures for chair design [5]

Average analysis of body exposure risk

Table 92.2 Measurement of anthropometric data by using dependent variable parameters of workstation

Dependent Variables

S.N.	Height of operater (Kg)	Weight of operater (Kg)	BMI = Wt./Ht.^{2}	Desk Height (In)	Chair Height (In)	Trunk In clination Forward/ Backward (degree)	Sitting knee height	sitting Height Errect	Length Between Eye And Needle (cm)	Distance Between operator and the machine (cm)
1	1.778	76	24.04086	31	18	21	24	25	25	11
2	1.7272	65	21.78854	31	18	16	21	23	20	13
3	1.7272	67	22.45895	31	18	25	23	22	23	11
4	1.778	74	23.40821	31	18	20	25	25	21	12
5	1.7526	64	20.836	31	18	15	26	24	22	12
6	1.6256	60	22.70512	31	18	19	20	22	20	11
7	1.7018	70	24.17024	31	18	23	23	20	26	13
8	1.6	57	22.26563	31	18	20	21	26	26	11
9	1.778	78	24.67352	31	18	25	24	23	23	13
10	1.6	55	21.48438	31	18	17	20	20	20	13

Source: Author

Questionnaire also helps to find exposure of risk from working and operational conditions which is contributed to mental and physical problems in various jobs among the workers [3]. Due to high frequency noise, Operators suffered from increment in body temperature, dizziness, poor vision and ear defects [4–6]. MSD problems causes the injury and disability in operators due to the handling of materials and machinery with repetitive work [7]. Shoulders and the lumber regions shows more dynamic actions causes more musculoskeletal complaints while with rapid upper limb assessment the upper limbs were most affected region [8]. Figure 92.2 shows the average analysis of exposure during the survey in clothing manufacturing company. Table 92.3 shows measurement of independent variable parameters of operators in workstation.

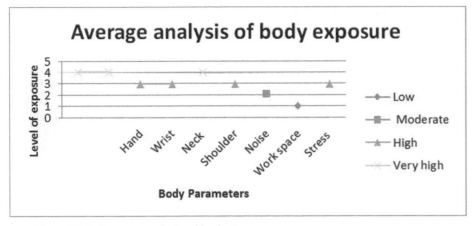

Figure 92.2 Average analysis of body Exposure
Source: Author

Table 92.3 Measurement of independent variable parameters of operators in workstation

Independent Variables

Work task	Working hrs.	Muscular pain before employment	Muscular pain in present work	Type of musculo-skeletal pain	Medical consultation due to pain	Age	Job Experience in yrs.
Sewing	8	No	Yes	Joint Pain	No	45	15
Sewing	8	Yes	Yes	Back Pain	Yes	28	5
Sewing	8	Yes	Yes	Knee Pain	Yes	34	10
Hemming	8	No	Yes	Neck & Legs	Yes	25	2
Hemming	8	No	Yes	Back & Neck	No	30	11
Sewing	8	No	Yes	Head & Back	No	32	12
Sewing	8	No	Yes	Back, Neck, Hand & Legs	No	40	16
Sewing	8	Yes	Yes	Back & Legs	Yes	34	10
Sewing	8	No	Yes	Neck Pain	No	28	9
Hemming	8	No	Yes	Knee & Joint Pain	No	42	12

Source: Author

Relation between operational and ergonomic factors

In any manufacturing industry there are factors to be affected on work, production, manufacturing process etc. These factors are operational factors causes ergonomical issues which are further related to directly or indirectly on operators working, comfort ability and productivity. From the survey some operational and some ergonomical factors are found out. Design of workstation, planning and control, inventory management, capacity, maintenance, and continuous improvement are the operational factors. And repetition of work, awkward posture, forceful motion, stationary position, direct pressure, vibration, extreme temperature, noise, and work stress are the ergonomical factors. Figure 92.3 shows the link between operational and ergonomical factors.

Design of new work station with CAD Tool

The new design is totally based on ergonomic conditions of the workstation which are suitable for workers. The chair dimensions are decided from anthropometric data collected during the survey. Therefore, by considering operator's suitability the new chair dimensions are taken for an average weight of 80 kg. To reduce the musculoskeletal risks of the sewing machinists there are a number of modifications could be done with simple or complicated ways. Some of the modifications may help to prevent the operators from their panic working conditions in day-to-day life. The operators have repetitive work for all day with same seating arrangement. If there are provisions to change the height of the chair while working, then workers may adjust it as per their convenience. It gives smooth working and reduces complaints about muscular

Operational factors   **Ergonomic factors**

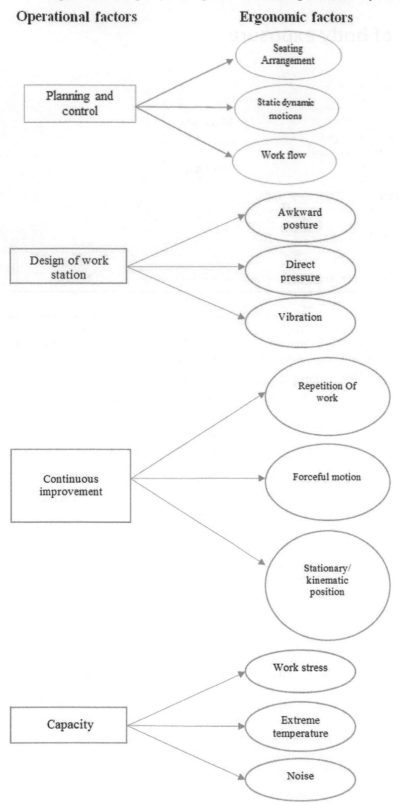

Figure 92.3 Link between operational and ergonomical factors
Source: Author

pains. The change in design of the chair is based on postures as well as the ergonomic considerations having less number of calculations. The main design parameter is the height of the chair. The model is made
with UG NX 10. A simple geared DC motor is used to operate the chair. When the operator seating on a
chair, the horizontal platform of the chair automatically adjusts as per the height of the operator. Auxiliary
power supply needs to operate the chair. Feedback switches provided to the horizontal platform to confirm
the true adjustment of the height of chair according to ergonomics. Also the middle spindle is revolving

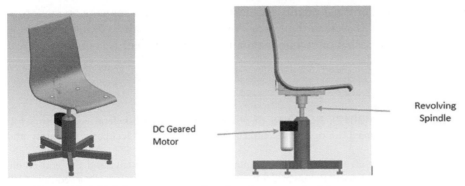

Figure 92.4 Link between operational and ergonomical factors
Source: Author

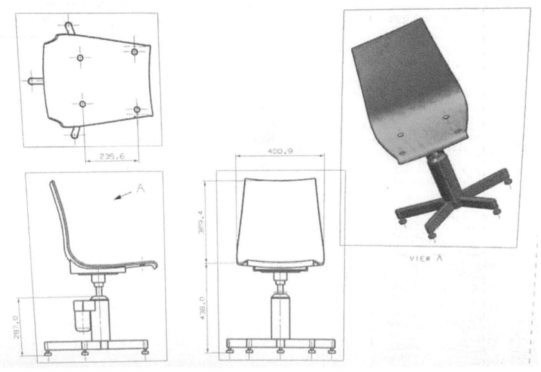

Figure 92.5 Design of chair with dimensions (mm)
Source: Author

to change the direction of the operator's position if required. The Isometric and side view of new seating arrangement is shown in Figure 92.5. The design of chair is shown in Figure 92.5.

Conclusion

The data collected from the various Garment manufacturing units were analysed and the observations taken during the survey were tabulated in different ways. The dimensions are taken from previous working chairs for modification of height of the workplace. To understand the actual situation of the operators in clothing industries where they were suffering from the different MSD problems. With the same chairs or stools it is difficult for every worker to manage their suitable posture during working as they all have different height and body structure. The exposure risk gives the idea of pain in body for continuous work task. The measured anthropometric data during survey helps in designing the new seating structure for the workers satisfaction. This paper puts the focus on ergonomic issues where the clothing industry workers were suffered. Due to that, jobs are affected as well as operators have dis-satisfaction while doing their work. The study and new design were used to reduce the musculoskeletal disorders and to prevent the work from dis-continuity. The new design is done with the CAD tool UG NX 10 with change in previous dimensions of the seating arrangements in surveyed industries. The DC geared motor, rotating spindle and the feedback switches are provided to new design chair. In the comparison of reduction in sick leaves, operators comfort ability, increased productivity and product quality etc. cost of the chair may manage by mass production. It may help the factory operators as well as the factory owner with smooth working in clothing industries. The validation of the design is carried out while the manufacturing of chair is completed now it is only at static analysis for the further study.

References

[1] Bulduk, S., Bulduk, E. O., and Süren, T. (2017). Reduction of work-related musculoskeletal risk factors following ergonomics education of sewing machine operators. International Journal of Occupational Safety and Ergonomics.

[2] Sheta, A., Abou-Ali, M., El-Gholmy, S. et al. (2018). Anthropometric body measurements and the ergonomic design of the sewing machine workstation. Textile Bioengineering and Informatics Symposium Proceedings, 11th Textile Bioengineering and Informatics Symposium, TBIS, pp. 519–524.

[3] Tarafder, N. (2019). Study of ergonomics in textile industry. Journal of Mechanical Robotics. 4(3) .

[4] Ahmad, A., Javed, I., and Abrar, U. et al. (2021). Investigation of ergonomic working conditions of sewing and cutting machine operators of clothing industry. Industrial Textile. 72(3) .

[5] Mahato, M., and Das, A. K. (2021). Design of a low cost adjustable seating system for comfort of weaver in 'de Sign Loom: A semi-automatic handloom. In Smart Innovation Systems and Technologies, (vol. 221). Singapore: Springer.

[6] Rana, K. S., Dr Randhawa, J. S., and Dr. Kalra, P. (2021). Healthy workplace healthy workers: balance for productivity enhancement –A case study of garment manufacturing units in North India. Turkish Journal of Computer and Mathematics Education. 12(6), 4744–55.

[7] Quiroz, J., Aquino, D., Rodriguez, E. et al. (2021). Redesign of workspace through an ergo-lean model to reduce musculoskeletal disorders in SMEs in the clothing accessories sector. International Journal of Engineering Trends and Technology. 69(12), 163–74.

[8] Barbosa, J., Carneiro, P., and Colim, A. (2022). Ergonomic assessment on a Twisting Workstation in a Textile Industry. Springer Science and Business Media Deutschland GmbH. pp. 411–419.

Chapter 93

Experimental and numerical investigation of high-density CNTs and EPOXY-BNNT composites

Shubham Kumar[1,a], Gaurav Arora[2,b] and Mohit Kumar[2,c]

[1]Final year student, M.tech(M.E) Chandigarh University, India

[2]Assistant Professor, M.E., Chandigarh University, India

Abstract

The use of the microwave process has a new approach to the process interaction for the composite materials and resin. This process also aims to reduce the production cycle, cost of tool material, remove bottlenecks, reduce part processing time, and reduction of energy consumption. In this paper, a predeveloped workpiece of LDPE, HDPE, HDPE+LDPE, and epoxy resin was used for non-destructive testing (NDT) to find changes in the physical and mechanical properties of microwave-processed samples with 20% CNTs and BNNT-based composite. The study shows data variation in HDPE at several points where the value of HRC was almost similar at indent points 1-. Value was around 30-40 HRC and at indent points 7, 8 shows 25 HRC and 40 HRC. Similarly, the HRC value of HDPE+LDPE was significantly higher on indent point 2 around 26 HRC but almost similar on every other point between 17-21 HRC, LDPE shows a relatively lower value at indent points 3, 24HRC as compared to other points however indent point values between 1, 2, 4 and 5 shows HRC value 28.7, 32.95, 31.7 and 29.7 HRC, respectively. Additionally, EPOXY-BNNT shows plain side 40.36 HRC, and top side 48.03 HRC, also data show no variation in HRC values during graph plots for HDPE+LDPE, LDPE, and EPOXY-BNNT.

Keywords: EPOXY-BNNT, HDPE, HDPE+LDPE, LDPE, MICROWAVE, nanocomposite

Introduction

The main problem that arises in nanocomposite material is: (i) Improving the composite material properties based on the requirement such as mechanical, thermal, and chemical properties. (ii) To optimise the processes involved in the development of nanomaterials: to reduce time and cost in the processing parameters, depending on the nanocomposite materials processing in terms of economy, quality, and productivity. (iii) Last one is the ecological requirements, since environmental and social challenges are major concerns, both composite material and the process must be environmentally friendly and must reduce the energy consumption required during the nanocomposite synthesis or transformation [1]. Since heating is the main parameter in the processing of the nanocomposite, the heating process can play an essential role. In this regard, microwave heating seems to be the solution since it provides equal heating. There has been a lot of interest in advanced polymer nanotube composite materials that have good mechanical strength, are lightweight, and have excellent high-temperature characteristics. These materials are widely used in a variety of commercial applications, including the aerospace, automotive, sporting goods, energy, infrastructural, and marine industries. One of the key causes of advancing composite technology is the enormous need for lightweight, extremely durable materials with improved electrical and thermal properties. The nanocomposite is increasingly frequently used as an advanced composite due to its low weight, exceptional adhesive quality, environmental resistance, and great thermal and electrical properties. Based on these and other criteria, the following set of polymer systems was selected as an initial epoxy resin, polyamide (PA), polyurethane (PU), ppolyethylene (PE), and polycarbonate (PC).

Recently, microwave heating is gaining popularity, since heat must be transferred to the interior of the material by conduction which provides uneven heating. However, microwave heat material on a molecular level, microwave heating is more efficient, quicker, and more selective than traditional heating, using 10-100 times less energy [2]. Material is heated using a microwave by being exposed to electromagnetic energy. The wavelengths of microwave frequencies range from 1 m to 1 mm, or from 300 MHz to 300 GHz. The use of these waves for industrial, medical, and scientific purposes is controlled to prevent interferences with radars and radio communication; the 2.45 GHz frequency is the most frequently used [3]. Microwaves can interact with material objects, just like other electromagnetic fields can [11]. Testing the nanocomposite is crucial to determining whether the material is appropriate for the application for which it is intended. Numerous techniques, including the mechanical test, can be used to measure the material. The Rockwell hardness test and tensile strength test will be our main topics in this research paper testing the hardness and tensile strength of nanocomposites is crucial to determining whether the material is appropriate for the application for which it is intended. Recently, many authors talked about the use of microwave

[a]cu.16bme1262@gmail.com, [b]gaurav.e11608@cumail.com, [c]mohit.e11538@cumail.com

DOI: 10.1201/9781003450252-93

for polymer, and epoxy composite curing because it affects mechanical and chemical properties. Feher talks about the fabrication of high-quality carbon-fiber-reinforced plastics via microwave curing. They also study the benefits such as quality curing and cost-effectiveness [4]. In a different study by Pal et al. talks high-frequency curing method of epoxy nanocomposites using carbon black powders with a sample size of 15–65 nm [5]. They also study the effect of carbon powder when added to epoxy as filler which shows significant improvement on 15 nm CB with 20% wt having 20-23 MPa tensile strength. Abu-Saleem et al. talks about the use of microwave radiation to cure recycled plastic to improve its strength of the recycled plastic [6]. It is widely known that recycled plastic loses its bonding strength depending on how many times it has been processed [7].

Studies done by Xu et al. talk about a cycle time reduction of nearly 63% compared to thermal processing when using microwave curing. The authors then talks about the mechanical and physical properties of microwave-cured samples to are superior to those cured by conventional means [8]. This consequently tells us the modern importance of microwave curing and the need for higher mechanical properties. In another study, author Sabet talks about LDPE-carbon nanotubes (CNTs) 10 wt% composite where the author uses 3-10% CNT content as filler. They find significant improvement in both mechanical, thermal, and chemical properties, in studies, it is noticed that CNT-added filler shows significate improvement up to 96% and 60 in young's modulus and tensile strength. In present studies, the primary objective is to use CNT as filler and to use the microwave-cured technique for nanocomposite. The details changes in mechanical properties of HDPE-CNT, Pure HDPE, HDPE+LDPE, and Epoxy-BNNT nanocomposites.

Characteristics of nanotube composite

Carbon nanotubes (CNTs) have unique properties that are enabled by their physicochemical properties in recent years. They are designed as valuable elements of polymer systems in the industry. By adding CNTs, manufacturers can control material properties. Conductivity, strength, flexibility/flow, thermal stability/flame resistance, static properties, weight, and other load requirements. Despite ongoing research into this technique, generally accepted methods for evaluating the release of CNTs from polymer systems are lacking. Investigating the warning potential of CNT emissions is an important aspect of safety potential assessment in the context of environmental, health, and safety risk assessment. Evaluation of the most useful methods for measuring CNTs released from polymer systems is also commercially important and is important for materials with multiple properties. To evaluate the most useful methods for measuring CNT emissions from polymer systems, it is important that the materials are commercially suitable and exhibits a variety of properties. The main industrial users of CNTs are electronics, data storage, defense, aerospace, and energy, which are early adopters of new materials and technologies. CNTs are also increasing penetration in various consumer goods markets such as sporting goods, packaging, and textiles. With cost reduction and the continued development of production scale, the number and variety of products and technologies containing CNTs will continue to increase, and the CNT industry will be able to compete with the carbon fiber industry. It becomes an important additive in the production of polymer composites. Although the number of products and applications that can benefit from CNTs is virtually endless, taking advantage of their simple properties, electrical conductivity, and mechanical strength were most exploited in early commercial applications.

Development of nanocomposites

A CNT polymer composite was produced using a commercial microwave applicator with a maximum power of 1.1 kW. Microwave applicators, HDPE-CNT and PP-CNT pellets, and microwave polymerisable composites. The development and characterisation of polymer-based CNT composites are briefly described in the next section. Microwave energy heats materials through interactions between microwaves and materials. When microwaves interact with materials, heat is generated in the fabric as a major function in the microwave frequency range. The absorption of microwave heat, which exhibits time-harmonic electric and magnetic fields, is caused by the conductivity and permeability of the material. 8A set of correlations known as Maxwell's equations defines these properties and fields. The combination of materials and microwaves is described by Maxwell's equations. The ability of a molten material to dissipate microwave energy and the dielectric constant dictate the amount of energy stored in the heated material.

Composite conductivity is the most important parameter in characterising the exothermic level of non-magnetic materials during microwave interaction. At the operating frequency of 2.45 GHz, the dielectric constant of the polymer ranges from 2.1-6.0 and the dielectric loss factor ranges from 0.0008-1.20. A polymer matrix composite (PMC) with a dielectric constant of 10.45 and a dielectric loss factor of 3.75 shows interesting facts with carbon black powder as a filler. The allotropes of carbon, or graphene, can also be used as fillers in microwave polymers. CNTs, graphene structures capable of absorbing microwave energy. The CNTs phase acts as a high-loss insulator, while the polymer acts as a low-loss insulator. According to this principle, CNTs are heated due to local energy conversion, and CNT transfer energy to the polymer by conduction. Thermoplastic materials, especially HDPE and PP, are difficult to heat to room temperature due to their low dielectric loss factor.

This thermoplastic is therefore reinforced with multi-walled CNTs using a compression molding technique that produces 1 mm thick particles in a die-cast part. Pellets were weighed and placed in an aluminum oxide mold. The board is then placed in a mold to ensure the pellets are sealed, then place the entire unit in the microwave, set the heater to the desired temperature or power, and preheat the unit. The microwave begins to fuse the CNTs and the CNTs heat up. CNTs as high-loss insulators start losing heat to the polymer. The polymer heats up and transfers heat to the nearby pellets through conduction and gets heat from the adjacent pellets [9].

As a result, all the grains begin to melt and bind together. The solidification of the pellets continues until the temperature of the pellets reaches the glass transition temperature and remains constant for a certain time to ensure good adhesion of the molten pellets. The molten composition is then cooled to room temperature in a microwave applicator. Carbon nanotubes, which are very absolute insulators, begin to dissipate heat from the polymer. Polymers are heated by conduction, transferring, and receiving heat from adjacent grains. In this way, all the granules dissolve and begin to connect. Freezing of the pellets was continued until the temperature of the pellets reached the glass transition temperature and was held constant for some time to ensure good adhesion of the melted pellets.

The molten compound was then allowed to cool to room temperature on a microwave rod. The temperature was monitored during processing with an infrared pyrometer mounted on rods through a 0.7 mm opening above the pressure plate. the processing conditions used to produce composites from pellets. The temperature of the particles increases with the exposure time. The glass transition temperatures expected for HDPE-CNT and PP-CNT particles are respectively 131°C and 16°C. A lower value of the glass transition temperature means less time for the particles to solidify. In the resulting discussion section, consistent temperature increases and decreases were observed in HDPE-CNT composites compared to PP-CNT composites. CNT-integrated microwaves absorb microwave energy and dissipate heat to cure the composite. A constant temperature was observed for some time after the temperature reached the glass transition temperature of the individual particles, indicating treatment with the compound. As a result of the experiment, it was observed that due to the uniform heating properties of the polymer, the polymer solidified within 100-125 seconds after reaching the glass transition temperature, and the polymerization of all treated samples was uniform. microwave energy. To confirm the presence of CNT, SEM, DSC, and TGA characterisations were performed on the pellet samples. Characterisation showed that the CNTs were randomly distributed throughout the polymer.

Commercial resin epoxy resin 3501 6 is widely used in the aerospace industry, and polyfunctional amines have been used in the experimental analysis of machining tools. For security. It is known to be more organic than diglycidyl esters of bisphenol A. Once cured as recommended, mix 10% hardener with 100% resin. BNNT epoxy composites were prepared using a centrifugal mixer. The mixture was stirred for 2 hours for better dispersion of the BNNTs, then the bath was sonicated for 1 hour. The dispersed BNNTs were evaporated in acetone by passing nitrogen through the mixture at 60°C. The residual acetone was then dried in a vacuum oven at 80°C for 10 hours. The residue was cooled from room temperature to 80°C in the presence of nitrogen. After keeping the mixture in a glassy state at room temperature for 8 hours, it is baked in an oven at –120°C for 1 hour conventionally. It was then allowed to cool to room temperature. After that, the finished composition was removed from post-processing and subjected to mechanical testing. Aluminum spheres with a length of 0.5 mm and a diameter of 0.1 mm were ground in a "machine shop" into a spherical powder with a diameter of 10 μm. The BNNTs were dispersed in acetone with the same temperature and time parameters described for the epoxy samples [10]. The samples were prepared with BNNT, 2% aluminum powder, and 3% Vf. The aluminum powder was cold pressed at 180 MPa for 15 min and then annealed at 580°C for 1 hour. Cool the sample to room temperature at a constant rate (i. e., 5°C/min). Also, uniform curing has been achieved for all the processed samples due to the uniform heating characteristics of microwave energy.

Results and Discussion

Mechanical analyses, jointly with the nondestructive testing (NDT) technics and methodologies, have been, are, and will be fundamental for the development and qualification of materials and components. The main techniques will be presented and briefly discussed [11]. Mechanical analyses are crucial during the manufacturing processes and must be carried out step by step to guarantee the final material's standards and requirements [12]. These analyses are to determine the properties of materials that are fundamental for the composite material elements and are becoming more and more crucial to identify the correct content of impurities and trace elements. Material identification, characterisation, and verification are essential for

Figure 93.1 Nanocomposite material: (a) HDPE and (b) HDPE+LDPE
Source: Author

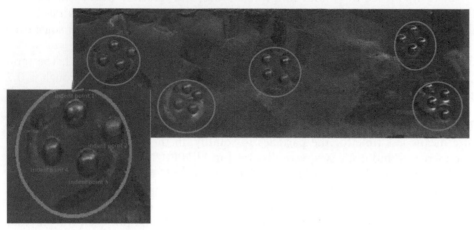

Figure 93.2 Nanocomposite material: (c) LDPE and (d) EPOXY-BNNT
Source: Author

Figure 93.3 Material with indent point markings on HDPE+LDPE
Source: Author

businesses to ensure their materials or products have been manufactured to the correct composite grade and conform to standards [13]. Different nanocomposite materials are shown in Figures 93.1 and 93.2.

Hardness test of composites

Rockwell hardness was measured using a manual analog Rockwell hardness tester. All tested values are an average of four composite samples which were reproduced as explained above. It has a calibration value of 58.3 HRC and an error value of –1.3 HRC, the Rockwell C hardness scale was used (indent diameter 6.35 mm and load 60 kg). The test procedure is divided into two key steps. The possibility of errors resulting from flaws in the material's surface and inaccurate measuring due to backlash from the depth of the created indentation are eliminated by conducting the test. hardness is the resistance against indentation by a harder body. Hardness testing is comparatively simple, quick, and efficient and is called non-destructive.

1. The force (9.8N) at which the steel ball (indenter) is pressed into the plastic material is referred to as preload. Applying pressure to the indenter causes it to indent to a predetermined depth.
2. The indent Points were used to measure the average HRC value due to the nature of the workpiece finding the average value is an important task since the workpiece also shows some defects in some points also four indent point has been taken to make sure HRC data on that particular point is correct.
3. Five different test points were taken on the surface of the workpiece, and four indent test points were taken from the hardness test. Similarly, every test piece is followed by the same experimental procedure example shown in Figure 93.3. However, in the case of HDPE nine different test points were taken since the workpiece has some defects and porosity or voids in the material also EPOXY-BNNT has two different test points, and three indent test points with two sides top and the plain side were taken for similar reasons to find the average value of hardness.

Investigating the effect of microwave curing in the HDPE and HDPE+LDPE sample are shown in Figure 93.4. HDPE shows inconsistent values during the hardness test, Indent Point 3 shows a higher than expected between 18 and 56 HRC. It was concluded, when nanocomposites were formulated in the microstructural

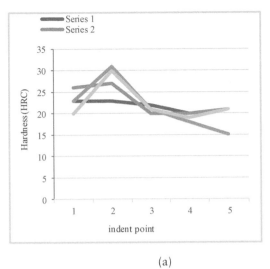

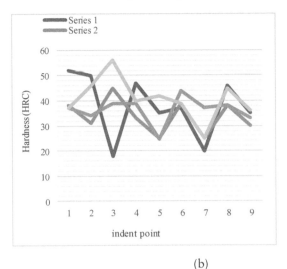

(a)

(b)

Figure 93.4 The graph between hardness value (HRC)-indent points: (a) HDPE and (b) HDPE+LDPE
Source: Author

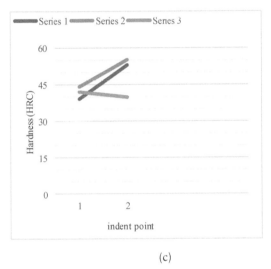

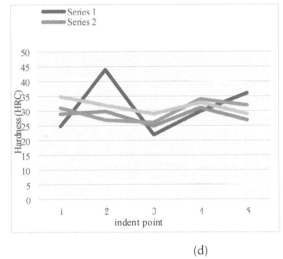

(c)

(d)

Figure 93.5 The graph between hardness value (HRC)-indent points: (c) LDPE and (d) EPOXY-BNNT
Source: Author

level sample have developed defects during the processing of the material. Some Indent Points Mainly Indent Point 1, Indent point 2, and Indent Point 3 show values above/similar to 50 HRC. Similarly, Indent Point 3 in series 1 shows a lower value of 18 HRC. When calculating the average value analysis shows a different trend in Indent Point 7 three hardness test values show almost similar HRC values meanwhile series 3 indent point 7 shows the lowest 26.7 HRC. In addition, Indent Point 8 with the highest 41.7 HRC test value. From this data, an accurate mechanical property can be deduced that there is non-homogeneous distribution in the matrix. Similarly, since a correlation between hardness and crystalline density results in increased hardness value, this result for HDPE shows hardness and crystalline density might have not been in the amorphous phase[12]. Investigating the effect of microwave curing in the LDPE and EPOXY-BNNT sample are shown in Figure 93.5. Studies by E. Alfredo Campo stated that a lower HRC value means better hardness. Similarly, a higher number means soft material which limits the application of the composite material [14]. The advantage of nanocomposites is their high strength-to-weight ratio which is a main factor of the hardness value.

In different studies published by Shimadzu corporation, the HRC value of HDPE is higher than the current studied value [15]. The epoxy test was done from two different sides top and plain even though fiber volume is not unequal this was done so that it can be identified as to which side shows better HRC value. The process is similar to the HDPE hardness test since the workpiece was small only two indent points can be taken as the measurement. In studies conducted by Jawad where author four different epoxy types of workpieces have been used the value of HRC is noted from 31 HRC to 42 HRC as discussed earlier higher value shows the material is soft while a lower HRC value shows the material is hard [11]. In comparison,

Table 93.1 EPOXY-BNNT hardness test

Indent point	Series 1	Series 2	Series 3	Side	Average value	Actual value = -1.3+ avg. value
1	39	44	42	Plain	41.66	40.36
2	53	55	40	Top	49.33	48.03

Source: Author

Table 93.2 HDPE hardness test

Indent point	Series 1	Series 2	Series 3	Series 4	Average value	Actual value = -1.3+ avg. value
1	52.1	38	37	37	41	39.7
2	50	31	34	46	40.2	38.9
3	18	45	39	56	39.5	38.2
4	47	33	39	40	39.7	38.4
5	35	25	24.5	42	31.6	30.3
6	37	39	44	39	39.7	38.4
7	20	25	37	25	26.7	25.4
8	46	38	38	45	41.7	40.4
9	35	30	33	36	33.5	32.3

Source: Author

Table 93.3 HDPE+LDPE hardness test

Indent point	Series 1	Series 2	Series 3	Series 4	Average value	Actual value = -1.3+ avg. value
1	23	26	23	20	23	21.7
2	23	27	31	30	27.7	26.4
3	22	20	21	21	21	19.7
4	20	20	18	19	19.25	17.95
5	21	21	15	21	19.5	18.2

Source: Author

Table 93.4 LDPE hardness test

Indent point	Series 1	Series 2	Series 3	Series 4	Average value	Actual value = -1.3+ Avg. value
1	25	29	31	35	30	28.7
2	44	30	27	32	33.25	32.95
3	22	25	26	29	25.5	24.2
4	30	31	34	33	32	31.7
5	36	27	32	29	31	29.7

Source: Author

the Epoxy- BNNT used in the current study has a 41.66 HRC to 49.33 HRC value which shows a higher value on the top side although the plain has almost the above/similar value as the author explained. This change in hardness value might be because the BNNT-reinforced composites exhibit transversely isotropic properties and are prone to the Composite defects that may form during the manufacturing process. Hardness test results are shown in Tables 93.1-93.4.

During the LDPE, HDPE+LDPE hardness testing phase, both workpieces show relatively higher HRC values than the other two workpieces. In case, for LDPE five indent points that were taken three value is much lower than expected, meanwhile, Indent Point 2 shows the highest value of HRC. The studies done by Guleria author show the average HRC value of LDPE is 40 HRC to 50 HRC which in comparison is much higher than expect it is noted that although some conditions are different [16]. It's been noted that the difference in structural bond and weight percentage fiber between LDPE and HDPE+LDPE makes this workpiece show much lower values in comparison to the rest of the workpiece. The mechanical characteristics of the CNT-reinforced polymers are significantly changed by microwave curing. The Young's modulus of the HDPE and HDPE composites, respectively, rose considerably by 295% for HDPE and 787.8% for HDPE + LDPE when compared to pure HDPE and HDPE + LDPE polymer. During HDPE + LDPE hardness testing workpiece shows material properties because of this behavior we can assume that LDPE is not bonded with HDPE on a structural level even though some carbon impurities were also introduced therefore issue might be due to the composite defects that they may have been formed during the manufacturing process few example Bonding defects, fiber defects, and fiber misalignment in the material. Mechanical inspection of composite materials helps ensure critical findings for the workpieces.

Tensile test of composites

The tensile strength of the microwave-treated specimen 85 mm 18 mm 4 mm was measured at room temperature using UTM (Tinius Olsen, UK (H50KS)). The microwave-treated composite was conditioned for at least 24 hours before tensile testing. The modulus of elasticity was determined using a clamp extensometer at a strain of 0.5 mm/min. Four samples were tested for each compound and the average was calculated. The measured length of the sample was 45 mm. Since the specimen does not require pre-test polishing, the thickness of the specimen remains the same. For the HDPE-CNT composite, the maximum tensile stress observed was 19.7 MPa with random CNTs with an increased modulus. Additionally, when comparing this study with existing data of epoxy-CNT based composite polymer shows 30-41 MPa mixed which was cured by a layer-by-layer manufacturing method which is higher. Due to LDPE-CNT brittleness compared to HDPE after microwave treatment, the LDPE-CNT composite's tensile strength is 2.53 Mpa. In studies by Sabet author prepared five samples with LDPE, LDPE-CNT (1 wt%, 3wt%, 5wt% &, 10wt%) showing tensile values between 2-16 MPa [17]. Table 93.5 shows tensile test results.

Conclusion

This study dealt with carbon nanotubes (CNTs)-based and BNNT-based polymer composite which revealed that 20 wt% CNTs addition to composite material increased their mechanical properties such as hardness and tensile strength. Additionally, as discussed earlier mechanical behavior of HDPE, LDPE and HDPE+LDPE show remarkable increment when compared to other similar studies from different authors. However, EPOXY-BNNT shows slightly lower mechanical properties on the plain side as explained but the top side exhibits not many changes when data were compared. It is noted that the fractured surface of HDPE+LDPE and LDPE-based polymer composites reveal CNTs are uniformly distributed and due to the curving and coiling characteristics of nanotubes, the superior CNTs property of high tensile modulus has not been fully utilized in composites. Unlike pure HDPE, the mechanical properties of the composites improved when enough CNTs were present. The results of the study indicated that a key factor in improving the toughness of HDPE/CNT, LDPE/CNT, and HDPE+LDPE/CNT composites were a homogeneous distribution of interfacial CNTs in the matrix.

References

[1] Belkhir, K., Riquet, G., and Becquart, F. (2022). Polymer processing under microwaves. In Advances in Polymer Technology. Hindawi Limited. doi: 10.1155/2022/3961233.

[2] Mahmoud Abu-Saleem, Yan Zhuge, Reza Hassanli, Mark Ellis, Md Mizanur Rahman, Peter Levett, Microwave radiation treatment to improve the strength of recycled plastic aggregate concrete, Case Studies in Construction Materials, 15, 2021, Page: e00728, ISSN 2214-5095, https://doi.org/10.1016/j.cscm.2021.e00728. (https://www.sciencedirect.com/science/article/pii/S2214509521002436)

[3] Belkhir, K., Riquet, G., and Becquart, F. (2022). Polymer processing under microwaves. Advances in Polymer Technology. 2022, 1–21. doi: 10.1155/2022/3961233.

[4] Feher, L. E. and Thumm, M. K. (2004). Microwave innovation for industrial composite fabrication - The HEPHAISTOS technology. IEEE Transactions on Plasma Science. 32(1). 73–79. doi: 10.1109/TPS.2004.823983.

[5] Pal, R., Jha, A. K., Akhtar, M. J., Kar, K. K., Kumar, R., and Nayak, D. (2017). Enhanced microwave processing of epoxy nanocomposites using carbon black powders. Advanced Powder Technology 28(4), 1281–90. doi: 10.1016/j.apt.2017.02.016.

[6] Mahmoud Abu-Saleem, Yan Zhuge, Reza Hassanli, Mark Ellis, Md Mizanur Rahman, Peter Levett, (2021). Microwave radiation treatment to improve the strength of recycled plastic aggregate concrete, Case Studies in Construction Materials, 15, Page: e00728, ISSN 2214-5095, https://doi.org/10.1016/j.cscm.2021.e00728. (https://www.sciencedirect.com/science/article/pii/S2214509521002436)

[7] Tesfaw, S., Fatoba, O., and Mulatie, T. (2022). Evaluation of tensile and flexural strength properties of virgin and recycled high-density polyethylene (HDPE) for pipe fitting application. Mater Today Proceedings. 62, 3103–13. doi: 10.1016/j.matpr.2022.03.385.

[8] Xu, X., Wang, X., Wei, R., and Du, S. (2016). Effect of microwave curing process on the flexural strength and interlaminar shear strength of carbon fiber/bismaleimide composites. Composites Science and Technology. 123, 10–16. doi: 10.1016/j.compscitech.2015.11.030.

[9] Arora, G., Pathak, H. and Zafar, S. (2019). Fabrication and characterization of microwave cured high-density polyethylene/carbon nanotube and polypropylene/carbon nanotube composites. Journal of Composite Materials. 53(15), 2091–2104. doi: 10.1177/0021998318822705.

[10] Padmanabhan, S., Gupta, A., Arora, G., Pathak, H., Burela, R. G., and Bhatnagar, A. S. (2021). Meso–macro-scale computational analysis of boron nitride nanotube-reinforced aluminium and epoxy nanocomposites: A case study on crack propagation. Proceedings of the Institution of Mechanical Engineers, Part L: Journal of Materials: Design and Applications. 235(2), 293–308. doi: 10.1177/1464420720961426.

[11] Jawad, D. K. and Ramesh, D. A. Mechanical properties of tensile, flexural, impact, hardness and water absorption tests on epoxy composite with glass fibre, black granite powder, white granite powder and stone powder. [Online]. Available from: www.ijert.org

[12] Koch, T. Bierögel, C., and Seidler, S. (2014). Conventional hardness values - data. In polymer solids and polymer melts–mechanical and thermomechanical properties of polymers. pp. 431–444. Springer: Berlin Heidelbergdoi: 10.1007/978-3-642-55166-6-72.

[13] Ibrahim, N. H., Zafir Romli, A., and Nik Ibrahim, N. N. I. (2020). Hardness optimization of epoxy filled PTFE with/without TiO2filler. in IOP Conference Series: Materials Science and Engineering. 839(1). doi: 10.1088/1757-899X/839/1/012002.

[14] E. Alfredo Campo, 5 - Physical Properties of Polymeric Materials, Editor(s): E. Alfredo Campo, In Plastics Design Library, Selection of Polymeric Materials, William Andrew Publishing, 2008, 175–203, ISBN 9780815515517, https://doi.org/10.1016/B978-081551551-7.50007-3. (https://www.sciencedirect.com/science/article/pii/B9780815515517500073)

[15] Shimadzu (2019). ei274 Hardness Test of Plastic Materials (ISO/TS 19278:2019), Shimadzu Corporation, November 13, 2019, https://www.shimadzu.com/an/sites/shimadzu.com.an/files/pim/pim_document_file/applica-tions/application_note/13781/jpi619014.pdf

[16] Guleria, A., Singha, A. S., and Rana, R. K. (2018). Mechanical, thermal, morphological, and biodegradable studies of okra cellulosic fiber reinforced starch-based biocomposites. Advances in Polymer Technology. 37(1), 104–12. doi: 10.1002/ADV.21646.

[17] Sabet, M. and Soleimani, H. (2014). Mechanical and electrical properties of low density polyethylene filled with carbon nanotubes. In IOP Conference Series: Materials Science and Engineering. 64(1). doi: 10.1088/1757-899X/64/1/012001.

Chapter 94

Improving reverse osmosis process in zero liquid discharge using Taguchi optimization method in alcohol production industry

Padmanabh Arun Gadge^a and Shubhangi Gurway^b

G H Raisoni University, Saikheda, India Priyadarshini Bhagwati College of Engineering, Nagpur, India

Abstract

With the growing revolution in industrialisation, alcohol and its derived products right from acetic acid, ethyl acetate to perfumes, food and paint are gaining popularity and demand day by day. The importance of alcohol industries is hence increasing steadily. India is the third largest alcohol production country having estimated value of about \$35 billion. With an average of about 319 distilleries in India producing about 3.25×10^9 L of alcohol and 40.4×10^9 L of wastewater annually getting higher environmental potentials, global warming potential and higher effect rate on toxicity. Alcohol industries are among the Top in "Red Category" in the list of Central pollution control board and hence need to be taken into consideration by governing bodies. Reverse osmosis processes are the process in zero liquid discharge (ZLD) are the choice of most of the industries in India having major advantage including generation of biogas and recovery of clean water.

The present paper showcase the application of Taguchi optimization method in reverse osmosis process for achieving ZLD in alcohol manufacturing company. As RO process is the foremost need of any industry to avoid pollution it's very necessary to find the processing parameters which affect its optimality. In present paper the experiment was conducted to investigate the effect of RO process on water quality after treatment which can best be defined using water permeate. For Taguchi design of experiment (DOE), input parameters like Potential of Hydrogen (PH), operating pressure (OP), Anti Scalant Agent (ASA) and oxidation reduction potential (ORP) used in prescribed way as it gives the better value of permeate and other quality parameter. Temperature is used to represent environmental factor as a noise factor. L27 orthogonal array have been designed and experiment have been conducted collecting permeate samples at different temperature of feed from lagoon. It is observed that the optimal setting computed by Taguchi loss function is the best as it gives the higher value of total signal to noise ratio for all responses. ANOVA revealed that potential of hydrogen is the most significant RO process parameters in affecting the permeate which is the main target process for optimal result followed by OP, ORP, and ASA. The plotted data of results in main effect plot of S/N ratio and response table shows that potential of hydrogen (PH) contribute 62% followed by 14% OP, ORP with 13% and ASA with 14%. The obtained results are then validated by conducting the confirmatory test for the optimal levels of combinations of the input parameters

Keywords: Reverse osmosis process, taguchi method, design of experiment, anova, orthogonal array

Introduction

Reverse osmosis process is one of the most important processes for achieving zero liquid discharge (ZLD) in different industries. It separate water from the effluent which is of very harmful for environment and maintaining the P.C.B norm. Being the basic operation in all the industries, it's very necessary to find the processing parameters which affect its optimality. The operating parameters namely, potential of hydrogen (PH), operating pressure (OP), Anti Scalant Agent (ASA), and oxidation reduction potential (ORP) may put a major influence. Similarly, environmental factors like feed salinity, temperature, procedure, type of membranes used, and clearer procedure used also play a major role. In order to increase recovery and quality of permeate i.e. chemical oxygen demand (COD), conductivity, total solids (TS) and hardness of permeate to be minimise during reverse osmosis operation it is very necessary to control these parameters as the product with desired quality characteristics are function of these parameter. Poor control on the desired responses generates poor quality permeate and results in decrease in recovery of reverse osmosis (RO) process and loss of productivity due to shut down and failure of machine. Present paper emphasise on optimization of above parameters to get the optimal machining condition [1, 2].

Design of Experiment

Designing of experiment is the foremost challenge for any experimentation process which can be best achieved if the optimal results are found with minimum number of test trials. Taguchi optimisation method is one of those tools which prove to be best for reducing the series of test perform. So, in our design of experiment we have used Taguchi method where PH, OP, ASA and ORP considered as an input parameters

---

^a padmanabh.gadge@ghru.edu.in, ^b shubhangi.pbcoe@gmail.com

DOI: 10.1201/9781003450252-94

which are the basic performance parameters which are shown in Table 94.1. with their subsequent levels. Permeate is one of the major quality concern of the RO process so we have decided to first concentrate on quality of permeate and optimize the same using single optimization method. Various experiments have been conducted considering the test combinations obtained during foundation of OA, to find the effect of all the selected process parameters on the permeate quality [3, 4].

Finding Main Function and Determining Its Side Effect

- Function: Optimising reverse osmosis process of ZLD in distillery industry by OP, PH, ORP and anti-scaling agent on permeate to maximise the ZLD process.
- Side effects: Effect of OP, PH, ORP and anti-scaling agent on permeate.

Determining Various Test Conditions

- Testing condition: The external source which might affect the performance of RO process is the environment in which the experiment is being conducted. Feed salinity, human error while operating the setup, cleaning process and temperature are one of those external factors.
- Deciding operating pressure and range: Operating parameter is one of the significant role playing element in maximising permeate and its quality of RO process. But the main point of concern is that the OP can affect the RO process positively at some specific level only. Achieving good quality permeated with increasing value of OP leads to membrane deterioration and loss in integrity of the system. So, the operating pressure range would be selected as per the specification given by the manufacturer for particular setup [5, 6].
- Deciding PH: Number of studies on RO process reveals that PH put a major influence on the output of RO on membrane efficiency. Hence for successful process and membrane maintenance, proper control of PH is required.
- Deciding ORP: ORP is one of those parameter which negatively impact on the quality of permeate. The negative the ORP the better will be the value of permeate but ORP is the major role player in efficiency and flawless operation of RO process.
- Deciding quantity of ASA: Permeate and its quality is again majorly affected by fouling and scaling which is the major loss in any Reverse osmosis process. Similarly in order to avoid machine failure, facedown of anti-scaling agent should be given according to feed [7, 8].

Deciding Performance Parameters

- Quality of RO permeate.

Deciding Objective Function

Optimising the process parameters of RO process to achieve high permeate.
    In order to optiprocess,e process we have S/N ratios with three conditions namely-

1. Smaller the Better
2. Larger the Better
3. Nominal the Better

In Taguchi design of experiment objective functions which are on positive side of the optimality like production rate, quality should be consider as "Larger the Better" similarly if some characteristics should be sufficient enough to achieve the objective function, then, "Nominal the Best" SN ratio can be a best possible option and for objective function like cost of production, should be "Smaller The Better" [9]. In our case our objective function is high permeate so we have selected "Larger the Better".

Identifying Process Parameters and their Level

Table 94.1 shows the process parameters with their optimum levels.

Formation of Orthogonal Array

For deciding the OA from standard OA, three things i.e., number of factors, their level and degree of freedom must be strictly taken into consideration. Table below shows the details of all the three particulars:
    Based on above calculation shown in Table 94.2. It was found that minimum 20 number of experiment should be conducted to get the optimal result. While using the Minitab-14 software we have two option for formation of OA which are L09 and L27, but as per the calculation done above in Table 94.2, it is observed

Table 94.1 Process parameters with their optimum level

Process Parameters	Abbreviation	Code	Range	Level 1	Level 2	Level 3
Operating pressures	op	A	35-45	35	40	45
ph	ph	B	5.9-6.9 0.35	5.9	6.4	6.9
ORP value	mV	C	30-330	-30	-180	-330
Anti-scaling agent(lit)	ROHIB	D	0.5-1.5	0.5	1	1.5
Noise factor						
Temperature (m/sec2)	tem	NF	22-32	22	27.5	32

Source: Author

Table 94.2 Formation of orthogonal array

No. of control factor	No of level	DOF of each factor (No. of level-1)	DOF of interaction A*B	A*C	A*D	Total DOF of control factor	Total DOF of interaction	Total DOF	Min no of experiment to be performed (Total DOF+1)
4	3	2	2*2=4	2*2=4	2*2=4	4*2=8	4+4+4=12	8+12	21

Source: Author

that 21 experiment can be best conducted only using 27 experiment which is almost near to 21 and hence L27 OA have been selected for experiment. Table 94.3 represent the OA with coded value of matrix for experimentation and subsequent result of permeate for given values of parameter settings similarly the Signal to Noise Ratio have also been shown.

Experimental Result

The experiments are conducted as per Taguchi method and experimental data is collected for permeate. Permeate is measured for each level of temperature and average value is taken as the final value which is shown in Table 94.3. As higher value of permeate leads to optimality of the RO process we have decided Larger the Better S/N ratio equation of which is shown below.

$$S/N = -10*\log (\Sigma (1/Y2)/n) \tag{1}$$

Where
n - number of experiments and
y - measured variable values.

Analysis and Discussion

The test was conducted followed by finding out S/N ratio for individual level of experiment. The data were plotted as shown in Figure 94.1 (a, b). Similarly Table 94.5 shows the responses to various SN Ratios which reflects that PH have the highest contribution to affect Permeate which is also depicted in Figure 94.2. Table 94.4 shows the response table showing mean S/N ratio for permeate.

Analysis of Variance

The investigation of the effect of selected design parameter on the objective function can be best possible by ANOVA. In our work we have used ANOVA to find out which parameters among the four majorly affect permeate. ANOVA at 95% of confidence level is done with F-test to decide the significant factor affecting the RO process. Table 94.5 represents the calculation of ANOVA for Signal to Noise ratio. Abbreviations used in the formulas mentioned in the table are shown below tables in order to get clear view of the method.

Similarly, mathematical calculation of permeate is done and shown in TABLE VI shown below.

From Table 94.6, it was found that p-value for OP, PH, ORP and ANTI AGENT is less than 0.05 i.e. (<0.05) hence all the selected parameters are interacting well with permeate whereas the interaction between two subsequent parameter was greater than 0.05 i.e (>0.05), hence they are not significant. Results witnessed the highest contribution of PH with 62% inn affecting the process and optimal result.

Table 94.3 Orthogonal array (L27) with coded value and experimental output with S/N ratio

Run	Operating pressure (Bar)	Ph	OR P (Oxygen reduction potential) Mv	Anti -Scaling Agent (Kg/M³)	Permeate	S/N
1	35	5.9	-30	0.5	54.557	34.737
2	35	5.9	-180	1.0	70.298	36.9388
3	35	5.9	-330	1.5	97.355	39.7671
4	35	6.4	-30	1.0	179.986	45.1048
5	35	6.4	-180	1.5	190.225	45.5853
6	35	6.4	-330	0.5	188.141	45.4897
7	35	6.9	-30	1.5	91.784	39.2553
8	35	6.9	-180	0.5	82.436	38.3223
9	35	6.9	-330	1.0	112.411	41.0162
10	40	5.9	-30	0.5	87.757	38.8656
11	40	5.9	-180	1.0	104.228	40.3597
12	40	5.9	-330	1.5	132.495	42.444
13	40	6.4	-30	1.0	212.556	46.5495
14	40	6.4	-180	1.5	221.465	46.9061
15	40	6.4	-330	0.5	217.861	46.7636
16	40	6.9	-30	1.5	121.294	41.6768
17	40	6.9	-180	0.5	113.506	41.1004
18	40	6.9	-330	1.0	145.231	43.2412
19	45	5.9	-30	0.5	120.957	41.6526
20	45	5.9	-180	1.0	138.158	42.8075
21	45	5.9	-330	1.5	167.635	44.4873
22	45	6.4	-30	1.0	240.126	47.6088
23	45	6.4	-180	1.5	247.705	47.8787
24	45	6.4	-330	0.5	242.581	47.6971
25	45	6.9	-30	1.5	150.804	43.5682
26	45	6.9	-180	0.5	144.576	43.2019
27	45	6.9	-330	1.0	178.051	45.0109

Source: Author

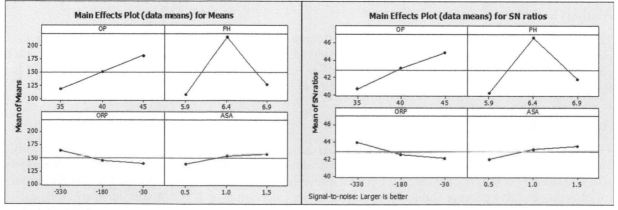

Figuer 94.1 (a) Main effect plots for data means permeate, (b) Main effect plots for sn ratios permeate
Source: Author

Validation of Experiment

As ANOVA result reveals the important contribution of Ph in RO process followed by OP, ORP and Anti scaling agent. For validation of the result the decided set of combination of parameters are OP at level 3

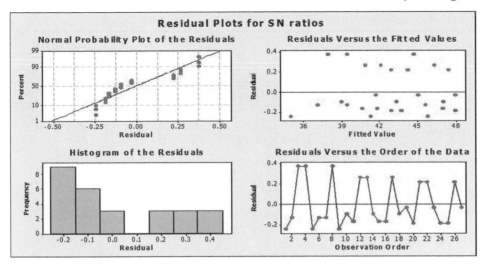

Figure 94.2 Residual plots for SN ratios permeate
Source: Author

Table 94.4 Response table showing mean s/n ratio for permeate

Level	OP	PH	ORP	ASA
1	40.49	39.94	42.99	42.43
2	42.99	46.62	46.19	43.38
3	44.81	41.72	42.11	42.48
Delta	4.32	6.67	1.08	0.9
Rank	2	1	3	4

Source: Author

(A3 = 45), PH at level 2 (B2 = 6.4), ORP at level1 (C1 =-30) and ANTI AGENT at level 3 (D3 = 1.5) or A3B2C1D3.

So, in this stage we have validated our test trial with parameter combinations which are found best suitable for optimum level of permeate as per calculation done before. Table 94.7 shows the different process parameter level at optimum level of permeate. Table 94.8 represent the values of those selected process parameters. Similarly, average value of response of which at optimum level is shown in Table 94.9.

A. Confidence Interval

Confidence interval at 95%

$$CI = \sqrt{\frac{F_\alpha(1, f_e) V_e}{n_{eff}}}$$

Where $F_\alpha(1, f_e)$ = The F ratio at the confidence level of (1-α) against '1' and error of freedom f_e.

$$n_{eff} = \frac{N}{1+\left[DOF \ associated \ in \ the \ estimate \ of \ mean \ response\right]}$$

=81/ (1+8) =9,

N= Total number of results = 27*3 = 81
f_e = error DOF = 4 (ANOVA table mean)
$F_{0.005}(1, 6)$ = 5.99 (Tabulated F value (Ross, 1996))
confidence interval at 95% is: [μ – CI] < μ < [μ + CI]
Where, μ- predicted mean of response characteristic.
Predicted values of permit at optimal setting A3B2C3D3
μSR = YSR + (A3– YSR) + (B2– YSR) + (C1– YSR) + (D3– YSR)

Table 94.5 Calculation of SS Adj, MS,F and % contribution for ANOVA

SS Adj	MS	F	% contribution
			Calculated values
Sum of square for factors	A	$SSA = a\sum_{i=1}^{a}(\bar{x}-\bar{\bar{x}})^2$	79.7238743
	B	$SSB = b\sum_{j=1}^{b}(\bar{x}-\bar{\bar{x}})^2$	199.43419
	C	$SSC = c\sum_{k=1}^{c}(\bar{x}-\bar{\bar{x}})^2$	17.4567687
	D	$SSD = d\sum_{l=1}^{d}(\bar{x}-\bar{\bar{x}})^2$	6.43863702
Total sum of squares	SST	$SST = \sum_{i=1}^{a}\sum_{j=1}^{b}\sum_{k=1}^{c}\sum_{l=1}^{d}\sum_{m}^{a}(x_{ijklm}-\bar{\bar{x}})^2$	319.8628
Sum of squares for interaction of factors	A x B	$SSAB = \sum_{i=1}^{a}\sum_{j=1}^{b}(x_{ij_-}\bar{x_i}-\bar{x_j}+\bar{\bar{x}})^2$	6.423
	A x C	$SSAC = \sum_{i=1}^{a}\sum_{k=1}^{c}(x_{ik_-}\bar{x_i}-\bar{x_k}+\bar{\bar{x}})^2$	0.735
	A x D	$SSAC = \sum_{i=1}^{a}\sum_{k=1}^{c}(x_{ik_-}\bar{x_i}-\bar{x_k}+\bar{\bar{x}})^2$	0.627
Mean sum of square (sum of square/ degree of freedom)	$MSA = SSA\,/\,dof_A$	$MSA = SSA\,/\,dof_A$ =79.7238743/2	39.8619371
	$MSB = SSB\,/\,dof_B$	$MSB = SSB\,/\,dof_B$ =199.43419/2	99.7170948
	$MSC = SSC\,/\,dof_C$	$MSC = SSC\,/\,dof_C$ =17.4567687/2	8.72838435
	$MSD = SSD\,/\,dof_D$	$MSD = SSD\,/\,dof_D$ =11.814356/2	5.90717798
	$MSAB = SSAB\,/\,dof_{A*B}$	$MSAB = SSAB\,/\,dof_{A*B}$ =9.386/2	2.3464
	$MSAC = SSAC\,/\,dof_{A*C}$	$MSAC = SSAC\,/\,dof_{A*C}$ =0.735/4	0.1568
	$MSAD = SSAD\,/\,dof_{A*D}$	$MSAD = SSAD\,/\,dof_{A*D}$ =0.627/4	0.2042
Fisher ratio	$F_A = MSA\,/\,MSerror$	39.8619371/0.2042	195.21027
	$F_B = MSB\,/\,MSerror$	99.7170948/0.2042	488.330533
	$F_C = MSC\,/\,MSerror$	8.72838435/0.2042	42.7442916
	$F_D = MSD\,/\,MSerror$	5.90717798/0.2042	28.9283937
	$F_{AB} = MSAB\,/\,MSerror$	5.90717798/0.2042	11.49
	$F_{AC} = MSAC\,/\,MSerror$	0.1568/0.2042	0.90
	$F_{AD} = MSAD\,/\,MSerror$	0.2042/0.2042	0.77
Percentage contribution (%) : (sum of square/total sum of square)*100	$PC_A = SSA\,/\,SST$	79.7238743/319.862	0.249244%
	$PC_B = SSB\,/\,SST$	199.43419/319.8628	0.6234%
	$PC_C = SSC\,/\,SST$	17.4567687/319.8628	0.054554%
	$PC_D = SSD\,/\,SST$	11.814356/319.8628	0.0369357%
	$PC_{AB} = SSAB\,/\,SST$	9.386/319.8628	0.005%
	$PC_{AC} = SSAC\,/\,SST$	0.735 /319.8628	0.115 %
	$PC_{AD} = SSAD\,/\,SST$	0.627/319.8628	0.033%

Source: Author

Table 94.6 ANOVA for mean of permeate

Source	DoF	Seq SS	Adj MS	F	P
OP	2	17638.6	8819.3	393.48	0.000
PH	2	59412.8	29706.4	1325.39	0.000
ORP	2	2987.4	1493.7	66.64	0.000
ASA	2	1721.8	860.9	38.41	0.000
OP+PH	4	96.5	24.1	1.08	0.445
OP+ORP	4	1.9	0.5	0.02	0.999
OP*ASA	4	9.7	2.4	0.11	0.975
Residual Error	6	134.5	22.4		
Total	26	82003.2			

Source: Author

Table 94.7 Process parameter level at optimum level of permeate

Parameters	Code	Levels		
		1	2	3
OP	A	35	40	45
PH	B	5.9	6.4	6.9
ORP	C	-30	-180	-330
Anti-Agent	D	0.5	1	1.5

Source: Author

Table 94.8 Average value of response at optimum level of permeate

Process parameters	SR	S/N_SR
Y	272.0614	42.80867
A3	181 177	44.87922
B2	215.6273	46.6204
C1	164.6401	43.99079
D3	157.8624	43.50764

Source: Author

Table 94.9 Confirmatory experiment for permeate

Permeate (M3/DAY)

Sample	NF1 220C	NF2 27.50C	NF3 320C
1	203.459	223.459	369.137
2	211.038	231.038	362.909
3	205.914	225.914	396.384
Avg	206.803667	226.8037	376.1433
Total Avg	269.9168889		

Source: Author

$\mu SR = 149.0818 + (181.177 - 149.0818) + (215.6273 - 149.0818) + (164.6401 - 149.0818) + (157.8624 - 149.0818) = 272.724$

Error variance $Ve = 22.4$ (Table 4)

$CI = \pm 6.091$

CI at 95% is: $[\mu SR - CI] < \mu SR < [\mu SR + CI]$

$266.633 < \mathbf{272.724} < 278.815$

Error variance Ve =22.4 (Table 7)

CI =+-6.091

CI of population at 95% is: [µSR– CI] <µSR< [µSR+ CI]

266.633<**272.724**<278.815

Conclusion

The optimal setting of input parameters like potential of hydrogen, oxidation reduction potential and Anti Scaling Agent (ASA) for permeate have been done and test trials with the designed L27 orthogonal array using Taguchi optimisation method have been conducted. The results of the experiments for selected process parameters were then feeded to MiniTab-14 software where the calculation of S/N ratio (For Larger the Better) shows that potential of hydrogen (PH) is the most significant parameter that contribute about 62% in affecting the value of permeate. Based on the graphs of S/N ratio for selected parameter settings OP at level 3 (A3 = 45), PH at level 2 (B2= 6.4), oxidation reduction potential (ORP) at level1 (C1 =-30) and ANTI AGENT at level 3 (D3 = 1.5) have been selected and confirmatory experiment have been performed considering temperature as a noise factor results and calculations of which were deeply described in former part of the work. ANOVA revealed that potential of hydrogen is the most significant reverse osmosis (RO) process parameters in affecting permeate which is the main target process for optimal result followed by operating pressure (OP), ORP, and ASA. The plotted data of results in main effect plot of S/N ratio and response table shows that PH contribute 62% followed by 14% OP, ORP with 13% and ASA with 14%. The obtained results are then validated by conducting the confirmatory test for the optimal levels of combinations of the input parameters.

References

[1] Krishnamoorthy, S., Premalatha, and M., Vijayasekaran, M. (2017). Characterization of distillery wastewater – an approach to retrofit existing effluent treatment plant operation with phycoremediation. Journal of Cleaner Production. 148, 735–50. doi:10.1016/j.jclepro.2017.02.045

[2] https://amritt.com/industries/india-consumer-packaged-goods-market/alcohol-industry-in-india/

[3] Shubhangi, G. and Padmanabh, G. (2022). Optimizing processing parameters of stone crushers through Taguchi method. 65, 3512-18. https://doi.org/10.1016/j.matpr.2022.06.088

[4] Foster, W.T. (2000). Basic Taguchi design of experiments. National Association of Industrial Technology Conference, Pittsburgh, PA.

[5] Hung, L. and Shingjiang, J. L. (2014). Effect of operating condition on reverse osmosis performance for high salinity wastewater.

[6] Hoang, T., Stevens, G.W., and Kentish, S. E. (2012). The influence of feed pH on the performance of a reverse osmosis membrane during alginate fouling. Desalination and Water Treatment 50, 220–25.

[7] L. Y. Dudley, L. Y. and J. S. Baker, J. S. (). The Role of antiscalants and cleaningchemicals to control membrane fouling by permacare.

[8] Al-Saleh, S. A. and Khan, A. R. (1994). Comparative study of two anti-scale agents Belgard EVN and Belgard EV 2000 in multi-stage flashdistillation plants in Kuwait. Desalination. 97, 97-107.

[9] Introduction to Taguchi Method. https://www.ee.iitb.ac.in.

Chapter 95

Experimental investigation of shear strength of adhesively bonded lap joint for dissimilar material and its validation using Taguchi and ANN approach

Pratik Lande[1,a], Neeraj Sunheriya[2,b], Rajkumar Chadge[1,c], Jayant Giri[3,d], Chetan Mahatme[1,e] and Akshaykumar V Kutty[3,f]

[1]Yeshwantrao Chavan College of Engineering Nagpur, Maharashtra, India

[2]National Institute of Technology, Raipur, India

[3]New York University, United States, NY

Abstract

Adhesively bonded joints offer a great advantage in joining complex structures or dissimilar materials. The failure of adhesively bonded joints is largely dependent upon the length of lap, adhesive thickness, and surface preparation. The substrate of dissimilar material having a single lap joint was made of aluminum 8011, ABS (Acrylonitrile Butadiene Styrene), and Nylon. An epoxy-based adhesive Huntsman Araldite AW 106 and hardener HV 953 was used. The rectangular specimen of 100 × 25 × 5 mm was used for tensile lap shear testing. Taguchi L9 experimental matrix array representing different control factors are employed to know the failure of the adhesive-bonded single lap joint. Levenberg-Marquardt algorithm trained neural network was used to predict the shear failure. It is observed that the overlap length of the joint affects the joint shear strength. For Al-ABS & Al-Nylon adhesive joint 20 mm overlap length gives better shear strength whereas for ABS-Nylon 25 mm overlap length gives better shear strength. From the MINITAB software, it is observed that the most affecting parameter on the shear strength of the joint was surface roughness on the overlap area of the substrates.

Keywords: Adhesive bond, dissimilar material, lap joint, Nylon, shear strength

Introduction

Industries such as automobiles, aerospace, and shipbuilding are constantly seeking and investigating new materials that provide lighter and stronger structures that allow higher fuel efficiency. This growing need increases the requirement to manufacture lightweight structures, for example body of the aircrafts and body frames of the vehicle etc.

The wide use of such materials led to the development of innovative techniques to join similar or dissimilar materials. The adhesive bond joint is one such modern way of joining similar or dissimilar material which has almost all the advantages of the conventional joining process. It is lighter in weight, waterproof, galvanic corrosion-free and even stress distribution around the joint is minimal. Adhesive joints provide several benefits over other traditional joining methods, particularly when joining different materials (e.g., diffusion bonding, riveting, welding, bolting, etc). The structural reliability of adhesive joints largely depends upon the joint geometry, mechanical properties of adherent and adhesive, bonding surfaces and loading condition. Stress analysis provides the mechanical performance characteristics of the adhesive joints.

For a single lap, joint optimum use of design parameters increases the joint strength. The failure load increases considerably as the overlap length of the joint grows, but after a certain point, the increase in overlap length has little effect on the strength of the joints. Increased overlap length enhances shear stiffness while decreasing longitudinal stiffness. According to the technical data sheet of adhesives, the standard proportion of resin and hardener by volume is 1:1. The Taguchi method is one of the best methods used to calculate the design of an experiment to optimize the number of experiments which need to be carried out. Further, this paper includes Taguchi Method to optimize the experimentation and ANN to validate the obtained results.

Literature Review

A vast number of researchers have investigated adhesively bonded joints as well as the materials used in them. Some researchers have conducted experiments on the strength of such joints, such as [1], who evaluated the adhesively bonded joints between two different materials and checked the influence of thickness of bond on the fractural characteristics of an interfacial fracture. They used ANSYS software to numerically model the problem. They discovered that increasing the bond thickness of a joint reduces the strength of

[a]pratiklande@gmail.com, [b]neeraj.sunheriya@gmail.com, [c]rbchadge@rediffmail.com, [d]jayantpgiri@gmail.com, [e]sachu_gm@yahoo.com, [f]avk322@nyu.edu

DOI: 10.1201/9781003450252-95

adhesively bonded joints without fractures. da Costa Mattos et al. checked the impacts of load rate on the single lap adhesively bonded joints between ASTM A36 steel plates under two distinct conditions: strain controlled quasi static rupture test and oscillating condition [5]. In quasi-static tensile testing, they discovered that these joints exhibit quasi-brittle behavior with fixed stiffness and little inelastic deformation. Structural adhesive joints are also evaluated by Arenas et al. and Reina et al under different non-favorable conditions like motor oil and water immersion [3, 8]. They examined joint deterioration (mechanical characteristics) as a result of ageing. They came to the conclusion that water degrades the adhesive faster than motor oil. The impact of water and engine oil on the mechanical properties of polyurethane adhesive joints are minimum.

Anyfantis, used FEA to analyses composite-to-metal joints utilising mixed mode law to define elasto-plastic loading and fracture response of the adhesive layer [2]. They found that the cohesive characteristics of a joint varied from the adhesive properties. They also found that the DZT and EPZ techniques underestimate experimental failure loads for DLJs involving thick metal substrates while overestimating experimental failure loads for DLJs involving thin metal substrates. The parametric analysis produced the well-known failure load plateau as the overlap length increased. DLJs with thin metal substrates outperform joints with thick metal substrates by giving 18% higher failure loads. Hazimeh et al. used FEA to investigate double lap adhesively bonded joints under impact stress [6]. They investigated the geometrical and material impacts of dynamic in-plane loadings on composite double lap joints (DLJ). They found that the stress is caused by structural or wave propagation factors. Time has little effect on structural or geometrical heterogeneity. Its contribution remains constant from the quasi-static to the impact case. They also found that increasing the adhesive shear stiffness and substrate longitudinal stiffness improves the average shear stress, but increasing the adhesive and substrate shear stiffness increases the stress heterogeneity.

A study by He, reviewed numerical simulations of several adhesively bonded joints [7]. He observed that an appropriate FEA model of an adhesive bonded joint must be capable of predicting failure in the adhesive and at the adhesive– adherent contacts, as well as accounting for complete nonlinear material behavior. It is also critical to compare the anticipated dynamic behavior of adhesive bonding structures from FEA to experimental test findings. A new approach for optimization of process parameter in improving surface roughness using Taguchi method is suggested by Athreya and Venkatesh [4]. They illustrated the procedure using Taguchi Method in facing operation of lathe machine. Variance analysis, S/N ratio and orthogonal array are the main key factors to check the performance parameters of facing operation.

Objectives of the study and its novelty

The present work focuses especially on the adhesively bonded joints and its analysis. There are various methods available for joining the dissimilar materials but some suitable methods are selected. Many of the researchers have done a number of researches in joining of similar and dissimilar materials. Some researchers also done investigations on grinding and joining operations by Taguchi and ANN approach. In the present study experimentation and analysis is covered for such type of joints.

The main objectives of study are:

- Optimisation of different control factors in adhesively bonded lap joints in dissimilar materials.
- Comparison of strength of adhesively bonded lap joints in different types of dissimilar materials.

Materials and Methods

The lap shear strength of a single lap adhesively bonded joint is investigated in this work. ASTM D1002 is the standard used for sample preparation by the American Society for Material and Testing. The universal testing machine was used to perform the lap shear test. The UTM results are verified using an artificial neural network.

Substrate and adhesive material

Aluminum 8011, ABS (Acrylonitrile Butadiene Styrene), and nylon were the substrates of the different materials of the single lap joint. Materials used for this investigation were adequately described to understand their behavior and analyze the experimental data obtained in this work. The stated strength of aluminum 8011 is 180MPa, which is obtained artificially. Previously, this aluminum 8011 was characterized using the ASTM- D1002 specimen standard. ABS has a tensile strength of 40 MPa. The nylon material utilized has a strength of 60 MPa. Tensile lap shear testing was performed on rectangular specimens with dimensions of $100 \times 25 \times 5$ mm3. To conduct tensile lap shear stress, the specimens were bonded for a single lap joint. Huntsman Araldite Resin AW106 and Hardener HV 953 IN were used as the adhesive for this experiment. This adhesive is epoxy based adhesive. Properties of adhesive are as follows:

- Mix ratio 1:1 by volume
- Curing time 10 -12 Hrs at normal temp.
- Yield strength 36.3 MPa
- Poisson's ratio 0.37
- Young's modules 1.8 GPa

Design of experiments

The single-lap joint (SLJ) is the simplest structure used to predict the strength and stability of adhesively bonded joints. As a result, it is the most studied in literature. However, the manufacturing of a bonded joint is influenced by numerous factors that complicate its design, including substrate surface preparation, geometrical parameters, and the proportion of resin and hardeners. Experiment design employing Full factorial design necessitates a greater number of trials. If the number of parameters increases, it becomes highly difficult and time-consuming. Taguchi proposed an especially calculated approach called the use of an orthogonal array to examine the whole parameter space through a limited number of trials to be carried out to avoid this issue. As a result, Taguchi proposes using the loss function to quantify the performance characteristics that deviate from the preferred goal value. This loss function's value is then modified as the signal-to-noise (S/N) ratio. To assess the S/N ratio, three categories of performance characteristics are often used. They are as follows: Smaller-is-better, Larger-is-best, and nominal-is-better. The single-lap joint (SLJ) is used to predict the strength and stability of adhesively bonded joints.

larger-the-better is used for the objective function and S/N ratio is as:

$$\eta = -10 \log 10 \left(\frac{1}{n} = \sum_{i=1}^{n} \frac{1}{y2} \right)$$

The parameters and their values were selected for experimenting, based on a literature review, and also taking into account the guidelines provided in the Adhesive manufacturer's handbook. Table 95.1 shows the variables and their levels.

Surface roughness for each paper grade is different for each material which is given as Table 95.2.

The degrees of freedom must be determined to pick an appropriate orthogonal array to accompany the tests. The following is an example:

D.O.F. = 1 - Mean Value

Total number of D.O.F. is taken as 9.

The L9 orthogonal array, as indicated in Table 95.3, is the best choice for research. As a result, nine experiments have been carried out.

Sample preparation

The samples were prepared using the settings provided in Figure 95.1 and computed using orthogonal array Table 95.3. De-greasing, abrasion, and cleaning are the three primary stages. The surface preparation is defined by this sequence. Degreasing was accomplished using a solvent-based technique. Acetone, a widely used organic solvent in industry, was employed as the organic solvent.

Table 95.1 Different levels and factors

Factors	Levels		
	A	B	C
Surface roughness	PG 1	PG 2	PG 3
Length of lap	20 mm	25 mm	30 mm
Resin and hardener ratio	1:1	2:3	3:2

Source: Author

Table 95.2 Surface roughness of adherent with paper grade

Material	PG0	PG60	PG80
Aluminum	0.02	1.4	3.9
ABS	0.04	6.2	4.9
Nylon	0.68	4.50	5.24

Source: Author

Table 95.3 Relationship between experiment number and control factors (Orthogonal Array)

Experiment No.	Relative Control Factor		
	P	Q	R
1	P	P	P
2	P	Q	Q
3	P	R	R
4	Q	P	R
5	Q	Q	P
6	Q	R	Q
7	R	P	Q
8	R	Q	R
9	R	R	P

Source: Author

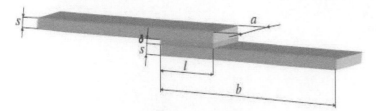

Figure 95.1 Schematic of the single lap joint
Source: Author

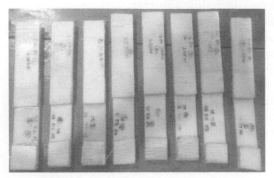

Figure 95.2 ABS-nylon joints
Source: Author

Figure 95.3 Aluminum-nylon joints
Source: Author

a = 25 mm, b = 100 mm, s = 5 mm, l = 20, 25, 30 mm

To clean the surfaces of each adhering surface exposed to the bonding region, lint-free cloths soaked in the organic solvent were employed. The treated surfaces were then left for one minute to allow the solvent to completely evaporate. The surfaces were then cleaned with a clean hot water rinse. The form of the poured water on the surfaces shows a clean, grease-free surface, whereas a drop-shape suggests an unevenly degreased surface. A drier with a stream of hot air to the surface was used to dry the fabric. After degreasing, mechanical abrasion was carried out manually by sanding the substrate surfaces involved in the bonding junction using sandpapers. Each work piece was abraded until no sign of surface shine was evident. Three distinct roughness are obtained by using one as the manufacturer's roughness and the other two using two Sandpaper grit sizes P60 and P80. It should be noted that a lower grit number produces a rougher surface preparation. As shown in Table 95.2, the use of different grit sizes of sandpaper allows for the study of the influence of surface roughness on joint resistance. The surface roughness of each treated surface was assessed using a surface roughness tester with different mesh sandpaper sizes. Table 95.2 shows the average roughness measured from three places at 2.5X magnification. A sand scratch direction perpendicular to the stress applied in the lap shear testing on a single lap joint was chosen to execute the abrasion treatment. ABS-nylon joints sample is shown in Figure 95.2 while aluminum-nylon joints sample is shown in Figure 95.3.

Validations of results

Prepared joints were tested on universal testing machine for lap shear. The method of tensile testing of the rod is used to test the lap shear test of the single lap joint. An ASI 40 tone load frame electromechanical universal testing equipment was used to determine the static shear strength. Single lap joint held by the automatic self-adjusting V-clamp. Once the machine is started it begins to apply load on a given single lap joint.

These values shown in Table 95.4 are used for Taguchi optimisation i.e, to calculate S/N ratio for prediction of optimum parameters for different materials.

Validation of measured values of AL-ABS SLJ using ANN

Measured tensile shear strength values by using universal testing machine are validated by using the artificial neural network using the MATLAB Software. The inputs of the configuration are overlap length, ratio of resin and hardener and surface roughness and output is shear strength. Topologies for every single lap joint are calculated and the best of these is considered. Levenberg-Marquardt (trainlm) is the training algorithm was used. For the above joint measured data values of regression and mean square error are given in Table 95.5.

The value of R is greater for 3-4-1 among all the above topologies. The best curve fitting is given by this topology. For this topology 3-4-1 ANN results are given below in Figures 95.4 and 95.5.

The blue bars, green bars and red bars correspond to training data, validation data, and testing data respectively (Figures 95.6 and 95.7. The histogram can -show you outliers, which are data points where the fit is considerably worse than the rest of the data. In this example, we can observe that most inaccuracies are between -0.919 and 0.67. The regression plot depicts the connection between target and the network output. In each graph, the dashed line indicates the ideal outcome - outputs = goals. The solid line is the best-fit linear regression line between the outputs and the objectives. The solid line is the best-fit linear regression line between the outputs and the objectives.

Table 95.4 Measured tensile shear strength

Experiment No.	Mean tensile shear stress (MPa)		
	Aluminum-ABS	ABS-Nylon	Aluminum-Nylon
1	5.12	1.64	4.71
2	7.70	2.65	7.26
3	6.50	2.45	6.91
4	7.04	2.56	6.17
5	4.29	1.80	4.88
6	6.62	2.95	6.14
7	5.49	2.09	5.29
8	5.06	2.58	6.19
9	3.40	1.93	5.00

Source: Author

Table 95.5 Regression and mean square error for AL-ABS

Topology	MSE	R-value	R_{Total}
3-4-1	$2.3239 e^{-1}$	$0.387 e^{-1}$	0.9552
	$3.701 e^{-2}$	$9.9190 e^{-1}$	
	$6.371 e^{-2}$	$9.760 e^{-1}$	
3-3-1	$1.739 e^{-1}$	$9.508 e^{-1}$	0.9481
	$1.860 e^{-1}$	$9.699 e^{-1}$	
	$9.196 e^{-2}$	$9.9846 e^{-1}$	
3-2-1	$2.108 e^{-1}$	$9.367 e^{-1}$	0.9551
	$2.113 e^{-1}$	$9.822 e^{-1}$	
	$1.835 e^{-1}$	$9.873 e^{-1}$	
3-1-1	$3.239 e^{-1}$	$9.101 e^{-1}$	0.8963
	$1.112 e^{-1}$	$9.940 e^{-1}$	
	$2.739 e^{-1}$	$9.6851 e^{-1}$	

Source: Author

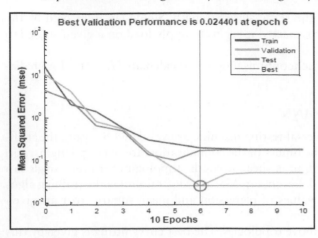

Figure 95.4 Graph of MSE vs Epochs for AL-ABS
Source: Author

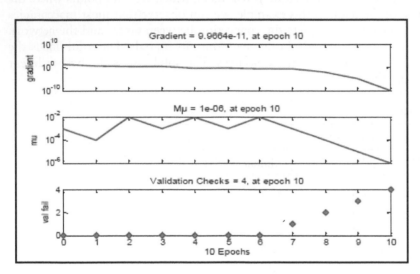

Figure 95.5 Graph of gradient and validation check for AL-ABS
Source: Author

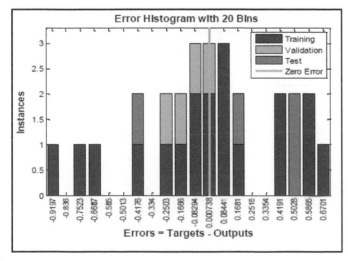

Figure 95.6 Error histogram with 20 Bins for AL-ABS
Source: Author

If R = 1, it means that the outputs and targets have a perfect linear connection. There is no linear relationship between outputs and targets if R is 0. For this R = 0.9552 means the precise relationship between outputs and targets.

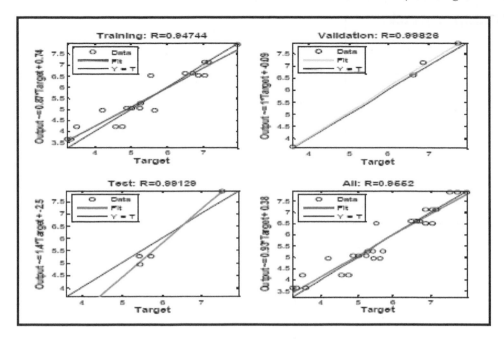

Figure 95.7 Regression plot for AL-ABS
Source: Author

Table 95.6 Regression and mean square error for ABS-Nylon

Topology	MSE	R-value	RTotal
3-4-1	1.983 e^{-2}	9.677 e^{-1}	0.8994
	1.586 e^{-2}	9.833 e^{-1}	
	3.1223 e^{-1}	8.853 e^{-1}	
3-3-1	1.686 e^{-2}	9.558 e^{-1}	0.8794
	1.438 e^{-1}	9.506 e^{-1}	
	1.206 e^{-1}	9.829 e^{-1}	
3-2-1	4.3941 e^{-2}	8.833 e^{-1}	0.8831
	1.862 e^{-2}	9.9865 e^{-1}	
	1.873 e^{-2}	9.641 e^{-1}	
3-1-1	5.422 e^{-2}	8.551 e^{-1}	0.8472
	1.059 e^{-2}	9.737 e^{-1}	
	1.3991 e^{-1}	8.556 e^{-1}	

Source: Author

Validation of measured values of ABS-Nylon SLJ using ANN

For ABS-nylon joint measured data values of regression and mean square error are given below in Table 95.6:

For topology 3-4-1 regression value 0.8994 is higher. Therefore, the relationship between output and target for this topology is taken into consideration and ANN results are as follows in Figures 95.8, 95.9, 95.10 and 95.12.

Validation of measured values of AL-nylon SLJ using ANN

For Al-Nylon joint measured data values of regression and mean square error are given below in Table 95.7. Also, the relationship between output and target for this topology is taken into consideration and ANN results are as follows in Figures 95.12-95.15.

Results and Discussions

After experimentation and calculations the following results are obtained for different types of joints.

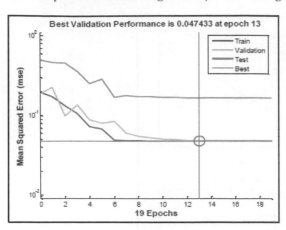

Figure 95.8 Graph of MSE vs Epochs for ABS-Nylon
Source: Author

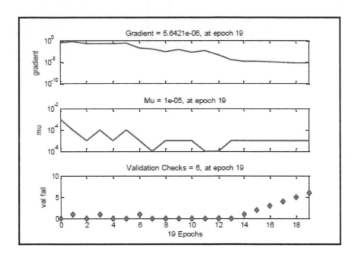

Figure 95.9 Graph of gradient and validation check for ABS-Nylon
Source: Author

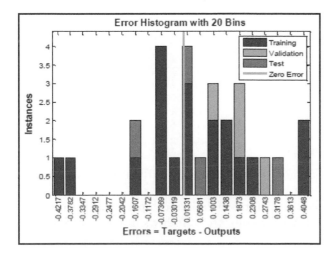

Figure 95.10 Error histogram with 20 Bins for ABS-Nylon
Source: Author

ANN Results for AL-ABS materials adhesive single lap joint

For this configuration signal to noise ratio, Delta value, Rank of the parameters and results are as follows:

Figure 95.16 shows the Main Effect plot, it is observed that maximum shear strength is found at 20 mm at which the S/N ratio is greater than the other two levels which are 16.07db. As the increase in the overlap length means of the S/N ratio decreases which means that the shear strength decreases with an increase

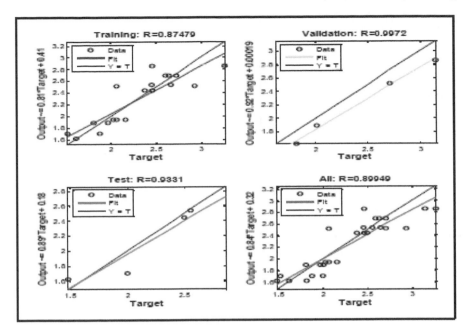

Figure 95.11   Regression plot for ABS-Nylon
Source: Author

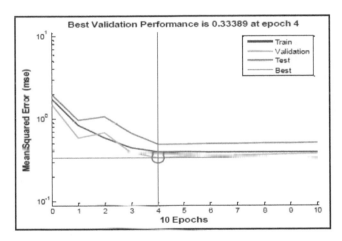

Figure 95.12   Graph of MSE vs Epochs for Al-nylon
Source: Author

Table 95.7   Regression and mean square error for Al-Nylon

Topology	MSE	R-value	RTotal
3-4-1	$1.154\ e^{-1}$	$9.507\ e^{-1}$	0.8114
	$1.455\ e^{-0}$	$8.506\ e^{-1}$	
	$1.022\ e^{-0}$	$9.085\ e^{-1}$	
3-3-1	$3.81\ e^{-1}$	$8.176\ e^{-1}$	0.8077
	$4.085\ e^{-1}$	$9.760\ e^{-1}$	
	$4.6739\ e^{-1}$	$8.459\ e^{-1}$	
3-2-1	$4.361\ e^{-1}$	$8.110\ e^{-1}$	0.8071
	$5.020\ e^{-1}$	$9.122\ e^{-1}$	
	$3.072\ e^{-1}$	$8.016\ e^{-1}$	
3-1-1	$3.633\ e^{-1}$	$7.8695\ e^{-1}$	0.8033
	$4.589\ e^{-1}$	$9.141\ e^{-1}$	
	$4.769\ e^{-1}$	$7.932\ e^{-1}$	

Source: Author

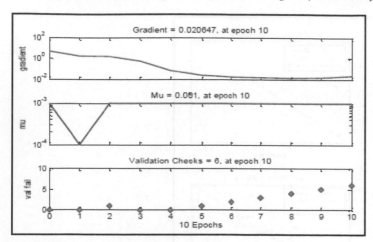

Figure 95.13 Graph of gradient and validation check for Al-Nylon
Source: Author

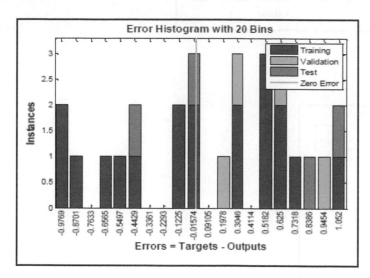

Figure 95.14 Error histogram with 20 bins for Al-nylon
Source: Author

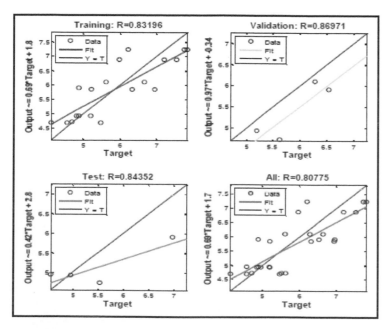

Figure 95.15 Regression plot for Al-nylon
Source: Author

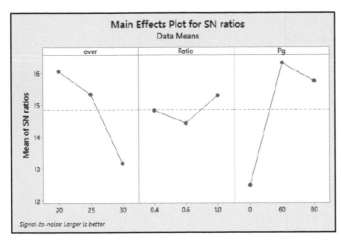

Figure 95.16 Main effect plot for AL-ABS adhesive SLJ
Source: Author

Table 95.8 Response for signal to noise ratios

Level	Overlap length	Ratio	PG
1	16.07	14.83	12.49
2	15.34	14.44	16.33
3	13.17	15.31	15.76
Delta	2.90	0.87	3.83
Rank	2	3	1

Source: Author

Table 95.9 Response for signal to noise ratios ABS Nylon SLJ

Level	Overlap length	Ratio	PG
1	6.848	7.267	5.038
2	7.555	7.630	8.090
3	6.774	6.280	8.049
Delta	0.781	1.350	3.053
Rank	3	2	1

Source: Author

in overlap length. Similarly, maximum shear strength is found at 1:1 proportion at which the S/N ratio is greater than the other two levels which are 14.83db and 16.33db respectively. As the variation in the proportion from 1:1, it decreases the shear strength for AL-ABS adhesive lap joint. And S/N ratio is also greater for Paper grade PG60 than others. From this plot, it is observed that the S/N ratio is less for PG0 & PG80 which means that for higher surface roughness shear strength is higher.

Table 95.8 show the responses for signal to noise ratios for different overlap lengths.

Results for ABS -nylon materials adhesive single lap joint

Table 95.9 shows the response for the signal-to-noise ratio for each parameter.

From Figure 95.17, maximum shear strength is found at 25 mm overlap length at which S/N ratio is greater than other two levels which are 7.555db. Similarly, maximum shear strength is found at proportion 2:3 and paper grade PG60 at which S/N ratio is greater than the other two levels which are 7.63db & 8.09db respectively. It is also observed that as the Overlap length increases from 20 mm to 25 mm S/N ratio increases while it decreases from 25 mm to 30 mm which means that shear strength increases with the increase in overlap length up to 25 mm, onwards it decreases. For a proportion of resin and hardener 3:2, shear strength is high. For Paper grade PG60 which gives a higher roughness value has higher shear strength.

Results for AL-nylon materials adhesive single lap joint

For this arrangement signal to noise ratio and results from Minitab Software are calculated and these results are shown in Table 95.10.

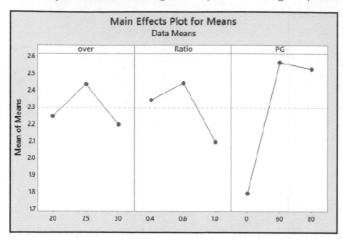

Figure 95.17 Main effect plot for ABS -Nylon adhesive SLJ
Source: Author

Table 95.10 Response for signal to noise ratios AL-Nylon SLJ

Level	Overlap length	Ratio	PG
1	15.82	15.61	13.73
2	15.11	15.51	15.82
3	17.76	14.59	16.14
Delta	1.06	1.03	2.41
Rank	2	3	1

Source: Author

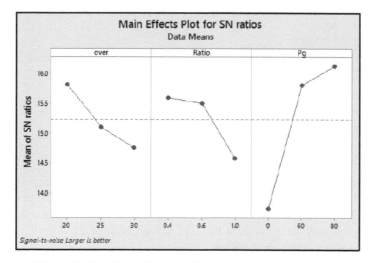

Figure 95.18 Main effect plot for AL-Nylon adhesive SLJ
Source: Author

Figure 95.18 shows that maximum shear strength is found at 20 mm at which the S/N ratio is higher than the other two levels which are 15.82db. Similarly, maximum shear strength is found at 1:1 proportions And Paper grade PG80 at which the S/N ratio is greater than the other two levels which are 15.61db & 16.14db respectively.

Conclusions

Finally, a failure mode analysis was performed to either confirm or find the optimal Lap shear strength. The following conclusions may be made from the results of the tensile Lap shear testing and failure mechanism analysis:

For a given 5 mm adherent thickness, a critical value of surface roughness of Aluminum, ABS, and Nylon was found to be equal to 1.4 μm, 6.2 μm, and 5.24 μm respectively. Surface preparation on the overlap area affects the lap shear strength. The failure mode analysis indicated an increase in the proportion of cohesive failure for fractured surfaces treated with various grit sizes of sandpaper. It demonstrates that surface roughness improves adhesion between adherents and glue. In contrast, a larger percentage of cohesive failure indicates a smaller number of cracks initiating at the adherent/adhesive junction. Overlap length of the joint also affects the joint shear strength. For Al-ABS and Al-Nylon adhesive joint 20 mm overlap length gives better shear strength whereas for ABS-Nylon 25 mm overlap length gives better shear strength. It was discovered that increasing the surface roughness of the adhering from low to high increased the lap joint shear strength. Shear stress concentration at the extremities of the overlap area increases as surface roughness increases. These findings were confirmed for both experiments involving the impact of surface roughness on lap shear strength. As the lap shear strength is better for Al-ABS at 1:1, AL-Nylon at 2:3, and for ABS-Nylon at 3:2. It is concluded that the Lap shear strength is maximum for different joint configurations with different proportions of resin and hardener. The artificial neural network reveals a nearly perfect relationship between the input parameters and the output parameter. Analysing the results obtained from MINITAB, it is observed that the most affecting parameter on the shear strength of the joint was surface roughness on the overlap area of the substrates.

References

[1] Afendi, M. and Teramoto, T. (2009). Effect of bond thickness on fracture behavior of interfacial crack in adhesive joint of dissimilar materials. Journal of the Adhesion Society of Japan. 45(12), 471–7. https://doi.org/10.11618/adhesion.45.471.

[2] Anyfantis, K. N. (2012). Finite element predictions of composite-to-metal bonded joints with ductile adhesive materials. Composite Structures. 94(8), 2632–9. https://doi.org/10.1016/j.compstruct.2012.03.002.

[3] Arenas, J. M., Alía, C., Ocaña, R. and Narbón, J. J. (2013). Degradation of adhesive joints for joining composite material with aluminum under immersion in water and motor oil. Procedia Engineering. 63, 287–94. https://doi.org/10.1016/j.proeng.2013.08.218.

[4] Athreya, S. and Venkatesh, Y. D. (2012). Application of Taguchi method for optimization of process parameters in improving the surface roughness of lathe facing operation. International Refereed Journal of Engineering and Science. 1(3), 13–19.

[5] da Costa Mattos, H., da Silva Nunes, L. C., and Monteiro, A. H. (2016). Load rate effects in adhesive single lap joints bonded with epoxy/ceramic composites. Latin American Journal of Solids and Structures. 13(10), 1878–92. https://doi.org/10.1590/1679-78252818.

[6] Hazimeh, R., Challita, G., Khalil, K., and Othman, R. (2015). Finite element analysis of adhesively bonded composite joints subjected to impact loadings. International Journal of Adhesion and Adhesives. 56, 24–31. https://doi.org/10.1016/j.ijadhadh.2011.07.012.

[7] He, X. (2011). A review of finite element analysis of adhesively bonded joints. International Journal of Adhesion and Adhesives. 31(4), 248–64. https://doi.org/10.1016/j.ijadhadh.2011.01.006.

[8] Reina, J. M. A., López, R. O., García, C. A., and Prieto, J. J. N. (2014). Technical evaluation of structural adhesive joints under adverse operation conditions. Materials Science Forum. 797, 169–74. https://doi.org/10.4028/www.scientific.net/MSF.

Chapter 96

Measurement of flank wears using edge detection method

Shrikant Jachak[1,a], Sahil Gaurkar[2,b], Avinash Thakre[3,c], Malhaar Jachak[4,d], Neeraj Sunheriya[1,e] and Prashant Kamble[1,f]

[1]Assistant Professor, Department of Mechanical Engineering, Yeshwantrao Chavan College of Engineering Nagpur, Maharashtra, India

[2]Assistant Professor, University of Glasgow, Scotland, United Kingdom

[3]Assistant Professor, Mechanical Department, Visvesvaraya National Institute of Technology Nagpur, Maharashtra, India

[4]Assistant Professor, Manipal Institute of Technology, Bengaluru, India

Abstract

In machining of parts, the rate at which inserts wear plays a crucial role in terms of economics of the cutting tools and hence understanding the inserts wear is of utmost importance. Turning tool wear monitoring is vital in metal cutting industry in terms of dimensional accuracy, shape deviations and overall surface quality. This paper reports a novel edge detection technique which is based on canny algorithm, for the measurement of flank wear of bonded and brazed tools in turning operation. A Two level four factor full factorial experimentation has been designed to analyse the experimentally observed data. The photograph of the used tools was taken and edge detection technique was used for finding the boundaries within these images. The edge detection algorithm Canny was used for this purpose. Input to the programme was in the form of photograph. The input indexed image was converted to Gray scale. MATLAB 14.0 was used for the measurement of the flank wear of the tools by successfully implementing the canny algorithm. The results of the tool wear obtained from the edge detection technique are experimentally validated with those obtained from the regression models. This study shows that edge detection technique can be effectively used for the wear measurement of tools. This technique can defiantly be used in the manufacturing industries for tool wear monitoring and measurement.

Introduction

Turning operation is one of the most important and widely used subtractive manufacturing methods. In machining of parts, the rate at which inserts wear plays a crucial role in terms of economics of the cutting tools and hence understanding the inserts wear is of utmost importance [1]. Turning tool wear monitoring is vital in metal cutting industry in terms of dimensional accuracy, shape deviations and overall surface quality. Traditional wear measurement technique includes, indirect and direct wear measurement. Direct wear measurement technique applies direct approach of tool wear quantity. Offline technique like optical microscope is used for direct measurement of tool wear. Normally, most of the offline tool wears measurement techniques are time consuming [2]. Cutting tool monitoring review indicates that tool wear condition can be successfully monitored by many direct measurement techniques such as machine vision [3–7]. Deep learning models can be effectively used for image processing provided sufficient data is available for the training purposes. White light interferometry can be used for measuring both types of wear [8–10]. Indirect tool measurement involves measurement of various process parameters which have a correlation with tool wear. Indirect wear measurement includes correlating wear with the changes in the parameters such as temperature, surface finish and force [11]. The surface irregularities due to tool wear can also be a criterion of determination of possible tool wear. Decimated wavelet transform decomposes the surface image of the given work piece to the sub- mages, which ultimately indicates the cutting tool wear [11]. The change in the original boundary indicates wear of the inserts. This worn boundary of the inserts helps find the probability functions which can accurately relate the worn boundary with original boundary with the aid of Bayesian inference [11, 12]. Cutting tools wear can also be measured with the aid of 2D matrices feature in MATLAB software, thus the method helps in monitoring any change in the insert shape [13, 14]. In turning operation, turning tool is subjected to two types of wear: crater wear and flank wear [11]. The severity of the cutting tool can easily be quantified with flank wear as compared to crater wear of the tool. Another tool that can be used for measuring the flank wear is unitary methodology [15]. The intent of this work is to develop a reliable method for predicting the flank wear of the turning tool. In image processing images taken are treated as 2D signals by applying already set signal processing methods. Flank wear of the turning tools can be effectively measured by using edge detection methods. In edge detection technique the boundaries of objects within images are detected. Boundaries of the images are detected by

[a]jshrikant8@gmail.com, [b]sahilgaurkar@gmail.com, [c]aathakare@mec.vnit.ac.in, [d]77msjachak@gmail.com, [e]neeraj.sunheriya@gmail.com, [f]mrpdkamble@gmail.com

DOI: 10.1201/9781003450252-96

detecting discontinuities in the image boundaries due to brightness. Edge detection is specifically used for image segmentation and data extraction in areas such as image processing, computer vision, and machine vision. Many edge detection algorithms are used. The algorithms that can be used are Sobel, Canny, Prewitt, Roberts, and fuzzy logic methods. Canny algorithm which is rarely used earlier has been used for wear measurement in this experimentation.

Methodology

Today, image processing system is the rapidly growing technologies, with its applications in many areas in the field of engineering and computer sciences. The methodology adopted in this work is explained here.

Experimental planning

The aim of the study was to measure the insert wear during the turning operation. Trials were conducted on a HMT TL 20 lathe. Aluminium 6061workpieces was used. Aluminium 6061t has good mechanical properties, good weldability and is used for general-purpose machining. The dimensions of aluminium 6061 alloy work pieces include 250 mm in length with an external diameter of 50 mm. The shank material was mild steel with a cross section 16 mm square and having triangular shaped carbide inserts. Carbide insert used are of 16 mm side with internal sides of 14 mm. The inserts TPKN2204P30 was used and has got the specification as: inscribed circle (IC) of 9525 mm, Insert thickness of 3 mm and nose radius of 0.8 mm. The cutting-edge condition is of negative land and shape of the selected inserts was equilateral triangle. The tool geometry designated was of side rake angle – 5° and back rake angle of 5° The end clearance angles of 5° with Side clearance angle of 5° and side or end cutting edge angle was of 15°. The turning process parameters were selected by a comprehensive literature survey. Pilot experiments were conducted for selecting the appropriate levels of cutting parameters like depth of cut, cutting speed, feed rate and the coolant flow rate required in this experimentation [16, 17]. The selected cutting conditions for performing experimentation are specified in the Table 96.1.

Experimental design

The experimental design was planned by using design of experimentation (DOE) [15]. The methodology has been used to analyse the experimentally observed data. A standard 2^4 full factorial experimentation was designed. The independent parameters viz. cutting speed, Feed rate, depth of cut (X3) and coolant flow rate were selected. The conventional brazed and bonded tools were used for this experimentation and measurement of wear of these tools was taken into consideration. The edge detection technique used for tool wear measurement has been explained in subsequent topics.

Tool wear measurement

Here, wear measurement of the tools used in this experimentation was carried out through image processing. It is a method of converting the images of wear into its digital form and then performing some kind of operations on it, for getting an enhanced image. This is basically done for extracting some important information. The method adapted is type of signal dispensation. Here input is a digital image and output is also an image or characteristics that are associated with the input image. In this type of signal processing system, an input image was treated as two dimensional signals. Set signal processing methods was applied to these images. Edge detection technique was used for measurements of the flank wear of the cutting tools. The photograph of the used tools was taken, and edge detection technique was used for finding the boundaries within these images. The edge detection algorithm Canny was used for this purpose. The advantage of using canny algorithm is that it reduces the amount of data to be processed. Input to the programme was in the form of photograph as shown in the Figure 96.1. Gray scale is usually the preferred format for image processing. The input indexed image was converted to Gray scale. MATLAB 14.0 was used for the

Table 96.1 Cutting conditions for Aluminium 6061 work pieces

Input factors	Mean level	Upper level	Lower level
Cutting speed, m/min	150	188	112
Feed rate, mm/rev	0.10	0.15	0.05
Depth of cut, mm	0.475	0.70	0.25
Coolant flow rate, ml/sec	34	48	20

Source: Author

Figure 96.1 Input image of flank wear
Source: Author

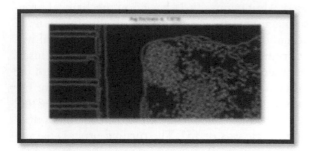

Figure 96.2 Flank wears of turning tools
Source: Author

Table 96.2 Composition of Aluminium 6061 alloys

Si %	Fe%	Mn%	Cu %	Mg %	Cr %	Zi %	Ni %	Ti %	Aluminium
0.40-0.8	0.7	0.15	0.15-0.40	0.8-1.2	0.04-0.35	0.25	-	0.15	Balance

Source: Author

Table 96.3 Flank wear for adhesive bonded tools for Aluminum - 6061

Tool No.	Cutting speed m/min	Feed rate mm/rev	Depth of cut mm	Coolant flow rate ml/sec	Wear (mm) adhesive bonded tools					
					Replication 1			Replication 2		
					1	2	3	1	2	3
1	112	0.05	0.25	20	0.19	0.19	0.18	0.19	0.18	0.19
2	188	0.05	0.25	20	0.24	0.25	0.25	0.23	0.26	0.25
3	112	0.15	0.25	20	0.38	0.34	0.35	0.37	0.34	0.36
4	188	0.15	0.25	20	0.53	0.56	0.53	0.53	0.55	0.54
5	112	0.05	0.7	20	0.48	0.41	0.43	0.44	0.45	0.43
6	188	0.05	0.7	20	0.69	0.66	0.64	0.63	0.68	0.68
7	112	0.15	0.7	20	0.98	0.94	0.97	0.96	0.96	0.97
8	188	0.15	0.7	20	1.66	1.59	1.63	1.65	1.61	1.64
9	112	0.05	0.25	48	0.18	0.19	0.18	0.17	0.19	0.19
10	188	0.05	0.25	48	0.22	0.22	0.23	0.23	0.21	0.23
11	112	0.15	0.25	48	0.33	0.34	0.33	0.35	0.34	0.31
12	188	0.15	0.25	48	0.57	0.49	0.48	0.55	0.49	0.50
13	112	0.05	0.7	48	0.35	0.35	0.36	0.36	0.35	0.37
14	188	0.05	0.7	48	0.54	0.58	0.59	0.55	0.57	0.59
15	112	0.15	0.7	48	0.71	0.63	0.65	0.70	0.64	0.65
16	188	0.15	0.7	48	1.59	1.61	1.56	1.58	1.62	1.56

Source: Author

Table 96.4 Flank wear for brazed tools for Aluminium - 6061

Tool No.	Cutting speed m/min	Feed rate mm/rev	Depth of Cut mm	Coolant flow rate, ml/sec	Wear (mm) brazed tools					
					Replication 1			Replication 2		
					1	2	3	1	2	3
1	112	0.05	0.25	20	0.31	0.28	0.28	0.29	0.27	0.31
2	188	0.05	0.25	20	0.32	0.39	0.41	0.35	0.39	0.38
3	112	0.15	0.25	20	0.39	0.45	0.39	0.39	0.42	0.42
4	188	0.15	0.25	20	0.70	0.73	0.66	0.72	0.71	0.66
5	112	0.05	0.7	20	0.63	0.59	0.61	0.62	0.62	0.62
6	188	0.05	0.7	20	0.79	0.80	0.79	0.80	0.78	0.80
7	112	0.15	0.7	20	1.45	1.60	1.47	1.48	1.58	1.46
8	188	0.15	0.7	20	1.80	1.82	1.78	1.79	1.81	1.80
9	112	0.05	0.25	48	0.26	0.24	0.25	0.24	0.26	0.25
10	188	0.05	0.25	48	0.32	0.33	0.31	0.34	0.33	0.29
11	112	0.15	0.25	48	0.38	0.45	0.37	0.40	0.42	0.38
12	188	0.15	0.25	48	0.58	0.56	0.50	0.57	0.55	0.52
13	112	0.05	0.7	48	0.46	0.47	0.46	0.46	0.46	0.47
14	188	0.05	0.7	48	0.73	0.72	0.78	0.76	0.73	0.74
15	112	0.15	0.7	48	0.95	1.05	1.05	0.98	1.02	1.05
16	188	0.15	0.7	48	1.68	1.62	1.58	1.63	1.62	1.63

Source: Author

Table 96.5 Optimisation of regression model for bonded tools

Regression model for bonded tools

Term	Min	Max	Coefficient	Optimal solution
Constant	-	-	0.5906	-
Cutting speed (m/min)	112	188	0.1557	188
Feed rate (mm/rev)	0.05	0.15	0.2323	0.05
Depth of cut (mm)	0.25	0.7	0.2677	0.25
Coolant flow rate (ml/sec)	20	48	-	-
Tool wear (mm)	0.146 mm			

Source: Author

measurement of the wear. Output being the last stage represented an altered image or final report of the image analysis. The sample results of the measurement of flank wear of turning tools for some trials are shown in the Figure 96.2.

Results

The results of wear measurement by using edge detection technique are presented here. The primary aim of this study was to measure the tool wear during the turning operation. Carbide inserts TPKN2204P30 was used. Aluminium 6061 alloy work pieces having 50 mm diameter were used. Chemical composition of the Aluminium 6061 alloy work pieces are shown in the Table 96.2. Fresh carbide inserts were used for every turning operation in order to ensure similar working conditions. Machining time for all the specimen was around 5 minutes with the input cutting conditions represented in the Tables 96.3 and 96.4. Conventional brazed tools and adhesively bonded tools were used in this experimentation. Total 16 trials on each turning tools with two replications and three repetitions were conducted. The edge detection canny algorithm was used for this purpose. The input indexed image was converted to Gray scale. MATLAB 14.0 was used for the measurement of the wear. The results of measurements of inserts wear for adhesively bonded tools and brazed tools respectively were shown in Tables 96.3 and 96.4. This study shows, application of edge detection for the measurement of cutting tools wear in the turning operations.

Table 96.6 Optimisation of regression model for brazed tools

Regression model for brazed tools				
Term	Min	Max	Coefficient	Optimal solution
Constant	-	-	0.74	-
Cutting speed (m/min)	112	188	0.1212	112
Feed rate (mm/rev)	0.05	0.15	0.2612	0.05
Depth of cut (mm)	0.25	0.7	0.3288	0.25
Coolant flow rate (ml/sec)	20	48	-	-
Tool wear (mm)	0.186 mm			

Source: Author

Table 96.7 Tool wear measurement with edge detection technique

Sr. No.	Wear zone captured by camera	Flank wear zone image	Total wear area of inserts (mm^2)
1			0.942
2			0.961
3			0.312
4			0.354
5			0.716

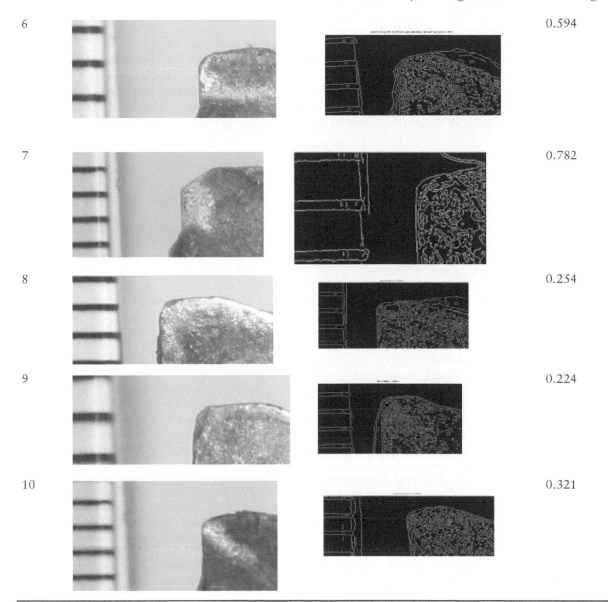

6		0.594
7		0.782
8		0.254
9		0.224
10		0.321

Source: Author

Formulation of regression analysis model

Regression analysis, a statistical and analytical modelling method was used for evaluating the performance of bonded and brazed tools in terms of tool wear. Their performances on tool wear were studied by varying cutting speed, feed rate, and depth of cut and coolant flow rate. The observed optimal values of tool wear calculated from model were compared with the measured values of tool wear. Tables 96.5 and 96.6 shows the regression analysis models of tool wear for bonded tools and brazed tools.

The regression analysis model for the tool wear of bonded tools is:

$$Twbonded = 0.5906 + 0.1557 *X_1 + 0.2323*X_2 + 0.2677*X_3 +0.0882*X_1*X_2 +0.0978 *X_1*X_3 +0.1194*X_2*X_3$$

Where, X_1 = Cutting speed (m/min)

$\quad\quad X_2$ = Feed rate (mm/rev)

$\quad\quad X_3$ = Depth of cut (mm)

$\quad\quad X_4$ = Coolant flow rate (ml/sec)

$\quad\quad$ Twbonded = Tool wear of adhesive bonded tools (mm)

The coefficient of correlation [R] between the flank tool wear observed and predicted by the regression model is 0.982 with R^2 = 96.55 and $R^2_{adjusted}$ = 94.25

Also the regression analysis model for the tool wear of brazed tools (Twbrazed) is:

$$Twbrazed = 0.74 + 0.1212 *X_1 + 0.2612*X_2 + 0.3288*X_3 + 0.1575*X_2*X_3$$

Here the coefficient of correlation [R] between the flank tool wear observed and predicted by the regression model is 0.977 with R^2 = 95.54 and $R^2_{adjusted}$ = 93.32. The results indicate better correlation between the experimental and predicted values of the flank wear of these tools.

Conclusions

In the wear measurement of the bonded tools and conventional brazed tools automatic calibration system was implemented successfully. Total cutting tools wear area of inserts (mm^2) is shown in the Table 96.7. Most of the measured values of flank wear for the bonded tools and brazed tools are found have close agreement with the results obtained from the regression models. This study shows that edge detection technique can be effectively used for the wear measurement. It also helps in understanding the overall condition of the tools. This technique can defiantly be used in the manufacturing industries for tool wear monitoring and measurement.

References

[1] Choudhury, S. K. and Kishore, K.K. (2000). Tool wear measurement in turning using force ratio. International Journal of Machine Tools & Manufacture. 40, 899–909.

[2] Bagga, P. J., Makhesana, M. A., Patel, K., and Patel K. M. (2021). Tool wear monitoring in turning using image processing techniques. Materials Today Proceedings. 44(Part 1), 771–5.

[3] Sortino, M. (2003). Application of statistical filtering for optical detection of tool wear. International Journal of Machine Tools and Manufacture. 43(5), 493–7.

[4] Castejo'n, M., Alegre, E., Barreiro, J. and Herna'ndez, L. K. (2007). On- line tool wear monitoring using geometric descriptors from digital images. International Journal of Machine Tools and Manufacture. 47(12-13), 1847–53.

[5] Danesh, M. and Khalili, K. (2015). Determination of tool wear in turning process using undecimated wavelet transform and textural features. Procedia Technology. 19, 98–105.

[6] Yu, X., Lin, X., Dai, Y., and Zhu, K. (2017). Image edge detection based tool condition monitoring with morphological component analysis. ISA Transactions. 69, 315–22.

[7] D'Addona, D. M. and Teti, R. (2013). Image data processing via neural networks for tool wear prediction. Procedia CIRP. 12, 252–7.

[8] Dawson, T. G. and Kurfess, T. R. (2005). Quantification of tool wear using white light interferometry and three-dimensional computational metrology. International Journal of Machine Tools and Manufacture. 45(4-5), 591–6.

[9] Devillez, A., Lesko, S. and Mozer, W. (2004). Cutting tool crater wear measurement with white light interferometry. Wear. 256(1-2), 56–65.

[10] Xiong, G., Liu, J. and Avila, A. (2011). Cutting tool wear measure- ment by using active contour model based image processing. In Proceedings of 2011 International Conference on Mecha- tronics and Automation (ICMA), (pp. 670–675), IEEE, Beijing, China, August 2011.

[11] Li, Y., Mou, W., Li, J., Liu, C. and Gao, J. (2021). An automatic and accurate method for tool wear inspection using grayscale image probability algorithm based on Bayesian Inference. Robotics and computer integrated manufacturing.

[12] Xiong, G., Liu, J. and Avila, A. Cutting tool wear measure- ment by using active contour model based image processing. In Proceedings of 2011 International Conference on Mechatronics and Automation (ICMA), (pp. 670–675), IEEE, Beijing, China, August 2011.

[13] Danesh, M. and Khalili, K. (2015). Determination of tool wear in turning process using undecimated wavelet transform and textural features. Procedia Technology. 19, 98–105.

[14] Arévalo-Ruedas, J. H. (2015). Wear analysis in cutting tools by the technique of image processing with the application of two dimensional matrices. Journal of Physics, 19, 98–105.

[15] Daicu, R. and Oancea, G. (2022). Methodology for measuring the cutting inserts wear. MDPI. 14, 469.

[16] Kamble, P. D., Waghmare, A. C. and Askhedkar, R. D. (2021). Application of hybrid Taguchi-Grey relational analysis (HTGRA) multi-optimization technique to minimize surface roughness and tool wear in turning AISI4340 steel. Journal of Physics: Conference Series 1913 (1), 012142

[17] Meng, Q., Arsecularatne, J. A., and Matthew, P. (2000). Calculation of optimum cutting conditions for turning operations using a Machining theory. International Journal of Machine Tools and Manufacturing. 40, 1709–33.

Chapter 97

Design of energy storage and transfer system for solar cooking

Tharesh Gawande[a] and Dilip Ingole[b]

Professor, Department of Mechanical Engineering, Ram Meghe Institute of Technology & Research, Badnera, India

Abstract

Continuous availability of solar energy is very essential for solar energy appliances, and to get the solar energy after the sunshine, it requires to be stored and transferred to the appliance. A lot of research is going on the use of solar energy for cooking, but still, some extensive techniques need to be developed for easily usable systems which will be versatile for all conditions. This paper presents the design of heat storage and transfer systems for solar cooking with energy storage and transfer system. In this paper, the key aspects like the methodology to develop the heat storage system, materials used in these systems, and design procedure has been presented. Also, the importance of heat transfer rate, cooking rate, and heat loss rate has been discussed. The effect of temperature on the energy storage system and its effect on cooking have been discussed.

Keyword: Design, energy storage system, PCM, solar cooking

Introduction

Solar cookers have been used since the 17th century. A lot of research has been done by various researchers. Most of the cookers available nowadays can cook the food in daytimes/sunshine hours only. This arises the need for an energy storage and transfer system which will allow the cooking after sunshine hours also. Scheffler cookers are widely adopted for communitive cooking as solar cookers in which solar radiations are used to heat the water and convert it into steam. Then the steam is used for cooking. This method only helps in accelerating cooking but doesn't allow all types of cooking. This problem can be solved by using energy storage systems. This system stores the energy, and it is available i.e., at the time of sunshine hours and releases when energy is not available i.e., after sunshine hours. Heat transfer fluid (HTF) is used to carry/transfer energy. This energy storage system may be a sensible storage system or latent heat storage system which depends on the material. In the presented work, phase change material is used for storing the energy along with HTF. This makes it a hybrid system. The energy storing capacity depends on the type of material used like types of PCM, types of heat transfer fluid, and material used for manufacturing of storage system. Also, while designing such a system energy loss is an important factor for which insulation plays a significant role. Also, the shape and size decide the energy storing and retrieving capability of a system, which in turn, decides cooking capacity. This paper deals with the design of such an efficient energy storage system that can be used for cooking food at any time.

Methodology

As mentioned in the previous section, HTF is used to transfer energy, and PCM is used for energy storage along with HTF. Hence, an energy storage system that can easily store and retrieve energy and transfer it to cooking applications with optimum value, has been designed by considering various factors. This system will be versatile for all conditions. The proposed system is shown in the following Figure 97.1. One of the important factors which is heat loss occurs during the heat transfer can be minimised by using proper insulations [1].

For designing this system, the following steps are important-

i.   **Study of previously developed systems**
     By thorough literature survey, one can come across the selection of the energy storage system.
ii. **Selection of the type of shape and geometry of the system**
     This requires knowledge of heat exchangers. Depending on suitability, the type of exchanger and its shape are selected. The shell type (cylindrical) heat exchanger has less heat loss compared to a plate or cuboidal type. Hence, here cylindrical shell and tube structure is used.
iii. **Selection of material**
     Material selection is also one of the main aspects of the designing of a system.
     Material selection should be done for the following component of the system:

---

[a]tharesh01@gmail.com, [b]dsingole@rediffmaiil.com

DOI: 10.1201/9781003450252-97

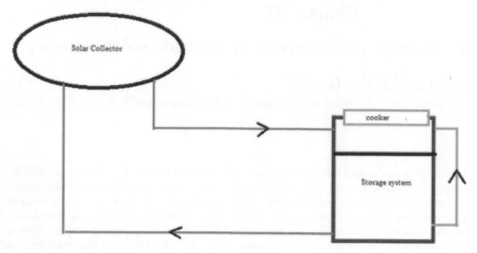

Figure 97.1 Proposed system of energy storage and transfer for cooking
Source: Author

iv. **Material for heat exchanger**

Heat storage and cooking systems are the heat exchangers. Hence its main function is to transfer the heat from one medium to another. So, it is always preferable to use a material having higher thermal conductivity. Generally, tubes may be made up of copper and shell from the still sheets [2]. And same are used in this work i.e., copper has been used for tube and stainless steel is used for the shell.

v. **Heat transfer fluid**

Heat must be transferred from the solar collector to the heat storage system to cookers and then back to the collector. This fluid should have good stability, low vapor pressure, low freeze point, low viscosity, and higher mass flow rate [3–8].

Hytherm 600 and Therminol 66 are the most suitable fluid and here Hytherm 600 is used as an HTF.

vi. **Material for storage (PCM)**

For latent heat storage systems, the selection of phase change material is very crucial.

Phase change material can be classified as: -

*   Organic: Paraffin and non-paraffin (fatty acids)
*   Non-organic: Salt hydrates and metallic
*   Eutectics: It's a mixture of two or more components [9]

There are lots of PCM available nowadays, but its selection is a very crucial thing [10]. It depends on the availability and applicability of the PCM for the selected purpose.

Paraffin wax [11–13] and HDPE [14] seem to be more promising PCM; hence these two are selected as PCM while designing this system.

vii. **Thermal design**

It is a very important step to find the capacity of the system and to validate its suitability for the considered application.

Design of energy storage system

While designing solar thermal energy storage system, following design criteria is very important [15]
  a. Technical criteria, b. Economic criteria and c. Environmental criteria

Types of heat exchangers

In general, heat exchangers can be classified according to construction, transfer processes, flow and arrangements, phase of the process fluids, and heat transfer mechanisms [15].

Thermal design procedure

The overall design procedure of a shell and tube heat exchanger is quite lengthy, and hence it is necessary to break down this procedure into steps [15]:

1. Approximate sizing of shell and tubs heat exchanger
2. Evaluation of geometric parameters also known as auxiliary calculations

3. Correction factors for heat transfer and pressure drop
4. Evaluation of the design, i.e., comparison of the results with the design specification

Design of storage system

The performance of the system can be evaluated based on heat supplied, that is cooking power supplied for the number of people. So, it becomes difficult to design the system accordingly by calculating the heat required and converting it into volumes. Hence it is preferred to design it volume-wise [16]. The temperature range for the design is selected from 200°-100°C as the usable energy for the cooking observed above 100°C. Also it is calculated that heat required for the cooking of 200 grams rice for four persons is 627 kJ as describes in 3.2.7 section.

Depending on the above data and based on the application, the coiled type heat exchanger is selected for the energy storage system.

It is made up of the number of coiled tubes which are inside the shell.

Dimensioning of the heat exchanger

Selection of tube size 3/8" (inner diameter 3/8" = 9.5 mm and outside diameter 12.5 mm) – Standard size [15]
These tubes are coiled with inclination of 30°

Distance between two tubes = 1.25 times outside diameter of tubes [17]   1

Clearance of first coil from shell = 15.625 + 15.625

= 31.25 clearance from shell (i.e. 15.625 mm radial clearance)   2

Diameter of second coil = Inner diameter of first coil – Clearance between tubes = 203.75 ≈ 200 mm
Similarly
Inner diameter of 2nd coil   = 175 mm
Diameter of third coil   = 140 mm
Inner diameter of 3rd coil   = 115 mm
Diameter of forth coil   = 80 mm
Inner diameter of 4th coil   = 55 mm
Therefore, four coils will be used for the heat storage unit which is shown in Figure 97.2.

Number of turns

Clearan   ce between two turns of a coils = Dia. of tube * 1.25 = 1.25 * 12.5 = 15.625

Number of turns in a coil (n) = $\dfrac{\text{Height of coil}}{\textit{distance coverd by 1 turn}}$ = 8 turns

Therefore, total turns = 8+1 = 9 turns

Length of the coils,

Length of first coil, $l_1 = 2\pi r_1 n$ (where, n is total no. of turns) = 7.35 m
Therefore, Total length of four coils $L = l_1 + l_2 + l_3 + l_4$

Volume of shell

For hollow shell (without tubes),
Volume of shell, $V_1 = \pi R^2 h_1$

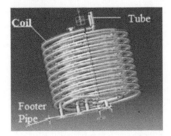

Figure 97.2 Coils inside the shell
Source: Author

Volume of tubes inside the shell,
$$V_2 = \pi \, (r_{to})^2 \, L = \pi \, \text{x} 0.0125^2 \, \text{x} \, 20 \, m^3$$
Volume of footer pipe inside the shell,
$$V_3 = \pi \, (r_f)^2 \, L = \pi \, \text{x} 0.019^2 \, \text{x} \, 0.270 \, m^3 \text{ Therefore,}$$
Volume of shell with tubes,
$$V = V_1 - (V_2 + V_3)$$
Therefore, volume available for PCM = volume of shell with tubes = 11 litre
Volume of oil (HTF) = $\pi \, (r_{ti})^2 \, L \approx 5$ litre

Also, a reservoir is used above shell of same diameter but height = 15 cm which is shown in Figure 97.3. This helps in heat transfer from oil to PCM continuously without pump also i.e., by natural convection (thermosiphon effect).

Therefore,
Volume of HTF reservoir = $\pi \, R^2 \, h_2$ = 10 L

Calculation of convective heat transfer coefficient of HTF

Therminol 66 is used as HTF. It is important to know the properties of HTF and as it varies with temperature. With the help of temperature properties of HTF like density, heat capacity and thermal conductivity (ρ, Cp, k) can be found out using Equations 3, 4 and 5 in SI units [18].

$$\rho_{htf} = \text{Density (kg/m}^3\text{)} = 1020.62 - 0.614254 * T - 0.00321 * T^2 \qquad 3$$

$$\text{Heat capacity (kJ/kg.K)} = 1.496005 + 33.13*10^{-4} * T + 8.97 * 10^{-7} * T^2 \qquad 4$$

$$\text{Thermal conductivity (W/mK)} = 0.118294 - 33*10^{-4} * T + 1.5 * 10^{-7} * T^2 \qquad 5$$

$$\text{Kinematic viscosity (mm}^2\text{/s)} = e^{\frac{(586.375 - 2.2809)}{T+62.5}} \qquad 6$$

$$\text{Vapor pressure (kPa)} = e^{\frac{(-9094.51 + 17.6371)}{T+340}} \qquad 7$$

From the available data of Therminol 66

At temperature 200°C

Density = 885.1 kg/m³,	Thermal conductivity = 0.106 W/mK,
Heat capacity= 2.195 kJ/kgK,	Dynamic viscosity=0.86 mPaS = 0.86*10⁻³ kg/ms
Kinematic viscosity = 0.97 mm²/s,	Vapor pressure = 2.23 kPa

At temperature 150°C

Density = 920 kg/m³,	Thermal conductivity = 0.110 W/mK,
Heat capacity= 2.014 kJ/kgK,	Dynamic viscosity = 1.52 mPa.s,
Kinematic viscosity = 1.65 mm²/s,	Vapor pressure = 0.40 kPa

Velocity V through pipe or coil can be calculated using Equation 8 as cross section area; density and mass flow rate are known. Also, Reynolds number (Re) is calculated, which is a ratio of inertial force to viscous force. It can be computed from Equation 9 [19].

$$V_{htf} = \text{mass flow rate/(density x area)} = \mathring{m}/(\rho A) = 4\mathring{m}/(\rho \pi d^2) \qquad 8$$

$$Re_{htf} = \text{inertial force/viscous force} = \text{vel*dist/kin viscosity} = \rho * d/\mu/V = \rho V d/\mu \qquad 9$$

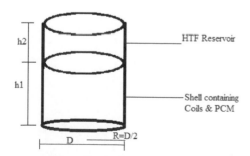

Figure 97.3 Outline of the shell of storage system
Source: Author

Prandlt number (Pr) is function of viscosity, specific heat and conductivity of HTF as shown in Equation 10. Nusselt number (Nu), for HTF flowing in a pipe is a function of Reynolds number, Prandlt number, diameter and length of pipe. These can be calculated from Equation 11 [19].

$$Pr = \mu * Cp/k \qquad\qquad 10$$

$$Nu = 3.66 + \frac{0.0668 * \frac{d}{l} * Pr * Re}{\left(1 + 0.04 * \frac{d}{l} * Pr * Re\right)0.66} \quad \text{(for } 3000 < Re < 5*10^6 \text{ and } 0.5 < Pr < 2000) \qquad 11$$

Nusselt numbers indicate the ratio of convective heat transfer to conductive heat transfer. With the help of Nusselt number coefficient of convective heat transfer (h), can be determined as shown in Equation 12.

$$h = \frac{k * Nu}{d} \text{ (since } Nu = hd/k) \qquad\qquad 12$$

Therefore,
For HTF temperature = 200°C and

$$Nu = 3.66 + \frac{0.0668 \frac{d}{l} * Pr * Re}{\left(1 + 0.04 * \frac{d}{l} * Pr * Re\right)0.66} = 6.75$$

Convective heat transfer coefficient,
$h_{htf} = k * Nu/d = 75.40 \text{ W/m}^2\text{K}$

Similarly, for **T = 150°C** and mass flow rate $\overset{\circ}{m}$ = 10 LPM
Vel = 2.35 m/s, Reynold number, Re = 13526.95, Prandlt number Pr = 27.82,
   Nusselt number, Nu = 6.65, Convective heat transfer coefficient, h_{htf} = 77.0 W/ m²K, atmospheric temperature= Ts = 30°C.

Heat transfer rate (capacity) through four coils

Initial temperature of PCM = atmospheric temperature= Ts = 30°C,
   Temperature of HTF; T_f = 200°C, k_c = k_{copper} = 374 W/mK
   $Qc = Q_1 + Q_2 + Q_3 + Q_4 + Qconv.$

$$= \frac{2\pi k l1(Tf - Ts)}{In\left(\frac{r2}{r1}\right)} + \frac{2\pi k l2(Tf - Ts)}{In\left(\frac{r2}{r1}\right)} + \frac{2\pi k l3(Tf - Ts)}{In\left(\frac{r2}{r1}\right)} + \frac{2\pi k l4(Tf - Ts)}{In\left(\frac{r2}{r1}\right)}$$

$$+ \text{hoil} * A(Tf\text{-}Ts) * \text{hpcm} * A(Tf\text{-}Ts)$$

$$= \frac{(Tf - Ts)}{\frac{1}{2\pi L}\left(\frac{1}{hoil*r1} + \frac{In\left(\frac{r2}{r1}\right)}{k} + \frac{1}{hpcm*r2}\right)} = \frac{(Tf - Ts)}{Rrest1}$$

Where, R_{rest1} = $\frac{1}{2\pi L}\left(\frac{1}{hoil*r1} + \frac{In\left(\frac{r2}{r1}\right)}{k} + \frac{1}{hpcm*r2}\right)$

But, temperature is not constant throughout the length as it drops along the length. Therefore, heat is calculated for the infinitesimal small length of dx and drop of temperature is dT. Hence heat transfer rate is given by Equation 13 [19]:

$$dQ = \frac{dx * (T - Ts)}{\frac{1}{2\pi}\left(\frac{1}{hoil*r1} + \frac{In\left(\frac{r2}{r1}\right)}{k}\right)} \qquad\qquad 13$$

This is equal to the heat transfer given by Equation 14

$$dQ = \overset{\circ}{m} Cp \, dT \qquad\qquad 14$$

Comparing Equation 13 and 14

$$\frac{dx * (T - Ts)}{\frac{1}{2\pi}\left(\frac{1}{hoil*r1} + \frac{In\left(\frac{r2}{r1}\right)}{k}\right)} = \overset{\circ}{m} Cp \, dT$$

$$\frac{dx}{\overset{\circ}{m}Cp * \frac{1}{2\pi}\left(\frac{1}{hoil*r1} + \frac{In\left(\frac{r2}{r1}\right)}{k}\right)} = \frac{dT}{(T - Ts)}$$

$$\frac{dT}{(T-Ts)} = \frac{dx}{Rt2} \qquad 15$$

Where R_{t2} is a thermal resistance of coil per unit length

$$R_{t2} = \overset{\circ}{m}\, Cp* \frac{1}{2\pi} \left(\frac{1}{hoil*r1} + \frac{In\left(\frac{r2}{r1}\right)}{k} \right)$$

Integrating equation 4.15 within temperature limits of T_{if} and T_{ff} where T_{if} and T_{ff} are initial and final temperature of HTF in coil.

$$\int_{Tfc}^{Tic} \frac{dT}{(T-Ts)} = \int_{0}^{L} \frac{dx}{Rt2}$$

Integrating and applying limits

$$\int_{Tff}^{Tif} \frac{dT}{(T-Ts)} = (-1)* [\log(T\text{-}Ts) - \log(T\text{-}Ts)]_{Tff}^{Tif} = -[\log (T_{if}\text{-}T_s) - \log (T_{ff}\text{-}T_s)]$$

$$= - \log \left[\frac{(Tif-Ts)}{(Tff-Ts)}\right]$$

And $\int_{0}^{L} \frac{dx}{Rt2} = [x/R_{t2}]_{0}^{L} = \frac{L}{Rt2}$

Therefore,

$$-\log \left[\frac{(Tif-Ts)}{(Tff-Ts)}\right] = \frac{L}{Rt2}$$

$$\log \left[\frac{(Tif-Ts)}{(Tff-Ts)}\right] = \frac{Rt2}{L}$$

$$\frac{(Tif-Ts)}{(Tff-Ts)} = \exp (Rt_2/L) \qquad 16$$

$$T_{ff} = T_s + \frac{(Tif-Ts)}{\exp(Rt2/L)}$$

Heat transfer across length of the coil causes temperature drop which is given by Equation 17
$dQ = \overset{\circ}{m}\, Cp\, (T_{ff}\text{-}T_{if})$ putting value of T_{ff}

$$dQ = \overset{\circ}{m}\, Cp\, (\, Ts - Tfi) * \left(1 - \frac{1}{\exp\left(\frac{Rt2}{L}\right)}\right) \qquad 17$$

Therefore,

Heat transfer rate, $Q_c = \overset{\circ}{m}\, Cp(Ts\text{-}Tfi) \left(1 - \frac{1}{\exp\left(\frac{Rt2}{L}\right)}\right)$

But lengths of four coils are different,
Therefore,

$$Q_{c1} = \overset{\circ}{m}\, Cp\, (Ts\text{-}Tfi) \left(1 - \frac{1}{\exp\left(\frac{Rt2}{L1}\right)}\right)$$

$Q_{c1} = Q_{c2} = Q_{c3} = Q_{c4} = 55.04$ kW

Therefore, total heat transfers by four coils = $Q_{c1} + Q_{c2} + Q_{c3} + Q_{c4}\, Q_c = 220.01$ kW

Heat content by HTF in tubes at 200°C; and atmospheric temperature as 30°C (for static condition)

Heat content by the HTF which is present in the tube at static condition (5 L)
$\qquad Q_{f1} = m\, Cp\, \Delta T$
Hence, total heat contained by HTF at 150°C
$\qquad Q_f = Q_{f1} + Q_{f2}$

Heat stored in PCM,

Mass of PCM
Mass = density x volume
But, volume should be ¾ of total volume only because of phase change of the PCM
Hence, heat stored in paraffin wax
$\qquad Q_{wax1} = m_c \{C_{pc} (T_{cm} + T_{ci}) + \Delta h_f + C_{pc} (T_{co} - T_{cm})\}$–4.18
For high density polyethylene (HDPE)
Density, $\rho = 970$ kg/m^3, Cps = 1.9 kJ/kgK, Cpl = 2.3 kJ/kgK, melting temperature, Tm = 128°C,
Therefore, total heat contained by the storage system at 200°C
For paraffin wax

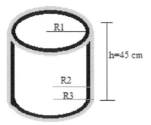

Figure 97.4 Outline of energy storage system with insulating radius
Source: Author

$Q_{wax1} + Q_f$
For HDPE
$Q_{hdpe1} + Q_f$

Heat loss through system (Heat loss rate)

The storage tank is made up of a GI sheet insulated by glass wool. The storage tank is cylindrical in dimension as shown in Figure 97.4. Heat loss to the atmosphere takes place in the radial and axial directions of a cylinder. D and h are the diameter and height of the cylinder. tcy and tcyin are the thickness of the cylinder and insulation of the cylinder respectively. Radial heat loss is given in eq.19 including convective as well as conductive loss [19].

$$Q_{radial} = (T_{pcm} - T_a)/ R_{resist} \qquad 19$$

$$\text{Resistance, } R_{resist} = \frac{1}{2\pi h} \left[\frac{In\left(\frac{R2}{R1}\right)}{k1} + \frac{In\left(\frac{R3}{R2}\right)}{ki} + \frac{1}{hoR3} \right]$$

Where,
h- height of shell (cylinder)
R_1 - inner radius of shell
R_2 - outer radius of shell = R_1 + 1 mm
$R_3 = R_2$ + thickness of insulation
Thickness of insulation

As mentioned earlier, various insulating materials are available but depending on application mineral wool has been selected which has thermal conductivity, ki = 0.04 W/mK critical radius of insulation,

Rc = ki/ho
ho – mean convective heat transfer coefficient of air
ho = 1.42 $(\Delta T/L)^{1/4}$ [15]

This radius is less than the radius of shell; hence further addition of insulation will decrease heat loss only.

The available mineral wool sheet thickness is tcyin = 7 cm.
So, it is wrapped once,
$R_3 = R_2$ + 7cm= 22.1 cm
$Q_{radial} = (T_{pcm} - T_a)/ R_{resist}$

$$R_{resist} = \frac{1}{2\pi h} \left[\frac{In\left(\frac{R2}{R1}\right)}{k1} + \frac{In\left(\frac{R3}{R2}\right)}{ki} + \frac{1}{hoR3} \right]$$

When, R_3= 15.1+ 14 = 29.1
$Q_{radial} = (T_{pcm} - T_a)/ R_{resist}$

$$R_{resist} = \frac{(Ts-Ta)}{1/\pi\left[\frac{tc}{R2^2 k1} + \frac{tcyin}{R3^2 ki} + \frac{1}{hoR3^2}\right]}$$

Heat loss through storage tank (shell) in axial direction is given by Equation 20

$$Q_{axial} = \frac{(Ts-Ta)}{1/\pi\left[\frac{tc}{R2^2 k1} + \frac{tcyin}{R3^2 ki} + \frac{1}{hoR3^2}\right]} \qquad 20$$

Therefore, Total heat loss $Q_L = Q_{radial} + Q_{axial}$

Heat loss rate without any cooking (for thickness of insulation 14 cm)

For paraffin wax
  Time to loss this heat upto 30°C
        = Heat stored in storage system/heat loss rate
  Time to loss this heat upto 100°C without cooking
        = Heat storage from 100-200°C/heat loss rate
For HDPE
  Time to loss this heat upto 30°C
        = Heat stored in storage system/heat loss rate
  Time to loss this heat upto 100°C without cooking
        = Heat storage from 100-200° C/heat loss rate
Heat storage at various temperature and heat loss rate is shown in Figure 97.5 however Figure 97.6 depicts the energy loss and balance with respect to time considering ambient temperature as 30°C.

Cooking parameter

The cooking of rice is considered for research purposes. For that quantity of rice is stockpiled to keep rice quality the same throughout the experiment. Use of pressure cooker of 2 liters is used for conventional as well for cooking by using stored energy. The standard method of cooking is used for cooking as cooking is supposed to be completed after blowing off steam thrice [31]. 200 grams of rice is used for each iteration.
  For heat calculation, Prestige PIC 20, 1200W Induction cooktop is used.
  The cooking is done at 1100W; Time taken to completely cook the rice is 9 min and 30 sec (570 sec)
  Therefore,
  Heat consumed for cooking of rice = 1100 * 570 = 627000 J = 627 kJ
  Therefore,
  Considering heat stored up to 200°C at 5 pm and cooking must be done at 9 pm (after 4 hours)
  Then,
  Heat in system = 7529.27 kJ for paraffin wax (thickness of insulation = 14 cm)
  Heat available after cooking at 10 pm
        = Heat available system – heat required for cooking

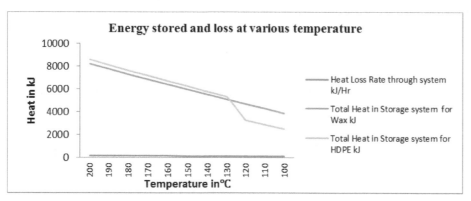

Figure 97.5 Energy stored and loss rate at various temperatures for paraffin wax and HDPE
Source: Author

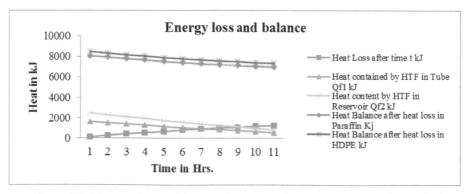

Figure 97.6 Energy loss and balance in system after time 'T' for insulation thickness 14 cm
Source: Author

Heat in system = 7910.11 kJ for HDPE (thickness of insulation = 14 cm)
Heat available after cooking at 9 pm
      = Heat available system – heat required for cooking

Heat transfer through pipe from solar collector

The collector will be placed on roof at the height of 10 feet from storage system where HTF is heated at 200°C so to transfer heat from solar collector to storage system heat loss is takes place which is shown in Figure 97.7.

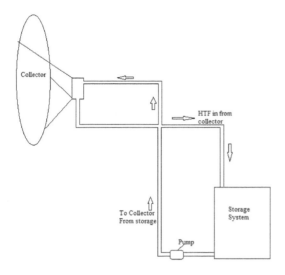

Figure 97.7 Heat transfer system
Source: Author

Table 97.1 Details of designed system

Parameter	Dimensions/quantity	Remark
Tube (inner) diameter	9.5 mm	
Tube (outer) diameter	12.5 mm	
Number of coils	4	Changes according to energy demand
Number of turns	9	----//-----
Length of tube	20 m	----//-----
Thickness of insulation for shell	14 cm	
Diameter of shell	300 mm	----//-----
Diameter of (inner) pipe	12.5 mm	
Diameter of (outer) pipe	19 mm	
Thickness of insulation for pipe	2 cm	
Height of shell	300 mm	----//-----
Height of reservoir	150 mm	----//-----
Volume of shell for PCM	11 L	
Volume of tubes for HTF	5 L	
Volume of reservoir for HTF	10 L	
Heat loss rate through pipe	43.14 J/s	Varies as per size
Heat loss rate through shell	40.89 J/s	----//-----
Heat transfer rate by HTF @10LPM	220.01 kJ/s	
Heat stored in paraffin wax at 200°C	4097.56 kJ	Varies as per size
Heat stored in HDPE at 200°C	4478.4 kJ	Varies as per size
Heat stored in HTF @200°C in reservoir	4119.27 kJ	Varies as per size

Source: Author

Critical radius of insulation,

Rt = ki/ho

ho – mean convective heat transfer coefficient of air

ho = 1.42 $(\Delta T/L)^{1/4}$ ---- [15]

As Rt < ro,

∴ Thickness of Insulation = 2 cm

∴ Radius = ro + 2 cm = 19 mm + 20 mm = 39 mm

Material of pipe is GI

Heat Loss through Pipe

Q_{LP} = (Tfi-Ta)/R_{esist}

$$R_{t3} = \left\{ \frac{1}{2\pi} \left[\frac{In\left(\frac{ro}{ri}\right)}{k1} + \frac{In\left(\frac{r3}{ro}\right)}{ki} + \frac{1}{hiri} + \frac{1}{hor3} \right] \right\}$$

The Designed system can be briefly described as in the Table 97.1.

Conclusion

From the literature survey, the need for energy storage is figured out. The design presented in this study reveals that the cooker with an energy storage system enables the cooking after sunshine hours also. Presented work illustrates that the energy storage system with helical coils can store the energy for a longer time which helps in cooking after sunshine hours. This energy system enables cooking 4 to 8 hours after the sunshine also. Without performing cooking system can store usable heat for more than 12 hours which suggest cooking can be done on next day upto next sunshine hours. The selected insulation signifies the importance of minimizing the heat loss rate and improving the effectiveness of energy storage systems. The presented study enables the designing of energy systems for any requirement as well as for any type of cooking i.e., household as well community cooking. This system is specifically designed for cooking applications, and it can be easily scaled and can be used for any energy storage requirement. This may aid in increasing the usage of solar cookers and it may assist in saving comparable conventional energy.

References

[1] Masatin , V,. Volkova, A., Hlebnikov, A., and Latosov, E. (2017). Improvement of district heating network energy efficiency by pipe insulation renovation with PUR foam shells. Energy Procedia. 11, 265–9.

[2] Permatasari, R. and Yusuf1, A. M. (1977). Material selection for shell and tube heat exchanger using. In AIP Conference.

[3] Solé, A., Fontanet , X., Barreneche, C., Martorell, I., Fernández, A. I., and Cabeza, L. F. (2012). Parameters to take into account when developing a new thermochemical energy storage system. Energy Procedia. 30, 380–87.

[4] Lin , W. and Ievers, S. (2009). Numerical simulation of three-dimensional flow dynamics in a hot water storage tank. Applied Energy. 86, 2604–14.

[5] Yang, Z., Garimella, S. V. and Flueckiger, S. M. (2013). Design of Molten-Salt Thermocline Tanks for Solar Thermal Energy Storage. pp. 1–50. CTRC Research Publications..

[6] Venkataramaiah , P. Lokesh, T. R. and Reddy, K. D. (2014). Parametric study on phase change material based thermal energy storage system. Energy and Power Engineering, Scientific Research. 6, 537–49.

[7] Malhotra, R. K., Kaushik, S. C., and Srivastva, U. (2015). Recent developments in heat transfer fluids used for solar thermal energy applications. Journal of Fundamental of Renewable Energy and application. 5(6), 1–11.

[8] Chaudhari, J. and Singh, S. K. (2016). Design and optimization of thermal storage tank using CFD. International Journal on Recent and Innovation Trends in Computing and Communication. 4(4), 126–31.

[9] White, M., and Qiu, S. (2013). Phase change material thermal energy storage system design and optimization. In Proceeding of ASME 2013, 7th Int. Conf. on Energy Sustainability & 11th fuel Cell Science, Engg & Tech Conference, ESfuel cell 2013, Minneapolis, MN, USA, 2013.

[10] Magin, R. L. (1961). Transition temperatures of the Hydrates of Na2S04, Na2HP04 and KF as Fixed Points in biomedical thermometry. Journal of Research the National Bureau of Standards. 86(2), 181–92.

[11] Buddhi, D . and Sharma, A. (2009). Review on thermal energy storage with phase change materials and applications. Renewable and Sustainable Energy Reviews. 13, 318–45.

[12] Silakhori, M. (2013). Accelerated thermal cycling test of microencapsulated paraffin wax/polyaniline made by simple preparation method for solar thermal energy storage. Materials, MDPI Journals. 1608-20.

[13] W., and Sun, W . (2017). Thermal analysis of a thermal energy storage unit to enhance a workshop heating system driven by industrial residual water. Energies, MDPI. 10, 1–19.

[14] Gasia, J. (2017). Phase change material selection for thermal processes working under partial load operating conditions in the temperature range between 120 and 200 C. Applied Sciences. MDPI. 7, 722.

[15] Thulukkanam, K. (2013). Heat Exchanger Design Handbook. Boca Raton: CRC Press.

[16] Gawande, T. K. and Ingole , D. S. (2019). Comparative study of heat storage and transfer system for solar cooking. SN Applied Sciences A Springer Nature Journal. 1.

[17] Thermopedia, B. R. J. (2011) . Thermopedia. Standards of the Tubular Exchanger Manufacturers Association, (TEMA), [Online]. Available from: http://dx.doi.org/10.1615/AtoZ.s.shell_and_tube_heat_exchangers.

[18] Heat Transfer Fluid by SOLUTIA, Sollutia, [Online]. Available: http://twt.mpei.ac.ru/TTHB/HEDH/HTF-66.PDF.

[19] Holman, J. P. (2010). Heat transfer. Ney York: McGraw Hill Companies, Inc.

[20] Lakshmi, S. and Chakkaravarthi, A. (2007). Energy consumption in microwave cooking of rice and its comparison with other domestic appliance. Journal of Food Engineering. 78, 715–22.

[21] Hytherm 500 by HP Lubricants. Hindustan Petroleum, 2019. [Online]. Available from: https://www.hplubricants.in/products/specialties/thermic- fluids/hytherm-500-and-600-thermic-fluid-oil.

[22] Therminol 66 Heat Transfer Fluid by EASTMAN. https://www.therminol.com/product/71093438

[23] Solé, A. and Neumann, H. (2014). Thermal stability test of sugar alcohols as phase change materials for medium temperature energy storage application. Energy Procedia. 48, 436–9.

[24] Xu, H . (2017). Selection of phase change material for thermal energy storage in solar air conditioning systems. Energy Procedia, 105, 4281–8.

[25] Lecuona , A. (2013). Solar cooker of the portable parabolic type incorporating heat storage based on PCM. Applied Energy, 111, 1136–46.

[26] Sebarchievici, C. and Sarbu , I. (2018). A comprehensive review of thermal energy storage. Sustainability, MDPI. 10, 1–32.

[27] Beemkumar, N. (2015). Heat transfer analysis of latent heat storage system using D-Sorbitol as PCM. ARPN Journal of Engineering and Applied Sciences, 10(no. 11), 5017–21.

[28] Diarce, G . (2015). Eutectic mixtures of sugar alcohols fo rthermal energy storage. Solar Energy Materials & Solar Cells. 134, 2115–226.

[29] Tian, Y. (2013). A review of solar collectors and thermal energy storage in solar thermal. Applied Energy, 104, 538–53.

[30] Bauer, T. (2009). Sodium nitrate for high temperature latent heat storage. In The 11th International Conference on Thermal Energy Storage – Effstock, Stockholm, Sweden.

[31] Gawande, T. K. and Ingole, D. S. (2021). Performance analysis of energy storage and transfer system for solar cooking using HDPE as a PCM. IRJET. 8(6), 461–463.

Chapter 98

Numerical investigations for smoke movement in industrial shed with C shaped collection channel

Ashvin Amale[1,a], Neeraj Sunheriya[2,b], Jayant Giri[1,c], Rajkumar Chadge[1,d], Sachin Mahakalkar[1,e], Pratik Lande[1,f] and Abhiram Dapkeg[3g]

[1]Yeshwantrao Chavan College of Engineering Nagpur, Maharashtra, India

[2]National Institute of Technology, Raipur, India

[3]University of Maryland, College Park, United States

Abstract

The present work deals with exhaust systems having two channel of C section type for a fireplace workstation. It is the optimisation of removing exhaust gases or smoke from workplace using two C type channel rather than using direct exhaust pipe coming out from workplace and connected to exhaust blowers. Shed is used to guide the smoke. At fire place station, there is accumulation of smoke inside the space, where workers work. Smoke circulates inside the fireplace's free space and becomes difficult to go outside. So, fans or blowers are needed to overcome this problem. Exhaust Pipes, attached with blowers, can be used directly to remove smoke, connected to fireplace's walls directly or a C channel can be used attached with this wall and after that the exhaust pipes, connected to fans or blowers to improve the exhaust system. C channel helps and guides the smoke flow to go outside of the fireplace workstation. For finding the path-line of smoke flow inside the fireplace workstation, computational fluid dynamics (CFD) is used. Using ANSYS FLUENT software, we find the flow pattern of the smoke flow and variation in temperature with the height inside the fireplace workstation. So simplified work geometry of fireplace workstation, is made using ICEM tool and meshing is also done using this software. This geometry has two C type channels attached to exhaust on each side. Then exhaust pipes are used to connect the C channel exhaust system and blowers. Solution and simulation is done using ANSYS Fluent software. The turbulence k Epsilon model is used to solve the problem. A further study includes the estimation of capacity of the blowers that are used for exhaust system and the best suitable blowers on the basis of the mass handling capacity of the blowers and temperature of the exhaust gases to be handled.

Keywords: Industrial shed, computational fluid dynamics, exhaust, C channel

Introduction

In industrial area like manufacturing, automobile, heat treatment of materials, forging etc. fuel like coal, oil etc. burn and smoke is generated. This smoke may contain soot particles, harmful gasses, fumes, toxic gasses etc. which may cause serious health problem to people working in that environment. So, this smoke and harmful gases need adequate exhaust system so that it can flow in a manner that it cannot accumulate inside working environment. For this exhaust fans and blowers of adequate capacity, are needed as per requirement. Fans and blowers are the power consuming devices. Therefore, design of the geometry model and ducts layout should be optimum, so that frictional and other losses can be reduced, and power consumption would be less. The capacity of exhaust fans or blowers should calculate first before ordering or purchasing new fans and blowers. Capacity of exhaust fans depend on the amount of smoke or flue gasses generated and smoke flow path, duct design and rate of removing of smoke and flue gasses and some losses like leakage loss, friction losses etc. Using many fans and blowers is good idea rather than using only one fan or blower.

For calculation of capacity of fans or blowers' relevant data like smoke generation per second (kg/s), temperature of smoke and flue gasses, static pressure and velocity pressure at inlet and outlet of blower, friction losses in duct, smoke velocity, smoke flow rate (m³/s), is needed. Operating temperature of fans or blowers is most important factor, when deciding the fans or blowers. Type of blowers or fan is decided on the basis of different factors like flow rate of smoke (m³/s) handled, suction pressure and delivery pressure required.

After calculation of capacity of blowers or fans, motor size is calculated depending on the basis of several factors like efficiency of fans or blowers, mechanical losses, and other factors. For smoke flow patterns computational fluid dynamics (CFD) tool is useful which provides graphical representation of smoke movement inside the working environment which is impossible to find experimentally. CFD tool is used to estimate capacity of blowers or fans based on boundary conditions like total smoke generated

[a]ashvin.amale@gmail.com, [b]neeraj.sunheriya@gmail.com, [c]jayantpgiri@gmail.com, [d]rbchadge@rediffmail.com, [e]sachu_gm@yahoo.com, [f]pratiklande@gmail.com, [g]abhiramdapke162@gmail.com

DOI: 10.1201/9781003450252-98

(kg/s), temperature and pressure of smoke generated, smoke and gasses handled per second by the blowers (m³/s), static and velocity pressure at inlet and outlet of blowers, friction losses in duct etc. Temperature variation, velocity profile, path-line of smoke or flue gasses can be shown using ANSYS software. Modelling of geometry is done using ANSYS ICEM software. And meshing of the geometry is also done using ICEM software. Then the FLUENT software is used for solving the problem by giving boundary conditions to the model. After adequate iteration, solution is converged, and we get the result. CFD-POST software is used for the result analysis and animation and simulation.

The objective of the study

- 3D CFD analysis of smoke movement in industrial shed with C shaped collection channel,
- geometry modelling and CFD simulation of smoke flow through C channel system on ANSYS Software (fluent),
- analysis of system with different exhaust pressure at outlet to C channel, and
- finding the best suitable capacity of blower on the basis of path-line generated of smoke and different exhaust pressure values.

Literature Review

The ventilation for smoke in different industrial sector has been carried out and CFD is done for different condition and parameter. Some of the important findings are mentioned in subsequent proceedings.

Chew et al. [1] examine the Atrium smoke movement with the use of computational fluid dynamics. They investigated fire-induced air flow, temperature, and smoke concentration fluctuate in an atrium (CFD). Two numerical studies were carried out in two atriam of equal size, fan placement, and fire type but distinct height. An actual copy in an atrium was also investigated, with varying fan positions and sizes but the same fire type. In the physical model, smoke was caught at the atrium ceiling and carried into the stores on the same level, with the fire source positioned at the base of a 50 m high atrium. The extraction fans at the smoke reservoir were unable to remove the smoke in the region adequately. The simulation findings highlight the significance of atrium height and fan position in terms of smoke extraction effectiveness. Atria bigger than 30 m in height with exhaust fans at the ends may demand special caution to adequately remove and manage the smoke. An investigation on station fire conditions in the Buenos Aires subway system by Li et al. [2] employed CFD to report the results of a study in which CFD was used as a numerical method to analyse ventilation performance at Buones Aires metro subway stations. Methods of natural and mechanical ventilation were examined. The investigation discovered that, while natural ventilation meets the temperature standards, it offers a possible egress concern by enabling hot gases to depart through the entrance ways. The research utilised a train fire release rate of 1.8 MW, which equated to a train fire involving mainly the combustible items under the car. They opted to deploy jet fans roughly 50 metres from each station in the tunnel. The location of the fire within the station would dictate the operation mode of the jet fan. The jet fan system was found to be a practical and cost-effective solution that satisfied the stipulated emergency ventilation standards in existing tunnels and stations. In an enclosure fire simulation, Xue et al. [3] examined several combustion models. The volumetric heat source model, the eddy break-up model, and the supposed probability density function model were all examined in enclosure fire simulation in this work. The computations were carried out using fluent, a commercial CFD package. The governing equations were solved using the finite volume method in a staggered grid arrangement. SIMPLEC was the algorithm employed. A power-law method is used in the numerical simulation. The prediction performance of three typical enclosure fires, a room fire, a shopping mall fire, and a tunnel fire, was compared and analysed. The high Reynolds number turbulence k- model with buoyancy adjustment and the discrete transfer radiation model were utilised in the simulation. For validation, we used matching experimental data from the literature. The most fundamental combustion model was the volumetric heat model. In the computational realm, the fire was represented as a volumetric heat source. In this model, the direct contribution of combustion species is neglected. The eddy break-up model was built on the solution of species transport equations for reactant and product concentrations. The mechanism of the reaction had to be specified specifically, and the reactions may be simple or multi-stage. In order to improve system safety, Willemann et al. [4] reviewed some of the computer modelling approaches and studies utilised to design a tunnel ventilation fan plant for the New York City Subway. Three fire intensity levels were considered: 44 W (low intensity), representing a low heat smoky tunnel environment, 1.8 kW (middle intensity), representing an under-car train fire, and 14.7 kW (high intensity), representing a completely engulfed car-train. The significance of the fire model in CFD simulations was highlighted. It was necessary to develop a simpler combustion method for determining heat generation rates and fire product production rates. They used the SES results to provide system boundary conditions under certain situations. When the velocity was less than the critical velocity, they used CFD analysis. This study showed how computer modelling and analysis contributed to the design of tunnel ventilation fan plants at New York City Transit. Shahcheraghi et al. [4] investigated the effect of fan start time on the operation of a subway station emergency ventilation system. A transient computational fluid dynamics simulation of a train fire in a subway station included a time-dependent fire progression. This study examined 46 different emergency ventilation fan start times (0, 60, 180 and 420 seconds). The 10 MW fire reached its peak in 960 seconds. Summer circumstances were taken into account since the temperature of the buoyancy forces was higher than that of the station's interior. As a result, buoyancy forces acted in opposition to the emergency ventilation system. Three boundary requirements were tunnel apertures, fans, and mezzanine connections to the surface. The tunnel apertures were used to create a mass flow barrier, with flow rates predicted using the SES fire simulation. The original tunnel boundary conditions were generated using SES simulations. The fans were viewed as constant volume flow rate boundaries, and the mezzanine exit was designated as an access point to the outside ambient conditions. Walls were modelled as having no sliding restrictions and roughness values of 0.25 cm. Gobeau et al. [5] conducted a research effort that used CFD modelling and tests in their evaluation of CFD to forecast smoke flow in complicated enclosed settings. The project was divided into three main parts. Phase 1 included CFD estimates for three genuine "complex areas." An underground station, an accommodation module on an off-shore platform, and a high-rise structure under construction were among them. Different CFD modelling methodologies were utilised to assess their impact on smoke movement prediction. Phase 2 resulted in a "benchmark" collection of experimental observations of hot smoke flow in basic small-scale constructions. A variety of basic shapes were built, instrumented, and tested, each targeting a different element of smoke layer physical behaviour. The third step involved a thorough study of CFD performance in simulating the Phase 2 benchmark trials. The computational grid and the discretisation technique were changed as part of the modelling procedure. Sakai et al. [6] have investigated the ventilation system of a lavatory inside an office building for the improvement in the amenities. In comparison to others, the concentration of indoor air pollutants might be kept low. It was also demonstrated that the use of this technology might result in energy savings due to the reduction in ventilatory volume. Zajicek et al. [7] simulated the broiler house for the improvement in the air flow circulation. Different cases of different design configurations were studied. Trokhaniak et al. [8] have performed the CFD simulation for the poultry farm to investigate the influence the fresh air through valves. The found that the valves place at 200 mm height gives better results compared to the valves situated at 400 mm height.

Description of the Problem and Geometry

Description of the Problem

An Industrial shed with C shaped collecting channel, having 22.5 m length, 10 m width and 10 m height is the dimension under consideration. There are two C channels at both sides of shed on top of the wall (at height of 10 m). There are two openings at outlet side and three at inlet side to c channel. In the center, the smoke is generated inside the two furnaces. The side windows are open for fresh air. There are six side windows which are fully open for geometry.

Smoke is generated in furnace and coming out from upper face of furnace. Then smoke is collected by the two C channels and smoke is sucked out from the C channel through ducts each. These ducts are connected to blower's inlet. Then smoke is carried out to atmosphere with help of chimney.

Objective of this analysis is to find the path line for smoke movement inside the industrial shed and in C channel and finding best capacity of blowers and number of blowers based on path line, vector profile and temperature profile. For this case, first geometry modeling is done using ICEM software. Then meshing of the model is also done using ICEM software. And finally, it is solved using ANSYS FLUENT by providing boundary conditions. Contour of velocity and temperature and pressure and path-line of smoke movement are shown by FLUENT software.

Capacity of blower is calculated using static pressure and velocity pressure head at suction duct connected to exhaust to C channel and the blower inlet, the friction loss in duct and the pressure head required for maximum discharge through chimney. Based on total head required for exhaust of smoke, capacity of blowers is calculated. And the power required for blower is calculated using efficiency of blowers and motor. Simulation of problem is done using CFD-post software, where smoke movement inside geometry is shown in animation form.

Geometry

There are two C channels attached to the industrial shed at the top of the wall for collecting smoke. Geometry is modeled with 10 m width, 10 m height and 22.5 m length. C channel is of 1 × 1 m cross section. The open side of C is towards inside of geometry and closed side is toward to outside of geometry for exhaust. It has two outlets connecting ducts of 1 × 1 m size and these ducts are connected with the inlet to blowers.

There are two furnaces of size 4 × 4 m placed at center of floor of geometry and smoke is coming out from furnace only from upper side of the furnaces. Six windows are fully open of size 2 × 5 m are present for fresh air. Chimney is used for smoke exhaust to atmosphere at height of 30 m. Specification of geometry model is given below (Table 98.1).

C channel is modeled using ICEM Software as shown in figure 98.1, 98.2, 98.3 and 98.4. It has two outlets at the back side and full opening at inlet for suction of smoke from the geometry model.
Table 98.2 shows the total number of elements selected for the mesh generation. Various boundary conditions are shown in Table 98.3. Figure 98.5 shows the convergence of solution.

Boundary Conditions and Convergence of Solution

The output is checked for convergence (that is, the mass imbalance is extremely near to zero), and if it is not within the necessary range, the programme returns to the beginning, using the freshly computed pressure, velocity, and other scalar variables as the next starting guess. The technique is repeated until convergence occurs (iteration).

Table 98.1 Specification of geometry

Outlet:	1 × 1 m (Four outlets)
C Channel:	1 × 1 × 22.5 m (Two outlets and three inlets Figure 98.I)
Furnace:	4 × 4 × 2 m (Figure 98.3)
Inlet.1:	4 × 4 m (Two furnaces)
Inlet.2:	2 × 5 m (Four windows)
Ducts:	1 × 1 m

Source: Author

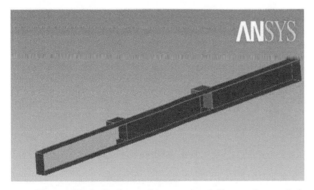

Figure 98.1 C channel geometry full opening at Inlet
Source: Author

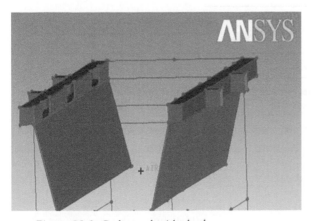

Figure 98.2 C channel with shed
Source: Author

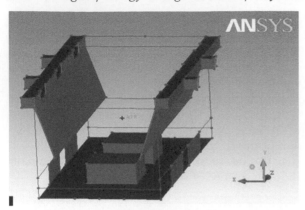

Figure 98.3 Furnace location inside the geometry
Source: Author

Figure 98.4 Surface meshing for the geometry model
Source: Author

Table 98.2 Number of element for mesh

Elements type		Elements parts		Total elements	
Node	328	Air	427313	Total Elements	526975
Line_2	5590	Geom.	66414	Total Nodes	107351
Tetra_4	427313	Inlet.1	1629		
Tri_3	93744	Inlet.2	1898		
		Outlet	207		
		Shed	29514		

Source: Author

Table 98.3 Boundary conditions

CASE	INLET				OULET
	Inlet.1		Inlet.2 Inlet pressure (Gauge pressure)		Exhaust pressure at c channel outlet (gauge pressure)
	Velocity (m/s)	Temperature (K)			
Case 1	1	400	0 atm.		-0.005 atm.
Case 2	1	400	0 atm.		-0.01 atm.
Case 3	1	400	0 atm.		-0.02 atm.
Case 4	1	400	0 atm.		-0.03 atm.
Case 5	1	400	0 atm.		-0.04 atm.

Source: Author

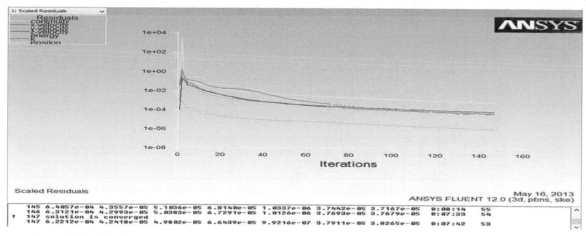

Figure 98.5 Solution convergence
Source: Author

CASE-1. At suction pressure -0.005 atmosphere

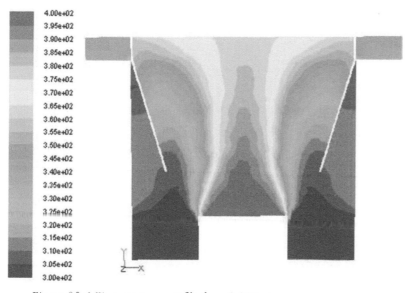

Figure 98.6 Temperature profile for - 0.005 atm. pressure
Source: Author

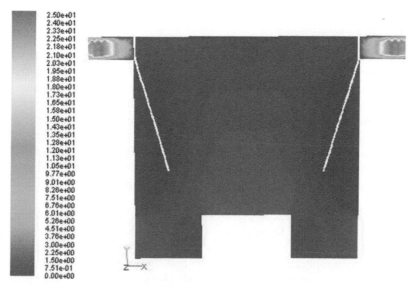

Figure 98.7 Velocity profile for -0.005 atm. pressure
Source: Author

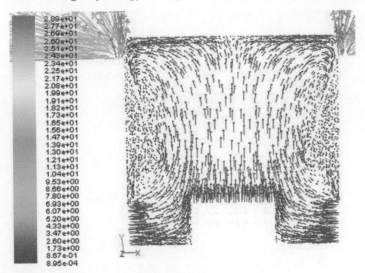

Figure 98.8 Vector profile for -0.005 atm. pressure
Source: Author

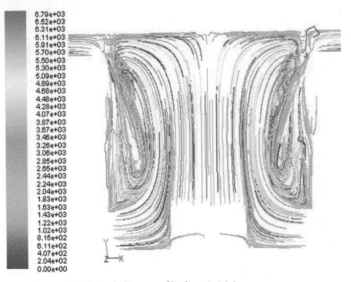

Figure 98.9 Path line profile for -0.005 atm. pressure
Source: Author

Results

Solution is calculated for all the cases (case 1, case 2, case 3, case 3, case 4, case 5) in FLUENT solver by giving boundary condition according to the cases listed in Table 3. After calculation contour for different variables like temperature, velocity, path-line and vector profile are obtained for different pressures (Figure 98.5-98.7).

It is observed that as we increase the pressure at outlet path-lines are goods and less eddy of smoke at bottom level. From temperature profile it is observed that temperature at low height is less as exhaust pressure increases. Vector profile shows the vector of smoke movement inside the geometry model. From these figures it is observed that at exhaust pressure -0.01 atmospheric (gauge pressure) path-line, vector and temperature profile are best and gives the optimum smoke movement of smoke. So, this pressure can be taken for further analysis and calculations.

The schematic diagram of temperature profile at exhaust pressure -0.01 atmospheric pressure for different planes -plane1 (z = 4.5 m), plane2 (z = 7.5 m), plane3 (z = 13 m), plane 4 (z = 21) is shown in Figure 98.13.

From Figure 98.13 it is observed that at plane1 (z = 4 m) temperature at low level is more due to smoke coming from furnace and at plane4 (z = 21 m) temperature is less due to less smoke is going there. CFD analysis is done to find out the path-line profile, vector flow profile, velocity profile and temperature profile for smoke movement inside the geometry model. After CFD analysis of geometry model at different exhaust suction pressures (-0.005 atm, -0.01 atm, -0.02 atm, -0.03 atm and -0.04 atm, gauge pressure),

CASE-2: At suction pressure -0.01 atmospheric

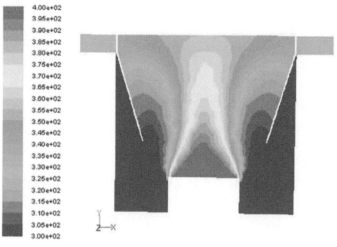

Figure 98.10 Temperature profile for - 0.001 atm. pressure
Source: Author

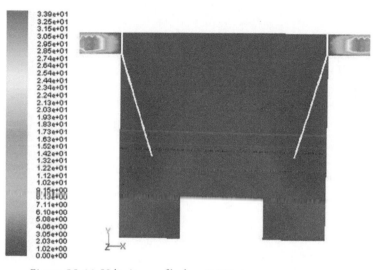

Figure 98.11 Velocity profile for -0.001atm. pressure
Source: Author

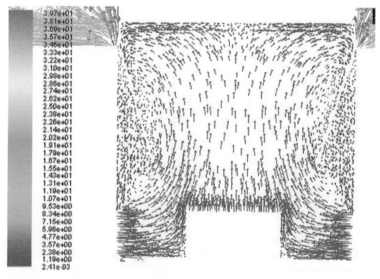

Figure 98.12 Vector profile for -0.001 atm. pressure
Source: Author

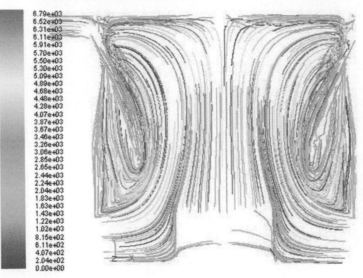

Figure 98.13 Path line profile for -0.001 atm. pressure
Source: Author

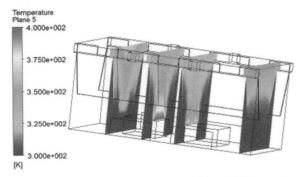

Figure 98.14 Temperature profile at different planes for -0.01 atm. pressure
Source: Author

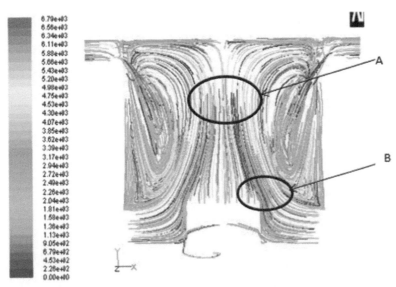

Figure 98.15 Path-line for smoke flow on a plane
Source: Author

It is observed that the suction pressure -0.01 atmospheric pressure (- 1013.25 N/m² gauge pressure) gives the optimum path-line profile, temperature profile and vector flow profile of smoke movement inside the geometry model.

Path-line of smoke movement is shown in Figure 98.14. This figure shows that the path-lines are smooth, and the smoke is surrounded and guided by shed and fresh air. This is shown in region-A in Figure 98.14.

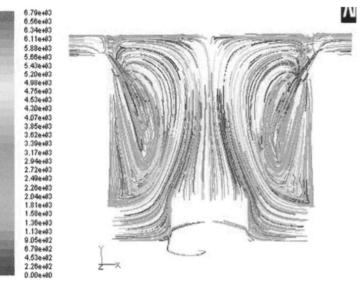

Figure 98.16 Path-line of smoke flow with shed model
Source: Author

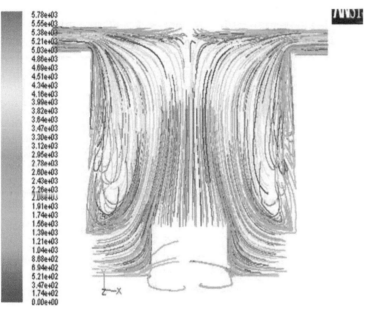

Figure 98.17 Path-line of smoke flow without shed model
Source: Author

All the smoke path-line is covered by the fresh air path line so less smoke spread out when smoke comes out from furnace. It is clearly visible in region-B that the smoke does not split out at corner of furnace as it comes out from the furnace. Fresh air helps and guide to flow of smoke and makes the environment good for workers at lower level of working station.

In this work CFD analysis is done for smoke movement in industrial shed with C shaped collection. For this smoke is treated as air means smoke properties is assumed to be same as air properties. But in practical case smoke properties may differ from air properties because smoke may have soot particles, fly ash particles etc. So, for finding the capacity of blowers and selection of types of blower, these factors should be considered.

Geometry has windows for fresh air. These windows are fully open in the CFD analysis. The size of the window gives the fresh air intake as the blower sucks air as well as smoke. The fresh air coming from windows helps to guide the smoke and prevent smoke from coming downwards. Therefore, at floor environment is free from smoke up to working height for workers.

C channel has two outlets and three inlets. These three inlets are distributed over the length of geometry which cover all smoke generated throughout the furnace. It is the optimization form of layout of exhaust ducts which carries the smoke. It is better to use more outlets for exhaust systems for lesser accumulation of smoke inside the working place. But the number of blowers or blower's capacity also increases for handling large smoke at outlet. Therefore the solution is optimised using CFD analysis and simulation for the problem.

Figures 98.16 and 98.17 show the comparison of path-line of smoke flow for geometry with shed and without shed. It is clearly shown in the figure that there is less eddy formation at lower side of geometry when we use shed. If shed is not used, smoke eddy form at lower height and it is not desirable. When shed is added to exhaust system eddy form at upper portion or at higher level, thus workers are safe due to no accumulation of smoke at lower level. A shed is used to guide the smoke generated, from the furnace to the exhaust outlet. From CFD analysis it is observed that smoke is guided by the shed and prevents eddies formation at lower level where fresh air enters inside the house. From the path-line it is clear that path line of smoke.

Conclusions

The objective of the work was to develop a computational fluid dynamics (CFD) simulation model for smoke movement inside the industrial shed with C shaped smoke collection channel. The purpose was to find the optimum path-line of smoke movement inside the geometry model and to estimate the power required for blowers on the basis of suction head required at exhaust and capacity of smoke handled per second (m³/s), using CFD calculation for design of industrial shed instead of expensive experimental testing and prototype production.

In this work, CFD analysis is done by varying suction pressure at outlet to c channel. It is observed that as the suction pressure at exhaust increases, smoke moves to c channel having smoother path-line and accumulation of smoke inside the geometry is less. Less smoke returns to the floor of geometry model.

For flow of smoke inside the workplace, shed design is good idea for better exhaust system. This increases the initial expenditure, but is necessary for better environment condition. From CFD analysis it observed that shed guides the smoke and less eddy of smoke is formed at lower level. All the smoke generated is collected at upper level and then drawn out by exhaust blowers.

Adding friction loss in ducts, head required for maximum mass flow rate of chimney, static pressure head and velocity pressure head at outlet to C channel gives the total head required for exhaust blower. Total power required for blower after calculation and considering blower efficiency, is 398.96 kW. There are four outlets to c channel so four blowers can be used of 99.65 kW theoretically. According to the available blowers in market best suited blower can be purchase.

References

[1] Chew, M. Y. L. and Liew, P. H. (2000). Smoke movement in atrium buildings. International Journal on Engineering Performance-Based Fire Codes. 2(2), 68-76.

[2] Li, Silas K. L., and Kennedy, W. D. (1999). A CFD analysis of station fire conditions in the buones aires subway system. ASHRAE Transactions. 105, 410-13.

[3] Xue, H., Ho, J. C., and Cheng Y. M. (2001). Comparison of different combustion models in enclosure fire simulation. Fire Safety Journal. 36, 37-54.

[4] Willemann, D. and Sanchez, J. G. (2002). Computer modeling techniques and analysis used in design of tunnel ventilation fan plants for the New York City Subway. Proceedings of the 2002 ASME/IEEE Joint Rail Conference.

[5] Shahcheraghi, N., McKinney, D., and Miclea, P. (2002). The effect of emergency fan start time on controlling the heat and smoke from a growing stationfire a transient CFD study. Tunnel Fires 4th International Conference, 2-4 December 2002.

[6] Gobeau, N. and Zhou, X. X. (2004). Evaluation of CFD to predict smoke movement in complex enclosed spaces. Research report. 255.

[7] Sakai, K., Murata, Y., Kubo, R., and Kajiya, R. (2007). A CFD analysis of ventilation system of lavatory in office building. Proceedings: Building Simulation.

[8] Brooks, R., White, P., and Logan, C. (2010). Design of smoke ventilation systems for loading bays & coach parks. Federation of Environmental Trade Associations.

Chapter 99

A study on mobile based NBFC catering towards rural Indian customers special reference to Igatpuri district in Maharashtra

Abdul Khalid[1,a] and Dr. H Sankaran[2,b]

[1]Research Scholar, mechanical engineering department, Sarvepalli Radhakrishnan University, Bhopal, India

[2]Assistant professor, mechanical engineering department, Sarvepalli Radhakrishnan University, Bhopal, India

Abstract

This research is an effort to find the impact and usage of mobile application on financial services, need of rural customers, and economic development depends on the fact that in emerging nations, people have a predisposition to think of a finance as a fundamental right, comparable to access to water, health, etc. While having access to finance may be essential for many households to lift themselves out of poverty, it does not always require a connection to a financial institution help of technology, data science and machine learning based rule engine platform tightly coupled with mobile application. This study is to show how mobile based applications are helping in rural Indian despite of those infrastructural constraints like electricity, transportation and infrastructure Becomes hurdle for getting financial needs for rural India. Major public sector banks are not present in rural India, in this scenario Non-Banking Financial Corporation (NBFC) must serve to financial needs due to electricity and power systems are having challenges to serve to customers, alternate solution to this problem is mobility. Stack of mobility comes into picture which plays a vital role becomes bridge between NBFC and customer.

Keywords: Data science, machine learning, NBFC, rule engine m-mobile platform

Introduction

A company that is engaged in the business of loan and advances, acquisition of shares, stock, bonds, hire-purchase insurance business or chit-funds business but does not include any institution whose principal business includes agriculture, industrial activities, etc. is defined as one that is registered under the Indian Company Act of 1956. RBI oversees and governs all of these business operations. The activities of Non-Banking Financial Corporation (NBFC) are similar to those of banks since they lend money and make investments, but they do not accept demand deposits or any of the other services that banks offer in exchange for them. As with banks, NBFCs are subject to Reserve Bank of India regulation (RBI). Moreover, the central bank makes public a report on the results of NBFC stress tests and their connections to system of finance. Although there is a justification for enhancing the NBFCs' asset-liability structure, there is already enough information available.

The RBI categorises NBFCs into a dozen different groups based on their activities. Asset financing, loan companies, investment companies, infrastructure, MFIs, and peer to peer (P2P) are among the categories. P2P lending was the most recent innovation when this new type of fintech lending began internationally. The RBI took an active role in bringing the P2P sector under its control by implementing new laws for it. This demonstrates that there is a distinct division between the activities of NBFCs and that all of them adhere to regulations, keep required capital, and fulfill other requirements. Liabilities for the present-day systemically significant NBFCs total 22.76 lakh crore rupees. In 2017–18, CP's share is approximately 6.17 percent, whereas the bank fund is approximately 17.74%. This demonstrates that NBFCs' liability profile is well known. It is simple to see how the liability profile has changed over time. In 2017–18, NBFCs' non–performing assets (NPAs) were around 5.8%. In contrast, the commercial bank's NPAs account for around 11.6% of its loan book. This demonstrates how less NPA there is relative to banks. In actuality, Subramanian, recently proposed a former CEA that NBFCs implement an asset quality review (AQR) similar to what banks do. This is necessary to assess the health of NBFCs and to plan for stressed loans. Retail loans made by NBFCs are expanding quickly. The assets increased by 46.2% in 2017–18. Growth from the prior year was 21.6%. Vehicle loans are the main contributor to this enormous growth, followed by consumer durables and mortgages. Such rapid growth and the source of it should be monitored by the government or regulators. In 2017–18, the NBFC industry generated total revenues of 51 lakh crore rupees and 38,600 billion rupees in net income. NBFCs' financial standing is also well-known. This information is essential for determining whether the NBFCs are operating their homes profitably or at a loss. S C Garg, Secretary of Economic Affairs, recently noted that data points on non-banking financial organisations are extremely lacking (NBFCs). He also discussed the gaps in the rules. Garg crafted this remark in reference to

[a]abdulkhalid06@outlook.com, [b]nikhil.rathod2008@gmail.com

DOI: 10.1201/9781003450252-99

the IL&FS situation because everyone was caught off guard when it suddenly became a problem. What initially appeared to be a single large NBFC called leasing for infrastructure and financial services defaulting on its debt has grown to such enormous proportions that it now threatens to bring down the entire industry and jeopardies several businesses that provide asset financing and personal loans. While the fight between the union government and the Reserve Bank of India (RBI) over several issues, including refinancing for NBFCs, is headed for a climactic showdown at the central bank's next board meeting on November 19, the crisis in India's NBFCs, caused by the IL&FS fiasco, has taken centre stage in the economic debate. The country has over 11,000 NBFCs registered with the RBI, which have boosted India's consumer economy as retailers from all industries jumped on the EMI bandwagon to entice customers. There is already talk that the entire sector may collapse unless drastic measures are taken to keep it afloat after IL&FS defaulted on payments to lenders and caused panic in the markets. The sustainability of this shadow banking sector, which has grown at a rate of about 20% and currently has a sizable aggregate book of Rs 26 lakh crore, has been questioned in a number of ways. As a result, it has grown to play a significant role in the gross domestic product of our country, with NBFCs alone accounting for a 12.5% increase. The majority of individuals favor NBFCs over banks because they think they can provide their financial needs quickly, efficiently, and safely. Additionally, a variety of loan products are offered, and their services are flexible and transparent.

Review of Literatures

By demonstrating its various perspectives, the survey of literature gives an understanding of the present topic and its vast scope. The evaluation of readily available literature reveals the scope, emphasis, depth, boundaries, and depth of analysis already performed on the research issue. As a result, the reviews are helpful in creating the hypothetical framework, identifying the research gaps and reflecting the potential research fields, as well as in defining the research objectives and more precisely putting together the research methods for the current study. The fundamental investigation is shaped and guided by the evaluation of the available research studies. This chapter has been broken down into four main sections and is organised chronologically. Studies that focused on the peculiar characteristics of Indian NBFCs included mounting risk appraisal models, developing models to forecast financial sustainability, analysing the effects of mergers and acquisitions, examining the financial connections between banking and non-banking entities, as well as setting up and pricing strategies. Final thoughts on the literature review are made at the end.

Acharya et al. [1] investigate the factors that led to the significant growth among non-bank financial institutions that accept deposits in India during the past ten years. These NBFCs are regarded as systemically significant by the Reserve Bank of India. We NBFC liabilities are shown to be significantly influenced by bank NBFC lending, which (i) accounts for a sizeable part of the liabilities of NBFCs; (ii) varies according to the allocation of banks to priority lending industries; (iii) declines as banks open more locations in rural areas compared to urban areas; however (iv) is essentially nonexistent for State Bank of India (SBI), the largest state-owned bank, and its affiliates, which have a sizable rural wing Bbeginning with the fall 2008 financial emergency, there was a permanent contraction shock in bank lending to NBFCs because of the change. These bank-NBFC relationships are predominantly existent for, and have an impact on, those NBFCs that finance assets or loans but not investment firms. In general, the results suggest that contrary to popular belief, lending to NBFCs in India is seen by banks as a replacement for direct lending in the non-urban sectors of the Indian economy. However, this replacement is hampered by irregularities in bank deposit flows because of the apparent differences in the level of government backing for certain banking groupings.

The current study provides a thorough grasp of how behavioral biases can be reduced by robo-advisory from an expert's point of view. These professionals work in India's information technology, finance technology, NBFCs, and Banking and Financial Services Industry (BFSI) as members of the top management level, the quality control team, middleware, or the product development team [2]. The researchers used an in-depth interviewing technique to learn more about the phenomena and gain understanding of automated advice services. Every interview was captured on tape and verbatim transcribed. Following the completion of the literature review, a detailed examination of the transcript material was performed using various categories. The findings are not meant to be generalised because the field of robotic advice is still in its infancy.

The 2007–2008 global financial crisis made it clear that there are some established financial principles that need to be reviewed [3]. The current paradigm has been questioned across the globe, which has led to an acceptance that modern financial structures and models must arise in order to take into take into consideration the variety of socioeconomic conditions that exist worldwide. In this backdrop, the requirement for a new financial system in India is pretty obvious, and the panel discussion that follows raises certain challenges unique to India that the various parties participating in this endeavor need to explore.

A new, structured product called securitisation has arisen to suit the funding needs of microfinance firms. This essay provides background information on the MFI industry and its funding sources. Interviews with senior executives from two microfinance institutions about their securitisation transactions follow the note [4]. In our note, we contend that in order to achieve the larger objective of financial inclusion, the microfinance industry must be revitalised. To do this, cooperation between banks and MFIs is necessary. Banks must assist MFIs in making the switch to low cost financing by providing cash advances or creative solutions like securitisation. Business banks must use MFIs to expand their small business origination and recovery capabilities.

The relative significance of microfinance organisations at the macro and micro levels is examined in this essay. We see a big impact on the majority of fronts. Participation of MFIs improves the economy's overall allocation of credit and savings. By lowering income disparity and poverty, their involvement improves economic welfare. Additionally, by putting traditional commercial banks under more competition and encouraging greater efficiency, their active presence helps to discipline them [5].

Islamic microfinance institutions were founded to offer Qard Hasan in fostering commercial activities and the growth of entrepreneurship among microloan borrowers [6]. The current practices that the IMFIs in the areas of: 1) social ideals of an Islamic perspective that can contribute favorably to Islamic microfinance, 2) the use of bank using a mobile device capacity to increase availability of financial services, and 3) the introduction of mobile banking in aiding mechanism for recording financial transactions are critically discussed in this paper. The project aims to make key discoveries that will lay a solid foundation for the effective model of IMFIs.

We took quotes from the literature on banking financial products and used them in 100 interviews with elderly banking customers [7]. We discovered that the genre's four key components— the narrative, the scene, the narrative, and the story—all failed to engage the target audience. By creating a template that authors of the genre can follow to fix this issue, we contribute to practice. We also advocate the use of metaphors and synonyms and contend that as part of a quality control procedure, draughts should be subjected to reader feedback. After that, the genre live up to its audience's expectations and prudent financial choices for everyone's financial future to be better of us will be made.

The procedure for establishing regulations for the European Union's structural banking reform, where it was suggested to restrict the ability of universal banks to conduct both trading and deposit-taking operations, was the subject of this study. However, no regulation was ultimately implemented [8]. We examine this instance of international non-regulation by utilising the conceptual idea a law's endogenisation, which emphasises the influence of the controlled in forming the perspectives of the regulators. In order to clarify the endogenisation procedure, we deploy institutional maintenance efforts to pinpoint the tactics employed by the regulated and their allies. We see that upkeep work includes mythologising the achievements of the global banking concept, criticising the reform ideas. The study of the study presented in this paper to know the impact, usage and awareness of mobile based banking services offered by NBFC special preference to rural India (gatpuri district).

Testing of research hypotheses

The regression analysis was to evaluate the research hypotheses with the help of SPSS version 21.0, a regression analysis was carried out, and the significance of the coefficients and t-values were assessed in order to accept or reject the hypotheses. When the estimated value of t exceeds the table value of $t = 2.58$ based on a two-tailed test with a p-value of less than 0.05, the t-value is significant. Linear regression is the most elementary and frequently used predictive analysis. When one variable is regarded as an explanatory variable while the other is regarded as a dependent variable.

Hypotheses of the study

Ho1   Regarding mobile-based services, NBFCs don't significantly differ from one another in their procedures.

Ha1   There is significant difference in NBFCs practices regarding mobile based services.

Ho2   There is less awareness about mobile-based services offered by non-banking financial companies amongst customers.

Ha2   There is much awareness about mobile-based services offered by non-banking financial companies amongst customers.

Ho3   Demographic factors do not play an important role in customers' perception about mobile based practices in Igatpuri.

Ha3   Demographic factors play an important role in customers' perception about mobile based practices in Igatpuri.

Ho4   NBFCs better practices do not contribute to improving an organisation's brand value.

Ha4   NBFCs better practices contribute to improving an organisation's brand value.

Reliability of data

The reliability analysis was conducted with the help of data collected from the respondents (Table 99.1). Cronbach's alpha value for the views and perception of respondents and mobile based practices offered by selected NBFCs in Igatpuri is 0.819. It shows the higher degree of reliability and suitable for research study and regression analysis to get the proper results (Table 99.2).

Testing of first hypothesis

Ho1   Regarding mobile-based services, NBFCs don't significantly differ from one another in their procedures.

Ha1   Regarding mobile-based services, NBFCs' practices differ significantly.

Table 99.3 presents the regression estimation for assessing the significant difference in mobile based service and practices offered by selected NBFCs in Igatpuri district. The study considered account details (6), next EMI dues status (8), top-up facility (9), interest payment (10) and part payment (11) as independent variables and significant difference in mobile based practices of selected NBFCs (12) as a result of the survey's questions, as a dependent variable. The coding of a question in a questionnaire is indicated by the numbers used with variables in brackets. The statement, as shown in Table 99.1, is statistically significant because the calculated value of t is higher than the table value (t-value = 5.021, p 0.01). In this way, we reject the null hypothesis and accept the alternative hypothesis, which is that there is significant difference in mobile based services and practices of selected NBFCs. Among the five independent variables, account details (.77) have a greater influence about the differences in mobile based practices of selected NBFCs followed by next EMI due status (.50) and top-up facility (.45).

Testing of second hypothesis

Ho2   There is less awareness about mobile-based services offered by non-banking financial companies amongst customers.

Table 99.1 Reliability of data

NBFCs and their mobile based service and practices	Cronbach's Alpha	Cronbach's Alpha based on standardised items	No of variables
Mobile Based Services and Practices Offered by NBFCs	.819	o.819	33

Source: Reliability analysis calculated by SPSS

Table 99.2 Some important factors considered for study: Mobile based services

Labels	Question No	Name of factor	Construct/Statement	Factor loading
1	10	Mobile based services	Interest payment	0.821
	11		Part payment	0.748
	09		Top-up facility	0.732
	07		Repayment schedule	0.714
	05		Wallet facility	0.702
	15		Telephone calls	0.699
	16		Television Advertisement	0.693
	17		Emails	0.688
	18		Agents/executive	0.672
	19		Social-media	0.659
	23		Income	0.592
	24		Education	0.563
	26		Gender	0.548
	27		Work experience	0.539
	29		Prompt services	0.519
	30		Positive response	0.501
	32		Less costly	0.585

Source: Component Matrix calculated by SPSS

Ha2 There is much awareness about mobile-based services offered by non-banking financial companies amongst customers.

Table 99.4 presents the regression estimation for awareness about mobile based services and practices. The study considered television advertisement (16), agent/executive (18), social media (19), print media (20) as independent variables and awareness about mobile based services amongst stakeholders (21) from the survey questionnaire as a dependent variable. As seen in Table 99.4, estimated value of t is less than the table value (t-value = 0.38, $p > 0.01$), the assertion is not statistically significant. Thus, the alternative hypothesis is rejected and the null hypothesis is accepted. There is less awareness about mobile based services amongst stakeholders.

Testing of third hypothesis

Ho3 demographic influences do not play an important part in customers' perception about mobile based practices in Igatpuri.
Ha3 Demographic factors play an important role in customers' perception about mobile based practices in Igatpuri.

Table 99.5 presents the regression estimation for demographic factors have important role in customers' perception about Mobile based services and practices. The study considered age (22), occupation (23), education (24), gender (26), and work experience (27) as independent variables and important role in stakeholders 'perception (28) as a dependent variable from the questionnaire for the survey. Table 99.3 shows that the statement is statistically significant because the calculated value of t is higher than the table value

Table 99.3 Regression estimation for significant difference in mobile based services and practices

Independent variables	Standardized coefficients (Beta)	t-value	Sig.
Constant		5.021*	.000
Account details (6)	.77		
Next EMI due status(8)	.50,		
Top-up facility(9)	.45		
Interest payment (10)	.21		
Part payment (11)	.15		

Note: *t-value is significant for $p < 0.01$

Table 99.4 Regression estimation for awareness about mobile based practices

Independent variables	Standardised coefficients (Beta)	t-value	Sig.
Constant		0.38**	.14
Television advertisement (16)	.06		
Social media (19)	-.05		
Print media (20)	-.32		
Agent/executive (18)	.12		

Note that for p 0.01 the t-value is not significant.

Table 99.5 estimation of regression using demographic variables play important role in customers' perception

Independent variables	Standardised coefficients (Beta)	t-value	Sig.
Constant		25.484*	.000
Gender (26)	.44		
Age (22)	.09		
Education (24)	.61		
Occupation (23)	.52		
Work experience (27)	.58		

Note: *t-value is significant for p < 0.01

(t-value = 25.484, *p* = 0.01). Consequently, we disregard the null hypothesis and favor the alternative one i.e., demographic factors play an important role in customers' perception about mobile based services and practices in Igatpuri district. Education (.61) among the independent factors has a significant effect on the stakeholders' perception, followed by work experience (.58), income (.52) and gender (.44).

Testing of forth hypothesis

Ho4   NBFCs better practices do not help to enhance the brand value of organisation
Ha4   NBFCs better practices help to enhance the brand value of organisation

Table 99.6 presents the regression estimation for better services and brand value of the NBFCs. The study considered prompt services (29), positive response (30), timely services (31) less costly (32) as independent variables and NBFCs better services and practices enhance brand value (33) based on the survey questionnaire's dependent variable. The assertion is statistically significant, as can be seen in the table, because the estimated value of t is higher than the table value (t-value = 3.216, *p* 0.01). Thus, we reject the null hypothesis and accept the alternative one i.e., NBFCs better practices help to enhance the brand value of organisations. One of the independent factors is, prompt services (.63) have a significant impact on increasing the value of brands and followed by positive response (.27). Table 99.7 shows the results of the testing of hypotheses.

Findings based on research problem, hypotheses and objectives of the study

- The present research tried to assess and indicate the status of the respondents who have smart phone and use it. The study found that 534, (89%) respondents have admitted that they have smart phones in Igatpuri district and remaining 66, (11%) have denied to the same.
- The study revealed and narrated the status of the respondents who deal with NBFCs and mobile based services offered by NBFCs. 345, (57.5%) respondents have admitted that they deal with NBFCs and use the mobile based service in Igatpuri district and the remaining 255, (42.5%) have said no to it.
- It is found under the research study about the non-banking financial companies and beneficiaries associated with the mobile based services offered by NBFCs in Igatpuri district including three wards. Most beneficiaries are with the Muthoot micro fin Ltd. With 102, (17%) respondents out of 600. HDFC investment limited got 96, (16%) after that Mahindra and Mahindra financial services got 81, (13.5) and found third position as per the responses from the beneficiaries. BoB financial solution got 72, (12%) with forth position, Bajaj finance Ltd got 69, (11.5%) with fifth position, Axis finance Ltd.

Table 99.6 Regression estimation for better services and practices help to enhance the brand value

Independent variables	Standardised coefficients (Beta)	t-value	Sig.
Constant		3.216*	.000
Prompt services (29)	.63		
Positive response (30)	.27		
Timely services (31)	.18		
Less costly (32)	.13		

Note that t-values are significant when *p* = 0.01

Table 99.7 Results of the testing of hypotheses

Hypotheses (H)	t-value	p-value	Results
Ha1: There is significant difference in NBFCs practices regarding mobile based services.	5.021*	.000	Accepted
Ha2: There is much awareness about mobile based services offered by non-banking financial companies amongst customers.	0.38**	.14	Rejected
Ha3: Demographic factors play important role in customers' perception about mobile based practices in Igatpuri.	3.216*	.000	Accepted
Ha4: NBFCs better practices help to enhance the brand value of organisation.	25.484*	.000	Accepted

Note: *indicates that the t-value is significant for *p* 0.01 and **indicates that it is not significant.

got 66, (11%) sixth position, Edelweiss finance and investment Limited 60, (10%) seventh position, and eight position booked by Anand Rathi global finance ltd. with 54, (9%) respondent out of 600 respondents.

- Research study tried to examine and found about the level of awareness among the respondents to concern with mobile base services and practices offered by NBFCs in Igatpuri district. The 345, (57.5%) beneficiaries admitted that they are much aware about the practices and services and remaining 255, (42.5%) beneficiaries were not aware about the same.

- The research study made attempts to know the views of the respondents about the various mobile based services offered by NBFCs in Igatpuri district. In which study found that 336, (56%) respondents admitted that NBFCs perform these activities and the remaining 264, (46%) respondents were not convinced with the services.

- The research study found and specifically narrated about the fields and areas covered by NBFCs towards mobile based services and practices in Igatpuri district. Respondents have booked their views separately. The details of each and every area and field covered in this study is as follow: transfer of money got first position with 540 responses out of 600 at highest among all having 90%, remittance of saving got fifth position with 480, 80% responses, loans and advances got 498, 83% responses with forth position, financial education got, 463, 77% responses with sixth position among all, insurance got 516, 86% responses with second position, bills and tax payment services got 507, 84% responses with third position, receiving pension schemes got 234, 41.5% responses with tenth position, government grants got 249, 41.5% with ninth position, technology enabled transaction got 441, 73.5% respondents with eight position and advisory and financial counselling services got 459, 76.5% responses with seventh position among all above fields and areas towards mobile based services.

- Research study tried to examine the different forces behind performing the various mobile based practice for the customers by NBFCs. Respondents have been sharing their diverse views while focusing the different variable and constructs. The preference of respondents is as follow: Govt. policies got 147, 24.5% responses out of 600, image building 102, 17%, increase business 198, 33%, uplift the society 75, 12%, automated services 66, 11%, and other reasons got 12 responses, 2% respectively.

- Research study analysed the views and opinion about the status of physical visit to the branches and offices of NBFCs by the customers for services among different groups of respondents. When it was enquired about the role of NBFCs and their services and programs which caused to the less visit of the customers, 390 respondents admitted yes NBFCs are making efforts to minimise the physical visit and rest of the 210 respondents said no to it.

- This study tried to find out the views and opinion of the respondent about the role of regulatory bodies like RBI to concern with the forces behind having mobile based services. A total of 309 respondents have admitted that regulators force NBFCs to offer mobile based services to the customers and remaining 291 respondents were differing from the statement.

- Research study found the status of the respondents, views towards the mutual benefits with the help of various mobile based services offered by NBFCs. In this regard 378, (63%) of the respondents have admitted and the rest of the 222, (37%), have responded no to it.

- It is found under the present research study about one of the significant variables to concern with making better image of NBFCs while offering various programs for the mobile based services. There were two options, and it was asked to the respondents, in response of that 405, (67.5%) respondents said yes whereas 195, (32.5) respondents said no to as their answer.

- Research study indicated towards the services and practices offered by NBFCs for awareness and attentiveness through mobile based services in Igatpuri district. A total of 318, (53%) respondents have admitted that NBFCs offer practices and services for awareness and attentiveness among the customers and the remaining 282, (47%) were viewed as no to it.

- It is found about the status of the respondents and their views about financial education to customers with the help of mobile-based services offered by NBFCs. In this regard 339, (56.5%) of the respondents have admitted yes and the rest of the 258, (43.5%), have negative views about the same.

- Research study found the status of respondents and their views about better financial services and practices offered by NBFCs for customers in Igatpuri district. A total of 357, (59.5%) respondents have admitted that they are convinced with the practices and services and the remaining beneficiaries 243, (40.5%) were not.

- The study tried to indicate towards the financial literacy as services and practices offered by NBFCs for the customers in Igatpuri district. A total of 438 respondents have admitted that they get these practices and services whereas the remaining 162, (27%) respondents did not think so.

- A research study found the status of the respondents and their views towards the time and money saving through mobile based services offered by NBFCs. In this regard 322, (52%) of the respondents have admitted yes and the rest of the 288, (48%), have responded no to it.

- Table no. describes the status of respondents and their views about merits of mobile based services as services and practices offered by NBFCs for customers in Igatpuri district. The 357, (59.5%) customers have admitted that they are having merits and are convinced with the practices/services remaining respondents 243, (40.5%) are not convinced with the statement.
- Research study found the regression estimation for assessing the significant difference in mobile based service and practices offered by selected NBFCs in Igatpuri district. The study considered account details (6), next EMI dues status (8), top-up facility (9), interest payment (10) and part payment (11) as independent variables and significant difference in mobile based practices of selected NBFCs. (12as a result of the survey's questionnaire, as a dependent variable. The coding of a question in a questionnaire is indicated by the numbers used with variables in brackets. The statement is statistically significant because the calculated t value (t-value = 5.021, p = 0.01) is higher than the table value. As a result, we accept the alternative hypothesis—that there are notable differences in the mobile-based services and practices of a subset of NBFCs—and reject the null hypothesis.
- It is presented under the study and regression estimation for awareness about mobile based services and practices. The study considered television advertisement (16), agent/executive (18), social media (19), print media (20) as independent variables and awareness about mobile based services amongst stakeholders (21) as an outcome of the survey questionnaire. Given that the calculated value of t is less than the value in the table (t-value = 0.38, p > 0.01), the statement is not statistically significant. Thus, we reject the alternative hypothesis and accept the null one i.e., there is less awareness about mobile based services amongst stakeholders.
- Research study found about the regression estimation for demographic factors have important role in customers' perception about mobile based services and practices. The study considered age (22), occupation (23), education (24), gender (26), and work experience (27) as independent variables and important role in stakeholders' perception (28) derived from the survey questionnaire as a dependent variable. The statement is statistically significant because the calculated t value (t-value = 25.484, p = 0.01) is higher than the table value. Consequently, we disregard the null hypothesis and embrace the alternative explanation, which is the demographic factor plays important role in customers' perception about mobile based services and practices in Igatpuri district.
- Research study found about the regression estimation for better services and brand value of the NBFCs. The study considered prompt services (29), positive response (30), timely services (31) less costly (32) as independent variables and NBFCs better services and practices enhance brand value (33) as an outcome of the survey questionnaire. The assertion is statistically significant because the estimated value of t is higher than the table value (t-value = 3.216, p = 0.01). As a result, we accept the alternative hypothesis rather than the null hypothesis, which is NBFCs better practices help to enhance the brand value of organisations. One of the independent factors was, prompt services (.63) have major effect on increasing the value of brands and proceeded by positive response (.27).

Conclusion

- This study is restricted to the key determinants of non-banking financial companies and their services offered by them in Igatpuri Maharashtra. Under the present research study, we have analysed the different practices with special reference to fields and areas to concern with financial services. These services and practices were evaluated based on the information obtained from the available secondary information presented as annual reports, literature, and previous research. On the other hand, fresh information, inputs, views, and ideas have been captured from the responders using a carefully constructed survey and dedicated questionnaire to the research problem. The current study denotes the present NBFCs and mobile based practices in the areas and parameters which are the most effective and enable greater scalability in assessing it, in terms of views and perception about it. This research study has very vast academic and research scope for the future research scholars in the field of NBFCs and their services and the perception about the same. This will give great help to the respective and distinguished research scholars, academicians, administrators, and understand mobile based application and impact on rural India regarding NBFCs operations and practices. This study's findings will be very beneficial for the organisation and authorities to change, adapt, and amend processes to maintain strong management system based on the findings of this research.

References

[1] Acharya, V. V., Khandwala, H., and Sabri Öncü, T. (2013). The growth of a shadow banking system in emerging markets: Evidence from India. Journal of International Money and Finance. 39, 207–30.

[2] Bhatia, A., Chandani, A., and Chhateja, J. (2020). Robo advisory and its potential in addressing the behavioral biases of investors — A qualitative study in Indian context. Journal of Behavioral and Experimental Finance. 25, 10028.

[3] Basu, S. (2016). India emerging: New financial architecture. IIMB Management Review. 28, 170–8.

[4] Jayadev, M. and Rao, R. N. (2012). Financial resources of the microfinance sector: Securitisation deals e-Issues and challenges Interview with the MFIs Grameen Koota and Equitas. IIMB Management Review. 24, 28e39.

[5] Abra, A., Hasan , I. and Kabir, R. (2021). Finance-growth nexus and banking efficiency: The impact of microfinance institutions. *The Journal of Economics and Business*. 114, 105975.

[6] Amran, A. M., Rahman, R. A., Yusof, S. N. S. and Mohamed, I. S. (2014). The current practice of Islamic microfinance institutions' accounting information system via the implementation of mobile banking. Procedia - Social and Behavioral Sciences 145, 81–87.

[7] Helliar, C. V., Lowies, B., Suryawathy, I. G. A., Whait, R., and Lushington, K. (2022). The genre of banking financial product information: The characters, the setting, the plot and the story. The British Accounting Review. 54(5), 101131.

[8] Munzer, M . and Pelger, C. (2022). Maintaining the universal banking model – A study of institutional work in the endogenization of a failed transnational post-crisis financial market reform. Critical Perspectives on Accounting. 95.

Chapter 100

Effect of learning rate on the performance of VGG-16 for prediction of cardiomegaly in chest X-ray images

Dr K Prasanna[1,a], Dr S Ravi Kumar Raju[1,b] and Dr J Krishna[2,c]

[1]G. Narayanamma Institute of Technology and Science, Hyderabad, India

[2]Annacharya Institute of Technology and Sciences, Rajampet, India

Abstract

Computer aided diagnostics (CAD) helps radiologists in medical examinations that use imaging technology like computed tomography, MRI, and X-rays to diagnose illnesses and treat injuries. Convolutional neural networks (CNN) have demonstrated favorable and extraordinary outcomes in fields like image identification in the field of neural networks and deep learning techniques. The technological improvements over CNN made CAD easier in medical diagnosis. A CNN model's performance is essential to a successful CAD system. Finding a high-performing CNN model is a difficult undertaking because it heavily depends on selecting the proper hyper-parameters and having sufficient training data. This research's objective was to assess the impact of various hyper-parameter settings on a particular VGG-16 model performance. In the study, a CNN-based VGG-16 was employed to identify cardiomegaly, a condition in which a patient has an enlarged heart. Investigated hyper-parameters including learning rate, batch size, and several optimisers. The model was trained with a variety of various values of the hyper-parameters during each training session on a dataset of X-ray images. The learning rate, which may be regarded as the most significant hyper-parameter based on the results, has shown a remarkable impact on the performance of the VGG-16 model. The optimiser option had minimal impact on the model's performance and obtained an equivalent level of accuracy when the batch size and learning rate were set to the correct values. Additionally, it was discovered that lower learning rate values often produced better performance compared to higher learning rates that combining a small batch size with a lower learning rate would produce the best model performance.

Keywords: Learning rate, optimiser, batch size, VGG-16, X-ray image, cardiomegaly

Introduction

Due to recent advances in deep learning, computer-aided diagnosis (CAD) research is growing in popularity [1]. CAD systems are used to give doctors a second look when they examine medical images to diagnose various disorders [2]. Convolutional neural networks (CNNs), one type of deep learning image classification technique, can be used to further increase the accuracy and efficiency of CAD systems. Despite being created more than 30 years ago, CNN's were not put into operation until the mid-2000s due to a lack of labeled data and computing power [3]. As observed from the recent study, CNNs are an effective technique that can be utilized by the CAD sector and has generated effective results showing promising in the diagnosis of medical images [4].

The VGG-16 model is one of several CNN architectures currently in use that offer results for image classification problems that are reasonably accurate. The x-rays that will be used in this work are one of the primary forms of medical imaging used in such research. These photos are used to train a CNN model to find patterns and characteristics associated with a particular disease to make a diagnosis. Radiologists use chest X-rays to examine patients and learn more about their health and possible chest diseases. Cardiomegaly, a condition where a patient has an abnormally large heart, is an example of a condition that may be diagnosed early with this imaging technique is a disease [5]. However, selecting the best hyperparameters is a difficult and time-consuming operation, and the success of CNN greatly depends on it [6]. In addition, due to the increase in demand for medical examinations and advances in imaging technologies, radiologists have difficulty in the manual interpretation of medical images, which results in inaccurate or indecorous diagnoses [7].

In CNN architectures, due to limited computing power, the need for adequate and reliably labeled data points, and the difficulty of determining the ideal configuration of hyperparameters, training a model is a difficult task. The purpose of the paper is to assess the effectiveness of categorizing chest X-ray images when trained using different hyperparameter settings. Learning rate, batch size, and various optimizers will be investigated as hyperparameters.

Different hyperparameter settings are expected to affect model performance. Some hyperparameters can affect performance more than others, either increasing or decreasing. The choices of hyperparameters are

---

[a]prasanna.k642@gmail.com, [b]ravisayyaparaju@gmail.com, [c]krishna.j.jk@gmail.com

DOI: 10.1201/9781003450252-100

also affected by the model architecture, and to solve this problem we focused on the VGG-16 model. The hyperparameters associated with the training phase were used in the present study and parameters related to the architecture were ignored in the present research.

The performance of the models is also affected by the lack of a quantitative and qualitatively sufficient, reliable data set with accurate labeling. Although the data set used in this study is balanced, it is not ideal because it is a small data set. To deal with this, we tried to reduce the impact of a small dataset using transfer learning and data augmentation. Because it offers more control over CNN behavior than using hyperparameter tuning algorithms, human trial and error is used to identify the best hyperparameter settings.

The second section deals with the theoretical basis and related work, which are used in this work. A description of the architecture designed and implemented is described in section 3. The method of experimentation and the measured results on the networks can be found in section 4 and the conclusion is given in section 5.

Basic Preliminaries and Related Work

Cardiomegaly

Cardiomegaly is a general word for defining medical conditions where the heart is enlarged and causes illness in the patient. Despite not being a disease, it nonetheless has a high death rate since heart failure is more likely to occur. Cardiomegaly is diagnosed using the cardiothoracic ratio (CTR) on medical images including computed tomography (CT) scan images and X-ray images of the patient's chest [5]. In general, for healthy patients, the CTR value will be equal to or less than 0.5. If the patient's CTR is greater than 0.5 then it is diagnosed as cardiomegaly. The CTR is evaluated as the ratio between the maximum heart diameter and length between the inner edges and the ribs of the chest X-ray, both seen in Figures 100.1 and 100.2 [8].

X-rays and Radiology

Modern medicine has relied heavily on X-ray imaging [7]. Radiology is a branch of medicine that employs image processing techniques and produces medical images that encompass critical health information. This information is further used to map the source of symptoms, diagnose diseases and treat injuries. These radiographic images can be produced by a variety of imaging techniques, including X-rays, magnetic resonance imaging (MRI), computed tomography (CT) scan images, and X-ray images. X-ray images are the

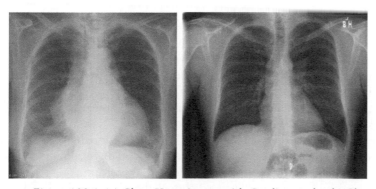

Figure 100.1 (a) Chest X-ray image with Cardiomegaly, (b) Chest X-ray image with cardiomegaly [8]

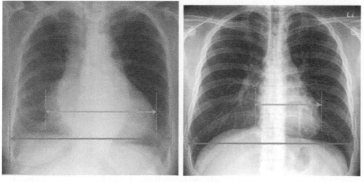

Figure 100.2 Biomarkers for measuring the CTR from the chest X-ray: Left image with Biomarkers of Cardiomegaly and Right image with Biomarkers without cardiomegaly [8]

most employed imaging techniques because it is affordable, simple to use, and painless for the patient. A chest X-ray, which produces a 2 D digital image, is used with CTR measurement to diagnose cardiomegaly. CTR is evaluated as X/Y, Where X is the maximum heart diameter (orange line in Figure 100.2) and Y is the length between ribs and inner edges [9]. Compared to other procedures, it takes less time, and an X-ray image can often be obtained in 30 minutes or less [10]. In its most basic form, the procedure is that radiologists see an image and analyse what they observe using their cognitive and perceptual abilities [11].

Problems with the Interpretation of Medical Images

Since the developments in imaging technologies, the generation and analysis of medical images are becoming a notch in disease diagnosis and the manual interpretation of medical imaging has gotten more difficult [7]. One billion radiologic imaging exams are reportedly performed worldwide each year, with radiologists interpreting a significant portion of them. The norm in X-ray and CT scan exams is this manual interpretation. Writing reports and doing a second exam with another radiologist are time-consuming steps in the process. Moreover, it increases the impact of human error on the procedure [12]. According to [11] estimated error rate for radiological diagnosis is between 3% and 5% where the 3% error rate translated into approximately 30 million errors [13].

Perceptual and cognitive errors fall under two types of radiological errors. Perceptual errors, which account for 60-80% of radiologists' errors, are the most common of the two types of errors. When a radiologist ignores the significance of an anomaly he has identified from an image, he has made a cognitive error. While a perceptual error occurs when the radiologist first misses an anomaly that can be seen on subsequent review and examination of the images [14].

According to WHO, the availability of radiologists for medical analysis is very tedious and there is a tremendous requirement by 2030 when the number of radiologists required will be doubled [10]. Depending on the skill level of the radiologist, the time it takes to review chest X-rays varies. Because of the less contrast and overlap with other organs and tissues, many medical images are difficult to interpret and require considerable, specialised training as well as a degree of skill on the part of the radiologist. A skilled and experienced radiologist may need a few minutes to review the image and create a report. Many radiologists must work overtime due to the dearth of radiologists and the increasing workload, which raises the risk of radiographic misdiagnosis, and risks to the patient will increase [10]. It is thought that misdiagnosis and delayed diagnosis are issues that need to be addressed since they can cause patient damage and prolonged response times for treatment.

Convolutional Neural Network

A type of DNN called CNN is used for image pattern recognition [15]. Each neuron receives stimulation in a limited area of the visual field known as its receptive field. The entire visual field is made up of all receptive fields. The input data is compressed in a CNN and transmitted among receptive fields among neurons rather than to its neighbor. A significant portion of the input in a process known as sparse interaction is selected where a kernel is smaller than the input. This results in the early detection of simpler patterns and later in the processing of more complex patterns. Fully connected layers, convolutional layers, and pooling layers are the three types of layers that make up a CNN. With this kind of topology, CNNs can be trained quickly and with a network that has several layers. The digital image is fed to a CNN for image classification, and the network analyses a sequence of pixels organised as a grid form of the digital image data. The dot product between the kernel matrix and a subset of pixels is performed by convolutional layers. To reduce the dimension of a subset, a filter is used by the pooling layer to choose the pixel within the filter dimension with the largest value. As a result, as the image representation is compressed, the number of weights (parameters) and necessary calculations decreases. A fully connected layer, as its name suggests, has all its neurons connected to both the input of the layer below it and the output of the layer above it, allowing for matrix multiplication. The next layer uses convolutional layers, which combine linear and nonlinear layers. The operations on nonlinear layers are done using Sigmoid and ReLU activation functions [16].

Hyper-parameters

Despite the positive results of CNN research, finding a high-performing model is still challenging as it requires years of experience, deep knowledge, and network training. The choice of hyperparameters has a significant impact on the performance of the network during training because they determine how the network behaves [6]. It can be challenging to determine ideal model hyperparameters. Today, there are several ways to determine the best hyperparameters using manual testing and tuning algorithms on different settings. The tuning techniques require a lot of effort and time when performed using grid search and random search [19].

Hyper-parameters and parameters are two different categories of variables found in CNN networks. The parameters are dynamic variables that are altered as the model processes the data. Hyper-parameters are static variables that are pre-set before training the model and the parameter values do not change throughout training. Many hyper-parameters are available, including optimizer type, batch size, learning rate, number of epochs. How rapidly the network updates the optimizer's settings depends on its learning rate. The number of epochs is determined based on the size of the training set and the number of times the training set to processed by the model. An epoch is defined as a complete pass over the training data points. The batch size is the number of samples from the dataset that are processed in each epoch. The optimiser will be used to specify which optimisation algorithms are suitable for the model to enhance accuracy and performance. The Adam algorithm and stochastic gradient descent are two examples of the various types of optimisers [15].

Related work

Increased computer performance and technical development have made deep learning techniques for the categorization of medical images that show promise. Research has been done to assess and demonstrate the possibilities of employing CNN to medical images for the prevalence of the disease diagnosis. Consequently, choosing the best hyper-parameter configuration for higher-performing networks is a challenging task. Both a study of hyper-parameters in neural networks and research developing models for the use of the diagnostic of chest ailments using X-rays are provided.

Bougias et al. [6] provided a summary of the difficulties in determining the ideal hyper-parameters for different neural networks. The performance and training effects of various hyper-parameter settings on several neural network architectures were compared using a dataset of plant communities. It was shown that utilising the Adamax optimiser, a variation of the Adam, the learning rate's optimal value existed in the range of 0.1-0.001, indicating that lower learning rates are desirable to prevent unstable training.

DNNs are often employed to categorize a variety of chest illnesses [20]. The study suggests an enhanced architecture that outperforms current architectures, particularly when applied to medical imagery. Given that it is a multiclass classification, the model's average accuracy achieved 89,77 of the various illnesses, which is an incredibly excellent performance.

Cardiomegaly was diagnosed using the ResNet model and an explainable feature map model [21]. The model was implemented on NIH chest X-ray data using the optimisers Adam, AdaGrad, SGD, and RMSProp are used in evaluating the model accuracy, and it resulted in a near 80% accuracy rating [24].

Effect of Learning Parameters on VGG-16 for the Prediction of Cardiomegaly

Dataset

The NIH chest X-ray dataset at Kaggle [22] will be used for testing in this paper. It was taken from the State Institute of Health. The dataset has 30,000 different 1024 x 1024 patient-quality chest X-ray images, for a total of 112,000. Each image is marked with "no findings" or a list of one or more diseases. There are 14 separate diseases and 15 classifications in total that are in the class "no findings". The study of hyperparameters is the focus of this research, so it would not make sense to train a deep-learning VGG-16 on so many different diseases. Consequently, the VGG-16 will only be trained on the previously specified cardiomegaly disease.

We use the Pandas package to get the required data points for the experiment. For projects like this that involve data analysis, this library can be used. The first step in the procedure was to filter solely classified as cardiomegaly data points and remove the data points that have other conditions indicated under multiple labels. These criteria were satisfied by 1093 data points in total. The same process was subsequently used to get a total of 1093 data points labeled as "no finds." Where the identities for the 14 diseases were not listed, however, the label "no findings" does not exclude the presence of other diseases.

VGG-16

VGG is one of the best-known architectures used in image data classification and analysis [18]. Layers alternate in it, which are divided into blocks. Each block consists of a two-dimensional Convolution layer, a batch normalization, a ReLU activation function, and a maximum pooling layer. This combination is repeated twice in each block as shown in Figure 100.3. There are several blocks in a row, followed by one or more linear layers, which form a vector at the output, indicating the evaluation of the probability that the image belongs to the given category.

Image classification and facial recognition applications use CNN architectures from the VGG family, which is a popular and promising set of architectures. Finding out how the network's depth (number of

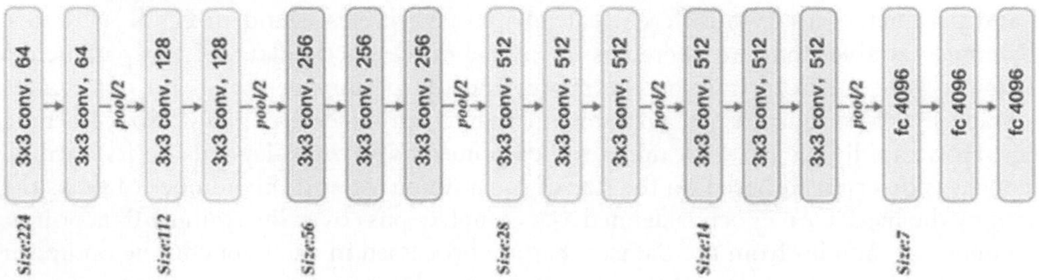

Figure 100.3 Graphical representation of the VGG-16 model [18]

layers) affects accuracy was the primary objective of the VGG study [17]. The 16 layers of VGG-16 have 13 layers of 3 × 3 Convolutions and three fully connected layers. The network's depth is boosted with a 3 × 3 convolutional filter applied to all 13 layers, followed by maximum pooling after each set of convolutional layers. The input dimension is condensed and flattened along the model when it reaches the output end [18].

As already mentioned, the experiment will use the CNN-based VGG-16 architecture. We chose PyTorch as the deep learning framework for model building. This framework offers many useful techniques as well as implementations of the most popular CNN architectures. The data were divided into three groups before building the model: training data made up 70% of the total, validation data 20%, and test data 10%. To ensure the best performance, the data is distributed category-wise.

The pre-trained VGG-16 model of the Pytorch framework will be the model used in the experiment. The performance of the model is already pre-trained with the well-known ImageNet model. A transfer-learning model would often have parts of its layers frozen to speed computation when training on large data sets. Since the data set used is not very large, it is not necessary to freeze the layers. In PyTorch, the last layer of the model is built to classify images into ten categories, although only two are required in this case. Consequently, we change the output of the last layer to two to make it suitable for binary classification.

In addition, during the experiments, the Pytorch imported and used a cross-entropy loss function to evaluate the loss over the accuracy and error rate. The following variables were varied during the experiment to determine their relative effects:

- Batch size
- Learning rate
- Optimiser Type

Our research work is mainly focused on establishing a relationship between learning rate and optimiser type. Due to their interconnection between the learning rate and optimiser type, they will be tested together. We carried out an extensive implementation to analyse the accuracy of the model with different optimisers integrated with multiple learning rates to observe how each affects model performance as the learning rate is fused as the input to the optimiser. Different learning rate values have different effects on the three optimisers. As a result, for each optimiser, different learning rate values are used by manual tuning. Batch size and different learning rates will be investigated as subsequent hyperparameters to provide more precise findings and explore their interrelationships. This hyperparameter will be tested using the top optimiser from previous tests. To take full advantage of GPU processing, all batch sizes that will be examined are powers of two.

Performance Measure

A metric is a computation used to assess model accuracy. One of the simplest metrics is classification accuracy, which is determined by the division of correctly predicted instances by the total instances predicted. Similarly, the error rate is the percentage of predicted incorrectly with a total number of predictions.

$$\text{Accuracy} = \frac{Predicted\ Correctly}{Total\ number\ of\ predictions}$$

$$\text{Error rate} = \frac{Predicted\ Incorrectly}{Total\ number\ of\ predictions}$$

There are situations in which using this statistical analysis does not yield better results and where the data is not evenly distributed over several categories [23]. However, given there are an equal number of data points in each category in this experiment, the measure should be enough. For each parameter set that is tested, a total of 1000 epochs were used during model training. Each epoch includes measurements of training accuracy and loss. The VGG-16 is trained on the lowest validation loss so. It may be utilized to determine the test accuracy for the next epoch. Figure 100.4-100.7 demonstrates the accuracy of the VGG-16 model in predicting cardiomegaly from X-ray images.

Empirical Analysis

Throughout the experiment, we calculate each hyperparameter's accuracy, we present Table 1 demonstrating the VGG-16s accuracy for several hyperparameters. The performance of the VGG-16 on the part of the data set that was not used in the training is represented by the precision of the test. Several plots are also provided for some of the hyperparameters to demonstrate how the value of each parameter may be selected to achieve high accuracy.

Optimizer Type and Learning Rate

The impact of various learning rates on optimiser performance is shown in Table 1. In each example in the table, a batch size value of 64 was used. The table shows that for all types of optimisers, a learning rate that is too high or too low makes the model perform worse. In Table 100.1, the entry in test accuracy "-", indicates the model outcome which is not during the training. A learning rate that is too high compels the optimiser to move more slowly in the direction of the loss function which is either a local or global minimum. As a result, the optimiser becomes unstable and either fails to reach a minimum or rapidly converges to a local minimum. It takes longer to train the VGG-16 because learning rates that are too low cause the optimiser to converge very slowly. As a result, the optimizer stays at only a local minimum. It is important to note that the optimizers could have achieved higher accuracy if more than 1000 epochs had been used. During an epoch, the VGG-16 is trained with the training data for one cycle. The training data and its associated information will be used once within an epoch. The database is portioned into one or more batches and the model is trained for each epoch using the portioned batch size. The epoch refers to the progression of going through the training instances of each batch. The size of the batch determines how many epochs there will be. For each epoch, the amount of data points is simply determined by multiplying the required iterations by the batch size. We can experiment using different epochs. In this case, the neural network is fed the same data several times. In other words, it neither fits too well nor too poorly. Before training begins, we

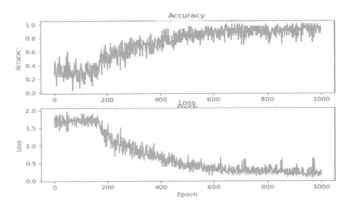

Figure 100.4 The VGG-16 performance with Adam optimiser when high learning rate = 0.00001
Source: Author

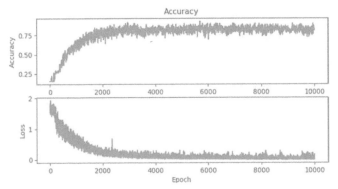

Figure 100.5 The VGG-16 performance while utilising the SGD optimiser on low learning rate = 0.00001
Source: Author

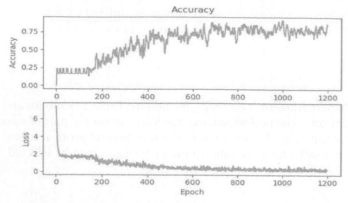

Figure 100.6 The VGG-16 performance with SGD optimiser when learning rate = 0.0001
Source: Author

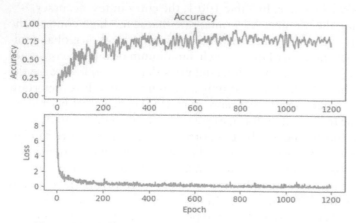

Figure 100.7 The VGG-16 performance using RMSProp's optimiser on a high learning rate = 0.00001
Source: Author

Table 100.1 Test the accuracy of the VGG-16 model using different optimizers learning rates

Optimiser	Learning rate	Test accuracy (%)
Adam	0.001	64.83
Adam	0.0001	80.56
Adam	0.00001	79.47
SGD	0.001	79.83
SGD	0.0001	80.20
SGD	0.00001	73.47
RMSProp	0.0001	-
RMSProp	0.00001	80.47
RMSProp	0.00005	58.56

Source: Author

Table 100.2 VGG-16 Test accuracy on different batch sizes and learning rates

Batch size	Learning rate = 0.00001	Learning rate =0.001
	Test accuracy (%)	Test accuracy (%)
16	78.29	34.15
32	77.73	56.10
64	79.37	64.82
128	77.28	78.99
256	76.91	79.73

Source: Author

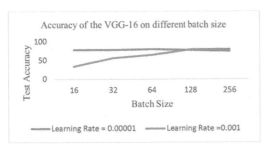

Figure 100.8 The VGG-16 Accuracy on various batch sizes and learning rates
Source: Author

set the number of epoch parameters while creating a model. The primary elusive to us is to fix the number of epochs based on the batch size. With the intricate design of the neural network and the data collected, we must decide the point at which the weights converge.

The graphs in Figure 100.4 demonstrate the VGG-16 model performance on learning rates when the Adam optimizer is used. Figure 100.4 shows that using a higher learning rate resulted in faster learning of the optimiser in a short period. Although Adam's optimiser learns faster with a higher learning rate, the results are erratic and unpredictable.

The effectiveness of the VGG-16 using SGD optimisation with various learning rates is shown in the graphs in Figures 100.5 and 100.6. The graphs show that the VGG-16 model learns more slowly at low learning rates. This shows that a low accuracy score does not necessarily mean a lack of improvement; rather, they indicate the need for additional learning time, and the graphs show that the accuracy continues to increase. The solution to this issue is to increase the number of epochs, but this will require additional computation time. Figure 100.6 shows that while SGD is less stable at higher learning rates, it is faster on the loss function when it is minimum.

The graphs in Figure 100.7 illustrate how the VGG-16 performs on the RMSProp optimiser pragmatic on learning rates. The outcome can be seen in the graphs with different optimisers. The VGG-16s output becomes unpredictable with a greater learning rate. The optimizer becomes stable at a lower learning rate, but the model also learns more slowly and takes longer to compute.

Table 100.2 presents the accuracy of the VGG-16 on different learning rates and batch sizes. When learning rates are higher, smaller batch sizes perform worse, as Table 100.2 clearly shows. VGG-16 performance is no worse than it would be for smaller batch sizes because larger batch sizes reduce the impact of changes in learning rate values. Figure 100.8 shows the performance graph.

The VGG-16 performs badly with lower batch sizes where a few images are selected for processing for each epoch. As a result, the VGG-16 performs worse because it has access to fewer training images. When smaller lot sizes are used, the VGG-16 performs better due to a lower learning rate. This means that learning from a small number of images has a negative impact that is minimised by a lower learning rate.

Discussion

The VGG-16 model's classification accuracy is significantly influenced by the learning rate, thus picking the right learning rate is essential to improve the performance in terms of accuracy rate. The performance of the VGG-16 model is negatively affected by using a learning rate value that is too high or too low. The results also showed that under the right conditions, such as a suitable learning rate and batch size, various types of optimisers were able to achieve comparable levels of precision. This means that the performance of the model was not fundamentally affected by the choice of the optimiser. To get the best results from small batches, it is important to use a lower learning rate.

It was crucial to manually tune the hyperparameters to see how they affected the model and each other. A major problem in this investigation was the considerable computing power required for each training session, which made the procedure time-consuming. Consequently, the search for the best settings for the hyperparameters should be automated if that is the intended result. Due to the short data set used in our research, it was difficult to attain a high level of accuracy in the classification. To solve this problem, we used transfer learning, where a model was pre-trained on data sets known as ImageNet. No conclusions can be drawn about the impact of hyperparameters on a dataset with even more data points because we trained on a small dataset.

Conclusions

The study focused on the importance of learning rate among the three hyperparameters examined, batch size, optimiser, and other hyperparameters. Compared to batch size and optimiser type, it has the biggest impact on VGG-16 performance. Regarding the selection of the best optimiser, the results demonstrated that all the examined optimisers could produce excellent accuracy. The given parameters and the type of optimiser is not significantly affected the performance of the VGG-16. These results also preserved how important it is to choose a learning rate, particularly since there is a restriction on the batch size if you want to improve the VGG-16s accuracy rate. In further research, additional hyperparameters beyond those presented here could be studied to learn more about how they affect model performance. The hyperparameters include dropout rates and experimentation with different activation strategies at each convolutional layer. Future enhancements could include the adoption of more comprehensive training on different deep learning architectures with a larger number of epochs.

References

[1]   Lee, M. S. et al. (2021). Evaluation of the feasibility of explainable computer-aided detection of cardiomegaly on chest radiographs using deep learning. Scientific Reports. doi: 10.1038/s41598-021-96433-1.

[2]   Arsalan, M., Owais, M., Mahmood, T., Choi, J., and Park, K. R. (2020). Artificial intelligence-based diagnosis of cardiac and related diseases. Journal of Clinical Medicine. doi: 10.3390/jcm9030871.

[3]   Sarpotdar, S. S. (2022). Cardiomegaly detection using deep convolutional neural network with U-net. 10.48550/arXiv.2205.11515.

[4]   Xu, W., He, J., Shu, Y., and Jiuxiang, H. J. (2020). Advances in Convolutional Neural Networks. In Advances and Applications in Deep Learning. doi: 10.5772/intechopen.93512

[5]   Candemir, S., Rajaraman, S., Antani, S., and Thoma, G. (2018). Deep Learning for grading cardiomegaly severity in chest X-rays: an investigation. Proceedings IEEE Life Sciences Conference (LSC 2018). pp. 109-113. 10.1109/LSC.2018.8572113.

[6]   Bougias, H., Georgiadou, E., Malamateniou, C., and Stogiannos, N. (2021). Identifying cardiomegaly in chest X-rays: a cross-sectional study of evaluation and comparison between different transfer learning methods. Acta Radiology. doi: 10.1177/0284185120973630

[7]   Buturache, A. and Stancu, S. (2020). A study of artificial neural networks hyperparameter tuning for data-driven decision systems. 19th International Conference on Informatics in Economy. Education, Research and Business Technologies. pp. 183-98. 10.24818/ie2020.04.03.

[8]   Zhou, B., Gao, J., Jiang, Q., and Chen, D. (2019). Convolutional neural networks for computer-aided detection or diagnosis in medical image analysis: an overview. Mathematical Biosciences and engineering.16(6), 6536-61. doi: 10.3934/mbe.2019326.

[9]   Salem Saeed Alghamdi et al. (2020). Study of cardiomegaly using chest x-ray. Journal of Radiation Research and Applied Sciences. 13(1), 460–67.

[10] Ga´c, P., Truszkiewicz K., and Poreba R. (2021). Radiological Cardiothoracic Ratio in Evidence-Based Medicine. Journal of Clinical Medicine, MDPI (2021), pp. 1–3.

[11] Zhangand, J., Xu, W., Yuxi, D. and Pan, Y. (2017). Learning to read chest X-ray images from 16000 + examples using CNN. Institute for Interdisciplinary Information Sciences, Tsinghua University, and Beijing.

[12] Krupinski, E. A. (2003). The future of image perception in radiology. Academic radiology. 10, 1–3.

[13] Berlin, L. (2007). Radiologic errors and malpractice: a blurry distinction. American Roentgen Ray Society. 189(3).

[14] Bruno, M. A., Walker, E. A., and Abujudeh, H. H. (2015). Understanding and confronting our mistakes: the epidemiology of error in radiology and strategies for error reduction. Radiographics. 35(6), 1668-76.

[15] Nielsen, M. (2015). Neural Networks and Deep Learning. Determination Press. 16. Mishra, M. Convolutional Neural Networks. In: Towards Data Science.

[16] Nepal, P. VGGNet Architecture Explained. In: Towards Data Science. https://medium.com/analytics-vidhya/vggnet-architecture-explained-e5c7318aa5b6

[17] Adhiyaman Manickam, Jianmin Jiang, Yu Zhou, Abhinav Sagar, Rajkumar Soundrapandiyan and R. Dinesh Jackson Samuel. (2021). Automated pneumonia detection on chest X-ray images: A deep learning approach with different optimizers and transfer learning architectures Measurement, 184, 109953, ISSN 0263-2241, https://doi.org/10.1016/j.measurement.2021.109953

[18] Bergstra, J. and Bengio, Y. (2012). Random search for hyper-parameter optimization. Journal of Machine Learning Research. 13(10), 281–305.

[19] Chaudhary, A., Hazra, A., and Chaudhary, P. (2019). Diagnosis of chest diseases in X-ray images using deep convolutional neural network. 2019 10th International Conference. 21. Yoo, H., Han, S., and Chung, K. (2021). Diagnosis support model of cardiomegaly based on CNN using ResNet and explainable feature map. IEEE 9 (2021).

[20] Alghamdi, S. S., Abdelaziz, I., Albadri, M., Alyanbaawi, S., Aljondi, R., and Tajaldeen, A. (2020). Study of cardiomegaly using chest x-ray. Journal of Radiation Research and Applied Sciences. 13(1), 23. Brownlee, J. (2022). Failure of accuracy for imbalanced class distributions. Machine learning mastery V.

[21] Yoo, H., Han, S., and Chung, K. (2021). Diagnosis support model of cardiomegaly based on CNN using ResNet and explainable feature map. IEEE Access. doi: 10.1109/ACCESS.2021.3068597.

[22] Saiviroonporn, P. (2021). Cardiothoracic ratio measurement using artificial intelligence: observer and method validation studies. BMC Medical Imaging. 21(1), 1–11. doi: 10.1186/s12880-021-00625-0.

Chapter 101

A comprehensive review on effective selection of metallic, polymeric and ceramic crown in dentistry

Shiv Ranjan Kumar[1,a], Rishika Mehani[2,b] and Gourav Bhansali[3,c]

[1]Bhagalpur College of Engineering, India

[2]Christian dental college, India

[3]JECRC university, India

Abstract

Application of mechanical design and manufacturing has provided a huge support for the research and development in the field of dental implant. Therefore, dental implants such as dental crown, jaw, and composite are designed and developed in such a way to completely satisfy patient requirements in terms of aesthetic look and mechanical parameters like strength and durability. A dental crown is a metallic or polymeric cap used to cover or restore broken or damaged teeth. This review aims to critically evaluate the need of metallic, ceramic or polymeric materials and to select among various available material which is rarely analysed either experimentally or clinically. Hence, selection of crown material is one of the tough decisions for patient as well as dentist. The comparative assessment is performed in the prospective of physical properties, mechanical properties, biological properties, wear resistance properties and corrosion resistance properties. Therefore, detail material property, performance and requirement chart should be prepared and communication between the dental manufacturing unit, the dentist or clinical person and the patient must be clear for effective, successful and long-lasting dental restoration.

Keywords: Ceramic crown, dental crown, metal crown, polymeric dental material

Introduction

In order to study in detail about the need/performance of materials for the replacement of human teeth, the effect of various factors like physical, chemical and mechanical properties of human teeth, various forces during mastication or chewing, environmental condition such as temperature, acidic, alkaline effect, position of teeth in mouth, patient history etc. must be assessed.

Composition of human teeth

Teeth are made of both organic and inorganic minerals, which are distributed between hard and softer layers of teeth. The dentin, enamel, and cementum are the harder portions of teeth, enamel being the hardest, whereas the pulp region is softer. These layers vary in material composition and nature. The outer layer being hard is calcium and phosphate rich to protect it from the abrasive wear due to food and the chemical action of various fluids. The various layers and their chemical composition have been presented in Table 101.1 [1, 2]. Table 101.1 indicates the presence of calcium hydroxyapatite fiber and collagen resin in the teeth in the form of composite.

Oral environment

The human digestive system is a combination of smaller subsystems involving the oral cavity, salivary glands, pharynx, esophagus, liver, gallbladder, stomach, pancreas, small and large intestines, appendix, rectum and anus [3]. The food goes into the human body by means of the oral cavity which has the 32 teeth, tongue, taste buds, and other components to aid the digestive system. The chewing process ensures breaking down the bite into smaller ingestible portions and the saliva, mixed with the help of the tongue, ensures the food is volatile. These teeth first appear when the child is 6-months-old, and are called Deciduous Teeth, which are 20. They shed at their respective times, last ones being second molars - 10-12 years and are replaced by permanent teeth. The name and number of each tooth in an arch (mandible/maxilla) are: Incisors (4 in numbers), canine (2 in numbers), premolars (4 in numbers), and molars (6 in numbers).

Prominent teeth ailment requiring counterfeit

The enamel is hard and biocompatible, yet the continuous action of solid food and fluids result in slow but steady decay of the layers. Alongside, the shape of molars is such that food gets stuck either between them and the gums area or on them. Due to this, the salivary glands secrete excessive saliva resulting in cavitation. Sugar from snacks and drinks, and bacteria due to poor dental hygiene are diagnosed as principal reasons

[a]ranjan.shiv@gmail.com, [b]rishikamehani1410@gmail.com, [c]gouravjain332@gmail.com

DOI: 10.1201/9781003450252-101

Table 101.1 Chemical composition of human teeth [1, 2]

S. No	Components of human teeth	Composition
1	Enamel	96% inorganic components (calcium hydroxyapatite), 4% organic components
2	Dentin	70% inorganic components (calcium hydroxyapatite), 20% organic components (90% collagen type1, non-collagenous matrix proteins and lipids
3	Cementum	45% inorganic components, 55% organic components
4	Pulp	

Table 101.2 Description of dental caries

S. No	Name of dental caries and issues	Description
1	Tooth decay or cavities	Cavities are formed when a layer of bacteria, called plaque, is formed on the teeth. The bacteria convert the sugars in the cavities into acids which decay the teeth. The degradation continues to damage the layers until precautionary steps are taken. They may be spotted as dark spots or holes on the teeth, besides pain and smell.
2	Gum disease	Gum disease is also known as Gingivitis which is associated with the gums. Prominent red alerts include red gums, bleeding and pain while chewing food. This may be an attribute of improper brushing, tobacco, pregnancy, et cetera. If untreated, it may lead to Periodontitis which is a more severe form of gum disease which may lead to loss of permanent teeth and inflammatory response throughout the body.
3	Bad breath	Bad breath or halitosis may be due to several medical or general issues including gingivitis, periodontitis, tooth decay, formation of plaque, medication, dry mouth, and cancer are prominent among others.
4	Sensitive teeth	The pulp is the conductor that passes and receives signals like hot and cold. When the enamel and dentin layer disintegrate, the impulses are directly passed on to the pulp leading to sensitivity. The ultimate treatment for infection of the pulp is removing it: root canal treatment.
5	Cracked or broken teeth	Any trauma to the tooth can lead to the development of cracks or its breakage. With the advances in dentistry, it is now possible to reestablish the lost tooth structure with the tooth-colored restorative material like GIC
6	Receding gums	Also known as Gingival recession is the movement of the margins of the gingiva covering the tooth, towards the apex of the tooth. This leads to exposure of the roots of teeth. Spaces are created between the two teeth where food debris and bacteria get accumulated leading to sensitivity, caries, and ultimately tooth loss.
7	Root infection	It occurs because of accumulation of pus in the tooth, also known as tooth abscess. The treatment to this is drainage of the pus
8	Enamel erosion	Forceful or faulty tooth brushing leads to wearing away of the enamel layer of the crown of the tooth. This further expose dentin, therefore giving yellowish appearance to the tooth. In severe cases, it leads to sensitivity.

Source: Author

for tooth decay [4]. Bacterial infection is responsible for the decay of both the organic and inorganic constituents of the teeth [5]. Prominent ailments of various types of teeth are presented in Table 101.2.

History and modes of tooth wear

Tooth wear has been studied for centuries due to its direct correlation with the health of a person. It was also correlated with the age of a person, meaning worn out teeth indicates an old person with poor oral and physical health. The focus on developing methods to minimise the loss of teeth has been a priority for people. Ancient romans used brushes made of horse's hair, Indian used bark of Azadirachta indica also known as Neem tree to ensure that the teeth are clean, and similarly Egyptians used toothpaste made of rock salt, dried iris flowers and mint [6]. With modern technology, these methods have taken shape into toothpastes and brushes offering cleaning and protection from all kinds of oral ailments.

The digestion process starts with the action of teeth to break food into smaller bits so that it may be efficiently mixed by the tongue. The action of saliva makes it a paste which makes it easy to pass through the esophagus. The brief working has been shown below:

- Incisors cuts the food into a short bite and with canines separate it from the whole piece.

- Molars and premolars work in conjunction with tongue and saliva to make a paste of the bite.
- The paste is then swallowed and passed to stomach via the esophagus.
- This process involves abrasive action of saliva, food particles and the teeth themselves, which causes the top layer to wear out. The steady wear is responsible for ailments requiring for complicated and long procedures.

To simplify the process of understanding tooth wear modes [7], following modes have been identified:

- Attrition: This mode is correlated with the wear of teeth when they are in direct contact with each other. Absence of foreign particles is an important identifier for this mode.
- Abrasion: This mode involves the action of foreign particles which are being introduced to the tooth continuously. Due to particles of different shapes and sizes, the force distribution changes implying that the tooth suffers wear at different wear rates. Abrasion can be categorised as two body or three body abrasion. Two body abrasions exist during the direct motion between two teeth or two different materials. Three body abrasions exist when external agent participates in the wear process between two materials. Many research have been conducted to evaluate two body and three body abrasive wear of dental materials [8–12].
- Erosion: Erosion is a painless process of tooth wear which may be correlated with pathologic, chronic or localised chemical reactants. It is the chemical etching of the outer layer involving chemicals and bacteria.
- Resorption: This mode of wear happens with introduction of substances which are being assimilated but due to their position, teeth become victims as well.
- Demastication: It refers to the wear due to the presence of bolus of food between opposing teeth.

Treatment for dental caries

Permanent teeth cannot be repaired by the body, unlike other organs and tissues in our body. This has created a huge demand for solutions such as dental crowns to counter tooth decay due to enzymes continuously reacting with the tooth enamel. The decay is usually identified by its black colored appearance on the tooth. This decay is irreversible and painful due to the veins in the tooth gums. When the ends of these veins come in contact with fluids at high or low temperatures, they send shivering signals to the nervous system. Thus, activities like eating or drinking become painful. Each tooth is a combination of 2 parts: Crown and Root.

The enamel forms a very thin layer on the surface, hence erosion due to the action of enzymes on salts and sugar is inevitable. Thus, as the crown gets worn out, the Dentin is exposed. Dentin contains a yellowish cement-like filling which provides strength to the tooth. But exposure of dentin to the oncoming enzymes acting rigorously on solid food stuck in teeth, salts, and sugars presents a threat to the gums of the oral structure which are linked to blood vessels. These blood vessels are extremely tender, and any interaction with the temperatures of higher or lower range exposes them to irritation and pain [13]. Dental crowns are, thus, installed to prevent the spread of enzymes to the Apical Foramen, which contains blood vessels and arteries.

After thorough cleaning of the decayed tooth area using methods like laser beam machining, acid decay, and dental drill to remove the cavity. Dental drills are a faster alternative, but an oral anesthetic becomes an essential consideration besides expertise in using the drill [14]. The general procedure for installation of dental crowns as per the RCT or Root Canal Treatment consists of many stages as depicted in Figure 101.1.

Material Fabrication Methods

The manufacturing of dental crowns is based on the type of ceramic available. They may be in block form or powder form. Processes like chemical vapor deposition, molding, 3-D printing, and other advanced manufacturing methods are now regularly used [15]. These methods provide high purity and high strength dental crowns for a long lasting and effective cure. The manufacturing methods have seen a boom with emerging technologies. The crowns made are customised and high precision. Some of major required/ desirable properties have been highlighted and presented in Table 101.3.

Upon fulfillment of these criteria for a counterfeit dental crown, a number of factors are considered before designing and manufacturing of the crown. Figure 101.2 indicates the detail procedure to be followed and factor to be considered for the manufacturing and designing of dental materials.

Data collection and detail of patient

It involves the collection and analysis of complete data such as detail prescription of dentist regarding the age of teeth, condition of dental caries, history of patient, medication, position of teeth, amount of natural teeth remaining, condition of gum and surrounding teeth.

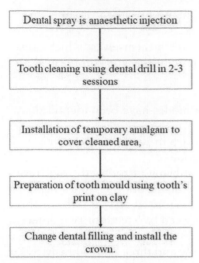

Figure 101.1 General procedure for installation of dental crowns as per root canal treatment
Source: Author

Table 101.3 Required properties of dental materials

S. No.	Required properties	Description of criteria
1	Bio-compatibility	For artificial implants staying in direct contact with the human body, biocompatibility is an important criterion. Certain prerequisites are tested and the behavior is observed under specialised conditions. (Add about the tests done)
2	Fracture strength/resistance	Teeth are subjected to high impact forces combined with regular fatigue developed while eating various foods. To ensure that the material is safe to use, these tests are done.
3	Toughness and hardness	Teeth should resist indentation and impact of different food items and injuries. This is important to check as materials with low hardness will lose the dimensional confinements. Also, due to deformed shape, food may trap in the teeth and result in degradation of other teeth.
4	Compressive strength, flexural strength and diametral tensile strength	During mastication process of hard or soft food particle, dental material should have strength to withstand compressive load, bending load and lateral expansion due to compressive load.
4	Water sorption and solubility	As the dental materials remain mostly in water environment. Hence, this property must be evaluated.
5	Translucency	To mimic, natural teeth, the developed material should have translucency like natural teeth
6	Wear resistance	To increase the life and efficiency of dental materials, their wear resistance must be evaluated under various medium such as dry, distilled water, food slurry medium etc.

Source: Author

Selection of crown material

Selection of material for crown also depends upon the prescription of dentist about the patient. Many factors like the age of patient, medical history, tobacco or smoking history, sensitivity issue for a particular material is also considered.

Due to involvement of the cap material with the human body, it is necessary to test against its biocompatibility and life cycle. Both non-destructive and destructive methods are utilised to ensure that the caps are crowns possess no harm to the user by the virtue of inert chemical nature. The structural properties are analysed based on appropriate hardness and strength. It has been observed that teeth are highly wear resistant, but poor hygiene practice and irregular eating habits contribute to the degradation [16]. Following criteria are critical for the evaluation of biocompatibility:

• The material for cap should not diffuse due to action of enzymes and fluids,
• The material should not possess toxic substances such as lead or any chemically reactive substance,
• The material used should not be carcinogenic in nature,

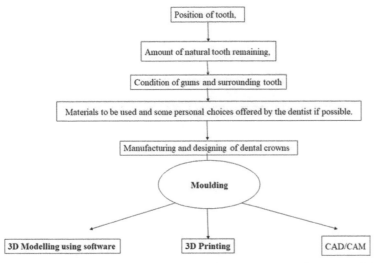

Figure 101.2 Detail procedures for design and manufacturing of dental crown
Source: Author

- The material should not possess any harm to either soft or hard tissues of the oral cavity.
- Based on these criteria testing is done in two ways which are In Vivo and In Vitro testing methods.
- These methods have been proposed as ISO standards due to high efficacy and failure prevention. Thus following testing and evaluation methods are used:
- General Biocompatibility testing is done by evaluating the reaction of material at cellular level,
- Immunological Testing involves tests like Clifford Materials Reactivity Tests in which if a material tests for NS, i.e., Not Suitable the material should not be used, and Kinesiologic test is also performed for critical evaluation [17].
- For energetic evaluation of the materials, Bio Energetic Tests are performed which involves Electro dermal Screening and Applied Kinesiologic tests [18].
- Besides these, trials are also carried out for further investigation and evaluation of these materials. A general procedure involves the following steps:
- Non-specific toxicity trials are carried in non-simulated environment in the labs to understand cellular level behavior,
- Specific toxicity trials are executed in a simulated clinical environment where humans may be test subjects,
- Clinical trials on humans is the final stage where precautions are also taken during the testing and critical evaluation is done.
- Based on Langelands' proposal for a testing sequence in 1984, the ISO Technical Report 7405 was adopted [19, 20] which had the following methodology:
- Initial Test: This involves Cytotoxicity and Mutagenicity Tests are used,
- Secondary Test: This involves sensitization, implantation tests, and mucosal irritation test,
- Usage Test.

Types of dental Crowns

Dental caps vary in material and points of application. Thus, various materials are used as per requirement and the tooth where it is to be placed. Generally three major types of crown materials are used, which are metallic crown, porcelain fused to metal crown, ceramic crown and polymeric crown.

The detailed comparative analysis of crown materials are presented in Table 101.4. Table 101.4 indicates that metal, ceramic and composites are mostly used to restore teeth as crown materials.

Metallic Crown: These crowns use metals infused during the preparation of the material powder itself. The metal provides strength, hardness, and longer life. They are made using additive manufacturing and casting methods. Advanced manufacturing methods also use laser melting to develop required contours. Ceramic metal interface enhances the strength while also staying food grade and effective. Another advantage of metallic crown is that they may be utilised more effectively for molars and premolars which undergo majority pressure and fatigue loading regularly. Popular choices are gold, platinum, and PFMDC. While these materials promise great performance, one disadvantage associated with them is that due to high strength and hardness, the teeth on the other side may wear out faster. The major characteristics of metal crown is its excellent strength, corrosion resistance and wear resistance which results in longer life of crown and teeth. The crown can be made of metal such as stainless steel, titanium, and other metal alloy

Table 101.4 Detain comparative analysis of crown materials

| Functions | Crown materials | | | |
	Gold	Porcelain	Porcelain fused to zirconia/ full zirconia	Dental composite
Strength	Gold is extremely durable, apply less abrasion on opposing teeth,	Very hard, durable, colour matching with teeth, less sensitive to hot and cold	Very hard but not fracture easily compared to porcelain, least sensitive, high degree of translucency and colour completely matched with natural teeth	Least strength and durable but is available in tooth coloured options. Best for those who have sensitivity issue with metal and ceramics
biocompatibility	Less reactive, can be reactive to some people, good for molar teeth. Biocompatible	Less reactive, suitable for front teeth	Less reactive	Least reactive
drawbacks	Not matching with natural teeth Less wear resistance compared to ceramics so can wear within a year if opposite teeth is made of ceramic or patient is having habit of grinding or clinching. Sensitive to hot and hold	Very hard but so can damage enamel and fracture itself easily.	Full zirconia is stronger than porcelain but as far as cosmetic point of view, it is not as attractive as porcelain	Weakest restoration hence, it should be changed regularly in small interval
Limitation (may not use)	Patients with specific metal allergies or sensitivities (testing is available); front teeth; patients with extreme sensitivities to hot and cold	Patient having clinching habit should not be recommended to use.	Suitable for full crown restoration not for partial restoration. Not suitable for patient having allergy from zirconia	If anyone of other alternatives is suitable
Cost	Most expensive	Least expansive	Less expensive	Least expensive

Source: Author

[21]. However, the color of metallic crown doesn't match with the human teeth. A dark line is visible at the gum-line below the teeth. Therefore, for the molar teeth which are on back side, metal crowns are widely used for capping the damaged teeth. In order to match the color of natural teeth, porcelain fused to metal is used as coating on metallic crown. Metals used for the crown are stainless steel, gold, titanium, and other metal alloy. Stainless steel crowns are used for temporary caps. Stainless steel crowns protects the tooth till the permanent crown is fabricated.

Other metal crown such as gold, nickel, chromium, titanium, cobalt alloy is expensive. All metal crowns are excellent in terms of strength and wear resistance. Hence, they can sustain biting and masticatory forces easily [22–27].

Porcelain-fused-to-metal

For the cap matching the color of human teeth and having excellent mechanical, corrosion and wear resistance, porcelain-fused-to-metal crowns are used. Several layer of porcelain are used to coat metallic crowns. Both the disadvantages of metallic crown such as mismatch of color and dark line at the gum-line. However, a single layer of porcelain coated on stainless steel doesn't have sufficient strength and hence the surface is prone to fracture or abrasion after some time. Multi-layers of porcelain or combination of porcelain with zirconia is generally preferred by dentist. However, compatibility of all the layer material is a crucial factor deciding the long lasting of crown materials.

Ceramic crown: Ceramic crowns are inexpensive, durable, and true to the original color of our teeth. Although they are not as strong as the metallic crowns, they still have a life of around 15 years even though they may last up to 25 years with proper care. They are widely used for the incisors due to their clear appearance and strength. Pressed ceramic dental crowns combined porcelain caps are a popular choice due to the better color match. While metal crowns may wear out the opposite teeth, ceramic crowns are a safer option. They resist chipping and bending during everyday use unlike metallic crowns. The only issue prevailing is the relatively lower strength which may cause the crown to break if excess force is applied. Popular examples include materials like zirconia, silicate, and so on.

Ceramic composites and metal additives are one such option which has been explored. Prime examples are porcelain, lithium disilicate, layered zirconia, and leucite-reinforced glass ceramic. These materials are a combination of kaolinite-glass-bone ash-alabaster and so on, lithium-oxygen-silicon ($Li_2O_5Si_2$), zirconia-oxide (ZrO^2 -95%) mixed with stabiliser yttrium oxide (Y^2O^3 -5%), and leucite is mainly silicon-oxide mixed with aluminum, potassium and sodium oxides (SiO^2 -63%, $Al2O^3$ -19%, K^2O -11%, Na^2O -4%) respectively. All the materials can withstand stresses and action of enzymes, but it has been observed that Zirconia is better. All the materials are biocompatible, hence, safe for use.

When a dental restoration is made of ceramic only, it is called all-ceramic restoration. The advantages of all-ceramic restoration is good strength, long life, corrosion resistance and aesthetic look. The disadvantage of all-ceramic restoration is its extreme hardness which can erode other teeth in contact during mastication or other processes. All-ceramic restoration may be single or multi-layer. In case of multilayer, ceramic core material is covered with a ceramic veneer [28, 29].

Polymeric Crown: Polymeric crowns are made from polymers which are long chains of monomers. These are made by using polymerization process where monomers are joined together. The properties of these materials can be modified by forming cross linking thermosetting plastics, adding plasticisers and using copolymerisation. An emerging material is PEEK or Polyetheretherketone used in prosthodontics. The primary advantage associated with these is the ability to withstand different chemicals working in the mouth, biocompatibility and strength besides potential for high volume manufacturing. One of the disadvantages associated with these materials is the excessive shrinkage and poor adhesion with teeth are being tested against continuously. Other materials include PMMA, PEMA, and Urethane among many others. Polymeric crowns are generally used for making dentures, soft lining material and artificial teeth. Apart from that, dental polymer are widely used for dental restoration and implant purpose [30–33]. Polymeric dental composite are used to repair cavities and cracks in the teeth.

The manufacturing methods also vary as per use but molding and 3D printing are a popular choice due to simplicity and efficiency both monetarily and manufacturing vise. Besides, with the help of emerging technologies it has become possible to do FEA analysis to understand the working and performance of the crowns. FEA tools have made it easier to understand material behavior in different load and working conditions which helps in selecting materials and use them in a better and safer way.

Conclusion

The complete details about the patient medical history, prescription, location of teeth, consumption diet etc. must be analysed by the dentist to offer the right dental crown for restoration. The patient must be aware of the advantages, disadvantages and limitation of each and every type of dental crown materials. Metal crown is excellent in mechanical, corrosion and wear resistance properties but lacks in aesthetic look. Multi-layer zirconia, porcelain fused to metal are considered as the best alternative in terms of dental requirement. All ceramic restorations are extreme in hardness and can cause wear to other teeth. Therefore, detail material property, performance and requirement chart should be prepared and effect communication between the dental manufacturing unit, dentist or clinical person and patient must be maintained for effective dental restoration.

References

[1] Nanci, A. (2008). Ten Cate's Oral Histology: Development, Structure, and Function. St. Louis, MO: Mosby Elsevier.

[2] Yahyazadehfar, M., Ivancik, J., Majd, H., An, B., Zhang, D., and Arola, D. (2014). On the mechanics of fatigue and fracture in teeth. Applied Mechanics Reviews. 66, 0308031–3080319. doi: 10.1115/1.4027431.

[3] Sensoy, I. (2021). A review on the food digestion in the digestive tract and the used in vitro models. Current Research in Food Science. 4, 308–19. https://doi.org/10.1016/j.crfs.2021.04.004.

[4] Dülgergil, Ç., Dalli, M., Hamidi, M., and Çolak, H. (2013). Early childhood caries update: A review of causes, diagnoses, and treatments. Journal of Natural Science, Biology and Medicine. 4, 29. https://doi.org/10.4103/0976-9668.107257.

[5] Saini, R., Saini, S., and Sharma, S. (2011). Biofilm: A dental microbial infection. Journal of Natural Science, Biology, and Medicine. 2, 71. https://doi.org/10.4103/0976-9668.82317.

[6] Sharma, A., Sankhla, B., Parkar, S.M., Hongal, S., Thanveer, K.5., Ajithkrishnan, C.G. (2014). Effect of traditionally used neem and babool chewing stick (Datun) on streptococcus mutans: An in–vitro study. Journal of Clinical and Diagnostic Research (JCDR). 8(7), ZC15-ZC17.

[7] Kumar, S. R., Patnaik, A., and Bhat, I. K. (2020). Factors influencing mechanical and wear performance of dental composite: A review. Materialwiss Werkstofftech. 51, 96–108. https://doi.org/10.1002/mawe.201900029.

[8] Kumar, S. R., and Patnaik, A. (2023). Mineral trioxide aggregate (MTA) as a filler in dental composite: Evaluation of micro-hardness and wear properties. Proceedings of the Institution of Mechanical Engineers, Part E: Journal of Process Mechanical Engineering, 237(5), 1988–1998. https://doi.org/10.1177/09544089221131473

[9] Kumar, S. R., Patnaik, A., and Bhat, I. K. (2016). The in vitro wear behavior of nanozirconia-filled dental composite in food slurry condition. Proceedings of the Institution of Mechanical Engineers, Part J: Journal of Engineering Tribology. 231. 23–40. https://doi.org/10.1177/1350650116641329.

[10] Kumar, S. R., Patnaik, A., and Bhat, I. K. (2018). Wear behavior of light-cured dental composite reinforced with silane-treated nanosilica filler. Polymers for Advanced Technologies. 29, 1394–403. https://doi.org/10.1002/pat.4251.

[11] Kumar, S. R., Patnaik, A., and Bhat, I. K. (2018). Abrasive wear behavior of silane treated nanoalumina filled dental composite under food slurry and distilled water condition. Science and Engineering of Composite Materials. 25, 541–53. https://doi.org/10.1515/secm-2016-0175.

[12] Sonal, Patnaik, A., Kumar, S. R., and Godara, M. (2019). Investigating influence of low fraction of polytetrafluoroethylene filler on mechanical and wear behavior of light-cured dental composite. Materials Research Express. 6, 085403. https://doi.org/10.1088/2053-1591/ab209a.

[13] Evans, S. S., Repasky, E. A., and Fisher, D. T. (2015). Fever and the thermal regulation of immunity: the immune system feels the heat. Nature Reviews Immunology. 15, 335–49. https://doi.org/10.1038/nri3843.

[14] Becker, D. E., and Reed, K. L. (2012). Local anesthetics: review of pharmacological considerations. Anesthesia Progress. 59, 90–102. https://doi.org/10.2344/0003-3006-59.2.90.

[15] Pillai, S., Upadhyay, A., Khayambashi, P., Farooq, I., Sabri, H., and Tarar, M., Lee, K.T., Harb, I., Zhou, S., Wang, Y., Tran, S.D. (2021). Dental 3D-Printing: Transferring art from the laboratories to the clinics. Polymers. 13, 157. https://doi.org/10.3390/polym13010157.

[16] Piemjai, M., and Santiwarapan, P. (2022). An enamel based biopolymer prosthesis for dental treatment with the proper bond strength and hardness and biosafety. Polymers. 14, 538. https://doi.org/10.3390/polym14030538.

[17] What is Clifford Materials Reactivity Testing? 2022, accessed 21 July 2022, https://www.rpmdentistry.com/articles/what-is-clifford-materials-reactivity-testing/.

[18] Colbert, A. P., Spaulding, K., Larsen, A., Ahn, A. C., and Cutro, J. A. (2011). Electrodermal activity at acupoints: Literature review and recommendations for reporting clinical trials. Journal of Acupuncture and Meridian Studies, 4, 5–13. https://doi.org/10.1016/s2005-2901(11)60002-2.

[19] ISO 7405:2019-03 Dentistry - Evaluation of biocompatibility of medical devices used in dentistry (ISO 7405:2018, Corrected version 2018-12.

[20] Schmalz, G. (1998). Concepts in biocompatibility testing of dental restorative materials. Clinical Oral Investigations. 1, 154–62. https://doi.org/10.1007/s007840050027.

[21] Randall, R. C. (2002). Preformed metal crowns for primary and permanent molar teeth: review of the literature. Pediatric Dentistry. 24(5), 489–500.

[22] Jabbari, Y. S. A. (2014). Physico-mechanical properties and prosthodontic applications of Co-Cr dental alloys: a review of the literature. The Journal of Advanced Prosthodontics, 6, 138–45.

[23] Roberts, H. W., Berzins, D. W., Moore, B. K., and Charlton, D. G. (2009). Metal-ceramic alloys in dentistry: a review. Journal of Prosthodontics. 18, 188–94.

[24] Viennot, S., Dalard, F., lissac, M., and grosgogeat, B. (2005). Corrosion resistance of cobalt-chromium and palladium-silver alloys used in fixed prosthetic restorations. European Journal of Oral Sciences. 113, 90–5.

[25] Okazaki, Y., and Gotoh, E. (2005). Comparison of metal release from various metallic biomaterials in vitro. Biomaterials, 26, 11–21.

[26] Serra-Prat, J., Cano-Batalla, J., Cabratosa-Termes, J., and Figueras-Àlvarez, O. (2014). Adhesion of dental porcelain to cast, milled, and laser-sintered cobalt-chromium alloys: shear bond strength and sensitivity to thermocycling. Journal of Prosthetic Dentistry, 112, 600–5.

[27] Svanborg, P., Längström, L., Lundh, R. M., Bjerkstig, G., and Ortorp, A. (2013). A 5-year retrospective study of cobalt-chromium-based fixed dental prostheses. The International Journal of Prosthodontics, 26, 343–9.

[28] Beuer, F., Schweiger, J., Eichberger, M., Kappert, H. F., Gernet, W., and Edelhoff, D. (2009). High-strength CAD/CAM-fabricated veneering material sintered to zirconia copings–a new fabrication mode for all-ceramic restorations. Dental Materials 25, 121–8.

[29] Sharkey, S. (2011). Metal-ceramic versus all-ceramic restorations: part III. Journal of the Irish Dental Association. 57, 110–3.

[30] Kumar, S. R., Patnaik, A., and Bhat, I. K. (2020). Factors influencing mechanical and wear performance of dental composite: A review. Materialwiss Werkstofftech, 51, 96–108. https://doi.org/10.1002/mawe.201900029.

[31] Verma, R., Azam, M. S., Kumar, S. R., and Patnaik, A. (2018). Mechanical and thermo-mechanical characterization of glass ionomer-filled dental composite. Polymer Composites. 40, 3361–7. https://doi.org/10.1002/pc.25197.

[32] Sharma, A., Alam, S., Sharma, C., Patnaik, A., and Kumar, S. R. (2017). Static and dynamic mechanical behavior of microcapsule-reinforced dental composite. Proceedings of the Institution of Mechanical Engineers, Part L: Journal of Materials: Design and Applications. 233, 1184–90. https://doi.org/10.1177/1464420717733770.

[33] Mahna, S., Singh, H., Tomar, S., Bhagat, D., Patnaik, A., and Kumar, S. R. (2019). Dynamic mechanical behavior of nano-ZnO reinforced dental composite. Nanotechnology Reviews. 8, 90–9. https://doi.org/10.1515/ntrev-2019-0008.

Chapter 102

Robust control of inverted pendulum on inclined rail at different angle of inclinations via proportional-integral-derivative controller

*Dr. Ashwani Kharola[1,a], Shalu Garg[2,b], Varun Pokhriyal[3,c],
Anuj Raturi[1,d], Syed Farrukh Rasheed[1,e] and Ajay Kumar[4,f]*

[1]Department of Mechanical Engineering, Assistant Professor, Graphic Era Deemed to be University, Dehradun, India

[2]College of Social Sciences, PG Student, University of Birmingham, Birmingham, United Kingdom

[3]Co-founder, Scabula Labs Private Limited, Dehradun, India

[4]Department of Petroleum Engineering, Assistant Professor, Graphic Era Deemed to be University, Dehradun, India

Abstract

This study investigates control of highly nonlinear inverted pendulum system climbing on an inclined plane inclined at different angles using a proportional-integral-derivative (PID) closed-loop control. Stabilisation of inverted pendulum on cart is a challenging control problem. Further the control can be made more challenging while stabilising the inverted pendulum system on an inclined plane. Hence in this study control of inverted pendulum has been made more challenging by varying the angle of inclination of the plane as 10°, 15°, 20°, 25° and 30° respectively. A mathematical model of the proposed system has been derived and simulated in Matlab/Simulink. The performance of the proposed controller has been measured in terms of settling time, overshoot responses and steady state error. The results indicate highly robust behavior of PID controller even at higher angle of inclination of the plane.

Keywords: Inclined rail, inverted pendulum, Matlab, nonlinear, PID, simulink

Introduction

Inverted pendulum is a popular robotic system acting as a benchmark for testing various control algorithms. Its inherent nonlinear dynamics makes it suitable for testing various control laws. The system finds numerous practical and industrial applications. The system has always been a source of attraction for researchers working in the domain of control theory and robotics. A number of approaches have been adopted for their adequate control. For an instance two separate proportional-integral-derivative (PID) controllers tuned with multi-objective genetic algorithm and adaptive particle swarm optimisation for control of inverted pendulum system has been proposed in [1]. The results highlight that later controller leads to less chattering, noise and faster settling time compared to former one. A PID stabilisation of inverted pendulum system has been suggested in [2]. The governing equations of the system have been derived using Lagrangian mechanics. The performance of the proposed system was further monitored in terms of settling time and pendulum angle offset.

Further, an output feedback linearisation method for control of inverted pendulum on wheel has been considered in [3]. The study considered stabilisation of pendulum angle and wheel position simultaneously. The results highlight that when pendulum angle was considered as output an unstable zero dynamics appeared whereas when sum of pendulum angle and wheel rotation were considered as output a stable zero dynamics appeared in the system. Moreover, an observer-free feedback controller having variable parameters for stabilisation of inverted pendulum has been proposed in [4]. The controller architecture was obtained by converting linear state-feedback controller. Numerical simulations were performed for validity of proposed controller. A study on four different control approaches namely PID, linear quadratic tracker (LQT), linear model predictive control (LMPC), and nonlinear model predictive control (NMPC) for balance control of inverted pendulum system is highlighted in [5]. These techniques were tested in Matlab/Simulink. The results indicate superior control performance using NMPC approach.

More recently, a novel sliding-mode-control approach for nonlinear models has been proposed [6]. The control law was developed through interruption estimator. The proposed technique effectively counteracted effect of noises encountered during modelling processes. Finally, simulations were performed which verified the effectiveness of proposed control law on inverted pendulum system. Various other strategies have also been successfully adopted for their adequate stabilisation [7, 8]. This study considers control of

[a]ashwanidaa@gmail.com, [b]shalugarg.1412@gmail.com, [c]varun1997.pokhriyal@gmail.com, [d]anujraturi@geu.ac.in, [e]syedfrasheed.me@geu.ac.in, [f]ajaykumar@geu.ac.in

DOI: 10.1201/9781003450252-102

inverted pendulum on inclined rail using PID control strategy. The governing equations for the system have been derived and a Simulink model has been developed in Matlab platform. The inputs considered for the system includes wheel position, wheel velocity, pendulum angle and pendulum angular velocity. The study can be considered as an extension of previous work done by author in [9]. In the previous study control of inverted pendulum on inclined planes have been done at an angle of 10° using PID and fuzzy controllers. Therefore, in this study a more robust PID controller has been designed which can control the inverted pendulum system at different angle of inclination of the plane i.e. 10°, 15°, 20°, 25°, and 30° respectively. The performance of the controller has been further monitored in terms of settle time, steady state error and overshoots responses.

Mathematical model of inverted pendulum on inclined rail

The system includes a pendulum of mass ($m = 0.5$ kg) and length ($l = 0.1$ m) mounted on a cart of mass ($m = 1$ kg). The pendulum is inclined at an angle (θ) and the cart can move freely on a slope surface at an angle (γ) as shown in Figure 102.1 [9].

Let, x1 be the distance from reference point to the ground and x2 be the distance from ground to pendulum axle. Then total displacement is given as in (1),

$$x = x_1 + x_2 \tag{1}$$

Further, forces acting on pendulum is indicated in Figure 102.2. In the figure R_x and R_y are the contact forces acting along x and y directions.

Summing the forces along horizontal-axis is given as in (2),

$$\Sigma F_x = m\ddot{x} \tag{2}$$

The interaction forces in x-direction is given as in (3),

$$R_x = m\frac{d^2}{dt^2}(x - l_c cos\theta) \tag{3}$$

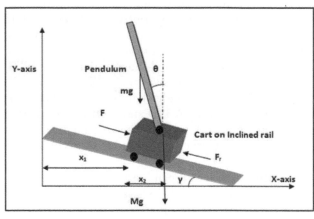

Figure 102.1 Inverted pendulum on inclined rail
Source: Author

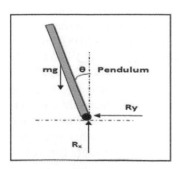

Figure 102.2 Forces on pendulum sub-system
Source: Author

Equation (2) and (3) gives value of R_x as in (4),

$$R_x = -m\ddot{x} + ml_c\ddot{\theta}cos\theta - ml_c\theta^2 sin\theta \tag{4}$$

Summation of forces along vertical-axis indicated as in (5),

$$- \Sigma F_Y = m\ddot{y} \tag{5}$$

The interaction forces in y-direction is given as in (6),

$$-R_y = m\frac{d^2}{dt^2}(y - l_c cos\theta) \tag{6}$$

Equation (5) and (6) gives value of R_y as indicated in (7),

$$R_y = mg + m\ddot{y} - (\ddot{\theta}sin\theta + \theta^2 cos\theta) \tag{7}$$

Summation of torque acting onp indicated as in (8),

$$\Sigma r_p = I\ddot{\theta} \tag{8}$$

Balancing the torque against interaction force indicated as in (9),

$$-Ryl_c sin\theta - Rxl_c cos\theta = I\ddot{\theta} \tag{9}$$

Forces acting on cart is highlighted in Figure 102.3. In the figure F and *Fr* represents the control force and friction force acting on the cart during its motion. A vertically downward force of magnitude (*Mg*) is also acted on the system due to gravity.

Summation of forces acting on cart along horizontal-axis is given as in (10),

$$Fcos\gamma - b\dot{x}cos\gamma + R_x = M\ddot{x} \tag{10}$$

Summation of forces acting on cart along vertical-axis is given as in (11),

$$-Fsin\gamma + b\dot{x}sin\gamma + R_y = M\ddot{y} \tag{11}$$

Summation of torque acting on the cart is given as in (12),

$$\Sigma r_c = I\ddot{\theta} \tag{12}$$

Equation (11) & (12) were solved to get value as in (13),

$$b\dot{x} - Rysin\gamma + Rxcos\gamma = 0 \tag{13}$$

In the above equations variables x and y are mutually dependent on each other. Therefore, eliminating one variables a simpler equation as in (14) is obtained,

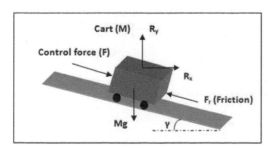

Figure 102.3 Forces acting on cart sub-system
Source: Author

$$x2 = \frac{t}{\sin \gamma} \tag{14}$$

Applying Equation (14) in Equation (1),

$$x = x1 + x2 = x1 = \frac{t}{\sin \gamma} \tag{15}$$

Also angle of inclination can be represented as in (16),

$$\tan \gamma = \frac{y}{x1} \tag{16}$$

Equation (16) when differentiated once yields (17),

$$\dot{y} = \dot{x} \tan \gamma \tag{17}$$

Equation (16) when differentiated twice yields (18),

$$\ddot{y} = \ddot{x} \tan \gamma \tag{18}$$

Equation (18) can be replaced in (7) and (11) to yield values as in (19) and (20),

$$R_y + mg + m\ddot{x}\tan\gamma = ml_c\ddot{\theta}\sin\theta + ml_c\dot{\theta}^2\cos\theta \tag{19}$$

$$-F\sin\gamma + b\dot{x}\sin\gamma + R_y = M\ddot{x}\tan\gamma \tag{20}$$

Equation (4) and (19) is substituted in (9) solved to yield value as in (21),

$$\ddot{\theta}(I + ml^2) - mgl_c\sin\theta = \ddot{x}ml_c(\cos\theta + \tan\gamma\sin\theta) \tag{21}$$

Equation (4) is substituted in (10) to yield value as in (22),

$$F\cos\gamma = (M + m)\ddot{y} + b\dot{x}\cos\gamma + ml_c\dot{\theta}^2\sin\gamma - ml_c\ddot{\theta}\cos\theta \tag{22}$$

The Equation (21) and (22) were adopted for building a Simulink model as given in Figure 102.4.

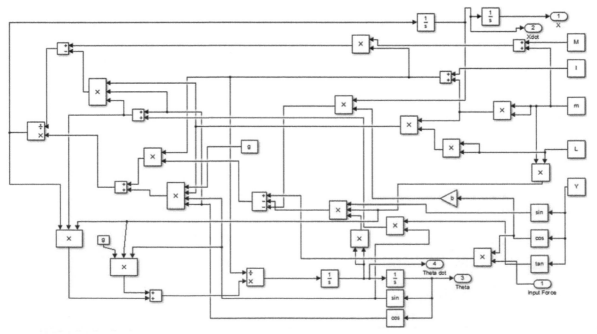

Figure 102.4 Simulink model of inverted pendulum on inclined rail
Source: Author

Proportional-integral-derivative control of inverted pendulum on different inclined surfaces

PID is a closed-loop controller device extensively usedP in numerous industrial applications [10]. It evaluates a difference value (*t*) between the desired value and output value and further applies a correction based on proportional (k_p), integral (*ki*) and derivative (k_d) terms [11, 12]. In this study two separate PID controllers were developed one for control of wheel position and wheel velocity and second one for control of pendulum angle and angular velocity. Further, the gains of the PID controller were tuned using auto-tuning function [13, 14]. The tuned PID gains for both the controllers are shown with the help of Table 102.1. The masked sub-system of proposed system along with PID controllers is shown in Figure 102.5.

Table 102.1 PID gains obtained after tuning

Sub- system	PID gains		
	k_p	k_i	k
Wheel	-851.83	-1142.02	-60.81
Pendulum	37.48	0.41	22.34

Source: Author

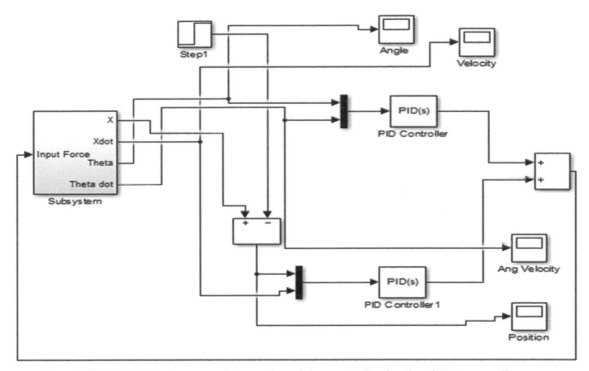

Figure 102.5 Masked sub-system of inverted pendulum on inclined rail with PID controllers
Source: Author

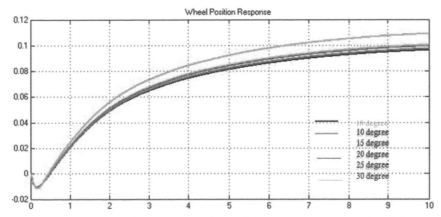

Figure 102.6 Simulation results obtained for 'wheel position' using different angles of inclined plane
Source: Author

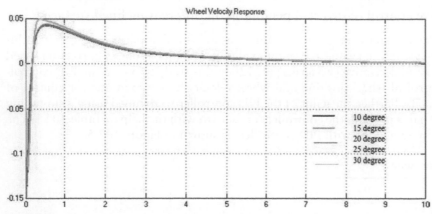

Figure 102.7 Simulation results obtained for 'Wheel velocity' using different angles of inclined plane
Source: Author

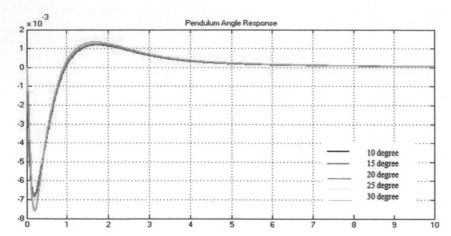

Figure 102.8 Simulation results obtained for 'pendulum angle' using different angles of inclined plane
Source: Author

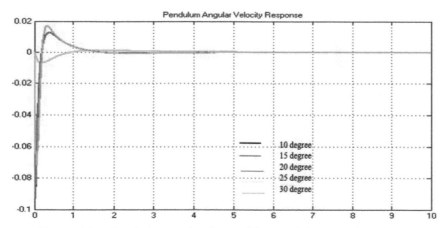

Figure 102.9 Simulation results obtained for 'pendulum angular velocity' using different angles of inclined plane
Source: Author

Table 102.2 Simulation results for output 'wheel position'

Angle of inclined rail	Overshoot ranges	Settling time	Steady state error
10°	-0.01 to 0.095	7.0 sec	0.004
15°	-0.01 to 0.096	7.0 sec	0.003
20°	-0.01 to 0.098	7.0 sec	0.002
25°	-0.01 to 0.1	7.0 sec	0.001
30°	-0.01 to 0.11	7.0 sec	0.05

Source: Author

Table 102.3 Simulation results for output 'wheel velocity'

Angle of inclined rail	Overshoot ranges	Settling time	Steady State Error
10°	-0.15 to 0.04	4.0 sec	0.005
15°	-0.15 to 0.04	4.0 sec	0.005
20°	-0.15 to 0.04	4.0 sec	0.005
25°	-0.15 to 0.04	4.0 sec	0.005
30°	-0.15 to 0.05	4.0 sec	0.005

Source: Author

Table 102.4 Simulation results for output 'pendulum angle'

Angle of inclined rail	Overshoot ranges	Settling time	Steady state error
10°	-6.8×10^{-3} to 1.3×10^{-3}	4.0 sec	0
15°	-6.8×10^{-3} to 1.3×10^{-3}	4.0 sec	0
20°	-6.8×10^{-3} to 1.3×10^{-3}	4.0 sec	0
25°	-6.8×10^{-3} to 1.3×10^{-3}	4.0 sec	0
30°	-7.6×10^{-3} to 1.4×10^{-3}	4.0 sec	0

Source: Author

Table 102.5 Simulation results for output 'pendulum angular velocity'

Angle of inclined rail	Overshoot ranges	Settling time	Steady state error
10°	-0.085 to 0.016	1.5 sec	0
15°	-0.082 to 0.016	1.5 sec	0
20°	-0.09 to 0.016	1.5 sec	0
25°	-0.01 to 0.001	1.0 sec	0
30°	-0.07 to 0.018	1.5 sec	0

Source: Author

The simulation results obtained for attributes wheel position wheel velocity, pendulum angle and pendulum angular velocity are shown with the help of Figures 102.6-102.9. respectively. Also the same data is represented in Tables 102.2, 102.3, and 102.5 for respective attributes.

Conclusion

The study considered control of highly nonlinear and multivariable inverted pendulum system climbing on an inclined rail using a proportional-integral-derivative (PID) controller. The PID controller gains were tuned using auto-tuned function in Matlab. In order to make the problem more challenging the angle of inclination of the plane has been kept as 10°, 15°, 20°, 25° and 30° respectively. A mathematical model of the proposed system has been developed and simulated in Matlab/Simulink platform. The results indicate highly robust behavior of PID controller even at higher angle of inclinations. The settling time required by PID controller to stabilise the system at different angle of inclinations were wheel position (7.0 sec), wheel velocity (4.0 sec), pendulum angle (4.0 sec) and pendulum angular velocity (1.5 sec). Further, minimal overshoot responses and almost zero steady state error were obtained through proposed controller. As an extension for future work a real-time model of the proposed can be developed and considered for control using PID controller.

References

[1] Valluru, S. K. and Singh, M. (2017). Stabilization of nonlinear inverted pendulum system using MOGA and APSO tuned nonlinear PID controller. Cogent Engineering. 4(1), 1–15.
[2] Shreedharan, S., Ravikumar, V., and Mahadevan, S. K. (2021). Design and control of real-time inverted pendulum system with force-voltage parameter correlation. International Journal of Dynamics and Control. 9, 1672–80.
[3] Rapoport, L. B. and Generalov, A. A. (2022). Control of an inverted pendulum on a wheel. Automation and Remote Control. 83, 1151–71.

[4]   Messikh, L., Guechi, E. H., Bourahala, F., and Blazic, S. (2022). Stabilization of the cart-inverted-pendulum system using trivial state-feedback to output-feedback control conversion. Journal for Control, Measurement, Electronics, Computing and Communications. 63(4), 640–55.

[5]   Tamimi, J. (2021). A comparative study for balancing and positioning of an inverted pendulum robot using model-based controllers. International Journal of Mechatronics and Automation. 8(3), 135–41.

[6]   Patil, Y. P. and Patel, H. G. (2021). Sliding mode control design for nonlinear perturbed systems for tracking performance. International Journal of Systems, Control and Communications. 12(4), 364–79.

[7]   Saini, P. and Thakur, P. (2022). H-Infinity based robust temperature controller design for a non-linear systems. Wireless Personal Communications. 126(1), 305–33.

[8]   Sikander, A. and Prasad, R. (2019). Reduced order modelling based control of two wheeled mobile robots. Journal of Intelligent Manufacturing. 30(3), 1057–67.

[9]   Kharola, A., Patil, P., Raiwani, S., and Rajput, D. (2016). A comparison study for control and tabilization of inverted pendulum on inclined surface (IPIS) using PID and fuzzy controllers. Perspectives in Science. 8, 187–90.

[10] Belwal, N., Juneja, P. K., Sunori, S. K., Jethi, G. S., and Maurya, S. (2023). Modeling and control of FOPDT modeled processes-a review. Lecture Notes in Networks and Systems. 467, 255–60.

[11] Prasad, L. B., Gupta, H. O., and Tyagi, B. (2018). Intelligent control of nonlinear inverted pendulum system using Mamdani and TSK fuzzy inference systems: a performance analysis without and with disturbance input. International Journal of Intelligent Systems Design and Computing. 2(3), 313–34.

[12] Chheda, A. M., Mandava, R. K., and Pandu, R. V. (2021). Design and development of two-wheeled self-balancing robot and its controller. International Journal of Mechatronics and Automation. 8(1), 1–8.

[13] Wang, J. J. (2011). Simulation studies of inverted pendulum based on PID controllers. Simulation Modelling Practice and Theory. 19(1), 440–9.

[14] Magdy, M., Marhomy, A. I., and Attia, M. A. (2019). Modeling of inverted pendulum system with gravitational search algorithm optimized controller. Ain Shams Engineering Journal. 10(1), 129–49.

Chapter 103

A review on polymer matrix composites used in automobile applications

Mahboob Amin[1,a] and Chetan Swaroop[2,b]

[1]Department of Mechanical Engineering, Shivaji Institute of Technology and Management, Prayagraj, India

[2]Department of Mechanical Engineering, United College of Engineering and Research, Prayagraj, India

Abstract

In the search of lightweight, high performance, low cost, energy saver, and low environmental impact materials for automobiles, worldwide researchers are switching their focuses over monolithic to composite materials. Polymer matrix composites (PMCs) are leading the metals, alloys and other matrix composite materials by their several good properties such as better mechanical strength, light weight, cost effectiveness, low energy consumption, and recyclability. A vehicle made of composite material parts and accessories becomes light in weight, fuel efficient and reduces the emission of greenhouse gases. Accessories made of natural fibers are more eco-friendly. This research paper is focused upon PMCs utilised in automobiles, their comparative properties with conventional metallic parts, application of various types of fiber as well as natural fiber reinforcements with different polymer matrix in automobiles, influence of nanoparticles on mechanical properties of PMCs, possible applications in automobiles, and spotlights the advantages of PMCs over monolithic materials.

Keywords: Polymer composite, carbon fiber, natural fiber, glass fiber, automobile

Introduction

The popularity of fiber-reinforced polymer composite (FRPC) is growing in high technology applications like aerospace and also in low technology industrial applications like sanitaryware, etc. because of their high stiffness as well as specific strength, tremendous chemical resistance, good fatigue endurance, low density, flexible design, and superior manufacturability [1, 2]. Applications of polymer matrix composite materials are expanding on a large scale in diverse engineering fields like automobiles, aerospace, marine, military, energy, sports, civil and infrastructure [3, 4]. The application field of composites is expanding as the properties of composites are improving by introducing natural fibers. Many researchers in the greenhouse technology area are involved in reducing greenhouse gases by producing fuel-efficient and lightweight vehicles by using polymer composites [5]. The automobile sector is most benefitted from composite materials. Composite materials are replacing the metallic parts of vehicles day by day, which results in reducing the weight of vehicles and making them fuel-efficient. More energy is required to produce conventional materials than polymer composite materials, whereas natural fibers require much lower energy than synthetic fibers [6]. Biofiber composites are capable to fulfill the daily demand for environmentally safe composite materials; they are the prominent alternative to monolithic materials [7]. In automobiles, polypropylene is greatly replacing engineering plastics because of its excellent mechanical properties, economically cheap, lightweight, viable, ease of processing, and recyclability [8].

Composite Material and Polymer-matrix composites

A composite material is multi-component and multi-phase system that consists of a minimum two components having dissimilar properties and forms through a compounding process. One component is a continuous phase, which is known as a matrix; it includes metal, non-metal, or polymer; the other component is a dispersed phase, which is known as a reinforcing material; like natural fiber, glass fiber, carbon fiber, and so on. A composite material possesses different properties apart from its components [9]. Composite materials have superior properties as compared to conventional materials used in production processes. They have good stiffness and strength, high corrosion, and fatigue, resistant, are lighter in weight, have electrical conductivity, thermal expansion, and are less visible on radar than metals. Composite made with the thermoplastic matrix can be joined by heating. The scrap of composite materials can be remolded or reused. Composites can be classified on the basis of matrix material as metal-matrix composite, ceramic-matrix composite, and polymer-matrix composite [10].

Long or short both types of fiber are inserted into a polymer matrix that keeps the fibers together to create FRPCs [11]. The mechanical load is supported by polymer-matrix composites (PMCs) reinforced fiber. The goal of the matrix is to keep fibers together, transmit loads from fiber to fiber and keep the fiber safe from environmental and/or mechanical changes [10]. Composites containing polymer matrix and

[a]mahboobamin7@gmail.com, [b]chetanswaroop019@gmail.com

DOI: 10.1201/9781003450252-103

natural fiber maintain their chemical and mechanical uniqueness by holding good interfacial bonding and resistance. PMCs are easier in production along with fabrication than metal and ceramic matrix composites. Thermoplastic polymers are much more popular as they are lower in cost, have excellent chemical resistivity, and have good mechanical properties compared to metal matrix composites. Although the main disadvantage of polymer matrix composites is that they are not biodegradable like synthetic plastics, large accumulations of PMCs in the environment are creating problems of pollution [12]. To a certain extent, natural fiber biocomposites are key alternatives to reduce environmental harm and maintain the demand for composite materials [13].

Using as matrix material thermosetting plastics and thermoplastics both have some advantages and disadvantages. At the leading edge, the properties of PMCs such as toughness, stiffness, strength, and density can be optimized to certain stiffness to density and strength to density ratios to fulfil the need for structural material. Table 103.1 lists the properties and commonly used polymers in polymer matrix composites [14–16].

The PMCs have many benefits like low cost and density, less abrasiveness, etc. We can make it economic, reduce the weight and improve the strength by changing its constituents. Figure 103.1 presents the fabrication processes of polymer matrix fiber reinforced composites [17].

Resin-transfer, compression, and injection molding processes are widely used to manufacture automobile parts. Filament winding process is applied to manufacture the crashbox of vehicles which absorbs the impact energy of collision at the time of accident [18]. Most parts of automobile like dashboard, storage tanks, and engine hood, etc. are produced by vacuum aided resin transfer molding [19].

Enhancement in Properties of Polymer Matrix Composites

Devaraju et al. [20] conducted impact, tensile and flexural tests on the specimens made of 30wt% of palm fiber and epoxy matrix with and without ZnO nanoparticle addition. The surface of palm fiber was treated with NaOH. Three different wt% 0.1, 0.3, and 0.5 of ZnO were used to make the specimens. The 0.5wt%

Table 103.1 Properties of thermoset and thermoplastic matrices and commonly used matrices [14-16]

Properties	Thermoset matrices	Thermoplastic matrices
Toughness	Moderate	High
Modulus	High	Moderate
Processing temperature	Low/room temperature	High
Application temperature	High	Moderate
Viscosity	Low	High
Recyclability	Average	Good
Commonly used matrices	Epoxy, polyester,	Natural rubber, acrylonitrile butadiene rubber,
	Unsaturated polyester,	Low-density polyethylene,
	Phenolic, vinyl-acetate,	High-density polyethylene,
	Ethylene covinyl-acetate	Polyvinylchloride, polystyrene,
	Phenol formaldehyde	Polypropylene, polymethyl meta crylate

Source: https://docs.google.com/document/d/1zC6aXU_XvVARf2G3RaE17YHK1-RmYHIJ/edit?usp=sharing&ouid=10306725452 0897941389&rtpof=true&sd=true

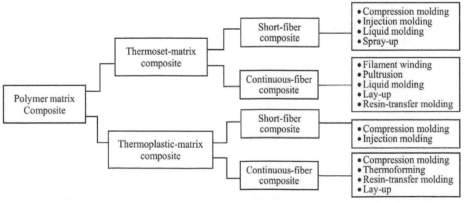

Figure 103.1 Fabrication processes of polymer matrix fiber reinforced composites [17]
Source: https://docs.google.com/document/d/1zC6aXU_XvVARf2G3RaE17YHK1-RmYHIJ/edit?usp=sharing&ouid=1030672 54520897941389&rtpof=true&sd=true

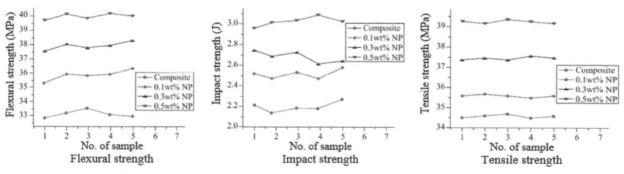

Figure 103.2 Mechanism of enhancement in properties on ZnO nanoparticle addition in PMC [20]

Source: https://docs.google.com/document/d/1zC6aXU_XvVARf2G3RaE17YHK1-RmYHIJ/edit?usp=sharing&ouid=1030672
54520897941389&rtpof=true&sd=true

ZnO nanoparticles composite demonstrated greater flexural, tensile, and impact strength. They concluded it is a green composite, non-hazardous, easily available, and low cost in contrast with Kevlar fiber, glass fiber, etc., This composite is better replacement material for automobile components. Figure 103.2 shows the mechanism of enhancement in properties on ZnO nanoparticle addition in PMC.

The mechanism of properties enhancement on ZnO nanoparticle addition in PMC are shown in Figure 103.2 [20].

Dubey et al. [21] carried out a study on polymer matrix composites made from fibers of agriculture, industrial, and other sources of waste. They found that waste filler composites increase the mechanical, physical, and wear properties while also lowering the cost of the composite. Treatment of fiber boosts the adhesion properties of fiber-to-matrix material than untreated fibers. They suggested it for the application in insulation boards, lightweight structural members, blocks, etc.

Chandrakar et al. [22] made six samples of composite material with crushed 75 micron Pistachio shell microparticulate and epoxy resin. The microparticulate wt% was 5, 10, 15, 20, 25, and 30. They found that the tensile strength improves with the rise of weight percent of microparticulate. Tensile strength increases by 23.7% with 5wt% filler, it was increased by 44.2% with 30wt% filler microparticle. Firstly, the compressive strength decreases rapidly up to 10wt% filler because the brittleness of composite increases with the increase of filler, after 10-30wt% the compressive strength increases up to a maximum level of 159.8MPa. The behavior of flexural strength was similar in nature to the tensile strength.

Girge et al. [23] studied the research on red mud, blast furnace slag and Linz Donawitz slag-filled polymer composites. Red mud filler provides higher durability and mechanical strength; it may be used in the production of lightweight automobile accessories. Fly ash in hybrid composite improves the compressive strength and K_2O improves the fire resistance and mechanical strength. Rather fiber and matrix properties; the property of PMC also depends on some factors like filler material, surface treatment of fiber, hybridisation and orientation of fibers, microparticle fillers, etc. Impact and tensile strength of PMCs rises up on inclusion of ZnO nanoparticles up to a certain wt%. Filler microparticle material improves tensile and flexural strength. Similarly, blast furnace slag also rises-up mechanical properties and hardness of composite material.

Figure 103.3 shows comparative peak load achieved, yield, and flexural strength on loading the epoxy-matrix composite material strengthened with glass, carbon, and carbon-glass fiber [24].

Pmc Used in Automobile Industries

Composite materials made with polymer matrix and fiber or whisker reinforcement are commonly used in the automobile field. In June 30, 1953 Chevrolet Corvette firstly used a fiberglass body made of PMC for their automobile [27]. Polymer matrix composites are more advantageous than metallic materials such as good specific strength and 20-40% weight saving, low thermal expansion, excellent fatigue resistant, fracture-resistant, dimensional stability, and potential for fast processing cycles. Globally 50% of thermoplastic and 24% of thermosets are being used in the automobile sector.

To reduce the emission levels and energy cost, automotive sector industries need advanced lightweight composite materials with better mechanical and chemical properties. Figure 103.4 illustrates the uses of PMC in automobile [25, 26].

Fiber Reinforcement in Polymer Matrix Used for Automotive Field

Carbon fiber

Carbon fibers with a diameter of 5 to 10 micrometers that are more than 90% carbonised are commonly used in composites. Carbon fiber is much popular in several engineering applications like automobiles,

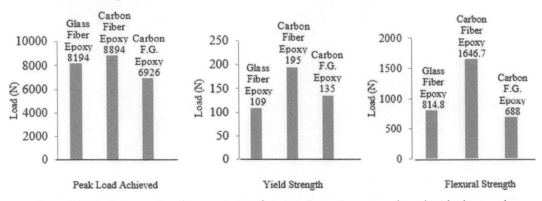

Figure 103.3 Comparative characteristics of epoxy composites strengthened with glass, carbon, and carbon-glass fiber [24]

Source: https://docs.google.com/document/d/1zC6aXU_XvVARf2G3RaE17YHK1-RmYHIJ/edit?usp=sharing&ouid=1030672 54520897941389&rtpof=true&sd=true

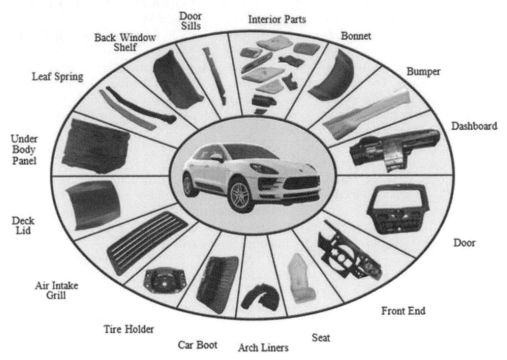

Figure 103.4 PMC used in automobile [25, 26]

Source: https://docs.google.com/document/d/1zC6aXU_XvVARf2G3RaE17YHK1-RmYHIJ/edit?usp=sharing&ouid=1030672 54520897941389&rtpof=true&sd=true

aerospace, marine, civil engineering and medical field by virtue of its important characteristics like lightweight, very good tensile strength and stiffness, self-lubrication, fatigue resistance, good vibration damping, low coefficient of thermal expansion, high stability at elevated temperature, high corrosion/chemical resistance, organic inertness, good electrical conductivity and electromagnetic properties [28]. The only down side of carbon fiber is its expensiveness as compared to glass fibers, natural fibers, and plastic fibers [29]. Sources of carbon fiber are polyacrylonitrile (PAN), pitch, and rayon. PAN contributes around 90% production of carbon fiber. Carbon fibers produced from PAN have reasonably good mechanical properties, low in density than carbon fiber produced from pitch. The diameter of carbon fiber pitch ranges from 10 to 11 micrometer having good tensile modulus about, of 170 to 980 GPa, and good coefficient of thermal expansion in axial direction but the density is higher than other carbon fiber produced from PAN and rayon. Using carbon fiber polymer composite, the weight of the vehicle can be significantly reduced by 60% [30].

Glass fiber

Of all the fibers used in composite materials, glass fiber may be the most common, because glass fiber is the least expensive one. Glass fiber is produced from silicon dioxide (SiO_2) in the form of very fine fibers of glass. Glass fibers have two main advantages; low cost and high strength, while this have some other array of properties such as dimensional stability, low thermal conductivity, good electrical insulator, incombustibility, durability and dielectric permeability. Silicon dioxide (SiO_2), lime stone and alumina (Al_2O_3) are used as raw material in production of glass fiber, various other compositions are being used commercially to achieve different properties. S-glass and E-glass are two more prevalent types of glass fibres. E-glass is also known as electrical glass and more commonly used fiber (about 99%) with polymer matrix in industrial applications, marine, and aerospace, etc [31].

The right orientation and composition of glass fiber makes desired functional characteristics and properties of polymer composites that are similar to steel, its specific gravity is ¼ times that of steel, and have more stiffness than aluminum [32]. Under laboratory conditions, the tensile stresses of glass fiber can achieved at up to 7000 N/mm², whereas commercial glass fiber can gain 2800-4800 N/mm² [33]. GFRP composites have been largely used for manufacturing automobile parts like engine covers, bumpers, body panels, door panels, and seat cover panels [34].

Natural fiber

In nature, natural fibers are widely distributed. They are found from various parts and species of plants like jute, bamboo, kenaf, sugarcane plants etc. [35].

Thermoplastics like polypropylene, polyvinyl chloride are well suited binding materials for natural fibers. Epoxy, polyester and phenolic are also good thermosetting matrix which is being used with natural fibers [36]. Different kinds of natural fibers could be used in various automotive applications.

Hemp fiber is found from bast of cannabis family. It has high Young's modulus and mechanical strength. Marcedes Benz and some automakers are using it to manufacture of spare wheel [37].

Flax fibers have various good qualities such as light weight, high strength, stiffness, toughness, specific tensile property and biodegradability. It has gained high interest in the automobile sector and can be a substitute of glass fiber. It is used to manufacture door panels, body panels and rear and trim shelf by Marcedes Benz and different automotive manufacturer [38].

Bamboo plant pulp is the source of bamboo fibers. It is used to manufacture floor mat, luggage compartment, speakers etc. [39].

Coir fibers are found from coconut fruit, it has high hardness, non toxic and highly durable due to high lignin present in it. It is used automobile manufacturer like Fiet for exterior and interior trims and Marcedes Benz for backrest, etc. [40].

- Sisal fibers are hard in nature, easy to cultivate. It has high resistance to alkali, acid, salty water and corrosion. It is used by BMW and other automakers to manufacture underbody panels and interior door panels.
- Pineapple leaf fibers are the by-product of pineapple. Automobile manufacturers have keen interest in it to manufacture natural fiber polymer composite for a variety of applications in automobile [41].
- Kenaf fibers have high mechanical properties, low density, nonabrasive and biodegradable. It can serve as good reinforcement for polymer composite which can be used in automobile industry [42].

In automotive sector, the vital scope of polymer matrices are racing brakes, clutch plate, body & parts of electrical vehicles, heavy trailer and truck springs, rocker arms, engine shrouds, fuels tank with filament

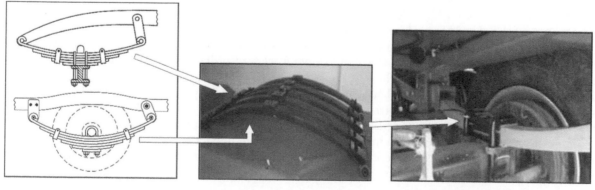

Metal laminated spring   Composite laminated spring   Composite material laminated spring

Figure 103.5 Metal laminated spring system replaced with composite material leaf spring system [59]
Source: https://docs.google.com/document/d/1zC6aXU_XvVARf2G3RaE17YHK1-RmYHIJ/edit?usp=sharing&ouid=10306725 4520897941389&rtpof=true&sd=true

wound, wheel and suspension arms [25, 43]. Although body and parts made with polymer composites makes a better vehicle in terms of light weight, durability, crashworthiness, eco-friendliness, fuel saver and reducing carbon emissions [44].

In several engineering applications, hybrid composite materials are being used because of their light weight, low cost, good strength, ease to development, and strength to weight ratio [45].

Main Parts of Automotive Manufactured by Using Polymer Matrix Composites

Brake Pad

A brake pad converts a vehicle's kinetic energy to heat energy. It has a facing frictional lining bonded to a steel back plate [46]. The performance and effectiveness of a braking system rely upon the frictional material and its fabrication process [47]. Kumar et al. [48] developed wear mechanism maps of phenolic matrix composites reinforced with flax-fiber (FFRC), basalt fiber (BFRC) and flax/basalt hybrid fiber (HFRC). The maximum wear was found for BFRC, HFRC, and FFRC as 1.6 μm, 6.4 μm, and 3.5 μm respectively. Due to the improved wear resistance of BFRC, good thermal properties and bonding qualities were found. The wear debris acted as a shielding barrier, which improved wear resistance. BFRC was found to be the most effective material for brake pads.

Ravikumar et al. [49] By employing epoxy resin and fibers from the palm kernel shell (PKSF) in compression molding, asbestos-free brake pads were developed. They added carbon, aluminum oxide, and calcium carbonate as abrasive filler element. As PKSF has low density, the hardness of PKSF pad decreased by 21% on increase in PKSF percentage from 25 to 35. The composite was highly stable between the temperature range of 0-200°C. Further weight loss was recorded by 25-60% at 400°C and nearly 80% at 800°C. Thus, the mechanism of wear is affected by temperature of interface, composition of friction pad composite, debonding phenomenon, and operating parameters.

Clutch Plate

A clutch is a mechanism that engages and dis-engages the engine shaft to the gear box. It transmits engine power to wheels of a vehicle by frictional force.

Prakash et al. [50] found the jute fiber to be much stronger than asbestos, and the jute fiber can be used to replace asbestos for clutch plates.

Kumar et al. [51] worked on clutch plate development using coconut shell char flax fiber(CSC FF)/epoxy resin and coconut shell char glass fiber(CSC GF)/epoxy resin. During experiments they found better results from CSC FF/epoxy composite than from CSC GF/epoxy resin for the application of automobile clutch plates.

Sundarapandian et al. [52] conducted experiments to replace the asbestos facing of a single plate clutch with coconut shell powder/coir/epoxy composites. They found improved mechanical properties in NaOH treated fibers and heat-treated shell powder. Eco-friendly coir/coconut shell powder/epoxy composite was the best material to replace the asbestos clutch plate material.

Bumper

A bumper is a safety device which protects the vehicle and its parts during a collision. It absorbs the impact energy and reduces the effect of impact. It is mounted on the front and rear ends of the automobile.

Selwyn et al. [53] fabricated aramid fiber layered composites with different orientations in second and fourth layers (A-90/45/90/45/90, B-90/30/90/30/90, and C-90/60/90/60/90. Specimen A reported excellent mechanical properties. The UTS and impact load were recorded as 147 MPa and 20 Joule, which was more than light weight steel with 120MPa and 9 Joule. The ductility and toughness were reasonably high. The ultimate stress and maximum withstand load of sample A were recorded as 86.8MPa and 13.46kN which were higher than those of stainless steel.

Ramasubbu et al. [54] manufactured sisal 21% /kenaf 9% fiber epoxy 70% hybrid composite by compression molding machine. They compared its mechanical properties with sisal 30% fiber epoxy 70% and kenaf 30% fiber epoxy 70% composites; and found that the hybrid composite was better than kenaf composite in tensile strength. The hardness, moisture absorption, and impact energy of hybrid composite were greater than the sisal-epoxy and kenaf-epoxy composites. It was investigated to be a good suitable material for light-weight automobile applications.

Leaf Spring

Leaf springs are semi-elliptical shaped strips which are used to absorb the energy of shocks and jerks from road and releases it slowly, which makes a safe and comfortable ride. The main factor in the choice of material for leaf springs is strain energy [55, 56]. It is used in the form of laminated bundle, held together by 'U' clamp bolts and clips.

The Chevrolet Corvette pursued to apply light weight epoxy polymer composite, reinforce with glass fiber to leaf spring and investigated fatigue life of this composite was five times more than of steel. This composite material leaf spring showed quicker response to stresses developed by poor road conditions and gave a smoother ride than leaf springs made of steel. Besides, composite material leaf spring offers less chance of disastrous disappointment and amazing properties of corrosion resistance [57].

Oztoprak et al. [43] worked on replacing the leaf spring made of EN45 material with an epoxy composite spring by static FEA analysis in Abaqus 6.12-1 software. They found that weight can be reduced by 80% with good mechanical properties by the use of carbon/epoxy leaf springs in place of EN45 steel leaf springs. It was found to be much stiffer than other leaf springs and had good load-displacement response. The hybrid system was better and can be used to manufacture leaf springs.

Mehul et al. [58] conducted static FEA on steel as well as composite material leaf springs for stiffness, capacity, and weight reduction. The weight could be reduced by 79.617% by using the same number of composite leaf springs, for the same strength, the weight could be reduced by 90.09% by using a mono-leaf composite material spring.

Guduru et. al. [24] worked on developing a monocomposite leaf spring for vehicles to put back traditional steel multi-leaf spring. Mechanical properties of mono-leaf springs were tested to search for the most suitable fabric of composite. Resulted that carbon-fiber/epoxy as well as glass-fiber/epoxy spring presents high flexural and tensile strength, carbon fiber epoxy spring showed less deflection than fiber glass/epoxy spring. More weight can be supported by carbon fiber/epoxy than by fiber glass/epoxy spring; weights were reduced by 75% and 69.4% respectively for carbon fiber and glass fiber springs. Figure 5 Shows metal laminated spring system replaced with composite material laminated spring system [59].

Result and Discussions

It is identified from this literature review that polymer composites have a significant role in automobiles. In the investigation of several researchers, it was concluded that:

- Polymer matrix served as a good matrix material for Glass, Carbon, and Natural fibers which are being used in automobiles.
- Inclusion of ZnO nanoparticles in palm fiber-epoxy matrix increased the tensile and impact strength. Agricultural, industrial waste and Blast furnace slag served as good filler and improved the composites' mechanical and also physical attributes.
- Surface treated natural fiber providing better bonding with matrix, that can serve as good mechanical properties for accessories manufactured.
- Fly ash improved the compressive strength whereas K2O improved the mechanical strength and fire resistant properties.
- Glass fiber/epoxy as well as carbon fiber/epoxy composites were the most effective cheaper, light in weight material with good shock energy absorbing capacity when used as leaf spring.
- BFRC had good wear resistant, thermal, and bonding properties that could serve as effective material for brake pads.
- NaOH treated coir/coconut shell powder/epoxy composite was better material to replace the asbestos in clutch plate. Natural fibers are abundantly available, lighter, cheap, biodegradable, and holds good mechanical behaviour. The automotive industries are widely using natural fiber composites.

The polymer matrix composite may be a goal changer in the automobile field because of its superior mechanical behaviour, weight reduction of vehicles, and increased fuel efficiency that will drop the greenhouse gas emissions. Carbon fiber, carbon epoxy, graphite epoxy, e-glass epoxy, vinylester, pineapple fiber, Flakes, etc. Composite materials are being vitally used in automobile industries.

Conclusion

Several researchers are doing large scale research on polymer matrix composites, by choosing different fibers and polymers. They are trying to improve the properties. This run-through review illuminates the research accomplished to improve the mechanical properties of PMCs such as stiffness, strength, and impact strength etc. The polymer matrix material is more advantageous over other matrix composites such as low costs, low density, and lesser abrasive properties. Carbon fiber/epoxy and glass fiber/epoxy composites are good for leaf springs. Carbon fiber/epoxy composite holds more load than glass fiber/epoxy composite; this makes it most suitable for leaf spring. Abundantly available natural fibers are best suited to the thermoplastic matrix. Researchers have been greatly focused on it. They are being used to manufacture several accessories for automobiles.

Broad and deep research is needed to explore the wide scope of polymer matrix composites. This is why researchers must concentrate on developing a low cost and potential manufacturing technique for composite materials having a balanced structure and properties.

One major limitation of polymer matrix composite is its non-biodegradability; in the future researchers could achieve its biodegradable properties.

Declaration of Competing Interest

The authors report no conflict of interest.

References

[1] Chen, Q., Zhang, L., Zhao, Y., Wu, X. F., and Fong, H. (2012). Hybrid multi-scale composites developed from glass microfiber fabrics and nano-epoxy resins containing electrospun glass nanofibers. Composites Part B: Engineering. 43(2), 309–16. doi: 10.1016/j.compositesb.2011.08.044.

[2] Beckermann, G. W. and Pickering, K. L. (2015). Mode I and Mode II interlaminar fracture toughness of composite laminates interleaved with electrospun nanofibre veils. Composites Part A: Applied Science and Manufacturing. 72, 11–21. doi: 10.1016/j.compositesa.2015.01.028.

[3] Lakhdar, M., Mohammed, D., Boudjemâa, L., Rabiâ, A., andBachir, M. (2013). Damages detection in a composite structure by vibration analysis. Energy Procedia. 36, 888–97. doi: 10.1016/j.egypro.2013.07.102.

[4] Rafiee, R. (2016). On the mechanical performance of glass-fibre-reinforced thermosetting-resin pipes: a review. Composite Structures. 143, 151–64. doi: 10.1016/j.compstruct.2016.02.037.

[5] Miller, L., Soulliere, K., Sawyer-Beaulieu, S., Tseng, S., and Tam, E. (2014). Challenges and alternatives to plastics recycling in the automotive sector. Materials (Basel). 7(8), 5883–5902. doi: 10.3390/ma7085883.

[6] Kim, W. et al. (2012). High strain-rate behavior of natural fiber-reinforced polymer composites. Journal of Composite Materialsvol. 46(9), 1051–65. doi: 10.1177/0021998311414946.

[7] Sahu, P. and Gupta, M. K. (2017). Sisal (Agave sisalana) fibre and its polymer-based composites: a review on current developments. Journal of Reinforced Plastics and Composites. 36(24), 1759–80. doi: 10.1177/0731684417725584.

[8] Agarwal, J., Sahoo, S., Mohanty, S., and Nayak, S. K. (2020). Progress of novel techniques for lightweight automobile applications through innovative eco-friendly composite materials: a review. Journal of Thermoplastic Composite Materials. 33(7), 978–1013. doi: 10.1177/0892705718815530.

[9] Wang, R.-M., Zheng, S.-R., and Zheng, Y.-P. G. (2011). Polymer Matrix Composites and Technology. Elsevier.

[10] Clyne, T. W and Hull, D. (209\21. An introduction to Composite Materials. Cambridge University Press.

[11] Teng, J. G., Yu, T., and Fernando, D. (2012). Strengthening of steel structures with fiber-reinforced polymer composites. Journal of Constructional Steel Research. 78, 131–43. doi: 10.1016/j.jcsr.2012.06.011.

[12] Fakhrul, T. and Islam, M. A. (2013). Degradation behavior of natural fiber reinforced polymer matrix composites. Procedia Engineering. 56, 795–800. doi: 10.1016/j.proeng.2013.03.198.

[13] Akampumuza, O., Wambua, P. M., Ahmed, A., Li, W., and Qin, X. (2017). Review of the applications of biocomposites in the automotive industry. Polymer Composites. 38(11), 2553–69.

[14] Plummer, C. J. G., Bourban, P.-E., and Månson, J.-A. (2016). Polymer Matrix composites: matrices and processing. Reference Module in Materials Science and Materials Engineering. pp. 1–9, doi: 10.1016/b978-0-12-803581-8.02386-9.

[15] Mahesh, V., Joladarashi, S., andKulkarni, S. M. (22021). A comprehensive review on material selection for polymer matrix composites subjected to impact load. Defence Technologyoly. 17(1), 257–77. doi: https://doi.org/10.1016/j.dt.2020.04.002.

[16] Charrier, J.-M. (1991). Polymeric Materials and Processing: Plastics. Elastomers Compos. Munich, New York: Hanser Publications.

[17] Campbell, F. C. (Ed.) (2010). Introduction to composite materials. Structural Composite Materials. ASM International. doi: 10.31399/asm.tb.scm.t52870001.

[18] Friedrich, K. and Almajid, A. A. (2013). Manufacturing aspects of advanced polymer composites for automotive applications. Applied Composite Materials. 20(2), 107–28. doi: 10.1007/s10443-012-9258-7.

[19] Kong, C., Lee, H., and Park, H. (2016). Design and manufacturing of automobile hood using natural composite structure. Composites Part B: Engineering. 91, 18–26.

[20] Devaraju, A., Sivasamy, P., and Loganathan, G. B. (2020). Mechanical properties of polymer composites with ZnO nano-particle. Materials Today: Proceedings. 22, 531–34.

[21] Dubey, S. C., Mishra, V., and Sharma, A. (2021). A review on polymer composite with waste material as reinforcement. Materials Today: Proceedings. Materials Today: Proceedings. 47, 2846–51.

[22] Chandrakar, S., Agrawal A., Prakash, P., Khan, I. A., and Sharma, A. (2021). Physical and mechanical properties of epoxy reinforced with pistachio shell particulates. AIP Conference Proceedings. 2341. doi: 10.1063/5.0049949.

[23] Girge, A. et al. (2021). Industrial waste filled polymer composites–a review. Materials Today: Proceedings. 47, 2852–63.

[24] Guduru, R. K. R., Shaik, S. H., Tuniki, H. P., and Domeika, A. (2021). Development of mono leaf spring with composite material and investigating its mechanical properties. Materials Today: Proceedings. 45, 556–61. doi: https://doi.org/10.1016/j.matpr.2020.02.289.

[25] Muhammad, A., Rahman, M. R., Baini, R., andBin Bakri, M. K. (2021). Applications of sustainable polymer composites in automobile and aerospace industry. Advances in Sustainable Polymer Composites. Woodhead Publishing. 185-207. doi: 10.1016/b978-0-12-820338-5.00008-4.

[26] Oladele, I. O., Omotosho, T. F., and Adediran, A. A. (2020). Polymer-based composites: an indispensable material for present and future applications. International Journal of Polymer Science. 8834518. doi: 10.1155/2020/8834518.

[27] Mueller, M. and Woods, B. (2004). Corvette. MotorBooks International.

[28] Roth, S., Stoll, M., Weidenmann, K. A., Coutandin, S., and Fleischer, J. (2019. A new process route for the manufacturing of highly formed fiber-metal-laminates with elastomer interlayers (FMEL). International Journal of Advanced Manufacturing Technology. 104(1), 1293–1301. doi: 10.1007/s00170-019-04103-4.

[29] MatWeb. Overview of materials for epoxy/carbon fiber composite. https://www.matweb.com/search/datasheet. aspx?matguid=39e40851fc164b6c9bda29d798bf3726 (accessed Aug. 04, 2022).

[30] D. Isaiah, D. (2016). Carbon fibre: the fabric of the future. Automotive World Megatrends Mag. 1.

[31] "15 Different Types of Fiberglass - Home Stratosphere." https://www.homestratosphere.com/types-of-fiberglass/ (accessed Aug. 04, 2022).

[32] Ramzan, E. and Ehsan, E. (2009). Effect of various forms of glass fiber reinforcements on tensile properties of polyester matrix composite. Faculty of Engineering & Technology. 16, 33–39.

[33] Rosato, D. V. and Rosato, D. V. (2004). Reinforced Plastics Handbook. Elsevier.

[34] Sathishkumar, T. P., Satheeshkumar, S., andNaveen, J. (2014). Glass fiber-reinforced polymer composites: a review. Journal of Reinforced Plastics and Composites. 33(13), 1258–75. doi: 10.1177/0731684414530790.

[35] Samuel, O. D., Agbo, S., and Adekanye, T. A. (2012). Assessing mechanical properties of natural fibre reinforced composites for engineering applications. Journal of Minerals and Materials Characterization and Engineering. 11(8), 780–84. doi: 10.4236/jmmce.2012.118066.

[36] Facca, A. G., Kortschot, M. T., andYan, N. (2006). Predicting the elastic modulus of natural fibre reinforced thermoplastics. Composites Part A: Applied Science and Manufacturing. 37(10), 1660–71.

[37] Badji, C., Beigbeder, J., Garay, H., Bergeret, A., Bénézet, J. C., and Desauziers, V. (2018). Exterior and under glass natural weathering of hemp fibers reinforced polypropylene biocomposites. Impact on mechanical, chemical, microstructural and visual aspect properties. 148. Elsevier Ltd. doi: 10.1016/j.polymdegradstab.2017.12.015.

[38] Huang, K., Tran, L. Q. N., Kureemun, U., Teo, W. S., and Lee, H. P. (2019). Vibroacoustic behavior and noise control of flax fiber-reinforced polypropylene composites. Journal of Natural Fibers. 16(5), 729–43. doi: 10.1080/15440478.2018.1433096.

[39] Hu, G., Cai, S., Zhou, Y., Zhang, N., and Ren, J. (2018). Enhanced mechanical and thermal properties of poly (lactic acid)/bamboo fiber composites via surface modification. Journal of Reinforced Plastics and Composites. 37(12), 841–52. doi: 10.1177/0731684418765085.

[40] Arrakhiz, F. Z., Malha, M., Bouhfid, R., Benmoussa, K., and Qaiss, A. (2013). Tensile, flexural and torsional properties of chemically treated alfa, coir and bagasse reinforced polypropylene. Composites Part B: Engineering. 47, 35–41. doi: 10.1016/j.compositesb.2012.10.046.

[41] Dai, H., Ou, S., Huang, Y., and Huang, H. (2018). Utilization of pineapple peel for production of nanocellulose and film application. Cellulose. 25(3), 1743–56. doi: 10.1007/s10570-018-1671-0.

[42] Mastura, M. T., Sapuan, S. M., Mansor, M. R., and Nuraini, A. A. (2017). Environmentally conscious hybrid bio-composite material selection for automotive anti-roll bar. International Journal of Advanced Manufacturing Technology. 89(5–8), 2203–19. doi: 10.1007/s00170-016-9217-9.

[43] Oztoprak, N. et al. (2018). Developing polymer composite-based leaf spring systems for automotive industry. Science and Engineering of Composite Materials. 25(6), 1167–76. doi: 10.1515/secm-2016-0335.

[44] Seydibeyoğlu, M. Ö., Doğru, A., Kandemir, M. B., and Aksoy, Ö. (2020). Lightweight composite materials in transport structures. Lightweight Polymer Composite Structures. 103–130, 2020, doi: 10.1201/9780429244087-5.

[45] Ravishankar, B., Nayak, S. K., and Kader, M. A. (2019). Hybrid composites for automotive applications: a review. Journal of Reinforced Plastics and Composites. 38(18), 835–45. doi: 10.1177/0731684419849708.

[46] Idris, U. D., Aigbodion, V. S., Abubakar, I. J., and Nwoye, C. I. (2015). Eco-friendly asbestos free brake-pad: Using banana peels. Journal of King Saud University Engineering Sciences. 27(2), 185–92. doi: 10.1016/j.jksues.2013.06.006.

[47] Nagesh, S. N., Siddaraju, C., Prakash, S. V., and Ramesh, M. R. (2014). Characterization of brake pads by variation in composition of friction materials. Procedia Material Science. 5, 295–302. doi: 10.1016/j.mspro.2014.07.270.

[48] Ilanko, A. K. and Vijayaraghavan, S. (2016). Wear behavior of asbestos-free eco-friendly composites for automobile brake materials. Friction. 4(2), 144–52. doi: 10.1007/s40544-016-0111-0.

[49] Ravikumar, K. and Pridhar, T. (2019). Evaluation on properties and characterization of asbestos free palm kernel shell fibre (PKSF)/polymer composites for brake pads. Material Resource Express. 6(11), 1165d2.

[50] Prakash, R. S., Kumar, N. M., Sabareesh, J., Sadeesh, C., and Sanjay C. (). Experimental study on behaviour of clutch plate lining using jute fibre. Engineering, Materials Science.

[51] Kumar, J. G. K. and Murugan, M. A. (2018). Mechanical and tribological behaviour of coconut shell char, flax and glass fiber reinforced composite material for automobile clutch plate. International Journal of Mechanical Engineering Research. 6(3), 15–18.

[52] Sundarapandian, G. and Arunachalam, K. (2020). Investigating suitability of natural fibre-based composite as an alternative to asbestos clutch facing material in dry friction clutch of automobiles. IOP Conference Series: Materials Science and Engineering. 912(5). doi: 10.1088/1757-899X/912/5/052017.

[53] Selwyn, T. S. (2021). Formation, characterization and suitability analysis of polymer matrix composite materials for automotive bumper. Material Today Proceedings. 43, 1197–1203.

[54] Ramasubbu, R. and Madasamy, S. (2022). Fabrication of automobile component using hybrid natural fiber reinforced polymer composite. Journal of Natural Fibers. 19(2), 736–46. doi: 10.1080/15440478.2020.1761927.

[55] Shukoor, S. A. and Rao, N. A. N. (2017). Fabrication, design optimization, and stress analysis of mono leaf spring using metal-matrix composites. International Journal of Innovative Technology and Research. 5(3), 6073-77

[56] Kabanur, B., Patil, P. S. (2017). Improve The design of leaf spring by reducing the frictional stress. International Research Journal of Engineering and Technology. 4,(8), 1363–70.

[57] Pruez, J., Shoukry, G. Williams, S., and Shoukry, M. (2013). Lightweight composite materials for heavy duty vehicles. pp. 2.

[58] Mehul, S., Shah, D. B., and Bhojawala, V. (2012). Analysis of composite leaf spring using FEA for light vehicle mini truck. Journl of Information, Knowledge. Research Mechenical Eng.ineering. 2(2), 424–28.

[59] Ma, L. et al. (2021). Structure design of gfrp composite leaf spring: An experimental and finite element analysis. Polymers (Basel)., 13(8), 1–22. doi: 10.3390/polym13081193.

Chapter 104

Nonlinear autoregressive neural network based multistep prediction of specific enthalpy of steam

Dr. Ashwani Kharola[1,a], Dr. Rahul Rahul[2,b], Vishwjeet Choudhary[3,c] and Anurag Bahuguna[4,d]

[1]Department of Mechanical Engineering, Assistant Professor, Graphic Era Deemed to be University, Dehradun, India

[2]Department of Electronics, Assistant Professor, Sri Venkateshwara College, University of Delhi, Delhi, India

[3]Department of Mechanical Engineering, Research Scholar, BITS-Pilani, Pilani, India

[4]Department of Mechanical Engineering, Assistant Professor, Tula's Institute, Dehradun, India

Abstract

This study investigates the application of nonlinear autoregressive (NAR) neural networks for multistep prediction of specific enthalpy of steam. Real-time data has been considered for precise prediction of specific enthalpy of liquid and specific enthalpy of vapor. The network has been trained using Levenberg-Marquardt learning algorithm. The multistep prediction technique has been analyzed by considering consecutive prediction of outputs for different samples sizes i.e., 5, 10, 15, 20, 25, and 30 respectively. The performance of NAR neural network model has been analyzed in terms of RMSE, MAE and Bivariate correlation coefficient. The results indicate that both RMSE and MAE increase with increase in sample size of multistep predictions. Further, the output specific enthalpy of liquid and specific enthalpy of vapor has been predicted with a maximum error less than 1% and 2% respectively. The results also indicate a very strong Bivariate correlation coefficient in order 0.99 obtained for different sample sizes for both the output properties of steam.

Keywords: MAE, multistep prediction, NAR, neural network, RMSE, steam, thermodynamic property

Introduction

Nonlinear autoregressive (NAR) neural network is a popular machine learning algorithm for series prediction. It is a closed-loop network which can perform multistep predictions [1]. NAR neural network has the capability to predict under situations when external feedback is not available by using internal feedback only. These networks are widely used for various industrial applications involving multistep predictions [2–4]. For instance, a NAR neural network combined with discrete wavelet transform to predict oil-dissolved gas absorption in a transformer has been proposed in [5]. The results indicate that the technique can precisely forecast several values of gas concentration thus making it suitable for transformer fault diagnosis. Further, a NAR neural network for forecasting inside air-temperature of a smart building is suggested in [6]. The study adopted a novel approach using Internet-of-Things (IoT) to collect real-time data of the building for training of NAR neural network model. Experimental results demonstrated high precision of the proposed approach. Moreover, a NAR neural network model for forecasting energy consumption in South African Universities considering their previous three years daily energy consumption data has been proposed in [7]. The acquired data was further filtered through singular spectrum technique. The results highlighted that high prediction accuracies can be obtained by adopting data filtering technique in forecasting.

Furthermore, two different neural network modelling techniques i.e., NAR and NAR-exogenous (NARX) to forecast wind speed at different locations in Malaysia for a period of 30 days has been suggested in [8]. The study further compared tangent sigmoid and log sigmoid transfer function for predicting wind speed. The results indicate promising performance of proposed neural network models with minimal errors between actual and predicted outputs. A NAR neural network model to predict extraction of lanolin from raw wool at temperature above its melting point is presented in [9]. The experimental data for temperature, pressure, solvent mass flow rate, wool packing density and time were collected and used for training of proposed network using Weibull statistical function. The results indicate that the proposed model can predict the lanolin breakthrough with an error of about ± 0.42%. Additionally, a novel algorithm developed by combining fuzzy c-means technique and NARX neural network for predicting temperature and silicon content of hot metal in a blast furnace is proposed in [10]. The efficiency of the proposed algorithm was evaluated by simulating it for the industrial manufacturing process. The results indicate superior performance of NARX neural network model designed using fuzzy c-means technique when compared to conventional NARX model.

[a]ashwanidaa@gmail.com, [b]rvermaresearch@gmail.com, [c]vishwjeet.c547@gmail.com, [d]anurag.bahuguna@tulas.edu.in

DOI: 10.1201/9781003450252-104

A NAR neural network model based on radial basis function for solving of nonlinear systems of minute ionised gas particles has been proposed in [11]. The proposed model was implemented using Van der Pol-Methiew Equation. The results indicate that excellent root mean square errors (RMSE) upto 1e-38 were obtained using proposed models. Recently, a study to compare NAR neural network with gray-box and black-box models for forecasting inside temperature of a building in Montreal, Canada has been presented in [12]. The outcomes indicate that the NAR neural network model can forecast the thermal behavior of the building more precisely compared to other two models. Thus, the review of past few year literatures suggests that NAR neural networks can be effectively used for multistep prediction of various categories of data but till date no such model has been developed for prediction of thermodynamic properties of steam. Therefore, in this study a novel NAR neural network model has been adopted for prediction of specific enthalpy of steam. The real-time data from Perry's Chemical Engineering Handbook has been considered for multi-step prediction of specific enthalpy of liquid and specific enthalpy of vapor. Examining data from the steam table is a tedious task especially when interpolated data is required hence the proposed NAR model can aid in accurate data prediction in minimal time for better validation. The NAR neural network model has been trained using Levenberg-Marquardt training algorithm. The accuracy of the model has been further measured in terms of root mean square error (RMSE), mean absolute error (MAE) and Bivariate correlation coefficient.

Nonlinear autoregressive neural network based prediction of specific enthalpy of steam

In NAR neural network predictions, the forthcoming values of a particular sequence are forecasted considering the previous data of the same sequence [13]. The network is initially generated in an open-loop and then transformed into a closed-loop for multi-step prediction [14]. In this study two separate NAR neural network models have been designed using 30 and one neurons in the hidden and output layer respectively for multistep prediction of specific enthalpy of liquid and specific enthalpy of vapor respectively. Further, the feedback delays which signifies the data needed for multistep forecasting were kept as two as shown in Figure 104.1. A set of 200 consecutive series of data for both of the above-mentioned thermodynamic properties of steam were collected from Perry's Chemical Engineering Handbook and used for training of NAR neural network model for multistep prediction of these properties with different step/sample sizes i.e. 5, 10, 15, 20, 25 and 30 respectively.

The mean squared error (MSE) response for training, validation and testing samples are indicated in Figure 104.2. The MSE decreases with the increase in number of epochs. Further, the finest validation

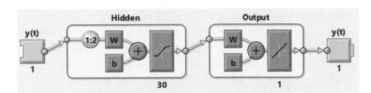

Figure 104.1 NAR neural network architecture considered for multistep prediction
Source: Author

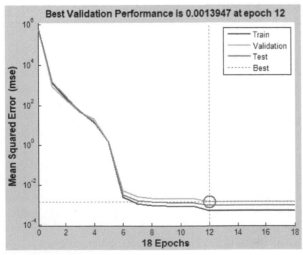

Figure 104.2 Best validation performance obtained at 12 epochs for NAR neural network model
Source: Author

performance is 0.0013947 at 12 epochs. The MSE values are close to zero, which indicates a minimal error between the desired output and predicted outputs.

The graphs obtained for gradient, *Mu* and validation checks after 18 epochs are shown in Figure 104.3. The gradient indicates the slope of the function and helps in predicting the effect of taking a small step from a particular point in any direction. After 18 epochs, the value obtained for gradient is 0.33506. Further, *Mu* is a control parameter for Levenberg-Marquardt algorithm and effects the error convergence rate. It is also used for controlling the weights of the neurons during training of NAR neural network model. After training, the value obtained for *Mu* is 0.001 at 18 epochs. The validation checks help in terminating the learning of network when the error stops improving after six consecutive epochs.

The errors obtained for training, validation and test samples represented in form of histogram is shown in Figure 104.4. The entire error array has been separated into twenty columns along x-axis. The samples collected from the data set were represented along y- axis. It is also clear from the results that the zero-error point lies in the column with centre -0.00121.

The regression plots obtained for different samples of proposed network is shown in Figure 104.5. The results indicate that excellent regression values of one have been obtained for all the category of samples which clearly highlight a strong correlation between desired output and predicted values.

The responses obtained for output, target, and error values with respect to time is shown with the help of Figure 104.6. The responses also highlight the time points considered for training, testing and validation process. Further, the autocorrelation plots indicate how the prediction errors are interrelated in time. It can be observed from Figure 104.7, that the correlations fall nearly within 90% of the confidence limit. Hence, the proposed model can be considered appropriate for forecasting.

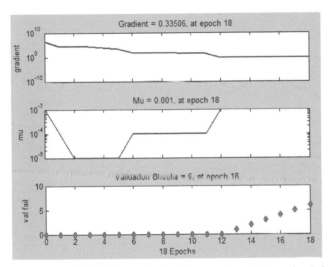

Figure 104.3 Plots obtained for gradient, Mu and validation checks after 18 epochs
Source: Author

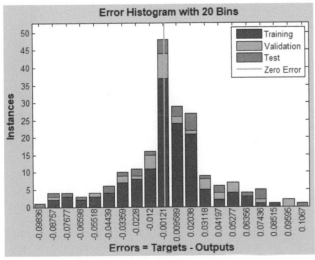

Figure 104.4 Error histogram obtained for NAR neural network model
Source: Author

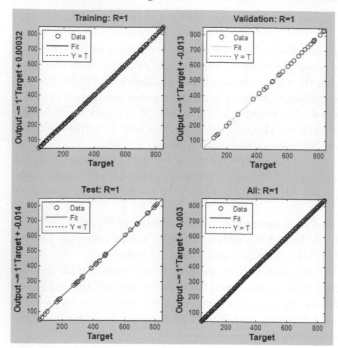

Figure 104.5 Regression responses obtained for NAR neural network model
Source: Author

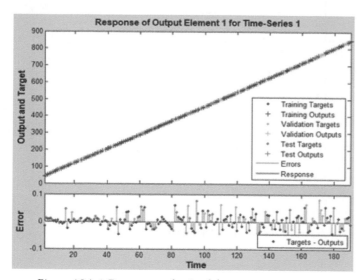

Figure 104.6 Responses obtained for output, target and error values with respect to time
Source: Author

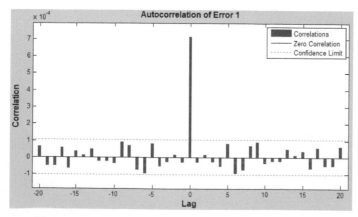

Figure 104.7 Autocorrelation plots obtained for NAR neural network model
Source: Author

Results and Discussion

The performance of NAR neural network has been examined in terms of RMSE, MAE and Bivariate correlation coefficient for different size of predicted samples using multistep prediction. The respective errors obtained for outputs 'specific enthalpy of liquid' and 'specific enthalpy of vapor' for different sample sizes are shown with the help of Figures 105.8 and 105.9 respectively.

It can be clearly observed from the above results that both the output thermodynamic properties of steam have been predicted within a minimal RMSE and MAE respectively. Further, it can be noticed that both RMSE and MAE increases with increase in sample size of multistep predictions. The output 'specific

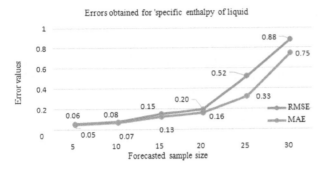

Figure 104.8 RMSE and MAE obtained for output 'specific enthalpy of liquid'
Source: Author

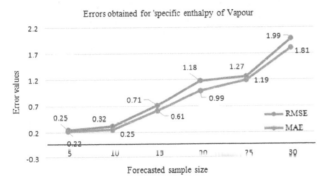

Figure 104.9 RMSE and MAE obtained for output 'specific enthalpy of vapor'
Source: Author

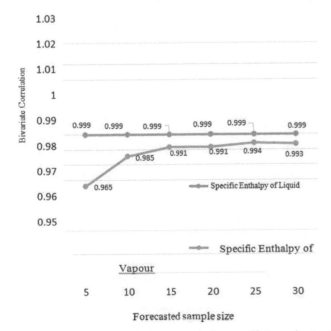

Figure 104.10 Bivariate correlation coefficient obtained for outputs with different sample sizes
Source: Author

enthalpy of liquid' has been predicted with a maximum RMSE and MAE of 0.88 and 0.75 respectively for a maximum sample size of 30 multistep predictions. Whereas the output 'specific enthalpy of vapor' has been predicted with a maximum RMSE and MAE of 1.99 and 1.81 respectively for a similar maximum sample size of multistep predictions. Further, the Bivariate correlation coefficient obtained for different sample sizes for both the output properties of steam are shown with the Figure 104.10. The results clearly indicate a very strong correlation between desired outputs and predicted outputs obtained through proposed NAR neural network model.

Conclusion

The study considered a novel multistep prediction approach based on NAR neural network for precise prediction of specific enthalpies of steam i.e., specific enthalpy of liquid and specific enthalpy of vapor. The NAR neural network has been trained using Levenberg-Marquardt learning algorithm. The real-time data on specific enthalpy of steam was collected and used for training of proposed network. The multistep prediction technique has been analyzed by considering consecutive prediction of outputs for different samples sizes i.e., 5, 10, 15, 20, 25 and 30 respectively. The performance of NAR neural network model has been analysed in terms of RMSE, MAE and Bivariate correlation coefficient between the desired outputs and predicted outputs. The results indicate that both RMSE and MAE increase with increase in sample size of multistep predictions. Further, the output 'specific enthalpy of liquid' has been predicted with a maximum RMSE (0.88) and MAE (0.75) for a maximum sample size of 30 multistep predictions. Whereas, the output 'specific enthalpy of vapour' has been predicted with a maximum RMSE (1.99) and MAE (1.81) for a maximum sample size of 30 multistep predictions. Lastly, the results also indicate a very strong Bivariate correlation coefficient in order 0.99 obtained for different sample sizes for both the output properties of steam. As a scope for future work various other machine learning algorithms can also be considered and compared for precise prediction of various thermodynamic properties of steam. Alternatively, the shape of the transfer function can also be varied to examine its effect on performance of NAR model.

References

[1] Tealab, A., Hefny, H., and Badr, A. (2017). Forecasting of nonlinear time series using ANN. Future Computing and Informatics Journal. 2(1), 39–47.

[2] Katic, K., Li, R., Verhaart, J., and Zeiler, W. (2018) . Neural network based predictive control of personalized heating systems. Energy & Buildings. 174, 199–213.

[3] Saini, P. and Thakur, P. (2022). H-Infinity based robust temperature controller design for a non-linear systems. Wireless Personal Communications. 126(1), 305–33.

[4] Belwal, N., Juneja, P. K., Sunori, S. K., Jethi, G. S., and Maurya, S. (2023). Modeling and control of FOPDT modeled processes-A review. Lecture Notes in Networks and Systems. 467, 255–60.

[5] Pereira, F. H., Bezerra, F. E., Junior, S., Santos, J., Chabu, I., De Souza, G. F. M., Micerino, F. and Nabeta, S. I. (2018). Nonlinear autoregressive neural network models for prediction of transformer oil-dissolved gas concentrations. Energies. 11(7), 1–12.

[6] Aliberti, A., Bottaccioli, L., Macii, E., Cataldo, S. D., Acquaviva, A., and Patti, E. (2019) . A non-linear autoregressive model for indoor air-temperature predictions in smart buildings. Electronics. 8(9), 1–17.

[7] Adedeji, P. A., Akinlabi, S., Ajayi, O., and Madushele, N. (2019). Non-linear autoregressive neural network (NARNET) with SSA filtering for a university energy consumption forecast. Procedia Manufacturing. 33, 176–83.

[8] Sarkar, R., Julai, S., Hossain, S., Chong, W. T., and Rahman, M. (2019). A comparative study of activation Functions of NAR and NARX neural network for long-term wind speed forecasting in Malaysia. Mathematical Problems in Engineering. 1–14. Article ID 6403081

[9] Valverde, A., Alvarez-Florez, J., and Recasens, F. (2020). Hybrid nonlinear autoregressive neural network - Weibull statistical model applied to the supercritical extraction of lanolin from raw wool. SN Applied Sciences. 2(160), 1651.

[10] Fontes, D. O. L., Vasconcelos, L. G. S., and Brito, R. P. (2020). Blast furnace hot metal temperature and silicon content prediction using soft sensor based on fuzzy C-means and exogenous nonlinear autoregressive models. Computers & Chemical Engineering. 141, 1–10.

[11] Bukhari, A. H., Sulaiman, M., Raja, M. A. Z., Islam, S., Shoaib, M. and Kumam, P. (2020). Design of a hybrid NAR-RBFs neural network for nonlinear dusty plasma system. Alexandria Engineering Journal. 59(5), 3325–45.

[12] Delcroix, B., Ny, J. L., Bernier, M. Azam, M., Qu, B., and Venne, J. S. (2021). Autoregressive neural networks with exogenous variables for indoor temperature prediction in buildings. Building Simulation. 14, 165–78.

[13] Kumar, D. A. and Murugan, S. (2018). Performance analysis of NARX neural network backpropagation algorithm by various training functions for time series data. International Journal of Data Science. 4(4), 308–25.

[14] Khan, F. M. and Gupta, R. (2020). ARIMA and NAR based prediction model for time series analysis of COVID-19 cases in India. Journal of Safety Science and Resilience. 1(1), 12–18.

Chapter 105

Research on knowledge based tutoring system of engineering graphics learning management

Ripon Roy[1,a], Md. Habibur Rahman[2,b] and Md. Helal Miah[3,c]

[1]School of Economics and Management, University of Chinese Academy of Science, Beijing, China

[2]Applied Chemistry & Chemical Engineering, Islamic University, Kushtia, Bangladesh

[3]Department of Mechanical Engineering, Chandigarh University, Chandigarh, India

Abstract

Regarding the salient features and advantages of knowledge based tutoring system (KBTS), this research will illustrate the learning management of basic concepts in engineering drawing through KBTS. The aims of the research include identification of the concepts in engineering drawing that students find challenging to learn, developing a KBTS for teaching those concepts, determining the effectiveness of the developed KBTS in achieving the instructional objectives, as well as studying technical and management issues related to the implementation of the KBTS and arriving at appropriate solutions to address such issues. Firstly, the model for the development of KBTS is optimised based on the modular structure of KBTS. Secondly, the study was conducted in five phases to investigate the efficiency of the KBTS optimised model. The validation of the KBTS was done using a single group pretest and posttest experimental design on a sample of 267 undergraduate engineering students of different programmers. Test scores were collected by administering the Pre-test and Post-test for comparison to determine the effectiveness of KBTS in achieving the stated instructional objectives. It is evident that the satisfaction score percentages ranged from 83.20 to 90.80 for different factors of the KBTS, with a mean value of 85.05. These values point to a high level of satisfaction for students in learning through KBTS. The research has practical value in developing a knowledge-based tutoring system regarding learning efficiency and flexibility.

Keywords: KBTS, engineering drawing, learning management, online pre-test, online post-test

Introduction

This Engineering drawing is one of the introductory courses in engineering education that intends to mentally develop an ability to visualize objects in various orientations. Real-world objects are essentially 3D (3 dimensional), and these objects are represented on paper using 2D (2-dimensional) drawings. Generating the necessary 2D graphics of the 3D objects and visualizing the actual 3D object from its 2D drawings requires specific abilities that are difficult to develop by conventional methods [1]. The research findings into the development of spatial skills in students have thrown new light into the process of content arrangement and instructional approaches concerning engineering graphics; consequently, both the content and curricula have moved in more unique ways. The ability on the part of the users to mentally manipulate the geometry and a level of spatial reasoning has become "conditio sine qua non" for efficient 3D modelling as we move into the arena of computer-aided design (CAD) [2]. Even a cursory look at any large group of learners will show vast heterogeneity and very significant differences in their learning styles. These diversities lead to differences in learning preferences, which necessitates an alignment of the learning preferences of the learners with the strategies and techniques followed in the instructional process [3]. Along with this, technology has brought about a sea change in education by opening new avenues of imparting instructions in more effective and "user-friendly" ways. Consequently, the focus is increasing on individual learners and adapting the entire teaching-learning process to suit the needs of individual learners. Flexible approaches to instruction are adopted to achieve this individualsation of learning, which overcomes the barriers of space and time as well as materials and method [4].

The main aim of educational technology is to design learning environments keeping in view instruction objectives and bringing out the best means of accomplishing them. In essence, the instruction process necessarily implies that a planned set of learning experiences are provided to each learner for achieving specific learning objectives. Sometimes, it may be necessary to modify the learning environment by re-arranging learning activities and presenting the content in alternate ways [5]. The multimedia content, a planned mixture of audio, animations, video, text, and graphics, forms the bulk of the educational content. This mix will vary with the type of content and instructional objectives to be achieved [6]. "Engineering drawing," also called "engineering graphics," is a particular course dealing with real-world objects and representing them on paper through different types of sketches [7]. Due to its extensive use in engineering practice, it is sometimes called the "language of engineers." The real world is essentially 3D, but people need to represent

[a]riponroystar1993@gmail.com, [b]habib.hk5722@gmail.com, [c]helal.sau.12030704@gmail.com

DOI: 10.1201/9781003450252-105

real-world objects on paper, which is essentially 2D [8]. The art and science of making this representation through multiple related views are the essences of "engineering drawing. Graphically depict real-world things in engineering drawings using different standards and strategies. To accurately portray the projections of the objects across many connected viewpoints in drawings, one must be able to picture the things in various orientations mentally. Many students struggle with this depiction, failing to fully comprehend the topics covered in the "engineering drawing" course. [9].

The key to engineering drawing is to visualise the thing in many positions and draw its projections; a computer-aided learning system can be employed more successfully [10]. Students can see the object in a digital 3D environment to facilitate simple visualisation. The best option in this situation is to deploy a knowledge-based tutoring system (KBTS). By enabling the student to adjust the position and orientation of the items and then see how the projected views have changed. As a result, this system can give the learner plenty of exploring options [11]. Contero (2005) developed a system to improve visualization skills in engineering education and enhance their spatial abilities using a twin approach of Web-based graphics applications and a sketch-based modeling system. By combining these two approaches, he could capture students' attention and foster two essential engineering skills, (i) freehand sketching and (ii) an understanding of the relationship between orthographic and axonometric views [12]. Along the same lines, Violante and Vezzetti (2015) used a web-based 3D interactive concept mapping model to teach abstract and challenging concepts in engineering drawing with a high degree of success [13].

KBTS can be designed for different domains. The research aim is to improve the efficiency of KBTS and efficiency will be measured with existing literature review. The research shows that such an intelligent tutoring system enables students to develop strategic flexibility to achieve the desired outcome in many domains [14]. In addition, KBTS also achieves a high level of transferability of learning across domains (Roll, 2011). The KBTS will work based on a database that stores a large amount of learning content, including resources in various multimedia formats like graphics and text, along with animation, video, and audio elements. Appropriate content will be derived from the database depending on the context and interactions between the learner and the system. Because of the above, it can be concluded that there is a need to develop a KBTS to facilitate the effective learning of basic concepts in engineering drawing [15]. Considering the salient features and advantages of KBTS, this research will illustrate the learning management of basic concepts in engineering drawing through KBTS. Engineering drawing, as a course, intends to mentally develop an ability to visualize objects in various orientations. KBTS is expected to be efficient in developing the required level of competency in the students concerning preparing engineering drawings. The Chandigarh University syllabus for the computer aided engineering drawing (CAED) course of the bachelor's degree programmer in engineering was considered for developing the KBTS. It is a common course for all programs offered by the university in the first year, including civil engineering, mechanical engineering, electrical and electronics engineering, electronics and communication engineering, and computer science and engineering.

Description of KBTS

The KBTS on basic engineering drawing concepts was designed as a web-based system. Learners can use the system independently to learn the course online at their preferred and most comfortable pace [16]. The KBTS was designed to teach three basic concepts of the CAED course to the learners, viz., (i) Orthographic projections, (ii) Projection of points, and (iii) Projection of lines [17]. These concepts were presented as three distinct modules in the KBTS. Instructional objectives and teaching points were identified for each of the three modules. The teaching points were presented in the KBTS in a logical sequence based on pedagogical principles. Embedded self-tests were included at appropriate places in the KBTS. These embedded test items perform the role of formative evaluation [18]. The learners were provided with necessary feedback and guidance based on the formative evaluation. The final summative evaluation was done on completion of the course to assess the students' learning, which indicates the effectiveness of the KBTS in achieving the stated instructional objectives [19].

Structure of KBTS

The KBTS was developed in terms of modules at the top level, a specific number of instructional objectives under each module, and several teaching points under each instructional objective. The organization of the content of KBTS in its modular structure is shown in Figure 105.1.

The KBTS was designed to teach three significant concepts. Three modules represent these concepts at the top level. For each of the concepts, instructional objectives were formulated. The instructional objectives were further divided into specific sets of teaching points to achieve the instructional objective when these points were successfully learnt.

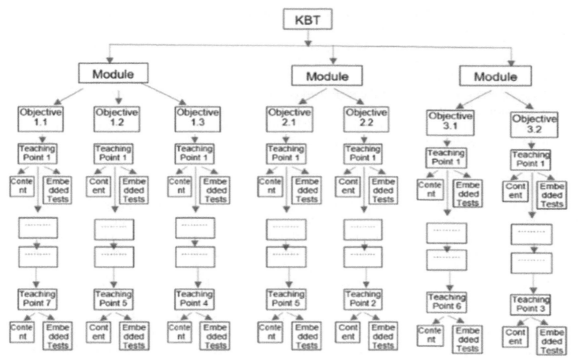

Figure 105.1 Modular structure of KBTS
Source: Author

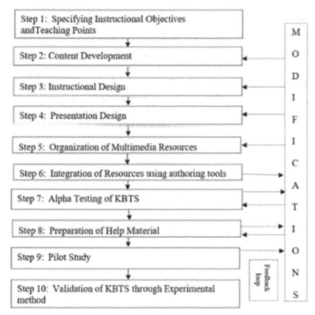

Figure 105.2 Model for development of KBTS
Source: Author

Development of KBTS

The development of KBTS involves several steps, from specifying the instructional objectives and teaching points to the final validation of the developed KBTS. A step-by-step procedure was followed for developing the KBTS as per the model shown in Figure 105.2.

Research Methodology

According to the problem statement mentioned in the introduction section, the main aims of the research include identification of the concepts in engineering drawing that students find challenging to learn; developing a KBTS for teaching those concepts; determining the effectiveness of the developed KBTS in achieving the instructional objectives; as well as studying technical and management issues related to the implementation of the KBTS and arriving at appropriate solutions to address such issues. Based on these objectives, the study was conducted in five phases.

- Phase 1: Identifying basic concepts in the engineering drawing course using the survey method that students find difficult to learn.
- Phase 2: Development of a KBTS for the identified concepts.
- Phase 3: Validation of the KBTS using a single group experimental design.
- Phase 4: Study the relative effectiveness of the KBTS in achieving the various instructional objectives and its usability, as well as the opinion of the students concerning their satisfaction while learning through KBTS.
- Phase 5: Focus group discussion to identify the management and technical issues involved in delivering the instruction through KBTS.
- The preliminary study on the validation of the KBTS was done using a single group pretest-treatment-post-test experimental design on a sample of 267 undergraduate engineering students of different programmers. Demographic details of students in the sample were collected through the registration process. Test scores were collected by administering the pre-test and post-test for comparison to determine the effectiveness of KBTS in achieving the stated instructional objectives.
- The students who were studying in the first year of the undergraduate program in engineering at 17 departments affiliated with the Chandigarh University in the Punjab state of India during the years 2022 constituted the population for the study. The data for the research study was collected from the students of three engineering departments. The departments were selected using purposive sampling. A stratified random sampling technique was used for selecting the sample of students for the main study. The hypotheses formulated for the study involved studying the differences in achievement through KBTS between different categories of students, and accordingly, these categories were taken as different strata. To study the difference between core engineering branches and IT-related engineering branches, students from three core engineering branches of Electrical & Electronics Engineering, Civil Engineering, and Mechanical Engineering, as well as students from two IT-related branches of engineering, viz., Communication Engineering and Computer Science & Engineering were included in the sample. A total of 267 students were included in the sample for the main study, 135 were from core branches of engineering, and the remaining 132 were from IT-related branches of engineering. Additionally, gender, place of residence (urban or rural), and medium of instruction up to the secondary level of education were the other stratification variables considered for selecting the sample.
- The sample (N = 267) comprised 50.56% (N = 135) of students of core engineering programs and 49.44% (N = 132) of students of IT-related engineering programs. Female students constituted 35.96% (N = 96) and male students constituted 64.04% (N = 171). 79.03% (N =211) of students had studied their secondary education through the English medium, and 20.97% (N =56) of students had pursued their secondary education through a non-English medium. Urban students constituted 49.06% (N = 131) and rural students constituted 50.94% (N = 136) of the sample.

The author visited the selected engineering departments to collect data for the main study. The steps followed in collecting the data are presented below:

- Step 1: Providing infra structural facilities comprising of computer systems having internet access and earphones along with proper arrangement.
- Step 2: General briefing and providing the URL of the website.
- Step 3: Registration to the site by the student by providing demographic data.
- Step 4: Student taking the online pre-test → providing pre-test data.
- Step 5: Student using the three modules (Module 1: Orthographic projections, Module 2 Projection of points, Module 3 Projection of lines) of the KBTS through online mode for learning basics concepts of engineering drawing.
- Step 6: Diagnostic test for students to assess understanding of the concepts.
- Step 7: System generates a customised learning package for each student by assessing his/her performance in the diagnostic test.
- Step 8: Student studies the customised learning package presented to him/her.
- Step 9: Student takes the online post-test → Providing post-test data.
- Step 10: Student filling online satisfaction survey questionnaire through google form → Providing satisfaction survey data.

As the administration of the complete package, including pre-test, post-test, course content, and satisfaction survey, was done online, the provision for appropriate infrastructural facilities was imperative for the study. For this purpose, the researcher communicated with departments to provide the necessary number of computers for the study along with broadband internet connectivity. As the multimedia content included extensive usage of audio, provision for using individual audio devices by each student was made.

The focus group discussion method was adopted to study the management and technical issues associated with instruction delivery through KBTS. A focus group comprising 12 members was formed by including six students, four teachers, and two multimedia specialists. The objectives of the focus group discussion were (i) to identify the factors affecting instruction through KBTS and (ii) the possible steps to increase the effectiveness of KBTS, (iii) identifying critical technical issues which affect the utility of KBTS and the ways to address those issues to enhance the effectiveness of KBTS. The qualitative data gathered from focus group discussions were subjected to content analysis.

Data Analysis and Result

Overall Pre-test and Post-test Scores

Descriptive statistics related to the pre-test and post-test scores is given in Table 105.1.

It can be observed from Table 105.1 that the minimum score in the pre-test is 10 (this is because answers are evaluated by assigning negative scores for wrong answers) and the maximum score is 80.50 with mean score of 17.40 and standard deviation of 13.65. The minimum score in post-test is 16 and maximum score is 110 with the mean value of 82.55 and standard deviation of 24.47. Further analysis in terms of gain is done in the following sections. The visual presentation of the distribution of overall scores in pre-test and post-test is presented in the Box Plot (Whisker Plot) shown in Figure 105.3. The Box Plot summarises the data in terms of five numbers. These five numbers correspond to the values of minimum, first quartile, median, third quartile, and maximum [20]. As can be seen from the plot, the whiskers go from first quartile to the minimum and from third quartile to the maximum. The box covers the range of values form the first quartile to the third quartile whereas the vertical line inside the box represents the median [21]. The box plot shows a significant gain in Post-test scores compared to the pre-test scores. The box plot of pre-test scores shows 8 outliers referring to students with IDs 31, 35, 95, 188, 206, 216, 241 and 259. The

Table 105.1 Details of overall scores in pre-test and post-test

Test	N	Maximum Possible score	Minimum score obtained*	Maximum score obtained	Mean score obtained	Std. Deviation	% score
Pre Test	267	110	-10.00	80.50	17.40	13.65	15.85
Post Test	267	110	16.00	110.00	82.55	24.47	75.05

*The test uses negative scoring for wrong answers and hence total score can be negative.

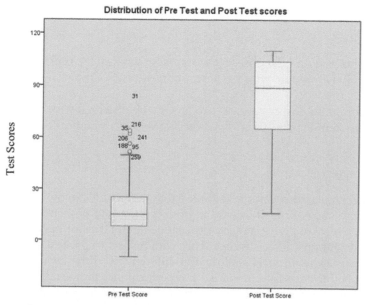

Figure 105.3 Box plot showing distribution of overall pre-test and post-test scores
Source: Author

Table 105.2 Descriptive statistics regarding the module-wise pre-test and post-test scores

Module	Test score	Maximum possible score*	Minimum score obtained**	Maximum score obtained	Mean score obtained	S.D.	Score
Module 1: Orthographic projections	Pre-test	40	-8.75	34.25	8.66	6.40	21.65
	Post-Test	40	5.00	40.00	30.20	7.92	75.40
Module 2 Projection of points	Pre-test	36	-8.50	28.50	5.39	7.20	14.98
	Post-Test	36	1.00	36.00	27.63	9.04	76.76
Module 3 Projection of lines	Pre-test	34	-7.75	25.25	3.35	5.98	9.85
	Post-Test	34	-3.50	34.00	24.73	10.86	72.81

*Indicates the maximum possible score for a student with respect to all the items testing the contents of a particular module.

**The total scores can be negative as the test uses the negative marking scheme for wrong answers.

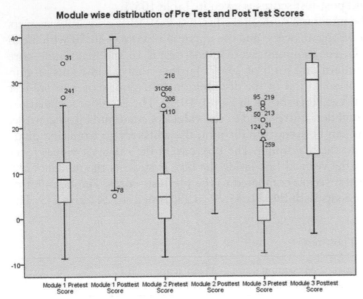

Figure 105.4 Box plot representing pre-test and post-test scores with respect to the three modules

Source: Author

corresponding post-test scores of these outlier cases indicate that even high achievers have benefitted substantially from the KBTS.

Module-wise Performance in Pre-test and Post-test

As there are three modules in the course, the pre-test and the post-test scores were separately tabulated for the three modules. The descriptive statistics of the module-wise pre-test and post-test scores are listed in Table 105.2.

A significant difference between pre-test scores and post-test scores can be visually observed with respect to all the three modules from the box plot shown in Figure 105.4. The box plot shows a significant gain in post-test scores compared to the pre-test scores.

Objective-wise Performance in Pre- and Post-test

A total of seven instructional objectives were formulated covering the three Modules included in the course. The performance of the students with respect to each of the objectives in both pre-test and post-test is discussed in this section. The instructional objective-wise performance of the students in terms of maximum score, minimum score, mean score, standard deviation and percentage score with respect to pre-test as well as post-test are shown in Table 105.3. As the maximum marks allotted is different for different instructional objectives, percentage score provides a better basis for comparison of performance in both the tests.

The data in Table 105.3 shows that the mean of percentage scores in pre-test for the seven Instructional Objectives range from 7.5-32.21% and that of post-test ranges from 70.71-80.52. These data show a marked improvement in post-test scores over pre-test scores.

Table 105.3 Descriptive statistics of instructional objective-wise pre-test and post-test scores

Instructional objective	Test score	Maximum possible score*	Minimum score obtained**	Maximum score obtained	Mean score obtained	S.D.	% Score
Objective 1	Pre-test	16	-4.00	14.75	5.15	3.48	32.1
	Post-test	16	2.25	16.00	12.24	3.27	76.53
Objective 2	Pre-test	10	-2.50	8.00	2.09	2.53	20.86
	Post-test	10	-1.00	11.00	8.05	2.76	80.52
Objective 3	Pre-test	14	-3.50	14.00	1.42	3.39	10.14
	Post-test	14	-3.50	14.00	9.89	4.38	70.71
Objective 4	Pre-test	15	-3.75	15.0	3.97	4.37	26.44
	Post-test	15	-1.25	18.00	12.55	3.61	83.67
Objective 5	Pre-test	21	-5.25	18.50	1.43	4.62	6.80
	Post-test	21	-1.50	21.00	15.08	6.73	71.87
Objective 6	Pre-test	14	-3.50	11.50	1.05	3.15	7.50
	Post-test	14	-3.50	18.00	10.19	4.96	72.84
Objective 7	Pre-test	20	-5.00	17.50	2.30	4.63	11.49
	Post-test	20	-0.50	20.00	14.53	6.67	72.64

*Maximum marks refer to the total marks allotted for all items testing an objective.

**The obtained scores can be negative as the test uses the negative marking system for wrong answers.

Table 105.4 Module-wise percentage gain in scores

Module no.	Title of the module	Mean of percentage gain	Std. deviation of percentage gain
1	Basics of orthographic Projections.	67.81	2.56
2	Projection of points.	71.14	32.03
3	Projection of lines	68.28	38.53

Source: Author

Effectiveness of KBTS in Terms of Gain Percentage

The gain percentage for each student was calculated in order to estimate the effectiveness of the course in terms of gain achieved by the students by learning through KBTS. The procedure followed for calculating the gain percentage is explained below.

$$Actual\ gain = Post\ test\ score - Pre\ test\ score \tag{1}$$

$$Maximum\ possible\ Gain = Maximum\ possible\ score\ in\ the\ Post\ test - Actual\ score\ in\ the\ Pre\ test \tag{2}$$

$$Percentage\ gain = \frac{Actual\ gain}{Maximum\ possible\ Gain} \times 100 \tag{3}$$

It may be noted that the maximum possible gain of the learner through the course is limited by his pre-test score. If the pre-test score is high, the scope for possible improvement in terms of actual absolute score becomes limited. But maximum possible gain is taken as the difference between the "maximum marks allotted" and the "marks scored in pre-test". As this maximum possible gain is taken for calculating the percentage gain, the effect of pre-test score is effectively controlled.

Module-wise Gain

Both pre-test and post-test were designed to cover items under three different modules. The gain in the score and percentage gain were separately calculated for the three modules. The results are shown in Table 105.4.

The data in Table 105.4 shows a significant positive gain percentage with respect to all the three Modules. The percentage gain for the modules ranges from 67.81-71.14%. With a mean of 69.08. Module

Table 105.5 Details of objective-wise percentage gain in scores

Objective no.	Instructional objective	Mean of percentage gain	Std. Deviation of percentage gain
1	Explain Principle of orthographic projections.	61.89	37.72
2	Distinguish between first angle and third angle projection systems.	72.25	45.40
3	Draw orthographic projections of simple objects.	65.79	37.63
4	Identify the quadrants in which points lie with reference to their position relative to principal planes.	75.12	36.68
5	Draw orthographic projections of points lying in various positions with respect to principal planes.	68.41	37.21
6	Draw projection of lines when the line is parallel to at least one of the principal planes.	67.97	42.50
7	Draw projection of lines inclined all the principal planes.	66.12	43.79

Source: Author

Table 105.6 Overall percentage gain of kbts

Gain	N	Minimum gain	Maximum gain	Mean gain	S.D.
Overall Percentage gain	267	-3.64	100.00	69.84	26.90

Source: Author

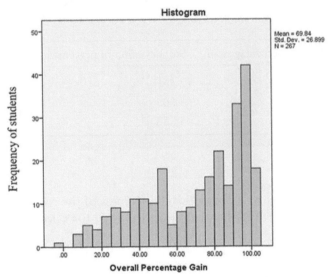

Figure 105.5 Histogram representing the frequency of distribution of gain
Source: Author

2 (projection of points) recorded comparatively higher gain whereas Module 1 (orthographic projections) showed comparatively lower gain, but the difference is very small.

Objective-wise Percentage Gain

The percentage gain was separately calculated for each of the seven instructional objectives to ascertain the effectiveness of the KBTS with respect to the identified instructional objectives. The distribution of the gain with respect to the seven instructional objectives is shown in Table 105.5.

The percentage gain for different instructional objectives ranges from 61.89 (Objective 1) to 75.12 (Objective 4) with a mean of 68.22. The data shows that mean percentage gain is more than 60% for all the seven instructional objectives which implies that the KBTS is effective in facilitating learning with respect to each of the seven instructional objectives.

Overall gain in the Scores

The overall gain of the KBTS package was calculated by using the formula specified in 1. Table 105.6 lists the overall gain achieved by the learners through the KBTS.

The minimum gain is -3.64% (as negative marking is used) and maximum gain is 100% with a mean gain of 69.84% and a standard deviation of 26.90. The mean percentage gain of 69.84 is significantly high which indicates the overall effectiveness of KBTS in facilitating learning.

The distribution of gain is visually presented in Figure 105.5. It can be observed from the figure that most of the students have achieved high gain because of the KBTS.

Discussion

Data obtained through the satisfaction survey from students showed that the satisfaction score percentages ranged from 83.20-90.80 for different factors of the KBTS, with a mean value of 85.05. These values point to a high level of satisfaction for students in learning through KBTS. Has reported student satisfaction scores ranging from 85.68-91.21, with a mean of 87.94, concerning his CBMMLP for learning concepts of electromagnetism [22]. The intelligent tutoring system for learning advanced topics in information security, developed by Mahdi et al. (2016), showed student satisfaction scores ranging from 89-91% for different factors, as per student feedback [23]. In the study by Kasapbasi (2014) on modelling and evaluating the WebCoach-based approach for learning content management system (LCMS), a student opinion survey revealed that 86% of the students were satisfied with the WebCoach system [24]. These findings support the results of the present study concerning the satisfaction experienced by students in learning through KBTS. A focus group meeting was conducted with 12 participants (six students, four teachers, and two multimedia specialists) to identify instruction's managerial and technical issues through KBTS and formulate strategies for solving them. In addition to the quality of content, interest value, proper sequencing of content, and a high level of interactivity were identified as major factors influencing the usage of a KBTS. It was observed that even though the KBTS was designed as an independent system, it can also be integrated with regular classroom teaching by providing a structured schedule for the students to access specific topics of KBTS and then having related follow-up activities in subsequent classes. It was observed that the KBTS could be used as an important resource for implementing the flipped classroom concept. It was also observed that providing an online tutor to a set of students would be effective for coordinating and managing the learning through KBTS.

Conclusions

As an outcome of the study, a validated knowledge-based tutoring system (KBTS), a web-enabled system for the management of learning basic concepts of engineering drawing for students' use, has been developed. Students can register on the website and freely access the study materials, tests, and diagnostic content for individualised learning. This study evaluated the effectiveness of the KBTS in learning selected concepts of engineering drawing. The outcomes of the study establish the effectiveness of KBTS in facilitating effective learning of the concepts on the part of the students. It is evident from the overall, module-wise and instructional objective-wise percentage gains and the gains at different cognitive levels.

Moreover, as an outcome of the study, the following conclusions were arrived at:

* The KBTS was equally effective with students pursuing engineering programs in core engineering branches and with students pursuing engineering programs in IT-related departments.
* The KBTS was equally effective for both boys and girls in learning the concepts.
* KBTS was equally effective with students who have pursued their secondary education in English and non-English mediums.

References

[1] Kudale, P. and Buktar, R. (2022). Investigation of the impact of augmented reality technology on interactive teaching learning process. International Journal of Virtual and Personal Learning Environments. 12(1), 1–16. doi: 10.4018/IJVPLE.285594:

[2] Kosmadoudi, Z., Lim, T., Ritchie, J., Louchart, S., Liu, Y., and Sung, R. (2013). Engineering design using game-enhanced CAD: The potential to augment the user experience with game elements. Computer-Aided Design. 45(3), 777–95. doi: 10.1016/J.CAD.2012.08.001.

[3] Almasri, F. (2022). Simulations to teach science subjects: connections among students' engagement, self-confidence, satisfaction, and learning styles. Education and Information Technologies. 1–21. doi: 10.1007/S10639-022-10940-W/TABLES/5.

[4] Simamora, R. M. (2020). The challenges of online learning during the COVID-19 pandemic: an essay analysis of performing arts education students. Studies in Learning and Teaching. 1(2), 86–103. doi: 10.46627/SILET.V1I2.38.

[5] Shieh, C. J. and Yu, L. (2016). A Study on information technology integrated guided discovery instruction towards students' learning achievement and learning retention. Eurasia Journal of Mathematics, Science and Technology Education. 12(4), 833–42. doi: 10.12973/EURASIA.2015.1554A.

[6] Abdulrahaman, M. D. et al. (2020). Multimedia tools in the teaching and learning processes: A systematic review. Heliyon. 6(11), e05312. doi: 10.1016/J.HELIYON.2020.E05312.

[7] Tumkor, S. (2018). Personalization of engineering education with the mixed reality mobile applications. Computer Applications in Engineering Education. 26(5), 1734–41. doi: 10.1002/CAE.21942.

[8] Oxman, R. (2017). Thinking difference: theories and models of parametric design thinking. Design Studies. 52, 4–39. doi: 10.1016/J.DESTUD.2017.06.001.

[9] Fujita, T., Kondo, Y., Kumakura, H., and Kunimune, S. (2017). Students' geometric thinking with cube representations: Assessment framework and empirical evidence. The Journal of Mathematical Behavior. 46, 96–111. doi: 10.1016/J.JMATHB.2017.03.003.

[10] Séquin, C. H. (2005). CAD tools for aesthetic engineering. Computer-Aided Design. 37(7), 737–50. doi: 10.1016/J.CAD.2004.08.011.

[11] Streitz, N. A. (1988). Mental models and metaphors: implications for the design of adaptive user-system interfaces. Learning Issues for Intelligent Tutoring Systems. 164–86. doi: 10.1007/978-1-4684-6350-7_8.

[12] Contero, M., Naya, F., Company, P., Saorín, J. L., and Conesa, J. (2005). Improving visualization skills in engineering education. IEEE Computer Graphics and Applications. 25(5), 24–31. doi: 10.1109/MCG.2005.107.

[13] Violante, M. G. and Vezzetti, E. (2015). Design of web-based interactive 3D concept maps: a preliminary study for an engineering drawing course. Computer Applications in Engineering Education. 23(3), 403–11. doi: 10.1002/CAE.21610.

[14] Aleven, V. et al. (2016). Example-tracing tutors: intelligent tutor development for non-programmers. International Journal of Artificial Intelligence in Education. 26(1), 224–69. doi: 10.1007/S40593-015-0088-2/FIGURES/1.

[15] Zheng, J. (2011). Teaching of engineering drawing in the 21st century. 2011 2nd International Conference on Mechanic Automation and Control Engineering, MACE 2011 – Proceedings. 1713–15. 10.1109/MACE.2011.5987287.

[16] Wang, M., Shen, R., Novak D.,, and Pan, X. (2009). The impact of mobile learning on students' learning behaviours and performance: report from a large blended classroom. British Journal of Educational Technology. 40(4), 673–95. doi: 10.1111/J.1467-8535.2008.00846. X.

[17] Kok, P. J. and Bayaga A. (2019). Enhancing graphic communication and design student teachers' spatial visualisation skills through 3D solid computer modelling. African Journal of Research in Mathematics, Science and Technology Education. 23(1), 52–63. doi: 10.1080/18117295.2019.1587249.

[18] Perry, J., Paas, L., Arreola, M. E., Santer, E., Sharma, N., and Bellali, J. Promoting E-governance through capacity development for the global environment. 980–1010. doi: 10.4018/978-1-60960-472-1.CH423.https://services.igi-global.com/resolvedoi/resolve.aspx?doi=10.4018/978-1-60960-472-1.ch423.

[19] Chao, K. J., Hung, I. C., and Chen, N. S. (2012). On the design of online synchronous assessments in a synchronous cyber classroom. Journal of Computer Assisted Learning. 28(4), 379–95. doi: 10.1111/J.1365-2729.2011.00463.X.

[20] Alwan, F. et al. (2017). Presenting ADAMTS13 antibody and antigen levels predict prognosis in immune-mediated thrombotic thrombocytopenic purpura. Blood. 130(4), 466–71. doi: 10.1182/BLOOD-2016-12-758656.

[21] DeCoursey, W. (2022). Statistics and probability for engineering applications. 2003. [Online]. Available from https://books.google.com/books?hl=en&lr=&id=1hvpJiK6nrAC&oi=fnd&pg=PP1&dq=The+box+covers+the+range+of+values+form+the+first+quartile+to+the+third+quartile+whereas+the+vertical+line+inside+the+box+represents+the+median+(DeCoursey,+2003)+&ots=n6GYhhKyEC&sig=4TJaSOyLuyL_ok3gEKMSaIXEjxc

[22] Shah, H., S., Attiq, S. (2016). Impact of technology quality, perceived ease of use and perceived usefulness in the formation of consumer's satisfaction in the context of e-learning. Abasyn Journal of Social Sciences. 9(1).Ngoc, P. et al. (2021). The impact of extrinsic work factors on job satisfaction and organizational commitment at higher education institutions in Vietnam. The Journal of Asian Finance, Economics and Business. 8(8), 259–70. doi: 10.13106/JAFEB.2021.VOL8.NO8.0259.

[23] Kasapbasi, M. C. (2014). Knowledge management integrated web based course tutoring system. procedia-social and behavioral sciences. 116, 3709–15. doi: 10.1016/J.SBSPRO.2014.01.828.

Chapter 106

Comparative study of different coil spring using finite element analysis

Ankur Rai[a], Brahma N and Agrawal[b]

Department of Mechanical Engineering, Galgotias University, Greater Noida, India

Abstract

The suspension system is used to monitor vibrations caused by shock loads due to road surface abnormalities. The primary purpose of an automotive suspension system is to smooth out and dampen shock impulses. It also consumes or dissipates energy so that the suspension system provides comfort and safety for passengers and vehicles. In this article, the cylindrical, conical, and barrel coil springs are designed in SolidWorks with precise dimensions, and the analysis is performed in ANSYS software. The material used for cylindrical, conical, and barrel coil springs is structural steel. The total deformation, maximum shear strain, equivalent stress and strain energy are analyzed under static conditions having a load of 2750 N and fatigue life and safety factor is also analysed for 1000000 number of cycles. In this paper, the comparisons of the cylindrical, conical, and barrel coil springs are done based on the finite element analysis (FEA) process. The results revealed that the total deformation of the barrel coil spring is greater than that of the cylindrical and conical coil spring. The barrel coil springs provide higher variable stiffness in a passenger vehicle than conical and cylindrical coil spring. The equivalent stress is also more in a barrel coil spring, so it will yield more in its shape. A coil spring is used to store mechanical energy. It can be twisted or stretched by some force and can return to its original shape when the force is released.

Keywords: Alloy steel, barrel coil spring, conical coil spring, cylindrical coil spring

Introduction

The system of springs, shock absorbers, axles, ball joints, and arms that link a vehicle to its wheels is referred as suspension. It maintains a high standard of ride handling under all driving conditions. During cornering, bumping, and braking – high stiffness and damping are required to supply the good handling properties of the suspension system. It is primarily responsible for dissipating kinetic energy and shock control. Light coil springs are used in most modern automobiles. Coil springs in the front and leaf springs in the back are common in passenger vehicles, which are smaller than commercial vehicles. The axle connects either side of the vehicle's wheels, allowing the wheels to travel independently of one another and thereby reducing body movement. In most new light vehicles, coil springs are used in the front suspension. The spring is used to store mechanical energy due to its elastic behavior. On applying force, it can be twisted, pulled, and stretched, when the force is removed, they revert to their original form. The spring's loading capacity is determined by the wire diameter, the spring's diameter, its shape, and thus the coil spacing. Failure occurs in coil springs due to poor material properties of the spring materials and high cyclic fatigue, where the induced stress should be held below the yield strength.

Lavanya et al. concluded that in comparison to chromo vanadium steel products used in automobiles, carbon structural steel is better suited for the construction of helical springs [1]. Alsahlani and Khashan studied for all load values, carbon composite springs exhibited the least amount of overall deformation. As compared to steel, deformation in carbon composites was reduced by 15%, while total deformation in copper alloys was reduced by around 54% [2]. Singh analysed that throughout their service lives, springs are subjected to varying loading. ANSYS, SolidWorks, Pro-E, CATIA, Autodesk Inventor, and other design software have also been used to conduct mechanical spring stress analysis [3]. Pawar et al. studied the helical coil spring used in the front suspension of a three-wheeled vehicle was subjected to stress analysis [4]. Siddharth et al. performed an investigating study to design the spring suspension system of a vehicle. Considering two basic models of spring i.e., variable pitch and cylindrical designs a 3D model is generated in SolidWorks and imported to Ansys for further material analysis on alloy steel and high carbon steel. Equivalent stress and deformation are analysed for different application of load and concluded that designed spring successfully fulfilled the motives of basic spring design in single combined spring structure [5]. Seunghyeon et al. performed finite element analysis (FEA) to test the effect of end coil angular position as well as centerline shape on the force axis [6]. Rahim et al. implements reconstructive signals of discrete wavelet transform (DWT) to assess the FEA of 14 decomposition signals of an automobile coil spring [7].

Sreenivasan et al. performed the material's deformation, equivalent stress, and strain concerning the use of helical spring suspension [8]. Gunaki, the LOTUS software was used to simulate the double-wishbone suspension mechanism and per-formed the analysis on the ANSYS software [9]. Arvind and Tripathy

[a]ankur.rai282001@gmail.com, [b]brahma.agrawal@galgotiasuniversity.edu.in

DOI: 10.1201/9781003450252-106

studied the geometries of the helical coil spring, as well as the geometries of a suspension spring for an automobile, to determine the static state of the helical spring. Used three different geometries of the helical spring which results in the maximum stress and deformation [10]. Win and Tun have presented and analysed the stress analysis of the Landcruiser Prado's rear suspension helical spring. Since the shear stress and deformation formed in the spring during loading are within acceptable limits, it is considered stable [11]. Mulla and Kadam discussed the elasticity and stress analysis of front-end automotive suspension springs [12]. Vukelic and Brcic explained the fracturing behavior of coil springs in automobile suspension systems and to in-crease spring stability to minimise failures [13]. Dolas and Jagtap studied the stress values of coil springs that were measured and found to be less than the yield stress, indicating that the design is stable. The results were compared to the original design and updated design, with the updated design having lower stress and displacement values [14]. Sagarsingh et al. stated that automobiles face various problems while driving on bumps on road and improper road conditions. To optimise the performance of coil spring suspension of an automobile of hatchback segment finite element method is used and validated using analytical method, includes modelling, meshing and post-processing of front suspension spring. Based on the results of static analysis and analytical calculations a comparison is made for optimized results which satisfy all the necessary conditions [15]. Felipe et al. discussed coil spring failure analysis using optical emission spectroscopy, hardness testing, optical & scanning electron microscopy and metallography. Below protective paint of spring wire helical pattern was identified [16]. Qingang et al. analysed the effects of 21 different configurations of conical and fusiform turbulator placed in DPHE and simulated in same with circular and rectangular tube configurations at RE 4000, RE 7000, RE 10000 and RE 13000 on heat trans-portation and turbulent flow designs [17].

The main objective is to design the model and study the analysis of coil springs having different geometrical modifications such as open-end cylindrical coil spring, open-end conical coil spring and open-end barrel coil spring for use in vehicle suspension systems made of structural steel. Coil springs are designed with precise dimensions in SolidWorks 2016 and analyzed with ANSYS. With a load of 2750 N, the directional deformation, equal elastic strain, equivalent stress, and overall deformation are calculated numerically.

Methodology

In this section, coil springs with three different geometrical modifications are taken for the passenger vehicle. Coil springs with various geometries have been designed and analyzed. The materials used are structural steel; the specifications, properties, modeling, meshing, flow diagram, and analysis are as follows in Table 106.2. The parameters selected for the design of coil spring as given below in Table 106.1.

Table 106.1 Design parameters of coil spring [2]

Specifications	Values
Wire diameter (d)	9.50 mm
Outer diameter of cylindrical coil (D)	58 mm
Outer diameter of conical and barrel coil (D) (One end)	58 mm
Outer diameter of conical and barrel coil (Other end)	48 mm
Free height of coil (H)	154 mm
No. of active coil (N)	8
No. of active coil at upper and lower Section of barrel coil	4
Test load (F)	2750 N
Pitch (P)	mm

Table 106.2 Mechanical properties of structural steel [2]

Properties	Values
Young's modulus	2.07E + 11 Pa
Density	7850 Kg/m^3
Ultimate tensile strength	2.5E + 08 Pa
Poisson's ratio	0.3

Coil spring CAD model

The CAD model for open-end cylindrical coil spring, open-end conical coil spring and open-end barrel coil spring is shown in Figure 106.1. These models are created by the use of designing software Solid Works 2016 [18]. These models have eight active coils, coil free height of 154 mm, and the diameter of wire is 9.50 mm. For analysis, the 3D model of coil springs was imported to the ANSYS software.

Pre-processing (meshing and model setup)

Meshing is done in ANSYS. As the structural analysis is being done on the part so it is preferred to use quad mesh element on the design (Figure 106.2). 3D meshing is required because of the solid structure of the coil spring. In cylindrical coil spring meshing, the nodes used are 2,72,651 and elements are 62,604, in conical coil spring meshing, the nodes used are 2,32,451 and elements are 52,362 and in barrel coil spring meshing, the nodes used are 2,00,521 and elements are 43,708. Meshing data is shown in Table 106.3.

One end of the spring is kept fixed, and the other end of the spring is loaded axially downward as shown in Figure 106.3. For static structural analysis the spring geometry is analysed along the y-axis for the total deformation, maximum principal strain, equivalent stress and strain energy. For Fatigue analysis the same spring is analysed along y-axis for fatigue life and factor of safety.

Here the upper end of the cylinder coil is fixed (movement in direction X, Y, Z = 0). And force is being applied in the downward direction (negative "- Y") from the bottom end to analyse the design study.

Results and Discussion

The model of the coil spring was designed using SolidWorks 2016 software. After that, the model will be imported to ANSYS 19.0 for analysis. In FEA, Mesh is a complex structure that consists of nodes (points)

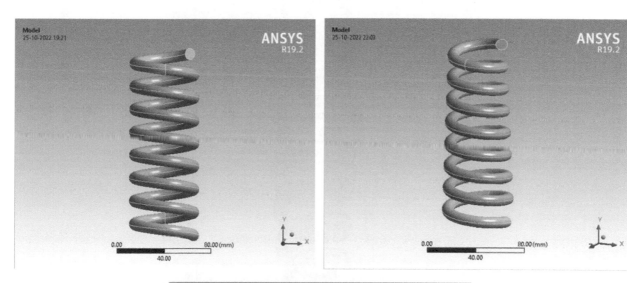

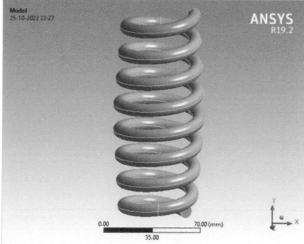

Figure 106.1 Geometrical design of coil spring in SolidWorks
Source: Author

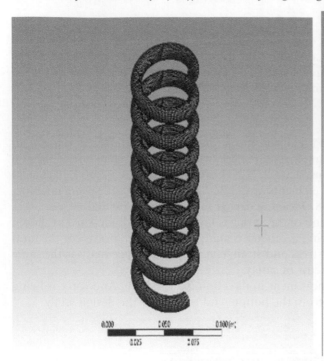
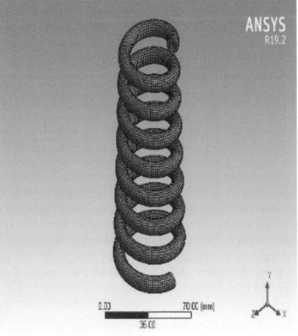
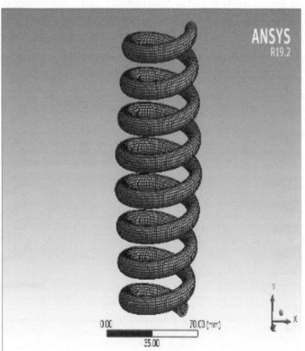

Figure 106.2 Meshing of coil spring geometries in ANSYS 19.0
Source: Author

Table 106.3 Meshing data

Coil Spring geometries	Number of elements	Number of nodes
Open-end cylindrical coil spring	62604	2,72,651
Open-end conical coil spring	52362	2,32,451
Open-end barrel coil spring	43708	2,00,521

Source: Author

that form a grid. Nodes and Elements are a very important part of FEA and will use them in every analysis process. This mesh is made up of cells and points that have material as well as structural properties, which determine the output of the structure under loading conditions. In Node, the degree of freedom shows the

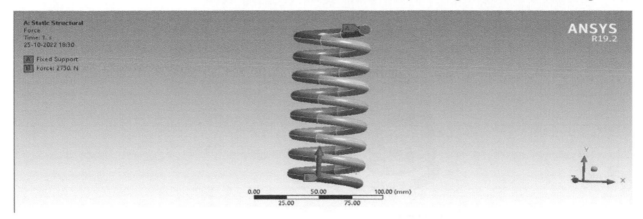

Figure 106.3 Boundary condition on the coil spring
Source: Author

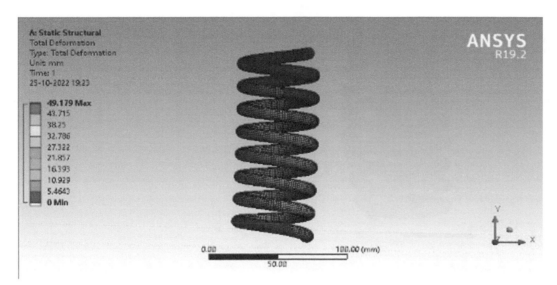

Figure 106.4 Cylindrical coil spring
Source: Author

possible movement of the point due to the loading of the structure. This model includes static analysis of an open-end cylindrical coil spring, open-end barrel spring and an open-end conical coil spring.

Structural analysis

The coil spring parameters, material properties, meshing, and loading boundary conditions are used in the static analysis. Under a static load of 2750 N the total deformation, maximum shear strain, equivalent stress and strain energy are investigated, as shown in the following figures.

Total deformation

It represents all the deformation in a model along all three coordinates (x, y, z). The value of total deformation for open-end cylindrical coil spring, open-end barrel coil spring and open-end conical coil spring is 17.343 mm, 13.576 mm and 22.109 respectively. From Figures 106.4, 106.5 and 106.6, conical barrel spring show least deformation as compared to cylindrical and conical coil spring. As the stiffness is inversely proportional to deflection so conical coil spring is less stiff and barrel spring is most stiff.

Maximum shear equivalent strain

It is the maximum amount of stress concentrate in a very small area. The value of maximum shear elastic strain for open-end cylindrical coil spring, open-end conical coil spring and open-end barrel coil spring is 0.0031, 0.0019 and 0.0034 respectively. Figures 106.7, 106.8 and 106.9 represents shear strain contour of cylindrical, conical, and barrel coil spring which shows that elastic shear strain for cylindrical spring and barrel spring is nearly same and maximum while for conical spring is least.

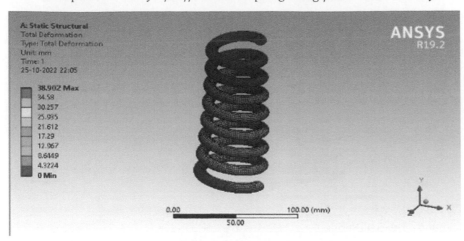

Figure 106.5 Conical coil spring
Source: Author

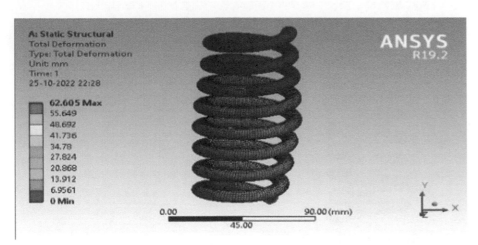

Figure 106.6 Barrel coil spring
Source: Author

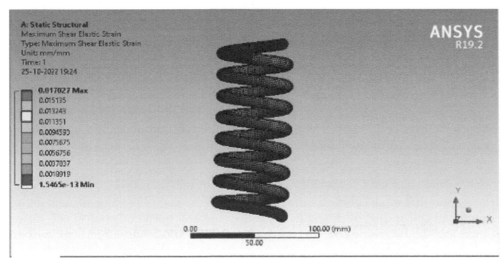

Figure 106.7 Cylindrical coil spring
Source: Author

Equivalent stress

Under multiaxial loading condition it predicts the yielding point of material using simple uniaxial tests. The value of equivalent stress for open-end cylindrical coil spring, open-end conical coil spring and open-end barrel coil spring is 423.66 MPa, 378.94 Mpa and 462.7 Mpa respectively. It can be seen in above Figures

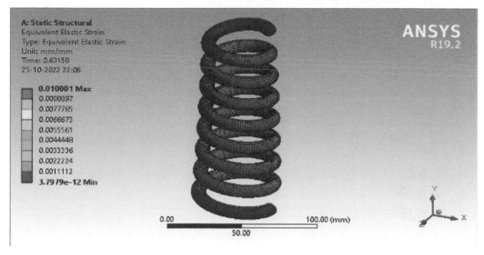

Figure 106.8 Conical coil spring
Source: Author

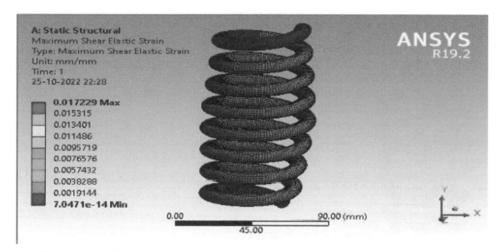

Figure 106.9 Barrel coil spring
Source: Author

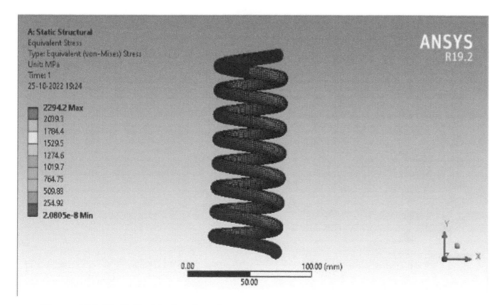

Figure 106.10 Cylindrical coil spring
Source: Author

106.10, 106.11 and 106.12 that the equivalent stress for conical coil spring is minimum as compared to cylindrical and barrel coil spring.

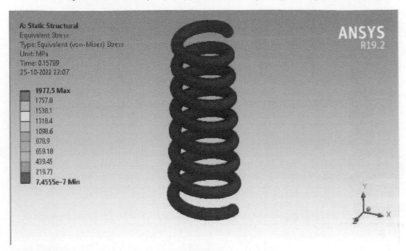

Figure 106.11 Conical coil spring
Source: Author

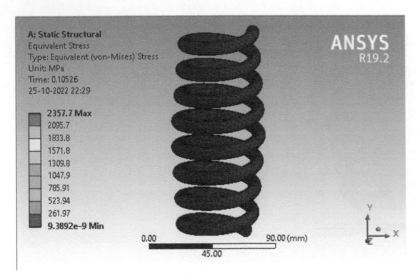

Figure 106.12 Barrel coil spring
Source: Author

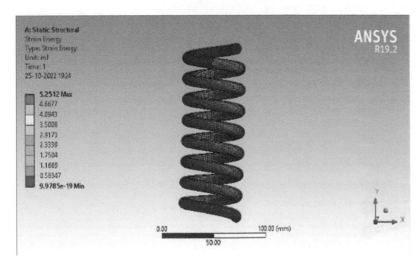

Figure 106.13 Cylindrical coil spring
Source: Author

Strain energy

It is the energy stored inside a body to restrict the strain formation on the body. The value of strain energy for open-end cylindrical coil spring, open-end conical coil spring and open-end barrel coil spring is 5.2512

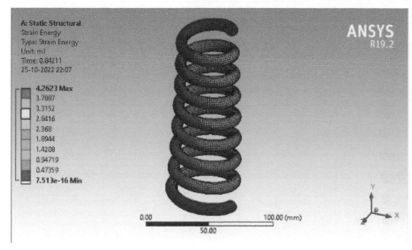

Figure 106.14 Conical coil spring
Source: Author

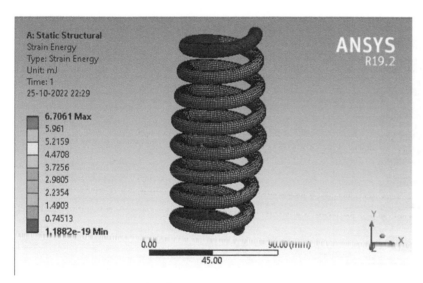

Figure 106.15 Barrel coil spring
Source: Author

mJ, 4.2623 mJ and 6.7001 mJ respectively. A strain energy is directly proportional to the deformation, Figures 106.13, 106.14 and 106.15 represents that conical coil spring contains least energy strain and barrel coil spring have maximum energy strain.

Fatigue analysis

When components are subjected to the cyclic load and the tendency of failure is analyzed, it is known as fatigue analysis. For this study, different coil springs are subjected to a load of 2750 N for 1000000 number of cycles and their fatigue life and safety factor is determined as shown in respective figures.

Fatigue life

It is the predicted time that a component or material will last before falling due to stress compression and strain formation. The value of fatigue life for open-end cylindrical coil spring, open-end conical coil spring and open-end barrel coil spring is 1.6×10^5 cycles, 1.4×10^5 cycles and 1.8×10^5 cycles respectively. Ascending order of fatigue life as represented in above Figures 106.16, 106.17 and 106.18 is Barrel coil spring > cylindrical coil spring > conical coil spring.

Fatigue of safety

The value of safety factor for open-end cylindrical coil spring, open-end conical coil spring and open-end barrel coil spring shown in Figures 106.19, 106.20 and 106.21 is 1.9, 1.3 and 2 respectively. Conical coil

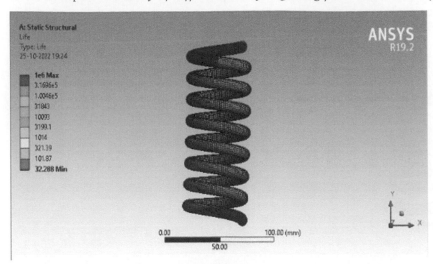

Figure 106.16 Cylindrical coil spring
Source: Author

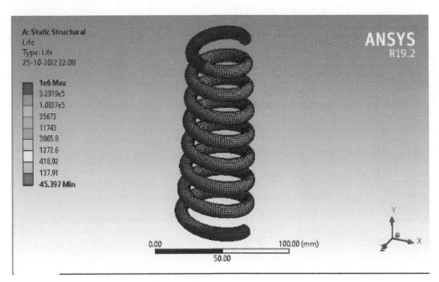

Figure 106.17 Conical coil spring
Source: Author

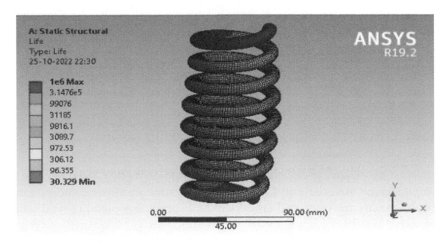

Figure 106.18 Barrel coil spring
Source: Author

spring least safety factor while cylindrical and barrel coil spring have same and highest safety factor, but all three types of coil springs have factor of safety greater than 1.

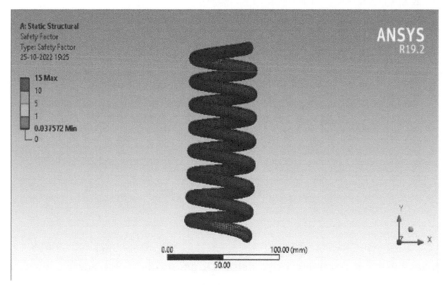

Figure 106.19 Cylindrical coil spring
Source: Author

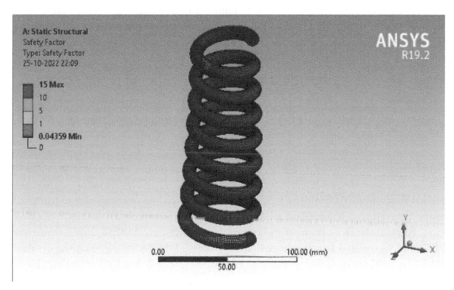

Figure 106.20 Conical coil spring
Source: Author

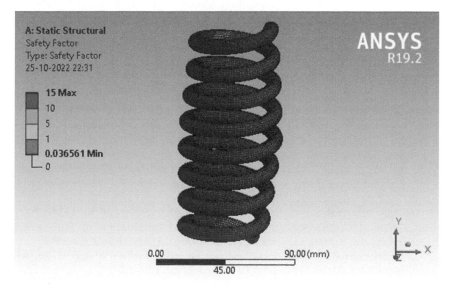

Figure 106.21 Barrel coil spring
Source: Author

Table 106.4 Analysis results

S.No	Parameters	Cylindrical coil spring	Conical coil spring	Barrel coil spring
1.	Total Deformation (mm)	17.343	22.109	13.576
2.	Maximum shear elastic strain	0.0031	0.0019	0.0034
3.	Equivalent stress (Mpa)	423.66	378.94	462.7
4.	Strain energy (mJ)	5.2512	4.2623	6.7001
5.	Fatigue life (cycles)	1.6×10^5	1.4×10^5	1.8×10^5
6.	Factor of safety	1.9	1.3	2

Source: Author

FEA results comparison

The comparison of three geometries based on parameters like deformation, strain, stress, energy, life and safety of factor is shown in Table 106.4. Total deformation, maximum equivalent shear stress, equivalent stress and energy strain is determined under static structural analysis and fatigue life and factor of safety is determined under fatigue analysis.

From the above data it can be evaluated that, cylindrical coil spring less 22.74% more deflection, have 63.1% more shear strain, 11.8% more stress and store 23% more strain energy whereas barrel coil spring displays 62.8% less deflection, 78.4% more elastic strain, 22.10 more stress and store 57.19% more strain energy than conical coil spring.

Conclusion

This paper aims to develop and analyse open-end cylindrical coil springs, open-end conical coil springs, and open-end barrel coil spring for passenger vehicles. In which the coil spring of structural steel is subjected to static analysis such as total deformation, maximum shear elastic strain, equivalent stress and strain energy, also for fatigue analysis such as fatigue life. From the analysis of the study, the following conclusions have been drawn.

Barrel coil spring shows least total deformation without buckling as compared to conical and cylindrical coil springs. The barrel coil springs provide high variable stiffness in a passenger vehicle as compared to conical and cylindrical coil springs. Elastic strain is highest in barrel coil spring i.e., 0.0034. For cylindrical coil spring equivalent stress is 11.78 % more and for barrel coil spring it is 22.10 % more than the conical coil spring. Barrel coil spring stores 6.7 mJ strain energy which is highest as compared to cylindrical and conical coil spring. Fatigue life and factor of safety for barrel coil spring is highest among all three designed coils springs.

Based on the modelling and finite element analysis (FEA), we conclude that the bar-rel coil spring made up of structural steel material is most suitable for passenger vehicle as compared to cylindrical and conical coil spring and these results can be further verified with such comparison in experimental application.

References

[1] Pastorcic, D., Vukelic, G., and Bozic, Z. (2019). Coil spring failure and fatigue analysis. Engineering Failure Analysis. 99, 310–18. https://doi.org/10.1016/j.engfailanal.2019.02.017.

[2] Alsahlani, A., Khashan, M. K., and Khaleel, H. H. (2018). Design and analysis of coil spring in vehicles using finite elements method. International Journal of Mechanical and Production Engineering Research and Development. 8,(4), 615–24. doi: 10.24247/ijmperdaug201864.

[3] Zhou, C., Chi, M., Wen, Z., Wu, X., Cai, W., Dai, L., Zhang, H., Qiu, W., He, X., and Li, M. (2020). An investigation of abnormal vibration – induced coil spring failure in metro vehicles. Engineering Failure Analysis. 108, 104238. 1350-6307. https://doi.org/10.1016/j.engfailanal.2019.104238.

[4] Patel, H. and Zhou, H. (2021) . Analysis and synthesis of conical coil springs. Proceedings of the ASME 2021 International Design Engineering Technical Conferences and Computers and Information in Engineering Conference. Volume 8A: 45th Mechanisms and Robotics Conference (MR) doi: 10.1115/DETC2021-69971.

[5] Yadav, S. D. and Lata, S. (2021). Design development and analysis of cylindrical spring with variable pitch for two wheelers. Materials Today: Proceedings. 47(40), 3105–11. doi: 10.1016/j.matpr.2021.06.130.

[6] Cho, S., Yeon, H., Kim, H., and Kim, C. W. (2021). Design of end coil angular position and centerline shape of C-type side load coil spring for reducing side load of MacPherson strut suspension. Journal of Mechanical Science and Technology. 35(3), 1153–1160. doi: 10.1007/s12206-021-0228-6.

[7] Rahim, A. A. A., Abdullah, S., Singh, S. S. K., and Nuawi, M. Z. (2021). Fatigue strain signal reconstruction technique based on selected wavelet decomposition levels of an automobile coil spring. Engineering Failure Analysis. 125, 105434. doi: 10.1016/j.engfailanal.2021.105434.

[8] Sreenivasan, M. et al. (2020). Finite element analysis of coil spring of a motorcycle suspension system using different fibre materials. Materials Today: Proceedings. 33(40), 275–9. doi: 10.1016/j.matpr.2020.04.051.

[9] Vivek, D., Praveen, R., Krishnan, A. S., Sabapathy, Y. K., Ebenezer, D. and Selvaraj, M. (2021). Development and analysis of GFRP conical springs BT. Trends in Manufacturing and Engineering Management. 409–418.

[10] Katyayn, A. and Tripathy, C. R. (2020). Design and analysis of helical coil spring to improve its strength through geometrical modification. International Journal of Technical Research & Science. 5(5), 20–25, doi: 10.30780/ijtrs.v05.i05.003.

[11] Chen, H., Lv, Z., Yang, X., Liu, X., and Li, P. (2021). Characteristics of a novel cylindrical arm spiral spring for linear compressor. International Journal of Refrigeration. 131, 1000–9, doi: 10.1016/j.ijrefrig.2021.07.006.

[12] Wang, K., Yang, Y., Yang, Y., Xu, M., Chen, S., Ling, L., Xiao, F., Zhang, K., and Ma, C. (2022). An experimental investigation of the mechanism and mitigation measures for the coil spring fracture of a locomotive. Engineering Failure Analysis. 135, 106157. https://doi.org/10.1016/j.engfailanal.2022.106157.

[13] Manouchehrynia, R., Abdullah, S., and Singh, S. S. K.(2022). Fatigue-based reliability in assessing the failure of an automobile coil spring under random vibration loadings. Engineering Failure Analysis. 131, 105808. https://doi.org/10.1016/j.engfailanal.2021.105808.

[14] Wei, X., Wang, G., Chen, X., Jiang, H., and Smith, L. M. (2022). Natural bamboo coil springs with high cyclic-compression durability fabricated via a hydrothermal-molding-fixing method. Industrial Crops and Products. 184, 115055. https://doi.org/10.1016/j.indcrop.2022.115055.

[15] Kushwah, S., Parekh, S., and Mangrola, M. (2020). Optimization of coil spring by finite element analysis method of automobile suspension system using different materials. Materials Today: Proceedings. 42, 827–31. doi: 10.1016/j.matpr.2020.11.415.

[16] Bergh, F., Silva, G. C., Silva, C., and Paiva, P. (2021). Analysis of an automotive coil spring fracture. Engineering Failure Analysis. 129, 105679. doi: 10.1016/j.engfailanal.2021.105679.

[17] Xiong, Q., Izadi, M., Rad, M. S., Shehzad, S. A., and Mohammed, H. A. (2021). 3D Numerical study of conical and fusiform turbulators for heat transfer improvement in a double-pipe heat exchanger. International Journal of Heat and Mass Transfer. 170, 120–995. doi: 10.1016/j.ijheatmasstransfer.2021.120995.

[18] Tan, Y., Lu, G., Cong, M., Wang, X., and Ren, L. (2021). Scavenging energy from wind-induced power transmission line vibration using an omnidirectional harvester in smart grids. Energy Conversion and Management. 238. 114–73. doi: 10.1016/j.enconman.2021.114173.

Chapter 107

Fea based camparison of different Go-Kart chassis

Ankur Rai[a], Kapil Rajput[b], Shah Abdul Aleem[c] and Fakidul Azam[d]

[1]Department of Mechanical Engineering, Galgotias University, Greater Noida, India

Abstract

A Go-Kart is a motorsport vehicle defined as a self-propelled vehicle. In this vehicle no suspension and differential are needed and it is fitted with four-wheel driving. The aim of the design is to make a featherweight, drive-friendly, and heavy-duty kart. To make the go-kart chassis highly rigid and provide a safe driver cockpit, we have used the principle of triangulation to avoid damage in case of an accident. A tabular cross-section seamless tube of 1.25 inches diameter and 3 mm thickness of AISI 1020 grade is used to fabricate the chassis. The front and rear impact test were simulated to the two different CAD Model of chassis having difference in engine mounting area, Model 1 is rear mounted vehicle and Model 2 is vehicle with right engine mounting. Results for the total deformation, maximum elastic shear strain, equivalent stress and strain energy is determined using Finite Element Analysis and compared for the both models of chassis. Several frame couplings are added at the critical location to brace the weak pipe and members who can experience the high load. Front and Rear bumpers are allocated in the chassis to protect and ensure less damage and safe driving. It can be evaluated that the total deformation for Model 2 is 67% and 50% more than Model 1 for front and rear impact respectively, while equivalent stress for front impact is approx. 100% and for rear impact 20% more and stored energy strain is highest in Model 2 in both cases.

Keywords: CAD Modelling, go-kart, impact analysis, static analysis

Introduction

A Go-Kart is a simple ride-height, motor-driven, light weight, and closed-packed vehicle. Due to handy and low-level karts these vehicles are used for racing events. It consists of various parts like chassis, axles, steering mechanism, wheels, pumpers, and tires. It has no suspension system due to low ground clearance and it is used in track racing [1]. Frame is the most important part of a Go-Kart vehicle; it bears all the load and force on it. Due to this kart structure should have toughness and steadiness to carry high load and have minimise displacement [2]. Go-Kart is an enterprising and generous sporting vehicle to the persons who are interested in the sport race because it has economical, plain fabric and secured racing. It can have both indoor or outdoor racing track [3,4]. In this project, we have aimed to design a chassis of Go-Kart and perform several analyses, construction is done using circulation cross-section pipes. SolidWorks software is used to perform modelling and analysis.

Nowadays, various types of Go-karts are used for racing named as internal combustion engine driven, electrically powered, hybrid and solar driven. Internal combustion Go-Kart got converted into electric kart by changing the design of mounting units using SolidWorks. Both model of IC engine and Electric kart is analysed using Ansys Modal on various impact tests like front, rear and side, their results are compared accordingly [5]. Go-Kart vehicles have no suspension and differential in it as it is raced in concrete or bituminous track during various events. An electric kart is designed using SolidWorks CAD software keeping an aim to obtain low weight and high performing vehicle. For testing and analyses AISI 1018 material is considered. Ansys Modal software is used to obtain deformation, Factor of Safety and Von-Mises Stress of rear, side, and front impact test [6]. Change in vibration and stress pattern can occur due to design changes. Research is performed to evaluate the vibration and structural strength properties of three designed kart. These three karts of different design are modelled according to national karting or racing events rulebook using CAD software. Modal analysis is performed to examine the difference in dis-placement, stress distribution, natural frequency, and impact test of all three karts. For FEA using Ansys AISI 1018 is considered as material and front, side and rear impact test is performed. Results show good structural and vibration properties with high factor of safety [7]. The chassis design is modelled to enhance the kart roll case strength as per SAE standard. Three different materials i.e., AISI 1018, AISI 1020 and AISI 4130 and different pipe thickness 1.0, 1.25, 1.5 mm is considered for Finite Element Analysis using modal software to determine the factor of safety, equivalent stress for all the above arrangements and compare their results. This will provide a basic optimisation strategy to build or manufacture a Go-Kart [8]. Frames must sprightful enough to hold static and dynamic loads generated by different mechanical components fitted on the frame to run vehicle. Go-Kart geometrical modification can lead to improvement in its structural strength. Several changes and variation have been done in a kart frame and it is reanalysed torsional rigidity and

[a]ankur.rai282001@gmail.com, [b]kapil.rajput@galgotiasuniversity.edu.in, [c]abdulaleem9990@gmail.com, [d]fakidul08@gmail.com

DOI: 10.1201/9781003450252-107

impact strength. Results illustrate that modified design provide better value in im-pact testing and torsional analysis [9].

One of the basic and important components of a vehicle is its frame, it acts as a support for assembling components like suspension, steering, braking and power-train on a vehicle and also provides a structural safety. Factors which should be kept in mind while design and analysis of vehicle is its weight aspect, performance, and safety. These factors can be achieved by considering various parameters like tube cross section material selection and shock or force absorption [10]. Various Go-Karting events companies like IKR, GKDC and NGKC offer their rulebook to manufacture the kart. CAD modelling and static analysis of a go kart is performed using SolidWorks and Ansys according NCKC rulebook. Results of FEA and analytical calculation are compared, and it is observed that they nearly have same values [11]. Front, rear and side impact test results of go kart chassis performed on four different materials is compared, Chassis is designed using Catia V5 and analysed using Ansys Workbench. Von-mises stress is used to define design and failure criteria of the different materials. Various considerations have been taken into consideration such as reliability, material selection and strength, easy manufacturing, absorption of energy and rigidity to investigate the perfect material and safety factor of low ground clearance kart [12]. In today's worlds, in development and innovation dynamics changes which are related as cycles plays an important role. A study aims to determine, characterise and examine these cycles of innovation process along with their behavior. They developed an electrically driven kart project to acquire detailed information of execution process. These results lead to the development of a model on which specified tasks like cycle description and analysis related aspects are performed. The results demonstrate the importance of different aspects in any process or function [13]. Professional drivers and high-speed vehicles are not required in several go karting events. A study aimed to design and manufacture a sound kart which offers greater fuel economy and high driver comfort with compact chassis size. NKRC rulebook is followed for design and analyzing using SolidWorks and Ansys respectively [14]. In a smaller and compact vehicle, major issue occurs related strength and supporting structure of the chassis. A systematic methodology is presented to perform the design and analysing of the kart. Structural strength analysis is performed using Workbench while considering an overload parameter which can be experienced by the driver during collision. Boundary conditions are given on the front longitudinal axis and frame lateral truss. Results shows a small difference in the experimental and analysed data [15].

It can be concluded by the above literature review that various studies have been taken place to examine the structural strength and rigidity of a go kart using finite element analysis. But, in these rescarch's none of them have compared or examined the best area to mount a go kart engine. In continuation of these studies, this paper is aimed to compare the different mounting locations for a go kart engine using finite element analysis. Following is the objective of this study:

- Perform the required analytical calculation of chassis.
- Deign two models of Go-Kart chassis (rear mounted engine and right mounted engine).
- Select the best material for the frame.
- Perform the static structural analysis to determine deformation, strain, stress and energy.
- Compare the simulation results for both models and conclude the most suitable model.

To develop the go-kart chassis cad model SOLIDWORKS 2016 software is used and the model file is converted to the. step or .igs file type. This converted file is inserted in the Ansys 19.2 Workbench static structural analysis module and further actions is performed accordingly.

Methodology

A Research is something which should be planned and executed very efficiently to determine the desired outcome, therefore Figure 107.1 illustrates method of approach taken into account in this paper.

Figure 107.1 Methodology for go-kart chassis FEA analysis
Source: Author

Go-Kart chassis 3D model

Rulebook of Indian Karting Race Season 7 is considered and analysed for designing the two models of go-kart chassis using SOLIDWORKS 2016 CAD software. At first stage, 2D sketch of the kart base is designed in top view followed by the 3-D sketching of front and rear body parts respectively. This combine sketch is given a 3D view by applying weldments of dimensions 30.01 mm outer diameter and 3.2 mm thickness [16]. The isometric view of designed models can be seen in Figures 107.2 and 107.3. The designs are differentiated from each other by their engine mounting location and driver sitting area. Model 1 has rear engine mounting while Model 2 have right hand side engine mounting. Tables 107.1 and 107.2 represent dimensions of model 1 and 2.

Material selection

The template is used to format your paper and style the text. All margins, column widths, line spaces, and text fonts are prescribed; please do not alter them. You may note peculiarities. For example, the head margin in this template measures proportionately more than is customary. This measurement and others are deliberate, using specifications that anticipate your paper as one part of the entire proceedings, and not as an independent document. Please do not revise any of the current designations. Tables 107.3 and 107.4 represents the various properties of material used.

Analytical calulation

It is the formulation of a given problem in a proper and well-defined manner and solve it using desired formulae and equations. For a go kart vehicle, kerb-weight of 75 kg and driver of 70–80 kg is considered [18].

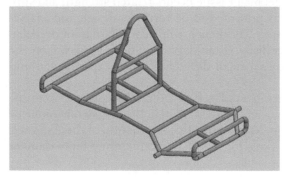

Figure 107.2 Model 1 (rear engine mounting)
Source: Author

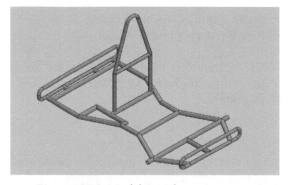

Figure 107.3 Model 2 (right engine mounting)
Source: Author

Table 107.1 Dimensions of model 1

Variable	Value
Track-width	1300 mm
Wheelbase	1150 mm
Kart Width	800 mm

Source: Author

Table 107.2 Dimesnions of model 2

Variable	Value
Track-width	1400 mm
Wheelbase	1250 mm
Kart width	700 mm

Source: Author

Table 107.3 Mechanical properties [17]	
Properties	Values
Tensile strength	440 MPa
Bulk modulus	140 GPa
Yield strength	370 MPa
Shear modulus	80 GPa
Poisson's ratio	0.29
Young modulus	205 GPa
Ultimate strength	394.72 MPa
Brinell hardness	126

Table 107.4 Chemical and thermal properties [17]	
Properties	Values
Thermal conductivity	51.9 W/mk
Chemical properties	
Element	**Content**
Manganese, Mn	0.60 – 0.90
Carbon, C	0.15 – 0.20
Sulphur, S	0.05 (max)
Phosphorous, P	0.04 (max)
Iron, Fe	Balance

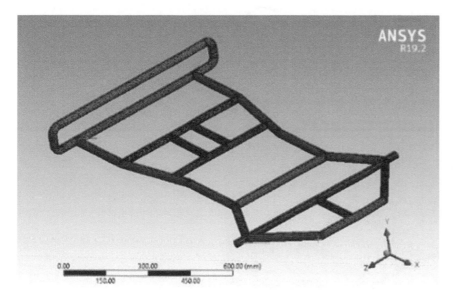

Figure 107.4 Model 1 meshed view
Source: Author

Load on kart with driver = 75 kg + 80 kg = 155 kg
Consider, 45 kg load as a safety load for analysis.
Total load on the kart will be = 155 kg + 45 kg = 200 kg
Let us consider, acceleration due to gravity to be 10 m/s².
Weight of the kart = Total Load × Acceleration due to gravity
Weight of the kart = 200 kg × 10 m/s² = 2000 N

Pre-processing

It is the model development for the calculations and simulations, it includes meshing, model setup, applying loads and constraints. First and most essential part in analysis is meshing, a good mesh and bring a drastic change in our results. Meshing is the process of dividing whole body into small nodes and elements on which the results are calculated [19].

A fine mesh is generated for both the models in Ansys 19.2 software, for mod-el 1 no. of nodes are 103240 and no. of elements are 54883 and for model 2 no. of nodes are 142458 and no. of elements are 85698. Type of mesh used is of quadrilateral type and elements size is of 10 mm (Figures 107.4 and 107.5).

Simulation techniques used for analysing the displacement, factor of safety, flexibility and other physical aspects taking place on structure or frame on applying force/pressure on it known as "finite element analysis" [20]. ANSYS 2019 R2 is used for the analysis of designed chassis. Following test have been performed to analyse the kart frame:

1. Front impact test: Rear end is fixed and force is applied from front face or node shown in Figure 107.6.
2. Rear Impact test: Figure 107.7 represents that front end of the chassis is kept fixed and force is applied from rear or back side of the chassis.

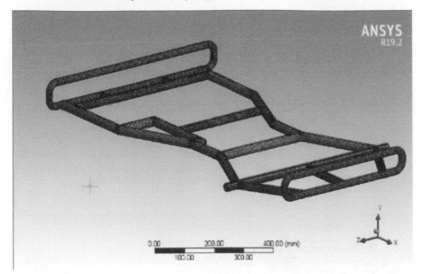

Figure 107.5 Model 2 meshed view
Source: Author

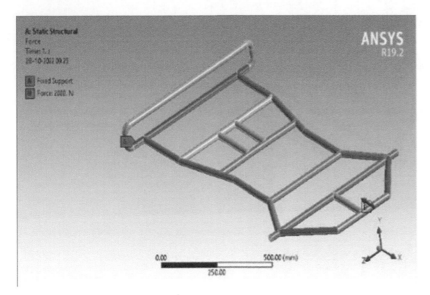

Figure 107.6 Front impact setup
Source: Author

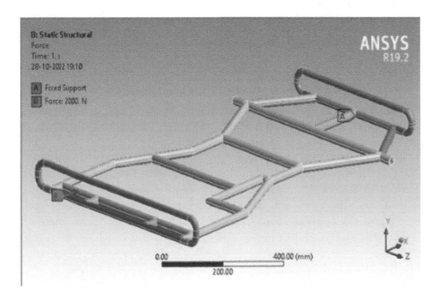

Figure 107.7 Rear impact setup
Source: Author

Results and Disussion

Impact testing is defined as the calculation of changes occurs in a body or vehicle due to its collision or impact. A heavy amount of force is exerted on the vehicle during collision or impact testing which may results in vehicle deformation or break-age. In impact analysis of the chassis, the one end of the chassis is kept fixed and force of 2000 N is applied on other end, the total deformation, maximum shear strain, equivalent stress and strain energy of the chassis is determined and compared for both models.

Front impact test

In any case of accident due to driver error, the front part of kart gets collide first. To analyse this front impact, four nodes of the rear end get fixed, and 2000 N load is applied at two front nodes and below mentioned parameters are analysed using static structural analysis for both models.

Total deformation

It is the calculation of the amount of deflection shown by the body due to axial force. Figures 107.8 and 107.9 display the total deformation formed in the Model 1 and Model 2 respectively. It can be seen that in model 1 total deformation is 0.17313 mm and for model 2 it is 0.10349 mm, so due to less deflection the stiffness in model 2 will be more. The stiffness value of Model 2 199325.5 N/mm and for Model 1 it is 11552.01 N/mm.

Maximum shear equivalent strain

Shear elastic strain is the change in the angle between the adjacent lines of a body from its original right angle. For Model 1 maximum shear elastic strain is 2.31 and for model 2 it is 5. Figures 107.10 and 107.11

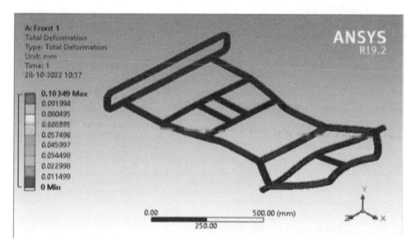

Figure 107.8 Model 1 deformation contour
Source: Author

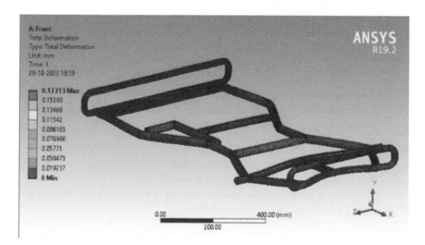

Figure 107.9 Model 2 deformation contour
Source: Author

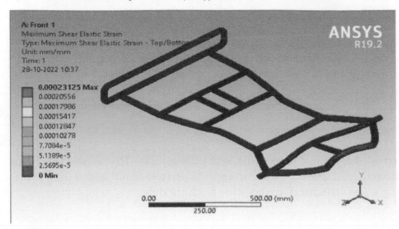

Figure 107.10 Model 1 elastic strain
Source: Author

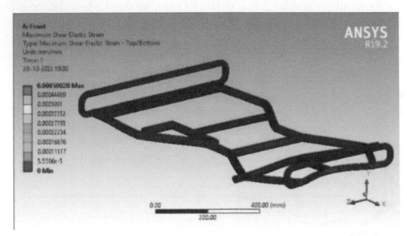

Figure 107.11 Model 2 elastic strain
Source: Author

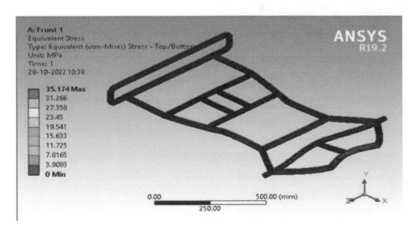

Figure 107.12 Equivalent stress model 1
Source: Author

display the contour image of maximum shear elastic strain for both the models. Model 2 has more elastic strain value which determine that for given deflection of deformation it generates large amount of stress to nullify the knuckle or fracture condition.

Equivalent stress

It is also known as von-misses stress. It is used to find out whether a body will break down or yield the applied force or pressure. Figures 107.12 and 107.13 represent the equivalent stress for Model 1 and Model 2 viz, 35.174 MPa and 78.21 MPa respectively. Model 2 develops more equivalent stress as compared to

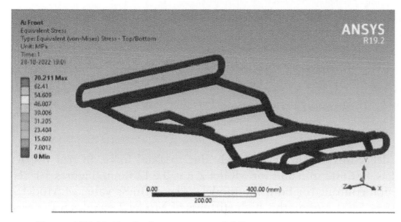

Figure 107.13 Equivalent stress model 2
Source: Author

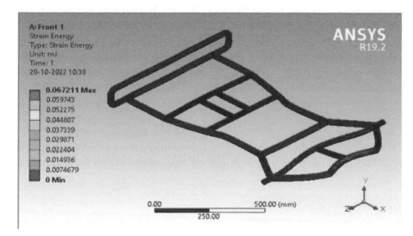

Figure 107.14 Strain energy model 1
Source: Author

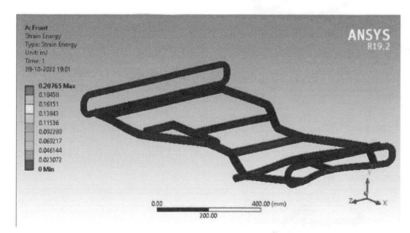

Figure 107.15 Strain energy model 2
Source: Author

Model 1 due to high amount deflection, this high amount of stress helps the design to restrict breaking or fracture due to high value force or load.

Strain energy

Energy stores or absorbs by the body when a force or a load is applied on it, is known as strain energy. From Figures 107.14 and 107.15 it can be accessed that Model 2 stores more amount of strain energy than Model 1. The value of strain energy for Model 1 and Model 2 is 66.26 mJ and 124.78 mJ respectively. The right engine mounted chassis stores high amount of potential or strain energy which makes it capable to stand against a high amount of force or high velocity impact.

Rear impact test

In rear impact analyses, the front end is kept fixed, and a force of 2000 N is applied form rear end to evaluate the total deformation, maximum shear strain, equivalent stress and strain energy at rear section.

Total deformation

It is the calculation of the amount of deflection shown by the body due to axial force. In rear impact analysis, total deformation for Model 1 is 0.36602 mm and for Model 2 it is 0.2428 mm. Figures 107.16 and 107.17 displays the total deformation contour for both the models. The stiffness value of Model 2 8237.23 N/mm and for Model 1 it is 5464.1822 N/mm, this high stiffness of Model 2 enables it to carry high amount of load.

Maximum shear equivalent strain

Shear elastic strain is the change in the angle between the adjacent lines of a body from its original right angle. Figures 107.18 and 107.19 displays the value of maximum shear elastic strain for Model 1 and Model 2 viz, 4.66×10^{-4} and 5.53×10^{-4} respectively. Model 2 generates more amount of elastic strain as compared to Model 1 due to high amount of stiffness. Model 2 has more elastic strain value which determine that for given deflection of deformation it generates large amount of stress to cancel out the knuckle or fracture condition.

Equivalent stress

It is also known as von-misses stress. It is used to find out that a body will breakdown or yield the applied force or pressure. Same as front impact, Model 2 generate more amount of equivalent stress as compared

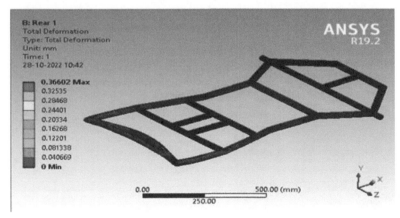

Figure 107.16 Model 1 rear impact deformation contour
Source: Author

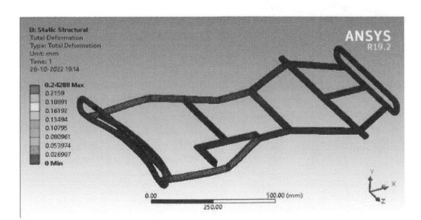

Figure 107.17 Model 2 rear impact deformation contour
Source: Author

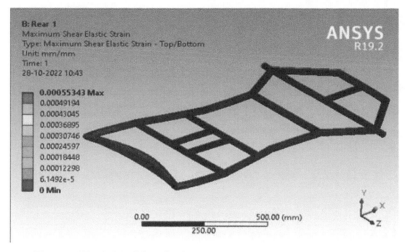

Figure 107.18 Model 1 elastic strain
Source: Author

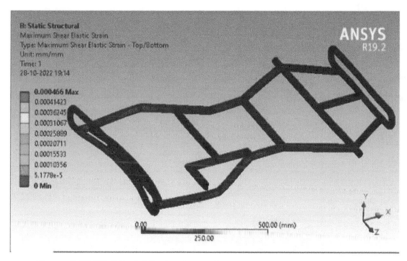

Figure 107.19 Model 2 elastic strain
Source: Author

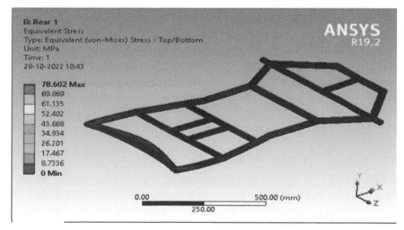

Figure 107.20 Equivalent stress model 1
Source: Author

to Model 1 shown in Figures 107.20 and 107.21. The equivalent stress for Model 1 and Model 2 is 65.251 MPa and 78.602 MPa. Model 2 develops more equivalent stress as compared to Model 1 due to high amount deflection, this high amount of stress helps the design to restrict breaking or fracture due to high value force or load.

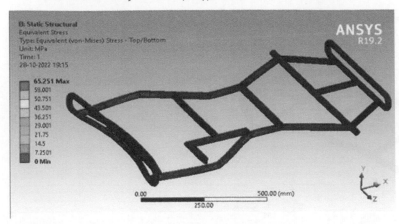

Figure 107.21 Equivalent stress model 2
Source: Author

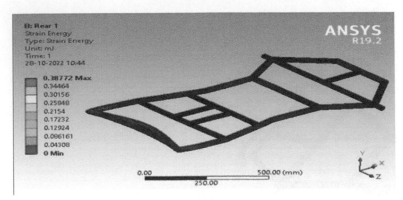

Figure 107.22 Strain energy model 1
Source: Author

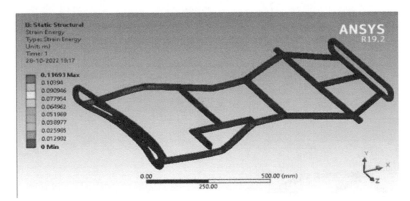

Figure 107.23 Strain energy model 2
Source: Author

Strain Energy

Energy stores or absorbs by the body when a force or a load is applied on it, is known as strain energy. As mentioned earlier that Model 2 produces more amount of strain than Model 1, it is resulted in higher energy storage in Model 2 as compared to Model 1, Figures 107.22 and 107.23 represents the energy strain value of Model 1 and Model 2 viz, 151.43 mJ and 203.45 mJ respectively. Right engine mounted chassis stores high amount of potential or strain energy which makes it capable to stand against a high amount of force or high velocity impact.

Comparison of results for front and rear impact test.

At above two sections, the total deformation, maximum shear elastic strain, equivalent stress and strain energy for front and rear impact test is obtained under static structural module of Ansys. All the derived data for each parameter is tabled in Table 107.5 and compared with each other.

Table 107.5 Results camparison

S. No		Variables	Total deformation (mm)	Maximum shear elastic strain	Equivalent Stress (MPa)	Strain energy (mJ)
1	Front Impact	Model 1	0.10349	2.31×10^{-4}	35.174	66.236
2		Model 2	0.17313	5.0×10^{-4}	70.211	124.78
3	Rear Impact	Model 1	0.2428	4.66×10^{-4}	65.251	151.43
4		Model 2	0.36602	5.53×10^{-4}	78.602	203.5

Source: Author

From the Table 107.5, it can be evaluated that the total deformation for Model 2 is 67% and 50% more than Model 1 for front and rear impact respectively, while equivalent stress for front impact is approx. 100% and for rear impact 20% more and stored energy strain is highest in Model 2 in both cases.

Conclusion

The root essentials of a Go-Kart vehicle are having a low weight to toughness ratio, with less ground clearance. The frame designing for Go-Kart is helpful in identifying the durability and deficiency of build and design. CAD model of the go-kart frame is developed according to the Indian Kart Racing rulebook. Further, Finite Element Analysis is performed using AISI 1020 carbon steel as chassis material and static structural analysis is performed using Ansys software and following results are concluded from the study:

- For both tests, total deformation formed in Model 2 is much large than in Model 1. Model 2 has more stiffer body as compared to another model. The maximum shear elastic strain for Model 2 is have large value than of Model 1 in both front and rear impact test. Equivalent stress for Model 1 is very low as compared to Model 2 which have 70.21 MPa and 78.602 MPa equivalent stress in front and rear impact respectively. Strain energy absorbed by the Model 2 is very high as compared to Model 1 which makes it capable to withstand the applied load for longer time.

After this detailed finite element analysis of the total deformation, maximum shear clastic strain, equivalent stress and strain energy, it can be concluded that Model 2 which have engine mounting location at right hand side of the driver is best capable or suitable for manufacturing a Go-Kart for any national or inter national event.

References

[1] Thavai, R., Shahezad, Q., Shahrukh, M., Arman, M., and Imran, K. (2015). Static analysis of go-kart chassis by analytical and solid works simulation. International Journal of Mechanical and Industrial Technology. 3(2), 73–78.

[2] Kiran, R. S. (2020). Design and analysis of go - kart chassis using distinctive materials. International Journal for Research in Applied Science and Engineering Technology. 8(7), 294–307. doi: 10.22214/ijraset.2020.7050.

[3] Attarde, R., Chougule, A., and Magdum, R. (2019). Material based FEA analysis of a go-kart chassis: A comparative atudy. Advanced Engineering Forum. 31, 10–25. doi: 10.4028/www.scientific.net/aef.31.10.

[4] Saini, N. K., Panwar, M. S., Maurya, A., and Singh, P. A. (2019). Design and vibration analysis of go-kart chassis. International Journal of Applied Engineering Research. 14(9), 53–57. [Online]. Available: http://www.ripublication.com

[5] Chandramohan, N. K., Shanmugam, M., Sathiyamurthy, S., Prabakaran, S. T., Saravanakumar, S., and Shaisundaram, V. S. (2020). Comparison of chassis frame design of go-kart vehicle powered by internal combustion engine and electric motor. Materials Today: Proceedings. 37(Part 2), 2058–62. doi: 10.1016/j.matpr.2020.07.504.

[6] Krishnamoorthi, S., Prabhu, L., Shadan, M. D., Raj, H., and Akram, N. (2020). Design and analysis of electric go-kart. Materials Today: Proceedings. 45, 5997–6005. doi: 10.1016/j.matpr.2020.09.413.

[7] Srivastava, J. P., Reddy, G. G., and Teja, K. S. (2020). Numerical investigation on vibration characteristics and structural behaviour of different go-kart chassis configuration. Materials Today: Proceedings. 39(xxxx), 176–82. doi: 10.1016/j.matpr.2020.06.488.

[8] Jay Prakash Srivastava, B. Krishna Chaithanya, K. Sai Teja, B. Venugopal, S. Vineeth, M. Rajkumar and Habeeb Khan. (2020). Numerical study on strength optimization of go-kart roll-cage using different materials and pipe thickness. Materials Today: Proceedings. 39(40), 488–92. doi: 10.1016/j.matpr.2020.08.217.

[9] Mithlesh, S., Tantray, Z. A., Bansal, M., Kumar, K. V. K. P., Kurakula, V. S., and Singh, M. (2021). Improvement in performance of vented disc brake by geometrical modification of rotor. Materials Today: Proceedings. 47(40), 6054–59. doi: 10.1016/j.matpr.2021.05.006.

[10] Prashant Thakare, Rishikesh Mishra, Kartik Kannav, Nikunj Vitalkar and Shreyas Patil. (2016). Design and analysis of tubular chassis of go-kart. International Journal of Research in Engineering and Technology. 5(10), 156–60. doi: 10.15623/ijret.2016.0510025.

[11] Pewekar, M., and Sandye, P. P. (2018). Design of subsystems of go-kart vehicle. International Journal of Science, Engineering and Technology Research (IJSETR). 7(1), January 2018.

[12] Saini, N. K., Rana, R., Hassan, M. N., and Goswami, K. (2019). Design and impact analysis of go-kart chassis. International Journal of Applied Engineering Research. 14(9), 46–52.

[13] Langer, S., Knoblinger, C., and Lindemann, U. (2010). Analysis of dynamic changes and iterations in the development process of an electrically powered go-kart. 11th International Design Conference. 2010, pp. 307–318.

[14] Patil, A. S. D. H. D., and Bhange, S. S. (2016). Design and analysis of go-kart using finite element method. Design and Analysis of Go-Kart using Finite Element Method. 3(1), 551–9.

[15] Bobrowskii, A. V., Zotov, A. V., Rastorguev, D. A., Gorokhova, D. A., and Ugarova, L. A. (2019). Analysis of the frame design of the subcompact racing car of go-kart class. IOP Conference Series: Materials Science and Engineering. 537(3), doi: 10.1088/1757-899X/537/3/032078.

[16] Indian Karting Race 7th Season Rulebook. (2021). 7, 1–50.

[17] Nitin, Jain Ankit and Pravin, Patil Prathamesh and Giten, Shet and Marachakkanavar, Mailareppa and Marachakkanavar, Mailareppa, Design and Analysis of Engine Mount Bracket for Go-Kart (March 14, 2019). Proceedings of International Conference on Sustainable Computing in Science, Technology and Management (SUSCOM), Amity University Rajasthan, Jaipur - India, February 26–28, Available at SSRN: https://ssrn.com/abstract=3352396 or http://dx.doi.org/10.2139/ssrn.3352396

[18] Muzzupappa, M. (2006). Methods for the evaluation of the Go-Kart vehicle dynamic performance by the integration of Cad/Cae techniques. Available: http://www.ingegraf.es/XVIII/PDF/Comunicacion17219.pdf

[19] Park, K. M. (2015). Design and fabrication of electric go-kart using 3D printing. ASEE Annu. Conf. Expo. Conf. Proc., vol. 122nd ASEE, no. 122nd ASEE Annual Conference and Exposition: Making Value for Society, doi: 10.18260/p.23789.

[20] K.S. Karthi Vinith, P. Sathiamurthi, Design And Fabrication Of Adaptive Spoiler For Go - Kart Vehicles, International Journal of Scientific & Technology Research. 9(3), March 2016, ISSN 2277-8616

Chapter 108

Finite element analysis of composite leaf spring for TATA-ACE

Ankur Rai[a], Dr. Sudhir Kumar Singh[b], Akash Raj Tyagi[c] and Husain Abbas[d]

Galgotias University, Greater Noida, India

Abstract

A suspension system in an automobile offers smooth and comfortable rides to passengers. In light motor vehicles, sedan and sport utility vehicle helical spring and soccer-type suspension system are present whereas in load-carrying vehicles leaf spring-type suspension system exists. In India, the most common load-carrying vehicle is Tata Ace. In this research, a comparative study of two leaf springs having a different arrangement of materials is carried out. Analytical calculations are performed to find the total load, bending stress, and deflection. Computer-aided design (CAD) models of the leaf spring of Tata Ace are designed using SolidWorks software. The analysis is carried out on the leaf spring made up of carbon steel and a combination of four materials. Finite element analysis of developed CAD models is evaluated using ANSYS software to study the static structural and fatigue life parameters. The finite element analysis results are compared with the analytical results. A comparison of finite element analysis and analytical results shows that in Model 1 stress varies by 11% and deflection varies by 13% while in Model 2 stress and deflection vary only 5% from analytical data. Leaf spring of composite material offers more attractive and effective results for structural and fatigue analysis as compared to leaf spring of single material.

Keywords: Composite material, deflection, fatigue life, leaf spring, static structural, Tata Ace

Introduction

In a car assembly, a suspension system acts as a protective layer for absorbing shock using different springs. These shocks are caused by irregular roads, road bumps, and off-road driving. Road wearing and irregularities are caused by the road infrastructure, changing weather conditions, etc. According to reports, about 20% of road accidents are caused due to the poor road conditions. Several factors like roadblocks, holes, cracks, differences in road level, and distinguished friction level cause accidents due to emergency braking or stoppage [1].

The need of suspension in a car or an automobile is to provide a comfortable and cushioned ride, a stable vehicle. Transmitting vehicle load to wheels and absorbing shocks and vibration. Modern-day vehicles are equipped with different types of suspension systems from spring type, helical type to leaf spring type. The vehicle which is used to transport heavy goods or carry loads from one place to another is equipped or fitted with a suspension system of leaf spring type to absorb energy, vibration, or shock [2]. A leaf spring suspension system is a case of layers of a leaf that are jointed together using U-bolts and rebounded clip at the center and both ends respectively. These types of suspension system have mainly used for vehicles that are used to carry a load, transport goods, etc. as leaf springs absorb, store and release the energy in a simultaneous process, unlike other springs. The uppermost leaf on its assembly is known as the "master leaf" and its ends are known as eyes which are connected to the vehicle body using connecting hardware in elliptical form as shown in Figure 108.1 [3].

Nowadays, automobile industries are working on weight reduction and low density for their vehicle. These parameters can be achieved by using materials of low specific density and high tensile and yield stress. For this, composite materials are being used by several manufacturing units for several automotive components like suspension springs, axles, body parts, etc. [4]. Specific strain energy is given by-

$$U = \frac{\sigma^2}{\rho E}$$

Where,

σ = Maximum Permissible Stress

ρ = Density of Material

of $\in$ = Modulus of Elasticity

[a]ankur.rai282001@gmail.com, [b]sudhirkumar.singh@galgotiasuniversity.edu.in, [c]akashtyagi7717@gmail.com, [d]husainabbad501@gmail.com

DOI: 10.1201/9781003450252-108

From the above equation, it can be observed that those materials which offer a low factor of density and modulus of elasticity results in high specific strain energy. Composite materials have this tendency of low weight and density ratio without comprising the strength, load capacity, and stiffness.

In the suspension system of heavy-duty vehicles leaf spring is used. A detailed analysis of a parabolic leaf spring whose thickness varies from center to both ends is performed using finite element and mathematical analysis. Finite element analysis (FEA) concludes that maximum stress is developed in the red area close to the surface which may fail in extreme conditions [5]. A material of high yield strength and modulus of elasticity can be preferred to manufacture a leaf spring. A comparative study of different composite materials is performed to study their characteristics and behavior in different operating conditions. Tensile stress and hardness test are considered for evaluation as per the standard of IS 1139:15 and Society of Automobile Engineers. It shows that composite-based leaf springs have low density after high strength [6]. Automobile trends have shown the replacement of steel springs with composite materials which offer a high strength-to-wear ratio and less corrosion. All leaf of the leaf spring is made up of different materials compared with one which has single material as a glass fiber for the whole leaf spring. Comparison based on results of Brinell hardness test, compression test, and impact strength. To replace steel springs, composite and glass fiber area potential candidates for a leaf spring without comprising strength and weight ratio [7]. When a composite material is used despite steel to manufacture a leaf spring it provides a high reduction in weight and offers high strength which works as a useful measure for energy conversation and reduction in overall fuel consumption of mini trucks [8]. Considering a light motor vehicle (LMV) or passenger car, its suspension should offer high ride quality and work on the basic principle of absorbing, storing, and releasing energy whenever required. In these vehicles, steel mono-leaf is preferred for manufacturing which is compared with composite-type springs consisting of carbon and flex fiber as primary and secondary materials respectively. The high factor of safety, strain energy, and rapid increase in frequency can be seen in leaf springs of composite type [9].

Nowadays, composite material is replacing convectional springs due to its better properties, environmental awareness, and application in different areas. The hand-layup fabrication technique is used to evaluate the composite leaf spring made up of composite matrix and E-glass fiber. Semi electron microscope (SEM) analysis is also performed to find out the microstructure of composite material leaf spring (CMLS). The strength, weight, and strain of CMLS are more than that of CSLS [10]. Before deciding on any material for analysis its mechanical property should be examined carefully. It depends on the material type, method of fabrication, and orientation of its structure. E-glass epoxy and carbon fiber are materials that show the best suitable mechanical properties to be used as leaf springs. It provides a weight reduction of 70% and 7% respectively [11]. The aluminum alloy also has the potential to be used as a material for leaf springs due to its low density, high strength, and non-significant deforming in different loading conditions. For evaluating the above, computer-aided engineering (CAE) analysis is performed using ANSYS which concludes weight reduction and reduction in stress at various loading conditions [12]. Various research has been done to describe or evaluate the static and fatigue analysis of different composite materials. These studies are conducted to find the performance of all these composite materials under different operating conditions. The standard and parameters decided for a Tata Ace mini truck are used in the above FEA [13]. In this paper, a comparative study is performed using FEA static and fatigue analysis to find the economical material for the leaf spring of a Tata Ace vehicle without comprising the strength and fatigue life [14].

The literature review reflects that; several studies have been done to evaluate the static analysis of the leaf spring using the composite of two materials whereas no study has been done yet to examine the fatigue and static analysis of four composite materials for leaf spring. The objectives of the current study are as follows:

- Develop the CAD model of leaf spring for single material and composite material.
- Investigate the static and fatigue analysis results for leaf spring.
- Compare the analytical results with FEA results.

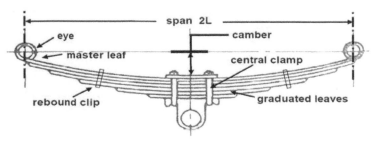

Figure 108.1 Leaf spring
Source: Author

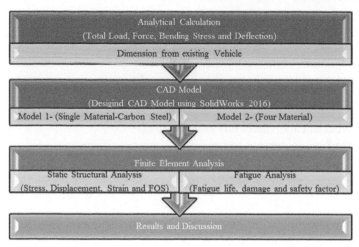

Figure 108.2 Methodology for FEA of mini-truck leaf spring
Source: Author

Table 108.1 Mechanical properties for different materials [11]

Material name/properties	1023 carbon steel	Cast alloy steel	Cast carbon steel	Chrome stainless steel
Yield strength (N/m²)	2.82 e+008	2.41 e+008	2.48 e+008	1.72 e+008
Tensile strength (N/m²)	4.25 e+008	4.48 e+008	4.82 e+008	4.13 e+00
Elastic modulus (N/m²)	2.5 e+011	1.9 e+011	2e+011	2 e+011
Poisson's ratio	0.29	0.26	0.32	0.28
Mass density (kg/m³)	7858	7300	7800	7800
Shear modulus (N/m²)	8 e+010	7.8 e+010	7.6 e+010	7.7 e+010

Methodology

Research is something that should be planned and executed very efficiently to determine the desired outcome. Figure 108.2 illustrates the steps involved in carrying out the study. The methodology provides a brief idea of workflow. In this research paper FEA method is used to determine the results of static and fatigue analysis for developed CAD Models. Results of FEA are validated by the means of analytical results [15]. For CAD modelling, SolidWorks software is used. The developed model is imported into Ansys to perform the simulation work.

Material Selection

A leaf spring material has low density, high tensile, and yield strength, and absorbs large no. of natural frequency. In this regard, the following material candidates are considered for the spring AISI 1023 carbon Steel, cast alloy steel, cast carbon steel, and chrome stainless steel. The properties of the following materials are described in Table 108.1 [16].

Two types of material combinations are used for analysing leaf spring performance-

Model 1 (single material leaf spring) - AISI 1023 carbon steel

Model 2 (composite material leaf spring) – Cast alloy steel, cast carbon steel, chrome stainless steel, and AISI 1023 carbon steels

Analytical Calculation

For a Tata Ace vehicle total load capacity defined by the company is 850 kg and the weight of the vehicle is 900 kg. Also, it is fitted with equipment or accessories not specified by the company which leads to considering a safety weight of 250 kg [17].

1. Total weight of the empty vehicle (Vw) = 900 kg
2. Maximum load capacity (L_c) = 850 kg
3. Total load on the vehicle = $V_w + L_c$ = 900 + 850 = 1750 kg

Considering all other loads, driver and passenger, and safety weight –

4.   Total load will become = 2000 kg

Assuming acceleration due to gravity is *g=9.81 m/s²*

5.   Weight = Total load × acceleration due to gravity
6.   Weight = 2000 × 9.81 = 19620 N

Consider factor of safety 2. Tata Ace has four wheels; each wheel has a leaf spring.
   Force = ¼ of total weight = ¼ × 19620 = 4905 N
   Safe loaf (F) = Calculated force/factor of safety
   2F = 4905
   F = 2452.5
   Span length (2L) = 930 mm

$$L = \frac{930}{2} = 465 \; mm$$

No. of the leaf (n) = 4
Width of each leaf (b) = 80 mm
The thickness of leaf (t) = 10 mm

1.   Maximum bending stress:

$$\sigma_b = \frac{6FL}{nbt^2} = \frac{6 \times 2452.5 \times 465}{4 \times 80 \times 10^2} = \frac{6842475}{32000}$$

2.   Total deflection:

$$\delta_{max1} = \frac{6FL^3}{Enbt^3} = \frac{6 \times 2452.5 \times 465^3}{206000 \times 4 \times 80 \times 10^3} = \frac{1.4795 \times 10^{12}}{6.592 \times 10^{10}}$$

$$\delta_{max2} = \frac{6FL^3}{Enbt^3} = \frac{6 \times 2452.5 \times 465^3}{200000 \times 4 \times 80 \times 10^3} = \frac{1.4795 \times 10^{12}}{6.4 \times 10^{10}}$$

Pre-Processing

CAD Model

A leaf spring of Tata Ace vehicle is designed using SolidWorks CAD Software, standard dimension of the vehicle is used, these dimensions are measured from a Tata Ace vehicle itself. Each part of a leaf spring, leaf spring, center plate, rebounded clip, U-Bolts, nut, and bolts is designed separately and assembled (assembly shown in Figure 108.3) using SolidWorks Assembly [18].

Meshing

Dividing the whole body into small elements and nodes which helps in getting better and qualified results during finite element analysis is known as meshing. In this analysis, mechanical type mesh is being used and the element is considered as 0.1 mm Figure 108.4. All other necessary details are described in the given Table 108.2 [19].

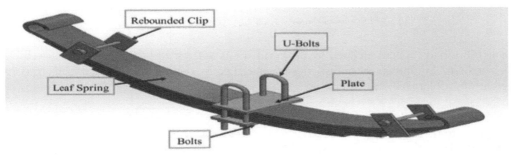

Figure 108.3 Assembly of leaf spring
Source: Author

Figure *108.4* Meshed view of the leaf spring
Source: Author

Table *108.2* Meshing details

Total nodes	12401
Total elements	7352
Maximum aspect ratio	6.0091
% of elements with an aspect Ratio < 3	96.6

Source: Author

Static Analysis Model 1- AISI 1023 Carbon Steel Material

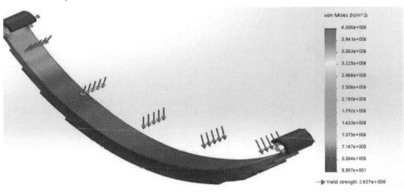

Figure *108.5* Stress development
Source: Author

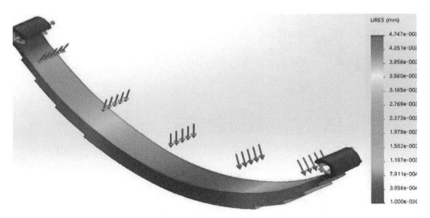

Figure *108.6* Resultant displacement
Source: Author

Load Calculation

To perform the static analysis of the designed leaf spring, SolidWorks simulation is used. Load calculation is done below:

The weight of the vehicle is 900 kg and the load carries a capacity of the Tata Ace vehicle is 850 kg. Hence the total load of the vehicle is 1750 kg.

Adding all loads (driver, passengers, and security weight): The total load is considered 2000 kg.

Taking acceleration due to gravity $g = 10$ m/s^2

Total weight of the vehicle = Total load × acceleration due to gravity = 2000 kg × 10 m/s^2 = 20000 N

Tata Ace has four wheels, each wheel will have a leaf spring, hence the resultant load on each leaf spring will be.1/4 of the total weight the of vehicle i.e., 5000 N,
   F= 5000 N

Results and Discussion

In static analysis force of 5000 N is exerted on the master leaf keeping the end clips fixed for both the material composition (defined previously). Aim of this study is to compare the stress development, resultant displacement, strain energy, and factor of safety of both models (Figures 108.5–108.12)

Fatigue Analysis

Fatigue analysis is the study of the probability that a material may fail or not on applying certain performance criteria like vibration, and force. It provides an estimate of the no. of cycles that a product or

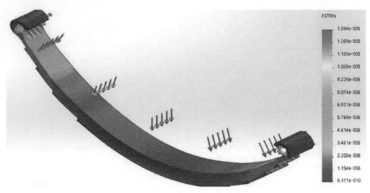

Figure 108.7 Equivalent strain
Source: Author

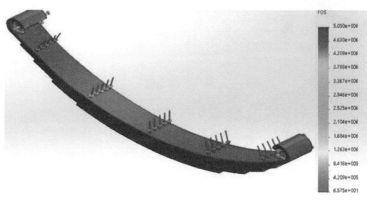

Figure 108.8 Factor of safety
Source: Author

Static Analysis Model 2- Composite Material

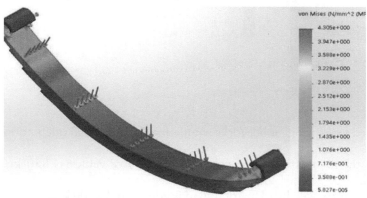

Figure 108.9 Stress development
Source: Author

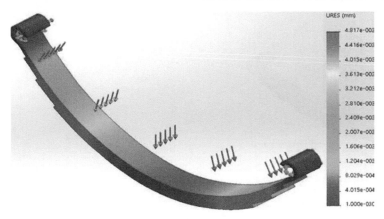

Figure 108.10 Resultant displacement
Source: Author

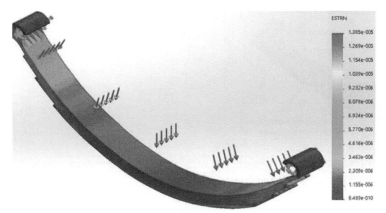

Figure 108.11 Equivalent strain
Source: Author

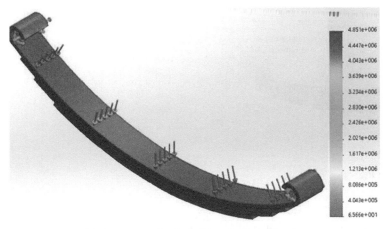

Figure 108.12 Factor of safety
Source: Author

material can withstand without getting failed. It provides a rough idea about crack initiation, its propagation, or its tendency to fail for a specific material. Fatigue analysis provides data for "fatigue damage", "fatigue life" and "fatigue load factor". Simple fatigue analysis can be performed using SolidWorks simulation software, before starting with the fatigue analysis analysts must perform static analysis because fatigue analysis requires several data which will be derived from the static analysis of that body [20]. The point of concentration for both models will be-

Fatigue Damage

Fatigue damage is a criterion that deals with crack initiation and its growth when a body is subjected to the condition of corrosion, fatigue loading, or tensile loading. For Model 1 Figure 108.13 shows the fatigue

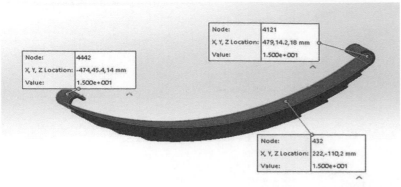

Figure 108.13 Model 1: Contour- fatigue damage
Source: Author

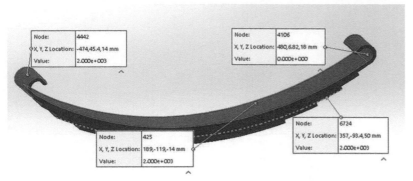

Figure 108.14 Model 2: Contour- fatigue damage
Source: Author

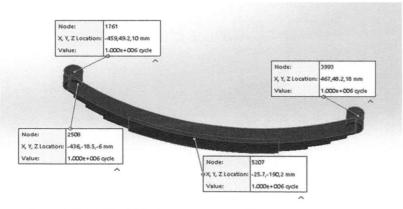

Figure 108.15 Model 1: Contour- fatigue life
Source: Author

damage contour at the nodes 432, 4121, and 4442 for the end clip, master leaf, and center plate, the fatigue damage is of value 1500 which is the maximum and at the bottom it minimum. Whereas Figure 108.14 shows the same for Model 2, which has a maximum fatigue damage value of 2000.

Fatigue Life

The finite no. of cycles i.e., loading stress that a body offers without failure due to natural causes is known as the fatigue life of a body. Figures 108.15 and 108.16 shows the fatigue life contour for Model 1 and Model 2 at the nodes 1761, 2508, and 5207 named as end clip, master leaf, and center plate, the maximum fatigue life shown by the leaf spring is 1000000 cycles in both models.

Fatigue Load Factor

The fatigue behavior of the adhesive joints is affected by the load factors which include amplitude, mean, frequency, and stresses. I mean stress or amplitude of stress increases, which results in a decrease in fatigue

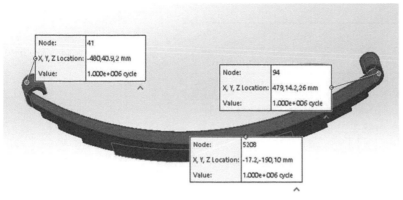

Figure 108.16 Model 2: Contour-fatigue life
Source: Author

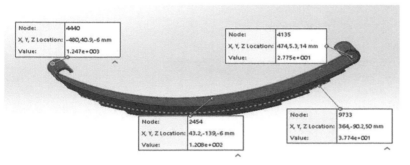

Figure 108.17 Model 1: Contour- load factor
Source: Author

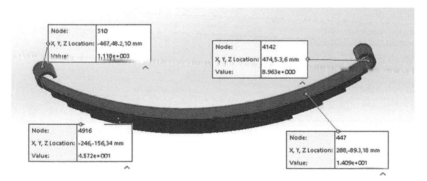

Figure 108.18 Model 2: Contour- load factor
Source: Author

life. Figure 108.17 shows the fatigue load factor contour at the nodes 2454, 4135, and 9733 i.e., end clip, master leaf, and joints of leaves, the min load factor is 37.74 in between leaves and min load factor of 1247 at the outer surface of end clip. While for Model 2, Figure 108.18 shows the contour for fatigue load factor respectively and its minimum value is 8.963 and maximum is 1180 at the same location.

Results Comparison

For a Tata Ace vehicle, the total load acting in it is 1750 kg which is taken as 2000 kg (after considering safety load). The amount of normal force acting on the master leaf is 2452.5 N. Leaf springs have a span length of 950 mm, 4 no. of leaves, a thickness of 10 mm, and a width of 80 mm. Analytical calculation results in maximum bending stress of 213.82 MPa and total deflection of 22.44 mm and 23.11 mm for Model 1 and Model 2 leaf springs respectively. In static analysis, a load of 5000 N is applied which results in stress development, resultant displacement, strain energy, and factor of safety of 237.7 MPa, 25.6 mm, 138.2 mm, and 6.7 respectively for Model 1 and 225.24 MPa, 24.3 mm, 137.5 mm and 8.2 respectively for Model 2. Table 108.3 shows the comparison of all values of evaluated parameters of Model 1 and Model 2 for better understanding.

Table 108.3 Static analysis comparison

Property	Model 1		Model 2	
	MIN	MAX	MIN	MAX
Stress (MPa)	5.597×10^{-5}	237.7	5.827×10^{-5}	225.24
Displacement (mm)	0	25.6	0	24.3
Strain	0	138.2	0	137.5
FOS	6.7		8.2	

Source: Author

Table 108.4 Fatigue analysis comparison

Property	Model 1	Model 2
Fatigue Damage	1500	2000
Fatigue Life	1000000	1000000
Load Factor	Min- 37.74 and Max- 1247	Min- 8.693 Max- 1180

Source: Author

Table 108.5 Experimental vs Model 1 static analysis

Property	CAE	Analytical	% Error
STRESS (MPa)	237.7 MPa	213.2 MPa	11%
STRAIN	23.6 mm	22.44 mm	13%

Source: Author

Table 108.6 Experimental vs Model 2 static analysis

Property	CAE	Analytical	% Error
Stress (MPa)	225.4 MPa	213.2 MPa	5%
Strain	24.3 mm	23.1 mm	4.9%

Source: Author

Table 108.5 and Table 108.6 illustrates the comparison of CAE and Analytical results for Model 1 and Model 2 respectively.

Considering fatigue analysis which determine the no. of cycles that a body performs before getting failed. Fatigue analysis evaluates fatigue damage, fatigue life, and load factor. For Model 1 maximum fatigue damage is 1500 for a life of 1000000 cycles having min load factor of 37.74 and a maximum load factor of 1247 between leaves and end clip outer surface respectively. Table 108.4 shows the comparison of fatigue analysis for Model 1 and Model 2. Model 2 has maximum fatigue damage of 2000 for 1000000 cycles with a maximum load factor of 1180. FEA static structural results are compared with the analytical results with the mean of percentage error. To obtain the percentage, the ratio of the difference in analytical and CAE results to analytical is calculated and multiplied by 100.

For static analysis results- In Model 1, analytical bending stress is 237.7 MPa and deflection is 25.6 mm whereas CAE stress is 213.2 MPa and deflection is 22.44 respectively providing a percentage error of 11% and 13% for both.

In Model 2, we have a percentage error of approx. 5% for both the bending stress and deflection as the analytical values are 225.4 MPa and 24.3 mm and the CAE value is 213.2 and 23.1 mm for stress and deflection respectively.

Conclusion

In this research, a comparative study of two types of leaf spring material is being performed using the computer-aided engineering (CAE) tool. The results of finite element analysis are verified with the analytical results to observe the variations in different parameters under study. From finite element analysis (FEA) and analytical results are summarised as follows,

- Static analysis results show that Model 1 offers high stress of magnitude 237.7 MPa but a large amount of this gets transferred in between the leaves of the spring. On the contrary side, Model 2 which is made up of composite material shows a very small amount of stress acts in between leaves, and maximum stress is induced to resist the applied load.
- Model 2 depicts minimum deflection as compared to Model 1 for the same amount of load (5000 N) transferred.
- Model 2 portrays a high factor of safety (8.2) while Model 1 shows a low factor of safety (6.7).
- Comparison of FEA and analytical results illustrates that in Model 1 stress varies by 11% and deflection varies 13st% while in Model 2 stress and deflection vary only 5% from analytical data.
- Model 2 has more fatigue damage life (2000) as compared to Model 1 have a fatigue damage life of 1500.

Finally, the current study concludes that a composite material leaf spring with four material combinations has a high tendency to replace the conventional material leaf spring for a Tata Ace mini truck. Composite material has a low density as mentioned in Table which leads to reducing the vehicle weight results without affecting the vehicle body strength, stiffness, and loading carrying capacity. Therefore, the suggested leaf spring in the vehicle offers a smooth, cushioned, and comfortable ride to the passengers by absorbing a large amount of vibration and shocks.

References

[1] IS 1135 (1995): Springs - Leaf Springs Assembly for Automobiles [TED 21: Spring], Fifth Revision, BUREAU OF INDIAN STANDARDS.
[2] Kumar, P. and Matawale, C. R. (2020). Analysis and optimization of mono parabolic leaf spring material using ANSYS. Materials Today: Proceedings. 33, 5757–64. doi: 10.1016/j.matpr.2020.06.605.
[3] Umanath, K., Prabhu, M. K., Yuvaraj, A., and Devika, D. (2020). Fabrication and analysis of master leaf spring plate using carbon fiber and pineapple leaf fiber as natural composite materials. Materials Today: Proceedings. 33(P1), 183–8. doi: 10.1016/j.matpr.2020.03.790.
[4] Ashwini, K. and Rao, C. V. M. (2018). Design and analysis of leaf spring using various composites-An overview. Materials Today: Proceedings. 5(2), 5716–21. doi: 10.1016/j.matpr.2017.12.166.
[5] Varma, N., Ahuja, R., Vijayakumar, T., and Kannan, C. (2021). Design and analysis of composite mono leaf spring for passenger cars. Materials Today: Proceedings. 46, 7090 8. doi: 10.1016/j.matpr.2020.10.073.
[6] Chavhan, G. R., and Wankhade, L. N. (2019). Experimental analysis of E-glass fiber/epoxy composite-material leaf spring used in automotive. Materials Today: Proceedings. 26, 373–7. doi: 10.1016/j.matpr.2019.12.058.
[7] Guduru, R. K. R., Shaik, S. H., Tuniki, H. P., and Domeika, A. (2021). Development of mono leaf spring with composite material and investigating its mechanical properties. Materials Today: Proceedings 45, 556–61. doi: 10.1016/j.matpr.2020.02.289.
[8] Mallesh, B., Gupta, B., Kumar, S. K., and Jani, S. P. (2021). Modeling and analysis of leaf spring with a different type of materials. Materials Today: Proceedings 45, 1945–9. doi: 10.1016/j.matpr.2020.09.223.
[9] Kamboj, M., Chetry, A., Kurien, C., and Srivastava, A. K. (2020). Computational study on the potential of aluminum alloy as a candidate material in automotive leaf spring. Australian Journal of Mechanical Engineering. 00(00), 1–12. doi: 10.1080/14484846.2020.1842617.
[10] Venkatesh, A. P., Padmanabhan, S., Allen Rufus, C., Lukmaan, H. M., and Rahman, A. R. A. (2021). Exploration of fatigue and modal analysis on mono leaf suspension made by natural composite materials. Materials Today: Proceedings. 47, 4262–7. doi: 10.1016/j.matpr.2021.04.568.
[11] Kong, Y. S., Abdullah, S., Omar, M. Z., and Haris, S. M. (2016). Failure assessment of a leaf spring eye design under various load cases. Engineering Failure Analysis. 63, 146–59. doi: 10.1016/j.engfailanal.2016.02.017.
[12] Arora, V. K., Bhushan, G., and Aggarwal, M. L. (2017). Enhancement of fatigue life of multi-leaf spring by parameter optimization using RSM. Journal of the Brazilian Society of Mechanical Sciences and Engineering. 39(4), 1333–49. doi: 10.1007/s40430-016-0638-z.
[13] Arora, V. K., Bhushan, G., and Aggarwal, M. L. (2014). Fatigue life assessment of 65Si7 leaf springs: a comparative study. International Scholarly Research Notices. 2014, Article ID 607272, 11. http://dx.doi.org/10.1155/2014/607272
[14] Kumar, K. and Aggarwal, M. L. (2015). Fatigue life prediction: a comparative study for a three-layer en45a parabolic leaf spring. Engineering Solid Mechanics. 3(3), 157–66. doi: 10.5267/j.esm.2015.5.003.
[15] Nadargi, Y. G., Gaikwad, D. R., and Sulakhe, U. D. (2014). A performance evaluation of leaf spring replacing with composite leaf spring. International Journal of Mechanical and Industrial Engineering. 3(3), 157–60. doi: 10.47893/ijmie.2014.1147.
[16] Kumar, M. S. and Vijayarangan, S. (2007). Static analysis and fatigue life prediction of steel and composite leaf spring for light passenger vehicles. Journal of Scientific & Industrial Research. 66(2), 128–34.

[17] Pawar , S. B. and Ghadge, R. R. (2015). Design & analysis of multi steel leaf spring. International Engineering Research Journal (IERJ) Special Issue 2. 4468–4474, 2015, ISSN 2395-1621

[18] Verma, A., University , S. G., Dhanbad, I. S. M. and Indore, S. (2016). Design and simulation of leaf spring for TATA-ACE mini loader truck using FEM. Imperial journal of interdisciplinary research. 2(10), 1421–7.

[19] Singh, H. and Brar, G. S. (2018). Characterization and investigation of mechanical properties of composite materials used for leaf spring. Materials Today: Proceedings. 5(2), 5857–63. doi: 10.1016/j.matpr.2017.12.183.

[20] Khaleel, H. H. and Sahlani , A. A. (2018). Modeling and analysis of leaf spring. International Journal of Mechanical Engineering and Technology (IJMET). 9(6), June 2018, 48–56, Article ID: IJMET_09_06_007

Chapter 109

Assessment of a LM procedure using the ISM and AHP models in indian MSMEs

Neha Gupta[a] and Dr. Apurva Anand[b]

Mechanical Engineering Department, Babu Banarasi Das Institute of Technology and Management, Lucknow, India

Mechanical Engineering Department, Maharana Pratap Engineering College, Kanpur, India

Abstract

Lean manufacturing (LM) is a technique that is frequently employed in manufacturing businesses to minimise waste and increase productivity. It is crucial to measure the changes with accurate measures after the lean principles have been adopted in an organisation. Therefore, decision-makers need a methodology that they can use to assess how well the LM philosophy has been applied. There have been many methods employed in the past to evaluate the application of lean practices. One of these techniques is the lean radar chart, which shows the discrepancy between the actual and desired performances graphically. To foster a healthy competitive atmosphere, numerous organisations evaluate the work of various departments by rewarding them. We describe a unique analytic hierarchy process (AHP)-based LM process. The process is hierarchical and iterative. The application of LM principles in Indian MSMEs is examined in this article. It specifically looks at how much lean can be adopted considering the numerous financial restrictions that Indian enterprises face in the current business climate.

Here, a novel approach was utilised to assess how well MSMEs implemented lean using interpretative structural modeling (ISM) and AHP models. These were ranked after determining the weights for that criterion. From a group of six MSMEs, the best lean implementation candidate was selected using the AHP model. Additionally, fuzzy was employed to compare those criteria weights in order to support the AHP results. This paper offers a novel approach to investigate the potential effects of a collection of 11 criteria on the success of lean adoption. The industry gains insight into how lean implementation may be made more responsive from the analysis of the AHP and ISM approaches.

Keywords: Analytical hierarchy process (AHP), fuzzy AHP, India, interpretive structural modeling (ISM), lean implementation, lean manufacturing (LM), MSMEs

Introduction

In current research, a society of manufacturing engineers (SME) that deals with the production of bicycle parts has its lean performance evaluated. The concerned SME has four main manufacturing cells, each of which is devoted to the production of a certain family of products and produces five different product variations. The SME has only been operating for the last six years, making it relatively new. Lean manufacturing (LM) has been implemented in a number of ways. Lean radial charts are sometimes used in businesses to determine the discrepancy between benchmark and actual performance. In the current research, a SME that deals with the production of bicycle parts has its lean performance evaluated. The concerned SME has four main manufacturing cells, each of which is devoted to the production of a certain family of products and produces five different product variations. The SME has only been operating for the last six years, making it relatively new. LM has been implemented in a number of ways. Lean radial charts are sometimes used in businesses to determine the discrepancy between benchmark and actual performance.

LM, also known as lean production, is the most well-known method for making such an improvement and it was developed from the Toyota production system. In order to build an efficient, high-quality system that produces finished goods at the rate of customer demand with little to no waste, lean manufacturing integrates just-in-time, quality systems, supplier management, and many other management strategies [10].

The following are the most famous lean manufacturing tools:

1. Cellular manufacturing essentially divides processes into units with all required resources.
2. Just-In-Time (JIT), which "pulls" manufacturing when a consumer originates a demand and transmits it backward.
3. The signal system known as Kanbans is utilised to implement JIT production.
4. Total Preventive Maintenance (TPM), in which staff members perform routine equipment maintenance to avoid breakdowns.
5. System engineers keep striving to speed up machine setup times through setup time reduction.
6. Total Quality Management (TQM), a system of ongoing improvement that emphasises customer's requirements and makes use of participative management.
7. 5S, which emphasises standardised operating procedures and good workplace administration.

[a]er.mech.neha@bbdnitm.ac.in, [b]apurva2050@yahoo.co.in

DOI: 10.1201/9781003450252-109

Basically, the main objective of LM tools is cost reduction through waste elimination. LM has identified seven waste sources: overproduction, waiting times, over processing, inventory, motion, and scrap [6]. Logistics, which can be defined as process planning, organization, and management aimed at optimizing the flow of material and information within and outside the company to maximise profit, plays a crucial role in the "lean" change even though businesses are involved in it throughout all their various departments, from marketing to control [3]. Internal logistics in discrete manufacturing are the main topic of this essay. Value stream mapping (VSM) [1] has distinguished itself among the literature's available LM technologies as a straightforward and efficient method for locating waste and reducing it while enhancing overall operational management.

However, it has been demonstrated that linear flow manufacturing systems can benefit from the VSM method [2]. On the other hand, it is difficult to map a complex production system and identify its key problems [4]. In this research, we demonstrate that the VSM may be successfully supported by the unified modeling language (UML) formalism [5] and the analytic hierarchy process (AHP) decision making approach [9] in order to evaluate the macroscopically identified improvement activities.

We describe a six-step LM process in particular.

- In the first step, UML is used to completely define the manufacturing system.
- In the second step, value stream analysis (VSM) is used to visualise the total process activities and discover system inconsistencies. if any anomalies are discovered.
- In the third stage, AHP evaluates the most effective method to eliminate of them.
- In the fourth step, the VSM approach is utilised to develop the view of the future of the value stream, with a reduced influence of non-value added activity, based on the results of the AHP.
- In the fifth step, UML is used to redesign the manufacturing system and then get rid of the major system flaws.
- The planned adjustments are finally put into action. The process is iterative to ensure that development is made consistently.

The proposed method provides a number of advantages over the conventional VSM method, including:

1. Complex production systems can use it.
2. It concurrently results in the process activities being detailed (due to the usage of the UML formalism) and the total value stream being presented in a high-level picture (using the VSM conciseness).
3. It allows you to quantify systemic irregularities (thanks to the AHP methodology).

The proposed method is used in a case study to examine the production process of OM Carrelli Elevatori SpA, a top producer of forklift trucks based in Bari's mechanical industrial sector (Italy). The flow of partially finished pieces that enter the forklift truck assembly line after being painted is referred to as the chosen production process. Internal logistics currently produces a lot of waste, has a lot of missing parts, and does not fully adhere to ergonomic standards in the working environment. The company's three major performance metrics have improved as a result of using the suggested technique in this case study, demonstrating the efficacy of the strategy.

Toyota created the idea of LM in the 1950s to compete with American productivity. The goal was to find and get rid of everything wasteful. Each process's inefficiencies and extraneous components are sought for and isolated by LM. This is simply accomplished by keeping careful tabs on the workers, tools, supplies, and rate of production. Anything that is wasteful merely raises costs and reduces consumer pleasure; it does nothing to generate value.

As a result, strict redundancy minimization reduces costs. As strategic elements of lean that businesses may use in the effective implementation of lean, Maike [28] regarded leadership commitment, employee autonomy, greater transparency, cultural fit, and successful implementation of practices to support the operational and tactical aspects of lean. Nash and poling are adamant proponents of bringing the community together and giving them the power to make and implement choices (2007).

Lean is a strategy that many businesses use to cut costs and boost production by removing waste from their operations. Manufacturing flexibility is crucial for agility and can be increased with the right application of LM techniques [30]. Where MSMEs are still in their infancy, lean implementation on a big scale can be adequate and successful. Due to its inherent advantages, such as cheap capital requirements, high employment generation, decentralization of industrial activity, exploitation of local resources, and expansion of the entrepreneurial base, MSMEs play a significant role in the socioeconomic growth of the Indian economy.

MSMEs hold a prominent position in the Indian economy. The majority of SMEs, however, struggle with long cycle times and inadequate systems for forecasting and planning production, which results in problematic inventory control systems and expensive costs that restrict the expansion of competitiveness levels [36]. This is due to the oversupply of outdated inventory in warehouses and the subpar customer service provided [23].

Since the bulk of research on this subject which has been published in the literature have centered on advanced economies, Singh [36, 37] addressed the need to study, analyse, and discuss the competitiveness of SMEs operating in developing countries e.g., [31, 20]. Large-sized firms are now trying to apply LM globally in order to retain their competitiveness. These days, a company's performance has a significant influence on the success of any firm. The best performance can be reached by utilizing LM. It is essential to assess the company's adoption of the critical success factors revealed by lean titans. Consequently, it is a multi-criteria decision-making challenge (MCDM).

In this study, the effectiveness of MSMEs' lean implementation is assessed using combinations of ISM and the analytical hierarchy process (AHP). In 1974, Warfield created ISM. ISM sought to organise system-related components in a hierarchical relationship [42]. There were identified eleven criteria, and weights were obtained. Using AHP, industries were ordered according to the weights. By identifying numerous enabling elements that can increase the agility of LM, this study aims to create a model. The approach used to create such a model entails a case study, field study, conversations with consultants and experts in the subject, and a thorough literature analysis. The study's results are analysed using two techniques, ISM and AHP. The new method for assessing the impact of key success elements for lean is presented in this study.

Literature Review

In the 1950s, W. Edwards and Frederick Taylor, two Toyota industrial engineers, created the LM tenets. Toyota improved and expanded this idea further to produce the Toyota Production System (TPS). The manufacturing industry had historically been looking for processes that may achieve better results, such as more throughput, superior quality, shorter cycle times, less firefighting, extra efficient operations, and lower operational expenses (Pagatheodrou, 2005). The Japanese manufacturing industry struggled in the years after the Second World War due to a lack of labor, financial, and raw material resources. Eiji Toyoda and Taiichii Ohno of the Toyota Motor Company in Japan created the concept of TPS or LM to maximise and effectively use existing resources (in the USA).

This idea centered on reducing waste, which was described as any number of resources spent in producing the finished product but not adding to the benefits to the client. In other words, LM relates to producing things at the right time, or when they are most needed (Womack et al., 1990). It reduces resource and time waste so that labor and shop floor efficiency are maximised and higher-quality manufacturing is accomplished at a lesser cost (Todd, 2000). Any business that wants to dominate the global market must transform to keep up with emerging technologies (Ohno, 1997; Monden, 1998). The practice that is most appropriate for your organization must be chosen in order to successfully implement lean (Wen and Chen, 2009). In a situation when numerous, various decision-making or selection problems must be resolved, the multi attribute decision making (MADM) method may be helpful.

A well-defined approach called interpretive structural modeling (ISM) helps a business to comprehend the intricate connections between the various components present in complex circumstances. The objective is to utilise ISM to rank the main criteria and sub-criteria for industry selection as well as the relationships between them. ISM is a tried-and-true strategy for determining connections between particular components that characterise a problem or an issue [25]. It is utilised to pinpoint complex and individualised issues. For the purpose of increasing the performance of the supply chain, many researchers have adopted the ISM technique [21, 39, 34] (Sage, 1977). A number of factors were created by Singh et al. (2007) to increase the competitiveness of MSMEs.

Even if the measurement is made at a nominal scale, the evaluation criteria still need to be quantifiable [27]. The prioritization of the criteria and examination of the interconnections between them were done using modified interpretative structural modeling (MISM). Understanding and ranking the links between various risk factors is challenging. As a result, they were grouped together, and an ISM computer-assisted learning technique was used to create a structural relation and illustrate how risks interacted with one another. This technique works by imposing order and direction on the difficulty of correlations between variables [38].

ISM is used to create a summary of the intricate relationships between parts for situations requiring complicated decision-making [35, 22] (Warfield, 1974). In order to better comprehend a company's total risk profile, Gorvett and Liu [22] applied ISM methodologies. This method was used by Attri [17] to identify and examine how the enablers for the implementation of total productive maintenance interact with one another (TPM).

For the purpose of identifying and analyzing the obstacles to TPM implementation, Attri [17] used the ISM technique. Raj [33] used the ISM approach to analyse how the variables impacting FMS flexibility are related to one another. These models assist in identifying the crucial element involved in a problem or issue. A strategy for handling the issue may be devised after the key factor or component has been identified. ISM is also easy to use and can reach a wider audience.

The most popular MADM technique, known as the analytical hierarchy process (AHP), entails creating a set of options and a set of shared goals. The industries are ranked so that the best industry can be chosen. ISM and fuzzy AHP were both employed by Parthiban [32] and Yang [41], with the former used to determine the relationship between sub-criteria and the latter to determine the relative weights of each criterion. The ability of an option to maximise the achievement of the goals established determines which is the best choice? AHP's compensating strategy and inability to handle linguistic variations are overcome by using a form of AHP called fuzzy AHP [26].

Research Gap

This study should try to close this gap. According to the current literature review,

- The adoption of LM using analytical hierarchy process methods has been the subject of numerous studies in the automotive, electrical, and electronic industries, but only a small number of studies have focused on its application in small- and medium-sized manufacturing industries.
- The effect of lean approaches on performance in production improvement, smaller and medium-sized manufacturing industries throughout Uttar Pradesh state is not well supported by research.
- Almost no research has been done in this crucial field of production philosophy. The small and medium sized manufacturing industries within Uttar Pradesh have yet to investigate and adopt this lean approach employing AHP techniques.
- Only a small number of research focused on the issues of inadequate knowledge of lean manufacturing employing AHP methods, non-compliant business owners about modern manufacturing concepts and cutting-edge technology, and their resistance to upgrade to contemporary and new technologies.

The current study will thoroughly and methodically examine sustainable LM for small and medium-scale manufacturing industries throughout Uttar Pradesh employing AHP methods and provide an assessment of their improving performance.

LM Procedure

This proposed LM procedure is hierarchical and based on six steps, reported in Figure 109.1.

- First, a comprehensive UML formalism explanation of the manufacturing process behavioral model and its related activities is provided. Because UML is a unified and well-known formalism that can be easily integrated into any simulation software, it was chosen to generate an effective model of the process under investigation using standard, pertinent, and effective diagrams.
- The second stage of the VSM approach is to integrate the flow of materials and information of the specified value stream while identifying system anomalies, or activities that don't add to the production process's value.
- Third, the optimal plan of action for solving the identified system anomalies is identified using the AHP decision-making approach.
- Fourth, the VSM approach is used to construct the future vision of the value stream, with a diminished impact of non-value-added operations, based on the results of the AHP procedure.
- Fifth step, the manufacturing system's comprehensive activity diagram is modified in the fifth step using UML formulation to fix the major anomalies.
- Sixth and final step of the procedure, the planned enhancements in the manufacturing process are performed in the sixth and final step of the process.

For continuous improvement, the process is iterative and can be performed recursively at least once a year. The proposed lean methodology would ultimately produce a stable and balanced flow that can more effectively meet final customer needs without compromising output. In the paragraphs that follow, we further describe the technique phases.

Step 1: The existing manufacturing system's UML model

UML [5], a graphic and textual modeling formalism helpful for analysing and defining organizations from both structural and behavioral aspects, forms the core of the first step of the proposed LM process. In

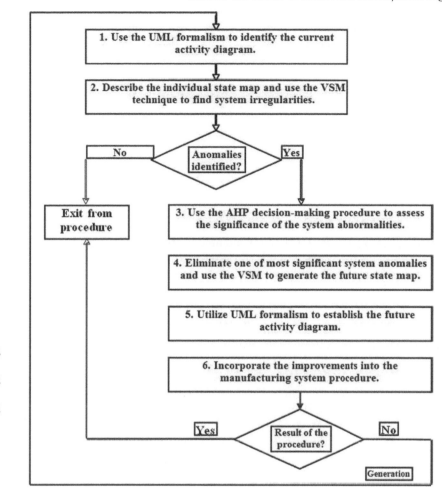

Figure 109.1 An illustration of the suggested LM process
Source: Author

particular, UML enables the standard and thorough description of the process activities and the associated players. A system can be described in UML using activity diagrams that give a broad picture of the system dynamics from the behavioral point of view.

The initial activity is represented by a solid circle, the final activity is represented by a bull's-eye symbol, and other activities are represented by a rectangle with rounded edges. Arcs are used to represent flows connecting activities. Forks and joins are used to represent concurrent actions that start and end at the same time. Decisions are used to represent alternative flows and are represented by a horizontal split.

Moreover, different system members may be involved in an activity. As a result, the diagram is divided into columns or swim lanes and partitions or swim lanes are utilised to indicate which actor is accountable for which activities.

Step 2: Evaluation of the current manufacturing system's value stream

The second phase of the technique involves getting a broad overview of the system to pinpoint the non-value-adding activities after the previous step has documented the specific production system operations in detail. For strategic/tactical planning, we use VSM [1], a widely used LM approach that entails a visual mapping of all processes and activities that go into producing a piece or part. The detection of waste is, in essence, the purpose of the VSM analysis. The VSM methodology helps businesses visualise the whole flow of a single product and identify any areas where the production process is slowed down. In the related studies, there are various instances of VSM applications to discrete manufacturing techniques (for instance, see [8]).

The following are typical components of the VSM method [7]:

- Finding the product or service to map involves determining the start and end points of the value stream that must be enhanced.
- The value stream's present position (state), containing all relevant information, is described in the current state map, which is created (tasks, costs, time for each task, delays in between stages of the process,

Table 109.1 AHP scale of comparisons

Intensity of importance	Definition
1	Equal significance
3	Not terribly important
5	Strong relevance
7	Extremely important
9	Really significant
2, 4, 6, 8	Values that fall in the middle of adjacent judgments

Source: Author

etc.). A team watches the production processes to create the current state map and records information (cycle times, buffer sizes, staffing needs, etc.) that is then represented in the map with standardised icons.

• Anomalies in the system are found by qualitatively analyzing the existing state map to see if each process activity is adding value. Bomb-style icons are employed to indicate the flow irregularities that have been found.

Step 3: Analytic hierarchy process evaluation of the production system anomalies

We now outline the third stage of the LM technique, which allows for a quantitative evaluation of the state map considering the system anomalies found in step one. We use AHP, a well-known multi-objective decision-making technique [9] for rating a variety of alternatives in accordance with a set of competing criteria of varying degrees of relevance, to carry out this assessment. We use AHP to isolate the most significant system anomalies from those identified at the preceding level. AHP does, in fact, show improved accuracy over other decision-making procedures because it is based on alternatives pair-wise comparison. The following components make up the AHP technique [9].

1. *Arranging the decision-making process in a hierarchy.* For the first AHP level, "Effectiveness" is chosen as the overarching goal. As a result, the second AHP level, which is composed of the n contributing criteria, is created. The selected n criteria, or Key Performance Measures (KPI), are useful metrics to assess the discovered defects in the information and material flow, including lead time, inventory, etc. The third AHP level is also based on the number of system abnormalities that have been identified and that we wish to compare to the benchmarks or KPIs defined in the second AHP level.

2. *Generating a collection of comparison matrices for pairs.* The contribution of all the upcoming layer selections to the current level must be estimated for every level of the AHP hierarchy (from bottom to top). To this aim, a pair wise comparison matrix **CM0** of dimension $n*n$ is required evaluating the importance of the n criteria of the second level in reaching the top objective and n additional matrices **CMi** of dimension $m*m$ with $i=1,...,n$ express each how important the removal of each identified anomaly of the third level is with respect to the i-th criterion of the second and higher level. Each element **CMi** *(j k)* of **CMi** with $i=0,...,n$ represents the relative importance of the j-th anomaly compared to the k-th one and is determined interviewing the decision maker and associating to such an importance an integer from 1 to 9 (see Table 109.1). Less important criteria are defined by reciprocals, so that

   CMi *(j, k) = 1 /* CMi *(k, j)*

3. *Determining priorities and normalised performances from comparisons.* For each comparison matrix **CMi** with $i=0,...,n$ determine the maximum eigen value and the corresponding eigenvector v_i with $i=0,...,n$. Obviously, v_0 includes n elements, while all the other eigenvectors v_i with $i=1,...,n$ include m elements. Compute the priorities vector normalizing eigenvector v_0 of **CM0** as follows:

   $$\mathbf{P} = v_0 / \Sigma_{j=1}^{n} v0 = [p_1......p_n]^T$$

   Where each element p_i with $i=1,..., n$ of **P** represents the normalised importance degree of the i-th criterion. Similarly, compute the normalised performance values of the alternatives against each i-th criterion as follows:

   $$CRIT_i = v_i / \Sigma_{i=1}^{n} v0 = [CRIT_1.........CRIT_j]^n \text{ with } i=1.........n$$

Establishing the decision-making model. Find the weighted sum of the standardised performance rating against each criteria weighted by the appropriate priority for each discovered system anomaly j with j=1,...,m:

$$PI_j \_ AHP = \sum_{i=1}^{n} Pi\ CRITij$$

The impact or significance level of the j-th anomaly with regard to the set of variables or KPI is quantified by PIj AHP.

5. *Ordering the options.* The system anomalies that were found are rated according to their overall AHP index, or PIj AHP. The anomaly with the maximum index PIj AHP, obtained by is obviously the top anomaly (3). Hence, the need for an early resolution to the problem increases with the value of such an index, or the higher the ranking of the anomaly position.

Step 4: Redesigning the manufacturing system, steps 4-5 through 6

Once the manufacturing system anomalies have been evaluated by the previous stage, the technique's next step is to restructure the overall system picture and reduce (some of the top ranked) non-value-adding operations. We use the VSM method to produce a future state map in order to accomplish this. The ideal future condition of the production system is illustrated on a map in the fourth step of the process (Figure 109.1). The future state map shows how the stream's process should be in its ideal state when the (top) crucial points have indeed been eliminated, ranked, and identified in earlier phases of the LM process.

In the fifth phase of the process (Figure 109.1), the sketched vision of the value stream is expressed using the UML formalism, and a unique activity diagram of the manufacturing system is described, outlining all the activities in the updated system. The envisioned manufacturing system's UML model is produced as a result. As a result, the manufacturing system's behavior model has been changed using the UML framework. In the sixth stage of the procedure (see Figure 109.1), the chosen improvement plans are now implemented in the manufacturing system, allowing for the achievement of the anticipated future state. Naturally, the proposed design and manufacturing process is iterated in order to ensure ongoing development and to validate the deployment approaches.

Methodology

Interpretive structural modeling

J. Warfield originally put forth ISM in 1973 to study the intricate social and economic systems. Individuals or groups can establish a road map for understanding the various links between the criteria found in challenging situations by using a computer-based learning process. The core idea is to use expertise and practical experience to solve a complicated system with many subsystem components before creating a multilayer structural model. A reach ability matrix is essentially created by choosing specified criteria, comparing them in a predetermined binary relationship, and then using the results of the comparison.

When analyzing ISM, the group determines whether and how the system's elements are connected. Ahuja [16]. It is organised and built on the basis of the connection, and the ultimate structure is derived from a complicated collection of system variables. It is also a model because a directed graph model is used to illustrate the ultimate relationship.

Steps for solving ISM technique are

1. Determining the essential components using the expert survey.
2. Creating a connection between the paired items that will be used to test them.
3. Building an element-based structural self-interaction matrix (SSIM) that shows the relationships between the elements pair-wise.
4. By employing binary relations and transitivity, an ISM tenet that states that "If element A is related to element B and element B is related to element C, then element A is related to element C," it is possible to evolve a reach ability matrix from the SSIM.
5. The matrix is used to split the elements into levels.
6. Incorporating items into a hierarchy to transform the resulting levels into an ISM-based model.
7. Examining the model to look for errors and figuring out the elements' normalised weights.

Analytical hierarchy process

According to Ravi et al., AHP is a well-liked multiple criterion decision-making method that integrates both qualitative and quantitative criteria [34]. It rates potential providers using a hierarchical system [21, 39]. It is also a method for group decision-making that aids in determining the best criteria for reaching a goal.

So, by creating a sequence of one-to-one comparisons, it offers the finest method to decrease qualitative and quantitative complex limitations. In multiple sourcing, the AHP is widely used to assess supplier portfolios [19]. It explains why decision-makers chose certain options in addition to assisting them in choosing the better ones [24]. The approach which was undertaken to apply AHP in order to compare and rank the various manufacturing cells based on the LM assessment is described here.

- *Identification of Parameters for Lean Assessment-* The three parameters were considered in order to compare and rank the four manufacturing cells on the basis of lean assessment: material flow, visual control, and metrics. Each parameter for lean assessment is compared against every other parameter. Now consider the equation $[A. x = \lambda max. x]$ where, A is the comparison matrix of size n×n, for n criteria, also called the priority matrix. X is the Eigenvector of size n × 1, also called priority vector. λmax is the Eigen value. Checking for Consistency Further, we need to find the ranking of priorities, i.e., the Eigenvector. The first step is to normalise the column entries by dividing each entry by the sum of the respective columns. Then we take the overall row average.
- *Checking for Consistency-* The assessments are then compared to substantial samples of wholly arbitrary judgments in order to determine their consistency, which is measured by the consistency ratio (CR). The rationality of the decision maker is assumed in AHP evaluations, therefore if A is chosen over B and B is preferred over C, then A is also preferred over C. The judgments are unreliable and the exercise must be redone if the CR value is larger than 0.1 since they are too close to randomness to be considered reliable.
- *Calculation of Consistency Ratio-* The next stage is to calculate λmax so as to lead to the consistency index and the consistency ratio. Consider $[A. x = \lambda max. x]$ where, x is the Eigen vector.

Consistency index (C.I) is found by the formula-

C.I. = (Principal Eigen Value – Size of Matrix) / (Size of Matrix-1) = $(\lambda_{max} - n) / (n-1)$

Consistency Ration (C.R.) = (Consistency Index) / (Random Index)

AHP calculations can be done in three stages:

- The hierarchy's construction.
- Analyses of matrices used for pair wise comparisons.
- The computation of the pair-wise comparison tables' relative priority weights.

The next stage is to compare each parameter of the lean performance assessment pair by pair across the company's four manufacturing cells. AHP can be utilised with individual lean radar charts as a multi-attribute decision-making model to satisfy the fundamental goal of comparing manufacturing cells on some shared uniform criteria.

The ranking of manufacturing cells and the final overall weights are shown below. According to current assessments, the fourth manufacturing cell is the best, while the third manufacturing cell is the worst and needs the most development.

Steps for solving AHP technique are

1. Decide on the objective and the standards that will help you achieve it.
2. Determine the suggestions provided as options to achieve the objective. All levels that come after the criterion levels are included in this.
3. The relevant weights for each comparison table should be found. For example, to compare the relative importance of one criterion to another, a scale from 1 to 5 is employed.
4. Create numerous square matrices for each criterion by comparing each feasible pair of alternatives with their ranks under each criterion.
5. Calculate the composite relative priority for each alternative and the steps that are illustrated in table.

Fuzzy AHP

The AHP has been used to tackle issues requiring numerous decision-making criteria [40]. However, due to differences in the decision maker's response, a straightforward and pair-wise comparison using a conventional AHP would not be able to fully capture the correct choice. Fuzzy logic is consequently added to the pair-wise comparison stage of the conventional AHP.

Steps for solving Fuzzy AHP technique are

In order to create three triangular fuzzy numbers, the evaluation of the success factors must first be modified. The triangular fuzzy numbers are then used to construct the AHP comparison matrices based on pair wise comparison technique. The major weights of the success factors and industries can be found using fuzzy AHP. Triangular fuzzy numbers (0-9) are used to represent arbitrary pair-wise assessments of success variables and to take ambiguity into consideration (fuzziness).

1. Evaluate the performance rating.
2. Create a success factor to help you determine the weights significance.
3. Create the matrix of fuzzy comparisons.
4. Determine the number of ambiguous elements based on the rankings.
5. Determine the crisp weight.
6. Determine the weight's normalised value.
7. Conducting the ranking.

Result and Discussion

In order to reduce waste in discrete production systems, the study offers a revolutionary internal logistics improvement LM process. It is feasible to identify the system processes that are actually essential and quantitatively evaluate them in order to determine the best course of action for achieving the desired improvement by combining these approaches. Future study will combine the suggested LM method with simulation tools to test potential enhancements.

The current research effort gives a thorough strategy for comparing and ranking the lean performances of four production cells in a SME using the AHP methodology. As can be shown, AHP performs flawlessly as a multi-attribute decision-making tool when comparing various alternatives based on various criteria. Lean performance evaluation of manufacturing cells allows for cell comparison and ranking according to standardised standards. Finally, it is suggested to frequently rank compare the cells at predefined intervals. In order to foster healthy competition and inspire other manufacturing cells to pursue constant development, the best manufacturing cell would also get rewards.

Thus, the driving force behind LM is to offer clients excellent quality at a lower cost, thereby enhancing customer happiness. People must employ the best methods (like ISM and AHP) to contribute to this adaptability. As a result, the results of these strategies enable us to be more useful. ISM and AHP approaches must be used for this. Because of the interplay between the criteria and sector experts, this is more practical. Although the performance elements are only shown to be meaningful when they interact with the criteria, the ISM improves its application by providing ideal interrelationships between them. AHP highlights its significance by including non-quantifiable factors like social, political, and economic. A useful decision-making process is also made possible by fuzzy AHP, which helps to lessen environmental and social impacts. These combined techniques could encourage increased research interest and assist in more effective lean deployment.

References

[1] Koç , E., and Burhan, H. A. (2015). An application of analytic hierarchy process (AHP) in a real world problem of store location selection. Advances in Management and Applied Economics. 5, 41–50.

[2] Dalalah, D., AL-Oqla, F., and Hayajneh, M. (2010). Application of the analytic hierarchy process (AHP) in multi criteria analysis of the selection of cranes. Jordan Journal of Mechanical and Industrial Engineering. 4, 567–578.

[3] Saaty, T. L. (1995). Transport planning with multiple criteria: The analytic hierarchy process applications and progress review. 29, 81–126.

[4] Vaidya, O., and Kumar, S. (2006). Analytic hierarchy process: An overview of applications. European Journal of Operational Research. 169, 1–29.

[5] Kumru, M., and Kumru P. Y. (2014). Analytic hierarchy process application in selecting the mode of transport for a logistics company. Journal of advanced transportation. 974–99.

[6] Al-Harbi K. M. A. S. (2001). Application of the AHP in project management. International Journal of Project Management. 19, 19–27.

[7] Ratna, S., and Prasad, D. (2018). Decision support systems in the metal casting industry: An academic review of research articles. Materials Today: Proceedings. 5, 1298–1312.

[8] Cebeci, U. (2009). Fuzzy AHP-based decision support system for selecting ERP systems in textile industry by using balanced score card. Expert Systems with Applications. 36, 8900–9.

[9] Saaty, T. L. (2008). Decision making with the analytic hierarchy process. International Journal Services Sciences. 1, 83–89.

[10] Abdulmalek, A. A., and Rajgopal, J. (2007). Analyzing the benefits of lean manufacturing and value stream mapping via simulation: A process sector case study. International Journal of Production Economics. 107, 223–236.

[11] Ahuja, V., Yang, J., and Shankar, R. (2009). Benefits of collaborative ICT adoption for building project management. Construction Innovation: Information, Process, Management. 9(3), 323–40.

[12] Attri, R., Grover, S., Dev, N., and Kumar, D. (2012). An ISM approach for modeling the enablers in the implementation of total productive maintenance (TPM). International Journal System Assurance Engineering and Management. 4(4), 313–26.

[13] Benyoucef, M., and Canbolat, M. (2007). Fuzzy AHP-based supplier selection in e-procurement. International Journal of Services and Operations Management. 3(2), 172–92.

[14] Crick, D. (2009). Managers of UK based SMEs' perception of their overseas performance and competitiveness: a study of regular and sporadic internationalizing firms. Journal of Strategic Marketing. 17(5), 397–410.

[15] Faisal, M. N., Banwet, D. K., and Shankar, R. (2006). Supply chain risk mitigation: modeling the enablers. Business Process Management Journal. 12(4), 535–52.

[16] Gorvett, R., and Liu, N. (2006). Interpretive structural modeling of interactive risks'. In Proceedings of the Enterprise Risk Management Symposium, (April 23–26), Chicago, IL, USA, pp. 1–12.

[17] Gunasekaran, A., Forker, L., and Kobu, B. (2000). Improving operation performance in small company: a case study. International Journal of Operations and Production Management. 20(3), 316–36.

[18] Ho, W., Xu, X., and Dey, P. K. (2010). Multi criteria decision making approaches for supplier evaluation and selection: a literature review. European Journal of Operational Research. 202(1), 16–24.

[19] Jharkharia, S., and Shankar, R. (2004). IT enablement of supply chains: modeling the enablers. International Journal of Productivity and Performance Management. 53(8), 700–712.

[20] Kabir, M. A., Latif, H. H., and Sarker, S. (2013). A multi-criteria decision-making model to increase productivity: AHP and fuzzy AHP approach. International Journal of Intelligent System Technologies and Applications. 12(Nos. 3/4), 207–29.

[21] Khakbaz, M. H., Ghapanchi, A. H., and Tavana, M. (2010). A multicriteria decision model for supplier selection in portfolios with interactions. International Journal of Services and Operations Management. 7(3), 351–77.

[22] Maike, S. R., Bogle, T. A., and Deflorin, P. (2010). Lean takes two! Deflections from the second attempt at lean implementation. Business Horizons. 52(1), 79–88.

[23] Ngamsirijit, W. (2011). Manufacturing flexibility improvement and resource-based view: cases of automotive firms. International Journal of Agile Systems and Management. 4(4), 319–341.

[24] Parhizkar, O., Smith, R. L., and Miller, C. R. (2009). Comparison of important competitiveness factor for small-to medium-sized forest enterprises. Forest Product Journal. 59(5), 81–86.

[25] Parthiban, P., Zubar, H., and Garge, C. P. (2012). A multi criteria decision making approach for suppliers selection. Procedia Engineering. 38, 2312–28.

[26] Raj, T., Attri, R., and Jain, V. (2012). Modeling the factor affecting flexibility in FMS. International Journal of Industrial and System Engineering. 11(4), 350–74.

[27] Ravi, V., Shankar, R., and Tiwari, M. K. (2005). Productivity improvement of a computer hardware supply chain. International Journal of Productivity and Performance Management. 54(4), 239–55.

[28] Sarmah, S. P., Acharya, D., and Goyal, S. K. (2006). Some models on value of information sharing in supply chain management. ICFAI Journal of Operation Management. V(2), 7–23.

[29] Singh, R. K., Garg, S. K., and Deshmukh, S. G. (2008). Strategic development by SMEs for competitiveness: a review. Benchmarking: An International Journal. 15(5), 525–47.

[30] Singh, R. K., Garg, S. K., and Deshmukh, S. G. (2010). The competitiveness of SMEs in a globalised economy: observations from China and India. Management Research Review. 33(1), 54–65.

[31] Steuer, R. E., and Na, P. (2003). Multiple criteria decision making combined with finance: a categorised bibliographic study. European Journal of Operational Research. 150(3), 496–515.

[32] Thakkar, J., Kanda, A., and Deshmukh, S. G. (2008). Interpretive structural modeling (ISM) of IT-enablers for Indian manufacturing SMEs. Information Management and Computer Security. 16(2), 113–36.

[33] Wang, Y. M., and Chin, K. S. (2011). Fuzzy analytic hierarchy process: a logarithmic fuzzy preference programming methodology. International Journal of Approximate Reasoning. 52(4), 541–53.

[34] Yang, J. L., Chiu, H. N., Tzeng, G. H., and Yeh, R. H. (2008). Vendor selection by integrated fuzzy MCDM techniques with independent and interdependent relationships. Information Sciences. 178(21), 4166–83.

[35] Yih, J. M., and Lin, Y. H. (2007). An integration of fuzzy theory and ISM for concept structure analysis with application of learning MATLAB. Intelligent Information Hiding and Multimedia Signal Processing, IIHMSP, Third International Conference on. 2, 187, 190, (pp. 26–28).

[36] Kurdve, M., and Bellgran, M. (2021). Green lean operationalisation of the circular economy concept on production shop floor level. Journal of Cleaner Production. 278, 1–11. https://doi.org/10.1016/j.jclepro.2020.123223.

[37] Ramkrishna S. Bharsakade, Padmanava Acharya, L. Ganapathy and Manoj K. Tiwari. (2021). A lean approach to healthcare management using multicriteria decision making. Springer, Operational Research Society of India. https://doi.org/10.1007/s12597-020-00490-5

[38] Krzysztof Ejsmont, Bartlomiej Gladysz, Donatella Corti, Fernando Castaño, Wael M. Mohammed and Jose L. Martinez Lastra (2020). Towards 'Lean Industry 4.0' – Current trends and future perspectives'. Cogent Business and Management. 7, 1–32. https://doi.org/10.1080/23311975.2020.1781995

[39] Handfield, R. B., Graham, G., and Burns, L. (2020). Coronavirus, tariffs, trade wars and supply chain evolutionary design. International Journal of Operations and Production Management. 40 (10), 1649–60.

Chapter 110

2D-thermal exchange assessment past backward-facing step inside a rectangular channel using LBM-MRT method

El Bachir Lahmer[a], Jaouad Benhamou[b], Mohammed Amine Moussaoui[c] and Ahmed Mezrhab[d]

Mechanics & Energy Laboratory, Faculty of Sciences, Mohammed First University, Morocco

Abstract

The technology of electronic devices has made significant progress, but their performance can still be affected by rising temperatures. This can lead to the failure of elements like processors, capacitors, and so on. At this stage, scientific researchers orientated their works to find efficient solutions to reduce the overheating of electronic units to maintain flawless operation. In this research, a numerical code based on the lattice Boltzmann method associated with the multiple relaxation times (LBM-MRT) was used to study the enhancement of heat transfer in a partitioned channel featuring a backward-facing step. The numerical results showed that the introduction of the vertical partitions has a substantial effect on the thermal exchange between the cooled air and the heated wall after the step. In addition, this study also looked at the impact of parameters such as Reynolds and Prandtl numbers on thermal exchange rates. These numerical findings helped to improve the quality of heat transfer of electronic components in regions with a low thermal exchange.

Keywords: Thermal management, heat transfer, fluid flow, backward-facing step, lattice Boltzmann method

Introduction

The study of fluid flow and convective heat transfer through a backward-facing step (BFS) in a partitioned channel is an important and widely researched area with numerous applications in a variety of industrial and engineering contexts [1–6]. There is a significant body of literature examining how different geometries can impact heat transfer and fluid flow processes in this type of system. A 2-D numerical study was conducted by Kanna et al. [7] to examine the evolution of the steady-state thermal exchange in a laminar incompressible wall jet past a backward-facing step. The results of the study indicated that the average Nusselt number increases with an increase in the Reynolds number and Prandtl number. Additionally, an increase in the step height and Prandtl number also resulted in an improvement in heat transfer, as evidenced by a higher average Nusselt number. Conversely, the average Nusselt number was found to decrease as the step length increased. Ayad et al. [8] conducted an experimental study to evaluate the enhancement of heat transfer and pressure drop in a heated rectangular duct with a backward-facing step using hybrid nanofluids (Al_2O_3-TiO_2). The results showed that an increase in the mass fraction and Reynolds number increased the Nusselt number. This, in turn, led to an improvement in heat transfer quality and a 23% increase in pressure drop compared to the results obtained using pure water.

In a study led by Kumar et al. [9], the impact of an adiabatic circular cylinder placed downstream of a step-side wall within a channel containing a BFS system on separated forced convection was examined using 2-D fluent software. The implementation of the circular cylinder at various Reynolds numbers grows the local Nusselt number, with the maximum value reaching 155% for the constant temperature case. Additionally, this arrangement resulted in an overall enhancement of heat transfer.

Other researchers have applied the lattice Boltzmann method (LBM) to study the BFS. Chen et al. [10] employed this method to investigate convective heat transfer with the field synergy principle in their study. The single relaxation time scheme (SRT) was utilised to simulate convective heat transfer and fluid flow, and the results demonstrated good agreement with existing experimental and numerical literature on the BFS. It was found that the inclusion of a square blockage increased the convective heat transfer rate when considering the field synergy principle.

The focus of our current study is to conduct a comprehensive assessment of the airflow and thermal exchange on a BFS system with the inclusion of four vertical partitions. To accomplish this, a simulation code based on the Boltzmann lattice method (LBM-DMRT) is being utilised to evaluate the performance of the mechanisms under examination in the present configuration [11,12]. The numerical results obtained in this study contributed to the enhancement of the heat transfer efficiency of electronic components in areas characterised by low thermal conductivity.

[a]lahmerelbachir@gmail.com, [b]jaouad1994benhamou@gmail.com, [c]moussaoui.amine@gmail.com, [d]amezrhab@yahoo.fr

DOI: 10.1201/9781003450252-110

Numerical Approach

The development of the LBM method over the past few years has enabled the modeling of temperature distribution and fluid flow in a more practical manner. This method involves a combination of the particle dynamics and Navier-Stokes equations at microscopic and macroscopic scales, respectively. It is characterized by its ability to model the fluid at a precisely defined level (mesoscopic level), as opposed to traditional methods [13].

The simulation of convective heat transfer and fluid flow focuses on the solution of the equation of lattice Boltzmann (LBE), which defines the propagation and collision mechanism of fluid particles employing the MRT model, as demonstrated by the equation below:

$$f_i(r + e_i \times \Delta t, t + \Delta t) - f_i(r, t) = -M^{-1} \times S_i \times (m_i(r, t) - m_i^{eq}(r, t)), \ i = 0, ..., 8 \tag{1}$$

The distributive function f_i determines the likelihood of locating a fluid particle at the location r at time t. The diagonal relaxation time and the transformation matrix are denoted by the symbols s_i and M, respectively. The moment vector and the equilibrium moment are referred to as m_i and m_i^{eq}, respectively [14].

The numerical computation of the LBE equation allows the characterisation of the fluid flow pattern in terms of the macro-quantities of density and motion, as demonstrated by the equations below:

$$\rho = \sum_{i=0}^{8} f_i$$

$$\rho u = \sum_{i=0}^{8} f_i \, e_i \tag{2}$$

Regarding the temperature field study, the thermal lattice Boltzmann equation (TLBE) with the introduction of the multiple relaxation time scheme (MRT) was used to simulate the thermal distribution g_i, as represented in the following expression:

$$g_i(x + e_i \times \Delta t, t + \Delta t) - g_i(x, t) = -N^{-1} \times Q_i \times (n_i(x, t) - n_i^{eq}(x, t)), \ i = 0, ..., 4 \tag{3}$$

where the diagonal relaxation time and the transformation matrix are represented by the symbols Q_i and N, respectively. The thermal momentum and the thermal equilibrium moment are denoted by n_i and n_i, respectively [13,15].

By utilizing a numerical method to solve the TLBE equation, it is possible to accurately calculate the thermal field at the macroscopic scale using the following equation:

$$T = \sum_{i=0}^{4} g_i \tag{4}$$

Throughout this research, the D2Q9 and D2Q5 schemes were utilised to model the airflow behavior and thermal exchange process, respectively. These schemes are depicted in Figure 110.1.

The other parameters and quantities used to solve the Boltzmann equation are thoroughly discussed in the literature [11, 13, 14, 16, 17].

Configuration System and Boundary Conditions

Studied System Problem

The present study examines a two-dimensional rectangular channel of length L, with distinct heights designated as the height of the upstream channel (he) and the height of the downstream channel (H). The backward-facing step configuration has a channel expansion ratio ($ER = \frac{H}{he}$) of 1.5. The configuration under investigation for this study comprises four vertical partitions with a length of $lp = \frac{H}{4}$ installed on the top wall of the channel. The partitions are separated by a distance of $D = S$. At the inlet, the airflow has a velocity of $u_0 = 0.1$ and a cold temperature of $\theta_{in} = 0$, which are featured by a parabolic profile. the maximum velocity of the airflow is located in the middle of the upstream channel domain, as determined according to the following formula:

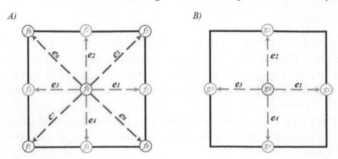

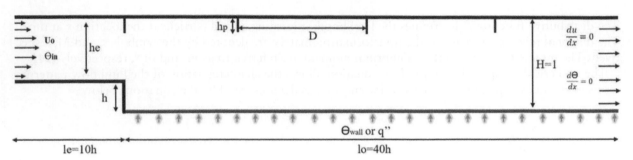

Figure 110.1 Modeling of fluid flow behavior (D2Q9) and heat transfer process (D2Q5) in 2D partitioned channel
Source: Author

Figure 110.2 Schematic illustration of the studied system under investigation
Source: Author

$$u(y) = \frac{3}{2} \times uo \times (1 - y/H) \times \left(y/H - \frac{h}{H}\right) \tag{5}$$

The Reynolds number, represented in the following equation, plays a crucial role in the analysis of fluid flow behavior. It is a dimensionless quantity that is used to describe the ratio of the inertial forces to the viscous forces in a fluid flow and is often used to predict the transition between laminar and turbulent flow regimes. By using the Reynolds number, it is possible to understand the characteristics of the fluid flow and how it may change under different conditions.

$$Re = \frac{2u_0 \times h}{3v} \tag{6}$$

Knowing that the step height of the BFS (h) was used to define the characteristic length in Equation 6. The symbols v and u_0 represent the kinematic viscosity and the maximum velocity rate of flow at the channel inlet, respectively. It was stipulated that the velocity and temperature gradients at the outflow of the channel were both equal to zero (Figure 110.2).

In the analysis of thermal issues, it is assumed that the temperature of all walls of the channel is adiabatic, excluding the lower wall on the downstream side of the step, which is maintained at a constant temperature of $\theta_i = 1$, as depicted in Figure 110.2.

Boundary Conditions

The boundary conditions of Zou and He [18] were applied at the entrance and exit of the channel domain. While the Boundary conditions for a rebound were used for the solid wall and vertical partitions, as shown in Figure 110.3 [14]. In regard to the thermal flow, Equation 4 can be used to define the unknown distribution function based on the known distribution functions at the entrance. The extrapolation of the second-order was employed to define the unknown distribution function at the channel exit [19].

Verification of the Code Accuracy

The present simulation code, which is established on the LBM-DMRT method, has been validated by comparing the results obtained with those from other references to assess the accuracy and reliability of fluid flow and heat transfer simulation.

The numerical results found by the MRT-LBM method are compared with the experimental data of reference [20]. The along-channel velocity profile was examined numerically against previous experimental findings provided by Fearn et al. [20]. As depicted in Table 110.1, the values of the velocity ratio achieved

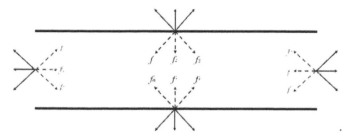

Figure 110.3 Illustration of the known and unknown distribution functions on the channel border
Source: Author

Table 110.1 Numerical and experimental measurements of the velocity ratio at the middle of the channel ($Re = 26$)

	u/U_0	
x/h	Experimental results of Fearn et al. [20]	Our numerical results
1.25	0.81	0.83
2.5	0.63	0.6
5	0.41	0.36
10	0.344	0.334

Source: Author

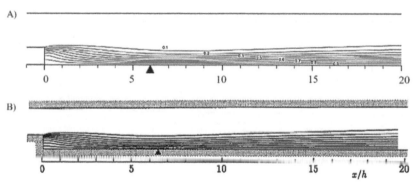

Figure 110.4 Comparison of isotherms distribution at Re= 100, ER= 1.5: A) Current results, B) Literature results [21]

from the current study are consistent with the experimental findings at the middle of the channel for $Re = 26$. Therefore, the numerical method currently used is well consistent with the literature [20].

The numerical study of the thermal area in a rectangular channel within a backward-facing step was investigated concerning the temperature field. The present results and the numerical results of the literature gave the same outcomes, as illustrated in Figure 110.4.

The numerical study of the thermal area in a rectangular channel within a backward-facing step was investigated concerning the temperature field. The present results and the numerical results of the literature gave the same outcomes, as illustrated in Figure 110.4.

In this research, we examine the local Nusselt number to understand the thermal exchange process within the channel region. The local Nusselt number is calculated using the following equation:

$$Nu = \frac{\partial \theta}{\partial y}\bigg|_{y=0} \tag{7}$$

The evolution of the local Nusselt number at $Re = 100$ and $ER = 1.5$ is plotted in Figure 110.5. Our numerical results are in good agreement with those reported in reference [21], demonstrating the validity of our DMRT-LBM-based code for simulating fluid flow and convective heat transfer.

Numerical Results and Discussion

A structural analysis was performed to identify the most suitable mesh size for accurately simulating fluid flow and heat transfer in the proposed system. The aim is to avoid the issue of mesh independence. The

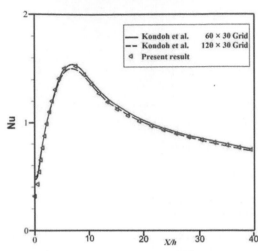

Figure 110.5 Evolution of the local Nusselt number at $Re = 100$, $ER = 1.5$
Source: Author

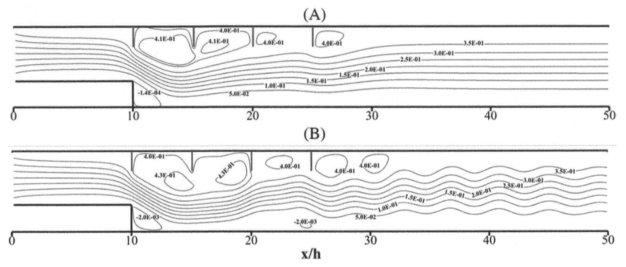

Figure 110.6 Fluid flow streamlines for $Pr = 0.71$: A) $Re = 70$, B) $Re = 140$
Source: Author

results indicated that a mesh size of 2000 × 120 has the most reliable results for all computational simulations conducted.

In this work, the impact of the Reynolds number and Prandtl number (Air with two different thermal properties) was considered to model the airflow and convective thermal exchange around the BFS system, and all other parameters were assumed to be fixed during the simulation.

The streamlines of the airflow for two different Reynolds numbers ($Re = 70$ and $Re = 140$) at a fixed Prandtl number ($Pr = 0.71$) are illustrated in Figure 110.6. The results show that vortices of various shapes appear downstream of the step and behind each partition at the upper wall. The genesis of these recirculation zones is attributed to the velocity difference between the mainstream air and the motion of the air behind each partition. The merging of two recirculation zones can be observed near the first two partitions. Once the Reynolds number rises, the vortices areas become more distinct in form. The airflow starts wavering by the influence of the vertical partitions. It can be concluded that the inertial effects dominate against the viscous effects.

Figure 110.7 displays the temperature distribution for Reynolds numbers equal to 70 and 140. The results achieved show that the thermal layers become compressed close to the bottom wall downstream of the step for specific values of the Reynolds number at the laminar state. A narrow thermal boundary layer is established near the bottom wall after the step. When the Reynolds number equals 140, the temperature distribution has a sinusoidal shape with a decrease and increase in the tightness of the temperature layers directly after the last partition.

The present study evaluates the thermal exchange rate in a rectangular channel consists of four vertical partitions and investigates the local Nusselt number under the effect of Reynolds numbers, as depicted in Figure 110.8. The results indicate a marked enhancement in the local Nusselt number in the vicinity of

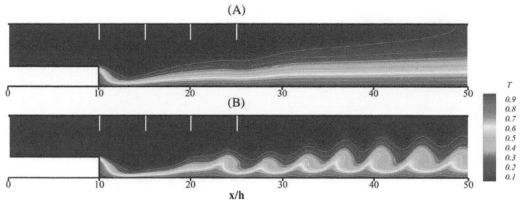

Figure 110.7 Thermal distribution for $Pr = 0.71$: A) $Re = 70$, B) $Re = 140$
Source: Author

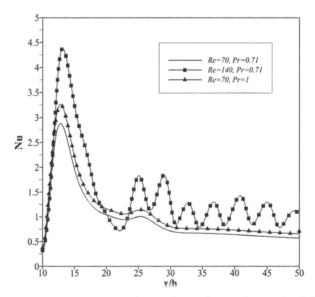

Figure 110.8 Local Nusselt number evolution for different cases
Source: Author

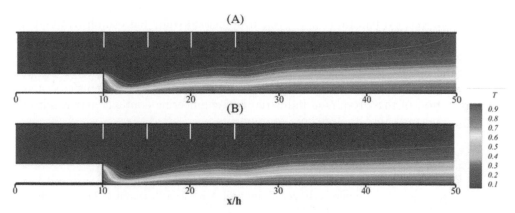

Figure 110.9 Temperature field for $Re = 70$: A) $Pr = 0.71$, B) $Pr = 1$
Source: Author

the bottom wall downstream of the step upon increasing the Reynolds number. The local Nusselt number exhibits a critical peak near the channel step for a Reynolds number of 70. This augmentation is a result of the influence of the four vertical partitions on the orientation of the airflow. This leads to an enhancement in the heat transfer rate. At a Reynolds number of 140, the primary peak near the step attains a value of 4.38. In addition, the high Reynolds number value leads to a disturbance in the fluid flow, resulting in the appearance of seven critical peaks downstream of the last partition. It can be interpreted that raising the Reynolds number leads to an improvement in the quality of heat transfer at a fixed Prandtl number.

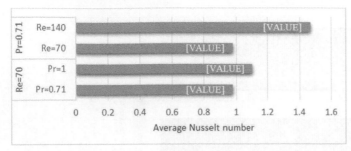

Figure 110.10 Average Nusselt number for different values of Prandtl number and Reynolds number
Source: Author

This research also examines the effect of the Prandtl number (*Pr*) on heat transfer. The Prandtl number is a dimensionless quantity that represents the relative rate of aerodynamic/hydrodynamic and thermal properties in a fluid. An increase in the Prandtl number results in a reduction of the thermal boundary layer thickness in the channel regions (Figure 110.9). Thus, hydrodynamic phenomena predominate ahead of the thermal phenomena of the fluid. In other words, the fluid flow impacts the temperature lines where the isotherms are pressed toward the bottom wall (Figure 110.9).

Regarding the evolution of the local Nusselt number, Figure 110.8 also illustrates the evolution of Nu for two different values of the Prandtl number. The thermal exchange rate rises when the Prandtl number changes from *Pr* = 0.71 (Air at *T* = 0°Cand atmospheric pressure) to *Pr* = 1 (air at *T* = 133.2°C and *P* = 20 *bar*) for Reynolds number equals 70, which means that the heat transfer from the solid wall to the fluid increases due to the increase of the hydrodynamic effects over the thermal effects of the studied fluid.

Conclusion

This work conducts numerical research on a 2D rectangular channel containing a backward-facing step and four vertical partitions. The actual code based on the lattice Boltzmann method associated with the double multiple relaxation times shows good accordance with the previous experimental and numerical literature to provide accuracy when modeling the heat transfer and fluid flow phenomena. This paper reports numerical results in terms of streamlines, isotherms, and local and average Nusselt number evolution. To recap the results achieved, Figure 110.10 presents the mean Nusselt number under the effect of the Reynolds number and Prandtl number. As the Reynolds number increases from *Re* = 70 to *Re* = 140, the thermal exchange rate rises by about 50% for *Pr* = 0.71. By raising the Prandtl number from *Pr* = 0.71 to *Pr* = 1 at a fixed Reynolds number (*Re* = 70), the heat exchange increases by 13%. It can be deduced that the rise in the Prandtl number and Reynolds number improves the thermal exchange rate appropriately with the involvement of the four vertical partitions. The present numerical study has established that the incorporation of inclined partitions in cooling systems leads to a notable enhancement in heat exchange at a low flow rate. Furthermore, it is observed that even though the effect is relatively small at a low flow rate. At a high flow rate, the thermal exchange becomes substantially improved. These numerical findings provide strong evidence for the potential of inclined partitions as a design feature in heat transfer systems. Therefore, the practical application of these results in industrial and engineering contexts may result in a substantial improvement in the performance of electronic components.

References

[1] Eleiwi, M. A., Tahseen, T. A., and Hameed, A. F. (2020). Numerical study of fluid flow and heat transfer in a backward facing step with three adiabatic circular cylinder. Journal of Advanced Research in Fluid Mechanics and Thermal Sciences. 72, 80–93. https://doi.org/10.37934/arfmts.72.1.8093.

[2] Moosavi, R., Moltafet, R., and Shekari, Y. (2021). Analysis of viscoelastic non-Newtonian fluid over a vertical forward-facing step using the Maxwell fractional model. Applied Mathematics and Computation. 401, 126119. https://doi.org/10.1016/j.amc.2021.126119.

[3] Talib, A. R. A., and Hilo, A. K. (2021). Fluid flow and heat transfer over corrugated backward facing step channel. Case Studies in Thermal Engineering. 24, 100862. https://doi.org/10.1016/j.csite.2021.100862.

[4] Abuldrazzaq, T., Togun, H., Alsulami H., Goodarzi, M., Safaei, M. R. (2020). Heat transfer improvement in a double backward-facing expanding channel using different working fluids. Symmetry (Basel). 12, 1088. https://doi.org/10.3390/sym12071088.

[5] B. Geridönmez, B. P. and Öztop, H. F. (2021). Effects of inlet velocity profiles of hybrid nanofluid flow on mixed convection through a backward facing step channel under partial magnetic field. Chemical Physics. 540 (202. https://doi.org/10.1016/j.chemphys.2020.111010.

[6] Lahmer, E. B., Benhamou, J., Admi, Mohammed, A. M., Jami, M., Mezrhab, A., and Phanden, R. K. (2022). Assessment of conjugate and convective heat transfer performance over a partitioned channel within backward-facing step using the Lattice Boltzmann method. Journal of Enhanced Heat Transfer. 29, 51–77. https://doi.org/10.1615/jenhheattransf.2022040357.

[7] Kanna, P. R. and Das, M. K. (2006). Heat transfer study of two-dimensional laminar incompressible wall jet over backward-facing step. Numerical Heat Transfer, Part A: Applications. 50, 165–87. https://doi.org/10.1080/10407780500506857.

[8] Abedalh, A. S., Shaalan, Z. A., and Saleh Yassien, H. N. (2021). Mixed convective of hybrid nanofluids flow in a backward-facing step, Case Stud. Therm. Eng. 25, 100868. https://doi.org/10.1016/j.csite.2021.100868.

[9] Kumar, A. and Dhiman, A. K. (2021). Effect of a circular cylinder on separated forced convection at a backward-facing step. International Journal of Thermal Sciences. 52, 176–85. https://doi.org/10.1016/j.ijthermalsci.2011.09.014.

[10] Chen, C. K., Yen, T. S., and Yang, Y. T. (2006). Lattice Boltzmann method simulation of backward-facing step on convective heat transfer with field synergy principle. International Journal of Heat Mass Transfer. 49, 1195–1204. https://doi.org/10.1016/j.ijheatmasstransfer.2005.08.027.

[11] Lahmer, E. B., Moussaoui, M. A., Mezrha, A., Botton, V., Henry, D. (2019). Double constricted channel for laminar flow and heat transfer based on double MRT-LBM, in: 24ème Congrès Français de Mécanique, semantic scholar. Brest. 16. https://www.semanticscholar.org/paper/Double-constricted-channel-for-laminar-flow-and-on-Lahmer-Moussaoui/ecb20516ca855705828f2abe69bb29138fe10ae0#extracted.

[12] Moussaoui, M. A., Lahmer, E. B., Admi, Y., and Mezrhab, A. (2019). Natural convection heat transfer in a square enclosure with an inside hot block. International Conference on Embedded Wireless Systems. 1–6. https://doi.org/10.1109/WITS.2019.8723863.

[13] A. A. Mohamad, Lattice Boltzmann Method: Fundamentals and Engineering Applications with Computer Codes, 2011. https://doi.org/10.2514/1.J051744.

[14] E. L. B. Lahmer, M.A. Moussaoui, A. Mezrhab, Investigation of laminar flow and convective heat transfer in a constricted channel based on double MRT-LBM, in: Int. Conf. Wirel. Technol. Embed. Intell. Syst. WITS 2019, 2019: pp. 1–6. https://doi.org/10.1109/WITS.2019.8723820.

[15] A. Mezrhab, M. Amine Moussaoui, M. Jami, H. Naji, M. Bouzidi, Double MRT thermal lattice Boltzmann method for simulating convective flows, Phys. Lett. A. 374 (2010) 3499–3507. https://doi.org/10.1016/j.physleta.2010.06.059.

[16] Benhamou, J., Jami, M., Mezrhab, A., Botton, V., and Henry, D. (2020). Numerical study of natural convection and acoustic waves using the lattice Boltzmann method. Heat Transfer. 493779–3796. https://doi.org/10.1002/htj.21800.

[17] Admi, Y., Moussaoui, M. A., and Mezrhab, A. (2020). Effect of a flat plate on heat transfer and flow past a three side-by-side square cylinders using double MRT-lattice Boltzmann method. 2020 IEEE 2nd International Conference on Electronics, Control, Optimization and Computer Science. pp. 1–5. https://doi.org/10.1109/ICECOCS50124.2020.9314506.

[18] Zou, Q. and He, X. (1997). On pressure and velocity boundary conditions for the lattice Boltzmann BGK model. Physics of Fluids. 9, 1591–98. https://doi.org/10.1063/1.869307.

[19] Lahmer,   E. B., Admi, Y., Moussaoui, M. A., and Mezrhab, A. (2022). Improvement of the heat transfer quality by air cooling of three-heated obstacles in a horizontal channel using the lattice Boltzmann method. Heat Transfer. 51, 3869–91. https://doi.org/10.1002/htj.22481.

[20] Fearn, R. M., Mullin, T., and Cliffe, A. K. (1990). Nonlinear flow phenomena in a symmetric sudden expansion. Journal of Fluid Mechanism. 211, 595–608. https://doi.org/10.1017/S0022112090001707.

[21] Kondoh, T., Nagano, Y., and Tsuji, T. (1993). Computational study of laminar heat transfer downstream of a backward-facing step. International Journal of Heat Mass Transfer. 36, 577–91. https://doi.org/10.1016/0017-9310(93)80033-Q.

Chapter 111

Power enhancement in solar PV configurations under partial irradiation scenarios using Futo-Sumdoku method

Jagadeesh K[a] and Chengaiah Ch[b]

Sri Venkateswara University College of Engineering, Tirupati, India

Abstract

In recent years solar photovoltaics (SPVs) is considered to be the most promising type of renewable energy source. One of the most critical factors that can affect the power production and energy efficiency of a photovoltaic (PV) system is the partial shading conditions (PSC). Under this condition, the panels of a PV system receive varying levels of radiation which effects power output of an SPV system in terms of mismatch losses and wired losses, and decreases its performance. Shade dispersion is one of the promising techniques to mitigate the losses in a PV system. Various conventional and hybrid PV array topologies and methods have been proposed in the literature to address this issue. This paper discusses three such static reconfiguration techniques (SRT) which are analysed with corner, square, horizontal and vertical shading patterns and are compared with the proposed) reconfiguration technique. In this paper a 5 × 5 PV array is simulated in MATLAB/Simulink environment to evaluate the various metrics such as global maximum power point (GMPP), mismatch losses (ML), fill factor (FF).

Keywords: Solar PV system, partial shading condition, global maximum power point (GMPP), mismatch losses (ML), fill factor (FF)

Introduction

In the present scenario, power utilisation has drastically increased which not only demands more power but also quality of power, while the accessibility to conventional resources (such as coal and petroleum) has deteriorated dramatically, creating a challenging future for power generation. Furthermore, traditional resource-based power generation emits greenhouse gases. Because of these factors, renewable energy-based power generation has gained an increased attention [1]. Among all renewable energy technologies, on the basis of availability, transportability and accountability PV-based energy is the right choice with low operational and maintenance costs. Photovoltaic (PV) generation systems are becoming increasingly important for both stand-alone and grid-connected systems [2]. At present most of the residential and semi residential complexes, industrial and farming industry are using solar as an alternative source of energy by establishing rooftops and small and medium solar plant to meet their energy requirements. The major factor that influences the output of a solar plant is the non-availability of sunlight [3]. This may be caused by low irradiance in winter and rainy season or by any opaque object in between sun and solar panels which decreases the generated power [6]. The researchers have proposed many reconfiguration techniques that reduce the losses and increases the power generation either by physical reallocation or by using metaheuristic algorithms.

To obtain the required power the panels are connected in series, parallel, series parallel, total cross tied (TCT), bridge linked (BL), and honeycomb (HC). In which TCT has shown better performance than the other configurations [9]. The static reconfiguration techniques are puzzle-based techniques which use physical reallocation of panels without altering the internal wiring to maximise the power output under shading conditions. complex algorithms, switching matrices, sensors, and auxiliary circuits are not required in static interconnection scheme [8]. Different reconfiguration techniques are studied and are analysed based on number of interconnections, mismatch losses (ML) and fill factor (FF) etc.

In this work, Sudoku, Magic Square, and Skyscraper puzzle arrangements were analysed. One common limitation of these techniques is that they cannot be extended to irregular array sizes. To address this issue, a new reconfiguration technique is proposed in this paper.

Model and Configurations of PV Modules

PV Cell Internal Diagram

The best results were obtained using the basic equivalent circuit model of a PV cell consisting of a single diode [5] with the lowest series resistance and diode unity value. The single diode model depicted in Figure 111.1 is the simplest and most widely used. Where:

---

[a]svujagadeeshk@gmail.com, [b]chintapudisvu@gmail.com

DOI: 10.1201/9781003450252-111

I_p = photo generated controlled current source

I_d = diode saturation current

R_{sh} = Shunt resistance

R_s = series resistance

From the equivalent circuit by applying KCL we get [19].

$I_p = I_d + IR_{sh} + I_{pv}$

The current produced by the PV panel is given as

$I_{pv} = I_p - I_d - IR_{sh}$

$$I_{pv} = I_p - I_o \left(e^{\frac{V_{pv}+(I_{pv}*R_s)}{V_T}} - 1 \right) - \frac{V_{pv} + (I_{pv} * R_s)}{R_{sh}} \qquad (1)$$

Where, $I_o = KT^m e^{\frac{-V_{Go}}{nV_T}}$ (Saturation current); $V_T = kT/q$ = Junction thermal voltage
Where k = Boltzmann's constant and T = temperature.

The PV and IV characteristics of a PV module is shown in Figure 111.2.

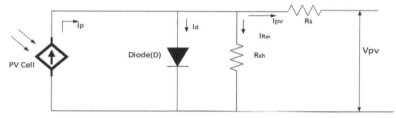

Figure 111.1 Basic equivalent circuit of a PV cell [5]

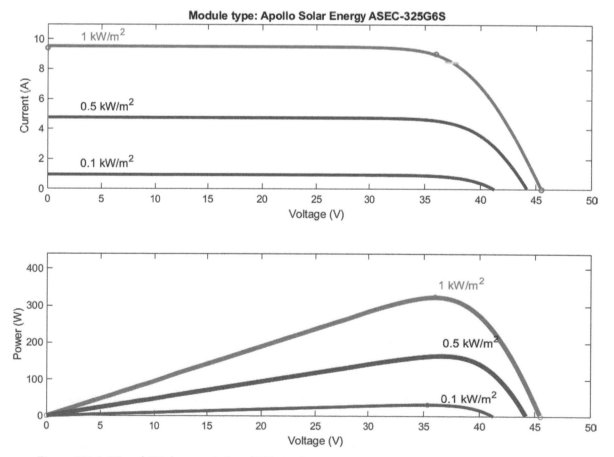

Figure 111.2 IV and PV characteristics of PV panel
Source: Author

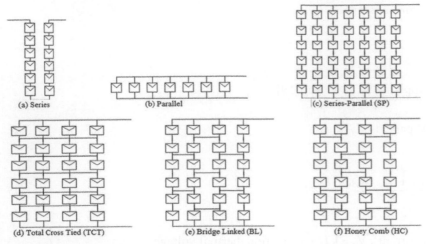

Figure 111.3 Different SPVA configurations
Source: Author

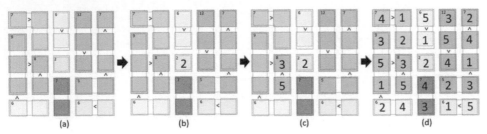

Figure 111.4 (a) Formation of 5 × 5 FSD puzzle (b) Arithmetic operation (c) Logical operation (d) Completed puzzle
Source: Author

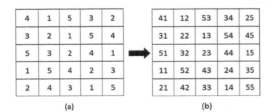

Figure 111.5 (a) FSD puzzle (b) FSD puzzle arrangement in PV array
Source: Author

In order to get the required output, the PV panels are arranged in series, parallel, series-parallel (SP), total cross tied (TCT), BL, HC configurations [9] as depicted in Figure 111.4. In which TCT shows improved output and BL is a bridge linked where the panels are connected to form a bridge to boost the power output. The benefits of TCT and BL are joined to form a (HC configuration shown in Figure 111.4 (f). Ties in TCT are more and less in BL, we must choose the ties carefully and adequately in HC. This can be achieved by joining ties in variants of two, four, and six modules.

Methodology

Futo-Sumdoku Puzzle Formation and Arrangement

Futo-Sum Doku (FSD) is a combination of Futoshiki and SumDoku puzzle techniques which belongs to Sudoku family. The FSD puzzle is a m × n grid with separate cage or cluster of cage that belong to different rows and columns. In this no numbers will be filled in at the start, each cell is also a part of a cage, which is indicated by different colors. Each of these cages has a sum, and you must determine which numbers to put in each cell based on the cage sum by obeying the greater than and less than symbols. Each cell in a cage is added up and must equal the cage's total. In this paper a 5 × 5 PV array system is considered for evaluation. Numbers 1 to 5 must be placed in each cell of the age such that the sum of the digits in the cage must me equal and should satisfy the logical relation between the cells in the cage if given. Each number should not be repeated in column or row. The Figure 111.4(a) shows 5 × 5 puzzle used in the work. The following are the steps to be followed for solving the puzzle.

i)   Third row and third column shown in the puzzle consists a single cage with a value 2. So, the value in that cell is 2 as shown in Figure 111.4 (b).

ii) In the third row second column the sum of the cage should be 8 and it is also given that the bottom digit in that cage is greater than the upper digit with this clue the digits 5 and 3 can be placed in that cage which gives the sum of 8 and the ambiguity of placing which number in which cell can be cleared by seeing at the logical operator placed in between the digits. It shows that the bottom digit in that cage is greater than the upper digit which means the number 5 is placed in bottom and number 3 is placed in upper cell of the cage as shown in Figure 111.4(c).

iii) Similarly, the next numbers should be placed in the cage by satisfying all the conditions in the puzzle as shown in Figure 111.4(d).

The advantage of this is that the puzzle has a unique solution and it can be implemented on the PV system.
   The FSD puzzle is rearranged as shown in Figure 111.5 for analysis purpose.
   After FSD puzzle arrangement the output power from the arrangement is analysed for five shading patterns namely single corner cell shading (SCS), square corner cell shading (SQCS), vertical shading (VS), horizontal shading (HS) and diagonal shading (DS) when subjected to 1000 W/m²,500 W/m² irradiances as shown in Figure 111.6. The results are discussed below.

Implementation of Proposed Model

Simulation of Proposed PV Topologies

A FSD puzzle pattern-based reconfiguration technique is proposed for a 5 × 5 PV system to enhance the GMPP under thepartial shading conditi (PSCs). FF, ML, power loss (PL), execution ratio (ER), and performance enhancement ratio (PER) are used to evaluate the performance of the PV configurations. The subsequent subsections provide more information on performance parameters and their analysis. The performance of the FSD reconfiguration technique is compared with other reconfiguration methods such as Sudoku, Magic Square and Skyscraper in order to confirm its efficacy.
   The sub-system model of the SP, TCT, BL, and HC connections are connected to variable DC sources and is simulated as illustrated in Figure 111.7. The Simulink model for each type of topology connection is grouped into a sub-system. The proposed system is implemented by considering Apollo solar energy ASEC – 325G6S panel and the corresponding solar PV panel ratings are given in Table 111.1.

Simulation of Partial Shaded Conditions

In this section, the following partial shading patterns are analyzed for a 5 × 5 PV array topology.
   They are as follows

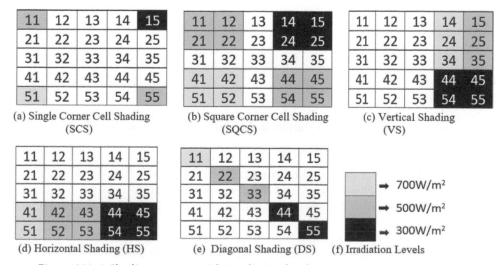

Figure 111.6 Shading patterns with irradiation levels
Source: Author

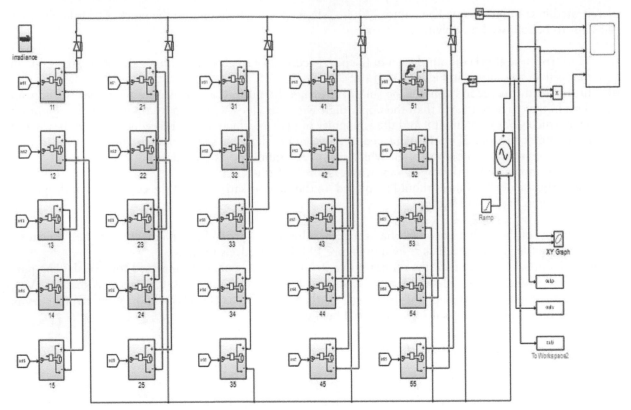

Figure 111.7 Simulink model of the proposed FSD method
Source: Author

Table 111.1 Solar PV panel ratings

Electrical parameters	Rating
Maximum power (Pmax)	325 W
Open circuit voltage (Voc)	45.48 V
Voltage at maximum power (Vmpp)	35.99 V
Short circuit current (Isc)	9.39 A
Current at maximum power (Impp)	9.03 A
Number of cells connected in series (Ns)	72

Source: Author

a)   Single corner cell shading (SCS)
b)   Square corner cell shading (SQCS)
c)   Vertical shading (VS)
d)   Horizontal shading (HS)
e)   Diagonal Shading

The shading patterns for all the considered reconfiguration techniques are shown with their irradiation levels in Figure 111.6 and dispersions in Figure 111.8.

a)   Single corner cell shading: In this type of shading the corner cells are selected and are partially shaded with different irradiance as shown in Figure 111.6 and the corresponding shade dispersion is shown in Figure 111.8(a). The PV characteristics of single corner cell shading pattern in TCT, BL and HC are shown in Figures 111.9-111.11.

b)   Square corner cell shading: In this type of shading the corner squares are selected and are partially shaded with different irradiance. The third row and third column are unshaded and all the other cells are partially shaded as shown in Figure 111.6. The PV characteristics of SQCS pattern in TCT, BL and HC are shown in Figures 111.12-111.14.

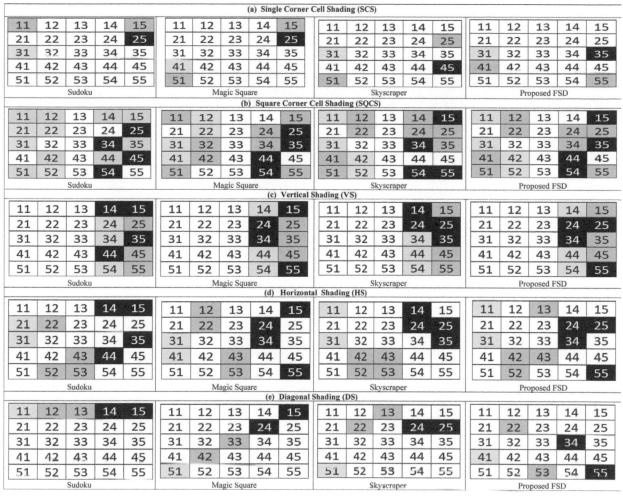

Figure 111.8 Shade dispersion patterns in different methods
Source: Author

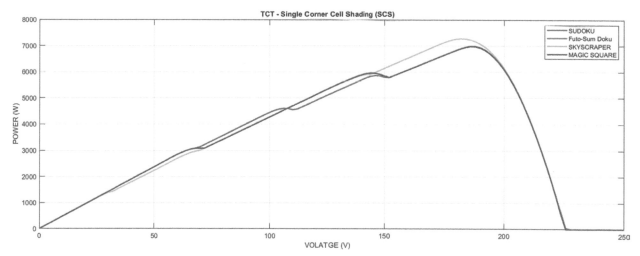

Figure 111.9 PV Characteristics of TCT for SCS shading
Source: Author

c) Vertical shading: In this type of shading last two columns selected and are partially shaded with different irradiance and all the other cells are un shaded as shown in Figure 111.6. The PV characteristics of vertical shading pattern in TCT, BL, and HC are shown in Figures 111.15-111.16.

d) Horizontal shading: In this type of shading bottom two columns are selected and are partially shaded with different irradiance and all the other cells are un shaded as shown in Figure 111.6. The PV characteristics of vertical shading pattern in TCT, BL, and HC are shown in Figures 111.18-111.20.

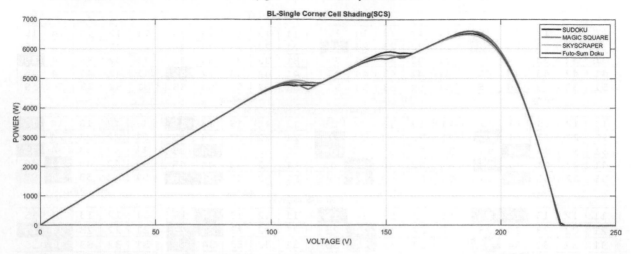

Figure 111.10 PV Characteristics of BL for SCS shading
Source: Author

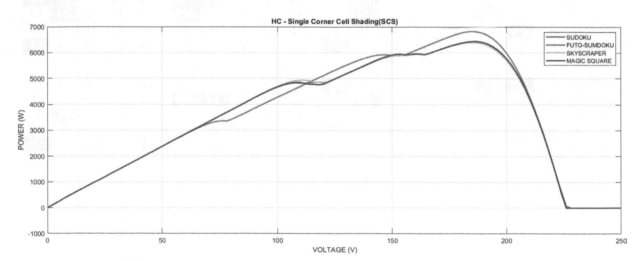

Figure 111.11 PV Characteristics of HC for SCS shading
Source: Author

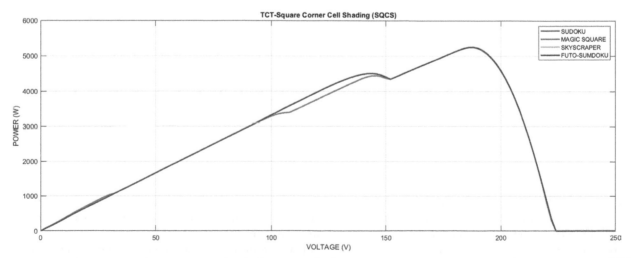

Figure 111.12 PV Characteristics of TCT for SQCS shading
Source: Author

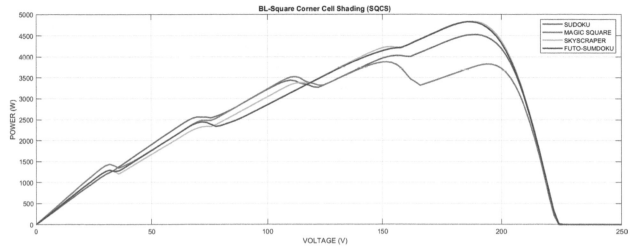

Figure 111.13 PV Characteristics of BL for SQCS shading
Source: Author

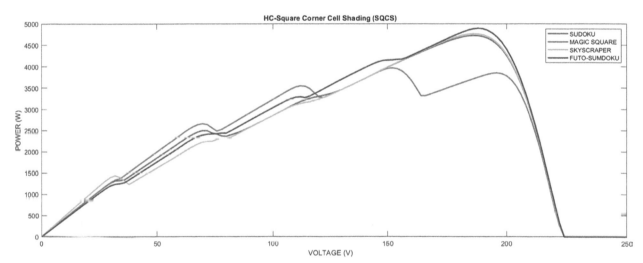

Figure 111.14 PV Characteristics of HC for SQCS shading
Source: Author

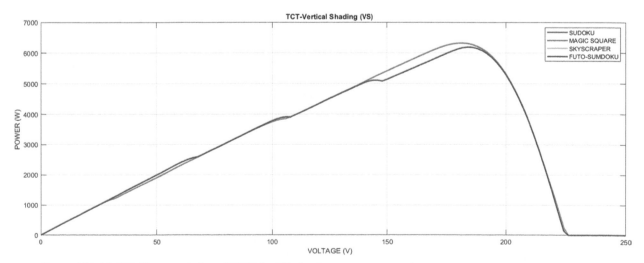

Figure 111.15 PV Characteristics of TCT for VS shading
Source: Author

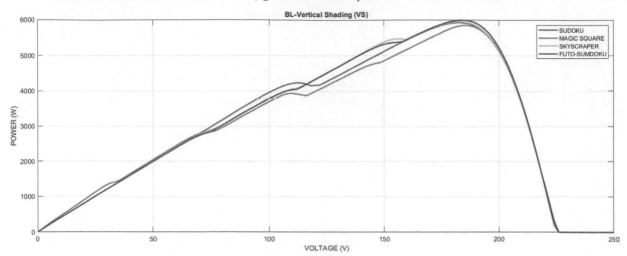

Figure 111.16 PV Characteristics of BL for VS shading
Source: Author

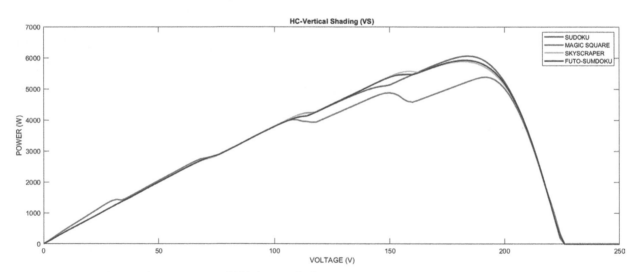

Figure 111.17 PV Characteristics of HC for VS shading
Source: Author

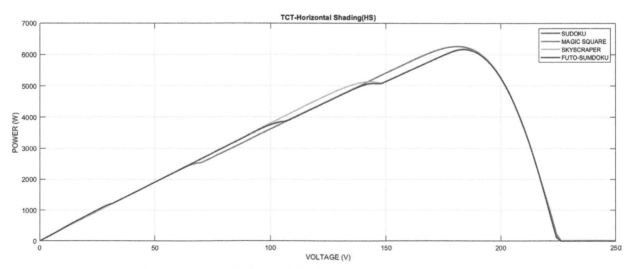

Figure 111.18 PV Characteristics of TCT for HS shading
Source: Author

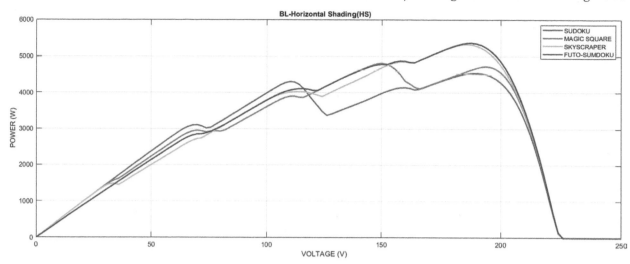

Figure 111.19 PV Characteristics of BL for HS shading
Source: Author

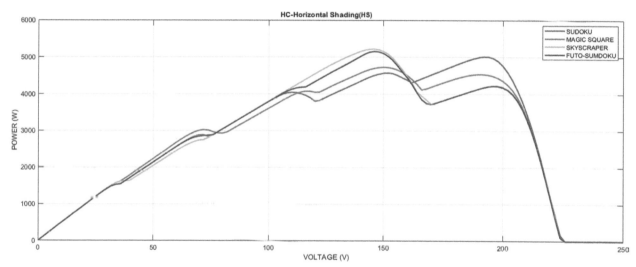

Figure 111.20 PV Characteristics of HC for HS shading
Source: Author

e)   Diagonal shading: In this type of shading the diagonal elements are selected and are partially shaded with different irradiance and all the other cells are un shaded as shown in Figure 111.6. The PV characteristics of vertical shading pattern in TCT, BL, and HC are shown in Figures 111.21-111.23.

Performance Analysis of Different Reconfigurations

This section gives the comparative analysis of different reconfiguration techniques with the proposed technique for 5 × 5 size PV array under uniform irradiance of 1000w/m². PV characteristics for five partial shading cases are analysed and suggested best performance topology for the PV system applications. The performance parameters are elaborated in the subsequent subsections.

a)   Fill factor (FF): Fill factor is the is the product of Voltage and current at maxima point of PSC to the product of Voc and Isc at the given STC [20].

$$FF = \frac{V_{mp}I_{mp}(at\ PSC)}{V_{oc}I_{sc}}$$

(2)

b)   Mismatch losses: The difference between the performance of a single PV panel and an entire system is referred to as the partial shading. This condition leads to the GMP and LMP point attributes.

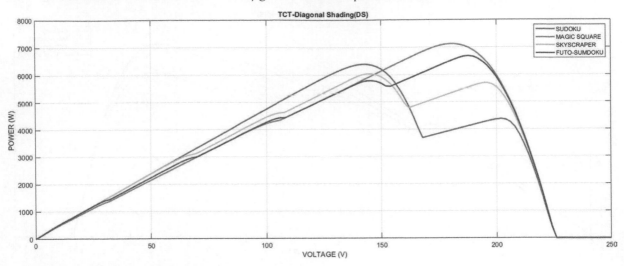

Figure 111.21 PV Characteristics of TCT for DS shading
Source: Author

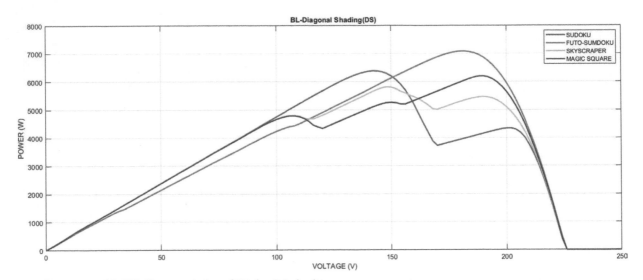

Figure 111.22 PV Characteristics of BL for DS shading
Source: Author

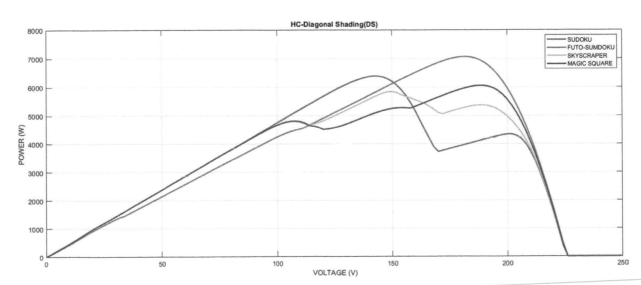

Figure 111.23 PV characteristics of HC for DS shading
Source: Author

Table 111.2 Performance metrics for TCT reconfiguration

Type of shading	Reconfiguration techniques	P_0 (W)	FF (%)	ML (W)	PL (%)	BEST
SCS	Sudoku	6431	60.23	1604	19.96	FSD
	Magic square	6441	60.33	1594	19.84	
	Skyscraper	6401	59.95	1634	20.34	
	Proposed FSD	6824	63.91	1211	15.07	
SQCS	Sudoku	4726	44.26	3309	41.18	FSD
	Magic square	3967	37.15	4068	50.63	
	Skyscraper	4764	44.62	3271	40.71	
	Proposed FSD	4892	45.82	3143	39.12	
VS	Sudoku	6062	56.78	1973	24.56	Sudoku
	Magic Square	5384	50.43	2651	32.99	
	Skyscraper	5885	55.12	2150	26.76	
	Proposed FSD	5926	55.50	2109	26.25	
HS	Sudoku	5018	47.00	3017	37.55	FSD
	Magic Square	4729	44.29	3306	41.14	
	Skyscraper	5158	48.31	2877	35.81	
	Proposed FSD	5220	48.89	2815	35.03	
DS	Sudoku	6051	56.67	1984	24.69	FSD
	Magic Square	6384	59.79	1651	20.55	
	Skyscraper	5842	54.72	2193	27.29	
	Proposed FSD	7062	66.14	973	12.11	

Source: Author

Table 111.3 Performance metrics for BL reconfiguration

Type of shading	Reconfiguration techniques	P_0 (W)	FF (%)	ML (W)	PL (%)	BEST
SCS	Sudoku	6977	65.35	1058	13.17	FSD
	Magic square	6986	65.43	1049	13.06	
	Skyscraper	7269	68.08	766	9.53	
	Proposed FSD	7272	68.11	763	9.50	
SQCS	Sudoku	5245	49.12	2790	34.72	SS/FSD
	Magic Square	5245	49.12	2790	34.72	
	Skyscraper	5254	49.21	2781	34.61	
	Proposed FSD	5254	49.21	2781	34.61	
VS	Sudoku	6193	58.00	1842	22.92	MS
	Magic Square	6323	59.22	1712	21.31	
	Skyscraper	6193	58.00	1842	22.92	
	Proposed FSD	6193	58.00	1842	22.92	
HS	Sudoku	6168	57.77	1867	23.24	FSD
	Magic Square	6182	57.90	1853	23.06	
	Skyscraper	6168	57.77	1867	23.24	
	Proposed FSD	6262	58.65	1773	22.07	
DS	Sudoku	6384	59.79	1651	20.55	MS
	Magic Square	7119	66.68	916	11.40	
	Skyscraper	6032	56.50	2003	24.93	
	Proposed FSD	6683	62.59	1352	16.83	

Source: Author

$$ML = P_{m\,(at\,STC)} - P_{m(at\,PSC)} \qquad (3)$$

c) Power loss: It is the difference between power maxima at STC and PSC to that of maximum power at STC [20].

$$\% \text{ Power loss (PL)} = \frac{P_{m\,(at\,STC)} - P_{m(at\,PSC)}}{P_{m\,(at\,STC)}} \times 100 \qquad (4)$$

The reconfiguration techniques are evaluated in terms of ML, FF, P0, and percentage power losses (PL). The arrangement of PV modules in row-wise and column-wise for Sudoku, Magic Square, Skyscraper and FSD techniques are shown in Figure 111.8. Table 111.2 shows the detailed analysis of various reconfiguration techniques and also gives the best reconfiguration technique.

Table 111.4 Performance metrics for HC reconfiguration

Types of shades	Reconfiguration techniques	P_0 (W)	FF (%)	ML (W)	PL (%)	BEST
SCS	Sudoku	6517	61.04	1518	18.89	FSD
	Magic square	6594	61.76	1441	17.93	
	Skyscraper	6474	60.64	1561	19.43	
	Proposed FSD	6596	61.78	1439	**17.91**	
SQCS	Sudoku	4523	42.36	3512	43.71	FSD
	Magic Square	3875	36.29	4160	51.77	
	Skyscraper	4832	45.26	3203	39.86	
	Proposed FSD	4823	45.17	3212	**39.78**	
VS	Sudoku	5924	55.48	2111	26.27	SS
	Magic Square	5846	54.75	2189	27.24	
	Skyscraper	5945	55.68	2090	**26.01**	
	Proposed FSD	5993	56.13	2042	25.41	
HS	Sudoku	4543	42.55	3492	43.46	FSD
	Magic Square	4810	45.05	3225	40.14	
	Skyscraper	5328	49.90	2707	33.69	
	Proposed FSD	5368	50.28	2667	**33.19**	
DS	Sudoku	6384	59.79	1651	20.55	MS
	Magic Square	7079	66.30	956	**11.90**	
	Skyscraper	5808	54.40	2227	27.72	
	Proposed FSD	6196	58.03	1839	22.89	

Source: Author

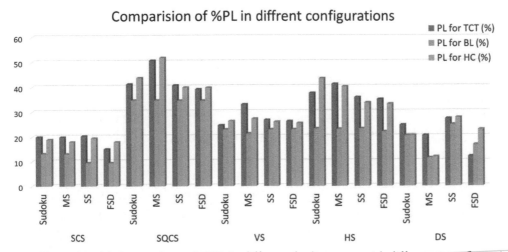

Figure 111.24 Comparison of %PL in different shading cases with different reconfiguration methods
Source: Author

From Table 111.2, it can be observed that the proposed FSD shows better performance than the existing reconfiguration techniques in terms of mismatch and power losses.

It is evident for SCS, SQCS, and HS shading the proposed FSD shows better performance than the existing reconfiguration techniques in terms of mismatch and PL and for VS and HS Magic square technique is showing better performance (Table 111.3).

- From Table 111.4, it is evident for SCS, SQCS and HS shading the proposed FSD shows better performance than the existing reconfiguration techniques in terms of mismatch and power losses and for VS skyscraper and DS Magic square technique is showing better performance.
- The comparison of power loss of different reconfiguration techniques with different shading patterns are shown in Figure 111.24. from the chart it can be observed for all the considered shading patterns the proposed FSD method gives better performance in most of the cases.

Conclusion

In order to enhance the global maximum power point (GMPP) under PSCs for a 5×5 photovoltaics (PV) array, an Futo - Sumdoku (FSD) puzzle-based reconfiguration technique is put forth in this study. Without changing the electrical connections, the FSD configuration moves the module's column-wise location inside the PV array. MATLAB/Simulink is used to validate the responses obtained from the reconfiguration techniques. In terms of FF, ML, and PL, the FSD configuration's performance is compared to that of the total cross tied (TCT), bridge linked (BL), honeycomb (HC), and other existing configurations including those for Sudoku, Magic Square, and Skyscraper techniques.

The observations made from simulation results are as follows:

Five different shading patterns are applied on Sudoku, Magic Square, Skyscraper and FSD techniques in TCT, BL and HC configurations. It is observed that for TCT configuration FSD has shown better results in single corner cell shading (SCS), square corner cell shading (SQCS), horizontal shading (HS), and diagonal shading (DS) by 15.07%, 39.12%, 35.03%, and 12.11% when compared to other methods. It is observed that for BL configuration FSD has shown better results in SCS, SQCS and HS by 9.50%, 34.61%, 22.07% when compared to other methods. Whereas in HC configuration FSD has shown better performance in SCS, SQCS, and HS by 17.91%, 39.78%, 33.19% when compared to other methods.

References

[1] CANRE, Solar Energy Technologies and Applications. Avaliable from: www.canren.gc.ca/techappl/index.asp?caid=5 & pgid=121, Sept. 4th, 2011

[2] John, G. (2011). Photovoltaic: solar electricity and solar cells in theory and practice. Available from: http://www.solarserver.com

[3] Gil, K. (2004). How do Photovoltaic Work? Science News. 1(1), 4.

[4] Pachauri, R. K., Mahela, O. P., Sharma, A., bai, J., Chauhan, Y. K., Khan, B., Haies, H. (2020). Impact of partial shading on various pv array configurations and different modeling approaches: a comprehensive review. IEEE Access. 10.1109/ACCESS.2020.3028473.

[5] Azzouzi, M., Dumitru, P., and Bouchahdane, M. (2016). Modeling of electrical characteristics of photovoltaic cell considering single-diode model. Journal of Clean Energy Technologies. 4, 414-20. 10.18178/JOCET.2016.4.6.323.

[6] N. Singh, N., Chauhan, Y. K., Singh, S. P., and Srivastava, A. K. (2018). Shading issues in photo-voltaic modules and its performance assessment. International Conference on Power Energy, Environment and Intelligent Control. 280-87. doi: 10.1109/PEEIC.2018.8665573.

[7] Pachauri, R. K., Singh, R., Gehlot, A., Samakaria, R., and Choudhury, S. (2019). Experimental analysis to extract maximum power from PV array reconfiguration under partial shading conditions. Engineering Science and Technology, an International Journal. 22(1), 109-30.

[8] Chandrakant, C. V. andMikkeli, S. (2020). A typical review on static reconfiguration strategies in a photovoltaic array under non-uniform shading conditions. CSEE Journal of Power and Energy Systems. 1-33. doi: 10.17775/CSEEJPES.2020.02520.

[9] Rani, B., Ilango, G., and Nagamani, C. (2013). Enhanced power generation from PV array under partial shading conditions by shade dispersion using Su Do Ku configuration. Sustainable Energy, IEEE Transactions. on. 4, 594-601. doi: 10.1109/TSTE.2012.2230033.

[10] Vijayalekshmy, S., Bindu, G. R., and Iyer, S. (2015). Performance improvement of partially shaded photovoltaic arrays under moving shadow conditions through shade dispersion. Journal of the Institution of Engineers (India): Series B. 97. doi:10.1007/s40031-015-0199-z.

[11] Manimegalai, D., Karthikeyan, M., and Vijayakumar, S. C. (2018). Maximizing the power output of partially shaded photovoltaic arrays using the SuDoKu configuration. ARPN Journal of Engineering and Applied Sciences. 13, 124-33.

[12] Winston, D. P., et al. (2020). Performance improvement of solar PV array topologies during various partial shading conditions. Solar Energy. 196, 228-42.

[13] Kandipati, R. and Ramesh, T. (2020). Maximum power enhancement under partial shadings using modified Sudoku reconfiguration. CSEE Journal of Power and Energy Systems (2020). doi: 10.17775/CSEEJPES.2020.01100.

[14] Horoufiany, M. and Ghandehari, R. (2018). Optimization of the Sudoku-based reconfiguration technique for PV arrays power enhancement under mutual shading conditions. Solar Energy. 159, 1037-46.

[15] Moger, T. (2020). Static Reconfiguration approach for photovoltaic array to improve maximum power. 2020 International Conference on Electrical and Electronics Engineering (ICE3). IEEE.

[16] Nihanth, M. S. S. et al. (2019). Enhanced power production in PV arrays using a new skyscraper puzzle-based one-time reconfiguration procedure under partial shade conditions (PSCs). Solar Energy. 194, 209-24.

[17] Meerimatha, G. and Rao, B. L. (2020). Novel reconfiguration approach to reduce line losses of the photovoltaic array under various shading conditions. Energy 196, 117120.

[18] Chao, K. -H., Lai, P. -L., and Liao, B. -J. (2015). The optimal configuration of photovoltaic module arrays based on adaptive switching controls. Energy Conversion and Management. 100, 157-67.

[19] Jagadeesh, K. and Chengaiah, C. H. (2022). A comprehensive review-partial shading issues in PV system. AIP Conference Proceedings. 2640(1). doi: https://doi.org/10.1063/5.0112913

[20] Palpandian, M., Winston, D. P., Kumar, B. P., Kumar, C. H., Thanikanti, S. B., and Hassan, H. A. (2021). A new ken-ken puzzle pattern based reconfiguration technique for maximum power extraction in partial shaded solar PV array. IEEE Access. pp. 1-1. doi:10.1109/ACCESS.2021.3076608.

[21] Natarajan, B., Murugesan, P., Udugula, M., Gurusamy, M., and Subramaniam, S. (2020). A fixed interconnection technique of photovoltaic modules using a sensorless approach for maximum power enhancement in solar plants. Energy Sources, Part A: Recovery, Utilization, and Environmental Effects. 1-23. 10.1080/15567036.2020.1831655.

[22] Nasiruddin, I., Khatoon, S., Jalil, M., and Bansal, R. (2019). Shade diffusion of partial shaded PV array by using odd-even structure. Solar Energy. 181. 519-529. 10.1016/j.solener.2019.01.076.

[23] Pachauri, R., Yadav, A., Chauhan, Y., Sharma, A., and Kumar, V. (2018). Shade dispersion-based photovoltaic array configurations for performance enhancement under partial shading conditions. International Transactions on Electrical Energy Systems. 28. e2556. 10.1002/etep.2556.

[24] Srinivasan, A., Devakirubakaran, S., and Sundaram, M. (2020). Mitigation of mismatch losses in solar PV system-two-step reconfiguration approach. Solar Energy. 206, 640-54. 10.1016/j.solener.2020.06.004.

Chapter 112

Design, analysis and simulation of a hydro-pneumatic suspension system for a quarter car model with two degrees of freedom

K Kedar[a], Prem Narayan Vishwakarma[b] and Ajay Sharma[c]

Department of Mechanical Engineering, Amity University, Greater Noida, India

Abstract

Hydro pneumatic suspension enables smooth suspension as the hydraulic fluid to the bladder containing the dinitrogen gas transmits the force due to uneven road. In the present work, design has been developed in such a manner that the isolation is possible to maximum extent, as four chambers of hydro pneumatic systems have been incorporated for each wheel. The dimensions have been fixed with respect to the already existing accumulator. The spacing of four chambers is done keeping in mind the same dimension so that more space is not required. Chambers of hydro pneumatic systems have been incorporated for each wheel for which the piston rod has been designed such that the impact of uneven road will be shared among the four chambers equally for individual wheel and increasing the smoothness of the suspension. Following the development of the mathematical modelling, the differential equations of motion are obtained from the model. The simulation for the quarter automobile model has two degrees of freedom. The findings illustrate how the displacement of both the sprung and the unsprung mass varies with regard to the passage of time.

Keywords: Accumulator, hydro-pneumatic suspension, quarter car model, sprung mass, unsprung mass

Introduction

When it comes to a car, the suspension system is the most significant component since it dampens the vibrations caused by the road. It makes a substantial contribution to the vehicle's stable operation, safety, and control. The accumulators, cylinders, flow resisters, lines, and fittings make up the majority of the components that make up a hydro-pneumatic suspension system.

There is a need to develop a smooth and even suspension system due to lack of which, the passengers can develop various spine problems and the automobile can be damaged. When compared with spring suspension systems, hydro pneumatic suspension systems would provide more smoother isolation from the sudden forces due uneven road as the force is transmitted via piston to hydraulic fluid and further to bladder containing dinitrogen gas. The bladder would be made of desmopan, a material with high wear resistance as it needs to withstand the compression due to the force transmitted by the piston due to uneven road and simultaneously the reaction force of dinitrogen as a result of compression. The hydraulic fluid to be used is liquid hydraulic mineral (LHM) with a density of 891 Kg/m^3 that transmits piston displacement pressure. Hydro-pneumatic suspension has advantages for military vehicles: progressive elasticity gives outstanding driving comfort with a modest load and allows the vehicle to carry a large weight (the requirements are mutually exclusive in classical suspension). At static suspension load, compressed nitrogen, encapsulated in the sphere with an elastic diaphragm, is six times more flexible than steel springs, providing exceptional driving comfort. A potential to decrease the vehicle's height by adjusting the road clearance, which increases aerial transport availability, a possibility to enhance road clearance for better cross-country driving. The suspension may self-level, compact hydro pneumatic column casing. It is especially important in the event of heavy vehicles, possible horizontal positioning of the hydro pneumatic column in the rear suspension of the vehicle (a saving of space that may be allocated building in some other special gear, or it may enlarge the loading or assault space capacity) [1–5]. Hydro pneumatic suspension system (HPSS) consists of double acting cylinder, two oil chambers, damping valve and one gas accumulator. The two chambers are connected through a flow control valve. The accumulator is connected to the chamber through a damping valve. The results showed that the proposed HPSS gives improvement over the passive suspension respectively, the acceleration by over 50%, the suspension working space by 12% and the dynamic tyre deflection by 3% when compared to the passive suspension system [6–10].

Design of Hydro-Pneumatic Suspension System

HPSS works by transmitting the force resulting because of an uneven road to the desmopan bladder containing dinitrogen via a hydraulic fluid (LHM). The hydraulic fluid is incompressible while Dinitrogen is compressible.

---

[a]k.kedar@s.amity.edu, [b]pnvishwakarma@amity.edu, [c]asharma3@amity.edu

DOI: 10.1201/9781003450252-112

Components Design

It houses the four accumulators, the piston cylinder and the piston rod. The designed piston rod central component will be inserted into this slot and simultaneously the piston assembly for the four chambers is aligned as per the requirement. The four accumulators containing the desmopan bladder will be inserted into these slots in such a manner that they are in the same level and the impact due to uneven road is transmitted equally among the four chambers equally. Figure 112.1 shows the design of cylindrical accumulators along with their dimensions.

The accumulator has to bear high pressure as the hydraulic fluid exerts the force on the bladder and at the same time, there is some reaction forces acting on the walls as well [7–10].

This piston rod is designed in such a manner that the force due to the uneven road gets distributed among the four pistons and similarly the hydraulic fluid gets pushed into the four accumulators at a time equally (Figure 112.2). The piston should be made of Al-Si alloy.

Assembly Design

The individual accumulators need to be assembled in the main body (Figure 112.3). Four slots are present into which the accumulators are inserted. The accumulators have been fixed into the slots present on the main body.

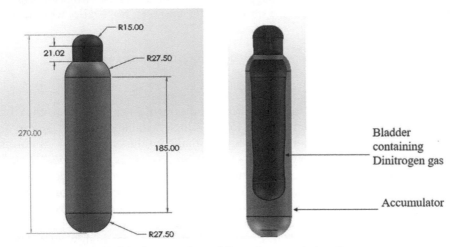

Figure 112.1 Cylindrical accumulator (dimensions are in 'mm')
Source: Author

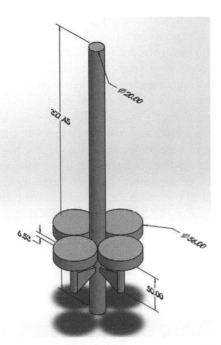

Figure 112.2 Design of piston rod along with dimensions (Isometric View) dimensions are in 'mm'
Source: Author

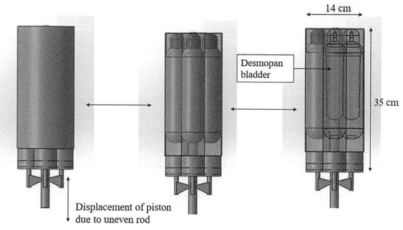

Figure 112.3 Complete assembly
Source: Author

Table 112.1 Parameters of suspension system

Parameter	Value
Mass of vehicle(M)	1300Kg
Static suspension force(F1)	797.25 N
Pressure(Po)	3 Bar
Volume(Vo)	0.25L
Polytropic constant(n)	1.3
Stiffness	56792.316N/m

Source: Author

Spring and Damping Characteristics of Suspension Systems

The system offers suspension after installing all pieces and changing hydraulic pressure by charging or discharging hydraulic fluid. The piston rod alters the accumulator's liquid volume and pressure (p1 → p2). The spring rate k is determined by the piston rod force and position change. The piston moves when the applied force reaches F to F*, transferring hydraulic fluid to the accumulator. This change continues until the accumulator pressure (and piston active surface) balances the system. This force balance underpins suspension system functioning and knowledge. The following computations will utilise it.

Calculation of Stiffness(k)

The gas present inside the accumulator is responsible for the elasticity of the entire installation. Its properties are primarily important for the behavior of the entire suspension system because it performs the most important work (spring stiffness).

Vo is accumulator volume without hydraulic pressure while Po is initial charging pressure. The manufacturing pre-pressure is 293.15K (about 20°C). Parameters for suspension system are given in Table 112.1.

Static suspension force will act when the mass M is suspended and it will compress the suspension to P_1 and V_1

Force on piston is given by (by using thermodynamic relations):

$$F_k = P_1 V_1^n / (PoVo/P1 - Ax - Ax)^n * A_k \tag{1}$$

Differentiating the Equation 1 w.r.t. x to get the stiffness,

$$K = dF_k / dx \tag{2}$$

After Differentiating Equation 1 and putting the value of P_1 and V_1, we get

$$K = F_1 n \{P_o V_o / F_1\}^n / \{P_o V_o / F_1 - x\}^{n+1} \tag{3}$$

After putting all the values of variable in Equation 10 we get the value of

K = 56792 N/m

Calculation of Damping Coefficient

The hydraulic fluid in the suspension system is utilised as the source to transmit the pressure on the surface from the piston to the accumulator. This occurs in the opposite direction of how the pressure is normally transferred. The energy that is transmitted to the suspension as a result of external stimulation needs to be decreased in order to achieve the desired outcome of lower vibration amplitude and to avoid the undesirable result of increased amplitude. The application of a braking force during the motion of the hanging components will, in most situations, result in the conversion of kinetic energy to heat. In most cases, the principle of friction serves as the foundation for this delayed damping force.

Throttle

The flow cross-delayed section's wide-to-narrow-to-wide transition slows flow. An purposefully tiny opening in the fluid channel between the cylinder and the accumulator gives a specific throttle valve for specified damping a circular cross-section. Hydraulic fluid flows fast due to the tiny cross-section. Due to the high flow velocity gradient from the center of flow to the inner wall of the borehole, high shear forces and therefore high-pressure loss. Parameters for orifice is given in Table 112.2.

Flow through the valve is consider as laminar and pressure loss can be calculated as:

$$\Delta p = V \, \upsilon \rho K_D \tag{4}$$

Δp is pressure difference across the flow through valve
V is volume flow rate
KD is constant
ρ is density of flowing fluid
υ is viscosity
K_D is calculated using the geometry and dimensions of the valve used

$$K_D = \frac{128 l_D}{\pi d_D^4} \tag{5}$$

$$F_{D,hyd} = \Delta p A_K \tag{6}$$

F_D is damping force, A_K is piston Area, Δp is pressure difference across the flow through valve
Comparing damping force with its formula i.e.

$$F_D = c^*(velocity)$$

Putting all the values to get the value of C as C=35609 Ns/m

Dynamic Behaviors Analysis

Suspension system analysis is performed using quarter car model to predict the behavior of the system. As the name indicates quarter car model consider one suspension unit in a four-wheel vehicle. It is two DOF

Table 112.2 Orifice parameters [6]

Parameter	Value
L_D	4cm
d_D	0.2cm
Density(ρ)	840Kg/m3
Kinematic viscosity	0.0000179m2/s
Mass flow rate	21.69551kg/s
Velocity of flow	10m/s

model (Figure 112.4). So there are two masses, entire mass of vehicle is supported on the suspension system is sprung mass (M) and unsprung mass is basically a mass of wheel.

Now we will drive the governing equation for the system. As the system 2 DOF hence 2nd degree differential equation will form. We will divide the above quarter car model into two system i.e. sprung mass and unsprung mass. We will use FBD for depicting the forces acting on the suspension (Figure 112.5).

X1 is the input given by the road which will make oscillate the suspension system. X3 movement of the vehicle due to the impact of X1.

Sprung Mass

There will be three types of force acting on this sprung mass i.e., spring force, damped force and inertia force which acts in opposite direction to the motion. All these forces will act in downward direction.

Using 2ⁿᵈ law

$$M\ddot{X}_3 + K_s(X_3 - X_2) + C_s(\dot{X}_3 - \dot{X}_2) = 0$$

$$\ddot{X}_3 = \frac{1}{M}\left(K_s(X_2 - X_3) + C_s(\dot{X}_2 - \dot{X}_3)\right)$$

Unsprung Mass

From above FBD (Figure 112.6) we can see there will 4 different forces act and in this mass there will be influence of both of the spring.

Using second law:

$$m\ddot{X}_2 + K_t(X_2 - X_{1)} = K_s(X_1 - X_2) + C_s(\dot{X}_1 - \dot{X}_2)$$

$$m\ddot{X}_2 + K_t(X_2 - X_1) - K_s(X_2 - X_3) - C_s(\dot{X}_2 - \dot{X}_3) = 0$$

$$\ddot{X}_2 = \frac{1}{m}(K_t(X_1 - X_2) + K_s(X_2 - X_3) + C_s(\dot{X}_2 - \dot{X}_3))$$

To solve these differential equation we will use the MATLAB/Simulink where we give X1 as input in the form of STEP function of speed bump of 10 cm.

As an input signal a Bump of height 10 cm is consider. It assumed to be step function which will remain constant with time. Simulink will show the result in the form of displacement of sprung and unsprung

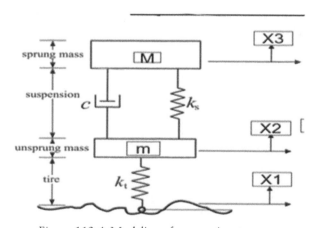

Figure 112.4 Modeling of suspension system
Source: Author

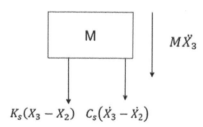

Figure 112.5 FBD of sprung mass
Source: Author

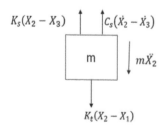

Figure 112.6 FBD of unsprung mass
Source: Author

masses. It will also show the settling time along with this we can also calculate the peak overshoot which will be helpful in predicting the stability.

Modelling on Simulink

Simulink model (Figure 112.7) for suspension system is presented for the differential equations relating parameters shown in "Table 112.3 Simulink parameters".

Result and Discussion

The parameters considered for the suspension system are equivalent to the volume of the accumulators that have been designed. Corresponding to those parameters, the stiffness and damping coefficient have been calculated. The Simulink model makes use of those parameters and assists us to simulate the working of the designed suspension system.

The inputs shown below (X_1, X_2 and X_3) refer to the displacement of the components shown in the mmodelling of suspension system.

Input (X_1) is shown in Figure 112.8. Step input value is taken as 10 cm.

We have considered a bump of height 10 cm for the simulation. The above plot shows the parameters related impulse response of the system. It states that the height of bump has been assumed to be constant with time.

Displacement (X_2)

The displacement for the unsprung mass is plotted w.r.t time (Figure 112.9). The displacement of wheel at t = 1 sec is touching the ground. The settling time is found to be at t = 3.737 sec. The highest displacement is found to be 0.1787 cm at t = 1.41 sec.

This result gives the empirical reaction of the suspension system.

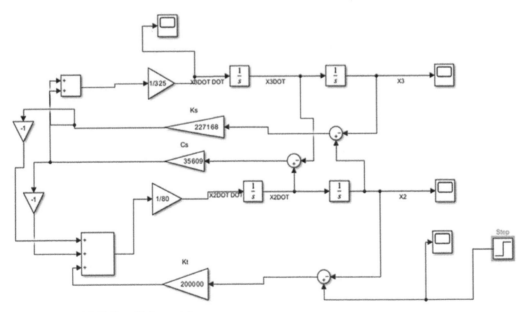

Figure 112.7 Simulink model
Source: Author

Table 112.3 Simulink parameters [3]

Parameter	Values
Sprung mass(M)	325 Kg
Unsprung mass(m)	80Kg
Suspension spring stiffness(Ks)	227168 N/m
Suspension damping coefficient(CS)	35609 N-s/m
Tire stiffness(Kt)	200000 N/m
Speed bump height(X3)	10 cm

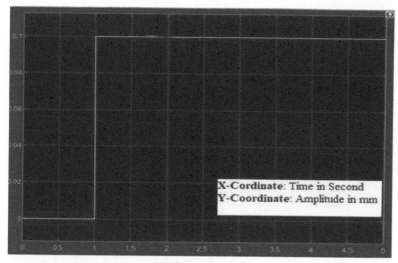

Figure 112.8 Plot for step input value (X_1)
Source: Author

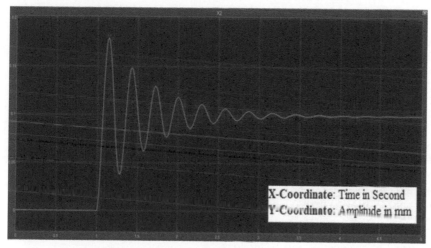

Figure 112.9 Plot for displacement (X_2)
Source: Author

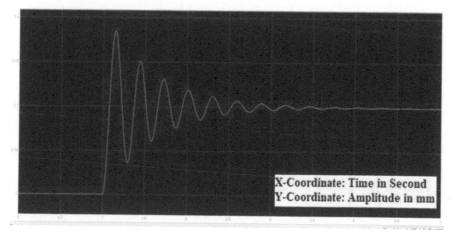

Figure 112.10 Plot for displacement (X_2)
Source: Author

Displacement (X_3)

The displacement for the sprung mass is plotted w.r.t time (Figure 112.10). The displacement of wheel at t = 1 sec is touching the ground. The settling time is found to be at t = 3.75 sec. The highest displacement is found to be 0.1839 cm at t = 1.152 sec.

This result gives the empirical reaction of the vehicle to which the suspension system has been installed.

Conclusion

The design has been developed in such a manner that the isolation is possible to maximum extent as four chambers of hydro pneumatic systems have been incorporated for each wheel. The dimensions have been fixed with respect to the already existing accumulator. The spacing of four chambers is done keeping in mind the same dimension so that more space is not required. Chambers of hydro pneumatic systems have been incorporated for each wheel for which the piston rod has been designed such that the impact of uneven road will be shared among the four chambers equally for individual wheel and increasing the smoothness of the suspension.

The spring and damping characteristics of the suspension system have been determined successfully. The suspension system has been modelled with the free body diagram and the subsequent differential equations have been obtained. The obtained differential equations have modelled in Simulink and have been helpful in the determination of displacement of sprung and unsprung masses.

References

[1] Ghazaly, N. M. and Moaaz, A. O. (2020). Hydro-pneumatic passive suspension system performance analysis using amesim software. International Journal of Vehicle Structures & Systems. 12(1), 9–12.

[2] Jonjo, R. E. and Nyalloma, S. T. (2020). Modeling the effect of road excitation on vehicle suspension system. International Journal of Engineering Materials and Manufacture. 5(1), 19–28.

[3] Reddy, V., Shankapal, S.R., and Monish, M. (2014). MODELLING AND SIMULATION OF HYDROPNEUMATIC SUSPENSION FOR A CAR.

[4] Badway, I. A., Sokar, M. I., and Raboo, A. (2017). Simulation and Control of a Hydro-pneumatic Suspension system. International Journal of Scientific & Engineering Research, 8(9).

[5] Sağlam, F., and Ünlüsoy, Y. S. (2014). State Dependent Riccati Equation Control of an Active Hydro-Pneumatic Suspension System. Journal of Automation and Control Research.

[6] Ali, G., and Emeritus, H. (2015). Car Dynamics using Quarter Model and Passive Suspension; Part V: Frequency Response Considering Driver-seat. In International Journal of Scientific Research Engineering & Technology (IJSRET) (Vol. 4).

[7] Hassaan, G. A. (n.d.). Car Dynamics using Quarter Model and Passive Suspension, Part VI: Sprung-mass Step Response. 17(2), 65–74. https://doi.org/10.9790/0661-17216574

[8] Mitra, A. et al. (2018). Development and validation of a simulation model of automotive suspension system using MSC-ADAMS. Materials Today: Proceedings. 5, 4327–434.

[9] Dirbas, W., Diken, H., and Alnefaie, K. (2022). Vibrational Behaviour of a Quarter Car Travelling over Road Humps with Different Suspension Systems.

[10] Tiwari, A., and Pathak, A. (2017). Design and Performance Analysis of Mechanical Hydro Pneumatic Suspension System. International Research Journal of Engineering and Technology.

Chapter 113

Force transmissibility analysis of a magneto-rheological damper

Prashant Narwade[1,a], Ravindra Deshmukh[2] and Mahesh Nagarkar[3]

[1]Department of Mechanical Engineering, D.V.V.P.C.O.E, Ahmednagar, Maharashtra, India

[2]Department of Mechanical Engineering, J.N.E.C, Aurangabad, Maharashtra, India

[3]Associate Professor, Department of Automation and Robotics, Pravara Rural Engg College, Loni, Ahmednagar

Abstract

Suspension is one of the crucial components of a car to isolate the oscillation of the vehicle body and provide ride comfort to the passengers. The performance of a magneto-rheological (MR) damper under vertical vibration transmitted to passengers from uneven road profiles is investigated in this article. MR dampers have an electromagnetic piston fitted in a cylinder filled with MR fluid. A smart material called MR fluid that is employed to control vibration can change its nature from a Newtonian fluid to a semi-solid substance by changing the magnetic field. The advantage of the MR damper over the conventional damper is that it can change damping force quickly in response to external magnetic field strength. They feature convenient construction, minimal power, and semi-active control. Comparative experimental trials are performed for MR damper and passive damper at various loads and speeds. The transmissibility ratio of the MR damper is calculated and compared with the conventional damper. It has been reported that the MR damper's viscosity increases with increasing current, decreasing the transmission of motion to the structure. The findings demonstrate that, at a 45 kg load, the MR damper's transmissibility is more than 10% higher compared to that of passive suspension. Similarly at 55 kg load transmissibility is more than 39% compared to that of passive suspension. This improvement in transmissibility effectively increases vehicle ride comfort.

Keywords: Magneto-rheological fluid, magnetic flux density, transmissibility, viscosity

Introduction

To isolate unpleasant vibrations produced about by different road profiles, a passive suspension system with fixed spring stiffness and damping parameters has been extensively employed in cars. These shock absorbers depend on mechanical means to convert vibration energy into heat. Passive dampers work on the principle of oil flow restriction by the piston in oil filled cylinder. A pressure difference between a compression chamber and an expanded chamber produced due to the fluid restriction exerts an opposing force on the vibrating body. Semi-active damper technology uses externally controlled technology or a powered actuator for dissipation of vibration energy. A damper and a spring that can simultaneously store and release energy make up a passive suspension system. Passive dampers have limitations in providing an adequate performance of isolation and stability. If the system is designed to optimize the road handling and stability of the vehicle, passengers are often subjected to a large amount of vibration. In contrary, if the suspension is soft, the vehicle's vibrations are reduced however its stability and road handling are substandard. They cannot balance the conflicting characteristics of the shock absorber. Semi-active damper like magneto-rheological damper has reduced the need for compromise between achieving the best performance in isolation and stability (Gillespie, 1992). While the MR damper produces optimal damping force for both low and high velocity, the linear viscous damper produces damping force proportional to damper velocity.

For human comfort, acceptable vibration acceleration values are between 0.53.5 m/s². Most of the time when travelling on highways, automobile vibrations and shocks are of a small scale. Off-road vehicle occupants, however, may experience intense and frequent stress, which can have an adverse effect on their health (ISO 2631-1:1997). Mehdi Ahmadian et al. examined the effectiveness of the skyhook, ground hook, and hybrid controls while observing the performance of three semi-active control strategies. The hybrid control test findings show that it can be relied upon to deliver a semi-active control that can be customized to the driving situation and vehicle dynamics for increased vehicle stability and ride comfort (Ahmadian and Pare, 2000). To evaluate the effectiveness of vibration control, Lee et al. created and integrated the skyhook controller with the HILS system. The suggested semi-active MR suspension system can significantly increase steering stability, according to bump and random road testing for ride quality (Lee and Choi, 2000). The topic of linear suspension systems subjected to deterministic road excitation has been covered by authors. However, the random nature of the frequency spectrum's descriptions of road abnormalities causes nonlinearity in the system. Comfortable ride quality is provided by semi-active suspensions with MR dampers, which help to reduce nonlinear vibrations (Verros et al., 2005). In order to adjust the appropriate input voltage to the MR damper, Lam proposed to use a sliding mode controller for the system (Lam and Liao,

[a]narwadeprashant@gmail.com

DOI: 10.1201/9781003450252-113

2004). A modified skyhook control algorithm can be used as the controller to reduce road disturbances in simulation (Bakar et al., 2008). In the preceding ears, comfort improvement has seen a lot of research activity, and to reduce vibrations, a wide range of control schemes and sensors are being explored (Savaresi et al., 2011). When the vehicle body is believed to be robust, the MR damper outperforms the passive suspension and exhibits a performance comparable to that of the active ones. But depending on the road condition and the damper properties, a performance drop may be seen in flexible structures (Stutz and Rochinha, 2011). Quantitative feedback theory (QFT), initially proposed by Zapateiro et al. for linear systems, is indeed applicable to nonlinear systems. This can be achieved by transforming the nonlinear dynamics into a linear system with uncertainty that accurately captures the desired plant's real behavior (Zapateiro et al., 2012). In order to help with the development of ride comfort, Jayachandran et al. developed the "sky-hook" control model. They reported that the body acceleration and wheel deflection of the suggested controller were superior to those of the passive system in terms of weighted seat acceleration values (Jayachandran and Krishnapillai, 2012). By creating an integrated seat suspension model for active seat suspension design, the authors showed how a correctly designed static output controller can provide higher ride comfort performance (Du et al., 2013). Weber et al. (2014) developed a semi-active vibration absorber with a real-time regulated magneto-rheological damper to enhance harmonic structure vibrations. At structural resonance frequency, they stated that it outperforms the passive vibration absorber by at least 12.4% and as much as 60.0%. Shiao et al. (2015) suggested an MR damper with multiple poles and demonstrated that the compression force increases 7.66 times at a test speed of 0.03 m/sec when the applied current rises from 0 A to 1.6 A. Dev Anand et al. (2016) made the observation that the magneto-rheological fluid's viscosity varies depending on the magnetic field generated inside the piston cylinder using the COMSOL program.By taking into account configurable damper delay time, five potential control strategies are identified. The largest performance range was discovered to be achieved by modified skyhook hybrid control, which also outperformed other strategies when the delay period was more than 10 ms. Ride comfort was substantially more sensitive to variations in delay times in the frequency range of 4–8 Hz than road handling (Qin et al., 2017). Authors have designed a stiffness-adjustable tuned mass damper to reduce the vibration of civil engineering structures using inerter-induced negative stiffness and magnetic-force-induced positive/negative stiffness systems (Wang et al., 2018). Yuan et al. (2019) concluded in their review that; dampers have been widely adopted in different diverse fields of vibration. In an MR damper, the formation of a magnetic field is a significant phenomenon. The copper coil windings may provide magnetic flux density up to 0.61 T to activate MR fluid. Thus, it is possible to offer changeable damping force semi-actively and improve the ride comfort of the car by altering the applied magnetic field (Narwade, 2022). Semi-active suspension has gained a lot of attention recently since they offer the adjustability of active suspension without requiring a lot of electricity. The performance of a traditional passive viscous damper is compared to that of a MR damper in the current work, and its transmissibility is assessed on a damper testing machine. Transmitted force at different frequencies and loads is studied.

The following is the outline of this article. The proposed semi-active MR damper is first described, followed by an explanation of the quarter-car experimental setup. The testing procedure for MR and passive dampers is shown in Section III. Section IV presents the force transmitted and transmissibility data for 45 kg and 55 kg of load. The final section provides a summary of the vibration acceleration graphs produced from the FFT analyzer.

Development of MR Damper and Experimental Setup

A piston, an excitation coil, and MR fluid are all enclosed inside the housing cylinder of the prototype MR damper that has been created for this research. MR damper is a distinctive mono tube damper created to provide a respectable semi-active control effect. The piston in Figure 113.1 contains slots for MR fluid movement and a groove for a copper winding to fit in. For the purpose of producing the magnetic effect, 330 turns of 30SWG wire gauge enameled copper wire are wound around the piston. A cylinder containing MR fluid and an electromagnetic piston that requires a DC current supply ranging from 0.1 Ampere to 2 Ampere constitute the MR damper. Since the magnetic loop only arises in magnetic materials, the magnetic leakage is disregarded. The designed damper has an extended length of about 300 mm and a 75 mm stroke length.

The MR fluid used in the damper is supplemented by pure iron particles or carbonyl iron particles suspended in a low viscosity carrier fluid like paraffin oil, silicone oil, or synthetic oil. If carbonyl iron particles are coated with guar gum and mixed with paraffin oil, the results are better than those of other mixtures. As illustrated in Figure 113.2, the quarter car model setup with a cam follower-type exciter is created to test the MR damper. Cam (exciter) rotary motion is transformed into linear vibratory motion. The damper is tested using devices such as a DC regulated controller kit with the necessary power supply, an accelerometer, an MR Damper, and an FFT analyzer. An accelerometer is built inside the FFT analyzer to monitor the acceleration of sprung mass. The top plate of the setup has a specified mounting point for the accelerometer as well as a facility for load attachment. Trials are done using the passive suspension as a reference object

Figure 113.1 Piston with copper winding
Source: Author

Figure 113.2 Experimental setup [10]

Table 113.1 Passive damper test at 45 kg load

Voltage (V)	Current (A)	Motor speed (RPM)	Displacement (mm)	Compression (m)
25	1.55	30	20	0.02
32	1.55	40	19	0.019
49	1.55	60	17	0.017
54	1.55	70	15	0.015
64	1.55	80	14	0.014
76	1.55	90	12	0.012

Source: Author

in order to assess the efficacy of the proposed MR Damper for the semi-active suspension. A FFT analyzer, which can produce several forms of displacement, velocity, and acceleration spectrum signals, was used to carry out the laboratory tests (Narwade, 2022). Testing of passive and MR damper

Shock absorber is tested for a number of repetitive cycles using cam follower mechanism. A shock absorber with a fixed load applied over it is a part of the setup. The test damper is connected to a follower that rests on a cam and is powered by an electric motor via a reduction gearbox. The motor speed is controlled using a thyristor (adjustable DC voltage) drive. In this situation a sinusoidal force is applied to the mass. With the increase in motor speed, displacement of the system reduces as cyclic time of exciter cam profile is reduced. Force transmitted is calculated from the compression of spring. Output displacement of passive damper is presented in Table 113.1, where voltage and current are obtained to find out power supplied to the system. Motor speed is changed to obtain results at different frequencies.

Under braking, the entire vehicle slows down as a single unit represented as one lumped mass located at its center of gravity. For quarter car model, 1/4th mass is considered for testing of damper, hence initially 45 kg load is applied on test rig for observations. Table 113.2 represents the observations of MR damper without magnetic field working like passive damper. For comparison all values of voltage, current and rpm are kept similar to find displacement, velocity and acceleration of passive and MR damper.

Displacement of MR damper at 2 Ampere current is shown in Table 113.3 for various rpm values. It can be observed that all values of displacement of MR damper are less than passive damper values at same load of 45 kg. Similarly displacements are observed by increasing load up to 55 kg, as shown in Table 113.4 and Table 113.5.

Results and Transmissibility of Passive and MR Damper

Transmissibility is the ratio of the force transmitted by the system (reaction force) to the force applied on the system. It can be also determined using ratio of output displacement of the system to the input displacement. Maximum transmissibility occurs at the natural frequency or at point before natural frequency when system is undamped or under damped. As damping ratio increases, maximum transmissibility which occurs at natural frequency decreases. In this article, transmissibility is calculated for the trials of damper on experimental set up with the help of following formulae.

1. Compression of spring = compression $\times 10^{-3}$ (in m)
2. Force transmitted (F_{Tr}) in N = Compression x spring stiffness, where spring stiffness = 12,732 N/m

Table 113.2 MR damper test at 45 kg load without magnetic field

Voltage (V)	Current (A)	Motor speed (RPM)	Displacement (mm)	Compression (m)
25	1.55	30	19	0.019
32	1.55	40	18	0.018
49	1.55	60	16	0.016
54	1.55	70	14	0.014
64	1.55	80	13	0.013
76	1.55	90	11	0.011

Source: Author

Table 113.3 MR damper test at 45 kg load with 2 Ampere current

Voltage (V)	Current (A)	Motor speed (RPM)	Displacement (mm)	Compression (m)
25	1.55	30	17	0.017
32	1.55	40	16	0.016
49	1.55	60	14	0.014
54	1.55	70	12	0.012
64	1.55	80	11	0.011
76	1.55	90	10	0.010

Source: Author

Table 113.4 MR damper test at 55 kg load without magnetic field

Voltage (V)	Current (A)	Motor speed (RPM)	Displacement (mm)	Compression (m)
25	1.55	30	17	0.017
32	1.55	40	15	0.015
49	1.55	60	12	0.012
54	1.55	70	11	0.011
64	1.55	80	10	0.010
76	1.55	90	9	0.009

Source: Author

Table 113.5 MR damper test at 55 kg load with 2 Ampere current

Voltage (V)	Current (A)	Motor speed (RPM)	Displacement (mm)	Compression (m)
25	1.55	30	15	0.015
32	1.55	40	13	0.013
49	1.55	60	11	0.011
54	1.55	70	10	0.010
64	1.55	80	9	0.009
76	1.55	90	8	0.008

Source: Author

3.   Power input (P) in watt $P = V \times A$
4.   Input Torque (T) in N-m

where N = motor rpm   (1)

5.   Force applied (F_a) in N, $F_a = T/R$ (2)

   Where R = Radius of Cam with lobe = 0.05 m

6.   Transmissibility = F_{Tr}/F_a (3)

7. Cam speed = Motor speed/7.5   (4)

As damping coefficient increases maximum transmissibility decreases and the maximum transmissibility occurs before natural frequency range. Passive damper and MR damper test result values are used to calculate the force transmitted and transmissibility.

Force Transmitted and Transmissibility at 45 kg Load

The approaches employed to examine the proposed transmissibility relationships from an experimental perspective are presented by the authors in this section. System is considered as single degree of freedom and gross 45 kg load is applied for the trial of passive damper as shown in Table 113.6. To obtain high torque at input excitation, motor speed is reduced at cam follower mechanism by 7.5 times using reduction gear box. All observations are written with respect to the cam follower speed.

Output velocity of Passive damper at 45 kg load increases with the increase in excitation speed from 4.51 mm/s to 8.84 mm/s. Similarly output acceleration increases from 0.274 m/s² to 2.02 m/s². Force transmitted calculated from the deflection of shock absorber, decreases with the increase in speed from 360 N to 216 N. Since damping force and velocity are proportional, damping force rises as velocity increases. In the same way, transmissibility reduces from 1.45 to 0.864, as it is directly proportional to force transmitted.

Table 113.6 Passive damper results at 45 kg load

Cam speed (RPM)	Compression (m)	Velocity (mm/s)	Acceleration (m/s²)	Force transmitted (N)	Power I/P (W)	I/P Torque (N-m)	Force applied (N)	Transmissibility
4.0	0.02	4.51	0.274	360	38.75	12.341	246.815	1.459
5.3	0.019	4.97	0.612	342	49.6	11.847	236.943	1.443
8.0	0.017	5.69	0.377	306	75.95	12.094	241.879	1.265
9.3	0.015	6.26	0.426	270	83.7	11.424	228.480	1.182
10.6	0.014	7.74	1.74	252	99.2	11.847	236.943	1.064
12.0	0.012	8.84	2.02	216	117.8	12.505	250.106	0.864

Source: Author

Table 113.7 MR damper results at 45 kg load without magnetic field

Cam speed (RPM)	Compression (m)	Velocity (mm/s)	Acceleration (m/s2)	Force transmitted (N)	Power I/P (W)	I/P Torque (N-m)	Force applied (N)	Transmissibility
4.0	0.019	4.02	0.269	389.5	38.75	12.341	246.81	1.578
5.3	0.018	3.79	0.724	369	49.6	11.847	236.94	1.557
8.0	0.016	5.29	0.406	328	75.95	12.094	241.87	1.356
9.3	0.014	6.22	0.945	287	83.7	11.424	228.48	1.256
10.6	0.013	7.49	1.568	266.5	99.2	11.847	236.94	1.125
12.0	0.011	8.41	2.274	225.5	117.8	12.505	250.10	0.902

Source: Author

Table 113.8 MR damper results at 45 kg load with 2 Ampere electric current

Cam speed (RPM)	Compression (m)	Velocity (mm/s)	Acceleration (m/s²)	Force transmitted (N)	Power I/P (W)	I/P torque (N-m)	Force applied (N)	Transmissibility
4.0	0.017	4.51	0.288	348.5	38.75	12.340	246.81	1.4119
5.3	0.016	4.97	0.29	328	49.6	11.847	236.94	1.3843
8.0	0.014	5.69	0.377	287	75.95	12.093	241.87	1.1865
9.3	0.012	6.26	0.456	246	83.7	11.424	228.48	1.0766
10.6	0.011	7.74	0.564	225.5	99.2	11.847	236.94	0.9517
12.0	0.01	8.84	0.719	205	117.8	12.505	250.10	0.8196

Source: Author

Parallel trend is observed for the MR damper as shown in Tables 113.7 and 113.8. For the same 45 kg load condition MR damper is tested without magnetic field and with magnetic field at 2 Ampere current. Vibration acceleration of MR damper without magnetic field is more than that with magnetic field. MR fluid responds to the magnetic field and changes its properties when a magnetic field is presented. The effect of force transmitted and transmissibility is shown in graphs. Force transmitted to the foundation is compared for all three dampers as shown in Figure 113.3. It is observed that the force transmitted at 2 Ampere is largely reduced than the force transmitted at 0 Ampere. Passive damper performance is in between 0 Ampere current and 2 Ampere current. The authors have derived and validated a relationship to obtain the force transmissibility of passive damper and MR damper. This relationship shows that when no magnetic field is produced, MR damper produces less resistance force than passive damper and when 2 Ampere electric current is supplied, MR damper's resistance force is more than passive damper's resistance or damping force. Transmitted force of the system goes on decreasing when we increase the amount of current to electromagnet. With increase in excitation speed from 4 rpm to 12 rpm, force transmitted to the system by all dampers reduces.

The graph shown in Figure 4 represents transmissibility of system for various damping conditions at 45 kg load. Transmissibility of MR damper is more effective than passive damper and MR damper without magnetic field. This indicates that transmissibility decreases with the increase in damping force.

Figure 113.4 Transmissibility at 45 Kg Load

Force Transmitted and Transmissibility at 55 kg Load

Semi- active suspension is also compared at different current values by conducting trials at 55 kg load. As with the increase in magnetic field MR damper performed better than passive damper, authors performed test on only MR damper with and without magnetic field. Results of MR damper without magnetic field is shown in Table 113.9 and results of MR damper with 2 Ampere current supplied to electromagnet is shown in Table 113.10. Force transmitted to the system is largely reduced by MR damper using high magnetic field produced by supplying 2 Ampere current as shown in Figure 113.5. Without magnetic field MR damper behaves as passive damper. Range of force transmitted by MR damper without magnetic field is from 348.5 N to 184.5 N, whereas for MR damper with magnetic field is from 307.5 N to 164 N. At low speed of vibration excitation large amount of damping force is produced by the MR damper with magnetic field. In the same way transmissibility of system is largely reduced by MR damper using high magnetic

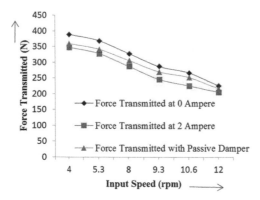

Figure 113.3 Force transmitted at 45 kg load
Source: Author

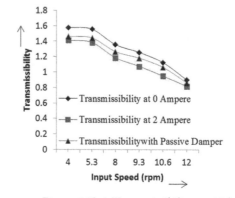

Figure 113.4 Transmissibility at 45 kg load
Source: Author

Table 113.9 MR damper results at 55 kg load without magnetic field

Motor speed (RPM)	Compression (m)	Velocity (mm/s)	Acceleration (m/s²)	Force transmitted (N)	Power I/P (W)	I/P Torque (N-m)	Force applied (N)	Transmissibility
4.0	0.017	5.625	0.289	348.5	38.75	12.340	246.81	1.4119
5.3	0.015	7.256	0.445	307.5	49.6	11.847	236.94	1.2977
8.0	0.012	15.16	0.46	246	75.95	12.093	241.87	1.0170
9.3	0.011	17.203	1.32	225.5	83.7	11.424	228.48	0.9869
10.6	0.01	18.911	2.04	205	99.2	11.847	236.94	0.8651
12.0	0.009	25.848	2.95	184.5	117.8	12.505	250.10	0.7376

Source: Author

Table 113.10 MR damper results at 55 kg load with 2 Ampere current

Motor speed (RPM)	Compression (m)	Velocity (mm/s)	Acceleration (m/s²)	Force transmitted (N)	Power I/P (W)	I/P Torque (N-m)	Force applied (N)	Transmissibility
4.0	0.015	5.9	0.274	307.5	38.75	18.107	362.14	0.8490
5.3	0.013	6.7	0.304	266.5	49.6	17.383	347.66	0.7665
8.0	0.011	14.31	0.554	225.5	75.95	17.745	354.90	0.6353
9.3	0.01	19.58	0.909	205	83.7	16.762	335.24	0.6114
10.6	0.009	24.52	1.262	184.5	99.2	17.383	347.66	0.5306
12.0	0.008	29.03	1.778	164	117.8	18.348	366.97	0.4468

Source: Author

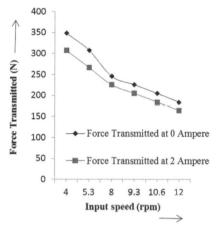

Figure 113.5 Force transmitted at 55 kg load
Source: Author

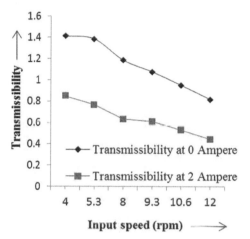

Figure 113.6 Transmissibility at 55 kg load
Source: Author

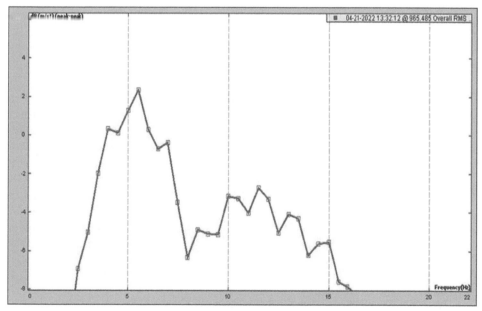

Figure 113.7 Acceleration of MR damper without magnetic field at 55 kg load

Source: Graph obtained from FFT Analyzer model COCO 80 available in Mechanical Department of Dr. Vithalrao Vikhe Patil College of Engineering

field produced by supplying 2 Ampere current as shown in Figure 111.6. Range of transmissibility by MR damper without magnetic field is from 1.41 to 0.73, whereas for MR damper with magnetic field is from 0.840.44. These values represent that more amount of isolation is observed in case of MR damper with large magnetic field at 2 Ampere current. When the transmissibility value is low, it means little vibration can be transferred from the system.

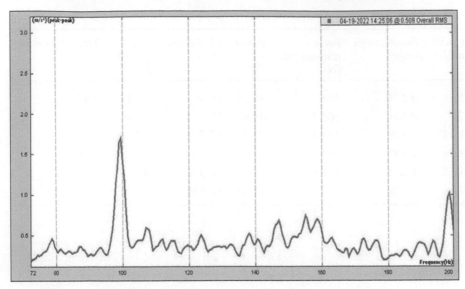

Figure 113.8 Acceleration of MR damper with magnetic field (2 Am current) at 55 kg load
Source: Graph obtained from FFT Analyzer model COCO 80 available in Mechanical Department of Dr. Vithalrao Vikhe Patil College of Engineering

Vibration Acceleration

The performance of the suspension system can be evaluated using the acceleration concept in transmissibility. CoCo-80/90 real time four channel FFT analyzer is used to collect the signals of displacement, velocity and acceleration amplitude. Output acceleration of damper obtained with the help of FFT analyzer is plotted in the Figures 113.7 and 113.8. It can be observed that acceleration 2.95 m/s^2 of MR damper without magnetic field at 55 kg is more than acceleration 1.77 m/s^2 of MR damper with 2 Ampere current magnetic field.

In passive suspension or MR damper without magnetic field damping force is proportional to the acceleration at resonance and increases with the acceleration. But MR damper with magnetic field produces high damping force by increasing current thus reducing the acceleration. Reducing acceleration value in vehicle suspension system provides comfort to the passengers.

Conclusion

Since viscous damping is a linear function of velocity, damping force resisting vibrations increases with increase in vibration velocity. But at low velocity and low-frequency vibrations, modest damping force is generated in Passive dampers. MR damper with fluids capable of varying their viscosity as a function of magnetic field produces required damping force at any frequency and velocity of vibrations. The primary feature of an MR damper suspension system is its ability to generate the necessary damping force at frequencies significantly lower than the resonant frequency. According to experimental findings, the MR damper system will improve transmissibility by 10% at 45 kg of load and 39% at 55 kg as compared to a typical fluid passive shock absorber. The acceleration amplitude of the MR damper is also improved by reducing acceleration by 40% in the case of 55 kg load condition. Reduced transmissibility and lower acceleration amplitude are helpful in increasing the passenger comfort at low frequency vibrations.

References

[1] Ahmadian, M. and Pare, C. A. (2000). A quarter-car experimental analysis of alternative semiactive methods. *Journal of Intelligent Materials Systems and Structures*, 11(8), 604–612.
[2] Bakar, S. A., Jamaluddin, H., and Rahman, R. A. (2008). Vehicle ride performance with semi-active suspension system using modified Skyhook algorithm and current generator model. *International Journal of Vehicle Autonomous Systems*, 6(3/4).
[3] Dev Anand, M., Janardhanan, K. A., Gopu, P., and Kinslin, D. (2016). Design modelling and study of magneto-rheological dampers in supension system. *Journal of Chemical and Pharmaceutical Sciences*, 9(1), 347–350.
[4] Du, H., Li, W. and Zhang, N. (2013). Vibration control of vehicle seat integrating with chassis suspension and driver body model. *Advances in Structural Engineering*. 16(1), 1–9.
[5] Gillespie, T. D. (1992). Fundamentals of vehicle dynamics. Society of Automotive Engineers, Warrendale, 2030.
[6] ISO 2631-1:1997: Available from https://www.iso.org/standard/7612.html (1997). Accessed 2 September 2022.

[7] Jayachandran, R. and Krishnapillai, S. (2012). Modeling and optimization of passive and semi-active suspension systems for passenger cars to improve ride comfort and isolate engine vibration. *Journal of Vibration and Control*, 19(10), 1471–1479.

[8] Lam, A. H. -F. and Liao, W. -II. (2004). Semi-active control of automotive suspension systems with Magneto-rheological dampers. *International Journal of Vehicle Design*, 33(13), 5075.

[9] Lee, H. S. and Choi, S. B. (2000). Control and response characteristics of a magneto-rheological fluid damper for passenger vehicles. *Journal of Intelligent Materials Systems and Structures*, 11(1), 80–87.

[10] Narwade, P., Deshmukh, R., and Nagarkar, M. (2022). Improvement of ride comfort in a passenger car using Magneto Rheological (MR) damper. *Journal of Vibration Engineering & Technologies*. https://doi.org/10.1007/s42417-022-00512-0

[11] Purandare, S., Zambare H., Razban, A. (2019). Analysis of magnetic flux in magneto-rheological damper. *Journal of Physics Communications*. 3(075012). https://doi.org/10.1088/2399-6528/ab33d7

[12] Qin, Y., Zhao, F., and Wang, Z. (2017). Comprehensive analysis for influence of controllable damper time delay on semi-active Suspension Control Strategies. *Journal of Vibration and Acoustics*. 139(3). https://doi.org/10.1115/1.4035700

[13] Savaresi, S. M., Poussot-Vassal, C., and Spelta, C. (2011). Semi-active suspension control design for vehicles. 1 st ed., Butterworth-Heinemann.

[14] Shiao, Y. J., Jow, M. L., and Kuo, W. H. (2015). Design and experiment of the magnetorheological damper with multiple Poles. *Applied Mechanics and Materials*. 764, 223227.

[15] Stutz, L. T. and Rochinha F. A. (2011). Synthesis of a magneto-rheological vehicle suspension system built on the variable structure control approach. *Journal of the Brazilian Society of Mechanical Sciences and Engineering*, 33(4), 445–458.

[16] Verros, G., Natsiavas, S., and Papadimitriou, C. (2005). Design optimization of quarter-car models with passive and semi-active suspensions under Random Road Excitation. *Journal of Vibration and Control*. 11(5), 581–606.

[17] Wang, Z., Gao, H., and Wang, H. (2018). Development of stiffness-adjustable tuned mass dampers for frequency retuning. *Advances in Structural Engineering*. 22(2), 473–485. https://doi.org/10.1177/1369433218791356

[18] Weber, F. (2014). Semi-active vibration absorber based on real-time controlled mr damper. *Mechanical Systems and Signal Processing*. 46(2), 272–288.

[19] Yuan, X., Tian, T., and Ling H. (2019). A review on structural development of magnetorheological fluid damper. *Shock and Vibration*. 1–33. https://doi.org/10.1155/2019/1498962

[20] Zapateiro, M., Pozo, F., and Karimi, H. R. (2012). Semiactive control methodologies for suspension control with magnetorheological dampers. IEEE/ASME Transactions on Mechatronics. 17(2), 370–380.